ADVANCES IN NEUROLOGY
VOLUME 39

Advances in Neurology

INTERNATIONAL ADVISORY BOARD

Konrad Akert, *Zurich*
Julius Axelrod, *Bethesda*
André Barbeau, *Montreal*
Günter Baumgartner, *Zurich*
Donald B. Calne, *Bethesda*
Macdonald Critchley, *London*
Roger C. Duvoisin, *New Jersey*
Stanley Fahn, *New York*
Robert A. Fishman, *San Francisco*
Gilbert H. Glaser, *New Haven*
Rolf Hassler, *Frankfurt*
Herbert H. Jasper, *Montreal*
Bryan Jennett, *Glasgow*
Richard T. Johnson, *Baltimore*
Harold L. Klawans, *Chicago*
C. David Marsden, *London*
J. Kiffin Penry, *Winston-Salem*
Alfred Pope, *Boston*
Dominick P. Purpura, *Stanford*
Derek Richter, *London*
Lewis P. Rowland, *New York*
Arnold B. Scheibel, *Los Angeles*
Peritz Scheinberg, *Miami*
Richard P. Schmidt, *Syracuse*
Donald B. Tower, *Bethesda*
T. Tsubaki, *Tokyo*
John N. Walton, *Newcastle upon Tyne*
Arthur A. Ward, Jr., *Seattle*
Melvin D. Yahr, *New York*

Advances in Neurology
Volume 39

Motor Control Mechanisms in Health and Disease

Editor

John E. Desmedt, M.D.
Brain Research Unit
University of Brussels
Brussels, Belgium

Raven Press ■ New York

Raven Press, 1140 Avenue of the Americas, New York, New York 10036

©1983 by Raven Press Books, Ltd. All rights reserved. This book is protected by copyright. No part of it may be reproduced, stored in a retrieval system, or transmitted, in any form or by any means, electronic, mechanical, photocopying, recording, or otherwise, without the prior written permission of the publisher.

Library of Congress Cataloging in Publication Data

Motor control mechanisms in health and disease.

   (Advances in neurology; v. 39)
   Includes bibliographical references and index.
   1. Movement disorders—Addresses, essays, lectures.
2. Motor cortex—Diseases—Addresses, essays, lectures.
3. Efferent pathways—Addresses, essays, lectures.
4. Biological control systems—Addresses, essays, lectures.
I. Desmedt, John E. II. Series. [DNLM: 1. Motor activity —Physiology. 2. Motor cortex—Physiology. 3. Motor cortex—Physiopathology. 4. Movement disorders—Physiopathology. 5. Nervous system—Physiology. 6. Nervous system—Physiopathology. W1 AD684H v.39 / WE 103 M918]
RC321.A276 vol. 39   616.8s [616.7'4]   83-13964
[RC925.5]
ISBN 0-89004-723-5

Made in the United States of America

   The material contained in this volume was submitted as previously unpublished material, except in the instances in which credit has been given to the source from which some of the illustrative material was derived.
   Great care has been taken to maintain the accuracy of the information contained in the volume. However, Raven Press cannot be held responsible for errors or for any consequences arising from the use of the information contained herein.

Advances in Neurology Series

Vol. 39: Motor Control Mechanisms in Health and Disease
J.E. Desmedt, editor. 1224 pp., 1983.

Vol. 38: The Dementias
R. Mayeux and W.G. Rosen, editors. 288 pp., 1983.

Vol. 37: Experimental Therapeutics of Movement Disorders
S. Fahn, D.B. Calne, and I. Shoulson, editors. 339 pp., 1983.

Vol. 36: Human Motor Neuron Diseases
L.P. Rowland, editor. 592 pp., 1982.

Vol. 35: Gilles de la Tourette Syndrome
A.J. Friedhoff and T.N. Chase, editors. 478 pp., 1982.

Vol. 34: Status Epilepticus: Mechanisms of Brain Damage and Treatment
A.V. Delgado-Escueta, C.G. Wasterlain, D.M. Treiman, and R.J. Porter, editors. 551 pp., 1982

Vol. 33: Headache: Physiopathological and Clinical Concepts
M. Critchley, A. Friedman, S. Gorini, and F. Sicuteri, editors. 417 pp., 1982

Vol. 32: Clinical Applications of Evoked Potentials in Neurology
J. Courjon, F. Mauguiere, and M. Revol, editors. 592 pp., 1982.

Vol. 31: Demyelinating Diseases: Basic and Clinical Electrophysiology
S. Waxman and J. Murdoch Ritchie, editors. 544 pp., 1981.

Vol. 30: Diagnosis and Treatment of Brain Ischemia
A.L. Carney and E.M. Anderson, editors. 424 pp., 1981.

Vol. 29: Neurofibromatosis
V.M. Riccardi and J.J. Mulvihill, editors. 288 pp., 1981.

Vol. 28: Brain Edema
J. Cervós-Navarro and R. Ferszt, editors. 539 pp., 1980.

Vol. 27: Antiepileptic Drugs: Mechanisms of Action
G.H. Glaser, J.K. Penry, and D.M. Woodbury, editors. 728 pp., 1980.

Vol. 26: Cerebral Hypoxia and Its Consequences
S. Fahn, J.N. Davis, and L.P. Rowland, editors. 454 pp., 1979.

Vol. 25: Cerebrovascular Disorders and Stroke
M. Goldstein, L. Bolis, C. Fieschi, S. Gorini, and C.H. Millikan, editors. 412 pp., 1979.

Vol. 24: The Extrapyramidal System and Its Disorders
L.J. Poirier, T.L. Sourkes, and P. Bédard, editors. 552 pp., 1979.

Vol. 23: Huntington's Chorea
T.N. Chase, N.S. Wexler, and A. Barbeau, editors. 864 pp., 1979.

Vol. 22: Complications of Nervous System Trauma
R.A. Thompson and J.R. Green, editors. 454 pp., 1979.

ADVANCES IN NEUROLOGY SERIES

Vol. 21: The Inherited Ataxias: Biochemical, Viral, and Pathological Studies
R.A. Kark, R. Rosenberg, and L. Schut, editors. 450 pp., 1978.

Vol. 20: Pathology of Cerebrospinal Microcirculation
J. Cervós-Navarro, E. Betz, G. Ebhardt, R. Ferszt, and R. Wüllenweber, editors. 636 pp., 1978.

Vol. 19: Neurological Epidemiology: Principles and Clinical Applications
Bruce S. Schoenberg, editor. 672 pp., 1978.

Vol. 18: Hemi-Inattention and Hemisphere Specialization
E.A. Weinstein and R.P. Friedland, editors. 176 pp., 1977.

Vol. 17: Treatment of Neuromuscular Diseases
R.C. Griggs and R.T. Moxley, editors. 370 pp., 1977.

Vol. 16: Stroke
R.A. Thompson and J.R. Green, editors. 250 pp., 1977.

Vol. 15: Neoplasia in the Central Nervous System
R.A. Thompson and J.R. Green, editors. 393 pp., 1976.

Vol. 14: Dystonia
R. Eldridge and S. Fahn, editors. 509 pp., 1976.

Vol. 13: Current Reviews
W.J. Friedlander, editor. 400 pp., 1975.

Vol. 12: Physiology and Pathology of Dendrites
G.W. Kreutzberg, editor. 524 pp., 1975.

Vol. 11: Complex Partial Seizures and Their Treatment
J.K. Penry and D.D. Daly, editors. 486 pp., 1975.

Vol. 10: Primate Models of Neurological Disorders
B.S. Meldrum and C.D. Marsden, editors. 270 pp., 1975.

Vol. 9: Dopaminergic Mechanisms
D.B. Calne, T.N. Chase, and A. Barbeau, editors. 452 pp., 1975.

Vol. 8: Neurosurgical Management of the Epilepsies
D.P. Purpura, J.K. Penry, and R.D. Walter, editors. 369 pp., 1975.

Vol. 7: Current Reviews of Higher Nervous System Dysfunction
W.J. Friedlander, editor. 202 pp., 1975.

Vol. 6: Infectious Diseases of the Central Nervous System
R.A. Thompson and J.R. Green, editors. 401 pp., 1974.

Vol. 5: Second Canadian-American Conference on Parkinson's Disease
F. McDowell and A. Barbeau, editors. 525 pp., 1974.

Vol. 4: International Symposium on Pain
J.J. Bonica, editor. 858 pp., 1974.

Vol. 3: Progress in the Treatment of Parkinsonism
D.B. Calne, editor. 402 pp., 1973.

Vol. 2: The Treatment of Parkinsonism—The Role of DOPA Decarboxylase Inhibitors
M.D. Yahr, editor. 303 pp., 1973.

Vol. 1: Huntington's Chorea, 1872–1972.
A. Barbeau, T.N. Chase, and G.W. Paulson, editors. 901 pp., 1973.

Preface

This book presents an expanding new field of neurology in combination with physiology and explores the neural mechanisms underlying movement and posture control in man. This book provides a basic neurologic framework for thinking in a more structured manner regarding the variety of clinical manifestations that result from various types of motor commands dyscontrol in man.

Over the years, membrane phenomena in individual neurons have not always been meaningfully related to the brain mechanisms underlying normal and abnormal motor behavior. On the other hand, research on humans has for some time been hampered by the constraint that only noninvasive methods can be used. These difficulties recently have been circumvented in two ways. Progress in computer averaging and other electronic devices have brought within the range of analysis the small potential changes recorded in man at some distance from their actual neural generators. Moreover, the enhanced motivation for acquiring data and for testing ideas directly on patients (not only in animals) encouraged the design of new methods whereby a tremendous potential for understanding human motor control or dyscontrol has increased. Briefly put, instead of asking whether a set of animal results may be helpful for understanding some neurological issue, it is now possible in many cases to clarify the issues through direct research on man. Intelligent cooperation makes man a preferred subject for many studies of voluntary commands. As a matter of fact, the uncovering of neural mechanisms specific to man (even at spinal cord level) can bring new insight to the basic neurosciences.

This book is a collection of chapters by over sixty internationally recognized experts. Personal viewpoints of these key experts add strength to the discussions. Concurrent presentations of some debatable issues by researchers in other chapters also give the reader additional exposure to concepts from various viewpoints.

This book captures the major segment of current knowledge on neural mechanisms of normal motor control or motor dyscontrol in man. It conveys the excitement of charting a variety of clinical phenomena in more precise pathophysiological terms. In order to resolve difficulties that may arise in a rapidly expanding field, the book discusses the aims and relevance of feasible procedures and points to snags that must be guarded against. It provides practical guidelines regarding non-invasive procedures in man and seeks to resolve specific issues on a wide variety of problems.

The nervous system has an enormous capacity for adjusting motor behavior in the face of external challenges, and when its own operational capabilities are either distorted or focally disrupted through disease or metabolic insults. For example, during normal aging or at preclinical stages of a neurological disease, a person may present subtle behavioral failures that may go unnoticed for some time, but which later acquire retrospective significance. Quantitative analysis of such problems has become feasible and has become pertinent in athletic training and sports medicine as well as in the evaluation of patients in geriatrics and neurology, and in revalidation.

PREFACE

The increasing demands of our population for maintaining efficient bodily functions and optimal lifestyles continue to reinforce trends for preventive medicine and the identification of disorders at their early functional (rather than later pathological) stage.

Besides diagnostic uses in clinical neurology, the methods discussed in this book for assessing motor control mechanisms have a wide range of applications. These applications include monitoring dynamic changes in patients undergoing surgical or rehabilitation procedures (restorative neurology), rating residual capabilities after accidents with body damage (forensic medicine), quantitative screening of either therapeutic or unwanted effects of new drugs during their development and subsequent licensing (clinical pharmacology, pain research), and surveillance of persons exposed to environmental hazards or specific work overloads (occupational medicine, deep sea diving, space missions). The book delves into the specifics of many such applications of motor dyscontrol and anticipates practical problems that will be encountered. Motor control studies in normal man also elicit much interest in physiology and psychology, and students studying brain mechanisms also see the importance of their studies.

This book is an essential reference source for the neurologist in view of the current breakthrough in motor dyscontrol mechanisms in disease.

John E. Desmedt
Editor

Acknowledgment

I would like to thank the authors who contributed the many new results and who helped assemble into this single volume so much critical information that provides an up-to-date account of this exciting field.

Contents

Concepts and Issues in Motor Control

1 Program Disorders of Movement
 Robert J. Grimm

13 Visual Control of Reaching Movements in Man
 M. Jeannerod and C. Prablanc

31 Posture Control and Trajectory Formation in Single-or Multi-Joint Arm Movements
 Emilio Bizzi and William Abend

47 Regulatory Role of Proprioceptive Input in Motor Control of Phasic or Maintained Voluntary Contractions in Man
 Jerome Sanes and Edward V. Evarts

61 Sensory Motor Processing of Target Movements in Motor Cortex
 C. Ghez, D. Vicario, J.H. Martin, and H. Yumiya

Microneurography: Fusimotor and Spindle Afferents Activities

93 Muscle Afferent Function and Its Significance for Motor Control Mechanisms During Voluntary Movements in Cat, Monkey and Man
 A. Prochazka and M. Hulliger

133 Critical Examination of the Case for or Against Fusimotor Involvement in Disorders of Muscle Tone
 David Burke

Muscle Sense

151 Muscle Sense and Effort: Motor Commands and Judgments About Muscular Contractions
 D. Ian McCloskey, Simon Gandevia, Erica K. Potter, and James G. Colebatch

Muscle Fatigue

169 "Muscular Wisdom" that Minimizes Fatigue during Prolonged Effort in Man: Peak Rates of Motoneuron Discharge and Slowing of Discharge During Fatigue
 C.D. Marsden, J.C. Meadows, and P.A. Merton

Force Gradation and Motoneuron Recruitment

213 Functional Properties of Spinal Motoneurons and Gradation of Muscle Force
Daniel Kernell

227 Size Principle of Motoneuron Recruitment and the Calibration of Muscle Force and Speed in Man
John E. Desmedt

253 Cutaneous Facilitation of Large Motor Units and Motor Control of Human Fingers in Precision Grip
Kenro Kanda and John E. Desmedt

Connectivities of Motor Cortex

263 The Nature of the Afferent Pathways Conveying Short-Latency Inputs to Primate Motor Cortex
E.G. Jones

Functional Organization of Motor Cortex

287 Intracortical Organization of Arousal as a Model of Dynamic Neuronal Processes That May Involve a Set for Movements
Tomokazu Oshima

301 Functional Organization of the Motor Cortex
R. Porter

321 Input–Output Organization of the Primate Motor Cortex
Peter L. Strick and James B. Preston

329 Contrasting Properties of Pyramidal Tract Neurons Located in the Precentral or Postcentral Areas and of Corticorubral Neurons in the Behaving Monkey
Christoph Fromm

347 Separate Cortical Systems for Control of Joint Movement and Joint Stiffness: Reciprocal Activation and Coactivation of Antagonist Muscles
Donald R. Humphrey and Dwain J. Reed

373 Interaction Between Motor Commands and Somatosensory Afferents in the Control of Prehension
Allan M. Smith, Robert C. Frysinger, and Daniel Bourbonnais

387 Direct Electrical Stimulation of Corticospinal Pathways Through the Intact Scalp in Human Subjects
C.D. Marsden, P.A. Merton, and H.B. Morton

Premotor and Supplementary Motor Cortex

393 Supplementary Motor Area and Premotor Area of the Monkey Cerebral Cortex: Functional Organization and Activities of Single Neurons During Performance of a Learned Movement
Cobie Brinkman and Robert Porter

421 Functional Organization of the Supplementary Motor Area
Jun Tanji and Kiyoshi Kurata

Segmental Mechanisms in Motor Control

433 Reciprocal Ia Inhibitory Pathway in Normal Man and in Patients with Motor Disorders
Reisaku Tanaka

443 Functional Organization of Recurrent Inhibition in Man: Changes Preceding and Accompanying Voluntary Movements
E. Pierrot-Deseilligny, R. Katz, and H. Hultborn

459 Recurrent Inhibition of Motoneurons During the Silent Period in Man
Jun Kimura

Long-Latency Myotatic Responses and Transcortical Feedback Control

467 Segmental Versus Suprasegmental Contributions to the Long-Latency Stretch Responses in Man
Christina W.Y. Chan

489 Long-Latency Myotatic Reflexes in Man: Mechanisms, Functional Significance, and Changes in Patients with Parkinson's Disease or Hemiplegia
Robert G. Lee, John T. Murphy, and William G. Tatton

509 Long-Latency Automatic Responses to Muscle Stretch in Man: Origin and Function
C.D. Marsden, J.C. Rothwell, and B.L. Day

541 Dissociated Changes of Short- and Long-Latency Myotatic Responses Prior to a Brisk Voluntary Movement in Normals, in Karate Experts, and in Parkinsonian Patients
James A. Mortimer and David D. Webster

555 Muscular Contractions Elicited by Passive Shortening
Ronald W. Angel

Vestibulo-Ocular Control and Nystagmus

565 Neuronal Organization of the Premotor System Controlling Horizontal Conjugate Eye Movements and Vestibular Nystagmus
Hiroshi Shimazu

Balance Control and Postural Regulations

589 Dynamic Characteristics of Vestibular and Visual Control of Rapid Postural Adjustments
M. Lacour, P.P. Vidal, and C. Xerri

607 Analysis of Movement Control in Man Using the Movable Platform
Lewis M. Nashner

621 Tonic Labyrinthine Reflex Control of Limb Posture: Reexamination of the Classical Concept
Christina W.Y. Chan

633 Patterns and Mechanisms of Postural Instability in Patients with Cerebellar Lesions
Johannes Dichgans and Karl-Heinz Mauritz

645 Rapid Postural Reactions to Mechanical Displacement of the Hand in Man
C.D. Marsden, P.A. Merton, and H.B. Morton

661 The Jendrassik Maneuver: Quantitative Analysis of Reflex Reinforcement by Remote Voluntary Muscle Contraction
P.J. Delwaide and P. Toulouse

Vibration-Induced Inhibition and Excitation in Man

671 Mechanisms of Vibration-Induced Inhibition or Potentiation: Tonic Vibration Reflex and Vibration Paradox in Man
John E. Desmedt

685 Motor Dyscontrol as a Hazard in Massive Body Vibration in Man
Gabriel M. Gauthier, Jean Pierre Roll, Maurice Hugon, and Bernard Martin

Human Locomotion and Gait Disorders

699 Reflex Control of Bipedal Gait in Man
E. Pierrot-Deseilligny, C. Bergego, and L. Mazieres

717 Pathophysiological Aspects of Human Locomotion
B. Conrad, R. Benecke, J. Carnehl, J. Höhne, and H.M. Meinck

Brainstem Reflexes and Their Diagnostic Uses

727 Anatomical and Functional Organization of Reflexes Involving the Trigeminal System in Man: Jaw Reflex, Blink Reflex, Corneal Reflex, and Exteroceptive Suppression
B.W. Ongerboer de Visser

739 Human Jaw Reflexes
J.P. Lund, Y. Lamarre, G. Lavigne, and G. Duquet

757 Comparative Study of Corneal and Blink Reflex Latencies in Patients with Segmental or with Cerebral Lesions
B.W. Ongerboer de Visser

773 Clinical Uses of the Electrically Elicited Blink Reflex
Jun Kimura

Exteroceptive Reflexes

787 Cutaneomuscular (Flexor) Reflex Organization in Normal Man and in Patients with Motor Disorders
H.M. Meinck, R. Benecke, S. Küster, and B. Conrad

797 Exteroceptive Influences on Lower Limb Motoneurons in Man: Spinal and Supraspinal Contributions
P.J. Delwaide and P. Crenna

Pain Research and Nociceptive Reflexes in Man

809 Nociceptive Flexion Reflexes as a Tool for Pain Research in Man
Jean Claude Willer

Deep Sea Diving

829 Deep Sea Diving: Human Performance and Motor Control Under Hyperbaric Conditions with Inert Gas
Maurice Hugon, Laurent Fagni, and Kunihiro Seki

Central Movement Disorders

851 Pathophysiology of Dystonias
J.C. Rothwell, J.A. Obeso, B.L. Day, and C.D. Marsden

865 Clinical Neurophysiology of Muscle Jerks: Myoclonus, Chorea, and Tics
C.D. Marsden, J.A. Obeso, and J.C. Rothwell

883 Visuomotor Control of Leg Tracking in Patients with Parkinson's Disease or Chorea
Nobuo Yanagisawa, Sadakazu Fujimoto, and Reisaku Tanaka

889 Slow Visuomotor Tracking in Normal Man and in Patients with Cerebellar Ataxia
Hirokuni Beppu, Minami Suda, and Reisaku Tanaka

897 Motor Unit Control in Movement Disorders
Jack H. Petajan

907 Analysis of Abnormal Voluntary and Involuntary Movements with Surface Electromyography
Mark Hallett

Spinal Cord Lesions and Paraplegia

915 Motor Control in Man After Partial or Complete Spinal Cord Injury
M.R. Dimitrijevic, J. Faganel, D. Lehmkuhl, and A. Sherwood

Diagnostic Uses of Human Reflexes in Peripheral Neuropathies

927 The Use of Monosynaptic Reflex Responses in Man for Assessing the Different Types of Peripheral Neuropathies
P. Guiheneuc

951 H Reflexes in Muscles of the Lower and Upper Limbs in Man: Identification and Clinical Significance
J. Deschuytere, C. DeKeyser, M. Deschuttere, and N. Rosselle

961 F-Wave Determination in Nerve Conduction Studies
Jun Kimura

Motor Control Titration of Drug Effects in Clinical Pharmacology

977 Central Actions of Neurotropic Drugs Assessed by Reflex Studies in Man
P.J. Delwaide, J. Schoenen, and L. Burton

997 Quantification of the Effects of Muscle Relaxant Drugs in Man by Tonic Stretch Reflex
Manuel Meyer and Csaba Adorjáni

1013 Analysis of Gait and Isokinetic Movements for Evaluation of Antispastic Drugs or Physical Therapies
Evert Knutsson

1035 Electromyographic Analysis of Bicycling on an Ergometer for Evaluation of Spasticity of Lower Limbs in Man
R. Benecke, B. Conrad, H.M. Meinck, and J. Höhne

Neuroplasticity and Restorative Neurology

1047 Neurobionomics of Adaptive Plasticity: Integrating Sensorimotor Function with Environmental Demands
G. Melvill Jones and G. Mandl

1073 Axonal Sprouting and Recovery of Function After Brain Damage
Nakaakira Tsukahara and Fugio Murakami

1085 Rehabilitation Versus Passive Recovery of Motor Control Following Central Nervous System Lesions
Paul Bach-y-Rita

1093 Bioelectric Control of Powered Limbs for Amputees
R.B. Stein, D. Charles, and M. Walley

1109 *References*

1181 *Subject Index*

Contributors

Abend, W, 31–45
Adorjani, C, 997–1011
Angel, RW, 555–563
Bach-y-Rita, P, 1085–1092
Benecke, R, 717–726,
  787–796, 1035–1046
Beppu, H, 889–895
Bergego, C, 699–716
Bizzi, E, 31–45
Bourbonnais, D, 373–385
Brinkman, C, 393–420
Burke, D, 133–150
Burton, L, 992–996
Carnehl, J, 717–726
Chan, WY, 467–487, 621–632
Charles, D, 1093–1108
Colebatch, JG, 151–167
Conrad, B, 717–726,
  787–796, 1035–1046
Crenna, P, 797–807
Day, BL, 509–539, 851–863
DeKeyser, C, 951–960
Delwaide, PJ, 661–669,
  797–807, 977–996
Deschuttere, M, 951–960
Deschuytere, J, 951–960
Desmedt, JE, 227–251,
  253–261, 671–683
Dichgans, J, 633–643
Dimitrijevic, MR, 915–926
Duquet, G, 739–755
Evarts, V, 47–59
Faganel, J, 915–926
Fagni, L, 829–849
Fromm, C, 329–345
Frysinger, RC, 373–385
Fujimoto, S, 883–888
Gandevia, S, 151–167
Gauthier, M, 685–697
Ghez, C, 61–92
Grimm, RJ, 1–11
Guiheneuc, P, 927–949
Hallet, M, 907–914

Höhne, J, 717–726,
  1035–1046
Hugon, M, 685–697, 829–849
Hulliger, M, 93–132
Hultborn, H, 443–457
Humphrey, DR, 347–372
Jeannerod, M, 13–29
Jones, EG, 263–285
Jones, GM, 1047–1071
Kanda, K, 253–261
Katz, R, 443–457
Kernell, D, 213–226
Kimura, J, 459–465, 773–786,
  961–975
Knutsson, E, 1013–1034
Kurata, K, 421–431
Küster, S, 787–796
Lacour, M, 589–605
Lamarre, Y, 739–755
Lavigne, G, 739–755
Lee, RG, 489–508
Lehmkuhl, D, 915–926
Lund, JP, 739–755
Mandl, G, 1047–1071
Marsden, CD, 169–211,
  387–391, 509–539, 645–659,
  851–863, 865–881
Martin, B, 685–697
Martin, JH, 61–92
Mauritz, KH, 633–643
Mazieres, L, 699–716
McCloskey, DI, 151–167
Meadows, JC, 169–211
Meinck, HM, 717–726,
  787–796, 1035–1046
Merton, PA, 169–211,
  387–391, 645–659
Meyer, M, 997–1011
Mortimer, JA, 541–554
Morton, HB, 387–391,
  645–659
Murakami, F, 1073–1084
Murphy, JT, 489–508

Nashner, LM, 607–619
Obeso, JA, 851–863, 865–881
Ongerboer de Visser, BW,
  727–738, 757–772
Oshima, T, 287–300
Petajan, JH, 897–905
Pierrot-Deseilligny, E,
  443–457, 699–716
Porter, R, 301–319, 393–420
Potter, EK, 151–167
Prablanc, C, 13–29
Preston, JB, 321–327
Prochazka, A, 93–132
Reed, DJ, 347–372
Roll, JP, 685–697
Rosselle, N, 951–960
Rothwell, JC, 509–539,
  851–863, 865–881
Sanes, J, 47–59
Schoenen, J, 977–996
Seki, K, 829–849
Sherwood, A, 915–926
Shimazu, H, 565–588
Smith, AM, 373–385
Stein, RB, 1093–1108
Strick, PL, 321–327
Suda, M, 889–895
Tanaka, R, 433–441, 883–888,
  889–895
Tanji, J, 421–431
Tatton, WG, 489–508
Toulouse, P, 661–669
Tsukahara, N, 1073–1084
Vidal, PP, 589–605
Vivario, D, 61–92
Walley, M, 1093–1108
Webster, DD, 541–554
Willer, JC, 809–827
Xerri, C, 589–605
Yanagisawa, N, 883–888
Yumiya, H, 61–92

Abbreviations Used in This Book

AEP: Auditory Evoked Potential
AHP: Afterhyperpolarization of Neuron After the Action Potential
ALS: Amyotrophic Lateral Sclerosis
ATPase: Adenosine Triphosphatase
BP: Bereitschaftpotential That Precedes a Voluntary Movement
CFF: Critical Fusion Frequency
CNS: Central Nervous System
CRN: Cortico-Rubral Neuron
CT: Computerized Tomography
CV: Conduction Velocity
CVA: Cerebrovascular Accident
EEG: Electroencephalography
EMG: Electromyography
EPSP: Excitatory Postsynaptic Potential
FF: Fast Fatigable Muscle Fiber Type
FR: Fast Fatigue-Resistant Muscle Fiber Type
FSR: Functional Stretch Response
GBS: Guillain-Barré Syndrome (Polyradiculopathy)
HD: Huntington Disease
H-Reflex: Monosynaptic Reflex (Hoffmann) Elicited by a Single Electrical Stimulus to Low-Threshold Group I Afferents in a Peripheral Nerve
Hmax: H-Reflex of Maximum Amplitude (See Recruitment Curve)
HRP: Horseradish Peroxidase
Ia: Group I Afferent Nerve Fiber Innervating the Muscle Spindle
Ib: Group I Afferent Nerve Fiber Innervating the Golgi Tendon Organ
IPSP: Inhibitory Postsynaptic Potential
IS: Initial Segment Where the Axon Arises from Its Parent Neuron
LED: Light-Emitting Diode
MC: Motor Cortex
MN: Motoneuron
M-Response: Direct Electrical Response Elicited in Muscle by Electrical Stimulation of a Peripheral Nerve (At Shorter Latency Than the H-Reflex)

MS: Multiple Sclerosis
MSR: Monosynaptic Stretch Reflex
MSW: Meters of Sea Water (in Deep Sea Diving)
NS: Non Significant (in Statistical Study)
PM: Premotor Area of Cerebral Cortex
PT: Pyramidal Tract
PTN: Pyramidal Tract Neuron
PSTH: Post-Stimulus Time Histogram
rCBF: Regional Cerebral Blood Flow
RL: Resting Length of Muscle
RN: Red Nucleus
S: Slow Muscle Fiber Type
SBS: Spino-Bulbo-Spinal Reflex Pathway
SD: Soma-Dendrite Region of a Neuron
SEP: Somatosensory Evoked Potential
SMA: Supplementary Motor Area in the Cerebral Cortex
SMU: Single Motor Unit
SP: Silent Period Elicited by Nerve Stimulation in the EMG of an Active Muscle
SR: Shortening Response of Muscle
STA: Stike-Triggered Averaging
TSR: Tonic Stretch Reflex
TVR: Tonic Vibration Reflex
VEP: Visual Evoked Potential
VOR: Vestibulo-Ocular Reflex

Abbreviated Names of Muscles
Bi: Biceps Muscle
FCR: Flexor Carpi Radialis Muscle
FCU: Flexor Carpi Ulnaris Muscle
FDI: First Dorsal Interosseous Muscle
G: Gastrocnemius Muscle
GS: Gastrocnemius-Soleus Muscle
PL: Palmaris Longus Muscle
Q: Quadriceps Muscle
S or Sol: Soleus Muscle
TA: Tibialis Anterior Muscle
TFL: Tensor Fasciae Latae Muscle

Program Disorders of Movement

*Robert J. Grimm

Neurological Sciences Institute, Good Samaritan Hospital and Medical Center, Portland, Oregon 97210

Recent information on the genesis and control of human movement calls for a new neurology with fundamental changes in our thinking about disorders of movement, their classification, and how they are to be studied and treated.

Nosology

As suggested elsewhere (Grimm and Nashner, 1978) it is useful to divide movement disorders into two groups (Table 1). Those represented by structural alterations above the level of motor units and beyond the central terminations of primary afferents are termed program disorders. The peripheral machinery for movement is normal, but a faulty performance is given as a consequence of instructional errors. Where the central nervous system (CNS) is normal, with deficits existing in the motor apparatus, they are called *nonprogram disorders*. Movements possess a muscle structure or assemblage of muscles actually deployed. Movements are also characterized by certain measures, or metrics, as velocity, force, acceleration, amplitude, frequency, electromyography (EMG) characteristics, etc. Disordered movements can thus be characterized by their structure and metrics during the performance of a task.

Elucidation of the structure and metrics of a disordered movement is a prerequisite to understanding and therapeutic advance.

Phases

In Figure 1, a curve illustrates the development of a program disorder such as Parkinson's disease, cerebellar ataxia, or spasticity of multiple sclerosis (MS) origin. Early in the initial phase (I) there are no symptoms. As the disorder worsens, a vague uncertainty presides to make the unsuspecting victim forego or not change previously easy tasks, such as jumping over puddles or walking down a stairway without holding on. Ordinary movements such as turning over in bed are met with difficulty (as in Parkinson's disease, for example). Fatigue, a sedentary life, or aging may be offered as excuses. It is as if there exists an internal surveillance mechanism, possibly at the level of efference copy, which apprehends, evaluates, and vetoes performances where a postural setup prerequisite to success has become unreliable.

Symptoms (phase II) are eventually recognized, especially when they persist; signs (phase III) of the disease follow. If there is a phase III, phase II is obtained by history and phase I, possibly, by reflection.

In phase I, the disease is initially concealed by the inherent "noise" of a normally distributed task performance in a population of heterogeneous genetic makeup, age,

---

*2311 N.W. Northrup (Suite 202), Portland, Oregon 97210.

TABLE 1. *Program and nonprogram movement disorders*

Program disorders	Nonprogram disorders
Huntington's chorea	Amputations
Stroke	Painful extremities
Subacute cerebellar degeneration	Sprains
	Myopathies
Wilson's disease	Myasthenia gravis
Hydrocephalus	Neuropathies
Cerebral palsy	Botulism
Parkinson's disease	Tick paralysis
Progressive supranuclear palsy	Spinal muscle atrophy
Multiple sclerosis	Periodic paralysis
Epileptic fits	Guillain-Barré syndrome
Myoclonus	
Tardive dyskinesia	
Blindness	
Hysteria	
Labyrinthitis[a]	
Malingering	

[a] Labyrinthitis is placed here on grounds that this sensory disturbance alters coactivation.

drafted for program disorders of movement; by appropriate scaling techniques, the slope of the process can be given.

Finally, assume the existence of an inherent redundancy in the motor system performance, whereby losses within a single system between interacting systems must be substantial before real trouble is noticed. We speak here not of a single and strategically placed lenticulostriate capsular infarct, but, rather, of a more diffuse or multisite phase I process that exists in the brain prior to recognition of the fact. Unexpected timing errors in an Olympic gymnast, if repeated, would command serious and immediate attention. However, subtle changes in motor performance of such an athlete in the routines of daily life might proceed a great distance before difficulties were appreciated.

The study of system redundancy thus emphasizes the circumstance and context of performance. As in renal failure, 85% of functioning glomeruli can be lost before blood urea nitrogen (BUN) level rises. In the case of both renal and CNS disease, unscheduled fatigue may be an early signature of trouble. We intercept central disorders of movement late, often after the inherent redundancy of interacting systems has fallen to the point where performance in some way becomes awkward and inefficient. When redundancy in performance falls below some previously set criterion, performance variability increases with uncertainty, hesitation, anxiety, or fear, in its wake. Where the risk of falling is possible, as evaluated by this unconscious surveillance, the task may be cancelled.

experience, and will. Symptoms intensify, interfere with peace of mind, are recognized, and eventually lead to medical consultation. Early in phase III, the individual's worst fears are borne out as signs appear. Assume a serial progression from one phase to another, however steady, remitting or sporadic. These phases can be

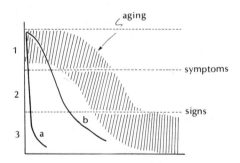

FIG. 1. Development of a program disorder of movement. Phase 1; no symptoms; phase 2; symptoms; phase 3; signs. The *shaded curve* depicts aging effects on competence. **a**: Acute onset disorder; **b**: chronic progressive disorder. *Abscissa:* Variable time course. *Ordinate:* Functional measures of phases; the width of shading represents redundancy.

Symptoms are what we experience and recall. Signs are physical defects, such as a short leg, abnormalities ordinarily beyond voluntary control, such as an extensor plantar response, reflex deficits in the pupil or involuntary errors in speech, swallowing or balance as detected by a trained eye. Where signs are absent, symptoms may be dismissed. All of us from time to time have awkward movements.

If there is merit in detecting trouble early, we must become deadly serious about the study of symptoms despite the inherent difficulties involved. When signs emerge, it can be argued that limited cycle stability of interacting systems is lost and that a system has moved to a more stable, however dysfunctional, position. From this point of view, decline in function is a relatively late event. The decline has exceeded some set point of performance control and could well be characterized by the catastrophe theory (René Thom, 1975). Longitudinal studies of families with spinocerebellar degeneration, studies of patients with a high probability of developing iatrogenic movement disorders, e.g., phenothiazine-induced tardive dyskinesia, or of patients with retrobulbar neuritis carrying a tissue haplotype of HLA B7 and DW2 antigens (a high probability for MS) are potential starting points for examining the ontogenesis of a movement disorder in its earliest phase.

Assumptions and Axioms

Students of movement disorders rely on certain ideas about the organization of the nervous system in order to understand and account for its decline when pressed by disease. The study of movement in normal and impaired subjects currently proceeds by three assumptions. The first is that movements are composed of a myriad of irreducible subsets of motor units which, by structure and metric design, are of rigid or fixed inconstancy in any motor performance regardless of complexity. It is the manipulation of such elemental units into various patterns that produces complexity.

The second assumption is that the principle of a hierarchically organized motor system, developed out of a generation of earlier studies on invertebrates, fish, amphibians, and birds by zoologists and ethologists can be applied to humans. The assumption comes in two parts: Complex movements are constructed from elemental units; the complexity itself is limited by certain inflexibilities, e.g., fixed ratios between limb movements regardless of task. Second, movements can be ordered in a hierarchy from elemental twitches and muscle stretch reflexes to complex movements of cortical design and control. Sophistication in current CNS theory dictates the existence of complex interactions or loops between supernuclear structures and spinal cord gray at Rexed layer IX, the site of motor neuron subsets. The prevailing and functional axis runs between cortex and cord. The hierarchical model provides a powerful conceptual tool for the analysis of motor control and dyscontrol. Whether or not it is correct is another matter as will be considered later.

Third, the hierarchical design of the CNS has, at all levels, and in all circumstances, a fundamental priority: Before voluntary movement of any kind, stabilization of posture comes first.

Axioms

Movements portray a hierarchically organized and interactive function in the nervous system which can be viewed as a program or performance; faces of the same phenomena.

Complexity implies a cortex-to-spinal cord axis with programs emerging from transactions between different levels of the CNS. Integrations are summarized in brainstem and spinal cord.

A program for a specific movement can be characterized by its muscle structure and metrics.

For convenience and by using common sense, human movements can be designated as either voluntary or involuntary, based on whether such performances are governed by a preceding conscious act of will. This distinction is artificial, however, and should not be trusted.

Performances summarize CNS interactions as expressed by a motor apparatus consisting of first-order elements, receptors, and skelctal-muscle linkages.

Performances are a normally distributed function of age, genetics, experience, training, will, and health as constrained by physical environment, cultural norms and expectations. Performance context is a crucial variable.

Even with practice, no two motor performances are identical.

The motor system is so arranged that posture and axial stabilization precede voluntary movement.

No sharp dividing line separates normal from abnormal performances.

Compensation for CNS or peripheral injury with normal or near-normal performance is a measure of redundancy. As expressed by learning, adaptation, repertoire, flexibility, execution, and accuracy under various circumstances, redundancy is a measure of health.

Motor dyscontrol is the reciprocal of redundancy.

Redundancy is the reciprocal of fatigue.

Fatigue is a function of unreliability, inefficiency, redundancy losses, and in certain cases, a defective machinery.

Program disorders of movement are those imposed on a normal motor apparatus; nonprogram disorders arise from a disabled peripheral motor apparatus.

Performance by recall implies the existence of mechanisms for storage, assemblage, execution, modification, etc. Where such hypothetical components are defective, they lead to higher order deficits which present as behavioral disorganization.

Neurological signs are generally unreliable in their ability to proscribe or predict performance changes possible with therapy.

In general, neurological signs arise relatively late in the course of the disease.

Similar motor deficits, e.g., ataxia can have strikingly different program errors of structure and metrics.

In program disorders, fatigue rises in proportion to the degree of attention given to supervise the performance of a previously semiautomatic subroutine, e.g., walking. Fatigue declines with learning, therapeutic repair of program errors or nonprogram elements, or by intrinsic recovery mechanisms, e.g., sprouting.

Therapy for dyscontrol syndromes begins with analysis.

Parkinson's Disease

In idiopathic Parkinson's disease, we encounter a set of discrete errors of movement, which are progressively admixed to produce a dyscontrol picture unique to each patient. In its late stages, diminished speech volume, bradykinesia, long latencies in response or movement changes, a resting tremor, involuntary rigidity imposed on axial, facial, and limb musculature, and superimposed age-induced and disuse changes in muscles and articular connections all conspire to produce a classical picture of a stooped, shuffling, whispering sufferer with poor balance, delays in response to questions, and various cognitive defects. Based on a now well-recognized neurochemical pathology of the nigrostriatal striatonigral system, patients with Parkinson's disease are now being treated with L-DOPA, dopaminergic agents, anticholinesterases, or catechol analogs on the theory that as in insulin-dependent diabetes, replacement of substances critical to, but missing from, the operation of normal movement, will alleviate symptoms or signs. Bradykinesia and rigidity are usually remarkably improved; tremor less so. Although such treatment improves the quality of life for many, it does not alter the course of the disease. It is ineffective in some patients and can have serious side effects. There is a curious "on-off" feature of patients with Parkinson's disease under drug treatment, in which there is an alarming and swift development of a virtually complete freezing of motion lasting for many minutes or longer despite relatively constant drug levels. In some ways, this is the reverse of the curious *kinesia paradoxica* phenomena observed in the pre-L-DOPA era. Inert, bedridden, and mute parkinsonism patients, when exposed to great stress, such as a fire or being caught on a bedpan by a student nurse, may produce prompt, facile, and organized movements, such as jumping out of bed and running out of a burning building or shouting "Get out of here and close the door!" It is as if strong emotion overrides the block to permit performance to occur, and one which, however brief, may illustrate relatively normal metrics and structure, before the reassertion of control by the disease state. A comparison between motor performances during periods of kinesia paradoxica and those dominated by parkinsonism signs would be instructive. It is a common observation that if a ball is unexpectedly tossed to a bradykinetic parkinsonism patient with the command "catch," it is properly caught. An experiment designed to study how the patient with Parkinson's disease carries out this "mini" kinesia paradoxica task would be useful.

The nature of the program disorder in parkinsonism continues to attract attention despite a current emphasis on therapeutics. Various technical approaches have provided new insights about the nature of this disease. Flowers (1976, 1978) examined upper-limb tracking responses of patients with Parkinson's disease under various target conditions. He demonstrated that patients lose their ability to produce open-loop ballistic-type movements and are thus left with a clumsiness imposed by slowness and the jitter of an impaired motor control system limited to a discrete, serial-step paradigm. Cognitive defects of attention, shifts in set, and rigidity analogous to the movement aspects also occur in Parkinson's disease patients and are concomitants of this process (R. Rafal, *personal communication*). Such insights underscore earlier work illustrative of defective program organization, e.g., sequencing inability to change or correct programs (Angel et al, 1970) and failure at simultaneous performances.

Recently, Hallett and Khoshbin (1980) using a torque motor and EMG setup, required parkinsonism patients to make various limb movements and then subjected them to different loads and perturbations. The patients were unable to modulate burst duration of ramp and ballistic motions, movements of a fundamentally different character. Rather, the patient with Parkinson's disease deploys a series of bursts in proportion to movement length, a finding in concert with Flowers' observation on tracking. Hallett and Khoshbin (1980) also demonstrated that rigidity and bradykinesia arise from different mechanisms and that parkinsonism slowness is not due to the cocontraction of agonist-antagonist pairs.

Rigidity in the limb afflicted with Parkinson's disease, independent of gain changes in muscle stretch reflex, has been the subject of studies using various approaches. Using microneuronography, Hagbarth et al (1975c) and Wallin and Hagbarth (1978) found aberrations in fusimotor control of spindle afferents in the Parkinson-affected limb. Errors are also demonstrable in motor unit recruitment (Grimby and Hannerz, 1974; Milner-Brown et al, 1979), and there exists an abnormally large amplitude, long-latency reflex in the upper limb of the patient with Parkinson's disease (Lee and Tatton, 1978), a potential contributor to limb rigidity. Under more explicit conditions, Mortimer and Webster (1978) demonstrated a degree of control over this reflex in Parkinson's disease patients. In preliminary platform studies of patients with Parkinson's disease, the metrics and structure of long-latency postural reflex responses in the legs are, with the exception of minor metrical changes, essentially normal (L. W. Nashner, *personal communication*).

The question can be asked whether or not bradykinesia is the outcome of a system preoccupied with postural control either because programs for open-loop control are abolished, or, if they do exist, are either untrustworthy or available only in specific circumstances, e.g., kinesia paradoxica. Although no one would hold that patients with Parkinson's disease are clumsy and slow by choice, if there exists a subconscious mechanism which vetoes risk-taking, e.g., open-loop movements, its pathological enhancement could effect the same functional end. Table 2 summarizes program errors now recognized in patients with Parkinson's disease.

Spasticity

Spasticity remains a dreary outcome of spinal cord insults from progressive MS lesions or dysfunction. In the legs it produces a stiff, fatiguing, weak, unstable gait carried on exaggerated muscle stretch reflexes. Such reflex changes are set up by Ia afferent capture of motoneuron membrane denervated by long tract lesions. Stretch-triggered clonus, alterations in cutaneous reflexes (Babinski response) and velocity-dependent muscle-stretch responses to passive displacement can be demonstrated at the bedside. Diminished movement speed, balance difficulties, fatigue, muscle cramps, and lack of bladder control provide a demoralizing clinical background to this condition often called spastic paraparesis.

Figure 2 illustrates platform responses of a free-standing, normal subject to provide here a background for findings to be described in cord-spastic patients. If a controlled platform is abruptly rotated downward through the axis of the ankle joint, it produces a brief stretch of tibialis anterior

TABLE 2. *Recognized program disorders in patients with Parkinson's disease*

Bradykinesia, rigidity, cogwheeling, and intention tremor
Suppression of tremor with voluntary action
Loss of rigidity with sleep; enhancement of rigidity with arousal
Inability to modulate myotatic reflex gain
Loss of fusimotor muscle spindle control
Motor unit recruitment pattern errors
Inability to modulate EMG burst durations
Relatively normal metrics and structure of postural responses; small errors in timing and amplitude modulation of long-latency stretch reflexes
Cognitive errors of attention and performance analogous to bradykinesia and rigidity; subcortical dementia
Deficits in sequencing, modification, and simultaneity of action

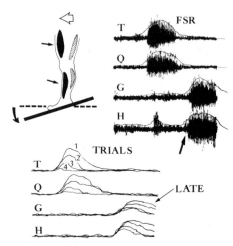

FIG. 2. Platform paradigm illustrating the metrics and structure of a postural response to a sudden forward rotation (6°/sec) stretching tibialis anterior, drawing the individual forward. EMG tracings of the response of tibialis (T), quadriceps (Q), gastroenemius (G), hamstring (H) to this maneuver. FSR, functional stretch response (long-latency stretch response) at 100 msec; a structural response (simultaneous onset T, Q) with indicated metrics of latency amplitude with later compensatory (vertibular) response structure in G, H at 200 msec *(arrow)*. Lower traces reveal serial response to 4 repeated platform rotations invoking adaptation (suppression) of the long-latency stretch response.

muscle. At about 100 msec, there is a simultaneous contraction of both the tibialis (T) and quadriceps (Q) constituting a structured response termed the functional stretch reflex (FSR) in the past, but here referred to simply as a long-latency response, a more neutral term. The body is then inappropriately pitched forward, triggering at 200 msec a compensating response of vestibular origin in the gastrocnemius (G) and hamstrings (H) muscle, restoring balance. Latencies, amplitude, and duration of the integrated EMG responses provide the metrics of this response structure. At the bottom of Fig. 2, line tracings of the long-latency stretch response show that in subsequent trials, this response is serially attenuated (adaptation) as the subject has learned to control body sway in this novel circumstance. Ordinarily, in normal persons myotatic leg reflexes are not perturbed by slow stretches of 6° per second. The absence of the myotatic reflex, a vigorous and prompt long-latency stretch response capable of adaptation when appropriate, and late compensatory responses (vestibular) are normal for this platform paradigm.

In a series of patients with cord spasticity, the platform has been used to elucidate quantitative characteristics of spasticity. Table 3 summarizes our findings in 8 patients, the majority with clinically definite MS with gait and neurological dysfunction (Kurtzke scale scores of 4 to 6). Such patients give platform-derived responses which differ considerably from normals. Myotatic stretch reflexes are of prominent amplitude, with asymmetries of structure and amplitude in the long-latency responses obtained from the legs. Adaptation is normal, but switching responses (not illustrated) are clearly abnormal. By "switching" we mean a situation in which a slit platform is used: One platform is suddenly lowered and the other elevated, forcing a quick change in postural strategy. Spastic subjects are not able to change such structural synergies rapidly due to the fact that they possess an enhanced degree of coactivation of antagonists that is not seen in normal subjects. The structure of the late-occurring compensatory response is normal, but the onset of this response in spastics began at variable times and at longer latencies.

Several years ago, baclofen, a fluorinated analog of gamma aminobutyric acid (GABA)—an inhibitory transmitter present in cord gray matter—was introduced in the belief that gait would improve by diminishing spasticity. The erstwhile hope of separating the effects of spasticity from muscle strength has led all neurologists astray at one time or another. In general, as spas-

TABLE 3. *Platform studies of spasticity*[a]

Stretch reflex (MSR)	Present[b]; amplitude increased
Long-latency stretch response structure	Abnormal; amplitude decreased
Long-latency stretch response metrics	Abnormal; latency increased[c]
Adaptation	Normal
Switching	Abnormal: prominent coactivation of antagonists[d]
Vestibular structure	Normal
Vestibular metrics	Abnormal: latency often increased

[a] From Grimm, Nashner, and Wollacott, *unpublished studies* on 8 patients. Responses of free-standing, spastic subjects on a controlled, split platform subjected to unexpected perturbations to elicit long-latency response structure and metrics.
[b] Usually absent in normal subjects.
[c] Functional stretch reflex (FSR); see text.
[d] Not illustrated: Smooth, complex shifts in structure as one leg on a split platform is rapidly lowered, the other elevated *viz* "switching."

ticity is drug-suppressed, there is, unfortunately, a corresponding reduction in strength. However, in two of our MS patients with similar problems, baclofen produced opposite effects. Patient RF walked better and reported that the drug was helping; BK became more unsteady, reported increased weakness, and stopped taking the drug.

As illustrated in Figure 3, functional difficulties appeared identical, but were the outcome of somewhat different program errors. BK presented with a prominent myotatic and long-latency response but a compensatory response of diminished amplitude. Baclofen abolished the myotatic reflex, depressed the amplitude of the long-latency response, and augmented the amplitude of the compensatory response. It effectively destroyed her "back-up" system, e.g., the enhanced myotatic reflex structure, suppressed her impaired long-latency reflex mechanism, and so amplified the late response that her balance became too poor to walk.

In patient RF, with initially prominent myotatic reflexes and long-latency responses, the addition of baclofen similarly abolished the myotatic reflex but only diminished the long-latency response amplitude. Similar to BK, the drug augmented the late-occurring compensatory response. The outcome here was that the loss of stiffness in his limbs did not progress to the point of destabilization, as long-latency reflex responses survived. Abolition of myotatic reflexes (as tested by platform stretches) resulted in a subjective improvement of gait and a lessening of cramping. The principal difference between the responses of BK and RF to baclofen was that the drug did not compromise the long-latency stretch response structure in RF, but did remove the myotatic reflex influence. Platform study (Table 4) of spasticity in free-standing subjects opens this dyscontrol program to fundamental analysis as well as providing a technique for quantifying and differentiating responses to therapy.

Cerebellar Ataxia

In a previous study (Nashner and Grimm, 1978), platform analysis of cerebellar ataxia provided a powerful insight into the nature of this classical program disorder. A complete disorganization of both the structural and metrical descriptions of movement was found, a counterpart of venerable clinical observations of difficulty with coordination, timing, and gait. Moreover, two unexpected findings emerged in that study. The

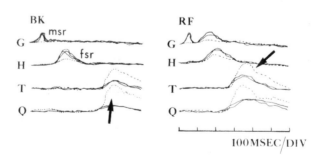

FIG. 3. Comparison of two similar cord-spastic patients, BK and RF, to baclofen. Rectified EMG traces: G, gastrocnemius; H, hamstring; T, tibialis anterior; and Q, quadriceps. *Line traces:* Controls (before baclofen); *dotted lines:* baclofen. msr, monosynaptic stretch reflex; fsr, functional stretch response (long-latency stretch responses). *Arrows* indicate augmentation of long-latency compensatory responses by baclofen.

TABLE 4. *Platform studies of the effect of physostigmine on cerebral ataxia*[a]

	Physostigmine	
Response	Before	After
Myotatic stretch reflex	Absent	Absent
Long-latency stretch response structure	Abnormal	No change
Long-latency stretch response metrics	Abnormal	No change
Adaptation	Abnormal	No change
Switching	Abnormal	No change
Vestibular structure	Normal	Normal
Vestibular metrics	Abnormal	Latency decreased[b]

[a] From Grimm, Nashner, and Wollacott, *unpublished studies* on 5 patients. (Physostigmine 2 mg, t.i.d.)

[b] The only significant change was a striking 50 to 75% decrease of the latency of the abnormally late compensatory response (vestibular) after the drug was given.

first was that cerebellar patients were unable to inhibit their long-latency stretch responses to inappropriate movements. Also unexpected was the curious finding of a long delay in the occurrence of the compensatory vestibular response. Normally, such responses occur between 200 and 300 msec, but in the cerebellar patients, they occurred at 800 msec or more. Thus, there appeared to be not only the loss of long-latency response adaptation, but when falls were induced, the hapless victim could not make a response quickly enough to prevent a stagger or fall.

Others observed that the anticholinesterase physostigmine improved the gait in familial spinocerebellar degeneration. A subsequent platform pilot study of cerebellar patients and their responses to physostigmine was carried out. Findings in this study were of considerable interest. First, it appeared that to one degree or another all patients on physostigmine improved in gait manifested by less ataxia, better turns, and gave subjective reports of improved balance. Platform studies confirmed this. Surprisingly, as shown in Table 5, a single and obvious parameter change was detected as a consequence of the drug. A striking reduction occurs in the latency of the compensatory response to vestibular origin. Reductions averaged between 50 and 75% of predrug latencies (each patient serving as his or her own control). Tracings (Fig. 4) are made of one such patient with reduction from a previous vestibular response latency of 1,000 msec to a value of about 450 msec, sufficient to preclude platform vestibular falls which occurred prior to the use of the drug.

TABLE 5. *Ataxia and late compensatory platform response to physostigmine (N = 5)*

		Compensatory (vestibular) response (msec)	
	Ataxia[a]	Before drug	After drug
Patient A	−5	Falls[b]	No falls (275)
Patient B	−2	Falls[b]	No falls (300)
Patient C	3	Falls (1,000)	No falls (450)
Patient D	−5	Falls (600)	No falls (200)
Patient E	−5	Falls (400)	No falls (300)

Blindfolded, free-standing patients were balanced on the platform (A–P sway, servomode) and asked not to fall.

[a] Ataxia scale (Nashner and Grimm, 1978). Scale: 0—normal, 9—bedridden; normal latency compensatory response 190 to 250 msec. After controls, patients placed on drug (2 mg p.o. t.i.d.) for 2 weeks, before being remeasured. Studies obtained 1 to 2 h after last dose.

[b] Unmeasurable.

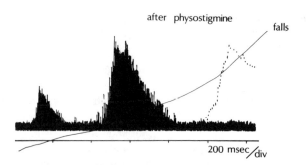

FIG. 4. Physostigmine effect on cerebellar ataxia. Initial discharge is the long-latency stretch reflex; the second response is the compensatory (vestibular) response. The *dotted* response is the compensatory response at 1,000 msec before physostigmine. After physostigmine, the compensatory response is reduced in latency to 550 msec. The *fine diagonal line* crossing the illustration indicates body attitude (and fall) before physostigmine.

The use of the platform in this situation served as a technique to study the effect of a drug on cerebellar ataxia. It revealed that a drug capable of crossing the blood-brain barrier and one with an acceptable risk-benefit ratio could have a specific effect on a single program metric. The fact that one portion of a faulty program could be improved not only gives promise to the therapeutics of cerebellar ataxia, but indirectly illustrates that the various components of structure and metrics may be under independent control. Although the mechanism by which physostigmine achieves this effect is conjectural, such studies emphasize that a functional analysis of a movement disorder has a place in the evolution of therapeutics.

Future Work

In practice, the study of program disorders of movement begins with the realization that most writing on the subject is obsolete. Views of movement disorders are principally derived from lesion studies on mammals, subhuman primates, or in humans whose problems (when studied at autopsy) are dominated by focal lesion models, unrelieved by serious reflection on the significance of new data on interacting systems. Information on movement dyscontrol is scattered, seldom verified, and focused on poorly characterized systems. If accumulated, data would fall into what Lide called class B data, akin to environmental data that is dependent on time and place and cannot be easily replicated. We must do better than this.

In the new neurology, we need a fresh, precise, and powerful nosology so designed to accommodate new insights and language from the study of program errors, redundancy, adaptation, learning and cognitive process as well as the nature, principles, and axioms that proscribe the structures and metrics of movement. Finally, there is an urgent need to overhaul completely our theoretical constructs about the organization of human movement so as not to waste our time with simplistic notions. Small, coherent, and fixed elemental units of movement, which are built up into increasingly complex assemblies does satisfy a hierarchical model and preserves a historical and biological relationship of human physiology to other vertebrates. Preservation, however, comes at a price. Earlier in this chapter, a pragmatic reliance on the hierarchical model provided a framework for thinking about motor control. There is no problem with the notion of hierarchy however designed, or its heuristic utility. The problem lies in its definition, its relevance, and its premise. Those who use it must simply keep these issues in mind. In the last analysis, such models leave us with an advanced robot with movement capabilities similar to humans. There is no assurance whatsoever that such assumptions (brilliantly developed by zoologists, ethologists, and physiologists), lately deployed for analytical purposes in humans, have much to do with the organization underlying the noisy, comprehensive, and facile flexibilities of human movement. Humans and, certainly, the great apes, demonstrate an effortless ability to shift control between voluntary and involuntary motions, are dominated by learning and adaptation, and possess an extraordinary repertoire of performance, the organizational principles of which are but dimly understood.

There is little documentation for this assumption that human movement is composed of complex montages of elemental motor units selected, orchestrated, and smoothed in such a way to produce versatile performances of precision, predictability, and error self-correction. It is true that humans can be trained to perform simple movements, the analysis of which provides data in line with this interpretation. In the strict sense, however, is the hierarchy and elemental movement composition idea correct?

There is no quarrel with the tautology that at final levels of resolution hard-wired motor units represent an irreducible element of performance. But how rigid is the composition of structure or the metrics of a learned performance? There is no assurance whatsoever from contemporary data, except in the most rigidly prescribed, stereotyped context that postural or manipulandum tasks adequately characterize human motor activity or give evidence for an elemental movement construct. Where studies of performance variability are undertaken, a good deal of trial-to-trial variation is seen in both metrics and structure, but the task is accomplished. One may reach a button at all sorts of odd angles and meet various difficulties in getting there.

What is needed are alternatives to the old but convenient hierarchical, elemental unit model. Out of the inherent, nonlinear aspects of human performance from interactive and noisy components, we can get a relatively stable, predictable, and analyzable function to satisfy the dictates of laboratory study and quantitative science. But linearization in the construction of human performance in health or disease probably gives the least significant insight about how the motor control system is organized. A new neurology must soon address these issues.

Visual Control of Reaching Movements in Man

*M. Jeannerod and C. Prablanc

Laboratoire de Neuropsychologie Expérimentale, INSERM U-94, Bron, France

Different aspects of normal goal-directed movements can be controlled more or less directly by visual input. In most cases, the goal of the movement is defined by visual parameters, such as its location in visual space, or its shape or size. These parameters are the basis for building up the motor program that determines the appropriate muscular commands. Second, the execution of the movement itself, as characterized by its velocity profile, its duration, and its terminal precision can also be subjected to direct visual control intervening during the trajectory.

Current theories have put different degrees of emphasis on central or peripheral aspects of motor control (see Kelso and Stelmach, 1976). In fact, none of these concepts alone could account for the experimental results obtained with various types of movements. Delays in visual feedback make the amount of visual control different for movements of different durations. In the experiments by Keele and Posner (1968), subjects were required to hit targets in a paced movement procedure, with or without visual feedback. For movements lasting less than about 200 msec, presence or absence of visual feedback was without influence on subjects' accuracy, although movements of longer duration were improved by vision. Thus, it takes 200 to 250 msec to process the feedback, and movements lasting less than 200 to 250 msec, as well as the first 200 msec of any movement, should be considered as ballistic events. However, the term ballistic appears rather misleading when used in the restricted sense that merely postulates the lack of a continuous visual control of the (brisk) movement. Ideally, ballistic means that a single initial motor command "package" can release the full extent of the movement (e.g., Keele, 1968; Desmedt and Godaux, 1978a). This mode of functioning is probably of limited importance when natural movements requiring a high degree of precision are considered. Furthermore, the intrinsic visuomotor delays only excludes the use of the visual feedback loop under certain time conditions, but does not preclude visual control of the movement exerted in other ways.

One possibility would be that initial pickup of visual information at movement onset exerts an open-loop visual control on its execution. The initial information could be stored as an internal visual "map" of the target to be reached, and used to direct the limb at the proper location via a feedforward mechanism. Such a regulation would clearly depart from the classical conception of ballistic movements, where visual control can be exerted only by a new

---

*To whom correspondence should be addressed: Laboratorie de Neuropsychologie Expérimentale, INSERM Unité no. 94, 16 Avenue du Doyen Lépine, 69500 Bron, France.

pickup of external information. Open-loop visual control would bypass input stages of visual information processing, and thus ensure a faster control of the movement parameters (Vince and Welford, 1967). Open-loop visual control of reaching movements could also be exerted indirectly by cues based on visual fixation of the target. In that case, the feed-forward mechanism would be derived not from the visual map, but from a central monitoring of the gaze axis position in the orbit, itself related to the target position in space (see Jeannerod and Prablanc, 1978; Jeannerod et al., 1980).

These different hypotheses are not mutually exclusive. Furthermore, these postulated modes of visual control are assumed to coexist with the other, widely admitted, mechanism by which vision can influence movements, i.e., closed-loop control by reafferent feedback. A number of authors (e.g., Megaw, 1974; Glencross, 1977) have postulated a two-level organization of visuomotor mechanisms. A higher level of motor programming would determine the main open-loop features of the movement, while a lower level operating under external feedback would allow a modulation of the duration and frequency of response of the motor units recruited by the program. In the experiment by Megaw (1974), subjects had to perform rapid movements of their unseen arm toward a visual target. The target was displaced by a few degrees with a variable delay after the onset of the arm movement. When the delay was below 100 msec, the program of the ongoing movement could be modified and the subject was accurate. This indicates that an external feedback giving an updated information on goal position may improve or refresh the open-loop control system.

Our own conception, based on the study of reaching movements, also implies several levels for visuomotor control. Schematically, the higher level (level 1) corresponds to a central programming of the movement based on visual input. An intermediate level (level 2) is represented by a center feed-forward modulation of the ongoing program. This stage corresponds to the open-loop visual control. It first implies that visual spatial cues about target location (e.g., a stored visual "map" and/or oculomotor signals) will be available; but it also requires the availability of informations about limb position with respect to the body prior to the movement. This information is assumed to match the visual map, which is also body-centered. Finally, a third level (level 3) involves detection and correction of the terminal error by a peripheral feedback loop. This level implies that the position of the target in space, and the position of the limb with respect to the target can be compared at some point of the trajectory of the movement. Admittedly, this model is still largely speculative. Particularly, level 2 will be hardly either demonstrated or disproved. At least, it may represent a plausible basis for the explanation of certain pathological states where visuomotor control is disrupted.

METHODOLOGICAL CONSIDERATIONS

Most authors have used a standard type of movement; namely, reaching at a target in front of the subject. Whatever the precise requirement (e.g., hitting with a stylus, finger pointing, moving a joystick, grasping) this type of movement usually involves several joints. A few attempts have been made at carefully controlling the number of joints or the degrees of freedom involved. The main problem, however, is the definition of movement speed. One way of dealing with this problem is to require the subject to move the arm at the maximum velocity and to measure the accuracy when target distance and target size are manipulated. The relationship between speed and accuracy is formalized by the classical "Fitts' law" (Fitts, 1954),

$$MT = a + b \log_2 (2A/W)$$

where MT is movement time, A is movement amplitude (distance from the target) and W, target width. This law predicts that, if movement amplitude is increased, or target size is decreased, the movement will slow down. In other words, if movement amplitude is doubled, the movement time does not change if width of target is also doubled (see reviews in Keele, 1968 and Welford, 1968).

Another way of dealing with the relationship between movement speed and amount of visual control (which determines accuracy) is to pace movement time by varying the frequency of ticks from a metronome (in that case, target size and distance must remain constant). Under this condition, the accuracy is clearly decreased when the movement speed increases (see Keele and Posner, 1968; Howarth et al., 1971). Practice may decrease the overall amount of error, although the speed/accuracy relationshp is preserved (Beggs and Howarth, 1972). In studies of natural reaching movements, emphasis is on accuracy rather than on speed. In that case, movement time can be studied as a dependent variable (see below).

The second methodological aspect of visual control studies is the manipulation of the available visual information provided by the view of the target, the stationary and/or moving hand, and of the hand position relative to the target. A number of techniques have been used to manipulate visual control, such as turning off the target at the onset of the movement, or having the subject execute his movement in the dark. The technique of Held and Gottlieb (1958) allows a separate manipulation of the view of the target and of the view of the limb. In this discussion, the terms closed-loop and open-loop conditions will be used to indicate, respectively, availability or nonavailability of visual cues related to the hand. Visual cues from the target can also be manipulated independently or in combination with those from the hand.

Another neglected aspect of visual motor control is the role of eye movements and/or position in directing the hand at the target. In experiments in which the head and gaze positions of the subject are not monitored, it is assumed that the subject fixates the target during his task. In fact, it can be of interest (a) to know precisely what the eyes are doing during hand movements at targets or, more generally, during reaching behavior; and (b) to determine whether eye position in the orbit may provide control cues for directing the hand movement. We have investigated this problem by requesting the subjects either to move their eyes freely while they move their hand or, conversely, to keep their gaze fixated at some neutral position (e.g., midline) during their reaching movement.

REACHING AT NATURAL OBJECTS

Mere observation of reaching and grasping movements in closed-loop and open-loop conditions provides useful information about the initiation of the movement and its trajectory, i.e., about those aspects which are under control of the program.

Subjects were seated in front of a cubicle placed on a table top on which three-dimensional target objects were displayed. The cubicle was separated horizontally into two equal compartments by a semireflecting mirror. Subjects looked through a window in the upper compartment, and placed their right hand in the lower compartment. In the first situation, the target-object was placed on the table in the lower compartment, so that the subject could see it directly through the semireflecting mirror. During movements directed at the target the subject could see his hand (closed-loop condition). In the second situation, a mask was inserted below the mirror so that the lower compartment was no longer visible. In this case, target-objects were displayed from the top of the upper compartment. Since the mirror was placed half-way between the target display and the table, subjects could see in the mirror a virtual image of the object projecting at the level of the table. Another object identical to that displayed in the mirror was placed directly on the table coinciding exactly with the virtual image seen in the mirror. This set-up resulted in a situation where the subject reached for the virtual image below the mirror without seeing his hand and met the second real object at the expected location. This situation, with no visual reafferences from the moving hand represented a typical open-loop condition. The targets were simple three-dimensional objects, the shape and orientation of which were chosen in order to minimize wrist rotation. They were displayed in the frontal medial plane at a variable distance from the body (maximum distance, 40 cm).

Subjects were required to perform rapid and accurate movements and to grasp the objects precisely and lift them. At the beginning of each trial, they had to place their hand on a stand near their body axis in the prone position with the fingers semiflexed. While a new object was displayed, subjects kept their eyes closed. At an acoustic signal, they opened their eyes but had to wait between 2 and 10 sec until a small light in front of them was turned off before performing the reaching movement. Movements were recorded on 16-mm cinefilm, by using a camera running at 50 frames/sec. Film data were processed by projecting single frames on a screen with a one-to-one magnification. The position in space of a small mark placed on the subject's wrist was plotted across successive frames. For each frame, the distance between the tip of the thumb and the tip of the index finger was measured. Finally, typical hand postures were drawn for each movement directly from the screen.

The movement of the hand as a whole, from the resting position to the contact with the target can be analyzed first. The duration of this component varied across the 5 subjects, but variation was less marked within individual subjects. For movements of an amplitude of 40-cm, mean duration ranged between 718.6 msec (102, 1 SD) in the fastest subject and 1176.6 msec (124, 7 SD) in the slowest. Standard deviation ranged between 10 and 17% of the mean except for one subject, where it reached 25%.

The position-against-time profile of the transportation component consistently fitted an asymmetrical S-shaped curve. Such a pattern is commonly observed for this type of movement, since its first recognition by Woodworth (1899). The movement first existed in a fast distance-covering phase, followed by a deceleration which tended to become very slow at the vicinity of the target. Duration of this low-velocity phase was between 150 and 250 msec, and never exceeded one-third of movement duration (Fig. 1).

Maximum velocity also varied from subject to subject, from about 80 cm/sec up to about 135 cm/sec for 40 cm movements. When the distance of the target relative to the body was varied, maximum velocity was found to increase almost linearly ($r = 0.84$; $p < 0.001$) with movement amplitude. On the overall sample, maximum velocity was poorly correlated ($r = 0.31$; $p < 0.05$) to movement duration. This can be explained by the fact that duration was little (if at all) influenced by movement amplitude ($r = 0.09$, NS) (Fig. 2). Of course, this result differs from what is observed with paced movements where duration, amplitude, and velocity are mutually correlated (see Beggs and Howarth, 1972). In the present case, the independence of duration with respect to the other parameters could also be a special feature of movements involving coordination between several limb seg-

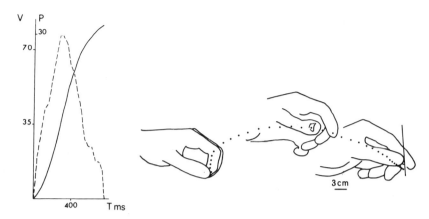

FIG. 1. Dynamic aspect of reaching movements. **Right:** Processing of a movement directed at a small object in the open-loop condition (the subject does not see his hand during the movement). Each dot represents the position of the hand on successive frames of the film. Higher density of dots at the end of the trajectory corresponds to the low-velocity phase. Note persistence of the low-velocity phase in the open-loop condition. Outlines of the hand position are redrawn from typical single frames. Note accurate finger posturing in the absence of direct visual control. **Left:** Velocity (V, in centimeters per second, *dotted line*) and position (P, in centimeters, *solid line*) profiles corresponding to the reaching movement displayed on the right. Tms: time in msec. [From Jeannerod (1981) with permission.]

ments, like prehension movements. This point will be discussed further below.

These dynamic characteristics of reaching movements were not affected by whether the visual feedback loop was open or closed. Duration and maximum velocity were similar in both conditions. The low-velocity phase at the end of the movement was present both when the hand was visible during the movement and when it was not.

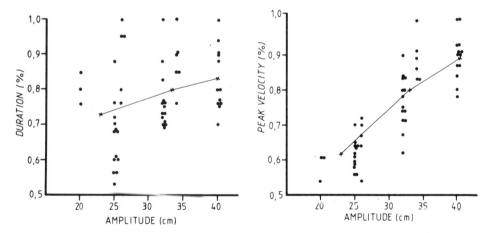

FIG. 2. Relationship of dynamic aspects of reaching movements to distance of the target. Data plot from 53 reaching movements in 3 different subjects. Target-objects are placed at a variable distance (20–40 cm) from the body. Note a good correlation between peak velocity and movement amplitude ($r = 0.84$) contrasting with a lack of correlation between duration and movement amplitude ($r = 0.09$). Ordinates have been expressed in percent of the maximum value in order to avoid intersubject differences.

The only significant difference in the two situations was the frequent occurrence of a small undershoot of hand with respect to target, when the visual loop was open.

Film analysis of reaching movements also provided a description of the finger grip movement occurring in anticipation to the grasp, during the transportation component. The first obvious feature of the "manipulation" component was that the fingers opened up to a maximum grip aperture, which was a function of the anticipated object's size although always greater than the actual size of the object. After the maximum grip aperture, the fingers began to close in anticipation of contact with the object. Position in time of maximal aperture was independent of the size of the aperture itself, but was strongly correlated with the position on the time-axis of the low-velocity phase of the transportation component. This feature is the basis for intersegmental coordination during prehension (see Jeannerod, 1981). It is exemplified in Fig. 3A, where the position and velocity profiles of the two components of an individual movement have been superimposed. Figure 3B represents a further attempt at visualizing the same time relationship in a larger sample of movements. Closing or opening the visual loop did not influence the timing of the manipulation component, nor the grip pattern itself. Anticipated finger posturing with respect to object size and shape appears to be remarkably preserved in the open-loop condition (Fig. 1B).

This study of natural reaching movements shows a remarkable independence of movement parameters with respect to visual control. This suggests that these parameters are mainly determined at the programming level, prior to movement initiation, and do not require further verification at the execution level. Visual control via peripheral feedback loops is thus probably less important than previously thought in specifying the temporal characteristics of this type of movement. It remains, however, that the accuracy of reaching is poor under the open-loop condition. Hence, one of the main questions addressed in the next section will be that of the precise role of feedback control for terminal accuracy, and of the relevant cues for this control.

THE EFFECT OF VARYING FEEDBACK CONDITIONS ON OPTIMAL CONTROL OF HAND MOVEMENTS

In order to vary widely the feedback conditions in which the reaching movements are executed, a more conventional task of hitting a small target was used. The studied variables were latency, duration, and accuracy of the hand movement.

Conditions of visual feedback used alternatively or in combination were as follows:

1. The hand could be visible throughout. This corresponds to the previously defined closed-loop condition,
2. The hand was never visible. This corresponds to the previously defined open-loop condition,
3. The hand could be visible only at rest before the movement was initiated, but not during the movement itself: static closed-loop condition,
4. The hand could be visible only during the movement, i.e., after the movement had been initiated, but not in the resting position prior to the movements: dynamic closed-loop condition,
5. The eyes were free to move. This corresponds to the visual condition where the subject "automatically" tracks the target at which he has to reach. This condition ensures foveal fixation of the target during the reaching movement: foveal vision condition,
6. No eye movements are allowed. The subject is required to keep his or her gaze fixated at the midline position. The target and the hand (if the visuomotor

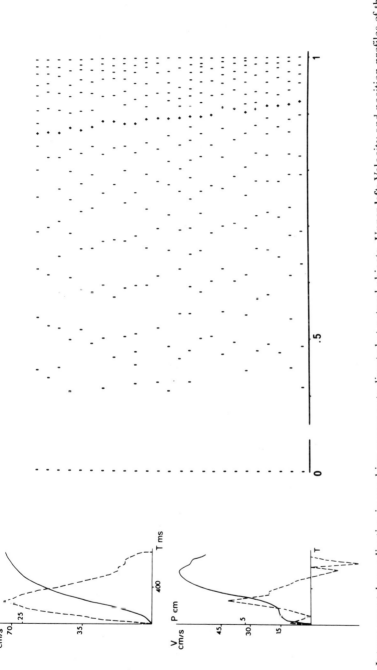

FIG. 3. Intersegmental coordination in reaching movements directed at natural objects. **Upper left:** Velocity and position profiles of the transportation component of a reaching movement. **Lower left:** Velocity and position profiles of the manipulation component of the same movement. Abbreviations as in Fig. 1. Note that the distance between the fingers (*solid line*) increases (finger grip opens) up to a maximum, and then decreases (finger grip closes in anticipation to contact with the object). The two diagrams have been superimposed to show that the point of maximum grip aperture is synchronous with the beginning of the low-velocity phase in the transportation component. **Right:** Hand position. Transportation components from 25 reaching movements have been classified from the upper to the lower part of the diagram. Each movement begins at position 0 on the left and ends at position 1 on the right. For each movement, position of the hand (as from film frames) has been sampled every 40 msec. Larger concentration of dots on the right corresponds to the slowing down of the hand near the target (low-velocity phase). A mark corresponding to the point of maximum grip aperture has been placed at the corresponding position for each movement. Note close correspondence between the mark and the beginning of the low-velocity phase. [From Jeannerod (1981) with permission.]

loop is closed) are only seen by the peripheral visual field: peripheral vision condition,
7. The eyes are free to move, but the target disappears at the onset of the saccadic eye movement toward it: restricted vision condition.

These conditions represent various degrees of availability for the visual signals assumed to be relevant for visual control of hand movements. By using different conditions in combination, it was possible to test systematically the relative contribution of the following factors: visual position and/or motion signals from the hand (conditions 1 through 4); visual error signals between hand position and target position (conditions 1, 2, and 7); oculomotor signals (conditions 5 and 6); signals arising from foveal fixation of the target (conditions 5 through 7).

Methods are fully described by Prablanc et al. (1979). The apparatus is shown in Fig. 4. Targets were presented through a matrix of red (600 nm wavelength) emitting diodes which could be lit randomly. The subject was sitting with his head fixed, in front of the surface R. He could see binocularly the virtual image E' of the target E, through a semireflecting mirror on the plane Q. The space between surfaces Q and R could be either illuminated, allowing the subject to see his hand (closed loop), or made completely dark (open loop). An electronic shutter fed by a logic pulse with a time response of 5 msec, was used to cut off the light. Binocular eye movements were recorded by an electrooculography (EOG) technique, using Beckman cup electrodes. A logic pulse was generated at the onset of the ocular saccade: It could be used to cut off the target at which the saccade was directed (condition 7).

Hand position was recorded by a thimble attached to the subject's forefinger, which indicated its XY coordinates when it was in contact with the surface R. This surface was covered with an isotropic resistive paper, fed by a current alternately switched along X and Y axes (Bauer et al., 1969; Prablanc and Jeannerod, 1973). The hand position signal was lost at the onset of the movement (when the hand left the surface) and a logic pulse was generated at this time. Another logic pulse indicated the end of the movement when the forefinger touched the surface. Targets were presented as step stimuli along the frontal plane. In each trial, the target stepped from position C' at the center of the display to a randomly selected E' position (10, 20, 30, 40 cm on the right side, 10, 20 cm on the left side). In condition 6, the subject had to fixate continuously another target located 2 mm ahead of C'.

Eight adult subjects participated in the experiment. Under binocular vision, they were instructed to point toward the targets as quickly and as accurately as possible with their right hand. At the beginning of each trial, the subject had to keep his gaze and his hand pointed at the central target C'. When the target jumped to a peripheral position, he had to point to it, without any particular instruction on the eye and hand response sequence (except for condition 6); then he had to keep his hand and gaze on this target until it jumped back toward the center C' (or reappeared at the center, after the peripheral target had been cut off as in condition 7). When the target came back to C' the subject was instructed to look and point back to it, and to keep there his eye and finger, until a new target reappeared at the periphery. The interstimulus interval was 5 sec, with a random variation of 1 sec, since the sequence of positions was also randomized. Where and when a target would appear was not predictable (except for the return target C'). The program allowing random stimulation, data acquisition, and interactive display of eye and hand responses ran on line with a PDP-8 DEClab with 8K memory and two DECtapes (DEC). At the beginning of each experi-

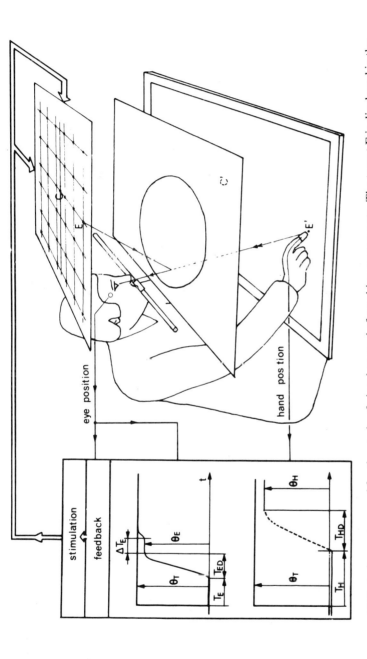

FIG. 4. Apparatus used for the study of visual control of reaching movements. The target E is displayed in the matrix C. After reflection in the semireflecting mirror C', its virtual image E' projects on the hand position recording surface. Note rigid head fixation. Horizontal eye movements are recorded by an electrooculographic technique, and hand position by a thimble attached to the finger; when the thimble is on the surface (fed by a current) it transmits its coordinates. Eye and hand data acquisition are performed on line; each elementary response is displayed onto an interactive terminal for some seconds, allowing the experimenter to nullify artifacts. Eye velocity can be used to trigger a feedback stimulation. For example, the onset of a saccade can trigger off the target. For a large target eccentricity $\theta_T = C'E'$ along the horizontal axis, the initial saccade usually undershoots its goal and the eye waits for ΔT_E before correcting the error by a second saccade. T_E; T_{ED}; θ_E and T_H; T_{HD}; θ_H are, respectively, for eye and hand the latency, duration of movement, and position of the response. The *dotted line* on the hand response indicates the loss of contact with the recording table during the movement. (From Prablanc et al., 1979, with permission.)

ment, a calibration of eye and hand positions was performed for each target. During the experiment, eye and hand responses were sampled at 200 Hz during 1.5 sec after onset of stimulus and the following measures were performed: latency of the initial saccade T_E, its duration T_{ED}, position of the eye O_E after the initial saccade, latency of the hand T_H, duration of the hand movement T_{HD}, and position of the hand pointing O_H (Fig. 4).

RESULTS

Latency data from this experiment were mainly concerned with providing a description of the eye-hand sequence in a visuomotor task. This will be given for a single condition of visual feedback (condition 2, open-loop), then extended to the other conditions. Mean latency of the hand movements in the open-loop condition ranged between about 350 and 400 msec. Within this range, it was statistically dependent on the eccentricity of the target $F = 8.4; p < 0.001$), i.e., it was longer for more peripheral targets (Fig. 5). The mean latency of the corresponding eye movements (outside condition 6) was also correlated with target eccentricity (this is a commonly observed fact: Bartz, 1962; Payne, 1967; Prablanc and Jeannerod, 1974). Accordingly, when the whole sample was considered, the hand latency-eye latency difference remained constant across targets of different eccentricities. However, by considering individual pairs of eye and hand latencies, only a weak correlation (if any) was found between the two ($r = 0.45$ and usually much lower). This shows that although eye and hand latencies are related to the same event, their respective variations are not time-locked to each other (see below). The latency of hand movements was affected by

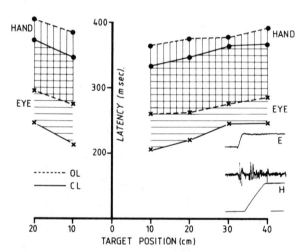

FIG. 5. Eye and hand latency as a function of target eccentricity. Targets are presented on either side of the midline in the peripheral visual field (target position, in centimeters). The latency of eye movements (x) and hand movements (●) have been plotted in two conditions: closed-loop, i.e., with full vision of the hand prior to, and during the movement (———, CL) and open-loop, i.e., with no vision of the hand (- - -, OL). Note longer latencies for both eye and hand movements in the open-loop condition. [From Prablanc et al. (1979), with permission.] *Insert, lower right:* Single example of a eye-hand sequence during reaching, in the open-loop condition. Note that the onset of the eye movement (E) precedes the onset of the hand movement (H) by about 120 msec. However, the EMG from biceps brachii (midline) shows an activation pattern which happens to be exactly synchronous with the eye movement. (From Biguer et al., 1982, with permission.)

varying the conditions. For instance, mean latency was always shorter in the closed-loop condition than in the open-loop condition (average difference, 16 msec, Fig. 5). In addition, mean latency was longer when no eye movements were allowed (condition 6) than when eye movements were present.

A few conclusions can be drawn from the latency study. First, the increase in hand latency with target eccentricity seems related to visual processing of target location by the peripheral retina, rather than to an increase in premotor time as a function of target distance. In fact, it is known that a simple increase in movement amplitude without changing retinal target location (e.g., by presenting targets at different distances in the sagittal plane) does not affect the latency of the movement (Brown and Slater-Hammel, 1949; review in Hayes and Marteniuk, 1976). Another conclusion can be drawn from the lack of correlation between eye and hand latencies when individual pairs are considered. To us, this means that motor commands responsible for eye and hand movements, respectively, are generated in parallel rather than in series. Accordingly, by considering the latency of arm muscles activation on the electromyogram (EMG) instead of that of the arm displacement, the hand latency gets very close to the eye latency (Fig. 5)(Jeannerod and Biguer, 1981; Biguer et al., 1982), thus indicating that there is no time allowed for cross-talk between eye movement and hand movement channels at the central level.

However, the arrangement of latencies of eye and hand displacements (eye latency systematically shorter than hand latency) may have some implications for movement control. For targets within 30 cm of eccentricity, the hand latency-eye latency difference allows the saccade to be completed (i.e., foveal fixation to be achieved) before the hand begins to move. On the contrary, beyond 30 cm of eccentricity, the saccade duration is longer than the hand-eye difference. This fact might explain the larger scatter of hand pointings whether they are directed at targets located within or beyond 30 cm from the midline, particularly in condition 3 when vision of the static hand position is the only available cue (see below, Fig. 7).

Finally, the difference in hand latency between open-loop and closed-loop conditions could be related to the processing of information concerning hand position at the beginning of the movement. It can logically be assumed that this processing takes a shorter time when both visual and proprioceptive information about limb position are matched, than when proprioceptive information is present alone. This hypothesis was not entirely verified, however. Indeed, as expected, latency was shorter in condition 3, where visual static cues are present before the movement begins, than in condition 2; but, it was shorter also in condition 4, which is not predicted by the theory.

Duration

In all conditions used, hand movement duration was clearly related to target eccentricity: The larger the extent of the movement, the longer its duration. This feature is in agreement with the result of studies using paced movements in a similar task of pointing at, or hitting the target, but it is in disagreement with our previous experiment where the task was to grasp three-dimensional objects. Our interpretation of this difference is that hand movements directed at objects that can be grasped imply the activation of multiple limb segments. The coordination of the segmental movements for the achievement of the terminal accuracy of the overall movement is a time-consuming process. In fact, duration of these reach-and-grasp movements executed at a "natural" rate (average duration ca. 900 msec) was longer than those of a comparable extent in the hitting task (average duration, ca. 450 msec). This difference may be due to the presence in the reach-and-grasp of a long low-velocity phase, which remains of a relatively constant duration whatever the movement amplitude. The low-velocity phase is much shorter in hitting movements. We have no direct measurement of the velocity profile of hand movements from our hitting experiment, but the curves shown by Beggs and Howarth (1972) clearly present evidence for a short low-velocity phase in hitting movements (see also Annet et al., 1958). Movement duration was uninfluenced by the different visual feedback conditions used. That was already the case for the other type of movements reported above.

Accuracy

Errors of the final position of the finger with respect to target location were measured in two ways. The mean error was computed by measuring the actual mismatch between the finger and each target, and by keeping in consideration its algebraic sign along the X axis ($+$, finger beyond the target, overshoot; $-$, finger within the target, undershoot). The mean absolute error was computed by taking account only of the finger-target distance on the X axis, whatever the sign of the error. These two measures first allow the description of a number of systematic trends in the observed errors. The mean error showed, in any condition where errors could be measured, a clear overshooting when the movement was directed at targets relatively close from the midline (10°, 20°). For more remote targets, undershooting was observed. From a similar experiment where subjects had to point in open-loop condition at targets located at different distances in the sagittal plane, Foley and Held (1972) observed only errors by overshooting, even for targets as distant as 40 cm from the body.

The absolute error appeared to increase with target eccentricity and, in addition, to be systematically larger in movements where the hand had to cross the midline to targets located on the side contralateral to the moving hand (Fig. 6). In closed-loop conditions, the movements were accurate only in the natural situation, when both the hand and the target were visible and when the eyes were free to move (combination of conditions 1 and 5). Errors appeared, with the above-mentioned characteristics, in situations where condition A was combined with either conditions 6 or 7. In these two cases the possibility of matching the final hand and gaze positions with target location is prevented. However, the similarity of the two curves corresponding to conditions 6 and 7, respectively, cannot allow us to determine lack of which factor (terminal error feedback signal, oculomotor signal . . .) is more critical for altering movement accuracy.

The absolute error of hand pointings was dramatically increased in the open-loop condition. Although the main error characteristics were still preserved (e.g., relationship with eccentricity, the open-loop errors distribution appears on Fig. 6 to represent a quite different population from those from the closed-loop experiment. Results obtained by combining condition 2 with either condition 5, 6, or 7 showed that the worst case was condition 2 plus 7; that is, a condition in which neither the hand nor the target are visible. In this case very few cues are available for hand movement control. The oculomotor signal which, in fact, may be the only one left seems to be of poor value as indicated by the comparison of conditions 5 and 6. Accuracy was even worse when the oculomotor signal was present (condition 5) than when it was not (condition 6).

The factors contributing to the increase in absolute error in the open-loop condition were investigated by using conditions where static or dynamic visual signals from

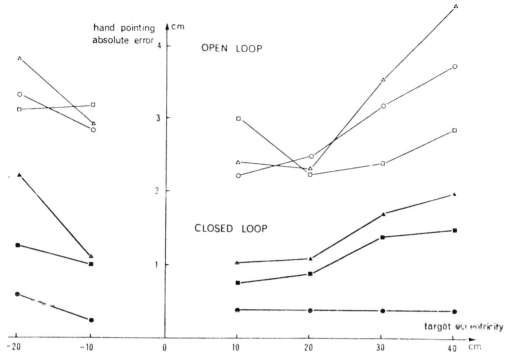

FIG. 6. Absolute error of the hand in different conditions of visual control. Errors (in centimeters) have been plotted as a function of target distance from the midline (eccentricity, in centimeters). *Circles:* Conditions with free gaze and foveal fixation; *squares:* condition without eye movements allowed (peripheral vision condition); *triangles;* target disappears at the onset of the eye movement (restricted vision condition). *Filled symbols:* Closed-loop conditions. *Open symbols:* Open-loop conditions. Note the global increase in absolute error in the open-loop condition. Data from 8 subjects. [From Prablanc et al., (1979), with permission.]

the hand could be selectively suppressed. When static signals were available prior to the movement (condition 3), the absolute error was consistently reduced with respect to the full open-loop condition. Greater accuracy of final finger pointing resulted from a reduction of both the overall amplitude of the error, and the variance with respect to the mean (Fig. 7). In this condition, with no information from the movement itself, the improvement in precision can only be due to a better calibration of the initial hand position with respect to the body and to the target. In condition 4, where the hand was only visible during the movement, the absolute error was also reduced, even to a greater extent than in condition 3. No conclusion should be drawn from this fact, however. Since condition 4 closely ressembles the normal, closed-loop condition (condition 1), it only demonstrates once more the usefulness of visual feedback signals for movement control.

DISCUSSION

Feedback signals generated by the movement itself seem to be a necessary condition for movement accuracy. Their role is amply demonstrated by the undifferentiated increase in absolute error which appears in any condition where their use is prevented. The problem is to determine which cues are fed into motor commands to improve final

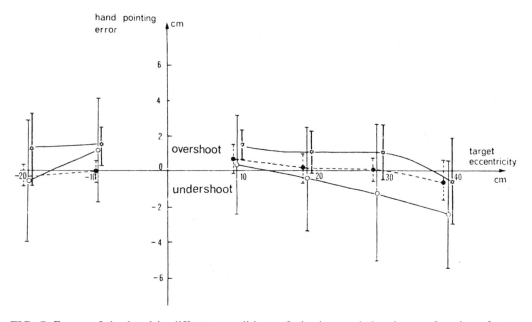

FIG. 7. Errors of the hand in different conditions of viewing one's hand, as a function of target distance. Errors have been plotted by keeping their algebraic sign. Positive values: overshoot; negative values: undershoot. In the normal open-loop situation *(open circles)* (condition 2) note large undershoots and wide scatter of hand pointings. Scatter is reduced in condition 3 (hand visible only prior to movement) (solid circles) and further reduced in condition 4 (hand visible only during the movement) *(stars)*. (From Prablanc et al., 1979, with permission.)

precision. The mentioned intrinsic delays for visuomotor processing imply that the relevant information must be picked up relatively early in the course of the movement. We also know (see condition 4) that a satisfactory control may still be exerted if the hand is made visible only after it has begun to move. Thus, visual motion signals from the hand can be used to determine if the planned velocity profile of the trajectory is concordant with the distance to be covered. In other words, the final position of the hand has to be extrapolated from the movement pattern and a possible error has to be anticipated in order to be corrected in time. Simple deductions as to this mode of control can be drawn from the analysis of reaching behavior: (a) due to the sequential organization of eye and hand movements directed at a target, it can be assumed that the signals relevant for error anticipation will be received by the peripheral retina,

since the eye is already fixating the target before the hand begins to move (Jeannerod and Biguer, 1981); and (b) because it is directly related to the movement amplitude (Fig. 2A) the peak velocity might represent the most relevant signal for controlling accuracy. In any case, the perceived signal (e.g., peak velocity) would have to be compared with the corresponding parameter stored in the program. Then the difference between the two would have to be added to, or subtracted from, the motor commands applied to the arm.

Another question is whether the late part of the trajectory, where the hand slows down when closing on the target, could also be a source for feedback signals. In that case, such cues as the hand-target distance, or the relative position of the two could be used to smooth the contact of the hand with the object and to improve accuracy. This view has been explicitly stated by Paillard

(1980), who thinks that central vision, "highly sensitive to position cues ... encodes the rate of change of position of the moving hand relative to the stationary target and allows corrective feedback to operate...." There is little evidence for such a mechanism, however. First, the slow phase at the end of the trajectory can be very short, particularly in hitting movements, where it is too short to allow the position signal to alter the motor commands. Second, the slow phase in itself is not due to some visual "braking" of the movement trajectory, since it persists virtually unchanged in open-loop conditions. The latter fact clearly indicates that the low-velocity phase is a built-in property of the movement sequence, and that it pertains to the program level rather than to a servo-controlled mechanism. Our conception of the low-velocity phase would be that of a target-acquisition phase (homing in on a target, Welford et al., 1969), i.e., a necessary constraint for movements requiring a high degree of precision, such as reaching at small or fragile objects. This view is consistent with out observation that the slowing down of the transportation component of reaching movements closely corresponds in time with the closure of the fingers (Fig. 3). We suggest that a certain state must be reached simultaneously by the different segments involved in the movement, a state which would be defined as the time-goal for the high-velocity part of all the movement components. It is a logical to speculate that for the sake of precision in reaching, the point in time where the velocity curves of the different components are phased must be different from the final stop at contact with the object (Jeannerod, 1981).

The fact that terminal error feedback cannot be used for controlling movement accuracy on-line is not incompatible with other possible functions. The terminal error at the end of a movement could be used to alter the motor program itself. In this way, the movement could become more and more precise on subsequent trials. This mechanism could represent a basis for the role of visual feedback in acquisition of skills or in adaptation to visuomotor conflicts. In prism adaptation for example, conditions involving the presence of a terminal error feedback are known to produce better adaptation and larger after-effects than when only the hand (but not the target) is visible (Freedman, 1968).

Open-loop visual control has been assumed to be another essential mode of movement control. This assumption is partly confirmed by the fact that vision of the stationary hand before the movement begins (as from our condition 3), is enough to bring movement accuracy close to the level of the normal condition, even though no visual feedback is available during the movement itself. In the complete open-loop situation (condition 2), the only signals generated by the hand position or movement are of proprioceptive origin. They are not sufficient by themselves to ensure a precise encoding of movement amplitude with respect to target location (e.g., Kelso et al., 1980). The position of the hand within the proprioceptive map (i.e., its relationship to the body coordinates) has to be calibrated with the help of other sensory cues. By "refreshing" the proprioceptive map, vision would ensure a better matching of hand-to-body position to the visual map in which the target-to-body position is encoded. This process would allow the hand movement to be performed adequately in the open-loop condition under kinesthetic control, through a comparison between the proprioceptive feedback and the visual goal stored in the internal visual map.

Arguments for visuomotor control in patients with optic ataxia can be drawn from pathological conditions in humans. Posterior parietal lesions are known to produce a specific impairment of visually directed reaching within the proximal extrapersonal space. Typically, the hand contralateral to the lesion misses targets within the corresponding part of the visual field. This syndrome (*optic ataxia*, Balint, 1909) has other variants: for instance, ataxia may affect the contralateral hand when reaching in all parts of the visual field, or both hands in the visual field contralateral to the lesion. Reaching movements in 3 such cases were studied by Vighetto and Perenin (1981) with the same procedure.

Comparison of the effects of various conditions shows that the 3 patients had virtually normal movements in the closed-loop condition with eye movements free. Opening the visuomotor loop (condition b) appeared to produce a very severe misreaching, (by undershooting) several times larger than in normal subjects in the same condition (Fig. 8A). Since the main effect of the open-loop condition is to suppress feedback signals, the observed deterioration of reaching in the patients means that they rely more (if not only) on feedback signals to direct their hand at target. This observation would thus indicate that patients with optic ataxia lack another mode of control which is normally present, even when the visuomotor loop is open.

The breakdown in movement control produced by the open-loop condition in optic ataxia could be due to poor programming of the movement, or to inability to use an open-loop mode of visual control. The latter hypothesis might be the correct one. In the 3 patients, vision of the stationary hand

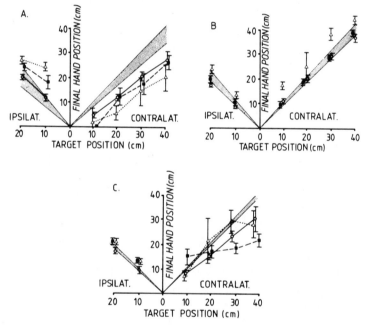

FIG. 8. Reaching errors in optic ataxia. Mean hand position and standard deviation have been plotted against target position in 3 patients suffering from optic ataxia due to unilateral posterior parietal lesions. Patients use their hand contralateral to the lesion to point at targets in a situation similar to that shown in Fig. 4. Contralateral: Pointings are directed toward the hemispace contralateral to the lesion, i.e., on the same side as the hand used for pointing. Ipsilateral: Pointings directed at the ipsilateral hemispace. *Dotted areas* represent the limits of the standard deviation of errors made by normal and control subjects. A: Open-loop condition with free gaze and foveal fixation. B: Condition 3, with hand visible prior to the movement. Note improvement of accuracy in all 3 patients with respect to A. C: Closed-loop condition with no eye movements allowed. Note deterioration of accuracy in all 3 patients. [From Vighetto and Perenin 1981, with permission.]

before the movement onset (condition 3) largely restored accuracy of pointing (Fig. 8B). Our explanation for this effect is that, in these patients, the absence of vision of the hand makes impossible its encoding within the proprioceptive map, a condition where open-loop visual control cannot operate. In normal subjects, this mechanism is assumed to be partly functional in the complete open-loop condition although it becomes fully accurate only when the hand can be seen (see Klein and Posner, 1974).

Signals related to the gaze position in the orbit, or to gaze fixation on the target may also represent cues for an open-loop control of hand movements. There are at least 2 basic assumptions subserving a role of oculomotor signals in reaching behavior: (1) Since the relationship of retinal coordinates to body axis does not hold constant (due to head and eye movements), the retinal locus where a target projects cannot be the only source of information about its egocentric localization. Signals derived from the eye-to-head position and from the head-to-body position must also be taken into account. (2) Oculomotor signals resulting from the process of matching the central retina with target are monitored centrally and can be used to guide the hand at the same target (Festinger and Canon, 1965).

By following up the same explanation as for the knowledge of hand position, we could assume that the visual process of fixating the target (foveation) is a prerequisite for encoding the eye position within the proprioceptive map and for relating the target position to the body coordinates. This "calibrated" eye position could thus represent an accurate landmark as to where to direct the hand. However, the experimental data reported here from normal subjects do not provide a basis for the use of such oculomotor signals in controlling the movements. Among the negative arguments are the lack of correlation between latency of the eye movement and latency of the hand movement when both are directed at the target; and the lack of effect on accuracy of whether the eye movement is present or not.

The role of oculomotor signals in directing the hand may become more apparent under pathological conditions when other cues cannot be used. This seems to be the case in optic ataxia. In the 3 patients reported by Vighetto and by Perenin and Vighetto, accuracy of pointing became very poor when eye movements were prevented (condition 6), even in the closed-loop situation (Fig. 8C). It should be stressed, however, that both normal subjects and in patients, the head was immobilized. Although this is required for a study of the role of oculomotor signals, it does not correspond to the normal physiological situation where the head is free to move. In that case, it can reasonably be assumed that the head-eye system behaves as a unit, the eye-to-head position maintained constant by the vestibulo-ocular reflex. Signals related to static head position could thus automatically ensure the referencing of the retinal map to body coordinates, and a monitoring of the eye position in the orbit would become unnecessary (Jeannerod et al., 1980). Another advantage of this mechanism is that the head-position signal produced by proprioceptive afferents is much stronger than the eye position signal. In fact, the recent work by Biguer et al. (1982) clearly demonstrates this point. Normal subjects tested in the open-loop situation as described here, but with the head free to move, perform accurate pointing movements with their unseen arm. The restoration of accuracy occurs only for those movements directed at targets located beyond 20 cm from the midline when head rotation is required for target fixation.

Motor Control Mechanisms in Health and Disease,
edited by J. E. Desmedt.
Raven Press, New York © 1983.

Posture Control and Trajectory Formation in Single- and Multi-Joint Arm Movements

*Emilio Bizzi and **William Abend

*Department of Psychology, Massachusetts Institute of Technology, Cambridge, Massachusetts 02139; and **Department of Neurology, Harvard Medical School, Boston, Massachusetts 02115

Understanding the way in which the motor system achieves the complex goal of controlling simultaneously the many degrees of freedom of movement presented by the arm has only infrequently been explored (Bernstein, 1967; Fukson et al., 1980; Georgopoulos et al., 1981; Gilman et al., 1976; Marey, 1901; Muybridge, 1901; Morasso, 1981; Soechting and Lacquaniti, 1981; Viviani and Terzuolo, 1980; and von Hofsten, 1979). This chapter presents experimental findings that reveal some of the control solutions that have evolved for generating coordinated multijoint movements. The results have been obtained through a kinematic approach rather than by circuitry analysis stated in terms of single-neuron activity. We first describe animal experiments dealing with one-degree-of-freedom movements and discuss the notions of final position control and of reference trajectories. This information serves as necessary background for another group of experiments which deals with the control of multijoint arm movements in humans.

OPEN-LOOP CONTROL OF FINAL POSITION

This section summarizes experiments directed at understanding some of the mechanisms subserving single-degree-of-freedom movements. As discussed below, the results of initial experiments on head movements were replicated in the context of elbow movements (Bizzi et al., 1976, 1978; Polit and Bizzi, 1979). The studies indicate that an animal can execute a simple pointing movement and maintain a new position in the absence of proprioceptive feedback. These findings have important implications regarding the functional relationship between descending commands and posture.

Monkeys were trained to make coordinated, horizontal eye-head movements directed at visual targets. In the intact animal, the unexpected application of a constant load was followed by an increase in electromyographic (EMG) activity of the neck muscles, presumably due to an increase in muscle spindle and tendon organ activity. As shown in Fig. 1, in spite of these changes in the flow of proprioceptive activity, the head reached its intended final position after the constant load was removed, a fact suggesting that the program for final position was maintained during load application and was not readjusted by proprioceptive

---

*To whom correspondence should be addressed: Department of Psychology, Massachusetts Institute of Technology, E10-238, Cambridge, Massachusetts 02142.

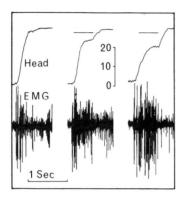

FIG. 1. Typical visually triggered horizontal head movements in chronically vestibulectomized monkey to appearance of target at 40°, but performed in total darkness. **Left:** An unloaded movement. **Center:** A constant-force load (315 gm·cm) was applied at the start of the movement resulting in an undershoot of final position relative to left, despite increase in EMG activity. **Right:** A constant-force load (726 gm·cm) was applied. Note head returns to same final position after removal of the load. Vertical calibration in degrees; time marker is 1 sec; EMG recorded from left splenius capitis. (From Bizzi, et al., 1976, with permission.)

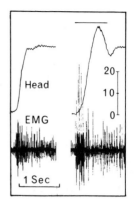

FIG. 2. Typical head responses of a chronically vestibulectomized monkey to sudden appearance of target at 40°. **Left:** An unloaded movement; **Right:** a load of approximately 6 times the inertia of the head was applied at the start of the movement, as indicated by the force record. Both movements were performed in total darkness, the light having been turned off by the increase in EMG (splenius capitis). Peak force exerted by the monkey is approximately 750 gm·cm; head calibration is in degrees; time marker is 1 sec. (From Bizzi et al., 1976, with permission.)

signals acting at segmental or suprasegmental levels. On the basis of this result, Bizzi et al., 1976, concluded that the central program establishing final position is not dependent on a readout of proprioceptive afferents generated during the movement but is preprogrammed. It should be stressed that the load disturbances were totally unexpected and that the monkeys were not trained to move their head to a certain position but chose to program a head movement together with an eye movement in order to perform a visual discrimination task (Bizzi et al., 1976).

In a second set of experiments, Bizzi et al. (1976) examined the effect of stimulating proprioceptors only during head movement. To this end, an inertial load was used to modify the trajectories without causing a steady-state disturbance (Fig. 2). As a result of the sudden and unexpected increase in inertia during centrally initiated head movements, a number of changes in head trajectory, relative to unloaded movements, were observed. An initial decrease in head speed, due to the inertial load, was followed by a relative increase in speed (which resulted from the kinetic energy acquired by the load being transmitted to the decelerating head). The increased speed led to an overshoot in head position, and this was followed by a return to the intended position (Fig. 2). The changes in head trajectory brought about by the sudden and unexpected increase in head inertia induced corresponding modifications in the length and tension of the neck muscles. The agonist muscles were, in fact, first subjected to increased tension because the load slowed the process of muscle shortening and then the shortening of the same muscles was facilitated during the overshoot phase of the head movement induced by the kinetic energy of the load. Such loading and unloading did, of course, provoke the classic muscle spindle response mediated by group Ia and group II afferent fibers which, in turn, affected the agonist EMG activity. Figure 2B shows that there was first a greater increase in motor unit discharge during muscle stretch than would have occurred if no load were applied, followed by a sudden decrease in activity at the beginning of the overshoot phase. Therefore, during a head movement, an unexpected inertial load induced a series of waxing and waning proprioceptive signals from muscle spindles, tendons, and joints, but the intended head position was eventually reached even in the absence of other sensory cues (visual and vestibular). This observation, together with those on the effect of constant-torque loads, suggests that the central program establishing final head position is not dependent on a readout of proprioceptive afferents generated during the movement but, instead, is preprogrammed.

To test further the hypothesis that the final head position is preprogrammed, Bizzi et al. (1976) investigated the attainment of final head position when monkeys were deprived of neck proprioceptive feedback. To ensure open loop conditions, animals were vestibulectomized 2 to 3 months prior to deafferentation [vestibulectomized monkeys recovered eye-head coordination (Dichgans et al., 1973)]. After deafferentation, the animals were still able to make accurate visually evoked responses (Bizzi et al., 1976). The experiment dealth with a constant torque applied during centrally initiated movements (Fig. 3). Just as with intact animals, when the load was applied unexpectedly at the beginning of a visually triggered movement, the posture attained by the head fell short of the intended final position. After removal of the constant torque, the head attained a position that was not significantly different from the one reached by the head in the no-load case (Fig. 3). These results can be explained by

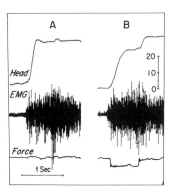

FIG. 3. Typical movements of a chronically vestibulectomized monkey with sectioned dorsal roots (C1-T3) made open loop (in total darkness). B: constant-force load (315 gm·cm) was applied at the start of movement, resulting in an undershoot while the load was on. Similarity of EMG pattern in 5^{Y}T,R5,¯(6pB shows lack of a stretch reflex. Peak force in B, approximately 315 gm·cm. Vertical calibration in degrees. (From Bizzi et al., 1976, with permission.)

modeling the neck muscles as opposing springs. The head will lie in a position at which the forces exerted by the agonist and antagonist springs are equal and opposite. Of course, if the head is moved to a new position by an external force, but the length-tension properties of the springs are not changed, then, when the external force is removed, the head will move back to the original position. This resting position could be more permanently modified by changing the stiffness (obtained by dividing the distance a spring is stretched into the magnitude of the resulting force increment) or the rest length of one or both of the springs, as described by Hooke's Law. Just as with springs, the force exerted by a muscle increases as the muscle is stretched; the spring characteristics of the muscle are adjusted by changing the neural input (Rack and Westbury, 1969). We therefore postulate that the motor program specifies, through a selection of a set of length-tension properties of agonist and antagonist muscles, an equilibrium point between these two sets of muscles that correctly position the head in relation to a visual target. In the absence of other forces, the final head position will be determined by that interaction of agonists and antagonists which results in a position where the tensions on the two sets of muscles are equal and opposite (Asatryan and Feldman, 1965; Rack and Westbury, 1969, Feldman, 1974a,b). Given this model, it is not surprising that the head overshoot during inertial loading is corrected with a return movement to the intended position, because the change in muscle length due to the inertial load will generate an increase in antagonist tension and, hence, a return head movement (Fig. 2B). By the same token, because head position is related to muscle-length and load, an undershoot is observed when a constant external opposing torque is applied (Figs. 1 and 3). The same hypothesis explains why the head moves to the intended final position when the constant torque is removed.

Thus, it seems that final head position in both intact and deafferented preparations should be viewed as an equilibrium point dependent on the firing rate of the alpha-motoneurons (MN) innervating agonists and antagonists, the length-tension properties of the muscles involved in maintaining the posture, the passive, elastic properties of the musculoskeletal apparatus, and the external load. In the intact animal, however, in parallel with this basic process, the proprioceptive system participates in the process of reaching final position by increasing muscle stiffness when a load disturbance is applied (Rack and Westbury, 1960). In fact, any stimulation of the proprioceptive apparatus, by virtue of its reflex connections, will modify the firing rate and the recruitment of alpha MNs and, therefore, force the selection of a new length-tension curve (Rack and Westbury, 1969).

In a complementary set of experiments involving forearm movements performed by rhesus monkeys, Polit and Bizzi (1979) extended the previously described findings on the final position control of the head. The monkey's forearm was fastened to an apparatus which permitted flexion and extension about the elbow in the horizontal plane (Fig. 4), and the monkey was trained to point at target lights with the forearm. Several target lights were spaced at 10° intervals along an arc which was centered on the axis of rotation of the elbow. The monkey was trained to move the forearm into a 12 to 15° wide target zone centered on the illuminated target light and hold that position for roughly one second. Experiments were conducted in a darkened room, so that the monkey saw only the target lights. A torque motor in series with elbow pivot was used to apply positional disturbances to the arm. On random pointing trials, the initial position of the forearm was displaced. In most cases, the positional disturbance was applied immediately after the appearance of the target light and was stopped just prior to the activation of the motor units in the agonist muscle. Hence, when the motor command specifying a given forearm movement reached the mus-

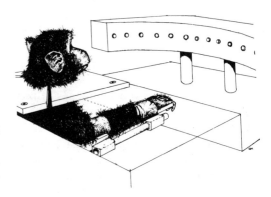

FIG. 4. Monkey set up in arm apparatus. Arm is strapped to splint, which pivots at elbow. Target lights are mounted in perimeter arc at 5° intervals. During experimental session, the monkey was not permitted to see its arm, and the room was darkened. (From Polit and Bizzi, 1979, with permission.)

cles, the positional disturbance had altered the length of the agonist and antagonist muscles, and the proprioceptive stimulation resulting from this disturbance had altered their state of activation. In spite of these changes, the intended final arm position was always attained; this was true whether the torque motor had displaced the forearm further away from, closer to, or even beyond the intended final position. There were no significant differences among the final positions achieved in these three conditions.

Naturally, the attainment of the intended arm position in this experiment could be explained by assuming that afferent proprioceptive information modified the original motor command. However, the results of previous work on final head position suggest an alternative theory: that the motor program underlying arm movement specifies, through the selection of a new set of length-tension curves, an equilibrium point between agonists and antagonists that correctly positions the arm in relation to the visual target.

To investigate this hypothesis, Polit and Bizzi (1979) retested the monkeys' pointing performance after they had undergone a bilateral C1-T3 dorsal rhizotomy (Taub et al., 1965, 1966, 1975). The animals were again required to produce pointing movements in an open-loop mode, as no proprioceptive activity could reach the spinal cord, and visual feedback of arm position was not present. Under these conditions, the animals could still produce pointing responses very soon after surgery (within 2 days in one case). The forearm was again displaced (at random times) immediately after the appearance of the target light and released just prior to the activation of motor units in the agonist muscles. For each target position, t-tests were performed for differences between the average final position of movements with undisturbed and disturbed initial positions. No significant differences were found. These observations suggest that, as in the case of head-position control, the final forearm position is directly programmed through alpha-MN activity, which selects the appropriate length-tension relationship for each of the muscles involved in the movement. The final limb position is reached when the tension generated by the agonists is equal and opposite to that generated by the agonists. This view implies that, for each limb position, particular levels of alpha-MN activity to the agonist muscles correspond to particular levels of activity to the antagonists.

A recent study of the EMG of agonist and antagonist muscles acting on the elbow and wrist joints in humans showed that a change from one posture to another involved a modulation of EMG activity in both flexors and extensors. Although the tonic EMG activity of flexors and extensors at each posture was variable, the ratio between the alpha-neuronal inputs to the two muscles was significantly less variable and no effect of the direction, amplitude, or velocity of the movement was detected (Lestienne et al., 1982). Once the final position was reached, a dynamic characteristic of the program underlying this agonist-antagonist innervation was noted. There was usually a progressive attenuation in the agonist and antagonist EMG activity without any change in final arm position (Polit and Bizzi, 1979, and Bizzi, Prablanc, and Hogan, *in preparation*). This finding indicates that the central programmer might gradually select a series of length-tension curves for agonists and antagonists, which may perhaps differ in slope, but which all specify the same final position.

It is tempting to speculate that this representation of posture as an equilibrium point between agonist and antagonist length-tension curves (see schematic representation of length-tension curves in Fig. 5) also has implications for movement (Feldman, 1974*a, b;* Cooke, 1979). If the central ner-

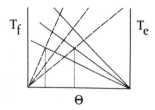

FIG. 5. Schematic representation of flexor (T_f) and extensor (T_e) length-tension curves. θ represents joint angle. (From Bizzi et al., 1982, with permission.)

vous system (CNS) were to abruptly specify new length–tension relationships (through a change in alpha-MN activity) for the muscles, movement would occur until a new equilibrium point was reached. Clearly, the suggestion that the CNS may control simple movements by specifying only final position is attractive because, in this way, a single process would subserve both posture and movement (Kelso and Holt, 1980; Sakitt, 1980). The details of the trajectory would, in fact, be determined only by the inertial and viscoelastic properties of the limbs and muscles. However, recent experimental findings reviewed in the next section, indicate that the CNS actively controls the trajectory in addition to the final position.

TRAJECTORY FORMATION IN SINGLE-DEGREE-OF-FREEDOM ARM MOVEMENTS

The goal of this series of experiments was to determine whether the final position control model is sufficient to account for all of the characteristics of elbow movements. This has been investigated by determining the time course of the neural signals executing the transition from an initial to a final position. If the transition to the final alpha-motoneuronal levels is achieved briskly (compared to the time course of the development of a motor unit twitch in response to a single action potential), then the speed of the movement could not be centrally controlled but would be determined only by the inertial and viscoelastic properties of the musculoskeletal apparatus. Alternatively, if the change to the final alpha-motoneuronal signal occurs slowly, this would indicate that there is active central control of the trajectory of the movement, in addition to control of the final position.

Monkeys performing a pointing task similar to the one described in the previous sections were studied (Fig. 4). Again, the main experimental procedure involved the use of force and positional disturbances which were applied with a torque motor coupled to the shaft of the pivot arm on which the elbow rested (Fig. 4). In some experiments the animal was prevented from detecting disturbances of forearm position by surgically interrupting sensory roots conveying afferent activity from the arm, neck, and upper torso (C1 to T3). After deafferentation, the forearm pointing responses, which had been learned in the preoperative state, could be easily evoked by presenting the targets (Bizzi et al., 1981, 1982). The movements were similar to those observed before deafferentation. As in the preoperative recordings, the EMG activity usually appeared as a moderate burst that gradually blended with the tonic activity characteristic of the holding phase. Although it is entirely possible that deafferentation, like any other CNS lesion, might have induced modifications in motor programming, the fact that we obtained similar results in intact and deafferented animals suggests that the same basic mechanism continued to control these simple movements.

Three experimental paradigms were used to determine the time course of the alpha-motoneuronal signals. The first two experimental manipulations relied on quickly forcing the forearm into the upcoming new final position. As the intact animal began its movement toward a new target position, a brief torque pulse (150 msec), the onset of which was triggered by the initial increase in EMG in the agonist muscles, drove the elbow quickly to the intended final position (Fig. 6B and C). However, instead of remaining in this position, the forearm returned to an intermediate point between the initial and final positions before reversing direction again and then continuing its trajectory toward the position specified by the target. Note that the return movement to the intermediate position was in the direction of extension, whereas the EMG activity was present in flexor muscles. This experiment suggests that these simple forearm movements do not result from rapid shifts in the equilibrium point. According to the hypothesis of final position control, we would expect the steady-state equilibrium position to be achieved after a delay due only to the dynamics of muscle contraction. Individual motor units, recruited at low levels of force, reach their peak force in 60 to 80 msec after the onset of the action potential in the muscle fiber (cf, Milner Brown et al., 1973; Collatos et al., 1977; Desmedt, 1981a). Our results indicate that for a movement of at least 600-msec duration (Fig. 6), the mechanical expression of the alpha-motoneuronal activity does not reach steady state until at least 450 msec have elapsed following the onset of action potentials in the muscle.

Similar conclusions can be drawn from a second experiment, which was performed in deafferent animals. The torque motor was used to displace suddenly and maintain the arm to what would be the location of the next target (Fig. 7A). The animals could not

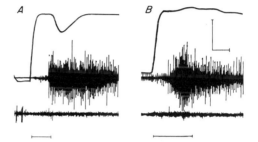

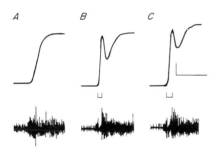

FIG. 6. Torque pulse assisting the natural movement of the arm toward the target position. *Upper trace:* Arm position with an elbow angle of 90° at the scale midpoint flexion upward. *Lower trace:* Flexor (biceps) EMG. *Bar* beneath the position trace indicates duration of the torque pulse. *A:* Control movement in the absence of a torque pulse. *B* and *C:* Two movements with assisting torque pulses. The arm reaches the intended target position early in the movement, transiently returns to an intermediate position, and then proceeds to the final equilibrium position. Note the unloading reflex in the EMG trace in *B* and *C*. Flx = flexion; Ext = extension.

FIG. 7. *A:* Forearm movements of a deafferented animal in the absence of visual feedback. Displacement of the forearm to a flexion position at which a target light was displayed. At the termination of the servo motor action, (indicated by *horizontal bar*), note the movement of the forearm toward extension and subsequent return to the position specified by the target. Flexor activity evoked by the target light is similar to that which was observed during undisturbed movements of same amplitude. Lower EMG corresponds to extensor muscles. There have been 200 instances of this behavior observed in 2 animals. *B:* Same animal. Displacement of the forearm to a flexion position at which a target light was displayed. No movement of the forearm at the termination of servo motor action. The EMG activity from flexors is triggered by the appearance of the light, and is similar to that which was observed during undisturbed movements. Lower EMG corresponds to extensor muscles. 220 instances such as the one displayed here have been observed in 2 monkeys. Time calibration, 550 msec. *Vertical bar* represents joint angle position, 20°. (From Bizzi et al., 1982, with permission.)

have expected a reward, as no new target was illuminated. In fact, because of the absence of any proprioceptive or visual information regarding arm position, the animals were unaware of the displacement. Now, with the arm still constrained by the servo motor, the target light corresponding to the new arm position was illuminated. To the trained monkeys, the appearance of this target light represented a signal to start the neural process involved in pointing to the target. This process became manifest through the appearance of EMG activity in the proper set of muscles, after the usual reaction time (Fig. 7A). After a predetermined time had elapsed following the onset of the EMG activity, the torque motor was turned off, releasing the arm. At this point, the arm was in exactly the correct position for receiving a reward. It is therefore remarkable that the forearm did not remain stationary. Instead, it moved toward the position from which it had originally been displaced, and then changed direction and returned to the position specified by the target light (Fig. 7A). While the to-and-fro movement took place, the agonist muscle developed an EMG pattern comparable to that observed during normal, undisturbed movements. Thus, in the presence of flexor muscle activity, we observed movement in an extensor direction. This remarkable finding cannot be explained if muscles are regarded purely as force generators, but is readily explained if the length dependence of muscle force is taken into account. It should be pointed out that, if alpha MN activity evoked by the target light had rapidly achieved levels appropriate for the new final position, then no return movement should have taken place (see schematic Fig. 5). The fact that a return movement did occur indicates that the control signal shifted slowly toward the final position. This conclusion is consistent with the observation that the amplitude of the movement toward the starting position decreased as the period of servo restraint of the arm was prolonged. Finally, when the servo action was maintained after the appearance of the evoked EMG activity for a period corresponding to the normal movement duration, then no significant movement of the arm was observed after it had been released by the servo (Fig. 7B). These findings suggest the existence of a gradually changing control signal during movement of the forearm from one position to the next and are not consistent with a view postulating a step-like shift to a final equilibrium point. Beyond this, these experiments indicate that the neural input to the muscles can be interpreted as specifying an equilibrium position plus a stiffness about the equilibrium position. Thus, in the transition from the initial position to the final position, the alpha-MN activity is defining intermediate equilibrium positions, which constitute a reference trajectory, whose end-point is the desired final position. As a result, following the cessation of the assist pulse (see Fig. 6), or at the cessation of torque motor action (as in Fig. 7), the limb heads for an intermediate position before reaching its final termination.

In a third set of experiments, further confirmation of the view indicating a gradual change in the control signal establishing the final equilibrium point was obtained. Both intact and deafferented animals were used and the following procedures adopted. Before the onset of visually triggered movements, the limb was clamped in its initial position; it was released at various times after the onset of evoked agonist EMG activity. The duration of the holding phase was varied randomly from 100 to 600 msec. Because of the deafferentation, there was neither an increase in EMG activity during the holding phase nor a pause in the agonist EMG activity after the sudden release. In intact monkeys, both events were routinely observed. The acceleration of the limb immediately after the release was measured and plotted as a function of the holding time, i.e., the time elapsed since the beginning of the EMG activity in the agonists. The plot of the acceleration in intact and deafferented monkeys showed a gradual increase for holding times up to 400 to 600 msec.

These findings are consistent with the notion of a reference trajectory which specifies a gradual shift in the equilibrium point. As the equilibrium point moves further away from the position at which the limb had been restrained, progressively larger torques are generated, resulting in progressively increasing values of acceleration following release and progressively faster movement trajectories. These findings are not in accord with the hypothesis postulating that arm trajectory is controlled by a simple rapid shift to a final equilibrium position. According to this hypothesis, the trajectory of the released arm should have been simply delayed, but its shape should not have been affected.

Physiologically, we do not know how the gradual shift in equilibrium point is programmed. Some of the factors responsible for the progressive increase in tension are the twitch contraction time of the muscle fibers (about 80 msec, see Collatos et al., 1977), the recruitment order, and the firing rate of arm muscle MNs (Henneman et al., 1965a,b; Tanji and Kato, 1973, 1981; Desmedt and Godaux, 1977a,b; Freund and Büdingen, 1978).

It should be emphasized that the results described here have been obtained by analyzing forearm movements performed at moderate speeds. In very fast movements, the shift in equilibrium point must be more abrupt and may even transiently involve a shift to a position beyond the intended equilibrium point. This would amount to a pulse step command of the type known to control eye movements and fast limb movements (Robinson, 1964; Desmedt and Godaux, 1978; Freund and Büdingen, 1978; Ghez and Vicario, 1978a,b). Thus, it is conceivable that, for simple one-degree-of-freedom movements such as those discussed in sections I and II, one of the roles of the reference trajectory is to specify movement speed. The notion of reference trajectory will be taken up again in the context of multijoint arm movements.

TRAJECTORY FORMATION IN MULTIJOINT ARM MOVEMENTS

The studies outlined in the first two sections have led to a theoretical framework dealing with the control of simple one-degree-of-freedom movements. We now wish to extend this framework to include the case of multijoint movements. It must be realized that multijoint movements do not involve just the same control problems as one-degree-of-freedom movements, the only difference being a compounding due to the several degrees of freedom. Rather, there are entirely new control issues which are not present in the case of one-joint movements. Therefore, we begin by discussing these new problems. Next, we describe some properties of human arm movements. These properties allow an insight into what multijoint control strategies are employed. We will then be in a position to discuss how the single-joint control concepts can be brought to bear on the problems of multijoint control.

Multijoint arm movements involve kinematic and dynamic issues which do not arise in the case of one-degree-of-freedom movements (Raibert, 1976; Horn and Raibert, 1977). The kinematic problem is that of trajectory formation. "Trajectory" refers to the path taken by the hand as it moves from one location in movement space to another and the speed with which the hand moves along the path. For a one-joint movement about the elbow with the wrist fixed, the path of the hand is mechanically constrained to an arc with a radius equal to the length of the forearm. Consequently, the speed and stiffness profiles of the movement must be controlled, but only the final position of the path need be specified. If, in addition, rotation about the shoulder joint is allowed, then the hand could approach a target along any of a vast number of paths. Now the controller must plan the entire path in addition to the speed and stiffness characteristics, and the path, speed, and stiffness of the hand will be the result of a complex interplay of the effects of the rotation of each of the involved joints.

Three differences between the mechanical properties of multijoint movements and those of one-degree-of-freedom movements are of particular interest. First, in multijoint movements, the torque required to move one joint is dependent on the position of the other joints. Second, there are joint interactional effects which result from muscles which span more than one joint (e.g., biceps brachii). Third, the dynamic formulations of multijoint movements contain interaction terms which are of purely mechanical origin. As a result of these cross-coupling effects, the rotation of one joint due to shortening of a muscle anatomically associated with that joint will effect movements of all of the other joints in the linkage. One interaction torque results from reactional forces, which are proportional to joint acceleration. As illustrated in Fig. 8A, if joint 1 is caused to rotate in the indicated direction, a reaction torque will act on joint 2. Similarly, joint 1 will rotate if a torque is produced at joint 2. Figure 8B illustrates the centripetal interaction torque. Here, line 2 is assumed to have been caused to rotate about joint 2 and is represented as a mass at the end of a rotating cable; the centripetal force acting on the mass establishes a reaction force (proportional to the square of the joint-2 angular velocity) which acts on joint 1 through level arm L. A movement at joint 1 would induce a centripetal torque at joint 2. A third interaction torque, referred to as the Coriolis torque, is proportional to the product of the angular velocity of the two joints and acts about joint 1. The magnitudes of all these interactional forces are affected by the particular trajectory of the arm and can often be quite large (Flash and Hollerbach, 1981). If there is no compensa-

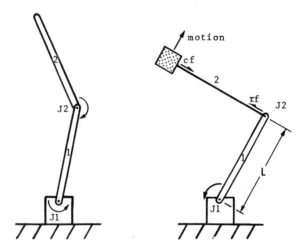

FIG. 8. Schematic illustration of interaction torques. **Left:** Horizontal two-link arm is fixed at one end to a wall by way of joint 1 (J1). Second degree of freedom provided by joint 2 (J2). If J1 is caused to rotate in indicated direction, a reaction torque will result in J2 motion. **Right:** Motion at J1 caused by centripetal interaction torque (see text). rf = Reaction force; cf = centripetal force.

tion for these coupling effects, motion about one joint would cause other joints to flail, so that errors in joint motion and hand motion would occur.

These issues in multijoint control raise many questions. Does the motor system first develop a trajectory plan and then determine those muscle stiffnesses which will generate appropriate joint movements? If so, what are the criteria used to select a particular trajectory? Is the trajectory plan specified in joint coordinates, or is the plan Cartesian, specifying the path of the hand and, therefore, requiring a transformation into joint coordinates? How are the necessary joint torques determined? Are they computed on a real-time basis, or is some memory approach used? What consideration is given to the joint interaction torques? If compensation is provided, are specific compensating torques injected, or is the stiffness of the linkage raised in order to overwhelm their effects? Finally, what is the role of final position control and of reference trajectories in the production of multijoint movements? Are there characteristics of these movements which, as in single-degree-of-freedom movements, indicate that factors other than final position control are operative? Does a reference trajectory specify speed, path, and stiffness?

In an effort to gain some insight into these questions, two-degree-of-freedom arm movements performed by normal adult humans were recorded. The wrist was braced, so that only movements about the shoulder and elbow joints were allowed. The subject, maintaining his arm in a horizontal plane passing through the shoulder (so that gravitational effects were constant), grasped the handle of a light-weight two-joint mechanical linkage and moved the handle to each of a series of visual targets which were distributed in the movement space (Fig. 9A). The geometry of the experimental arrangement was such that the signals from potentiometers of the two mechanical joints could be used to compute the hand and joint trajectories.

In an initial study using this apparatus, Morasso (1981) instructed subjects simply to move their hand from one target to another; there was no instruction regarding speed or accuracy. Data similar to Morasso's is presented in Fig. 9, where several movements are illustrated. Two findings are of interest. First, the path taken by the hand from one target to another was usually straight or only gently curved (Fig. 9B). A similar result has been obtained in the monkey (Gilman et al., 1976). This finding may be expected on the basis of casual observation of human behavior. It is of interest because the straight hand movements result from the combined effects of rotation of both joints; convoluted movements would result if the two degrees of freedom were not perfectly coordinated. Therefore, the tendency to produce straight hand paths suggests that path planning must occur, in addition to joint final position control.

A second finding reported by Morasso (1981) is illustrated in Fig. 9C, D, and E, where kinematic data are presented for three movements which were performed in different parts of the work space. The data show that the joint position and the joint velocity traces vary widely from movement to movement, while the speed of the hand is always roughly bell-shaped, even when the joint angular velocities are complex. The work space independence of the hand speed profile is consistent with, but of course does not prove the notion that the CNS plans a movement in terms of the hand kinematics and then transforms the plan into joint coordinates. This is in agreement with Bernstein's statement: "The hypothesis that there exist in the higher levels of the CNS projections of space, and not projections of joints and muscles, seems to me to be at present more probable than any

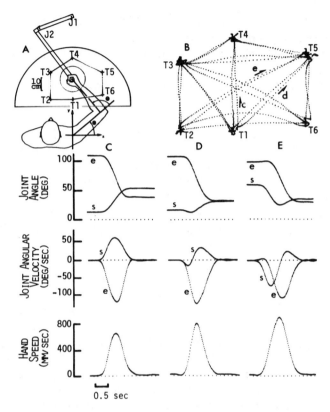

FIG. 9. A: Plan view of a seated subject grasping the handle of the two-joint hand-position transducer (designed by N. Hogan). The right arm was elevated to shoulder level and moved in a horizontal work space. Movement of the handle was measured with potentiometers located at the two mechanical joints of the apparatus (J1, J2). The subject was positioned so that J1 lay on the Y-axis. A horizontal semicircular plate located just above the handle carried the visual targets. Six visual target locations (T1 through T6) are illustrated as crosses. The digitized paths between targets and the curved path were obtained by moving the handle along a straight edge from one target to the next, and then along a circular path; movement paths were reliably reproduced. **B:** A series of digitized handle paths (sampling rate, 100 Hz) performed by one subject in different parts of the movement space. The subject moved his hand to the illuminated target and then waited for the appearance of a new target. Targets presented in random order. *Arrows* show direction of some of the hand movements. **C, D, E:** Kinematic data for three of the movements the paths of which are shown in **B**. Letters show correspondence (e.g., data under **C** are for path c in **B**). e = elbow joint; s = shoulder (angles measured as indicated in **A**).

other" (1967, p. 50). In Bernstein's view, such a scheme reduces the number of variables controlled by the CNS.

Since subjects tend to produce roughly straight hand paths when given no instruction other than to move the hand to a target, a group of 9 subjects was instructed to approach targets by way of curved paths (Abend et al., 1982). There was no instruction regarding accuracy or speed. The subjects first performed a series of movements in the absence of visual feedback of arm position and then repeated the experiment with feedback present. Two unexpected results are illustrated in Fig. 10 where, for each of six movements, the hand path and the time course of the hand speed and path curvature are presented. First, paths 3, 4,

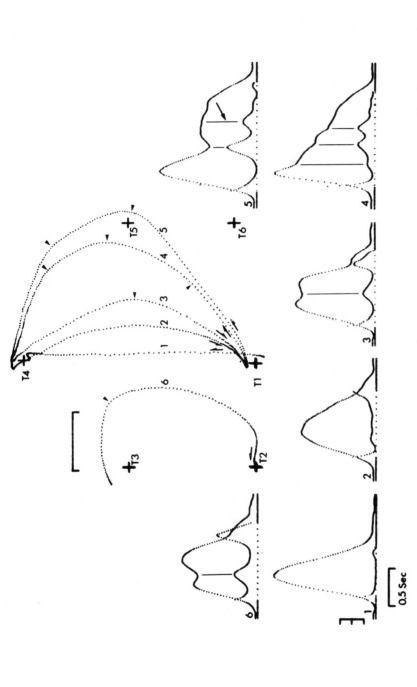

FIG. 10. Six movements performed by the same subject in the absence of visual feedback. Movement 1: Subject told only to move her hand to the target, T4. Movements 2, 3, 4, and 5: Subject told to move her hand between targets T1 and T4 by way of smoothly curved trajectories, ranging from shallow to highly curved ones. Movement 6: Subject told to make curved movement from T3 to T2. Targets same as in Fig. 9. The 6 movement paths are superimposed on one set of target symbols. *Arrows* indicate the direction of hand movement along the path. Numbers indicate correspondence between the hand path of a movement and the superimposed hand speed and path curvature profiles. In each case, the curvature profile is the more shallow of the two curves. The data are plotted from the time of illumination of the target light. Because subjects often began and ended movements with paths so highly curved that the curvature values are off scale, the curvature is plotted only for those times during the movement when the hand speed is greater than 150 mm/sec. *Bar* near hand paths = 10 cm. For the speed and curvature profiles, *full ordinate bar* = speed of 200 mm/sec; *half bar* = a curvature of 1/100 mm (inverse of radius of curvature). For paths 3, 4, 5, and 6 *arrowheads* = local curvature maxima along the trajectory; curvature maxima are also denoted by *vertical lines* over the curvature profiles.

and 5 appear to be composed of a series of gently curved segments, which meet at more highly curved regions. For example, arrowheads denoting local peaks in curvature appear to divide movement 4 into 4 segments. Second, the hand speed profiles are irregular compared to those associated with straight trajectories. There are valleys or inflections which are temporally associated with peaks in the path curvature; the hand tends to slow during the more highly curved parts of the trajectory. Curvature peaks with associated speed irregularities are typical of curved movements. Some unusual features have been noted, however. Occasionally, subjects produce curved trajectories with a relatively constant curvature and with a speed profile which is only vaguely irregular (movement 2). Infrequently, a local curvature peak is not reflected in the speed profile (Fig. 10, movement 5, arrow), or a speed irregularity may occur in the absence of a change in curvature; these phenomena are most often seen when a subject chooses to move his hand quite slowly, in which case even the hand speed profile of a straight trajectory may be irregular. The results were independent of whether visual feedback of arm position was available.

The path and speed discontinuities, rather than being the properties of a special kinematic pattern, characterize a wide variety of movements. This has been investigated in three ways (Abend et al., 1982). First it was shown that the characteristics are independent of the part of the horizontal work space employed (e.g., Fig. 10, movement 6). Second, the results were unchanged when the subject attempted to mimic constant-curvature, nonconstraining guide paths. Third, the wrist splint was removed, and the horizontal movement plane of the hand was lowered from shoulder to waist level. In this way, the work space of the arm was altered, a small vertical movement component was required, and the number of degrees of freedom was increased. Despite these changes, the character of the hand kinematics was unchanged.

There is no a priori reason why the hand should slow when the curvature is high if the hand speed is an expression of the rotation of each of the active joints. Therefore, the question arises as to whether the speed valleys might reflect inertial effects of one of the two active joints. Ordinarily, during the course of a curved movement, a joint comes to zero speed and reverses direction. However, two findings argue against the speed valleys being a reflection of a joint reversal. First, joint reversals were required for some straight movements but did not impart a hand speed valley (Fig. 9E). Second, subjects occasionally produced curved movements with bell-shaped speed profiles, even though a joint reversal was required. These points suggest that the speed valley is the result of factors imposed by the multijoint controller rather than events involving individual joints.

In summary, there appears to be a tendency to move the hand along roughly straight paths, and this tendency persists even when the subject is in the midst of producing curved trajectories. Usually, curved paths had a segmented appearance, as if the multijoint controller were approximating a curve with a series of low curvature segments. In addition, there are hand speed characteristics which appear to be related to the character of the path. The hand speed profile could be viewed as a series of peaks, each peak corresponding to a straight segment of the path. Discontinuities in movement characteristics have been noted previously (Brooks et al., 1973; Navas and Stark, 1968; von Hofsten, 1979; Viviani and Terzuolo, 1980). The basis for these findings is unknown, but, as discussed in the next section, we believe that they may serve as clues in understanding the organization of the central motor controller.

ISSUES IN MULTIJOINT CONTROL

The studies outlined in this chapter have been directed at understanding the neural mechanisms subserving arm trajectory formation and the maintenance of posture. We have discussed the concepts of final position control and reference trajectory in terms of single-degree-of-freedom movements. In this section, we will employ these ideas in discussing the characteristics of multijoint movements.

We wish to emphasize at the outset that the multijoint control problems requiring explanation have to do with trajectory formation, and not posture control. A particular arm posture is a function of the equilibrium position of each joint. There is not reason to expect that posture maintenance of a multijoint limb is managed by the CNS in any fundamentally different manner than that of a single-degree-of-freedom limb.

As in the case of elbow movements, there are several features of multijoint movements which indicate that the trajectory of the movement is controlled, in addition to the final position. These features include the tendency for the hand to move along straight paths, the segmented appearance of the curved movements, and the temporal correspondence between path curvature peaks and speed discontinuities. It is unlikely that such movement characteristics can be accounted for by assuming that movements are simply passive transitions to a new posture and governed only by the inertial and viscoelastic properties of the arm. Instead, they probably indicate the presence of specific trajectory strategies that are determined by a central reference signal which optimizes some variable such as the distance moved, the energy dissipated, or mechanical stress and wear.

For single-degree-of-freedom movements, a reference trajectory may determine a dynamic equilibrium in the interaction of the muscles which are active about a moving joint. The equilibrium point chosen would determine the stiffness of the joint. It is interesting to speculate that the trajectory (path and speed) of multijoint movements is also determined as a dynamic equilibrium state. Now, however, in order to control the conformation of the entire linkage, the reference signal would have to specify the interaction of all of the arm muscles. We have already speculated in section III that multijoint movements may be planned in terms of the hand. This notion could provide an organizing principle. It may be optimal to plan movements in terms of the stiffness of the endpoint of the linkage (the hand). In this way, the motion of the endpoint in response to an arbitrary external force input would be specified (Hogan, 1980). A hand stiffness plan may be implemented by transforming a hand signal into the necessary joint reference torques (Bernstein, 1967; Saltzman, 1979). Alternatively, the plan may be generated by a reference signal which codes the appropriate interaction of all of the arm muscles, in which case no explicit coordinate transformation would be required.

ACKNOWLEDGMENTS

This research was supported by National Institute of Neurological and Communicative Disorders and Stroke Research Grants NS09343 and NS06416, National Institute of Arthritis, Metabolism, and Digestive Diseases Grant AM26710, National Aeronautics and Space Administration Grant 22-009-798, and National Eye Institute Grant EY02621.

Motor Control Mechanisms in Health and Disease,
edited by J. E. Desmedt.
Raven Press, New York © 1983.

Regulatory Role of Proprioceptive Input in Motor Control of Phasic or Maintained Voluntary Contractions in Man

*Jerome Sanes and Edward V. Evarts

Laboratory of Neurophysiology, National Institute of Mental Health, Bethesda, Maryland 20205

The nature of the regulatory role of proprioceptive information in motor control has been increasingly questioned since Lashley's (1917) observations of the residual ability of a patient to move a deafferented lower extremity. More recently, several studies have shown a lack of major deficits in accurate head or limb positioning following surgical deafferentation in monkeys or ischemic deafferentation in humans (Bizzi et al., 1976, 1978; Polit and Bizzi, 1979; Kelso and Holt, 1980). There are, however, other experiments demonstrating the importance of afferent input in a variety of tasks performed by humans. For example, alterations of position sense and sense of effort were observed following limb deafferentation (Goodwin et al., 1972; Gandevia and McCloskey, 1977a, 1977b). Furthermore, performance of fine motor tasks, such as reproducing alphabetic characters (Laszlo and Bairstow, 1971), was disrupted when a subject's limb was deafferented by ischemia. And it is noteworthy that acute inactivation of the gamma loop in humans impaired the ability to activate motor units tonically (Fukushima et al., 1976), although phasic activation was not impaired.

Many of the previous studies aimed at understanding the role of afferent inputs in the course of active movement have investigated the effects of relatively large disturbances delivered during relatively large movements (Polit and Bizzi, 1979; Kelso and Holt, 1980; Schmidt and McGown, 1980). However, there is substantial evidence that cutaneous and muscle receptors have especially high sensitivity for small-amplitude signals (Matthews and Stein, 1969; Poppele and Bowman, 1970; Knibestol, 1975; Goodwin et al., 1978; Knibestol and Vallbo, 1980), and this suggests that the role of these receptors in voluntary movements might be more clearly determined if fine movements were studied. The present experiments were undertaken to determine whether peripheral disturbances delivered to the arm caused greater impairments of small as compared to large voluntary movements.

EXPERIMENT 1

A portion of a larger investigation on the effects of peripheral disturbances delivered during movements of the forearm is de-

---

*To whom correspondence should be addressed: Laboratory of Neurophysiology, Room 2D-10, Building 36, National Institute of Mental Health, 9000 Rockville Pike, Bethesda, Maryland 20205.

scribed here. We characterized the accuracy of large and small movements performed in the absence or presence of brief mechanical disturbances. We hypothesized that delivery of brief mechanical disturbances during movement would cause relatively greater impairments in small than in large movements.

Methods

Eleven subjects without neurologic abnormalities participated in this experiment. Subjects grasped a handle that could be rotated by pronation or supination of the forearm. The handle was coupled to the axle of a servocontrolled brushless DC torque motor (Aeroflex TQ-52). Each subject viewed two vertically oriented oscilloscope beams, one beam representing a target and the other beam corresponding to the orientation of the handle. Subjects were instructed to maintain alignment between the two beams and then to reestablish alignment when the target beam jumped to a new position. The magnitude of handle rotation necessary to establish realignment was 30° in certain blocks of trials, 10° in other blocks of trials, and 3° in still other blocks of trials. Prior to a given block of trials, subjects were given sufficient practice to calibrate themselves. For all blocks of trials, the target remained displaced from the center of the display until subjects moved the handle and reestablished realignment for 1.2 to 2.0 sec in a small zone (± 5% of the movement size) surrounding the target. At the end of the hold period, the target jumped to the center of the video tube. Subjects were instructed to move the handle as quickly and as accurately as possible to match the handle beam with the target beam. The final position for all movements was 0° of forearm rotation (i.e., vertical orientation of the handle).

Eight different trial types, presented in a pseudorandom sequence of 48 events, were performed by all subjects. The forearm was supinated for one-half of the movements and pronated for the remaining one-half. The display indicating handle orientation was blanked for half of the trials. The handle movement was stopped for 100 msec for 8 of the 48 trials. The disappearance of visual feedback of position and the obstruction of the forearm movement occurred when the handle was moved beyond the initial hold zone. The perturbation trials were selected so that in a sequence of 48 trials the forearm trajectory was briefly stopped two times each when there was or was not visual feedback for each direction of movement. Subjects first performed a series of 250 practice trials, and were encouraged to perform with ± 15% error. No perturbations were delivered during practice. On a subsequent occasion (1 to 5 days later) subjects performed an additional 250 test trials during which at least 10 perturbation trials of each type were presented. Data were analyzed with PDP-11 computer systems. Figure 1 illustrates the behavioral situation and the general scheme for data analysis. After onset of the movement to realign the handle beam with the centrally located target beam, relative error was measured when velocity first reached zero (dynamic error) and 500 msec after this point (static error). Relative errors in performance were computed by dividing the movement size (in degrees) into the angular difference between the actual handle position and the vertical handle position called for by the target and then multiplying this result by 100. Movement time (MT) was measured within 5-msec accuracy from the first change in velocity to the first zero-cross of velocity. The MT analysis was based on data obtained from 9 subjects; MT information was not recorded from 2 subjects. Dynamic and static relative error and MT were averaged across conditions and analyzed for significance with a repeated measures analysis of variance.

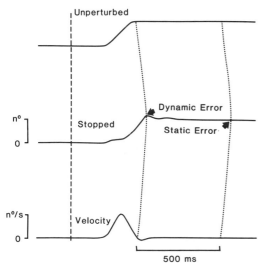

FIG. 1. Diagrammatic representation of the task performed in Experiment 1. The *upper 2 traces* show schematics of the handle position for an unobstructed and a stopped supination movement, respectively. Note in the *middle trace* that the handle is stopped for 100 msec nearly immediately after the onset of movement. The *lower trace* represents handle velocity for an unopposed movement. The left most vertical dashed line indicates when the target beam jumped to the middle of the video tube (i.e., the cue requesting the subject to begin a rapid movement to realign the handle position with the target beam). The other vertical dashed lines indicate when dynamic and static errors in performance were measured (see text for definition). The vertical scales to the left represent angular displacement of the handle (in degrees) and handle velocity (degrees/sec). The bars are scaled ambiguously, since in the analyses the movements were scaled in the ratio of 10:3:1 (for 3°, 10°, and 30°, respectively).

Results

Dynamic Effects

Unopposed supination movements of subject MD under visual open-loop (i.e., no visual feedback) conditions are illustrated in Fig. 2A and C. In this figure the 3°, 10°, and 30° movements have been scaled with ordinates being in the ratio of 3:10:30; put in another way, the scale has been selected so as to make the figures comparable in terms of relative error. After brief obstructions of MD's forearm movements relative dynamic error increased for all movement amplitudes (Fig. 2B). In addition, enhancement of dynamic error was greater when the forearm was stopped during small as compared to large movements.

The grouped data were similar to the observations seen for subject MD. The most significant observation for the group of subjects was the inverse relationship between relative dynamic error and movement size. As movement size decreased the dynamic error increased, $F(2, 20) = 28.0$, $p < 0.0001$. The increase in dynamic error was linear, $F(1, 10) = 109.09$, $k < 0.0001$. Figure 3 illustrates the dynamic error when the movement was unopposed (left sections of figure) or stopped (right sections of figure), respectively. When movements were obstructed the relative dynamic error increased, $F(1, 10) = 63.44, p < 0.0001$. Furthermore, there was substantially greater increase in dynamic error for the small in comparison to the large movements following brief obstructions of forearm movement, $F(2, 20) = 22.45, p < 0.0001$. Interestingly, dynamic error was smaller when visual feedback of the handle position was absent $F(1, 10) = 20.83, p < 0.001$. This main effect was significant because dynamic error was greater when perturbations were presented during closed-loop movements, [feedback—perturbation interaction, $F(1, 10) = 30.01, p < 0.001$]. There was no difference in dynamic error for movements of different directions ($p > 0.05$), an expected finding, since the final position was identical for both pronation and supination movements.

Static Error

Figure 4 depicts the grouped data for relative errors of position when subjects

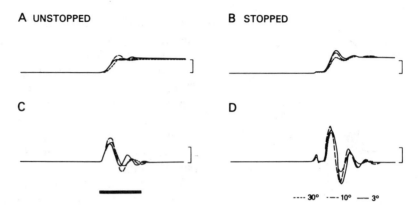

FIG. 2. Supination movements from one subject. Unobstructed handle position (**A**) and velocity (**C**) of movements of 3°, 10°, and 30° from subject MD. To the right are shown the corresponding handle trajectory (**B**) and velocity (**D**) when MD's movements were briefly stopped. Note the increase in dynamic error for all movement sizes and the inverse relationship between dynamic error and movement size when the handle was briefly stopped. *Calibrations:* upper traces (**A** and **B**) 3°, 10°, and 30° depending on movement size (see key at lower right). Lower traces (**C** and **D**) 10°/sec, 33°/sec, 100°/sec depending on movement size. *Horizontal scale* = 500 msec.

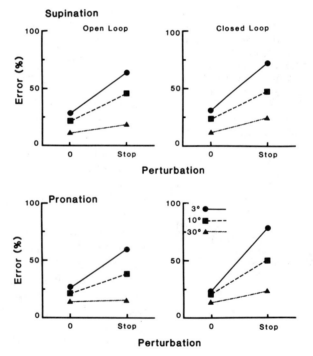

FIG. 3. Dynamic error for the group of subjects. As indicated in the text relative dynamic error (ordinate) increased as the movement size decreased. This was evident for both supination (**Top**) and pronation (**Bottom**) movements, when the handle displacement was unobstructed (0 on *abscissa*) or briefly stopped, and when subjects had visual feedback of handle position (closed-loop) or were unable to see the orientation (open-loop) of the handle position. When the handle was briefly stopped, note that the relative dynamic error increased a greater amount for the small (3°) in comparison to the large (30°) movements. This effect was evident for all movement conditions (supination or pronation; and open-loop or closed-loop feedback).

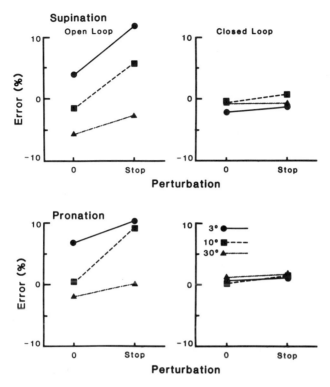

FIG. 4. Static error for the group of subjects. Relative static errors in performance for both supination and pronation were greater for the small (3°) than for the large (30°) movements only when subjects were not able to visually monitor (open-loop) handle position. Injection of the stop during movement increased static error, again only in the visually open-loop condition. When subjects saw the oscilloscope beam that represented handle position, static error (closed-loop) was similar for all movement sizes, and unaffected by delivery of the perturbation.

attained the final intended position. In the absence of visual feedback of the handle's position (open-loop), subjects depended on proprioceptive cues to position the forearm accurately. When both closed- and open-loop movements were included in the analysis of variance, it was found that relative static error increased as movement size decreased, $F(2, 20) = 13.59$, $p < 0.001$. The linear component for movement size was also reliable, $F(1, 10) = 35.61$, $p < 0.0001$. In addition, static error increased when forearm movements were obstructed, $F(1, 10) = 12.33$, $p < 0.01$. There was only a trend for the interaction between movement size and absence or presence of the stop to be significant ($p = 0.12$). Thus, although the static error for small movements qualitatively increased when perturbations were presented (see left portions of Fig. 4 top and 4 bottom), this observation was not reliable for this group of subjects. A separate analysis of variance based only on the static error obtained when subjects were deprived of visual guidance revealed reliability of all the main effects noted above (all p's < 0.01), but still failed to indicate an interaction between movement size and perturbation for static error ($p = 0.19$). When 3° movements were briefly stopped and there was no visual guidance planned comparisons of mean static error revealed that static error was larger for that condition than static error when perturbed visual closed-loop and both types of nonperturbed movements were performed ($p < 0.05$, Student's t-test, 2-tailed). Static error was relatively constant at ± 2% (right portion of Fig. 4) when subjects had visual feedback of the handle position. Decreases in movement size or delivery of perturbations did not alter accurate realignment of the handle position when subjects were able to monitor handle positions visually.

Movement Time

The MTs for all behavioral conditions are available in Table 1. Movement time decreased when the forearm encounterd an obstruction while moving toward the target, $F(1, 8) = 316.29$, $p < 0.0001$. An incidental observation on MT was that supination movements were performed faster than pronation movements, $F(1, 8) = 5.31$, $p = 0.05$. There was no significant effect of movement size or visual feedback condition on MT (p's > 0.05).

EXPERIMENT 2

The results of the previous experiment suggested that accurate performance of small movements is impaired by brief obstructions occurring soon after movement onset. The following experiment was an initial attempt to describe the mechanisms by which peripheral inputs modify motor performance.

Methods

Motor units were sampled from subjects without neurologic abnormalities. Tungsten microelectrodes (tip diameter of 2 to 4 μm, 125 μm shaft diameter) were inserted into the first dorsal interosseous (DI) muscle of the seated subjects while the forearm was pronated. The first metacarpophalangeal joint was positioned over the axle of the torque motor, the index finger was wedged into a holder that prevented finger movement, and the thumb was immobilized in a fully extended position. Ground and reference cup electrodes were placed on the dorsal surface of the hand. All tests were done in the isometric situation. Subjects were instructed to perform one of two tasks. In the first task, subjects sought to generate a single motor unit impulse, a task that required a brief period of practice for each new motor unit. Subjects received both visual and auditory feedback as to success or failure of the attempt to generate one impulse. For the second task, subjects maintained a sustained rate (about 10 Hz) of motor unit discharge for 5 to 10 min. The torque motor was used to periodically and randomly (4- to 7-sec intertrial interval) stretch or shorten the DI during both of these tasks. A 600-msec trapezoidal stimulus with 50-msec rise/fall time of either 3°, 6°, or 12° amplitude was used. A visual cue provided a signal calling on the subject to generate a single impulse from a motor unit and on certain trials finger displacements were produced either at a fixed delay (50 to 150 msec) after the visual cue or immediately after the occurrence of the first motor unit impulse following the cue. During maintenance of a sustained rate of discharge, displacements occurred either randomly or at a fixed interval in relation to the discharge of the motor unit.

Data were analyzed with a PDP-11/34 computer. The effects of finger displacements on generation of single impulses were evaluated by (a) comparing the proportion of successful generations of single spikes in the presence or absence of disturbances, and (b) comparing the number of spikes

TABLE 1. *Movement time (msec) grouped according to direction and size of movement*

	Supination		Pronation	
	No-stop	Stop	No-stop	Stop
3°	192.5	154.4	200.8	157.9
10°	199.6	158.6	204.0	171.2
30°	209.2	162.8	215.3	172.1

generated when the finger was displaced to the number of impulses that occurred during control trials. Effects of displacement on tonic discharge were determined by comparing the discharge rates of motor units prior to and following displacement of the index finger.

Results

Before studying displacement effects on motor unit discharge, there was determination of the isometric force level at which the motor unit was recruited. Thirty-four motor units from DI with recruitment thresholds ranging from 0.016 ft/lb to 0.512 ft/lb were studied.

Phasic Activation

Effects of finger displacement on activation of single motor unit spikes can be considered either in terms of success (i.e., generation of one impulse rather than more or less than one impulse) or in terms of the quantitative effects of the displacement or the number of impulses. The results of the first analytic approach (success or failure to generate a single impulse) are shown in Table 2. Each motor unit was placed in one of 5 discrete categories according to the percentage of trials that (for the particular unit) a subject successfully generated a single motor unit impulse. When the finger was not displaced (i.e., control trials), subjects successfully generated one impulse in most of the motor units on more than one-half of the attempts. Only 4/26 tests in 26 motor units showed a success rate of less than 40% for single-spike generation. In contrast, for 78 tests of the effects of displacements (of different direction and amplitudes) on single-spike generation, there were 52/78 cases in which success was less than 40%. Finger displacements significantly reduced the number of successful attempts to generate one spike when stretches of 3° and 6° and shortenings of 6° were delivered, $\chi^2 > 7.75, p < 0.01$. When both directions of finger displacement were considered, success in generating one action potential was impaired in 19 of the 26 motor units tested. Successful control of 5 motor units was selectively disrupted by passive DI stretch and reliable control of 4 motor units was selectively affected by passive shortening of DI. The successful generation of the remaining 10 single-spike motor units was disrupted by changes in DI length in both directions.

Examples of the results of the second approach (quantitative effects of the displacements on number of impulses) to determine the effects of displacement on single-spike generation are illustrated in Fig. 5. Motor units were classified as responsive to displacement if the number of impulses occurring following displacement was different from the number of impulses observed during control trials. For 22 of the 26 motor units, the number of impulses was changed by displacing the finger. For 11 of these responsive motor units firing rate was changed almost exclusively during the 600-msec displacing stimulus. An example of a motor unit of this type is shown in Fig. 5A.

TABLE 2. *Frequency distribution of successful control of single impulse*

	0–20%	21–40%	41–60%	61–80%	81–100%	Total
CONTROL	2	2	9	8	5	26
3° Stretch	10	6	1	3	1	21
3° Shorten	7	3	5	5	1	21
6° Stretch	11	3	1	3	0	18
6° Shorten	12	0	5	0	1	18

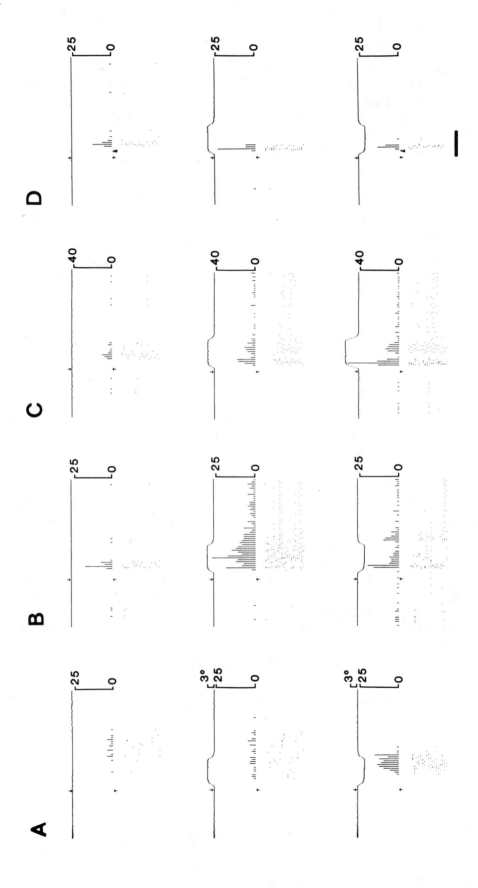

For this motor unit (threshold = 0.44 ft/lb), peripheral disturbances that shortened the DI evoked greater changes in discharge frequency than stretches of DI. The discharge frequency of another 11 motor units increased at short latency after ramp onset and the impulse rate then remained above zero for a variable period of time after DI was passively returned to the initial length. A motor unit of this type (threshold < 0.002 ft/lb) is illustrated in Fig. 5B. Two additional effects of delivering peripheral disturbances to the finger when subjects sought to generate one impulse were graded sensitivity to the amplitude/velocity characteristics of the disturbance (Fig. 5C) and the earlier occurrence (relative to the request) of impulses in 15 motor units when the DI was stretched (Fig. 5D). For these motor units, the latency of the first impulse following the request was shifted 40 to 120 msec sooner when the muscle was stretched. Disturbances that shortened the DI typically did not affect the latency of the phasic response.

Tonic Activation

In general, the effects of finger displacements on tonic maintenance of motor unit were weak. Nevertheless, the discharge frequency of 17 of the 26 units tested was phasically modulated by displacement of index finger during active maintenance of an 8- to 12-Hz frequency of DI motor unit discharge. For 12 of these 17 motor units, the effects on DI motor unit discharge were confined to the first 100 to 200 msec following onset of the stimulus. However, the modulation of motor unit discharge frequency of 5 motor units was often observed soon (< 150 msec) after the return of DI muscle to its initial length. Sustained changes in discharge rate enduring throughout the 600-msec stimulus and beyond, were rarely seen.

Relationship to Motor Threshold

Although the number of DI motor units sampled in this work was small, motor units

---

FIG. 5. Responses of DI motor units to ramp stimuli during attempts to generate single spikes. The 12 pairs of traces shown above represent average responses of DI motor units (lower trace of pair) recorded from 4 human subjects when the length of DI muscle (as shown in the upper trace of pair) was unperturbed, stretched (upward deflection), or shortened (downward deflection). The *full arrows* in each segment indicate the delivery of the cue requesting subjects to attempt to generate a single spike. In each segment of the illustration, the upper set of traces shows a subject's ability to generate single spikes when no disturbance was delivered. Note the intersubject variation in successful attempts concerning (1) the number of impulses and (2) the latency of the first spike following the cue. A: This motor unit (subject AL) was responsive during the 600-msec stimulus, especially during DI muscle shortening of 3° (lower set of traces). The scatter in motor unit impulses observed following muscle stretch was probably related more to the variability in latency to begin motor unit discharge following onset of the cue rather than increased responsiveness to the perturbing stimulus. B: The section shows a sustained increase in motor unit discharge rate (subject VJ) following either 3° stretch or shortening of DI muscle. The response was more vigorous when DI muscle was stretched (middle pair of traces). C: The amplitude/velocity sensitivity of a DI motor unit (subject EE) is shown here. Both 3° and 6° stretches of DI muscle increased the firing rate of the motor unit, with greater changes being observed for larger amplitudes of stretch. D: The latency shift of motor units is depicted here (subject MW). The bolder arrowheads are lined up with the first impulses when the DI muscle was stretched. Note the lag in latency for the series of control trials (upper set of traces) and when DI muscle was shortened (lower set of traces). *Calibration:* Histogram values in impulses/sec. Analog traces in degrees of angular rotation about the metacarpophalangeal joint. Time mark = 500 msec.

with a wide range of thresholds were studied. In agreement with previous reports (Buller et al, 1980) there appeared to be a tendency for low-threshold motor units to be more sensitive to passive changes in the DI length at the velocities used in these experiments, although there were several instances of low- and high-threshold motor units exhibiting nearly equivalent responsiveness to ramp stimuli.

DISCUSSION

A substantial amount of evidence suggests that small movements should be more dependent on peripheral inputs than large movements. Thus, a number of cutaneous and muscle receptors exhibit their greatest sensitivity to small amplitude input signals (Matthews and Stein, 1969; Poppele and Bowman, 1970; Knibestol, 1975; Knibestol and Vallbo, 1980). Furthermore, some reflexes have their greatest gain in response to small displacements (Marsden et al., 1978; Cooker et al., 1980; Hoffer and Andreassen, 1981), and changes in discharge frequency of muscle spindle afferents and motor units is greatest during the initial increases in tension during active movements (Tanji and Kato, 1973; Vallbo, 1973). Behavioral studies have also provided evidence that peripheral inputs contribute significantly to complex or fine motor acts. For example, the precise temporal patterning of EMG activity seen during locomotion is dramatically altered by peripheral disturbances (Forssberg, 1979; Wand et al., 1980) and the finer aspects of position sense, sense of effort, and accurate reproduction of alphabetic characters is severely impaired by removal of afferent input (Lazlo and Bairstow, 1971; Gandevia and McCloskey, 1977a,b; Goodwin et al., 1978).

The results of the present experiments indicate that peripheral disturbances influenced the final positioning of the limb and disrupted voluntary control of motor unit discharge frequency. This result might seem to differ from results of many earlier studies (e.g., Polit and Bizzi, 1979; Kelso and Holt, 1980). There were, however, significant differences between our experiments and those done by others. First, to address the issue of peripheral versus central control of movement we examined a variety of motor tasks. Movements requiring fine control, such as a 3° displacement and generation of a single motor unit impulse, were clearly disrupted by delivery of peripheral disturbances, thus agreeing with the finding of Laszlo and Bairstow (1971). In contrast, large movements and gross control of motor unit impulse rate were disrupted much less than small movements, thus agreeing with Polit and Bizzi (1979), Kelso and Holt (1980), and others. It would appear that peripheral inputs selectively impair performance of small, or precise, motor tasks.

A second difference between the present and many of the other studies was our use of intact subjects instead of deafferented preparations. Current methods of deafferentation surgically remove or functionally inactivate most of the peripheral return from muscles, joints and skin. Although it is true that the available techniques for deafferentation fail to eliminate motor functions that would appear to require the sense of kinesthesia, it is not necessarily correct to interpret residual function as evidence of the unimportance of peripheral input in regulating final limb position. Examples of retained function by deafferented subjects are accurate reaching, grasping and manipulation of objects (Taub et al., 1975) and positioning of limbs in space (Lashley, 1917; Kelso, 1977; Polit and Bizzi, 1979; Kelso and Holt, 1980). The basis for the retained function following deafferentation has been attributed to the extraordinary efficiency and accuracy of central motor commands. The present experiments cast doubt on such a notion, since we have demonstrated (1) a mismatch between intended and actual final position when peripheral receptors of intact humans are acutely stimulated during active movement and (2) precise control of motor unit discharge rate is also disrupted by brief changes in muscle length.

A possible explanation for the retained accuracy of motor performance following either acute or chronic disruption of muscle, joint and cutaneous afferent systems is that subjects rapidly adapt to the absence of peripheral inputs. The adaptive mechanism would necessarily be modification of the motor commands to the muscle and since there is no peripheral return from muscles, joints, or skin receptors adaptive control would be contingent on knowledge of results of each movement.

Preliminary experiments from our laboratory suggest that acute removal of afferent information can be partially overcome provided subjects are aware of the results of their efforts to move. We had hypothesized that removal of peripheral input would impair the ability of subjects to generate single DI impulses, but to our surprise, subjects performed this task accurately, provided visual and auditory feedback were available, until the neuromuscular junction was blocked. An explanation of this result may be found by considering the strategy employed by subjects to perform this task successfully. When attempting to generate a single impulse, a subject emits a central command that he believes will result in a single impulse. The success or failure of the attempt is revealed by visual and/or auditory feedback. If (on a given trial), a subject fails to generate any impulses, effort will be increased on the next trial, whereas if too many impulses occur, effort will be decreased on the next trial. Thus, subjects can recalibrate effort from one trial to the next.

In contrast to the successful ability of deafferented preparations to generate phasic output, tasks requiring maintained muscular effort are not performed reliably by deafferented subjects (Lashley, 1917; Provins, 1958; Fukushima et al., 1976; Gandevia and McCloskey, 1977a,b). In the experiments reported here, tonic discharge of motor units was modestly modified by peripheral inputs in the intact preparation, but when subjects were deafferented there was profound disruption of attempts to sustain DI discharge rates (*unpublished observation*). In this situation, when motor unit discharge and force feedback were available, subjects were able to increase motor unit frequency when requested. However, the frequency of discharge dropped off quickly and then, after a short period (< 500 msec), the discharge rate was increased. When both visual and auditory feedback of the motor unit discharge were unavailable, subjects increased the discharge rate of motor unit impulses only briefly. Gandevia and McCloskey (1977a,b) have discussed mechanisms that might be responsible for the inability of deafferented subjects to accurately sustain a muscular effort. Anesthesia of peripheral afferents may reduce, or remove, dynamic assistance by short- or long-latency stretch reflexes, and a tonic source of motoneuron facilitation.

The mechanism responsible for perturbations inducing more error for small as compared to large movements is uncertain. From the second experiment it was clear that greater, and often more prolonged, changes in motor unit discharge rates resulted from delivery of perturbations when subjects attempted to generate single impulses. In addition to increasing DI motor unit discharge rate for the single unit that the subject sought to activate with a single impulse, other motor units were often recruited when perturbations were delivered while subjects sought to discharge a single DI spike. The greater sensitivity of DI motor units to disturbances during phasic motor unit control may account for the observations of increased positional error for small in comparison to large movements. An issue of some significance is the source of excitation that recruits more motor units and increases discharge rate during small phasic movements. The evidence of the sensitivity of the segmental afferent and efferent systems to small signals (Matthews and Stein, 1969; Tanji and Kato, 1973; Goodwin et al., 1978) would suggest that spinal cord systems are responsible for the greater modulation of small outputs. However, there is direct evidence that the discharge rate of pyramidal tract neurons in the monkey are modulated more when perturbations are delivered while monkeys perform small precise movements in comparison to maintenance of postures (Fromm and Evarts, 1977, 1978). Modulation of tonic or large-motor events is less apparent when assayed behaviorally (these experiments) or physiologically (Fromm and Evarts, 1978).

A peculiar effect noted for several motor units was an increased response following finger displacements that shortened DI muscle in comparison to motor unit responses following muscle stretches. This was especially apparent when subjects maintained a relatively constant rate of motor unit discharge. This result would appear to be paradoxical, since unexpected muscle shortening typically silences EMG activity (Alston et al., 1967), presumably by unloading muscle spindles (Burke et al., 1978). However, externally produced muscle shortening in the isometric situation also passively decreases force, thereby decreasing Golgi tendon organ firing rates and probably changing the bias of cutaneous (Hulliger et al., 1979) and joint afferents (Grigg and Greenspan, 1977). The increased discharge in DI motor units observed following passive shortening of DI could best be explained by removal of the overwhelming inhibitory influences of IB afferents (Watt et al., 1976). Another factor that may have contributed to the failure to observe muscle silence when DI muscle was shortened may be gleaned from the observations of Kanosue et al., (1980). These workers suggest that the excitatory fusimotor bias is diminished when human subjects maintain force in an isometric task. Reduction of an already low level of muscle spindle discharge frequency, as in muscle shortening, would not be expected to cause as great an effect if, instead, the spindle firing rate were high, as when subjects regulate force in an isotonic task, before delivery of displacement changing muscle length. Quite aside from the mechanism responsible for increased DI motor unit firing observed following muscle shortening, other investigators have observed paradoxical effects on muscle activity following changes in muscle length when subjects maintained isometric force (Traub et al., 1980). Indirect unloading of the thumb muscle increased surface EMG activity only when subjects held the thumb against an immovable object: The expected decrease in EMG activity was observed following indirect unloading of thumb flexors when subjects maintained thumb position against moveable objects. In the present experiments we did not test DI responses while subjects regulated muscle length. It is fully expected that DI responses would conform to classic definitions of reflex responses if DI muscles length is changed while subjects regulate position.

When subjects attempted to generate single impulses from DI motor units, stretches of DI muscle that occurred shortly after the visual cue decreased the onset latency of most DI motor units. Disturbances that shortened DI length did not change the latency of impulse onset. From experiments studying reflex excitability during reaction time tasks, it has been demonstrated that alpha-MNs are hyperexcitable, as inferred from increased reflex amplitudes, in the period between delivery of the cue requesting movement and the onset of volitional muscle activity (Pierrot Deseilligny and Lacert, 1973; Michie et al., 1976; Kots, 1977). The changes in reflex amplitudes in the reaction interval are probably related to increased discharge rates of supraspinal, especially pyramidal tract, neurons that precede onset of voluntary movements (Evarts, 1974; Evarts and Tanji, 1976; Hamada and Kubota, 1979). If it is assumed that DI MNs are hyperexcitable (because of increased cortico-motoneuronal discharge) immediately preceding generation of a single spike, the increased muscle spindle discharge due to muscle stretch would summate with descending excitatory influences to DI MNs, thereby increasing the probability that DI MNs (and motor units) would discharge. The earlier latency of DI impulses observed following muscle stretch, compared to latencies on control trials, reflects the increased probability of discharge. The opposite effect on excitability of DI MNs (i.e., increased latency) might have been expected following DI shortening, but it did not, in fact, occur (see Fig. 5D, lower panel).

Sensory Motor Processing of Target Movements in Motor Cortex

*C. Ghez, D. Vicario, J. H. Martin, and H. Yumiya

Center for Neurobiology and Behavior, Columbia University, College of Physicians and Surgeons, New York, New York 10032

How the brain translates sensation into action represents a central and challenging question for motor physiology. The complex nature of the problem is evident from the multiple demands to be met whenever we rapidly move our limbs to a target. Information about the target controls the contraction of specific muscles among many possible alternatives. Thus, arbitrary components of a virtually infinite array of sensory inputs provide the spatial coordinates of the target- and control-selected elements of a second large array of muscles. Furthermore, accuracy requires not only the control of muscles acting as prime movers, but also of muscles that maintain postural stability. Appropriate sequencing mechanisms assure that commands reach certain muscles before they reach others. Finally, the nervous system must deal with the complex nonlinear properties of muscles and the dynamic changes in input signals arising in the course of movement. Changes in muscle length represent late and indirect consequences of neural signals, and changes in afferent activity can affect intervening control elements.

The major supraspinal structures that control the spinal cord have a discrete somatotopic organization on which an orderly arrangement of peripheral inputs is mapped. This input-output relation can be characterized as homotopic. Regions controlling a given body part receive sensory information from closely related body parts. For example, in the motor cortex there exists a clear relation between the muscles controlled by local aggregates of neurons and the sites in the periphery which provide sensory input to these neurons (see Asanuma, 1981). A similar input-output plan also applies to the red nucleus (Ghez, 1975; Larsen and Yumiya, 1980). Such homotopic input-output relations are unlikely to be responsible for the initiation of skilled voluntary responses where sensory information from arbitrary targets must be conveyed heterotopically across somatotopic boundaries. How such heterotopic information, arising from different sensory modalities and widely varying peripheral loci, is conveyed to the output representations in motor cortex or red nucleus is a central question for understanding the initiation of the skilled movement. Moreover, since such heterotopic input-output relations are subject to learning, this question is relevant to the more general problems of plasticity and memory which underlie the adaptive control of movement.

We have approached these problems through studies of tracking performance and correlated neural activity. The task re-

---

*To whom correspondence should be addressed: Center for Neurobiology and Behavior, College of Physicians and Surgeons, 630 West 168th Street, New York, New York 10032.

quires the animals to perform rapid, accurate motor responses with their forelimbs according to information provided by a display. Such a tracking task challenges the animal's ability to translate complex sensory information into appropriate motor commands. Changes in display properties and in the loads opposing the animal's movements have been used to probe the adaptive capacity of the intervening neural processes. This chapter is divided into three sections. The first section (A) examines the features of tracking performance to define general rules governing the processing of sensory inputs related to the target. From these results it appears that, as a first approximation, the behavior conforms to a closed-loop feedback model with continuous sampling of sensory events. The second section (B) examines the properties of neurons in the arm area of motor cortex and suggests that the arm area of motor cortex includes two separate regional subdivisions: one involved in response initiation and another more concerned with the control of ongoing movement. The third section (C) reports anatomical differences in projections to the two regions of motor cortex. We will discuss our observations and their possible implications from a very general perspective to provide a broad, albeit speculative, conceptual framework. We will attempt to define the minimal features of the translational processes between stimulus and response, and to distinguish these translational mechanisms from processes of a more adaptive nature (Houk and Rymer, 1981), which control the translational mechanisms themselves. This distinction is important because, during performance, temporal constraints are crucial for the first, but not for the second. Our data suggests that when learning the task, the animals generate complex internal models of both the display and their peripheral plant as well as a set of correspondence rules relating the two. Contextual cues allow the animal to select the appropriate models for processing sensory information from the display. We also suggest that neural mechanisms allow decision and response selection to antedate the stimulus.

Methods

Tracking performance and single-cell correlates of behavior were studied in cats, using a versatile and flexible tracking paradigm (Ghez and Vicario, 1978a; Ghez and Martin, 1982). The animal is restrained in a nylon sleeve. Its head and left humerus are rigidly fixed to an external frame which mechanically isolates the forearm from the body. The animal's forearm is strapped in a splint attached to a lever mounted on the axle of a servo-controlled torque motor equipped with a strain gauge, potentiometer, and tachometer (Fig. 1A). The lever can be rigidly fixed to record forces under near isometric conditions. When the lever is free to rotate, the servo-controlled torque motor can simulate inertial, viscous, or spring loads.

The animal faces a display (incorporating a retractable feeder), which can be moved horizontally from side to side by a second servo-operated torque motor. The angular position of the display is a function of a voltage difference (error voltage) between a criterion target level, under experimental control, and the output of one of the transducers in the manipulandum controlled by the animal. The display position can be made to reflect either an error in the force exerted by the animal on the fixed lever, or an error in lever position when it is free to rotate. The servo operates to bring the display in front of the animal when the force or position (produced by the animal's forelimb muscles acting on the lever) matches the criterion target level. The display moves away from the animal's mid-sagittal axis to the right or the left, reflecting the error voltage, when the force or position signal is greater or less than the target level.

Behavioral sessions are subdivided into

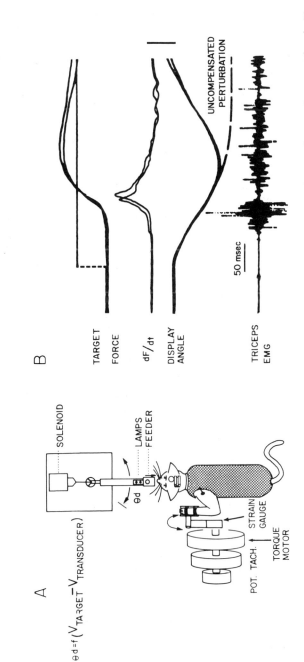

FIG. 1. **A:** Schematic diagram of experimental arrangement. **B:** Isometric responses to two equal step changes in target level. *Upper traces:* Target and force levels. *Second traces:* Corresponding changes in dF/dt. *Third traces:* Change in display angle corresponding to change in target level and in subsequent force applied to the lever. The *dashed line* indicates the shift of the display to the same target step when the animal's limb was removed from the lever (uncompensated perturbation). The *bottom trace* shows the triceps EMG. The correspondence between the burst of EMG activity and the rising phase of dF/dt should be noted. *Vertical calibration:* 0.5 Newtons, 17.4 Newtons/sec, $10°$ (display angle).

discrete trials separated by time-out intervals of a few seconds. At the beginning of each trial, the animal adjusts the force it applies to the fixed lever (or the lever position) so as to align the display with its midsagittal axis. Following a randomly varied interval (1 to 2 sec), the target level is abruptly changed. This target step creates an instantaneous error voltage and a display shift to which the animal responds by rapidly adjusting the force it exerts on the lever (or the lever position) to bring the display back to center. After a steady period of realignment, the animal is rewarded with food. Two features should be noted. First, a step change in target level requires about 200 msec to move the display to a new steady-state position. Inertia and internal friction cause the display to act as a low-pass device. Its properties were sufficiently linear that target steps of different sizes produced proportional angular displacements the peak acceleration and peak velocity of which are also proportional to the target step. As a result, information contained in the velocity and other derivatives of the initial display trajectory predict its final position and, therefore, the amplitude of the response required to bring the display back to center. Second, when driven by the target step, the display sweeps by the cat's face and the animal can use either visual cues or vibrissal deflection to sense its motion. Opaque contact lenses are applied to the animal's eyes to make it rely only on cutaneous information from vibrissal deflection. To make the animal rely on vision, the vibrissae are either shaved or retracted with a fitted mask.

Alterations in task conditions can be used to assess the adaptive capacity of processes intervening between stimulus and response. (1) Changes in gain of the display system, made by increasing or decreasing its angular response to a given value of the error signal, require the animal to alter its response strategy since a given amount of display motion now requires a different magnitude of output. (2) Changes in display polarity can be made. With the standard polarity of displayed error, display shifts to the right require extensor force adjustments, and to the left, flexor force adjustments. Requirements are reversed when the polarity of displayed error is inverted. Animals trained with both display polarities during training can learn both relationships and can be switched from one condition to the other without degrading their performance. In adapting to a sudden polarity change, the animals rely entirely on the effectiveness of their responses, since no additional cues are given. (3) The loads opposing lever movement can be changed to challenge the animal's adaptive capacity under anisometric conditions. (4) The response from the display to the animal's output can be delayed to deny the animal the information it would normally receive about the effectiveness of its response. In order to probe the animal's immediate response strategy, we have made extensive use of surprise or trick trials where behavior or unit activity is assessed when unexpected changes in task conditions are made from one trial to the next.

Single units were recorded with paralene-insulated, etched tungsten electrodes driven by a hydraulic microdrive fitted on an X-Y stage, which allows two-dimensional adjustment of penetration sites in the cortex. Concentric bipolar electrodes were implanted stereotaxically in the cerebral peduncles (where liminal stimuli produced contraction of forelimb muscles) for antidromic identification of cortical projection neurons. Control over the behavioral experiments and data acquisition was carried out with a general purpose computer PDP 11T03. An interactive language developed by P. Oratofsky provided raster displays and histograms of unit activity synchronized with particular events of interest. Data could be selected and sorted to detect trends and relations between cellular and behavioral events. An algorithm to determine automatically the time of onset of behavioral responses and their initial direction was programmed to exclude experimenter bias. The algorithm identified the peak of the first derivative of the force response to target shift, and the direction of the force change. Then, in a series of backward iterations, the first bin where the rate of change of force exceeded zero was located. The onsets of increase in unitary activity were scored as the middle of the first interspike interval after stimulus, which was less than half as long as the average interspike interval in the prestimulus period.

A. Tracking Performance Reflects the Operation of a Continuous Feedback System

General Features of Motor Responses: Pulse Step Control

During training, the cats learned to match accurately the force they applied to the lever, or the lever position, with the criterion target level according to task conditions. Examples of isometric responses are shown in Fig. 1B. Step changes in the target, shifting the display away from midline, elicited rapid adjustments in force or position which returned the display to midline. The responses accurately compensated for target steps of different sizes and directions. Under isometric conditions the peak force was a linear function of the peaks of the first and second derivatives of force (Ghez and Vicario, 1978b). Similarly, when limb motion was studied, the change in limb position was a linear function of the earlier peaks in velocity and acceleration (Ghez, 1979). The times taken to achieve the peak dF/dt or the peak acceleration, remained constant and were independent of the change in force or position. Similar rules have been found to apply to limb movements in humans (Taylor and Birmingham, 1948; Freund and Budingen, 1978; Lestienne, 1979) and monkeys (Lamarre et al., 1980). An additional feature of the behavior in the present task was that exposure to one size or to a limited range of display shifts was sufficient for the animals to make appropriate responses to a wide range of stimuli. Even for a novel stimulus size, the response derivatives remained scaled. Thus, the animals learned a general response strategy.

These observations suggest that the dynamic phase of the response trajectory is governed by an initial phasic control signal of variable amplitude and limited duration. The final force or position is governed by a tonic command which may be represented as a step (Ghez and Vicario, 1978b). Under constant loading conditions, the ratio of these two components has an approximately constant value. Both components are modulated in amplitude according to the requirements indicated to the animal by display shifts of different sizes.

The notion of an initial amplitude-modulated phasic command was also supported from electromyographic (EMG) recordings of agonist muscles during rapid responses (characterized by steep relations between the changes in force or displacement and their derivatives). Under conditions where both agonist and antagonist were silent at the initial resting force level, a single burst of activity is seen in the agonist coinciding with the rising phase of either dF/dt or acceleration. Although the integrated EMG activity in the burst correlates with the peak value of dF/dt (and in the anisometric case with the peaks of both acceleration and velocity), the duration of the burst does not, and it remains independent of the size of the response (see Desmedt and Godaux, 1977a, 1978b and Freund and Budingen, 1978 for comparable studies in humans). Similar EMG and kinematic results have recently been reported in both intact and deafferented monkeys (Lamarre et al., 1980). Those findings indicate that the phasic and tonic components of the motor strategy do not depend on sensory feedback.

Although the amplitude of the initial EMG burst is likely to reflect the magnitude of supraspinal control signals, the termination of this burst is likely to be strongly influenced by segmental mechanisms. Such mechanisms include post-spike hyperpolarization and recurrent inhibition in the motoneurons (MN) and other factors (Burke and Rudomin, 1978) that may contribute to burst termination. Thus, the initial force impulse can result from an excitatory supraspinal signal of limited duration, which is further sculpted by local negative feedback mechanisms. The resultant amplitude-modulated command appears to represent a feed-forward control signal to overcome the low-pass properties of muscle (Partridge, 1965), viscosity of muscle-tendon linkages, and limb inertia (Ghez, 1979; Partridge and Benton, 1981). Responses with these properties can be considered ballistic in that a major portion of the force or position trajectory is determined by an impulsive command limited to the accelerative phase (or the rising phase of dF/dt) (see Desmedt, 1981a).

Under constant loading conditions, the ratio of phasic and tonic components of the descending command appears to be closely regulated. This ratio is, however, adaptively altered (in the absence of any selective reinforcement procedure) when the configuration of opposing loads is changed (Ghez, 1979). When viscous loads which oppose movement are experimentally increased, humans (Taylor and Birmingham, 1948) and cats (Ghez, 1979) adapt by increasing initial phasic torques and thus increase the pulse-step ratio. When elastic loads are changed, the cats adapt by corresponding changes in both phasic and tonic torques (Ghez, 1979).

In experiments where inertial rather than viscous loads were imposed, adaptive increases in the pulse-step ratio are constrained because instability or overshoot of final position can result from a strategy which increases phasic torques to maintain high initial velocities. Thus, with a high moment of inertia, the peak velocity remains markedly reduced over controls, even with extensive training (see also Taylor and Birmingham, 1948). The EMG activity may also fail to show a well-formed initial burst in the agonist, and cocontraction of the antagonist becomes conspicuous. These alterations do not result from the added load opposing the initial acceleration but, rather, from changed mechanical requirements associated with accurate termination of movement. Indeed, had the peak velocity remained the same, the animal would have had to actively decelerate the limb. Otherwise, the momentum (mass × velocity) of the moving limb would have produced overshoot beyond the desired final position. We have verified this by comparing the trajectories of movements opposed by an inertial load (simulated by acceleration feedback) with ones where the forces opposing acceleration were the same but where the need for active deceleration was eliminated. This was achieved by half wave rectifying the acceleration signal used as a feedback controlling the torque motor. With this condition, the subjective feel of the lever resembled that of a bicycle wheel where a ratchet mechanism prevents the foot from being dragged when pedaling stops. Under this latter condition, which effectively eliminates momentum, the animal gradually increased initial acceleration to produce movements that were similar to both unloaded conditions and to those associated with viscous loads. Thus, increasing demands for control of the terminal phase of movement are normally met by decreasing the initial phasic command coupled with cocontraction, rather than by producing a precisely timed phasic burst in antagonist muscles to decelerate the limb. Cocontraction serves to increase the stiffness of the joint at the terminal equilibrium point (Polit and Bizzi, 1979; Hogan, 1980; Lestienne et al., 1981) and may be controlled by a specific population of cortical neurons (Humphrey and Reed, 1981, *this volume*).

These observations have the following implications: (1) Under isometric or under constant loading conditions, force or limb trajectories are directed and scaled from their inception according to earlier sensory events from the display. (2) The phasic torques which contribute to limb trajectory are under adaptive control to take into account both the magnitude of opposing loads and the mechanical consequences of rapid movement. Therefore, changes in steady-state commands producing rapid or gradual shifts in the mechanical equilibrium points (Feldman, 1967; Polit and Bizzi, 1977; Cooke, 1979) cannot be regarded as the sole determinant of limb trajectories.

The adaptive control of phasic and tonic components of the motor output suggests that performance of the task allows the cats to generate an internal model of the peripheral plant. This model includes the static and dynamic forces opposing movement of their limb as well as late reactive events (Bernstein, 1967) arising as a consequence of movement. This view is also compatible with the recent observations of Bizzi and Abend *(this volume)* on the control of limb trajectories in monkeys. Our results suggest that the pulse-step ratio could represent a principal variable under adaptive control. Additionally, we propose that the level of cocontraction is principally determined by the reliability of the model of the peripheral plant and to be under separate adaptive control. (See Ghez and Martin, 1982 for further discussion of this issue in relation to triphasic EMG patterns occurring in rapid limb movement.)

Factors Affecting Reaction Time: Role of Stimulus Uncertainty

The sudden shift of the display elicited appropriate motor responses with extraordinarily brief latencies. Response latency was dependent on the sensory modality and intensity, as well as on the magnitude of the motor output measured as the rate of change of force (Ghez and Vicario, 1978a). These relationships are attributable to modality-specific differences in sensory processing time and to spatial and temporal summation in central and segmental relays (Ghez and Vicario, 1978a). Response latency did not vary with the accuracy of the initial force or position response. However, a systematic decrease in latency occurred as the length of training increased. Cats trained for several months to 2 years have daily mean latencies (measured from target step to first change in dF/dt) of 50 to 70 msec (Ghez and Vicario, 1978a). Cats trained for 2 to 3 years reveal still lower daily mean latencies between 40 and 50 msec. Similar effects of extensive training have been reported in humans (Mowbray and Rhoades, 1959).

In all cats, the response latency was not dependent on uncertainty in the time, direction, or magnitude of the stimulus (Ghez and Vicario, 1978a) (Fig. 2). In Fig. 2A, the open histogram represents latencies of isometric responses of a given size to a stimulus the amplitude of which was maintained constant for over more than 50 trials. The shaded histogram represents the latencies to a stimulus of the same amplitude presented within a series of randomly varied stimulus amplitudes (eliciting responses of different sizes). In Fig. 2B the two histograms illustrate, in a second animal, latencies to target shifts of constant amplitude and direction (open histogram) and of responses under conditions of directional uncertainty (shaded histogram). The responses in the shaded histogram were made to target shifts of the same direction as those in the control series that were embedded as surprise trials (average: 1 in 10 trials) within a series of target shifts in the reverse direction. In both cases, the latency distributions overlap, with no significant differ-

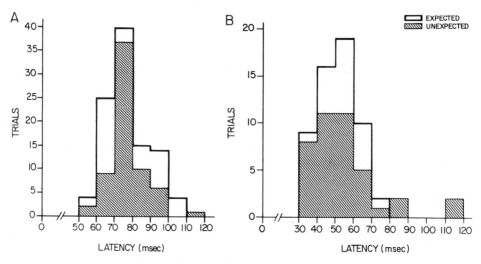

FIG. 2. Influence of uncertainty in amplitude and direction of target shift on response latency. **A:** *Hatched histogram* illustrates the distribution of response latencies to target shifts of set amplitude presented during a series of trials where perturbation amplitude was varied randomly (mean latency = 77.2 msec, SD = 10.7 msec, SE = 1.3, N = 65). *Open histogram* shows the distribution of response latencies measured during control trials where animal was presented target shifts of set amplitude only (mean latency: 77.2 SD = 11.9, SE = 1.2, N = 102). **B:** *Hatched histogram* shows the distribution of response latencies to target shifts eliciting extensor responses presented randomly in a series of trials requiring flexor responses (mean latency = 54.5 msec, SD = 19.1, SE = 3.0 msec, N = 40). *Open histogram* shows the distribution of extensor response latencies under conditions where the animal expected only target shifts requiring extensor response (mean latency = 51.4 msec., SD = 10.7, SE = 1.4, N = 56).

ences for conditions of either amplitude or direction uncertainty. In conclusion, the need to estimate either the magnitude or the direction of the response required by the stimulus does not add time to the processing of sensory events (see also recent findings in primates by Georgopoulos et al., 1981). Rather, response latency depends only on the sensory modality, stimulus, and response amplitudes, and the level of training. Furthermore, the time of onset of motor responses is not determined by the demands of specific information processing interposed between sensory events and the motor output commands.

Continuous Control of Motor Output by Newly Acquired Sensory Information

Another challenge to the neural processing of targeted movements occurs when a new stimulus appears before the response to a first target is completed. When human subjects are required to respond to two stimuli delivered in quick succession, the latency of the response to the second stimulus is prolonged (Telford, 1931) and occurs approximately a reaction time after the first response. This suggested intermittency in the processing of serial stimuli (Navas and Stark, 1968), as if stimuli-eliciting responses were processed by a unique channel of limited capacity (Welford, 1980). However, the observed lengthening in reaction time varies in different task conditions (Vince, 1948; Welford, 1952, 1980; Brebner, 1968; Megaw, 1972). This single-channel formulation implies that information about a second stimulus may be stored during the processing associated with the first stimulus.

These issues were addressed by examining the cat's responses to a second target shift when presented during the first reaction time or during the dynamic phase of the first response. Cats performed isometric force adjustments with their forearm in response to sequences of two-step changes of target levels presented at various interstimulus intervals (ISI) (double stimulation). In addition, the response of one animal to delayed feedback was examined. The double stimulation or delayed feedback was always introduced on surprise trials presented at random within a much larger series of single stimulus trials.

Double stimulation

When presented with two successive target steps, the animals always either aborted their responses or increased the force they exerted on the lever according to the direction of the second stimulus (Vicario et al., 1979) (Fig. 3). In Fig. 3A, averaged ($N = 12$) control and test trials are shown for surprise stimuli requiring the animal to abort

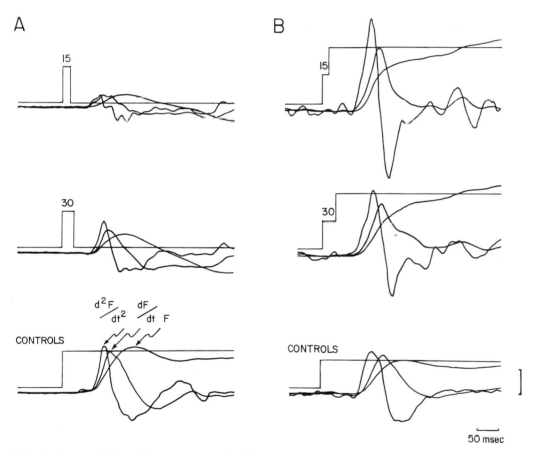

FIG. 3. Average ($N = 12$) responses to double target steps. A: Responses when the second step required the force to return to the baseline level. The animals force trajectory *(F)* and its first *(dF/dt)* and second derivatives *(d^2F/dt^2)* are shown together with the timing of target steps. Top two sets of traces show responses for interstimulus intervals of 15 and 30 msec. The bottom set shows control traces obtained when only one target step was delivered. B: Responses when the second step required an increase in the final force output. All traces as in A. *Vertical calibration:* 0.5 Newtons, 12.5 Newtons/sec, 500 Newtons sec². Averages synchronized with target steps.

its force response after ISIs of 15 and 30 msec. The first stimulus required extension and the second, a return to baseline. In contrast to what one might have expected had new target information been sampled intermittently or stored while awaiting the programming of the first response, the second stimulus causes the animal to abort the ongoing response immediately. Both the peak force and the first derivative of force are reduced when compared to the control trials (bottom of the figure). In these experiments, where the second stimulus called for termination of the response, both the magnitude and the duration of the initial EMG burst are reduced. In Fig. 3B, the second target step required an increase in the level of force at ISIs of 15 and 30 msec. Here again, the animal effectively updates its response by increasing both the first derivative of force and the final force achieved. The effectiveness of updating is dependent on the ISI. For ISIs of 15 and 30 msec shorter than the time from response onset to the peak dF/dt of the first response (i.e., during the impulsive phase of the response), both the resulting peak value of the second derivative of force, $d2F/dt2$, the peak dF/dt and the integrated EMG of the initial agonist burst were increased, but burst duration was not lengthened. For ISIs exceeding the time from onset to peak dF/dt, the force derivatives corresponding to the second response were reduced relative to controls. Response returned to control levels for ISIs of ≥ 250 msec. The EMG activity associated with the second response also varied with the ISI, showing a marked depression for values slightly longer than the time to peak dF/dt. Thus, for ISIs of 30 to 50 msec, the integrated EMG burst of the second response was reduced to only 30 to 40% of control values associated with the response to the first stimulus. Thereafter, a slow recovery took place with a return to control values for ISIs of >200 msec.

Cats are thus capable of using sensory information to update evolving motor responses in the present tracking task. For short ISIs this updating process is accomplished by amplitude modulation of the phasic output associated with the initial response. The effectiveness of such updating is limited by the time remaining within the phasic period. For longer ISIs, transient refractoriness is observed which does not delay the time of onset of the second response, but reduces its amplitude. Such refractoriness may be due to segmental mechanisms associated with the abrupt termination of the initial EMG burst, or to masking phenomena in various sensory nuclei conveying target information to central structures (Judge et al., 1980; Laskin and Spencer, 1980). The finding of continuous and effective response modulation for ISIs within the time to peak dF/dt may be analogous to response grouping (lack of separable responses) sometimes observed at short ISIs in human double-stimulation experiments (Welford, 1980). Such "grouping" is usually discussed in terms of the time course of single-channel loading. In our case this initial period corresponds to the interval during which phasic response initiation signals can gain access to MN pools before being significantly modified by the segmental consequences of MN excitation. In the more common case where positional responses to serial stimuli are studied, mechanical factors introduce significant delays between muscle contraction and a measurable response. Moreover, segmental influences and feedback effects make interpretation of late motor outputs problematic. Nonetheless, using a light-weight manipulandum, trained monkeys can modify their targeted movements with only a small increase in reaction time when a new stimulus is presented (Georgopoulos et al., 1981).

Delayed feedback

When feedback of the force error was delayed on random trials, the animal's initial force response returned the display to center only after a given delay interval. Feedback delays of 60 to 180 msec were presented at random. If sensory information from the display was sampled or used intermittently, the ensuing response should consist of a series of discrete steps in the form of a staircase. This was, however, never observed; rather, the animal's response showed a continuous increase in force until the display was affected by the initial response at the end of the delay period (Fig. 4). The upper traces of Fig. 4A show ensemble averages of control and test force responses. The lower traces show the associated control and test display trajectories. In Fig. 4B, the first derivatives of the response and of the display illustrate more clearly the method used to determine when divergence in the trajectories takes place. In addition, response and display derivatives are shown for a delayed feedback of 180 msec. Averages collected for trials at a feedback delay of 180 msec were used as control trajectories for determining the late stimulus and response events (S3 and R3). R3, at the turnaround of the force trajectory, occurs after the initial force response at R1 has altered the display trajectory at S3. The updated responses R2 and R3 to the effective stimuli in S2 and S3 had only slightly longer latencies (mean 20 msec) than those of the initial responses. These small differences are predictable from the reduced stimulus and response amplitudes associated with later stimuli (see Ghez and Vicario, 1978a).

In order to obtain the most accurate estimate of the point of divergence of the control and test response trajectories, a special procedure was used to select the responses serving as controls in the absence of delayed feedback. This procedure was based on our previous observation that the early components of the force trajectory (the first and second derivatives of force, in particular) were predictive of the ensuing force profile. The first 50 msec of the first derivative of the initial component of the force response in each test trial with delayed feedback was taken as a template to select among control responses the one that most closely approximated it. Using this procedure, the bin where the test and control responses diverged could be computed and the time of onset of the updated response determined.

In conclusion, the results support the concept that the responses are under continuous control by the display, and further indicate that sensory events produced by specific features of display motion cannot simply act to trigger a fixed, predetermined motor response (cf. Houk, 1978). If this were so, delaying feedback should not influence the cat's response, since the effect of the delay is to allow a display trajectory to unfold, which the cat normally experiences when its response latency is prolonged. Since the configuration and accuracy of short- and long-latency responses are identical (Ghez and Vicario, 1978a), the cat must take its response into account to interpret the stimulus. Thus, the effective neural signal determining the response is likely to be derived by the animal on the basis of a comparison between actual stimulus events and those predicted on the basis of its motor response. This implies that when learning the task, neural mechanisms generate an internal model of the display. We assume that parameters of such a model would be governed by an adaptive controller so that subsequently presented stimuli might be processed differently.

Discussion

A striking feature of the behavior elicited in our tracking task is the short, reflex-like latency of motor responses elicited by display shifts. The general rules governing targeted limb movements in cats indicate, however, that this behavior differs from both reflex actions and triggered voluntary

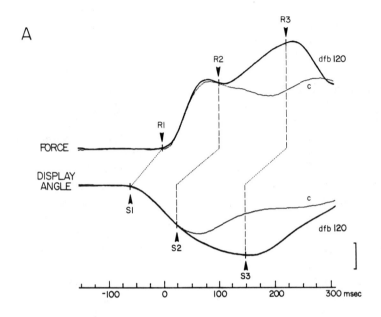
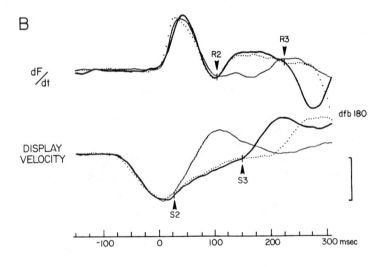

responses, described in humans while sharing some of the properties specific to both. As in many reflexes, the direction and intensity of the responses reflect those of particular sensory events. In our targeted response, these sensory events are produced by movement of the display in one or another direction. The tracking responses are altered in a fashion similar to what may be expected of a reflex when the stimulus is either curtailed or increased by double stimulation.

Tracking performance differs, however, from reflex behaviors in several respects. First, the behavior is learned, and adaptive changes in response configuration occur in relation to the achievement of an appetitive goal. Both the topography and gain of the stimulus-response relationship are rapidly changed when display properties (Ghez and Vicario, 1978b) or loading conditions are altered (Ghez, 1979). Adaptive changes in direction and gain of the vestibulo-ocular reflex have also been documented (Miles and Fuller, 1974; Gonshor and Melvill Jones, 1976a,b; Melvill Jones and Davies, 1976; Robinson, 1981; Melvill Jones and Mandl, *this volume*). In that case, however, the changes in direction are slow to develop and eventually constitute a new stable property of the system. The adaptive changes in phasic and tonic components of the motor program which regularly and rapidly follow changes in load in our paradigm suggest that response configuration is critically dependent on an internal representation of the complex properties of the peripheral plant. Although the nature of such a representation is at present uncertain (see Saltzman, 1979); it must be stored and regulated by mechanisms that are fundamentally different from those that determine the response. Second, the nature of the responses obtained using delayed feedback suggests that complex comparisons are required in sensory processing as well. We have proposed earlier that the neural signal determining response updating might be derived from a comparison of afferent input with an internal representation of display properties based on recently acquired information. A final difference between tracking performance in our task and reflex action is that the output configuration is not dependent on which sensory modality conveys information about the display. Similar rules ap-

FIG. 4. Responses obtained when force feedback was delayed 120 msec (DFB 120). **A:** The upper traces show superimposed ensemble averages of control *(light trace)* and test *(heavy trace)* responses obtained on trials when the display was unresponsive to the animal's output for a period of 120 msec following the target stimulus. The *lower traces* show superimposed averages of display trajectories for the control and test trials. Averages of 10 trials in each condition were synchronized by the onset of the force response. Control trials were selected by a matching procedure using the first 50 msec dF/dt of the force response as described in the text. *Arrows* S1, S2, S3 indicate the times of divergence (from the control trajectory) of the test display trajectory which elicited the three force responses at arrows R1, R2, R3. *Vertical calibration:* 0.5 Newtons, 10°. **B:** For the same trials as in A, the traces show superimposed averages of first-force derivative and display velocity of control *(light traces)* and DFB 120 msec delayed trials *(heavy traces)*. *Dotted traces* show averaged responses and display velocity for trials where the feedback delay was increased to 180 msec (DFB 180). These traces are superimposed to illustrate the algorithm used to determine the timing of late stimulus and response events (S2, S3, R2, and R3). S2 was determined by the divergence of control and test display velocities, R2 by the divergence of control and test dF/dt. S3 was determined by the divergence of display velocity of the DFB 120 and DFB 180 trials, since the third effective stimulus is the onset of the resumption of display feedback which begins to return the feeder to the center position. R3 was determined by the divergence of dF/dt for the DFB 120 and DGB 180 trials; i.e., when the animal stopped exerting increasing force because the display had become responsive. *Vertical calibration:* 15 Newtons/sec, 180 degrees/sec. All traces are averages ($N = 10$) synchronized with force onset (R_1).

ply when responses are elicited by visual, vibrissal, and combined inputs.

Both the short absolute latency and the factors by which it is influenced distinguish our tracking behavior from triggered voluntary responses commonly observed in man. Triggered responses are generally postulated to include a process that selects the category of response appropriate to the behavioral goal. The notion that response selection intervenes during the reaction time interval is derived from observations that the latency of motor responses increases progressively with the number of choices available (see Welford, 1980). Additionally, the translation of sensory information into "motor commands" is also assumed to be subject to a limited information handling capacity (Welford, 1952, 1980). These considerations have suggested that voluntary behaviors involve a more digital process than reflex actions, which more closely resemble analog transformations (Houk, 1978). Neither of these two constraints, however, applies in our paradigm, since correct responses are made to both predictable and unpredictable stimuli appearing singly or in pairs.

Under conditions applying in our task (see below), input-output processing is simplified to eliminate the need for response selection during the reaction time interval and a category of motor response emerges with the automatic (and analog) properties of reflexes and the flexibility of voluntary responses. We propose that the neural mechanisms involved implement a set of spatial rules which determine the topographic correspondences between sensory inputs from the display (target) and elements of the motor output representations. This set of rules, formally analogous to a map, would allow any member of an ordered array of sensory inputs providing direction and magnitude information to access specific output elements without the need for a selection or programming operation to intervene in the reaction time interval. Thus, prior to the occurrence of the display shift, contextual cues indicate to the animal the processing it will have to perform and thereby determine behavioral set. Then, appropriate elements process incoming and outgoing signals according to the specific expectations embodied in the models. Errors occurring during performance of the task and exposure to changing conditions are considered to act on adaptive control elements which alter internal models and maps.

In human subjects, a decrease in the susceptibility of reaction time to the number of choices available is known to occur with practice for highly compatible stimulus-response conditions. For example, when human subjects must move the finger to which a vibratory stimulus is applied, the increase in latency with increased numbers of possible stimuli is modest (Smith, 1979). In general, however, stimulus-response compatibility is a post hoc explanation given to explain differences in susceptibility ot choice effects (Welford, 1980). It implies the existence of a set of rules which may be used to connect ordered sets of stimuli (possibly even of a symbolic nature) and responses (Duncan, 1977). In certain cases, this notion is applied to the spatial demands of the task (Fitts and Deininger, 1954). It is, however, unclear to what extent "compatibility" differences reflect constraints relating to preferential stimulus-response linkages or rather to high levels of practice obtained from every-day experience with demands similar to those of the task. All these considerations may apply to our task and our highly practiced subjects. In addition, several other attributes of our task may simplify input-output requirements and encourage learning of a general response strategy rather than just individual stimulus response pairs. (1) The task itself can be viewed as requiring the transformation of a stimulus amplitude and direction into a response with similar dimensions and where only the direction and filter parameters need adjustment. (2) The compensatory nature of the tracking task requires the animal to attend only to a single input (rather than to both a target and a cursor, as in pursuit tasks), thus simplifying the spatial demands made on attentional mechanisms (Posner et al., 1980). (3) The lack of temporal and spatial discontinuities of the display and its linear properties allow the information provided by its trajectory to be highly redundant.

B. Roles of the Motor Cortex in the Initiation and Control of Voluntary Motor Responses

The extremely short latency in behavioral task suggested that only a few processing stations need be interposed between the arrival of target information and the generation of motor commands and that, within limits, new response requirements can be fairly directly relayed to motor structures. These considerations and the documented role of the primate motor cortex in movement (Phillips and Porter, 1977; Evarts and Fromm, 1978; 1981) led to the following questions: (1) Can the motor cortex contribute to the initiation of responses with such short reaction times? (2) Do the patterns of discharge of task-related neurons reflect the existence of a preset input-output relationship activated by the arrival of a signal derived from the stimulus? (3) What is the relationship between task-related neural activity and the specific homotopic input-output organization documented in acute physiological studies (Asanuma, 1981). To approach these questions, we recorded the activity of single neurons in the motor cortex of 6 cats trained to perform isometric responses in the compensatory tracking task described above. Differences in peripheral receptive fields and in task-related patterns of single unit activity were observed between pre- and postcruciate regions of motor cortex (area 4γ). A parallel series of experiments was conducted to determine the anatomical differences in projections to these two regions. (section C).

Functional Specialization Within Area 4: Regional Differences in Receptive Fields and Task Relations

Penetrations were made in the lateral halves of the anterior and posterior sigmoid gyri. Intracortical microstimulation (ICMS) was applied at 500-μm intervals and following the recording of each neuron to ascertain its location within the somatotopic representation in motor cortex. Single neurons were sampled in areas where ICMS, using currents of ≤ 20 μA, produced contraction of forelimb muscles active in the task (Ghez et al., 1982). Approximately 450 neurons were found to show a clear temporal relation between changes in their activity and behavioral events during task. Most neurons could also be driven by somatosensory stimuli to the forelimb while the animal was at rest and not performing the tracking task.

Receptive fields fell into two broad classes which we called simple and complex. Throughout the arm area of the motor cortex (defined by ICMS and by histologic reconstruction), neurons were found with discrete cutaneous or deep receptive fields (simple fields) (cf. Brooks et al., 1961a,b; Buser and Imbert, 1961; Asanuma et al., 1976). In addition, many neurons were observed whose forelimb receptive fields had more complex temporal or spatial characteristics (complex fields). The response of some of these neurons was temporally labile, being brisk at times, then diminishing or disappearing entirely within seconds or minutes, only to reappear a short time later. Such neurons have also been noted by others (Brooks et al., 1961a,b; Brooks and Levitt, 1964; Baker et al., 1971). Other neurons classified as complex were those with directional receptive fields on the skin, those that received convergent input from superficial and deep receptors, and those showing multiple excitatory and inhibitory foci which could be spatially discontinuous. Neurons with one or more of these complex properties were preferentially located in regions of area 4γ rostral to the cruciate sulcus, where they represented about 45% of neurons with peripheral receptive fields. In contrast, complex fields were only observed in about 10% of neurons caudal to the cruciate sulcus. Comparable rostrocaudal differences in receptive field properties have been reported in acute preparations (Buser and Imbert, 1961; Brooks et al., 1961a,b; Brooks and Levitt, 1964).

Dramatic differences were also observed between rostral and caudal portions of the arm area of motor cortex in the timing of neuronal discharge during task performance. Only units in rostral regions were modulated in advance of the animal's force response (lead cells), although task-related activity that lagged force production (lag cells) was observed in both rostral and caudal regions. The change in lag cells could often be attributed to stimulation of the neurons' receptive field which accompanied the behavioral response. For example, a neuron which was passively driven by stimuli to the hair on dorsal forearm was also phasically activated during elbow flexion, which produced shearing of the skin around the strap securing the forearm to the lever. However, this close association was not always seen and many neurons with exquisitely sensitive cutaneous or deep receptive fields in the responding limb or muscle, were not modulated during either the phasic or tonic portion of isometric or isotonic responses. The regional differences observed in all animals are schematically illustrated for the rostral (MCr) and caudal (MCc) compartments of area 4 (Fig. 5).

Although lead cells constituted only about 10% of all task-related units, they

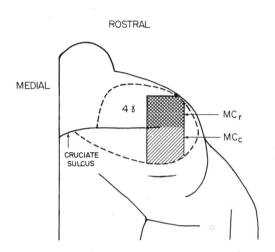

FIG. 5. Functional specialization in feline motor cortex. Motor cortex (area 4) is surrounded by *dashed line*. The area within the *solid line* schematically illustrates the regions where micro stimulation elicited forearm movements at ≤ 20μA in 6 cats. *Hatched area:* Region from which cells with simple receptive fields were recorded and where task-related activity lags force response (includes MCr and MCc). *Crosshatched area:* Approximate location of cells with complex receptive fields and whose task-related activity on average leads force response (MCr only).

were more extensively evaluated because only this population of neurons could have contributed to response initiation. Approximately 80% of the 45 lead cells examined had receptive fields of the complex type, suggesting that such neurons receive complex convergent patterns of peripheral input. Of the lead cells, 90% showed reciprocal response patterns with increased activity for forearm force in one direction (either flexor or extensor) and decreased or no modulation for force developed in the opposite direction. The modulation in unit activity of lead cells could be characterized as phasic (14%), tonic (22%) or, in the majority of cases, mixed (64%). In 80% of lead cells, the degree of modulation in activity varied with one or more of the following parameters: rate of change of force, steady-state force developed, or integrated EMG in agonist muscles. These discharge patterns and output properties are comparable to those reported for neurons in monkey motor cortex during conditioned arm movements (Evarts, 1966; 1968; Lamarre et al., 1978; Thach, 1978; Fetz and Cheney, 1980) and isometric responses (Smith et al., 1975; Hepp-Reymond et al., 1978). The differences in properties of single neurons described imply a functional subdivision of the cat motor cortex and may correspond to the dual representation of forelimb muscles documented by Pappas and Strick (1979, 1981a) using ICMS. We have regularly confirmed their finding that low-threshold effects in a given muscle can be obtained from discontinuous sectors within area 4 (see Anderson et al., 1975; Jankowska et al., 1975; Strick and Preston, 1978a; Kwan et al., 1978; for results in monkey). Furthermore, our observations show that neurons recorded in a zone from which low-threshold ICMS effects in a given muscle are obtained need not lead the contraction of that muscle during execution of a learned behavioral response.

Mapping of Target Related Sensory Inputs in Neurons of MCr

The finding that activity of lead cells in rostral motor cortex was not as tightly coupled to events arising in the periphery as traditional input-output mapping had suggested (Asanuma, 1975) raised the possibility that they receive other inputs. To determine whether sensory inputs associated with display motion might be coded in their activity, we first examined, trial-by-trial, the times of onset of changes in unit activity in relation to the stimulus and the force response. Next, to dissociate effects related to the magnitude and direction of stimulus from effects related to ensuing responses, we examined unit activity when the gain or polarity of the error signal controlling the display was changed.

Timing of lead cell activity: Stimulus Synchronization

Surprisingly, 85% of lead cells showed a consistent temporal relationship to the stimulus, a relationship that was largely independent of reaction time. To illustrate this, raster displays of unit activity recorded during single trials were sorted according to reaction time and aligned with the onset of the force change (Fig. 6). The trials with the shortest reaction time are at the top of the rasters. The small plus signs mark the time of occurrence of the stimulus and the zero on the abscissa is the onset of force change. In part A, unit activity is increased with extensor responses elicited by display movement in one direction, and in B, the activity of the same unit is suppressed with flexor responses elicited by display movement in the other direction. These rasters show the reciprocal character of the changes in unit activity. Both increased and decreased unit activity occurred with a consistent latency from stimulus rather than remaining aligned with the onset of the response. The amount of lead of unit activity relative to the onset of the force response varied as a function of reaction time. In Fig. 7, the time by which the change in unit activity leads onset of force response is plotted against the response latency for each trial (see legend). Note in part C that this relationship also applies when the range of reaction times is large. These figures also show that when reaction time is unusually short a unit's activity may actually lag the response. On these axes, a unit, the activity of which was exactly timed to the occurrence of the stimulus would be represented by a regression line with a slope of 1.0. The line for a unit timed to response onset would have a slope near zero. In 38 of 45 lead cells examined, the relationship between lead and reaction time showed a significant correlation coefficient. The slopes of corresponding lead-reaction time regression lines were distributed around a

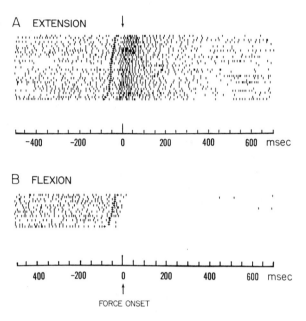

FIG. 6. Modulation of lead cell activity is time-locked to stimulus. A and B: Spike rasters synchronized by the onset of extensor and flexor force, respectively. Rasters are sorted by reaction time with shortest RT shown on top and longest on bottom. + signs mark the time of occurrence of the shift in display position (i.e., stimulus).

mean value of 0.92 (Fig. 8). Thus, the timing of onset of lead activity in rostral motor cortex is better correlated with the stimulus than with the ensuing response. Neurons discharging after the onset of the response (lag cells) typically exhibited lead-reaction time regression slopes which were close to zero (correlation coefficients not significant). In 10 of 13 such units, the variance in latency from stimulus to unit onset was greater than that from force onset to unit onset. This is compatible with lag cells being driven by stimulation of their receptive fields during the behavioral response.

Task-related units in forelimb motor cortex could not be driven passively outside of the context of task by visual, vibrissal, or other stimuli of the type associated with display movement. They were, however, often observed to fire during spontaneous movements of the forelimb. The task-related activity of both lead and lag cells was contingent on the occurrence of behavioral responses (Fig. 9). When reinforcement was withheld, behavioral responses extinguished and changes in unit activity no longer followed shifts in the display (B). In trials with incomplete or aborted responses (C), the perirespose time histograms show intermediate amounts of modulation. Thus target-related sensory information is conveyed to neurons in the rostral portion of the arm area of motor cortex in a contingent fashion, and confirms the relation between unit modulation and response magnitude. In addition, for this unit and 5 others that were fully analyzed, unit activity showed a relationship to the amplitude of the stimulus. Since, for a constant stimulus, these cells also exhibited a relation to the amplitude of the response, only a display gain manipulation could fully dissociate stimulus and response amplitudes. When the display gain was changed, the force required remained the same, but the amplitude of the stimulus was different. Fig. 9D shows perirespose time histograms of unit activity in trials following adaptation to a reduction in display gain to one half that of controls in A. For averaged force responses which are

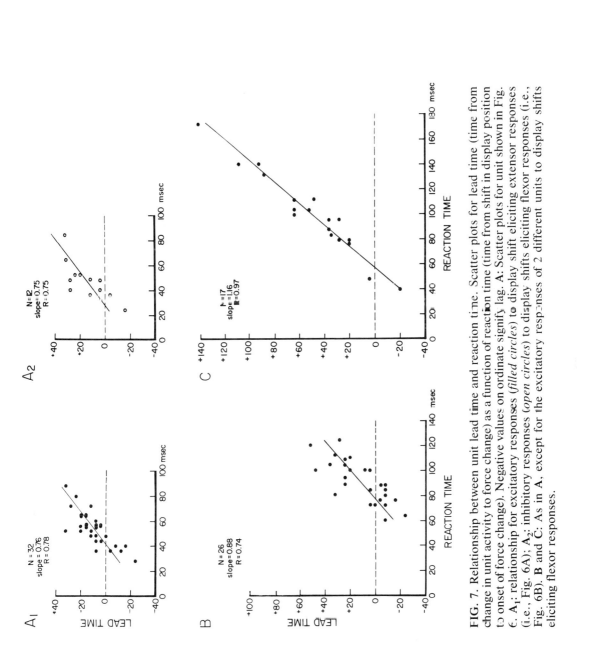

FIG. 7. Relationship between unit lead time and reaction time. Scatter plots for lead time (time from change in unit activity to force change) as a function of reaction time (time from shift in display position to onset of force change). Negative values on ordinate signify lag. A: Scatter plots for unit shown in Fig. 6. A_1: relationship for excitatory responses (*filled circles*) to display shift eliciting extensor responses (i.e., Fig. 6A); A_2: relationship for inhibitory responses (*open circles*) to display shifts eliciting flexor responses (i.e., Fig. 6B). B and C: As in A, except for the excitatory responses of 2 different units to display shifts eliciting flexor responses.

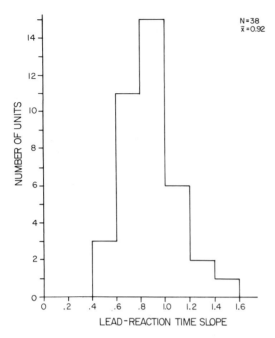

FIG. 8. Histogram of linear regression slope for lead-reaction time relationship. All units included in this histogram exhibited significant lead-reaction time relationships.

comparable to those illustrated in A, the unit activity is markedly reduced.

Changes in stimulus-response relations in lead cells

These observations suggest that the activity of lead cells reflects sensory information related to display movement, but is also contingent on a specific behavioral set. Two possible mechanisms may be envisaged to account for this phenomenon. First, as shown for neurons in monkey posterior parietal cortex and frontal eye fields, the behavioral set associated with task performance could uncover a receptive field in a fixed location which was not, however, detectable when the animal was examined under passive conditions (Wurtz and Mohler, 1976; Yin and Mountcastle, 1977; Robinson et al., 1978; Bushnell et al., 1981; Goldberg and Bushnell, 1981; Mohler et al., 1973; Motter and Mountcastle, 1981). Alternatively, the stimulus site producing excita-

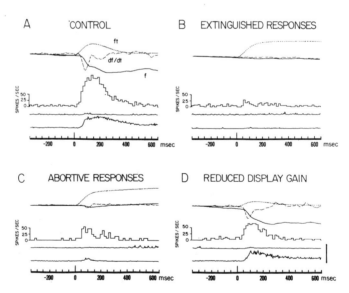

FIG. 9. Lead-cell activity is contingent on behavioral response. A: Control trials, averages ($N = 47$) synchronized by the onset of display movement; from top to bottom, display position (ft), df/dt, force, periresponse time histogram of unit activity, triceps EMG, and biceps EMG. B: Extinguished responses as in A, except for trials where reward was withheld and animal no longer responded ($N = 21$). C: Abortive responses, as in A, except for trials selected for incomplete responses ($N = 10$). D: Reduced display gain as in A, except for half control target display movement elicits comparable force response ($N = 22$). *Vertical calibration:* 1 Newton, 15 Newtons sec; 20°.

tion or inhibition in the neuron could vary according to the direction of the response required by the stimulus.

We could distinguish between these possibilities by dissociating the coding of the direction of display movement (i.e., stimulus) from coding of the response in task-related unit activity. Units were recorded in 3 animals trained to respond appropriately when the polarity of the displayed error was inverted (Martin et al., 1981) to examine neuronal activity associated with flexor and extensor responses elicited by display movements to right or left.

Two classes of lead cells were observed in rostral motor cortex. The first class showed reciprocal changes in activity (i.e., increases or decreases) depending on the direction of force production, but independent of the direction of display movement (Fig. 10). In parts A and B, average extensor and flexor responses are elicited, respectively, by display movement to the right or to the left. Rasters sorted by reaction time and synchronized by force onset, illustrate the tendency of this unit to modulate its activity with the onset of the stimulus rather than with the onset of the response. In part C, display polarity was inverted. Unit activity increases with the same response as in A, but is temporally related to the stimulus which in part B produced flexion. Part D shows the relation between lead time and reaction time under the two conditions in A and C, and demonstrates that stimulus onset determines tim-

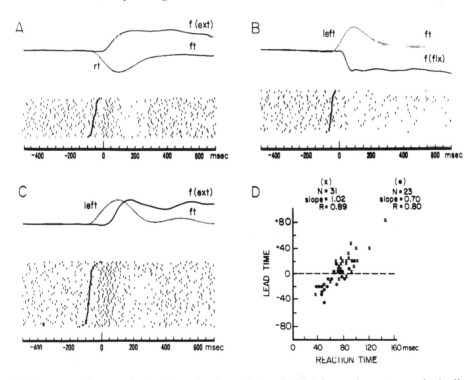

FIG. 10. Response characteristics of a unit, the activity of which is contingent on a single direction of force production. A: Extensor responses elicited by display movement to the right; top: average force and display position synchronized by force onset; bottom: spike raster, aligned with force onset and sorted in order of progressively increasing Rt (top to bottom). B: Flexor responses elicited by display movement to the left, otherwise similar to A. C: As in A, but extensor responses elicited by display movement to the left. D: Relationship between lead time (negative values indicate lag) and reaction time for control trials (filled circles, A) and trials with inverted polarity of displayed error (Xs, C).

ing of changes in unit activity. Thus, display-related afferent imput to these neurons can be selected according to behavioral set, possibly by gating mechanisms acting on neurons presynaptic to them.

A second class of lead cells (Fig. 11) showed changes in activity with a single direction of display movement and did not vary with the direction of force production by forelimb muscles. Modulation of activity in these cells, while timed to the stimulus, was contingent on an overt behavioral response as in other lead cells. This class of lead cells may be important in processing target-related stimulus information associated with a fixed effective stimulus site. Cells with similar properties have been observed in monkey posterior parietal cortex (Wurtz and Mohler, 1976; Yin and Mountcastle, 1977; Robinson et al., 1978; Bushnell et al., 1981) and frontal eye fields (Mohler et al., 1973; Goldberg and Bushnell, 1981), but these cells are also responsive to visual stimulation in the absence of a motor response. Alternatively, this group of cells may control one or more behavioral responses which follow the display shift in our paradigm. Two such associated responses were attempts at head rotation and eye movements toward the moving display. The direction of these responses was correlated with the direction of display movement, but was independent of both polarity of displayed error and direction of force produced by the forelimb muscles. Cells whose activity modulates with a single direction of display movement may be important in controlling proximal muscle synergies associated with postural adjustment and joint stabilization which precede the actions of muscles acting on more distal joints (El'ner, 1973). This is consistent with evidence of Asanuma et al. (1981) in pyramidectomized cats suggesting that efferent

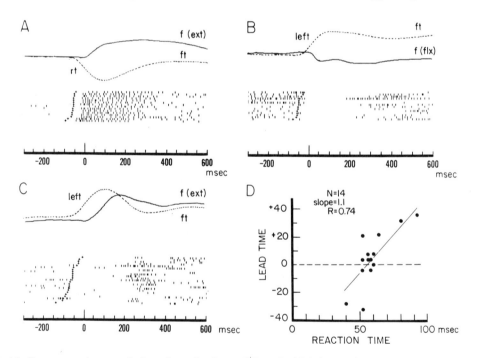

FIG. 11. Response characteristics of a unit, the activity of which is contingent on a single direction of display movement. A–C: As in Fig. 6. D: Scatter plot of the relationship between lead time and reaction time for trials illustrated in A.

zones controlling distal forelimb muscles also send impulses to proximal muscles.

Discussion

The striking conclusion emerging from these observations is that neuronal activity in rostral motor cortex encodes sensory events specifically related to target stimuli as well as the direction and magnitude of the ensuing behavioral response. The topographic features of stimulus representation (i.e., direction and magnitude of display movement encoded in the cell's response) was in many cells specific to the particular transformation required to produce the forelimb response. Our observations suggest that the conclusions of Evarts and Tanji (1974) for intended responses to proprioceptive stimuli may have more general applicability. Thus, in a task requiring spatially accurate responses to stimuli moving in one direction or another, we propose that central programs implement an isomorphic relation or map relating target and response dimensions. The resulting constraints to the flow of sensory information would provide for efficient transfer of a range of stimuli to the elements of motor cortex appropriate for controlling the required muscles. Such a map established through learning and implemented prior to the stimulus could account for the reflexlike properties of tracking behavior we have documented. Related concepts have been employed in models of prey-catching behavior in the frog and toad (Didday, 1976; Arbib, 1981) in which a visual sensory sheet is mapped directly onto a motor sheet in the optic tectum (Ingle, 1976; Ewert, 1976). Although the cat's behavior is the result of associative learning of an arbitrary, rather than a hardwired set of correspondences, once established, such a mapping process can reduce translational delay and output variability. To account for the animal's ability to adapt to a display polarity switch, it is also necessary to invoke a higher-order process capable of modifying or replacing one map with another. The neural mechanisms for the selective implementation of a specific map may be formally analogous to one's thought to control the reflex action of cutaneous inputs as a function of phase in the step cycle during locomotion (Pearson and Duysens, 1976; Forssberg et al., 1977; Rossignol and Gauthier, 1980).

Although aggregate lead activity in motor cortex could contribute to both the configuration of the response and its time of onset, cortical activity could also act as a gating signal applied to a lower-order station. However, reanalysis of the activity of red nucleus neurons, recorded under identical task conditions (Ghez and Kubota, 1977; Ghez and Vicario, 1978c), revealed that lead activity in this subcortical structure is also better timed to the stimulus and reflects its magnitude (Ghez and Martin, *unpublished observation*). These findings indicate that the variance in reaction time cannot be accounted for by the variance in timing of supraspinal signals originating in either motor cortex or red nucleus. Although this conclusion seems counterintuitive at first, our behavioral data also suggest that response latency is not under explicit control by the processing of target-related inputs. The variance in response latency thus appears as an emergent property of the system as a whole, probably resulting from the participation in varying degrees of several nonsynchronized descending pathways activated in parallel. Their signals must be integrated with afferent inputs by interneurons and/or MNs whose threshold properties introduce new variability. The timing of the commands ultimately delivered to individual muscles may thus be influenced by segmental mechanisms integrating phasic descending commands with changing peripheral signals. This peripheral input could provide information concerning initial conditions including limb position and body posture (El'ner, 1973; Ghez, 1979; Nashner, 1980; Soechting et al., 1981). The propriospinal system at C_3-C_4, which receives convergent descending and afferent inputs, may provide a critical link in the elaboration of the motor command and its integration with peripheral information (Alstermark et al., 1981; Lundberg, 1979).

Whereas our data suggest that presetting mechanisms acting on, or presynaptic to, neurons in motor cortex can control stimulus-response topography, the encoding of output magnitude remains uncertain. First, although the summed neural activity of both lead and lag cells was, in most cases, significantly related to one or another parameter of the ensuing force response, this finding must be interpreted with caution. Many lead and lag cells exhibited peripheral receptive fields in the responding limb which were likely to be stimulated after onset of muscle contraction. The activation of these homotopic inputs by early response events may have contributed to observed relations between unit activity and motor output. Such feedback might in fact be under central control (Ghez and Lenzi, 1971; Ghez and Pisa, 1972; Evarts and Fromm, 1978) and could be important in the control of tonic force. Second, our limited sample of lead cells subjected to changes in display gain reveals paradoxical features of task-related neuronal responses. For example, on a trial-by-trial basis, using a stimulus of constant size at a given display gain, unit activity of phasic cells covaried with the ensuing peak dF/dt of the response. However, when similar force responses were elicited with a different display gain, the relationship between unit activity and dF/dt changed to reflect the larger or smaller stimulus. This raises the possibility that the gain of the stimulus-response relation might be controlled by mechanisms separate from those controlling response topography. It also suggests that adaptation to changes in display gain (Ghez and Vicario, 1978b) does not involve the control of the transmission in sensory pathways afferent to motor cortex.

Our findings of stimulus-related properties in neurons of the cat motor cortex contrast with observations made in primates during arm movement. In the monkey, neurons of motor cortex have been reported to modulate their activity with a consistent lead over the motor response (Evarts, 1966; Porter and Lewis, 1975; Lemon and Porter, 1976; Meyer-Lohmann et al., 1977; Cheney and Fetz, 1980; cf., however, Lamarre et al., 1980). Recent work, however, indicates a functional subdivision of even the primate motor cortex into rostral region (MCr) and caudal regions (MCc). Neurons in rostral portions appear to discharge earlier than ones caudal in area 4, when monkeys perform a motor task (Lamour et al., 1980; Humphrey, *personal communication*). Moreover, during a visual tracking task, neurons in posterior arcuate cortex (situated rostral to MCc) show responses which are time-locked to the visual stimulus (Jennings et al., 1980). This suggests that the expansion of frontal cortex in primates relative to carnivores further subdivides sensory and motor properties possessed by single neurons in the cat rostral motor cortex.

C. Differential Cortical and Thalamic Projections to Rostral and Caudal Motor Cortex

The different properties of neurons in the rostral and caudal portions of the arm area of the cat motor cortex raise several questions. The first concerns the pathways relaying target information to the lead cells in motor cortex and simple homotopic input to lag cells. The second question concerns whether duplicate representations of other body parts may exist in separate regions of motor cortex. Finally, could the differences in projections to different subregions of motor cortex account for the functional differences observed. To address these issues, we have determined the topography of thalamocortical and corticocortical projections to area 4γ (Hassler and Muhs Clement, 1964) in the cat, using retrograde and anterograde tracer techniques. In a retrograde series, intracortical pressure injections of horseradish peroxidase (80 nl, 30% HRP, Sigma Type VI) were made in a different site area 4γ in each of 10 cats. These sites were distributed throughout the pericruciate region. After a survival period of 30 h, the animals were perfused through the heart with 0.9% saline followed by a mixture of 1% paraformaldehyde and 1.25% glutaraldehyde in 0.1 M phosphate buffer (pH 7.4) for 30 min, and subsequently perfused with cold sucrose-buffer (Rosene and Mesulam, 1978). Then the brains were blocked and cut into 50-μm thick frozen sections. Every third section was processed with tetramethyl benzidine (TMB) and counterstained with neutral red (Mesulam, 1978). The distribution of retrogradely labeled neurons was determined by examining sections under bright-field illumination. An X-Y plotter, electronically coupled to the microscope stage, was used to record locations of labeled neurons. Injections were restricted to cortical gray matter proper and did not extend into white matter.

Cortico-cortical Connections

Taken together, the HRP injections in pericruciate cortex retrogradely labeled neurons in areas 6, 3a, 1-2, 5a, 5b, and 2pri as well as in portions of area 4 surrounding the injection site. Scattered labeling was present in other areas including the banks of orbital and suprasylvian sulci and areas 4δ, 4sfu, 4fu, 3b, and 7. Systematic differences were seen in the locations of cells labeled after pre- and postcruciate injections (Fig. 12). Injection sites were targeted to regions where task-related units had been recorded in trained animals. Precruciate injections of HRP produced retrograde labeling mainly within irregular and sometimes patchy areas of 4γ surrounding the injection site (A). Labeling in other cortical areas was sparse. In contrast, the postcruciate injection (B) produced dense labeling in regions to cells labeled within area 4γ itself.

A comparison of different injection sites revealed a topographic arrangement of projections from somatic sensory cortex to postcruciate area 4. Figure 13 summarizes results from 6 injections. Thus, injections in the medial portion of postcruciate area 4, thought to control hindlimb muscles (Thompson and Fernandez, 1975; Nieoullon and Rispal-Padel, 1976; Larsen and Yumiya, 1979), received projections from portions of areas 1-2, 3a, and 2pri considered to receive sensory input from hindlimb (Woolsey, 1958; Levitt and Levitt, 1968; Landgren and Silfvenius, 1969; Haight, 1972; McKenna et al., 1981). More lateral regions of area 4γ thought to control forelimb muscles (Asanuma et al., 1968; Nieoullon and Rispal-Padel, 1976), received inputs from sensory forelimb areas (Oscarsson and Rosen, 1963; Haight, 1972; Iwamura and Tanaka, 1978; McKenna et al., 1981). The scattered labeling in area 2 and 5 which appeared after precruciate injections did not show apparent topographic organization. Whereas the projections of primary and secondary somatic sensory cortices dominate in postcruciate regions, the projection from somatic association cortex (areas 5a and 5b) was different. Both MCr and MCc injections labeled equal numbers of cells in this region. A series of anterograde experiments was carried out in 6 cats to confirm the preferential projection of areas 2 and 2pri to the posterior portions of area 4. Multiple injections of either HRP or tritiated amino acids (Lasek et al., 1968; Cowan et al., 1972; Jones et al., 1979; Mesulam and Mufson, 1980) were placed in areas 2 or 2pri in separate experiments (Fig.

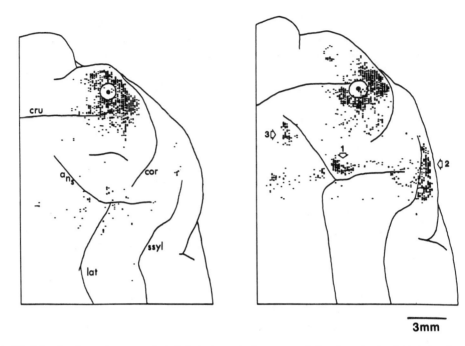

FIG. 12. Distribution of retrogradely labeled cortical neurons following single HRP injections. Surface reconstructions made from sagittal sections. **Left** and **Right:** Injections in the lateral part of precruciate and the lateral part of postcruciate area 4, respectively. Each *short bar* represents one labeled neuron. *Filled black circle* and *surrounding line* indicate sites of penetration of HRP-filled micropipette and extent of the reaction product at injection site, respectively. *Arrows 1 and 2* point, respectively, to clusters of retrogradely labeled cells in area 2 and in area 2pri. *Arrow 3* points to a cluster of labeled neurons in area 4δ in the depths of the cruciate sulcus. Cru, ans, cor, ssyl, and lat indicate cruciate, ansate, coronal, suprasylvian, and lateral sulci, respectively.

14). The terminal labeling is most dense in the postcruciate portion of area 4γ. Similar results were obtained with injections in area 2.

Thalamocortical Projections to Area 4γ

In the thalamus, retrogradely labeled cells were found in the ventrolateral (VL) nucleus, the shell zone at the ventrobasal complex (VB) border and in the central lateral (CL) nucleus. Within VL, clusters of labeled neurons formed lamellae in roughly parasagittal planes extending rostrocaudally which were topographically related to the cortical injection site. Medial parts of VL project to medial precruciate, lateral VL to medial postcruciate, and intermediate regions of VL to portions of precruciate cortex in between. In addition to neurons within VL proper, postcruciate injections labeled clusters of neurons in the shell zone at the VB border. The latter accounted for approximately 30% of labeled neurons in the thalamus following postcruciate injections, whereas only 5% of labeled neurons were in this region following precruciate injections. The strip of labeled neurons in VL lies in a more medial position following the precruciate injection in Fig. 15A than following the postcruciate injection in Fig. 15B. Also to be noted are the clusters of labeled neurons in the shell zone at the VB border following the postcruciate injection. The finding of labeled neurons in the shell zone at the VB border confirms earlier ob-

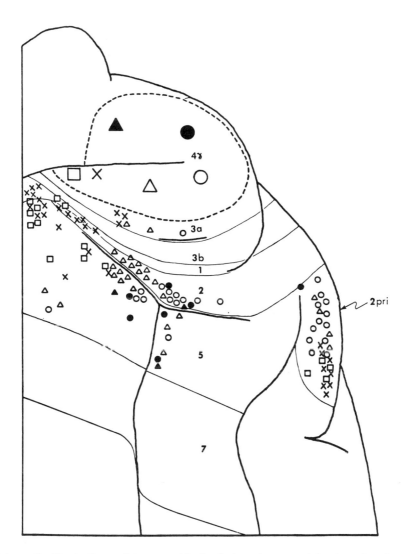

FIG. 13. Schematic illustrations of topographical relations between injection sites in area 4 and locations of labeled neurons in anterior part of ipsilateral cortex. The *large symbols* indicate the locations of single injection sites in different experiments. The corresponding *small symbols* indicate the locations of labeled neurons. Labeled neurons within area 4, area 6, and in the banks of orbital, coronal, and suprasylvian sulci are omitted for clarity.

servations (Larsen and Asanuma, 1979; Asanuma et al., 1979; Hendry et al., 1979). Our current data do not allow us to state that this group of neurons are functionally distinct from the other labeled cells in VL. Retrograde labeling of neurons in CL as reported by Strick (1975) was seen in the caudal portion of CL following each cortical injection; however, no obvious topographic arrangement of these labeled neurons was seen.

Discussion

These anatomical findings indicate clear differences in both corticocortical and thalamocortical projections reaching MCc and MCr. Projections from somatic sensory

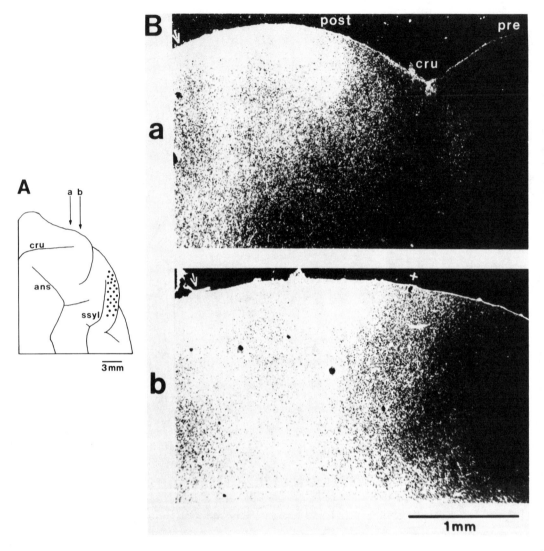

FIG. 14. Differential terminal labeling between MCc and MCr demonstrated by autoradiography after injections of tritiated amino acids into area 2pri. A: Drawing of cortex showing multiple injection sites *(dots)* in area 2pri and locations of the parasagittal sections (a,b) shown in B. B: Dark-field photomicrographs of parasagittal section through area 4. *Arrows* in B indicate the border between area 4 and area 3a. *Plus sign* indicates the anteroposterior level of the lateral extension of the cruciate sulcus.

cortex and from the shell zone at the VB border are preferentially distributed to MCc. Since neurons in both the shell zone (Asanuma et al., 1979) and somatic sensory cortex (McKenna et al., 1981) may exhibit simple receptive fields in the limbs, either of these inputs could account for the preferential distribution of simple fields in MCc (see Asanuma, 1981; Jones, *this volume*). In contrast, the major projection to MCr appears to arise in the VL nucleus itself.

Our observations unfortunately do not resolve the question of whether duplicate motor representations exist. Taken together, the following considerations tend, however, to support such a view. First, our

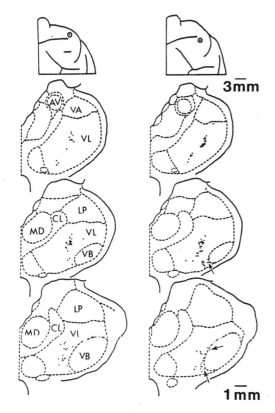

FIG. 15. Distribution of retrogradely labeled neurons in the thalamus following HRP injections into the lateral part of precruciate (**Left**), and the lateral part of postcruciate area 4 (**Right**). *Top row:* Injection sites. *Lower 3 rows:* Thalamic frontal sections. Each dot represents one labeled neuron. Tracings of thalamus from top to bottom represent, successively, more caudal sections. *Arrows* indicate the labeled neurons in the shell zone at the VB border.

anterograde and retrograde studies suggest that the postcruciate gyrus is somatotopically organized to receive input from the forelimb and hindlimb representations of area 2 and 2pri. Second, physiological observations (Sasaki et al., 1973; Y. Shinoda, *personal communication*) indicate that, in the cat, both fore- and hindlimb representations in the deep cerebellar nuclei focus their outputs (via VL) preferentially on regions rostral to cruciate sulcus. The presence of duplicate cortical representations may imply a similar duplication within VL proper since the entire projection from VL to MCr and MCc is topographically organized. Such a dual representation in VL has not, however, been documented. Alternatively, MCr might be specialized to accomplish synergic associations between head and forelimb movement which would be in line with convergent projections to MCr by forelimb and face regions of somatic sensory cortex. Preliminary experiments in progress, however, do not favor this hypothesis.

A second unresolved question concerns the pathways which convey target-related heterotopic information to lead cells in MCr. Two possibilities may be entertained. On the one hand it is possible that display-related sensory inputs are processed by cortical sensory mechanisms before giving rise to signals conveyed to the motor cortex and then to spinal levels. A more attractive alternative is that under the highly practiced conditions, input information may be processed by subcortical structures. The cerebellum may provide the proximate input signal determining lead activity in MCr via the VL nucleus of the thalamus. The latter is supported by observations showing both increased reaction time and a concomitant retardation of lead activity in motor cortex neurons in primates, following cooling or ablation of deep cerebellar nuclei (Brooks et al., 1973; Meyer-Lohmann et al., 1977; Lamarre et al., 1978; Conrad, 1978; Brooks, 1979). Increases in reaction time to exteroceptive stimuli in the cat have been reported to follow cooling of the ventrolateral nucleus of the thalamus (Benita et al., 1979).

AN OVERVIEW: TOWARDS A SYNTHESIS

We have examined a learned and highly skilled behavior of the cat under well-controlled conditions. The dominant feature of the behavior is the short latency with which accurate responses are made to a range of stimuli which vary unexpectedly in direction and amplitude. Extra time does not appear to be required for selection of specific response parameters during the reaction time. We have argued that the apparent simplification of processing demands during the reaction time could result from use of internalized models of target and peripheral plant and of a map of correspondences evoked by contextual cues prior to the stimulus. The behavioral data and correlated patterns of neuronal activity in motor cortex support our hypothesis that central motor programs determining response topographies may be represented as a map which relates target and response dimensions to each other. We have proposed that the behavioral set associated with preparation to respond in specific ways to a range of stimuli provides preferential pathways for the efficient transfer of behaviorally relevant information to the sectors of supraspinal systems controlling components of the motor synergy. Our data do not shed light on the role played by pyramidal neurons with simple receptive fields in the efferent zones of the caudal motor cortex; these were active only late during force development or during tonic force production. However, it seems likely that their output contributes to the composite descending signal ultimately reaching MNs. Whether they function in a feedback role, possibly dependent on changes associated with behavioral set, remains to be clarified. In any case, the specific input-output relations of different sectors of MCc provide for the parallel processing of inputs which could concurrently bias the MN pools of many different muscles.

The simple block diagram of Fig. 16 proposes that two processing networks underlie the stimulus-response transformations (S-R processors) in targeted movement. Each of these processors is assumed to comprise several functional components which perform specific operations on input information. Parameters of these operations are under adaptive control in order to optimally achieve higher order behavioral goals. The target S-R processor deals with stimuli specifically related to a target of interest, in our case, error signals arriving through visual or vibrissal channels. At an initial stage of this processor, incoming signals are compared with the ones predicted from a learned model of the display (delayed feedback experiments). The resultant error is used to modify ongoing input to constitute a derived forcing function. This derived input is then processed through a network which, perhaps by elements acting as neural gates, constrains the spatial flow of information according to a preset map. The signals derived from this spatial transform would be processed by a network producing phasic and tonic outputs. Both the ratio between phasic and tonic components and the overall gain of the input-output relation are governed by adaptive control elements which take into account an internal model of the peripheral plant. Rostral components of motor cortex are a component of this target S-R processor. The anatomical results suggest that spatially processed inputs initiating movement may reach this spatially processed inputs initiating movement may reach this region from the ventrolateral nucleus and thus possibly take origin in the cerebellum.

The parallel S-R processor receives input signals from the moving limb and body parts affected by motor commands and may, perhaps within a predicted range, act to compensate for deviations between the actual and the intended or reference trajectory. The parallel processor is assumed to include stretch reflexes and other spinal components of the motor servo (Houk and Rymer, 1981) as well as supraspinal elements, such as MCc. This parallel processor would be governed by adaptive controllers that modulate the gain of multiple input output channels with stable topographic relations.

The subdivision of function into signal

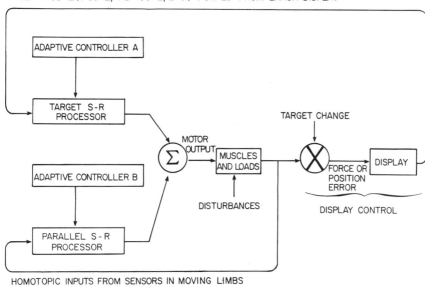

FIG. 16. Dual process model of targeted movement. Motor output signals are assumed to be generated by two processing networks. The Target S-R Processor is devoted exclusively to the processing of stimuli related to the target and operates as a feedback to minimize the displayed error. Output signals from the Target S-R Processor are summed with those from the Parallel S-R Processor to produce the final motor output. This parallel processor is assumed to be capable of processing inputs from somatic and other receptors acting in parallel through hardwired connections on a more spatially complex array of muscular elements than processor A. The Adaptive Controller B is assumed to modulate only the gain of the elements of the parallel processor. The simplified diagram as shown is not intended to exclude influences of the Target S-R Processor or the Parallel S-R Processor. (This diagram has been modified from Houk and Rymer, 1981, with permission.)

transformations carried out by multiple parallel pathways is a general organizational principle of the nervous system. Our results underscore that unified motor output emerges from multiple antecedent elements. This may mirror sensory processes which decompose a unitary event along several parallel channels. One resultant property is that the timing of behavioral events is not under the control of any single system. We thus propose that the temporal details of the overall synergy (i.e., the onset of contraction in different muscle groups) reflects the property of the total system rather than a variable controlled by a particular process. Similar ideas have emerged for neural circuits involved in locomotion (cf., Grillner, 1981).

The acquisition of skill, or motor learning, may thus be a process which simplifies the complexity of input-output requirements of targeted movements in the form of internalized maps of correspondence. To the extent that both input and output domains can be ordered or parameterized (cf. stimulus-response compatibility), all specific pairings may be equivalent and decision time effectively controlled. Such simplifications undoubtedly lead to a degree of stereotyping in motor behavior which is justified by the gains in time and reliability.

The present tracking task embeds the subject in a continuous loop of action and reaction and provides a clear window into the simplest form of sensory-motor association, that between a subject's action and its immediate consequences. Therefore, such tasks may provide an opportunity to study both the translational mechanisms which allow arbitrary target information to engage motor elements, and also adaptive modifications of these mechanisms when new circumstances require learning.

Our results emphasize that the classic distinctions drawn between the sensory or motor nature of particular brain regions, while of heuristic value, should not obscure the close interrelations between sensory and motor processing. Sensory input can provide cues to enable widespread changes in the processing of subsequent information and the adjustment of transformational programs. Additionally, sensory input provides ongoing information about the trajectory of targets of interest as well as signals used in local neural servo mechanisms. Similarly, in the context of a highly automatized task such as the one studied here, the concept of command does not seem readily applicable. In fact, once the nervous system is ready to respond to the stimulus and the changes in state alluded to above are implemented, the proximate "command" is the occurrence of the expected stimulus.

ACKNOWLEDGMENTS

It is pleasure to acknowledge the invaluable assistance of Paul Oratofsky for developing computer software, Kathrin Hilten for the illustrations, and Ruth Sharton for her devotion in typing multiple versions of this manuscript.

Motor Control Mechanisms in Health and Disease,
edited by J. E. Desmedt.
Raven Press, New York © 1983.

Muscle Afferent Function and Its Significance for Motor Control Mechanisms During Voluntary Movements in Cat, Monkey, and Man

*A. Prochazka and **M. Hulliger

*Sherrington School of Physiology, St. Thomas Hospital Medical School, London, United Kingdom and **Institut für Hirnforschung der Universität Zürich, Zürich, Switzerland

In mammals, the execution of movements is mediated by activity in α-motoneurons (MN), which serve as the final common pathways to muscles. For any given movement, the number of α-MNs involved is likely to be far outweighed by the number of proprioceptive afferent and fusimotor efferent fibers, which are also activated and whose discharge may contribute to the control of the movement. This chapter will critically assess recent afferent recording experiments in intact mammals, in relation to current theories of motor control. The data amassed in acute experiments on muscle afferents and fusimotor neurons have been reviewed elsewhere (Matthews, 1972; Stein, 1974; Murthy, 1978; Prochazka, 1980; Proske, 1981; Hulliger, 1981).

Of the various proprioceptors, only muscle spindle afferents have their firing properties directly influenced by central nervous efferents. Since fusimotor efferents have for technical reasons not been recorded in voluntary movements, the nature of spindle afferent activity during movement continues to attract a special interest. Soon after the first recordings from fusimotor neurons in acute preparations (Leksell, 1945), Hunt and Kuffler (1951) suggested that an important function of fusimotor neurons was to maintain spindle afferent discharge in the face of muscle shortening during active movements. This view was reformulated in more restricted terms, when Granit (1955) suggested that such maintained discharge might be provided by a rigid linkage between skeletomotor and fusimotor neurons. The concept of α-γ linkage has since come to dominate notions of how the central nervous system (CNS) controls these neurons, particularly after Valbo's (1971) findings on spindle firing during voluntary contractions in humans, which substantiated earlier reservations (Matthews, 1964) about the generality of the follow-up servo concept (Merton, 1951). The popularity of the α-γ linkage concept persisted in spite of the demonstration in 1962 by Matthews of the existence of functionally distinct types of fusimotor neurons (static and dynamic), and subsequent experiments showing that they could be selectively influenced from different CNS structures (Matthews, 1972; Appelberg, 1981). Since the advent in the last decade of afferent recordings in freely moving animals, early reservations about the generality of α-γ linkage have gathered momentum.

---

*To whom correspondence should be addressed: Sherrington School of Physiology, St. Thomas's Hospital Medical School, Lambeth Palace Road, London SE1 7EH, United Kingdom.

1. TECHNIQUES EMPLOYED TO STUDY AFFERENT ACTIVITY DURING VOLUNTARY MOVEMENT

1.1 Microneurography in Humans

Vallbo and Hagbarth, (1968) developed a microelectrode technique for single-fiber recording from peripheral nerves in man. This has since provided valuable information on the activity of spindle and Golgi afferents during voluntary movement. The main advantage of this technique is that the cooperation of human subjects and their capacity to learn quickly a range of motor tasks can be enlisted for the investigation of proprioceptive activity under natural conditions. At present, the main shortcomings are the difficulties with single-unit classification (see 2.2) and the limited stability of the recordings. The latter has restricted the speed and amplitude of the movements and the size of the contractions (loads) investigated. The details of the technique have been described elsewhere (Vallbo, 1973; Vallbo et al., 1979). Briefly, the nerve is first located by palpation and/or low-intensity electrical stimulation, with evoked low threshold proprioceptive sensations and/or muscle contractions providing clues as to how to direct the electrode. For a rough orientation this may be done first with percutaneous stimulation through surface electrodes. Thereafter, a more precise localization is achieved by juxtaneural stimulation through either the recording electrode or a separate stimulating electrode. In the latter case, the electrode is left *in situ* throughout the experiment to serve as a guide to the nerve. This is particularly useful when the recording site is deeply situated (up to 3 cm from the surface), as occasionally occurs with the radial nerve.

The recording electrode is an insulated (normally epoxy resin) tungsten needle with 200-μm shaft and 1 to 15-μm tip diameters, and with an impedance between 100 and 400 khm at 1 kHz (cf. Vallbo et al., 1979). It is manually inserted through the skin, normally with the aid of a pair of insulated forceps, and is then advanced through the underlying tissue. The search for the nerve is by trial and error and, if need be, by repeated recourse to electrical stimulation. The searching time may be between a few minutes and an hour, and sometimes the experiments must be discontinued because of unacceptable pain. On penetration of the perineurium, multiunit activity, often in the form of a short-lasting burst, is usually first encountered, either from a skin nerve fascicle or a muscle nerve fascicle. The appropriate muscle fascicle is then searched for by small adjustments of the insertion angle and depth of the electrode, with or without leaving and reentering the nerve. At this stage the experimenter is guided by the characteristic mass responses of muscle afferents to stretch and pressure applied to the appropriate muscle. Once the desired fascicle is located, single-unit activity may be obtained by further minute adjustments of the electrode position.

At the recording sites commonly used, the separation into distinct skin and muscle nerve fascicles is usually very clear. Thus, in the major nerves above the elbow or in the knee region, specific fascicles supplying individual or small groups of muscles may be recognized (Sunderland, 1945). This facilitates the identification of afferents (Vallbo et al., 1979) as either muscle or cutaneous fibers, although further differentiation poses appreciable problems (see 2.2).

When the subject is relaxed, the signal to noise ratio often is good enough to permit safe triggering of single-unit impulses by automatic level discriminators. However, when the subject is asked to contract the parent muscle of a particular afferent unit, the general noise level increases, largely due to two types of interfering signals. First, other nerve fibers, either afferents or efferents, presumably in the same fascicle, may be activated. Second, electromyographic (EMG) activity from muscles in the vicinity of the recording electrode may become manifest and interfere with the recording, since most subjects spontaneously cocontract a wide range of muscles, even when they are asked to move just a single finger. Both types of interference increase with contraction force, and can often limit microneurographic studies to low contraction levels. The recordings are most stable when the electrodes are left freely floating, supported only by the surrounding tissue. Thus, when movements occur at the recording site, the electrodes tend to follow, hopefully without displacement of the electrode tip. However, this is not always so, and sufficiently large movements transmitted from the parent muscle or from cocontracting muscles nearby can indeed displace the electrode tip. This tends to limit the duration of single-unit recordings to an average of about 15 min, but this may vary between a few minutes and several hours. These limitations have meant that the amplitude, speed, and force of contractions studied have been restricted to the lower end of their physiologic ranges.

Microneurographic studies in man must be conducted according to the Declaration of Helskinki (1964), with the subject's informed consent and knowledge that he or she may, at any time, ask for the experiment to be stopped. In practice, the usual reason for the occasional experiment being prematurely discontinued is that the experimenters realize that the procedures involve more than passing sensations of pain and prevent the subject from relaxing. Until now, no cases of irreversible damage caused by the nerve impalements seem to be known. Less serious side-effects may occur, however, most commonly paresthesias, lasting between a couple of days to a few weeks.

1.2 Midbrain Recordings in Cats and Monkeys

The unique anatomical separation of the cell bodies of tendon organ and spindle afferents in the brainstem (Szenthagothai, 1948; McIntyre, 1951) allows selective recordings from spindle afferents of the jaw-closing muscles during voluntary movement. Pilot studies were carried out in anesthetized cats (Davey and Taylor, 1967; Taylor and Davey, 1968; Cody, et al., 1972), and recordings in conscious animals were independently announced in June 1972 by Matsunami and Kubota in monkeys, and by Cody and Taylor in January 1973. Two other groups have obtained comparable data in conscious monkeys (Goodwin and Luschei, 1975; Lund et al., 1979), and more material has emanated for Taylor's laboratory (Taylor and Cody, 1974; Cody et al., 1975; Appenteng et al., 1980; Taylor and Appenteng, 1981). The recordings from the midbrain nucleus of the trigeminal nerve (MeN V) were obtained with microdrive chambers implanted chronically over trepanations rostral to the tentorium. Conventional metal microelectrodes could then be driven through to the midbrain during recording sessions. Efferent-stimulating electrodes to aid in afferent identification were implanted by Matsunami and Kubota (1972) and Lund et al. (1979), but these authors did not monitor muscle length. Taylor's group used a compliant, V-shaped, stain-gauge transducer to record the vertical displacement between the maxilla and mandible in a sagittal plane, arguing that lateral movements of the cat jaw were negligible. In the monkey, lateral movements are significant, and Goodwin and Luschei (1975) arranged to monitor both vertical and horizontal components of movements in the frontal plane. The stability of single-cell recordings is very variable, with the duration ranging from a few seconds to a maximum of about 15 min, if the animal's head is free to move.

The number of afferents recorded per animal depends on the successful positioning of the recording chamber. If one compares the average yield of spindles per animal in different studies, one finds that Matsunami and Kubota (1972) reported 2, Cody et al.

(1975) reported 6, Goodwin and Luschei (1975) reported 10, and Lund et al. (1979) reported 1. The variability in yield between individual animals may be very large (the figures published by Goodwin and Luschei (1975) show that of 39 afferents recorded in 4 animals, 35 were from 1 animal).

1.3 Dorsal Root Recordings in Cats and Monkeys

Attempts have been made to record from single neurons in spinal cord structures and peripheral nerves of freely moving animals (Wall et al., 1967; Kubota, et al., 1967; Marks, 1969; Brindley, 1972; Courtney and Fetz, 1972). The first such recordings from reliably identified spindle afferents in conscious animals were reported in May, 1975 by Prochazka et al. and by Loeb (1976).

Schieber and Thach (1980) have recently developed a spinal cord chamber for recording from dorsal root ganglion cells in awake, restrained monkeys. The techniques used by Prochazka and Loeb (1979) are similar. During one aseptic operation, fine wire electrodes are inserted into dorsal roots or dorsal root ganglia, and are anchored to nearby dural or vertebral structures. In Loeb's arrangement the wires lead directly to a socket on the cat's back, to which a preamplifier may be connected. In Prochazka's technique, connecting wires lead subcutaneously to a head-piece, to which a telemetry device may be clipped. Both groups monitor muscle length with implanted length gauges (Prochazka et al., 1976; Loeb et al., 1977), or, more recently, with external length gauges attached to implanted, percutaneous fixation wires (Prochazka et al., 1979). Electromyographs are recorded either with implanted wire electrodes, or with percutaneous needle electrodes (Prochazka et al., 1979). Routine video filming during recording sessions has been performed by Loeb's group since 1976 and by Prochazka's group since 1980.

Successful recordings rely on the chance juxtaposition of electrode tips and single afferent fibers or cell bodies. The stability of the recordings is remarkable and afferents are frequently held for at least 24 hrs. of normal activity, and occasionally for many days (Loeb and Duysens, 1979). The stability of recordings from dorsal root ganglia as opposed to dorsal root fascicles is slightly better, and the ganglion provides a more homogenous target during the electrode implantation. On the other hand, unitary recordings from fascicles are more likely to be those of large muscle afferents (Prochazka et al.). An adequate selectivity may be retained, while improving the chances of successful recordings, by using electrodes with surprisingly long uninsulated tips (up to 150 μm, using 17 μm-diameter wire). The average yield of identified spindle afferents per animal reported by Loeb and Duysens (1979) was 5, whereas Prochazka et al. have obtained an average yield of 3. Although there is some uncertainty about the reliability of classification (see 2.3), spindle afferents comprise 20 to 30% of the population sampled in the dorsal root ganglion, tendon organs comprise 15 to 20%, and the remainder are skin afferents, or afferents of unidentified origin.

Recording sessions involving afferents of interest have generally included periods in which the animals were anesthetized, during which the afferents could be identified, and movements of the limb could be imposed. Loeb and Duysens (1979) have normally used the anesthetic ketamine which does not necessarily reduce muscle tone. These authors have not published comparisons of spindle afferent discharge during voluntary movements and movements imposed during deep surgical anesthesia. Prochazka et al. (1979), using anesthetics which abolish muscle tone (propanidid, pentobarbital) and greatly reduce fusimotor activity, have routinely imposed movements intended to mimic the movements which occurred when the animal was awake.

Schieber and Thach (1980), using a modified Evarts recording chamber, have recorded from dorsal root ganglion afferents from the forearm of a monkey. For electrode stability the monkey was placed in a primate chair, which limited movement of the animal's trunk, head, shoulders, and arm. The animal moved a manipulandum by angulating its wrist, and the angular displacement rather than muscle length was monitored. Electromyographs were obtained with percutaneous wire electrodes. This technique is important, as it promises to complement the human data on spindle afferent discharge during skilled hand movements.

2. IDENTIFICATION OF AFFERENTS

2.1 Criteria Available to Distinguish Between Afferents

Matthews (1972) has provided a detailed description of the distinguishing functional characteristics of proprioceptive afferents. In the following, we draw together the characteristics which seem most useful in identifying afferents recorded in conscious animals. In acute experiments on anesthetized animals, it is customary to rely on two criteria for the characterization of a muscle afferent: the response of the afferent to an electrically evoked muscle twitch, and the conduction velocity (CV) of the afferent. Since the limb is normally denervated except for the muscle of interest, the possibility of recording from joint capsule afferents is limited to those few cases where joint afferents occur in the muscle nerve (e.g., quadriceps in cat). Muscle stretch is sometimes used to obtain a preliminary identification of an afferent (see below) and can assist in classification in those rare cases in which an afferent of group I or II CV is suspected of being neither a spindle nor a tendon organ afferent.

Before considering the tests relevant to human neurography and chronic animal preparations, it is worth recalling that even in acute experiments, an unambiguous association between a recorded afferent and a morphologically defined receptor ending can only be established for spindle afferents. These can be recognized beyond all doubt only when it is shown that the afferent at issue is excited by electrical stimulation of ventral root filaments containing functionally single γ fibers. This is based on the morphologic evidence that among the muscle receptors, only spindles are supplied by branches of the small fibers (Barker, 1974 review). Although Golgi tendon organs may be powerfully excited by stimulation of motor axons with CVs above the γ range, the excitation of an afferent during motor unit activation does not suffice for its unambiguous identification as a tendon organ afferent, given the widespread occurrence of β fibers (see 4.1.2). These supply both extrafusal and intrafusal muscle fibers, and may therefore activate both Golgi and spindle afferents.

However, although in acute preparations at least spindle afferents can be unmistakably identified, this is not as readily achieved in human neurography and chronic animal preparations. Selective activation of single γ fibers has not yet been achieved, and the pressure-block technique (Leksell, 1945) to permit the activation of γ but not α fibers during voluntary effort is uncomfortable for the subjects and difficult to perform when the nerve is not freely accessible. In conclusion, the reliable identification of afferents is a crucial and continuing problem in experimental work, both in freely moving animals and in human subjects.

Muscle Twitch

Tendon organs can usually be differentiated clearly from muscle spindles on the basis of the isotonic twitch response, providing that all α-MNs to the muscle are stimulated. Submaximal twitches may result in spindles in inactive portions of the muscle responding like tendon organs, and tendon organs like spindles (Stuart et al., 1972). Stimulation of efferent fibers proximal to their point of entry into the muscle is the least uncertain way of ensuring the excitation of all α-MNs. Stimulation with intramuscular electrodes carries the risk of incomplete excitation, and stimulation with electrodes applied to the surface of the muscle carries the risk of the stimulus spreading to surrounding muscles.

For technical reasons, reliable maximal twitch tests cannot always be performed in studies of chronically implanted animals or in humans. In the former, the resulting ambiguity can be largely overcome by the use of suxamethonium (see below). However, in humans this test cannot be routinely employed because of the risks and the time involved. Thus, the twitch test must be used, and particular care has to be taken to ensure its reliability.

Conduction Velocity (CV)

The main source of error in the estimate of CV of an afferent lies in the measurement of the length of nerve between the points of stimulation and recording. In acute experiments, this length is best measured after the nerve has been dissected free at the end of the experiment, with the measurements taken under standardized conditions, either *in situ* or after removal of the nerve. Even when considerable care is taken, however, the ranges of axonal CV measured in different laboratories can show systematic differences of as much as 35% (e.g., Stephens and Stuart, 1975; Proske and Waite, 1976). Extracellular semimicroelectrode recording, especially in conjunction with window discriminators, introduces the hazard of measuring responses and CVs of afferents other than the one of interest.

Vibration

In the cat, Brown et al. (1967b) showed that small-amplitude, longitudinal vibration of a deefferented muscle selectively excited spindle primary endings. Tendon organs and spindle secondaries were relatively insensitive, although tendon organ sensitivity could be greatly increased by the stimulation of α-MNs innervating the muscle. When vibration is applied laterally to the muscle belly, however, the selectivity, at least for spindle primary and secondary afferents, is far less satisfactory (Bianconi and VanderMeulen, 1963), and even tendon organs can be remarkably locally sensitive (U. Proske, *personal communication*). The use of vibration applied laterally to tendon or muscle belly therefore remains of unproven value for afferent identification. Indeed, given the variability in mechanical coupling and differences in sensitivity of tendon organs in different muscles even within a given species (Alnaes, 1967), it is doubtful that general criteria for differential responses to externally applied vibration could be obtained.

Dynamic Index

The dynamic index, defined as the decrease in discharge frequency within 0.5 sec after the end of a large ramp-and-hold stretch (amplitude 10 to 15% of the resting length of the muscle belly) is, on average, larger for primary than for secondary endings of deefferented spindles, but there is considerable overlap (Matthews, 1963; Koeze, 1968, 1973; Browne, 1975; Cheney and Preston, 1976a). This overlap may, in part, occur because a given external stretch to the muscle would not stretch all spindles by the same amount (Meyer-Lohmann et al., 1974). This is reflected in differences in static position responses, so that the ratio of dynamic index to position response (which we refer to as the dynamic ratio) might be a more general distinguishing characteristic (Jami and Petit, 1979; Banks et al., 1981).

The dynamic responsiveness of tendon organs to muscle stretch can be very marked in the presence of muscle tone (Houk and Henneman, 1967; Houk et al., 1980), and can be appreciable even in a passive muscle (Goslow et al., 1973). However, for the majority of low-threshold tendon organ afferents, the dynamic index seems to be of the same size as for the secondary spindle afferents, provided that the parent muscle is totally relaxed. Unfortunately, systematic comparisons of dynamic indices of spindle afferents and tendon organs of widely ranging thresholds have not been carried out. With these reservations in mind, it can be concluded that in passive muscles, the dynamic index and the dynamic ratio provide useful criteria for identifying primary spindle afferents but do not differentiate between secondary spindle and tendon organ afferents.

Acceleration Responses

At the beginning and end of the ramp phase of a ramp-and-hold stretch, spindle primary afferents show characteristic transient responses, here referred to as the initial burst and the deceleration response, respectively. The initial burst has also been observed in spindle secondary afferents and Golgi tendon organs (Alnaes, 1967; Proske and Gregory, 1977; Schäfer and Schäfer, 1973), and its size is dependent on the duration of preceding muscle rest and fusimotor action. It can, therefore, not be relied on as an identifying characteristic for spindle primaries. In contrast, the deceleration response, which manifests itself as a clear prolongation of the first (compared with the subsequent) interspike intervals at the end of the dynamic phase of a ramp-and-hold stretch (Matthews, 1972, p. 167; Cheney and Preston, 1976a) seems to be an uncontaminated manifestation of acceleration sensitivity. Linear analysis of responses to small-amplitude sinusoidal movements has shown that acceleration sensitivity is present with primary (Matthews and Stein, 1969a; Poppele and Bowman, 1970; Goodwin et al., 1975) but not with secondary (Matthews and Stein, 1969a; Poppele and Bowman, 1970; Cussons et al., 1977) nor with tendon organ afferents (Anderson, 1974). This is in accordance with the finding that deceleration responses are present with primary but not with secondary afferents (Brown et al., 1965; Koeze, 1968; Cheney and Preston, 1976a), in spite of the fact that the ramp stretches commonly used were of an amplitude in excess of that for which spindle behavior is linear. For Golgi tendon organs, deceleration responses have not been investigated systematically. However, from published responses to ramp-and-hold, triangular, or step changes in length, it appears that tendon organ afferents do not normally show deceleration responses (Jansen and Rudjord, 1964; Houk and Henneman, 1967; Houk, 1967; Stuart et al., 1970; Dutia and Ferrell, 1980).

Finally, both the dynamic index and the deceleration response can only be adequately assessed when properly controlled stretches are employed. Ramp-and-hold stretches with smoothed turning points result in drastically altered ramp responses of primary afferents and an abolition of their deceleration responses (Cussons et al., *unpublished observation*). Such undesirable smoothing occurs when stretching devices with poor frequency responses are used, or when the stretches are manually applied, as is often the case with tests in chronically implanted animals or in human subjects.

Suxamethonium

The effects on dynamic index of a dose of suxamethonium (100 to 200 µg/kg i.v.) is characteristically different in muscle spindle primary and secondary afferents (Rack and Westbury, 1966; Dutia, 1980). Although primary endings may be identified with the use of suxamethonium, the responses of some tendon organs are disturbingly similar to those of secondary endings (Dutia and Ferrell, 1980). However, if suxamethonium is administered at a level of anesthesia associated with detectable muscle tone, the ensuing rapid abolition of this tone causes a characteristic increase in the threshold of afferents shown separately to respond like tendon organs to electrically evoked muscle twitches *(unpublished observations)*. Nevertheless, spindle secondaries and tendon organs are difficult to distinguish on the basis of their responses to i.v. suxamethonium alone.

Variability of Discharge

Matthews and Stein (1969) showed that the coefficient of variation of discharge frequency was higher in primary than in secondary endings of deefferented soleus spindles in the cat. There was no overlap, except when the afferents had very low resting discharge rates (but afferents in the 60 to 80 m/sec CV range were not studied). Fusimotor action increased the coefficient of variation of firing in both types of afferents, a factor which limits the usefulness of this criterion in cases where the absence of fusimotor action cannot be guaranteed. Furthermore, although quantitative studies have not been reported, the discharge of tendon organ afferents to muscle stretch is normally very regular (Houk and Henneman, 1967; Houk and Simon, 1967; Houk et al., 1980; Proske, 1981). It seems unlikely, therefore, that spindle secondaries and tendon organs could be clearly differentiated on this basis. At any rate, any study on the variability of interspike intervals should be performed under strictly controlled isometric conditions, and the analysis should be supported by statistical tests of stationarity (Matthews and Stein, 1969).

Small-Amplitude Sensitivity

Spindle primaries respond much more sensitively than secondaries to rhythmic variations in muscle length of ≤ 0.01 resting lengths (RL) (Matthews and Stein, 1969a). However, this small-amplitude sensitivity depends on the mean muscle length (Matthew and Stein, 1969a; Hunt and Ottoson, 1977; Hunt and Wilkinson, 1980). This is more pronounced for primary than for secondary afferents so that the difference between these two types of afferent may only become fully manifest when the parent muscle can be stretched close to its maximal physiologic length (Goodwin et al., 1975; Hulliger, 1979). Moreover, static fusimotor action can reduce or abolish this difference (Goodwin et al., 1975; Cussons et al., 1977; Jami and Petit, 1981; see also Brown et al., 1967a).

The high sensitivity of some tendon organ afferents to stretching in the presence of skeletomotor activity also reduces the potential value of small-amplitude responsiveness as an identification criterion. Furthermore, it is not known whether the values of sensitivity obtained for cat soleus spindles are applicable to other muscles and other species (see 4.1, human data).

2.2 Reliability of Afferent Identification in Human Neurography

Human skeletal muscles seem to be as richly supplied with muscle spindles and Golgi tendon organs as those in the cat (see Barker, 1974). Human spindles possess primary and secondary endings (Cooper and Daniel, 1963; Swash and Fox, 1972), and spindles and tendon organs in humans show the same structural characteristics as in other mammals. These three types of receptor afferents are likely to be encountered during microneurographic recording and ought to be classified according to the criteria listed above. However, not all tests available in acute or chronically implanted animal preparations can readily be performed in human volunteers. In particular, tests with complete extrafusal and intrafusal relaxation cannot be routinely performed owing to the risks and the timescale involved with deep anesthesia or suxamethonium test.

Muscle Relaxation

Reliable recordings of EMG activity and muscle force are of obvious importance for measurements of the size of voluntary contractions and, especially, for the assessment of the degree of muscle relaxation.

Force records

These are difficult to interpret unless it is excluded that active force is contributed from other muscles, e.g., synergists of the parent muscle of the afferent under investigation. Selective force recordings from individual muscles have so far only been obtained for finger extensor muscles (Hulliger and Vallbo, 1979; Vallbo et al., 1981; Hulliger et al., 1982; Vallbo and Hulliger, 1982).

Intramuscular EMG recordings

Intramuscular EMGs run the risk of being sensitive to only a fraction of the activity in the whole muscle. With these recordings it is deemed important that the electrodes be placed in the immediate vicinity of the receptor (Vallbo, 1970a; Hagbarth et al., 1975b; Burke et al., 1976a, 1980c), yet this is only feasible for those receptors that can be precisely located by palpation. It is a disadvantage of intramuscular EMG recording that it may be painful and that the accurate placing of the electrodes may be time-consuming, particularly in cases of uncertainty about the precise localization of the receptor. Nobody would claim that local EMG silence can provide a guarantee of fusimotor silence in the whole muscle and in the spindle of interest, yet it may serve as a useful indicator, since the prevailing view is that manifest spindle excitation, attributable to fusimotor activity, does not normally occur in the absence of reliably monitored EMG activity (Hagbarth et al., 1975a; Burke et al., 1976b, 1980c,d; Vallbo et al., 1979; Burke, 1980, 1981). However, this evidence applies only for the specific motor paradigms investigated. Also, this view has not found general acclaim, and Struppler (1981) has recently reiterated the original claims (Burg et al., 1974; Szumski et al., 1974), that selective activation of fusimotor neurons unaccompanied by skeletomotor activity could occur during the Jendrassik maneuver and during contraction of remote muscles. In view of the controversy over this issue, the proposal that selective activation of fusimotor neurons was a specific feature of extensor muscles (Struppler, 1981) deserves further examination.

Surface EMG recording

This procedure is easier to achieve and is pain-free for the subject. However, the surface EMG electrodes are prone to pick up activity from muscles adjacent to the muscle of interest. Such contamination may be minimized by arranging, prior to the nerve impalement, a number of surface electrodes in suitable positions over muscle bellies so that by selection of suitable pairs of these electrodes, the EMG activity of the parent muscle of any afferent can be relatively selectively recorded (Hulliger et al., 1982).

Multiunit Recordings

Single-unit recordings may not be representative for the ensemble responses of muscle afferents during voluntary movements. The attempt, therefore, has repeatedly been made to characterize the ensemble responses by multiunit recordings from nerve fascicles supplying the muscle of interest (e.g., Hagbarth et al., 1975a,b, 1981; Burke et al., 1978a, 1979a, 1980a,d; Hagbarth and Young, 1979; Young and Hagbarth, 1980). It was argued that these recordings were dominated by Ia spindle afferents, since the responses to passive muscle stretch and release and to electrically induced muscle twitches were very similar to those of single Ia afferents (Hagbarth and Young, 1979; Burke et al., 1979a, 1980d; Hagbarth et al., 1981). However, even if such responses are dominated by large nerve fibers, it is difficult to accept that Aα axons or Ib afferent fibers from tendon organs do not seriously contaminate such recordings.

Recordings by Burke et al. (1979a, Fig. 2) illustrate these concerns. The multiunit response contains a conspicuous Ib component during the early twitch rising phase. Moreover, during a pressure block of α-fiber activity (EMG silence, Fig. 2) the multiunit neurogram during voluntary effort was strikingly reduced. Part of this reduction might well be accounted for by an abolition of some α contribution to the multiunit neurogram. It would seem very difficult in awake human subjects to demonstrate that α-MN activity cannot contribute to multiunit recordings. Silence in multiunit neurograms would have to be demonstrated during voluntary contractions and under conditions of guaranteed silence not only of spindle afferents but also of Golgi afferents. The claimed absence of a contribution to multiunit recordings from tendon organ afferents would only appear plausible if complete silence during the rising phase of twitch contractions could be shown, first for maximal twitch contractions and, second, repeatedly during each recording, since with multiunit recordings small intrafascicular displacements of the electrode are difficult to detect. Thus, silence during the rising phase of a maximal twitch observed at the beginning of a recording session does not guarantee that the same holds true a few minutes later, particularly when the loss of single-unit recording sites occurs regularly observed (Hagbarth et al., 1981). Therefore, it seems that for the time being there is very little to be gained from multiunit recordings during voluntary movements.

Single-Unit Recordings

The criteria commonly used to differentiate muscle spindle and tendon organ afferent units from cutaneous afferents, Pacinian corpuscle afferents from deeply situated structures (Hunt and McIntyre, 1960), or joint receptor afferents, may be considered safe (Vallbo et al., 1979). Thus, single units are taken to be muscle afferents when they give brisk and consistent responses to pressure applied to a confined portion of a muscle, when they respond to passive muscle stretch performed by lightly tapping the appropriate tendon or by passively moving the appropriate joint, and when they fail to respond both to light touch of the skin overlying the muscle and to pressure applied to neighboring joints.

These, and only these tests also permit a reliable identification of the receptor-bearing muscle. In contrast, the observation of afferent excitation during voluntary isometric contraction of the suspected parent muscle may be misleading, given most subjects' unintentional tendency to cocontract many other muscles, even during the simplest tasks (See 1.1).

Muscle Twitch

Undoubtedly, the most reliable method to elicit twitch contractions for the differentiation of spindle and tendon organ afferents is to stimulate all of the motor nerve fibers supplying the parent muscle of a single afferent unit (see 2.1). In practice, this is approximated by stimulating with electrodes placed over or within the muscle, or by intrafascicular stimulation through the neural recording electrode, when this rests at a single-unit recording site (Hagbarth et al., 1975a; McKeon and Burke, 1980). Due to the relatively specific subdivision of the limb nerves into fascicles supplying individual muscles (Sunderland, 1945), such intrafascicular stimulation may permit a relatively selective activation of the receptor-bearing muscle. It is doubtful, however, that stimulation in any of these ways permits a reliable activation of all of the motor units of the muscle in question. McKeon and Burke's (1980) assertion that muscle spindle afferents could be positively identified with "barely visible twitches" can hardly claim credibility in view of the paradoxical spindle-type responses of tendon organ afferents to twitch contractions of single motor units (Houk and Henneman, 1967; Stuart et al., 1972; see also 2.1).

Regrettably, in the past the differentiation between single tendon organ afferents and spindle afferents was not always attempted with rigorous and systematic twitch tests (Hagbarth and Vallbo, 1969; Vallbo, 1970a,b, 1971, 1974a,b; Burke and Eklund, 1977; Burg et al., 1974). Even in some of the recent reports it is not clear how many, or indeed whether any, single afferent units were classified on this basis (Hagbarth and Young, 1979; Young and Hagbarth, 1980; Hagbarth et al., 1981). However, twitch tests were performed in the following investigations: Hagbarth et al., 1975a,b; Burke et al., 1976a,b; 1980c,d; Hulliger and Vallbo, 1979; Vallbo and Hulliger, 1981, 1982; Vallbo et al., 1981; Hulliger et al., 1982. Normally, it is possible to deduce the proportion of units that was rigorously tested and, even in some cases, the size of the twitch contractions relative to the maximal voluntary contraction force (e.g., Hulliger et al., 1981; Vallbo and Hulliger, 1982). This latter piece of information is relevant, since it permits a rough estimate of the proportion of motor units activated in the test and thus of the reliability of the classification. On the other hand, it seems unrealistic, given the limited stability of the present microneurographic techniques to expect that maximum twitch contractions could routinely be performed.

It has been argued that some of the published recordings from single units that were not twitch-tested (see references above), but were, nevertheless, classified as primary spindle afferents, might, in fact, have been from tendon organ afferents with a high dynamic sensitivity for muscle stretching in a perhaps incompletely relaxed muscle (Taylor, 1981). This would at least partly account for the dearth of published recordings from tendon organ afferents.

It would also cast considerable doubt on the validity of the conclusion drawn in recent reviews (Hagbarth, 1979; Vallbo et al., 1979; Burke, 1980, 1981; Vallbo, 1981) about the generality of α-γ coactivation as a principle of organization of skeletomotor and fusimotor activity. The possibility of incorrect classification cannot be excluded in the absence of convincing twitch tests. However, evidence for α-γ coactivation has also been found in the sub-sample of twitch-tested and reliably classified spindle afferents (see 4.1). Furthermore, there are other important factors likely to contribute to the scarcity of data from tendon organ afferents, above all the fact that single units are invariably searched for under conditions which certainly do not favor the detection of tendon organ afferents: The subjects are always asked to relax as much as possible during the exploration of the nerve, the parent muscle normally is at short or intermediate length, well below the physiologic maximum, and the search for single units often is guided by explorative tests clearly favoring the detection of afferents with high dynamic sensitivity (see 1.1). Thus, as noted elsewhere (Vallbo, 1973, 1974a; Burke, 1981), the sample of single units is strongly biased in favor of primary afferents at the expense of secondary and tendon organ afferents.

In conclusion, the credibility of future

microneurographic investigations would greatly benefit from consistent classification of muscle afferents with near-maximal muscle twitches and from an unambiguous characterization of each sample of single units with respect to the reliability of the twitch tests (proportion of units so tested, size of the twitches in percent of maximal voluntary contraction force of the parent muscles).

Primary and Secondary Spindle Afferents

For reasons discussed above (Section 2.1), reliable estimates of the CV of muscle afferent fibers have not yet been achieved (see also Vallbo et al., 1979). Thus, the test normally relied on in acute studies to differentiate between group I and II afferents has not been available in human neurography, nor even in many of the chronic animal studies (see 2.3). Classification has therefore been based on an assessment of the dynamic sensitivity to muscle stretching or on a combination of several criteria, including dynamic sensitivity (Vallbo, 1970a, 1971, 1974a,b; Burg et al., 1974; Szumski et al., 1974). As regards the reliability of methods used, there are two main shortcomings. First, apart from the deceleration response to adequately controlled ramp-and-hold stretches (see above, acceleration sensitivity), which is a unique feature of primaries, none of the other criteria permits an unequivocal classification of all units. Second, for all response parameters used for spindle afferent classification (dynamic index, dynamic ratio, small amplitude sensitivity, variability of interspike intervals), the differences between primary and secondary afferents are of a quantitative nature. Thus far, however, published work on microneurography has relied in qualitative assessments. This is further aggravated by the fact that mechanical stimulation for classification tests was often performed manually (Hagbarth and Vallbo, 1968, 1969; Burg et al., 1974; Szumski et al., 1974; Burke and Eklund, 1977, Burke et al., 1978b, 1979b). Thus, in these studies, the test stretches probably had a smoothed sigmoid time-course, so that the dynamic index, let alone the deceleration response, could not have been assessed. Notable exceptions are the studies of Vallbo (1974a,b), Hagbarth et al. (1975a,b), Burke et al. (1976a,b), Hulliger and Vallbo (1979), Hulliger et al. (1982), Vallbo et al. (1981) and Vallbo and Hulliger (1982), in which adequately controlled stretching devices were used to apply ramp-and-hold stretches with the prerequisite sharp edges (see section 2.1).

It is regrettable that, in a number of studies, a subclassification of spindle afferents into primary and secondary units was never attempted (Vallbo, 1970b; Burke et al., 1978a,c; 1979a; 1980c,d; Hagbarth and Young, 1979; Young and Hagbarth, 1980; Hagbarth et al., 1981). In those studies where a differentiation into primary and secondary afferents was attempted (e.g., Burke at al., 1976a,b) an unknown fraction of spindle afferents may have been incorrectly subclassified, either because the stretches were inadequate or because the dynamic sensitivity was not assessed quantitatively. It is hoped that this situation will improve in view of the valuable information that might be obtained from studies with reliably classified secondary afferents, on the characteristics of static fusimotor activity during voluntary movement, assuming that the selective control of secondary afferents by static fusimotor fibers holds true in man (see 4.1).

Recent quantitative studies have confirmed that in man, as in cats and monkeys, neither the measurements of dynamic index or ratio (Hilliger et al., *unpublished observations*), nor of the variability of discharge (Burke et al., 1979b; Hulliger et al., *unpublished observations*) permit an unequivocal classification of all spindle afferents as either primary or secondary units, although in the study of Burke et al.

(1979b; some of this uncertainty might arise from the spindle afferent subclassification on the basis of abrupt application of pressure to the tendon of the receptor-bearing muscle. In conclusion, the identification of afferents in human neurography could be (and indeed is being) greatly improved by the routine use of (1) adequate ramp-and-hold servos which allow quantitative estimates of deceleration responses, dynamic indices, and dynamic ratios to be obtained, and (2) careful twitch tests in which attempts are made to elicit maximal muscle twitches. In view of the great importance of correct afferent identification, it is not unrealistic to expect the same emphasis to be placed on this aspect of a neurography experiment as was placed on obtaining and analyzing voluntary movement data.

2.3 Reliability of Afferent Identification in Chronic Animal Recordings

Midbrain Recordings in Cats and Monkeys

In the cat, the evidence argues against the presence of tendon organ afferents in the midbrain nucleus of the trigeminal nerve (Szenthagothai, 1948; Jerge, 1963a; Cody et al., 1972). Thus, units within this nucleus responding to muscle stretch and palpation will almost certainly be spindle afferents. The existence of interneurons responding like muscle spindles and tendon organs in the adjoining nucleus supratrigeminalis (Jerge, 1963b) means that contamination could occur if the recording electrodes were slightly inaccurately placed. Histologic verification of recording sites, as carried out by Matsunami and Kubota (1972), Goodwin and Luschei (1975), and Lund et al. (1979) therefore adds credibility to the data. It should be stressed that the anatomic separation of spindle and tendon afferents mentioned above has yet to be verified in monkeys.

The chances of Taylor's mistaking tendon organs for muscle spindles would therefore seem remote (see Taylor and Davey, 1968 and Cody et al., 1972). However, distinguishing spindle primaries from secondaries has been more troublesome, and these workers generally preferred to speak of high-frequency and low-frequency units. Cody et al. (1972) considered the measurement of CV not to be technically feasible. Apart from the difficulty of the postmortem dissection, there may be uncertainty about the exact site of stimulation, and a further complication is introduced by the probable reduction in afferent fiber diameter between the cell body and the trigeminal motor nucleus. The chances of mistaking a peridontal mechanoreceptor for a muscle spindle are probably minimal, as the former normally respond to pressure applied to teeth and gums only.

Matsunami and Kubota (1972) stimulated within the semilunar ganglion with a concentric needle electrode and characterized afferents in the conscious animal on the basis of their responses to submaximal muscle twitches and to muscle stretch, which does not reliably distinguish between spindle and tendon organ afferents (see 2.1). Judgment in this case depends on the assumption that the anatomic separation of these afferents seen in the cat also pertains to the monkey. The same applies to the study of Goodwin and Luschei (1975). No attempt was made in either study to classify spindle afferents as primaries or secondaries. Lund et al. (1979) recorded from 7 afferents adjacent to the trigeminal motor nucleus. The afferents were judged to be muscle spindles on the basis of their responses to imposed jaw movement, and their estimated CV.

Dorsal Root Recordings

In the cat studies of Loeb et al. (1980) and Prochazka (1980) afferent identification is carried out after the animals have been deeply anesthetized. Both groups apply muscle stretch, palpation, tendon taps and electrically evoked twitches as test stimuli, none of which, alone, allow unequivocal identification of afferents (see Section 2.1). Loeb and Duysens (1979) encountered technical difficulties with the use of suxamethonium, and preferred to rely on estimates of CV to distinguish between spindle primary and secondary afferents. These estimates were obtained by stimulating intramuscularly at a strength sufficient to excite the afferents directly. The conduction paths were measured during postmortem dissections (see section 2.1). Loeb and Duysens (1979) believed that their estimates were accurate to between 5 and 20%. In one puzzling case, an afferent with an estimated CV of 85 m/sec (which would normally be considered well within group I range) was classified as a secondary, but all other presumed secondaries had estimates of CV lower than 60 m/sec. In most cases, Loeb et al. (1980) relied on the responses to muscle twitches evoked with intramuscular electrodes, to distinguish spindle primaries from tendon organs. The problems associated with this procedure have been discussed above. Prochazka et al. (1979) routinely determine the dynamic index, albeit with manually applied ramp-and-hold stretches, the smooth turning points of which usually obscure deceleration responses. Nevertheless, since the values of dynamic index before and after i.v. suxamethonium are compared, the recognition of spindle primaries has presumably been reliable. Some uncertainty with regard to distinguishing between low-threshold tendon organs and spindle secondaries does, however, remain. The responses to suxamethonium administered in the presence of vestigial muscle tone (section 2.1) probably provide the least ambiguous criteria, short of CV estimates obtained after dissection with direct stimulation of the muscle nerve.[1]

In view of the continuing lack of agreement about the properties of joint capsule afferents (Burgess and Clark, 1969; Zalkind, 1971; Tracey, 1979; McIntyre et al., 1978; Ferrell, 1980) it is not possible at this stage to define unequivocal criteria for the identification of such afferents.

3. PROPERTIES OF TENDON ORGAN DISCHARGE DURING VOLUNTARY MOVEMENT

3.1 Human Data

We are aware of published recordings of only four presumed tendon organ afferents in man (Vallbo, 1970a, 1974a; Burg et al., 1973; Burke et al., 1976a; 1980b) (see section 2.2). Tendon organs have probably been encountered often enough to suggest that it is choice rather than necessity which has determined their low publication rate. It may have been thought that their discharge behavior can, in any case, be predicted from the properties of tendon organ afferents established in acute experiments (Houk et al., 1980; Proske, 1981) and that neurography merely confirms such predictions. Yet this is probably only true for weak isometric muscle contractions which form a minute part of an animal's repertoire of movements.

Vallbo (1970a) published records obtained during isometric muscle contractions from an afferent axon tentatively classified as a tendon organ. Since twitch tests were

---

[1] An ingenious new technique for estimating axonal CV in animals has recently been described by Hoffer et al., (1981). Recordings from two pairs of electrodes in a cuff enclosing a peripheral nerve are averaged, sweeps being triggered from the neural spikes recorded in a spinal root. Conduction velocity is derived from the difference in latency between the two resulting spike waveforms, the distance between electrodes being taken as the conduction length. Since the latter is limited by the dimensions of the neural cuff, the accuracy of the estimate presumably depends critically on the accuracy of the latency and distance measurements. Thus far, this method has only been applied to ventral root recordings. (see also Appenteng et al. (1980) for another variant of the spike-triggered averaging approach.)

not performed, the identification rested mainly on the observation that the two variables, active muscle force and (afferent) impulse frequency, agreed very well, and indeed this is the salient feature of the data. Smooth changes in force were accompanied by corresponding smooth changes in afferent firing rate. On the other hand, step changes in discharge rate during smoothly increasing isometric contractions are evident in a second tendon organ recording (Vallbo, 1974a). This pattern was reasonably interpreted as resulting from the progressive recruitment of motor units associated with the tendon organ receptor. The data of Burg et al. (1973) show a tendon organ afferent discharging during the muscular response evoked by a tendon tap. Burke et al. (1976a) showed a tendon organ afferent apparently responding subharmonically to muscle vibration in a fully relaxed muscle.

No useful conclusions can be drawn from this scanty data base. Moreover, an interesting point arises: How often does a human tendon organ afferent discharge relate to whole muscle force (as in Vallbo, 1970a) and how often is such a relationship obscured by "recruitment steps" (as in Vallbo, 1974a)?

3.2 Cat Data

The disclosure by Houk and Henneman (1967) that tendon organs are very sensitive to active muscle contractions gave rise to numerous investigations in which the characteristics of these receptors were reappraised (Houk et al., 1980; Proske, 1981). Chronic animal recordings from single afferents identified as tendon organs showed that discharge occurred mainly during active muscle contraction (Prochazka et al., 1976; Lewis et al., 1979a). In another study, Prochazka and Wand (1980a) reported that tendon organ afferents all fired during unobstructed active muscle shortening. The firing rates often exceeded 100 sec, even up to the highest shortening velocity observed, 1.8 resting-lengths/sec. In contrast, the firing rate of muscle spindles usually dropped to well below 100/sec when shortening velocities exceeded 0.2 resting-lengths/sec. The ratio of tendon organ afferents to spindle afferents (primary and secondary) in cat hindlimb muscle nerves ranges from 0.3 to 0.4 (Barker, 1962), so it is conceivable that during rapid, active muscle shortening, the net discharge from a muscle's tendon organs may outweigh the net discharge from its spindles.

Since afferent activity from antagonistic muscles might significantly contribute to the control of such movements (e.g., Capaday and Cooke, 1981), the balance between effective tendon organ and spindle afferent discharge might be more than redressed, when the total afferent influx associated with the movement of a particular joint is considered.

Loeb (1980) has published recordings from three hindlimb tendon organs during locomotion, and also found that the activity of the afferents suggested a close correlation with the timing of activity of the receptor-bearing muscle. Interestingly, one afferent showed little tendency to follow short-term irregularities in EMG amplitude. The suggestion was made that this afferent was responding to slow motor units which, once recruited, fluctuated little in their tension output. Another characteristic of this afferent was its abrupt onset of firing at a fairly steady frequency. The latter property is familiar from acute experiments, and has also been observed in man (see section 3.1) and in conscious cats by Prochazka and Wand *(unpublished observation)*.

As in human neurography, tendon organ afferents have, relative to spindle afferents, been neglected in the chronic animal studies due to the seemingly predictable nature of their discharge behavior. However, some important questions remain to be answered. For example, how do tendon organs respond when active muscle shortening is suddenly obstructed? Do tendon organs display bursts of discharge during rapid muscle stretching, as has been described for spindle afferents (Hagbarth et al., 1980, 1981; Prochazka and Wand, 1981b)? Is there a speed of unobstructed muscle shortening beyond which tendon organ afferents are silent and if so, does this limiting speed vary from muscle to muscle?

4. PROPERTIES OF MUSCLE SPINDLE DISCHARGE DURING VOLUNTARY MOVEMENT

4.1 Human Data

In view of the numerous reviews on spindle afferents in man (Hagbarth, 1979; Vallbo et al., 1979; Burke, 1980, 1981; Vallbo, 1981), this section is restricted to current developments and controversial issues. In several respects, the human data on spindle afferent responses during voluntary movement seem to be at variance with those in cats and monkeys.

4.1.1 Basic Spindle Properties in Man: Comparative Aspects

Morphology

The main conclusion from the majority of morphologic investigations of human muscle spindles is that the latter are very similar structurally to those of the cat (Cooper and Daniel, 1963; Kennedy, 1970; Swash and Fox, 1972; Barker, 1974; Kucera and Dorovini-Zis, 1979). Thus, the sensory innervation of human spindles normally consists of 1 or 2 Ia fibers supplying primary endings and of several group II fibers supplying secondary endings. Furthermore, human spindles usually are generously supplied with intrafusal muscle fibers (up to 14; Cooper and Daniel, 1963; Swash and Fox, 1979) which, on morphologic and histochemical grounds, have been differentiated into nuclear bag 1, nuclear bag 2, and nuclear chain fibers (Kucera and Dorovini-Zis, 1979). Furthermore, a rich motor supply to intrafusal muscle fibers has been described. Thus, as in the cat, three types of motor terminals have consistently been found: p1 and p2 plates and trail endings (Swash and Fox, 1972; Barker, 1974; Saito et al., 1977). The p1 plates, innervated by large motor axons, which sometimes were also seen to innervate extrafusal muscle fibers (Cooper and Daniel, 1963; Swash and Fox, 1979), were regularly found on bag fibers (Cooper and Daniel, 1963; Kennedy, 1970; Swash and Fox, 1972) but only rarely on chain fibers (Swash and Fox, 1972). The p2 and trail endings were found to be innervated by small, presumably γ axons (Swash and Fox, 1972). The p2 plates were more frequently seen on bag than on chain fibers (Kennedy, 1970; Swash and Fox, 1972), and trail endings were found predominantly but not exclusively on chain fibers (Kennedy, 1970; Swash and Fox, 1972). However, in none of these studies were bag 1 and bag 2 fibers differentiated (Ovalle and Smith, 1972; Barker et al., 1976; Banks et al., 1977). Therefore, the fusimotor innervation pattern emerging from cat studies (see Emonet Denand et al., 1980; Boyd, 1981), remains to be confirmed in man. The current evidence for the cat supports the view that skeletofusimotor β fibers supply p1 plates (Barker, et al., 1970) situated mainly on bag 1 fibers (Barker et al., 1980), that dynamic γ fibers innervate p2 plates also predominantly on bag 1 fibers (Barker et al., 1976; Barket et al., 1978) and that static γ fibers supply trail endings, mostly on bag 2 and chain fibers (Barker et al., 1973).

For now, the controversy about the motor innervation of the cat muscle spindle seems to have been settled to everybody's satisfaction (see Boyd, 1981). However, new developments are in sight, since nuclear chain fibers are no longer considered as a uniform group of muscle fibers. Long-chain fibers have been singled out on morphologic grounds (Barker et al., 1976) and it has become apparent that histochemically they differ from the majority of (short) chain fibers (NADH-TR reaction, but not adenosine triphosphatase (ATP) reaction, (Kucera, 1980). For human spindles two histochemical types of chain fibers have also been described (Saito et al., 1977). However, these fiber types were not characterized by their morphologic features (long- versus short-chain fibers), and their histochemical profile is the reverse of that described by Kucera (1980) in the cat, since the human chain fibers differed in their ATPase staining but not in the NADH-TR staining. This issue is of some importance for the question of skeletofusimotor innervation in man, since in the cat fast α axons, exerting static fusimotor action, have been found to innervate predominantly long-chain fibers (Harker et al., 1977; Jami et al., 1978, 1979; Emonet Denand et al., 1980).

Structural differences between human and other mammalian spindles have, however, also been described. Thus, Swash and Fox (1972), but apparently not Cooper and Daniel (1963), found that primary afferent fibers were preferentially distributed to nuclear bag fibers. This is at variance with the sensory innervation of cat and other mammalian spindles, where primary afferent terminals normally are distributed to bag 1, bag 2, and chain fibers (Barker, 1974; Banks et al., 1979, 1981; and *personal communication*). The human studies referred to were performed before the existence of these three types of intrafusal fiber was generally accepted (Barker et al., 1976; Banks et al., 1981). It remains an open question as to whether in man primary afferents might be more responsive to activation of bag 1 than bag 2 fibers and, compared with the cat, more susceptible to dynamic than static fusimotor activity. However, this speculation rests on the assumption that the now widely accepted specific innervation pattern of the cat, with dynamic axons acting on bag 1 fibers and static axons operating bag 2 and chain fibers (Bessou and Pages, 1975; Boyd et al., 1977; Banks et al., 1978; Banks et al., 1981) is also valid for human spindles.

Further, human spindles contain more intrafusal muscle fibers than other mammals (Cooper and Daniel, 1963; Barker, 1974). This is largely accounted for by larger numbers of nuclear chain fibers, since human spindles were never found to contain more than 3 or 4 bag fibers (Cooper and Daniel, 1963; Kucera and Dorovini-Zis, 1979), as is also typical for other mammalian species (Barker, 1974). Again, one might speculate that as a result, static fusimotor action in man, particularly on secondary afferents, might be more powerful than in cats.

In man a closer structural relationship than in cat may exist between spindles and extrafusal motor units. It has been reported that the capsule of human spindles can enclose neighboring extrafusal muscle fibers (Cooper and Daniel, 1963; Kucera and Dorovini-Zis, 1979). Thus, human spindle afferents might be even more responsive to contraction of individual motor units than spindle afferents in cat, where such responsiveness to twitches in single motor units has repeatedly been described (Binder et al., 1976; Windhorst and Meyer-Lohmann, 1977; Binder and Stuart, 1980). Although direct evidence to support this view is not yet available, the possibility ought to be borne in mind, when afferent recordings during voluntary contractions are interpreted (see recruitment order below).

In vitro *studies of human spindles*

Two studies have been performed on isolated spindles from normal human muscle (Poppele and Kennedy, 1974; Newsom Davis, 1975). In both cases, the properties of human spindle afferents were in good agreement with those of cat spindles studied *in vivo,* although the classification of the afferents had to rely partly on indirect criteria (see 1.1), since the axonal CV could not be measured. The spindle origin of the afferent responses was ascertained by direct visualization of the sense organs recorded from, and/or the performance of twitch tests (Newsom Davis, 1975). In both studies individual afferents were classified as primary or secondary on the basis of differences in their dynamic sensitivities to stretch.

Newsom Davis (1975) found that the absolute discharge rates and position sensitivity of human intercostal spindle afferents were the same as in cat (Andersson et al., 1968a), provided that allowance was made for the difference in resting length (RL) of the parent muscles, by relating changes in discharge rate to proportional changes in muscle length (in % RL). Thus, the same value of about 3 impulses/sec/% RL was found for human and feline intercostal spindle afferents (Andersson et al., 1968a; Newsom Davis, 1975) and for cat soleus afferents (assuming a muscle belly resting length of 60 mm, applied to data from Harvey and Matthews, 1961; Lennerstrand, 1968; Brown et al., 1969). When the dynamic sensitivity of the spindle afferents was assessed it emerged that, as in the cat (Andersson et al., 1968a; see also section 2.1), there was a group of intermediary afferents, for which only modest values of dynamic index were found, in accordance with *in vivo* recordings in man (Section 2.2).

Poppele and Kennedy (1974) found that the frequency response characteristics of spindle afferents from human extensor digitorum proprious were indistinguishable from those of cat tenuissimus spindle afferents (Poppele and Bowman, 1970). Regrettably, the authors gave no clues about the absolute values of the dynamic sensitivity of the primary and secondary spindle afferents, nor did they indicate whether, as in the cat, nonlinear properties (in particular a high sensitivity for small-amplitude movements) were encountered.

4.1.2 In Vivo *Recordings in Man*

Discharge rates

Thus far all microneurographic recordings from human spindles revealed remarkably low discharge rates of spindle afferents, both in relaxed muscles and during voluntary contractions (see Vallbo et al., 1979; Burke, 1981; Vallbo, 1981). At resting length, human spindle afferents often were silent while during stretch, tonic maintained rates of discharge rarely exceeded 25/sec. Peak discharge rates during the fastest stretches studied were normally below 100/sec, and often below 50/sec. This is in striking contrast to the data from acute or chronic recordings in cats and monkeys (see Cody et al., 1975; Goodwin and Luschei, 1975; Prochazka et al., 1976, 1979; Loeb and Duysens, 1979). Acute animal experiments do not necessarily provide relevant standards of reference, since muscles are often stretched to their physiologic limits, and fusimotor neurons tend to be stimulated at high, possibly nonphysiologic, rates. However, these reservations do not apply to data from chronically implanted animals.

It is unlikely that the difference in discharge rates is attributable to an unfortunate choice of muscles studied in man, since these, in fact, cover a respectable spectrum (see Vallbo et al., 1979), including the hindlimb muscles that have been extensively investigated by Loeb's and Prochazka's groups. Neither is there any compelling reason to expect such a discrepancy from the comparative studies of human and other mammalian spindles discussed above. A more likely explanation is that, owing to the limited stability of the recording technique (see section 1.1), human studies have been confined to a narrow range of experimental conditions (see also Taylor, 1981). Most studies have been performed at short or intermediate muscle lengths, the velocities of imposed or voluntary movements have rarely exceeded 0.1 RL/sec (e.g.,

Hulliger et al., 1981; Prochazka, 1981) which is far below the maximal velocities reached in either mammalian studies (Prochazka et al., 1979; Lewis et al., 1979a, and below), and the voluntary contractions rarely exceeded 10% of maximum voluntary force of the parent muscle (up to 30% in single cases, e.g., Hulliger and Vallbo, 1981). Thus, in man neither the afferents' responsiveness nor the fusimotor efferents' excitatory power have been explored to their full capacity.

Position sensitivity

In only two studies has the position sensitivity of human spindle afferents been quantitatively assessed (Vallbo, 1974a, for finger flexor muscles; Hulliger et al., 1981, for finger extensors). At first sight the figures reported for afferents from relaxed muscles seem very low indeed, as the mean values ranged between 0.18 (flexor primaries) and 0.28 (extensor primaries) impulses/sec/degree joint movement. However, when position responses are related to proportional length changes (Newsom Davis, 1975; Hulliger et al., 1981), figures of the same order of magnitude as in the cat (see above, *in vitro* recordings) are obtained. Thus, with a (length)/(joint angle) conversion factor of 0.33 mm/degree (Kaplan, 1965) and a muscle belly resting length of about 200 mm, the position sensitivity ranges between 1.2 and 1.8 impulses/sec/% RL.

Dynamic and small-amplitude sensitivity

Primary spindle afferents in the cat exhibit a marked dynamic sensitivity. This is particularly pronounced for small-amplitude movements. Thus, for small sinusoidal stretches (below 50/µm) in soleus the sensitivity may be 10 times as high for 1-mm sinusoidal stretches (Hulliger et al., 1977a). In man, dynamically sensitive spindle afferents have frequently been encountered. Their responses to properly controlled ramp-and-hold stretches (Section 1.1) applied to the relaxed parent muscle show a high dynamic index or ratio (Hulliger et al., to be published). Also, during voluntary movements such units tend to be sensitive to small disturbances or irrregularities of movement (Vallbo, 1973a, 1974b; Burke et al., 1978a,b; Hagbarth and Young, 1979; Young and Hagbarth, 1980). Since in some instances appreciable responses to very small movements (< 0.5 degree at the wrist or at finger joints; Hagbarth and Young, 1979; Young and Hagbarth, 1980) were recorded, it has been concluded that apparently primary afferents in man exhibited the same nonlinear characteristics and the same pronounced sensitivity to small-amplitude movements as in the cat (Hagbarth and Young, 1979). However, in the absence of a rigorous analysis, this conclusion seems premature, all the more so, since in the cat the high sensitivity to small movements was found only with well-stretched muscles (see Section 1.1). Caution is also indicated because of the finding of Swash and Fox (1972), that at the site of the primary endings human nuclear bag fibers may not be as devoid of myofilaments as feline bag fibers [the nonuniform distribution of myofilaments along the intrafusal fibers being taken to be largely responsible for the manifestation of a pronounced small-amplitude sensitivity (Matthews, 1972)].

Further quantitative studies are obviously necessary to clarify this issue. Attention should be paid to the amplitudes of movement chosen. If any cat data may be extrapolated, a small-amplitude sensitive range of approximately 0 to 0.001 RL (cat soleus) should be looked for. For human finger extensor muscles this corresponds to a 0.2-mm stretch or an angular movement of 0.6° at the metacarpophalangeal joint (cf. Kaplan, 1965), which is well within the range of movements readily studied in human neurography. The emphasis placed above on the need for quantitative studies is by no means to deny the published qualitative findings their possible significance for the control of movement (Hagbarth and Young, 1979; Vallbo et al., 1979; Young and Hagbarth, 1980; Burke, 1981).

Spontaneous fusimotor activity in resting muscles?

It is well known that, in spinal decerebrate or anesthetized cats, γ-MNs may be spontaneously active (e.g., Hunt, 1951; Hunt and Paintal, 1958; Diete-Spiff et al., 1962; Ellaway, 1971; Noth, 1971). Also, in alert cats there is indirect evidence for spontaneous activity of fusimotor neurons when there is EMG silence in the parent muscle. Prochazka *(unpublished observation)* found that spindle afferent sensitivity and bias can change in the absence of detectable joint movements, apparently in relation to the degree of alertness of the cat or the type of movement anticipated by the animal (see section 4.2). These findings are in clear contrast to the bulk of evidence obtained thus far in man. As forcefully argued in recent reviews (Vallbo et al., 1979; Burke, 1980, 1981), there is at present no convincing evidence to support the contention of Burg et al. (1974) and Struppler and Velho (1976) that fusimotor neurons in man may be activated without the activation of α-MNs. This does not, however, preclude the possibility that under specific, as yet unidentified, conditions, this may occur. The experiments of Burke et al. (1980c) are of particular interest in this context. On indirect evidence these authors suggested that the threshold of activation of fusimotor neurons could be altered by various maneuvers designed to alter presumed descending or segmental reflex inputs to these neurons (see below).

β-Innervation in man?

In cat, it would seem that the sole function of γ-MNs is fusimotion (Hunt and Kuffler, 1951; Kuffler and Hunt, 1952). On the other hand, α-MNs are always skeletomotor, sometimes skeletomotor and fusimotor (in which case they are termed β) but never exclusively fusimotor (Ellaway et al., 1972). After the first morphologic (Cooper and Daniel, 1963; Adal and Barker, 1962) and functional (Bessou et al., 1965) indications of the existence of such skeletofusimotor β innervation, it remained uncertain just how important quantitatively β innervation was. Due largely to the improvement of histologic and experimental techniques it has been accepted that β innervation in cat (Barker et al., 1970; Emonet-Denand et al., 1975; Emonet-Denand and Laporte, 1975; McWilliam, 1975; Laporte and Emonet-Denand, 1976; Harker et al., 1977) and rat (Andrew and Part, 1974; Barker, 1974) is widespread and functionally significant. Moreover, β-fusimotor action is not exclusively dynamic. There is now good evidence that fast β axons (above 85 m/sec in the cat) exert static fusimotor action, whereas slower β axons (60–85 m/sec) exert dynamic action (Harker et al., 1977; Jami et al., 1978, 1979; Laporte, 1979; Emonet-Denand et al., 1980).

In man, the morphologic evidence available indicates that β innervation is as widespread as it is in cat, since large myelinated axons have regularly been found to supply p1 plates on bag fibers (see above). There is, however, some residual uncertainty since, as in the cat, it is not clear whether *all* p1 plates are innervated by β axons, or whether some might be supplied by dynamic γ axons (Barker et al., 1980).

In unanesthetized human subjects, it is, for obvious reasons, extremely difficult to perform conclusive experiments to assess the proportion of fusimotor drive mediated by β and γ axons. A first attempt has been made by Burke et al. (1979a) using Leksell's (1945) pressure-block technique, which in their hands gave a satisfactory selective block of γ-MNs. In 5 of 10 observations (2 single-unit recordings and 3 multiunit recordings, for a critique of the latter, see section 2.1) presumed fusimotor effects persisted during a complete block of α, and thus also of β-MNs activity, when the subject made a voluntary effort to contract the parent muscle. The authors concluded that the fusimotor effects seen during a voluntary contraction are mediated by myelineated small-caliber fibers, which probably innervate intrafusal structures exclusively (γ-fusimotor fibers). It is not necessary to postulate that skeletofusimotor β-fibers are responsible for tight α-γ coactivation seen in man during voluntary contractions. Clearly, the only reasonable interpretation of the fusimotor action persisting during a complete block must be that it was mediated by small, probably γ-MNs. This does not imply, however, that all fusimotor action during nonblocked voluntary contractions was mediated by γ fibers, since in the absence of direct recording from γ fibers, the distinct possibility cannot be excluded that with increasing voluntary effort during the progressing block, activity also increased. Thus, an increased γ drive during the block could have compensated for the possibly substantial β contribution that may have been present before the block.

In the remaining 5 tests, the subject lost the ability to voluntarily activate spindle afferents when the block of α fibers was complete. The most straightforward interpretation of this finding would be an abolition of fusimotor action mediated by β fibers. The authors' rejection of this possibility is based on the observation that the abolition required a near-complete block. They concluded that only slow, i.e., dynamic β axons could have been involved, but that these were too weak to elicit spindle accelerations of the size observed during voluntary contractions. This seems untenable in view of the numerous examples of powerful β action (e.g., Bessou et al., 1965; Emonet-Denand et al., 1975, 1980; Laporte and Emonet-Denand, 1976).

In conclusion, Burke et al. (1979a) have demonstrated that some of the fusimotor effects observed during voluntary contractions in man are most likely mediated by γ-MNs. However, a contribution by β axons to normal fusimotor action is also to be expected, both on morphologic and physiologic grounds.

Static and dynamic fusimotor neurons in man?

Convincing evidence has been obtained that during voluntary contractions, human spindle afferents are subjected to fusimotor drive. Morphologic evidence further suggests that separate static and dynamic fusimotor fibers exist in man. However, the differentiation of static or dynamic fusimotor effects must rely on the observation of decreased or increased dynamic sensitivity (relative to passive controls) in responses to large stretches. With ramp-and-hold stretches, the amplitudes should be at least 0.1 RL (Matthews, 1962, 1972), and for sinusoidal stretches amplitudes > 0.04 RL at 1 Hz would appear safe (Hulliger et al., 1977a). Such quantitative sensitivity testing has not yet been performed in human microneurographic studies, although most reviewers (apart from Vallbo, 1981; Hulliger, 1981) seem to have come to the tacit agreement that static and dynamic fusimotor neurons are established entities in man, Appelberg et al. (1981) have recently described a method of quantitative sensitivity testing for reflex studies in fusimotor neurons, which relies on recordings of primary afferent responses to imposed test stretches, and similar methods could no doubt be useful in man (Hulliger, 1981). A first attempt in this direction has indeed been made by Burg et al. (1976), using sinusoidal test stretches, yet in the absence of a quantitative analysis (based, e.g., on averaged cycle histogram responses) an unambiguous assessment of their findings seems difficult.

In conclusion, the existence of functionally distinct static and dynamic fusimotor neurons remains to be established in man. In this context, it is pertinent to note, that static and dynamic fusimotor neurons can be differentiated not only in the cat but also in monkeys (Koeze, 1968; Cheney and Preston, 1976b,c).

α-γ coactivation

It has been a recurrent theme of recent reviews (Hagbarth, 1979; Vallbo et al., 1979; Burke, 1980, 1981; Hagbarth, 1981; Vallbo, 1981) and indeed of most human microneurographic studies that during voluntary contraction there is a fusimotor outflow which largely runs parallel with the skeletomotor outflow in space, time and intensity (Vallbo, 1973), a tight α-γ coactivation (Burke et al., 1979a), or α-γ coactivity and α-γ cosilence (Hagbarth, 1981). Figure 1 illustrates one type of finding normally taken as evidence for α-γ coactivation. The figure shows the response of a primary afferent during a load-bearing tracking movement performed with the index finger. The afferent was not silenced during the shortening movement (downward deflection in the joint angle record), in contrast to the pause in afferent discharge regularly seen when similar movements were imposed passively (not illustrated). Also, during the subsequent lengthening phase there was no conspicuous dynamic response, as was present during passive movements. Thus, modulated fusimotor activity just adequate to offset the afferent's length response probably accompanied the voluntary contraction seen in Fig. 1, and the time-course of the fusimotor outflow to this afferent may have been rather similar to that of the skeletomotor activity, as seen in the EMG record (see below). In the following points, attention is drawn to the considerable amount of important detail that is easily masked by the rather general formulation of the concept of α-γ coactivation.

1. The range of motor performance investigated thus far is rather limited. Thus, afferent responses during imposed resisted movements have, for instance, not as yet been recorded. In chronically implanted cats, this has turned out to be a paradigm in which predominant and apparently independent dynamic fusimotor activity was revealed (see section 4.2).

2. In view of the distinct possibility of β innervation in man, evidence in spindle afferent discharge of fusimotor action linked to skeletomotor activity is by no means sufficient to demonstrate α-γ coactivation. In most cases, the data may at best be compatible with coactivation of skeletomotor and fusimotor neurons (including β-motoneurons) rather than with strict α-γ coactivation. Indeed, the fact that during very slow muscle shortening, human spindle afferents could show reductions in firing rate (Burke et al., 1978b), indicates that the presumed skeletomotor-linked fusimotor action was modest and it could be argued that β-MN firing at normal motor unit firing rates (0–30/sec) would alone be capable of providing this amount of fusimotion.

3. The evidence adduced to support the concept of coactivation is based exclusively on afferent recordings. As detailed below (section 4.3) this virtually precludes any precise deductions concerning details of the time course and the intensity of fusimotor outflow. Thus, in most cases, little more can be concluded than that α and fusimotor neurons are active at the same time. Coactivity may therefore be a more adequate term than coactivation.

4. The overall fusimotor excitation which may be inferred from observations of increased afferent discharge (compared with responses at rest) could be due to activity in either static or dynamic fusimotor neurons (see above). As long as the functional properties of these types of neurons are not specified in man, observations of maintained afferent discharge during slow shortening movements (as in Fig. 1) may not safely be attributed to static fusimotor action. Even in the cat it is not known

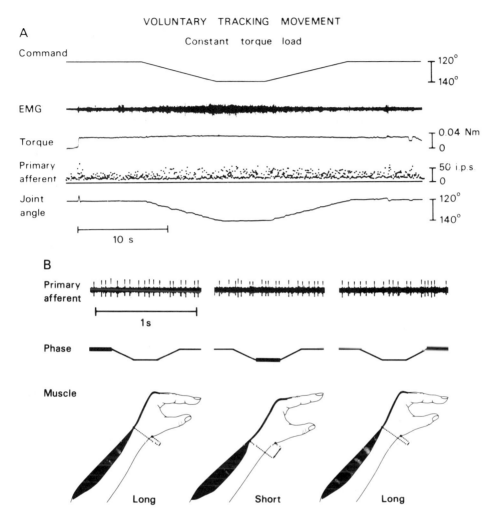

FIG. 1. Sample record and schematic illustration of a tracking movement as performed by a human subject with the fourth finger of the left hand.
A: From top to bottom, time-course of *command* signal with extension movements downwards; EMG recorded with surface electrodes; *torque* record (transducer signal); activity of a *primary spindle afferent* fron extensor digitorum communis (fourth finger portion) displayed as instantaneous rate of discharge; metacarpo-phalangeal *joint angle* of the fourth finger as provided by the transducer of the mechanical stimulator, which was used to generate loads of approximately constant torque. B: Sample records of the same afferent's discharge during 1-sec periods of the three separate hold phases, and schematic illustration of the M-P joint angle and muscle length of the finger extensors during these hold phases. Since finger extension (cf. *short*) was accompanied by muscle shortening, it was plotted downward. [From Hulliger, Nordh and Vallbo 1981, with permission.]

whether, for movements as slow as those in Fig. 1, dynamic fusimotor neurons may not be capable of compensating for spindle unloading (Hulliger, 1981). Circumstantial evidence for the participation of static fusimotor activity is, however, provided when, under the same conditions, secondary afferents are also excited (Vallbo and Hulliger, 1982). This rests on the expectation that the relatively selective excitation of secondary afferents by static axons in cat (Appelberg et al., 1966; Cheney and Preston, 1976b) can also be confirmed in man.

5. The afferent responses during alternating load-bearing movements, which have been listed as examples of presumed α-γ coactivation, in fact included a wide range of patterns, most likely determined by the balance between the length component and the fusimotor component of modulation which together shaped the responses (see 4.2). Thus, depending on the speed of movement, anything from full compensation of length responses by presumed modulated fusimotor drive (slow movements, Hulliger et al., 1981; Fig. 1) to largely length-dominated responses (fast movements, Burke et al., 1978b) has been found. Intermediate types of responses may exhibit double bursts of discharge during a single cycle of movement, when stretch responses during lengthening can be distinguished from fusimotor-evoked firing during shortening (Hagbarth et al., 1975b; Burke et al., 1978b). When slow movements are opposed by elastic loads the fusimotor component may apparently even prevail over the length component of the response, so that spindle afferent discharge may increase during muscle shortening and decrease during lengthening (Hulliger and Vallbo, 1979). The trend in these findings is in some ways similar to that described for spindle afferent recordings in freely moving cats, where the range of movement speeds was relatively unrestricted (Prochazka et al., 1979; Section 4.2). It should of course be stressed that the observation of an increasing and decreasing spindle afferent discharge with increasing and decreasing muscle length does not, of itself, rule out α-γ coactivation. Depending on the speeds involved, it may, however, raise questions regarding the strength, and therefore the functional significance of the presumed α-linked fusimotor action.

6. During slow load-bearing movements, or during active position-holding, the available human data indicate that under these conditions, the discharge rates of spindle primary and secondary afferents are not related to absolute muscle length (e.g., Fig. 1) This may be taken as indirect evidence for α-γ coactivation, but quite apart from this, there is the important ramification that these afferents may not normally provide direct position signals (Hulliger et al., 1981; Vallbo et al., 1981). However, since muscle spindle afferents have on other grounds been implicated in the mediation of position sense and kinesthesia (Goodwin et al., 1972; McCloskey, 1978, 1980), it seems possible that central mechanisms might be involved in the extraction of position information from the efferent and afferent discharge to and from spindles. On the other hand, the demonstration in chronic cat recordings of a significant static component of length sensitivity in both primary and secondary spindle afferents during voluntary movements (e.g., Prochazka, 1981, Figs. 5 to 8), indicates that more human data are needed to delineate the absence of position sensitivity in human spindle afferents during active movements.

Recruitment Order of Fusimotor Neurons?

Burke et al. (1978c) found that in isometric contractions individual spindle afferents were reproducibly activated at a fixed level of contraction strength specific for each unit. From this they drew the conclusion that fusimotor neurons in man had a fixed recruitment order, as do skeletomotor neurons (Burke and Edgerton, 1975; Buchthal and Schmalbruch, 1980; Desmedt, 1980a). Motor neurons are apparently recruited in order of cell size (Henneman et al., 1965; Henneman, 1981; Stein and Bertoldi, 1980). However, the question of whether this applies to γ-MNs, and indeed whether γ-MNs have a fixed recruitment order at all, is not clear, as illustrated by the recent findings that a fixed recruitment order among γ-MNs was not a general feature in the rabbit (Murphy, 1981).

Thus, a number of reservations arise vis-à-vis the interpretation by Burke et al., mainly because their evidence is rather indirect.

1. The authors were clearly aware that the activation of spindle afferents at discrete force levels could also reflect the afferents' responsiveness to local changes in length, possibly elicited by the recruitment of neighboring motor units. However, the assertion that "no correlation was found between the sensitivity of a spindle to external length changes and its ease of activation in a voluntary contraction" (p. 101) seems not to have been backed up by the desirable quantitative measurements. Furthermore, even rigorous data relating the threshold for voluntary activation to the threshold (rather than the sensitivity!) for activation by external stretch would be difficult to interpret, since the same external stretch applied to the whole muscle may cause widely varying local length changes (Meyer-Lohmann et al., 1974).

2. Given the likely occurrence of β innervation in man, the apparent order in the recruitment of spindle afferents may equally well be taken as a manifestation of the orderly recruitment of skeletomotor neurons, including skeletofusimotor neurons. If this were true, the results of Burke et al. (1978c) would indicate that the subpopulation of β-MNs followed the same rules of recruitment as the main population of α-MNs. This would, however, preclude firm conclusions on any general order of activation of γ-MNs. The authors reject this possibility on the grounds that "nerve block experiments in man using pressure indicate that the fusimotor effects are at least partly due to γ-MNs" (p. 111). However, such tests were not performed on the afferents of the recruitment study. Thus, unless it can be shown that fusimotor recruitment takes place at a fixed level of voluntary effort, and this persists during a selective block of α motor axons, the conclusion that such spindle activation is mediated by γ fibers must remain tentative (see above, β-*innervation*).

3. In cat, individual spindle afferents are commonly excited by a number of fusimotor neurons. The maximal numbers found in single-unit studies (5 in Hunt and Kuffler, 1951; 6 in Jami and Petit, 1978; 7 in Petit et al., *personal communication;* 8 in Emonet-Dénand and Laporte, 1975, Table 2; 10 in Emonet-Denand et al., 1977, Fig. 15) probably still underestimate the fusimotor supply to a spindle, given the technical difficulties of single-fiber isolation. In man, the fusimotor supply to spindles is, if anything, richer than that in the cat (see above). If the step-like activation of spindle afferents reported by Burke et al. (1978c) was indeed due to the consistent recruitment of one and the same fusimotor neuron, their conclusion would at best hold for a limited sample of fusimotor neurons (10 to 20%). Alternatively, the possibility cannot be excluded that on successive trials separate fusimotor neurons to the same spindle were activated at approximately the same contraction force.

4. Considerable care is also indicated with any generalization from the limited sample of Burke et al. (1978c). It seems quite possible that the effects described, provided they were indeed due to a recruitment of fusimotor neurons, were largely obtained from static neurons (if they exist in man, see above), since under isometric conditions any excitatory effect of dynamic fibers might be obscured by variability of discharge, given their much smaller excitatory strength compared with static fibers (Hulliger, 1979).

Finally, the afferents of Burke et al. (1978c) were classified only on the basis of twitch tests of unspecified size and, as discussed above, some doubt in distinguishing between muscle spindle and tendon organ afferents remains. This is worrying since most of the recordings showed afferent discharge patterns which could reasonably be expected of tendon organ afferents: See

Vallbo's (1974a) recording of a presumed tendon organ showing similar "recruitment" steps in discharge rate.

In conclusion, it seems exceedingly difficult, using the indirect methods available, to establish that the entire pool of fusimotor neurons (including γ-MNs) is governed by a recruitment order. Convincing evidence can probably only be obtained by direct recording from pairs of fusimotor neurons (Murphy, 1981), with the technique widely employed for the analysis of recruitment order of α-MNs (see Henneman et al., 1974; Henneman, 1981). In contrast, β-MNs being a particular class of α-MNs may reasonably be expected to reveal such orderly recruitment, given the wealth of evidence for its presence in α-MNs in general.

Independence of Skeletomotor and Fusimotor Activity?

Thus far, little evidence has been adduced in man to show that fusimotor activity may be controlled independently of skeletomotor activity. However, it would be premature to dismiss such independence, since the range of conditions hitherto investigated is still limited (see above, *discharge rates*). In fact, preliminary results from recent human investigations indicate that under circumscribed conditions of motor performance the balance between fusimotor and skeletomotor activity may not be rigidly fixed (Vallbo and Hulliger, 1979, 1981; Burke et al., 1980c).

The basic observation of Vallbo and Hulliger (1981) is illustrated in Fig. 2 which shows the responses of a primary spindle afferent from the index portion of extensor digitorum communis during slow load-bearing tracking movements. The afferent, identified by twitch testing, revealed a pronounced dynamic sensitivity to ramp-and-hold stretches (not illustrated). The task was performed with the index finger, and since the loads opposed finger extension, the parent muscle had to be activated (as verified with surface EMG). There was convincing, albeit indirect, evidence for fusimotor drive accompanying the skeletomotor activity, since the afferent was not silenced during the shortening movement (A, C, shortening plotted downward). In contrast, in the relaxed state, the afferent immediately fell silent when comparable passive movements were imposed (not illustrated). Also, the discharge rate during shortening (A, C) was at least as high as during lengthening (B, D) (Hulliger and Vallbo, 1979).

When the load opposing the movement was of constant torque (A, B), the afferent discharge throughout the movement was approximately constant although highly irregular (see Fig. 1). In contrast, with an elastic load (zero force in the resting position: 120° in Fig. 2), there were marked bursts of afferent discharge both at the beginning (C) and end (D) movement. The surface EMG from the receptor-bearing muscle did not show anything comparable (see Vallbo and Hulliger, 1981, Fig. 2), nor could the responses in C and D be accounted for by length or tension transients accompanying the initial and final phases of the movement. Thus, these disproportionally large onset and termination responses (C, D) may most simply be ascribed to a transient dissociation between fusimotor and skeletomotor activity. Burke et al. (1980c) estimated the balance between skeletomotor and fusimotor activity by determining the threshold of activation of presumed spindle afferents during slowly rising voluntary, isometric contractions of the receptor-bearing muscle, both when the trained subject was otherwise relaxed, and during a number of conditioning maneuvers.

The threshold of afferent activation was measured as the level of contraction force at which the first clear increase of afferent discharge was observed. Any change in activation threshold was interpreted as a change in the overall drive either to the entire pool of skeletomotor neurons or to the

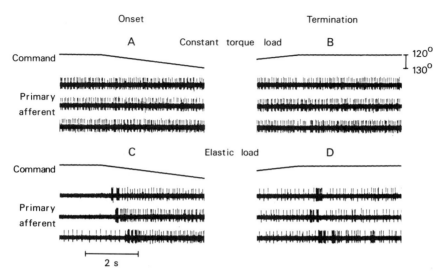

FIG. 2. Discharge from a muscle spindle primary ending in the index portion of the extensor digitorum communis muscle at the onset of voluntary shortening movements (A, C) and at the termination of lengthening movements (B, D) with two different kinds of load: Constant torque load (A, B), and elastic load with zero torque at the resting position (120°). The subject performed slow ramp movements (2.5°/sec) at the metacarpophalangeal joint of his left index finger in a visual tracking paradigm. The command signal at the top gives the desired joint position which the subject followed with some delay. Downward deflection indicates muscle shortening. In the 6 tracking sequences illustrated, the size of the load was, from above, 0.02, 0.04, 0.04 Newton m and 0.001, 0.002, 0.002 Newton m/°. Same unit as in Fig. 1 [From Vallbo and Hulliger, 1981 with permission.]

group of fusimotor neurons supplying the spindle under investigation. In either case, this overall drive would reflect the complex result of converging excitatory and inhibitory inputs from descending or segmental pathways. A reduction in the threshold of activation, interpreted as an increase in fusimotor action, was found with caloric vestibular stimulation (vestibulo-spinal influences) as well as with cutaneous or proprioceptive stimulation (segmental influences) from circumscribed receptive territories. The inverse pattern was found with cutaneous or proprioceptive stimulation from other receptive territories. The Jendrassik maneuver, on the other hand, consistently failed to alter the threshold of spindle afferent activation, indicating that it did not affect the α-γ balance during the isometric contractions investigated (see also Section 2.2). Thus, these negative results provided a valuable control. Obviously, both investigations provide only indirect evidence for α-fusimotor independence, and the data available do not permit a differentiation between β and γ, nor between static and dynamic fusimotor activity. Nevertheless they make an encouraging start in delimiting the extent of α-γ linkage in man.

4.2 Cat and Monkey Data

Cody and Taylor (1973) and Cody et al. (1975) found that in normal chewing movements, afferents identified as spindles of the jaw-closing muscles generally increased their firing during muscle lengthening and decreased it during shortening. In other sorts of movements, however, this simple relationship was not seen. It was argued that the results were consistent with static fusimotor action occurring phasically during muscle shortening, "approximately in parallel with the α-motor activity," while dynamic fusimotor action was tonically set at the beginning of a movement. Taylor and Appenteng (1981) have now expanded this idea, suggesting that dynamic fusimotor drive might be set to a fairly constant level to give appropriate spindle sensitivities to stretch, while static fusimotor drive is varied as a temporal template of the intended movement. This differs from Phillips' (1969) scheme in that the temporal template of static action would not lead to constant spindle firing during unobstructed movements.

Goodwin and Luschei (1975) concluded that fusimotor neurons to spindles in jaw-closing muscles in monkeys were coactivated along with extrafusal muscle, but that this did not seem capable of sustaining spindle discharge in the face of rapid shortening of the muscle. In fact, it is not clear why these authors favored coactivation, since a tonic activation of static and dynamic fusimotor neurons could equally explain most of their observations.

Dorsal Root Recordings

In recordings from dorsal root afferents identified with suxamethonium as being spindle primaries, Prochazka et al. (1975) observed modulations of firing rate during rapid voluntary movements, which corresponded to changes in length of the receptor-bearing muscles. In later studies, the generalization was proposed that for unobstructed movements involving velocities < 0.2 RL/sec (RL measured between muscle origin and insertion), fusimotor action could provide an effective, or even predominant modulatory influence on spindle afferent firing. In contrast, at higher velocities, the firing patterns of spindle afferents were dominated by their responses to variations in muscle length, leaving it open as to whether tonic or phasic fusimotor activity was present. This generalization was consistent with most relevant data available from human and animal recordings of spindle discharge during movement (Prochazka, 1981).

It is known that for low speeds (i.e., low repeat frequencies) of constant amplitude, alternating movements, static but not necessarily dynamic fusimotor activity modulated in phase with the movements can largely abolish the modulation of spindle afferent firing caused by the movements (Lennerstrand and Thoden, 1968c; Baumann and Hulliger, *unpublished observation*). However, as the speed (repetition frequency) increases, the capacity of phasic fusimotor activity to abolish spindle afferent modulation is likely to reach a limit. This is because with increasing frequency the depth of modulation in response to the constant amplitude movements increases (Matthews and Stein, 1969a; Poppele and Bowman, 1970; Rosenthal et al., 1970; Hulliger et al., 1977a), whereas the response to modulated fusimotor activity declines (Andersson et al., 1968b; Chen and Poppele, 1978). Generally, this limit is likely to be reached at lower repeat frequencies the larger the amplitude of movement, and for a given amplitude at lower frequencies with dynamic rather than static fusimotor action. This is because the excitatory strength of static fusimotor fibers is considerably higher than for dynamic fibers (Lennerstrand and Thoden, 1968a,b; Lewis and Proske, 1972; Cheney and Preston, 1976b; Emonet-Denand et al., 1977), particularly when the fusimotor activity is modulated (Hulliger, 1979). It would be interesting to know whether or not the critical speed of movement at which the limit is reached corresponded to 0.2 RL/sec. If the critical speed were shown to exceed 0.2 RL/sec, this would indicate that in voluntary unobstructed movements, the fusimotor system is not used at its full capacity to abolish afferent modulation. Loeb and Hoffer (1981) have shown examples of afferents not conforming to the generalization that modulations of firing rate closely correspond to variations in muscle length (> 0.2 RL/sec). We believe that the important data collected by Loeb and Hoffer would have been greatly strengthened if the responses of such afferents to i.v. suxamethonium had been observed. Since the afferents referred to seemed to show a behavior which might reasonably be expected of tendon organs, the absence of any records or individual description of identification tests makes judgment difficult.

In some recording sessions, spontaneous movements were occasionally obstructed in order to evaluate the modulatory strength on spindle firing of α-linked fusimotor action. Although evidence of such action has now frequently been adduced (Prochazka et al., 1976, 1977; Prochazka, 1980; Prochazka and Wand, 1981a), its modulatory strength proved surprisingly weak (Prochazka et al., 1979). In a more quantitative study, Prochazka and Wand (1980b, 1981a) used electronic filters to mimic the dynamic transducing properties of spindles. This procedure allowed the sensitivities of the afferents during voluntary movements to be compared with that during deep anesthesia. This analysis suggested that in unobstructed movements, there was evidence of steady, relatively low levels of both static and dynamic fusimotor action. A consistent finding was that in imposed movements involving slight to moderate resistance, spindle primary sensitivities were nearly always elevated, and this implied steady, predominantly dynamic fusimotor action. Figure 3 shows the remarkable change in sensitivity seen when comparing the responses of spindle primary afferents during unobstructed voluntary movements (A) and imposed, resisted movements (B). In A, the relatively high level of discharge of the afferent at short muscle length and its low sensitivity to muscle lengthening (during anesthesia in C, left) indicate predominant static fusimotor action, whereas the high sensitivity in B suggests predominant dynamic fusimotor action. There are indications that such increased dynamic sensitivity may often be independent of skeletomotor activation (Prochazka and Wand, 1980b, 1981a), although an element of α-γ_D or α-β_D coactivation cannot be ruled out in all cases. In powerful obstructed contractions, evidence for further activation of static and dynamic fusimotor neurons was sometimes seen. However, the overall impression was of a fusimotor system, the main modulatory influences on spindle firing of which were not phasically linked with α-MNs. The analytical technique just described clearly has its limitations. Since it relies on linear modeling, comparisons of sensitivity are restricted to movements of similar muscle velocities (see Houk et al., 1981a). Deductions about fusimotor action are made in a qualitative way, with reference to general properties of steady-static and dynamic fusimotor action on spindle firing established in acute animal experiments (see Section 4.3).

Spindle Responses to Rapid Stretch

In 1980, four groups independently published data suggesting that during rapid muscle stretching, spindle afferents fired segmented bursts of discharges, which in some cases seemed closely related to ensuing bursts of EMG response of the receptor-bearing muscle. These findings are significant for the understanding of stretch reflexes, since they provide indirect support for Ghez and Shinoda's (1978) suggestion that long-latency EMG responses to stretch are not necessarily exclusively mediated by long-loop pathways via the cerebral cortex (Tatton et al., 1975; 1978, Evarts and Fromm, 1978; 1981).

Tracy et al. (1980) confirmed in decerebrate and spinal monkeys the findings of Ghez and Shinoda (1978), and showed in decerebrate cats that spindle primary afferents fired in segmented bursts during rapid muscle stretching. Hagbarth et al.

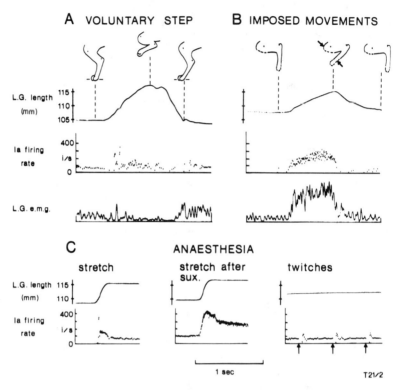

FIG. 3. Cat ankle extensor spindle primary afferent, contrasting the responses to length variations during an unobstructed voluntary movement (**A**) and an imposed, resisted movement (**B**) (top: video tracings). **C:** Responses of the afferent during deep pentobarbital anesthesia to imposed muscle stretch (**left**), imposed stretch ca. 30 sec after 200 μg/kg i.v. suxamethonium (**middle**) and to electrically evoked muscle twitches (**right**). The pattern of discharge in A is suggestive of static rather than dynamic fusimotor action, whereas the high sensitivity to stretch in B strongly implicates dynamic fusimotor action.

(1980, 1981) then published spindle-dominated multiunit recordings from human nerves in which periodicity was evident during muscle stretching. The periodicity was attributed to segmented bursts of spindle discharge, although in a prior study (Burke et al., 1978a), the long-latency responses in similar multiunit recordings were thought to be of efferent origin, on the grounds that the second volley was not seen in single afferent unit recordings. Prochazka and Wand (1981b) reported that in normal cats, single-spindle, primary afferents, identified with suxamethonium, fired in bursts during rapid muscle stretching and that these bursts were followed at short latencies by EMG responses. In stretches of very short duration, which resulted in only one spindle afferent burst, only one EMG response was seen. It was concluded that both the short- and long-latency EMG bursts depended, at least in part, on the prior occurrence of bursts of spindle discharge. It is interesting in retrospect to examine a recording of Prochazka et al. (1977, Fig. 5) showing the discharge of a spindle primary afferent recorded in a normal cat landing from a fall. The afferent showed clear bursts of firing during the resultant rapid muscle stretches. Finally, Loeb (1981) recently mentioned a similar phenomenon observed in two rectus femoris spindles during the rapid stretch phases of locomotion in normal cats. Thus, there is good evidence that muscle spindle primary afferents respond during rapid stretches with bursts of discharge with a periodicity of between 30 and 50 msec. The origin of these bursts may lie in mechanical oscillations within the muscle, short-range intrafusal stiffness, fusimotor reflexes, or a combination of these. Whatever the mechanism, the existence of these nonlinear responses inevitably raises serious difficulties in the interpretation of the many recent studies on stretch-evoked responses recorded from limb muscle and from CNS neurons, where long-latency components of the responses are often attributed to long-loop pathways (see Chan, *this volume;* Marsden et al., *this volume;* Lee et al., *this volume*).

Dorsal Root Recordings in Monkeys

The recordings from monkey dorsal root afferents of Schieber and Thach (1980) are potentially significant. Figure 5 shows the discharge behavior of a presumed spindle afferent from a wrist extensor muscle, recorded during slow hold-ramp-hold tracking movements. The identification tests for this afferent are shown in Fig. 4, A to D. Of the 8 criteria used for afferent identification, the most useful are probably the vibration response (Fig. 4B, showing one-to-one following at 128 Hz) and the twitch response (Fig. 4C). Strictly speaking, the tests used do not allow an unambiguous distinction between spindle and tendon organ afferents (see Section 2.1). However, taken together, the responses shown in Fig. 4A to D are indeed more suggestive of a spindle afferent than of a tendon organ afferent.

The discharge behavior of the afferent during voluntary movements was intriguing. Whereas in the slow tracking movements of Fig. 5, the afferent increased its firing during muscle lengthening (B and D) and at the onset of muscle shortening (A and C), in the more rapid voluntary movements of Fig. 4E, its firing was closely related to the variations in muscle length. We estimate the slow movements (28°/sec) to have involved muscle speeds of about 0.05 RL/sec, compared with maximal speeds of about 0.4 RL/sec for the faster movements (assuming a 15 to 20% muscle-length change for a 90° angular change at the wrist). Thus, the responses are in accord with the 0.2 RL/sec generalization discussed above. Schieber and Thach (1980) argued that the increases in discharge at the onset of muscle shortening in Fig. 5A were apparently poorly related to EMG, and that this suggested a dissociation of α and γ activity.

Although this conclusion is tenable for the afferent in question (and even in this case a component of α-linked fusimotor action cannot be ruled out), it would be premature to base any broad generalizations on the behavior of the 5 afferents observed in this study.

4.3 Simulation Experiments to Deduce Fusimotor Activity from Afferent Discharge

Little precise information is available on the activity of fusimotor neurons during voluntary movement. There are three main reasons for this. First, most of the fusimotor neurons are the small γ rather than the large β-MNs and are, therefore, probably more difficult to record from. This is perhaps the least significant reason, since the axon diameters of γ fibers and secondary spindle afferents are similar and recordings have indeed been obtained from the latter. Second, in ventral roots or in peripheral nerves, the identification of single units as efferent γ fibers poses considerable problems, as regards both identification of the target muscle and measurement of CV. Third, even if these problems were overcome, the classification of γ fibers as static or dynamic fusimotor neurons would probably be beyond present techniques. Thus, knowledge of fusimotor activity during natural movements is indirect and based on inferences from spindle afferent recordings, yet interpretation of spindle afferent data

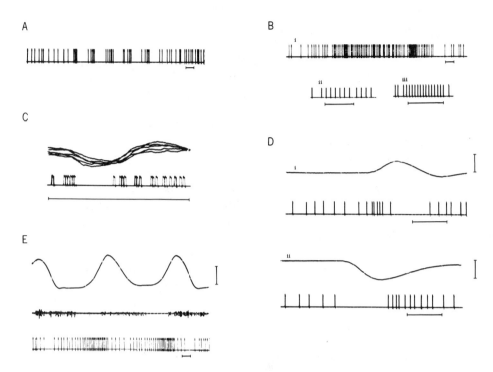

FIG. 4. Discharges of an extensor carpi radialis (ECR) afferent recorded from the dorsal root ganglion of a behaving monkey. A–D: Identification tests indicating that the afferent was probably a spindle primary. E: Voluntary, fast alternating movements. All traces: Horizontal calibrations: 100 msec, vertical calibrations: 25°, wrist flexion (muscle lengthening) upward. The unit's activity is shown as uniform pulses discriminated one-to-one from its action potentials. A: Five rhythmic taps were delivered manually over the receptive field in ECR. The unit responded to each press with a burst and to each release with a pause. B: A 128-Hz tuning fork applied manually to the receptive field elicited an increase in discharge (i), including periods of regular discharge at 64 Hz (ii), and occasionally at 128 Hz (iii). C: Wire electrodes inserted percutaneously on either side of the 1-cm receptive field were used to stimulate local muscle twitches (1 msec, 20 V, 0.5/sec stimulation) at the start of the trace. The position trace was doubly differentiated and amplified to detect small extensor deviations (downward) of the wrist resulting from muscle shortening, during which the unit paused (5 traces superimposed). D: 50-msec torque pulses passively stretching ECR caused the unit to burst and then to pause as the wrist reextended (i); those passively slackening ECR caused the unit to pause and then to discharge as the wrist reflexed (ii). E: During active alternating movements the unit was activated only with the flexion (stretch) phase (EMG recorded from ECR). [From Schieber and Thach 1980, with permission.]

from voluntary movements is difficult and often inconclusive due to a remarkable scarcity of experimental work on afferent responses during concomitant variations of muscle-length and fusimotor activity. Even approximate predictions of such responses are uncertain, as a result of the numerous nonlinear characteristics of the muscle spindle as a whole (see Matthews and Stein, 1969a; Goodwin et al., 1975; Hasan and Houk, 1975a,b; Hulliger et al., 1977a,b; Hulliger and Noth, 1979; Houk et al., 1981a,b; Baumann and Hulliger, 1981; Emonet-Denand and Laporte, 1981). These difficulties have recently been exacerbated by the finding of long-lasting (up to 40 sec) aftereffects of fusimotor activity on the sensitivity to stretching of primary spindle afferents (Baumann et al., 1981; and *unpublished observations*). These aftereffects manifest themselves during slow movements as an enhancement of dynamic sen-

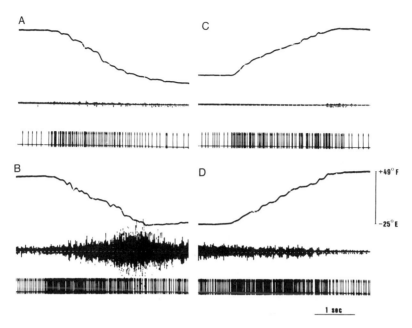

FIG. 5 Same afferent as in Fig. 4, recorded during hold–ramp–hold tracking movements. To the left are extension ramps (ECR shortening) and to the right flexions (ECR lengthening). The upper frames were performed under flexor load, the bottom under extensor load. Each frame shows an individual trial top: angular wrist position; middle: EMG recorded with wire electrodes inserted percutaneously on either side of the unit's receptive field in ECR; bottom: is the discriminated activity of the afferent unit. [From Schieber and Thach 1980, with permission.]

sitivity, not only after dynamic but, remarkably, also after static fusimotor stimulation.

We have recently attacked the issue of fusimotor activity during voluntary movement using a new simulation technique to determine, in acute preparations, the patterns of fusimotor activity that shaped the afferent responses during spontaneous motor activity in freely moving animals. Recordings during two types of movement in chronically implanted animals are necessary. First, active movements performed by the alert animal, such as evoked steps (Fig. 6, right), spontaneous stepping movements, or imposed movements resisted by the animal, and second, passive movements imposed by the experimenter, when the animal is temporarily anesthetized (Fig. 6, left). Such passive stretching is normally used for the classification of the afferent units recorded during active movements. In all cases, reliable records of the length changes of the parent muscle, in addition to the recordings of afferent activity, are indispensable.

In the simulation experiments in acute preparations, movements are reproduced from digitally stored segments of the original recordings (Frei et al. 1981). This technique permits a precise simulation not only of variations in length but also of any differences in mean muscle length between passive and active movements. The latter precaution is of considerable importance given the known dependence of spindle afferent sensitivity on the mean length of the receptor-bearing muscle (Matthews, 1963; Goodwin et al., 1975; and *unpublished observation*).

The passive movements imposed in the chronic preparation during anesthesia are used in the acute simulation experiments to select identified spindle afferent units on the basis of a close similarity between their

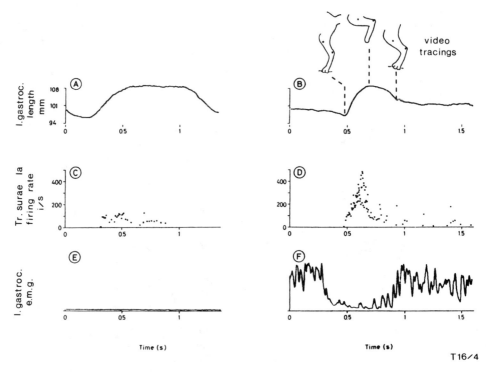

FIG. 6. Cat ankle extensor spindle primary afferent (identified with suxamethonium). **Left:** Passive movement. Ankle movements imposed during deep anesthesia. **Right:** Evoked step. Voluntary step evoked by touching plantar surface of foot. [*Abscissa* = time(s).]

responses and those of the original recordings. The purpose of this selection procedure is to ensure that the simulations of active responses are performed with afferent units with similar response characteristics to those of the chronic recordings.

Examples of simulations of the passive movement imposed during anesthesia and of the step evoked by light touch to the footpad of the chronically implanted cat shown in Fig. 6 are seen in Figs. 7 and 8. The same movements as in the original recordings (Fig. 6A and B) were repeatedly reproduced in the soleus muscle of an acute preparation (imposed passive movement in Fig. 7A, evoked steps in Figs. 7B and 8A and B). In Fig. 7 the responses of two separate primary afferents (C/D and E/F) to stretching in the absence of fusimotor stimulation are shown. It may be seen that with simulation of the passive movement (in C

and E) there was a fair agreement between simulated and original responses (Fig. 6B). In particular, the mean discharge rates of all three responses were around 100/sec. Thus, the response characteristics of the afferents used for simulation were approximately matched with those of the original recording.

In contrast, the responses of the same primary afferents to a simulated evoked step (Fig. 7D and F) failed to come anywhere near the original response (Fig. 6D). The peak rates of discharge in Fig. 7 were around 250/sec (D) and below 200/sec (F), whereas in the original response (Fig. 6D) a value of 500/sec was reached. However, with stimulation of single fusimotor fibers, responses more like the original recording could be obtained. This is illustrated in Fig. 8, where the same evoked step was simulated (A and B). During the length changes,

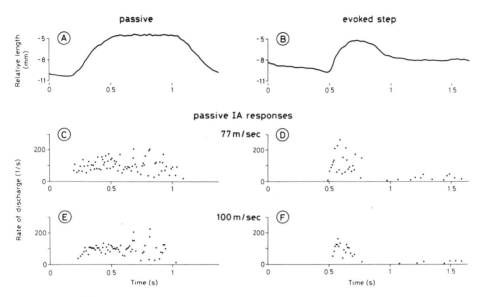

FIG. 7. Simulations without fusimotor stimulation. Discharge of two soleus spindle primary afferents C/D (CV 77 m/sec) and E/F (CV 100 M/sec) in response to muscle length variations (A and B) derived from the chronic recordings of Fig. 6 and reproduced by an electromagnetic servodevice in deeply anesthetized cats. Responses C and E approximately match those of Fig. 6C, whereas responses D and F are clearly smaller than the equivalent response seen in the conscious animal (Fig. 6D).

single fusimotor fibers supplying the same two primary afferents as in Fig. 7 were excited (C/D and E/F) with two different patterns of stimulation. In Fig. 8C and E the rate of stimulation was determined by the filtered EMG signal of Fig. 6F, which was digitally stored like the movement signals. This filtered signal (Fig. 8G) was fed into a pulse-frequency modulator driving a stimulator. In Fig. 8D and F the same fusimotor fibers were stimulated at constant rate (tonic pattern, H). The preliminary findings obtained for evoked steps may be summarized as follows:

(1) The original responses could not be reproduced with mere stretching in the absence of fusimotor stimulation (Fig. 7). In particular, the peak discharge rate of the original record (around 500/sec) far exceeded that in passive simulations (usually below 200/sec). In this analysis, attention was paid to an adequate simulation not only of the time course of movement but also of the mean muscle length at which it occurred originally.

(2) Neither tonic nor EMG-linked patterns of static fusimotor activity, concomitant with the simulated movement, could induce afferent responses that came anywhere near the pattern of the original responses, irrespective of the level of fusimotor activity. The stimulation rates investigated ranged between 20 and 150/sec (tonic levels, or peak rates with EMG-linked stimulation). The most pronounced static effects found are shown in Fig. 8C and D. Although static action provoked drastic effects on the afferent response (compare these with the simulated passive responses in Fig. 7D and F, and with the original response in Fig. 6D), it failed to imitate the marked response during the rapid stretching of triceps that accompanied ankle flexion. The

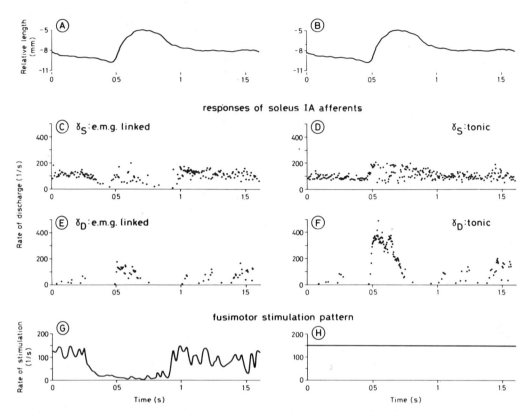

FIG. 8. Discharge of the same soleus spindle primary afferents as in Fig. 6, in response to the evoked-step length variations of Fig. 6B, but in the additional presence of static fusimotor stimulation (C and D) and dynamic fusimotor stimulation (E and F). Fusimotor stimulation in trials C and E was rate-modulated according to the waveform shown in G, derived from the EMG trace of Fig. 6F. These trials thus simulated α-γ-linkage. In trials D and F, fusimotor stimulation was at a constant rate of 150 Hz (as shown in H). The pattern of spindle afferent discharge in F was the most similar to that observed in the conscious animal (Fig. 6D).

potency of static action is further indicated by the finding that stimulation rates as low as 20/sec (not illustrated) altered the afferent response. Yet the type of effect elicited was the same as that seen with the high rates of stimulation illustrated in Fig. 8C and D.

(3) Dynamic fusimotor fibers also failed to give the desired imitation when the stimulation pattern was of the EMG-linked type (Fig. 8C and E), irrespective of the rate of stimulation used. However, with tonic stimulation, responses could be obtained with the same general features as those in the original record (compare Fig. 8F with Fig. 6D).

In conclusion, the simulation technique has opened up the possibility of directly testing hypotheses of the patterns of fusimotor activity occurring during natural movements, such as the concept of α-γ linkage.

The preliminary results with two examples of spontaneous movements in freely moving cats, indicate that the dominant component of fusimotor activity does not conform under all conditions to the concept of α-γ linkage (Granit, 1955) which we take to entail a close similarity between skeletomotor and fusimotor activity (cf. Hulliger, 1981), since in the present simulations, EMG-linked activity in both static and dynamic fusimotor fibers clearly failed to reproduce the original afferent responses.

CURRENT CONCEPTS OF FUSIMOTOR CONTROL AND PROPRIOCEPTIVE AFFERENT FUNCTION

5.1 Definition of Terms

One of the problems in understanding the relationships between skeletomotor activity, fusimotor activity and movement has been that descriptive terms such as α-γ linkage, α-γ coactivation, and α-γ independence (dissociation) have not been adequately defined. Thus, the term α-γ coactivation is taken by some to indicate that when α-MNs are activated and silenced, so are γ-MNs; at the other extreme, however, it is taken to indicate no more than simultaneous activity permitting separate activation and silencing). It may therefore be helpful to delineate some of these terms, whenever possible, in accordance with the original concepts and/or the prevailing usage.

α-γ Linkage

The original concept of Granit (1955, p. 254) was based on the finding that α and γ reflexes have proved to be linked, coexcited, and coinhibited, often with the γ reflexes leading. This was further supported by observations obtained with stimulation of central structures, and the basic idea has frequently been reiterated (Granit, 1979). Accordingly, with this, linkage is used here to describe the case where, apart from possible earlier recruitment, γ-MNs are only active when α-MNs are active and where the time-course of their activity is similar. The extreme case of linkage would be the one illustrated in Fig. 9A, namely strict proportionality between α-firing rates and γ-firing rates.

The effect on spindle primary discharge of linked α-γ activity would, of course, depend on the degree of mismatch between intrafusal and extrafusal length variations. Of the many possible resultant Ia firing patterns, two are illustrated in Fig. 9A. The first corresponds to Hunt and Kuffler's (1951) suggestion that spindle afferent firing is held constant in the face of variations in muscle length, by virtue of compensatory fusimotor action. The second corresponds to the scheme of Eldred et al. (1953) in which spindle afferent discharge increases during voluntary shortening.

α-γ-Coactivation

In current usage, coactivation is a looser term implying a less rigid relationship between α and γ activity, although the patterns of discharge still should show a certain degree of similarity. The range of possibilities is indicated by the shaded area in Fig. 9B. For example, a phasic component related to the concomitant α activity might be superimposed on a tonic component of γ activity.

From this view, α-γ linkage is seen as a special case of coactivation. Coactivation also forms the basis of the idea of servo-assistance in the control of movement (Fig. 9B), in which fusimotor activity keeps spindle afferent firing at a constant "bias level" in the face of changes in muscle length (Matthews, 1972; Stein, 1974; Phillips, 1969).

α-γ Independence or Dissociation

Thus far, these terms have been used to imply a lack of correlation between α and γ activity. A special case of independence, phasic α and tonic γ activity, is shown in Fig. 9D. Another special case, shown in

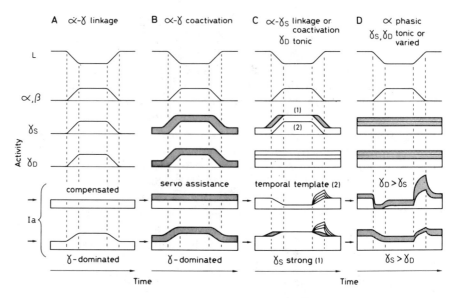

FIG. 9. Schema of α-, β-, γ-efferent and spindle primary afferent firing rates during idealized voluntary movements, according to the different prevailing views on fusimotor control. **Top to bottom:** Muscle length (L), shortening downwards; α, β, γ_S, γ_D-firing rates; Ia firing rates (upper assuming weak γ-action, lower assuming strong γ-action). **A:** α-γ-linkage (extreme case: strict proportionality between α- and γ-firing rates). **B:** α-γ-coactivation (*shaded areas* indicate the range of possibilities). **C:** α-γ_S-coactivation, γ_D-tonic (*1:* γ_S-coactivation strong; *2:* γ_S-modulation a "temporal template" of the length changes). Note increasing levels of tonic γ_D-firing rates, reflected in increasing Ia responses during muscle lengthening. **D:** Independence of both γ_S- and γ_D-activation from α- and β-activation (special case: constant γ_S- and γ_D-firing rates, phasic α- and β-activation).

Fig. 9C, is the possibility that only static γ activity is linked or coactivated with the phasic α activity, whereas dynamic γ activity is independent (in this instance, tonic). Such a scheme was suggested by Matthews (1972, p. 524) as being consistent with the evidence then available.

Further Possible Combinations

It should be stressed that the highly schematic drawings of Fig. 9 neither exhaust the possible combinations of α- and γ-firing patterns, nor do they adequately describe the range of primary afferent responses which might be expected to occur. Nevertheless, the essential elements of the currently favored theories of fusimotor control are represented, and the figure may provide a useful basis for hypothesis testing and for the development of more detailed theories.

5.2 Current Views and Future Directions

With the above definitions in mind, let us now summarize the current positions of the various groups involved in afferent recording during voluntary movement.

Human Neurography

Vallbo et al. (1979) stated that the balance of their evidence was ". . . against the view that the (normal human) subject can induce fusimotor activation of his spindle endings by any volitional maneuver without contracting the receptor-bearing muscle." Thus, "when the skeletomotor system is silent, so is the fusimotor system" (Burke et al., 1976a). Burke et al. (1978c) concluded that "the fusimotor drive to a muscle is proportional to the skeletomotor drive to the muscle, and that skeletomotor and fusimotor neurons are subjected to similar, if not identical, command signals." These interpretations are in line with the extreme case of α-γ-linkage as defined above. In the last 2 years, there has been a moderation of this view, and the term "flexibility in α-γ balance" has been introduced to describe apparent variations in the closeness of α-γ-linkage (Vallbo and Hulliger, 1979, 1981; Burke et al., 1980c). The "flexible α-γ-balance" envisaged by Burke et al. corresponds best to the "α-γ-coactivation" category in the above definitions, whereas Vallbo and Hulliger considered a higher degree of independence to be present, with transient periods of disproportionally enhanced fusimotor activity.

Recordings in Cats and Monkeys

On the basis of recordings from jaw muscle afferents, Taylor and Appenteng (1981) proposed that "in some natural movements dynamic fusimotor activity is set to a fairly constant level for a particular task, while static fusimotor drive fluctuates to form a 'temporal template' of the intended movement." This is a modification of Matthews' (1972, p. 524) suggestion of α-$γ_S$-coactivation with γ-independence, a special case of which is tonic $γ_D$-activation (Fig. 9C).

Loeb and Hoffer (1981) believed that their data were best explained by α-$γ_D$-coactivation in extensors and α-$γ_S$-coactivation in flexors. However, these authors also thought that it was unlikely that a single rule of fusimotor control could explain the full range of spindle behavior that they had observed.

The data of Prochazka and co-workers (1976, 1977, 1979) indicated that modulations in spindle-firing rates caused by α-linked fusimotor action are generally relatively small. Prochazka and Wand (1980b, 1981a) suggested that much of the functionally important fusimotor action occurring during normal movements (at least in cats) was independent of skeletomotor activity. This view is strengthened by the simulation experiments described above (Section 4.3). Furthermore, different classes of movements seemed to be associated with different levels of static and dynamic fusimotor action. In particular, dynamic action seemed to be consistently implicated in spindle primary afferent responses to imposed, resisted length changes (Prochazka and Wand, 1980b, 1981a; see also Fig. 3). These interpretations correspond to the α-γ-independence category in the above definitions. The afferent recordings in behaving monkeys of Schieber and Thach (1980) were also interpreted as favoring α-γ-independence or dissociation, although weak α-γ-linked fusimotor action can probably not be ruled out on the basis of these data (see Section 4.2).

Recognizing the Importance of Length Variations

Largely as a result of the chronic animal recordings, there has been a reemphasis of the importance of variations in muscle length in modulating spindle-afferent firing during voluntary movements. Whatever their favored hypothesis of fusimotor control, most groups now agree that for fast, unobstructed, alternating movements, spindle primary afferents increase their firing rates during muscle lengthening and decrease them during muscle shortening. Whether 0.2 RL/sec can, in this context, be considered a reasonable lower limit of the velocity of a fast movement remains to be seen. By the same token, much evidence exists that variations in fusimotor action, whether α-linked or not, can significantly obscure the dynamic relationship between spindle primary afferent firing and muscle length when muscle velocities are small or equal to zero (Hulliger et al., 1981).

Future Directions

The importance of reliable afferent identification is highlighted in this and other recent publications (Taylor and Prochazka, 1981). It is recognized that a full and systematic use of the available criteria discussed in section 2.1 is an area where improvements are possible and desirable.

It is becoming increasingly apparent that for accurate deductions about fusimotor action to be made from spindle-afferent recordings during movement, the nonlinearities of the afferent responses to concomitant fusimotor discharge and variations in muscle length must, in some way, be taken into account. The simulation studies described above seem, at present, to provide the most accessible approach to this problem. In order to develop analytical models, much more information is required about spindle-afferent responses to concomitant length variations of widely ranging amplitude and combined activity of the different fusimotor neurons innervating a given spindle. The existence of intermediate categories of fusimotor action (Emonet-Denand et al., 1977), and indeed, a possible functional division of static γ-MNs into two types (Gladden, 1981) must eventually be included in interpretations of the voluntary movement data.

ACKNOWLEGEMENTS

We thank Dr. A. Vallbo, and Drs. M.H. Schieber, and W.T. Thach for permission to reproduce Figs. 1 and 2 and 4 and 5, respectively. Our thanks are also extended to Drs. D. Barker and F. Emonet-Denand for valuable discussions, and to Drs. K. Appenteng, P.B.C. Matthews, U. Proske, and P. Wand for helpful criticism of the manuscript.

This work was supported by MRC grant G80–0040–2N and Swiss National Science Foundation, grant no. 3.225.77.

Critical Examination of the Case for or Against Fusimotor Involvement in Disorders of Muscle Tone

*David Burke

Department of Neurology, The Prince Henry Hospital, Sydney, Australia

The view that there is a primary defect of the fusimotor control of muscle spindle endings in different disorders of muscle tone has become so entrenched that to suggest otherwise often meets with resistance from both clinicians and physiologists. The evidence on which the fusimotor system has been implicated is indirect, inconclusive, and circumstantial. Conclusions have been repeated without attention to previously expressed reservations, so often that their repetition has conferred a validity on the data that they have not earned.

The first section of this chapter critically reviews the data that are believed to implicate the fusimotor system in disorders of muscle tone. The conclusion of this review is that the case is "not proven." In the second section evidence exonerating the fusimotor system is discussed.

GROUNDS FOR IMPLICATING THE FUSIMOTOR SYSTEM

1. Analogy with the Decerebrate Cat

The dependence of decerebrate rigidity in the cat on muscle spindle afferent activity, the subsidence of the rigidity during selective blockade of γ efferents with local anesthetic (Matthews and Rushworth, 1957a,b), and the decrease in muscle spindle discharge on severance of the ventral roots (see Matthews, 1972), all indicate that heightened fusimotor tonus is a necessary condition for the appearance of rigidity following intercollicular section of the midbrain. However, even in this preparation, fusimotor overactivity is not the sole factor. Decerebrate rigidity is accompanied by intense suppression of inhibitory reflex pathways (see for example, Eccles and Lundberg, 1959), and their release profoundly modifies the pattern of muscle tone (Grillner and Udo, 1970; Burke et al., 1972; Rymer et al., 1979).

The resemblance of human spasticity and rigidity to decerebrate rigidity in the cat was noted when analytical techniques available for use in man were in their infancy (Walshe, 1919). The fact that two disparate clinical syndromes could have been thought to resemble the decerebrate cat serves to emphasize that the resemblance is superficial: The conditions are similar only in that "muscle tone" is increased. In the cat, decerebrate rigidity is an acute reaction of the nervous system to an insult; it subsides with the passage of some days, even when bodily hemeostasis is maintained. Spasticity and rigidity are chronic responses; their development may occasionally be un-

*Department of Neurology, The Prince Henry Hospital, University of New South Wales, P. O. Box 1, Kensington, N.S.W. 2033 Australia.

expectedly rapid, but the conditions are then maintained, and plastic changes in synaptic organization can occur. To try to force a clinical syndrome to fit the pattern of the decerebrate cat may be a useful intellectual exercise, but the dissimilarities should not be neglected. The appropriate animal model for studying a human syndrome is one in which the model is intended to reproduce the human disorder, not vice versa. Studies using more appropriate animal models have failed to demonstrate a primary pathogenetic role for the fusimotor system (see later).

2. Selective Blockade of the Fusimotor System using Local Anesthetic

In the cat it is possible to block small γ efferents selectively with local anesthetics (Matthews and Rushworth, 1957a, 1957b). With this rationale, Rushworth (1960) injected dilute procaine into the motor points of spastic and rigid muscles of human patients. The resulting decreases in muscle tone were attributed to selective blockade of γ efferents, and it was suggested that spasticity and rigidity depended on overactivity of the fusimotor system. Subsequently, spasticity was attributed to a selective disorder of dynamic fusimotor neurons and rigidity to a selective disorder of static fusimotor neurons (Jansen, 1962; Rushworth, 1964; Dietrichson, 1971a, 1971b, 1973). Comparable experiments were performed by Landau et al. (1960), using intrathecal and epidural injections of local anesthetic, and by Gassel and Diamantopoulos (1964a), using a tibial nerve block, but they concluded that there was no evidence of excessive fusimotor activity in spasticity or Parkinsonian rigidity.

It is not surprising that injection of local anesthetic into or around a motor nerve decreases spasticity. Any reduction in the afferent traffic entering spinal cord circuits—even interruption of purely cutaneous nerves or nerves in the opposite limb—will reduce spasticity in remote muscles (Dimitrijević and Nathan, 1967). Since there are many afferents in the group II to IV range in muscle nerves, a selective nerve block that spared α efferents and group I afferents would still not affect γ efferents in isolation. In any case, it is debatable whether a selective local anesthetic block can be applied in man, even if it is possible to do so in the cat, where the motor nerve is of smaller diameter and can be dissected free and the agent can be applied directly to the appropriate fascicles. Experiments with microneurography have demonstrated that in man local anesthetics usually block small axons preferentially, all other things being equal (Hagbarth et al., 1970; Torebjörk and Hallin, 1973; Hagbarth et al., 1975; Mackenzie et al., 1975; Burke et al., 1976b; Hallin and Torebjörk, 1976), but the rate of development of the block is quite unpredictable, and occasionally large myelinated axons will be blocked before small ones. Local anesthetics produce a continuously developing and continuously recovering block, with different myelinated axons at different stages of block the same time, the effect on any one axon depending on its fiber diameter and on diffusion of the agent to its nodes of Ranvier. There is no stable stage during which all or most small myelinated axons will be blocked and all or most large myelinated axons spared.

In human studies, the local anesthetic has been applied directly around the muscle nerve, into the subarachnoid and extradural spaces, or into the motor point. Injection around the peripheral nerve or nerve roots ensures that the fibers most affected will be those to which the agent diffuses quickly: Small fibers in fascicles deep in the nerve trunk will be spared when more peripheral axons are totally blocked. Thus, the severity of the block will not be even over the cross-sectional area of the nerve, and cutaneous fascicles will be blocked as readily as fascicles innervating muscle. Injection into the motor point (or more diffusely into the muscle belly) does eliminate complications due to the involvement of cutaneous fascicles, but the "motor point" is a diffuse area of motor innervation, not a localized point. It is impossible to avoid uneven concentrations of the agent with consequent irregularity of effect. Furthermore, the tapering of nerve fibers at the motor point tends to reduce size differences, the very basis for a selective block.

Human studies have employed controls designed to show that the injected local anesthetic spared most large myelinated fibers, but these controls do not withstand critical inspection. The initial studies relied on the demonstration that voluntary power, as assessed clinically or with a dynamometer, was little affected. The inherent variability of such estimations, their subjective nature, dependent as they are on the maintained cooperation of the patient, the degree to which the observer offers encouragement and the degree to which synergistic muscles are brought into action to assist the prime mover, render such estimations unreliable as a measure of the extent of involvement of α efferents in a deliberately incomplete nerve block. In studies on the tibial nerve, the usual criterion of a successful block has been suppression or abolition of the ankle jerk with relative preservation of the M wave of triceps surae elicited by stimulation of the nerve trunk above the site of the block. This control assumes that the only afferents affecting the excitability of triceps surae motoneurons (MNs) are those from triceps surae. This is demonstrably not so. With injections around the main nerve trunk fascicles other than those innervating triceps surae, such as those from the sural nerve or the cutaneous and motor divisions of the posterior tibial nerve, could be affected. If so, this would produce changes in the excitability of triceps surae MNs in spastic patients (see Dimitrijević and Nathan, 1967; Bathien and Bourdarias, 1972) and in normal subjects (Fig. 1), (see Hagbarth, 1960; Delwaide, 1971; Hugon,

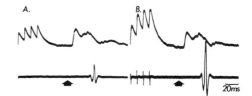

FIG. 1. The effect of stimulation of the posterior tibial nerve on the Achilles tendon jerk. In A and B the traces are, *upper trace*: multiunit muscle afferent activity recorded from the nerve fascicle innervating soleus (rectified and smoothed, time constant 0.01 sec); *lower trace*: EMG of soleus. Reproducible tendon taps were delivered (at 0.2 Hz) at the *vertical arrows*, producing a burst of muscle afferent activity (at a latency of 7 to 8 msec) and reflex EMG (at a latency of 35 msec). B: Tendon taps were conditioned by a train of 4 stimuli to the posterior tibial nerve at the ankle, the stimulus rate being 100 Hz and the intensity above threshold for a twitch of the intrinsic muscles of the foot. The first stimulus was delivered 80 msec before the tendon tap. A: Stimulator leads were disconnected so that despite the stimulus artifacts in the neural recording no stimuli were delivered. In both A and B, 16 responses have been averaged; control and test responses were alternated in an unpredictable sequence. The posterior tibial stimuli were insufficient by themselves to activate soleus skeletomotor neurons (no direct EMG response) or soleus fusimotor neurons (the afferent response to percussion was not enhanced), but must have depolarized the skeletomotor neurons so that the subsequent tap elicited a greatly enhanced tendon jerk. Hence, a local anesthetic block of the tibial nerve in the thigh or popliteal fossa could affect posterior tibial afferents, and so remove background activity increasing the excitability of the Achilles tendon reflex.

1973b). Even if the block happened to affect only those nerve fascicles innervating triceps surae and was selective, group II to IV afferents would also be involved, and it would be as logical to attribute any change in the ankle jerk to blockage of these afferents as to blockage of γ efferents. Additionally, this control may not really be adequate to detect mild involvement of α efferents (and, by implication, involvement of large afferents). Indeed, one might expect the M wave to be less affected than the tendon jerk in a block that was not selective at all. Tendon percussion sets up a high frequency, repetitive discharge in group Ia afferents. If a significant number of Ia afferents were partially blocked, it is conceivable that each afferent could transmit single but not multiple impulses due to Wedensky-like inhibition (Gassel and Diamantopoulos, 1964b; cf. Mackenzie et al., 1975), so that the afferent volley reaching the MN pool would be more dispersed and of lower amplitude. The same phenomenon could occur with α efferents but escape detection when the M wave is used as the control. Finally, there have been suggestions that α motor fibers are less susceptible than spindle afferents to a nerve block, be it ischemic (Magladery et al., 1950) or due to local anesthetic (Gassel and Diamantopoulos, 1964a), although the evidence for this is not strong.

> *Note:* The experiments of Dietrichson (1971a,b) are subject to the reservations made in the preceding paragraphs, but require specific mention, since they have been highly influential. Measurements of the H reflex and the M wave were made to demonstrate that conduction in muscle-spindle afferents and α efferents was not affected when the tendon jerk was abolished. There was only one site of stumulation, the popliteal fossa. The local anesthetic was injected "around the tibial nerve in the popliteal fossa"; it is not stated whether this was proximal to, distal to, or at the same site as the H-reflex stimulating electrodes. If proximal, the M wave cannot test conduction of α efferents across the block; if distal, the H wave cannot test conduction of spindle afferents across the block. If at the same site the reliability of the test is dubious. The effective site of stimulation of a nerve deep in the popliteal fossa is uncertain at best, and with high stimulus levels there would be longitudinal spread of current even if the nerve was close to the stimulating electrodes. Spindle afferents could have been stimulated above and α efferents below the block. Equally disturbing, however, is the efficacy reported for the "selective fusimotor blocks." For example, in all 8 spastic patients, a selective fusimotor block lasting 5 to 35 min (mean 10 min) followed single injections of 2.5 ml of 0.25% prilocaine (Citanest, an agent of potency identical to lidocaine) around the tibial nerve in the popliteal fossa. Considering the small volume injected and the relative remoteness of the nerve (even if it is located by electrical stimulation), it would be remarkable if the fascicles innervating triceps surae could be affected in isolation, and even more remarkable if a block of small nerve fibers innervating triceps surae could be maintained selectively for so long.

Gassel and Diamantopoulos (1964a) specifically studied the selectivity of local anesthetic blocks in man. In all subjects, the agent abolished the H reflex elicited by stimulation below the block but not that elicited by stimulation above the block. The suppression of the tendon jerk paralleled the disappearance of the H reflex elicited by stimulation below the block in most subjects. Motor fibers appeared more resistant to block than afferent fibers. They concluded that it is not possible to produce a selective block of small fibers in man using local anesthetics.

If a selective block of small myelinated fibers could be produced in human subjects using local anesthetics, there are, in the opinion of this author, no experimental controls that would demonstrate unequivocally that the block was selective. Even the controls of Gassel and Diamantopoulos (1964a) are not adequate. Furthermore, the demonstration that a block was selective would not guarantee that the accompanying changes in reflex excitability were due to denervation of muscle spindle endings rather than to blockage of small afferents of muscular or cutaneous origin. *The use of local anesthetics in man to produce a selective fusimotor block is fraught with so many unpredictable and uncontrollable factors in both the block itself and the interpretation of its results, that it should be discarded.*

3. The Jendrassik Maneuver

In the belief that performance of the Jendrassik maneuver activated fusimotor neurons selectively, Dietrichson (1971a) attributed the failure of the Jendrassik maneuver to potentiate the tendon jerk of spastic patients to an inability to increase futher an already high level of background fusimotor activity. Similar findings were reported by Buller (1957). However, other authors have found that the tendon jerk (Landau and Clare, 1964; Gassel and Diamantopoulos, 1964c; Bishop et al., 1968) and the tonic vibration reflex of spastic patients can be potentiated by performance of the Jendrassik maneuver (Burke et al., 1972; Dimitrijević et al., 1977). Failure to reinforce the tendon jerk can indicate a fusimotor system defect only if the fusimotor system is normally responsible for the reflex reinforcement produced by the Jendrassik maneuver. Such is not the case (Figs. 2 and 3). Some of the evidence has been reviewed by Burke et al., 1980a; more recent studies are discussed elsewhere (Burke, 1981a). It is now generally accepted that potentiation of

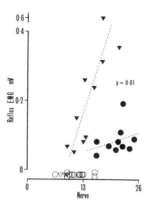

FIG. 3. Relationship between the amplitude of the rectified and smoothed soleus muscle afferent volley and the peak-to-peak amplitude of the resulting reflex EMG of soleus for the data of Fig. 2. The taps that failed to produce a tendon jerk are shown as *open symbols* next to the appropriate afferent volley size. The interrupted lines are linear regression lines for the taps that produced reflex EMG. The data obtained during reinforcement maneuvers (*filled triangles*) differ significantly from those obtained when relaxed (*filled circles*). Hence, reinforcement not only altered the threshold of central reflex mechanisms (cf. Fig. 2), it also altered their sensitivity. (From Burke et al., 1981a, with permission.)

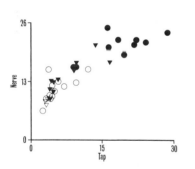

FIG. 2. Relationship between the intensity of percussion on the Achilles tendon (in arbitrary units) and amplitude of the rectified and smooth muscle afferent response from the soleus muscle (also in arbitrary units). *Circles:* Sequences when the subject was relaxed; *triangles:* Sequences for which the subject performed reinforcement maneuvers. *Filled symbols,* percussion elicited a tendon jerk; *open symbols,* there was no reflex response. Reinforcement lowered the threshold for a tendon jerk but did not alter the afferent response to tendon percussion; i.e., reinforcement did not sensitize spindle endings; it produced its effects on the reflex pathway through a mechanism independent of fusimotor neurons. (From Burke et al., 1981a, with permission.)

proprioceptive reflexes by the Jendrassik maneuver occurs through a central mechanism quite independent of the fusimotor system (Figs. 2 and 3; Burke et al., 1981a). In this light, Dietrichson's findings may actually be evidence that there was a defect of central gain-controlling mechanisms in his patients.

4. Comparison of the Tendon Jerk and the H Reflex

In clinical neurophysiology, it has become almost axiomatic that the tendon jerk and the H reflex are essentially identical reflexes which differ only in that the tendon jerk requires the integrity of the muscle spindle, and that differences in the behavior of the two reflexes accurately reflect the level of fusimotor drive. At best, this is an oversimplification. In spastic patients, it is commonly held that the tendon jerk is exaggerated more than the H reflex. The discrepancy in the degree of accentuation of these reflexes has been taken as evidence that fusimotor tone is elevated in spasticity. This conclusion assumes that (a) an increase in fusimotor drive would produce a parallel increase in the tendon jerk; (b) the level of fusimotor drive is the only significant factor affecting muscle spindle responsiveness; and (c) the only significant difference between the tendon jerk and the H reflex is the contribution of fusimotor activity to the former. Let us examine each of these assumptions.

(a) An Increase in Fusimotor Drive Would Produce a Parallel Increase in the Tendon Jerk Reflex

This assumption depends on two premises: First, the background fusimotor drive will potentiate the spindle response to tendon percussion; and second, a greater spindle discharge will generate a larger reflex response. The first premise has clear precedents in animal experiments, but attempts to demonstrate this in man have met with scant success thus far. However, the failure to do so could be due to methodological considerations rather than some species difference. The only reliable way of activating the fusimotor system in man involves the performance of a voluntary contraction, and that could change the transmission of the percussion wave to spindle endings in the muscle. The second premise has a sound basis; as in Figs. 2 and 3, the amplitude of the tendon jerk can be demonstrated to vary with the strength of tendon percussion (see also, Dietrichson, 1971a), and with the size of the afferent volley (Burke et al. 1981a). It can be concluded that assumption (a) is probably valid.

(b) The Level of Fusimotor Drive is the Only Significant Factor Affecting Muscle Spindle Responsiveness

It is well established that changes occur in extrafusal muscle with the development of spasticity and rigidity (Edström, 1970) and with disuse (for example, Davis and Montgomery, 1977), and there is recent evidence that changes in extrafusal muscle fibers "must be mainly responsible for the increased muscle tone" in spastic and rigid patients during locomotion (Dietz et al., 1981). There is a shift toward more slowly contracting motor units in the spastic patient (Hopf et al., 1974; Mayer and Young, 1979), and the paresis can lead to contracture formation (Herman, 1970). It is not unreasonable to expect some changes in intrafusal muscle, and it is unlikely that spindle responsiveness would remain the same, whether or not there was any alteration in resting fusimotor activity. Indeed, chronic immobility does affect muscle spindle sensitivity (Williams, 1980), and the extrafusal muscle atrophy due to treatment with triamcinolone heightens spindle discharge throughout the range of muscle lengthening (Botterman et al., 1981). After fusimotor stimulation and occasionally spontaneously, stretch of a spindle primary ending can elicit a high-frequency initial burst of impulses at the onset of movement, probably due to the formation of stable bonds between the actin and myosin filaments of the intrafusal fibers (Brown et al., 1969). If a similar phenomenon occurred in spastic paretic muscle, possibly as a result of relative immobility, tendon percussion could evoke a bigger spindle output and a larger reflex response than occur normally. Hence, if it were demonstrated that muscle spindle responsiveness was excessive in spasticity (and the available evidence suggests otherwise), one could not automatically conclude that fusimotor drive was excessive.

(c) The Only Significant Difference Between the Tendon Jerk and the H Reflex is the Contribution of Fusimotor Activity to the Former

This assumption is incorrect. The thought that the reflex arc of the tendon jerk may not be identical to that of the H-reflex is not new. This possibility was raised in 1952 by Teasdall et al. and their reservations have been amplified by other workers, including Homma and Kano (1962), Gassel and Diamantopoulos (1966), and Herman (1969). It cannot be maintained that changing fusimotor activity will have no effect on the H reflex, since the resulting change in background spindle activity will alter the excitability of the MN pool. However, fusimotor activity aside, the tendon jerk and the H-reflex differ in at least 6 respects, as follows:

(i) The distribution of the reflexes

The H-reflex is elicited most readily in triceps surae. It can be recorded in other limb muscles but, particularly in the upper limb, this usually requires some form of background excitation of the MN pool, for example, a weak contraction, performance of the Jendrassik maneuver, the use of a pair of stimuli to elicit the reflex, or the development of spasticity (see Deschuytere et al., this volume). In normal man the tendon jerk can be readily elicited from many limb muscles, including those of the upper limb, without the need for background excitation. To explain the elusiveness of the H-reflex in upper limb and other muscles it has been suggested that the conduction velocities (and electrical excitability) of group Ia afferents and α efferents are similar in the nerves innervating these muscles, so that any emerging reflex volley would be occluded by an antidromic volley in α efferents. However, the first MNs to be activated reflexly should be small, low-threshold MNs, which have small diameter axons of relatively high electrical threshold (see Desmedt, 1980a, 1981a). Occlusion should not occur until a reasonably strong electrical stimulus had recruited these axons into the M wave. Furthermore, the conduction velocities of Ia afferents and of α efferents do not change when a maneuver is performed to elicit the H-reflex in a muscle where none could previously be recorded.

(ii) The effect of ankle joint angle

Rotation of the ankle joint will affect the excitability of the soleus MN pool due to changes in length of the agonist, its synergists, and its antagonists, and to any effects arising from activation of cutaneous and joint receptors. Passive dorsiflexion of the ankle can result in inhibition of the H reflex of soleus in normal man (Mark et al., 1968; Delwaide, 1971), although this is not a consistent finding (Herman, 1969). However, the muscle stretch produced by dorsiflexion will alter the responsiveness of primary spindle endings in soleus so that percussion of the Achilles tendon could evoke a larger afferent volley. Hence, provided that the greater size of the afferent volley was sufficient to outweigh the inhibitory background activity, the tendon jerk of triceps surae would be potentiated in the dorsiflexed position. Herman (1969) has reported that this is the case, at least within the range from 30° plantar-flexion to the zero position.

The discrepancy in the effects of stretch on the two reflexes is greater in spastic patients (Herman, 1969). Within the stated range of movement, dorsiflexion facilitates the tendon jerk (Herman, 1969) but inhibits the H reflex (Burke et al., 1971), and the degree of inhibition is greater than in normal subjects (Herman, 1969), presumably due to release of transmission in flexor reflex afferent pathways in the spastic state (Burke and Lance, 1973). This factor alone could explain the finding that there is greater exaggeration of the tendon jerk than of the H-reflex in the spastic state: The more dorsiflexed the ankle, the greater the accentuation of the tendon jerk relative to the H reflex. In most studies that have compared the tendon jerk and H-reflex in normal subjects and spastic patients, the angle of the ankle joint was not specified, and it is not always clear that the angle was the same for the tendon jerk and the H-reflex tests. However, even if the angle was kept constant within and between subjects, this factor could still account for the discrepancy in the reflexes.

(iii) The afferent fibers activated by the stimuli

The H-reflex stimulus will activate group Ia and Ib fibers in roughly similar proportions, although, even in humans, a slightly lower threshold and slightly faster conduction velocity can be demonstrated for Ia afferents (Pierrot-Deseilligny et al., 1981). In humans tendon percussion can activate Golgi tendon organs (and secondary spindle endings), but not as intensely as primary spindle endings. Hence, the afferent volley for the H-reflex will involve significantly more group Ib afferents. When one considers the dispersion of the group Ia afferent volley by the time it reaches the MN pool (see below), the slightly slower conduction velocity for the Ib afferent and the slightly longer reflex pathway (classically disynaptic) are not enough to prevent overlapping Ia and Ib actions on the MN pool.

The tendon tap will activate predominantly muscle afferents in triceps surae but some spread of the percussion wave to the pretibial flexor muscles cannot be prevented. The H-reflex stimulus can be (but probably is not always) delivered solely to the tibial nerve, so that afferents from the antagonist are not activated, but it is difficult to prevent activation of afferents from the small muscles of the foot, innervated by the posterior tibial nerve *(unpublished observation)*. Hence, the percussion-evoked afferent volley will be contaminated by afferents from antagonistic muscles, while the electrically evoked volley will be contaminated by a significant number of afferents from the small muscles of the foot. The afferent volleys may also contain activity in cutaneous afferents, but their slower conduction velocity should ensure that they do not reach the MN pool during the composite group Ia excitatory postsynaptic potential (EPSP).

(iv) The pattern of activity in the activated afferent fibers

With the single electrical stimulus used for H-reflex testing, there will be only one discharge in each afferent fiber activated by the stimulus. With tendon percussion, primary spindle endings respond with a high-frequency burst of impulses, reaching instantaneous frequencies of 100 to 200 impulses/sec or higher for the second and subsequent impulses. Thus, with tendon percussion, the afferent volley will contain repetitive discharges in individual Ia afferent fibers, there being 5 to 10 msec or less between the consecutive discharges. Taking into account the dispersion of the group Ia afferent volley by the time it reaches the MN pool (see below), there is adequate time for the second discharges in fast Ia afferents to reach the MN pool at the same time as the initial discharges in slow Ia afferents.

(v) The dispersion of the afferent volley

With the H-reflex stimulus, afferent axons are activated synchronously at the level of the popliteal fossa. It seems safe to assume an afferent conduction time from popliteal fossa of 15 msec, a conduction velocity of 60 m/sec for the fastest Ia afferents (see Mano et al., 1976: Pierrot-Deseilligny et al., 1981), and 40 m/sec for the slowest Ia afferents. On this basis, EPSPs from the slowest conducting Ia afferents would be generated in the MN pool 7.5 msec after those from the fastest Ia afferents. This calculation is supported by recent studies on human subjects by Ashby and Labelle (1977). The rise-time of the composite group I EPSP of individual soleus MNs due to electrical stimulation in the popliteal fossa was determined as usually between 3.5 and 5.5 msec, with a mean duration of 3.6 msec. It should be noted that Ashby and Labelle (1977) used an electrical stimulus just subliminal for the H-reflex. This would have activated predominantly low-threshold fast-conducting group Ia afferents, and would have spared slower conducting afferents unless they were close to the stimulating electrodes. Hence, the values presented by Ashby and Labelle (1977) probably underestimate the rise time of EPSPs elicited by stimuli which are strong enough to produce an H-reflex.

Assuming a conduction velocity of 56 m/sec for the fastest Ib afferents from soleus (see Pierrot-Deseilligny et al., 1980), the group Ib effects on the MN pool would occur 1.5 to 2 msec after those due to the fastest Ia afferents (1 msec for the slightly slower conduction velocity, plus 0.5 to 1 msec for the interneuron in the classically disynaptic pathway). Hence, with an H-reflex stimulus, there will be significant overlap of the autogenetic group Ia excitatory effects and autogenetic group Ib inhibitory effects—a conclusion that is also apparent from the studies of Pierrot-Deseilligny et al. (1980). Indeed, their calculations suggest an even earlier onset of detectable Ib effects (about 0.8 msec after the onset of detectable Ia effects).

With tendon percussion, the dispersion of the group Ia volley by the time it reaches the spinal cord will be greater because the afferent pathway now includes time for propagation of the percussion wave to spindle receptors (probably 40 m/sec, cf. Brown et al., 1967), receptor-transducing mechanisms, and conduction from receptor to popliteal fossa. Because of differing positions of different spindles in the muscle, the additional time from percussion to popliteal fossa will differ for different Ia afferents. In human recordings, this additional time is 4 to 8 msec for the fastest afferents (*unpublished observation;* cf. Fig. 1). Hence, a latency of at least 20 msec can be assumed for the time from percussion to the onset of the effects due to the fastest Ia afferents on the MN pool, with those due to the slowest Ia afferents occurring about 10.0 msec later. These figures seem quite reasonable. Measurements have been made in man of the composite EPSP evoked in gastrocnemius MNs by tapping the Achilles tendon (Noguchi et al., 1979). The rise time of the tap-evoked composite EPSP was found to be 8.2 ± 1.7 msec (mean, $\pm$ standard deviation). The long duration of the tap-evoked EPSP in man is not unexpected. In the cat the rise time of the composite group Ia EPSP due to abrupt muscle stretch is approximately 50% longer than that of an electrically evoked EPSP of similar size (Stuart et al., 1971). Hence in man, the duration of the rise-time of the EPSP evoked by tendon percussion is such that second impulses in the fastest Ia afferents will reach the MN pool before or at about the same time as the initial impulses in slower Ia afferents. Even in the cat, where the effects of dispersion are minimal due to shorter distances and higher conduction velocities, the EPSP due the the second impulses in Ia afferents merges with the first, the delay being only 1.3 to 1.5 msec (Stuart et al., 1971).

Clearly, with the H-reflex stimulus, the more slowly conducting Ia afferents will reach a MN pool that has already been conditioned by the activity in more rapidly conducting Ia and Ib afferents. With tendon percussion, the MN pool will have been conditioned only by fast-conducting afferents, but second discharges in individual fast-conducting Ia afferents may contribute to the reflex discharge.

The situation is even more complicated. First, reflex activation of low-threshold (small) MNs can lead to efficient recurrent inhibition (Hultborn et al., 1979), and there is adequate time for undischarged MNs of high-threshold to receive Renshaw inhibition before the arrival of the slowest Ia afferents. Clearly, the longer duration of the percussion-evoked composite EPSP makes this more likely to occur with the tendon jerk than with the H-reflex. This factor could be important when considering spasticity, since it has been suggested that there is abnormal control of recurrent inhibitory pathways in patients with upper motor neuron lesions (Veale et al., 1973). This suggestion has, however, not been confirmed in more extensive studies using a different technique (Katz and Pierrot-Deseilligny, 1982). There appears to be a defect of the supraspinal control of Renshaw cells during voluntary or postural contractions in spastic patients (see Castaigne et al., 1978), but Renshaw cell activity at rest seems within the normal range. Second, the dispersion of the Ia afferent volley is such that there is adequate time for the fastest Ia afferents to exert effects through autogenetic disynaptic and trisynaptic pathways (Watt et al., 1976), and even through more highly polysynaptic spinal pathways (see Hultborn and Wigström, 1980), before the slowest Ia afferents begin to exert their effects. Again, the likelihood of this occurring increases with the degree of dispersion of the Ia afferent volley, being greater for the tendon jerk than for the H reflex.

It can be argued that the long duration of the group Ia excitation at the MN level would be detectable in a very dispersed EMG response if it were significant. However, low-threshold MNs have slowly conducting efferents, and high-threshold MNs have fast conducting efferents (see Desmedt, 1981a for review). The effects of these differences may well be enough to produce the reasonably synchronized EMG burst that is usually recorded. Interestingly, the duration of the reflex EMG potential is significantly longer with the tendon jerk than with the H reflex (Dietrichson, 1971a).

Note: In human reflex studies, the term monosynaptic reflex is used synonymously with tendon jerk or H-reflex. As indicated above, disynaptic, trisynaptic, and polysynaptic excitatory pathways could also contribute to producing the reflex discharge. There is no good proof that in man the tendon jerk and H-reflex are exclusively monosynaptic; indeed, there is no proof that they are even predominantly monosynaptic, although measurements of the minimal central latency have indicated a monosynaptic component (Magladery et al., 1951). The latency variability ("jitter") of individual motor units in consecutively elicited H-reflexes suggests that the discharge is oligosynaptic (cf. Brown and Rushworth, 1973) but has not established that it is monosynaptic. There is a much higher jitter with mechanically elicited reflexes than with those elicited electrically, even for the same motor units (Stålberg and Trontelj, 1979)—a phenomenon that could be due to greater dispersion of the mechanically evoked afferent volley, and possibly to greater participation of polysynaptic pathways.

If there is a significant disynaptic/trisynaptic contribution to human monosynaptic reflexes, the studies of Delwaide (1971, 1973) and some of those reviewed by Lance et al. (1973) and Ashby et al. (1980) are capable of reinterpretation. Vibration-induced inhibition of the H-reflex and the tendon jerk, although "premotoneuronal," need not be "presynaptic," and the suppression of this phenomenon in spastic humans could be due, not to a suppression of "presynaptic inhibition," but to some change in interneuron excitability. In normal humans, the degree of vibration-induced inhibition of the H-reflex is on average 66%, of which 25% can be attributed to spread of vibration to the anterolateral compartment (Ashby et al., 1980). Hence, it can be estimated that afferents in the posterior compartment are responsible for suppression of the discharge of 50% of the MNs active in the control H reflexes. Actually, according to the size principle less than 50% of the active MNs would have to be inhibited to reduce the EMG potential by 50% (Desmedt and Godaux, 1978a). Thus, it would require only that disynaptic and trisynaptic pathways *contributed* to the discharge of less than 50% of the MNs active in the control H-reflex for the vibration-induced inhibition to be completely explicable by postsynaptic inhibition at interneuronal level. As judged by the duration of the rise time of the composite EPSP discussed earlier, this explanation is a quite plausible alternative to presynaptic inhibition.

(vi) The MNs participating in the reflex contraction

As usually elicited, the H reflex of triceps surae involves predominantly soleus motor units, although the M wave appears more readily in gastrocnemius (Homma and Kano 1962; Levy, 1963; Buchthal and Schmalbruch, 1970; Hugon, 1973a). On the other hand, the tendon jerk can be recorded in gastrocnemius as well as soleus (Homma and Kano, 1962), although it appears at lowest threshold in soleus (Levy, 1963). It is possible to record selectively from soleus using surface EMG electrodes distal to the insertion of gastrocnemius into the Achilles tendon (Hugon, 1973a), but in this location a mechanical artifact may be recorded with tendon percussion. Hence, a more proximal and less selective site is likely to be chosen in comparative studies of the tendon jerk and H reflex, a compromise which could disadvantage the H-reflex potential.

The basis for the lower reflex threshold of soleus is probably that it is composed almost exclusively of low-threshold, slow-twitch motor units, whereas gastrocnemius contains both fast- and slow-twitch units, the former predominating in a ratio of 3:1 (see for example, R. E. Burke, 1967, 1973). With low-intensity electrical stimulation, only small low-threshold motor units would be activated reflexly. With stronger intensities, large motor units of higher threshold may well be activated reflexly but be unable to participate in the reflex EMG burst if, as is likely, their large-diameter axons were preferentially recruited in the M wave (Homma and Kano, 1962). It is common practice to use stimuli that produce a small M response, since if the M wave remains constant, experimentally induced changes in the H-reflex are probably not due to displacement of the stimulating electrodes. If there is an M wave, adjustment of stimulus levels so that the tendon jerk and H reflex produce reflex EMG bursts of identical amplitude could result in the H reflex stimulus activating a greater proportion of the MN pool than the tendon jerk stimulus. In addition, if there are differences in the MNs contributing to the reflex contraction in the tendon jerk and the H reflex, the changes in motor unit properties that occur both in spasticity (Edström, 1970; Hopf et al., 1974; Mayer and Young, 1979) and as a result of inactivity (Davis and Montgomery, 1977) may well be a significant factor complicating the interpretation of comparative reflex studies.

In summary, there are quite good theoretical grounds on which to maintain that fusimotor activity is not the only difference between the tendon jerk and the H-reflex. The importance of these other factors is unknown, but so too is the extent of the difference made by fusimotor activity. It is as logical to attribute a difference in the behavior of the two reflexes to a change in one of these other factors as to a change in fusimotor drive. It is safe to assume that a difference in the behavior of the tendon jerk and the H-reflex results from a change in fusimotor drive only if it can be shown that all of these factors are unimportant or that all remain constant under the particular experimental circumstances. With presently available techniques for use in man neither condition can be satisfied. It can be concluded that comparisons of the H reflex and tendon jerk are not reliable as measures of fusimotor activity. In any case, it would be difficult to devise a more indirect means of estimating the activity of fusimotor neurons than to compare the skeletomotor activity produced by two different reflexes.

EVIDENCE FROM RECORDINGS OF MUSCLE SPINDLE AFFERENT ACTIVITY

The technique of microneurography (cf. Hagbarth and Vallbo, 1968; Vallbo et al., 1979; Burke, 1981; Prochazka and Hulliger, this volume) offers a more reliable and certainly less indirect method of estimating fusimotor activity. From such recordings it has become apparent that, in normal man, the gain of proprioceptive reflexes can be regulated through mechanisms completely independent of the fusimotor system. Changes in reflex gain occur: (a) immediately before a voluntary contraction; (b) when the subject is given a warning of the need to make a voluntary contraction (anticipation); (c) during the build-up phase of the tonic vibration reflex; (d) when the tonic vibration reflex is suppressed voluntarily; (e) when joint position is maintained despite changes in load on the contracting muscles; and (f) as a result of performance of the Jendrassik maneuver. The fusimotor system is not involved in the alterations of reflex gain in any of these instances, at least under the particular experimental conditions (see Burke, 1981 for discussion; see also Figs. 2 and 3). The inhibition of tendon jerks and H reflexes by muscle vibration can also be considered a change in reflex gain and, although the precise mechanism remains to be established beyond all reasonable doubt (see earlier), it is certainly not due to suppression of fusimotor activity.

It is not denied that a greater afferent input would result in a greater reflex response, all other things being equal, but in each of the above instances the change in reflex gain occurs through a central mechanism—perhaps through an effect at interneuronal level on reflex transmission. These studies have established, not surprisingly, that the nervous system has mechanisms other than the fusimotor system for changing the gain of proprioceptive reflexes. Indeed, it has yet to be demonstrated that, during normal motor behavior, reflex gain can be altered by altering fusimotor drive. The biological purpose of the fusimotor system still must be elucidated, but there is no evidence to suggest that it lies in regulating the gain of proprioceptive reflexes.

Given that questions still surround the role of the fusimotor system as a gain-controlling mechanism and that the nervous system has adequate alternative mechanisms no matter what the true role of the fusimotor system, it is reasonable to question whether defective fusimotor activity could be responsible for disorders of "stretch reflex" gain. The limited data available from preliminary studies in patients support more extensive studies that have been made in animal models designed to reproduce the human disorder. The hypertonia of spasticity and rigidity do not appear to result from a primary defect of the fusimotor system (see later). Hypotonia has been found to be associated with hypoactivity of the fusimotor system (Gilman, 1969; Gilman et al., 1971, 1974). Whether this defect is relevant to clinically detectable hypotonia is considered below.

1. Disorders Associated with Hypotonia

A decrease in the resistance to passive stretch is a characteristic finding in spinal shock and is often found in cerebellar disease. With spinal shock there is generalized reflex suppression, involving tendon jerk reflexes and exteroceptive reflexes in addition to "muscle tone."

Based on recordings of the afferent response to abrupt muscle stretch, Hunt et al., (1963) concluded that withdrawal of fusimotor activity plays a minor role, if any, in spinal shock in the cat and the monkey. In man, however, it was concluded that depression of fusimotor function is one factor contributing to the hyporeflexia of spinal shock (Weaver et al., 1963). The human study was based on serial comparisons of the tendon jerk and H reflex and on the premise that primary spindle endings require background fusimotor activity to be sufficiently sensitive to generate an afferent volley capable of producing a tendon jerk. The validity of comparisons of the tendon jerk and the H-reflex as a measure of fusimotor activity has been discussed earlier. The premise rests mainly on the ability of selective fusimotor blockade using local anesthetic to abolish the tendon jerk, evidence which is of dubious validity.

In normal man, there is no evidence of background fusimotor drive to the muscle spindle endings of completely relaxed muscles, whether those muscles are flexors such as the pretibial muscles, or extensors, such as triceps surae (see Vallbo et al., 1979; Burke, 1981; Burke et al., 1981b). The dynamic sensitivity of human primary spindle endings is high in the relaxed state, and the afferent response to stretch or percussion does not decrease when a muscle is temporarily deefferented by a complete nerve block using either local anesthetic (Fig. 4) or pressure. Fusimotor activity does not appear to be a prerequisite for a tendon jerk (Berke et al., 1981b). Spinal shock may well be accompanied by hypoexcitability of fusimotor neurons as well as hypoexcitability of skeletomotor neurons, but the suppression of the tendon jerk cannot be attributed to the removal of a factor that is not normally necessary anyway. Unless there are extrafusal changes that diminish spindle responsiveness, the afferent response to tendon percussion should be normal in spinal shock in human patients, much as it is in the cat and monkey (Hunt et al., 1963).

For the hypotonia of cerebellar disease

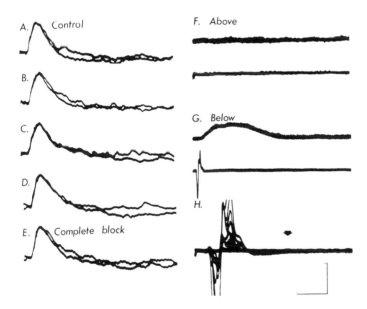

FIG. 4. The effect of a complete block of the sciatic nerve on the response of soleus muscle afferents to reproducible percussion on the Achilles tendon. Nerve block was produced by injection of lidocaine (10 ml, 2%) around the sciatic nerve in the thigh, proximal to the neural recording site. A–E. Consecutive averages of the muscle afferent response to tendon percussion (rectified and smooth, time constant 0.01 sec). Each trace consists of two successive averages (to demonstrate reproducibility at each stage), each of 32 responses. A: Control; B–D: during development of the block; E: when the block was clinically and electrically complete. F and G, contraction force at the ankle joint (*upper trace*) and EMG of soleus (*lower trace*) in response to maximal stimulation above the block (F) and to stimulation below the block (G). In F and G, 5 traces have been superimposed; the EMG amplification is 5 times greater in F than in G. H, soleus EMG responses to 10 gradually increasing stimuli delivered below the block, to demonstrate the absence of the H reflex (which should have appeared near the vertical arrow). *Vertical calibration:* 200 νV, F and H; 1 mV, G. *Horizontal calibration:* 20 msec, A–E; 100 msec, F, G; 10 msec, H. (From Burke et al., 1981b, with permission.)

and that of spinal shock to result from fusimotor withdrawal, one must assume that normal muscle tone is dependent on an active stretch reflex. However, muscle tone is tested clinically when the subject is completely relaxed or is as near to complete relaxation as possible. Under these circumstances, passive muscle stretch usually fails to produce a reflex contraction, unless the stretch is so abrupt that it elicits a tendon jerk. If there is no reflex component to the resistance to stretch, fusimotor activity can play no role, whether or not the fusimotor system can be shown to be hypoactive under other conditions. In completely relaxed normal subjects muscle tone is dependent solely on non-neural factors, such as the viscoelastic properties of muscle and tendon, joint stiffness and limb mechanics.

In agreement with this view, it has been found that general anesthesia and muscle relaxants do not reduce muscle tone of normal subjects below the level found in relaxed conscious subjects (Lakie et al., 1980).

> *Note:* The static responsiveness of primary spindle endings in gastrocnemius of monkeys has been reported to be reduced by cerebellar ablation (Gilman, 1969) and by medullary pyramidotomy (Gilman et al., 1971), both of which produce lasting hypotonia. It has also been reported during the stage of hypotonic hemiparesis following ablation of cortical areas 4 and 6 (Gilman et al., 1974). At face value, these findings suggest the presence of background fusimotor activity in the normal monkey and, therefore, appear to be at variance with the data from human spindle afferents referred to above (see Vallbo et al., 1979; Burke, 1981; Burke et al., 1981b). However, these animals were lightly anesthetized, rigidly fixed in a spinal frame, and subjected to extensive surgery, which, of necessity, included laminectomy. When lightly anesthetized but otherwise intact, slow, passive, stretch evoked reflex EMG activity on both sides, although the threshold was higher on the hypotonic side (Gilman et al., 1974). Clearly, this situation is quite unlike that existing with the relaxed human subject and cannot be described as normal. It is probable that the experimental invasions necessary in animal studies are responsible for the background fusimotor activity found in the neurologically intact animal. These studies do demonstrate that the "set point" and, therefore, the responsiveness of fusimotor neurons may be depressed by lesions which also result in clinically detectable hypotonia (cf. Gilman and Ebel, 1970). This finding is certainly relevant when one considers the total motor deficit of a hypotonic patient, but it is of limited importance to the state of tendon jerks and muscle tone in patients at rest.

Hypotonia is not necessarily a figment of the expectant clinician's imagination. Not all subjects relax readily to command, and clinicians become trained to accept a certain degree of failure of relaxation as within normal limits. The absence of resistance due to this "acceptable" degree of failure of relaxation may be interpreted as hypotonia when it occurs in the presence of an overt cerebellar deficit, even if it would be called normal in a different context. Patients with cerebellar lesions do not appear to manifest the same degree of failure of relaxation as normal subjects; indeed, their major deficits include the initiation, gradation, alteration, and cessation of voluntary movement (Holmes, 1917, 1922). On the side(s) of the cerebellar deficit, muscle tone can be expected to fall into the lower end of the normal range.

It is not meant to infer that stretch reflex hypoactivity does not occur with cerebellar lesions. However, it would take a background contraction for any such hypoactivity to become manifest, and only under these circumstances could fusimotor hypoactivity play any pathogenetic role. Perhaps the asthenia of the cerebellar patient, the greater sense of effort required to perform a task in the absence of paresis, is a manifestation of withdrawal of stretch reflex assistance to the contraction (Holmes, 1917, 1922; cf. Gandevia and McCloskey, 1977).

With chronic cerebellar lesions, a reduction in muscle tone below that seen even in completely relaxed normal subjects could result if the motor deficits led to ligamentous stretch, joint laxity, and other soft tissue changes. These phenomena could occur, for example, secondary to an absence of the normal reflex damping of voluntary movement (Neilson and Neilson, 1978; Neilson and Milder, 1981). If this is true, the hypotonia detectable at rest could be considered one of the sequelae of stretch reflex hypoactivity.

2. Disorders Associated with Hypertonia

Spasticity and rigidity have been attributed predominantly if not exclusively to fusimotor overactivity (Jansen, 1962; Rushworth, 1960, 1964), but this view has been contested (cf. Landau, 1969), and it has even been suggested that the prime defect in parkinsonism is fusimotor hypoactivity (cf. for example, Stern and Ward, 1960). There is good evidence for central changes in reflex transmission in both disorders. For example, in spasticity there is diminution of the vibration-induced suppression of phasic reflexes (Delwaide, 1971, 1973; Ashby et al., 1980); the onset of the tonic vibration reflex (TVR) is abrupt, and the spastic patient cannot suppress the reflex as can normal subjects (Hagbarth and Eklund, 1968; Burke et al., 1972; see Burke et al., 1976b); flexor reflexes become hyperactive and altered in pattern (Shahani and Young, 1971, 1973); there are abnormalities in the frequency response curves of the "action tonic stretch reflex" (Neilson and Lance, 1978); the incidence of F waves in single MNs is higher (see for example, Schiller and Stålberg, 1978); recurrent inhibition may be defective (Veale et al., 1973), at least during voluntary and postural contractions (Castaigne et al., 1978; Katz and Pierrot-Deseilligny, 1982); performance of the Jendrassik maneuver does not potentiate the tendon jerks of some spastic patients (Dietrichson, 1971a); the recovery curve of the H reflex following a conditioning stimulus is abnormal, but because it is similarly altered on the normal side of hemiplegics the precise meaning of this change is debatable (cf. for example, Magladery et al., 1952; Zander Olsen and Diamantopoulos, 1967; Yap, 1967; Sax and Johnson, 1980). In parkinsonian rigidity, spinal proprioceptive reflexes such as the TVR and the tendon jerk are relatively normal (Hagbarth and Eklund, 1968; Burke et al., 1972; Lee and Tatton, 1975; Mortimer and Webster, 1978; Chan et al., 1979), but polysynaptic exteroceptive reflexes contain consistent abnormalities (Delwaide et al., 1974), and there are abnormalities in long-loop reflexes (Lee and Tatton, 1975, 1978), which appear to correlate with the severity of rigidity (Mortimer and Webster, 1978, 1979, this volume; cf., however, Chan et al., 1979) and which are not present in nonrigid parkinsonian patients (Lee and Tatton, 1975; Thomas et al., 1977).

If the prime defect in spasticity and rigidity lay in the fusimotor system, the abnormality would be manifested as overactivity of muscle spindle afferents, and the characteristic features would be reproducible in normal subjects merely by making muscle spindles discharge excessively. With muscle vibration, primary spindle endings of normal subjects can be driven to discharge with an intensity that exceeds that seen when the muscles of patients are subjected to stretch (Burke et al., 1976a; see later). The normal subject does not become spastic or rigid when his muscles are vibrated, and neither do rats and mice kept in a vibrating environment for 2 weeks (Matthews, 1976). In normal man, a TVR may develop in the vibrated muscles, but its appearance is variable in different muscles and different subjects, it may require a background contraction of the vibrated muscle, and it can be suppressed at will by the subject. It seems that excessive muscle spindle activity is not adequate to produce spasticity and rigidity. Central defects are essential.

(a) Spasticity

A number of animal models of spasticity have been developed that have not suggested fusimotor overactivity in spasticity. In monkeys made hemiplegic following spinal cord hemisection or by higher lesions in cerebrum and brainstem, the afferent volleys elicited by tendon percussion were found to be similar on the spastic and normal sides when the tendon jerk amplitudes differed by a factor of 15 (Meltzer et al., 1963). Moreover, in these animals spontaneous variations in the amplitude of the tendon jerk were not accompanied by changes in the percussion-evoked afferent volley. Similar findings have been reported for cats (Fujimori et al., 1966) and monkeys (Aoki et al., 1976) rendered hemispastic by hemisection of the spinal cord. In the latter study, the reflex potential recorded from ventral roots on percussion of the patellar tendon was augmented by 30 to 40% on the hemisected side, even though the responsible afferent volley recorded in dorsal roots was smaller by 10 to 30%. The greater size of the afferent volley on the normal side could be due to a number of factors: atrophy of the quadriceps by 5 to 15% on the chronically hemisected side; the use of light barbiturate anesthesia in the acute experiment; the transection of the spinal cord performed in the acute experiment; and greater fusimotor responsiveness to experimental trauma on the normal side.

The activity of single group Ia afferents has been studied in animal models, but only the responses to static stretch have been analyzed. Ablation of cortical area 4 in monkeys has been reported to lead to an increase in the discharge rates of primary endings of the paretic gastrocnemius (Gilman and Van der Meulen, 1966). These animals had some features of spasticity, although their postural deformity was considered to be primarily of nonreflex (dystonic) origin (cf. Denny-Brown, 1966). In a subsequent more extensive analysis performed on monkeys in which cortical areas 4 and 6 had been ablated, Gilman et al., (1974) reported that the discharge rates of primary endings in the paretic gastrocnemius were low during the stage of "shock" (see earlier), and recovered to normal levels with the development of hypertonia, but at no stage exceeded normal. The reasons for the apparent discrepancy between these two studies are not immediately clear. The recordings of Gilman and Van der Meulen (1966) were made at only one muscle length, usually 5 mm beyond zero, which was defined as that length at which muscle stretch caused a detectable change in force as recorded by a sensitive myograph. Asymmetric extrafusal properties (for example, changes in extrafusal muscle resulting from hemiparesis; active contraction due either to the dystonic posture or to differences in reactivity associated with the deliberately light level of anesthesia) could have produced a different zero length for the normal and hemiplegic animals, and might have escaped detection, since EMG was not recorded.

Zero length was defined in the same manner by Gilman et al. (1974), but the resulting errors could have been minimized by determining spindle discharge frequency at six lengths rather than one and by measuring the degree of stretch required to produce a discharge from each ending. In both studies, the animals had hyperactive tendon jerks and increased muscle tone (although it is not clear that the disorder of tone was typically spastic), but in neither study were the spindle responses to abrupt stretch or to tendon percussion evaluated. Nevertheless, on the data presented, the conclusion of Gilman et al. (1974) that "the hypertonia of this experimental model is not based upon a tonic heightening of fusimotor activity" seems justified. The same conclusion has been reported for the hypertonia that develops 6 days after spinal cord transection in cats (Lieberman et al., 1979). Again the responses of gastrocnemius primary spindle endings to static stretch were lower during the hypotonic stage than in control animals. They recovered with the development of hypertonia, but did not exceed control levels.

Recordings have been made from single Ia afferents of spastic patients (Szumski et al., 1974; Hagbarth et al., 1975; Wallin and Hagbarth, 1978), but these recordings were primarily directed at elucidating the mechanisms of clonus. Some incidental findings (effects of the Jendrassik maneuver; the appearance of spindle discharges early on the falling phase of a twitch contraction) led Szumski et al. (1974) to consider the presence of greater dynamic fusimotor activity in their patients than in normal subjects. However, the Jendrassik maneuver does not operate through the fusimotor system (Figs. 2 and 3; see also Vallbo et al., 1979; Burke et al., 1981a; Burke, 1981), and in any case, spindle endings in normal subjects can discharge as early on the falling phase of a twitch contraction as was seen by Szumski et al. (1974) in spastic patients (Hagbarth et al., 1975; and *unpublished personal observations*). The only published study to have compared the discharge rates of primary spindle endings in spastic patients with those seen in normal subjects is a preliminary communication by Hagbarth et al. (1973), who found no difference in background activity or in the dynamic and static responses to passive stretch of endings in triceps surae. This study involved only 9 spastic and 12 normal spindle endings, but the sample has since been enlarged with identical findings. Until there are more data it would be imprudent to state uncategorically that there is no defect of fusimotor functions in spasticity. However, thus far there is no evidence of increased fusimotor drive, and this finding is in agreement with the earlier animal studies.

(b) Parkinsonian Rigidity

There are probably no completely satisfactory animal models of parkinsonian rigidity, but the models available do allow some conclusions to be made. Rigidity develops in the reserpine-intoxicated rat in the absence of fusimotor activity (see Steg, 1966). There are no fusimotor fibers in the neonatal kitten, and in the adult cat there is a stage of reinnervation following nerve section and resuture during which skeletomotor but no fusimotor activity has been reestablished. Both preparations develop rigidity following administration of haloperidol, even though both lack a functioning fusimotor system (Thulin and Blom, 1974). Biochemically these models may resemble parkinsonian rigidity in some respects; clinically, however, the resemblance is debatable. Nevertheless, the animal studies suggest that fusimotor activity may be irrelevant to the development of rigidity.

The study of Wallin et al. (1973) is often cited as demonstrating that there is excessive static fusimotor drive in parkinsonian rigidity and, by implication, that this is inappropriate and pathogenetic. These findings have recently been elaborated and clarified (Hagbarth et al., 1975; Burke et al., 1977; see also Vallbo et al., 1979). Fluctuations in rigidity are associated with parallel fluctuations in spindle activity and EMG, much as can occur in a normal subject who is not relaxed. The spindle response to stretch of rigid muscles looks similar to that seen in normal subjects who are performing a weak voluntary contraction. Thus, there is evidence of fusimotor activity in rigid patients who are trying to relax, and this activity is not seen in normal subjects who are relaxed. However, the rigid patients are not really at rest, and there may be similar fusimotor activity in normal subjects who cannot relax or who are performing a weak voluntary contraction. The findings can be explained if the excessive drive onto skeletomotor neurons came not through spinal reflex arcs but from supraspinal centers, and if this drive activated α- and γ-MNs together, much as occurs in a voluntary contraction. Hence, fusimotor overactivity accompanies but does not appear to cause the skeletomotor overactivity. There is no evidence of a primary defect of fusimotor function in parkinsonian rigidity (Burke et al., 1977). Similar conclusions have been expressed by Mano et al. (1979) in the only other published study of muscle spindle activity in patients with Parkinson's disease.

ACKNOWLEDGMENT

This work was supported by the National Health and Medical Research Council of Australia.

… Motor Control Mechanisms in Health and Disease, edited by J. E. Desmedt. Raven Press, New York © 1983.

Muscle Sense and Effort: Motor Commands and Judgments About Muscular Contractions

*D. Ian McCloskey, Simon Gandevia, Erica K. Potter, and James G. Colebatch

School of Physiology and Pharmacology, University of New South Wales, Sydney, Australia

This chapter deals with the perceptual effects of the actions of motor commands within the central nervous system (CNS). Many other possible sensory consequences, arising as a result of motor commands leading to actions and those actions in turn to stimulation of sensory nerves, are not considered here.

In a long and controversial history, the view that there might be perceptual consequences of the internal actions of motor commands has appeared in many forms (see McCloskey, 1980). In brief, the theory is that signals are generated within the CNS from, or together with, the commands for muscular contractions and that these signals influence sensory perception either by modifying the processing of incoming sensory information or by entering higher sensory areas so as to evoke, in their own right, sensations of various kinds. Acceptance of this theory does not compel rejection of a major role for conventional, peripherally arising sensory inputs. Possibly the most widely accepted version of the hypothesis of a sensory role for internal collaterals of motor commands concerns visual stability and the commands to the extraocular muscels—an hypothesis that is usually attributed to Helmholtz (1867), and which has come under serious challenge recently (see Miles and Evarts, 1979; McCloskey, 1980). Motor commands to eye muscles are not considered further in this chapter.

Terminology in this field has become somewhat confusing. Sperry (1950) introduced the term corollary discharge to describe supposed internal signals which arise from motor signals and which influence perception. The term has frequently been misused in referring to internal actions of command-related signals which have no demonstrated perceptual effects. Efference copy is another term that is often misused. It was introduced by von Holst (1954) to refer to internal, command-related signals which cancel so-called reafferent signals (sensory discharges set up purely as consequences of the motor signals), so as to leave unaffected exafferent signals (sensory discharges caused by external influences). Cancellation of a specific element of afference, not merely a general suppression of afference, is the key element of efference copy as von Holst defined it, and the term is applicable whether or not the surviving exafference is perceived. Various terms including sensation of innervation, sense of effort, and felt will, have been used to refer to sensations said to arise directly from the internal actions of motor commands.

---

* To whom correspondence should be addressed: School of Physiology and Pharmacology, University of New South Wales, P. O. Box 1, Kensington, N.S.W. 2033 Australia.

1. FAILURE OF MOTOR COMMANDS TO CAUSE SENSATIONS OF MOVEMENT

If a limb muscle is completely paralyzed either by injection of local anesthetic around the nerves that serve it (Melzack and Bromage, 1973), or by depriving it of its blood supply (Goodwin et al., 1972), or by local or systemic administration of a neuromuscular paralyzant (Campbell et al., 1980; McCloskey and Torda, 1975), no sensation of movement accompanies an attempt to move. A subject so affected may describe his immobile limb as feeling heavy, or as if stuck or pinned down. He is aware of making an attempt to move the limb and, indeed, efferent neural activity can be recorded in the motor nerves proximal to the blocked region (see below) in association with such attempts. Nevertheless, there is no sensation of movement or of altered position of the part to which that neural activity was directed. Even when there are peripheral sensory inputs compatible with success of the attempted movement, no movement is perceived (McCloskey and Torda, 1975). These observations provide strong evidence against the view that corollary motor discharges can provide sensations of limb movement.

In apparent conflict with this conclusion is the observation that amputees frequently report that a phantom limb moves in response to motor commands despatched to it. Here, only apparent movements which involve alteration of the relative positions of parts within the phantom are relevant to the question of corollary discharges and movement (Goodwin et al., 1972; McCloskey, 1978, 1980), and such movements have been found to occur only when twitching occurs in the muscles of the stump in response to attempts to move (Henderson and Smyth, 1948; Melzack and Bromage, 1973). Clearly, therefore, the apparent movements may be perceived through afferent signals from the stump, and corollary discharges need not be postulated to account for them.

Another apparently conflicting observation is the ability of experimental animals to relearn movements after transection of the dorsal spinal roots (Taub and Berman, 1968; Bossom, 1974). Without peripheral afferent signals relating to those movements, such animals must base their learning on other signals, and corollary discharges must be considered. Until reliance on visual and other signals (including, possibly, afferent signals traveling in the ventral spinal roots; Coggeshall, 1980) is shown to be wholly inadequate for performance, corollary discharges need not be regarded as necessary here, however. In any case, an involvement of corollary discharges need not be as providers of sensations of movement, a role for which the other evidence cited above makes them most unlikely. Perception of exerted, static muscular force or effort is more likely to depend on corollary discharges (see below), and it may well be that retraining of movements after deafferentation depends on association of perceived muscular efforts (via corollary discharges) and unfelt, but visually observed or otherwise rewarded, movements that ensue.

2. SENSATIONS OF EXERTED MUSCULAR FORCE OR EFFORT

(a) Qualitative Aspects

It is a common experience that muscular fatigue is associated with perception of an increased effort to produce a given tension. As the muscular tensions and cutaneous pressures involved in an effort can be the same in a fatigued state as in normal circumstances, it seems unlikely that the altered perception is based on the discharges of receptors signaling tensions or pressures. (The specific possibility of a contribution from muscle spindle afferents is discussed below.) However, the fatigued muscles are weak, and must require larger-than-normal, centrally generated commands to drive them to produce any given tension. This raises the possibility that the altered perceptions are based on corollaries of these increased motor commands. If so, the prediction must follow that sensations of increased effort or exerted muscular force will accompany conditions in which increased centrally generated motor commands are necessary to achieve any given muscular performance. This prediction has been thoroughly tested.

Objective measurement of perceived muscular effort can be made through various matching tasks. Thus, for example, a subject can be required to achieve a given tension by contracting a muscle group isometrically against a strain gauge so as to align a visual signal of achieved tension with a preset target. By arranging for the corresponding muscle group on the opposite side of the body to contract against a similar strain gauge, but without visual feedback, an objective indicator can be arranged. The subject is asked to make the required contraction on the reference side (with visual feedback), and to match it with a similar contraction of the muscle group on the opposite side of the body (Gandevia and McCloskey, 1977a,b). A particular aspect of the reference contraction might be nominated by the experimenter for matching. For example, achieved tension can be matched by most subjects when it is specifically nominated, and this is so even when the relation between command and achieved tension is experimentally disturbed (McCloskey et al., 1974; Roland, 1973). If tension is not specified, but instead the subject is left to decide for himself about the aspect of the reference contraction to be matched (as by instructing him only "to make both contractions the same"), the behavior usually elicited is for the subject to appear to be guided by the efforts made on both sides. This, too, is so even when the relation between command and achieved tension is experimentally disturbed. Thus, for example, if the contracting muscles on the reference side are weakened by fatigue (from a prior period of weight-bearing), or by local infusion of a neuromuscular paralyzant, the target tension is achieved only by an unusually large motor command and effort. That this is perceived and influences the matching process is indicated by a larger-than-normal matching contraction made with the unaffected indicator muscles on the opposite side (McCloskey et al., 1974; Gandevia and McCloskey, 1977b, c). An example of such behavior is shown in Fig. 1.

Many subjects find it difficult to align isometric tensions, so a more familiar form of contraction has been used in many otherwise similar tests. In these tests the subjects were asked to lift a weight by contracting a particular muscle group on the reference side and to match it by selecting a weight on the indicator side such that both were made "to feel the same" (Gandevia and McCloskey, 1977a–c; Gandevia et al., 1980). With the instruction framed in this way the subjects were left to decide for themselves which aspect of the reference contraction was to be matched. Again, the usual behavior is for the subject to appear to be guided by the effort made on the reference side in choosing the appropriate match.

It should be noted that subjects are not requested to match their subjective efforts in these tasks. Nevertheless, muscular effort does seem to guide the subjects' choices. When there is a specific request to match effort, subjects can do so (Roland, 1973; and see section 2d).

When muscle groups are used in which the opposite side of the body cannot be used as an indicator, subjective estimates of the magnitude of a contraction can be used. Thus, for contractions of the respiratory muscles, subjects can be asked to estimate

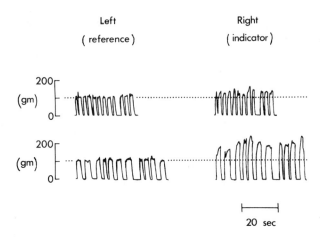

FIG. 1. Records taken from a subject matching isometric contractions of the extensors of the index fingers as a control (**top**) and during partial curarization of the left forearm and hand (**bottom**). The panels on the *left* show the reference tension for the left finger (100 g: displayed to the subject together with a target) on repeated contractions. The panels on the *right* show the tensions exerted by the right finger (no visual feedback). These were the subject's attempts to make contractions that were subjectively the same as those made on the reference side. During partial curarization of the reference side, the subject chose larger tensions than normal to match the same reference tension.

the magnitude of various resistive loads to breathing. After partial curarization, when the efforts required to breathe against any given resistance are higher, higher than usual subjective estimates are made of those resistances, even when air flow is reduced (Campbell et al., 1980; see Fig. 2). Again the behavior is consistent with subject being guided by the size of the motor command, or effort, involved (Gandevia et al., 1980).

In all experiments of these types subjects behave as if their judgments were based, at least in part, on the size of their centrally generated voluntary motor commands rather than on sensory signals related to the actual tensions and pressures achieved. Many conditions have been studied and the results are summarized diagrammatically in Fig. 3. In the first category (second panel, Fig. 3) are conditions in which the muscular response to a given motoneuron (MN) output is reduced. The conditions tested were fatigue and partial curarization (McCloskey, 1973; Gandevia and McCloskey, 1977b,c). In these, the diminished muscular response required an increased spinal MN outflow, which, in turn, was achieved through increased centrally generated motor command. In the second category (third panel, Fig. 3) were conditions in which MN excitability was depressed so that a normal spinal MN outflow was achieved only in response to increased centrally generated commands. Here the conditions studied have been cerebellar hypotonia (Holmes, 1917; Angel, 1980) and MN inhibition induced through excitation by vibration of muscle spindles in the antagonist of the contracting muscle (McCloskey et al., 1974). Again, given contractions were demonstrated to involve increases in perceived effort, or increases in the preceived heaviness of objects they lifted. In the final category were conditions in which muscular weakness resulted from intracranial motor lesions such as simple motor strokes (Gandevia and McCloskey, 1977c). In these objective observations were again made of increases in perceived effort or of increased apparent heaviness of lifted objects. This occurred even when the lesion produced only motor deficits, with no clinically detectable loss of peripheral sensation. In

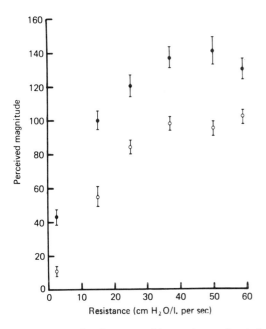

FIG. 2. Results from a subject who estimated the magnitude of added inspiratory resistive loads. Mean (±SEM) of 10 estimates of each resistance is plotted against the added resistance during the control run *(open circles)* and during partial curarization *(closed circles)*. At all levels of resistive load, perceived magnitude during curarization exceeded control estimates. (From Campbell et al., 1980, with permission.)

such conditions, the weakness results from damage to corticofugal pathways and possibly also to cortical structures, so that presumably, increased centrally generated traffic is required along unaffected pathways to achieve a normal input to spinal MNs. It is of interest that when a motor stroke is so severe that complete paralysis rather than weakness occurs, some patients feel no effort accompanying attempts to move (Mach, 1886; Gandevia, 1982). It has been suggested that this might be because the sensation of effort is mediated by collateral signals which diverge from the descending command signals at a point below the lesion (McCloskey, 1978). When the lesion is incomplete, the compensatory increase in traffic in the unaffected pathways is sampled below the lesion and leads to consciousness of increased effort. When the pathways are completely interrupted, no signals would exist to provide a sensation of effort. Study of localized intracranial lesions may provide opportunities to test this hypothesis.

It might be argued that corollary discharges have been taken too readily to be involved in the various phenomena described above and that signals from peripheral sensory receptors could still be responsible. Such an argument would now require specific proposal of the receptor type or types to be considered as alternatives and an explanation of how their discharges could account for the various phenomena summarized in Fig. 3. One such specific proposal, the muscle spindle, has been made by Granit (1972). He has pointed out that, because of α-γ linkage, "a spindle component in excitation is included in the expectations related to the accomplishment of motor acts commanded by the will," and that, "the periphery itself is 'corollarized' by α-γ linkage." Although it seems likely that increases in spindle discharge would accompany increases in motor command in muscle fatigue and also possibly during partial curarization (panel 2, Fig. 2), it seems quite unlikely that spindle firing and centrally generated command would run parallel in the other conditions in which there is a sense of increased effort, or of apparently increased heaviness of a lifted object (panels 3, 4, Fig. 2). In any case, Granit's proposal has been tested directly in experiments using high-frequency (100 Hz) vibration of muscle. Vibration applied over the tendon of a muscle powerfully excites the primary endings of the muscle spindles and also significantly increases the firing of spindle secondaries and Golgi tendon organs (Burke et al., 1976a,b, 1980b). An involuntary reflex contraction of the vibrated muscle occurs and has been attributed to excitation of the spindle endings (DeGail et al., 1966; Hagbarth and Eklund, 1966). Hagbarth and Eklund (1966) were the first to observe that during vibration "a sub-

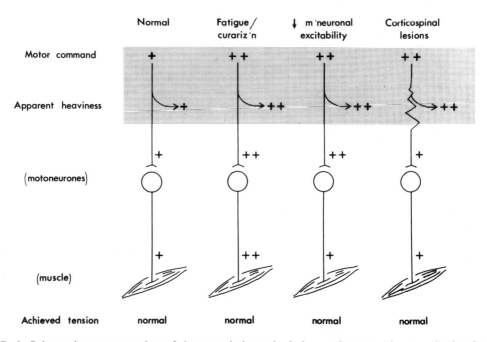

FIG. 3. Schematic representation of the association *(shaded area)* between the magnitude of centrally generated motor command and either the perceived muscular force exerted or the apparent heaviness of a lifted object. Each panel depicts a descending command to spinal MNs, the MN output to muscle, and the muscular tension achieved. An estimate of the relative magnitude of signals at each level is given by plus signs. **First panel:** Normal contraction; **second panel:** changes in the various signals when muscle responses are depressed, as by fatigue or partial curarization; **third panel:** changes when MN excitability is depressed, as by cerebellar lesions or by muscle-spindle-based inhibition evoked by vibration of the antagonist; **fourth panel:** changes when corticofugal pathways are partially interrupted by intracranial lesions, such as motor strokes.

ject gets a feeling of relief or lessening of tension," not a feeling of heaviness or increased effort, as Granit's proposal of a spindle origin of these sensations would predict. Indeed, it is the reduction in centrally generated command made possible by the involuntary, vibration-induced contraction, that seems responsible for the subjectively reduced effort, and for this effect when demonstrated in objective matching procedures (McCloskey et al., 1974).

(b) Heaviness

One cannot judge the weight of an object if a muscular effort cannot move or support it. This must be so whether judgment about weights is usually based on the discharges of peripheral sensory receptors or on corollary discharges. As one attempts to lift a heavy object, centrally generated commands increase through the range of the effort, and peripheral sensory signals related to the effort also vary through this range. Unless one knows which of the range of commands or sensory signals is associated with lifting or supporting the object, one cannot say how heavy an object is—only that it is heavier than some other object that can be lifted or supported. No can one discriminate between the heaviness of two objects when neither one of them can be lifted or supported.

Simple experiments have shown that for estimating weights it is sufficient for the nervous system to be provided with a quite crude signal to indicate which command in a range of motor commands is sufficient to move a given weight (Gandevia and McCloskey, 1978). Figure 4 summarizes these experiments, which were done as follows. Subjects flexed the distal joint of the thumb so as to press down on one end of a lever, at the other end of which variable weights were attached. Objective indications of perceived heaviness were obtained by the subject choosing weights of apparently equal heaviness with similar contractions of the thumb on the opposite side of the body. The thumbs on both sides were anesthetized by local injection of lignocaine at their bases so as to reduce peripheral afferent input. The lever arrangement on the reference side was then altered (see Fig. 4) so that no weight was actually lifted or supported at all. Instead, the thumb tip pressed isometrically against a strain gauge, and the tension signal from the gauge triggered a large electromagnetic vibrator to lift the lever away rapidly as soon as a certain force was attained. For the subject, it was as if the weight suddenly "disappeared" from under the thumb at the instant he would ordinarily have lifted it. In these circumstances, however, he did not accelerate the weight from its stop nor support it. Nevertheless, subjects matched weights "lifted" in this way with conventionally lifted weights on the opposite side of the same magnitude as those chosen when both sides lifted conventionally. The CNS could use the very crude signals associated with the mechanically unloaded contraction as some kind of event marker to identify the appropriate motor command in the range of commands employed. That it was the level of motor command so identified, rather than the absolute intramuscular tension, was indicated by the increased matching weights chosen when the thumb flexor on the reference side was fatigued by a prior period of weight bearing.

(c) Influences from Peripheral Sensory Fields on Perceived Heaviness

On the basis of the observations and experiments outlined above, the hypothesis can be advanced that corollary discharges contribute to the sense of muscular force or effort, and through this, to the sense of the heaviness of lifted objects. The hypothesis fits the many observations summarized in Fig. 3, and has the advantage that there are not satisfactory alternative hypotheses based on sensations coming from peripheral sensory receptors.

If sensations of heaviness depend on the magnitude of centrally generated motor commands, then objective observations of perceived heaviness can be used to study changes in such motor commands in various circumstances. This reasoning was the basis of a series of experiments (Gandevia and McCloskey, 1976, 1977a; Gandevia et al., 1980) in which heaviness was studied in conditions of altered peripheral sensation. For objects lifted by flexion of the terminal joint of the thumb, it was found that apparent heaviness was increased when the skin and joint of the thumb were anesthetized by digital nerve block (Fig. 5). Moreover, anesthesia of the adjacent index finger instead of the thumb had a similar, but smaller, effect (Gandevia and McCloskey, 1977a). For weights lifted by flexion of the elbow (Gandevia and McCloskey, 1976), anesthesia of the hand was similarly associated with increases in the apparent heaviness of lifted objects. Thus, it seemed that the motor commands delivered to perform a particular movement usually receive facilitation from a quite wide peripheral sensory field associated with that movement, so that removal of the tonic sensory input requires compensatory increases in centrally generated commands. This interpretation receives further support from the finding that electrical stimulation of the index finger causes reductions in apparent heaviness of

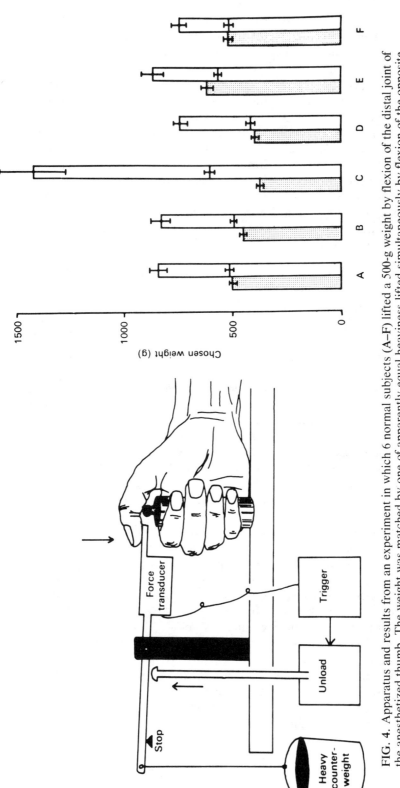

FIG. 4. Apparatus and results from an experiment in which 6 normal subjects (A–F) lifted a 500-g weight by flexion of the distal joint of the anesthetized thumb. The weight was matched by one of apparently equal heaviness lifted simultaneously by flexion of the opposite thumb (also anesthetized). In a control series both thumbs lifted the weights by depressing one end of the "see-saw" at the opposite end of which the weights were suspended. The mean weights (± SEM, N = 10) chosen by each subject to match 500 g are shown by the *shaded bars*. When, instead of lifting 500 g on the reference side, the subjects had only to exert an isometric tension of 500 g before triggering an electromechanical device, which rapidly carried the lever away from under the thumb (see apparatus, at left), further matches were made. In these, the thumb on the reference side did not lift or support the 500 g once the 500 g tension was achieved, although the thumb on the opposite (matching) side made its lifts as in the control series. The lower segments of the *unshaded bars* show the weights chosen. When the thumb flexor on the reference side was first fatigued by a period of weight bearing, the matching weights chosen were those shown by the full unshaded segments. Although the group of subjects showed no consistent change from the control series in matching weights chosen during rapid unloading on the reference side, fatiguing the lifting muscle significantly increased the matching weights in all subjects. See text for discussion. (From Gandevia and McCloskey, 1978, with permission.)

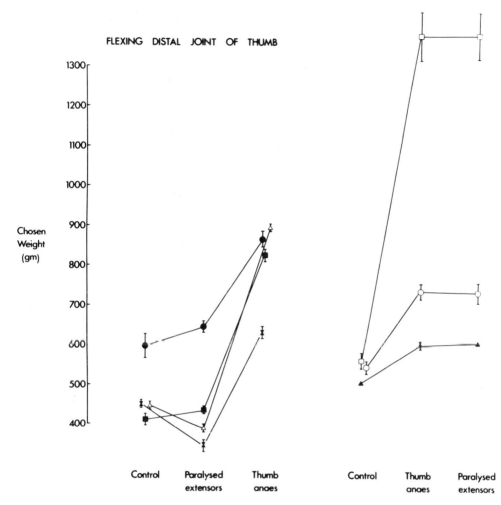

FIG. 5. Results from 7 subjects estimating the weight of 500 g lifted by the flexor of the thumb. The mean estimate (±SEM, 10 trials) made by each subject in different experimental conditions is shown. Data from the same subject are indicated with a distinct symbol, and are joined by lines. *Left:* Results obtained in 4 subjects when the extensor of the thumb was first paralyzed by radial nerve block and then, while extensor paralysis persisted, the thumb itself was anesthetized. Extensor paralysis caused no systematic alteration in heaviness, but subsequent anesthetization of the thumb increased heaviness. *Right:* Results obtained in 3 subjects when the thumb was first anesthetized and then, while digital anesthesia persisted, the extensor of the thumb was paralyzed. Anesthetization of the thumb increased heaviness, and no further alteration in heaviness occurred following extensor paralysis. (From Gandevia et al., 1980, with permission).

objects lifted by thumb flexion. However, for thumb extension, the role of the inputs from peripheral sensory fields is different, as anesthesia of the thumb causes reductions in the apparent heaviness of objects lifted by extending the thumb (Gandevia and McCloskey, 1977a).

The experiments just described were undertaken on the basis of reports (Marsden et al., 1978b) that anesthesia of the joint and skin of the thumb abolished the long-latency response to stretch of the flexor of the terminal joint of the thumb. If this response usually contributes a substantial element of reflex assistance to a muscular contraction, it was reasoned that removing it by anesthetizing the thumb would require centrally generated motor commands to increase so as to compensate for the loss, and the apparent heaviness of objects lifted by flexion of the anesthetic thumb would thereby increase. The results agreed with this logic, and were initially interpreted in these terms (Gandevia and McCloskey, 1977a). However, Marsden et al. (1979a) subsequently contradicted the initial reports of Marsden et al. (1977d) on the effects of anesthesia, finding that anesthetization of the thumb did not abolish the long-latency response, but only reduced it by an average of 30%. They pointed out that, for the very large changes in perceived heaviness seen during digital anesthesia to be due to compensation for such relatively small reductions in reflex assistance to movement, the long-latency reflex would normally have to contribute an exceptionally large and unlikely element of excitation. The initial and subsequent findings of Marsden and co-workers were obtained in subjects making rapid, visually guided isotonic contractions. For similar stretches imposed during simple isometric holding contractions, we find no effects of digital anesthesia on long-latency responses. During visually guided isometric force tracking, however, long-latency reflexes are reduced by digital anesthesia, and lose the modifiability by volitional set which exists for isometric holding contractions (C.K.C. Loo, D.I. McCloskey, and E.K. Potter, *unpublished observation*). It seems that the behavior of long-latency responses depends on the category of muscular contraction in which they are observed, so it would be unwise to come to conclusions about their contributions in simple lifting performances on the basis of their behavior in quite different conditions. Nevertheless, it now seems unlikely that changes in long-latency reflexes will account for the changes in preceived heaviness caused by altering sensory inputs from related sensory fields. Thus, the facilitatory and inhibitory effects previously thought to have acted through these reflexes must now be taken to be acting elsewhere to require the compensatory changes in motor commands reflected in the alterations in perceived heaviness.

A specific objection to the earlier findings in this series was raised by Marsden et al. (1979a), who suggested that digital anesthesia is accompanied by co-contraction of the antagonist of the thumb flexor, thus requiring a larger than usual flexor contraction to lift a given weight. Such an effect, they argued, meant that in the conditions of apparently increased heaviness there was, in fact, an increased intramuscular tension in the thumb flexor, and so corollary discharges need not have been held responsible. This objection cannot be sustained in the face of the subsequent finding (Gandevia et al., 1980) that the effects of digital anesthesia on heaviness persist even when the antagonist is selectively paralyzed by injection of local anesthetic around its motor nerve (Fig. 5).

(d) Quantitative Aspects and Problems

The case in favor of a contribution of corollary discharges to sensations of exerted muscular force, or heaviness, is based largely on qualitative considerations. Quantitative aspects of these sensations are, however, more complex. Changes in perceived force or heaviness usually do not bear a simple inverse relation to the changes in muscular strength which they accompany. Indeed, such quantitative considerations have been the basis of arguments against interpretations involving corollary discharges (Roland, 1978). During partial neuromuscular block, for example, perceived force and heaviness increase, but rarely exceed 2 to 4 times their control levels, even when tension voluntarily attainable is reduced to 10 to 20% (see Fig. 6). The situation appears to be similar in fatigue (Fig. 6) and after motor strokes. When the excitability of relevant motor pathways is altered, however, as with vibration-induced changes in muscle-spindle activity (McCloskey et al., 1974) or with altered inputs from related sensory fields (Gandevia and McCloskey, 1976, 1977a) maximal voluntary strength is rarely affected although forces exerted in submaximal contractions are perceived to be unusually high.

It is possible that the relation between motor command and exerted tension is not a linear one, so that some of the apparent discrepancies of proportionality between command and perceived force may reflect nonlinearities in this relation. However, in the specific case of the hindlimb muscles of the cat where the "command" signal to MNs from various spinal afferents was altered, linearity of the relation appeared to be a good first approximation (Henneman et al., 1965a). Of course, this need not be true for responses of human muscles to commands generated centrally. Furthermore, a changing level of reflex assistance to the voluntary motor command may occur at different output levels thus contributing to any nonlinearities. For example, Marsden et al. (1976a) described changes in the gain of load-compensating reflexes in the human thumb and related the gain to "the effort made by the subject rather than the actual pressure exerted by the thumb." Motor activity evoked through such vari-

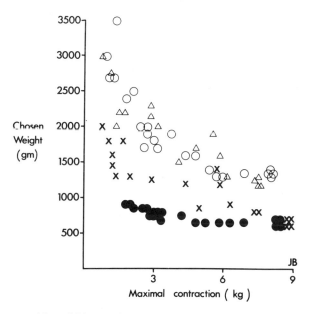

FIG. 6. Results from a subject lifting weights by flexion of the distal joints of the thumbs. The weights chosen on one side to match a reference weight of 500 g lifted on the other are shown. *Circles:* Results obtained during weakness induced by partial curarization of the forearm and hand on the reference side. *Closed circles:* Lifts when the reference thumb was unanesthetized. *Open circles:* Lifts when thumb was locally anesthetized by digital nerve block. *Crosses:* Results obtained when the thumb flexors on the reference side were fatigued, and hence weakened, by prolonged weight bearing (thumb unanesthetized) and by *open triangles* (thumb anesthetized). In all cases the perceived heaviness of the referenced 500 g (i.e., the weight chosen to match it), is plotted against the maximal voluntary contractile strength of the thumb flexor on the reference side in the prevailing experimental conditions. Full discussion in text.

able reflex mechanisms might contribute significantly to achieved force but not, as voluntarily generated activity does, to the perception of force.

Figure 6 illustrates results from an experiment in which some of these difficulties are apparent. A subject lifted a weight of 500 g by flexion of the thumb and gave an objective indication of its apparent heaviness by choosing a weight of apparently equal heaviness lifted by thumb flexion on the other side. When the reference thumb was weakened by either partial curarization (filled circles), or fatigue induced by prior weight-bearing (crosses), the apparent heaviness of the 500-g weight increased. Even when the maximal contractile strength of the muscle was only 1 to 2 kg, compared to its normal maximum of 9 kg, apparent heaviness was increased only about three to fourfold for fatigue, and less than twofold for curarization. At any level of muscular weakness, fatigue had a more pronounced effect on apparent heaviness than curarization (this may be partly due to selective decurarization in those motor units performing repeated contractions; see Gandevia and McCloskey, 1977b). When the experiments on fatigue and curarization were performed again, this time after locally anesthetizing the thumb on the reference side (open circles and triangles, respectively), the changes in perceived heaviness were more marked. Although digital anesthesia increased the apparent heaviness of the 500 g by only ~ 2.5 times when muscular strength was normal, it increased it by a greater proportion when muscular weakness meant the 500 g reference represented almost a maximal effort. Clearly, the effects of digital anesthesia and increased motor output were more marked when the two occurred together (Gandevia and McCloskey, 1977b). Nevertheless, lifting the reference weight with a muscle made so weak that a near-maximal contraction was required, whether with or without digital anesthesia, did not cause the apparent weight of the lifted object to approach the maximal contractile strength of the muscle in its unweakened state (9 kg). The subject shown in Fig. 6 (typical of the subjects studied in these experiments) chose only 3 to 3.5 kg as the apparent weight lifted when making near-maximal contractions with fatigued or curarized muscles during digital anesthesia.

These discrepancies are highlighted here to indicate that it is unlikely that perceived muscular force or apparent heaviness are signaled solely by a simple linear transformation of motor command through corollary discharges. However, an internal signal that is linearly related to the command may well be available, as Roland (1978) has shown that subjects can match their calculated efforts with good, quantitative precision when specifically instructed to do so. The results discussed here have shown that subjects probably choose neither the effort alone nor the tensions or pressures produced alone when making judgments about exerted muscular force or apparent heaviness. The same results do show, however, that the magnitude of the centrally generated command contributes strongly to these judgments, but other factors must be considered. Possibly, a signal of intramuscular tension could be used to scale a corollary discharge according to the size and strength of the muscle contracting, or possibly such a signal could interact with a corollary discharge to modulate its sensory effects. Mechanisms such as these deserve further study.

3. COROLLARY DISCHARGES AND KINESTHETIC AFFERENT INPUTS

For some years it was conventional physiological opinion that the discharges from intramuscular mechanoreceptors do not have access to consciousness. Proprioceptive sensations were said to depend on the activity of receptors in the joint capsules and ligaments. Later studies forced a revision of this view and it is now clear that muscle receptors contribute to kinesthetic sensibility (see Goodwin et al., 1972).

One argument against muscle spindles making any contribution to kinesthesia was that these receptors do not give a discharge which unambiguously signals muscle length or rate of change of muscle length. This is because muscle spindles fire not only in response to passive stretching of the whole muscle in which they lie, but also in response to contraction of their own striated muscular ends caused by discharge of fusimotor (γ-efferent) fibers. There is now, however, considerable evidence that muscle spindles do contribute to kinesthesia, and it seems likely that corollary discharges provide the mechanism by which a useful kinesthetic signal is extracted from their total discharge. The proposal of such a role for corollary discharges comes from experiments on vibration and loading of muscles in man.

When high frequency vibration is applied through the skin over a muscle or its tendon, the vibrated muscle contracts involuntarily in a tonic vibration reflex which is usually attributed to excitation of the primary endings of muscle spindles (DeGail et al., 1966; Hagbarth and Eklund, 1966). The subject whose muscle is vibrated experiences an illusion of movement at the joint about which the vibrated muscle operates (Goodwin et al., 1972), and this illusory movement occurs in the direction that would normally stretch the vibrated muscle. Illusory movements occur in opposite directions when vibration is applied to agonists and to antagonists, but none are experienced when the vibration is applied over the joint. This provides evidence that muscle receptors rather than joint receptors give the signals that cause the illusions.

Even when all the joints and skin of the hand are anesthetized, vibrations of the long flexor tendons within the anesthetized hand causes illusory sensations of extension of the fingers and thumb. Thus, although there is every likelihood that paciniform corpuscles and other mechanoreceptors within the skin and joints of the region normally would be excited by vibration, the kinesthetic illusions depend not on these but on the excitation of receptors within the muscles. Discharges seen in multifiber recordings from joint nerves during vibration (Millar, 1973) are likely to be due to the excitation of paciniform endings and, in any case, are not relevant to kinesthetic illusions. The vibration-induced illusions are predominantly illusions of movement—of velocity of joint rotation (Goodwin et al., 1972; McCloskey, 1973) and continue for as long as vibration is applied.

It seems virtually certain that the spindle primaries are involved because of their high sensitivity to vibration (Brown et al., 1967b; Burke et al., 1976a,b) and because of the appropriateness of illusory movement as a sensation arising from a receptor-type normally more sensitive to dynamic than to static stimuli. Nevertheless, considerable excitation of spindle secondaries and Golgi tendon organs is also caused by muscle vibration in man (Burke et al., 1976a,b; 1980a).

If activity of muscle spindles is responsible for vibration-induced kinesthetic illusions, why do not similar illusions occur when the same spindles fire in response to the activity of fusimotor fibers? It is known that fusimotor activity increases with increasing strength of isometric voluntary contraction, so that the level of fusimotor-induced spindle activity also increases (e.g., Vallbo et al., 1979; Prochazka and Hulliger, *this volume*); so why are there no illusions? An answer to these questions could be that corollary discharges are used within the nervous system to distinguish the spindle firing that is appropriate for any level of voluntary isometric contraction, from that which is inappropriate and therefore of kinesthetic significance.

One simple way for this to occur would be for the corollary discharges to act as an efference copy and for the fusimotor-induced activity of spindles to act as reafference (Goodwin et al., 1972). Figure 7 shows this proposal diagrammatically. A signal (A) related to the input to fusimotor neurons (or it could, theoretically, be a collateral of the fusimotor axons themselves) acts synaptically at a cell (S) to offset the fusimotor-induced spindle activity acting on the same cell through input B. Only spindle activity in excess of that induced through the fusimotor system would pass through cell S, and so would constitute a kinesthetic signal (or exafference, in von Holst's terminology).

The location of a station for such subtractive interaction (cell S) within the CNS is not known. However, there are experimental observations that strongly suggest the existence of such a system, or of a system somehow able to evaluate the kinesthetic significance of spindle inputs. Muscle vibration probably causes an increase of spindle firing from its preexisting level to some constant level—in the simplest case, by 1:1

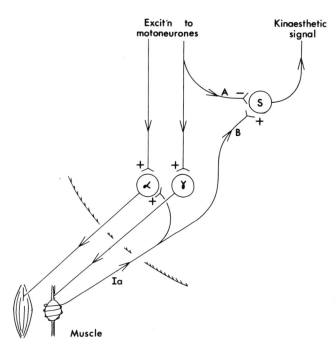

FIG. 7. Model showing how a kinesthetic signal might be extracted from the discharge of muscle spindle afferents. The diagram shows descending pathways activating spinal α- and γ-MNs. Signals, shown here in pathway **A**, and related to the input to the γ-MN, provide an inhibitory input to cell S. The afferent pathway from the muscle spindles (**Ia**), as well as providing excitation to α-MNs, is shown to provide an excitatory input to cell S along pathway **B**. Subtractive interaction between **A** and **B** at cell **S** could extract a signal of kinesthetic significance from the spindle input. In the terminology of von Holst (1954), pathway A carries an efference copy which cancels from the total afference in pathway B, the reafference due to fusimotor activation: A kinesthetic signal (exafference) results.

entrainment, to the vibration frequency (see Matthews, 1972). It is, therefore, expected that the kinesthetically significant proportion of vibration-evoked spindle firing would decrease as the force of muscular contraction (and, hence, fusimotor-induced spindle firing) increases. Such a process was proposed by Goodwin et al. (1972) to explain the failure of vibration to cause kinesthetic illusions when applied to a strongly contracting muscle. A similar argument applies to the findings illustrated in Fig. 8. When the velocity of a vibration-induced illusory movement is observed, it is found to decrease as the load borne by the vibrated muscle increases (McCloskey, 1973). This would be expected if only part of the spindle discharge—that part in excess of the level appropriate or expected for the prevailing contraction—were perceived. In a very strongly contracting muscle, vibration would be expected to add little or nothing to the prevailing level of spindle discharge, and so no illusions would occur. Results such as these (Goodwin et al., 1972; McCloskey, 1973) therefore indicate that some central process sorts and transmits only the kinesthetically significant portion of receptor firing.

4. SUBJECTIVE TIMING OF THE ONSET OF VOLUNTARY MUSCULAR CONTRACTIONS

With sensory inputs from muscles, joints, and skin available to signal positions

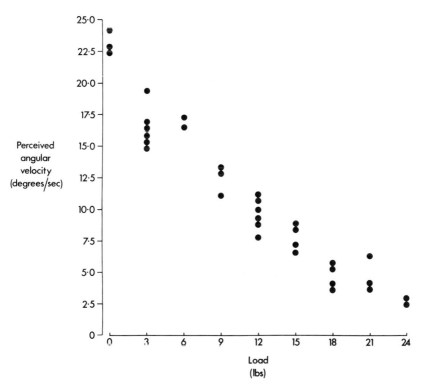

FIG. 8. Angular velocity of the illusory movement of elbow extension induced in a subject by vibration of biceps brachii at 100 Hz, is plotted against the load at the wrist bone by tensing the vibrated biceps. Biceps carried the loads only briefly and was not fatigued. These results fit the model presented in Fig. 7.

and movement, and a sense of effort or muscular force as well as these for signaling force or heaviness, there are clearly many sources of information on which one might rely when judging the timing of various muscular contractions.

As a simple first step in studying the perceived timing of muscular contractions, normal subjects were asked when they perceived an actual movement to have begun. To do this, the subjects were required to make a rapid voluntary movement with a specified muscle group at any instant they chose during a period (~1 sec) while a light was shown. At a variable time within the same period a brief electric shock, adjusted in intensity to feel like a sharp tap, was applied to the subject's ankle. It was found that EMG activity had to commence in the specified muscle well before the reference shock was given for the perceived movement and the shock to seem to coincide (Fig. 9). The same subjects were then asked to carry out further similar trials, but this time to attempt to disregard their actual movements, and to base their judgments on the moments at which their commands to move were issued. No strategies were suggested for this. Despite an initial attitude of dismay at even attempting the task, almost all subjects performed accurately, and similarly, and now usually found that the reference stimulus had to precede EMG activity by about 50 msec for command to move and the reference stimulus to appear to coincide (Fig. 9). Similar relative timing was chosen when the specified muscles were completely paralyzed, so that no actual movement occurred in response to a given command. In these experiments the efferent volleys in the muscle nerve were recorded instead of the EMG which, of course, was silent (D. Burke, J.G. Colebacth, D.I. McCloskey, and E.K. Potter, *unpublished observation*). Such observations indicate that we know the moment at which we tell a part to move separately from the moment at which we feel the actual movement that ensues.

These studies have been extended to include observations on two separate muscles in individual subjects, usually anatomically

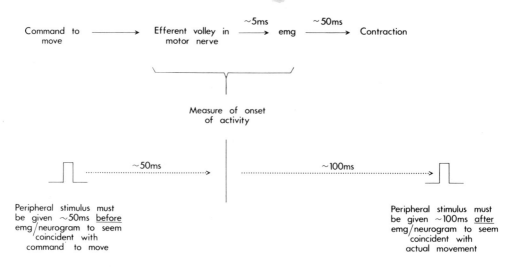

FIG. 9. This diagram shows the temporal relationships between the measured onset of motor activity (the EMG or, less frequently, the efferent volley in a motor nerve) and the placement of a peripheral reference stimulus which seems to coincide with either a command to move, or with an actual movement.

remote muscles in the jaw and the foot. The moments of felt movement and of command-to-move were measured against the same reference stimulus for each muscle. The subject were then asked to attempt to make simultaneous movements with the two muscles. Their judgments about apparent simultaneity were nearly always those expected from their aligning commands-to-move rather than actual felt movements.

Such a preference for command-related signals may be wise. For rapid voluntary movements of the type studied, sensory feedback from the moving part, no matter how accurate, can give information only about the past, and so must have severely limited value to the moving subject.

ACKNOWLEDGEMENTS

The research reported has been supported by the National Health and Medical Research Council of Australia.

"Muscular Wisdom" that Minimizes Fatigue during Prolonged Effort in man: Peak Rates of Motoneuron Discharge and Slowing of Discharge During Fatigue

C. D. Marsden, J. C. Meadows, and *P. A. Merton

Physiological Laboratory, University of Cambridge, Cambridge CB2 3EG, United Kingdom

Years ago, Naess and Storm-Mathisen (1955) reached the conclusion that electrically excited tetanic contractions fatigue human muscle more rapidly than voluntary contractions. At the Cambridge meeting of the Physiological Society in September 1968, under the title Muscular Wisdom, we demonstrated this fact and went on to show that an electrical tetanus is better maintained, almost as well maintained as a voluntary contraction, if the frequency of stimulation is progressively reduced from 60 to 20/sec over the course of the first minute—a procedure we called artificial wisdom (Marsden et al., 1969). These experiments and others that arose from them are described in the first part of this chapter. The records of Fig. 3 were those used to illustrate the demonstration. These findings have been confirmed and extended by Jones et al. (1979). Subsequently, an anatomic peculiarity (the Martin-Gruber anastomosis between ulnar and median nerves in the forearm) in two of the authors allowed to record from single motor units throughout maximal voluntary efforts starting abruptly and carried to fatigue (Marsden et al., 1971a). During a maximal contraction, as will be described, the frequency of firing of single units drops, matching the frequencies we had found gave the most judicious artificial wisdom in the earlier experiments. For the early part of a contraction these results were independently confirmed by an antidromic collision technique.

METHODS

Most of the experiments were done on the adductor of the left thumb. The hand was held in the splint illustrated in Fig. 1. There was a similar splint for the right hand, which was used in some of the single unit experiments. To start with, tension was recorded by a myograph employing an RCA 5734 mechanoelectric transducer valve (Merton, 1954, Fig. 1). In later experiments the strain gauge shown in Fig. 1 of this paper was used. It employed two silicon gauges (Ether Ltd, Type 3A-1A-350P) in a bridge. In both cases the transducer output was amplified by direct-coupled amplifiers. The transient response of the later pattern strain gauge was measured in two ways and found to be more than adequate. First, a 5-kg weight was hung from the thumb band by a piece of string. Records taken while

---

*To whom correspondence should be addressed: Physiological Laboratory, University of Cambridge, Downing Street, Cambridge CB2 3EG, United Kingdom.

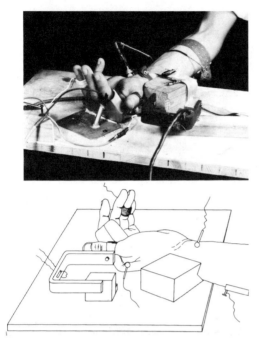

FIG. 1. The apparatus for recording electrical and mechanical responses from the adductor pollicis. The U-shaped member carries the strain gauges and receives the pull of the thumb via a brass band around the proximal phalanx. The recording electrodes are a lint-covered silver ring around the ring finger and a padded silver electrode over the muscle near the metacarpophalangeal joint of the thumb. The stimulating cathode is over the ulnar nerve at the wrist, with the indifferent anode plate higher up the forearm over the cutaneous border of the ulnar bone.

the string was suddenly cut through showed that the trace returned to the base line in 10 msec, afterward oscillating about the baseline at some 400/sec, the natural frequency of the U member. Second (to check the response during extension rather than shortening), the gauge was extended by a compliant force of 5 kg. A piece of string was then arranged round the ends of the U member as a Spanish windlass, and tightened until the output of the gauge returned to baseline. Cutting this string was equivalent to the sudden application of a 5-kg load. When this was done the rate of re-

sponse did not differ from that found by the first method.

Muscle temperature was measured with a homemade needle thermocouple (outside diameter 0.47 mm), which was inserted into the adductor pollicis from the dorsal side near the head of the first metacarpal, to a depth of about 1.5 cm from the skin surface. In order to standardize muscle temperature we always worked (unless otherwise stated) with a hot subject in a hot room. The subject had to be hot enough for the veins on the back of his hands to dilate; if necessary he put on outdoor clothes and ran up and down stairs. The hand to be used was then warmed in water at 44°C or so for a few minutes. Electric heaters raised the room temperature to 27°C or more. Under these conditions, as we have frequently observed, the temperature in the adductor pollicis is 35.5 to 36.5°C and remains at that level indefinitely, with the subject's hand under experiment in the splint. To work on cold muscle the room was cooled, if possible to 20°C or less, by turning off all heating and opening windows. The arm to be used was put into water at 15°C up to the elbow, for 5 min. This gives a temperature in the adductor of roughly 24°C, which changes only slowly provided the room temperature is low and the subject cool.

Action potentials were led off by surface electrodes, one behind the head of the first metacarpal near the insertion of the adductor pollicis, and one a strip of silver covered in lint wrapped round the ring finger. This latter gave much the same action potentials as an electrode on the palm over the origin of the adductor, and was more convenient. The leads are scarcely specific for the adductor pollicis, in particular the electrode at the base of the thumb will record from the first dorsal interosseous; but, as this muscle also adducts the thumb and is supplied by the ulnar nerve, this ought not to invalidate the records. In other experiments a fine bipolar needle electrode (outside diameter

0.45 mm) was thrust into the adductor either from the palmar or the dorsal surface of the hand, near the head of the first metacarpal. The outer shaft of the needle was earthed, a pair of wires inside, each 50 μm in diameter (34-gauge), forming the active electrodes.

When it was desired to limit voluntary contractions to the ulnar-supplied muscles acting on the thumb, the median nerve was blocked just above the elbow, usually with 1% lidocaine hydrochloride with norepinephrine 1:200,000 (Desmedt, 1973b). In the experiments on single units the ulnar nerve at the elbow was similarly blocked. As with all the experiments, the subjects were the authors. Two of the subjects, each of whom had two left ulnar blocks and one right ulnar block in the course of a week, noticed that their ulnar nerves remained tender to pressure and gave paresthesiae more readily than normal for several months. Five years later, however, there is no sign that they had suffered permanent damage.

In some experiments the median nerve was blocked by pressure. This proved unexpectedly easy, being completely successful at the first attempt. With the elbow fully extended to stretch the nerve, a stout rubber strap (made by cutting longitudinally into a strap for holding the plate electrodes of an electrocardiograph) was placed tightly around the elbow. The strap lay directly on the medial epicondyle and, behind it, on the olecranon, thereby avoiding pressure on the ulnar nerve in the groove between them. A discarded electroencephalograph (EEG) electrode under the strap in front was used to press on the median nerve where it can be palpated on the brachialis muscle, just medial to the brachial artery. This electrode also served to lift the strap clear of the brachial artery and the antecubital veins. The radial pulse remained at full volume with minimal congestion of the veins. The thenar muscles became paralyzed, and the cutaneous median distribution in the hand became anesthetic after about 40 min of pressure; however, even after 1 hr, there was still feeble movement in the distal phalanx of the thumb, moved by the long flexor in the forearm. This was virtually extinguished after 80 min in the key experiment. On releasing the strap, recovery was rapid and complete. When this type of block was used, the recording apparatus was raised to shoulder height so that the elbow was held extended. (In subject PAM it was noticed that full extension of the elbow with the arm in this position eventually caused median block without any externally applied pressure on the nerve; but this block dissolved after any transient accidental flexion.) Pressure block has the advantage over lidocaine block in that the hand can be cooled by immersion in water in the ordinary way. This is not possible after block with local anesthetic due to the vasodilatation that prevails. Block of the ulnar nerve by pressure is described in the results section.

The ulnar nerve was excited electrically through a cathode at the wrist, pressed onto the nerve by a tight rubber strap. In most experiments the indifferent anode was a plate 6 × 4 cm on the extensor surface of the upper forearm. The skin under the electrodes was not prepared in any way. Stimuli were delivered through an isolating transformer. In the earlier experiments they were condenser discharges with a time constant of about 50 μsec. In the later experiments they were 50 μsec square pulses from a dual high-voltage stimulator.

The above arrangement of electrodes (which has been used as standard for many years) probably gives as little pain as any other and is perfectly satisfactory except at the higher rates of tetanization. At frequencies much above 100/sec, high voltages are necessary to ensure that the stimuli remain supramaximal as the tetanus continues, and it was when we first tried such frequencies that we noticed a tendency for extensor muscles to contract under the anode. At the same time, the long extensor of the thumb could be felt to tighten up at the anatomic snuff box. This would clearly lead to an erroneous reduction in the recorded tension.

Considerable effort was expended in investigating this artifact. The relation between shock strength and tetanic tension was examined at various frequencies. Records were taken with 50-μsec stimuli increasing in strength in 10-V steps from 40 V to 100 or 120 V (the top steps being decidedly painful). In these tests tetani lasting 1 sec were given every 15 sec. At 100/sec, with a carefully placed cathode, the tension that developed in a tetanus would increase as the stimulus voltage increased from 40 to 60 or perhaps 70 V; it would then not increase any further as the stimulus was turned up to, say 100 V. This plateau in the response was taken to indicate that the response was supramaximal in that range. Above about 100 V the tension achieved in a tetanus might remain steady, but often it would either increase by a few percent (not often more than 5%) or decrease by a similar amount. This was attributed to a spread of excitation at these high voltages either to the muscles that adduct the wrist (which would tend to pull on the thumb and increase the recorded tension) or to the extensors of the thumb (which, as we have said, would reduce it). These effects appeared to be least when the anode near the elbow was centered over the subcutaneous border of the ulna with its proximal edge some 6 cm below the tip of the olecranon.

It should be emphasized that with a carefully placed cathode in a lean subject these artifacts only appeared at voltages above what would normally be used for stimulation at frequencies up to 100/sec; but for higher frequencies and for accurate comparisons at all frequencies it was believed important to eliminate them as far as possible. Moving the anode to various sites on the upper arm made matters worse; but a lint-covered sheet-lead anode some 7 cm by 4 cm wrapped around the lower end of the ulna behind and medial to the stimulating cathode was a distinct improvement, a very satisfactory plateaux being obtained up to 120 V, the highest voltage the subject chose to bear. With this wrist anode there was no visible or palpable contraction of the long muscles in the forearm, even at the highest voltage, and this remained true when the frequency of stimulation was raised, up to 1,000/sec.

The peak tension in a tetanus was less at frequencies of 250/sec and above than at the best frequency of about 150/sec (Desmedt, 1973*b*). At 500/sec the peak tension was 14% less than at 150/sec and was not well maintained, having fallen by about 20% at the end of a second. A technical point with these high frequencies is that tetani of slightly submaximal shocks paradoxically develop more peak tension. This tension is better maintained than in supramaximal tetani, presumably because the motor fibers do not respond to each shock; thus, the mean rate at which motor impulses reach the muscle is nearer to the best frequency than it is with true supramaximal stimulation. In cooled muscle these effects are seen at lower frequencies of stimulation. In muscle at 24°C a 50/sec tetanus is larger than a 100/sec tetanus.

The few experiments on the abductor of the little finger were done with the hand strapped flat on a board. Two surface electrodes were pressed against the muscle, one somewhat proximal to the midpoint and one near the distal end. In other experiments a needle electrode was used. Tension was recorded in a rough and ready manner, by clamping the strain gauge so that the middle phalanx of the little finger pressed directly against the end of the gauge when the muscle contracted. Contraction of the muscle gives abduction combined with flexion.

RESULTS

Comparison of the Peak Tension Achieved Voluntarily and by Electrical Stimulation in Brief Contractions

The rapid fatigue of electrically excited tetani discovered by Naess and Storm-Mathisen (1955) is of immediate relevance to the question of voluntary fatigue only if it is known that similar frequencies of motor discharge are involved in voluntary contractions. We begin, therefore, by determining the frequency of electrical stimulation needed to match the force that can be exerted at the start of a maximal voluntary effort. The comparison was made in the adductor pollicis by Merton (1954). He concluded that the peak voluntary force was equal to that in a maximal tetanus. He believed that a maximal tetanus was achieved by a rate of stimulation of 50/sec, the highest rate he could stand as a callow investigator. We have found, however, as has Desmedt (1973b), that a tetanus at 100/sec is about 10% larger than one at 50/sec; thus, the earlier conclusion must be scrutinized.

The question of whether a voluntary contraction can match the best electrical tetani is of interest in its own right, but, when pursued with the accuracy obtainable with the present instrument, it is somewhat complicated by the difficulty of discovering what can be achieved by electrical stimulation. In simple 1-sec tetani repeated every 15 sec, a frequency of 100/sec is not quite optimal. At 140/sec, a roughly 1% larger force is exerted, but only as an instantaneous peak in the first quarter of a second (Desmedt, 1973b). At 200/sec, the peak reached is lower again, about the same as at 100/sec or a little less. On the other hand, the muscle is presumably not fully activated even at 140/sec, because the maximum rate of rise of tension, in a differentiated record, is reached at 250/sec (compare Desmedt and Emeryk, 1968). It is of interest that repeated maximal tetani at 250/sec leave the muscles sore for a day, which is not noticed with 100/sec tetani or with voluntary contractions. Taking up the slack in the system, by means of a few stimuli at a slower rate preceding the tetanus (e.g., 4 at 50/sec), slightly raises the tension at 140/sec ($< 1\%$), by bringing the peak forward, but does not enable a 250/sec tetanus to show superiority, as had been hoped. The tension exerted thus depends critically on the regimen of stimulation, and there is no guarantee that the best regimen has been found. This topic is relevant to later sections of this chapter. In what immediately follows, a simple 100/sec tetanus is used as a standard for comparison, accepting that full activation, if that could be achieved, would realize a higher tension.

Another factor that has to be taken into account and guarded against in all accurate work is a warming-up effect in the muscle. Fully rested muscle gives a smaller tetanic tension than muscle that has recently been working. Thus, the first 1-sec long, 100/sec tetanus of the day is up to 7% smaller than the stable level reached after a dozen or so such contractions at 15-sec intervals. This again is not the largest tension obtainable at 100/sec, for if three tetani are dropped out of the sequence, the next tetanus (after a 1-min interval) is about 4% bigger than the stable level. Succeeding tetani at 15-sec intervals decay back to the stable level with a time constant of about 1 min. If the interruption of the sequence lasts longer than 1 min the increase in size of the subsequent tetanus is smaller; the warming up wears off. The effect is not thought to be due to actual change of temperature, for it occurs in muscle that has been fully heated in a water bath, as described in the Methods section. It has been found that the smaller size of tetani in rested muscle is not because it needs a higher rate of stimulation. Warming-up does not occur if the circulation is arrested. The effect has not been investigated in detail. It is guarded against by maintaining an unbroken rhythm of stimulation in an experiment. It is another fact which makes the idea of full activation, achieving A. V. Hill's P_o (the greatest force the muscle can exert when fully activated), elusive for tetanic stimulation, just as it was originally shown to be for the twitch by Desmedt and Hainaut (1968). It has not proved possible to drive the force up to a well-defined consistent ceiling.

Two long series of comparisons of electrically excited and voluntary contractions of adductor pollicis were carried out. There was a contraction every 15 sec; every 30 sec there was alternately a maximal tetanus at 50/sec and at 100/sec, each lasting 1 sec; halfway in between these, the subject made a maximal voluntary contraction, intended also to last 1 sec. The median nerve was blocked at the elbow (see Methods section) so that the muscle acting on the thumb available for voluntary activation, mainly the adductor pollicis, ought to have been the same as that reached by the electrical stimulus to the ulnar nerve at the wrist. The tensions exerted in the 50 and 100/sec tetani remained very constant. Accurate measurements were not made on 35-mm film but on wider traces from an ultraviolet (UV) recorder, on which 1 kg was roughly 1 cm. The peak tensions reached in the voluntary contractions were more variable and seemed to depend on the skill with which the subject exerted himself. Typical examples of the best contractions are shown in Fig. 2. In general, the peak force reached was greater than that in a 50/sec tetanus and slightly less than that in a 100/sec tetanus. The best voluntary contractions closely approached the 100/sec tetanus and occasionally appeared briefly to exceed it, by perhaps 1%; but the peak force was seldom if ever maintained as well as in the electrical tetanus. By the end of a second, the voluntary tension was often about equal to that in the 50/sec tetanus. A particularly well-maintained contraction is shown in Fig. 2D. The fastest rates of rise of tension achieved in a voluntary effort were also greater than in a 50/sec tetanus (as can also be seen in Merton, 1954, Fig. 2) but never in any of the 40 or so comparisons as fast in a 100/sec tetanus. These results suggest that the rate of activation in a voluntary effort is fastest, probably rising to about 100/sec, in the first few tenths of a second, and has usually fallen considerably within a second. The single unit results, as we shall show, bear out this supposition.

The fact that a voluntary contraction can achieve, or almost achieve, the maximum force of which a muscle excited through its nerve is capable, as shown by the above results, agrees with the conclusions of Merton (1954); but he must have reached the correct answer only by a happy cancellation of errors. He underestimated the tetanic force in his belief that it had reached its maximum at 50/sec. And his large ball-race device, intended to prevent the use of median-supplied muscles in the voluntary contractions by confining movements to the plane of the palm, may well have prevented the use of a fraction of the ulnar-supplied muscle as well and caused him to underestimate the voluntary strength. Thus, we find that if, with the median nerve not blocked, the pull of the thumb is voluntarily confined to adduction strictly in the plane of the palm, the force that can be exerted is roughly equal to that in a 50/sec tetanus.

It is of interest that in these experiments, done early in 1970, it was specifically noted that after median nerve block (which anesthetized the thumb) the subject found it more difficult to exert himself in a contraction of the adductor. Subsequently, Marsden et al. (1971b, 1977a) found that it needed a greater subjective effort to flex an an-

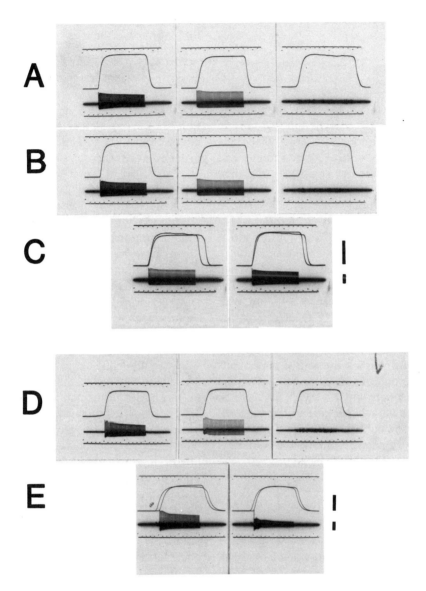

FIG. 2. Comparison of brief voluntary and electrically excited contractions in the adductor pollicis. **Above:** force records; **below:** action potentials. The recording sweep lasts just over 2 sec. There are time markers at the top and bottom of each frame, at 0.1-sec intervals; each tenth marker (at 1-sec intervals) is accentuated; there are less accentuated markers 0.5 sec after a 1-sec marker. A, B, and C: Median nerve blocked at the elbow by local anesthetic. A: A 1-sec tetanus at 100/sec, the same at 50/sec, and, finally, a voluntary contraction. B: The same. C: *Left:* a 50/sec tetanus and a voluntary contraction superimposed; *Right:* the same, but the tetanus at 100/sec. D and E: another experiment: Median nerve blocked at the elbow by pressure. D: as A and B. E, as C, but with the muscle cooled to about 27°C. *Calibration bars:* 5 kg and 10 mV.

esthetic thumb, and that servo action in the active muscle, flexor pollicis longus, was depressed.

Cooled Muscle

In muscle cooled by immersion of the arm in water at 15°C for 10 min to a muscle temperature of about 24°C, the 50/sec tetanus is larger than the 100/sec tetanus and the voluntary force realized in a maximal effort is also less than that in a 50/sec tetanus (Fig. 2E) being, in fact, closely the same as that in the 100/sec tetanus. This result is most simply explained if the rate of innervation in the voluntary contraction in cold muscle is the same as it is in warm muscle, starting around 50/sec and accelerating up to about 100/sec. Keeping with this, it is seen that the rate of rise of the voluntary contraction is, as before, about the same as in the 50/sec tetanus. This experiment, in which numerous comparisons were made on cold muscle, has only been performed once, but as the results are in accordance with expectations there seems no reason to doubt them. The view is that a maximal voluntary contraction in cold muscle is smaller than it might be, because the rate of innervation is too high. This interpretation conforms with various results in cold muscle given below, but the alternative, that the voluntary contraction is small because the rate of innervation slows in the cold, to below 50/sec on the plateau, although improbable, cannot be excluded on present evidence. The matter might be settled by Piper analysis (see Discussion Section).

Comparison of the Tensions Achieved Voluntarily and by Electrical Stimulation in Prolonged Contractions

Figure 3A is a record of the tension exerted in a maximal voluntary adduction of the thumb lasting about 95 sec. In this and in subsequent records in Fig. 3 the circulation was arrested for the duration of the contraction by a cuff on the upper arm, in order to avoid the complication of a return of blood flow in the contracting muscle, if the tension should fall sufficiently to permit this. The voluntary contraction was strictly limited to ulnar-supplied muscle in the hand by blocking the median nerve at the elbow, in this case by lidocaine with epinephrine. The subsequent tetani (Fig. 3B–D), obtained as before by supramaximal electrical stimulation of the ulnar nerve at the wrist, therefore call on the same motor units as the voluntary contraction and can be strictly compared with it. (The run of Fig. 3B was actually made before that shown in Fig. 3A. The other records in Fig. 3 were made in the order in which they are shown.) There was a rest interval of 15 min between each contraction in the experiment from which the records of Fig. 3 were taken. A longer interval would be desirable, but a limit was imposed by the likely duration of the lidocaine block. The conclusion we wish to draw from records A to D is that there is no single frequency of electrical stimulation which matches the voluntary contraction at all. The slower frequencies, 35 and 50/sec, give tetani which are well maintained, but which fall short at the start. A frequency of stimulation fast enough to lift the tension at the start into the volun-

FIG. 3. Comparison of voluntary and electrically excited contractions in the adductor pollicis. *Upper trace,* tension; *lower trace,* action potentials. A reference line is drawn in at a tension slightly greater than the tension at the start of the two voluntary contractions. Each contraction lasts for 95 sec or slightly more. A and F, maximal voluntary efforts. B, C, and D, electrically excited tetani at 100/, 50/, and 35/sec, respectively. In E (artificial wisdom), stimuli were given at 60/sec for 8 sec, 45/ sec for 17 sec, 30/sec for 15 sec, and 20/sec for 55 sec. *Calibration bars:* 5 kg and 10 mV.

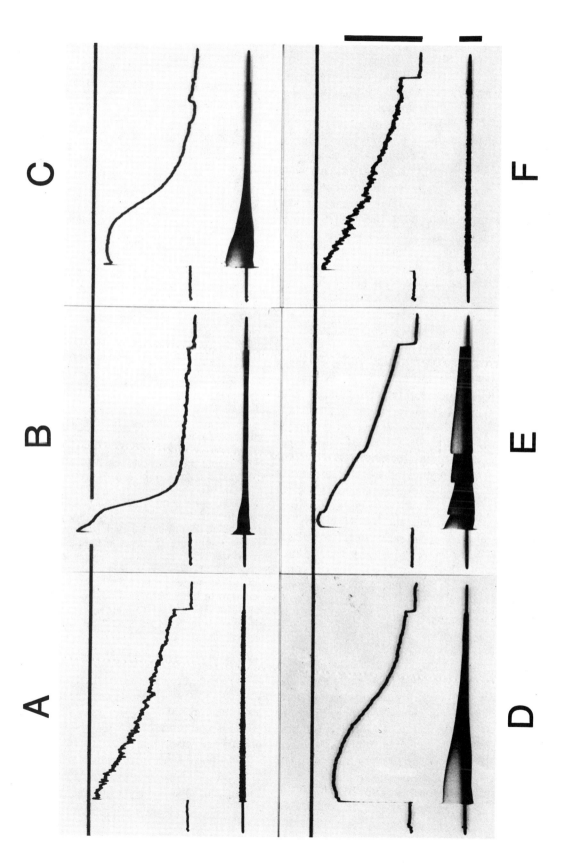

tary range, soon leads to a rapid decline, as Naess and Storm-Mathisen (1955) discovered and as confirmed by Jones et al. (1979). For the start of the contraction, the results agree approximately with the more careful comparisons of the previous section. Although there is always uncertainty about the rigor of all human experiments of this type, it is difficult to see in what way these records can be misleading, even though the nerve stimulus remains supramaximal throughout the faster tetani. This point was not checked during the experimental runs illustrated in Fig. 3; but it was checked, by turning up the stimulus strength after the tension had fallen, on other occasions when the records were essentially the same.

The natural inference from records A to D of Fig. 3 is that the rate of motor discharge in a maximal voluntary effort is high at the beginning, but soon drops. We therefore tried the effect of tetanizing the muscle with frequencies that fell in a series of steps. Figure 3E shows the result of applying such artificial wisdom. The final record, Fig. 3F, is a second voluntary contraction, which closely matches the first. During the artificial wisdom run, the tension is somewhat less, for most of the time, than in the voluntary runs, but the whole picture is much closer to a voluntary run than any of the single frequency tetani, as Jones et al. (1979) have also found. The frequencies and the durations for which they ran, during artificial wisdom, were determined by trial and error in exploratory experiments to minimize fatigue. The sequence could doubtless be improved to give an even closer match to a voluntary run, if more time were spent and if more steps were available (the present number being limited by the hand equipment). Other comparisons are shown in Fig. 4. A perfect match might, however, never be obtainable if, as seems likely from the single-unit recordings we show later, the optimal frequencies differ from motor unit to motor unit.

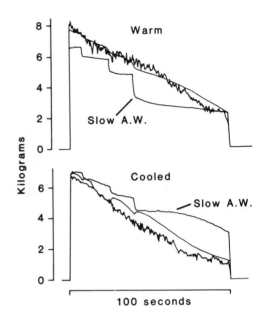

FIG. 4. Comparison of voluntary and electrically simulated voluntary contractions (artificial wisdom) in warm and in cooled muscle. In each case a maximal voluntary contraction lasting 100 sec was compared with a standard artificial wisdom run, using the same sequence of frequencies and durations as in Fig. 3E (but with the 20/sec period extended by 5 sec to give a total duration of 100 sec) and with a slow artificial wisdom run (marked Slow AW) in which the intervals were the same but the respective frequencies were 40/, 27/, 17/, and 10/sec. For the two voluntary contractions, the median nerve was blocked at the elbow by pressure. The hand was cooled separately for each run on cold muscle. In each case the muscle temperature was measured with a needle thermocouple and was close to 24°C. The circulation was arrested by a cuff on the upper arm for the duration of a contraction, in both warm and cold muscle. (Tracings from UV-recorder records.)

Cooled Muscle

In muscle cooled, as before, by 10 min in water at 15°C, a maximal contraction gives less tension than before; but the fatigue curve is still more or less matched by a run of artificial wisdom at the same frequencies as were used with warm muscle. However, artificial wisdom using lower frequencies throughout gives a bigger contraction all the time than the standard frequencies. This is shown in Fig. 4. In warm muscle, the slower frequencies are seen to give less tension throughout. A possible interpretation is that innervation do not change for a cold muscle and (like a standard artificial wisdom series) are too high to suit a cold muscle. In cold muscle lower frequencies do better. The same interpretation has already been applied to the results with brief contractions of cold muscle. We return to these topics later.

Ischemic Fatigue with Rest Periods

A few experiments were done in which the subject rested for periods during a maximal contraction, the circulation remaining occluded. Figure 5A and D show voluntary contractions interrupted by two, half-minute rests. After each rest the tension recovered somewhat, but then fell away clearly faster than it would have done had the original contraction persisted. In records B and C, the initial voluntary contraction is succeeded, after a rest interval, by two electrically excited tetani with a half-minute rest in between. In Fig. 5B, tetani at 20/sec do not match the corresponding voluntary contractions; in particular, they fail in the initial rapid rise of tension to a peak. In Fig. 5C, tetani at 35/sec nearly, but not quite, match the start of the rested voluntary contractions, but then decay more rapidly. These results suggest that, after a rest, the voluntary effort resumes at a frequency in excess of 35/sec, which then steeply falls to about 20/sec. The behavior of single units in similar circumstances is described later.

Optimal Frequencies in Brief Contractions

The phenomena demonstrated above on contractions lasting minutes make themselves felt even in contractions of 1-sec duration. As previously described, if the fastest rate of rise of force is required, fatigue, if it may so be called, comes on at the optimal frequency of stimulation for rate of rise (250/sec) even before peak tension is reached; so that the tetanic force is less than at 140/sec or 100/sec. The greatest tetanic force is obtained by accelerating rapidly up to 140/sec (i.e., by preceding the tetanus by a few impulses at a slower rate); but the peak is not held and a rapid fatigue sets in at once. The tension falls off by 2 to

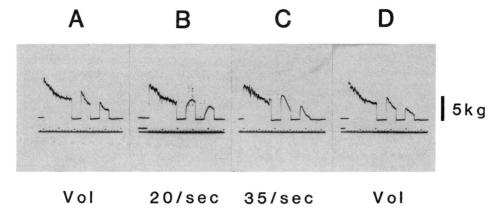

FIG. 5. Ischemic fatigue with rest periods. A and D: Force records during maximal voluntary contractions with two 0.5-min rest periods. The circulation was arrested throughout. B: As A and D, but, after each rest period, electrically excited tetani at 20/sec. C: Same as B, but the tetani at 35/sec. Time markers: 10 sec, 50 sec, and 100 sec.

3% by the end of the 1-sec tetanus. The tension is held better if the rate of stimulation is dropped to 100/sec after 0.25 sec at 140/sec. Tested in the usual stable series of 1-sec tetani at 15-sec intervals, the tetani in which the frequency was dropped to 100/sec were roughly 1% larger at the end of 1 sec. This meant that at the end of the mixed tetanus the tension was almost the same as that at the end of a simple 100/sec tetanus. A simple 1-sec long, 140/sec tetanus always ends up lower than a 100/sec tetanus. So, even in the first fraction of a second, slowing is advantageous. These experiments were done after we had seen that single units behave in just this sort of way, accelerating rapidly and then decelerating almost at once. We might comment in passing that, since it needs some experimental cunning to get the largest and best maintained electrically excited tetani lasting one second, it would not be surprising if skill had to be developed by the subject to match such tetani by voluntary effort. The fact that the brief voluntary contractions thus far recorded seldom match up to the best electrical tetani as regards peak tension and have never quite done so as regards rate of rise of tension is not, therefore, to be regarded as the last word on the subject.

Fatigue in Relation to the Number and Frequency of Motor Impulses

We have shown, in agreement with Jones et al. (1979), that the frequency of motor discharge has to fall in order to get the best out of the muscle. Bigland-Ritchie (1981) reaches the same conclusion. No complete explanation of this need for slowing can be offered. Two factors seem to be chiefly involved. First, repetitive excitation of the nerve induces some kind of failure of muscle activation, which might, for instance, be due to transmission block in terminal motor nerve twigs, to neuromuscular block, or to impairment of action potential-contraction coupling (Desmedt and Hainaut, 1968). For want of a better term, we refer to this phenomenon, whatever its true cause, as activation failure. The development of activation failure makes it desirable that the rate of voluntary innervation should be as slow as is consistent with activating the contractile mechanism. Second, during a prolonged contraction, the time scale of the contractile mechanism slows down, so that the rate of innervation needed to activate it optimally falls progressively as a contraction proceeds. The first factor puts a premium on slowing and the second factor allows it to happen without penalty.

The evidence for activation failure is that, with rapid tetani, when the muscle is somewhat fatigued, reducing the rate of stimulation slows the subsequent rate of fatigue, as seen in the artificial wisdom run of Fig. 4, or, if the change of frequency is left later, leads to an actual rise in the tension exerted, as illustrated in Jones et al. (1979, Fig. 3) and, in cold muscle, in Fig. 4 of this chapter. If the tension recovers when the frequency is reduced, the preceding rapid fatigue cannot have been due to wearing out the contractile mechanism itself, but must have been because of some failure in the chain of events leading to its activation. In another form of this experiment, the muscle is fatigued, with the circulation arrested, to, perhaps, one fifth of its initial tension, by a 30-sec tetanus at 100/sec. Immediately afterward (the circulation remaining occluded), single maximal twitches at a rate of 1/sec are found to be as large as control twitches obtained before the tetanus, and to have, if anything, a slightly faster rate of rise (as seen on a differentiated record). The accompanying action potentials are only slightly diminished. Hence, after fatigue at 100/sec, the contractility of the muscle seems unimpaired, so the preceding fatigue is to be attributed to activation failure.

The use of twitches to test the state of muscle is now, however, regarded with suspicion (Edwards et al., 1977). In brief tetani, activation failure is apparent at a frequency of 250/sec (which gives the fastest rate of rise) even before the peak tension is reached. This follows from the previously mentioned fact that a slower tetanus gives a larger tension. At the optimal frequency of 140/sec, the effects of activation failure can be demonstrated (by changing to 100/sec) within one-quarter sec. It is not improbable that activation failure sets in right from the start and that the maximum tension that can be reached by adjusting the distribution of the stimuli in time is a compromise between the rate at which tension builds up and the rate at which activation failure deepens.

The diminishing utility of rapid rates of stimulation during a long contraction can be demonstrated during a 20/sec tetanus. It begins, of course, well below the peak tensions obtainable at 100 or 140/sec. Thus, to start with and for the first 60 sec or so of the contraction, increasing the rate of stimulation causes a rise in tension. After that, increasing the rate, e.g., to 40/sec, gives a sharp decline of tension; the muscle, at that time, is optimally activated by 20 impulses/sec. We say optimally rather than fully because, as we have seen, the concept of full activation of the contractile machinery is elusive under our conditions. Optimal is intended merely to mean that neither raising nor lowering the frequency of stimulation increases the tension developed, without any implication that the machinery is fully engaged.

To start with, we imagined that in a voluntary contraction the rate of innervation fell off in such a way that activation failure did not develop, so that the contractile machinery was fully activated all the time, until it was exhausted. Certainly, after a long ischemic voluntary contraction, the contractility of the muscle tested by single twitches is much impaired, almost normal-sized action potentials triggering only small twitch responses (Merton, 1954, Fig. 6, and this chapter, Fig. 8). But this point of view had to be abandoned in the light of a new comparison.

The experiment was to record tetani over a wide range of frequencies, plotting the tension against the number of stimuli delivered, rather than against time (Fig. 6). From 200/sec down to 20/sec the curves are remarkably similar, when compared in this way, and there is a family likeness even at 10 and 5/sec. Thus, although at the start, force varies with frequency in the expected manner, after a time it becomes, to a first approximation and for frequencies of 20/sec and above, a function only of the number of impulses delivered and not of their frequency. This result has more than one implication. In the first place, if we are to agree with what was said above, that the rapid decline in force (in terms of time) at 200 and 80/sec is largely to be attributed to activation failure, it is not likely that fatigue at 20/sec which, as plotted in Fig. 6, follows such a similar course, is due to something else. It is presumably also due to activation failure. The 20/sec contraction lasts a long time only because activation failure develops that much more slowly. We must suspect that the same would be true of a maximal voluntary contraction; that the rate of innervation would slow in order to minimize and delay activation failure and could not avoid it wholly. We cannot prove this for a voluntary contraction; but we can do the next best thing and demonstrate its probable truth for an artificial wisdom run, which closely resembles a voluntary contraction.

To do this, an artificial wisdom run comprising a total of 3,600 impulses, starting at 60/sec and falling to 20/sec, was recorded and the tension plotted against the number of impulses delivered. The curve so obtained was compared with a similarly plotted curve (actually the mean of two) for a fixed frequency tetanus at 60/sec also containing 3,600 impulses (Fig. 7). The two curves are almost identical, although the

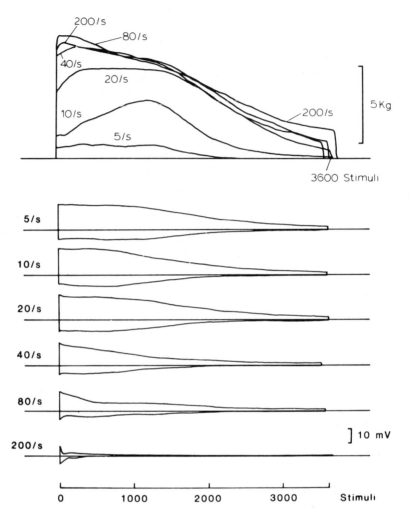

FIG. 6. Force and action potential records over a range of frequencies of stimulation from 5/sec to 200/sec, plotted against the number of stimuli delivered (intended to be 3,600 in each case). Tracings of the force records are superimposed at the top, with the envelopes of the action potentials traced separately below. The records were made by altering the speed of the recording paper in proportion to the rate of stimulation. The fastest record lasted 18 sec and the slowest 12 min.

duration of the artificial wisdom run was 2.3 times longer than the 60/sec tetanus. A preliminary account of this experiment has already appeared (Marsden et al., 1976). We argue, as before, that since activation failure certainly occurs in the 60/sec tetanus and must largely dictate its form, the same is likely to hold during artificial wisdom, which obeys the same relation between tension and number of impulses delivered. In the artificial wisdom run, the frequencies were chosen to be optimal, in the sense in which we use that word. Presumably, the same is true in a voluntary run, which closely matches it in time course. Hence, we conclude tentatively that activation failure occurs in a maximal voluntary effort, but that its incidence is minimized by a slowing of motor discharge, so arranged that the rate of discharge at any time is not faster than the optimal frequency. Viewed another way, we may say that since (under

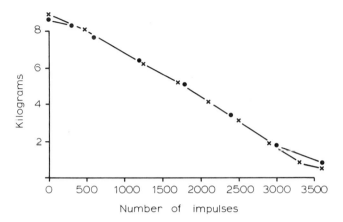

FIG. 7. A comparison of the force in a tetanus at 60/sec and in an artificial wisdom run, as a function of the number of stimuli delivered. The artificial wisdom run (x) started in the same way as that illustrated in Fig. 3E, i.e., with 60/sec for 8 sec, 45/sec for 17 sec, and 30/sec for 15 sec, but the 20/sec-period that followed was extended from 55 to 95 sec to bring the total number of stimuli to 3,600. The artificial wisdom run lasted 2 min 15 sec. The 60/sec tetani lasted 1 min and, thus, also contained 3,600 stimuli. The curve plotted is the mean of two runs (●) at 60/sec, one made before the artificial wisdom run and one afterwards.

the conditions we are concerned with) the tension depends for a first approximation on the total number of impulses delivered, and not on their frequency, the best-maintained and longest contraction will be obtained if the impulses are spread out as thinly at each stage of the contraction as is consistent with optimal activation at that stage. This is as far as we can carry an explanation for the slowing of motor discharge during a voluntary effort. The unsatisfactory feature is that we do not understand why the muscle obeys the particular relation between tension and number of impulses, seen in Figs. 6 and 7, which dictates the need for slowing; nor is it clear what advantage this relation offers the animal. It is quite possible, of course, that evolutionary pressures will not have tended to optimize performance in the type of task; namely, a long ischemic, isometric contraction, that we have set the muscle here.

Failure of Action Potentials

Thus far, there have been no conjecture as to the site of the activation failure, the existence of which we have inferred. The important discovery of Jones et al. (1979) that tetanic fatigue curves very similar to those illustrated here are seen in directly stimulated curarized mouse muscle, indicates that neuromuscular block is not likely to be an important factor, except, perhaps at the very highest frequencies (see also Bigland-Ritchie, et al., 1979). Recent work with direct stimulation of fatigued human muscle points firmly in the same direction (Merton et al., 1981). In the present experiments, the muscle action potentials always diminish during repetitive stimulation, at the slower rates after about 1,000 impulses have been delivered (Fig. 6). This does not necessarily mean that there is neuromuscular block, since, in single muscle fibers stimulated directly, the action potential, recorded intracellularly, diminishes greatly under repetitive stimulation (e.g., Lüttgau, 1965). Whatever the cause, the failure of action potentials observed here may well have something to do with the activation failure that develops.

The diminution of the action potential which comes on after about 1,000 impulses is not clearly seen at higher stimulation rates. This is thought to be because the phenomenon is obscured by another effect. The muscle action potential, as recorded with surface electrodes lasts, in resting muscle, for a total of some 25 msec, but it lengthens out during repetitive stimulation. Hence, successive action potentials soon run into each other and begin to cancel each other if the interval between stimuli is ≤ 25 msec, i.e., at rates of stimulation of ≥ 40/sec. This effect appears to be responsible for much of the slow decay during the 40/sec tetanus and for the rapid fall off at the start of the 200 and 80/sec tetani in Fig. 6, seen also in Desmedt (1973b, Fig. 4).

The similarity of the action potential records for 10 and 5/sec to that for 20/sec suggests that the fatigue of the contraction at 10 and 5/sec has the same cause as fatigue at 20/sec and higher frequencies and is also due to activation failure. Contrary to our original conceptions, it seems that there is no frequency of stimulation, however slow, that would exhaust the contractile mechanism but not cause activation failure. This is certainly the case for fused tetani. However, at 5/sec, the contraction consists of twitches with very little tetanic fusion, and the twitches (under ischemic conditions, of course) fade away rapidly after 1,500 impulses, at a time when the action potential is not greatly diminished. A similar state is found after the end of a long ischemic voluntary effort (see below). These observations were originally made much of, when we believed voluntary activation fatigued the muscle without activation failure; but as we have said, testing the contractile state of the muscle by means of twitches is now regarded with suspicion.

Another factor that encouraged the belief that activation failure does not occur in a voluntary effort was the claim by Merton (1954) that during voluntary effort the action potentials caused by interposed single maximal shocks to the nerve were not reduced in height. Opposite results were reported by Merton (1957) and by Stephens and Taylor (1972), and Merton's original view is clearly incompatible with the results of this chapter that action potentials always decay in a long tetanus (Fig. 6) and, in particular, during artificial wisdom (Fig. 3). We have reinvestigated the matter and obtained essentially the same results as Stephens and Taylor (1972). Two of several experiments of this kind that gave similar results are illustrated in Fig. 8. The abductor of the little finger was used, since with this muscle it is easier than with adductor pollicis to be certain that one is recording the action potentials of the muscle in question only. Paired maximal stimuli at an interval of 30 msec were used as test stimuli, applied every 10 sec. The response to the first stimulus may be reduced if it arrives at the muscle at about the same time as a burst of voluntary activity, but the antidromic volley from the first stimulus collides with any descending voluntary impulses and thus eliminates them, so that the second stimulus has a clear field. It is the second response that is measured. The open circles are from an experiment in which the circulation was arrested by a cuff during the maximal voluntary effort and for 2 min afterward. The voluntary force falls virtually to zero in 3 min. The test action potential stays up fairly well for 1 min and then falls to about 35% of its initial height by the time the contraction ends. After the subject relaxes it recovers to 75% of its initial height, but these action potentials evoke no detectable twitches. When the cuff is released at 5 min, the action potential gradually recovers the rest of the way, and the twitch returns. An experiment of this kind with a needle electrode in the adductor pollicis gave very similar results. The experiment on the little finger was repeated next day without arrest of the circulation (filled circles). The voluntary fatigue is still severe, but the test action potential is much less affected. After the voluntary effort ends, action potential and test twitch recover at once. The first

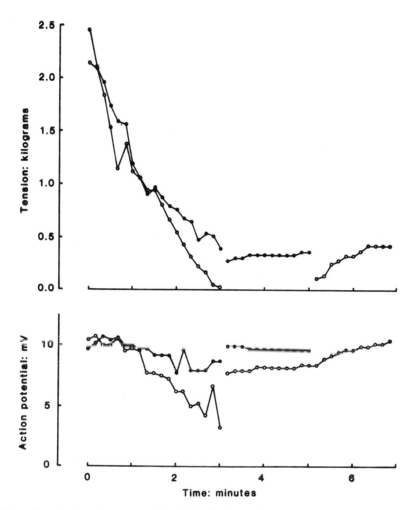

FIG. 8. Failure of evoked action potentials during fatigue of a maximal voluntary contraction in the abductor of the little finger. Every 10 sec, a pair of maximal nerve stimuli, 30 msec apart, was given at the wrist. Surface electrodes were applied 3 cm apart over the belly of the muscle. The height of the second evoked action potential of each pair is plotted below. Force exerted is plotted above. *Open circles:* A 3-min maximal effort with the circulation arrested for the duration of the contraction and for 2 min afterward. The points in the plot of force after 3 min are of twitch size. After the voluntary contraction ends, the test stimuli do not evoke a detectable twitch until the circulation is restored at 5 min. *Filled circles:* A similar run without occlusion of the circulation.

experiment shows that in 1954 Merton was misled, for reasons which cannot be determined at the present time. It may be noted, however, that this method of recording action potentials was not straightforward. The second experiment shows that substantial fatigue may be induced with only a modest diminution of electrical response. Whether the diminution observed in these experiments is due to neuromuscular block or to the action potential fatigue investigated by Lüttgau (1965) cannot be decided.

FIRING FREQUENCIES OF SINGLE MOTOR UNITS

The conclusion, to which the above experiments point, that the rate of firing of motor units slows progressively from a high initial value during a maximal effort, would be greatly strengthened if single units could be shown to behave in this manner. Numerous authors (e.g., Bigland and Lippold, 1954b; Norris and Gasteiger, 1955; Desmedt and Godaux, 1977a; Grimby and Hannerz, 1977; Grimby et al., 1981; Desmedt, 1981a) have recorded with needle electrodes during strong voluntary contractions. Although Desmedt and Godaux saw bursts at 120/sec in ballistic contractions and Norris and Gasteiger mention frequencies of up to 140/sec, we have not been convinced, and adequately long records of single units firing at 50 to 100/sec, which are the tetanic frequencies we believe must match the peak voluntary tension in the hand muscle; but the problem of isolating single motor units is extremely difficult during any vigorous effort, as we soon discovered. Therefore, like Bigland and Lippold (1954b), tried to thin down the population of active units in the muscle by a partial block of the ulnar nerve at the elbow with local anesthetic.

In the second subject (JCM), as the block developed, several units were found by probing with the electrode which stood out in better isolation; but when they were held their rhythm became irregular, as if impulses were dropping out in the region of the block. Finally, an excellent unit was found that did not deteriorate in this way, from which records were made during maximal efforts lasting a minute or more. In the very first record this unit started firing at 72/sec (measured over the first 10 interspike intervals) and reached a peak frequency of 150/sec, before slowing progressively. In a second run, these figures were 56/sec and 105/sec, respectively. It was then discovered that this unit could not be excited by an electrical stimulus to the ulnar nerve above the block at the elbow, but it could be reached by a stimulus to the median nerve at the elbow, but not from the median nerve at the wrist. It appears, therefore, that this unit in adductor pollicis was supplied by an aberrant median nerve fiber that passed in a connection between the median and ulnar nerves in the forearm. The behavior of this unit convinced us that the records from single units isolated by partial nerve block were quite untrustworthy (a possibility entertained by Bigland and Lippold, 1954b), and, henceforth, we only worked with aberrant median-supplied units. The type of abnormality responsible for their presence, which, we subsequently learned, is called the Martin-Gruber anastomosis. Iyer and Fenichel (1976) and Kimura et al. (1976) believe that it occurs in about 15% of the population. If this is so, we have been very fortunate, for we have found such units not only in the left hand of subject JCM, but in both hands of CDM as well. (None could be found in either hand of subject PAM; the right hand of JCM was not tried.)

Figure 9 shows two 1-sec long sections from a continuous 1-min record of such an aberrant median unit in the right adductor pollicis of CDM during a maximal voluntary effort. The frequencies during the complete run are plotted in Fig. 10. These are average frequencies for consecutive sets of 10 interspike intervals. The unit starts at 47/sec and slows to 15/sec. The peak frequency (the reciprocal of the shortest single interspike interval) is 65/sec. Figure 11 gives the evidence that identifies this unit as an aberrant median unit. Three other long records were made from this unit during maximal contractions, none of them notably different from the one illustrated.

Numerous records were also made from this unit during submaximal voluntary efforts. When the subject activated the unit he also contracted the median-supplied muscles of the thenar eminence, which exerted tension on the strain gauge. We found, as stated by Merton (1954), that the force exerted on the gauge, set to record from the adductor, by contraction of the median-supplied muscles was about 30% of the force exerted by the adductor. Although

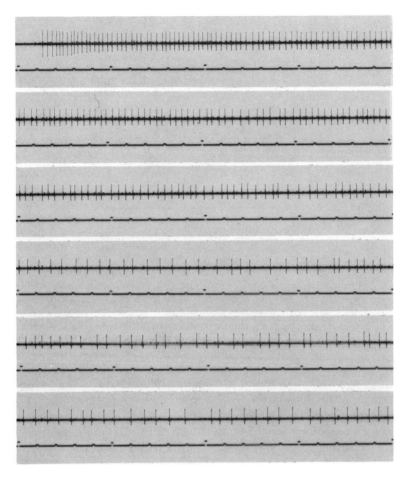

FIG. 9. Records from a single motor unit in adductor pollicis during a maximal voluntary effort lasting 1 min. This was an aberrant median-supplied unit found after blocking the ulnar nerve at the elbow. The first two strips are consecutive records of the start of the contraction. Subsequent strips are 2-sec sections cut from the continuous record, covering the intervals 8–10, 16–18, 38–40, and 58–60 sec, respectively. Subject CDM. Time markers 1, 0.5, and 0.1 sec. Voltage calibration as in Fig. 11.

there is no guarantee that the aberrant unit will behave in the same way as the units in these unparalyzed thenar muscles, we thought it sufficiently likely that it would make it meaningful to record their tension. The inset on Fig. 10 shows the tension plotted against the initial frequency in 16 runs on this unit. Similar graphs were obtained with 3 other single units. The lie of the points is similar to that found for tetanic stimulation of the whole adductor (Merton, 1954). Records of five submaximal contractions lasting roughly 5 sec each are shown in Fig. 12. The frequencies during these 5 runs are plotted in Fig. 13, together with the frequencies for the first 5 sec of the maximal run, already plotted on a slower time scale in Fig. 10. We see that the unit slows during submaximal contractions in the same manner as in the maximal effort. Frequencies much below 10/sec are not present. When the unit slows beyond that point it soon drops out altogether. If the frequency of firing is slow enough, the needle elec-

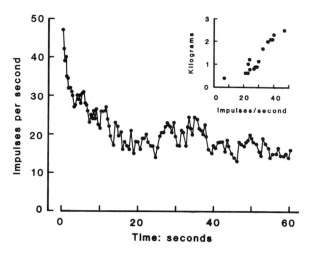

FIG. 10. The frequency of discharge of a single motor unit in adductor pollicis during a maximal voluntary effort lasting 1 min. A plot of the complete record from which sections are shown in Fig. 9. The values shown in this figure and in Figs. 13, 14, and 16 are the reciprocals of the period occupied by the first 10 interspike intervals, then the next 10 interspike intervals, and so on. Each point is plotted on the time axis at a position corresponding to the end of the 10-spike period to which it refers. Inset: The initial frequency of discharge of the same unit in 16 submaximal efforts, plotted against the force exerted by the thumb at the start of each effort.

trode can be seen to twitch in time with the sound of the spike in the loudspeaker.

In these first experiments in July 1969 we obtained prolonged recordings from 3 units in CDM's left hand, from 2 units in his right hand and from 1 unit in JCM's left hand. At least two long recordings were taken from each of these units. Every unit slowed conspicuously during a maximal (or a steady submaximal) effort. The behavior of the unit already illustrated was about average. Figure 14 shows a graph of the unit which slowed least over 1 min. This unit was exceptional mainly for the high frequencies maintained toward the end of the contraction. A record from a unit which reached a high peak frequency (100/sec) and exhibited greater slowing is shown in Fig. 15. Its behavior during the whole run from which the record of Fig. 15 was taken is plotted in Fig. 16. The frequencies reached toward the end of the 1-min contraction were in the range 15 to 20/sec, as with nearly all other units, the large slowing being a reflection of the high initial rate. A record from a still faster unit (150/sec) was given in the preliminary communication (Marsden et al., 1971a). The fastest recorded peak frequency was 190/sec (the 13th interval in Fig. 18C, measured with a microscope to give an accuracy of 1 or 2 impulses/sec). It is noteworthy that the discharge of motoneurons (MNs) is never regular, even at the highest frequencies. This is clearly to be seen in all the records illustrated (Figs. 9, 12, 15, 17, and 18). Occasionally, the more or less steady slowing is interrupted by a rapid burst (e.g., in Fig. 18C at 2.5 sec).

These single units thus amply confirmed the expectation that MNs would be found to slow during a prolonged voluntary effort. The actual frequencies observed at both the beginning and the end of a maximal contraction were remarkably similar to the rates of tetanization we had found best in the artificial wisdom experiments. The main discrepancy is that the single units slow down more rapidly in the first 20 sec or so than artificial wisdom suggested they would. Further data were obtained in a later series of experiments, described in the next section, the results of which are given in Tables 1 and 2.

Figure 16 also shows that the rate of discharge recovers somewhat after a period of rest with the muscle ischemic. In view of the recovery of tension that occurs under these circumstances (Fig. 5), this is not surprising. Properly controlled experiments of this type may throw light on the question of whether afferent signals from the muscle influence the slowing of discharge in a long contraction. In one pair of experiments, in which the comparison was made on the same unit, there was no evidence that after 5 sec rest the recovery in the firing frequency was any different whether or not the circulation was arrested.

The last technique for isolating aberrant units, which we have not yet fully explored, is to block the main ulnar pathway by high-frequency stimulation of the ulnar nerve at the elbow. After not many seconds of maximal stimulation at 100/sec, profound conduction block develops and there is little in the way of electrical response to the stimulation to be picked up by a bipolar needle electrode in the adductor pollicis and only a small mechanical contraction. With the main ulnar pathway out of action in this way, but the median to ulnar anastomosis unaffected, voluntary effort will bring in an aberrant unit in almost as good isolation as when the nerve is blocked at the elbow by local anesthetic or ischemia. The record in Fig. 17A was obtained in this way. The advantages of this method are that it is quicker than ischemic block, and less likely than

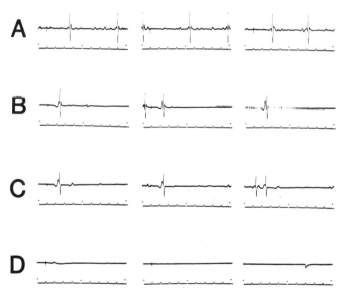

FIG 11. Evidence identifying the unit shown in Fig. 9 as an aberrant median-supplied unit in adductor pollicis. A: Three records of the unit discharging during moderate voluntary effort. B and C: The same, but with a stimulus to the median nerve at the elbow (shock artifact just visible at 10 msec). The action potential evoked was all-or-none with the stimulus and is, therefore, in a single unit. This action potential is the same as the voluntary potentials seen in A, at the start of the second frame of B and, most strikingly, at about 10 msec (in front of the electrically evoked potential) in the third frame of C. D: Muscle relaxed; first record, stimulus to median nerve at wrist; no response, showing that the unit is not an ordinary median nerve unit; second record, stimulus to ulnar nerve above the blocked region at the elbow; no response, showing that the block was complete and that unit is not an ordinary ulnar nerve unit; third record, 200 μV calibration. Time markers 10, 50, and 100 msec.

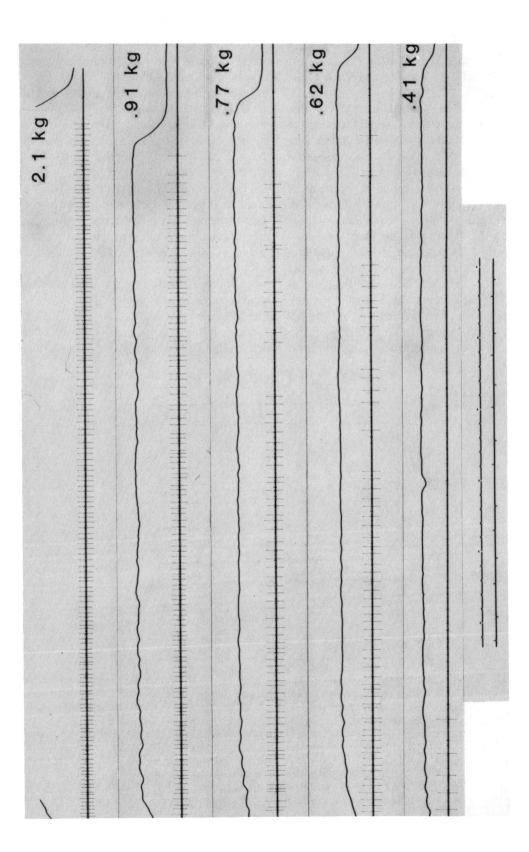

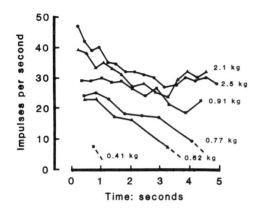

FIG. 13. A plot of the discharge frequencies during the submaximal contractions shown in Fig. 12, together with those in the same unit during the first 5 sec of a maximal contraction, which exerted 2.5 kg. (These latter points are those plotted, on a slower time scale, at the start of Fig. 10. The first points at all tensions are plotted on the inset on Fig. 10.) The position of each point on the time axis corresponds to the end of the period to which it refers. If it were plotted at a time corresponding to the middle of the relevant 10 interspike period the graphs, particularly at the lower tensions, it would fall somewhat more steeply.

block by local anesthetic to damage nerve fibers, when done repeatedly.

In practice the elbow stimulus is turned up gradually, giving time for each increment of contraction to decline before increasing the strength again. In this way pain is minimized and when the stimulus is left running at its final strength it is not unduly distressing. While records of the single unit are taken, the background of contraction and electrical activity due to the elbow stimulus is reduced if the circulation is arrested by a blood pressure cuff around the upper arm, but this is not essential.

The conduction block induced by continued 100/sec stimulation at the elbow is peripheral or, at any rate, predominantly so because interposed single shocks to the ulnar nerve at the wrist cause only a very small twitch, presumably due to the excitation of aberrant fibers (in these experiments we were recording tension at high gain). If an aberrant supplied unit is under observation, a wrist shock excites it in an all-or-none manner, thus confirming directly that the aberrant motor fiber in question has joined the ulnar nerve by the time it reaches the wrist; an observation that could not be made when the ulnar nerve is blocked by anesthetic or ischemia at the elbow, since in those circumstances all the motor units can respond to a wrist stimulus. Fig. 17B–D show the unit of Fig. 17A excited in this way at various frequencies. Measurement of the original records, on which the stimulus artifact from the wrist stimulus is just visible, gives a latency from stimulus to spike of about 3 msec.

On three occasions when stimulating this unit at the wrist, F waves (Magladery and McDougal, 1950; Dawson and Merton, 1956; McLeod and Wray, 1966) in this unit were seen. One of these is shown in Fig. 17C. In each case the separation of the two action potentials was 26 msec, giving a latency for the F response from the time of the wrist stimulus of some 29 msec. In this experiment the identity of the spikes during voluntary effort, or with all-or-none stimulation at the wrist, or in an F wave response was so striking that there seems little room for doubt that they were all from the same motor unit.

Stimulation of the ulnar nerve (during tetanic stimulation at the elbow) gives mechanical twitches, as seen in Fig. 17; but in

FIG. 12. Records from the same unit as Figs. 9, 10, and 11 during submaximal contractions, lasting roughly 5 sec, at the tensions shown. Top: Tension. The tension trace went off the oscilloscope during the 2.1-kg contraction but was recorded (and was reasonably steady) on the simultaneously run record on a SE 2005 UV recorder. Time markers at bottom, 1, 0.5, and 0.1 sec. The 100 μV calibrations are the small deflections below the time trace.

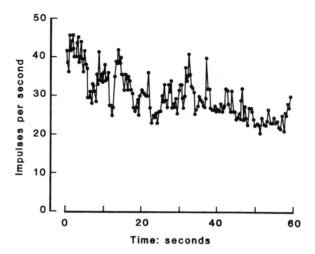

FIG. 14. The discharge frequencies of the single unit in adductor pollicis which showed the least slowing during a 1-min maximal effort. Subject CDM.

subject CDM (in whom the technique was tried), there must be several median units in the adductor, for the twitch is not all-or-none with the recorded spike. In Fig. 17E–G, from another experiment, a short length of voluntary run is compared with pieces of record from a period of stimulation at the wrist at 2/sec with the stimulus at threshold, so that not all the stimuli excited the single unit recorded from. The all-or-none nature of the spike is confirmed, but the change in the twitch when it is absent is small, so that there must be other aberrant units which are not recorded from by the electrode. Similarly, with voluntary activation (Fig. 17A), the ripples in the mechanical record are not strikingly correlated with the firing of the unit. In future experiments to resolve this difficulty and get the mechanical response of a single unit, subjects with only one aberrant unit might be sought, or spike-triggered averaging techniques applied.

As previously mentioned, stimulation at

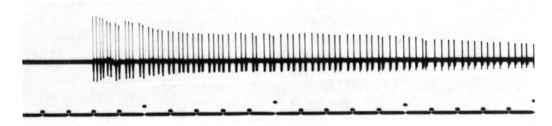

FIG. 15. A record of the start of a maximal contraction in a unit which showed a high initial rate of discharge. The shortest interval is 10 msec, between the third and fourth spikes, corresponding to an instantaneous peak frequency of 100/sec. Immediately before this run, the circulation to the hand was arrested by a cuff around the upper arm, but it is certain that this made no difference to the pattern of discharge, for an almost identical record was obtained from the same unit at the start of another maximal contraction some minutes before, when the circulation was free. On this occasion the peak frequency was again 100/sec, reached at the third interspike interval. Time markers, 1, 0.5, and 0.1 sec. Subject CDM.

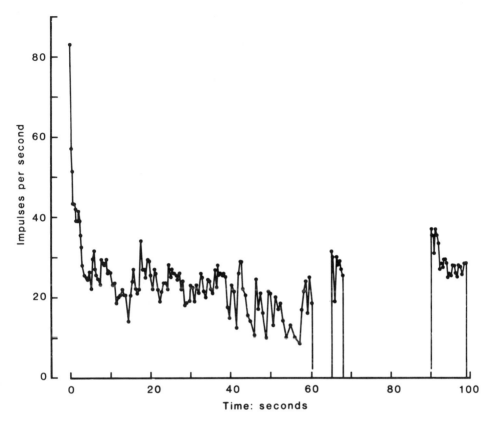

FIG. 16. Discharge frequencies during the entire run of which the record in Fig. 15 is the start. The subject first made a maximal contraction in the usual way, at 60 sec he relaxed, rested for 5 sec, and then made a 3-sec maximal effort; he then rested for a further 22 sec before a final maximal effort lasting 9 sec. Circulation was arrested throughout.

100/sec at the elbow cannot wholly abolish electrical responses from ulnar units, unless the circulation is occluded by a cuff on the upper arm. With the ulnar-supplied units on the edge of recovery and giving small and irregular responses, a brief interruption of tetanic excitation is followed by a larger response, which must not be confused with the potential due to an aberrant unit. Thus, a single stimulus at the wrist, at the timing relative to the elbow stimuli that we used, by sending an antidromic volley in ulnar fibers toward the elbow, cancelled out one descending volley of the tetanus and was followed by a gap longer than the usual 10 msec. This apparently allowed some perceptable recovery of ulnar units, so that following a wrist shock there was often a small potential (caused by an elbow volley) at an interval of about 15 msec from the time of the wrist shock, about 4 times longer than the latency of the spike of an aberrant unit. This potential is to be seen in Fig. 17B and again, marked by a black bar, in Fig. 17F and G. This response is independent of stimulation of the median unit under study, for it occurs when the latter is not excited by the wrist shock, as can be seen in Fig. 17F and G.

The Effect of Cooling the Muscle on the Firing Frequency of Single Units

A question already raised is whether the slowing of MN discharge during a voluntary

TABLE 1. *Rates, in impulses per second, of single units in warm and in cold muscle in subject CDM*

Date	Muscle temperature °C	Mean rate for 5 sec	First pair	Fastest pair	Rate over 10 intervals at start	Fastest over 10 intervals	Rate over 10 intervals at 0.5 sec	Rate over 10 intervals at 3 sec
Aug 5, 1970	36	35	60	115 (0.40)	52	68 (0.3)	58	48
	36	42	52	125 (0.24)	67	92 (0.3)	70	39
	36	42	51	115 (0.02)	51	61 (0.2)	48	39
	36	43	51	105 (0.16)	55	67 (0.2)	62	24
	28	—	60	135 (0.06)	85	85 (0.1)	—	—
	27	—	73	75 (0.06)	59	59 (0.1)	—	—
	27	—	22	72 (0.30)	36	50 (0.3)	—	—
	35	45	46	130 (0.30)	56	74 (0.2)	56	46
	35	40	36	98 (0.42)	42	61 (0.5)	56	35
Aug 6, 1970	37	47	44	145 (0.18)	70	90 (0.2)	73	44
	37	41	63	145 (0.16)	63	82 (0.3)	59	31
	37	42	53	145 (0.26)	79	81 (0.1)	60	32
	30	—	59	85 (0.12)	62	62 (0.1)	—	—
	30	—	59	125 (0.10)	92	92 (0.1)	—	—
	30	—	51	125 (0.03)	81	100 (0.2)	—	—
	22	—	64	—	—	—	—	—
	22	—	52	—	—	—	—	—
	23	—	55	—	—	—	—	—
	25	—	51	—	—	—	—	—
	26	—	29	—	—	—	—	—
	37	53	78	150 (0.24)	88	102 (0.3)	57	55
	37	43	72	125 (0.21)	84	88 (0.3)	48	29
	37	48	44	190 (0.16)	70	100 (0.3)	78	44
	37	48	44	150 (0.31)	80	96 (0.2)	59	40
	26	—	44	100 (0.08)	> 60	> 63 (0.1)	—	—
	25	—	68	90 (0.03)	> 68	> 68 (0.1)	—	—
	25	—	72	110 (0.06)	> 86	> 86 (0.1)	—	—
	27	—	51	100 (0.06)	67	67 (0.1)	—	—
	27	—	73	90 (0.02)	69	69 (0.1)	—	—
	27	—	56	95 (0.05)	64	64 (0.1)	—	—
	27	—	63	85 (0.04)	71	71 (0.1)	—	—

Each line is based on measurements of the record of a 15-sec maximal effort. The figures in brackets give the approximate time in seconds from the start of the discharge to the midpoint of the period in which the relevant frequency occurred. Where impulses were difficult to count, a conservative figure is given preceded by a > sign. Each day's work was on one single unit.

effort, which appears so nicely to match the requirements of the muscle, is an inherent property of the central nervous system (CNS), or whether it is controlled by feedback from the muscle itself, e.g., by a slowing of excitatory discharge from muscle spindles. With this in mind we looked specifically to see whether the rate of discharge was lower when the muscle was cooled, on the supposition that the optimum frequencies would be lower in a cold muscle. If the frequency of MN discharge fell when the muscle was cooled, that would establish a presumption that afferent feedback was influencing the MNs. It is not feasible to cool the hand after blocking the ulnar nerve with local anesthetic, due to the prevailing vasodilatation, from sympathetic block (cf. also Desmedt, 1973b). We therefore had to develop methods for blocking the nerve with

TABLE 2. *Rates, in impulses per second, of a single unit in warm and in cold muscle in subject JCM*

Muscle temperature °C	Mean rate for 5 sec	First pair	Fastest pair	Rate over 10 intervals at start	Fastest over 10 intervals	Rate over 10 intervals at 0.5 sec	Rate over 10 intervals at 3 sec
36	32	31	130 (0.12)	71	78 (0.1)	48	26
36	36	65	95 (0.17)	73	74 (0.2)	49	29
36	32	82	125 (0.03)	84	84 (0.1)	46	25
29	46	48	125 (0.10)	70	77 (0.2)	68	34
28	42	42	95 (0.17)	68	73 (0.2)	64	32
28	46	79	125 (0.03)	85	85 (0.1)	62	54
20	—	46	—	—	—	—	—
20	—	46	—	—	—	—	—
20	—	42	—	—	—	—	—
21	—	51	—	—	—	—	—
22	—	48	—	—	—	—	—
26	38	38	105 (0.15)	65	72 (0.1)	52	33
26	32	68	115 (0.04)	75	75 (0.1)	50	24
26	39	47	100 (0.04)	73	78 (0.1)	50	46
26	44	91	125 (0.07)	89	89 (0.1)	50	44
36	43	64	105 (0.04)	84	86 (0.1)	47	46
36	38	71	120 (0.20)	79	79 (0.1)	48	30
36	33	98	150 (0.07)	82	82 (0.1)	42	27

Each line is based on measurements of the record of a 15-sec maximal effort. The figures in brackets give the approximate time in seconds from the start of the discharge to the midpoint of the period in which the relevant frequency occurred. (Date: 9 August 1970.)

ischemia, which blocks the sympathetic nonmyelinated fibers late, instead of early. One successful experiment in May 1970 was done with ischemic block produced by pressing on the ulnar nerve at the elbow with a rubber bung. A strap around the arm held the bung firmly in place, while the brachial artery and the main veins in the antecubital fossa were relieved of pressure by a bridge, cut from expanded polystyrene, over which the strap passed.

In an improved method, used in subsequent experiments, pressure was applied by a rolled-up sphygmomanometer bag over the ulnar nerve behind the elbow. The elbow was held at a right angle and the deflated bag fixed over the nerve by long straps of zinc oxide surgical plaster running from the level of the wrist, up the forearm, over the bag, and down to the wrist again. When the bag was inflated the resultant pull was transferred by the straps to a large area of forearm skin. The circulation of the hand was not obviously embarrassed, and the radial pulse remaining full. Complete ischemic paralysis of ulnar motor fibers took about 30 min. Sometimes local pain under the cuff developed before or after this time. It was relieved by deflating the cuff for a minute or two and then reinflating. Not much time was lost in this way, as block came on more rapidly after the second inflation. After 50 to 60 min inflation the cuff was deflated for a few minutes. This sequence could be repeated several times, and in some experiments so repeated, block developed in 10–15 min after reinflation. If the needle electrode was recording from a single aberrant supplied motor unit at the time when the cuff was deflated, what appeared to be the same unit was picked up again when ulnar block had redeveloped,

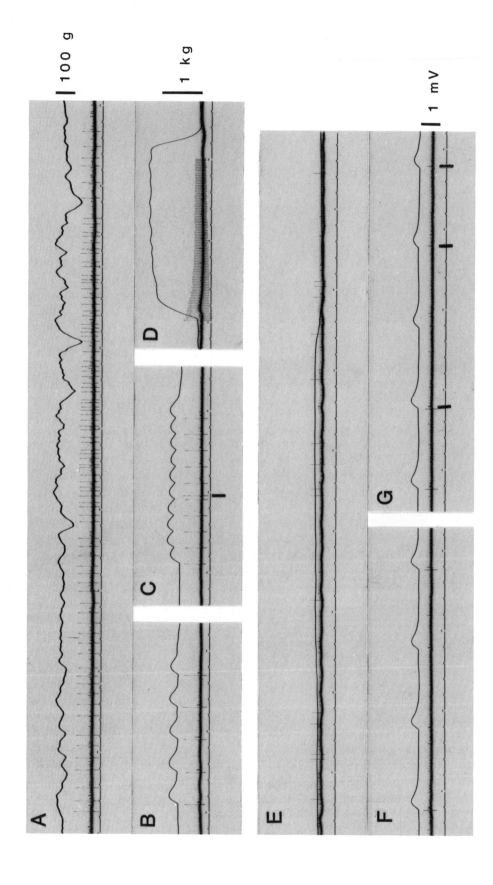

the subject, of course, taking care to keep still in the interim. The longest series of this kind lasted 5½ hr.

To alter temperature, the subject's arm was immersed in water to about 10 cm above the elbow. The forearm sloped upward so that the thenar eminence, where the needle electrode and the needle thermocouple were inserted into adductor pollicis, just broke surface. Cold water with ice floating in it was used to cool the hand. When it was desired to rewarm, the sink was emptied and filled with hot water at about 44°C. To be rid of alternating current (AC) interference, it was necessary to disconnect the lead sink trap from the waste pipe while recording.

There was little indication that cooling the muscle slowed MN discharge. The answer was not as clear-cut as was desired because, during a maximal voluntary contraction, the motor unit action potential breaks down in cold muscle and becomes uncountable. At muscle temperatures around 22°C only the first spike or two is distinguishable (Fig. 18A) the record soon becoming virtually flat. With submaximal contractions at this temperature good spikes continue for seconds, eventually becoming smaller and polyphasic (Fig. 18B). At around 27°C spikes may remain countable for a fifth of a second or so (Fig. 18D) before the record becomes chaotic, or may remain in good shape for 5 sec or longer. What goes wrong in the way of conduction block, desynchronization, slowing of action potentials in individual muscle fibers, etc., to wreck the spike during strong contractions in a cold motor unit can only be guessed. When the muscle is rewarmed the spike recovers completely.

In warm muscle the first pair of spikes at the start of a maximal effort are almost never at the highest frequency; usually they are at about half the peak frequency as shown in Tables 1 and 2. The fastest rates are often not reached for a fifth of a second or more in CDM, by which time, even in muscle cooled only to 27°C the spike has often failed. We are bound in such a case to find, therefore, that the peak frequencies that can be measured are lower in cold than in warm muscle, but if comparisons are made at similar times in a contraction the differences are unconvincing.

Three experiments were done, two on single units (which may or may not have been the same) in CDM, and one on a single unit in JCM (Figs. 18E and F). The results for each individual maximal contraction (none omitted) are given in Tables 1 and 2, and the averages in Table 3. The majority of contractions lasted just over 5 sec, and there was 2 min rest between each contraction in a set, usually of 3 or 4 contractions, at one temperature. Changing the temperature for the next set and reblocking the nerve took a half hour, or more. There was a reblocking without change of temperature near the end of the second experiment on CDM. Looking first at the results for muscle cooled to about 22°C, when only the instantaneous frequency for the first pair of spikes can be measured, it is seen that in

FIG. 17. Recording from a single aberrant-supplied unit in adductor pollicis during block of the main ulnar pathway by continuous maximal stimulation of the ulnar nerve at the elbow at 100/sec. Force recorded above, action potentials below, time markers; 0.1, 0.5, and 1 sec. A: voluntary activation of the aberrant-supplied unit. B: electrical stimulation of the ulnar nerve at the wrist at 5/sec (still during continuous 100/sec stimulation at the elbow). The same unit is excited. C: The same at 10/sec; an F wave, marked by a *black bar*, is seen in the single unit under observation. D: The same at 100/sec. E: A unit from another experiment, during a voluntary contraction. F and G: Threshold stimulation of this unit by a stimulus at the wrist at 2/sec; only some stimuli excite the unit. Short lines are placed under the small potentials mentioned in the text. Subject CDM. Voltage calibration bar against record G applies to all the records. Force calibration against D applies to all the records except A.

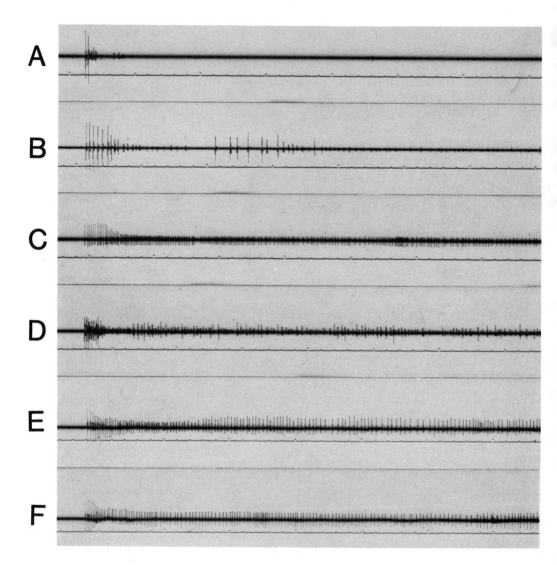

FIG. 18. Records from single units in cooled muscle. A: Subject CDM, record at the start of a maximal effort; muscle temperature 21.9°C. B: The same unit; a submaximal contraction a few minutes after the previous record, temperature 23.4°C. C: A maximal effort after rewarming to a muscle temperature of 37°C. D: The same after cooling to 25.6°C. E: Subject JCM, maximal effort, muscle at 26°C. F: The same after rewarming to 36°C. *Calibration bar:* 1 mV for records A, B, D, E, and F; 2.5 mV for record C.

the first run on CDM at this temperature, the rate (64/sec) was higher than for any preceding contraction on either day, although it was lower than in two of the runs when the muscle had been rewarmed. The average (Table 3) was 50/sec (pulled down by the final run with 29/sec), distinctly below the average for warm muscle that day 57/sec), but little less than the grand average for both days (53/sec). In JCM the rates at around 22°C (mean 47/sec) were also decidedly below those for warm muscle (68/sec). At temperatures around 27°C the frequencies for the first pair were higher on both

TABLE 3. *Average rates from Tables 1 and 2 of single units in warm and in cold muscle*

Subject and date	Muscle temperature °C	Rate for first 5 sec	First pair	Fastest pair	Rate over 10 intervals at start	Fastest over 10 intervals	Rate over 10 intervals at 0.5 sec	Rate over 10 intervals at 3 sec
CDM Aug 5, 1970	35–36	41	49	115	54	70	58 (83%)	38 (54%)
	27–28	—	52	94	60	65	—	—
Aug 6, 1970	37	46	57	150	76	91	62 (68%)	39 (43%)
	22–26	—	50	—	—	—	—	—
	25–27	—	61	96	69	70	—	—
	30	—	56	112	78	85	—	—
Aug 5 and 6, 1970	35–37	44	53	134	66	82	60 (73%)	39 (48%)
	25–28	—	58	95	66	68	—	—
JCM Aug 9, 1970	36	36	68	126	79	80	47 (59%)	30 (38%)
	20–22	—	47	—	—	—	—	—
	26–29	41	59	113	75	78	57 (73%)	38 (49%)

For CDM the averages for the 2 days are first given separately and then for the 2 days lumped together. The figures in brackets express the rates at 0.5 and 3 sec as a percentage of the fastest rate over 10 intervals.

days in CDM than in warm muscle (at 35–37°C), but again somewhat lower in JCM (59/sec as compared with 68/sec).

The comparison on which we place most weight is that between the mean frequency measured over the first 10 interspike intervals in warm muscle and in muscle at around 27°C. As can be seen from Tables 1, 2, and 3 there is very little difference, either in the spread among individual runs or in the means. Indeed, the grand mean frequencies for both days experiments on CDM are identical. If the highest rates for the 10 intervals are compared, warm muscle is faster in CDM. This is probably because, as previously mentioned, he reaches peak frequencies in warm muscle at times when his spikes cannot be counted in cool muscle. In JCM, who peaks earlier and whose spikes were better preserved, there is little difference between warm and cold muscle. In JCM it was possible to count spikes in muscle at about 27°C for the full 5 sec of each run. The mean rate over that period was, in fact, somewhat higher (41/sec) in cold than in warm muscle (36/sec).

The peak rates for the fastest pair in a run, were always higher, in the means, in warm muscle than in muscle at about 27°C. This is conspicuous in CDM (134/sec vs 95/sec) in whom the peak rate in warm muscle is usually late, but is also definite (126/sec vs 113/sec) in JCM, in whom this factor cannot be invoked. It is not improbable that this slowing is only apparent, for a pair of impulses at 126/sec, interval 7.9 msec, traveling down the median nerve in the upper arm, might well become slightly separated to 8.8 msec (113/sec) by traversing some 40 cm of cooled nerve in the forearm and hand, with a longer than normal relative refractory period, and hence reach the muscle at a lower instantaneous frequency than that at which they left the MN. Our conclusion from these three experiments is that cooling the muscle probably does not have any effect on the frequency of MN discharge in a maximal voluntary effort.

PEAK FREQUENCIES FROM ANTIDROMIC COLLISION

It can be objected to the single unit experiments that these solitary median-supplied units in the middle of an otherwise paralyzed muscle may behave abnormally. They are, for example, not subject to the reflex effects of the muscle afferents from this muscle, which presumably run in the blocked ulnar nerve. A check on the peak frequencies of firing in the unparalyzed muscle can be obtained in the following way. If a maximal shock is applied to the ulnar nerve at the wrist during a voluntary effort, a volley of motor impulses descends to the muscle and gives rise, after a latency of about 3 msec, to a synchronized muscle action potential, the M wave. At the same time, a similar volley travels antidromically toward the spinal cord. If, before it reaches the spinal cord, the antidromic impulse in any fiber meets a voluntary impulse descending, the two collide and cancel each other. Thus for a period after the shock no voluntary impulses get through to the muscle and, hence, the EMG immediately after the evoked action potential is silent.

The antidromic volley takes (in subject PAM) about 16 msec to reach the spinal cord. Immediately after it arrives, a few of the invaded MNs discharge an orthodromic impulse which travels down to the muscle (taking, in the fastest fibers, roughly $16 + 3$ msec) and gives rise to the F wave (Magladery and McDougal, 1950; Dawson and Merton, 1956; McLeod and Wray, 1966). The earliest F waves, in the fastest motor fibers, thus appear about $16 + 16 + 3 = 35$ msec after the original shock. These events are depicted in Fig. 19, the heavy lines representing the volleys due to the shock and the F wave volley. After the arrival of the antidromic volley, signaled by the discharge of the impulses that cause the F wave, the MNs enter a period of antidromic block and, in addition, are influenced to an unknown extent by the effects of the orthodromic volley in afferent fibers that the shock also sets up and which arrives at the cord at about the same time as the antidromic volley. Because of these complications, consideration of the part of the record after the F wave is deferred. For the moment, we direct our attention to the record before the F wave, because any voluntary activity which manages (for reasons that will emerge) to appear in front of the F wave must be due to impulses that left the spinal cord before the antidromic volley arrived and therefore cannot have been influenced by it.

Under ordinary circumstances the record remains silent after the main evoked potential (as labeled on Fig. 19) until the F wave. This is because any muscle action potential in this interval must be due to a motor impulse which reached the wrist after the shock (otherwise the action potential it causes would precede the main evoked potential) and which left the cord before the arrival of the antidromic volley (otherwise the action potential it causes would come after the F wave). Normally, all such impulses (e.g., the one labeled P in Fig. 19) will collide with the antidromic volley, as already described.

Voluntary action potentials can appear before the F wave, however, if the MNs are firing so fast that in the situation where one impulse in a fiber has just not reached the wrist when the shock goes in (e.g., shock Q in Fig. 19), the next impulse in that fiber is discharged from the MN before the antidromic volley arrives (e.g., shock R in Fig. 19). The second impulse then finds that the way through the antidromic volley is clear, because the antidromic impulse in that particular fiber has already been annihilated by colliding with the first descending impulse. Hence, the second descending impulse gets through to the muscle, and it arrives ahead of the F wave. How fast does a MN have to fire to get an action potential in front of the F wave? The most favorable situation, giving the slowest allowable rate of discharge for this to happen, will be when one impulse has just reached the wrist when the shock is given and the next impulse leaves the MN immediately before the antidromic volley arrives. If, for the moment, we consider only the most rapidly conducting fibers, the first impulse must have left the cord on the figures we have been using 16 msec before the shock; and the second one

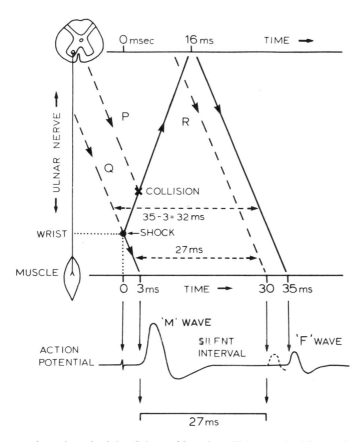

FIG. 19. Diagram to show the principle of the antidromic collision method for setting a lower limit to the rate of discharge in the most rapidly discharging motor fibers. A maximal shock at the wrist at time zero sends volleys *(heavy lines)* direct to the muscle (arriving at 3 msec), and antidromically to the spinal cord (arriving at 16 msec). The antidromic volley, on arrival, causes the discharge of a smaller volley *(heavy line)*, which gives rise to an F wave in the muscle at 35 msec. The *dashed lines* represent three descending voluntary impulses, P, Q, and R. P collides with and annihilates the antidromic motor impulse in its nerve fiber. Thus, a second voluntary impulse in this fiber can get through to the muscle, and, if it is early enough, arrive before the F wave. What is the greatest interval between these two voluntary impulses that will allow this to happen? P will not abolish the antidromic impulse if it is earlier than Q; so that sets the beginning of the interval. A response to the second impulse will not appear before the F wave if not before the F volley; that sets the end of the interval. The longest possible interval is thus the time shift between Q and F, which is 2 × 16 = 32 msec, or, looked at in another way, 35 − 3 = 32 msec. For the second impulse to cause a response that can be clearly recognized as in front of the F wave, it should be about 5 msec in front. Such an impulse is R, which is 27 msec after Q. Thus, 27 msec is the longest interval between two impulses in the same fiber, which will get a voluntary potential clearly in front of the F wave. It is equal, as can be seen, to the time interval between the start of the main synchronized action potential caused by the shock and the first voluntary potential to appear after the silent interval. It corresponds to a minimum frequency of 37 impulses/sec.

leaves 16 msec after the shock. The critical interval between successive impulses is, therefore, 32 msec and the frequency is 31/sec. All other situations are less favorable.

Thus, if the first impulse is 2 msec above the wrist at the time of the shock (having left the cord 14 msec earlier) the critical interval falls to 30 msec, because the second impulse must still leave within 16 msec after the shock. The same is true for the slower conducting fibers. Consider a fiber for which conduction time from cord to wrist is 18 msec. The first impulse must, therefore, leave the cord 18 msec before the shock in order to arrive at the wrist simultaneously with the shock. But the second impulse cannot wait, as the impulse in the faster fiber could wait until 16 msec after the shock, for if it does so it will be overtaken by the F wave in the fastest fibers before it gets to the muscle. It would have to leave at least 2 msec earlier in order not to be overtaken by the F wave before it gets to the wrist, and somewhat earlier still to keep ahead all the way to the muscle (which is some distance beyond the wrist). Hence, the permissible interval between impulses must be less than $18 + (16 - 2) = 32$ msec; i.e., less than in the fastest-conducting fibers. Thus, 31/sec is the absolute minimum frequency of discharge in fast- or slow-conducting fibers that will ever give action potentials in front of the F wave. In practice, action potentials will not be recognized as clearly in front of the F wave unless they are, say, 5 msec in front (27-msec interval, see Fig. 19), which raises the figure for the minimum frequency from 31 to 37 impulses/sec.

The figure of 16 msec for conduction time from wrist to spinal cord in subject PAM, used above, is obtained by working back from the observed latencies of the main evoked potential, the M wave, and the F wave (3 msec and 35 msec) assuming that the central latency for the F wave is negligible. For the second subject, CDM, the F wave latency is 30 msec, so that his MNs have to fire faster to get voluntary action potentials in front of the F wave. In subjects PAM and CDM, the distances from the seventh cervical spine to the proximal recording electrode, measured with the arm abducted to a right angle, are in the ratio 34:30, close to the ratio of their F wave latencies.

There are three potential sources of error that have been glossed over in the above account. First, the central latency of the F wave is about 1 msec (Renshaw, 1941). This means that the discharge of the MNs can be influenced by the antidromic volley for 1 msec before the F wave. Second, Dawson (1956) showed that there are sensory fibers in the human ulnar nerve that conduct some 10% faster than the fastest motor fibers. The ascending volley in these fibers would, therefore, reach the spinal cord nearly 2 msec before the antidromic volley, and could, therefore, influence the motor discharge for $(2 + 1) = 3$ msec before the F wave. Third, there is no guarantee that the F waves recorded in any particular experiment will include some in the fastest fibers, so the true position of the F wave in the fastest motor fibers (which is reference point in all the arguments) might be in front of the F waves actually recorded. All three factors operate in the same direction and are additive. Hence, we can only be certain that a motor discharge has not been influenced by the ascending volley if it precedes the F wave by a margin of a few msec. The 5 msec noted above ought to cover this margin.

The matter can be looked at in another way. Imagine one of the fastest motor fibers firing regularly at intervals just less than the critical interval. If a shock at the wrist is timed to coincide with the arrival of an impulse in this fiber, the muscle action potential caused by this impulse will start at the same time as the direct maximal evoked muscle action potential. The next action potential will, as we have so assigned, fall just before the time of the F wave. Hence, we see that the critical interval is just the time from the start of the maximal evoked potential to the start of the F wave. Likewise, if in an experiment an action potential appears in front of the F wave, the minimum frequency of discharge that can have been responsible for it is calculated from the time interval between the start of the main

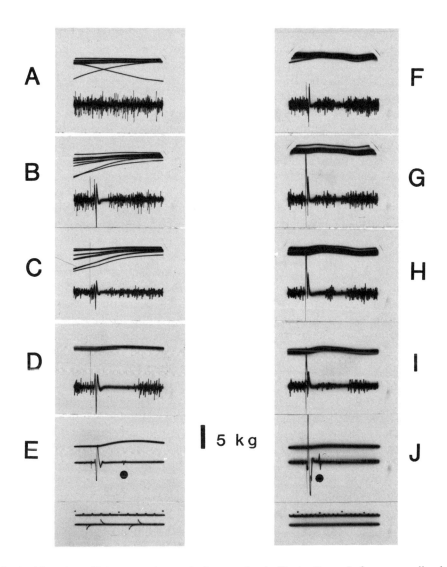

FIG. 20. Antidromic collision experiment during maximal efforts. Records from a needle electrode in the abductor of the little finger. Maximal stimuli at 1/sec to the ulnar nerve at the wrist. Ten records superimposed in each frame. A to E: Subject PAM. A: Control: Maximal contraction but no stimulus. B and C: Records taken at the starts of 10 maximal efforts. D: 10 sweeps at 1/sec starting 10 sec after the beginning of a maximal effort. E: Records with the muscle relaxed to give the time of the F wave (indicated by a *black dot* below the trace). Records F to J: On a slower time base, subject CDM. F: Records at the starts of 10 maximal efforts. G: The first 10 records at 1/sec in a 30 sec maximal effort; H: The next 10; I: The final 10. J: Muscle relaxed to give the time of the F wave (again indicated by a *black dot*). *Time markers:* 10, 50, and 100 msec. *Voltage calibration:* 1 mV.

evoked potential and the start of this action potential. This is illustrated by the line at the foot of Fig. 19 marked 27 msec.

The actual records obtained vary greatly. Even at the start of a maximal contraction no action potentials may appear in front of the F wave, presumably because the shock catches all the fibers with unfavorably placed impulses. It was necessary, therefore, to superimpose records of the start of numerous contractions. Such records are shown in Fig. 20. Records A to E are from subject PAM; records F to J (on a slower time scale) from CDM. Records B, C, and F are superimposed records of the start of maximal contractions, the subject attempting, by anticipating the stimulus, to be exerting himself fully at the time of the stimulus. Record A is a control, without the shock. Records E and J were made with the muscle relaxed to give the time of the F wave. It is clear that in B, C, and F, voluntary action potentials appear well in front of the position of the F wave. In record B the start of the voluntary potentials is particularly well defined; it is 16.4 msec after the time of start of the main evoked potential (seen most clearly in E). This corresponds to a rate of discharge of the responsible MNs of 61/sec, this, of course, being the minimum rate that could produce such a record. The rate derived from record C is 79/sec and from record F (subject CDM), 70/sec. Numerous other records gave similar figures (and did not suggest that there was any systematic difference in the peak rates of the two subjects). Thus, these results confirm the previous estimates of firing frequency early in a maximal effort from comparison of voluntary and electrically excited contractions and from single-unit recording.

The other records in Fig. 20 provide direct evidence that the rate of motor discharge slows early on in a maximal contraction, for, in records taken much after the start, there is rarely any sign of voluntary activity in front of the F wave. Thus, in record D, which was started 10 sec after the beginning of a maximal effort, 10 responses were recorded at 1-sec intervals. There is no voluntary activity in front of the position of the F wave, making it likely that the frequency in all units had fallen (according to the calculations above) below 31/sec. A single experiment of 10 trials cannot be held to establish this with certainty. There might have been faster firing units the impulses of which were unfavorably timed in relation to the shock on each occasion. But there were several other experiments on both subjects which gave similar results. In records G, H, and I with the other subject, stimuli were given every second during a 30 sec maximal contraction. The first 10 responses appear in G, the next 10 in H, and the final 10 in I. Record G is similar to F (superimposed starts), but slowing is manifest in H and I. In CDM the rate corresponding to the critical interval was 37/sec. Thus, it appears that in both subjects the frequency of discharge drops to about 30/sec after 10 sec. These records, therefore, not only demonstrate slowing, but confirm the conclusion from the single unit work that the rate of slowing is faster than was suggested by artificial wisdom, in which a frequency of 45/sec was held until 25 sec.

It is particularly valuable to have estimates of peak firing rates and this evidence of rapid slowing for subject PAM. This is because most of the artificial wisdom experiments and the estimates of peak frequency from comparison of brief voluntary and electrically excited contractions were done on him, and all the single unit work was done on the other subjects.

Considering the experiments in Fig. 20 critically, it appears that the only obvious way they could give a false answer would be if the wrist shock were not truly maximal. This is a very real danger, because, in a vigorous effort of adduction of the thumb, various forearm muscles, notably the flexor carpi ulnaris, contract and tend to lift the stimulating electrode. For this reason we always used shocks that were greatly supramaximal when the subject was relaxed. There are two pieces of evidence that the shocks did, in fact, remain supramaximal during voluntary effort. First, there was no obvious change in electrode position in the first 10 sec of a maximal effort; so the fact that no action potentials appeared in front of the F wave after 10 sec suggests that the shock (which was clearly maximal at 10 sec) was maximal throughout. Second, if action potentials were getting through before the F wave because the shock was submaximal, they ought to be distributed evenly along the record, and not bunched up toward the F wave as they invariably were.

We turn now to the part of the record after the F wave. Extending earlier arguments, it is clear that if every MN is firing at 31/sec or faster (again using figures from PAM), every antidromic impulse set up by the wrist shock will collide with a descending voluntary impulse before it can reach the spinal cord. Hence, there will be no antidromic block. It would be expected, as a result, that the level of voluntary activity immediately after the F wave would be the same as the initial level preceding the shock, and would continue at that level until the silent period, due to the pause in the firing of muscle spindles, developed (Merton, 1951). To a first approximation this seems to be what is happening in Fig. 20 records B and F. The silent period, better seen in records H and I, is inconspicuous, because in a maximal effort the superimposed twitch is necessarily small. Records B and F therefore give evidence that the majority of the units (or, at any rate, of those that make a substantial contribution to the recorded action potential) are firing faster than 31/sec in PAM and 37/sec in CDM. This interpretation is somewhat uncertain because it ignores the possible effects of the ascending volley in dorsal root fibers that the shock sets up. It seems unlikely, however, that the effect of this volley would be just to cancel out antidromic block and make it appear that there was not any, since this is what one would have to suppose its effect to be. The silence, or relative silence, of records D, H and I immediately after the F wave, only allows us to conclude, from considerations of antidromic block that, after 10 sec or more, not all the units are firing faster than the critical frequency for that subject. This contributes little to the stronger evidence for slowing already obtained from looking in front of the F wave in these records.

In a second series of experiments, single sweeps were recorded. These experiments were done on the abductor of the little finger, recording with a unipolar needle electrode in the belly of the muscle and stimulating the ulnar nerve at the wrist with supramaximal shocks. The subject made a maximal contraction every 30 sec; he held it for rather more than 10 sec. The sequence was paced by three warning clicks at 1-sec intervals, the subject contracting the muscle abruptly at the third click. Half a second later the first stimulus was delivered and a record taken; two further stimuli were given 3 and 10 sec after the start of the contraction. The three records and a time scale were recorded on a single camera frame. Five typical consecutive frames are shown in Fig. 21A, together with a frame without voluntary contraction to give the time of the F wave. In 4 runs on CDM in which a total of 100 frames were taken, action potentials preceded the time of the F wave in the 0.5-sec records in every case except one. At 3 sec in the majority of cases, and at 10 sec in the large majority of cases, they did not. Thus, there was clear evidence for rapid firing at the start of a maximal contraction, followed by rapid and progressive slowing. To measure these records, to get the times at which action potentials in front of the F wave occur, is a somewhat arbitrary business on account of the difficulty, using a needle electrode, of deciding how small an irregularity to count as an action potential. Attempting to maintain an even criterion, measurements were made, converted to minimum frequencies of discharge, and plotted in Fig. 22A. At the first half-second, most of the units have minimum frequencies in the range of 50 to 90/sec. The true rates are necessarily somewhat higher, it is impossible to estimate but how much higher without a knowledge of the number of units within electrical reach of the needle, the degree of synchrony in their discharge, and

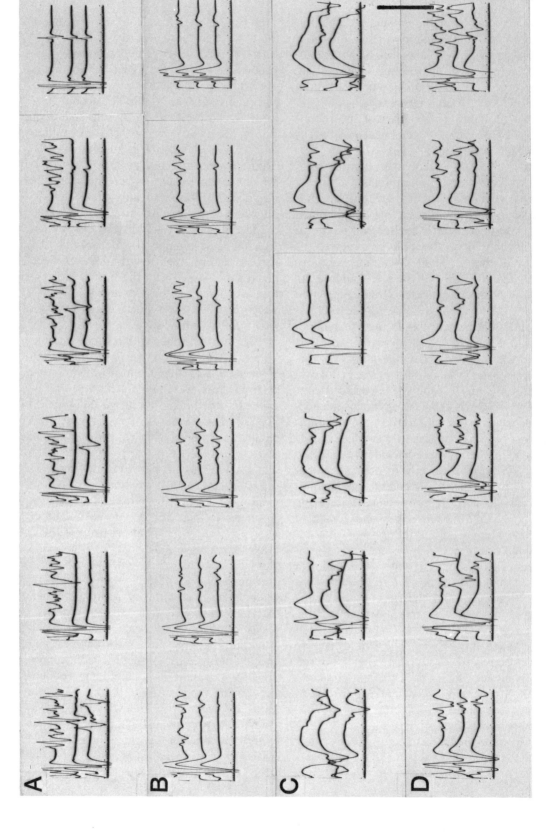

other data that we do not possess. The degree of slowing from 0.5 to 3 sec, in these experiments on CDM, is in accordance with the slowing of single units in his adductor pollicis over the same interval, shown in the last two columns of Tables 1 and 3. In two runs, a total of 50 frames, on PAM, the results of which are plotted in Fig. 22B, slowing during the first 3 sec was inconspicuous, but was definite, although smaller in degree than in CDM, after 10 sec. Specimen records are shown in Fig. 21B.

Frequencies from Antidromic Collision in Cooled Muscle

In the experiments just described, runs with the hand cooled were intermixed with those at normal temperature. When the deep muscle temperature was around 22°C, chaotic records, which could not be deciphered, were obtained, as with the cold single units described earlier (Fig. 18A). Examples from subject PAM are given in Fig. 21C. At about 27°C, although the main synchronous action potential was prolonged (Fig. 21D), measurements were possible with difficulty. The results at about this temperature for two runs on CDM and two runs on PAM are given in Fig. 22. In these runs the hand was not immersed in cold water and was slowly warming up. In CDM there is nothing to choose at either 0.5, 3 or 10 sec, between the hot and the cold results, the temperature difference of which was, on the average, about 9°C. In PAM the frequencies at 0.5 sec are, if anything, slightly higher in the cold (no doubt coincidentally) and the slowing by 10 sec, therefore more pronounced; but in this subject the temperature difference was only about 6°C, on the average. None of these results can be regarded as of the highest reliability and the degree of cooling is less than could be desired, but, like the results with cold single units, they do nothing to encourage a belief that MN discharge to cold muscle is slowed.

DISCUSSION

Three independent lines of evidence show that the MNs of the human adductor pollicis fire at rates of the order of 100/sec at the start of a maximal voluntary effort and then slow down progressively over the next minute to around 20/sec. Our peak rates, up to 190/sec, are in excess of anything previously reported. Desmedt (1981a) found 120/sec in strong brisk ballistic contractions. The mechanism of slowing in maintained contractions is not explained. It might be reflexly controlled. We looked for evidence of this by cooling the muscle. Our three independent methods all indicated that the MNs of the cold muscle fired as fast as ever, although it would apparently have been advantageous if they had slowed down. It may not be necessary to look outside the MN itself for the answer. Granit et al. (1963) found that the discharge rate of some MNs in the cat fell in much the same way we have found them to do in human voluntary effort, when they were excited by passing a constant current through them from a microelectrode. This line of investi-

FIG. 21. The antidromic collision experiment with warm and cold muscle. Records from a needle electrode in the abductor of the little finger. Every 30 sec the subject made a maximal effort lasting rather more than 10 sec. Shocks were delivered to the ulnar nerve at the wrist after 0.5, 3, and 10 sec. The action potentials resulting were recorded from above downwards on three traces of the oscilloscope. The fourth trace carried time markers at 1 msec (barely visible), 5 msec, and 10 msec; sweep duration 50 msec. A: Subject CDM; Five sample frames and one with the muscle relaxed to show the F wave; muscle not cooled. B: Subject PAM; otherwise as A. C: A continuation of the same experiment on PAM with the muscle cooled to 23°C. D: The same continued, after warming up to 27°C. *Voltage calibration bar:* 2.5 mV (applies to all records).

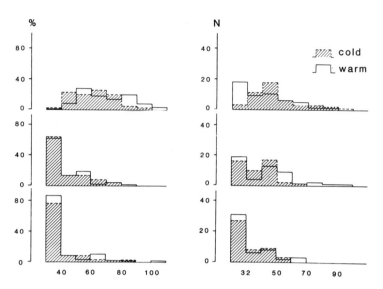

FIG. 22. Results of the antidromic collision experiment on warm and cold muscle, of which sample frames are shown in Fig. 21. *Abscissas* give the lower limit for the frequency of motor discharge (impulses per sec) that could account for an action potential in the position measured. **Left:** Values of 40 and below are plotted in the first block, values from 41 to 50 in the second block, 51 to 60 in the third block, etc. **Right:** The first block contains values of 32 and below, the next block 33 to 40, then 41 to 50, 51 to 60 as before. *Solid Lines:* Warm muscle near 36°C. *Dotted lines and hatching:* cold muscle. *Top graphs:* From the shocks at 0.5 sec; *middle:* from those at 3 sec; and *bottom:* from those at 10 sec. **Left:** Subject CDM; 101 frames with the muscle warm; 39 frames with the muscle near 27°C; the results plotted as percentages of the total to facilitate comparison of the warm and cold results. **Right:** Subject PAM; 50 frames with the muscle warm compared with 50 frames at a temperature around 30°C.

gation has been continued by Schwindt and Calvin (1972) and Schwindt (1973). How the discharge characteristics of MNs come to be matched to the needs of the muscle fibers they supply (or vice versa) is a question for the future. Insofar as the production of the right pattern of innervation for the muscle and for the task in hand is a matter of skill, as we have argued it is, it might be imposed on the MNs by the higher motor centers, for example, by the cerebellum.

A source of evidence on the frequency of MN discharge and its slowing, about which nothing has yet been said, is the Piper rhythm. The Piper rhythm is a striking but little-studied oscillatory potential of roughly 50/sec, which can be led off by surface electrodes over muscles in vigorous voluntary contraction (Piper, 1912). Its size at times approaches that of the maximal action potential, and it must, therefore, involve a synchronization of most of the motor units at the Piper frequency (Merton, 1957, 1981). It slows during a prolonged contraction. This appears to confirm that the predominant frequency of MN discharge slows too. But the Piper rhythm also slows if the muscle is cooled. Adrian (1925) remarks that this effect "is sufficiently obvious to be used as a class demonstration." In preliminary experiments with Hammond and Morton we have seen this phenomenon, which could be taken as evidence for slowing of MN discharge by cold. The apparent conflict with the results of the present chapter is, as yet, unresolved.

The need for slowing in strong contractions is dictated by unexplained properties of the muscle. There is first the curious relation we have described that, at physiological frequencies, the degree of fatigue is a function of the total number of impulses delivered, independent of frequency. This first property clearly makes it advantageous for motor discharge to slow. Second, there is the apparently separate property that, during any prolonged contraction, the frequency of stimulation needed to produce the greatest possible tension falls progressively. This second property allows the MNs to slow and still to get the best out of the muscle; indeed it compels them to slow if they are to. A maximal voluntary contraction does, indeed, seem to get the best out of the muscle, for although it can be mimicked by an electrically excited tetanus, the frequency of which falls progressively, it cannot be significantly bettered (except in cold muscle, in which wisdom apparently fails). These two properties of muscle that govern slowing (and which may be interrelated) are not understood.

We give reasons for thinking that the limitations set by activation failure (expressed in the first of the above two properties) largely determine the time course of all that type of fatigue that we have observed. Brown and Burns (1949) believed that neuromuscular block was important and served to protect the muscle fibers from excessive rates of excitation. The highly significant observation that typical fatigue curves can be obtained by direct stimulation of curarized muscle (Jones et al, 1979) indicates that neuromuscular block is not likely to be of prime importance. This is borne out by the recent observation (Fig. 23) that direct stimulation by massive pulses of current passed between plate electrodes on either side of the muscle cannot cause fatigued human muscle to contract, other than very feebly (Hill et al., 1980; Merton et al., 1981). The main factor in activation failure is presumably failure of excitation-contraction coupling; but Brown and Burns were right about the undesirability of fast maintained rates of stimulation. They did not, of course, know about action potential fatigue (see Lüttgau, 1965) and their results would require reinterpretation in light of it. In maximal contractions in our preparation, activation failure is apparently felt from the very beginning. A varying degree of activation failure, if that remains the appropriate term for it, may exist even in resting muscle and manifest its presence in the phenomenon of warming-up. This was originally shown for the twitch by Desmedt and Hainaut (1968) and may be related to calcium release from the sarcoplasmic reticulum (Desmedt and Hainaut, 1978). So it is not improbable that in all contractions there is an element of activation failure—that the muscle is never fully activated. Indeed, the idea of full activation has proved elusive. The force developed depends critically on the distribution in time of motor impulses and at any moment there is only one frequency which gives the largest force, that frequency being determined by the past history of innervation. There is no definite plateau of force over a range of frequencies of stimulation to correspond with the idea of full activation. Furthermore, at any frequency of stimulation which gives the largest forces, there is no plateau in time of the force exerted, which would indicate a constant level of activation. In the words in which we have already summed-up the position, it has not been possible to drive the force up to a well-defined ceiling.

If we are right, the force exerted continually depends on a compromise between activation of tension generation and a concomitant aggravation of activation failure. Performance would depend on a skillful juggling with these variables by appropriately timing the pattern of arrival of motor impulses. These subtleties of innervation are likely to be of real significance and not mere curiosities of the laboratory. Even a few percent of extra muscle power would be potent in natural selection as, indeed, it would be in present-day athletics. We must also envisage the possibility that people may

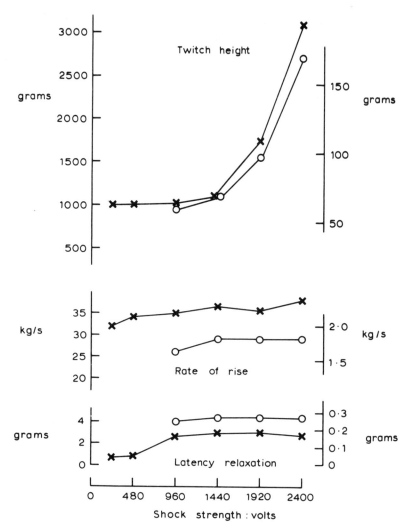

FIG. 23. Direct stimulation of fatigued and fresh muscle. The adductor pollicis was stimulated by condenser discharges of 50 μsec time constant at the voltages shown on the *horizontal axis*. Single stimuli were applied through plate electrodes on the palm over the muscle and on the back of the hand. The peak tension and maximum rate of rise of tension in the resultant twitches are plotted as a function of stimulating voltage. The size of the latency relaxation preceding the twitch was also measured. The *crosses* are for unfatigued muscle and the *open circles* (ordinate scales at the right, expanded 16 times) are for fatigued muscle. After the fatigue produced by 4 min of maximal voluntary contraction with the circulation arrested, the twitch, its rate of rise, and the latency relaxation are all reduced by a factor of about 16; but there is no sign that the electrical threshold of the fibers is much altered by fatigue, for the curves are the same shape as before. Below 960 V the latency of contraction (not plotted) is about 4 msec, corresponding to stimulation of intramuscular nerves. At 960 V and above, the latency jumps back to about 2.5 msec, indicating that muscle fibers are being excited directly; at the same time the brief latency relaxation increases sharply in size, reflecting the greater synchrony with direct excitation. Since the muscle shows striking weakness when it is undoubtedly being excited directly, fatigue cannot mainly be due to neuromuscular block or to any similar cause. (From Merton et al., *unpublished experiment*, with permission.)

vary in the skill with which they innervate their muscles. Those individuals of slight build who surprise by their ability to lift heavy objects or to unscrew recalcitrant jam jar lids, may succeed because they program their MN frequencies with more skill than their brawnier fellows have been obliged to acquire. In a wider sense, patterns of motor discharge clearly do differ from subject to subject, and are probably partly inherited, as evidenced, for example, by the way people walk or by their handwriting. The individuality of the written signature is the basis of commercial, legal, and many other aspects of modern life, and is taken so much for granted that we forget it is one of the glaring facts of motor physiology. But individuality may be apparent even in simple movements. In this chapter, we found that subject JCM tended to reach peak frequencies earlier than CDM, while from the antidromic cancellation experiments, PAM appears to slow less than the others over the first second. Dr. Rashbass informs us that, within a small group of subjects, records of saccadic eye movements can be assigned with some confidence to the subject from whom they were taken. There is not much scope in this movement for variation, other than in the change of frequency of motor discharge with time to the extraocular muscles.

ACKNOWLEDGMENTS

The sequences of stimuli used in artificial wisdom were programmed by a Devices Ltd. 5-dial Digitimer, with extra Logic Unit Type 3080, Gated Pulse Generator and Counter Unit Type 3250. This equipment was designed by H.B. Morton of the National Hospital, Queen Square.

Functional Properties of Spinal Motoneurons and Gradation of Muscle Force

*Daniel Kernell

Department of Neurophysiology, Jan Swammerdam Institute, University of Amsterdam, The Netherlands

For the production of an adequate motor behavior, it is essential that the CNS can control muscle force in a finely graded manner. In limb muscles of the adult mammal, each extrafusal muscle fiber is innervated by only one motoneuron (MN); each MN may often control between hundreds or a few thousand muscle fibers, and each muscle is typically controlled by several tens to some hundreds of MNs. The group of muscle fibers that are innervated by one MN is called a muscle unit (Burke, 1967), and the group of MNs controlling one muscle is referred to as a MN pool. In reflex or voluntary contractions, muscle force is graded by two mechanisms: by a variation in the number of active MNs (recruitment gradation) as well as by a variation in the discharge rate of the MNs (rate gradation). Use of these force-controlling mechanisms is complicated by the fact that, even within a single muscle, individual units may differ markedly from each other with respect to their contractile speed, endurance, and maximum strength (Wuerker et al., 1965; Burke et al., 1973). Much of this variation seems to be continuous in nature. For descriptive purposes it is, however, usually practical to group the muscle units into the categories or types of Burke: fast-twitch fatigue-sensitive (FF), fast-twitch fatigue-resistant (FR), and slow-twitch fatigue-resistant (S) (Burke et al., 1973). With respect to their maximum force, the FF units are typically the strongest ones by far, and the S units, on average, tend to be weaker than the FR units of the same muscle (e.g., Burke et al., 1973; Dum and Kennedy, 1980; McDonagh et al., 1980). The way in which the CNS uses the various units of a muscle depends, to an important extent, on the properties and organization of the interfacing device, the MN pool. This chapter deals with results from our own experimental work on cats.

THE FREQUENCY-CURRENT RELATION OF SPINAL MNs

In almost all kinds of voluntary or reflex motor activity, the muscle fibers are activated by repetitive MN discharges. In maintained contractions, these discharges are evoked by relatively steady, postsynaptic currents, which are produced as a result of the summation of a very great number of brief postsynaptic events. In a silent cell, such currents give rise to a maintained depolarizing shift of membrane potential, and in a discharging MN they produce an increase of impulse frequency (see Granit et al., 1966; Burke, 1968). The way in which

---

*Department of Neurophysiology, Jan Swammerdam Institute, University of Amsterdam, Eerste Constantyn Huygensstraat, 1054 BW Amsterdam, The Netherlands.

this signal-tranformation occurs is not easily investigated by means of synaptic activity, because postsynaptic currents are difficult to measure and control. Hence, the repetitive encoding properties of spinal MNs have largely been analyzed by aid of maintained stimulating currents that were injected via the tip of an intracellular microelectrode (e.g., Granit et al., 1963a,b; Kernell, 1965a–c; Schwindt, 1973; Baldissera and Gustafsson, 1974; Calvin, 1975; Baldissera et al., 1978).

The threshold current for maintained repetitive firing (rhythmic threshold) is, on an average, about 1.5 times as strong as the current required for eliciting a single spike (Granit et al., 1963a; Kernell, 1965a). The difference between these two kinds of threshold current is likely to be caused by the fact that a current step produces an initial overshoot of change in membrane potential (Ito and Oshima, 1965).

The relation between impulse frequency and the intensity of steady injected current (the f-I relation) is approximately linear over a considerable range of lower stimulus intensities from the rhythmic threshold upward. Firing within this range is referred to as firing within the primary range (Fig. 1). At higher discharge rates than those of the primary range, the cell is typically more sensitive to changes in stimulus intensity (steeper f-I relation). Firing at these higher rates has been referred to as firing within the secondary range (Fig. 1). At the very highest discharge frequencies the f-I curve often becomes less steep again and, finally, as current becomes too strong the cell stops firing altogether (inactivation). For unknown reasons, many MNs are unable to produce maintained discharges within the secondary range (Kernell, 1965b, 1979; Baldissera and Gustafsson, 1971; Schwindt, 1973). All cells can, however, fire within this high-grain range for a brief initial period after the abrupt onset of stimulation. It should be noted that primary range and secondary range are mainly to be regarded as convenient descriptive terms. There is

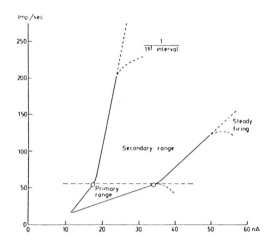

FIG. 1. Typical relation between impulse frequency (imp/s) and intensity of stimulating current (nA) for hindlimb MNs of the cat. Based on average data from experiments of Kernell (1965b). Separate curve drawn for first impulse interval and for firing following the initial phase of adaptation ('steady firing').

not necessarily any great difference in firing mechanisms between these two ranges of discharge (Kernell, 1968; Kernell and Sjoholm, 1973; Baldissera and Gustafsson, 1974; Baldissera et al., 1978).

After the sudden onset of activation of a MN, its firing rate declines even if stimulation is kept constant. This adaptation is particularly rapid during the first few impulse intervals, but it may go on for many seconds and, very slowly, even for several minutes (Fig. 2). For discussion, it is useful to distinguish between a rapid initial phase, taking place within the first second, and a subsequent late phase (Fig. 1). The initial phase of adaptation is commonly dominated by a pronounced drop in instantaneous impulse rate from the first to the second interval. The f-I relation has the same general shape initially in a discharge as later on, but the slope of the curve (f-I slope) declines rapidly during initial adaptation (Fig. 1): (Granit et al., 1963a; Kernell, 1965a,b). The transition from primary to secondary range typically occurs at about the same firing rate, but at a much weaker stimulus intensity, initially in a discharge compared to later

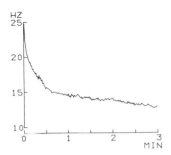

FIG. 2. Typical time course of late phase of adaptation. Average relation between impulse rate (Hz) and time (min) for 6 gastrocnemius MNs. All the cells were activated by a constant current of 5 nA above the rhythmic threshold. (Modified from Kernell and Monster, 1982a, with permission.)

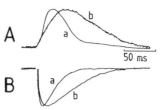

FIG. 3. Twitch (A) and afterhyperpolarization (B) from two MN–muscle unit conbinations (*a* and *b*, respectively) of one and the same gastrocnemius muscle. The records are averages of 30 to 40 sweeps each. Single spikes giving rise to twitches and AHPs were evoked by brief (1-msec) pulses of injected current. The spikes themselves are not shown. Same time calibration for all records. Normalized amplitudes. The slow twitch was 0.8 g and the fast one 4.2 g. The slow AHP was 4.3 mV and the fast one 3.9 mV. (From experiments of Kernell, 1979.)

on. For a given MN, the initial as well as the late phase of adaptation more pronounced at strong than at weaker intensities of stimulating current (Fig. 1). During the course of late adaptation, however, the sensitivity to *changes* in stimulus intensity remains nearly constant (Granit et al., 1963b; Kernell, 1965a; Kernell and Monster, 1982a).

Control experiments have demonstrated that, at least within the primary range, maintained postsynaptic inhibition and excitation usually have the same effect on the discharge rate as the addition or subtraction of a fixed amount of injected current (Granit et al., 1966). This is true also for synaptic activity that produces a considerable increase in the resting membrane conductance of the neuron (Kernell, 1969). Thus, postsynaptic conductance changes do not short-circuit the effects of currents on firing rate; in a discharging cell, different sets of somatic synapses may simply add their effects together without any mutual shunting effects. Such intuitively unexpected results are easily accounted for theoretically (Kernell, 1968; Kernell and Sjoholm, 1973).

The repetitive properties of spinal MNs are, to an important extent, dependent on the afterpotentials that succeed each single somadendrite spike. Of particular importance is the after-hyperpolarization (AHP) (see Fig. 3B), which arises because the spike is succeeded by a gradually declining increase of potassium conductance (Coombs et al., 1955). Recent studies have indicated that this potassium conductance mechanism is different from the one which is responsible for the rapid fall of the spike (Barrett et al., 1980). The AHP-conductance is apparently activated by calcium ions that enter the cell during the spike (Barrett and Barrett, 1976; Krnjevic et al., 1978). It is intuitively easy to understand that a prolonged AHP-conductance should be of importance for repetitive firing (Eccles, 1953). If a neuron is stimulated by a steady current of suprathreshold intensity, the stimulus will produce a spike as soon as the threshold voltage is reached. This initial spike is followed by its hyperpolarizing AHP current, which counteracts the stimulating current. The stronger the stimulating current, the sooner would it be expected to outbalance the gradually declining AHP current, and the higher would the discharge rate be expected to become. With a steady stimulation of just threshold intensity, no second spike would be expected to appear until the AHP of the first spike were over. Thus, when stimulated steadily at threshold intensity, a cell might be expected to fire repetitively at a minimum rate for which the impulse intervals are equal to the duration of AHP. Experiments have shown that in spinal MNs this is very nearly the case (Kernell, 1965c). In spinal MNs of the cat, the AHP lasts for about 50 to 200 msec (Eccles et al., 1958), and the minimum rates for maintained firing are between approximately 5–22/sec, different in different MNs (Kernell, 1965c, 1976, 1979). If a steady stimulating current is decreased a little below the strength needed for the minimum rate, firing becomes very erratic or stops completely.

During the initial impulse intervals after the abrupt onset of a steady stimulation, the decline in impulse rate seems mainly to be caused by a nonlinear summation of the AHPs of consecutive spikes (e.g., Ito and Oshima, 1962; Kernell, 1972; Kernell and Sjoholm, 1973; Baldissera and Gustafsson, 1974; Baldissera et al., 1978; Barrett et al., 1980). The postspike conductance changes associated with the AHP are also of great importance for the shape and slope of the f-I relation (Kernell, 1968; Kernell and Sjoholm, 1973; Baldissera and Gustafsson, 1974; Baldissera et al., 1978). The late adaptation seems to be caused by some cumlative after-effects of many consecutive spikes (Kernell and Monster, 1982a). We do not yet know the nature of these after effects in MNs; for a general discussion of adaptive mechanisms in neurons, see Jack et al. (1975).

THE TENSION-FREQUENCY RELATION OF MUSCLE UNITS

It is well known since the whole-muscle experiments of Cooper and Eccles (1930) that the relation between muscle tension and stimulus frequency (T-f relation) follows a sigmoid course, whereby the relationship is steeper over an intermediate range of stimulus rates than at lower or higher ones (Fig. 5). Within this steep region of the T-f curve, the muscle is maximally sensitive to variations in stimulus frequency; this is the optimal region for rate

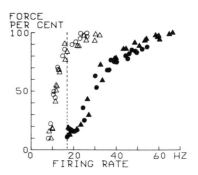

FIG. 4. Mean contractile force (%) produced by regular discharges at different rates within the primary range (Hz) in the two MN—muscle unit combinations of Fig. 3 (a = *filled symbols;* b = *open symbols*). Each cell was stimulated by a series of constant injected currents of 2.5 sec duration, and the plotted discharge rates and forces were measured at 0.5–1.0 sec (*triangles*) and at 2.0–2.5 sec (*circles*) after the onset of stimulation. In unfused contractions, the plotted values refer to *mean* force production. For each unit forces have been expressed as a percentage of the maximum force displayed in the graph. The maximum rate within the primary range in each one of the two units was capable of eliciting about 85% of the respective maximum tetanic tension. Vertical interrupted line drawn through minimum rate of fast unit. (From experiments of Kernell, 1979.)

modulation of muscle force (optimal T-f range). The optimal T-f range is situated at lower stimulus frequencies for a muscle or muscle unit with a slow twitch than for one whose twitch is faster (cf. Figs. 3 and 4; Cooper and Eccles, 1930; Kernell et al., 1975). It should be emphasized that stimulus rates corresponding to the optimal T-f range are functionally more relevant than stimulus rates referring to asymptotic phenomenon such as fusion frequency or frequency for maximum tetanic tension. The differences between the T-f relations of fast and slower muscle fibers are probably at least as much dependent on differences in the speed of relaxation as on differences in the speed of shortening. Furthermore, it should be noted that the optimal T-f range shifts to slower stimulus rates as a muscle is extended within its physiologic working range (Rack and Westbury, 1969). All ex-

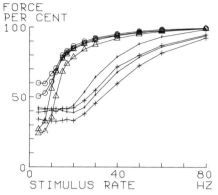

FIG. 5. Slowing effects of chronic stimulation on the tension-frequency relation of fast muscle. Isometric force (%) versus stimulus rate (Hz) for peroneus longus muscles that had previously been subjected to different rates of chronic stimulation (*triangles, circles*) and for their contralateral controls (*crosses*). The plotted values were obtained in each case by stimulating the muscle nerve with 1-sec bursts of brief (0.1 msec) supramaximal pulses at the rates displayed along the abscissa. Plotted force values are the peak forces generated by each such burst, expressed as a percentage of the maximum force of each muscle (stimulus rates up to 200 Hz tested). Prior to this acute experiment, chronic muscle nerve stimulation had, for each one of the 4 cats, been given during 8 weeks in a continuous sequence of 1-sec periods of activation alternating with 1-sec periods of rest. During the recurring activation periods, stimulus rate was slow for two of the muscles (10 Hz; *circles*) and fast for the two remaining ones (40 Hz; *triangles*). In the chronically stimulated muscles, the histochemical reaction for myofibrillar ATPase was about the same in all fibers, and it approached the reaction seen in the normal soleus muscle. All the chronically stimulated muscle nerves were deafferented, and none of the cats showed any behavioral reaction to the stimulation. In the absence of chronic stimulation, deafferented muscles did not become slower than their contralateral controls. (From experiments of Kernell et al., 1981.)

amples of the present article concern isometric contractions at the muscle length for which a twitch had its maximum amplitude (close to maximum physiologic length; cf. Rack and Westbury, 1969).

RHYTHMIC PROPERTIES OF MNS AND THE RATE MODULATION OF MUSCLE UNIT FORCE

In maintained reflex or voluntary contractions, MNs of slow-twitch units become active at lower minimum rates than those encountered among faster muscle units (Fig. 6B; Kernell and Sjoholm, 1975; Grimby et al., 1979). Other experiments have shown that no controlling synaptic machinery is needed for producing such differences between fast-twitch and slow-twitch MNs: also when activated by long-lasting injected currents, the minimum rate of maintained firing is significantly lower for MNs with a slow twitch than for MNs with a faster one (Fig. 6; Kernell, 1979). As previously mentioned, the intrinsic minimum rate of a MN is determined by the duration of its AHP (Kernell, 1965c), and this afterpotential is known to be more long-lasting for MNs of slow muscles and muscle units than for MNs innervating more rapidly contracting fibers (Eccles et al., 1958; Burke, 1967; Huizar et al., 1977; cf. Fig. 3). The matching between AHP duration and muscle unit speed is such that the minimum rate of a MN approximately corresponds to the lower end of the optimal T-f range of its muscle fibers (Fig. 4) (Kernell, 1965b, 1979; Milner-Brown et al., 1937b; Kernell and Sjoholm, 1975; Baldissera and Parmiggiani, 1975; Monster and Chan, 1977). Thus, the minimum rate of a MN is typically close to the frequency at which consecutive twitches of its muscle fibers start to sum; for a given MN—muscle unit combination, the AHP and the twitch are phenomena of rather similar durations. How does this match between muscle speed and MN frequency range come about?

It is known that a fast mammalian muscle becomes slower if it is innervated by the MNs of a slow muscle, and vice versa (Buller et al., 1960; Vrbova et al., 1978). Thus, MNs may have a profound effect on their muscle fibers' speed of shortening as well as

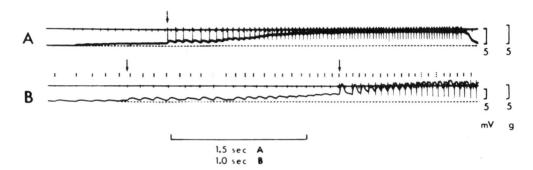

FIG. 6. Ascending-size-order of muscle unit recruitment in the first deep lumbrical muscle of the cat in response to (A) repetitive electrical stimulation of the contralateral motor cortex and to (B) pinching of the ipsilateral plantar cushion. A and B from different cats. In each case recording of EMG (*above*) and isometric tension (*below*). In B, the times of occurrence of the smallest spikes have been indicated by *vertical bars* above the record. *Time calibration:* 1.5 sec for A and 1.0 sec for B. (From Kernell and Sjoholm, 1975.)

of relaxation. With respect to the AHP of MNs, some interesting influences exist in the opposite direction as well (Czeh et al., 1978). There are, however, no known effects from a muscle onto its MNs that are specific for fast or slow muscle fibers (Buller et al., 1960; Kuno et al., 1974). A fast muscle may be slowed down by stimulating it for a number of weeks at rates characteristic for slow MNs (see Vrbova et al., 1978), and a slow muscle may become faster after prolonged activation by bursts of stimuli at high rates (Lømo et al., 1980). Thus, the contractile speed of a muscle might somehow be regulated by the average rate at which it was activated by its MNs during activity. This attractive explanation is unlikely to be valid in its simplest version, however, because a fast muscle may actually become slowed down to roughly the same extent by fast stimulus rates as by slow ones (Fig. 5) (Kernell et al., 1981; cf. Hudlicka and Tyler, 1980).

Not only the minimum rates but also the maximum ones are higher for a MN with a brief AHP and twitch than for cells with longer lasting AHP and twitch (Kernell, 1965c, 1979). The maximum rate within the primary range corresponds roughly to the upper end of the steep portion of the T-f curve (Fig. 4) (Kernell, 1965c, 1966b, 1979); thus, the primary range of a MN corresponds approximately to the optimal T-f range of its muscle fibers. For cells capable of prolonged firing within the secondary range, the absolute maximum rate is close to the stimulus frequency needed for eliciting a maximum tetanic tension from the respective muscle units (Kernell, 1965c, 1966b). It is likely that maintained MN activity usually takes place within the primary range. Firing within the secondary range would, however, be needed for producing the full tetanic tension of a unit. Within the secondary range, the steep slope of the f-I curve tends to compensate for the decreased slope of the T-f curve at high rates.

The time course of the AHP-conductance is not only of major importance for the minimum rate of a MN, but it is probably also one of the main factors determining the maximum rate within the primary range (i.e., the firing rate at which the f-I curve increases its slope; Kernell, 1968; Kernell and Sjoholm, 1973; Baldissera and Gustafsson, 1974; Baldissera et al., 1978). The AHP does not, however, explain why the maximum rate of maintained firing within the secondary range would tend to be lower for slow MNs than for faster MNs (Kernell, 1965c). The highest rate that a MN may sustain without becoming inactivated should mainly depend on its spike generating properties.

The maximum attainable discharge rate of a MN is very much higher for the initial impulse intervals than for the firing obtained during subsequent periods of constant stimulation (Granit et al., 1963a; Kernell, 1965b). Furthermore, thanks to the high gain of a MN at the onset of activation (Fig. 1), very high initial rates, far up into the secondary range, may be evoked by stimulus steps of comparatively moderate intensity. The ease with which a MN produces such high initial rates is of importance for the control of motor behavior, because very high initial frequencies would be needed for producing a high rate of rise of tension at the sudden onset of a contraction (Buller and Lewis, 1965; Baldissera and Parmiggiani, 1975). Furthermore, particularly in slow-twitch units, high initial firing rates may enhance the force production of a unit during a considerable time (seconds) by activating a catchlike mechanism in the muscle fibers (Burke et al., 1976).

RECRUITMENT ORDER AND THE SIZE PRINCIPLE

In mixed limb muscles of cats or man, individual muscle units vary considerably in force; in some muscles, the most powerful units may be over 100 times stronger than the weakest ones (e.g., Wuerker et al., 1965; Milner-Brown et al., 1973a; Jami and Petit, 1975; Desmedt, 1980). In voluntary or reflex contractions, weak units tend to become more easily activated than stronger ones (Fig. 6) (examples from animals: Burke, 1968; Kernell and Sjoholm, 1975; examples from man: Milner-Brown et al., 1973a; Monster and Chan, 1977; Stephens and Usherwood, 1977; Desmedt, 1981a, *this volume*). This tendency will be referred to as the ascending-size-order of recruitment, and it corresponds to an activation sequence according to the size principle of Henneman (1981; Henneman et al., 1965a; Henneman and Olson, 1965). Statistically speaking, an ascending-size-order of recruitment may remain valid even if various synaptic inputs activate a MN pool in a somewhat different sequence (Kernell and Sjoholm, 1975). Mechanically, the ascending-size-order of recruitment has obvious advantages because the relative force contributions of newly recruited units will tend to become as equal as possible for all units if use of the strong ones is reserved for contractions of great total force (Henneman and Olson, 1965; Milner-Brown et al., 1973a). How does the ascending-size-order of muscle unit recruitment come about?

Although the degree of correlation varies between different muscles, strong muscle units generally tend to be innervated by more rapidly conducting (i.e., thicker) axons than those of weaker units (e.g., Wuerker et al., 1965; Burke, 1967). Furthermore, compared to MNs with thin axons, MNs with thicker axons tend to have larger cell bodies (Fig. 7; Cullheim, 1978; Kernell and Zwaagstra, 1981), thicker and somewhat more numerous dendrites, and a larger total surface area (Barrett and Crill, 1974). Thus, statistically speaking, the ascending-size-order of muscle unit recruitment reflects an ascending-size-order of MNs and axon recruitment (Henneman et al., 1965a). Why would the CNS so commonly activate the small MNs of a pool more easily than the larger MNs?

In animal studies, stretch of a muscle leads to an ascending-size-order recruitment of its own MNs via the autogenetic excitation from muscle afferents (Henneman et al., 1965a; Burke, 1968). Part of this autogenetic excitation is caused by direct monosynaptic connections from muscle spindle afferents to MNs, and the monosynaptic excitatory postsynaptic potentials (EPSPs) are known to have a greater maximum amplitude for the most easily recruited MNs than for the other ones (Burke, 1968, 1979). Detailed studies of the central connections of single primary afferents from muscle spindles have shown that, within the homonymous MN pool, a large percentage of all the cells are excited by each single afferent (Mendell and Henneman, 1971). The widespread distribution of synapses means that an increase of excitation to one cell will automatically be accompanied by an increase of excitation to most other MNs of the same population as well. The order of recruitment will then depend on two factors: (1) quantitative differences in the distribution of active synapses to the various MNs of the pool, and (2) quantitative differences in electrical excitability among these MNs. The higher the electrical excitability, the less excitatory current would by definition be needed for recruiting a MN. One of the main factors of importance for the electrical excitability of a MN is its resting input conductance (reciprocal of input resistance). The smaller the input conductance, the less current would by definition be needed for producing a given amount of depolarization. This is, of course, valid for injected as well as for postsynaptic currents. Hence, a difference in EPSP size between various MNs does not necessarily indicate a corresponding difference in the number or density of synapses received by the respective cells. Knowledge concerning the input conductance of MNs is a prerequisite for the analysis. It is well known that slow-axon MNs have a smaller resting input conductance than that of MNs with faster axons (Kernell, 1966a; Burke, 1967, 1968; Barrett and Crill, 1974; Kernell and Zwaagstra, 1981). As one would expect, there is also a positive correlation between input conductance and soma size (Fig. 8; Kernell and Zwaagstra, 1981). The low input conductance of small cells is certainly partly caused by their minute dimensions. This is, however, not the whole truth; input conductance (or resistance) is not a simple measure of neuronal size.

For the total input conductance of a MN, its dendrites are likely to be of major importance. The input conductance of a dendrite would be expected to be proportional to its diameter raised to a power of 1.5 (see Jack et al., 1975). Thus, a doubling of the diame-

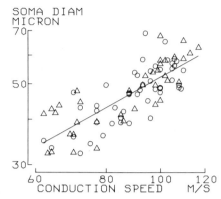

FIG. 7. Relation between diameter of cell body (μm) and conduction velocity of axon (m/sec) among spinal MNs. The cells were anatomically reconstructed after intracellular labeling with procion dye or HRP. Measurements were made of the cross-sectional soma area, and the plotted values for soma diameter were calculated from measured areas according to the formula for a circle. Logarithmic coordinates. Regression line calculated by method of least squares ($r = 0.78$, $N = 87$, $p < 0.001$). *Triangles:* Cells that were also used for the measurements of Fig. 9. (From Kernell and Zwaagstra, 1981, with permission.)

ter of a dendrite would be expected to increase its input conductance by a factor of only $2^{1.5} = 2.8$. A doubling of soma diameter would, of course, be expected to increase soma area by a factor 4. Hence, if all

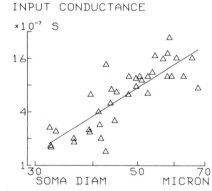

FIG. 8. Relation between input conductance (Siemens) and soma diameter (μm) among spinal MNs. Logarithmic coordinates. Regression line calculated by method of least squares ($r = 0.83$, $N = 37$, $p < 0.001$). Same series of experiments as Fig. 8. (From Kernell and Zwaagstra, 1981, with permission.)

TABLE 1. *Relation between input conductance and size in spinal MNs*[a]

Parameter	Cells with slow axons	Cells with fast axons	p
Axonal conduction velocity (m/sec)	< 90	≥ 90	
Soma diameter (μm)	39.5 ± 5.1	54.7 ± 6.9	< 0.001
Input conductance/soma area (mS/cm^{-2})	6.6 ± 3.1	12.8 ± 5.1	< 0.001
$\Sigma\ d^{3/2}$/soma area (cm$^{-1/2}$)	1.9 ± 0.6	1.9 ± 0.4	n.s.
N	17	20	

[a] Anatomical dimensions of MNs measured after intracellular labeling with procion dye or HRP. Soma area taken to be 4 times the measured cross-sectional area of the cell body. Soma diameter calculated from the measured cross-sectional area according to the formula for a circle; d = dendritic stem diameter. Statistical significance of differences tested by t-test; means ± SD. (p = probability; n.s. = not significant).

For comparisons to recruitment experiments, the cells have been grouped according to their axonal CV; the dividing line was arbitrarily chosen at 90 m/sec. From Kernell and Zwaagstra, 1981, with permission.

cells had the same specific membrane resistance and the same relative dimensions, the ratio between neuronal input conductance and soma area would be expected to be smaller for large than for small cells. In reality, the situation is quite the opposite: the ratio between input conductance and soma area is substantially larger for large fast-axon MNs than for the smaller MNs with slower axons (Table 1, Kernell and Zwaagstra, 1981). The unexpectedly high input conductance of large MNs was not caused by the presence of unexpectedly many and/or thick-stem dendrites among these cells (see values for $\Sigma\ d^{3/2}$/soma area in Table 1). Hence, the findings summarized in Table 1 indicate that the mean specific membrane resistance tended to be greater in the small spinal MNs than in the larger sized MNs. This means that, even if all the MNs of a pool were simultaneously activated by the same density of synapses, the small low-conductance MNs would show a larger postsynaptic depolarization and become more easily recruited than the larger MNs (see Kernell and Zwaagstra, 1981).

Thus, the most common order of MN recruitment would probably prevail even if the relevant synapses were randomly distributed among the cells of a pool. This does not, of course, exclude that other mechanisms may contribute as well to the ascending-size-order of recruitment. It has been suggested that, because of extensive branching, axons terminating on large cells might be less effectively invaded by presynaptic impulses than axons ending on smaller cells (Luscher et al., 1979). The overall density of synapses seems to be about the same for all ventral horn cells irrespective of their size (Gelfan and Rapisarda, 1964). There is as yet no evidence to indicate that small and large MNs would differ systematically with respect to the average amount of depolarization needed for spike initiation. When calculated for unpublished results of Kernell (1966a), the product of input resistance and rhythmic threshold-current (= equivalent voltage threshold) was not significantly correlated to axonal conduction velocity ($r = -0.31$, $N = 17$, $p > 0.1$; cf. also Stein and Bertoldi, 1981).

A different synapse distribution to MNs innervating different types of muscle units might conceivably be easily organized if the various cells were consistently situated in anatomically segregated regions. Within spinal MN pools, the craniocaudal distribution of MNs of different sizes (Burke et al., 1977) or muscle unit types (Kernell et al., 1979) is not always random. The degree of topographic type segregation seems, however, far too weak to form a useful basis for a type-specific distribution of synapses to various kinds of MNs (Burke and Tsairis, 1973; Burke et al., 1977; Kernell et al., 1979).

RECRUITMENT ORDER AND MUSCLE UNIT ENDURANCE

The most easily recruited units of a muscle would also be the ones most frequently used. Hence, it is appropriate that the smallest axons and MNs, which would usually be the ones most easily recruited, are provided with relatively slow muscle units of high endurance (Burke et al., 1973; Stephens and Usherwood, 1977). Among fast-twitch units, however, there is, in many muscles, no evident difference in axonal conduction velocity (CV) between MNs with fatigue-sensitive and fatigue-resistant muscle fibers (e.g., for the gastrocnemius in the cat, Burke et al., 1973). In spite of this apparent similarity in neuronal size, a tonic stretch reflex seems to recruit the fatigue-resistant FR units more easily than the stronger and more fatigue-sensitive FF units (experiments on the cat gastrocnemius; see Burke, 1979, 1981). Do FR and FF MNs commonly differ from each other with respect to their electrical excitability?

In our own recent experiments on gastrocnemius MNs of anesthetized cats, a repetitive discharge was elicited by a much weaker injected current in FR MNs than in the FF cells. The rhythmic threshold was 10.9 ± 4.7 nA (mean $\pm$ SD) for 11 FR MNs and 21.1 ± 8.3 nA for 15 FF MNs ($p < 0.005$; Kernell and Monster, 1981). With respect to the rheobase for a single spike, similar findings have been reported by Munson et al. (1980). These differences in electrical excitability seem mainly to be dependent on the fact that the neuronal input resistance is greater for FR than for FF MNs (Dum and Kennedy, 1980). Thus, even if all the fast-twitch MNs of a pool were simultaneously activated by the same amount of postsynaptic excitatory current, cells with fatigue-sensitive muscle fibers (FF) would still show a smaller postsynaptic depolarization and be less easily recruited than cells with fatigue-resistant fibers (FR). Among MNs of the same size, differences in input resistance would depend on differences in mean specific membrane resistance. Thus, the greater excitability of FR than of FF MNs might perhaps be caused by differences between these two kinds of cell with respect to their passive membrane properties.

It is well known that the fatigue-resistance of muscle fibers may be enhanced by training (e.g., Holloszy and Booth, 1976). The greater contractile endurance of FR than of FF muscle units might be a secondary result of a more extensive usage of FR than of FF MNs, caused by their difference in electrical excitability (see Edgerton and Cremer, 1981; Hainaut et al., 1981).

DURATION OF FIRING AND MUSCLE UNIT TYPE

As we have seen, the last recruited units of a muscle are likely to be the fast-twitch fatigue-sensitive ones, which would typically be expected to be used mainly for relatively brief force explosions (Walmsley et al., 1978). Are MNs of FF units then generally intrinsically incapable of tonic firing? When penetrated by a microelectrode and stimulated by steady current, many MNs refuse to deliver a discharge lasting for even ≥ 1 sec (Kernell, 1965a). Mishelevich (1969) found such phasic response patterns in many fast-twitch MNs but not in slow-twitch MNs. However, phasic discharge patterns may easily arise in a penetrated MN as a result of some kind of damage by the microelectrode (Kernell, 1965a; Schwindt, 1973). The mechanisms of these noxious effects are unknown (see p. 1301 of Schwindt and Crill, 1980). The emergence of a phasic response pattern is typically an early sign of cell deterioration, and this reaction may be seen in MNs with any kind of muscle unit. The findings of Mishelevich (1969) might, for example, mean that fast-twitch MNs tend to be more sensitive than the slow-twitch ones to the noxious effects of microelectrode penetration. It should be emphasized, however, that well-maintained discharges may be elicited by constant stimulation in MNs innervating any one of the 3 major types of muscle unit (FF, FR, S;

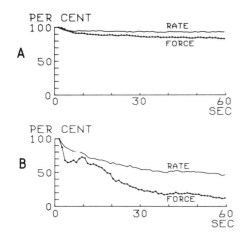

FIG. 9. Decline in firing rate and mean force for (A) slow-twitch and (B) fast-twitch muscle unit from m.gastrocnemius medialis during constant intracellular stimulation of the respective MNs at 5 nA above rhythmic threshold. All values given as percentage of the respective type of measurement at the second sec of discharge. For the unit of **A**, the first plotted rate (Second sec) was 16 Hz, the twitch contraction time was 53 msec, and the fatigue index (Burke et al., 1973; Kernell et al., 1975) was 0.97. For the unit of **B**, the corresponding values were 29 Hz, 25 msec, and 0.07. Despite the low fatigue index of the latter unit, most of the force-drop in **B** was probably caused by the drop in discharge rate (see text). It should be noted that the firing rates in B (from 29 to 13/sec) were substantially slower than those used for the determination of fatigue index (bursts of 40 Hz). (Modified from Kernell and Monster, 1982*b*.)

Kernell, 1979; Kernell and Monster, 1982*b*; cf. Kernell and Sjoholm, 1975; Freund et al., 1975). The FF MN of Fig. 9B did, for example, discharge continuously throughout our total test period of 4 min. There is no known physiologic class of MNs that consistently refuses to produce a tonic discharge in response to a steady stimulation of adequate intensity.

STABILITY OF FIRING RATE AND MUSCLE UNIT TYPE

In voluntary contractions, long-lasting steady discharges usually seem to be most easily produced in MNs, with a slow twitch and a high resistance to contractile fatigue (e.g., Stephens and Usherwood, 1977; Grimby et al., 1979). Cat experiments have suggested that in hindlimb muscles, normal posture is mainly maintained by activity in the very fatigue-resistant S units (Walmsley et al., 1978). Are the MNs of S units then in any way more suitable than those of FR and FF units for the production of a maintained postural contraction? With this question in mind, we have recently been comparing different types of gastrocnemius MNs with respect to their late phase of adaptation (Kernell and Monster, 1982*b*).

We estimated the amount of late adaptation by measuring the drop in firing rate from the 2nd to the 26th sec of a discharge, produced by constant intracellular stimulation at 5 nA above the rhythmic threshold. This drop in discharge rate was much less pronounced for slow-twitch MNs than for those innervating faster muscle fibers (correlation between frequency drop and twitch contraction time: $r = -0.76$, $N = 23$, $p < 0.001$). Among the fast-twitch cells, there were no significant differences between FF and FR MNs. Thanks to their low amount of late adaptation, slow-twitch MNs seem more suitable than the fast-twitch MNs for producing a maintained contraction in response to steady stimulation. Two examples are shown in Fig. 9. Both units were stimulated by the same amount of suprathreshold current. In the S unit there was little drop in either rate or force with time (Fig. 9A). In the FF unit, firing rate as well as force showed a conspicuous decrease during the first minute of constant activation (Fig. 9B). Even in the absence of fatiguing and/or potentiating processes in the periphery, the MN adaptation of Fig. 9B would have been expected to have caused a drop in force by about 80% over the time illustrated (calculated by aid of average T-f curve for gastrocnemius units of corresponding twitch speed).

Thus, at the moderate discharge rates of Fig. 9B, only a minor part of the total drop in force was apparently due to peripheral fatigue in the muscle unit. Initially in the discharge, on the other hand, the effects of MN adaptation were evidently counteracted by peripheral potentiation (see Kernell et al., 1975) at 5 to 10 sec after onset of activation, the force was increasing even while firing rate was continuously decreasing (Fig. 9B). For fast-twitch units that are activated by a constant excitatory drive, the tension produced in a maintained contraction depends on a complicated balance between the force effects of MN adaptation, peripheral potentiation, and peripheral fatigue. Why is the amount of late adaptation so much less marked for slow-twitch MNs than for the fast-twitch MNs? As noted, the late phase of adaptation is likely to be caused by cumulative aftereffects of many consecutive spikes in a discharge. Thus, the drop in rate over a given period of constant stimulation should be more pronounced the greater the number of spikes fired per unit time. Indeed, we found a very high correlation between the starting rate of such a discharge (measured at 2 sec) and the subsequent decline in discharge frequency (measured over 2 to 26 sec of constant stimulation; $r = 0.92$; 23 cells, $p < 0.001$). Thus, the small amount of late adaptation in slow-twitch MNs seems a consequence of the fact that when activated by a given amount of suprathreshold current, these cells start to discharge at slower rates than those of fast-twitch MNs. For cells of gastrocnemius muscle, such differences in starting rate would mainly reflect differences in minimum rate (Fig. 10; Kernell, 1979), which are caused by the differences between fast-twitch and slow-twitch MNs with respect to their duration of AHP (Eccles et al., 1958; Kernell, 1965c).

THE BALANCE BETWEEN RATE AND RECRUITMENT GRADATION OF MUSCLE CONTRACTION

If all the MNs of a pool were only just recruited without any further increase in rate, they would all fire at their minimum impulse frequency. At the minimum rate of a MN the twitches of its muscle fibers bare-

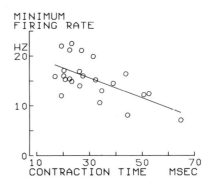

FIG. 10. Relation between minimum rate of maintained firing of gastrocnemius MNs (Hz) and the twitch contraction time of their muscle units (msec). Regression line calculated by method of least squares ($r = -0.62$, $N = 24$, $p < 0.005$). Contraction times measured from onset to peak of nonpotentiated twitches. (From experiments of Kernell, 1979, with permission.)

ly sum, and the mean force produced is typically only some 10 to 25% or less of the maximum tetanic tension (Kernell and Sjoholm, 1975; Kernell, 1979). Hence, most of the total force that a muscle might deliver would be unusable if there were no rate modulation of the MNs. In reality, the maximum tetanic force of a muscle can be mobilized in a voluntary contraction (Milner-Brown et al., 1973b; Desmedt, 1958, 1973b; Jones et al., 1979). Thus, when considering the whole working range of a muscle, rate modulation is indeed very important (Milner-Brown et al., 1973b; Kernell and Sjoholm, 1975).

In practically all known types of centrally evoked contractions, an increased recruitment is paralleled by an increased discharge rate among already active units (Fig. 6, Milner-Brown et al., 1973b; Kernall and Sjoholm 1975; Kernell, 1976). Such a simultaneous use of both mechanisms is very much to be expected: if synaptic inputs to a MN pool have a widespread distribution (cf. Mendell and Henneman, 1971), then an increase in excitation to one MN would automatically be accompanied by an increase in excitation to most of the other MNs as well.

How are the two parallel mechanisms for

force gradation balanced? That might, of course, vary considerably between different contractions depending on the distribution of synaptic excitation and inhibition among the MNs of the pool. The smallest and slowest MNs of a pool have a more narrow range of rate modulation than that of the fastest and largest MNs (Kernell, 1965c, 1979; cf. Table 2). To some extent, this difference in frequency range might be compensated for by the presence of a steeper f-I slope in large than in smaller cells (Kernell, 1966a). Such systematic differences in f-I slope are, however, not evident within all MNs pools (Kernell, 1979). How is it avoided that the smallest and most easily recruited MNs of a pool become saturated and/or inactivated before the largest and fastest ones have become maximally activated? This problem will be discussed by aid of a very simple pool model the main characteristics of which are summarized in Table 2. In this context, only firing within the primary range will be considered; firing within the secondary range seems less stable (Kernell, 1965b) and might perhaps be needed predominantly for maximal contractions of brief duration (e.g., a few sec or less). It is seen in Table 2 that, for the two versions of this model, the total stimulus intensity needed for producing a maximum discharge rate within the primary range is 2.1 to 3.4 times greater for the largest MNs than for the smallest MNs. Thus, if all the MNs of the pool were activated by the same absolute amount of excitatory current, the smallest MNs would become far too strongly excited before the largest ones reached their maximum rate. In general, however, all cells would not be expected to receive the same absolute amount of synapses, because the synaptic density appears to be about the same for large and smaller neurons of the ventral horn (Gelfan and Rapisarda, 1964). If the cells of the model of Table 2 were all activated by the same current density, then the absolute intensity of excitatory current would at all times be much greater for the largest cells than for the smallest ones. In such a case, simultaneously occurring changes of firing rate, as seen over a given range of total muscle

TABLE 2. *Properties of model of spinal MN pool*

Parameter	Smallest cells	Largest cells	Ratio largest/smallest
Recruitment data			
Membrane area (arbitrary units)	1	4	4
Input resistance (MΩ)	5	0.5	0.1
Threshold current (nA)	3	30	10
Rate data (primary range)			
AHP duration (msec)	200	50	0.25
Minimum rate (Hz)	6	21	3.5
Maximum rate (Hz)	31	79	2.5
Version 1			
f-I slope (Hz/nA)	1	2	2
Current for maximum rate (nA)	28	59	2.1
Version 2			
f-I slope (Hz/nA)	1.4	1.4	1
Current for maximum rate (nA)	21	71	3.4

AHP = afterhyperpolarization. f-I slope = slope of frequency-current relation. Current for maximum rate = threshold current + (maximum rate − minimum rate)/f-I slope. Membrane areas are considered to be proportional to the square of soma diameters; the largest α-MNs have about twice the soma diameter of the smallest ones (Zwaagstra and Kernell, 1981). Other data for the model are approximations based on experimental observations by: Kernell (1966a; input resistances, threshold currents, f-I slopes of version 1), Eccles et al. (1958; AHP durations), Kernell (1965c; relation between AHP duration and limits of firing rate), Kernell (1979; f-I slopes of version 2).

force, might be expected to be several times larger for the last recruited MNs than for the MNs recruited already in the very weakest contractions. Such differences in rate modulation have actually been observed in voluntary contractions (Gydikov and Kosarov, 1974; Monster and Chan, 1977; Monster, 1979). For approximate calculations on the model pool of Table 2, one may make the simplifying assumption that the absolute intensity of excitatory current is proportional to neuronal surface area (complications arising from cell geometry neglected). In such a case, a full activation of the largest cells would be associated with a net excitation of only 14.8 (59/4; version 1) or 17.8 (71/4; version 2) nA in the smallest MNs, which is less than the intensity these latter MNs would require for a maximum activation (28 and 21 nA for versions 1 and 2, respectively, Table 2). These approximate calculations suggest that, if all the MNs of a spinal pool were activated by about the same density of synapses, an adequate balance between recruitment and rate modulation might follow automatically from the distribution of MN sizes and inherent response properties (e.g., specific membrane resistance, frequency range, f-I slope). Special synaptic systems for the selective suppression of small cells would, for example, generally not seem to be needed to preserve activity in a whole pool throughout a broad range of force gradation.

Size Principle of Motoneuron Recruitment and the Calibration of Muscle Force and Speed in Man

*John E. Desmedt

Brain Research Unit, University of Brussels Faculty of Medicine, Brussels, Belgium

Electrophysiological studies of voluntary motor control in man involves several types of noninvasive methodologies: (1) computer averaging of scalp recorded cerebral potentials that precede or accompany voluntary movements (Bereitschaftspotential or readiness potential, motor potentials) (Deecke et al., 1969, 1977; Deecke and Kornhuber, 1977, 1978; Kutas and Donchin, 1977; Gerbrandt, 1977; Marsden et al., *this volume*); and (2) recording of muscle potentials [electromyography (EMG)] by surface or concentric needle electrode (Adrian and Bronk, 1929) which is a well-established method in clinical EMG (Kugelberg, 1947; Buchthal, 1961; Desmedt, 1981b; Milner-Brown et al., 1981), in kinesiology (Bouisset, 1973; Bouisset and Maton, 1973; Joseph, 1973; Petersen, 1973; Carlsoo et al., 1973; Vredenbregt and Rau, 1973; Basmajian, 1978), and in studies of motor control mechanisms. A unique feature of the needle EMG method is that it directly uncovers activity patterns of single spinal motoneurons (MNs): when an MN is synaptically activated, the action potential is conducted in all the ramifications of its motor axon, and it triggers (with a high safety factor) an action potential in all the muscle fibers innervated by that MN. Sherrington identified the functional unit of motor control as the motor unit (MU), which is a single MN and all the skeletal muscle fibers innervated by its axon (see Creed et al., 1932). Motoneurons represent the final common path through which all central nervous system (CNS) motor commands must be channeled. The degree of fractionation of muscle contraction that the CNS can achieve in any movement cannot go beyond the limit represented by the single MU. In fact, the size of MUs (in terms of number of its muscle fibers and amount of mechanical force it produces) differs characteristically in different muscles, in conjunction with the precision of movement control actually achieved by the muscle (see Buchthal and Schmalbruch, 1980).

Early EMG studies have emphasized that voluntary contractions of increasing force tend to recruit MUs with larger and larger action potentials. It is also known that mammalian muscles include muscle fibers of different biochemical types. Any MU is homogeneous in this respect, since it only includes muscle fibers of one specific type (Brandstater and Lambert, 1973; Kugelberg, 1973, 1981). This is in line with evidence that the type of motor innervation (hence of MN) is a major factor in the speci-

---

*Brain Research Unit, University of Brussels Faculty of Medicine, 115 boulevard de Waterloo, Brussels 1000, Belgium.

fication and subsequent maintenance of muscle fiber types (Buller et al., 1960; Vrbova et al., 1978; Lomo et al., 1980; Buchthal and Schmalbruch, 1980; Brown et al., 1981; Desmedt, 1981a). At least 3 main muscle fiber types with characteristic properties are commonly distinguished. The type IIb or fast glycolytic, fatigable (FF) MUs have a larger innervation ratio (more muscle fibers per MU) and muscle fibers of larger sectional area that produce a larger specific force output (Burke, 1973; Burke and Tsairis, 1973; Edgerton and Cremer, 1981). The FF MUs produce more mechanical force and are innervated by a larger MN that has more membrane surface area, smaller input resistance (Kernell, *this volume*) and a motor axon of larger diameter (Burke, 1973; Barrett and Crill, 1974; Cullheim, 1978). In contrast, the type I or S slow, oxidative) MUs develop less mechanical force in conjunction with a smaller innervation ratio (less muscle fibers per MU) and with muscle fibers of smaller sectional area that produce less specific force output than the FF (Burke and Tsairis, 1973; Burke et al., 1974). The S motor unit is innervated by a smaller MN with a smaller membrane surface area, a higher input resistance and a motor axon of smaller diameter. The type IIa or fast oxidative-glycolytic, fatigue-resistant (FR) MUs can be considered intermediate (see Burke and Edgerton, 1975; Edgerton and Cremer, 1981).

The MNs innervating one muscle share common connectivities and functional properties, and they are considered to form an MN pool (Creed et al., 1932; Mendell and Henneman, 1971; Burke et al., 1977). The patterning of motor commands in a given MN pool is generally discussed in terms of Henneman's size principle (Henneman et al., 1965a,b; 1974; Henneman, 1981) which has largely been substantiated by recent studies as a basic concept, even though specific exceptions have been documented (Burke, 1968, 1973, 1981; Burke and Edgerton, 1975; Edgerton and Cremer, 1981; Stein and Bartoldi, 1981; Desmedt, 1980a, 1981a, Desmedt and Godaux, 1981; Kernell, *this volume*). The size principle implies an orderly functional organization of the MN pool whereby small MNs are recruited before larger MNs. Motoneurons themselves cannot be studied in intact man, but these problems can be tackled through an analysis of single MUs.

The mere voltage amplitude of a recorded MU may bear some relationship to the MU actual size (in terms of number of muscle fibers) (Goldberg and Derfler, 1977), but this is somewhat unreliable. The recorded shape and voltage of a MU potential are much influenced by the geometric relationships between needle electrode tip and the muscle fibers that are scattered throughout an extended MU territory (cf. Buchthal and Schmalbruch, 1980; Desmedt, 1981b). A safer method for identifying MU (and thus MN) size is based on computer averaging of the single MU twitch by the method of spike-triggered averaging (STA) (Stein et al., 1972; see review by Kirkwood and Sears, 1980). Briefly put, the twitch of a selected MU the function of which is being studied in natural movements, is extracted by electronically averaging the analog force output of the whole muscle, while the duty cycle of the computer is triggered via a window trigger circuit by the action potentials of the chosen single MU (Fig. 1A).

The first dorsal interosseous muscle has fairly parallel fibers so that rather similar mechanical conditions prevail for the different MUs that all pull obliquely on the distal tendon at the base of the index finger. The twitch force extractated by STA for the different MUs can be considered as somewhat smaller than, but roughly proportional to, the actual force exerted by each of these MUs on the muscle tendon. The twitch forces of interosseous MUs range from 0.1 to 10 g (Fig. 2), while the twitch contraction times (CT) range from 25 to 85 msec. One point must be made about the fact that STA extractions are usually carried out on a MU

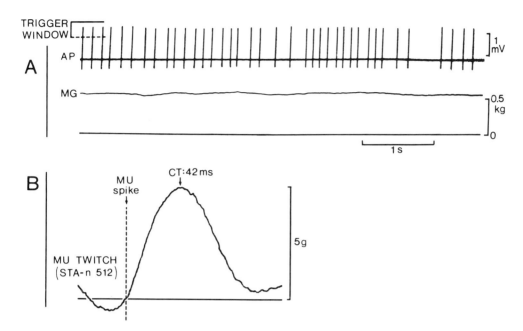

FIG. 1. The extraction of single motor unit twitch by spike-triggered averaging of the isometric muscle force. A: Recording of action potentials of a single MU with a highly selective bipolar needle electrode (2 insulated Nichrome wires 40 μm diameter stuck together a few micrometers apart in a hypodermic needle) (AP, upper trace), and of interosseous muscle isometric force (MG, lower trace). An electronic window trigger, set as indicated on the *left*, ensures that no potentials smaller or larger than that of the selected MU will trigger the duty cycle of the averaging computer. B: Single MU twitch extracted from 512 samples. The onset of MU twitch is taken from the time of the MU spike *(vertical interrupted line)*. (From Desmedt, 1981a, with permission.)

firing at 8 to 10/sec during the steady voluntary contractions. Such repetitive activation must produce some staircase effect whereby the MU twitch force increases and the MU twitch CT decreases with respect to control resting conditions (Desmedt and Hainaut, 1968). Such changes may affect slow MU types to a larger extent (P. Bawa, *personal communication),* leading possibly to an underevaluation of the true range of MUs forces or CTs. Studies of interosseous MUs activated by electrical microstimulation in the muscle at low discharge rates (2-sec intervals) indeed discloses a somewhat wider range of CTs, from 34 to 140 msec (Young and Mayer, 1981). However, the STA method (Fig. 1) is generally preferred, because it is difficult to relate the MUs studied by microstimulation to those recorded in natural voluntary movements.

MU RECRUITMENT THRESHOLD AND TWITCH FORCE

During steadily increasing voluntary contractions (slow ramp), single MUs are recruited in an orderly manner according to the force of their twitch (Fig. 2). The degree of ordering is remarkably high and correlation coefficients are greater than 0.8 and even 0.9 (Milner-Brown et al., 1973a, 1981; Tanji and Kato, 1973, 1981; Freund et al., 1975; Desmedt, 1981a). This extends Henneman's (1981) size principle based on animal experiments to slow voluntary contractions in man. This statement assumes that, as in mammals, the MU twitch force roughly indexes the MU size (see above), and hence also the size of the corresponding spinal MN.

The CTs of single MUs tend to be larger

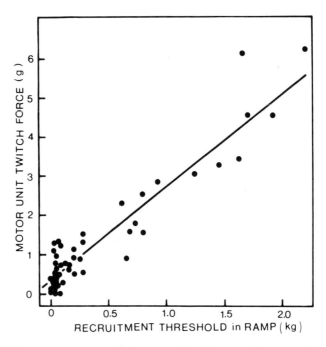

FIG. 2. First dorsal interosseous muscle of a normal adult man. The recruitment threshold of 50 single MUs in slow voluntary abduction of the index finger is plotted in the abscissa (kg). The twitch force of each MU (extracted by STA) is plotted in the ordinate (g). Linear correlation exceeds 0.9 between the two parameters. (From Desmedt, 1980, with permission.)

for MUs with lower recruitment thresholds, but the correlation coefficient is only about 0.6 (Milner-Brown et al., 1973a; Desmedt and Godaux, 1977b). The functional organization of the human hand may emphasize the MU force output differential as a more relevant parameter than the MU twitch CT (Desmedt, 1981a). Contractile properties of a single human MUs and their plasticity with usage should be analyzed further (Edgerton and Cremer, 1981; Hainaut et al., 1981).

Ballistic Versus Ramp Modes of Voluntary Motor Control

In view of the range and versatility of voluntary movements of hand muscles which receive a rich corticomotoneuronal innervation in primates (cf. Phillips, 1969; Kuypers, 1973; Phillips and Porter, 1977), it can be asked whether exceptions to the size principle might not occur in finger movements. In such studies it is important to contrast the motor commands and EMG patterns for movements carried out either in the ballistic mode or in the ramp mode (Woodworth, 1899; Stetson and McDill, 1923; Wachholder and Altenburger, 1926). Briefly, a movement carried out at voluntary speed (time-to-peak exceeding about 400 msec) is continuously controlled by sensory input from the moving body parts; in such movements the motor commands can be adjusted throughout by comparing the current with the intended position, as in the feedback control of a guided missile. In contrast, a ballistic movement is preprogrammed by the motor centers and triggered off as a unit whereby it has to run its full course virtually without the possibility of modification; this situation is somewhat like that of a projected missile for which no corrections can be made after the trigger has been pulled.

Electromyographic recordings provide evidence for a clear difference between such continuously controlled or ballistic movements (Wachholder and Altenburger, 1926; Wachholder, 1928). Figure 3A shows a brisk voluntary contraction of the first dorsal interosseous in normal man. A brief strong burst of muscle potentials precedes onset of ballistic abduction of the index finger by about 40 msec, and the time-to-peak of the angular movement of 25° is only about 80 msec. The EMG burst is completed when the angular displacement reaches about midamplitude, the ballistic movement being carried on by its own momentum until it is arrested against a block (as in typing, piano playing, etc.), or slows down and stops. Electromyographic bursts of similar pattern are recorded in brisk, isometric contractions with no external displacement (as when pressure controls are operated by fingers). On the other hand, when limb segments of sizable inertia are moved briskly, the momentum may carry the movement too far unless antagonistic muscle activity produces a decelerating force. A triple burst is then recorded involving first the agonist, second the antagonist, and then the agonist again (Wachholder, 1928; Angel, 1974, 1981; Hallett et al., 1975a; Lestienne, 1979; Hallett and Marsden, 1981; Ghez et al., *this volume*). The timing and strength of the antagonistic burst are centrally programmed so as to achieve ballistic displacements of appropriate size. However, the essential fea-

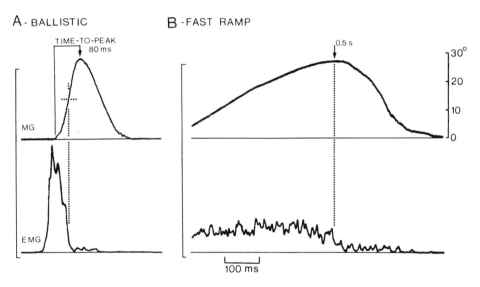

FIG. 3. Patterning of motor commands in ballistic (**A**) or fast ramp (**B**) voluntary contraction in man. First dorsal interosseous muscle. The index finger is attached to a lever fixed on the shaft of a torque motor (Printed Motors Ltd, type G 9M4H) producing a constant load except between trials when the lever rests against a stop. The abduction of index finger is carried out self-paced. The displacement *(upper trace)* of about 25° is recorded along with the rectified-integrated surface EMG *(lower trace),* and 16 trials are averaged on computer. The load is 7 Newtons/cm in **A** and the ballistic movement has a time-to-peak of 80 msec. The EMG burst starts before any movement is recorded due to excitation-contraction coupling delays and to load opposing the movement. The ballistic burst is virtually completed by the time the movement is about half-way to its peak *(vertical dotted line)*. **B**: A fast ramp (against a load of 17 Newtons/cm) carried at voluntary speed with a time-to-peak 500 msec; EMG is continuous and starts decreasing at about peak of movement only *(vertical dotted line)*. (From Desmedt, 1981a with permission.)

ture of the ballistic mode is the initial burst in agonist muscle (Fig. 3A). The delay between EMG burst and onset of movement depends on delay in excitation-contraction coupling and time required to generate enough force to overcome the load. It is remarkable that the duration of the ballistic surge of muscle force changes rather little for contractions of different sizes (see Fig. 9).

In a ramp contraction with the force rising smoothly at voluntary speed, the EMG presents a continuous graded activity that only starts declining at about ramp peak (Fig. 3B) (Wachholder, 1928), even in relatively fast ramp contractions that reach their peak force in 0.5 sec.

The rather simple dichotomy between ramp and ballistic contractions is supported by various data (see Desmedt and Godaux, 1978a; Evarts and Fromm, 1978; Cooke, 1980; Desmedt, 1981a), but it must be stressed that both modes of motor commands are blended in many skilled movements.

The two main modes of packaging of central motor control must be considered in conjunction with acquisition of bodily skills and with use of sensory cues and corollary discharge for the calibration of commands (see Desmedt, 1981a; Bizzi and Abend, *this volume*; Sanes and Evarts, *this volume*). In the simple skilled response, a ballistic pattern conceivably involves open-loop release of a brief package of motor commands that elicit the MN burst in the agonist MN pool; sensory feedback and knowledge of environmental results of the movement then allow recalibration of the commands package for subsequent trial, as skilled performance improves or adapts to changed conditions (Desmedt, 1980a, 1981a).

Recruitment of Single Motor Units in Fast Ramps or in Ballistic Contractions

The voluntary motor commands to MN pools achieve precise titration and timing of MUs activities in skilled mechanical performance. When voluntary ramp contractions are carried out with increasing speed, the force level at which a MU is recruited tends to drop as the time-to-peak of the ramp decreases from about 2 to 0.5 sec. (Tanji and Kato, 1973; 1981). Figure 4A documents a tibialis anterior MU with a high threshold of 10 kg in slow ramp (rate of rise of force 1.4 kg/sec). The MU becomes active at the same force level in the third ramp (from the right) carried out at 6 kg/sec. It is already recruited at 4 kg in the fifth ramp carried out at 24 kg/sec. Similar results were reported for small hand muscles for ramp rates of 2 kg/sec (Budingen and Freund, 1976). The data for fast ramps make sense on mechanical grounds. The excitation-contraction delay between action potential and force production in muscle fibers is no longer negligible when compared to the rate of force increase in these ramps, and earlier firing of MUs ensures that their twitch force contributes at about the same level of force output of the whole muscle (Budingen and Freund, 1976).

Ballistic contractions, however, reach peak in about 100 msec, and the rate of force development in tibialis anterior is as high as 100 kg/sec. The same MU now fires before the muscle produces any recordable force (Fig. 4A). This fits in with the gross EMG data showing the ballistic burst to precede the onset of force (Fig. 3A). The argument for fast ramps makes no sense in ballistic conditions (for example by saying that the firing of any MU would be timed so that the peak of its twitch would occur at the same level of muscle force), since the range of twitch CT of single MUs (see above) is very much wider than the intervals of only a few msec between the first spikes of different (slower and faster) MUs in the ballistic burst (Fig. 5D, G). An entirely different kind of motor control is, in fact, operating in ballistic contractions which involve a brisk surge of force followed by silence agonist muscle (Fig. 3).

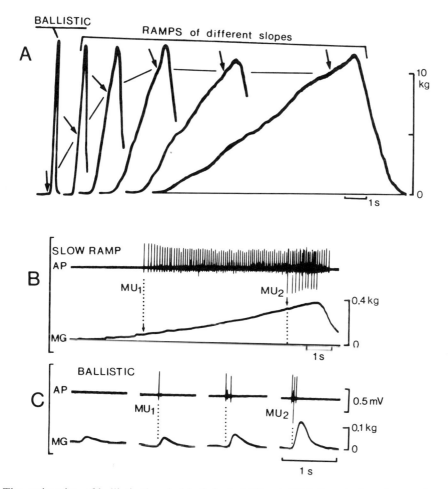

FIG. 4. The estimation of ballistic threshold of single MUs. A: Tibialis anterior of normal man. On brisk ballistic contraction of 12-kg peak force *(left)* and several ramp contractions of same force carried out at different voluntarily controlled speeds *(right)*. The *arrows* point to onset of firing of the same (high-threshold) single MU. The MU starts firing at 10 kg in slow ramps, but at smaller forces in faster ramps, and at "zero" force in the ballistic contraction. (From Desmedt, 1980a, with permission.) B,C: First dorsal interosseous muscle of normal man. Two single MUs are recorded in slow ramp contraction (B) and in 4 ballistic contractions of different peak force (C). Motor unit 1 (MU_1) has a ramp threshold of 40 g, and its twitch extracted by STA has a force of 0.3 g and a CT of 65 msec. Motor unit 2 (MU_2) has a ramp threshold of 310 g, and its twitch has a force of 1.5 g and a CT of 39 msec. The smaller, slower MU_1 is recruited before MU_2 both in ramp (B) and in ballistic contractions (C). MU_2 (spike identified by larger downward phase) only fires in the larger ballistic contraction *(right)*, while MU_1 fires 3 spikes in that contraction. The brief ballistic burst of MU spikes occurs before the muscle develops any significant force. AP: Action potential; MG: isometric myogram. (From Desmedt and Godaux, 1977b, with permission.)

Therefore, the recruitment force threshold of MNs in ballistic contractions cannot be estimated from the muscle force at the time of MU firing since this is virtually zero force for most MUs. There is thus a problem of experimental design, which has been resolved by using for titration the peak force of the smallest ballistic contraction in which that MU is activated (Desmedt and Godaux, 1977a,b). The subject carries out at easy intervals of a few seconds at least 50 ballistic contractions that achieve a wide range of peak forces from a few grams to near maximum muscle force to explore a wide range of ballistic forces. Any given MU fails to be activated in all the ballistic contractions which fail to reach a certain peak force; it discharges with increasing probability for slightly stronger contractions and fires in all the ballistic contractions that produce a larger force. The ballistic force threshold is taken as the mean between the maximum peak force for which the MU never discharges and the minimum peak force for which it always fires (Fig. 4B). The 2 MUs present clearly identifiable action potentials and their slow ramp thresholds are 40 and 310 g, respectively. Figure 4C compares ballistic contractions of different amplitudes. The smaller MU fires in all contractions exceeding 20 g peak force, and the larger MU is only recruited for ballistic contractions exceeding 110 g. The smaller MU fires 2 spikes (third trace from left) or 3 spikes (fourth trace) as the ballistic contraction increases above its own recruitment threshold.

Figure 5 provides another example with 3 single MUs simultaneously recorded from the same site and with consistent recruitment thresholds in slow ramp contractions (Fig. 5A). In this experiment the action potential of the second MU recruited was smaller than that of the first MU. The true size of a MU can, indeed, be more reliably estimated from its twitch extracted by STA. The same MUs are displayed in ballistic contractions of 3 selected peak forces in Fig. 5B–D. Several trials are shown for each of the 3 forces in order to show the uncertain MU firing close to ballistic threshold. MUs spikes are separated by intervals of only a few milliseconds in the ballistic bursts and their actual sequence is somewhat variable from trial to trial. Thus, in the 6 trials of Fig. 5F, MU_1 fires ahead of MU_2 twice, but thereafter 4 times. In the stronger ballistic contractions of Fig. 5G, MU_1 fires before MU_2 5 times in 6 trials, while MU_3 always fires after MUs 1 and 2.

Such variations in the temporal order of initial firing provide no evidence for or against the rank order of MUs. One reason is that these are spikes recorded in the muscle after different conduction delays along the motor axons. The illustration in Fig. 6 depicts 2 MNs of different size with their MUs. The smaller MN with an axon of smaller diameter (see Burke, 1973; Barrett and Crill, 1974; Cullheim, 1978) is recruited before the larger MN with a bigger axon (see Henneman, 1981; Stein and Bertoldi, 1981). Thus, of 2 MNs recruited in a slow ramp at different force levels, the second conducts its spike to the muscle faster than the first (Fig. 6B). This will not affect the EMG-recorded recruitment order in a slow ramp in which onsets of firing of different MUs are widely apart (Figs. 4B and 5A). However, when the same MNs fire at very short intervals in a ballistic burst, the faster axonal conduction of the second MN may result in an earlier MU potential (Fig. 6C), even if that MN was actually recruited slightly after the first MN. The range of alpha-axon conduction velocities in the cat covers a factor of 2 (Bessou et al., 1963). In man the conduction distance from spinal cord to hand is about 80 cm, and the maximum alpha motor axon velocity is about 65 m/sec. Thus, a misleading crossover of MU spikes recorded in the muscle can be expected when corresponding MNs fire at intervals < 5 to 10 msec. An additional factor is that, in strong ballistic contractions, the powerful motor commands to the MN pool activate many MNs of different size virtually simultaneously; under such conditions it is not surprising that background excitatory states and other factors may introduce a slight time variability over a few msec in the actual onsets of firing of different MNs. Such variability is unlikely to affect the rate of force production significantly.

The quantitative data on the slow ramp and ballistic force thresholds of each MU (Fig. 7) demonstrate a highly consistent

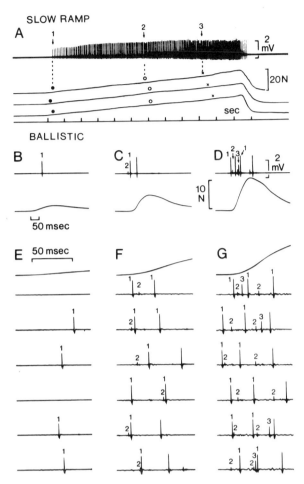

FIG. 5. Recruitment of 3 single MUs during slow ramp or ballistic voluntary contractions of the first dorsal interossous muscle in man. Three single MUs are clearly identified from this site. The MU_1 has the largest spike, MU_2 has the smallest spike, and MU_3 has a spike of intermediate size. Thresholds scatter over a wide range, since MU_1 is recruited at about 150 g, while MU_3 only fires at about 1.5 kg in slow ramp contractions (A). The peak voltage of MU_1 tends to increase somewhat during the ramp, but its waveform is preserved (as monitored on high-speed oscilloscope sweeps throughout) and leaves no ambiguity about the stability and reliability of MU identification. Three trials are shown in their isometric myograms and symbols (*dot* for MU_1, *circle* for MU_2, and *cross* for MU_3) indicate respective ramp thresholds in different trials. B–G: Brisk ballistic contractions of 3 different peak forces, namely, 160 g (B,E), 510 g (C,F), and 1.1 kg (D,G). The oscillograms of MU spikes *(upper traces)* and of myogram *(lower trace)* are shown at much higher-sweep speed in E, F, and G. In these columns only one myogram is displayed, but 6 traces of electrical activity document variability of MU discharges from trial to trial. With the smaller ballistic contraction of about 160 g (B,E), only MU_1 discharges, but not in all trials, since force is very close to its threshold of recruitment. With ballistic contractions of about 510 g (C,F), MU_1 fires securely 2 spikes and MU_2 fires 1 spike. It is only in the ballistic contractions > 1 kg that MU_3 is recruited (D,G) and it fires insecurely 5 times in the 7 trials displayed. Thus, each MU is activated only at a given peak ballistic force, and it never fires for contractions below that force. The temporal order of firing varies from one trial to the next in ballistic burst. (From Desmedt and Godaux, 1979a, with permission.)

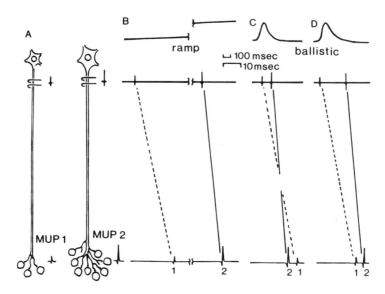

FIG. 6. Cartoon suggesting how differences in conduction velocities of the motor axons can influence the temporal order of MU spikes recorded in the muscle during a ballistic burst. The extracellularly recorded spikes are depicted as of larger size for the bigger motor axon and larger MU potential (MUP$_2$) in **A**. The times of occurrence of the 2 spikes in the proximal motor axons *(upper line)* and in the muscle *(lower line)* are consistent in their sequence in a slow ramp contraction (**B**) and sometimes in a ballistic contraction (**D**). However, they can be discordant in that the larger MU spike of MU$_2$ can occur before the smaller spike of MU$_1$ in the ballistic burst recorded in the muscle, even though the first MN is activated before the second, according to the size principle (**C**). The reason is that the larger motor axon conducts its spike at a faster rate whereby it can catch up the slower spike if the 2 MNs fire at close interval. No such ambiguity can occur in slow ramp contractions (**B**), where the different MNs are recruited at large intervals, since the muscle force rises slowly. (From Desmedt and Godaux, 1979a, with permission.)

proportional relationship between the force thresholds in either ramp or ballistic modes. The linear correlation coefficients between the two thresholds range from 0.91 to 0.97 in the different human muscles tested (Desmedt and Godaux, 1978b). The results are quite representative of the MUs in these muscles, since the high-threshold MUs of the samples were indeed only recruited at forces larger than half-maximum voluntary strength of the muscle. Thus, regardless of whether the MUs were recruited at either small or large slow ramp forces, they all exhibited fairly proportional reductions in threshold in the ballistic mode. The reduction is quite marked in interosseous, tibialis, and soleus muscles, where the coefficients were 0.38, 0.28 and 0.24, respectively (Fig. 7). This implies that a roughly 3 or 4 times larger fraction of the MN pool must be activated to produce a given ballistic peak force than for achieving the same level of force in a slow-ramp contraction.

The question of whether the size principle applies to ballistic contraction is important, and the answer is yes. This can be further documented by considering pairs of MUs simultaneously recorded (Desmedt and Godaux, 1978b). Of 193 pairs of single MUs studied, only 3 pairs reversed their order between the 2 contraction modes (these MUs had rather close ramp thresholds). The probability of a random distribution of thresholds in the pairs is < 0.005. The lack of preferential recruitment of the larger-faster MUs in human ballistic contractions indicates that, for a given movement, the voluntary motor commands are channeled through the same connectivities to the MNs of the pool, irrespective of the contraction speed or of the ballistic versus ramp mode of motor control. However, this chapter does not deal with recruitment of motor axons by graded electrical stimulation applied directly to peripheral nerves (Bergmans, 1973; Brown et al., 1981).

Functional Significance of Rank Ordering in the MN Pool

A rather fixed recruitment order makes motor control design considerably simpler by allowing the motor command packages to be assessed and possibly upgraded from one trial to the next through comparison of kinesthetic feedback with the environmental results of the skilled movement. Skilled motor behavior would be impossible to achieve if MN pool activation was versatile (see Henneman, 1981). This point can be elaborated by considering the coding of muscle force by Golgi tendon organs. Each Golgi receptor is responsive to the contraction of discrete MUs. About 10 single muscle fibers, each belonging to separate MUs with different contraction properties, attach to and activate one Golgi receptor whose firing threshold is not identical for these different MUs (Reinking et al., 1975). The mechanical coupling between single muscle

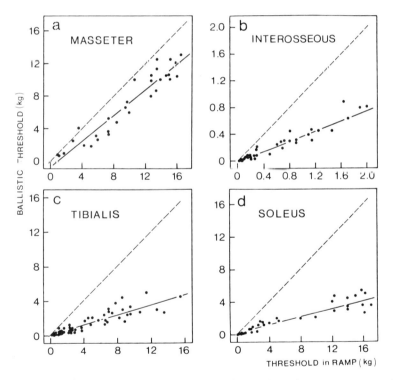

FIG. 7. Relations between the force thresholds for recruitment in a slow-ramp contraction reaching its peak force in about 10 sec *(abscissa)* and the ballistic force thresholds determined from a large series of ballistic contractions of different force. A: 31 single MUs recorded from masseter. The calculated linear regression is: $y = 0.80x - 0.81$ ($R_{xy} = 0.95$). B: 54 single MUs recorded from first dorsal interosseous. The linear regression is: $y = 0.38x + 0.01$ ($R_{xy} = 0.97$). C: 63 single MUs recorded from tibialis anterior. The linear regression is: $y = 0.28x + 0.07$ ($R_{xy} = 0.92$). D: 32 single MUs recorded from soleus. The linear regression is: $y = 0.24x + 0.33$ ($R_{xy} = 0.91$). Motor unit samples range from very low threshold to high threshold and are representative of muscle MU population. (From Desmedt and Godaux, 1978*b*, with permission.)

fibers and Golgi receptor capsule is such that smaller-slower MUs may exert an equal, and sometimes indeed a greater excitatory effect than a faster-larger MU acting on the same receptor. Therefore, the collective tendon organs discharges from an active muscle must present a unique nonlinear relationship to the muscle force at the tendon, when the single MUs are recruited in standard order. The Golgi receptors coding of muscle force would be completely distorted if MUs were to be recruited in random or reverse order (Desmedt, 1980a). For consistency of the force coding, the MN recruitment in the pool must be orderly. This appears as an essential factor of motor control design whereby, for example, ballistic skilled movements can be learned and recalibrated by comparison of kinesthetic feedback with environmental results.

Mechanical Design of Ballistic Contractions

Human muscles with different proportions of fast and slow muscle fibers produce ballistic voluntary contractions with rather similar time courses (Desmedt and Godaux, 1978b). Figure 8D illustrates the different CTs of isometric twitch elicited by a single maximal electrical stimulus delivered to the

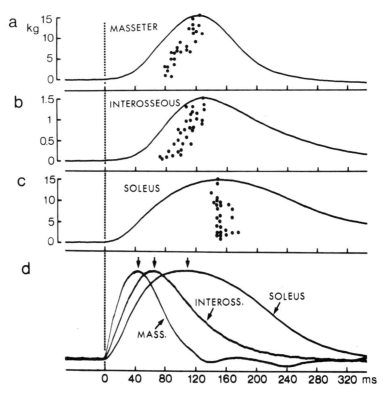

FIG. 8. Mechanical features of ballistic contractions in different muscles, compared to the single-twitch contraction times. A–C: Contraction times (abscissa, msec) of isometric ballistic voluntary contractions producing different peak forces (ordinate, kg) for jaw closing (A), abduction of the index finger (B), or extension of the ankle (involving soleus, since the knee is flexed at 110 degrees) (C). Only one myogram of a strong ballistic contraction is displayed for each muscle and the *dots* indicate the time to peak of a series of brisk contractions of weaker force. D: Isometric twitches of the 3 muscles elicited by a maximal electrical stimulus delivered to the motor nerve. The vertical amplification of the myograph analog signals has been adjusted for displaying twitches of equal size in the figure. (From Desmedt and Godaux, 1978b, with permission.)

motor nerve of masseter, interosseous, or soleus muscles (Table 1). For strong voluntary ballistic contractions that achieve over one half of maximum tetanic force, the times-to-peak present smaller differences, namely 130 msec in masseter and interosseous as compared to 155 msec in soleus. For ballistic voluntary contractions of smaller and smaller force, the time to peak shortens progressively to about 80 msec for masseter and interosseous but does not change significantly in soleus (Fig. 8A–C). The minimum time-to-peak is recorded for ballistic contractions achieving a force of one tenth or less of the maximal tetanic force. These data correlate with differences in firing patterns of single MUs in the 3 muscles. In soleus, each MU fires only one spike in most ballistic bursts, even up to peak forces of 8 kg, which largely exceed the ballistic threshold of the MU (Fig. 9E–H). In contrast, most MUs of masseter or interosseous muscles discharge several spikes when their ballistic threshold is exceeded (Fig. 4C, 9A–D). Repetitive discharges in the ballistic burst probably account for increasing time to peak of the stronger contractions (Fig. 8A,B; Table 1). The rate gradation is more conspicuous in masseter and interosseous, where the twitch CT (Fig. 8D) is much shorter than the time-to-peak of ballistic voluntary contractions.

The finding that most ballistic contractions have time-to-peak between 100 and 150 msec in muscles with a wide range of twitch (muscle contraction to single maximal nerve volley) contraction times (Table 1), suggests that brain processing times and other factors in motor control (such as inertia of moving parts) presumably set a lower limit to the useful range of speeds of brisk movements. It would seem that the histochemically defined faster versus slower muscle-fiber types may be much less significant than the patterning of the central motor commands for determining the actual speed of brisk movements in human muscles studied thus far (Desmedt, 1981a).

When considering how the brain uses the different muscles, it is necessary to distinguish: (1) the recruitment order of MNs within each pool, and (2) the possibility that one MN pool or another MN pool among synergist MN pools might be selectively activated. In other words, different MN pools might be engaged preferentially in certain movements, irrespective of whether or not the size principle applies within each MN pool. Gross EMG activities in soleus and gastrocnemius of the conscious cat executing natural movement disclosed pertinent data. Thus, the slower soleus is engaged before the faster gastrocnemius in all postural or locomotion activities, including brisk jumps (Smith et al., 1977; Walmsley et al., 1978). However, gastrocnemius is selectively activated, virtually without soleus, in the very rapid paw shakes that the cat makes to remove sticking tape from its paw or when stepping in water (Smith et al., 1980; Smith and Spector, 1981). Fast paw

TABLE 1. *Mechanical features of ballistic voluntary contractions and of muscle twitch elicited by a single maximal nerve shock in different human muscles*

Muscle	Time-to-peak of twitch elicited by nerve shock (msec)	Time-to-peak of ballistic contractions (msec)		Slope coefficient of ballistic threshold relative to ramp threshold[a]
		Weak	Strong	
Masseter	42	80	130	0.80
First dorsal interosseous	60	80	130	0.38
Tibialis anterior	84	90	130	0.28
Soleus	100	145	145	0.24

[a] See Fig. 7.

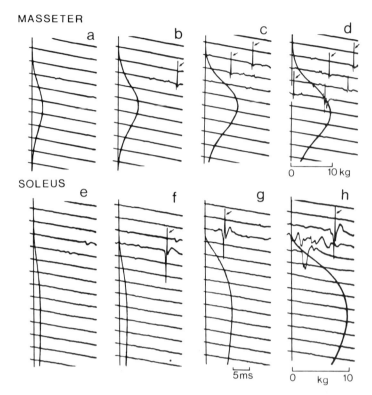

FIG. 9. Single MU discharges during ballistic voluntary contractions of increasing peak force in masseter (A–D) or soleus (E–H) muscles. The MU spikes are recorded on fast oscilloscope sweeps (free running) that are photographed on camera film moving from top to bottom. The isometric myograms are recorded by a stationary oscilloscope spot moving to left side (see calibration). The force threshold of the single masseter MU is reached for the contraction in B. The MU fires 2 spikes for the stronger contraction in C, and 3 spikes for an even stronger contraction in D (rate gradation). The spikes of the MU are indicated by small *arrows*. The single soleus MU fires one spike at threshold in F, and it does not fire more spikes in G–H when its threshold is widely exceeded. (From Desmedt and Godaux, 1978b, with permission.)

shaking is the only movement recorded in cats in which soleus remains virtually inactive while the faster (synergic) gastrocnemius is activated. This makes sense in motor control design, since soleus is actually too slow contracting to achieve such very fast shaking moves (Smith and Spector, 1981).

Ballistic Doublet Discharge and Rate Gradation of Motor Units

The contrast between ramp and ballistic modes of motor control is also documented by difference in rate gradation. The mean instantaneous frequency of discharge of single MUs at any chosen force level is expressed as the reciprocal of interval between 2 successive spikes of same MU at that force (Desmedt and Godaux, 1977a). Several successive trials must be pooled to obtain reliable mean values at any force level considered. Figure 10 illustrates typical findings for a tibialis anterior MU that is consistently recruited at 0.54 kg in slow ramp contractions. For a voluntary ramp reaching 12 kg in 5 sec, the MU discharges at 9/sec at onset of firing and accelerates progressively to about 30/sec as the ramp develops more force. The relation between

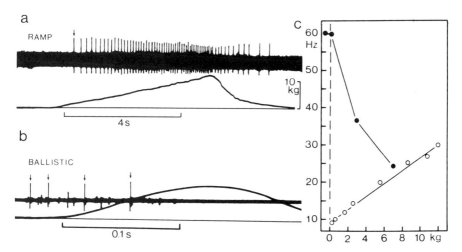

FIG. 10. Rate gradation of single tibialis MU in slow ramp or in ballistic voluntary contractions. **A:** Slow ramp reaching its peak of about 12 kg in 5 sec. A well-isolated single MU stands out of background noise, and its spike is intensified by Z-modulation for photography. The voltage of MU spike changes slightly during ramp, but the waveform monitored throughout on fast-running oscilloscope sweeps is constant. The MU accelerates its firing as the ramp develops more force, and it decelerates as the ramp relaxes. **B:** Brisk ballistic contraction reaching about 12 kg in only 110 msec, recorded on a much faster time base. The same single MU discharges 4 spikes in the ballistic burst *(arrows)*, which is initiated by doublet firing at short intervals. **C:** Instantaneous firing frequency (reciprocal of interspike interval; ordinate, Hz) as a function of simultaneously recorded isometric force of tibialis (abscissa, kg). In a ramp contraction reaching 12 kg in 5 sec (as in A), the MU firing frequency increases regularly *(circles)* from about 10 to 30 Hz. In strong ballistic contractions the same MU fires at a mean rate of 60 Hz (first interspike interval) before the muscle develops any significant force, and the instantaneous frequency decreases for subsequent firing while ballistic force increases. The data are mean value bases on several trials under identical conditions. (From Desmedt and Godaux, 1977a, with permission.)

instantaneous firing rate and muscle force is roughly linear (Fig. 10C). In contrast, in ballistic contractions of the same peak force of 12 kg, the mean instantaneous frequency of the same MU is as high as 60/sec for the first 2 intervals at onset of ballistic burst (Fig. 10B,C). The mean instantaneous frequency rapidly drops for subsequent discharges of the MU as the ballistic force rises. The mean instantaneous frequency ranged from 60 to 120/sec for 24 MUs studied at onset of ballistic bursts in tibialis, and there was always a drop in frequency later in the burst.

The interval between the first 2 spikes of a MU is generally shorter than between the second and third; hence, the usual doublet onset firing. Similar doublets are observed in masseter and first dorsal interosseous muscles, but not in soleus. For example the interval between the first 2 MU spikes in Fig. 9C,D is 12 msec, and corresponds to an instantaneous frequency of 83/sec at onset. Such modulations of MU firing go well beyond the steady frequencies classically recorded in ramps, and they can be observed for submaximal ballistic contractions that do not recruit all available MUs in the muscle. Therefore, this mechanism is called on not only after all MNs in the pool have been recruited, but it is also a characteristic feature of ballistic bursts (except in soleus) over an extensive range of forces and even at small force.

These results can be discussed in relation to two modes of firing analyzed in microphysiological studies of cat spinal MNs. For a steadily increasing sensory input (say, Ia fibers activated by muscle stretch), the MNs firing frequency increases smoothly and soon reaches a plateau of adapted firing rate which is about 20/sec for slow MNs and 37/sec for fast MNs (Granit, 1958; Burke, 1968). The MN behavior is rather similar to the one in slow ramp voluntary contractions in man (Figs. 4B, 5A, and 10A) (Milner-Brown et al., 1973c; Desmedt and Godaux, 1977a; Desmedt, 1981a). The cat MNs can fire much faster than this adapted firing rate, but only transiently under natural stimulation, and this has been called the secondary range of firing (Kernell, 1965f, this volume; Schwindt, 1973). The second range involves a regenerative firing mode in which the initial segment of MN is reactivated by delayed dendritic depolarization that persists after the trigger zone in the MN initial segment has repolarized following a first spike. Synaptic actions on MN dendrites may regulate the degree of dendritic delayed depolarization, and so control the regenerative activities of the MN. Furthermore, the delayed depolarization tends to become smaller with successive action potentials whereby the regenerative cycles responsible for the MN extra spikes are terminated (Calvin, 1974). Other possible factors are provided by the type of motor commands. Fast pyramidal tract neurons of the cat motor cortex also have the capability for extra-spike discharge (Calvin and Sypert, 1976), which would make for intense initial MN activation. In addition, the corticomotoneuronal excitatory postsynaptic potentials (EPSPs) of primates present marked facilitation when repeated at short intervals (Porter, 1970). This is no doubt relevant for ballistic motor commands, which involve a strong transient corticomotoneuronal activation of MNs dendrites that could result in regenerative firing of doublets or triplets.

In clinical EMG, doublets of MU potentials are not uncommon when the subject voluntarily starts contracting the muscle somewhat briskly, for example when he responds to a verbal order or to a sensory signal in a reaction-time paradigm (Partanen, 1978; Desmedt, 1981b, Fig. 8). Initial MU doublets at short interval also occur during normal locomotion in EMG bursts in flexors and extensors of the limb (Zajac, 1981). The occurrence of doublet or triplet MU firing at onset ballistic burst is functionally important since this enhances the surge of force that initiates the brisk movement. Double electrical stimulation of a motor nerve at such short intervals produces surplus force (that is, more than twice the force of a single volley) and considerably increases the rate of tension development in cat and human muscle fibers (Buller and Lewis, 1965; Desmedt and Hainaut, 1968). Studies on single giant muscle fibers of barnacle further show that the surplus force contributed by a second activation at short interval results from a considerably potentiated intracellular release of calcium ions, as disclosed by the photoprotein aequorin (Desmedt and Hainaut, 1977, 1978).

Another phenomenon that may increase the force output of the second muscle response, independently of any larger release of calcium ions into the myoplasm, is a potentiation of the actin-myosin mechanical interaction at crossbridges, thus a purely mechanical effect. In any case, the surplus force produced by double muscle fiber activation at short interval is an important and physiologically relevant process. In single MUs of the cat, a doublet elicited before a repetitive train causes a long-lasting enhancement of force output that has been termed *catch phenomenon*. (Burke et al., 1976). These data thus fit into a consistent pattern that is highly significant for the ballistic mode of motor control, whereby brisk productions of mechanical force are produced and titrated.

The Problem of Phasic Motor Units

Isometric ballistic contractions have been considered thus far, but similar results can be shown for isotonic conditions. Figure 11 shows 2 single MUs with different recruitment thresholds in a slow isometric ramp (Fig. 11A). The spike of the second MU is larger than that of the first MU in this example. In brisk isometric ballistic contractions the MUs are recruited in the same order (Fig. 11B); an isotonic displacement transducer with zero load is used (Desmedt and Godaux, 1979a). The MUs are also activated according to size principle under isotonic ramp or ballistic conditions (Fig. 11C,D). However, the question arises whether a separate population of alpha-MNs, called the phasic MNs (Granit et al., 1957) might not be uniquely involved in isotonic ballistic movements, a possibility for which claims have been made both in cats (Kernell and Sjoholm, 1975) and in man (Grimby and Hannerz, 1977, 1981). This problem has been examined by analyzing 33 single MUs in ballistic movements with

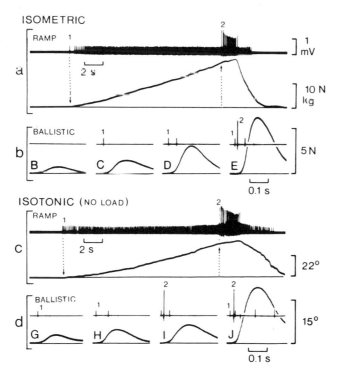

FIG. 11. Comparison of isometric and isotonic (zero load) ballistic contractions of the first dorsal interosseous muscle. Voluntary abduction of the index finger. *Upper trace:* Highly selective recording of 2 single MUs with the same electrode. MU_1 has a smaller spike and is clearly distinguished from MU_2. MU_2 presents a small decrease in peak voltage during ramp contractions, but the waveform monitored on fast oscilloscope sweeps is constant. The lower trace presents the isometric myogram (**A,B**) or the isotonic displacement under zero load (**C,D**) expressed as angular displacement in degrees at metacarpophalangeal joint of the index finger. The 2 single MUs are recruited in the same order throughout. When tested with ballistic contractions of different peak force or amplitude (only 4 trials are shown for each conditions), MU_1 is clearly recruited for smaller contractions than MU_2 (**B,D**). (From Desmedt and Godaux, 1979a, with permission.)

zero load. Each MU only started firing when the angular abduction of the index finger exceeded a certain amplitude (Fig. 12A–C), and its isotonic ballistic threshold was estimated from about 50 brisk movements of different size. When the same MUs were tested in slow isotonic ramps with angular velocity of about 3 degrees/sec, only 22 of the 33 MUs were apparently activated at reproducible movement amplitudes (Fig. 13A, dots). The remaining 11 MUs failed to discharge in slow ramps, even at maximum angular displacement (Figs. 12C, 13A, circles), and could indeed be called phasic, since they readily discharged in brisk isotonic movements (Fig. 12B,C). The mean ballistic threshold of the 11 phasic MUs was 9.5° ± 3.0°, thus larger than the mean ballistic threshold of 2.0° ± 2.4° of the other 22 MUs ($p < 0.001$) (Desmedt, 1981a).

For a correct interpretation of these data, it is, however, necessary to consider the force actually required to carry out such a ramp. For example, when a weight of 500 g is attached over a pulley to the index finger, 9 of the 11 phasic MUs are readily activated in slow ramp movements (Fig. 12E). The 2 remaining MUs (arrow in Fig. 13A) fail to discharge in isotonic ramps with a load of 500 g, but they do so when the loads is increased to 1,000 g. The addition of load considerably reduces the isotonic ramp threshold of all MUs that already fired in ramps with zero load (Fig. 13B dots). The phasic MU of Fig. 12 (star) that had not been recruited in ramps with zero load (Fig. 12D), discharges at a threshold displacement of 11.2 degrees in ramps with a load of 500 g (Fig. 12E) and at a threshold of only 1.7 degrees when the load is increased to 1,000 g (Fig. 12F) (Desmedt, 1981a). That MUs recruited in brisk movements with zero load might not fire in slow movements with zero load (Fig. 12D) is not surprising because the ballistic mode requires a large surge of muscle force to throw the body part into fast motion (force is the product of mass by acceleration). Such observations provide no evidence for a distinct population of so-called phasic MUs selectively engaged in brisk movements. These MUs indeed have a higher recruitment threshold also in isometric slow ramps: mean of 2.0 ± 0.6 kg as compared to the mean of 0.30 ± 0.45 kg for the 22 MUs recruited in the ramp with zero load ($p < 0.001$)

The correlation between the isotonic ballistic threshold of the 33 single MUs studied and their isometric ramp threshold (that correlated to single MU twitch force, Fig. 2) is $R = 0.92$. Thus, the thresholds of interosseous MUs for abduction of index finger

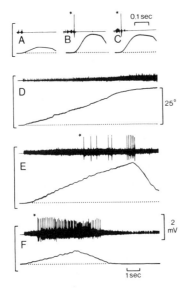

FIG. 12. Recruitment of a so-called phasic motor unit in the first dorsal interosseous muscle. Abduction of index finger under zero load in A–D. The angular displacement is indicated in degrees of rotation at metacarpophalangeal joint. A–C: Ballistic movements of 4.1° (A) or 12° (B–C). A single MU with a large spike (*star*) is recruited for large ballistic movements, but not in slow ramp movements of even maximal amplitude (24°) (D). Background activities are poorly isolated, low-threshold MUs. Notice that the time base is 10 times slower in D than in A–C. E: The same MU is activated for a slow movement of only 11.2 degrees, when a load of 500 g has been added to the isotonic myograph. F: With a load of 1,000 g the threshold drops further to 1.7°. (From Desmedt, 1981a, with permission.)

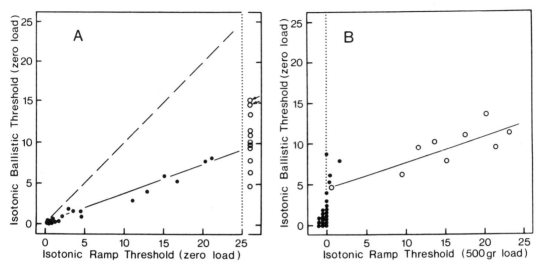

FIG. 13. Recruitment thresholds of 33 single MUs of the first dorsal interosseous muscle in ballistic abduction movements with zero load (ordinate, degrees angular displacement). The ordinate represents recruitment thresholds of the same MUs in slow ramp movements (degrees of angular displacement) carried out either with zero load (A) or with an isotonic load of 500 g (B). The single MUs represented by circles are not activated in ramps of maximum amplitude under zero load, but they discharge with a 500-g load. The 2 MUs indicated by an *arrow* in A *(right)* are only recruited in ramp movements with 1,000-g load. The MUs represented by *dots* are recruited by the slow ramp with zero load, and their angular threshold is considerably reduced when the load of 500 g is added in B. (From Desmedt, 1981a, with permission.)

are ranked according to size principle both in slow ramp and in ballistic contractions, irrespective of whether or not they are carried out under isotonic or isometric conditions.

Rank Deordering of Single MUs when the Muscle Functions as a Synergist

Ever since the stereotyped recruitment order of MNs in a pool was proposed (Henneman et al., 1965a), there have been interesting attempts to falsify the hypothesis. On the whole the latter has remarkably resisted such attempts. Most claims for a flexible or volitionally controlled rank deordering of MUs (cf. Basmajian, 1978; Grimby and Hannerz, 1977) have not been supported by any robust data.

For example, it is indeed not unexpected that MUs with close recruitment thresholds would occasionally fire in reverse order (cf, Henneman, 1981). In brisk movements, different MUs can fire a few milliseconds apart and exchange position in the ballistic burst (Figs. 5 and 6), but this is trivial and will not critically influence force production by the whole muscle. In certain studies, the use of poorly selective electrodes or unstable recording conditions may have confused interpretations by suggesting that another MU had been recruited when, in fact, the same MU potential merely changed voltage (Figs. 10A and 11A) (cf. Schmidt and Thomas, 1981). Monitoring waveforms of MUs action potential on fast oscilloscope sweeps ensures that the same MU is being studied in the various tests.

Highly consistent deordering of MU recruitment in slow ramps have now been observed in multifunction muscles (Thomas et al., 1978; Schmidt and Thomas, 1981; Desmedt, 1980a). This may offer a clue for explaining some of Basmajian's data. The first dorsal interosseous muscle has 2 *df*. It is the only muscle producing abduction of index

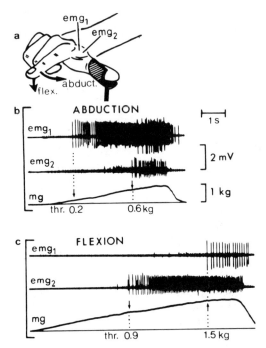

FIG. 14. Consistent deordering of 2 single MUs in relation to the direction of the movement. First dorsal interosseous muscle. A: Two single MUs are recorded from 2 muscle sites about 10 mm apart (emg_1 and emg_2). The index finger is moved either in abduction away from axis of the hand (prime mover function of interosseous), or in flexion (synergist function). B: For finger abduction, MU of emg_1 fires when the ramp reaches 0.2 kg, while the other MU is only recruited at 0.6 kg. The lower trace is the isometric myogram. C: For finger flexion, MU in emg_1 only fires when the ramp reaches 1.5 kg, while the MU in emg_2 is recruited at 0.9 kg. Thus, the recruitment thresholds of both MUs augments in flexion, but the threshold of the smaller slower MU increases more, whereby the 2 MUs are deordered. (From Desmedt, 1980a, with permission.)

finger away from the hand axis, but it also acts as a synergist to the long flexor muscle to produce flexion of index finger (Fig. 14A). An example of deordering of a MU pair is illustrated. The slower MU with a recruitment threshold of 0.2 kg in abduction (Fig. 14B) presents a much higher recruitment threshold of 1.5 kg in flexion (Fig. 14C), whereas the second MU (with a twitch of shorter CT) concomitantly increases its recruitment from 0.6 kg in abduction (Fig. 14B) to 0.9 kg in flexion (Fig. 14C). For this MU pair, deordering with movement direction is recorded consistently in successive trials involving either abduction or flexion of index in random sequence (Desmedt and Godaux, 1981). By changing direction of voluntary force, the subject could selectively activate either the first or the second MU in isolation. These effects relating rank reversals to precise levels of muscle force and movement directions contrast with the irregular and poorly reproducible effects of Grimby and Hannerz (1977). When a MU pair can be so deordered in ramp finger flexion, a similar deordering is also recorded for ballistic finger flexion. The phenomenon is indeed related to the movement in which the muscle functions, irrespective of speed of contraction.

The mean recruitment thresholds of any single MU, either in abduction or in flexion, disclose a rather wide scatter (Fig. 15A), and quite a few MUs with a low abduction threshold present a higher threshold in flexion (Desmedt, 1981a). Pairs of single MUs simultaneously recorded (Fig. 14A) are considered in Fig. 15B to evaluate the actual incidence of rank deordering. For the 142 MU pairs studied, the differences between mean abduction thresholds ranged from 0.02 to 1.9 kg (mean threshold of the larger MU minus mean threshold of the smaller MU). If the MU of higher abduction threshold also presents a higher flexion threshold than the other MU of the pair, the threshold difference in flexion will be positive and be plotted above the abscissa in Fig. 15B. This is the case for 126 MU pairs. However, 16 MU pairs present a negative threshold difference in flexion, of which 12 are considered statistically significant because the threshold difference exceeded 2 SDs for either directions. Thus, significant rank reversals occur in 12 of the 142 MU pairs with threshold differences exceeding 0.2 kg

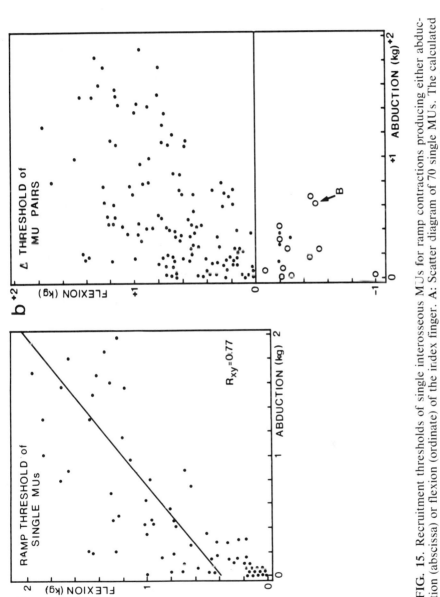

FIG. 15. Recruitment thresholds of single interosseous MUs for ramp contractions producing either abduction (abscissa) or flexion (ordinate) of the index finger. A: Scatter diagram of 70 single MUs. The calculated regression is: $y = 0.81x + 0.38$ ($R_{xy} = 0.77$). Motor units with a low threshold in abduction tend to have a higher threshold in flexion. B: Scatter diagram of the differences in recruitment thresholds for 142 pairs of interosseous MUs recorded simultaneously during either finger abduction or flexion. The MU pairs plotted below the zero ordinate are recruited in reverse order during finger flexion. The 12 MU pairs indicated by a *circle* show a threshold difference > 2 SD in either direction. An *arrow* points to the MU displayed in Fig. 14. (From Desmedt, 1980a, with permission.)

(Desmedt, 1981a). An incidence of 8.4% actual deordering is small but significant, and the mechanisms responsible have been investigated by STA extraction of single MU twitches either for finger abduction or for finger flexion. Each of the 32 interosseous MUs so studied contributed mechanical force to finger movements in either direction, and the twitches extracted for any MU have the same contraction time (as expected) and roughly proportional peak force (Fig. 16A,B). When the muscle acts as prime mover in finger abduction, the MUs with larger twitch force are consistently recruited after the smaller MUs, and a linear relation to the recruitment threshold is found (Fig. 16C; cf. Fig. 2). However, for finger flexion, the relation is less satisfactory (Fig. 16D), mainly because smaller MUs tend to present unusually large recruitment threshold for flexion. For example, the 11 interosseous MUs with twitch force < 1.2 g are all recruited at forces < 0.15 kg in finger abduction (Fig. 16C), whereas their thresholds rise up to 0.5 kg in finger flexion (Fig. 16D). Motor units with twitch > 7 g are never recruited at < 1.2 kg in either direction.

Such features were not shared by the long flexor muscle, which is the prime mover for index finger flexion. Motor units of this muscle are not activated in index finger abduction. For finger flexion, a highly significant correlation is found between twitch force and flexion threshold of single MUs of the long flexor (Fig. 16E). Therefore, the less satisfactory relation in Fig. 14D must be related, not to finger flexion per se, but to the interosseous acting as a synergist rather than as prime mover in flexion (Desmedt, 1981a). The intriguing feature of rank deordering with movement direction in a multifunction muscle is therefore related to the unexpected upward shift of recruitment threshold in flexion for the smaller MUs. Such shifts are not physiologically meaningful, since interosseous MUs contribute significant force to the flexion movement (Fig. 16B).

Size Principle and Connectivities in the MN Pool

The data presented do not support the hypothesis of selective volitional access to single MNs in a pool. Sizable rank reversals only occur in about 8% of MU pairs in conjunction with their use in another movement (Fig. 15B). This would be in line with some difference in the patterning of descending connections that terminate in synergist MN pools.

The consistent relations of MNs size both with susceptibility to discharge and with size and force of the corresponding MU have far-reaching consequences for the motor control of movements. For a standard input to the MN pool, the larger MNs present EPSPs of smaller voltage (Burke, 1968, 1973). The larger MNs have a smaller input resistance (Kernell, 1966; Burke, 1968, Barrett and Crill, 1974) so that a larger synaptic current will be necessary to produce a threshold depolarization and activate the larger cells. However, a mere proportional grading of input resistance with MN size would not explain the recruitment order nor the difference in synaptic potentials voltage if the larger cells received the same density of active synapses (whereby larger synaptic currents generated by the larger *number* of synapses would more or less make up for the smaller input resistance). If this was the case, larger cells would appear as just scaled-up versions of smaller cells (Zucker, 1973). In order to account for the smaller synaptic potentials voltage in larger MNs it could be postulated that each MN is influenced by roughly the same number (not the same density) of active excitatory synapses. On the other hand, the input resistance decreases much faster with increasing MNs size than would be expected from a size effect alone, be-

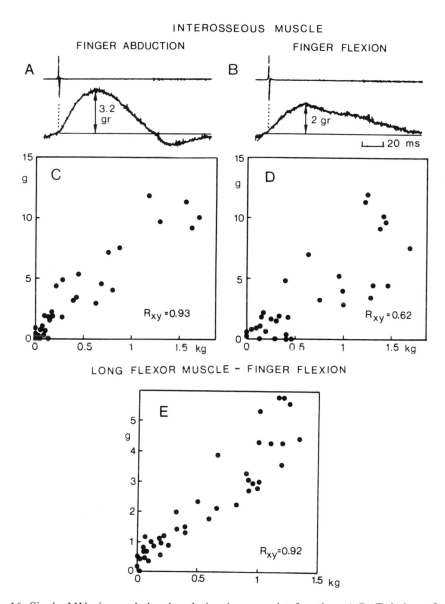

FIG. 16. Single MU size and the deordering in synergist function. A,B: Twitches of a single interosseous MU extracted by STA during voluntary abduction (A) or flexion (B) of the index finger. The triphasic spike of the MU is displayed on same time scale. C: Force threshold of interosseous MUs (*abscissa*, kg) in finger abduction plotted against MU twitch force (*ordinate*, g). D: Force threshold of same MUs in finger flexion (*abscissa*, kg) plotted against twitch force (*ordinate*, g). E: Force threshold of single MUs of the long flexor muscle during flexion of the index finger (*abscissa*, kg) plotted against MUs twitch force (*ordinate*, g). (From Desmedt, 1981a, with permission.)

cause larger MNs have a reduced specific membrane resistance whereby a given synaptic activation is less efficient on larger MNs (Kernell and Zwaagstra, 1981; Stein and Bertoldi, 1981; Kernell, *this volume*). Thus, cell size appears as a useful index of MN properties, but it would not by itself account for the recorded recruitment order.

The latter depends on the distribution in the pool of intrinsic MN properties somehow related to MN size and on active synaptic connectivities (Stein and Bertoldi, 1981). Dynamic changes in synaptic efficiency as produced by post-tetanic potentiation can influence differentially the relative recruitment thresholds of MNs (Luscher et al., 1979). The activation of sensory inputs with unequal distribution in the MN pool can also affect recruitment properties (Kanda et al., 1977; Smith and Spector, 1981; Burke, 1981; Kanda and Desmedt, *this volume*).

The data on rank deordering can now be considered in relation to the patterning for prime mover and synergist MN pools, respectively. Since the phenomenon appears related to upward shifts of recruitment threshold of some of the smaller interosseous MNs (with a high input resistance), it is proposed that such MNs receive a smaller than usual number of excitatory synapses from the voluntary flexion commands. By contrast, in the MN pool of the prime mover, synaptic terminals would be consistently distributed whether for abduction in interosseous pool (Fig. 17, right) or for flexion in long flexor pool (Fig. 17, left) (Desmedt, 1980a; Desmedt and Godaux, 1981). In a multifunction muscle, when the MN pool is activated as a synergist function rather than a prime mover, smaller MNs in spite of their higher input resistance, can thus be recruited later in a ramp, because they receive an unusually small share of active excitatory synapses (Fig. 17, right). In this situation, the larger MNs do not appear to be more easily recruited in the synergic pool (Fig. 15A, 16D), which makes sense in functional design. This is because a more efficient recruitment would require a higher than usual number of active excitatory synapses, in view of their similar input resistance, and this does not appear worthwhile nor meaningful to have.

In this discussion of deordering of 8.4% of MN pairs with movement direction, it must be stressed that the speed of movements is not a factor influencing the recruitment order. The motor commands are apparently channeled through the same connectivities for either flexion or abduction (Fig. 17), irrespective of the ramp versus ballistic mode of motor control. The state of affairs is perhaps somewhat different for very rapid paw shakes of the cat's hindlimb, which provide the single hitherto documented example of a natural movement that must be carried out faster than would be possible for the slow MUs of the synergic soleus muscle to accomplish (Smith and Spector, 1981).

The question must be asked whether the recorded deordering of the interosseous MN pool in finger flexion might not upset the kinesthetic feedback input as discussed above, but this actually raises no problem for two reasons: (1) the deordering in the

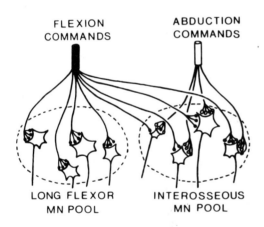

FIG. 17. Illustration suggesting different connectivities of voluntary motor commands in MN pools. Interneurons, propriospinal relays, and other details have been omitted. Active excitatory synapses are supposed to be fairly equally distributed throughout the MN pool of the prime mover, either in flexion *(left)* or in abduction *(right)*. Only 4 synaptic terminals are depicted per MN. In contrast, the active excitatory synapses for flexion commands are less numerous on some of the smaller MNs of the synergic interosseous pool. (From Desmedt, 1980a, with permission.)

synergist MN pool has a rather limited incidence, only 8.4% of MU pairs and (2) there is no concomitant deordering in the MN pool of the prime mover (Fig. 16E). It can indeed be supposed that the coding of actual force of any movement should rely primarily on kinesthetic feedback input from the prime mover muscle.

ACKNOWLEDGMENTS

The work reported has been supported by grants from the Fonds de la Recherche Scientifique Médicale and the Fonds National de la Recherche Scientifique (Belgium), and by the Muscular Dystrophy Association of America.

Motor Control Mechanisms in Health and Disease,
edited by J. E. Desmedt.
Raven Press, New York © 1983.

Cutaneous Facilitation of Large Motor Units and Motor Control of Human Fingers in Precision Grip

*,**Kenro Kanda and John E. Desmedt

*Brain Research Unit, University of Brussels Faculty of Medicine, Brussels, Belgium; and **Department of Physiology, Chiba University, Chiba, Japan*

Clinical observations on patients with disturbances of finger sensibility from peripheral nerve lesions and experiments involving local anesthesia of digital nerves have shown that input from cutaneous receptors important for the motor control of precise hand movements (see Goodwin et al., 1972; Gandevia and McCloskey, 1977; Marsden et al., 1978a; McCloskey, 1980; McCloskey et al., *this volume*; Smith et al, *this volume*). From the time of Sherrington (1906) the physiological role of exteroceptive afferents has been investigated, namely by recording flexion reflexes or other responses to cutaneous stimulation in animals and more recently in man (see Young, 1973; Hugon, 1973b; Meier-Ewert et al., 1983; Meinck et al, *this volume*; Delwaide and Crenna, *this volume*; Willer, *this volume*). The cutaneous input represents a significant part of the flexor reflex afferent (FRA) inputs which exert significant influences on spinal interneuron circuits and on ascending pathways (see Eccles and Lundberg, 1959; Lundberg, 1979).

Recent investigations have further disclosed that the motor cortex and its pyramidal tract neurons (PTNs) receive short-latency projections from exteroceptive inputs (see Wise and Tanji, 1981a,b; Jones, *this volume;* Strick and Preston, 1978a,b, *this volume*). In man such short-latency cutaneous projection to precentral cortex is indexed by component P22 of the averaged scalp-recorded somatosensory evoked potential (SEP) to electrical stimulation of contralateral fingers (Desmedt and Cheron, 1981; Mauguière et al., 1983). The detailed cortical organization and functional significance of the exteroceptive input for movement control are not yet clear. This chapter considers the effects of cutaneous input on the recruitment of different motoneurons (MNs) in a spinal MN pool, emphasizing the possible relevance of these interactions for control of the precision grip (Napier, 1956, 1962) between thumb and index finger in man.

Recruitment Patterns in Synergic Muscles

Ever since Ranvier's disclosure of different contraction properties in red or pale muscles, the question arose of whether reflex or voluntary motor commands might involve differentially the different muscles belonging to a given motor synergy, or more specifically the motor units of different types that make up these muscles. For ex-

---

*To whom correspondence should be addressed: Department of Physiology, Chiba University Medical School, 1-8-1 Inohana, Chiba-shi 280, Japan.

ample, early studies have compared the reflex activation of the different muscles making up the triceps of the cat. Denny Brown (1929) observed that the slower soleus muscle was more readily activated during stretch than the faster gastrocnemius and soleus (Sol) could, indeed, respond alone under appropriate conditions. As a rule, the chief extensor muscles of mammals possess a deep, slowly contracting component and a superficial faster contracting component: for example slower crureus and faster vastus lateralis in knee extensors; slower medial short head and faster lateral short head of triceps in elbow extensors; slower deep head and faster superficial head of supraspinatus in shoulder extensors. The physiological significance of this anatomical specialization of muscle fibers is not yet clear. Correlations between morphology, histochemistry, and physiology of single motor units (or muscle units) were analyzed in detail (see Wuerker et al., 1965; Close, 1972; Burke and Edgerton, 1975; Edgerton and Cremer, 1981; Henneman, 1981). Each single motor unit only has muscle fibers of one type (Kugelberg, 1973, 1981). In man, motor units of different fiber type present much less anatomical segregation in separate synergic muscles or muscle parts than is the case for cat, rabbit, or rat.

There is, at present, good evidence for Henneman's size principle whereby the different MNs in a spinal pool (corresponding to a given muscle) are recruited in a consistent order that roughly matches their cell size, smaller MNs being recruited before larger MNs (Henneman et al., 1965a,b; Henneman, 1981; Burke, 1981; Stein and Bertoldi, 1981; Kernell, *this volume*). The size principle also applies to human muscles executing ramp contractions (Milner-Brown et al., 1973a,b; Tanji and Kato, 1973, 1981; Freund and Budingen, 1976), and also brisk ballistic movements. In other words, the faster-stronger motor units that present a higher recruitment threshold in slow ramp contractions are not preferentially activated when the same muscle produces brisk movements (Desmedt and Godaux, 1977a,b, 1978b, 1981; Desmedt, 1980a, 1981a). The size principle appears widely applicable in human muscles carrying out undisturbed voluntary contractions, irrespective of the speed of force production (*preceding chapter*). The recruitment threshold of single motor units presents a high correlation with twitch force of the motor unit, which, in turn, appears to reflect the number of muscle fibers innervated by the corresponding motoneuron and presumably the cell size of this MN as well. Recruitment deordering appears to be rare in undisturbed voluntary contractions, and a clearly documented example of such deordering has only been recorded thus far in multifunctional muscle when used as a synergist rather than as a prime mover (Desmedt, 1980a, 1981a; Desmedt and Godaux, 1981).

Studies of freely moving cats with electrodes and transducers chronically implanted in limbs have compared electromyographic (EMG) activities and force production in the different synergic muscles making up the triceps surae. The data indicate that the slower soleus is always activated before the faster gastrocnemius in a wide range of movements from quiet quadrupedal standing to brisk vertical jumps up to 1 meter (Smith et al., 1977; Walmsley et al., 1978). The medial gastrocnemius adds only a small force to that produced by soleus in quadrupedal standing, and this fits in with the classical involvement of soleus in postural function (see Denny Brown, 1929). When larger forces must be generated in locomotion at various rates or in brisk jumping, the soleus is fully activated, and the gastrocnemius becomes more strongly involved. In none of these locomotory activities was the gastrocnemius active without the soleus being activated, and this is consistent with a stereotyped recruitment order in these natural activities (Walmsley et al., 1978).

An interesting exception was, however, recorded under the peculiar conditions associated with the execution of the very rapid paw shakes observed when a cat wants to get rid of water (after stepping into water) or of some adhesive that has been stuck onto its paw (Smith et al., 1980; Smith and Spector, 1981). These paw shakes involve very fast alternating movements with a mean shake cycle of 88 msec. This appears faster than would be possible for the (slow) soleus to achieve. In any case, the faster gastrocnemius is active, whereas the slower soleus is virtually silent during rapid paw shakes. This selective involvement of the faster synergist muscle in paw shakes thus makes sense functionally. Furthermore, fast paw shakes involve cutaneous stimulation of the paw (see below). It should be noted that this observation as such provides no argument for or against the size principle of MN recruitment within either soleus or gastrocnemius MN pool. Studies of single motor units or MNs (rather than whole-muscle EMGs) should test whether the rank ordering of MN recruitment in each MN pool is modified by the muscle switch in paw shakes.

Effects of Exteroceptive Input on Recruitment Patterns

The presence of some exteroceptive input may significantly affect the patterning of MN pool activation. For example, in approximately three-quarters of precollicular decerebrate unanesthetized cats, the repetitive electrical stimulation of sural nerve increased EMG and force production in the faster medial gastrocnemius (MG) while depressing the slower Sol (Kanda et al., 1977). Figure 1 shows this differential effect on muscle EMG and force production of MG and Sol. Superimposition of a weak repetitive stimulation at 100 Hz and at 5.6 times the threshold of the most excitable fibers in the nerve produces a great decrease in EMG and force activity in Sol. At the same

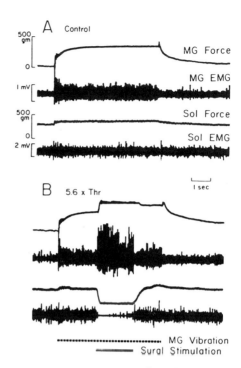

FIG. 1. Differential effect of sural nerve electrical stimulation (100 Hz) on the force output and EMG activity in medial gastrocnemius and soleus muscles when superimposed on TVR. Decerebrate cat. A: Control responses to vibration, without sural stimulation. B: Same vibration, but with a period of sural nerve stimulation at 5.6 times the threshold for the most excitable nerve fibers in the nerve. (From Kanda et al., 1977, with permission.)

time, there is an increase in the EMG and force output of MG. Thus, in this experiment on decerebrate cat, the exteroceptive input produced by sural nerve stimulation facilitated the tonic vibration response (TVR) in MG while depressing it in the synergic soleus. Similar effects can be obtained under such conditions by pinching the lateral ankle skin innervated by the sural nerve (Kanda et al., 1977).

The polysynaptic potentials produced by electrical stimulation of the ipsilateral sural nerve in MNs innervating slow twitch motor units (type-S motor units; see Burke and Tsairis, 1973; Burke, 1973, 1981; Burke and Edgerton, 1975; Edgerton and Cremer,

1981) in soleus are prodominated by late inhibitory postsynaptic potentials (IPSPs). In contrast, the gastrocnemius MNs innervating fast-twitch motor units received from the same input early excitatory effects, whereas IPSPs are much less pronounced (Burke et al., 1970). These results are in line with indirect evidence from Sherrington's laboratory (see Creed et al., 1932).

Cutaneous input indeed affects the recruitment of MNs in decerebrate cats (Kanda et al., 1977). Tonic discharges of motor units were elicited by muscle vibration (TVR; see Hagbarth, 1973; Lance et al., 1973; Desmedt, *this volume*) or by steady muscle stretch. Figure 2 illustrates an example of MN deordering by cutaneous input. Skin pinch was superimposed on stretch of MG and Sol. The response in whole MG muscle during pinch consisted of two bursts of increased activity, while inhibition clearly predominated in Sol (Fig. 2). The MG filament record (third trace) shows that a motor axon with a moderate action

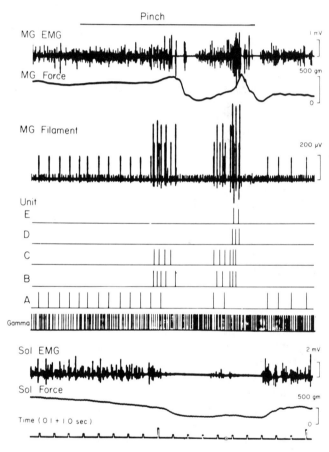

FIG. 2. Decerebrate cat. Records of EMG activity and force output of medial gastrocnemius *(two uppermost traces)* and soleus *(two lowermost traces)*, together with simultaneous record of action potentials of a small number of medial gastrocnemius motor axons [in a natural filament of the muscle nerve *(third trace)*]. Isolated activities of 5 identifiable gastrocnemius motor axons are shown in the *middle traces*. The effects of a strong pinch of ankle skin elicits marked excitation-inhibition sequences in the gastrocnemius records, and pure inhibition in soleus. The γ axon trace showed no inhibitory effect despite the transient suppression of gastrocnemius α motor axon activity. (From Kanda et al., 1977, with permission.)

potential amplitude fires slowly before skin pinch. During skin pinch, discharges were recruited in other MG motor axons, all with larger spike amplitude. Activities of 5 different MNs (sorted out by a spike discriminator device and checked against high-speed records) are labeled A through E in the rank order of their recruitment during muscle stretch, during repeated trials with skin pinch. This order corresponds closely to the rank order of the respective axonal conduction velocities (CV). In this record, the MG response to skin pinch includes two bursts of increased activity that are each followed by a transient inhibition (seen in both whole MG responses and in single units recorded from spinal motor root filaments). However, activity in axon A (with lowest recruitment threshold) ceases during the period when axons B to E (with higher recruitment threshold) are firing. In the second burst, the recruitment of axons D and E with the highest threshold only occurs after apparent complete suppression of the discharge in axon A. Note that the activity pattern of MG unit A during skin pinch is rather similar to that of whole Sol muscle. The units recorded in filaments in Fig. 2 were not identified in terms of type of their motor units. In any case, unit A is very likely of the slow type in view of its slow axonal CV and high sensitivity to muscle stretch, tendon tap, and tendon vibration. This experiment thus provides a rather clear example of differential control of recruitment thresholds of α-MNs in a MN pool.

Cutaneous afferents, together with group II and III muscle and joint afferents constitute the FRA (Eccles and Lundberg, 1959). The FRA elicit excitation in flexor muscles and inhibition in extensor muscles in spinal cats. They act together on many ascending pathways such as the ventral spinocerebellar tract and spinoreticulocerebellar tract (Lundberg, 1959; Lundberg and Oscarsson, 1962). It has been suggested that interneurons intercalated in reflex pathways from FRA act as lower motor centers that also receive inputs from several higher supraspinal structures. Therefore, ascending FRA pathways would signal activities that have been integrated in these lower motor centers, rather than merely passing along raw afferent input via FRA (Oscarsson, 1973; Lundberg, 1979). In addition to feeding onto the FRA ascending pathways, low-threshold cutaneous afferents act through separate polysynaptic reflex pathways on spinal MNs (Engberg, 1964; Hongo et al., 1969; Fedina and Hultborn, 1972; Hugon, 1973b; Illert et al., 1976; Lundberg et al., 1977; Meinck et al., *this volume*). There is no ready interpretation of the differential cutaneous input effects on low- and high-threshold MNs in conjunction with known data about the neuronal circuits related to FRA inputs. Finally, the exteroceptive inputs are conveyed via fast ascending pathways to the motor cortex where short-latency influences can be exerted onto PTNs projecting back to spinal interneurons and MNs, particularly those for distal muscles (see Jones, *this volume*; Strick and Preston, *this volume*). The possible role in motor control of cutaneous long loops is still obscure (see Delwaide and Crenna, *this volume*).

Cutaneous Input and MN Recruitment in Man

Noxious skin stimulation is known to exert inhibitory effects on various motor activities in man (Young, 1973; Shahani and Young, 1973b; Godaux and Desmedt, 1975b) and elicits withdrawal reflexes (Hagbarth, 1960; Willer, *this volume*). Mild skin stimulation eliciting tactile sensations can also influence motor control (Garnett and Stephens, 1980), and repetitive electrical stimulation (50 Hz at 4 times subjective threshold) of finger for 4 to 6 min affects slow ramp recruitment threshold of single motor units in first dorsal interosseous muscle in man (Garnett and Stephens, 1981): at short latency, the predominant effect is an increased probability of firing of faster higher-threshold motor units, and a reduction in firing for slower lower-threshold motor units. This is in line with the data for triceps in decerebrate cats (Kanda et al., 1977). The overall effect of prolonged electrical stimulation of index finger is to promote recruitment of higher-threshold motor units and to apparently delay the recruitment of the smaller lower-threshold motor units so that deorderings of recruitment order can occur (Garnett and Stephens, 1981).

We studied precision grip (see Smith et al., *this volume*) in human fingers and enquired whether, in first dorsal interosseous, single motor unit recruitment could be influenced by natural tactile stimulation. Highly selective bipolar tungsten wire electrodes fixed in an hypodermic needle were used to obtain reliable steady records of single motor units (see Desmedt and Godaux, 1975, 1977a). The hand rested with the palm downward into a molded plastic support. The first phalanx of the thumb was clamped in a steady position with a rubber mounted device. The fixation of hand and thumb ensured steadiness of the anatomic insertions of the interosseous muscle on the thumb, while permitting voluntary flexion of the distal phalanx of the thumb so as to achieve contact and active palpation with the palmar tip of the index.

The first phalanx of the index finger pressed against an isometric mechanical transducer (Flatline load cell ELF-1000-250), while the second and third phalanxes were kept fixed in semiflexed position, thus ready to meet the tip of the thumb easily when the latter was to be voluntarily flexed. Additional clamps were used in several experiments to make sure that thumb-index contacts and palpations did not mechanically influence the interosseous by changing the distance between the muscle's insertions. All the devices maintaining the hand or fingers were equipped with rubber sheets to avoid any pain, and they were placed so as not to interfere with the finger innervation and circulation. For each of the 45 single motor units studied in 3 subjects, the isometric twitch was extracted by spike-triggered-averaging (STA) with the usual precautions (see Milner-Brown et al., 1973a; Desmedt and Godaux, 1977b, 1978b; Desmedt, *this volume*). An example of STA-extracted single twitch is shown in Fig. 3A.

Figure 3B shows the recruitment in ramp voluntary contraction of two single-motor units. MU-1 with a twitch of 0.9-g force and 54-msec contraction time (CT) was recruited consistently when the interosseous index abduction force reached 90 g; this level is indicated by the horizontal interrupted line in B and C. MU-2 with a twitch force of 1.14 g and a CT of 47 msec was recruited at a higher ramp force of 180 g. The subject then maintained a steady abduction force of 70 to 90 g, and flexed the distal phalanx of thumb to achieve cutaneous contact between fingertips with slight to-and-fro palpation movements. As a result, MU-2 fired in isolation without MU-1 although the interosseous force remained quite steady at a level significantly below the one that had been required to elicit MU-2 firing in the absence of cutaneous stimulation of the index tip (Fig. 3C). The precision grip was then discontinued, and the index abduction was relaxed before being resumed (right side of Fig. 3C). In the absence of cutaneous stimulation, MU-1 was again found to fire when the index abduction force reached its previous threshold of 90 g (horizontal interrupted line).

Among the 45 single interosseous motor units studied in detail, 16 were influenced by natural stimulation of the index fingertip in precision grip. The recruitment threshold in ramp voluntary contractions was consistently studied in 9 of these units, and the reductions of threshold associated with cutaneous stimulation were quite significant (Fig. 4). All these motor units can be considered as higher-threshold units (see Des-

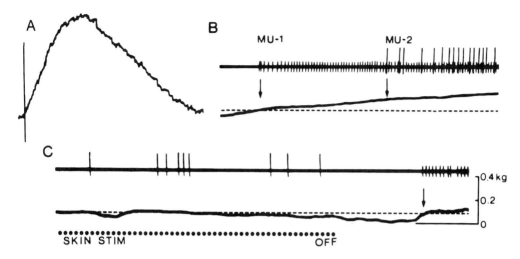

FIG. 3. Normal adult man, first dorsal interosseous muscle. **A:** Spike-triggered-averaged twitch of motor unit MU-2 (256 samples averaged); the contraction time from motor unit action potential to twitch peak is 47 msec and the twitch force is 1.14 g. The ramp recruitment threshold of MU-2 is 180 g. **B and C:** Action potential discharge *(upper trace)* of single motor units MU-1 and MU-2, recorded at a single muscle site with highly selective bipolar electrode. *Lower trace* represents isometric force for voluntary abduction of the index finger. **B:** Standard recruitment in a slow ramp contraction of MU-1 at 90 g, and of MU-2 at 180 g *(arrows)*. *Horizontal interrupted line* is drawn through the threshold of MU-1. **C:** Effect of contact of thumb tip with index tip in natural palpation *(horizontal dotted line)* facilitates the recruitment of MU-2 while the abduction force is barely at the threshold of MU-1 (70 to 90 g); after cessation of the skin stimulation by active touch, the muscle is voluntarily relaxed and then abducted, which recruits MU-1 at its usual undisturbed threshold *(arrow)*.

medt, 1980a, 1981a), since their ramp recruitment threshold was above 0.2 kg, and their twitch force 1 g. The remaining 29 interosseous motor units seemingly unaffected by cutaneous stimulation were lower-threshold units, since 26 of them had twitch forces < 1.5 g (and 21 of these had twitch forces below 0.9 g). In human muscles, the twitch contraction time of STA-extracted single motor units is a less consistent parameter than twitch force in conjunction with unit recruitment properties (Desmedt, *this volume*).

The motor units with higher recruitment threshold in undisturbed ramp did not all show such facilitation by cutaneous stimulation, and 9 of the units with threshold > 0.2 kg were apparently not influenced. The study did not document any increase of recruitment threshold of the lower-threshold motor units during natural finger tip stimulation. Data such as shown in Fig. 3C would not provide any robust evidence on that point. It is true that MU-2 fired in the absence of activity of MU-1, but the concomitantly recorded abduction force did not clearly exceed the MU-1 undisturbed recruitment threshold either.

Comments on Cutaneous Input and Motor Control of Fingers

Our results are in line with Garnett and Stephens (1980, 1981) in showing a facilitatory effect of cutaneous input on higher-threshold motor units of the first dorsal interosseous muscle in man. They extend their conclusions by showing that significant drops of the recruitment threshold occur during the natural stimulations involved in precision grip and active touch behavior. The pattern of glabrous skin contacts

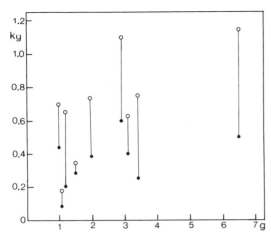

FIG. 4. Voluntary ramp abduction force of the first dorsal interosseous muscle in man necessary to recruit single motor units (*ordinate*, kg) plotted in relation to the force of STA twitch of the same motor unit (*abscissa*, g). Nine single motor units sensitive to skin stimulation of the index tip during active touch are shown. *Circles:* Recruitment threshold in the absence of skin contact. *Dots:* Recruitment threshold of the same motor units (joined by *vertical line*) in the presence of skin stimulation by contact with thumb.

achieved in our experiments was only standardized to a degree compatible with the execution of natural small-amplitude palpations. It is rather typical of palpation of objects under natural conditions.

The relevant factor is not the initial movement of the distal phalanx of thumb to reach index tip but, rather, the actual contact and slight friction between the glabrous skin of the fingers tips. The facilitatory effect on higher-threshold motor units appeared to be maintained throughout the period of contact and palpation between fingers tips. Therefore, the relevant sensory input must involve primarily the low-threshold mechanosensitive units that are present in the glabrous skin of the fingers (Johansson and Vallbo, 1979) rather than joint afferents.

Thus, the cutaneous input elicited through holding an object between index and thumb exerts an excitatory drive on interosseous MNs, whereby the precision grip is reinforced and confirmed. This motor behavior is of paramount importance in skilled manipulatory activities of the human hand, and it is functionally relevant that the precision grip be reinforced through facilitation of higher-threshold MNs as soon as the finger tips touch an object that must be held or manipulated. Loss of this facilitatory effect in patients with neuropathies or nerve lesions accounts at least partly for clinical deterioration of prehension and skilled uses of the hand. Experimental removal of cutaneous finger sensations by local anesthesia results in increased perceived heaviness of weights lifted by the fingers (Gandevia and McCloskey, 1977; Marsden et al., 1979; Gandevia et al., 1980; McCloskey, 1981; McCloskey et al., *this volume*). No matter whether short- or long-loop circuits are mainly involved in these effects (a question that we leave open at this stage), the net practical result is that prehension and manipulation are assisted and made more reliable, since relevant cutaneous contacts with the object elicit a surplus excitation in the MN pool that has been thrown into action by the descending motor commands which have initiated the prehension.

In this relation, it is more significant for higher-threshold MNs to be facilitated (whereby larger motor units add up significant contributions to the muscle force) than for lower-threshold MNs to be inhibited, since the latter effect can be rather counterproductive and of doubtful use since it would only remove a force contribution smaller than the one added by the recruited larger motor units. This reasoning holds if cutaneous contact primarily has to enhance the precision grip. Nevertheless, the fact that the detailed studies on cat's triceps revealed a somewhat complex series of events with transitions between facilitatory and inhibitory effects over time (Fig. 3) should elicit further elaboration of our current views. In the hand, it can indeed prove meaningful physiologically to blend inhibitory effects with facilitation along time, namely, when the surface characteristics of objects are being explored in active touch. Along this line, Lundberg et al. (1977) showed in the cat spinal cord that cutaneous afferents have disynaptic excitatory connections onto interneurons intercalated in disynaptic Ib inhibitory pathways to MNs. Interneuronal transmissions in Ib pathways can be inhibited by two systems originating in brainstem (Engberg et al., 1968), or facilitated from corticospinal and rubrospinal tracts (Illert et al., 1976, 1977). Thus, the brain has many possibilities for regulating transmission in Ib reflex pathways, for example in order to adjust the set of interneuronal transmissibility in conjunction with specific fine-movement control requirements. Parallel control of Ib pathways, not only from supraspinal levels, but also by cutaneous inputs suggests that, as stated by Lundberg et al. (1978), "In exploratory movements, this facilitated inhibition of agonist muscles gives a purposeful break on the movement when meeting an obstacle [during locomotion]." It remains to be explored whether this applies to the special features of motor control of small hand muscles, and no ready answer is available right now. The human hand developed an extraordinarily wide repertoire of adaptive manipulatory activities, which must involve a special neuronal circuitry for current guidance from active touch-related inputs. The analysis of mechanisms in control of skilled finger movements has barely begun.

ACKNOWLEDGMENTS

This research has been carried out at the Brain Research Unit of Brussels University where K. Kanda visited in July 1980. It was supported by grants from the Fonds de la recherche Scientifique Médicale and Fonds National de la recherche Scientifique.

# The Nature of the Afferent Pathways Conveying Short-Latency Inputs to Primate Motor Cortex

*E. G. Jones

James L. O'Leary Division of Experimental Neurology and Neurological Surgery, and McDonnell Center for the Study of Higher Brain Function, Departments of Neurology and Neurological Surgery, and Anatomy and Neurobiology, Washington University School of Medicine, St. Louis, Missouri 63110

It is an old belief that messages arising in the specialized sensory receptors of muscles and in receptors in other peripheral tissues may serve as guides to the activities of the motor cortex (Jones, 1972). This idea retains its popularity, and in recent years perhaps the greatest attention in the field has been directed toward examining the adaptive capacity of motor cortex neurons to compensate for peripherally imposed perturbations of movement performance that are presumably signaled to the motor cortex. As a consequence, the trained animal performing a stereotyped task with a manipulandum that is intermittently subjected to a perturbation (Evarts, 1973; Conrad et al., 1974; Porter and Rack, 1976; Evarts and Tanji, 1976; Evarts and Fromm, 1977, 1978), has become a common sight in the laboratory. In this type of experimentation it has been shown not only that corticomotoneuronal discharge may be modulated by perturbations of certain kinds of movements but also that the cortical response and the compensatory muscular contractions can vary depending on prior instructions to the animal. Comparable adaptive behavior in response to a perturbation has been described in humans performing a rapid, repetitive, flexion-extension movement of the thumb. In addition to having a latency rather longer than the well-known spinal tendon-jerk reflex, this behavior in man has been shown to depend on the integrity of the dorsal columns of the spinal cord (Marsden et al., 1973, 1977a).

In his clear treatment of the subject in 1969, Phillips spoke of the capacity of the nervous system to compute the difference between a real (i.e., perturbed) and an intended movement (by which he meant real and intended muscle length), and to modulate corticomotoneuronal output accordingly. Since then, many workers have focused their attention on the large-diameter muscle afferents from the muscle stretch receptors as the principal feedbacks to motor cortex. However, the kinds of perturbations mentioned in the experimentation referred to above may be expected to affect many kinds of sensory receptors. It has been known for many years that units in the motor cortex of man and experimental animals can have cutaneous and other deep— as well as muscle-related receptive fields (Albe-Fessard and Liebeskind, 1966; Welt et al., 1967; Asanuma et al., 1968; Fetz and Baker, 1969; Brooks and Stony, 1971; Gold-

---

*Department of Anatomy and Neurobiology, Washington University School of Medicine, 660 South Euclid Avenue, St. Louis, Missouri 63110.

ring and Ratcheson, 1972; Rosen and Asanuma, 1972, 1975; Lemon et al., 1976; Lemon and Porter, 1976; Wong et al., 1978; Fetz et al., 1980). Indeed, the direct relay of group I muscle afferents in the motor cortex, at least of the primate, has not been particularly well documented (Devananden and Heath, 1975; Lucier et al., 1975). The input to the motor cortex units with peripheral receptive fields, not only depends on the dorsal columns of the spinal cord (Brinkman et al., 1978; Asanuma et al., 1979), but also arrives at very short latency—as short as 5 to 10 msec in experiments in which electrical stimulation of peripheral nerves has been used. In man, data on averaged evoked potentials show a motor cortex response to a natural stimulus occurring with a latency very similar to that observed over the sensory cortex (Desmedt and Cheron, 1981).

These observations suggest that a variety of sensory inputs may play a role in motor cortex function and that the dorsal columns of the spinal cord are central characters in the whole motor performance. This cooperative interaction was hinted at by Wall (1970) when he wrote of the motor function of the dorsal columns of the spinal cord. In what follows, I shall attempt to define the anatomical pathways where these varied influences may reach the motor cortex of the primate and question to what extent they may be direct through the thalamus or indirect via the sensory cortex.

Group I Afferent Input to Cortex

Most workers who have described group I afferent input to the vicinity of the monkey motor cortex have recorded the shortest latency field potentials or unit responses, not in the motor cortex proper (area 4) but in a region in the floor of the central sulcus, area 3a (Fig. 1) (Phillips et

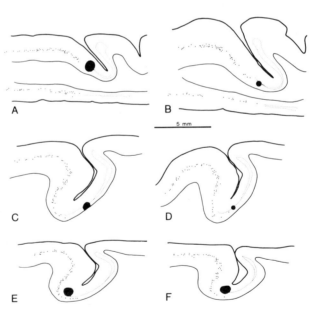

FIG. 1. Sites of maximal foci (*large dots*) of field potentials elicited by near-threshold electrical stimulation of group I afferents in hindlimb or forelimb nerves. From 4 different rhesus monkeys, A, B: horizontal sections in hindlimb representation; C to F: sagittal sections in forelimb representation. *Smaller dots:* Giant pyramidal cells of area 4; *interrupted lines:* granule cell layer of areas 3a and 3b. In each case, group I focus occupies attenuated granule cell zone characteristic of area 3a. (From Jones and Porter, 1980, with permission.)

al., 1971; Wiesendanger, 1973; Lucier et al., 1975; Heath et al., 1976; Hore et al., 1976; Jones and Porter, 1980). Reports of equally rapid group I inputs to area 4 of monkeys have been reported only in abstract form (Devanandan and Heath, 1975) or written with some degree of equivocation, the driving of motor cortex units being relatively weak or insecure (Lucier et al., 1975). Inputs from higher threshold muscle afferents, on the other hand, seem more assured (Wiesendanger, 1973; Hore et al., 1976).

Area 3a, after a period of uncertainty, can now be described in architectonic and connectional terms (Jones and Porter, 1980). In the hindlimb representation (Fig. 2), it is characterized by a zone of attenuated layer IV granule cells intervening between the thick granularity of area 3b and the most posteriorly situated giant pyramidal cell of area 4. In the forelimb representation, similar characteristics are visible (Fig. 3) but there is some degree of overlap of granule cells and giant cells at the most anterior end of the attenuated granule cell zone. This has led to a good deal of confusion, since it is only this small overlap region that has a structure similar to that of area 3a in other common experimental animals such as the cat (Jones and Porter, 1980). In the monkey, some workers have referred only to the overlap zone as area 3a, others only to the nonoverlapping part of the attenuated granule cell zone, and yet others to both. From the distribution of the short-latency, group I

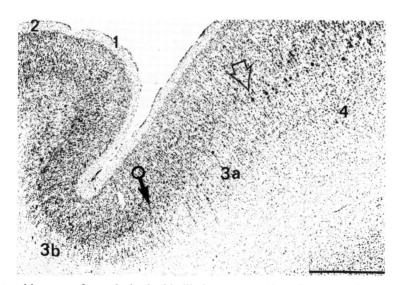

FIG. 2. Cytoarchitecture of area 3a in the hindlimb representation of a cynomolgus monkey (*horizontal section*). *Ringed arrow:* Point of attenuation of granule cell layer; *open arrow:* most posteriorly situated giant pyramidal cell. Between the two is area 3a. (From Friedman and Jones, 1981, with permission.)

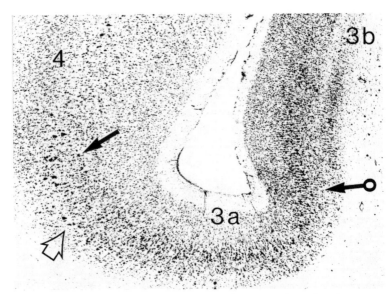

FIG. 3. Cytoarchitecture of area 3a in the forelimb representation of a rhesus monkey. Sagittal section. Note small degree of overlap of attenuated granule cell layer of area 3a *(straight arrow)* and most posteriorly situated giant pyramidal cell *(open arrow)*.

field potentials in the arm and leg representations, which always lie within the attenuated granule cell zone (Fig. 1), we prefer to adopt a definition of area 3a as the zone of granule cell attenuation ignoring, at least for practical purposes, the small degree of architectonic overlap at its anterior border in the hand representation. It will be shown that area 3a defined in this way is an integral part of the somatic sensory cortex.

Distribution of Sensory and Other Afferent Pathways in the Monkey Thalamus

Recent anatomical tracing studies have tended to show that the distributions of many sensory and motor-related pathways in the thalamus are far more restricted than was previously thought and that relatively few of these converge on the same thalamic relay nuclei. Figure 4 shows in a horizontal section, some of the principal ventral nuclei of the monkey thalamus. It has been known in general terms for some years that the oral division of the ventral posterior lateral nucleus (VPLo) of Olszewski (1952) (the Vim nucleus of European workers) receives cerebellar inputs and projects to motor cortex and that the caudal division of the ventral posterior lateral nucleus (VPLc) receives medial lemniscal inputs and projects to sensory cortex. It has sometimes been thought that a fairly substantial transition zone between the two nuclei should receive overlapping cerebellar and lemniscal afferents and project to area 3a. However, it is clear from the figure that there is a rather clear-cut border between the two nuclei. It can also be shown that there is no overlap in the terminal territories of cerebellar and lemniscal fibers at this border (Figs. 5 and 6). One source of past confusion lies in the fact that the VPLc nucleus and the lemniscal terminations undershoot the VPLo nucleus ventrally (Fig. 5), a feature that is not obvious in sections cut in the conventional frontal plane and illustrated in the standard atlases. Muscle afferent influences, which travel in the medial lemniscus (Phillips et al., 1971), thus relay in VPLc, not in VPLo.

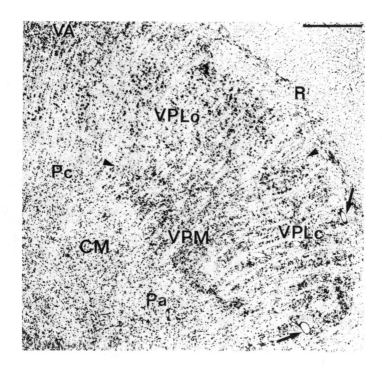

FIG. 4. Horizontal section through the thalamus of a cynomolgus monkey showing open neuropil and large cells of cerebellar relay nucleus to area 4, VPLo, and mixed cell population of lemniscal relay to areas 3a, 3b, 1, and 2 of sensory cortex, VPLc. *Arrow heads:* Line of junction between the two nuclei where lemniscal and cerebellar inputs abut but do not overlap. CM, centre médian nucleus; Pa, anterior pulvinar nucleus; Pc, paracentral nucleus; R, reticular nucleus; VPM, ventral posterior nucleus. (From Friedman and Jones, 1981, with permission.)

In contrast, fibers arising in the three deep cerebellar nuclei terminate in the VPLo nucleus (Fig. 6) but extend uninterruptedly from this into the overlying caudal ventral lateral (VLc) nucleus (Thach and Jones, 1979; Asanuma et al., 1980). This, together with the fact that VPLo and VLc nucleus both project to area 4 in the cortex, leads to the conclusion that the two form part of a common cerebellar relay nucleus to motor cortex (Jones et al., 1979).

Just as there is no overlap in the thalamic terminations of lemniscal and cerebellar pathways, so inputs to the thalamus from the globus pallidus (Fig. 7) also seem to be confined to an independent relay nucleus, the VLo, with no overlap into the cerebellar territory (Percheron, 1977; Tracey et al., 1980). The oral ventral lateral (VLo) nucleus projects to areas of cortex anterior to area 4. Past indications of terminations of cerebellar fibers in it and of its projection to area 4 stem from a failure to note its unique architecture, consisting of islands of cells that can lie almost completely surrounded by parts of the VLc nucleus.

The Thalamic Relay to Motor Cortex

The common cerebellar relay nucleus VPLo-VLc, including its extension, the so-called nucleus X of Olszewski (1952), projects only to area 4, but there are a number of refinements in the pattern of organization. The common nucleus is somatotopically organized with caudal parts of the dentate and interposed nuclei projecting medially, anterior parts laterally, and inter-

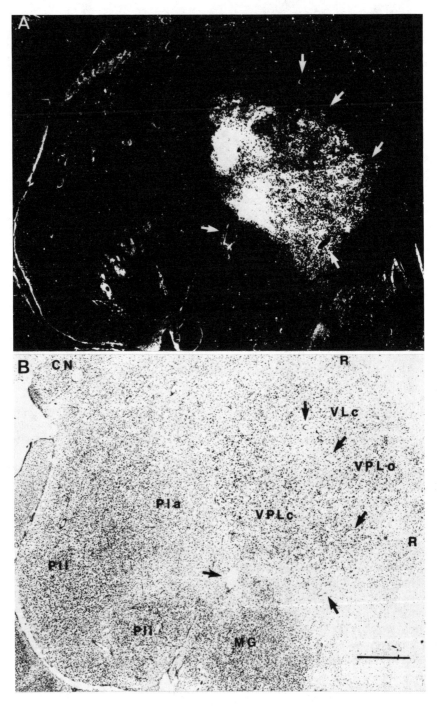

FIG. 5. Darkfield (**A**) and brightfield (**B**) photomicrographs from cynomolgus monkey at same magnification and from same sagittal section showing that terminal distribution of lemniscal fibers, as outlined by autoradiography following injection of ^{3}H-labeled amino acids in contralateral dorsal column nuclei, is confined to VPLc nucleus. Arrows indicate same blood vessels. MG, principal medial geniculate nucleus; Pla, anterior pulvinar nucleus; Pll, lateral pulvinar nucleus; Vlc, caudal ventral lateral nucleus. (From Tracey et al., 1980, with permission.)

mediate parts in between (Thach and Jones, 1979; Asanuma et al., 1980). Since these parts project, respectively, to face, leg, and arm representations in the motor cortex, it may be assumed that the pattern also reflects a basic somatotopy in at least two of the deep cerebellar nuclei (Perry and Thach, 1979). The fundamental plan of somatotopic organization in the common VPLo-VLc-X nucleus is one in which a major body part is represented by a curving lamella extending more or less through the dorsoventral and anteroposterior dimensions of the nucleus, although curving back over the dorsal aspect of the VPLc nucleus (Fig. 8). Each lamella then projects to an anteroposterior strip of area 4 (Jones et al., 1979) and multiple sensory respresentations in area 4 (Strick and Preston, 1978) are presumably associated with inputs mainly from different dorsoventrally situated parts of the lamellas.

One indication of differential afferent routes through the common cerebellar relay nucleus to area 4 comes from the observation that more dorsally situated (mainly VLc) parts of the nucleus project anteriorly in area 4, while more ventrally situated (mainly VPLo) parts project posteriorly, to parts of area 4 deep in the central sulcus (Strick, 1976a; Tracey et al., 1980; Friedman and Jones, 1981) (Figs. 6, 9, and 10). It is of interest that receptive fields recorded in anterior parts of area 4 are mostly situated in deep peripheral tissues, whereas those recorded in posterior parts are mainly cutaneous in character (Strick and Preston, 1978; Tanji and Wise, 1981). This immediately raises the question of differences in the nature of peripheral inputs to the VLc and to the VPLo parts of the thalamic cerebellar relay nucleus.

There is little indication of a differential distribution of cerebellothalamic fibers that might be relevant. It is clear that the input from a single deep nucleus is fractionated and distributed as bundles of axons and clustered terminations that need not overlap with similar bundles and clusters emanating from the other two nuclei (Thach and Jones, 1979; Asanuma et al., 1980). These clusters and the thalamocortical relay cells on which they terminate probably form the basis for the focal or columnar nature of the input to the motor cortex. But the overall distribution of clusters derived from each deep cerebellar nucleus extends through the dorsoventral extent of the common VPLo-VLc-X nucleus (Fig. 6).

Among other afferent inputs, only the spinothalamic and ascending vestibular pathways relay in the ventral, VPLo part of the common cerebellar relay nucleus (Tracey et al., 1980; Asanuma et al., 1980). This raises the question as to their role in the transmission of sensory information at short-latency directly to motor cortex. The VPLo nucleus seems to be the sole thalamic terminus for a rather modest vestibular projection (Lang et al., 1979), arising from ventral aspects of the vestibular nuclei (Tracey et al., 1980) in which spinovestibular fibers terminate (Mehler, 1969). The spinothalamic terminations extend into the VPLo nucleus from the VPLc nucleus (Fig. 11) and are more substantial than the vestibulothalamic contribution. But they are far less dense and their terminal ramifications much more widely spaced than those of the cerebellothalamic system in VPLo or of the lemniscal system in VPLc.

Several workers have reported the presence of units with deep or cutaneous receptive fields in the vicinity of the VPLo nucleus in both anesthetized (Horne and Tracey, 1979; Lemon and Van der Burg, 1979; Asanuma et al., 1979) and unanesthetized monkeys (Horne and Porter, 1980). Some of these units have been reported to be antidromically activated by stimulation of the motor cortex (Lemon and Van der Burg, 1979; Asanuma et al., 1979). In the unanesthetized animal the units seem to discharge at short latency only in re-

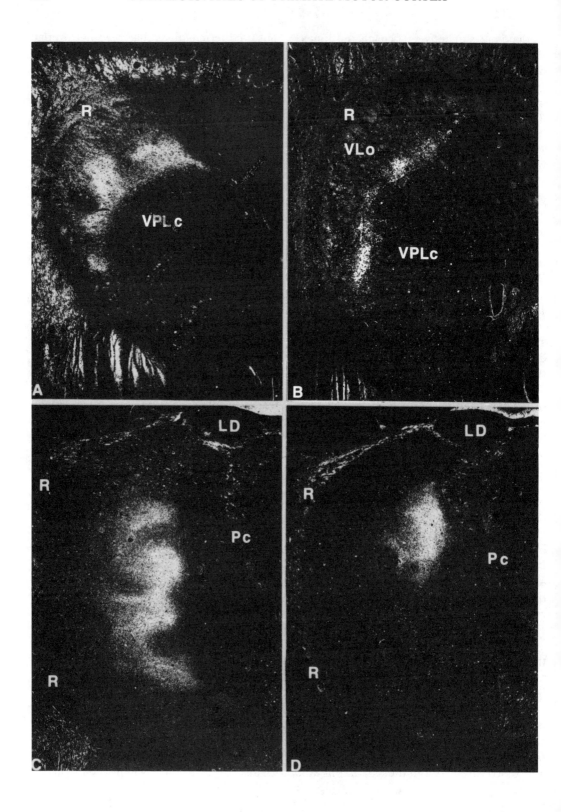

sponse to stimuli applied to a passive limb, and either not at all or much later than motor cortex units in response to perturbations of a manipulandum (Strick, 1976b; Horne and Porter, 1980; MacPherson et al., 1980). Units in VLc, which from the projection of the nucleus to anterior parts of area 4 might be expected to have deep receptive fields, do not seem to be driven readily by peripheral stimuli (Horne and Porter, 1980), or also respond only at rather long latencies (MacPherson et al., 1980).

It is also reported that the units in VPLo with passive peripheral receptive fields are discharged at very short latencies—as short as 3 msec—to electrical stimulation of peripheral nerves (Asanuma et al., 1979; Horne and Tracey, 1979; Lemon and Van der Burg, 1979). This has led some to predict that the VPLo zone receives a direct lemniscal input and projects directly to area 4, thus accounting for both the cutaneous receptive fields and the short latency of input to area 4. Seemingly in confirmation of this viewpoint, section of the dorsal columns of the spinal cord causes the receptive fields of units in area 4 to disappear (Brinkman et al., 1978) and the area 4 evoked potential is dramatically reduced (Asanuma et al., 1980).

However, from what has been described above, it is clear that there is no anatomical confirmation of lemniscal inputs to VPLo or VLc. We are left, therefore, with the spinothalamic or vestibulothalamic pathways as candidates for mediating the short-latency input. At the present time neither of these seems particularly appropriate. Although the parent cells of spinothalamic tract fibers terminating in VPLo lie in laminas of the spinal cord (Tracey et al., 1980) in which spinothalamic tract cells have both cutaneous and deep receptive fields (Applebaum et al., 1975, 1979; Willis et al., 1979), section of the dorsal columns should not interfere with their projection to the thalamus. Section of the dorsal columns might be expected to interfere with the spinovestibulothalamic projection but as pointed out above the vestibulothalamic component is rather small and, moreover, it is reported that the spinovestibular pathway (also small in the primate; Mehler, 1969) does not involve large diameter afferents (Wilson, 1972). The only alternative source of short-latency inputs to the motor cortex thalamic relay, the deep cerebellar nuclei, also seems unlikely in view of the long latency (up to 20 msec) and uncertain responses of units in these nuclei to peripheral stimuli (Eccles et al., 1974; Allen et al., 1977; Harvey et al., 1979), and also because cerebellar ablation does not appear to reduce the area 4 evoked potential (Malis et al., 1953).

We are faced then with something of a paradox: The known anatomical pathways do not match the reported results of physiological experimentation. Assuming that the anatomical methods are sufficiently sensitive, we must, therefore, conclude that some important clue has been missed in the

---

FIG. 6. Photomicrographs from 4 different cynomolgus monkey brains. A: Corticothalamic fiber labeling by autoradiography following a large injection of ^{3}H-labeled amino acids in area 4. The corticothalamic labeling in this sagittal section exactly reciprocates the thalamocortical projection and is restricted to the cerebellar relay nucleus (VPLo-VLc). B: Also from a sagittal section, cerebellar terminations in the same common nucleus following injection of one of the deep cerebellar nuclei. C and D from frontal sections of different brains at the same level show labeling mainly in ventral part (VPLo) of cerebellar relay nucleus after an injection of posterior parts of area 4(C) and in dorsal parts (VLc) after an injection of anterior parts of area 4(D). Punctate concentrations of label reflect focal nature of input to cortex (cf. Fig. 8). LD, lateral dorsal nucleus. (A, C, and D, from Jones et al., 1979, with permission.)

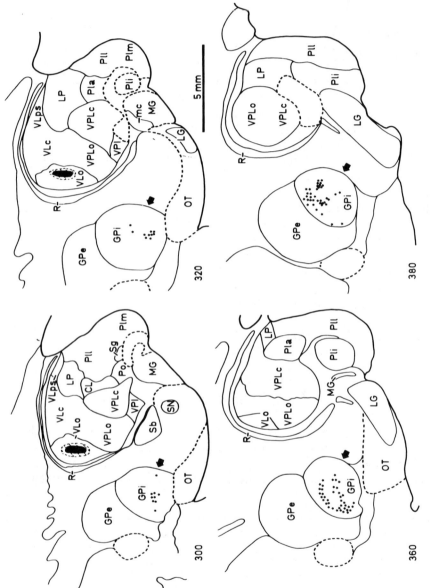

FIG. 7. Sagittal sections from a cynomolgus monkey brain showing restriction of retrograde cell labeling to internal segment of globus pallidus following injection (*black area*) of horseradish peroxidase in thalamic VLo nucleus. (From Tracey et al., 1980, with permission.) CL, central lateral nucleus; GPe, external division of globus pallidus; GPi, internal division of globus pallidus; LG, lateral geniculate nucleus; MC, magnocellular medial geniculate nucleus; OT, optic tract; Sb, subthalamic nucleus; SN, substantia nigra. (From Tracey et al., 1980, with permission.)

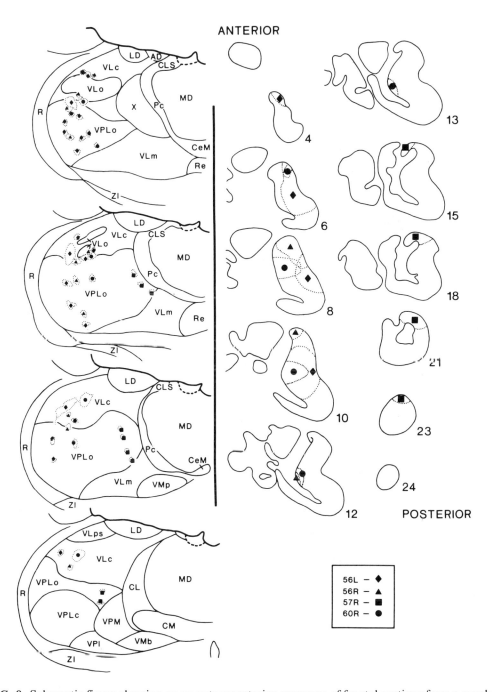

FIG. 8. Schematic figure showing on an anteroposterior sequence of frontal sections from a monkey thalamus (left), the distribution of inputs from different parts of the contralateral dentate nucleus (right). Comparable symbols indicate sites of injection of ^{3}H-labeled amino acids in dentate nucleus and axoplasmically transported label in the thalamus. Note that posterior parts of dentate nucleus project medially in common VPLo-VLc nucleus and anterior parts laterally. Each lamella of projections is further subdivided into elongated patches of terminal ramifications. CeM, central medial nucleus; CM, centre médian nucleus; VM, ventral medial nucleus; X, nucleus "X" of Olszewski (1952); ZI, zona incerta. (From Thach and Jones, 1979, with permission.)

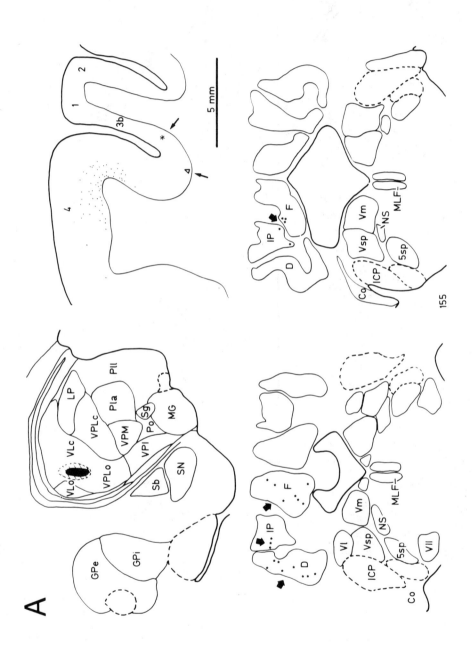

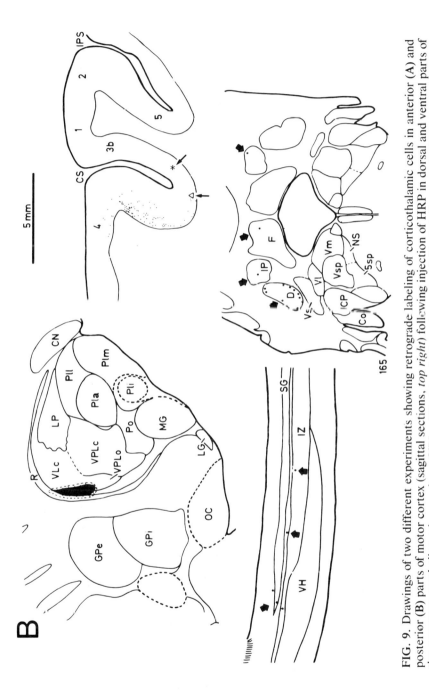

FIG. 9. Drawings of two different experiments showing retrograde labeling of corticothalamic cells in anterior (A) and posterior (B) parts of motor cortex (sagittal sections, *top right*) following injection of HRP in dorsal and ventral parts of the common cerebellar relay nucleus (VPLo-VLc) (sagittal sections, *top left*). Small arrows indicate area 3a. Each injection is associated with retrograde labeling in deep cerebellar nuclei (frontal sections, *bottom*) but only injection involving more ventral parts of the relay nucleus leads to retrograde labeling in spinal cord (B, *bottom left*). Co, cochlear nuclei; D, dentate nucleus; F, fastigial nucleus; ICP, inferior cerebellar peduncle; IP, interposed nucleus; OC, optic chiasm; 5sp, spinal trigeminal nucleus. (Modified from Tracey et al., 1980, with permission.)

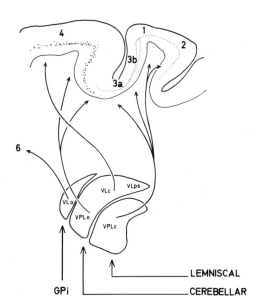 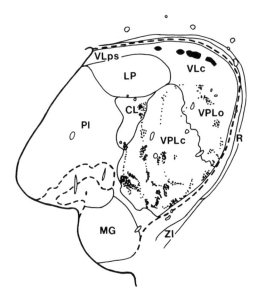

FIG. 10. Schematic figure on sagittal sections, summarizing preceding figures and showing independence of pallidal, cerebellar, and lemniscal relays in monkey ventral thalamic nuclei and projection of dorsal and ventral parts of the common cerebellar relay nucleus (VLc-VPLo) to anterior and posterior parts of area 4. GPi, internal division of globus pallidus.

FIG. 11. Drawing of a sagittal section showing distribution of spinothalamic fibers in monkey thalamus as labeled by degeneration (*fine dots*) following a hemicordotomy at the third cervical segment. Fibers spread from lemniscal relay (VPLc) only to ventral (VPLo) part of cerebellar relay to motor cortex. *Large dots*, dorsally placed, indicate terminations of fibers emanating from a small part of interposed nucleus, as labeled in same brain by autoradiography (cf. Fig. 5). (From *unpublished work* of C. Asanuma, W. T. Thach, and E. G. Jones, with permission.)

experimental situation or that the short-latency lemniscal input to motor cortex reaches it not directly via the thalamus but indirectly via the postcentral sensory cortex.

The Thalamic Relay to Sensory Cortex

Ever since work (Malis et al., 1953; Kruger, 1956) showing that ablation of the postcentral gyrus did not substantially alter the peripherally evoked potential in motor cortex, the sensory cortex has not been a popular route for relaying short-latency inputs to motor cortex, except in the case of area 3a which was thought to provide the muscle afferent input to area 4, thus seemingly closing a loop between working muscles and the corticomotoneuronal path. Recently, however, Asanuma et al. (1980), have shown that the motor cortex evoked potential can be reduced by as much as 75% following ablation of the sensory cortex. Furthermore, Marsden et al. (1977b) have found that human patients with localized destruction of the postcentral gyrus or involvement of the white matter underlying it, have difficulty in making the normal, rapid, automatic corrections of a thumb flexion movement in response to disturbances imposed during its performance. These observations lead us back to a consideration of the corticocortical route as a rapid pathway into motor cortex.

The dedicated lemniscal relay nucleus, VPLc, and its trigeminal counterpart, ventral posterior medial (VPM) nucleus project on cortical areas 3a, 3b, 1, and 2 in the postcentral gyrus and on the second somatic sensory area (SII) in the lateral sulcus

(Jones and Powell, 1970; Jones, 1975; Burton and Jones, 1976; Jones et al., 1979; Friedman, Jones and Burton, 1980; Friedman and Jones, 1981). If VPLc is mapped with microelectrodes introduced horizontally from behind (Friedman and Jones, 1981) (Fig. 12), a large proportion of the nucleus is found to be devoted to cutaneous representation: Most of the units encountered have light tactile receptive fields; but near the anterior surface of the nucleus, units are mainly activated by deep stimuli: pulling on tendons and manipulation of muscle bellies. This small region seems also to include the group I afferent relay (Wiesendanger et al., 1979). Injections of tritiated amino acids in the deep zone demonstrate a thalamocortical projection only to area 3a (Figs. 13 and 14); the terminal ramifications of the thalamocortical fibers stop abruptly opposite the most posteriorly situated giant pyramidal cell of area 4 (Fig. 14). A similar narrow zone in which units are mainly activated by deep stimuli can be demonstrated in the dorsal aspect of VPLc by microelectrodes introduced vertically from above, before they pass into the large cutaneous core of VPLc, as originally shown by Poggio and Mountcastle (1963). Injections of isotope in this zone demonstrate a projection not only to area 3a but also to area 2 (Fig. 13). This raises a question about the nature of the receptors providing input to area 2 and whether, like area 3a, there may be a substantial muscle receptor input. Area 2 has for long been regarded as receiving input principally from joint receptors (Powell and Mountcastle, 1959; Werner and Whitsel, 1968; Dreyer et al., 1974). However, a few units responding to muscle stretch have been described in area 2 (Burchfiel and Duffy, 1972). Considering the findings of Clark and Burgess (1975) that many peripheral nerve fibers activated by movement of a joint are connected with stretch receptors not in the joint capsule, but in muscles overlying the joint, large-diameter, muscle afferent inputs to area 2 may be more significant than previously supposed.

In contrast to the anterior and dorsal deep shell of VPLc, the larger cutaneous core can be shown by appropriately placed isotope injections to project only to areas 3b and 1, which are the known cutaneous receiving areas of the first somatic sensory cortex (SI) (Powell and Mountcastle, 1959; Werner and Whitsel, 1968; Dreyer et al., 1974; Kaas et al., 1979) (Fig. 13). Like in the projection from VPLo-VLc-X to area 4, small rodlike aggregations of VPLc cells receiving a cluster of lemniscal terminations project to small focal zones in the cortex (Friedman and Jones, 1980).

Corticocortical Connectivity of Sensory-Motor Cortex

From the last section, it would seem appropriate to have corticocortical connections joining the 4 fields of SI to the motor cortex and relaying the inputs characteristic of the four fields to the motor field. As pointed out above, although not popular as a route for all afferent influences to enter area 4, a projection from area 3a to area 4 was essential to theories concerning the nature of motor cortical reflex responses to peripheral perturbations of muscle length. The pattern of corticocortical connectivity linking the pre- and postcentral gyri is not, however, a simple, straightforward one.

The pattern of corticocortical connections based on autoradiography following injections of tritiated amino acids in the separate fields of the pre- and postcentral gyri is shown in a series of summary surface maps in Figs. 15 and 16 (Jones et al., 1978). The results can also be obtained using retrograde transport methods. The transported label following injections of area 3a shows that this field, rather surprisingly, projects backward to area 1, but not forward to area 4. Area 3b, the large cutaneous

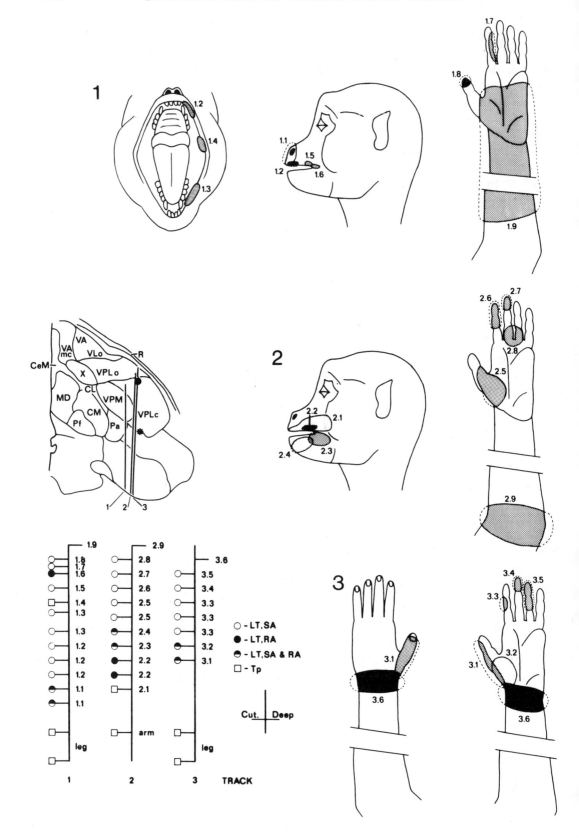

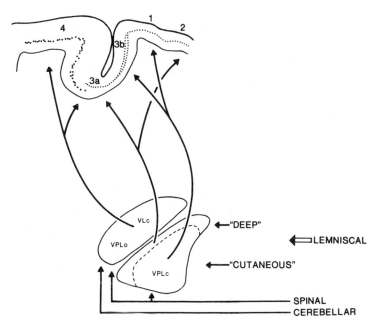

FIG. 13. Schematic figure summarizing on sagittal sections, the distributions of spinal, cerebellar and lemniscal terminations in the ventral thalamic nuclei, the cutaneous core and deep shell zones of the lemniscal relay nucleus and their separate projections to areas 3b-1 and 3a-2. (From Friedman and Jones, 1981, with permission.)

field, also projects mainly backward, to areas 1 and 2. Area 1, another mainly cutaneous field, projects backward to area 2 and forward principally to area 3a. Only area 2, the deep field, and a neighboring part of area 5 (Jones et al., 1978; Strick and Kim, 1978) project forward into area 4. The main output connections of area 4 in these regions are to areas 3a, 1, 2, and 5.

Thus, of the somatic sensory fields, only area 2 would seem to be an appropriate source of relatively direct peripheral inputs to motor cortex. It may be this connection that serves to maintain the amplitude of an area 4 evoked potential in the intact animal. But whether it is appropriate to provide the principal relay for deep and cutaneous input at short latency to area 4 is a far more difficult question that cannot be answered on the basis of present knowledge.

Subcortical Projections of the Somatic Sensory Areas

In seeking to close the loop at the cortical level between peripheral input and corticomotoneuronal output, some workers have focused on the direct corticospinal and cortico-brainstem projections of the postcentral cortical fields, particularly area 3a. After all, if such connections exist, why should they not form the efferent limb of a cortical reflex pathway that bypasses the motor cortex?

FIG. 12. Microelectrode tracks (1–3), entering monkey thalamus horizontally from behind, encounter neurons with cutaneous receptive fields *(bars left)* throughout greater part of lemniscal relay nucleus (VPLc, including trigeminal component, VPM). However, in anterior 300–500 μm of the nuclei, neurons with deep, particularly muscle-related, receptive fields *(bars right)* are encountered. *Large dot* on track 3 indicates localizing lesion made at beginning of zone of deep responses. *Figurines* show positions of cutaneous receptive fields. LT: Light touch; RA: rapidly adapting; SA: slowly adapting; Tp: tap. (From Friedman and Jones, 1981, with permission.)

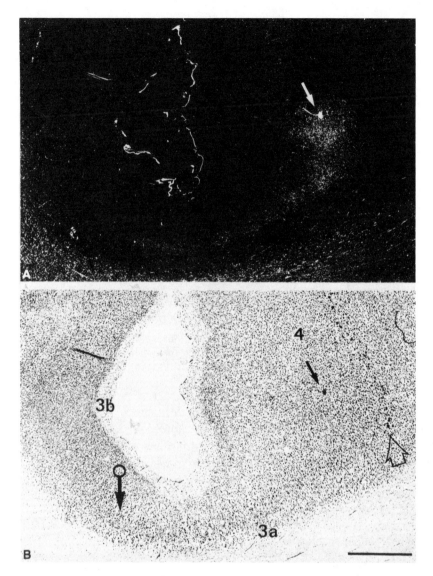

FIG. 14. Darkfield (A) and brightfield (B) photomicrographs at same magnification and from same sagittal section showing restriction of thalamocortical labeling to area 3a following small injection of ^{3}H-labeled amino acids in anterior part of deep shell zone of lemniscal relay nucleus (VPLc) in thalamus. (From Friedman and Jones, 1981, with permission.)

FIGS. 15 (top) and 16 (bottom). Surface maps of pre- and postcentral cortical fields drawn from sagittal or horizontal sections (15 top left) so as to unfold the cortex into a flat sheet. Each *short line* represents a focus of terminal ramifications of corticocortical fibers labeled by axoplasmic transport from injections of ^{3}H-labeled amino acids at sites indicated in *black*. Note that only areas 1, 2, and 5 can be shown to provide corticocortical inputs to area 4, and that area 4 projects back to all fields except area 3b. Label in upper part of area 6 in lower maps of Fig. 16 is in supplementary motor area. *Inset* to Fig. 15 shows autoradiographic labeling in areas 1 and 2 *(arrows)* after injection in area 3b. (From Jones et al., 1978, with permission.)

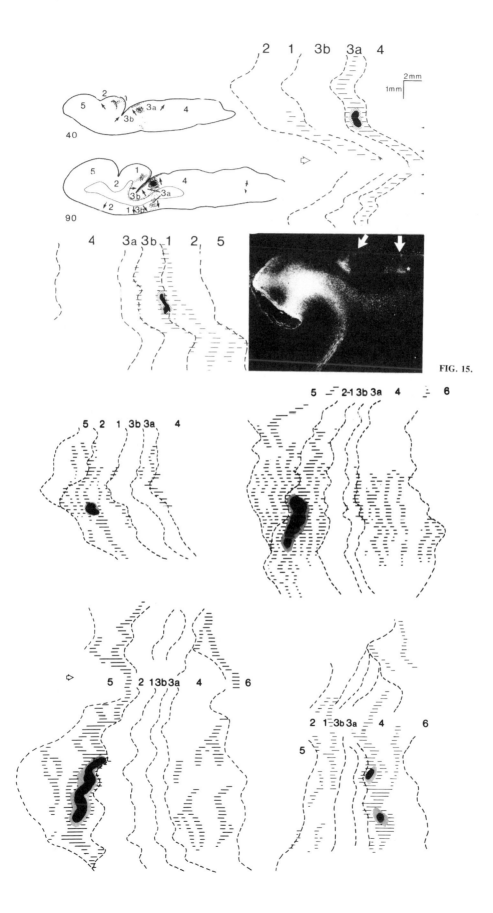

FIG. 15.

FIG. 16.

Although it is true that pyramidal cells of layer V in areas 3a, 3b, 1, and 2 give rise to axons that descend to the spinal cord (Jones and Wise, 1977), it is difficult to view them in the same light as the similar pathways descending from area 4. In the first place, the numbers of corticospinal neurons, in comparison with those in area 4, are quite small and the temporal realtionships of their discharges to peripheral perturbations far from clear. Second, the corticospinal fibers descending from the sensory fields are destined to terminate high in the dorsal horn, even in the substantia gelatinosa, unlike those from area 4 that terminate deep in the intermediate zone and in the ventral horn (Coulter and Jones, 1977). Reports that corticospinal fibers from area 3a may also terminate deep in the intermediate zone and ventral horn actually stem from experiments in which area 3a was designated as the zone of granule cell and giant pyramidal cell overlap in the hand representation. If, as we now know, area 3a should be regarded only as the attenuated granule cell zone, projections to the deep intermediate zone and ventral horn cannot be detected (*unpublished observation*). It seems unlikely that corticospinal fibers descending from the sensory fields and terminating high in the dorsal horn can have particularly direct access to motoneurons and, thus, it is difficult to see them as a route for rapid, corticomotoneuronal adaptive responses. A rapid return route from sensory cortex to motoneurons via the red nucleus also seems unlikely, since the only corticorubral neurons so far detected outside the motor areas are in the small anterior, overlap zone of area 3a in the hand representation (Jones and Wise, 1977). Such cells are probably best regarded as in the posterior part of area 4. Hence, the motor cortex itself still appears to be the most viable candidate for the efferent limb of our motor cortex-to-motoneuron adaptive loop, but the afferent limb of the loop still needs to be defined in anatomical terms.

In summary, a large body of recent work in primates indicates that under appropriate conditions, varied kinds of peripheral influences can modulate motor cortex output. These afferent influences reach the motor cortex at very short latencies and seem to be mediated by the dorsal columns of the spinal cord and the medial lemniscus.

Although there is evidence for short-latency cutaneous and deep inputs to the thalamic relay nucleus to motor cortex, there is no anatomical confirmation of a lemniscal input to this nucleus. However, in addition to cerebellar inputs, which are unlikely to relay short-latency inputs, spinothalamic and vestibulothalamic fibers are distributed to parts of the thalamic relay and the former might account for some of the short-latency inputs.

Lemniscal fibers, including those conveying impulses mediated by group I muscle afferents, are distributed with some spinothalamic fibers only to the thalamic relay to sensory cortex. This nucleus is composed of a large central core in which units respond mainly to cutaneous stimuli and which projects only to areas 3b and 1, the cutaneous fields of the first somatic sensory area. Over the central core, on the anterior and dorsal surfaces of the lemniscal relay nucleus, is a thin shell in which units respond mainly to deep stimuli. Anterior parts of the shell project only to the group I receiving area of sensory cortex (area 3a). Dorsal parts project to area 3a and to area 2, a further area dominated by deep inputs.

Although it is possible that short-latency lemniscal inputs may reach motor cortex via corticocortical pathways after a relay in the sensory cortex, anatomical studies indicate that few of the fields of the sensory cortex actually project to area 4. The principal field providing such input is area 2, not as has sometimes been postulated, area 3a.

Corticospinal fibers descending from the sensory cortex, because of their terminations high in the dorsal horn of the spinal cord, are unlikely to furnish the efferent limb of a rapid corticomotoneuronal response to a peripheral perturbation. Corticorubral fibers do not arise in the sensory fields.

All of these data lead to the conclusion that the known anatomical pathways for short-latency inputs leading to rapid adaptive responses in the corticomotoneuronal pathway do not readily match the results of other experimentation and further work is called for.

The Premotor Cortex of Monkeys

None of the subcortical afferent systems considered above seems to have direct connections with the premotor cortex. The premotor cortex, can be considered to consist of the agranular areas anterior to the giant (Betz) cell containing area 4, i.e., area 6 of Brodmann (1909), or the intermediate precentral area of Campbell (1905). Area 6 was later divided into subdivisions areas 6a alpha and 6a beta by Vogt and Vogt (1919) (Fig. 17). Area 6a alpha was a part of Brodmann's area 4; it includes areas 4s and 4a of Von Bonin (1949) and was considered by Woolsey et al. (1952) to be merely an extension of area 4 that held the representation of axial musculature. There seems now to be good evidence for contesting Woolsey's

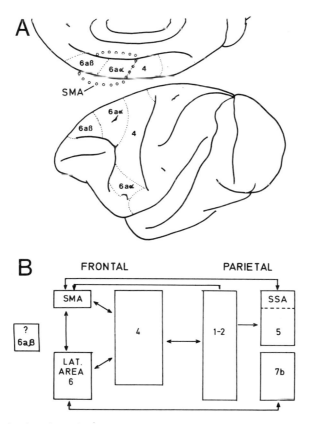

FIG. 17. A: Schematic drawing of the positions of the agranular areas of the frontal lobe after Brodmann (1909) and Campbell (1905), with the finer subdivisions, areas 6a alpha and 6a beta, delineated by Vogt and Vogt (1919), and superimposition of the supplementary motor area (Penfield and Welch, 1951; Woolsey et al., 1952). The supplementary motor area includes the medial part of area 6a alpha and possible parts of 6a beta. Lateral parts of 6a alpha appear to represent at least one other independent premotor field. No significant information is available about area 6a beta. B: Intracortical connectivity of the premotor areas. *SSA:* The supplementary sensory area, a part of area 5. (From Jones and Powell, 1969, and Murray and Coulter, 1981*b*, with permission.)

point of view on connectional and functional grounds. These tend to confirm the earlier cytoarchitectonic parcellations of the Vogts.

Supplementary motor area

The principal cortical inputs to the supplementary motor area emanate from areas 1, 2, and 5 of the somatic sensory and parietal cortex ipsilaterally (Jones et al., 1978; Murray and Coulter, 1981b) and from area 4 bilaterally (Jenny, 1979; Jones et al., 1978; Muakkassa and Strick, 1979). Its cortical output is back only to areas 4 and 5 ipsilaterally and possibly to lateral area 6 (Arikuni et al., 1980; Jones et al., 1978; Muakkassa and Strick, 1979). The supplementary motor area thus seems to be a region of considerable convergence that would exert some of its transcortical effects on area 4. It seems rather surprising, therefore, that few movement-related neurons in the supplementary motor area have obvious peripheral receptive fields (Brinkman and Porter, 1979; Wise and Tanji, 1981). It seems evident, however, that one important functional aspect of the supplementary motor area is that, in relation to many movements, it leads the motor cortex. In conscious monkeys, movement-related neurons, particularly if they are direction selective, tend to discharge in advance of those in area 4 (Brinkman and Porter, 1979; Tanji et al., 1980). In addition, movement-related neurons in the supplementary motor area, in contrast with those in area 4, usually discharge in relation to movements of both contralateral and ipsilateral limbs (Brinkman and Porter, 1979). Bilateral activation of the supplementary motor area is demonstrable (with unilateral activation of area 4) in studies of regional cerebral blood flow in man during movements of one hand (Roland et al., 1980). Perhaps even more interestingly, the supplementary motor area shows increased blood flow without an associated increase in area 4 when a subject simply thinks about the movement, without actually performing it (Roland et al., 1980).

It now seems necessary to consider area 6 to be composed of at least three parts: medial, lateral, and anterior (Fig. 1) discussed below.

The medial component. Largely on the medial surface of the hemisphere and invading the banks of the cingulate sulcus, is the supplementary motor area of Penfield and Welch (1951) and of Woolsey (1952). It occupies the part of area 6a alpha lying on the medial surface and may also incorporate some or all of the part of area 6a beta lying there.

It is distinguished by a coherent, full representation of the body musculature that is independent of that in area 4. The representation may be identified by intracortical microstimulation (Wise and Tanji, 1981) by the distribution of corticospinal neurons projecting to different regions of the spinal cord (Murray and Coulter, 1981a) and by the distribution of corticocortical fibers from different parts of the representation in the somatic sensory cortex (Jones and Powell, 1969).

Lateral area 6. The remaining parts of area 6, largely situated on the lateral surface of the hemisphere, can probably be divided into two components. One, lying dorsal to the upper limb of the arcuate sulcus corresponds approximately to area 6a beta. About this, we can say virtually nothing. The second part, corresponding approximately to lateral area 6a alpha, is an hourglass-shaped area with expanded upper and lower portions and a narrow connecting portion running down the posterior bank or floor of the convexity of the arcuate sulcus (Fig. 1). The principal difference between this lateral part of area 6 and the supplementary motor area is that, unlike the supplementary motor area, it does not receive inputs from areas 1 and 2 of the somatic sensory cortex (Chavis and Pandya, 1976; Jones et al., 1978; Jones and Powell, 1969; Jones and Powell, 1970). It is also doubtful that it projects to the spinal cord, again contrasting with the supplementary motor area (Murray and Coulter, 1981a). Other cortical inputs come from area 4 and from certain parietal areas: from area 5 (Jones and Powell, 1969; Jones and Powell, 1970) and from the anterior part of the inferior parietal lobule (Chavis and Pandya, 1976) (area 7b) which may be connectionally analogous to area 5 (Burton and Jones, 1976). Its cortical outputs are to area 4: dorsal parts, around the superior precentral dimple, project to leg and trunk representations and ventral parts, around the inferior precentral dimple, to arm and face representations (Arikuni et al., 1980; Muakkassa and Strick, 1979). Other outputs may go to the supplementary motor area and possibly back to areas 5 or 7 (Jones and Powell, 1970; Pandya and Vignolo, 1971).

Relatively little is known about the functional properties of the lateral component of area 6. No earlier lesion studies affected it in isolation. A recent study in conscious behaving monkeys (Wise and Weinrich, 1981), indicates that many neurons change their discharges in relation to a movement task involving projecting the arm toward a target in a delayed, two-cue situation and that these discharges can antecede those of motor cortex neurons. In a study in paralyzed monkeys, Rizzolatti et al. (1981a, 1981b) found that many neurons in a comparable area had complex receptive fields. These were commonly bilateral, usually on hand(s) and mouth, could have a visual component and often behaved as though related to movements and events that would bring the hands or objects in visual space toward the mouth (see also Kubota and Hamasa, 1978). Rizzolatti et al. remark that some of these features have similarities to neuronal behavior in area 7b, a source of input to lateral area 6.

These anatomical and physiological observations argue against lateral area 6 being simply the axial representation of area 4. They suggest far more complicated activities which, like those of the supplementary motor area, may lead area 4 in initiating movement. The functional relationships of other parts of area 6, such as area 6a beta and two other portions suggested by the work of Muakkassa and Strick (1979), remain unknown.

Thalamic afferents It seems remarkable that very little is known about the thalamic inputs to either the supplementary motor area or to lateral parts of area 6. From the early work of Walker (1949) we may assume that thalamic inputs to both arise predominately in the ventral lateral complex. Recent work on this complex, however, would tend to indicate that the large region receiving the cerebellothalamic projection and furnishing input to area 4, does not project to these other areas (Asanuma et al., 1982; Jones et al., 1979; Tracey et al., 1980). Attention, therefore, comes to focus on more anterior parts of the ventral lateral complex, especially the oral ventral lateral (VLo) nucleus, which receives pallidal inputs (Tracey et al., 1980). These are even further removed from the peripheral somatic sensory receptors than the cerebellum and its thalamic relay seem to be. But the cerebellocortical and pallidocortical circuitry now seem more discrete than previously thought and it is now necessary to ask what differential effects pallidum and cerebellum might exert on their separate cortical targets and whether and how these come together with one another and with the influences of the parietal lobe in the motor and premotor areas.

ACKNOWLEDGMENTS

Personal and collaborative work reported here was supported by grants Nos. NS 10526, NS 12777, F32-NS 05884, and T32-NS 07057 from the National Institutes of Health, United States Public Health Service. I thank Bertha McClure and Ronald Steiner for technical assistance. The help of Drs. C. Asanuma, D. P. Friedman, and W. T. Thach is also gratefully acknowledged.

Intracortical Organization of Arousal as a Model of Dynamic Neuronal Processes That May Involve a Set for Movements

*Tomokazu Oshima

Department of Neurobiology, Tokyo Metropolitan Institute for Neurosciences, Tokyo, Japan

Both behavioral arousal and the orienting reflex involve the earliest stage of initiating animal movement. According to Pavlov (1927), the orienting reflex interrupts the animal's ongoing activity to explore the surrounding space, thus becoming the starting point of a new behavior. General contribution of the cerebral cortex to the orienting reflex is demonstrated by the electroencephalograph (EEG) activation or arousal (Moruzzi and Magoun, 1949; Magoun, 1963; Moruzzi, 1972). Just as the orienting reflex is composed of the initial interruption of activity such as motor arrest and the following investigatory behavior, the EEG arousal can be divided into the initial transient or phasic phase and a later sustained or tonic phase (Sharpless and Jasper, 1956; Walsh and Cordeau, 1965; Courtois and Cordeau, 1969; Steriade, 1970; Steriade et al., 1974a,b; Steriade and Hobson, 1976; Inubushi et al., 1978b). The motor area of the cerebral cortex has become a suitable subject for analysis of the activity of single neurons during EEG arousal. This region exhibits a typical pattern of EEG arousal from synchronization to desynchronization, like the association cortex (cf. Phillips et al., 1972). In addition, the upper motor neurons such as fast and slow pyramidal tract (PT) cells can be identified antidromically, by stimulating their axons (Takahashi, 1965; Oshima, 1969), and studied in the earliest period of initiating movement through different phases of EEG arousal.

Unit discharges recorded extracellularly from various cortical regions change on EEG arousal as increases (excitation) in some neurons and decreases (suppression or inhibition) in others. Jasper (1958) summarized the situation of research about a decade after the Moruzzi and Magoun (1949) concept of ascending reticular activating system. He stated:

> The activating effect of the ascending reticular system on the cerebral cortex cannot be adequately described in terms of either gross excitation or inhibition, since both processes seem to be present to a more or less degree. Activation seems better described as a reorganization of temporal and spatial patterns of neuronal discharge in a matrix of more sustained excitatory and inhibitory patterns.

However, until recently it was beyond the resolving power of microphysiological techniques to substantiate the concept of the reorganization of neuronal activities during EEG arousal.

On the other hand, research on the upper motor neurons was initiated by two different approaches. One was the chronic experiment on monkeys, where the activities

---

*Department of Neurobiology, Tokyo Metropolitan Institute for Neurosciences, 2-6 Musashidai, Fuchu-City, Tokyo, Japan.

of fast and slow PT cells were recorded extracellularly during various states including the sleep-waking cycle (Evarts, 1965; Steriade, 1978; and others). The other was the acute experiment on cats lightly anesthetized or immobilized. The intracellular potentials were recorded by glass microelectrodes during high-frequency repetitive stimulation of the mesencephalic reticular formation (Klee et al., 1964; Akimoto and Saito, 1966; Klee, 1966). Both yielded many new findings: (1) On EEG arousal, fast PT cells decrease and slow PT cells increase their firing rates (Evarts, 1965). (2) During a later phase of EEG arousal, fast PT cells start to increase the firing rate, and slow PT cells maintain their increased firing rate tonically (Steriade et al., 1974a,b). (3) Presumed fast PT cells exhibit membrane hyperpolarization (Klee et al., 1964; Klee 1966) or depolarization during reticular stimulation (Akimoto and Saito, 1966). (4) Some interneuron-type neurons exhibit characteristic changes of discharge during EEG arousal (Steriade et al., 1974a; Steriade and Deschenes, 1974; Steriade, 1978). These results were encouraging, but it seemed necessary to bridge the gap between the chronic and acute experiments by adopting adequate preparations. We used the cat with its brain transected at the level of the pons (midpontine pretrigeminal preparation) or the cervical cord ("encéphale isolé" preparation of Bremer, 1935), immobilized by gallamine triethiodide under artificial respiration and unanesthetized during spontaneously occurring sleep-waking cycle as well as during stimulation of the reticular formation or application of natural stimuli (acoustic or visual) to induce EEG arousal. The analytic power of the intracellular technique was fully employed to qualify and quantify the neuronal activities altered during EEG arousal (Inbushi et al., 1978a,b; Ezure and Oshima, 1981a–c). These experiments made it possible (1) to detect the membrane potential changes during EEG arousal, as the index far more sensitive and accurate in timing than the firing rate changes; (2) to qualify the membrane potential changes in terms of postsynaptic excitation, inhibition, disfacilitation, and disinhibition by means of the intracellular current injection, and (3) to extend the sampling of neurons from all the cortical laminas through I to VI including fast and slow PT cells.

METHODOLOGICAL CONSIDERATIONS

The experiment was performed on the cat with head fixed in a stereotaxic apparatus and placed in a dimly lighted and sound-attenuated shielded room under the monitoring of systemic blood pressure, heart rate, body temperature, and CO_2 content of expired air. General care of the preparation was essential in maintaining the intracellular recording for a reasonably long period to repeat observations of EEG arousal. The experimental procedure and arrangement have been fully described elsewhere (Inubushi et al., 1978a,b; Ezure and Oshima, 1981a).

Reticular-Induced and Natural Episodes of EEG Arousal and Their Physiological Intensities

Earlier intracellular investigations were limited to PT cells with several provisos such as use of general anesthetics or lack of identification of fast and slow PT cell groups. The reticular stimulation caused membrane depolarization in some PT cells and hyperpolarization in some others, and the rate of occurrence of these two changes was viewed differently by two groups of investigators (Klee et al., 1964; Akimoto and Saito, 1966; Klee, 1966). This discrepancy has recently been explained by the following two factors: (1) one and the same fast PT cells respond with either (a) hyperpolarization, (b) hyperpolarization followed by depolarization, or (c) exclusive depolarization to EEG arousal according to its different physiological intensities (Ezure et al., 1980; Ezure and Oshima, 1981a). (2) Slow PT cells always respond with depolarization to EEG arousal (Inubushi et al., 1978a; Ezure and Oshima, 1981b). As a result of these investigations, reticular stimulation with various intensities became a useful means with which to induce EEG arousal to different degrees but raised another problem of possible effect of current spread to the structures neighboring the reticular formation. In these studies it was important to compare the reticular effects with those of various natural stimuli (acoustic, visual, etc.) or with arousal that occurred spontaneously. The depolarizing response of fast PT cells was assumed to occur during the most physiologically intense arousal, and was observed also during natural EEG arousal (Ezure and Oshima, 1981a). Thus, in reticular-induced EEG arousal, the cellular responses mimicked well the responses to natural episodes of EEG arousal. Attempts to define the physiological intensity of EEG arousal were made in terms of the intensity of stimuli used to induce EEG arousal as well as of the exerted effects on EEG such as the total length of aroused pattern and the degree of shift from synchronization to desynchronization (Ezure et al., 1980; Ezure and Oshima, 1981a).

Analysis of Intracellular Responses

Measurements of membrane potential and impedance enabled us to identify four basic types of cellular response in respective neurons during EEG arousal: (1) postsynaptic excitation (E), or increases of excitatory postsynaptic potentials (EPSPs), identified with membrane depolarization and accompanying decrease of the membrane impedance, (2) postsynaptic inhibition (I), or increases of inhibitory postsynaptic potentials (IPSPs), with membrane hyperpolarization and impedance decrease, (3) disfacilitation (DF), or decreases of EPSPs with membrane hyperpolarization and impedance increase, and (4) disinhibition (DI) or decreases of IPSPs, with membrane depolarization and impedance increase. Rates of occurrence of unitary EPSPs and IPSPs were also counted to support the identification of the four types of responses (Inubushi et al., 1978b). These methods may indicate only the net change of afferent synaptic bombardments onto a cell under test since different origins of the afferent inputs to the cell cannot be identified with these methods. Nevertheless, the results of identification are useful in relating the dominant cellular responses to the relevant neuronal connections, particularly for the cases of DF and DI in which the neuronal relays composed of at least three cells are the prerequisite (see below).

Sampling of Cells and Its Bias

To construct the distribution histogram of cell location with some ease, cells were sampled along the track of microelectrode penetrated perpendicularly to the cortical surface, the depth in recording of the IS-SD spike potential from the cell somata was read on a pulse-driven electrode-manipulator, and the direction of electrode penetration was histologically examined for correction, if necessary, after the experiment (Inubushi et al., 1978a). Ideal for the present analysis is obviously to sample the cortical neurons of all possible types through the whole cortical laminas. The recent intracellular technique has greatly been advanced for sampling both large- and small-sized neurons. However, no one can maintain that sampling bias is avoided. Rather, our procedure of sampling may inevitably yield some bias because of at least two factors: (1) the density of cell population and (2) the size of cell somata favorable for recording. For example, we can sample a good number of cells in lamina II in spite of their small size (high membrane resistances of 10–70 MΩ) (Inubushi et al., 1978b; Oshima, 1978) due to their dense population.

On the other hand, sampling of small-sized cells in laminas V and VI is very limited for the sparse population density. As a whole, the distribution in depths of intracellularly recorded cells shows clear tendency of sampling small cells in relatively superficial layers and large cells in deep layers (Fig. 1 of Inubushi et al., 1978b). We assume that this is a fairly good reflection of the properties of all types of cells in respective laminas.

With the microelectrodes filled with 2 M potassium citrate or 2 M KCH_3SO_4 solution, and having electrical resistances of 30 to 80 MΩ, sampling was made in the so-called carefree manner. Thus, the experimenters did not care what types of cells were collected during electrode penetration. Their concern was concentrated on obtaining only reasonably good recording conditions for repeated tests of EEG arousal. Selection of cells for analysis was made later according to the criterion that sufficient numbers of trials should be collected to ascertain the cellular responses to EEG arousal at different physiological intensities. Once the sample neurons are accepted as the natural representatives of respective laminas, their various activities related to the topography are integrated to create a cortical network. The carefree cell sampling was so rewarding in studies of the nonspecific events such as EEG arousal, that the structuralistic approach became much easier than had been presumed.

RESPONSE PATTERNS AND LAMINAR DISTRIBUTION OF CORTICAL NEURONS

Fast PT Cells and Their Family Neurons

The dual-activity pattern in fast PT cells, that is, initial decrease and subsequent increase of firing rates, were confirmed during behavioral arousal in monkeys (Evarts, 1965; Steriade et al., 1974a,b). These changes were also obtained using the transected brain preparations of cats (Steriade and Deschenes, 1974; Inubushi et al., 1978a). Dual activity changes were readily observed in fast PT cells in the acute experiment during the EEG arousal either occurring spontaneously or induced by acoustic, photic, or reticular stimuli. These changes are illustrated in Fig. 1A as an initial membrane hyperpolarization (marked in 4 with a horizontal bar a) and a subsequent depolarization (bar b) for the arousal indicated by the EEG in the motor cortex (2) and hippocampal activity (3) during reticular stimulation (1). The hyperpolarizing pulses are injected every second (5). The effective membrane resistance (R_M) is calculated from the potential difference (V in Fig. 1C) between the resting and hyperpolarized levels divided by the current intensity (I), and plotted in B to show an initial transient increase corresponding to the hyperpolarization, that is, DF response and a following decrease during the depolarization, i.e., E response.

The amplitudes of DF and E responses depended on the physiological intensity of EEG arousal (Fig. 2). It is interesting that a certain reciprocity is seen in the changes of these DF and E responses. The DF responses develop in full at the juxtathreshold stimuli and maintain their amplitude at approximately constant levels or tend to decrease to more intense stimuli. E responses are more graded with the increase of stimuli. In a case plotted with circles, DF responses (closed circles) are masked by E responses (open circles) when the stimuli are twice threshold or more. Thus, the exclusive DF and exclusive E responses can be observed as two extremes to the weakest and the most intense stimuli, respectively. In between DF followed by E (DF + E) responses are obtained. Thus, these three types of responses, DF, DF + E and E, are usually observed in fast PT cells regardless of whether EEG arousal is reticular-induced or occurs naturally. This raised intriguing arguments concerning the intensity

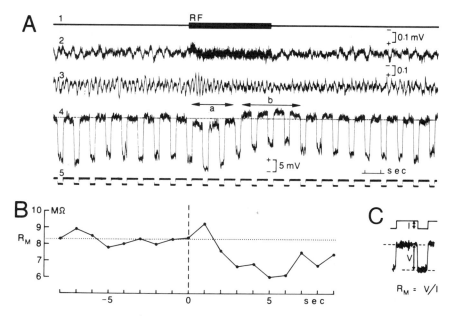

FIG. 1. DF + E response of a fast PT cell. In a pretrigeminal cat preparation this cell was recorded at a depth of 1,460 μm with latency of 0.5 msec of antidromic excitation elicited by stimulating cerebral peduncle. A: Reticular stimulation (40 A, 100 Hz) (1), causes EEG arousal in motor cortex (2), and hippocampus (3), and initial hyperpolarizing and late depolarizing responses in membrane potentials (4). Current steps of −2.0 nA (5) are applied every second to measure effective membrane resistance (R_M). In (4), a *dotted line* gives reference level of −55 mV, and spikes are cut off by using an electrical low-pass filter. B: R_M plotted during periods of control records and of induced EEG arousal after calculation from average potential shift (V) and used current (I) measured as illustrated in C. In this or each of following figures with plottings of R_M, a *vertical broken line* marks onset of EEG arousal, and a *horizontal dotted line* indicates the mean value of R_M during control before EEG arousal. (Fig. 1A from Oshima, 1981, cf. original records in Fig. 6 of Inubushi et al., 1978b, with permission.)

concept of arousal. To the question, can you measure the novelty of stimuli inducing the orienting reflex? you would say no, on the basis of DF response pattern, but say yes relying on the E response property. It is argued that the DF response is a biologically passive process or a general one set on receipt of the novelty factor of stimuli, whereas the E response represents an active process of encoding the intensity factor of stimuli. Many non-PT cells that exhibited a DF + E response pattern were found in the depths of 800 to 1,600 μm comparable to those of location of fast PT cells. Those cells showed the same response properties as fast PT cells, and were treated as the family neurons of the latter.

Slow PT Cells and Their Family Neurons

Clear response dichotomy such as the DF + E response pattern is not seen in slow PT cells. During relatively intense, long-lasting EEG arousal, slow PT cells exhibit slowly rising or tonically sustained depolarization accompanying R_M decrease, i.e., E response. We term this the E + E response, implying that the response involves two stages corresponding to the initial phasic and late tonic EEG arousal. Many non-PT cortical neurons can be classified with their E + E response as the family neurons of slow PT cells (Fig. 3). The motor EEG (A1) shows enduring desynchronization with little indications to divide the EEG arousal

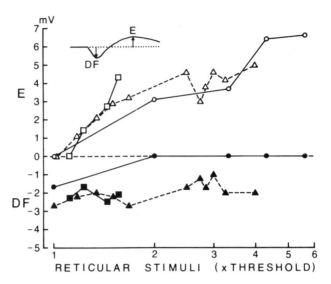

FIG. 2. Stimulus-response relations in DF and E responses to reticular stimuli. In 3 fast PT cells, maximal amplitudes of DF and E responses are measured separately *(inset)*, and plotted with different symbols against stimuli as times the threshold for eliciting EEG arousal. (From Ezure and Oshima, 1981*a* with permission.)

into phasic and tonic phases, whereas the visual EEG (A2) shows initial slow and later fast wave patterns, as marked with bars a and b, respectively, which may correspond to the two arousal phases. E + E responses in slow PT cells and their family neurons located in relatively deep cortical layers (depths, 500–1,500 μm) were continuous with no indications for two separate phases (cf. Fig. 3). However, separate surges of depolarization to intense EEG arousal were seen in the non-PT family neurons located in relatively superficial layers (depths, 100–900 μm).

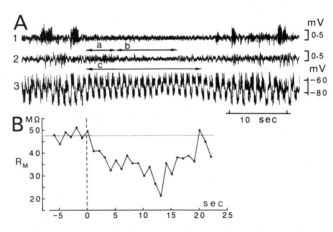

FIG. 3. E + E response in a non-PT cell (depth, 1,120 μm). Encéphale isolé cat. **A:** Motor *(1)* and visual EEG leads *(2)* are displayed with intracellular potentials *(3)* during spontaneous EEG arousal. Hyperpolarizing pulses of −0.5 nA are applied every second. Bars a–c indicate characteristic changes in respective records (see text). **B** plots R_M. (Modified from Ezure and Oshima, 1981*b*, with permission.)

Other Types of Cortical Neurons

In the most intense EEG arousal, responses in various neurons were of four types: E, I, DF, and DI, in either phasic or tonic phase of EEG arousal. Various combinations occurred. Including PT cells and their family neurons, a total of 164 cells were identified (Ezure and Oshima, 1981c) as the recipients of either E, I, DF, or DI during phasic EEG arousal (Table 1). Of these cells, 32 showed no appreciable activity changes during late tonic EEG arousal, and were termed late unresponsive. The other 102 cells responded with late E (+E), 16 with late I (+I), 12 with late DF (+DF) and 2 with late DI (+DI) (Table 1). Neurons thus identified were named by the initial and late responses such as DF + E cells for fast PT cells and their family neurons or as E + E cells for slow PT cells and their family neurons.

Figure 4 illustrates an I + E cell. On EEG arousal in motor cortex (A1) and hippocampus (2), this cell responds with an initial hyperpolarization (bar a in 3) followed by a long-lasting depolarization (bar b). Corresponding to these changes are continuous decreases in R_M as plotted in B, indicating that the responses are composed of I + E in this cell. Figure 5 shows an E + DF response. During long-lasting EEG arousal in motor cortex (A1) and hippocampus (2) the membrane potential is transiently depolarized (bar a in 3) with a concomitant initial decrease in R_M (A3, plotted in B, as indicated with a downward arrow), and thereafter tends to be hyperpolarized with a phase of increasing R_M (bars b and c in A3, plotted in B). This neuron is therefore identified as an E + DF cell. Figure 6 illustrates an I + DI response. Two separate phases of initial hyperpolarization (bar a in A3) and late depolarization (bar b) are demonstrated during long-lasting EEG arousal (1, 2). Corresponding R_M decrease and increase (Fig. 6B) indicate that this neuron shows an I + DI sequence.

The specimen records in Figs. 1 and 3 to 6 show that changes in the membrane potential and R_M to EEG arousal are generally small. In particular, when their control levels are fluctuant by the influence of spindle burst waves (cf. Figs. 5, 6), observations had to be repeated to assess reliability of responses. The response patterns of the 164 cells are summarized in Table 1. From this we learn: (1) All the cortical neurons respond initially with their specific activity patterns to phasic EEG arousal, but a certain population of these cells fails to respond later to tonic EEG arousal (late unresponsive cells, U in Table 1). (2) Excitation predominates other actions, particularly during tonic EEG arousal (incidence, 102/164). (3) When late responses occur, there are 9 types of combinations with the initial responses. It does not seem that all the theoretically possible combinations (4^2 = 16) do occur.

The Laminar Distribution

The depth profile discloses unique topographical patterns of response. For phasic EEG arousal (Fig. 7A) laminas I and II involve only E cells, lamina III$_a$ E and I cells and laminas III$_b$, V, and VI all the types of E, I, DF, and DI cells. The number of synaptic actions increases from 1 to 4 by the factor of 2 through the superficial to deep laminas. During tonic EEG arousal (Fig. 9B) laminas V and VI include only +E cells, lamina III$_b$ +E and +I cells, and laminas III$_a$ – I all the types of +E, +I,

TABLE 1. *Classification of neurons with responses to phasic and tonic EEG arousal*

	Cell type	Tonic phase				
		U	+E	+I	+DF	+DI	Total
Phasic phase	E	11	40	10	10	0	71
	I	17	22	6	2	2	49
	DF	4	34	0	0	0	38
	DI	0	6	0	0	0	6
	Total	32	102	16	12	2	164

U: late unresponsive.

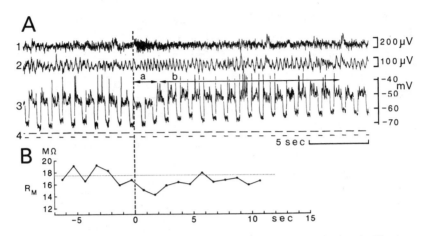

FIG. 4. I+E response in a non-PT cell (depth, 670 μm). Encéphale isolé cat. A: Illustrates (1) motor EEG, (2) hippocampal activity, (3) intracellular potentials, and (4) injected currents (−0.9 nA) during spontaneous EEG arousal. B plots R_M. (Modified from Ezure and Oshima, 1981c, with permission.)

+DF, and +DI cells, except for the late unresponsive cells, which are located through all the laminas with a tendency to increase from deep to superficial layers (Fig. 8A).

MODELS AND FUNCTIONAL IMPLICATIONS

Formulation of Arousal Circuit Models

We read the depth profile illustrated for phasic EEG arousal in Fig. 7A as follows according to a simple model circuit drawn in B and C. Eight parallel, 4-neuron relays are arranged through 4-layer divisions on a vertical plane of the cerebral cortex. These relays include all possible combinations of synaptic connections between excitatory (open circles with terminal circlets) and inhibitory neurons (filled circles with terminals) in the upper 3-layer divisions. The phasic arousal starts mainly with excitation of laminas I and II cells in response to the excitatory input form the nonspecific af-

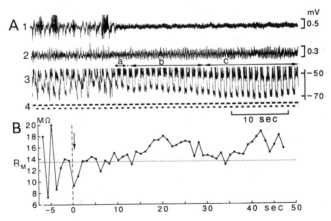

FIG. 5. E+DF response in a non-PT cell (depth, 320 μm). Encéphale isolé cat. Spontaneous EEG arousal is illustrated with the same conventions as in Fig. 4. Bars a and c mark E and DF responses, respectively, and bar b indicates a phase of transition from E to DF. (Modified from Ezure and Oshima, 1981c, with permission.)

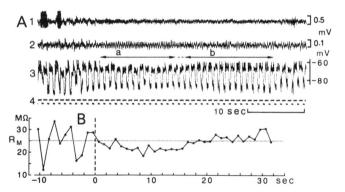

FIG. 6. I+DI response in a non-PT cell (depth, 660 μm). Encéphale isolé cat. Spontaneous EEG arousal is shown with the same conventions as in Figs. 4 and 5. (Modified from Ezure and Oshima, 1981c, with permission.)

ferent (a in C). These neurons transmit E and I to lamina III$_a$ cells, which in turn cause 4 types of responses in laminas III$_b$ and V and VI cells as the result of cascade transmissions through all combinations of excitatory and inhibitory actions. Theoretically, the excitatory nonspecific input (a) is directly received by all E cells, and its terminals are distributed as drawn in C in proportion to the increasing number of E cells toward the cortical surface. An inhibitory nonspecific input (b in C) might also contribute to phasic EEG arousal by providing I cells with direct inhibitory terminals.

In a circuit model proposed earlier (Inubushi et al., 1978b) we assumed two mechanisms to explain tonic maintenance of EEG arousal often outlasting the reticular stim-

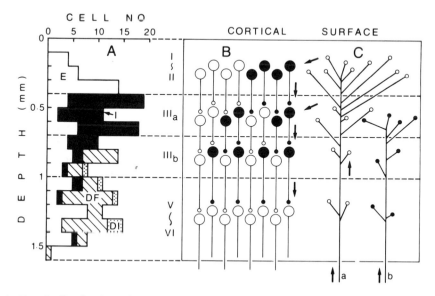

FIG. 7. A: Depth distribution of neurons with E (open columns), I (filled), DF (hatched), and DI responses (stippled). B: A cascade model of neuron relays for phasic EEG arousal. Open symbols: Excitatory neurons or synapses; filled symbols: inhibitory. C: Possible distribution of nonspecific afferents with excitatory (a) and inhibitory terminals (b). Arrows in B and C indicate direction of information flow. (From Oshima, 1981, with permission.)

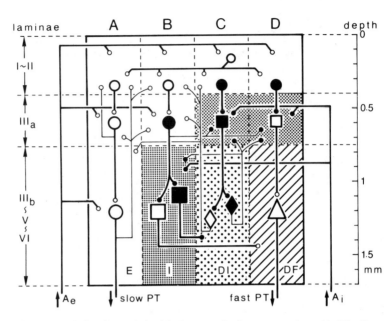

FIG. 8. An earlier arousal circuit model with four vertical neuron relays (A–D). Compartments are patterned as *open* for E, *finely stippled* for I, *roughly stippled* for DI, and *hatched* for DF. Excitatory (A_e) and inhibitory afferents (A_i) are illustrated at *left* and *right*, respectively. (Modified from Inubushi et al., 1978b, with permission.)

ulation. One is continuous inflow of afferent impulses. The other is the automatic recycling of intracortical impulses through positive feedback loops. In Fig. 8 these feedback loops are drawn with thin solid lines. Their actions are the reinforcement of the basic cascade pattern carried on by the relays drawn with thick lines. If these feedback loops actually operate, the vertical flow of intracortical impulses is bidirectional. Some horizontally spread impulses are also suggested in this figure. However, we take the downward cascade as an essential feature of the neuronal operation during phasic arousal, and view this dynamic circuit as downward directed.

The depth profile corresponding to tonic EEG arousal in Fig. 9A and B leads to the postulate of a circuit model drawn in C and D. Cascade transmissions occur then from the deepest layers, starting mainly with excitation there and being processed through excitation and inhibition in lamina III_b cells and finally bringing about indirect +E, +I, +DF, and +DI in laminas I and II and III_a cells. Theoretical distributions of excitatory afferent terminals (a) onto +E cells and of inhibitory terminals (b) onto +I cells are illustrated in Fig. 9D. In addition, the upward cascade transmission during tonic EEG arousal becomes spatially restricted and yields a fraction of unresponsive cells toward the cortical surface (Fig. 9A).

Positive feedback loops reinforcing the basic cascade pattern would also operate during tonic EEG arousal. Therefore, vertical flow of impulses through the cortical layers could be bidirectional, but we take the upward cascade as essential for this phase of EEG arousal, as depicted in a simple form in Fig. 9C.

The transition from phasic to tonic EEG arousal would be determined by a balance in synaptic bombardments from the two separate afferent systems (Figs. 7C and 9D) as well as by the intracortical mechanisms involving refractoriness and recycling of impulses in the neuron relays.

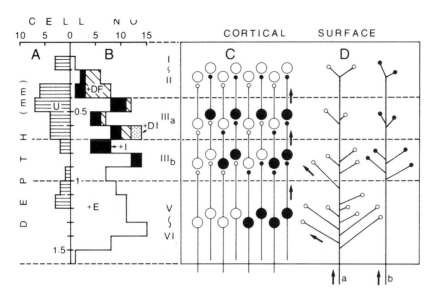

FIG. 9. A: Depth distribution of late unresponsive cells. B: Distribution of cells of late responses patterned with same conventions as in Fig. 7A. C: A cascade model of neurons for tonic EEG arousal. D: excitatory (a) and inhibitory (b) afferent terminal distributions participating in tonic EEG arousal. (From Oshima, 1981, with permission.)

A composite model for the phasic and tonic EEG arousal is drawn in a simplest form in Fig. 10A, where 64 cells are arranged in a matrix to construct 16, four-neuron relays in the vertical direction. Each layer division contains 8 excitatory and 8 inhibitory cells, and the 16 columns of relay represent all possible combinations of excitatory and inhibitory connections. The cell axons drawn with solid lines show downward connections, and those with interrupted lines give upward connections. The afferent systems (A_1 and A_2) are illustrated in Fig. 10A at left and right, respectively, similarly to those drawn in Figs. 7C and 9D. In Fig. 10B, 64 compartments are illustrated in correspondence with the cell topography in A with their response types during phasic EEG arousal: E (open compartments), I (filled), DF (hatched), and DI (stippled), with an arrow that directs the characteristic cascade transmissions. Figure 10C illustrates the response patterns during tonic EEG arousal with similarly patterned compartments and an upward arrow. The late unresponsive cell compartments in this figure are marked with crosses and demarcated with thick solid lines from other active compartments. In the composite model of Fig. 10A we assume the presence of the cells with vertically bifurcated axons, and further postulate the intracortical mechanisms which assure the direction of either downward or upward cascade transmissions corresponding to two different phases of EEG arousal.

Functional Implications

Behavior of fast and slow PT cells during EEG arousal is of particular interest because of their roles of carrying the main motor command signals (Evarts, 1965, 1966, 1968; Fetz and Baker, 1973; Fetz and Finocchio, 1975; Evarts and Fromm, 1977, 1978; Fromm and Evarts, 1977; Fetz and Cheney, 1980). The transient DF of fast PT cells during phasic EEG arousal may be interpreted as the reset of their activity without there being a powerful suppression due

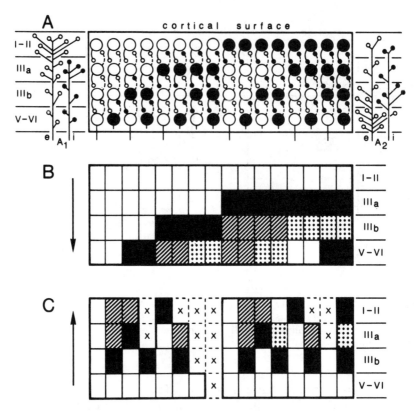

FIG. 10. Arousal circuit model. **A:** A simplified circuit constructed from 64 cells. **B** and **C:** Response patterns during phasic and tonic EEG arousal, respectively. *Arrows:* Directions of characteristic cascade transmission. Late unresponsive cell compartments marked with *crosses* in C are chosen arbitrarily in an approximate proportion to the population of observed response types. (From Ezure and Oshima, 1981c, with permission.)

to postsynaptic inhibition. Fast PT cells so reset would be in a state of readiness in that the increased R_M is effective to build up greater depolarization to the next excitatory drive and that elimination of the ongoing EPSPs improves the signal-to-noise ratio. Correspondingly, slowly developing excitation in slow PT cells would lead to raised excitabilities of lower motor neurons, which would present a favorable condition for initiation of behavior such as the postural adjustment set. Thus, we postulate that these fast and slow PT cells represent a readiness state with their respectively different activities corresponding to the initial stage of motor arrest in animals on receipt of a novel stimulus from its environment.

Other neuronal activities of E, I, DF, and DI would cooperate with each other to have the output neurons such as PT cells ready for action. All the cortical neurons thus contribute more or less to phasic EEG arousal or initial stage of orienting reflex.

The upward cascade pattern of transmissions during tonic EEG arousal resembles the intracortical pattern of information processing brought about by the specific afferent inputs disclosed by anatomical findings of the laminar distribution of some specific thalamocortical fibers (Strick and Sterling, 1974; Jones, 1975; Deschenes and Hammond, 1980) and physiological analyses of field potentials and single-unit responses (Marshall et al., 1943; Bishop and

Clare, 1952; Amassian, 1953; Amassian et al., 1955; Li et al., 1956; Toyama et al., 1977a,b; see also Towe et al., 1968; Doetsch and Towe, 1976a,b; Mountcastle, 1979; Singer, 1979).

Many deep-layer cells excited during tonic EEG arousal will project their axons to many target neurons in the subcortical structures such as the thalamus. These cells send the signals that may play a role of selective gating or resetting in both the sensory and motor systems in the subcortical nuclei (Phillips and Porter, 1977). Since some subcortical structures such as the thalamus send excitatory afferents to the cerebral cortex, it is conceivable that interactions occur during tonic EEG arousal between the cortex and the subcortical nuclei. These considerations would lead us to postulate that an animal reaches a stage of exploring its environment in a relatively specific manner.

Although the present work provided no direct evidence for the tangential spread of neuronal activities, the cerebral states during phasic and tonic EEG arousal can be summarized in Fig. 11. The neural event on phasic arousal (A) starts mainly in superficial layer cells (dots) and spreads in a cascade manner toward deeper layers as directed with arrows. The whole cortex is in action either by E, I, DF, or DI. When tonic EEG arousal starts with E in deep layer cells (dots in B), the neural information is processed upward within the cortex, as indicated with upward arrows. Simultaneously, it is also processed downward, as indicated with downward arrows in B, as the action of the cells projecting to the subcortical nuclei. As the result of interactions between the cortex and the subcortical nuclei, there appears to be an uneven distribution of active (open areas in Fig. 11B) and inactive zones (filled), the latter being occupied by late unresponsive cells. Active here means not only E but also either I, DF, or DI. From the developmental aspect of behavior, tonic EEG arousal or the cerebral state with the neural pattern drawn in Fig. 11B might be the prelude to the most specific patterns with which specified functional columns or modules of cell assembly (Mountcastle, 1957, 1979; Asanuma, 1975; Szentagothai, 1975, 1978; Phillips and Porter, 1977; Eccles, 1980) take over the nonspecific activites to create an adaptive behavior.

PERSPECTIVES

The arousal circuit model in Fig. 10 is the most simplified formula of the dynamic patterns in EEG arousal. This skeleton scheme undoubtedly needs to be tested and reformed by future research (Inubushi et al., 1978b). Nevertheless, the arousal circuit model provides a general aspect of neuronal integration in line with Jasper's (1958) view. This small measure of success might further allow us to discuss briefly some additional perspectives obtained through the present series of study.

Taxonomy of Cortical Neurons

If the arousal circuit model of Fig. 10A is treated rigidly, it requires 12 classes of neurons of which 9 have been identified by their initial and late responses. A question may arise whether the present classification could contribute to a workable polythetic scheme in the taxonomy of cortical neurons (Tyner, 1975; Mann, 1979). An advantage may be the full possibility in naming of all cortical neurons at least in the phasic phase

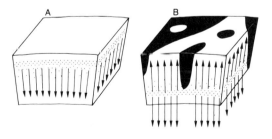

FIG. 11. Stereoscopic view of dynamic patterns in the cortex during phasic (A) and tonic arousal (B). (From Oshima, 1981, with permission.)

of EEG arousal. The argument that the essence-indicating names and the role-indicating names should be avoided in taxonomy (Rowe and Stone, 1977) is persuasive in the case of EEG arousal, which has been regarded as nonspecific in function. Many motor cortical neurons with their particular activity patterns during EEG arousal are not attention units, but may play other specific roles in other different behavioral contexts. Even in the EEG arousal, different responses occur in single neurons according to the phasic and tonic phases. This leads to the problem of the hierarchical levels in our understanding of the function of neurons.

Contextual Pluralism of Neuronal Function

The arousal circuit model may provide a good example of the cooperative system in which various constituent neurons show different activity patterns in a single behavioral context. Furthermore, each of these neurons may behave differently in different behavioral contexts. For example, most superficial layer cells are first recipients of nonspecific afferents on phasic EEG arousal, but later receive one of the four different actions with possibly some specific nature during tonic arousal. In a further processing of specific afferent inputs, finely organized patterns of activity may develop in the superficial layer cells, as suggested in anatomical (Szentagothai, 1978) and physiological studies (Toyama et al., 1977a,b). Also, superficial layers may be a candidate site of modifiable synapses related to learning (Eccles, 1978). Superficial and deep layer cells behave differently in the process of behavioral conditioning (e.g., Gabriel et al., 1980).

The context-dependent pluripotency of a single neuron has its basis on the convergent inputs from plural sources to and the divergent axonal trajectory from the neuron. Examples are easily obtained in many specific (e.g., the corticospinal tract, cf. Jankowska et al., 1975; Shinoda et al., 1976) and nonspecific systems (e.g., the brainstem reticular core, cf. Hobson and Scheibel, 1980). One specific function, as in cerebral cortex, may result from operation of a system to which many neurons of different species contribute with one particular activity pattern from their respective repertoire of response types. Changes in the activity pattern of the motor cortical neurons from EEG synchronization to desynchronization revealed the functional pluralism in the two phases of EEG arousal. Thus, single neurons are used differently in the plural contexts, which may suggest a *bricolage* (Lévi-Strauss, 1962) or tinkering in functional organization. The cellular response patterns underlying EEG arousal are congruent with the neuron theory (Eccles, 1973).

Theoretical Possibilities

A theoretical treatment would facilitate our understanding of the dynamic patterns processed in the aroused cortex. An earlier cooperative analogy with the compass needle model of ferromagnetism (Cragg and Temperley, 1954) is perhaps interesting in explaining the basic nature of the cortical arousal. Recent advances might lead us to better understanding, if the arousal circuit model be treated as the dissipative structures (Katchalsky et al., 1974; Nicolis and Prigogine, 1977). Katchalsky's three essentials for the system to exhibit dynamic patterns (pp. 54–55 in Katchalsky et al., 1974) would be fulfilled in this case, and it is hoped to stimulate further experimental quantification of the patterns.

ACKNOWLEDGMENTS

The author wishes to thank Professor Shizuo Torii, Mr. Shikio Inubushi, Dr. Toshinori Kobayashi, and Dr. Kazuhisa Ezure for their cooperation in this study. Thanks are also due to Miss Takako Sato for her skillful technical assistance. This work was supported in part by a grant-in-aid for scientific research from the Ministry of Education, Science and Culture of Japan.

Motor Control Mechanisms in Health and Disease,
edited by J. E. Desmedt.
Raven Press, New York © 1983.

Functional Organization of the Motor Cortex

*R. Porter

*The John Curtin School of Medical Research, Australian National University,
Canberra, Australia*

The discharges of corticospinal neurons provide one of the direct influences on the alpha-motoneurons (MNs) which activate the musculature of the limbs. In primates, a proportion of these corticospinal neurons make excitatory synapses on the surface of MNs through cortico-MN connections (Bernhard and Bohm, 1954; Preston and Whitlock, 1960, 1961; Landgren et al., 1962a). This means that the pyramidal cells of origin of these connections are only one synapse removed from the initiators of muscle contraction and that the impulses conducted to cortico-MN synapses could theoretically be as important for the control of MN activity and movement as are those that enter the spinal cord from primary endings of muscle spindles to produce monosynaptic reflex excitation.

Beginning with the initial work of Fritsch and Hitzig (1870); Ferrier (1875); Jackson (1897: see 1931); and Sherrington (1906), ample evidence has accumulated from stimulation studies to support the statement that corticospinal (and cortico-MN) somata are particularly concentrated in the precentral gyrus of the primate cerebral cortex (Woolsey et al., 1952; Landgren, et al., 1962b, Woolsey, 1964). Corticospinal fibers that arise from other regions such as the postcentral gyrus make connections with dorsal horn elements in the spinal cord and not with pools (Kuypers and Brinkman, 1970; Kuypers, 1973). In the precentral situation, a topographic organization exists in which the populations of corticospinal neurons to forelimb MNs are, in general, aggregated within the middle portion of the precentral gyrus between the territories which accommodate the "hindlimb" population and the "face" population (Fig. 1).

The colonies of cortico-MN somata which will provide excitation to a given single alpha-MN are aggregated within these zones and the projection areas that are occupied by the colonies connecting with different MNs overlap (Landgren et al., 1962b; Asanuma and Rosen, 1972; Andersen et al., 1975; Jankowska et al., 1975; Phillips and Porter, 1977). In spite of this overlap, the geographic organization of such efferent neurons within the precentral gyrus offers opportunities for selective study of neuron behavior that may be associated with the performance of particular motor functions, such as movements of the hand, elbow, or foot. Local populations may be stimulated by weak electric currents through small zones of the cerebral cortex (Phillips and Porter, 1962; Asanuma and Rosen, 1972; Andersen et al., 1975). Alternatively, microelectrodes may sample the discharges of individual neurons that are associated with the natural movements performed by conscious animals (Evarts, 1964, 1966, 1968; Porter, 1972; Lemon et al., 1976).

---

*The John Curtin School of Medical Research, Australian National University, P.O. Box 334, Canberra City, A.C.T. 2601, Australia.

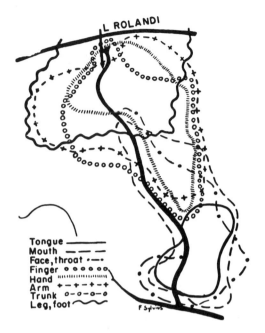

FIG. 1. Diagram of the right perirolandic cortex of the human brain. The overlapping outlines indicate the areas from which any motor responses of the indicated parts were ever evoked in different individuals in a series of neurosurgical explorations. Although the symbol for trunk appears in the key, no trunk area is shown on the map. It is clear that the cortical outflow and the corticospinal neurons which are capable of influencing the contralateral foot, hand and face occupy extensive overlapping territories. For the hand, fingers, and face these are predominantly precentral and there is very little extension behind the central sulcus regions. (From Penfield and Boldrey, 1937, with permission.)

It would, of course, be incorrect to consider that the electrical activation of cells within a limited cortical zone would cause discharge only of neurons belonging to a single class, such as those with a corticospinal destination. Activation of the intermingled corticostriate, corticothalamic, corticocortical, corticopontine, corticobulbar, and corticoreticular efferents from the same region must be anticipated. Electrical discharges in some of these (e.g., corticorubral or corticoreticular fibers) may well contribute, over indirect pathways, to motor responses evoked by the cortical stimulus. Moreover, particularly if repetitive stimulation of the local zone is used, neurons in outlying regions of cortex, perhaps far removed from the local zone but connected functionally with it, may also be influenced (Jankowska et al., 1975).

Similarly, when the discharges of a single cell, or group of cells, within a defined region of the cerebral cortex are recorded and analyzed, it cannot be assumed that the responses so documented specify the actions of a particular colony of cells. Even when the demonstration of antidromic activation of the neuron following a stimulus delivered to the medullary pyramidal tract allows the cell to be identified as a pyramidal tract neuron (PTN), its membership of a particular class of corticobulbar, corticoreticular, or corticospinal axons remains in doubt. Moreover, the similarities or differences in the natural behavior of adjacent neurons within a tiny region of cerebral cortex do not define functional associations and do not permit classification of the neurons into functional classes. Until both the definition of action under natural conditions and the anatomical connectivity for the same neuron are available, functional associations can only be presumed.

The objective of studies using stimulation and recording methods is ultimately to seek an understanding of organization and function of the human motor cortex. Enough observations have been made directly in man to make it clear that the experimental results of studies in subhuman primates have immediate relevance. Both the extensive studies of Penfield and Boldrey (1937), employing electrical stimulation of the cerebral cortex in conscious patients and the observations made with microelectrode penetrations of the human cerebral cortex (Li and Tew, 1964; Goldring and Ratcheson, 1972) indicate the parallel nature of the findings in man and monkeys.

Correlations between cortical neuronal activity and movement performance are found to be variable and flexible (Fetz and Finnocchio, 1972; Evarts and Tanji, 1974). The neuronal responses themselves are the outcome of the complex interactions between a very large number of convergent influences from a wide variety of sources. Hence, variation in neuronal discharge from one repetition of the same motor performance to another may be an indication of subtle changes in the balance of these many influences on the pyramidal cell. These variations may be related to changing contributions of the sample cell to the total population response which produces the movement or to other influences (e.g., attention or anticipation), which modify the relationship between cell discharge and movement. But even if a corticospinal neuron's responses are artificially paced, an immense capacity for variation exists within its destination in the MN pool of the spinal cord, where again, convergent influences from an enormous variety of excitatory and inhibitory inputs can determine the effectiveness of a particular active line. Even the powerful tendon-jerk responses to synchronous activation of the afferent line from primary endings of muscle spindles are dramatically modifiable in response to other inputs from cutaneous receptors or changes in the balance of descending signals to the spinal motor pools. The challenge to the physiologist is to understand the flexibility that exists in the relationship between cerebral cortical activity and movement performance and not to regard it simply as a series of deviations from a fixed, average relationship. In the search for an understanding of flexibility of association and the mechanisms which determine its nature may be uncovered secrets which contribute to knowledge about learning and memory, especially as they apply to the learning of the motor skills on which so much of human behavior and accomplishment depends.

STRUCTURE OF THE MOTOR CORTEX

Area 4 (Brodmann) is the thickest region of the cerebral cortex. Yet a narrow cylinder of cortex in this area contains, distributed through the total thickness from pia to white matter, the same number of neuronal cell bodies as do cylinders through other cortical zones (Rockel et al., 1980). Between these widely spaced cells is an extensive neuropil, which consists "predominantly of dendritic spines and axon terminals" (Sloper and Powell, 1979, p. 127). The gigantic capacity for synaptic connectivity, which is provided within this massive neuropil, is the most impressive feature of the motor cortex. Although a variety of cell types has been identified by many workers (see Jones, 1975a), Sloper et al. (1979) have calculated that the proportions of pyramidal cells and stellate cells (which they identify as the two major cell types in the cortex) was also similar in the motor area to that in other zones of cortex and that 72% of the cells were pyramidal cells.

Pyramidal cells are the output cells of the motor cortex. The recent use of retrogradely transported labels such as horseradish peroxidase (HRP) has added detail to knowledge obtained from earlier tracing studies and has revealed the stratification that exists for output elements with different destinations. The pyramidal cells of lamina III make corticocortical connections, including those with the other hemisphere (callosal). The pyramidal cells of lamina V provide the efferent axons for corticostriatal, corticorubral, corticothalamic (intralaminar), corticopontine, corticobulbar, and corticospinal connections. Lamina VI contains the smaller pyramidal cells that project to the relay nuclei of the thalamus as corticothalamic connections (Jones and Wise, 1977; Jones et al., 1977).

Corticospinal Neurons

The corticospinal neurons, which merit our most detailed study in this consideration of motor cortex function, have the largest somata and are, in general, more deeply situated within lamina V than those efferent neurons whose axons are destined to contact nearer targets (in the striatum, or the red nucleus or the pons). Jones and Wise (1977) measured the somata of corticospinal neurons that had been identified by retrograde transport of HRP from terminals in the cord. They reported an average diameter of 25.7 ±6.0 μm. But such measurements fail to emphasize other dramatic features of these cells, which could be so important for considerations of function. There is a wide range of sizes of corticospinal neurons, and the smaller cells with more slowly conducting axons make up 80 to 90% of the population. From the antidromic responses to lateral corticospinal tract stimulation of a very large number of individual neurons in the wrist zone of the monkey's motor cortex, Humphrey and Corrie (1978) estimated that the modal conduction velocity (CV) of the majority of corticospinal cells was 8 to 12 m/sec, whereas the smaller proportion of rapidly conducting axons had a modal CV of 50 to 55 m/sec.

Even though the somata are situated in the deeper parts of lamina V, the apical dendrites of these large cells ascend through the total thickness (2 to 3 mm) of the cortex, branch extensively, and are studded with synaptic spines. The apical dendrites traverse all laminas of the cortex and provide a vertical collector potentially for influences from all laminas. However, there are also prominent horizontal dendrites. These arise from the proximal part of the apical dendrite and range out within lamina V, where they extend for ≥ 500 μm. Horizontal collectors are also prominent at the base of the pyramidal cells. These are even longer, more prominent, covered with synaptic spines, and branch to extend into wide zones of lamina V and lamina VI, providing an extensive surface for synaptic influence (Fig. 2). The functional implications of this complexity could be very important. Moreover, if the recent results of Hamada et al. (1981), based on a limited number of 7 PTNs, are confirmed, and if it is found that the cells contributing slowly conducting axons to the pyramidal tract usually have several ascending apical dendritic branches and a differently organized basal dendritic aborization, with less prominent "tap-root" basal dendrites, than those of cells with fast axons, functional correlates of the different connectivities that could result may have to be examined.

Recurrent Collaterals

Before it leaves the cortical mantle to enter the white matter and descend to the spinal cord, the axon of a corticospinal neuron typically gives off a number of collateral branches within lamina VI (Cajal, 1911). Although there is physiological evidence for

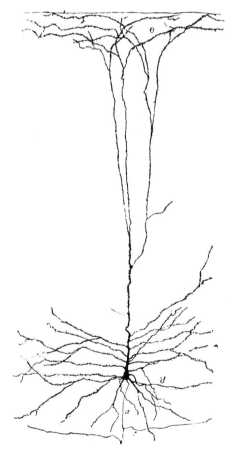

FIG. 2. Pyramidal cell from the human cortex. (From Cajal, 1911, with permission.)

influences exerted back on PTNs and other cortical cells by these recurrent collaterals (Phillips, 1959; Armstrong, 1965; Stefanis and Jasper, 1964; Tsukahara et al., 1968), the precise anatomical connectivity of these branches within the cortex has not been established, and the total extent of their distributed influence on other cells is in doubt, although it could potentially occur within a zone up to 3 mm in diameter (Szentagothai, 1976). Synapses made by recurrent collaterals would be expected to have an excitatory action, but interneurons producing inhibition of some PTNs are also influenced. It is possible that the recurrent collaterals provide the links whereby particular neurons of a given class (e.g., corticospinal neurons with destinations in a given motor pool) are made to respond similarly and to discharge together in association with a particular movement performance while the representatives of another class of cells closely intermingled within the same cortical zone and distributed beneath a surface area several millimeters in diameter will have their activities suppressed.

Afferents to the Cortex

The structural arrangement of the corticospinal neurons are only a minority within the total population of structures in the motor cortex. Emphasis must be placed on the enormously complex synaptic bulk, which makes up the neuropil and which presumably determines all the effects are relayed to spinous and other contacts on the surface of the pyramidal cell, where they are integrated in some manner and eventually reflected in the firing of action potentials along the axon of the efferent cell. Traveling in the neuropil are the axons of corticocortical and thalamocortical afferents. In the case of area 4 the former derive from only particular parts of ipsilateral and contralateral cortical zones.

In the ipsilateral hemisphere corticocortical afferents enter area 4 from area 2 in the parietal lobe and from area 6 and the supplementary motor area (Jones and Porter, 1980). Ipsilateral and callosal corticocortical synapses are distributed within superficial layers of the cerebral cortex. Electron microscopic evidence indicates that contacts are made with the dendrites of pyramidal cells (Jones and Powell, 1970a). But, although Jones (1980) has drawn the conclusion from double-labeling histochemical techniques that corticocortical connections are made with pyramidal cells which themselves provide reciprocal corticocortical projections, this remains to be demonstrated for individual identified neurons, and it will be essential to know whether other classes of neurons are also influenced. All the deeply situated pyramidal cells of lamina V have prominent apical dendrites which ascend through the termination zone of the corticocortical afferents in laminas II and III. Moreover, connections with nonpyramidal cells (interneurons) must remain a significant possibility.

Corticocortical connectivity is not uniform. For example, large parts of the hand and foot representation within area 4 lack callosal connections. Where contralateral corticocortical projections do exist, they are organized in zones or strips about 0.5 to 1 mm wide separated by less densely connected regions of similar size (Jones et al., 1979). This suggests that corticocortical influences are directed to some parts of a projection territory and not to others.

Thalamocortical Projections

Thalamocortical afferents, at least in the primary receiving areas of the cerebral cortex, are considered to end monosynaptically on the spiny stellate cells which are packed into the thick zone of lamina IV and which are considered to be the major recipients of thalamic projections (Szentagothai, 1975, 1976). Lamina IV is insignificant in the motor cortex. Nevertheless, Sloper et al. (1979) indicate that stellate cells exist within area 4 in similar proportions to those in other cortical regions. The thalamocortical afferents to area 4 of the motor cortex (from nucleus VPL_o) distribute terminals through lamina III and particularly within its deeper part (Jones, 1975a). It is likely that some of these endings are on spiny stellate cells. The spiny stellate cell sends axonal branches vertically as a tuft which parallels and may enclose and have an intimate association with the apical dendrites of pyramidal cells. Specific input-output coupling has been inferred from this anatomical description and columns of related neuronal elements have been inferred for the motor cortex as well as for somesthetic and visual areas (Asanuma, 1975). However, it is not known whether the synapses of the spiny stellate cell are with dendrites of a particular class of output cells. In addition, thalamocortical contacts directly with pyramidal neurons have been described in electron micrographs (Sloper, 1973; Strick and Sterling, 1974). Jones (1980) refers to the additional possibility that thalamocortical synapses at the border of lamina V and VI may engage the prominent basal dendrites of pyramidal cells or even particular corticothalamic projections back to the thalamic relay nucleus (Fig. 3).

Cortical Interneurons

The spiny stellate cell is only one of the classes of intracortical interneurons which could be engaged by afferent fibers and could then provide synapses which might influence efferent pyramidal cells. As identified by electron microscopy, the spiny stellate synapses on the spines of apical dendrites of pyramidal cells would be deduced to be excitatory. Basket cells have also been described (see Jones 1975a; Szentagothai, 1976). In the precentral gyrus, they are described with dendritic branches extending through a flattened disk at right angles to the length of the gyrus. Their axons provide horizontal branches which extend for long distances (1 to 2 mm) in the same plane and in most layers of the cortex. The morphologic features of their synapses on pyramidal cell somata indicate that these interneurons can produce inhibitory influences on the pyramidal cells contacted within this strip across the gyrus.

Other interneurons with presumed inhibitory or excitatory influences on pyramidal cells (or other interneurons) have been described. Their effects add complexity to the organization within a limited zone of cerebral cortex. But the opportunity exists, with the availability of reliable methods for precise anatomic description of the dendritic architecture and axonal terminations of individual cells, the physiologic responses of which have been examined with an intracellular microelectrode, to begin to describe the circuit diagrams which allow the "knitting together" (Sherrington, 1906) within the motor cortex of all the complex synaptic influences which result in outputs from particular identified pyramidal cells.

INFLUENCES OF CORTICOSPINAL DISCHARGE

Intracellular recording of the responses within alpha MNs produced by weak stimulation of appropriate small regions of cerebral cortex revealed clearly that convergence of excitatory influences resulted, with contributions from a number of cortico-MN units making up a colony of such cells distributed within a region of cerebral cortex (Landgren et al., 1962b; Phillips and Porter, 1964; Phillips and Porter, 1977). Fetz and Cheney (1980) correlated the changes in electromyographic (EMG) activity in forelimb muscles associated with a simple wrist flexion-extension task with the individual action potentials generated by sample neurons in the motor cortex, the firing of which changed with the movement task. Their technique of examining postspike facilitation of EMG indicated that for slightly less than one-third of cortical neurons that

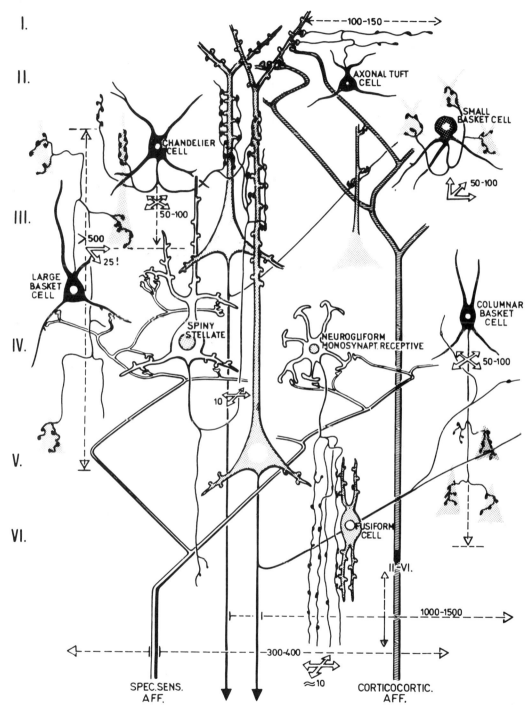

FIG. 3. Generalized diagram of the cell types and the connectivity of afferent systems within the cerebral cortex. (From Szentagothai, 1975, with permission.)

were task related, postspike facilitation of forearm EMG activity occurred with a short latency (mean 6.7 msec to onset) and a timecourse reminiscent of that of a monosynaptic cortico-MN excitatory postsynaptic potential (EPSP). Moreover, in about half the cases, postspike facilitation was observed in the EMG of more than one muscle in correlation with the discharges of the one cortical cell. This indication of the divergence of the influences of a given cortico-MN fiber to affect the MNs of more than one muscle requires most serious consideration in connection with possible functional implications of cortico-MN firing.

Divergence of influence is also suggested by Shinoda et al. (1981), who injected HRP into single identified corticospinal axons to trace their terminations in the spinal cord of the monkey. Although it is not possible to trace all the connections to their exact cell of termination, particularly those with the more distant dendrites of MNs, the branches of corticospinal axons are so extensive that they support communication with the MNs of several motor nuclei (see also Asanuma et al., 1979). This important observation deserves full attention. Shinoda et al. (1981) set out to examine the intraspinal morphology of electrophysiologically identified corticospinal axons originating from the hand area of the monkey's motor cortex. Corticospinal axons, which could be directly activated by weak stimulation of the contralateral motor cortex, were impaled with HRP-filled glass micropipettes, and HRP was injected into the axon iontophoretically with periodic checks of the impalement and the preservation of the action potential produced by cortical stimulation.

Three days before the experiment to record from and label corticospinal axons, HRP had been injected into a number of forelimb nerves on both sides and the nerves were damaged at the injection site. This was to permit the labeling, by retrograde transport of HRP, of the MN pools in the spinal cord contributing axons to these forelimb nerves.

Histologic study of transverse frozen sections of the spinal cord, after staining for HRP using the diaminobenzidine method, allowed the reconstruction of the branching pattern of the labeled corticospinal axons. Axon collaterals departing from stem axons in the lateral funiculus ran medioventrally and entered the cervical gray matter at the level of laminas V, VI, and VII of Rexed. These collaterals bifurcated and spread out in the intermediate zone and laterally in lamina IX. Terminal branches were prominent in lamina IX. In this respect, the pattern of the monkey's corticospinal axon termination differed significantly from that described in the cat, where terminations were not found among MNs (Futami et al., 1979). The terminal aborizations were distributed to circumscribed areas within lamina IX. Both bouton terminaux and bouton en passage were identified, but the target structures could not be visualized and described. It was reasoned that these endings were with distal dendrites of MNs, and there is other electrophysiologic evidence to suggest that this must be the case (Porter and Hore, 1969). Even some of the endings in the intermediate zone of the spinal cord could, of course, be on distal dendrites of MNs. Some synaptic boutons made apparent contacts with proximal or distal dendrites of HRP-labeled-MNs.

Axon collaterals could project to a number of different columns of motor nuclei. Since these motor columns could be identified as sending axons into quite different peripheral nerves (e.g., median nerve and radial nerve), the conclusion was drawn that a given corticospianl axon made synaptic contact with MNs of quite different muscles. Divergence of corticospinal influence was therefore demonstrated. However, it will be important to show the exact terminations to individual MNs, because the dendrites of cells within one region may extend into zones that are occupied by other terminal branches of the same axon. It will require additional experiments to show whether the population of MNs so engaged act synergistically (under some conditions) and to demonstrate what is the physiologic consequence of the divergent influence. Are some MN groupings more powerfully influenced than others when the divergent effects of a single corticospinal axon are examined quantitatively?

Lemon and Muir *(personal communication)* have examined the postspike facilitation of the EMG activity of particular small muscles of the hand (such as the first dorsal interosseous muscle) associated with the simultaneously recorded discharges of cortico-MN somata. The short-latency effects seen in these experiments appear to be larger than those that have been observed by Fetz and Cheney (1980) and may be limited to influences observable only in a single muscle. Clough et al. (1968) described the larger monosynaptic cortico-MN effects that were observed in the MNs innervating the most distally acting muscles of the forelimb.

Some Properties of Cortico-Motoneuronal Synapses

Landgren et al. (1962a) described the increasing effectiveness in alpha-MNs of successive pyramidal tract volleys produced by a high-frequency repetitive train of stimuli delivered to the cerebral cortex. This facilitation of cortico-MN synapses by preceding activity in the cortico-MN axon was confirmed by Kernell and Wu (1976b) and by Shapovalov (1975). Attempts to describe the facilitation quantitatively and to analyze its time course could not be confined to the study of the synapses made by a single pyramidal tract axon. Minimal cortico-MN EPSPs were studied. They were always small in amplitude (≤ 200 μV), and they were consistently demonstrated to possess the characteristic increasing effectiveness of successive volleys in a train of cortico-MN impulses (Porter and Hore, 1969; Porter, 1970; Muir and Porter, 1973). Since, under natural conditions during movement performance, corticospinal neurons discharge bursts of impulses often at high frequency for brief periods, this temporal facilitation could have physiologic significance in more effectively depolarizing MNs to threshold for discharge (Porter and Muir, 1971) (Fig. 4).

The cross-correlation studies of Fetz and Cheney (1980) do allow examination of the influences on EMG activity, in particular muscles, of the discharges of individual single corticospinal neurons. In addition to the evidence for divergence of effectiveness and to confirmation of stronger excitatory influences on elbow, wrist, and finger extensors than flexors (see Preston et al., 1967), these experiments also produced indications of increased effectiveness of brief interspike intervals in the discharges of cortico-MN neurons affecting the EMG of some muscles. This finding for the effects of discharges of single cortico-MN cells makes the possibility of a physiological role for temporal facilitation even more commanding.

DISCHARGES DURING MOVEMENT OF CORTICOSPINAL NEURONS

Development of methods for examining the impulse activity of identified PTNs in conscious, freely moving monkeys allowed analysis of the signals transmitted along efferent fibers from the cortex in relation to normal behavioral responses of the animal (Evarts, 1964, 1965a,b). Such studies tend to be dominated by observations of the activities of the largest neurons with the fastest axonal conduction velocities. The studies by Evarts concentrated on the natural functional correlates, in relation to a limited and defined movement performance, of those neurons whose axons could be demonstrated to enter the pyramidal tract in the pons and medulla. As previously mentioned, this classification of cells as PTNs

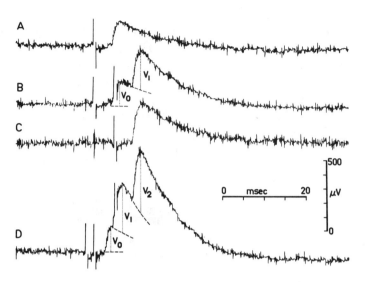

FIG. 4. The increasing synaptic effectiveness (facilitation) evident in minimal cortico-MN excitatory postsynaptic potentials following successive stimuli at physiologic frequencies for the discharge of cortical cells. V_0, V_1, and V_2 are the successive voltage changes produced by 3 stimuli imitating a brief cortico-MN burst of impulses. (From Muir and Porter, 1973, with permission.)

still leaves open the definition of axonal destination and does not necessarily identify a uniform class of cortical efferent neurons.

Careful and informative studies of Lawrence and Kuypers (1968a,b) revealed identifiable contributions that must be made to movement performance of monkeys by neuronal activity in the pyramidal tracts. They sectioned the pyramidal tracts bilaterally and then allowed maximum compensation to occur over time and with prolonged retraining. It became clear that intactness of pyramidal tract function was uniquely necessary for the independent finger movements which were required if the monkey was to winkle out tiny morsels of food from small holes in a test board. The animals had lost the ability to fractionate the use of the flexor and extensor muscles of the fingers and could not, for example, extend the index finger and move it while keeping the remaining digits flexed. The fingers could be used together to close the hand around a larger piece of food and bring it to the mouth, but could not perform a "precision grip" using only opposition of the tips of the thumb and index finger. The animal might display difficulty in relaxing the grasp on food brought to the mouth, even though no such difficulty in releasing the grip was seen in climbing or clinging to the cage (see also Tower, 1940; Kuypers, 1973). In general, it was found that monkeys with bilateral pyramidal tract lesions lacked full speed, precision, and skill in many of their movements. Infant monkeys subjected to bilateral pyramidotomy within a few weeks of birth, before full myelination of the pyramidal tracts and before formation of cortico-MN synapses had occurred, never learned to make precise, independent finger movements (Lawrence and Hopkins, 1976).

Bucy et al. (1966) cut the pyramidal tracts bilaterally at the level of the cerebral peduncles. They reported that monkeys subjected to this operation could use the thumb and index finger alone to pick up small seeds. This ability may have depended on the sparing of some pyramidal tract axons. Sparing of a small number of pyramidal tract projections may also account for the remarkable observation reported by Bucy et al. (1964) that a patient in whom the extent of histologic loss of pyramidal tract axons following a surgical lesion within the peduncles was observed by deMeyer to be 83% of the total tract, recovered fine, individual movements of the fingers (see also Jane et al., 1968). Pyramidal tract axons provide only one of many efferent projections from the motor cortex, and deficits of movement performance following lesions of the cortex or damage to the internal capsule are much more dramatic than those that follow section of the pyramidal tracts (Bucy, 1957; Brinkman and Porter, *this volume*). Denny-Brown (1966) concluded that damage to the cortex of the pre- and postcentral gyrus, caused "a severe and permanent defect in all delicate spatial adjustments of movements of the hand, and to a lesser extent of the foot and mouth. The small eversions, abductions, rotations that enable precise palpation and exploration or withdrawal, depend on the integrity of pre- and postcentral gyrus."

Recordings have been made from pyramidal tract and/or other neurons within the precentral gyrus, which have explored how the natural behaviors of these cells might be involved in the production of skill and precision in movement performance, particularly in the execution of movements using the forearm and hand. A number of experimental strategies have been used. Very few of these have addressed the relationship between PTN responsiveness and fractionation of muscle contraction or the precision grip. Most have attempted to confine the movement performance to a stereotyped repetition of alternating wrist or elbow extension/flexion movements. These can be performed against changing loads (requiring changing force development) or can be subjected to sudden perturbations which must be compensated. In a small number of experiments, the animals have been permitted to perform natural self-paced, free-range movements and attempts have been made to describe the particular relationships between neuronal activity and identifiable elements of this total movement repertoire (e.g., finger flexion no matter what the context within which this movement is performed). Each of these approaches has allowed description of significant information about motor cortex neuron involvement in the total control machinery for movement performance.

Evarts (1966) trained monkeys to perform a stereotyped conditioned wrist extension movement in response to a light flash. The monkey held its extended fingers on a telegraph key until the light flash signal was presented. It was rewarded with delivery of juice into its mouth for carrying out a brisk wrist extension movement to release the key within a response-time of less than 350 msec. The EMG activities in flexor and extensor muscles of the wrist were recorded through the skin, and the discharges of PTNs in the contralateral precentral gyrus, whose responses changed with the movement performance, were sampled. Closely associated activity in PTN was found to occur consistently in advance of extensor muscle activation. A burst of neuronal discharge preceded the onset of extensor muscle activation by 50 to 80 msec. This was not a response to the light signal per se because the discharge did not occur in the absence of a movement response. The latency of the cell's discharge following the conditioning light stimulus covaried with the reaction time for muscle contraction, and these two measures were strongly positively correlated. Hence, it was clearly established that PTNs changed their firing in advance of a conditioned wrist-movement response and could have been concerned in its initiation (Fig. 5).

Different PTNs change their firing at different times in relation to a natural movement performance using the contralateral arm and hand. The population of active cells is seen to be continuously changing

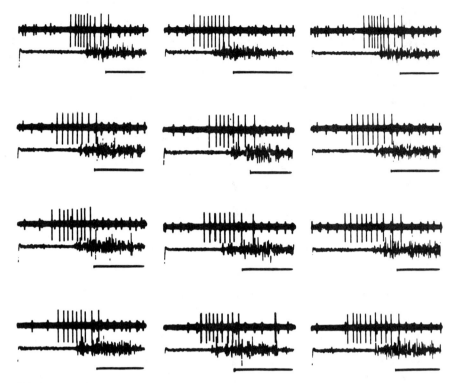

FIG. 5. Discharges of a PTN in the motor cortex (*upper trace* in each group) always precede a conditioned wrist extension movement (EMG in second trace) in a reaction-time experiment. A light signal at the onset of each trace is followed by a brisk wrist extension movement and the opening of a Morse-key contact which causes the third trace to appear on the oscilloscope screen. Reaction time is measured between the beginning of the traces and the onset of this third black line. (From Evarts, 1966, with permission.)

with time during the performance of a total movement sequence. It has been suggested that this changing pattern of active neurons is related to the modulation of contraction of particular muscles in the limb. Particular aggregations of PTN have a characteristic temporal association with the recruitment of their associated muscle or muscles to the movement sequence (Porter and Lewis, 1975).

Evarts (1968) trained monkeys to perform repetitive wrist flexion/extension movements to move a lever to which loads could be added using a system of pulleys (Fig. 6). This allowed a dissociation of the position achieved by the wrist from the force required to be developed in flexor or extensor muscles to maintain that position. Recordings from a given PTN differed depending on the muscular force that had to be generated using a particular muscle group. For the same position of the wrist, a large amount of PTN discharge occurred when that position was held against a great load while little or no PTN discharge occurred when the load aided the holding of the position and little or no muscular force was required (Fig. 7). This relationship between PTN activity and the development of force in particular groups of muscles has been confirmed by others in different experiments (Lewis and Porter, 1974; Smith et al., 1975; Hepp-Reymond et al., 1978; Thach, 1978).

Humphrey et al. (1970) recognized that a very large number of cortical neurons could cells. Humphrey (1972) described the optimum responses of the control system that were needed to produce the most accuracy in matching theoretical to actual performance.

FLEXIBILITY OF THE RELATIONSHIP BETWEEN PYRAMIDAL TRACT DISCHARGE AND MUSCULAR CONTRACTION

Fetz and Finocchio (1971, 1972) trained a monkey to produce isometric contractions of particular forelimb muscles while the arm was held stationary in a cast. The EMG records from each of 4 muscles were used as reinforcement criteria in the training to teach the monkey to produce isolated isometric contractions of one of these muscles while suppressing, more or less completely, the EMG responses in the other 3. Recordings were then made from neurons in the precentral gyrus. Some cortical neurons were found, the discharges of which occurred in association with contractions of only one of the 4 muscles; they began to fire about 70 msec before the onset of muscle contraction and reached peak frequency at the time of maximum muscle activity. The animal was then given food reinforcement for producing bursts of cortical cellular activity in these neurons while suppressing all muscle contraction, including that in the closely related muscle. After many repetitions, the monkey could produce bursts of cortical cell firing without any measurable EMG response in the sample muscles. Thus, the relationship between cortical cellular discharge and muscle contraction was flexible. Other connections with the spinal final common path could have been so modified during the training that they prevented motor output in response to corticospinal discharge.

Schmidt et al. (1974) also argued for a plastic relationship between cellular discharge in the motor cortex and the force to be generated in a wrist flexion/extension

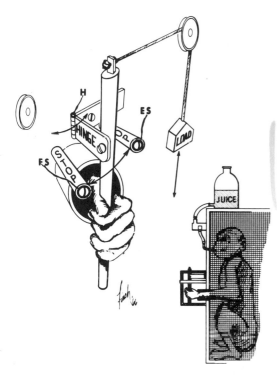

FIG. 6. Diagram of the method used for dissociating wrist displacement from the muscular force required to produce that displacement in the experiments of Evarts (1968).

be influencing the development of force in a complex group of fixating and prime-moving agonist muscles even in such an apparently simple task as wrist flexion. They succeeded in the technically difficult task of making simultaneous recordings from a number of cortical neurons, all of which were associated in their discharge with a given movement performance. They found that when account was taken of the slow time course of muscle contraction and the resultant slow movements of a limb, and when a delay of 50 to 100 msec was introduced between impulse activity in the cortex and the development of torque, it was theoretically possible to match the actual force and the displacement of a limb manipulating a joystick with the output of a linear control system, the input to which was the firing rate of a small population of cortical

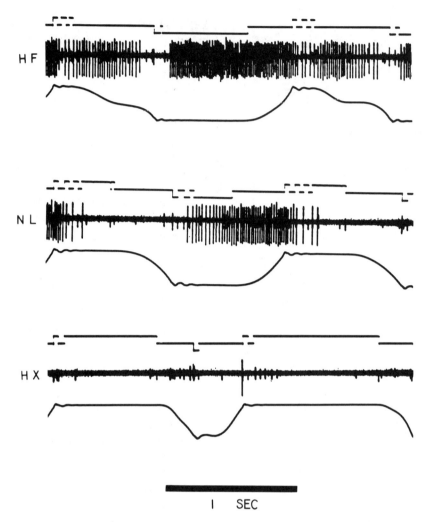

FIG. 7. Repetitions of a wrist flexion-extension task carried out by a monkey causes the third trace in each set to move from the flexor position (highest level) to the extensor position (lowest level). In the *upper tracings,* when a large load opposes the flexion movement, the PTN from which recordings were made produced maximum discharge which extended into the extensor phase as the load was paid out and the flexor muscles were still contracting. Without any load on the lever, the *middle records* were obtained for the same cell. When the load aided flexion and no muscular force was needed to produce that position, the cortical cell became essentially silent. (From Evarts, 1968, with permission.)

task against a simulated spring load. They considered that the relationship between the firing of a cortical cell and the load to be moved had some dependence on the animal's previous experience of load changes. Tanji and Evarts (1976) have examined some of the factors, other than force to be developed, which can influence the firing of PTNs. In certain situations, when an animal is required to move the forelimb in one of two directions (to push or to pull a lever), depending on the instruction given with a colored light cue, many PTNs increase their firing well in advance of the performance of the movement. This increased firing while the animal was prepared for the movement (having received the instruction about the direction of movement) but before the signal to move had been given, was more characteristic of PTNs in the precentral gyrus than it was of other cortical neurons in this area or of PTNs in the postcentral gyrus. Moreover, the effects of the prior instruction were differential. Those cells, whose bursts of discharge would precede and accompany a push movement when it was signaled to be performed, would have their firing rate increased by the instruction that informed of a push direction for the future movement; the same signal would depress the firing of cells whose natural bursts accompanied pull movements.

ELECTROPHYSIOLOGICAL STUDIES IN MAN

Availability of convenient electronic averaging computers enabled several groups of workers to seek correlates, in the electroencephalograms of human subjects, of the neuronal changes occurring in the motor cortex in association with the performance of voluntary movements, even though Jasper and Penfield (1949) had detected no characteristic potential changes other than suppression of the beta waves over the exposed precentral gyrus before the performance of movements. Many of these averaging studies indicate the occurrence of a readiness potential which may be detected bilaterally and over wide regions of the scalp beginning up to 1 sec before the onset of movement (Gilden et al., 1966; Deecke et al., 1969; Kato and Tanji, 1972). This early change has been reported by some as giving way to a pre-motion positivity (Deecke et al., 1969, 1977; Deecke and Kornhuber, 1977) also recorded bilaterally and then a true motor potential restricted in its maximum amplitude to the zone over the contralateral precentral representation of the moving limb and beginning 50 to 100 msec before the onset of movement.

Brindley and Craggs (1972) recognized the difficulties involved in averaging the potential changes associated with a very large number of repetitions of a movement and, in animal experiments, sought an alternative method of detecting cortical signals associated with movement performance. They used filtered recordings from the cortex to detect a signal envelope (processed mapped voluntary movement signal), which, even in single trials, was a reliable event associated with activity in a particular muscle group. The rising phase of this processed signal preceded the occurrence of movement by 100 msec or so and was independent of afferent signals ascending to the cortex from the spinal cord (Craggs, 1975).

In a few cases, the discharges of individual cortical neurons have been recorded in the exposed precentral gyrus of conscious human subjects undergoing surgery for epilepsy or Parkinson's disease. Li and Tew (1964) reported changes in discharge with voluntary contraction and relaxation of the finger. Goldring and Ratcheson (1972) extended these observations. A few cells which discharged whenever the fingers were flexed actively to make a fist were also caused to discharge by passive flexion of the fingers. Extension of the fingers either voluntarily or produced passively by the surgeon was associated with suppression of the firing of these same cells. Since, in the small sample examined by these authors, cortical neurons in the precentral gyrus did not respond to light tactile stimulation of the skin or to clicks, it was concluded that there is a selected kinesthetic input to cells in the motor cortex. These findings agree with many of those discovered in monkeys and emphasize the relevance of the work with monkeys to understanding the functional machinery of the human motor cortex. Indeed, Goldring and Ratcheson (1972) confirmed that a small proportion of cells in the primate precentral gyrus are associated in their discharge with the motor systems of both the ipsilateral and the contralateral hand and can receive a kinesthetic input from both (Lemon et al., 1976) even though strictly contralateral associations are predominant.

AFFERENT PROJECTIONS TO PRECENTRAL NEURONS FROM RECEPTORS IN THE MOVING LIMB

Early studies of Woolsey et al. made it clear that evoked potentials in response to peripheral stimuli could be recorded within the motor area of the precentral gyrus as well as in the postcentral sensory cortex. Hence, the concept of motosensory (Ms) and sensorimotor (Sm) areas was developed, with the motor (M) component dominating in the precentral gyrus. Malis et al. (1953) stimulated the sciatic nerve of anesthetized monkeys and recorded evoked potentials before and after ablation of regions which could have been concerned in the relay of nervous influences to precentral cortex. They found that the responses in the leg area of motor cortex to sciatic nerve shocks were unaltered by resection of the adjacent poscentral cortex, by removing both pre- and postcentral cortex of the other hemisphere or by resection of the cerebellum. They deduced that the precentral motor area received its own projection from peripheral afferents, independent of afferent projections to these other regions. They found that cutaneous as well as deep stimuli were capable of causing evoked responses in precentral cortex and concluded that projections from both superficial and deep receptors must converge on this region.

Albe-Fessard and Liebeskind (1966) examined the responses of single neurons in regions of the motor cortex from which evoked potentials could be recorded in chloralose-anesthetized monkeys. They found that the inputs to cells in area 4 were topographically separated to the extent that neurons in the arm area were influenced mainly from the arm and those in the leg area mainly from the leg. In contrast to neurons in the postcentral receiving cortex of the same animal, whose discharges were almost exclusively influenced by tactile stimuli, the majority of cells in the motor cortex were excited by stimulation of deep structures and particularly by movement of limbs. Adey et al. (1954) also demonstrated, with evoked potential recordings, the input from deep structures to the precentral gyrus. The shortest latencies of the responses recorded by Adey et al. were often found in the depths of the central sulcus and the earliest effects of a peripheral stimulus could be detected in the depths of the sulcus as little as 5 msec after a stimulus to the hand. Some of these very early responses could have been from cells in area 3a. Phillips et al. (1971) delivered electric shocks to the deep radial nerve and to the deep palmar branch of the ulnar nerve in anesthetized baboons. When weak shocks, capable of exciting only low-threshold group I afferents were delivered, evoked responses were limited to area 3a. These responses were abolished by section of the dorsal columns. The majority of the responses of single cells within area 3a were obtained when group I muscle afferents were stimulated. They were also influenced by brisk movement of joints, which could have activated muscle, tendon, and/or joint and fascial receptors, but they could not be caused to discharge by gentle stroking of the skin (see Jones and Porter, 1980).

Even so, evidence has accumulated which indicates that short-latency projections exist from muscle spindle receptors in the periphery to neurons located within area 4. Wiesendanger (1973) demonstrated that PTNs of area 4 in anesthetized monkeys received influences from muscle spindle receptors. He deduced that the effects of secondary endings were more potent than those of the primary endings of the muscle spindle. But, using controlled stretches to activate the dynamically sensitive primary endings as well as to maintain the static firing of secondary endings, Hore et al. (1976) and Lucier et al. (1975) concluded that the neurons in both the arm and the leg representation of area 4 were in receipt of influences from both the velocity-sensitive (primary) endings of muscle spindles and the length-sensitive (secondary) endings.

Rosen and Asanuma (1972) examined the relationship between the input region which, when stimulated, could cause discharge of cortical neurons in area 4 of tranquilized monkeys and the muscle group caused to contract by highly localized intracortical microstimulation in the vicinity of the same cortical neurons. Strong evidence was accumulated for a close association between the peripheral zone whose receptors projected influences to the neurons in area 4 and the motor field, which was activated by those same neurons. Hence, cells whose function could be deduced to promote flexion of the fingers were activated by peripheral receptors that would have been excited by this movement (e.g., cutaneous receptors on the tips of the digits) (see Doetsch and Gardner, 1972).

There is now clear evidence that even small doses of anesthetic agents modify the recordings that can be obtained from cells in the cerebral cortex (Baker et al., 1971; Lemon and Porter, 1976b). Fetz and Baker (1969) found that 85% of a large sample of single neurons from which they made recordings in the leg area of the motor cortex of unanesthetized, freely moving monkeys received a clear and reproducible input from passive movement of one or more joints of the contralateral leg. A similar high proportion of cells in the arm area are also influenced by joint movement, and fewer are affected by cutaneous stimulation (Fetz et al., 1974; Lemon et al., 1976; Lemon and Porter, 1976). Frequently, the effective stimulus to produce a response in a given cortical neuron was found to be limited to a movement of a single joint in a single direction (e.g., flexion of the metacarpophalangeal joint of the middle finger) while isolated movement of adjacent joints (e.g., interphalangeal joints) or other movements at the same joint (e.g., ulnar or radial deviation of the extended metacarpophalangeal joint) were ineffective (Fig. 8).

Experiments on conscious, freely moving monkeys have a disadvantage because it is not possible to analyze precisely which receptor populations are causing the effect on the receiving cell in area 4. In spite of this limitation, it has now been clearly established that some influences are produced by muscle spindle afferents, that other deep receptors probably combine their effects with these influences to produce inputs to many cortical neurons and that the contribution of skin receptors to the projection is relatively small (Porter, 1978). It remains possible (as has been suggested by Fetz et al., 1974) that influences from receptors in joint capsules and from afferents innervating tendons and fascia are of special importance in providing signals about movement to cells in the motor cortex which influence self-initiated movement performance.

There is also strong evidence for the oligosynaptic nature of the pathway from peripheral receptors to the cells of the motor cortex. Lemon and Porter (1976) found some responses of cortical neurons in conscious monkeys occurring less than 10 msec after a peripheral stimulus which caused natural activation of muscle receptors or skin receptors in the forelimb. It is likely that the relay for these projections is within the oral portion of the ventral posterior lateral (VPLo) nucleus of the thalamus of the monkey (Strick, 1976; Loe et al., 1977; Horne and Tracey, 1979; Lemon and van der Burg, 1979). Horne and Porter (1980) recorded the discharges of cells in the ventrolateral parts of the thalamus in conscious monkeys. Neurons in VPLo, which discharged in association with the performance of movements of the contralateral forelimb were also in receipt of influences from deep receptors (and, for some, from cutaneous receptors) in that same limb. Other parts of the ventrolateral thalamus contained neurons whose dis-

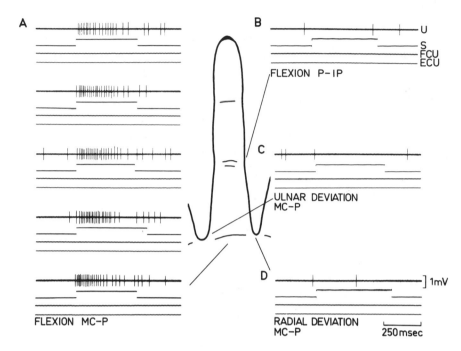

FIG. 8. Responses of a precentral neuron to passive movements of a single joint [flexion of the metacarpophalangeal (MC-P) joint of the middle finger]. This movement caused a response in the cell whenever it was produced by the experimenter. Other directions of movement at the same joint, and movement of neighboring joints (interphalangeal joints of the same finger) caused no such responses. (From Lemon and Porter, 1976, with permission.)

charges accompanied movement performance but did not receive short-latency inputs from peripheral receptors (in VLo). The majority of cells in the caudal portion of the ventral posterior lateral (VPLc) nucleus which discharged in association with movement performance were in receipt of projections from cutaneous receptors. The latencies of the responses to natural stimulation (as short as 5 or 6 msec) were consistent with a relay of information from deep receptors sensitive to limb movements in VPLo, en route to the motor cortex. Other projections could also reach the motor cortex from area 2 of the sensory cortex and from more remote cortical regions in receipt of information about the execution of the movement (Fig. 9).

The relevance of the peripheral receptors in the moving limb for the discharges of neurons in area 4, which are associated with the movement performance, was tested by Evarts (1973). Cortical neurons that were discharging in relation to the performance of a holding task with the contralateral forelimb were caused to change their firing 10 to 20 msec after a sudden displacement from the hold position was imposed. In a different task Porter and Rack (1976) also found that a sudden displacement of the fingers caused extra discharge of finger-flexing cortical neurons within 20 msec of the imposition of a displacement. Conrad et al. (1974) interpreted the very similar observations made by them as implicating this oligosynaptic feedback in a load-compensation function of the motor cortex.

Short-latency influences produced in the motor cortex by activation of peripheral receptors are dependent on the intactness of the dorsal column pathways (Brinkman et al., 1978). The fibers that convey the information from the receptors may be second-order spinothalamic afferents whose course

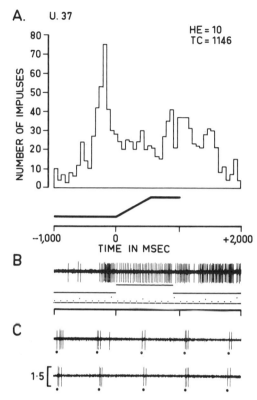

FIG. 9. **A**: Responses of a PTN summed over 10 repetitions of a self-paced, lever-pulling task to obtain a food reward. The natural activity of this cell was always associated with finger flexion both before grasping the lever (preceding time 0) and before flexing the fingers round the small morsel of food (at about time +1,000 msec). **B**: Discharges during a single performance of the task. A natural afferent input caused responses of this cell when the monkey was relaxed and at rest. Discharges caused by brisk taps applied to the palmar pad at the base of the index finger caused responses of the cell with a minimum latency to the first impulse of 7 to 10 msec. (From Brinkman et al., 1978, with permission.)

takes them through the dorsal columns, since Tracey et al. (1980) could produce no conclusive evidence that neurons in the dorsal column nuclei projected axons to nucleus VPLo of the thalamus. If primary afferent fibers in the dorsal columns are responsible for the effects in area 4, they must produce these effects by relay through VPLc and by corticocortical projections to the motor cortex.

Significance of the short-latency projections from peripheral receptors to the motor cortex has been reevaluated by Brinkman (1981) and by Jones *(this volume)*. Surgical section of the dorsal columns removes the short latency inputs to the motor cortex and abolishes (in man) the automatic compensation for sudden imposed disturbances of voluntary muscle contraction (Marsden et al., 1973, 1977*d*). After recovery from the surgical lesion, and after a period of retraining, even discrete movements of the fingers were well executed after complete section of the cuneate fasciculus in monkeys. In retrieving a raisin from a small well, both before and after dorsal column section, monkeys extended the index finger and opposed the thumb while keeping all other fingers flexed and out of the way. Thus, although the short-latency influence on cells in the motor cortex could be important during the learning of independent control of the fingers, it seems not to be essential for performance. It may be that the feedback about movement from peripheral receptors in the moving limb achieves its real significance in control of learned movement performance at lower levels of the central nervous system where the matching of intended movement against achieved result could be conducted.

The short-latency responses of cortical cells to perturbations of the moving limb are also modifiable depending on the instruction given to a monkey about the direction of a movement to be made following the perturbation. Evarts and Tanji (1974) found that the bursts of impulses in cortical cells 20 to 40 msec after a sudden disturbance of the wrist could be enhanced or suppressed depending on the prior instruction given to the animal about the direction of the movement response to follow the disturbance. Thus, the afferent limb of the cortical loop could be made operative (closed) or inoperative (open) depending on other influences conveying aspects of an instruction for movement performance to the PTNs of the motor cortex.

# Input–Output Organization of the Primate Motor Cortex

*Peter L. Strick and James B. Preston

Research Service of the Veterans Administration Medical Center and Departments of Neurosurgery and Physiology, State University of New York at Syracuse, New York 13210

There are several major features which characterize traditional views concerning input-output organization in the motor cortex. First, the output organization of the motor cortex has been described as a single, distorted map of the body in which each part is represented only once (Penfield and Rasmussen, 1950; Woolsey et al., 1952; Welker et al., 1957). Second, a larger area of motor cortex is allotted in these maps to the representation of distal musculature (e.g., Penfield and Rasmussen, 1950; Woolsey et al., 1952; Welker et al., 1957). Third, the density of direct connections to spinal motoneurons (MN) is greatest from regions of motor cortex representing the distal limb (Preston et al., 1967; Phillips, 1969; Kuypers, 1973; Jankowska et al., 1975). Finally, afferents from muscle and/or joint receptors have been viewed as the dominant somatosensory input, although cutaneous inputs also have been demonstrated.

Our preliminary studies (Strick and Preston, 1978a, b) and more recent experiments (Strick and Preston, 1982a, b) suggested an entirely different pattern of input-output organization for the hand representation in the motor cortex. We have demonstrated two spatially separate representations of the hand in area 4 of the squirrel monkey. These representations receive different concentrations of deep and cutaneous peripheral afferent input.

RESULTS

We have mapped the motor cortex of squirrel monkeys because prior studies (Welker et al., 1957; Sanides, 1968) have demonstrated that, unlike many other primates, none of the forelimb representation in area 4 is buried in a sulcus. This feature greatly facilitates accurate mapping. Maps of the distal forelimb representation were constructed by monitoring the movements and/or electromyographic (EMG) activity evoked by intracortical stimulation (Asanuma et al., 1976) at multiple closely spaced sites with area 4. In Fig. 1, the full extent of the digit and wrist representation is indicated by the black rectangle shown on the lateral view of a squirrel monkey's cerebral cortex. The map of the distal forelimb representation of this animal is illustrated in the bottom half of Fig. 1. We have defined the motor response at each site as that muscle or joint movement that was evoked at the lowest intensity of stimulation.

Several features characterize this map and the maps of 8 other animals examined. The most important feature is that the representation of the distal forelimb consisted

---

*To whom correspondence should be addressed: SUNY Upstate Medical Center, Department of Physiology, 766 Irving Avenue, Syracuse, New York 13210.

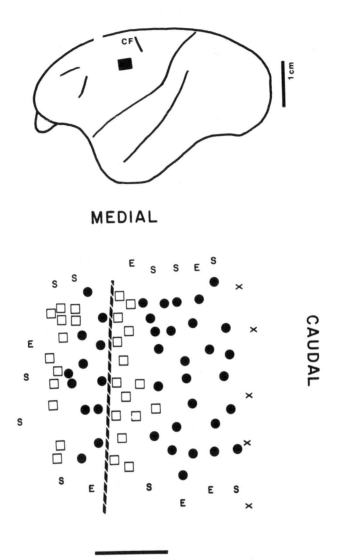

FIG. 1. Motor representation of the distal forelimb in area 4 of one squirrel monkey. **Top:** Lateral view of the left hemisphere. *Black rectangle* on the surface indicates the size and location of the distal forelimb representation. CF: central fissure. **Bottom:** Motor map of the distal forelimb representation. Symbols indicate sites of microelectrode penetrations and the movements evoked at each site. *Filled circles:* Thumb or fingers; *squares:* wrist or radioulnar joint; E: elbow; S: shoulder; X: no response to stimulation up to 30 μA. *Dashed line* separates the rostral and the caudal hand motor representations. Gaps where penetrations are absent were regions occupied by surface blood vessels. Scale: 1 mm.

of alternating bands of digit and wrist responses. One band of digit responses (filled circles) was located in caudal area 4 and a band of wrist responses (squares) was located just anterior to this band of digit responses. A second band of digit responses was located just rostral to the wrist band. Finally, a second band of wrist responses was located immediately rostral to the second band of digit responses. We have termed the caudal bands of digit and wrist responses the caudal zone of hand representation and the rostral bands the rostral zone of hand representation. The two zones of hand representation were bordered medially, laterally, and rostrally by regions from which stimulation evoked responses from more proximal joints. Caudal to the area of hand representation was a region from which stimulation (≤ 30 μA) evoked no movements. Histologic analysis showed that both the rostral and caudal zones were located entirely within area 4. The sites where stimulation failed to evoke movements were located in area 3a. In other animals, stimulation (≤ 20 μA) failed to evoke movement in 90% of the penetrations subsequently shown to be in area 3a. This is in stark contrast to area 4, where stimulation < 10 μA routinely evoked movement.

Although two spatially separate representations of the digits and wrist were observed in all animals, the relative size of these representations varied. In all animals, the band of digit responses in the caudal zone was larger than the band in the rostral zone. In addition, the location of the two hand representations relative to the central fissure was also variable.

We have observed no differences in the types of movements evoked by stimulation in the caudal and rostral zones. For example, in the animal presented in Fig. 1, 6 different finger and thumb movements were evoked. All of these movements could be evoked from both zones. Not only the same movements, but also the same muscles, were represented in the rostral and caudal zones. We determined muscle representation by monitoring EMG responses to intracortical stimulation. In most cases, EMG activity evoked at threshold was confined to a single muscle, although small increases in stimulus intensity above threshold often evoked activity in additional muscles. We defined the muscle most represented at each site as the muscle responding to the lowest intensity of stimulation. Two animals were mapped extensively with intracortical stimulation while EMG was monitored from 8 different muscles. For 6 of 8 muscles in one animal and 7 of 8 muscles in the other, we could find at least one site in each of the zones where they were most represented. Furthermore, each zone contained the representation of both extrinsic and intrinsic muscles of the hand. The separation between the most represented sites for a muscle in the caudal zone and the sites in the rostral zone was at least 1.2 mm and in 8 cases > 2 mm. Based on the calculations of others (Jankowska et al., 1975; Asanuma et al., 1976), these separations indicate that the response at a site in one zone could not have resulted from either current spread or collateral monosynaptic activation of a site in the other zone. The foregoing results lead us to view the outputs from the two separate representations as parallel pathways with comparable influences on the same MN pools.

Given the similarity in outputs from the two representations, we thought that a clue to their function might come from an analysis of their afferent inputs. In another set of experiments, single units were recorded in the rostral and caudal zones of area 4 and in adjacent portions of area 3a. The adequate somatosensory stimulus and receptive field was determined for each of 463 neurons recorded in 189 penetrations in the two hand-wrist representations of area 4 as well as adjacent parts of area 3a. Neurons were classified into two categories: cutaneous or deep. Of the 463 neurons, 106 were located in the rostral zone and 267 in the caudal zone. In the rostral zone, 104 neurons were classified as deep and only 2 as cutaneous. In contrast, 258 neurons in the caudal zone were classified as cutaneous and only 9 as deep. Thus, single neurons activated by cutaneous input were clearly concentrated in the caudal hand representation, whereas neurons responding to deep inputs were concentrated in the rostral hand representation. We concentrated our analysis to the caudal zone because the region comparable to our caudal zone is buried in the central sulcus in other primates, and has not been sampled extensively in prior studies.

Although area 3a was not the primary focus of this study, 90 neurons were recorded from the part of this area adjacent to the caudal zone of area 4. The distribution of neurons responding to cutaneous and deep inputs was similar to that seen in the rostral zone of area 4; 82 neurons were classified as deep and only 6 neurons as cutaneous.

The differential distribution of deep and cutaneous inputs to the two motor representations of area 4 is illustrated by maps derived from multiple penetrations in individual animals (Fig. 2). The filled symbols indicate tracks in which all the isolated units were classified as cutaneous. The unfilled symbols indicate tracks in which all the isolated units were classified as deep. Circles mark tracks in which the response to intracortical stimulation was movement of the digits (fingers or thumb). Squares mark tracks in which the response to stimulation was movement of the wrist (wrist or radioulnar joint). Stars mark tracks in which stimulation (up to 25 µA) evoked no motor response. All of the units recorded in the tracks marked by stars were classified as deep. The two motor representations of area 4 are clearly evident in the map of the animal illustrated in Fig. 2. The boundary between the rostral digit representation (unfilled circles) and the caudal wrist representation (filled squares) is indicated by the dashed line. Histologic analysis showed that the area 3a-4 border was located just caudal to the penetrations marked by filled circles. This border is indicated by the solid black line. In this animal, only units classified as deep were observed in the rostral

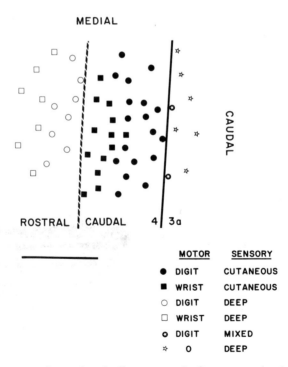

FIG. 2. Sensory-motor map of area 4 and adjacent area 3a from one animal. Symbols are placed at sites of microelectrode penetrations and indicate the motor output and the somatosensory input of each site. Symbols defined in lower right corner. *Dashed line* separates rostral and caudal motor representations. The boundary between area 4 and area 3a was determined by histologic analysis and is indicated by the *solid line*. Scale: 1 mm.

zone. In contrast, only units classified as cutaneous were found in the caudal zone. Both deep and cutaneous units were recorded in the two tracks (each indicated by the star inside of a circle) located just caudal to the area 3a-4 border.

Figure 3 shows examples of cutaneous receptive fields recorded in the caudal zone of the same animal. Each receptive field illustrated was recorded from the neuron nearest the site in each penetration where movement was evoked at the lowest stimulus intensity. A receptive field chosen in this manner was representative of the fields of other neurons recorded in the same track. The motor response evoked by intracortical stimulation at each recording site is indicated by the letters next to each figurine. There were no obvious differences, in this or other animals, between the receptive fields of neurons in the digit and wrist representations of the caudal zone. Neurons whose receptive fields were confined to the tips of the digits were recorded in both the digit and wrist representations, as were neurons whose receptive fields included the palm of the hand. Another feature that is obvious from this figure, and which was characteristic of all animals, was the predominance of receptive fields confined to the volar surface of the hand.

Figure 3 also illustrates our observation

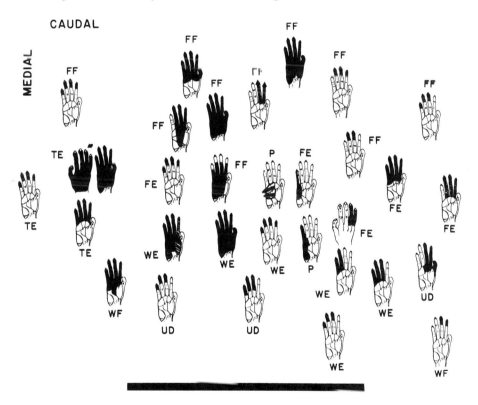

FIG. 3. Cutaneous receptive fields and motor responses observed at each caudal zone site illustrated in Fig. 2. The cutaneous receptive field for one neuron in each track is indicated by the shading on a figurine. Each figurine is placed at the site of the penetration in which it was recorded. Skin areas where stimulation gave the best response are shown in black. Skin areas where stimulation gave additional weak responses are marked with diagonal lines. Motor response evoked in each penetration is indicated by the letters adjacent to each figurine. FF: Finger flexion; FE: finger extension; TE: thumb extension; WF: wrist flexion; WE: wrist extension; P: pronation; UD: ulnar deviation. Scale: 1 mm.

that neither the size nor the location of a cutaneous receptive field correlated with the movement evoked by stimulation in the same track. For example, those neurons with receptive fields on the volar surface of the hand were recorded in tracks where stimulation evoked finger or wrist extension, as well as in tracks where stimulation evoked finger or wrist flexion. Illustrated in this figure are two neurons with receptive fields that included the dorsal surface of the hand. These neurons were recorded in tracks where stimulation evoked extension movements. In other animals, flexion movements also were associated with neurons having receptive fields on the dorsal surface of the hand.

In the rostral zone, there was a correlation between receptive field location and the movement evoked by stimulation. Both the receptive fields and the motor response were confined to the same body part for over 75% of the penetrations in the rostral zone. For example, when stimulation evoked a finger movement then passive movement of the fingers was, in most cases, the adequate stimulus for neurons recorded in the same track. However, the direction of passive movement which excited neurons did not correlate with the direction of movement evoked by intracortical stimulation. The sample was equally divided between tracks where the direction of passive and evoked movement was the same and tracks where they were opposite.

DISCUSSION AND CONCLUSIONS

Our results demonstrate a pattern of input-output organization for the primate motor cortex that has not been previously recognized. The forelimb area in the motor cortex of the squirrel monkey has two spatially separate representations of the digits and wrist. Stimulation in the two representations evokes the same movements and activates the same muscles. Furthermore, each representation can be characterized by the type of somatosensory afferent input it receives. The caudal digit-wrist representation receives input predominately from cutaneous afferents, whereas the rostral digit-wrist representation receives input predominately from muscle and/or joint afferents.

Our findings raise an important question: Is the pattern of double representation in the squirrel monkey species-specific, or does it represent a more general pattern of motor cortex organization? Two spatially separate representations of distal forelimb musculature also exist in the motor cortex of the cat (Pappas and Strick, 1981a,b). There is anatomical evidence that a double representation of the distal forelimb may exist in monkeys with a deep central sulcus (Zant and Strick, 1978). Furthermore, motor representation of the distal forelimb in the stump-tailed monkey apparently has two spatially separate regions where only finger movements were evoked (see Kwan et al., 1978, Fig. 3C). Thus, double representation appears to be a generalized pattern of organization for the representation of the forelimb in the motor cortex of the cat and New and Old World monkeys.

There is also evidence that the pattern of afferent segregation which we have observed in the squirrel monkey is present in the motor cortex of macaques. Tanji and Wise (1981) mapped the somatosensory input to the hindlimb area in the motor cortex of unanesthetized macaques and found that cutaneous afferent input was concentrated in the caudal part of area 4, whereas noncutaneous input was concentrated more rostrally. Recently, similar observations also have been reported for the forelimb region of the macaque motor cortex (Lamour et al., 1980). Clearly, area 4 can no longer be considered as a homogeneous structure which receives predominately kinesthetic input. There is significant cutaneous input to primate motor cortex and cutaneous and kinesthetic inputs are concentrated in separate regions of area 4.

Based on our results we have proposed that "the double representation reflects two motor control systems within area 4 that deal with different components of motor behavior" (Strick and Preston, 1978a,b). Although it would be surprising for these systems to be completely independent, they may have been anatomically separated to facilitate differential control over their inputs and/or outputs. Classical studies implicate the motor cortex in a number of reflex-like motor responses (see Denny-Brown, 1960; Phillips, 1969). The concentration of deep inputs in the rostral motor representation suggests that it may be designed to control movements that use kinesthetic feedback for their execution and guidance, as for example in load compensation. On the other hand, the concentration of cutaneous inputs in the caudal motor representation suggests that this cortical zone may be designed to control movements, which use tactile feedback for their execution and guidance. Thus, we suggest that the rostral and caudal zones of motor cortex are differentially involved in control of motor behaviors that depend on deep or on cutaneous inputs, respectively.

ACKNOWLEDGMENTS

This study was supported in part by funds from the research and development Service of the Veterans Administration and USPHS Grant NS 02957. We acknowledge the assistance of Ms. Agnes Helcz in this research.

Motor Control Mechanisms in Health and Disease,
edited by J. E. Desmedt.
Raven Press, New York © 1983.

Contrasting Properties of Pyramidal Tract Neurons Located in the Precentral or Postcentral Areas and of Corticorubral Neurons in the Behaving Monkey

*Christoph Fromm

Laboratory of Neurophysiology, National Institute of Mental Health, Bethesda, Maryland

Until recently, our knowledge of the movement-related activity of corticofugal neurons has been restricted to pyramidal tract neurons (PTNs) of the primary motor cortex (area 4). However, about half of the PTNs in the primate brain are located in the somatosensory cortex (SI) of postcentral gyrus comprising Brodmann's areas 3, 1, 2, and in the adjacent area 5 (Russel and DeMyer, 1961; Murray and Coulter, 1981). This chapter presents results on the activity of postcentral PTNs during voluntary movement and compares PTNs with a different class of motor cortex output neurons, corticorubral neurons. The corticospinal and corticorubrospinal systems represent the two major parallel routes available to the motor cortex for control of forelimb motoneurons (MNs) (Kuypers and Lawrence, 1967; Shapovalov, 1975) and either system is able to subserve voluntary movement (Lawrence and Kuypers, 1968a, b). For the cat, the similarity in anatomical and functional organization of both descending systems has often been stressed (Endo et al., 1973; Ghez, 1975; Jankowska, 1978; Larsen and Yumiya, 1980). For the

monkey, however, there is mounting evidence from anatomical (Kuypers and Lawrence, 1967; Catsman-Berrovoets et al., 1979; Hartmann-von Monakow et al., 1979) and electrophysiological work (Humphrey and Rietz, 1976; Humphrey and Corrie, 1978) that PTNs with collaterals to red nucleus (RN) and corticorubral cells are sufficiently separate in their cellular origin and subcortical destination within the RN to suggest their differential role in motor control.

Although having a common origin from cortical layer V, PTNs in area 4 and in the subdivisions of SI and area 5 differ in size as well as terminations (Kuypers, 1960; Liu and Chambers, 1964; Coulter and Jones, 1977; Jones and Wise, 1977; Murray and Coulter, 1981), suggesting specialization of PTNs depending on their origin and destination. Indeed, the different fields of origin, different afferent inputs to these fields (Mountcastle and Powell, 1959; Powell and Mountcastle, 1959; Paul et al., 1972; Jones and Porter, 1980), and different projection patterns of PTNs may mean that there is no common denominator for PTNs of different cortical areas except the bundling together of their axons as they descend.

In the present study these various cortical output neurons were compared with re-

---

*Max-Planck-Institut für Biophysikalische Chemie, Abt. Neurobiologie, Postfach 968, 3400 Göttingen, Federal Republic of Germany.

spect to (1) temporal relations to onset of volitional movement, (2) relationship to graded muscle forces, and (3) responses to afferent input. In connection with afferent responsiveness, we will focus on the effects of alteration of feedback caused by stopping an active movement just as it was starting. The response to arrest of an ongoing movement should provide data relevant to the hypothesis (Evarts and Fromm, 1978) that the PTN is a summing point for central programs and afferent feedback. According to this hypothesis, PTN output should signal discrepancies between the actual movement (as revealed by feedback) and the intended movement (corresponding to the central program).

It is possible that this concept may also be extended to postcentral PTNs, and we sought to obtain evidence for central inputs carrying information of the impending movement into primary sensory receiving areas and thereby generating corollary discharge. Previous studies on postcentral neurons (Evarts, 1972, 1974; Bioulac and Lamarre, 1979; Soso and Fetz, 1980) did not involve identification of PTNs, which send their efferent fibers to subcortical target neurons and provide them with corollary signals.

METHODS

Paradigm

Data were obtained from two monkeys (A and B). For monkey A, a visual pursuit tracking task was used to lead to performance of high velocity, large supination–pronation movements. Cortical responses to afferent input were elicited by torque pulses. Details of paradigm and data processing for monkey A have been described (Fromm and Evarts, 1977, 1978). The main data are from monkey B, in which we recorded from RN cells (Fromm et al., 1981) as well as in the cortex. The task was compensatory rather than pursuit tracking. The monkey viewed a single row of 31 light-emitting diodes (LEDs) of which the central LED was red and all the rest were green. Only one of the LEDs was illuminated at any one time; its position depended on the summation of two signals: (1) a voltage corresponding to the angular position of the handle grasped by the monkey and (2) a "set-point" voltage, which was automatically switched between three possible values according to a predetermined pseudorandom schedule. When the sum of these two voltages was sufficiently close to zero, the central LED was illuminated. Reward was delivered when the monkey had positioned the handle so as to cause uninterrupted illumination of the central LED for 2 to 5 sec. A handle movement of 1.5° corresponded to a shift from one LED to the next, so that reward depended on rather accurate ($\pm 0.75°$) postural stability. There were three different set-points requiring the handle to be held in three different positions: 0° vertical, 15° supination, and 15° pronation. The large movements (triggered by a set-point shift) were of lower velocity than large movements in the previous paradigm, because the monkey had to terminate movement accurately within the small target zone.

A torque motor coupled to the handle allowed variation of steady-state loads in either direction including a zero-load condition; the monkey was merely required to keep the handle in the vertical position. The torque motor also provided a means of imposing pronating and supinating ramp-and-bold displacements (15° amplitude) with varying ramp velocities for a total duration of 200 msec during which the handle was under positional servocontrol. Occasionally, the monkey's movements were brought to a halt (immediately after detection of movement onset) by electronically clamping the handle in a fixed position for a period of 200 msec. For successive cycles of the paradigm, active and passive displacements as well as interspersed stops were varied according to a pseudorandom order.

Antidromic Identification

Following microelectrode recordings from the RN, three pairs of bipolar stimulating electrodes were permanently implanted: one within the pyramidal tract (PT) at the level of the caudal one-third of the inferior olive; the other was placed at the caudal pole of the magnocellular RN; the third was situated at the posterior parvocellular RN. The following abbreviations are used throughout the text: PT-CRNs are cortical cells which responded antidromically to stimulation of both PT and RN; CRNs were antidromically driven only by RN stimulation, and PTNs were excited solely by PT stimulation. Cortical neurons failing to respond to any stimulation site are called non-PTNs. The antidromic latency of each neuron was noted for threshold and just suprathreshold stimulation, and was measured from stimulus onset to first deflection of the antidromically evoked spike. For identification, the antidromic response had to follow each shock of a short, high-frequency train with constant latency and had to be absent following a spontaneously occurring spike for a period of twice the antidromic latency plus the axonal refractory time (spike-collision test). Each cell that was antidromically driven was tested for interaction between stimulation of PT and RN, and spike occlusion was observed for neurons classified as PT-CRNs. A number of criteria were observed for the existence of axon collaterals to RN rather than current spread to the parent axons in the case of PT-CRNs (cf. Fromm et al, 1981). Single-unit recording was supplemented by intracortical microstimulation (< 30 μA, 100-msec train of 300 Hz) usually in the vicinity of recording from PTNs.

Histology

Toward the end of the experiments, a number of electrolytic microlesions were made through the recording electrode at selected penetration sites in order to aid subsequent histologic reconstruction of recording tracks from sagittal, Nissl-stained sections. For cytoarchitectonic delineations we followed the descriptions of Jones and co-workers, particularly regarding the definition of area 3a (Jones et al., 1978; Jones and Porter, 1980; Friedman and Jones, 1981; Jones, *this volume*). A number of neurons with distinct physiological properties were focally located in a junctional region between areas 2 and 5, where a clear-cut border could not be determined (Mountcastle et al., 1975; Jones et al., 1978). This region appeared rather as a transitional zone, on cytoarchitectonic grounds, and also on the basis of its thalamic connectivity. A small injection of tritiated amino acids restricted to this site led to labeling of the ventroposterior nucleus (caudal part, as is typical for area 2) and of the lateral posterior nucleus and anterior pulvinar nucleus (as described for area 5) (cf. Jones et al., 1978).

RESULTS

Location and Electrophysiological Data

A total of 208 PTNs, 27 PT-CRNs, and 22 CRNs were recorded from the motor cortex. Both types of corticorubral projection neurons were found over large parts of area 4 except for penetrations through the wall of the precentral gyrus. Antidromic responses to RN stimulation were not observed in postcentral regions.

Note: Ninety-two PTNs were identified in postcentral areas with the majority ($N = 41$) located in area 2; 12 were found to lie in area 3a, 15 in areas 3b and 1, 14 near the area 2/5 boundary (see Methods section), and 10 clearly in area 5. The fact that the minimum antidromic latency of postcentral

PTNs was 1.1 msec (as compared to 0.6 msec for precentral PTNs) confirms anatomical data on lack of large PTNs in the postcentral gyrus (Jones and Wise, 1977; Murray and Coulter, 1981). The axonal conduction velocities of CRNs ranged from 8 to 20 m/sec, an observation consistent with the relatively small size of these cells. Although the proportion of CRNs in our sample was considerably smaller than the proportion in a study using finer, more selective microelectrodes in anesthetized monkeys (Humphrey and Rietz, 1976), our proportion (12%) of the total PTN population that sent collaterals to RN (termed PT-CRNs) corresponds well with the branching percentage of 10% in that study (Humphrey and Rietz, 1976; Humphrey and Corrie, 1978). Most of our PT-CRNs belonged to the fast subset of the PTN population (50–60 m/sec) and were usually antidromically activated by stimulation of the caudal electrode pair placed near the magnocellular pole of RN, whereas CRNs were more often antidromically driven by the rostral electrodes situated at the parvocellular RN. The PT-CRNs were found somewhat deeper than the CRNs in any given cortical penetration, often intermingled with other PTNs. These observations are in agreement with recent anatomical and electrophysiological findings (Humphrey and Rietz, 1976; Catsman-Berrevoets et al., 1979; Hartmann-von Monakow et al., 1979) demonstrating the differential projections to either subdivision of RN from two sets of cortical cells.

Temporal Relation to Active Movement

In order to compare CRNs, PT-CRNs, and PTNs in the same cortical loci, analyses were restricted to those PTNs that had been recorded in the same penetrations with CRNs and PT-CRNs. Ideally, one would like to focus on simultaneously recorded CRNs and PTNs, but this was a rare event. Figure 1 shows two pairs of simultaneously recorded units; both PTNs discharged vigorously prior to movement onset, while discharge of either CRN and PT-CRN occurred at the onset of movement rather than before, with discharge frequency rising during the course of movement. The rather slight changes of discharge frequency in CRNs were observed even in penetrations where all PTNs were intensely modulated even with smallest pronation-supination movements (about 1°) and where intracortical microstimulation produced supination or pronation of the arm. In contrast to CRNs, pre- and postcentral PTNs had activity that peaked immediately before or after beginning of movement. Furthermore, the activity of the muscles involved in these movements peaked around movement onset. The more delayed activity of CRNs and of a number of PT-CRNs tended to buildup gradually and to peak near the termination of movement. This time course was also observed in RN neurons (Fromm et al., 1981) and is illustrated in Fig. 2, where discharge of several corticorubral neurons has been aligned on termination of movement.

Figure 3 shows the onset times of neural discharge with respect to movement onset. The PTN population preceded the PT-CRNs by about 60 msec ($p < 0.001$), and the PT-CRNs, in turn, were earlier ($p < 0.005$) than the CRN group. Distributions of CRNs and RN neurons overlapped and coincided mostly with the EMG onsets. These pooled data confirmed the results obtained by comparing timing of neurons within each individual penetration.

Figure 4 includes all PTNs from the motor cortex to compare their timing to that of postcentral neurons. For area 3a of SI, all PTNs but none of the non-PTNs discharged before the earliest electromyograph (EMG) changes of any arm muscle. Discharge before earliest EMG onset occurred less frequently in other postcentral areas, with such prior discharge occurring in 55% of PTNs recorded in area 2/5 junctional region and in 20% of the PTN population in areas 2 and 1. There were no significant timing differences for PTNs and non-PTNs in these SI regions. For area 4, 80% of the PTNs discharged in advance of EMG onset. With regard to mean onset times, the total PTN population in area 4 lagged behind that of area 3a by 23 msec ($p < 0.025$); this timing difference became greater (48 msec) and

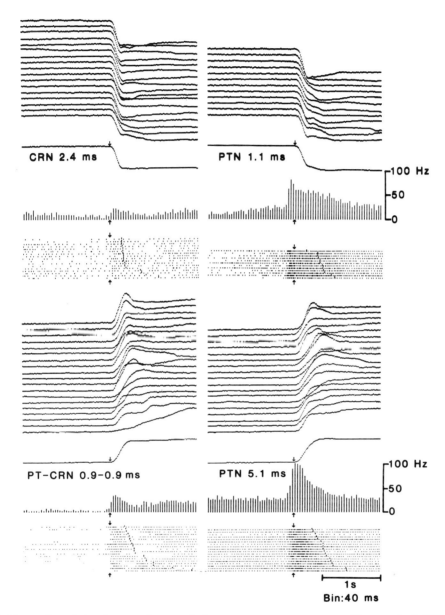

FIG. 1. Two pairs of simultaneously recorded motor cortex neurons are illustrated: A CRN (antidromic latency: 2.4 msec) together with a PTN (antidromic latency: 1.1 msec) active during supination movements in the upper displays and a second pair (PT-CRN and PTN) discharging with pronation movements in the lower half of the figure. In each display, the individual traces indicate handle position which correspond from top to bottom with the raster rows of unit discharge. The averaged position signal is directly above the histogram showing average discharge frequency in Hz, calculated for 40-msec bins and corresponding to the raster below. All raster have been aligned on the first detectable change of velocity of handle movement *(small arrows)*. Displays have been reordered according to increasing movement duration *(top to bottom)*, with completion of the 15° movements being represented by a *heavy dot* in each raster row. Numbers in msec refer to antidromic response latency; thus, the PT-CRN responded within 0.9 msec to both PT and RN stimulation.

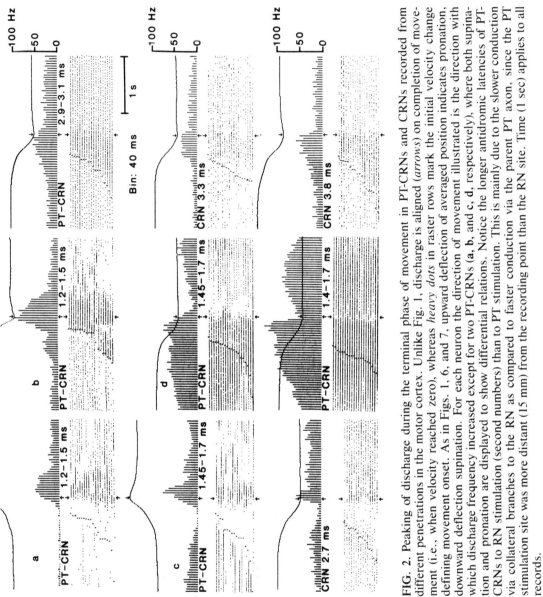

FIG. 2. Peaking of discharge during the terminal phase of movement in PT-CRNs and CRNs recorded from different penetrations in the motor cortex. Unlike Fig. 1, discharge is aligned (*arrows*) on completion of movement (i.e., when velocity reached zero), whereas *heavy dots* in raster rows mark the initial velocity change defining movement onset. As in Figs. 1, 6, and 7, upward deflection of averaged position indicates pronation, downward deflection supination. For each neuron the direction of movement illustrated is the direction with which discharge frequency increased except for two PT-CRNs (a, b, and c, d, respectively), where both supination and pronation are displayed to show differential relations. Notice the longer antidromic latencies of PT-CRNs to RN stimulation (second numbers) than to PT stimulation. This is mainly due to the slower conduction via collateral branches to the RN as compared to faster conduction via the parent PT axon, since the PT stimulation site was more distant (15 mm) from the recording point than the RN site. Time (1 sec) applies to all records.

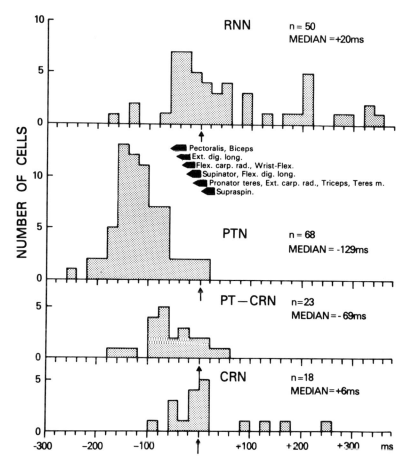

FIG. 3. Onset times of discharge of RN neurons of PTNs recorded in the same penetrations in area 4 together with PT-CRNs and CRNs, and of muscles in relation to movement onset (zero time). For each unit and muscle, the earliest clear increase of activity with any movement (supination or pronation) was measured from the histogram-averages with respect to first detectable change in velocity of handle movement (*arrow*). All data from monkey B.

highly significant ($p < 0.001$) if we compared mean onset time of all area 3a PTNs (122 msec) with onset time of a subgroup of 33 motor cortex PTNs (mean: 74 msec) that were located in the rostral bank of central sulcus adjacent to area 3a. These later discharging PTNs belonged mostly to the finger-hand representation of motor cortex as revealed by intracortical microstimulation.

Relation to Graded Steady-State Forces

A relationship between discharge frequency and steady-state force was observed for both pre- and postcentral PTNs but not observed in corticorubral neurons. About 70% of 132 PTNs in area 4, an equal proportion of 39 PTNs in areas 2 and 2/5, and almost all (6 of 7) PTNs in area 3a were found to have different discharge frequencies depending on steady-state load. In contrast, steady-state firing rates of CRNs did not vary with different levels of muscle activity due to different static loads or associated with different joint positions; in this respect CRNs also resembled RN neurons (Fromm et al., 1981). Again, this difference between CRNs and PTNs is of particular

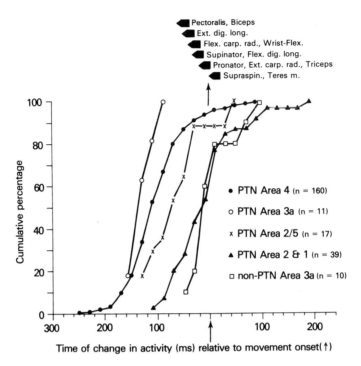

FIG. 4. Neural onset times are plotted as cumulative percentage of the total number of neurons in each group, as indicated in parentheses. For each unit the earliest clear change of frequency (decrease or increase), was measured. Median values of onset times prior to supination–pronation movements were 123 msec for area 3a PTNs, 106 msec for area 4 PTNs, 65 msec for PTNs in area 2/5 (including 13 PTNs in a boundary zone between areas 2 and 5 and 4 PTNs clearly in area 5), 12 msec for PTNs of area 2 and 1 (consisting of 36 PTNs in area 2 and 3 in area 1), and 7 msec for non-PTNs in area 3a. Except for the two latter populations, the distributions were significantly different from each other (t-test) ranging from $p < 0.005$ between area 2/5 PTNs and PTNs in area 2 and 1 to $p < 0.025$ for the timing difference between PTNs in area 3a and area 4. All data from monkey B. (From Fromm and Evarts, 1982, with permission.)

significance for penetrations in which both kinds of neurons were recorded simultaneously or at least in close proximity (Fig. 5, left diagram). Of 23 PT-CRNs, 12 were not influenced by load changes, whereas the remaining 11 units exhibited nonmonotonic relations such as shown in Fig. 5 (open circles, right diagram) or increased in several steps over the entire load range. The PTNs usually showed the most marked changes of frequency for relatively small load changes close to the shift point of torque direction, i.e., near zero load. By finely grading load near the point of reversing load direction, it was possible to demonstrate a sigmoid-shaped relationship of PTN activity to load, this relation being similar to that seen in motor units (Bigland and Lippold, 1954b; Clamann, 1970; Milner-Brown et al., 1973c; Desmedt and Godaux, 1977a). The PTNs in sensorimotor cortex were found to have a roughly proportional increase of firing rate over a rather limited load range, discharge rate commonly reaching a plateau with further increase of loads (cf. Fig. 5 left, Fig. 6A, Fig. 7B). It is concluded that the PTN outputs are especially important in controlling, besides other target neurons, the early recruited portion of MNs for fine adjustments in the range of low forces.

The findings may help to reconcile some apparent discrepancies in previous studies

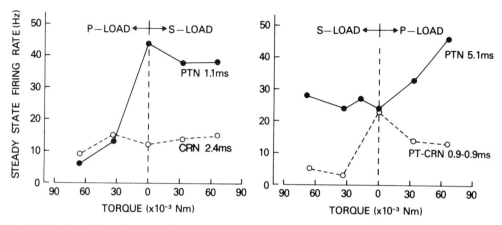

FIG. 5. Relation of steady-state firing rate to static load for the same two pairs of simultaneously recorded neurons as in Fig. 1. Monkey was required to keep the handle in vertical position during variation of torque. Each point represents averaged firing rate over 2-sec periods calculated from at least 15 trials and measured 2 sec after load change. P-load means torque opposing pronation; S-load requires net force in the supinators.

(Evarts, 1969; Schmidt et al., 1975; Smith et al., 1975). Hepp-Reymond et al. (1978) demonstrated that it was essential to use small variations of static force to reveal a linear relation between force and discharge frequency over a restricted range of loads. In the earlier reports, quite large steps of loads were used so that the units rapidly entered the saturation range (Evarts, 1969) and appeared as "on-off" cells when the direction of force shifted (Schmidt et al., 1975). A monotonic relation to load was found in 6 of 7 PTNs in area 3a, in about half of PTNs in area 4 and in junctional areas 2/5, and in 28% of PTNs in area 2. The incidence of such force relations does not necessarily reflect differences across these cortical subdivisions. For motor cortex PTNs the occurrence of a monotonic relation to torque was found to correlate with directionally specific changes of discharge frequency with the active supination-pronation movements and, furthermore, to depend on the extent to which a unit was related to the active movement. In this context, the findings of Cheney and Fetz (1980) are most revealing. Neurons for which the peripheral destinations ("muscle fields") could be established; i.e., corticomotoneuronal cells, showed monotonic relations to magnitude of load. There is, however, no evidence that postcentral PTNs have monosynaptic connections with MNs (see Discussion section). Nevertheless, their load dependence and sensitivity were comparable to that of wrist flexion-extension related corticomotoneuronal cells which ranged from 250 to 480 Hz/nm (Cheney and Fetz, 1980). We measured the increment in firing rate per increment in static load over the linear range and found the steepest rate-torque slopes in PTNs of area 3a (mean: 560 Hz/nm), followed by PTNs in the motor cortex and junctional area 2/5, which had similar mean slopes in common (about 300 Hz/nm). PTNs in area 2 had the lowest rate-torque slopes (mean: 170 Hz/nm). These load sensitivities, therefore, might indicate some gradual differences between cortical areas. For motor cortex PTNs, there was a correlation between recruitment threshold for tonic firing and antidromic latency. Slower conducting PTNs tended to maintain a certain, although decreased firing rate at loads opposing their "antagonist" muscles at which the faster conducting PTNs tended to be silenced (cf. both PTNs in Fig. 5). Higher recruitment thresholds for the

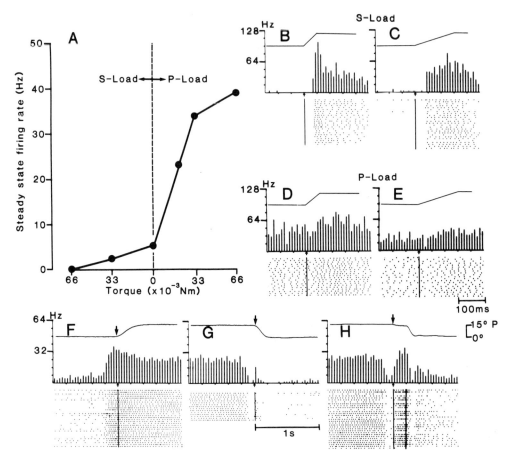

FIG. 6. For a PTN (antidromic latency: 1.8 msec) in area 3a, this figure illustrates relation to magnitude and direction of torque (A), responses to passive ramp displacements (B–E), discharge with active pronation (F), and unperturbed (G) and stopped (H) supination movements. Same arrangements of displays as in previous figures, with alignment of unit discharge on movement onset (*center lines* and *arrows*). B–E: Pronating ramp-and-hold displacements (15°) with different ramp velocities (300°/sec in B and D; 100°/sec in C and E) were imposed on different static loads: 0.066 nm torque loading the supinators (S-load) in B and C, and at loads requiring pronator activity in D (0.066 nm) and E (0.033 nm). Thus, the dynamic–static response to stretching the supinators (with the dynamic component depending on ramp velocity) was greatest when force was being exerted by the supinators and almost abolished for the other load direction. In parallel, the response to stopped supinations (H, stop-period of 200 msec marked by *two vertical bars*), which was recorded at a small torque (0.018 nm) loading supinators was greatly diminished when the pronators were loaded (not shown). Notice histogram ordinates, bin widths (10 msec), and time (100 msec) in B–E being different from F–H (bins: 40 msec, time: 1 sec). (From Fromm and Evarts, 1982, with permission.)

larger PTNs in a given penetration were particularly obvious when clusters of adjacent PTNs with different antidromic latencies could be recorded. This would support the view that static muscle force is encoded by gradation of frequency as well as by recruitment of PTNs according to cell size.

The EMGs of prime movers increased with torques loading them and decreased with torques unloading them over the range of forces tested. The EMG also varied with the isotonic movements performed with a small constant torque (0.018 nm) and as a function of steady-state joint position. In

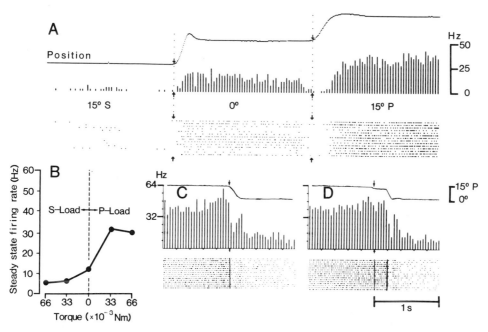

FIG. 7. Relationship of a PTN (antidromic latency 6.4 msec) in an area 2/5 border region to the three joint positions (A) and to load (B). Same arrangements of displays and conventions as used in the preceding figures. Time and bin widths (40 msec) apply to all displays. Peculiar to neurons located in this region was that the early reciprocal changes of activity (in this unit, decrease prior to pronation in A and increase prior to supination in C were opposite to the later changes occurring with attainment and maintenance of joint position; this unit shows a congruent relation for position and force. Other units of this area could show a discordant relation; for example, early increase prior to supination, decreased discharge frequency with supinated joint position, but enhanced activity when supinators were loaded. Notice that the later decrease of frequency (C) was prevented for the period of movement-stop (D).

accordance with the length-tension properties of muscle, even with external load held constant, EMG activity was thus greater for a position in which the muscle was shorter. Similarly, the area 3a PTN (Fig. 6) discharged phasically before and with pronation (F), maintained tonic discharge for the 15° pronated joint position (F, left part of G, and H) and increased with loads requiring pronator force during maintenance of 0°-vertical position (A). In a reciprocal manner, this unit became silent either at the supinated position or with increasing loading of the supinators at the vertical handle position. This congruent relation to load and position consistent with the muscle activity pattern, was the rule for reciprocally related PTNs in motor cortex (PTN 1.1 msec in Figs. 1 and 5), area 3a, and also for the majority of PTNs in area 2.

Pyramidal tract neurons in the area 2/5 junction differed from PTNs in other areas in showing the opposite sign of frequency change prior to and during the initial active movement as compared to the subsequent discharge associated with acquisition and maintenance of joint position (Fig. 7A, C). The unit appeared to provide information on the joint position, having decreased discharge frequency for the supinated and increased activity for the pronated position. This late position-related discharge is shown to depend on afferent input, since it does not occur as long as the arm is being prevented from moving (cf. C and D). The relation to load and position could then be either congruent (seen in 25% of these units, see PTN in Fig. 7), or discordant (seen in 75%, see legend Fig. 7).

Responses to Passive Displacement and to Stop of Active Movement

To assess afferent feedback to PTNs during movement execution, movements were sometimes brought to a brief halt. In addition, we compared neuronal responses to afferent input caused by pronating and supinating displacements of the handle during holding. For the muscles, arrest of movement exaggerated the changes of activity that would have occurred with both directions of the unperturbed movement. Thus, when active shortening of a muscle was obstructed there was an enhancement of EMG activity beginning at 20 msec and continuing throughout the stop period. Conversely, when a movement involving lengthening of a relaxed muscle was brought to a halt, the stop caused an even further decrease of EMG. For the majority of area 4 PTNs, the responses to stop were analogous to those in muscle. Increased discharge frequency associated with a particular direction of movement was further enhanced by stopping that movement.

In contrast, the PTN illustrated in Fig. 6 exhibited decreased activity with the control movement (G) but was caused to have increased activity during arrest of movement with a buildup of discharge lasting throughout the stop period (H). In general, these responses to halt of movement were directionally specific, correlated with the response to passive ramp displacement, and depended on the background force. For example, the excitatory response when shortening of supinators was being hindered (Fig. 6H) correlated with the reflex excitation following passive stretch of supinators (by pronating ramp, B). This unit was neither influenced by the supinating ramp nor by stop of pronation movements. In addition, response of PTNs to the halted supination was greater for supinator than pronator loads in parallel with the changed response to the ramp (cf. Fig. 6, B and D). Analysis of trial-to-trial variability (as seen in H) revealed that the intensity of these responses was positively correlated with the amount of effort of the impeded movement (speed and force development). It is important to note that the stop response of the 3a PTN shown in Fig. 6 differs from those in muscles and also in the majority of PTNs in area 4. This difference might be due to a different input-output relation, implying that this 3a PTN received feedback from antagonist muscle stretch receptors. Such input-output relation and directional pattern of stop response was found in half of area 3a PTNs but only in a minority of PTNs in other areas.

A fundamentally different type of stop effect was found exclusively in postcentral neurons. In the unit of Fig. 7, the late decrease of firing rate after onset of the control movement (C) was delayed for the period of stop (D). Such delay of late discharge until removal of stop was observed in PTNs and non-PTNs chiefly located in areas 1 and 2 and in non-PTNs of area 3a. This delay effect of stop is consistent with the idea that these neurons were driven by activation of kinesthetic receptors in the moving limb, an idea that is also supported by the sameness of their activity with active and passive (ramp) movements. In contrast, arrest of movement did not lead to this delayed discharge in PTNs of motor cortex or area 3a.

For area 3a, response latencies to afferent input produced by the ramp displacement were inversely related to the timing of the same neurons in relation to active movements (see Fig. 4). Latencies to ramp displacement were shortest in non-PTNs of area 3a (mean: 12 msec), which was significantly earlier than for PTNs of the same area (mean: 20 msec). There were no clear latency differences between different subgroups of SI PTNs, but response latencies of postcentral PTNs as a group (mean: 22

msec) were shorter than those of precentral PTNs (mean: 26 msec). Dynamic ramp responses increasing with velocity of ramp and depending on background load were most pronounced in area 3a, but also occurred to a lesser extent in motor cortex and posterior SI. Excitatory ramp responses usually increased with increasing firing rates due to load, but the reverse also occurred, especially in area 3a PTNs (Fig. 6, B–E). Across these cortical subdivisions there was a significant tendency for large fast conducting PTNs to show dynamic ramp responses, while dynamic-static and purely static responses prevailed in slower conducting PTNs (Table 1). Directionally specific changes of frequency with passive and active movements predominated in PTNs of areas 4, 3a, and posterior SI,

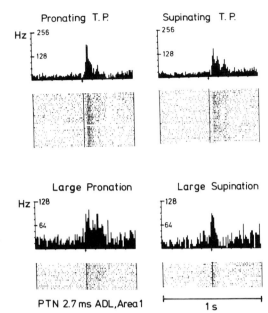

FIG. 8. Directionally nonspecific, late discharge of a PTN in area 1 with fast large (20°) supination-pronation movements (monkey A). Unit was excited for both directions of torque pulse applied during steady holding of the handle. *Top center line* displays indicate onset of torque pulse (TP, 50 msec-rectangular pulse, 5° amplitude); rasters and histograms (bin: 10 msec, notice different ordinate scale) of bottom displays were aligned on onset of movement.

TABLE 1. *Relation between antidromic latency and type of response to passive ramp-and-hold displacement*

Antidromic latency	Ramp response			
	Dynamic	Dynamic–static	Static	Total
Precentral PTNs				
< 2.0 msec	64	23	5	92
≥ 2.0 msec	14	29	9	52
Postcentral PTNs				
< 2.0 msec	9	2	0	11
≥ 2.0 msec	18	22	9	49

The 144 precentral and 60 postcentral PTNs have been arbitrarily categorized as rapidly (antidromic latency < 2.0 msec) and slowly conducting (antidromic latency ≥ 2.0 msec). Units that responded only to the ramp phase were classified as dynamic, whereas units with higher discharge rates during ramp than hold phase of displacement were classified as dynamic-static. Units were classified as having static responses if their firing rates during the ramp were less than or equal to those associated with maintained position. Besides showing a correlation between antidromic latency and type of response to passive displacement, these data indicate the preponderance of pure dynamic responses (total of 105 PTNs) as compared to dynamic-static ($N = 76$) and static ($N = 23$) responses and the relative lack of fast-conducting PTNs in postcentral areas.

whereas in more rostral parts of SI (3b and 1), directionally nonspecific responses prevailed. Figure 8 illustrates the typical late, phasic discharge with both directions of large movement and the bidirectional excitation caused by the perturbation. Phasic ramp responses in these rostral areas of SI were usually not correlated with ramp velocity. The incidence of short (< 50 msec) latency responses to passive displacements was lower in CRNs (23%) as compared to PTNs (62%) recorded in the same penetrations, and the CRN responses were rather weak. Activity of CRNs did not change by stopping the movement. However, afferent responses of PT-CRNs did not appear to differ from those of motor cortex PTNs and occurred in about the same proportions.

DISCUSSION

The similarities between PTNs in motor cortex and SI transcend the differences that might have been expected on the basis of their location in areas as functionally different as the primary motor and somatosensory cortex. In contrast, the differences between motor cortex CRNs and PTNs (and partly also PT-CRNs) were abundantly clear, even for neurons recorded in the same penetrations. These differences between CRNs and PTNs were seen in time course of discharge with respect to onset of voluntary movement, relation to steady-state force and to angular joint position, short-latency reflex responses to passive limb displacements, and responses to arrest of movement. These numerous differences were sufficiently clear to warrant a substantially different role for the CRN and PTN outputs from motor cortex. In fact, the corticorubral projection neurons seem to have features more in common with their target RN neurons (Fromm et al., 1981) than with the PTNs recorded in their immediate neighborhood. Our results fail to provide evidence for any strong coupling of CRNs and RN neurons with the periphery, for we did not observe many cases of short-latency somatosensory feedback. We also did not find a relationship between CRN and RN neuron discharge and steady-state posture, load, or concurrent EMG activity. It seems possible that CRNs and RN neurons may encode signals relevant to the termination of voluntary movement, an idea that is consistent with their late peaking during the terminal phase of movement. In contrast, many pre- and postcentral PTNs are undoubtedly involved in the initiation phase of movement, with the initial burst of activity from the corticomotoneuronal cells within the PTN population representing the decisive descending signal to start the movement. The PTNs receive short-latency afferent feedback, which is of particular importance during control of ongoing movement. Moreover, PTNs are closely coupled with EMG activity, thus being related to changes in muscle activity necessary to maintain a given level of force at a given joint position.

We have noted elsewhere (Fromm et al., 1981) that a key to understanding the role of the corticorubral system may lie in a differential phylogenetic development in the primate of RN and its interconnected structures and in the increasing importance of inputs to RN from nonprimary cortical area 6 at the expense of direct inputs from the periphery (see also Allen et al., 1977). The collateral influences on the magnocellular part of RN (deriving strictly from PTNs of ipsilateral area 4) appear to recede quantitatively in favor of the CRN projection (Humphrey and Corrie, 1978), which is mainly directed toward the parvocellular RN and originates not only from area 4, but also from premotor area 6 and bilaterally from the supplementary motor area (SMA) (Catsman-Berrevoets et al., 1979; Hartmann-von Monakow et al., 1979). Interestingly, the recent findings on SMA single units in the awake monkey produced a number of negative results, i.e., a lack of any clear relation to muscle force and a sparseness of afferent inputs (Brinkman and Porter, 1979, this volume; Smith, 1979; Wise and Tanji, 1981a; Tanji, this volume). These features of SMA neurons are reminiscent of the features in CRNs of area 4.

Our findings point to the need for caution in equating the function of the motor cortex with that of its PTNs. We emphasize this point because the sampling bias toward large cells in chronic preparations (Towe and Harding, 1970; Humphrey and Corrie, 1978) means that the behavior of the large-spike neurons (and thus of large PTNs) has tended to dominate the picture of neuronal activity recorded from motor cortex. We remain relatively ignorant of the many small-sized cells, and there is a need for future studies to identify other corticofugal neurons.

The differences, but also the striking similarities between PTNs in area 4, area 3a, posterior SI, and area 5 with respect to responses to ramp displacement and movement-stop and possibly also with respect to certain features of their dual load and position sensitivity might, in part, be attributed to the submodality distributions within these cortical areas. All of these postcentral subdivisions and the motor cortex are known to receive inputs from muscle receptors but to a different degree and with varying contributions from group I and II muscle receptors and with varying convergence from joint receptors (Mountcastle and Powell, 1959; Powell and Mountcastle, 1959; Phillips et al., 1971; Burchfiel and Duffy, 1972; Schwarz et al., 1973; Mountcastle et al., 1975; Hore et al., 1976; Lemon and Porter, 1976; Fetz et al., 1980; Soso and Fetz, 1980). Pyramidal tract neurons in the rostral part of SI [areas 3b and 1, which receive cutaneous inputs; (cf. Powell and Mountcastle, 1959; Paul et al., 1972; Tanji and Wise, 1981)] seem to constitute the most distinct group. This does not necessarily imply a general difference from motor cortex PTNs. Recent work points to foci for different modalities within the precentral gyrus, with the caudal part receiving cutaneous inputs and the rostral part of motor cortex receiving "deep" inputs (Strick and Preston, 1978, this volume; Lemon, 1981; Tanji and Wise, 1981); most of our recordings were obtained from the rostral part of the motor cortex. Furthermore, cutaneously activated neurons in caudal motor cortex and areas 3b-1 were more likely to be directionally nonspecific in relation to active and passive movements and to show phasic discharge that did not correlate well with velocity of displacement, all in contrast to neurons in rostral motor cortex (Lamour et al., 1980; Wise and Tanji, 1981b).

Discharge prior to voluntary movement cannot, *a priori*, be expected to be related to modality segregation in different cortical fields. Early discharge, at least that occurring prior to onset of EMG activity, is commonly thought to originate from centrally programmed input, whereas the later discharge, once the movement is in progress, might reflect interaction of both peripheral and central inputs. Early discharge appears to be a feature shared by PTNs in area 4 with their counterparts in area 3a and junctional area 2/5. Besides other possible sources, thalamocortical connections or corticocortical channels from motor cortex (Jones and Powell, 1969; Jones et al., 1978; Vogt and Pandya, 1978; Friedman and Jones, 1981) could provide such a central input by which PTNs in the respective postcentral areas may be informed of a forthcoming movement, thus adjusting their target neurons in advance of movement. Incidentally, PTNs in these areas have in common a mismatch response to altered afferent feedback caused by the stop of movement. In contrast, the later discharge with active movement observed in the remaining large parts of SI (Fig. 4), together with the evidence that this change of frequency did not occur as long as the limb was prevented from moving, suggests that the vast majority of PTNs in areas 1 and 2 can be regarded as pure "sensory" neurons. Inputs from joint and skin receptors may completely account for their activity during voluntary movement.

In one area (3a) at least, we found a clear functional differentiation between non-PTNs and PTNs. The non-PTNs of area 3a seemed to function as sensory receiving neurons; they were influenced at the shortest latency by afferent stimulation but were the last to be activated in the course of voluntary movement. The behavior of PTNs in area 3a can be best understood by assuming that they receive a central input in addition to an input from the periphery. Evidence for such dual inputs has been given in the two other reports on single unit recordings from area 3a in the behaving monkey, but these reports did not provide information with respect to timing and identified PTNs (Yumiya et al., 1974; Tanji, 1975). Our observations gave no clear evidence for differences between PTNs and non-PTNs in other fields of SI; this failure might be in part due to the fact that a number of the so-called non-PTNs may have been PTNs evading antidromic identification.

The concepts of corollary discharge and efference copy (Sperry, 1950; von Holst and Mittelstädt, 1950) involve the notion of centrally programmed discharges into primary sensory centers such that during a self-generated movement the activity in these centers will result from the summation of peripheral and central inputs. The behavior of some postcentral PTNs could be interpreted by these concepts (see Fig. 7), but regardless of the question of central versus peripheral factors, the dual relation of many SI PTNs to load and position, for example, indicates that the CNS must interpret the joint position signals of these neurons in the light of associated levels of muscle tension. One should also note Granit's (1972) statement: "All these terms (sense of effort, corollary discharges, etc.) were invented before it was known that the periphery itself is 'corollarized' by alpha-gamma linkage to one of our most highly developed sense organs (the muscle spindle) which also projects to the cortex."

The theory of servo-assistance of movement (Phillips, 1969; Matthews, 1972; Granit, 1975) has provided a framework that fits with the type of cortical stop response that was similar to the muscle response. During movement, a balanced coactivation of alpha-MNs and gamma-MNs (innervating the spindle organs) means that fusimotor action can compensate for the unloading effect on spindle due to extrafusal shortening; if an unexpected halt to shortening occurs, the unloading effect will be reduced and spindle afferents will signal misalignment by an increase of discharge rate because of the persisting fusimotor action. As further speculated by Phillips (1969), this signal should not only reach the lower but also the upper motoneuron, the PTN. Regarding the behavior of spindle afferents in man and awake animals, this concept has been supported (Vallbo, 1973; Burke et al., 1978a, b; Hulliger and Vallbo, 1979; Prochazka et al., 1979). Although our results are in agreement with Marsden et al. (1972, 1976a, *this volume*), who showed that the gain of the resulting servo action depends on the background load and the effort of movement, some words of caution are in order here, especially with respect to the input–output pattern shown in Fig. 6. As to the afferent side of the loop, involvement of Golgi tendon organs cannot be excluded. Furthermore, the sign of any such transcortical feedback loop depends crucially on the efferent limb and type of target neuron. Finally, the role of transcortical reflexes in load compensation [even in those PTNs exciting alpha-MNs, for example Sakai and Preston (1978)] has been questioned (Ghez and Shinoda, 1978; Miller and Brooks, 1981), and it has been suggested that such reflexes are for corrections of small internally generated irregularities during precise movements rather than for responses to external disturbances (Fromm and Evarts, 1978; Evarts and Fromm, 1981).

Anatomical data favor the view that signals descending from SI PTNs modulate transmission of ascending somatosensory information, whereas motor cortex PTNs have relatively more direct access to spinal MNs (Kuypers, 1960, 1973; Liu and Chambers, 1964, Coulter and Jones, 1977). However, data from fiber tracing (Coulter and Jones, 1977) suggest that corticofugal fibers from area 3a terminate near the spinal motor nuclei. The results of Koeze et al. (1968) point to a certain independence of motocortical control of alpha- and gamma-MNs, but unusually high currents were necessary to drive gamma-MNs by cortical surface stimulation (Clough et al., 1971). These findings lead to the speculation that the major corticofugal control of gamma-MNs might actually be exerted by the region buried in the floor of the central sulcus, the very area that receives the main feedback information from spindle afferents. The scanty, often absent muscular effects by intracortical microstimulation in the vicinity of PTNs recorded from area 3a (a common feature in all postcentral areas) contrasted to the abundance of low-threshold microstimulation effects evoked in the adjacent area 4 [see also Wiesendanger and Sessle (1980)]. This suggests that area 3a must exert its effects on structures other than alpha-MNs, and gamma-MNs would seem to be a likely alternative. Most important, human muscle spindle activity has been shown to reflect isometric force as well as the time course of load-bearing movements (Vallbo, 1973; Burke et al., 1978b, c; Hulliger and Vallbo, 1979), both of which are reflected in the descending signals of PTNs in area 3a (cf. Fig. 6). The finding that the firing rates of human spindle afferents did not vary with different joint positions during active holding against a load or were even somewhat higher at the shorter muscle length (Vallbo et al., 1981) becomes plausible in view of the behavior of PTNs in area 3a, if indeed they drive fusimotor neurons. Area 3a may then be able to extract information on steady joint position by evaluating spindle discharge and command signal to the fusimotor neurons.

Again, however, we must avoid advancing any model of a transcortical feedback loop involving these PTNs, because the significance of such feedback depends primarily on the peripheral destination of the PTN's axon, and this information is not available. Thus, it is premature at present to discuss the relative importance of sensory as compared to motor roles of SI PTNs, since the problem of their subcortical projections is far from being settled.

ACKNOWLEDGMENTS

I am most grateful to Dr. E. V. Evarts for his invaluable assistance and advice in all stages of this work. The author received a Heisenberg-fellowship from the Deutsche Forschungsgemeinschaft.

Separate Cortical Systems for Control of Joint Movement and Joint Stiffness: Reciprocal Activation and Coactivation of Antagonist Muscles

*Donald R. Humphrey and Dwain J. Reed

Laboratory of Neurophysiology, Emory University School of Medicine, Atlanta, Georgia 30322

In most studies of the voluntary control of the arm and hand, attention has been focused upon simple flexion and extension movements about single joints (e.g., Evarts, 1966; Stark, 1968; Humphrey et al, 1970; Thach, 1970; Lee and Tatton, 1975; Polit and Bizzi, 1979; Fetz and Cheney, 1980). Typically, such movements involve a pattern of activation of agonist muscles, with simultaneous suppression of the antagonists. Thus, the antagonist motoneuron (MN) pools that control the position of the joint are activated (or inhibited) reciprocally, in the manner widely presumed to occur during normal movement since the pioneering studies of reciprocal innervation by Sherrington (1906).

It is clear from simple observation, however, that there are also many normal forms of motor activity in which antagonist muscles are activated simultaneously. In many of these cases, antagonist co-contraction serves to increase the apparent mechanical stiffness or impedance of the controlled joint, thus fixing its posture or stabilizing its course of movement in the presence of external force perturbations. The existence of antagonist co-contraction in normal motor activities was clearly recognized by some early investigators (e.g., Tilney and Pike, 1925), but over the past several decades it has received remarkably little experimental attention (see review by Smith, 1981). More recently, however, a small number of investigators have begun to recognize that this pattern of muscle activation is both more pervasive and significant than has previously been assumed. Recent human behavioral studies, for example, have suggested that a wide range of normal motor activities are generated by a weighted combination of control signals from two partly independent central systems: one organized for reciprocal activation of antagonist muscles, and another for their co-activation (Feldman, 1980a,b).

This chapter presents the first direct neurophysiological evidence for the existence of these separate central systems, and discusses the major features that such a form of organization provides the mammalian motor system. In monkeys trained to control the position of the wrist in the presence of perturbing forces, we have noted that two different modes of control are used, depending on the speed or frequency of the perturbation. With slow perturbations,

---

*To whom correspondence should be addressed: Department of Physiology, Emory University School of Medicine, Woodruff Medical Center, Atlanta, Georgia 30322.

wrist position is controlled by graded activity in the particular set of muscles whose action opposes in direction the applied force; when the applied force alternates in direction, reciprocal activity occurs in the antagonist muscles which act about the joint. With rapid perturbations, however, the joint is also tonically stiffened by antagonist co-contraction. Thus, by changing the frequency of the applied perturbation, we have been able to change the balance of strategies used by the animal to control the position of his wrist, and to thus examine the properties of the reciprocal and coactivation systems. Simultaneously, we have observed the discharge of single neural elements both at the level of the MN pool and within the contralateral motor cortex. Our results suggest the important new concept that separate neuronal systems exist within the motor cortex for the generation of these two major patterns of muscle activity, with their output signals converging onto common elements at the level of the MN pool.

METHODS AND RESULTS

1. Features of the Motor Task

The major features of our behavioral task are shown in Fig. 1. Small monkeys, seated in restraint chairs, are trained with operant conditioning procedures to grasp a small, vertically oriented handle, that is free to rotate about the axis of wrist flexion and extension. To obtain a fruit juice reward, the animal must maintain his wrist within ± 2.5 to 4.0° of a fixed position, despite torque perturbations that are applied to the handle in the direction of wrist flexion or extension by an electronically controlled torque motor. Because the animal must sense, track, and null an imposed torque, we refer to this behavioral paradigm as a torque tracking task.

Extraneous movements of the animal's fingers and upper arm are minimized during task performance by two methods. First,

FIG. 1. A trained monkey performing the torque tracking task. The torque motor used to perturb the wrist is enclosed in the black housing with angular position scale. Attached to the upper shaft, enclosed in a transparent plastic housing, is the potentiometer used to measure angular wrist displacement. The net flexor-extensor torque exerted by the animal is detected by balanced strain gauges, located on the horizontal shaft of the handle assembly. The animal's gaze is fixed on a small light display, which indicates to the animal the correct position of its wrist.

the animal must firmly grip the handle in order to compress an enclosed, spring-loaded microswitch, and thus initiate the perturbation and reinforcement cycles. Second, the arm is restrained in a padded, contoured rest, which flexes the elbow to 90° and aligns the forearm and wrist with the handle assembly. Together, these procedures promote a firm and reproducible grip—minimizing finger movements—and an upper arm posture that minimizes activity in non-task-related muscles. Electromyograph (EMG) recordings from biceps, triceps, and deltoid muscles have verified that modulation of their activities is not related to cyclic variations in wrist torque, in the well-trained animal.

In the experiments summarized here, two basic torque perturbations were used: (1) a series of constant torques (5–10-sec duration), ranging from 250 to 1,250 g/cm in the flexor and in the extensor directions; and (2) a sinusoidal or triangular waveform, repeated at a frequency of 0.1 to 1.5 Hz, and varying from 1,250 g/cm opposing flexion to 1,250 g/cm opposing extension in peak-to-peak amplitude. These torque amplitudes are approximately 60 to 80% of the torques that monkeys weighing between 4 and 7 kg can exert steadily about the wrist for a period of 1–2 sec. To the human observer performing the same task, the frequency band used ranges from a very slow perturbation, which is easily sensed and nulled by conscious effort in a graded, continuous manner, to one that exceeds the upper limit for precise, voluntary tracking. Figure 2 shows examples of various motor performance variables recorded from one monkey during task performance.

Mechanical Impedance of the Wrist Under Passive Conditions and During Active Joint Position Control

Since the animal must maintain a nearly constant wrist position in the presence of perturbing forces, we, in effect, require it to maximize the apparent mechanical impedance of its wrist over a range of perturba-

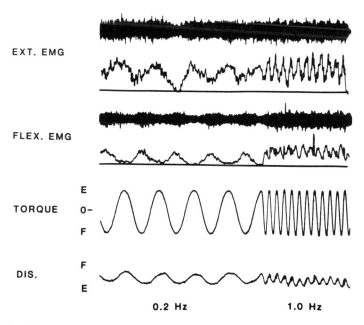

FIG. 2. Examples of motor performance variables recorded during task performance. Each double row of EMG traces shows the unprocessed recording obtained with surface electrodes overlying the major flexor and extensor muscles of the wrist, and immediately below this a rectified and low-pass filtered version of the same recording (second-order filter, critically damped, 50-msec time constant). The net wrist torque generated by the animal is also shown, with net extensor torque (E) upward. The applied torque was sinusoidal, ranging from 1,000 g/cm opposing flexion to an equal value opposing extension. The wrist displacements (DIS) produced by the applied torque are shown in *bottom trace;* peak-to-peak displacement at 0.2 Hz is ± 3°. Note the decrease in wrist displacement, indicating an increase in wrist stiffness, when the perturbation frequency is suddenly increased from 0.2 to 1.0 Hz. This increased stiffness is produced by an abrupt increase in tonic coactivation of both flexor and extensor muscles, as can be seen in the filtered traces.

tion frequencies. It is clear, however, that a fraction of this impedance will be due to passive physical factors: the inertia of the handle assembly and the animal's hand, and the viscoelastic properties of its wrist and its controlling muscles. In order to know the relative importance of neural input to the muscles in the control of wrist position, we must know, therefore, how both the total impedance of the joint and its passive component vary over the range of perturbation frequencies used. The contribution of neurally generated muscle activity may then be estimated by subtraction. We obtained this information in the following way.

First, we required the alert monkey to track a sinusoidally varying torque perturbation, while maintaining a displacement error of $< \pm 3$ degrees. The perturbation varied in amplitude from 1,000 gcm opposing flexion to 1,000 gcm opposing extension, and over the frequency range from 0.2 to 2.0 Hz. From measures of peak-to-peak wrist torque (ΔT) and angular joint displacement ($\Delta 0$), we computed the in-phase component of wrist impedance, i.e., wrist stiffness ($= \Delta T/\Delta 0$), as a function of the applied torque frequency. This computed function, which defines total wrist stiffness as a function of perturbation frequency, is shown in the top curves of the upper panel in Fig. 3. The phase angle between wrist torque and displacement was also computed for one experiment, and is shown by the curve labeled active in the lower panel.

Next, we tranquilized the animal with ketamine hydrochloride (10 mg/kg) or sodium pentobarbital (10–15 mg/kg), taped his

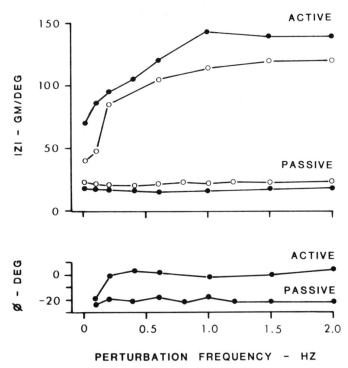

FIG. 3. Mechanical impedance of the wrist in the tranquilized and actively tracking monkey. Results are shown from two different experiments with 2 monkeys (*open circles,* experiment 1; *closed circles,* experiment 2). Top: magnitude of wrist impedance for the flexion-extension axis. Bottom: phase angle between wrist torque and displacement; a negative angle indicates that displacement lags torque. Note the increasing magnitude of wrist impedance in the actively tracking monkey as the perturbation frequency is increased. In contrast, the magnitude of wrist impedance is constant from 0.2 to 2.0 Hz in the passive or tranquilized monkey.

hand in griplike posture to the handle, and perturbed his wrist with a sinusoidal torque just sufficient in amplitude (100–200 gcm) to produce displacements similar to those seen in the alert and actively tracking state. The EMGs of the wrist muscles were essentially flat in this case, indicating that true passive conditions prevailed. From these torque-displacement measures, we then computed the passive component of wrist stiffness as a function of frequency; the amplitude of this component is shown in the bottom curves of the top panel in Fig. 3. The phase angle between torque and displacement was also computed (passive curve, lower panel in Fig. 3), and was found to be essentially constant over the frequency range studied. An apparent displacement phase lag of 20 degrees was in this case artifactual, and due to a departure of the displacement waveform from a sinusoidal shape.

From these and similar data, we drew the following conclusions. (1) Wrist stiffness, the reciprocal of wrist compliance, is the dominant contributor to total wrist impedance, accounting for 90 to 95% of total joint impedance under both active and passive conditions over the frequency range studied. Thus, the inertia of the animal's hand and the viscoelastic properties of the joint and its muscular attachments contribute negligibly to joint position control over this frequency band. (2) Only 15 to 20% of the apparent stiffness of the wrist in the actively tracking is due to passive factors; moreover, the magnitude of this component remains constant as the perturbation frequency increases up to 2.0 Hz. The major contribution to wrist stiffness comes instead from neural activation of the flexor and extensor muscles, an activation which rises progressively between 0.1 and 1.5 Hz. (3) This increase in stiffness is paralleled by an increase in response speed, as indicated by the decrease in the phase lag between wrist displacement and the compensatory torque applied by the animal (shown in Fig. 3, lower panel, active curve). This increase

in response speed could be due, in turn, to any one or a combination of the following factors. (1) An increased predictive response by the animal, so that the contractions of his wrist muscles were more effectively synchronized with the applied perturbations as they increased in speed. Such prediction could arise, for example, from a more effective activation of central motor structures by velocity-sensitive receptors in the wrist and hand, as they were excited more intensely in turn by the increasing velocity of wrist displacement. (2) An increase in the response speed of the muscular elements activated; for example, a recruitment into activity of faster-twitch motor units. Or (3), a tonic co-contraction of flexor and extensor muscles. By tonically stiffening the joint, the imposed torque would be opposed with essentially no phase lag.

We turn now to evidence from EMG recordings, which suggested that each of the factors 1 through 3 was involved over at least part of the range of perturbation frequencies in the animal's attempt to control the position of his wrist.

Firing Patterns of Wrist Flexor and Extensor Motor Units

Figure 4 shows examples of the processed EMGs recorded from the major flexor and extensor muscles of an alert monkey's wrist, as he tracked an applied torque, which varied sinusoidally at 0.2 to 1.5 Hz from a peak opposing flexion to a peak opposing extension. Examination of such records revealed that a number of motor strategies were used over this range of perturbation frequencies, in the animal's attempt to control the position of his wrist.[1]

With slow perturbations, wrist position was maintained by a graded, near-recipro-

---

[1]The terms motor strategy are used here to denote a mechanism used by the motor system to control a particular set of muscles, or the position of a joint; they do not imply that such mechanisms are consciously activated by the animal.

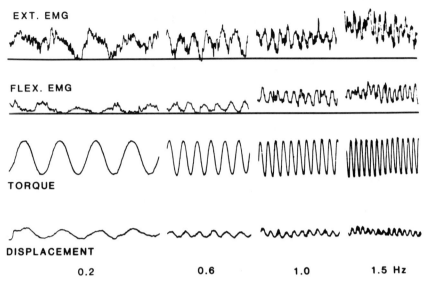

FIG. 4. Levels of reciprocal activation and coactivation of wrist flexor and extensor muscles as a function of perturbation frequency. The records are from an actively tracking monkey whose wrist displacements were kept below ± 4 degrees. Rows 1 and 2: Rectified and filtered EMGs obtained with intramuscular leads from the extensor and flexor carpi radialis muscles, at tracking frequencies that are indicated at the bottom of each column of traces. Row 3: Net wrist torque. Net extensor torque is upward. The torque ranges from 1,250 g/cm in the extensor direction to an equal peak in the flexor direction. Row 4: Angular displacements of the wrist. The peak-to-peak displacement at 0.2 Hz is ± 3.5 to 4.0 degrees.

cal activation of the flexor and extensor muscles (Fig. 4, EMG patterns at 0.2 and 0.6 Hz). As the perturbation frequency was increased from 0.2 to 0.6 Hz, the peak levels of muscle activation did not change; however, joint displacement decreased by 30 to 40% (cf. Fig. 4), indicating that wrist stiffness increased significantly. Clearly, this increase in joint stiffness must have resulted from a better synchronization of the animal's opposing muscle activity with the applied torque. With faster torque perturbations (0.6 to 1.5 Hz), two additional strategies were brought into play. First, the flexor and extensor muscles were co-activated tonically, as is shown in Fig. 4 by the increasing levels of the EMG traces above the reference baselines at 1.0 and 1.5 Hz. This co-activation contributed to a further increase in joint stiffness of about 15 to 20%. Second, the peak-to-peak amplitudes of the EMG modulations also increased, indicating that motor unit discharge was more synchronized, with unit potentials summing more effectively, or that motor units with larger spikes (and faster twitch tensions) were recruited into activity.

To examine more closely these frequency-dependent shifts in MN activity, we turned next to a study of the behavior of single motor units during task performance.

Recordings were obtained from 128 motor units in the flexor ($N=22$) and the extensor ($N=106$) carpi radialis muscles of 3 monkeys. The recordings were in sets of 2 to 4 units, with intramuscular, bipolar leads (1 to 2-mm tip separation, 25-µm tips); unit spike trains were separated by voltage-window discriminators. With 11 sets of 3 units each, we used spike-triggered averaging of the wrist torque record to obtain an estimate of each unit's twitch tension (cf. Mendell and Henneman, 1971; Stein et al., 1972; Desmedt and Godaux, 1977b; Kirkwood and Sears, 1980; Desmedt, 1981a). To characterize the properties of these units, a plot of relative twitch amplitude as a function of twitch duration is shown for the unit with the largest, the intermediate, and the smallest spike potential (Fig. 5). As described for other muscles in other species (Goldberg and Derfler, 1977; Tanji and Kato, 1973; Desmedt, 1981a), motor units with larger spikes tend to generate larger and more rapid twitches. Since staining of the wrist flexor and extensor muscles of the monkey for myofibrillar ATPase reveals at least two histochemically different classes of muscle fiber *(unpublished observations)*, the fast and slow-twitch units that we describe here may be analogous to the type FF and S units, respectively, described by Burke et al. (1973).

An an example of the task-related discharge of these fast and slow-twitch motor units, consider first the behavior of the set, the spike potentials and twitch tensions of which are shown in Fig. 6. During the generation of steady extensor torques of different amplitude, the order of unit recruitment was from small, slow-twitch unit to larger, faster-twitch unit; indeed, this order of recruitment was observed in 32 of 34 sets of recorded units, and was expected from the work of other investigators (Henneman et al., 1965; Milner-Brown et al., 1973; Freund et al., 1975; Desmedt and Godaux, 1977b, 1978c; Desmedt, 1981a). Somewhat unexpected, however, was evidence of a similar order of unit recruitment as a function of the speed or frequency of the torque perturbation, even though its amplitude was held constant.

An example of this type of frequency-dependent recruitment is shown in Fig. 7. At a perturbation frequency of 0.2 Hz, the firing

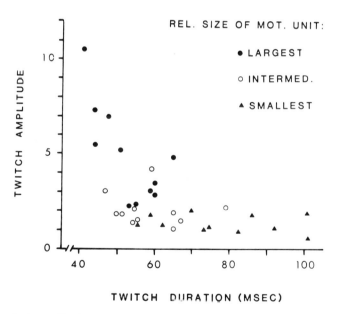

FIG. 5. Relative twitch amplitudes and durations for 33 single motor units with the extensor carpi radialis muscle. The data are from 11 sets of 3 simultaneously recorded units each. Measures of twitch amplitude are in arbitrary but linear units, with a value of 10 being approximaely 3 g. Amplitudes were measured from averages of 200 twitches for each motor unit. Twitch durations were measured between 10% peak amplitude points on the rising and falling phases of the averaged twitch. Separate symbols are used for twitches generated by the motor units with the largest, intermediate, and smallest spike potentials in each set. Note the 2:1 range in twitch duration, and the 10:1 range in amplitude.

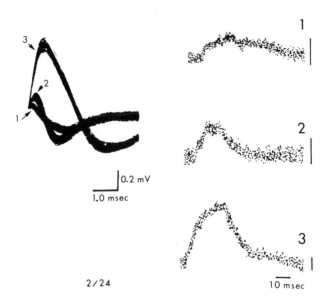

FIG. 6. Spike potentials and twitch tensions of three motor units recorded simultaneously from the extensor carpi radialis muscle. The spike potential of each motor unit (20 superimposed traces each) and its average twitch tension are designated by the same number. Twitches were averaged from the separated spikes of each unit during generation of a steady wrist torque; each tension record shown is the average of 200 consecutive twitches. Note the reduced gain for the twitch of unit 3; its amplitude is approximately 5 times that of the twitch of the smallest unit (number 1). *Vertical calibration:* 0.2 g.

rate of the slow-twitch unit (number 1 in Fig. 6) is modulated maximally: the rate declines to zero during peak antagonist (flexor) force, and reaches a maximum of 60 spikes/sec during peak activation of the parent extensor muscle (unit 1, 0.2-Hz column in Fig. 7). The large, fast-twitch unit fires only an occasional spike, however, and is essentially inactive at this perturbation frequency (unit 3, 0.2-Hz column). As the perturbation frequency increases from 0.2 toward 1.5 Hz, a significant change occurs in the pattern of discharge across the motor unit set. The slow-twitch unit begins to fire more tonically, with less rate modulation, and the fast-twitch unit is recruited progressively into a cyclic, modulated discharge. The decrease in discharge rate modulation in the slow unit appears to result from an increasing, tonic excitatory drive on the parent MN, which drives it toward firing-rate saturation. This same drive appears also to impinge on the fast-twitch unit, bringing it closer to threshold and allowing the modulated input to it to now be evident in its discharge pattern.

The generality of these frequency-dependent changes in motor unit activity is illustrated in Fig. 8, which summarizes the changes in firing-rate modulation as a function of perturbation frequency seen in the fast and slow-twitch units of the 11 sets (cf. Fig. 5). In these graphs, the extent of the firing-rate modulation during each torque cycle is indicated by the difference between the upper and lower curves. Note that rate modulation is maximal in the slow-twitch units at a perturbation frequency of only 0.2 Hz, and that it is eroded progressively by an increasing minimum or tonic rate (lower curve) as the perturbation frequency rises (Fig. 8, left). In contrast, rate modulation increases in relation to perturbation speed in the fast-twitch units (Fig. 8, right). When the units in the remaining 23 sets were ranked according to spike amplitude, a sim-

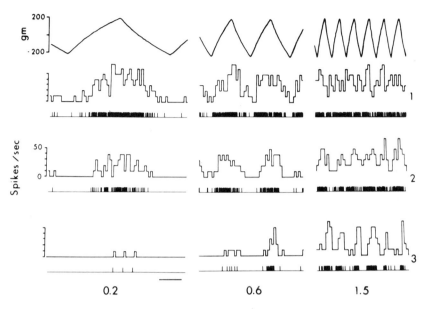

FIG. 7. Discharge patterns of slow and fast-twitch motor units at three perturbation frequencies. Records are shown for single torque cycles at perturbation frequencies of 0.2, 0.6, and 1.5 Hz. Net extensor torque is upward (peak of 200 g or 1,000 g/cm), and net flexor torque is downward (peak of −200 g). Each of the next three rows of two traces each show the spike train and the simultaneously computed firing rate of the unit whose number is shown to the right. The spike trains are those of units with the corresponding numbers in Fig. 6. The record of discharge rate lags the spike train by 100 msec, and is thus displaced slightly to the right. Perturbation frequencies are indicated at the bottom of each column of simultaneously recorded traces. *Horizontal bar* at 0.2 Hz: 1 sec.

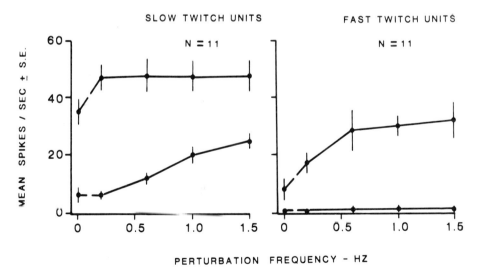

FIG. 8. Peak-to-peak modulations in discharge rate of slow- and fast-twitch motor units as a function of tracking frequency. The units are those whose twitch durations and amplitudes are plotted in Fig. 5. Each dot shows the average maximum discharge rate *(upper curve)* and average minimum rate *(lower curve)* seen at peak extensor and peak flexor force, respectively, during 20 consecutive torque cycles at the perturbation frequency indicated. *Vertical bars:* ± 1 SEM.

ilar pattern of activity was observed across presumed slow and fast-twitch units in 21 sets; thus, the findings summarized in Fig. 8 were quite consistent throughout the sample of motor units studied.

It is clear from these findings, therefore, that at different frequencies of joint perturbation, the control of joint position is accomplished by different patterns of MN discharge; these patterns and their contributions to the control of joint position are summarized in Table 1. In particular, the control signals that impinge on the MN pool at low and at higher frequencies of perturbation are quantitatively if not qualitatively different. At low frequencies, the input is graded and specific for the particular MN pools whose muscles oppose the direction of the applied force; as this applied force alternates, smoothly modulated activity occurs in a reciprocal manner in flexor and extensor MNs. At frequencies that begin to exceed those that are easily sensed and tracked voluntarily, an additional control signal is added, a tonic coactivation drive on both flexor and extensor pools. This tonic drive is evident most in the discharge of slow-twitch units, whose more sustained activity now contributes to a tonic stiffening of the joint. In addition, the dynamic stiffness of the joint is increased by a shift in the focus of modulated activity toward the fast-twitch units, whose faster twitch speeds decrease the response time of the tracking system and allow opposition of applied forces with less phase lag. The results also suggest that the modulated input to the MN pool may also increase slightly with increasing perturbation speed.

It was not clear from these data, however, whether these two general control signals impinged on the MN pool from separate sources, or whether they converged at some higher point in the motor system and arrived at the MN pool over a common set of pathways. To gain further information in

TABLE 1. *Summary of wrist flexor (F) and extensor (E) motor unit activity and its contribution to the control of wrist position at different perturbation frequencies*

Perturbation frequency (Hz)	Apparent wrist stiffness	Major patterns of motor unit activity	Major mode of control
0.1–0.2	50–80 g/degree; displacement lags the applied torque by 20–30 degrees.	Smooth modulation in the firing rates of slow-twitch motor units; unit populations in F and E muscles are activated reciprocally.	Graded, reciprocal. Wrist position controlled by moment-to-moment sensing and nulling of the imposed torque.
0.4–0.6	100–110 g/degree; phase lag decreases to near 0 degrees.	Same. The increase in wrist stiffness and the decrease in phase lag appear to be due to better synchronization of unit activity with the imposed torque.	Graded, reciprocal
0.6–1.5	120–140 g/degree. Phase lag remains near 0 degrees despite a threefold increase in perturbation frequency.	(a) Slow-twitch units decrease firing rate modulation and begin to fire more tonically in both F and E muscles. This partial coactivation tonically stiffens the wrist and aids in keeping the displacement phase lag near 0 degrees.	(a) Tonic co-contraction
		(b) Fast-twitch units are recruited into activity, so that the modulated output of the muscle is maintained. The more rapid twitch speeds of these units allows a more effective tracking of the higher frequency perturbations.	(b) Graded, reciprocal activation, but perhaps by a different mechanism than that involved at 0.2 Hz.

this regard, we turned next to a study of the discharge of motor cortex cells during task performance.

Neuronal Discharge Patterns in the Wrist Control Area of the Precentral Motor Cortex

After each animal was trained in the motor task, a cortical recording chamber was implanted over the arm area of the motor cortex contralateral to the trained arm. Stimulating electrodes were also implanted with tips at the dorsal borders of the medullary pyramidal tract (PT) and the red nucleus (RN), for antidromic identification of PT or corticorubral (CR) neurons, respectively. The wrist control areas of the precentral gyrus were then located by stimulating the cortex through an epidural electrode (surface anode; 10-pulse train, 200 pps; 0.2 msec, 0.5–1.5 mA pulses), while recording evoked EMGs from the forearm muscles. The areas were then defined more precisely with intracortical microstimulation (cf. Asanuma and Sakata, 1967; Asanuma and Rosen, 1972), applied at the end of each recording session and during detailed mapping experiments just before killing the animal.

During daily recording sessions, microelectrodes were inserted into these stimulation-defined areas, and the following data were obtained from neurons that were judged to be task related: (1) the discharge of the unit during task performance; (2) responses to passive rotation of the joints of the arm and hand, or to light tactile and pressure stimuli applied to the skin; and (3) responses to stimulation of the PT and of the RN. To date, we have recorded from 213 task-related cells in 2 monkeys. Complete observations were obtained for 108 neurons, whose discharge patterns have been analyzed quantitatively and in detail. The findings summarized here are based principally on this smaller sample, but are supported fully by less complete data from the remaining cells. Of these 108 neurons, 42% were rapidly conducting PT cells (cf. Humphrey and Corrie, 1978), 13% were slowly conducting PT cells, 11% were CR neurons, and 34% were not excited antidromically at the moderate stimulus intensities that were tolerated comfortably by our alert animals. As long as they lay in the same cortical zone, we have not found these anatomically distinguished classes of cells to behave, as populations, in ways that were functionally different with respect to reciprocal or coactivation of antagonist muscles; therefore, we will not distinguish among them further in the summary that follows.

Our major finding is that these task-related cells fell into two major functional groups on the basis of discharge patterns. The two groups overlapped in spatial distribution, but were centered at slightly different zones within the precentral wrist area. Moreover, the patterns of muscular response evoked by stimulation within these two zones differed significantly. These points are illustrated in Figs. 9 through 14.

The contour maps of Fig. 9 (left) show the low and intermediate threshold zones of the precentral wrist flexor-extensor area in one monkey, as defined by intracortical microstimulation and EMG techniques. Let us consider first the low-threshold extensor and flexor area (regions enclosed by the inner, continuous, and dashed contour lines, respectively). Together, these low-threshold foci comprise the first of the two functional zones mentioned above. Microstimulation at low intensity (10, 0.2-msec, 2- to 10-μA pulses at 300/sec) within the medial part of the extensor focus evoked activity only in the wrist and finger extensor muscles (extensor carpi radialis and ulnaris, extensor digitorum communis). An example of the EMG evoked from this region is shown in Fig. 10A. Stimulation at < 10 μA within the flexor focus, which is enclosed in the extensor focus, evoked activity in both flex-

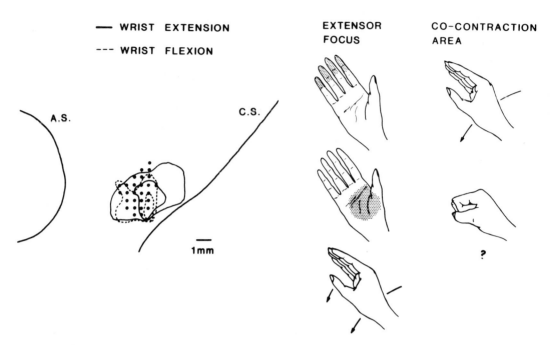

FIG. 9. Locations of the precentral wrist zones explored in the present study, and the receptive field properties of units within them. **Far left:** a scale drawing of the precentral arm area. Anterior is to the **left**, medial toward the **top**; A.S. = the arcuate sulcus, C.S. = the central sulcus. Superimposed on this drawing are contour lines, which enclose regions from which activation of wrist extensor *solid lines*) or wrist flexor muscles *(dashed line)* was obtained at particular stimulus intensities. The inner enclosed regions are those from which EMG responses were evoked at current intensities < 10 μA, whereas the surrounding enclosed regions are those from which responses were evoked at stimulus intensities between 10 and 25 μA. The inner, low-threshold extensor zone, which contains within it a low-threshold flexor zone, is referred to here as the extensor or the flexor–extensor (F–E) focus. The region anterior to and partly encircling this focus is referred to as the co-contraction zone, for reasons outlined in the text. The locations and types of somatosensory receptive fields for task-related units in these two zones are indicated schematically at the right.

or and extensor muscles, but principally in the former (Fig. 10B).

Of the task-related neurons in these low-threshold foci, 70% responded briskly to stimuli delivered to the joints or the skin of the forearm and hand. Effective cutaneous stimuli were localized pressure or a tactile probe swept quickly across the unit's receptive field. Cutaneous receptive fields were most frequently localized to the palmar surface of the fingertips or the proximal third of the palm, regions that would be stimulated intensely by manual exploration, or an attempt by the animal to control a manipulandum grasped by the hand. Effective proprioceptive stimuli consisted of passive flexion-extension of the wrist or, less frequently, of the fingers. Units that responded to passive wrist movement were excited by movements opposite in direction to those produced actively, and with which the discharge of the cell was correlated. These receptive field properties are summarized in Fig. 9 (extensor focus column).

The task-related cells in these low-threshold foci tended to fire in relation to activity in either wrist flexor or wrist extensor muscles, but rarely both. On the basis of discharge patterns during task performance, the units were classified into two subgroups. The first subgroup consists of cells (M_L units, $N = 59$) whose discharge was

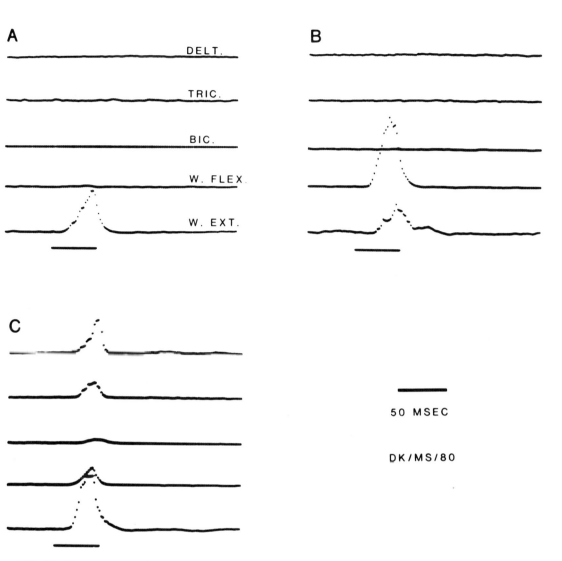

FIG. 10. Electromyograph responses evoked in five muscle groups in the arm by microstimulation in the F–E focus and in the adjacent co-contraction zone. Stimulus parameters: 10 pulse train, 0.2-msec pulses, 300/sec. The EMGs have been rectified and low-pass filtered. Stimulation begins at the onset of the *horizontal bar* shown beneath each column of simultaneously recorded traces. Each trace is the average of 20 evoked responses. A: Stimulation at 5 μA in the medial part of the extensor focus evokes activity in only the wrist extensors. B: Stimulation at 9 μA in the lateral part of the extensor focus evokes activity also the wrist flexor muscles. C: Stimulation in the anterior part of the co-contraction zone evokes activity in all five muscle groups at approximately the same threshold intensity (20–22 μA), and with approximately the same latency. W. ext. = Wrist extensor, W. Flex = wrist flexor, Bic = biceps, Tric. = triceps, and Delt. = deltoid muscles.

clearly modulated (M) in phase with modulations in wrist muscle or torque activity at low (L) perturbation frequencies (0.1–0.2 Hz). An example of the task-related discharge of one of these cells is shown in Fig. 11. Peak levels of unit discharge led those of related muscle activity or wrist torque by up to 300 msec; thus, although sensitive to somatosensory stimuli, it is unlikely that these cells were driven only by sensory

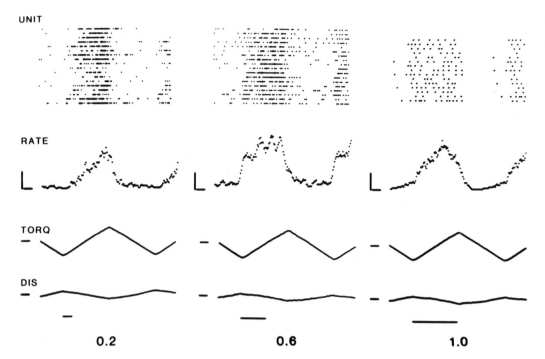

FIG. 11. Discharge pattern of a cortical M_L unit at tracking frequencies of 0.2, 0.6, and 1.0 Hz. Each column of traces shows, from above downwards, the following variables. *(1)* A raster display of the neuron's discharge pattern during 15 to 18 consecutive torque perturbation cycles. *(2)* The unit's average discharge rate during a cycle. *(3)* The average wrist torque generated by the animal during a cycle. The torque (TORQ) varies from 1,000 g/cm in the extensor (upward deflection) to 1,000 g/cm in the flexor direction. *(4)* The average displacement of the animal's wrist (DIS). Flexion is upwards, and peak-to-peak displacement at 0.2 Hz is approximately ± 3 degrees. All traces in each column are aligned with respect to time. Note that the time base has been expanded successively at 0.6 and 1.0 Hz, so that phase relations can be easily compared at each perturbation frequency. *horizontal bars* at the bottom of each column represent 0.5 sec. *Horizontal bars* to the left of rows *2* through *4* represent the zero levels for the variables indicated. *Vertical bars* in the discharge rate averages represent 10 spikes/sec. Note that cell discharge leads generation of an extensor muscle torque. (PT cell, F–E focus, 14 m/sec.)

feedback from the arm and hand. Thirty-nine of the cells increased their discharge in relation to activity of wrist extensor muscles, and 20 in relation to wrist flexor activity. If we define the variable percent modulation in firing rate (PMFR) by the formula

PMFR = 100 × change in firing rate during the torque cycle/ maximum rate during the cycle,

then PMFRs ranged from 30 to 100% at 0.2 Hz for these cells, with a median of 58%. As the perturbation frequency increased, the PMFR also increased in 29 of 59 cells, remained essentially unchanged in 21 of 59 cells, and decreased in 9 of 59 cells. It is quite likely that these cells contributed to the reciprocal, modulated form of activity seen in wrist flexor and extensor motor units at 0.1 to 0.6 Hz (Table 1).

The second subgroup of task-related cells in the flexor and extensor foci differed only quantitatively from those in the first. This subgroup consists of neurons (M_H units, $N = 17$) that showed no clear modulation (M) in firing rate at 0.2 Hz, but did show these modulations at higher (H) perturbation frequencies. A striking example of one such unit, which might have been classed also as an M_L cell, is shown in Fig. 12. The PMFR of this unit increased from approximately 30% at 0.2 Hz to almost 80% at 1.0 Hz. Cells of this type may well have contributed strongly to the increasing, modulated drive on MNs seen at higher perturbation frequencies, and most clearly in the discharge of the fast-twitch motor units. In summary, then, the outputs from the flexor and extensor foci are muscle specific, they are modulated reciprocally, and their depth of modulation during each torque cycle is increased with rising perturbation frequency.

Let us now consider the second of the overlapping functional zones mentioned above. The center of this region forms a partial belt, which surrounds all but the posterior parts of the low-threshold, flexor and extensor foci. Microstimulation in particular in the anterior part of this belt evokes a coactivation of wrist flexor and extensor muscles at thresholds ranging from 12 to 20 μA; indeed, at some loci, stimulation at only 20 μA coactivated muscles acting about the wrist, elbow, and shoulder,

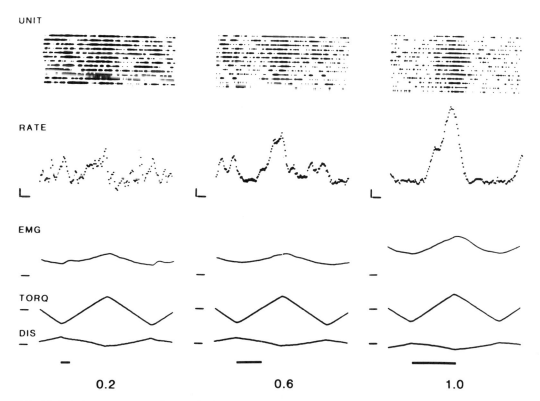

FIG. 12. Discharge pattern of a cortical M_H unit at three tracking frequencies. Format as in Fig. 11, except that the average rectified and filtered EMG from the wrist extensor muscles is shown also. Note that as the tracking frequency increases from 0.2 to 1.0 Hz, the baseline (tonic) level of EMG activity also rises (tonic coactivation). A similar change does not occur, however, in the background or tonic discharge rate of the cortical cell. Such changes were, in fact, seen in only 15% of the M-class units, with the majority of the cells discharging only in relation to the modulated component of the wrist flexor and extensor EMGs. The increase in the modulated discharge rate of this cell parallels that seen in the EMG record, as perturbation frequency rises. Note that the unit discharge leads the peak in extensor EMG activity. (PT cell, F-E focus, 42 m/sec.)

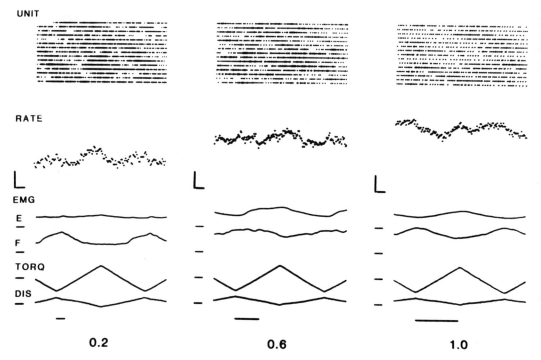

FIG. 13. Discharge pattern of a cortical S^Δ cell as three tracking frequencies. Format as in Figs. 11 and 12, except that averaged EMG traces are shown for both wrist extensor (E) and flexor (F) muscles. Zero levels of EMG activity are shown by the *horizontal bars* below and to the left of each filtered EMG trace. Note that tonic levels of activity increase in both muscles (co-contraction) as the perturbation frequency rises from 0.2 to 1.0 Hz. This increase is paralleled by an increase in the steady firing rate of the cortical unit. (CR neuron, 14 m/sec, co-contraction zone.)

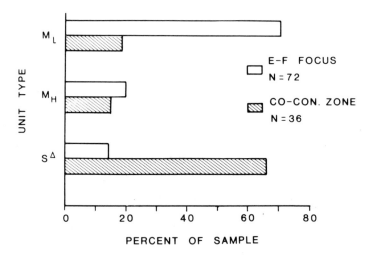

FIG. 14. Percentages of unit samples from the extensor-flexor (E-F) foci and from the adjacent co-contraction zone that were classified into M_L, M_H, and S^Δ categories. The number of cells in each sample is indicated.

with threshold currents being approximately equal for all muscles (see Fig. 10C). Such activation of several muscles, including antagonists, did not appear to be due to spread of stimulating current to adjacent, but separate foci for each muscle group. Rather, the coactivation appeared to be due to stimulation of some intrinsic, higher-threshold system—or of an afferent input to it—that was specifically organized for simultaneous activation of antagonist muscles at one or more joints in the arm. Such responses might easily have been missed had we observed evoked movements rather than evoked EMGs. Indeed, in some cases, coactivation of one or more sets of antagonists—at approximately the same threshold current—was accompanied by no clearly defined movement at any single joint.

The neurons within this anterior, coactivation zone were relatively insensitive to stimulation of the arm and hand. In those few cases (less than 20%) in which a task-related unit did respond to somatic stimuli, the response was weak, and the effective input was a rapid flexion-extension of the wrist through an arc of 30° or higher. Yet, when the animal voluntarily cocontracted his wrist muscles, as in stabilization of the wrist or tightening of a grip on the handle, these cells discharged in a brisk and tonic manner (these properties are summarized in Fig. 9, right). In brief, the task-related units in this zone appeared to be controlled principally by central structures, with little influence exerted on them by the somatosensory periphery.

The most frequent, task-related discharge pattern in the coactivation zone differed significantly from those seen with greatest frequency in the flexor and extensor foci. When observed at only one perturbation frequency, unit discharge appeared to be poorly related or even unrelated to the motor task; the cells fired at a steady frequency, with little modulation in relation to the torque cycle (PMFRs < 20%). Yet, as the perturbation frequency was raised, the tonic discharge rates increased ($N=22$) or decreased ($N=10$) progressively, in a manner that, in general, paralleled the tonic, co-contraction component in the discharge of slow-twitch motor units. An example of the discharge pattern of one of these neurons is shown in Fig. 13. We refer to such units as S^Δ cells, as the major change seen in their discharge rates with increasing perturbation frequency was a progressive shift (Δ) in steady (S) or tonic background level.

How spatially segregated in intracortical space were these two functional classes of task-related units? The bar graphs in Fig. 14 show the percentages of M_L-M_H and S^Δ cells observed in the two microstimulation-defined zones. Clearly, M_L-M_H neurons were detected most frequently in the low-threshold, extensor-flexor foci, whereas S^Δ cells were seen most frequently in the co-contraction zone. These differences in frequency of unit observation in each zone are highly significant statistically ($\chi^2 = 66.4$, $df = 2$, $p < 0.001$). It should be emphasized, however, that the true spatial intermingling may be more extensive than our current sample of units suggests. For example, we were not attuned to the existence of S^Δ units during the initial phases of our study, when the extensor and flexor foci were explored most heavily with microelectrodes. Thus, S^Δ units may exist in greater density in these foci than our present results suggest.

DISCUSSION

These results provide the first direct neurophysiological evidence for the existence in the primate brain of two partially independent motor control systems of the type described in the beginning of this chapter. Before discussing the functional significance of this form of organization, however, it will be appropriate to consider first how a co-contraction of antagonist muscles provides for a graded control of joint stiffness.

Comment is also required about the manner in which the signals from these two control systems add at the MN pool, and about the relationship of our cortical unit results to previous studies of motor cortex organization and neuronal behavior.

Control of Joint Stiffness by Antagonist Co-contraction

Studies of length-tension (L-T) relations in muscles with intact stretch reflexes have shown that the apparent stiffness of the muscle is held roughly constant over a considerable range of muscle length by reflex regulation (Nichols and Houk, 1976; Crago et al., 1976; Houk, 1976; Feldman, 1980a). With increasing levels of tonic excitatory drive on the MN pool, the apparent stiffness of the muscle (i.e., the slope of its L-T curve) is increased by only 1.5 to 2.0 times (Feldman, 1980a; Hoffer and Andreassen, 1981). Instead, an increased level of MN bombardment simply shifts the L-T curve toward a shorter muscle length, without significantly changing its slope (Feldman, 1980a). In the absence of any major change in the stiffness of individual muscles, then, one might well ask how a tonic co-activation of antagonists can alter the apparent stiffness of a joint over a range of 4 to 7 times, as suggested by the results of the present study.

An answer to this question is best given with the aid of Fig. 15. In this figure, net joint torque is shown on the ordinate, and

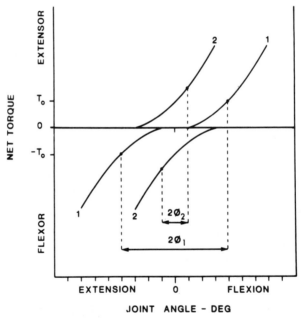

FIG. 15. Illustration of how a tonic co-contraction of antagonist muscles may lead to a significant increase in joint stiffness. Reflex-modified, length–tension curves for the extensor muscles are shown above the zero torque (or tension) line, and those for the flexor muscles below it. Increasing extensor length (joint flexion) is to the **right**, and increasing flexor muscle length (joint extension) is to the **left**. Two L-T curves are shown for each muscle, corresponding to two levels of tonic MN activation. Level 2 is greater than level 1. The increase in muscle activation does not change the slopes of the L-T curves, but instead shifts them toward shorter muscle lengths, or lower levels of joint displacement. The amplitude of the imposed torque disturbance is assumed to vary from $+ T_o$ to $- T_o$, as indicated on the *ordinate*. The displacements $2\phi_1$ and $2\phi_2$ are those resulting from this applied torque at levels of antagonist muscle coactivation 1 and 2, respectively.

joint displacement in the flexed or extended directions on the abscissa. Hypothetical L-T curves are shown above the zero torque line for an extensor muscle, and below this line for a flexor muscle. Since the tensions exerted by these muscles are opposite in direction, as are their length changes, the flexor L-T curves are reflected 180° about both axes from the extensor muscle curves.

Consider now the angular displacement that would be produced by the *external* application of a net torque of magnitude $\pm T_o$, about the axis of joint flexion-extension. Assume that the coactivation drive on the two muscles is at level 1, so that both are in the relaxed state, and the joint is at the zero angle position. As the applied torque rotates the joint in the flexed direction (to the right), the extensor muscle is stretched. When the angle 0_1 is reached, sufficient passive and reflex tension has been generated in the extensor muscle to null the imposed torque; joint movement then ceases. As the applied torque now reverses direction and passes through zero, the joint returns to the zero angle position and moves in the direction of extension. When the joint reaches the angle -0_1, the net tension exerted by the stretching flexor muscle nulls the imposed torque, and joint movement again ceases. Thus, with this alternating torque of magnitude $2T_o$, the joint moves through the total angle 20_1, and the apparent stiffness (S) of the joint is $S = 2T_o/20_1 = T_o/0_1$.

Assume now that the tonic *co*-activation of the two muscles is raised to level 2. The muscles are now co-contracting, and their L-T curves shift toward and overlap at the zero angle joint position. No joint movement occurs, because the torques exerted by the two muscles add algebraically to zero. However, when the alternating torque $\pm T_o$ is applied to the joint, it moves only through the angles $\pm 0_2$ before the algebraic sum of the torques exerted by the muscles is sufficient to null the imposed torque. In this case, $0_2 = 0_1/4$, so that the apparent stiffness of the joint is 4 times that in the previous example. This fourfold increase is achieved despite the fact that the slope of the L-T curve for each muscle varies by only 1.5 to 2.0 times over its entire range, and is not increased significantly by muscle activation. Thus, through co-activation of antagonists, joint stiffness can theoretically be varied over a considerable range despite a more limited variation in the stiffness of individual muscles.

It should also be noted that a tonic coactivation may also increase the apparent dynamic stiffness of a joint, i.e., the speed and effectiveness of neurally generated opposition to an applied disturbance. At moderate levels of coactivation, for example, slow-twitch units fire tonically and fast-twitch units are brought close to threshold. With a rapid stretch of the muscle, therefore, the most significant increment in discharge will occur in fast-twitch motor units; because of their faster mechanical responses, compensatory reactions occur more quickly, and the joint movement is more effectively damped.

Summation of Control Signals at the MN Pool

At higher frequencies of joint perturbation, the co-contraction signal is most evident in the discharge of slow-twitch motor units. In contrast, the increasingly modulated, reciprocal drive on flexor and extensor MNs is best seen in the discharge of fast-twitch units. One might ask, then, if these two major control signals are distributed preferentially to slow and to fast-twitch MNs as perturbation speeds rise? We do not believe that this is the case. Indeed, our results can be explained more parsimoniously on the basis of known properties of MN recruitment, and intrinsic limitations in firing rate.

At low perturbation frequencies, the principal input to the MN pool is modulated in phase with the applied torque. At the moderate torque levels imposed in the present study, only slow-twitch units are recruited into activity by these slow perturbations. Nonetheless, the slow units reach a peak firing level during each torque cycle that is close to their physiological limit (40–60 spikes/sec). As the speed of joint perturbation rises, the modulated input to the MN pools is increased and a tonic coactivation signal is added. The latter drives slow-twitch units toward their maximum firing rate; as a result of this rate saturation, the increasing modulated drive is not seen in their discharge pattern. In contrast, the coactivation signal brings fast-twitch units closer to threshold, allowing the increased modulated input to produce greater changes in discharge rate during each torque cycle. Thus, an explanation of our observed MN firing patterns requires no assumption of selectivity of descending inputs to motor units of different type, nor of any recruitment order different from that established in previous studies (Henneman et al., 1965a; Milner-Brown et al., 1973b; Freund et al., 1975; Desmedt and Godaux, 1977b; Desmedt, 1980a, 1981a). Both control signals appear to converge on both fast and slow-twitch MNs.

Relation to Previous Studies of Motor Cortex Organization

Our results indicate that there are two general output systems within the stimulation-defined wrist area of the precentral motor cortex. The first is centered within the posterior part of the wrist area, and is organized for the control of single muscles or groups of synergists acting about the wrist; it receives somatotopically precise sensory input from joint and cutaneous receptors in the wrist and hand. The second output system occupies overlapping cortical space, but is centered a bit more anteriorly; it co-activates antagonist muscles at the wrist, and perhaps also at adjacent joints in the more proximal arm. This system receives significantly less somatosensory input than the first, and unit receptive fields are more complex and diffuse. How do these findings compare with previous studies of motor cortex organization, and of the afferent input to it?

We wish to stress that these results do not conflict with the elegant mapping studies of the precentral hand area by Asanuma and Rosen (1972), whose results suggested that each intrinsic muscle of the hand was controlled by a separate, radially oriented, cortical column. We did not study the hand area. Moreover, Asanuma and Rosen also noted that stimulation within the precentral wrist area often evoked a co-contraction of flexor and extensor muscles. Thus, our results are compatible. In addition, more recent studies of the organization of precentral areas controlling movements about wrist, elbow, and shoulder have indicated that extensive spatial overlap exists in the neurons that control muscles of the forearm and proximal limb; moreover, each joint control zone covered a relatively wide expanse of the precentral arm area (Kwan et al., 1978). These findings are also compatible with our results.

With respect to an anterior-posterior partitioning of somatosensory inputs, our results are also compatible with previous studies. Strick and Preston (1978a,b) have discovered two hand-wrist representations in the motor cortex of the squirrel monkey; the anterior area receives input principally from joint receptors, the posterior hand-wrist zone from cutaneous receptors. We did not detect a similar separation of joint and cutaneous inputs in this study, but we did not explore fully the rostral bank of the central fissure, where cutaneous inputs may impinge more selectively. Finally, in studies of the macaque precentral gyrus, Wong et al. (1978) also noted that neurons within the more anterior part of the stimulation-defined wrist area were less responsive to somatosensory stimuli than those lying more posteriorly, with more complex receptive fields.

There is one major aspect of our data, however, which does suggest a form of motor cortex organization that has not been clearly stressed in previous studies. In most of these studies, maps have been obtained with observations of joint movement; EMG recordings were used to supplement these observations, but they were not conducted systematically (Asanuma and Rosen, 1972; Anderson et al., 1975; Strick and Preston, 1978a; Kwan et al, 1978). In contrast, we relied principally on EMG recordings. And with these, we were able to detect coactivation of muscles within the arm at stimulus intensities that in some cases evoked no clearly defined movement at any joint. In addition, we routinely plotted stimulus threshold contours, instead of mapping only those regions that activated a muscle at the lowest current intensity. As a result, our maps of precentral muscle control areas suggest that these zones are of larger extent and overlap more extensively than is suggested by the results of previous studies

One such microstimulation map of the precentral arm area is shown in Fig. 16. Evoked EMGs were recorded from bipolar surface leads (1-cm separation) overlying wrist-finger extensor, wrist-finger flexor, biceps, triceps, and deltoid muscles (virtually identical maps have been obtained with chronically implanted, intramuscular leads). The animal was alert during the mapping procedures, and stimuli consisted of 10 cathodal pulses at 300/sec, with 0.2-msec, 2- to 40-μA pulses. Two major features are seen in this ensemble of maps for the various muscle groups indicated. First, there are multiple foci from which a particular muscle can be activated; this is particularly true when considering the regions of intermediate threshold (10–25 μA regions). Note, however, that the low-threshold foci (< 10 μA) are fewer in number. The *second*

feature is that the zones for activating a particular muscle overlap extensively; indeed, at almost any point a stimulus of 10 to 30 μA intensity may evoke activity in two or more of the muscle groups indicated. It should be stressed that this overlap is *not* due simply to a spread of stimulating current to more discrete foci for the control of single muscles, as is indicated by the asymmetrical shapes of the threshold contours.

On the basis of these maps and the data summarized in this chapter, we propose the following, comparatively new view of the organization of the precentral arm area. Namely, that each major muscle group is controlled by one or two low-threshold regions, which, in turn, are imbedded in an output system that is capable of evoking activity in the antagonist muscles at the same joint, and perhaps also in sets of antagonists acting at adjacent joints. What is the functional significance of this form of somatotopic organization? By virtue of such topography, spatially restricted inputs to the precentral gyrus could: (1) evoke specific movements about a particular joint; (2) with increased intensity, evoke coactivation of antagonists and stabilization of the course of movement; and (3) evoke the muscle activity at adjacent joints that is necessary for postural support of the movement. It should be noted that previous investigators have suggested, on the basis of somewhat different data from cats, that a given cortical zone may evoke both limb movement and the postural supporting reactions that must accompany that movement (Gahery and Nieoullon, 1978; Massion, 1979).

If this hypothesis is correct, however, does it not question our assumption that the majority of single neurons observed in this study were in some way related to control of wrist muscles? Might not some of these cells have been involved in control of muscles acting about the fingers, the elbow, or even the shoulder? With respect to any sin-

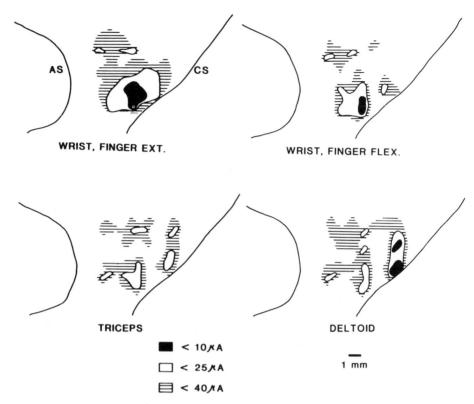

FIG. 16. Current-threshold contour maps in the precentral motor cortex of the monkey, for activation of various muscle groups in the contralateral arm. Muscle activation was measured electromyographically, with surface leads overlying the indicated muscle group (1-cm electrode separation). The biceps map is similar to the triceps map and is not shown. All maps are drawn to scale, with the positions of the arcuate (AS) and central sulci (CS) being indicated. Shown by different shadings are the cortical regions from which activity was evoked in each particular muscle group at current thresholds < 10 μA, from 10 to 25 μA, and from 25 to 40 μA. The lowest threshold observed along each transcortical penetration was used in constructing the map for each muscle group. Note that each set of muscles has multiple foci for activation by currents in the 10- to 25-μA range. Note also that a superposition of these maps, all from the same animal, would reveal a tremendous overlap in the locations of intermediate threshold foci for the various muscle groups. The irregular shapes of the contours indicate that this overlap is not a simple consequence of current spread to spatially contiguous, but separate, low-threshold foci. The zone explored with recording electrodes in this animal is coextensive with the < 10 and 10- to 25-μA zones in the wrist extensor map.

gle neuron, we cannot answer this question definitely. We can, however, argue that as a population it is highly likely that the M_L and M_H cells were involved in control of wrist flexor and extensor muscles, for the following reasons. (1) The majority of these cells were identified as output neurons (PT and CR cells); (2) they were located in the lowest threshold zone for activation of wrist flexor and extensor muscles; (3) the discharge of these cells preceded activity in the wrist flexor-extensor muscles by an appropriate interval (50–300 msec); (4) the temporal correlations between unit discharge and wrist muscle EMG activity were high (Figs. 11 and 12); (5) as a set, the units were highly responsive to somatosensory inputs from the wrist and hand, as described for units in precentral, wrist control zones (Lemon and Porter, 1976; Fetz et al.,

1980); and finally, (6) EMG recordings from several muscle groups in the arm during task performance showed that activity modulated in relation to the torque cycle occurred only in the wrist flexor-extensor muscles.

With respect to the S^Δ or putative co-contraction cells, however, we must relax this confidence, since an increasing level of tonic coactivation of muscles occurred in biceps, triceps, and deltoid muscles as the perturbation frequency rose, as well as in the wrist flexors and extensors. These changes in muscles acting at other joints suggest that our co-contraction units may have been part of a more general network for producing postural stabilization throughout the limb.

Sources of Afferent Drive on Task-Related Cortical Cells

Our view of the general sources of afferent drive on task-related cortical cells is summarized in Fig. 17. M-class units are shown as two subgroups, each of which controls independently a flexor or extensor MN pool. S^Δ units are shown as a separate population of co-contraction cells, whose outputs act by way of unknown networks to co-activate flexor and extensor MNs.

Consider first the operation of this system at low frequencies of joint perturbation (0.1–0.2 Hz). A human observer tends to track such frequencies in a conscious and continuous manner; i.e., the applied torque and small wrist displacements are continuously sensed, and graded opposing torques are applied voluntarily. The graded, reciprocal EMGs of wrist flexor and extensor muscles in the monkey, when tracking at these low frequencies, suggest that the animal is performing the task in a similar manner. Thus, the major mode of control is by the pathway shown as number 1 (dashed line) in the flow diagram. Sensory input data from the perturbation and proportional motor commands are processed voluntarily by higher central structures whose outputs drive M-class neurons. Although more direct pathways exist from hand and wrist receptors to these cortical cells (Lemon and Porter, 1976; Evarts and Fromm, 1978; Fetz et al., 1980; Wolpaw, 1980), we do not believe this source of afferent drive is a major one at the low joint movement velocities occurring at 0.2 Hz. At this low perturbation frequency, the central drive on the putative co-contraction cells is also weak.

Consider now a perturbation frequency of 1.5 Hz. At this frequency, the modulated output signal from the motor cortex is maintained or even increased in amplitude. The shapes of the discharge rate curves during each torque cycle suggest, however, that M-class cells are now driven significantly by a signal arriving over short-latency pathways from velocity-sensitive pressure or joint receptors in the hand or wrist, respectively (cf. Figs. 12 and 13). That is, pathway 2 in Fig. 17 is now more significant that at low frequencies in the total drive on M-class cells. In contrast, S^Δ or co-contraction cells, which receive little input from the periphery, are now driven strongly by centrally originating inputs. In this connection, it is of interest that the major sense of effort reported by a human during torque tracking at 1.5 Hz is one of tonic clamping of the wrist, i.e., a tonic co-activation of flexor and extensor muscles. Thus, the conscious mode of voluntary control has switched from a graded reciprocal form at 0.2 Hz to one of tonic joint stiffening at 1.5 Hz.

This summary is, of course, largely hypothetical, but it is consistent with our data and with the reports of human performers of the same motor task. It should also be noted that the major pathway from our presumed coactivation cells to flexor and extensor MNs is largely unknown. Although some PT neurons fell into this category, the major coactivation effects could be exerted by way of brainstem or even cerebellar-brainstem pathways (see, for example, Tilney and Pike, 1925; Smith, 1981).

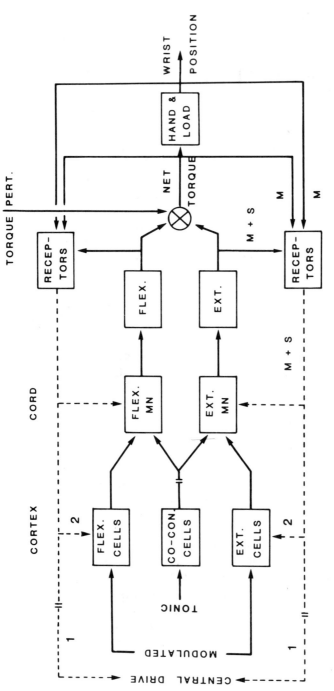

FIG. 17. Summary hypothesis of the afferent drive on cortical cells that control separately the reciprocal and the coactivation of antagonist muscles. Somatosensory input or feedback pathways are indicated by *dashed lines*. The letters M and M + S are used to indicate whether the task-related input over various somatosensory pathways would be both modulated (M) or have both modulated and steady components (M + S) during the torque perturbation cycle. Only feedback from tendon organs, activated by muscle contraction, would be expected to have a steady component. Thus, the tonic drive on coactivation cells must originate centrally. The interrupted output from the putative co-contraction (S) cells implies an uncertainty concerning the circuits by which their effects are exerted on spinal MNs. The flexor (FLEX) and extensor (EXT) cells at the level of the cortex are M-class neurons. The numbers 1 and 2 refer to the major routes of somatosensory influence on cortical cell activity during the tracking of low and higher frequency perturbations, respectively.

Functional Significance of Reciprocal and Co-Contraction Modes of Control

What are the normal roles played by the two systems described here in the control of movement and posture? The single muscle or reciprocal activation system is most likely used to produce smooth, controlled joint movements, under the precise control of somatosensory feedback. M-class cells, for example, discharged in relation to flexion or extension movements of the wrist. In contrast, the coactivation (co-contraction) system appears to provide for a principally central control of joint stiffness. Note that this view is not equivalent to the view that movement and posture are controlled independently. Single or reciprocal activation of antagonist muscles may be used either to produce movement, or, as in the motor task used here, to maintain a constant joint posture in the face of slow perturbations. Conversely, coactivation of antagonist muscles during the movement of a joint may stabilize the course of the movement (see below).

But what is the functional significance of this form of organization? Why have a separate central system for the control of joint stiffness? We believe that the coactivation system performs two major functions, both of which complement those of the reciprocal activation system. First, a co-contraction of antagonists may be used to fix those joints which must provide a postural base of support for movements occurring at other joints. Second, through a separate control of joint stiffness, the posture or the movement of a joint may be buffered from the effects of unpredictable force (or torque) disturbances. A stiffening of joints in the arm by antagonist co-contraction is familiar to all, for example, who have controlled the movement of a power tool as it cut through material of varying composition and unknown resistance, or who have grappled with an opponent in a sport such as wrestling. In such cases, the magnitude and the direction of sudden opposing forces are unknown prior to their occurrence. In order to maintain control of joint position, the motor system must respond to these disturbances with minimum phase lag. We have already seen how this might be accomplished through a tonic coactivation of antagonist MN pools: (1) antagonist muscles co-contract, the joint is tonically stiffened, and external forces are opposed with no time delay; and (2) the MN pools are shifted into the recruitment range for fast-twitch motor units, so that dynamic compensatory responses occur more rapidly. In this connection, it should be noted that such a buffering action is perhaps best performed by a system that is principally under central rather than continuous, peripheral sensory control, for in this way the effects of unpredictable sensory input are minimized.

As a corollary, it should be noted that this buffering action need not be confined to externally imposed forces. Kinesiologists, for example, know that a co-contraction of antagonist muscles may be prominent in the early phases of the learning of skilled movement. As the movement is learned, the level of co-contraction is reduced, and it is replaced by a more economical form of distributed activity in single muscles. Thus, during motor learning, the co-contraction system may be used to reduce the magnitude of movement errors that are produced by inappropriate commands from the joint-movement control system. We have observed this use of antagonist co-contraction in our trained monkeys on many occasions. For example, when first grasping the handle after a period of rest, the animal did not know the speed or exact form of the perturbation that would be imposed about his wrist. In such cases, joint position is initially controlled by a high level of antagonist coactivation, which then gives way gradually to a lower level, more economical form of graded, reciprocal activity in the

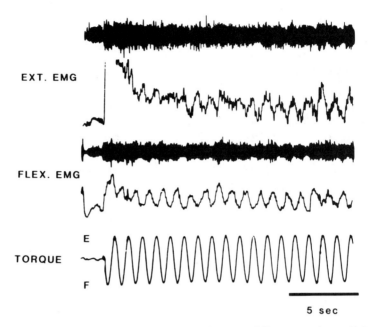

FIG. 18. Example of the use of antagonist co-contraction to stabilize the wrist until the nature of an applied perturbation is learned. Both unprocessed and rectified-filtered EMGs are shown for the wrist flexor and extensor muscles. At the onset of the sinusoidal variations in the torque trace, the animal first grasps the handle. At this time, it does not know the shape and the frequency of the perturbation that is applied to its wrist. Through a co-contraction of the flexor and extensor muscles, however, the animal maintains its wrist in approximately the same position. As the nature of the perturbation is sensed and learned, the co-contraction attenuates (filtered records), and is partially replaced by a lower level of reciprocally modulated activity in the two sets of antagonist muscles.

members of the antagonist pairs. An example of this type of motor activity is shown in Fig. 18.

Our final point concerns the implications of a separate co-contraction control system for our understanding of various motor pathologies. It is well known, for example that the resting or reflex stiffness of a joint may be increased (or decreased) tonically from normal levels in a variety of cerebral motor disorders; the rigid limb of a patient with Parkinson's disease is only one familiar example. Might not some malfunction of a co-contraction system of the general type envisioned here be involved in some of these forms of motor pathology?

ACKNOWLEDGMENTS

We thank Mr. R. Rader and Mr. P. Pelliterri for their technical assistance in these experiments. We also thank Drs. J. Houk and T.R. Nichols for helpful discussions, and Dr. W.D. Letbetter for careful reading of the manuscript. This research was supported by grant No. NS 10183 from the National Institute of Neurological and Communicative Disorders and Stroke.

Interaction Between Motor Commands and Somatosensory Afferents in the Control of Prehension

*Allan M. Smith, Robert C. Frysinger, and Daniel Bourbonnais

Centre de Recherches en Sciences Neurologiques, Département de Physiologie, Université de Montréal, Montréal, Québec, Canada

Of the 9 evolutionary trends with which LeGros-Clark (1959) characterized the order of primates, the first 3 applied particularly to the changing structure of the forelimb and the hand. These features can be summarized as involving the development of the pentadactyl hand, with enhanced mobility of the digits, especially the thumb, the replacement of claws by flat nails, and the development of highly sensitive tactile pads on the digits. Figure 1 shows the changes in the structure of the evolving hand from a prosimian to a catarhine monkey, leading to the eventual diversity seen between three different species of anthropoid apes. In general, most primate hands are capable of some degree of prehensile function, which has been defined by Napier and Napier (1967) as a flexion and adduction of the fingers in which "the digits approximate in such a manner that an object may be grasped and held securely by one hand against the effects of external influences." The variety of hands among the different primate species, as these authors have noted, is largely related to the degree of divergence and opposability of the thumb.

For example, the elongated hand of the gibbon shown in Fig. 1 is well suited to a hooking form of prehension used in brachiation in which the thumb is kept well adducted and flexed. However, the thumb of this species is relatively short compared to the other fingers and not well adapted to thumb-finger opposability and precision handling. Similarly, the hand and forearm of the chimpanzee and gorilla are not only adapted to brachiation, but they are also further modified to accommodate knuckle-walking, a terrestrial quadripedal form of locomotion in which the fingers are flexed against the palm, and the weight is born by the dorsal surface of the phalanges (Tuttle, 1967).

Napier and Napier (1967) have attempted to establish an index of opposability based on measurements of the ratio of thumb length to index ray length for various species of catarrhine primates. The results of this survey show that, apart from humans, the highest opposability quotients are obtained by the baboons and macaque monkeys. How much the prehensile ability of ancestral humans contributed to their intellectual development is a fascinating subject on which to speculate. The importance of this contribution is suggested by the partially reconstructed hands of early hominid remains approximately 1.75 million years

---

* To whom correspondence should be addressed: Centre de Recherches en Sciences Neurologiques, Faculté de Médecine, Université de Montréal, C. P. 6128 (Succersale A), Montréal, Québec, H3C 3J7, Canada.

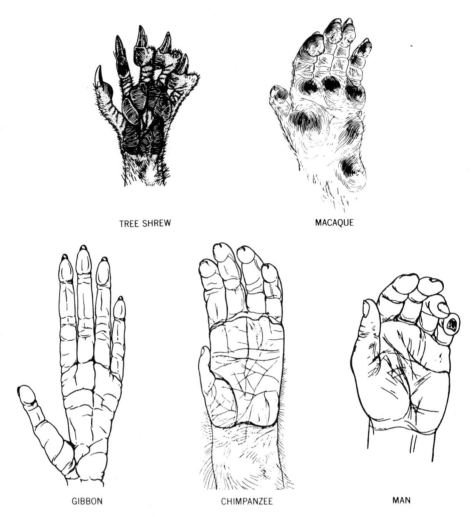

FIG. 1. A comparison of the hands of different primates species: a prosimian, the tree shrew, and a catarrhine or Old World Monkey (macaque), are shown above; below, are the hands of 3 anthropoid apes. The hand of the tree shrew was taken from Le Gros Clark (1959), the macaque from Napier and Napier (1967), the gibbon from Schultz (1949), and the chimpanzee and man from Napier (1960).

old from the Olduvai Gorge found in association with primitive stone tools. The fragments of these finger bones suggest a primate with a particularly powerful grip, but with a hand structure that is poorly adapted to precision handling. It would seem that as Napier (1976) commented, "the tools of *homo habilis* were as good as the hands that made them." However, the expanding brain and increased neuronal control probably contributed considerably to the improved tool-making techniques.

The Form and Control of Prehension

Napier (1956, 1962) has suggested that all prehensile movements of the hand can be divided into two categories: power grips and precision grips. These two operations are distinguished both on an anatomical as well as a functional basis. In power grips, prehensile strength is the primary objective and the thumb is adducted for moderate gripping force. However, to obtain maximum power, the thumb is abducted in what

Napier (1956) refers to as the "coal hammer grip" (Fig. 2). In precision grips, the fine control of prehensile force is the primary objective, although one of the principal purposes of this grasp is to facilitate the adroit manipulation of an object held by the fingers and thumb. Landsmeer (1962) has suggested the term "precision handling" to describe this dynamic maneuvering of objects by the fingers and to distinguish it from the precision grip (i.e., pinching), which is an isometric clasp between the thumb and any other finger. This pinch or precision grip can also be applied with varying degrees of force, and in contrast to the power grip, the thumb is abducted for delicate pressure and adducted for maximum force in the lateral pinch of the thumb and forefinger as seen in Fig. 2. When near-maximum forces are exerted either in the lateral pinch or in the power grip, the antagonist muscles of the wrist, thumb, and fingers co-contract to provide stiff postural support of the hand as well as to make each finger a rigid buttress capable of resisting the clamping forces of the thumb. Prehension therefore, is one of the rare instances in motor control, where antagonist muscles co-contract under isometric conditions.

The extent to which learning contributes to the control of prehension is a complex question, and it is likely that voluntary grasping integrates a variety of cutaneous and muscle reflexes with motor commands of central origin. From birth, neonatal primate infants display a reflex grasping to tactile stimulation of the palm. Gradually, this clasping loses its stereotyped character and, in an orderly sequence of steps, normal prehension develops postnatally (Erhardt, 1974). It is of some interest to note that monkeys that have experienced reflex grasping in infancy recover prehensile function when the limb is later surgically deprived of all sensation by dorsal rhizotomy, (Taub, 1976, 1977). However, monkeys that have had an arm surgically deafferented *in utero* never develop prehen-

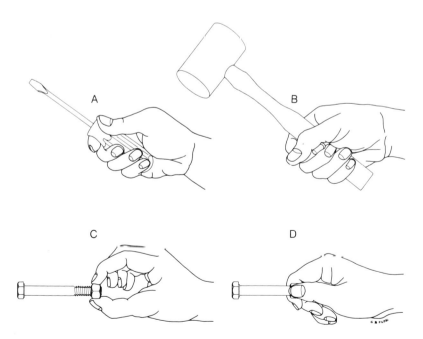

FIG. 2. A: Power grip with the thumb adducted for moderate strength. B: Thumb abducted for maximum prehensile force. C: Precision grip with the thumb abducted for precision handling. D: adducted for maximum pinching force.

sion spontaneously, although these same animals may be trained to make grasping movements by rewarding successive approximations of the desired gesture (Taub, 1981). The implication is that the neonatal instinctive or reflex grasping may be essential for the development of normal prehension.

We had the opportunity to study monkeys in various stages of learning controlled lateral pinch using the thumb and forefinger. These animals were trained to exert and maintain a specific force on a strain gauge for a fruit juice reward. A tone indicated to the animal that its finger pressure was between an upper and lower limit. Figure 3 illustrates the stages of training in the normal monkey. At first each animal was rewarded for a brief pinch of the transducer, and gradually the animal was required to maintain a specific constant force between an upper and lower limit. Well-trained animals were capable of controlling the forces for one second. Initially, the monkey applied a hard pinch very rapidly and then allowed the force to decline. With experience, the monkey learned to apply force more gradually and to maintain it at lower peak levels than during the early stages of learning.

In Fig. 3, it can be seen from the surface activity of the flexors and extensors that these antagonist groups are coactivated from the outset and throughout the various stages of learning. In fact, during a maintained lateral pinch of moderate force (about 200 g), most of the extrinsic as well as the intrinsic muscles of the hand were simultaneously active as shown by the electromyograms (EMGs) recorded in each of the wrist and finger muscles during 20 precision grips in 5 monkeys (Smith, 1981). Using a computer detection algorithm, the time when each muscle began contracting was compared to the onset of prehensile force (Fig. 4). The muscles are recruited over a period from about 300 msec before force onset to approximately 200 msec afterward. The earliest muscle to begin contracting is the extensor digitorum communis (EDC), which opens the hand prior to grasping. This particular behavior may serve first to stretch the long finger flexors before their contraction in order to increase prehensile force and may be somewhat similar to the crouching that precedes jumping and wide jaw opening prior to hard biting. Some of the muscles showing early activity are involved in positioning the wrist and fingers on the manipulandum. The position of the wrist, for example, is particularly important in determining prehensile strength, because the tendons of the finger extensor muscles are too short to allow maximum grasping force when the wrist is fully flexed. The optimum posture for forceful prehension is with the wrist in slight extension at an angle of about 35° from the forearm axis. The onset of contraction in adductor pollicis (ADP) and flexor pollicis brevis (FPB) are both tightly locked to the time of force onset, suggesting that they function as "prime movers" in this isometric precision grip. It is likely that the first dorsal interosseous is also a primary contributor to the lateral pinch, but because of its double abducting and flexing function (Desmedt, 1981a), the onset of activity in relation to the force change is more variable. Finally, a third group of muscles is recruited simulta-

FIG. 3. A series of steps in the learned maintained precision grip. Twenty force traces and the associated surface EMG activity. Each sample was taken during training as the animal increased prehension time from 250 to 1,000 msec. Approximately 2 weeks of daily training separate the first sample of 1-sec prehension and that taken at the end of training. **Lower right corner:** Schematic display of the constraints for the maintained precision grip.

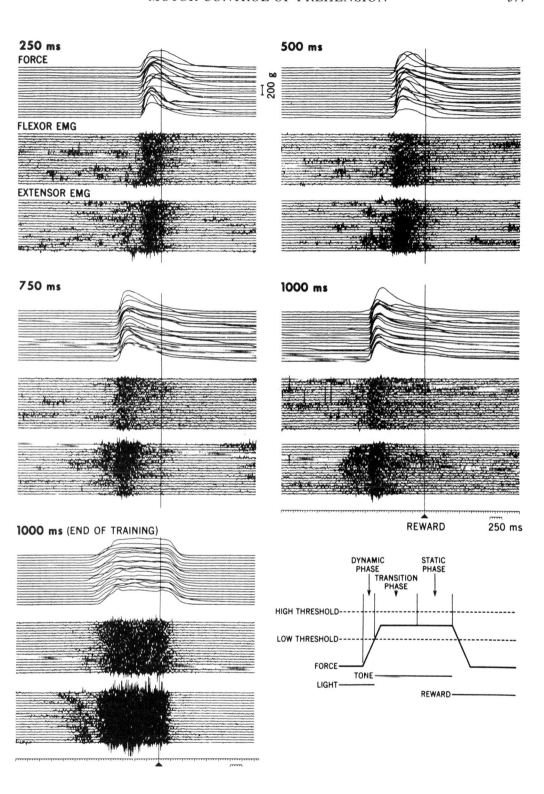

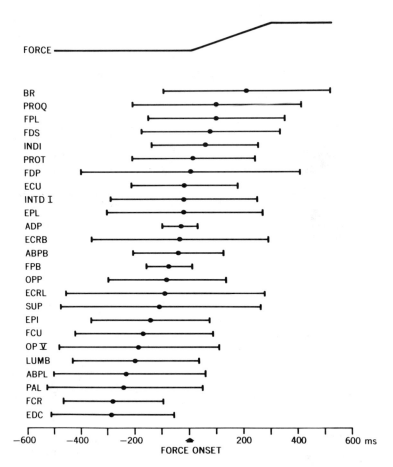

FIG. 4. Mean onset of activity in each forearm muscle ± 1 SD is shown in relation to the onset of prehensile force. Each mean was derived from approximately 100 prehensions in 4 or 5 monkeys. Abbreviations used: ABPB: abductor pollicis brevis, ABPL: abductor pollicis longus, ADP: adductor pollicis, BR: brachioradialis, ECRB: extensor carpi radialis brevis, ECRL: extensor carpi radialis longus, ECU: extensor carpi ulnaris, EDC: extensor digitorum communis, EPI: extensor proprius indicis, EPL: extensor pollicis longus, FCR: flexor carpi radialis, FCU: flexor carpi ulnaris, FDP: flexor digitorum profundis, FDS: flexor digitorum superficialis, FPB: flexor pollicis brevis, FPL: flexor pollicis longus, INDI: indicis, INTD 1: first dorsal interosseous, LUMB: first lumbical, OPP: opponens pollicis, OP V: oppenens digiti V, PAL: palmaris longus, PROQ: pronator quadratus, PROT: pronator teres, SUP: supinator.

neously or slightly after force onset and may provide the increased rigidity needed to sustain moderate prehensile forces.

Motor Cortex

Probably more than any other single region of the brain, the motor cortex and its efferent pathway, the pyramidal or corticospinal tract has been associated with motor control of the hand in primates (Phillips and Porter, 1977). The functional and anatomical studies of Kuypers (1973; cf, also Lawrence and Kuypers, 1968) have demonstrated the existence of direct cortical projections to motoneurons (MNs) of the distal limb musculature in the cervical enlargement as well as the unique contribution of the pyramidal tract (PT) to the control of independent finger movements. Phillips and colleagues (Landgren et al., 1962; Phillips and Porter 1964; Koeze et al., 1968) established that some of these cortical connections are monosynaptic to MNs of the wrist and finger muscles. Moreover, the average amplitude of the cortical monosynaptic potentials is greatest for the intrinsic muscles of the hand, all of which are active during prehension. Only EDC of the extrinsic muscles has a comparable cortical input (Clough et al., 1968; Phillips, 1969). This muscle is particularly important in prehension because it initiates grasping by opening the hand in addition to providing finger rigidity during clasping, and terminating the sequence by releasing held objects from the hand. Lawrence and Hopkins (1976) made observations on finger control in neonatal primates before the corticomotoneuronal connections are established, and following neonatal PT transection. These experiments confirmed that corticomotoneuronal connections are an essential neuroanatomical substrate for the development of independent finger movements and precise manipulation with the hand. However, the size of the PT as a whole does not appear to correlate with either the ability to control a precision grip or manual adroitness in precise handling. From our own observations, the gibbon has an enormous bulbar pyramid and executes a rapid and proficient power grip in brachiation but, nevertheless, performs poorly on a test of independent finger movement similar to that used by Lawrence and Hopkins (1976).

Transection of PT at the level of the inferior olive does not abolish grasping with either the power or the precision grip. Hepp-Reymond et al. (1972, 1974) have established that the peak strength of the precision grip returns after bilateral destruction of the PT. However, a residual deficit was noted in both the reaction time and the force-building time after unilateral pyramidotomy, an effect which was generally more pronounced in monkeys with bilateral lesions. It also appears that the monkey has more difficulty in maintaining a stable constant force precision grip after bilateral pyramidotomy.

Recordings from single neurons within the hand area of the motor cortex have shown that the discharge frequency of some neurons is correlated linearly with both the force and rate of force change in isometric prehension (Smith et al., 1975). This confirmed earlier reports of a monotonic relation between the discharge frequency of pyramidal tract neurons (PTNs) and the force exerted by wrist muscles during movement and maintained position (Evarts, 1968; 1969; Humphrey et al., 1970). Nevertheless, some debate did develop over the functional significance of the small correlation coefficients found in such studies (Schmidt et al., 1975) and the plasticity of the relationship between neuronal discharge frequency, muscle activity, and movement (Fetz and Finnochio, 1971, 1975; see Fromm, *this volume*). To some extent these disagreements may be explained by the differences between the motor control of limb movement versus isometric prehension: (1) In isometric prehension nearly all muscles of the forearm and hand are simultaneously active. This enhances the probability of finding coincident and spurious changes in neuronal activity. (2) Even muscles not directly contributing to prehensile force, such as the pronator and supinator muscles of the forearm, increase their synergistic activity in proportion to the prehensile force. (3) A number of covarying parameters such as joint angle and muscle length are constant under isometric as opposed to isotonic conditions. In isometric situations the activity of motor cortical neurons may more clearly indicate muscle tension and developed force. However, it may be that the discharge of some motor cortical neurons reflect an encoding of more than one motor parameter. (4) There is evidence from the activity of trigeminal gamma motoneurons and spindle afferents (Lund et al., 1978) and from the spindle activity in thumb muscles (Vallbo, 1971, 1973) to indicate that these muscle afferents are under gamma excitation during voluntary isometric contractions and that they may accurately signal muscle tension and the resultant exerted force. Finally, the hand is densely innervated by specialized mechanoreceptors and clearly fulfills sensory as well as motor functions. In addition, the sensitivity of many of the rapidly adapting receptors is increased by moving the fingertips over a surface to be explored. Johansson and co-workers (Johansson and Vallbo, 1979; Johansson and Westling, 1981; and Westling and Johansson, 1980) have shown that these afferents are likely to be particularly activated during prehension; in return they exert a powerful excitatory influence over neurons of the motor cortex (Wiesendanger, 1973; Murphy et al., 1978; Strick and Preston, 1978, *this volume*). The net result is that during isometric prehension, cutaneous afferents of the hand and PTNs of the motor cortex represent two parts in a positive feedback loop. Moreover, this returning signal from cutaneous and proprioceptive afferents is not limited to the motor cortex, but probably includes all the motor centers of the cerebello-thalamo-cortico-brainstem system.

More recent physiological and morphological studies have altered and refined our understanding about how the corticospinal neurons influence the distal forelimb muscles in the primate. Fetz et al. (1980) have used the action potentials of PTNs to trigger repeated averages of distal limb muscle activity. A postspike facilitation with a mean latency of about 10 msec was observed for many neurons suggesting that these cells make monosynaptic connections with the MNs. In general, the discharge frequency of the cortico-motoneuronal units could be correlated with force and rate of force change under both isometric and auxotonic (i.e., elastic) conditions (Cheney and Fetz, 1980). A surprising finding of these studies was that the synchronous facilitation often occurred between a single PTN and activity in more than one muscle of the wrist and fingers (Fetz and Cheney, 1978, 1979; Fetz and Finnochio, 1975). In general this facilitation was observed most frequently among synergist muscles such as the extensors of the wrist and fingers (Fetz and Cheney, 1978), but occasionally facilitation of two antagonist muscles such as the biceps and triceps (Fetz and Finnocchio, 1975) could also be demonstrated. Shinoda et al. (1981) have labeled the terminals of corticomotoneuronal cells through the intraaxonal injection of horseradish peroxidase (HRP) together with the specific subgroups of ventral horn MNs in the cervical enlargement to which they project. The pictures in these studies show not only multiple terminations of a single corticospinal tract fiber with a MN pool, but also extensive ramification and terminals on other MNs as well. These terminal arborizations provide a morphologic explanation for extensive postspike facilitation between the muscles of the hand and wrist. The action of these neurons may have important physiologic implications. For example, the fixed and ordered recruitment of synergistic muscles may account for many species-specific behavior patterns. A second implication is that separate populations of corticomotoneuronal cells may control particularly incompatible patterns of muscle contraction such as the reciprocal or the coactive contraction of antagonist muscles.

Supplementary Motor Area

The supplementary motor area (SMA) occupies the medial part of the area 6 (see Brinkman and Porter, *this volume;* Tanji and Kurata, *this volume*).

It has frequently been observed in both humans and monkeys that, after frontal lesions involving SMA, there is a forced grasping of objects that make contact with the palm and fingers of the hand and questions are still pending about the nature of this mechanism (cf. Brinkman and Porter, *this volume*).

The discharge of single neurons in the SMA has been studied in monkeys trained to perform a maintained precision grip with either hand (Smith, 1979). An arm area of the SMA can be identified and distinguished from the surrounding tissue on the basis of the neuronal responses to somatic stimulation of the forelimb and the increased discharge of single units during movement of the upper extremity. Receptive fields were often very wide and units appeared to discharge in association with movement about several joints. Sixty-one of 134 recorded neurons increased their discharge frequency during a maintained prehension with the contralateral hand. No neurons were related to gripping with the ipsilateral hand. This observation apparently conflicts with a report by Brinkman and Porter (1979, and *this volume*) that the activity of many SMA units could be related to the movement of either the contralateral or the ipsilateral hand, but the critical difference between these two studies may be the motor task itself. In Brinkman and Porter's study the monkey is required to pull with the arm as well as pinch with the fingers. It may be that the widefield SMA neurons are bilaterally driven when proximal and axial muscles are active but not when only forearm muscles contract. This interpretation is strongly supported by the recent study of Tanji and Kurata (1981). During prehension, only a few neurons increased activity before the onset of activity in the forearm muscles, whereas most units became active about 100 msec afterward (Smith, 1979). The discharge frequency of only 2 of the 61 recorded neurons significantly increased when greater prehensile force was applied and no apparent modulation of activity related to the rate of force change *(dF/dt)* could be found.

In addition to lesions of the SMA of the previously trained primate, there is a characteristic inability to release hand-held objects (i.e., forced grasping), and there is a return to the rapid application of prehensile force seen in the early stages of learned grasping (Smith et al., 1981). Figure 5 illustrates the changes in the maintained pinch that occur after a unilateral ablation of the SMA. There is a significant increase in the rate of force change during the dynamic period and the mean static force is greater as well. Apart from the inability to release the transducer, the grasping resembles the early stages of learning the maintained precision grip (compare Figs. 3 and 5). The ratio of activity of the forearm flexors to the extensors also appears to be increased after the lesion.

The SMA appears to be present only in animals in which postural prehension is well developed such as the raccoon, porcupine, and various primates (Jameson et al., 1968; Lende and Woolsey, 1956; Penfield and Welch, 1951; Woolsey et al., 1950). The forced grasping seen after SMA lesions suggests that this cortical area may serve a special function in tonic or postural grasping. Since both cutaneous and kinesthetic afferents exert excitatory feedback on PTNs controlling prehension, it may be that the SMA is needed to interrupt this loop in order to suppress the grasp reflex and allow precision handling of objects.

Cerebellum

The cerebellum also plays an important role in controlling movements of the hand and wrist. Thach (1970*a,b*) demonstrated that both the Purkinje cells of the cerebellar cortex and the neurons of the dentate and interpositus nuclei become active prior to and during quick movements of the wrist. Subsequently, Thach (1975) found that the earliest changes in neuronal discharge frequency occurred in the dentate nucleus, followed by the motor cortex and interpositus nucleus. That the dentate is particularly important in the initiation of move-

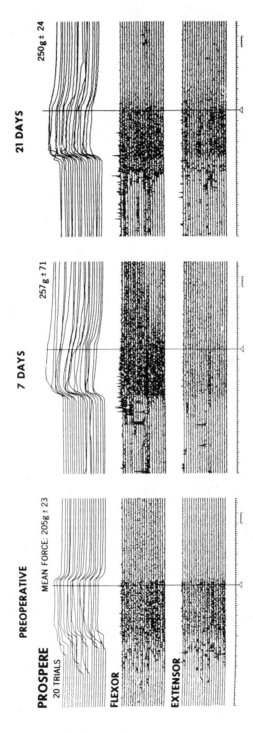

FIG. 5. Increase in both rate of prehensile force application and grasping force after a contralateral lesion of the SMA. The postoperative force traces (*middle top*) show the failure to release the transducer characteristic of forces grasping. *Lower traces* show simultaneous surface EMG from the forearm flexor and extensor muscles. *Time calibration*: 250 msec.

ment has been supported by evidence that electrolytic lesions or reversible cooling of the dentate cause a lengthening of reaction times (Lamarre et al., 1981; Vilis and Hore, 1980). More recently, Schieber and Thach (1980) has suggested that the interpositus nucleus, acting through the rubrospinal tract, is responsible for activation of gamma-motoneurons (see Fromm, *this volume*). Nevertheless, we believe it is probably that the rubrospinal pathway does not project exclusively to the gamma-motoneuron system, since in the monkey these neurons have also been shown to make monosynaptic termination on alpha-motoneurons (Shapovalov et al., 1971, 1974).

Experiments in our laboratory have shown that during isometric prehension the majority of cerebellar Purkinje cells, recorded in the forelimb area of the anterior lobe and identified by the conspicuous climbing fiber potentials, decrease their discharge frequency during isometric prehension. In contrast, the activity of other neurons, presumed to be inhibitory cells of the cerebellar cortex, increases during maintained grasping (Smith and Bourbonnais, 1981). Figure 6 illustrates an example of both Purkinje cell and non-Purkinje cell performance. A study of forearm muscles made during the same precision grip revealed that all the muscles of the hand and wrist were simultaneously active. An apparent paradox exists between the studies of reciprocal flexion and extension wrist movements in which Purkinje cells are excited (Thach, 1970*b*; Mano and Yamamoto, 1980), and isometric prehension, in which the majority of Purkinje neurons appear to decrease their activity (Smith and Bourbonnais, 1981). An hypothesis, based on observations by Tilney and Pike (1925) proposes that the cerebellar cortex plays a role in switching control from coactive to recipro-

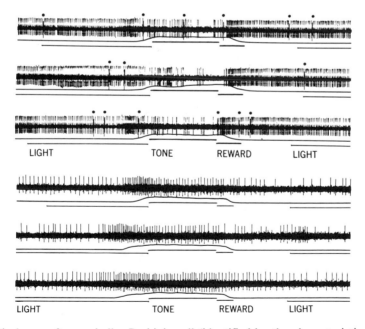

FIG. 6. Top: Discharge of a cerebellar Purkinje cell (identified by the characteristic climbing fiber discharge shown by *black dots*) on three consecutive prehensions. Bottom: Discharge of an unidentified cerebellar neuron (thought to be either a basket, stellate, or Golgi II cell) on three successive prehensions.

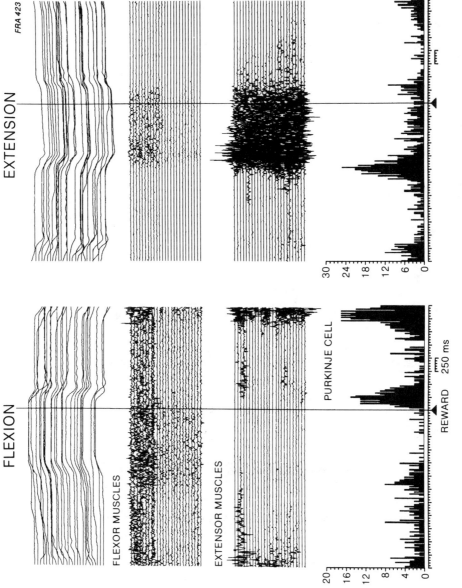

FIG. 7. **Left**: A series of wrist flexion displacements maintained for a 1-sec period (**top**), the surface EMG activity from the forearm flexor and extensor muscles (**middle**), and the histogram of summed Purkinje cell activity (**bottom**). A similar arrangement is shown for extension on the **right**. Upward deflection indicates flexion and downward extension.

cal muscular activity and vice versa (Smith, 1981). The central premise of this hypothesis is that the inhibition of cerebellar nuclear cells by Purkinje cells is ultimately translated into a relaxation of antagonist muscles. However, Mano and Yamamoto (1980) raise a serious objection to this proposition since, in general, cerebellar Purkinje cells increase discharge frequency during wrist movement regardless of whether the movement is in flexion or extension. However, it is important to know if all the synergist and antagonist muscles are reciprocally active in this task. We have studied this in monkeys required to flex and extend the open hand against a static load similar to the behavioral task used by Cheney and Fetz (1980). Our results demonstrate that, although the 3 flexor and 3 extensor muscles of the wrist contract reciprocally, some other synergist muscles of the forearm are active in both flexion and extension movements. However, all muscles behave reciprocally in isometric flexion or extension. Under these conditions it is difficult to affirm that any neuron discharging in a non-reciprocal way during movement, is related to the reciprocal prime mover muscles. It might be, as Mano and Yamamoto (1980) have argued, that the discharge is related simply to movement regardless of direction. Figure 7 illustrates an example of a Purkinje cell demonstrating a clear reciprocal discharge pattern with excitation during wrist extension and deactivation during wrist flexion. It is possible that the discharge of this particular Purkinje neuron acts to inhibit the flexor muscles of the wrist. Further investigations will attempt to examine the discharge of the same Purkinje neurons during both the reciprocal contraction and the coactivation of antagonist muscles.

SUMMARY

In 1931, Hughlings Jackson made a distinction between voluntary and automatic movements. According to Jackson the former are varied and purposeful gestures, whereas the latter include stereotyped movements largely under immediate sensory control. In this regard it may be said that the motor function of the primate hand includes both voluntary as well as automatic responses. Phillips and Porter (1977) noted that injury to the nervous system may often disrupt tactile exploration and skilled manipulation, although the same lesion may fail to interfere with automatic prehension used in posture and locomotion. The effects of selective brain destruction as well as the results from single-cell recordings from moving animals have indicated some of the different contributions made by various parts of the cerebello-thalamo-cortical motor system. Prehension offers a unique opportunity in which the interactions between motor commands and somatosensory afferents may be studied.

ACKNOWLEDGMENTS

The technical assistance of Mr. Gilles Blanchette and Mr. Jean Jodoin greatly contributed to the experiments conducted in our laboratory. We are grateful to Dr. James P. Lund for his constructive criticism of the manuscript. RF is an H. H. Jasper Postdoctoral Fellow.

Direct Electrical Stimulation of Corticospinal Pathways Through the Intact Scalp in Human Subjects

C. D. Marsden, *P. A. Merton, and H. B. Morton

National Hospital, Queen Square, London; Physiological Laboratory, Cambridge; and University Department of Neurology, Institute of Psychiatry, London, United Kingdom

Stimulation of the exposed cerebral cortex in conscious, unanesthetized humans presents unique opportunities that have long been exploited, notably by Foerster and by Penfield during brain surgery. However, the scope of these experiments was necessarily restricted with regard to number, duration, and elaboration. Hence, a technique that allows the use of healthy subjects without even breaking the skin, greatly widens the field of investigation. The present procedure arose out of experiments with D. K. Hill in which a method was developed that used high-voltage pulses of current for direct stimulation of human muscles (Hill et al., 1980). In January 1980, the same method was applied, with immediate success, to human motor cortex and visual cortex (Merton and Morton, 1980a,b,).

Earlier known attempts to stimulate the brain through the skull in man (Gualtierotti and Paterson, 1954; Merton, 1947, *unpublished observation;* Brindley, 1965, *unpublished observation*) used repetitive stimuli (either alternating current or pulses). The intensity of such stimuli that can be used is limited by pain. Merton and Brindley do not claim to have succeeded in stimulating. Gualtierotti and Paterson may have done so (their paper is mainly about anesthetized baboons, with only one human experiment described); but their method was not continued. The present technique limits, and often completely avoids pain by using single, brief pulses. Some discomfort—a sort of jolt—is inevitable; but is less severe than is often felt with maximal stimulation of peripheral nerves.

The stimulator closely resembles an old-fashioned condenser-thyratron device except that it discharges the condenser through one or more voltage-controlled rectifiers (thyristors). To start with, a 0.1 μF condenser charged to about 1,000 V was used; it was discharged through 100 Ω in shunt with the subject to keep the time constant down to 10 μsec. Later a 1 μF condenser at 350 to 500 V with a time constant of discharge of about 100 μsec was found to be as effective as a stimulus and not obviously more painful. For safety, and to be free of shock artifact when recording, the stimulus is passed through a pulse transformer. A low-output impedance is necessary, since the interelectrode resistance is highly nonlinear and falls to low values when several hundred volts are applied. It

---
*To whom correspondence should be addressed: The Physiological Laboratory, University of Cambridge, Downing Street, Cambridge CB2 3EG, United Kingdom.

was for this reason that a conventional square-pulse nerve stimulator that we tried, although it gave adequate open-circuit voltage, was only barely able to stimulate, and then only with a pulse length that accentuated pain.

Electrodes were either ordinary stuck-on silver-cup electroencephalograph (EEG) electrodes, filled with jelly, or a pair of saline-soaked pad electrodes about 6 cm apart on a handle. To stimulate the hand and arm, the electrode that went positive during the stimulus was placed over the surface marking for the arm area of the motor cortex (halfway down a line from 1 cm behind the vertex to 2 cm above the superior-anterior insertion of the pinna). The negative-going electrode was several centimeters forwards. This placement was successful at the first attempt in all 20 subjects who have been tried. However, to get the lowest threshold may require adjustment; the position of the best point is critical to within 1 or 2 cm on the scalp.

With the stimulus on the hand area, single shocks in a relaxed subject cause twitchlike movements of the opposite fingers and wrist. It was noticed at once that the threshold for exciting a muscle was greatly lowered if, instead of relaxing, a voluntary contraction of that muscle was made. This phenomenon (apparently newly observed) makes it possible to stimulate a particular muscle selectively. So generally true is this that when records were to be made from adductor pollicis (see below), the full experiment was set up without first looking to see that adductor pollicis would contract to cortical stimuli. Cortical stimulation is unique; you simply decide what muscles you want to excite and it excites them, within limits, that is; and providing the subject cooperates. At any rate, with the muscle contracting voluntarily to start with, the threshold for cortical stimulation is not obviously lower for pairs of stimuli, or for short bursts at intervals of 2 to 10 msec; which is not, perhaps, what might be expected from animal experiments (Phillips and Porter, 1977).

When electrical records are taken from a contracting muscle with surface electrodes, it is found that each shock causes a typical synchronous action potential with a sharp latency (Fig. 1). The duration of the action potential (15–20 msec) is not much greater than that from stimulation of the motor nerve, so the cortical volley must be fairly synchronous. The latencies observed are so brief that fast corticospinal fibers [probably making monosynaptic connection with the spinal motoneurons (MNs)] must be involved. Latencies for a tall subject, from the cortical stimulus to the start of the evoked muscle action potential are 16 msec for forearm muscles, 22 to 25 msec for hand muscles, and 34 msec for tibialis anterior. This very easily performed measurement of conduction time in the motor pathway has obvious clinical applications.

Another important quantitative relation has been observed. Due to the dispersion (slight though it is) of the cortical volley, it

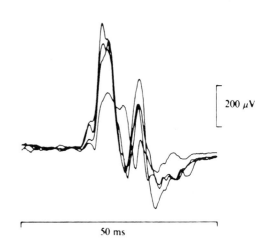

FIG. 1. Stimulation of the arm area of the motor cortex in subject P.A.M. The records shown are of action potentials from the contracting muscles. Stimulation is at the start of the sweep. Four records are superimposed. The latency of the responses was 16 msec. (From Merton and Morton, 1980a, with permission.)

is not possible to say what fraction of the MNs are involved in a cortically evoked motor action potential by measuring its peak height, because dispersion is bound to lower the peak height of the action potential compared to one from stimulation of the motor nerve near the muscle. We resolved this problem by measuring the mechanical force produced in a cortically evoked twitch. A convenient movement is adduction of the thumb, which is carried out mainly by the ulnar-supplied adductor pollicis but also by median-supplied muscle in the thenar eminence. To lower the cortical threshold (as described above) so that the adductor muscles were activated selectively, the subject adducted his thumb against a strain gauge, keeping the force constant at 4 kg (about half maximal). The force developed in a cortically evoked twitch was then compared with the force in a twitch elicited by simultaneous maximal stimulation of the ulnar and median nerves at the wrist (with a similar background force of 4 kg). Figure 2 shows that a cortical stimulus (which was quite moderate) causes a twitch that closely matches the electrically elicited maximal twitch (Marsden et al., 1981a). It thus appears that a single cortical shock has access to all the motor units in the adductor muscle and can cause a maximal twitch.

A possible alternative interpretation is that the cortical shock excites only a proportion of all the motor units, but excites them more than once, thus producing a large twitch. This possibility is more or less ruled out by double stimulation at the cortex and wrist together, which gives the third superimposed twitch in Fig. 2. This twitch is no bigger than the others. Since the wrist stimuli will only block by collision the first impulse in each fiber of a cortical discharge, second and later impulses, if there were any, would reach the muscle and cause a twitch larger than maximal for a single volley. This does not occur with cortical stimuli of the size in question; so multiple firing of MNs as a result of the single cortical stimulus cannot be significant here. With larger cortical shocks there does appear to be multiple firing.

Cortical excitation is followed by depression. With a background of voluntary contraction this shows up as a silent period in the EMG lasting ≥ 100 msec, and starting immediately after the cortically evoked action potential (the beginning of the silent period can be seen in Fig. 3). A presumably related depression during repetitive stimulation of the exposed cortex was described by Penfield (1958). In a small series of experiments on flexor pollicis longus, the threshold for a silent period appeared to be close to the same as the threshold for excitation of the muscle. This suggests that refractoriness and Renshaw inhibition of MNs might be an important factor at the start of the silent period. But there seems to be more to it than that. With big cortical

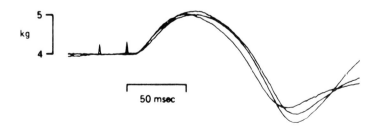

FIG. 2. Three superimposed twitches from the adductors of the thumb. Initial force of 4 kg. The twitch from stimulation at the wrist is caused to start at the same moment as the cortically evoked twitch by delaying it by 22 msec. The third twitch is from double stimulation at cortex and wrist. (From Marsden et al., 1981a, with permission.)

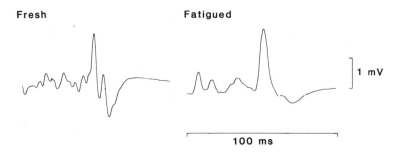

FIG. 3. Action potentials from the adductor pollicis with stimulation of the opposite motor cortex. The stimulus was delivered 20 msec after the start of the sweep and caused a response about 25 msec later. Each trace plotted is the average of 4 sweeps. Four minutes of severely fatiguing contractions between the two traces has little effect. (From Merton et al., 1981, with permission.)

stimuli, the silent period is prolonged, completely interrupting the background voluntary force, so that the downward overshoot after the twitch seen in Fig. 2 continues downwards, sometimes to the baseline. This is not seen with the silent period following a stimulus to the motor nerve (Merton, 1951).

During voluntary fatigue there is no sign that the cortical threshold changes, although the method would not reveal small alterations (Merton et al., 1981). The muscle used for experiments on fatigue was again the adductor pollicis, with surface electrodes. The subject pulled against a strain gauge with his maximum effort. The scalp electrodes for this experiment were the stick-on type for stability. To get a baseline, 4 single cortical shocks of moderate strength were delivered during brief maximal contractions, with 1 min of rest in between each contraction. The average of the evoked muscle action potentials is on the left in Fig. 3. Then the subject made a maximal effort lasting 4 min, during which the force exerted decreased to about a quarter of its initial value. The circulation to the arm was not arrested. Toward the end of this contraction, 4 cortical stimuli, of the same strength as before, were delivered at 10-sec intervals. The average response to these 4 stimuli is shown on the right in Fig. 3. It is not smaller than response in fresh muscle. Although size of recorded muscle action potential may be influenced by changes in the synchrony of the motor volley, duration of the individual fiber action potentials contributing to it, or size of the individual action potentials, it seems improbable that these factors are conspiring to conceal any large change in the size of the cortical volley. The results, at any rate, give no evidence that the cortical pathway fatigues during such a contraction. There is also, of course, no evidence that any other link in the pathway to the muscle, e.g., the MN or the neuromuscular junction, has begun to fail under these conditions.

The experiments described above are only the beginning of work on the human motor cortex. Muscles all over the body (face, neck, shoulder, arm, and leg) have been excited from the cortex when their threshold is lowered by voluntary contraction. Excitation is not always the first event seen; but these experiments are in a preliminary stage.

The other part of the brain that has been reached by stimulation through the scalp is the visual area (Merton and Morton, 1980a,b). The observations here are wholly subjective. The stimuli must be larger than for the motor cortex (1,700 V), 10 μsec time constant, in our original experiments; and are apt to be somewhat unpleasant, jerking the head by spread of current to the neck muscles and also causing some pain. Only two subjects have been successfully stimulated out of about half a dozen tried with various degrees of determination. With pairs of electrodes over the occipital region a single stimulus causes a brief phosphene in the visual field. The phosphene subtends about 5°, and may be roughly circular or, with a horizontal pair of electrodes, it may be extended into a horizontal band. It has a faint, colorless and indefinitely structured appearance, with a few wavy bright lines, rather difficult to see, as the phosphene is very brief. It may be superimposed on a generalized phosphene due to spread of current to the retina. The cortical phosphene is confirmed as such because it moves in the correct way in the visual field when stimulating electrodes are moved. With electrodes at the level of the occipital pole the phosphene (in one subject) was referred to a level 1 or 2 degrees below the fixation point. When the electrodes were moved upward on the scalp, the phosphene moved downward, and vice versa. Similarly, when the electrodes were moved laterally, to the right or to the left, the phosphene moved in the field in the opposite direction. Vision was not obviously interrupted by the largest cortical stimuli we used. No gap could be detected in the appearance of a cathode-ray tube sweep.

Supplementary Motor Area and Premotor Area of Monkey Cerebral Cortex: Functional Organization and Activities of Single Neurons During Performance of a Learned Movement

*Cobie Brinkman and Robert Porter

Experimental Neurology Unit, The John Curtin School of Medical Research, The Australian National University, Canberra, Australia

Studies of movement control have traditionally examined connectivity of brain and spinal cord, using anatomical methods to trace pathways (Kuypers, 1980; Jones and Burton, *this volume*) and physiological stimulation and recording in anesthetized preparations (Phillips and Porter, 1977). A wealth of functional data have also been collected in behavioral studies of patients with damage to motor centers (Denny-Brown, 1966) and of animals with selective lesions in brain motor regions. Recently, the natural discharges of neurons were studied in the brains of intact, behaving animals trained to perform a variety of movement tasks (Evarts, 1966; Porter et al., 1971). The connectivity of neurons can be tested through stimulation of sites remote from the recording area and through natural stimulation in limbs of animals trained to stay passive and relaxed to allow manipulation (Lemon and Porter, 1976; Lemon, 1981). In this way, it is possible to gain direct information about functional relationships of a normal nervous system, uninfluenced by anesthetics and without damage from disease or lesions. Although recordings in conscious animals have been made in the cerebellum (Harvey et al., 1977; Thach, 1978a; Harvey et al., 1979), basal ganglia (DeLong and Georgopoulos, 1979), and ventrolateral thalamus (Horne and Porter, 1980; Strick, 1976), the majority of the observations has come from the primary motor area (MI) of the precentral gyrus of the cerebral cortex (Evarts and Tanji, 1976; Lemon and Porter, 1976; Tanji and Evarts, 1976; Evarts and Fromm, 1978; Lemon, 1981; Porter, *this volume*), which is the origin of the majority of direct and indirect pathways to the interneurons and motoneurons (MNs) of the spinal cord (Jones and Burton, *this volume*). Much less attention has been paid to cortical areas which lie upstream from the precentral motor cortex. Studies of the premotor (PM) areas should shed light on the way in which the motor cortex itself may be instructed. Anatomically, one of the most important projections to the motor area (Brodmann's area 4) is derived from a belt of cortex immediately anterior to it, Brodmann's area 6 (Fig. 1). On the basis of anatomical and functional data (see below) this cytoarchitectonic homogenous area may be divided

---

*To whom correspondence should be addressed: The John Curtin School of Medical Research, Australian National University, P.O. Box 334, Canberra City, A.C.T. 2601 Australia.

in a medial part called the supplementary motor area (SMA, labeled M II in Fig. 1; after Woolsey et al., 1958) and a lateral part, the PM. This chapter deals with recordings from neurons within these areas in conscious monkeys.

I. SMA

Historical Background

The term SMA was first used by Penfield and Welch (1951) who distinguished in man and monkeys between a primary motor area, which included area 4 and part of area 6 on the convexity of the hemisphere and a secondary or supplementary motor area situated in the superior frontal gyrus and made up of mainly area 6 on the medial wall of the hemisphere above the cingulate sulcus. Focal stimulation of the primary motor area resulted in localized movements of the contralateral half of the body, whereas SMA stimulation (usually with more prolonged and stronger stimuli) elicited complex synergistic movements of the contralateral upper extremity and body and occasionally, of the ipsilateral side as well. A somatotopic arrangement was found, and it was possible to outline a distorted figurine of the contralateral body half in both the primary and the supplementary motor area. In Fig. 1, these figurines are shown for the monkey brain (simiusculi) as described by Woolsey et al. (1952). This arrangement of motor areas has been found in a wide variety of species, including rat (Woolsey, 1958), dog (Gorska, 1974), rabbit (Woolsey, 1958), raccoon (Jameson et al., 1968), and squirrel monkey (Welker et al., 1957). The difference in movements obtained from stimulation of the two motor areas is similar across species and suggests that SMA is involved in the control of movement in a manner different from the primary motor area.

Possible Functions of SMA in Movement Control

Only the anatomical connections of SMA have been worked out in some detail (see

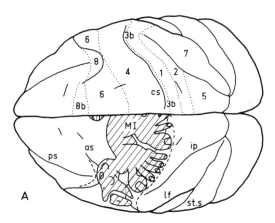

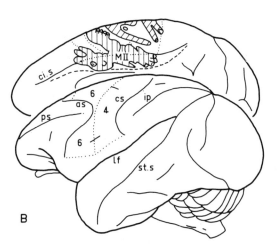

FIG. 1. Some cytoarchitectonic subdivisions and motor maps of the macaque brain. **A**: Superior aspect, **B**: lateral aspect and mirror view of part of the medial surface of the cerebral hemisphere. Cortical fields are numbered according to Broadmann (Roberts and Akert, 1963). The somatotopic arrangements found in stimulation studies of the frontal lobe have been outlined as distorted monkey figurines (simiusculi) and are redrawn after Woolsey et al. (1952). The area normally buried within sulci has been folded out and is indicated by an *interrupted line*. MI, in **A** and partly extending onto the medial surface as shown in **B**, forms the primary motor area. MII, in **B**, is normally almost entirely hidden in the median fissure and forms the SMA. Note that MI exceeds the boundary of cytoarchitectonic area 4 and extends rostrally into area 6, and that MII seems to lie partially in area 4. *Abbreviations:* as, arcuate sulcus; ci.s, cingulate sulcus; cs, central sulcus; ip, intraparietal sulcus; lf, lateral fissure; ps, principal sulcus; st.s, superior temporal sulcus.

Jones and Burton, *this volume*). Very few physiological studies in anesthetized animals have been done, and the role of SMA in motor control, as revealed by deficits after SMA lesions, remains uncertain. Nevertheless, two theories have been suggested: (1) SMA is especially involved in the control of posture, and (2) SMA is a center of sensorimotor integration, where somatosensory feedback from wide areas of the body is put together and then passed on to primary motor cortex (Wiesendanger et al., 1973). To these may be added a third hypothesis, which attributes to SMA a more direct role in the programming of movements, in view of its position upstream of MI; i.e., SMA would act as a feed-forward center rather than one of feedback (Tanji and Taniguchi, 1978; Brinkman and Porter, 1979; Roland et al., 1980; Tanji et al., 1980).

Is SMA involved in the control of posture? Stimulation of SMA in anesthetized animals elicits complex movements of the extremities, often at more than one joint, and only rarely have discrete distal extremity movements been found (Woolsey et al., 1952; Penfield and Welch, 1951; Welker et al., 1957). After ablation of MI (but not after section of the pyramidal tract, Woolsey, 1975), these responses were abolished and, as a pure SMA effect, only some gross proximal movements were observed (Wiesendanger et al., 1973). Unilateral ablation of SMA produced as a lasting deficit a bilateral hypertonia at the shoulders, while bilateral lesions resulted in increased tone especially of flexor muscles of the elbow and shoulder (Travis, 1955; but see Coxe and Landau, 1965). The lack of distal extremity movements obtainable from stimulation of SMA and the mainly proximal effects of SMA lesions are taken to imply that SMA is concerned mainly with the control of body posture, and not individual extremity movements.

Is SMA a feedback or a feed-forward area in the control of movement? In anesthetized monkeys, neurons in SMA were influenced by stimulation of more than one peripheral nerve, and of nerves in both the fore and hind limbs. This led to the idea that SMA is an area of sensorimotor integration, important in providing feedback from the moving limbs to MI (Wiesendanger et al., 1973). This notion is strengthened when the afferent connections of SMA are considered. It receives afferents from a large part of ipsilateral primary somatosensory cortex (areas 1 and 2) as well as from secondary somatosensory area (SII); reciprocal connections exist with area 3a, and with area 5 of the posterior parietal association cortex (Pandya and Vignolo, 1971; Chavis and Pandya, 1976; Jones et al., 1978; Vogt and Pandya, 1978; Jones and Burton, *this volume*). Its thalamic afferents are derived from a number of nuclei of the ventrolateral thalamus such as the nucleus ventralis anterior (VA), the nucleus ventralis lateralis, both pars oralis (VLo), and caudalis (VLc), the nucleus X, the centrum medianum and other parts of the intralaminar nuclei (Kievit and Kuypers, 1977; Bowker et al., 1979). Some of these nuclei are also recipients of projections from cerebellar nuclei (Mehler and Nauta, 1974; Kievit, 1979; Thach and Jones, 1979; Kalil, 1981) and basal ganglia (Mehler and Nauta, 1974; Kim et al., 1976). Thus, SMA can rightly be considered as an area of sensory convergence. Its thalamic afferents point to it as a major site of reentry of corticosubcortical circuits through the cerebellum and basal ganglia. Integration of somatosensory information reaching SMA from all these sources could result in a modulation of MI cortical activity through changes in the tonic inhibitory influence said to be exerted by SMA on MI neurons (Wiesendanger et al., 1975).

On the other hand, on the basis of its efferent connections, SMA could be expected to play a much more direct role in the organization and control of movement. Its cortical efferents seem to be preferentially directed to the ipsilateral cortical motor areas (areas 4 and lateral 6) and, through the corpus callosum, with these same areas and SMA in the opposite hemisphere (DeVito and Smith, 1959; Pandya and Vignolo, 1971). Likewise, its subcortical projections are all directed to structures considered to be parts of the motor system. Its thalamic projection is reciprocal (Brinkman, *unpublished observation*). Bilateral projections exist to the striatum: caudate nucleus, putamen, and claustrum (Kunzle, 1978), and to the red nucleus (Kuypers and Lawrence, 1967; Kunzle, 1978; Hartmann-von Monakow and Akert, 1979); projections to pontine nuclei are ipsilateral (DeVito and Smith, 1959; Brodal, 1978; Kunzle, 1978). Although long a point of contention (Bertrand, 1956; DeVito and Smith, 1959), a spinal projection from SMA has now been firmly established (Coulter et al., 1975; Jones and Wise, 1977; Biber et al., 1978; Palmer et al., 1981). The pattern of these efferent connections would allow SMA to be involved in the programming of movements, influencing the primary motor areas as well as the subcortical motor centers directly.

It is in this framework that the recording studies in conscious monkeys described below were initiated. In such a preparation, not only would it be possible to investigate motor aspects of SMA function to determine whether the area was especially involved with postural adjustments or proximal movements, but also to examine whether SMA neurons received, or were influenced directly by somatosensory information from the moving limb or other parts of the body.

Natural Activity of SMA Neurons in Conscious Monkeys Performing a Learned Motor Task

The major study in this area is that by Brinkman and Porter (1979). Since the same techniques as for MI (Lemon et al., 1976; Lemon, 1981) were used, direct comparison of the characteristics of neurons in SMA with those of MI was possible. Monkeys were trained to pull a horizontal lever, spring-loaded, or electromagnetically controlled, into a target zone, reaching of which was indicated by the onset of a 600-Hz tone. This signaled the monkey to release the lever and retrieve a small food reward presented in a standard position. Since SMA might influence movements bilaterally (Travis, 1955) the animals were taught to do this task with either hand. No restraint was used on the other extremity as the animals, after a period of training, tended to leave this arm quite relaxed as evidenced by the absence of electromyographic (EMG) activity. In order to investigate peripheral afferent input into SMA neurons, the animals were also trained to allow manipulation of joints and cutaneous stimulation without struggling, staying completely relaxed. In view of the convergence of inputs found in earlier experiments, this exploration included not only the arms but the entire body surface. In addition to recording the discharges of neurons in SMA, EMG data were collected simultaneously from representative muscles in both upper limbs, and stimulating electrodes were placed into the corticospinal (pyramidal) tract of some animals for identification of SMA cells sending their axons into this tract. Since only the animal's head was fixed, leaving the monkey otherwise free to move (see Fig. 2), whenever a neuron was found with its discharge modulated in relation to the movement, attempts could be made to establish the particular aspect of movement with which it was associated (e.g., proximal or distal, flexion or extension). By presenting the food reward in a variety of planned positions, the animal was required to repeat a particular movement. This allowed changes in elbow and shoulder movements to be explored; distal movements could be studied using other strategies such as holding the forearm in a fixed position and presenting food in a manner that required wrist flexion or wrist extension, for instance, for successful retrieval. Discrete movements of the digits were achieved by using special test boards (Brinkman and Kuypers, 1972) from which food could be recovered only by using opposition of index finger and thumb in a precision grip. Movie films were taken of all animals and careful histology with serial sections through the recording site was done on completion of the experiments.

Identification of SMA in the Conscious Animal

The movements evoked by stimulation of SMA in the anesthetized animal are particularly sensitive to the level of anesthesia. Microstimulation of the medial wall of the hemisphere in conscious monkey using the same parameters as for MI (where localized muscle contractions are readily obtained) never resulted in responses (Smith, 1979; Wise and Tanji, 1980; Palmer et al., 1981; but see Macpherson et al., 1982). In the recording experiments the medial wall of the hemisphere was systematically explored for neurons responsive during movements of the upper extremity, from a point a few millimeters anterior to the medial tip of central sulcus (see Fig. 1) to a point close to the rostral pole of the hemisphere. It was found that neurons whose activity was modulated during the movement task were concentrated mostly in an area above the cingulate sulcus on the medial surface of the hemisphere but extending for a few millimeters on to the lateral surface as well, and located at the anteroposterior level of the arcuate

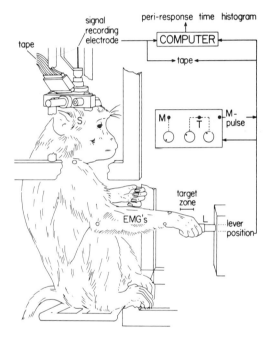

FIG. 2. Experimental design for recording of neuronal activity in the supplementary SMA and PM areas in the conscious monkey as employed by Brinkman and Porter (1979a,b), and drawn from a single frame of a movie film. A metal frame is attached to the monkey's skull over a suitably placed craniotomy and the head can be firmly fixed by bars attached to this frame and the cage. Mounted on the headpiece is an electrode carrier and a multipin socket connected to EMG recording electrodes placed in representative muscles and stimulating electrodes implanted into the pyramidal tract(s). The monkey has been trained to pull a horizontal lever (L) into a target zone. Lever position is measured and from this signal a reference pulse (M) is derived which indicates the start of the lever movement. This pulse and the lever position signal are used by a computer to construct a histogram indicating the relationship between lever movement and neuronal discharge during a number of pulls. These periresponse-time histograms are displayed on a screen in the laboratory and also printed out. Also shown in the figure and used during filming is a set of lights indicating the start of lever movement (M), reaching of the target zone (T), and overshoot (third light, not marked). Measurement of joint angles is facilitated by paint dots on shoulder, elbow and wrist joints. Only the animal's head is fixed and panels can be removed from around the body, thus leaving the animal free to carry out a wide range of movements and allowing the experimenter to test afferent inputs into neurons from peripheral receptors in skin, joints, and muscles from the entire body.

sulcus (see Fig. 1). More laterally on the convexity, responsive neurons were found only infrequently, and encountered more often only at the concavity of the arcuate sulcus; such cells were considered to be part of the premotor area (see below). The responsive zone on the medial wall was found to occupy area 6, and to coincide to a large extent with the forelimb representation area of the classical SMA motor map (Woolsey, 1958). Electrode penetrations posterior to this area rarely yielded any neurons with discharges modulated during forelimb movements. Anteriorly, a few neurons were found related in their activity to some aspects of vision or eye movements. Still more anteriorly, no clear modulation was found in relation to movement performance.

Movement-Related Neurons in SMA

In 3 monkeys, the activity of 271 neurons was recorded while they performed the task first with the contralateral and then with the ipsilateral extremity. When 14 cells that seemed to be related to visual events associated with the task were excluded, 214/257 cells (79%) were related to some aspect of movement performance; 216/229 (94%) changed their discharge regardless of whether the contralateral or ipsilateral limb was used.

The movement-related neurons of SMA could be divided into three groups: those related to proximal movements (92/215, 43%), those related to distal movements (108/215, 50%), and those neurons which showed changes in firing throughout the execution of the motor task and not during only one aspect of it (15/215, 7%). Most neurons, both those related to proximal and those related to distal movements seemed to increase their discharge before the onset of a particular movement and their activity decreased subsequently, but some neurons maintained a high discharge rate while a given position was maintained. Few neurons showed activity related to a maintained posture only.

Figures 3 and 4 show the relationship between neuronal activity and movement for a cell associated with proximal and distal movement, respectively. Proximal movements occurring during the task performance were flexion, extension, and adduction at the shoulder and flexion and extension at the elbow when reaching for the lever or food or when pulling the lever, and supination during food consumption. Distal movements were those of the wrist associated with positioning of the hand on the lever before pull and during lever release, mainly flexion and extension, and ulnar deviation during pull (see figurines in Figs. 3 and 4); extension and flexion of fingers to release or grasp the lever, and opposition of index and thumb (precision grip) for food retrieval. Figure 3 shows a neuron whose activity was always related to adduction of the arm at the shoulder. The impulses occurring during a period of 1 sec before and 2 sec after the start of the lever movement (indicated by 0) during 20 successive pulls were stored by computer and displayed as a periresponse time histogram (see Fig. 2). The 3-sec timespan allowed for complete execution of the movement sequence of reaching for the lever, pulling and releasing it, reaching for the food reward, and retrieval and consumption of the food. The top histogram is that obtained when the contralateral limb was used, the bottom one is that for the ipsilateral extremity. The traces below each histogram are an indication of the averaged lever position signal (see Fig. 2).

The neuron in Fig. 3, associated with adduction, showed an increase in its discharge whenever adduction occurred. First, when the animal pulled the lever toward its body (Fig. 3, top figurine on the right), second, during the reach for food when the animal had to project the limb across the body mid-

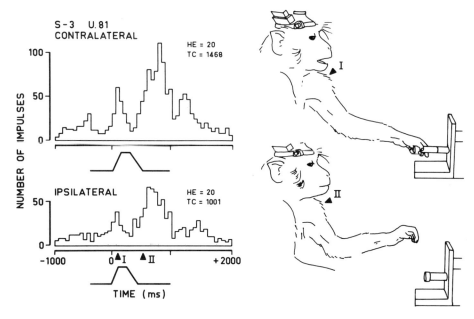

FIG. 3. Periresponse-time histograms of a neuron in SMA always associated with a proximal movement, adduction of the arm at the shoulder. The *ordinate* shows the number of discharges and the *abscissa* the time before and after the start of lever movement at time 0. For this neuron, these times were 1,000 and 2,000 msec, respectively. The average lever excursion during a number of successive pulls (histogram events, HE, in this case 20) is indicated by the trace below each histogram. TC: Total number of impulses occurring during these pulls. The neuron showed increases in its discharge whenever adduction occurred. The figurines on the right were taken from single frames of a video tape when adduction could first be detected. The times the frames were taken are indicated by the *triangles* below the histograms. The neuron was modulated with movements of both the contralateral and ipsilateral extremities and the shape of the histograms is quite similar. This was a common finding for SMA neurons. Neuronal activity also seems to precede the movement.

line (Fig. 3, bottom figurine on right), and third, when the food was placed into the mouth. Each time, the increased discharge of the neuron is reflected in a peak of the histogram. The shapes of the histograms for task performance with each arm are similar; slight differences in timing occur which correspond with a difference in speed of performance. The animal normally preferred to use its ipsilateral (left) hand, and movements of the preferred hand were faster. Similarity of histogram shapes for individual neurons obtained with movements of either limb was a common finding among animals, especially for those neurons associated with proximal movements. The number of impulses recorded could be almost the same for either extremity as well.

Neurons related to distal movements tended to show somewhat more differentiation in histogram shapes depending on which hand was used for the task. This was the case particularly for neurons related to movements of fingers and thumb. In normal monkeys, the preferred hand shows a clearer differentiation and fractionation of distal movements than does the nonpreferred hand. The neuron in Fig. 4 was always associated with a distal movement, extension of wrist. Analysis of the movie film showed that extension of the contralateral wrist occurred at the start of lever release (Fig. 4,

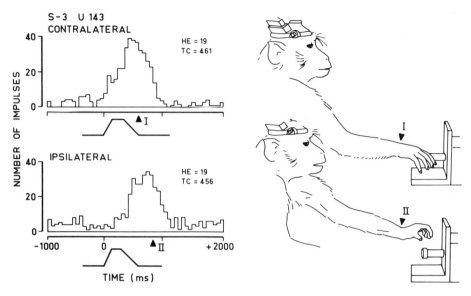

FIG. 4. Periresponse-time histograms of a neuron in SMA always associated with a distal movement, extension of the wrist as shown in the figurines on the right. Such distal neurons were found as frequently in SMA as were proximal ones. The slight delay in changes of activity observed for the ipsilateral hand was one reflected in the movement performance. Ipsilateral wrist extension occurred somewhat later.

top figurine on right), but extension of the ipsilateral wrist occurred during the reach for the food after the lever pull (Fig. 4, bottom figurine on right); consequently, the maximum increase of neuronal activity was shifted in time. There was a tendency for distal neurons to have a smaller number of discharges associated with movement of the ipsilateral hand than the contralateral. Nevertheless, perhaps their most striking property was that, like the proximal cells, they displayed modulation with both contralateral and ipsilateral movements. Distal extremity movements are controlled by pathways originating in the contralateral half of the brain only (Brinkman and Kuypers, 1973), and a bilateral control of SMA over such movements might have to be through the corpus callosum or through effects on subcorticospinal brainstem descending systems.

For some neurons, it was possible to compare the modulation in their activity with that of the EMG recorded simultaneously in representative muscles. The EMG recorded during the same set of pulls was analyzed by computer in the same way as neuronal activity had been (see Fig. 11). It was found that the neuronal activity tended to lead changes in EMG, usually by 120 to 180 msec. Similar time lags have been reported by Tanji and Kurata (1979, *this volume*). These findings suggest a role for SMA in the initiation of movement.

Pyramidal Tract Neurons in SMA

It is now established, both by anatomical (Coulter et al., 1975; Jones and Wise, 1977; Biber et al., 1978; Brinkman, *unpublished observations*) and physiological means (Bertrand, 1956; Palmer et al., 1981) that SMA has an efferent projection to spinal cord presumably through the pyramidal tract (PT) (Kunzle, 1978), but only Brinkman and Porter (1979) have used antidromic stimulation of the PT in recording experiments in an attempt to identify SMA neurons as PT neurons (PTNs). Only 8/159 (5%) of neurons tested were found to be PTNs, with antidromic latencies ranging from 1.2 to 3 msec (cf. Palmer et al., 1981). As a group, their behavior did not seem to be different from that observed in non-PTNs within SMA but, until a larger sample has been collected such a comparison must remain tentative. The most salient feature of the PTNs was that they too were associated with similar movements carried out using either limb. An example of an SMA—PTN is shown in Fig. 5.

Afferent Input into SMA Neurons

The problem regarding afferent inputs from peripheral receptors into SMA is two-fold: (1) Can such inputs be demonstrated in the passive, relaxed animal? and (2) Does somatosensory information generated during movement, such as that elicited by imposing a disturbance on the movement, influence SMA neurons? The first question was investigated by Brinkman and Porter (1979). Their monkeys had been trained to allow natural stimulation of peripheral receptors of limbs, body, and head while they stayed passive and relaxed as indicated by complete absence of EMG activity. Only 24/175 (14%) of neurons tested could be activated in this fashion. The most effective stimulus was joint movement (21/24) in one direction only (19/21) but more than half of this group could be driven by movements of more than one joint of the contralateral arm, and sometimes by stimulation of the contralateral leg or the ipsilateral extremities as well. Cells responding to skin or hair receptor stimulation, or to both cutaneous stimulation and joint rotation were found infrequently. Latencies for responses to cutaneous or muscle palpation, as determined with a mechanoelectric probe, ranged from < 20 to 30 msec. However, responses of a given cell were often not faithful and became difficult to elicit with repeated stimulation. Taken together, these data suggest convergence of afferent inputs onto individual SMA neurons (cf. Wiesendanger et al., 1973). However, these inputs can be found only with difficulty, and only for a limited number of cells, in a passive and relaxed animal. Such results contrast with Smith (1979), who reported that 13/24 (61%) of SMA neurons responded to stimulation of contralateral arm in experiments with no simultaneous recording of EMGs and without extensive training of the monkey to remain relaxed, so that startle type responses were elicited from many neurons. It is likely that the low proportion of responsive cells found by Brinkman and Porter (1979) reflects a true property of SMA neurons.

Responses of SMA neurons to an im-

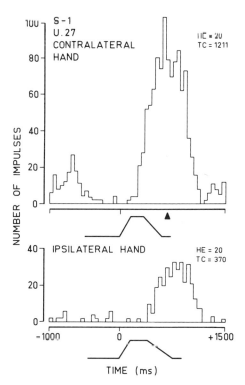

FIG. 5. Discharge pattern of a PTN in SMA. The neuron was always associated with finger extension such as occurred during lever release. The triangle refers to the moment this movement could first be detected on the video tape (figurine in Fig. 11 shows this movement for the animal).

posed disturbance have been sought by Wise and Tanji (1980, 1981) in the hindleg representation area, and by Tanji et al. (1980) and Brinkman (*unpublished observation*) in the forelimb area. Only a small proportion of neurons in the hindlimb area (4 to 8% of those related to the movement, up to 10% of the total sample) (Wise and Tanji, 1980) were found to be responsive to the peripheral disturbance. In the forelimb area, 0 (Brinkman) to 5% (Tanji et al., 1980) could be influenced. Thus, the significance of the abundant connections from somatosensory areas with SMA remains unclear. Information reaching these areas does not seem to get through to many cells in either the passive or the moving animal (see Tanji and Kurata, *this volume*).

Somatotopy in SMA

Stimulation of SMA in anesthetized animals resulted in complex movements which tended to reveal some somatotopic pattern, less strict than that found in MI, with face representation located most rostrally, hindlimb area most caudally, and forelimb and trunk area in between (see Fig. 1, MII simiusculus). After ablation of the precentral motor area, this somatotopy was lost, and it was claimed that stimulus spread from SMA to MI was responsible for the pattern seen in intact preparations (Wiesendanger et al., 1973). In stimulation experiments in man, somatotopy has not been very obvious, as vocalization was found to be the predominant response, and extremity movements were less common (Penfield and Welch, 1951). However, careful inspection of the individual cases suggests some clustering of points from which similar responses could be obtained.

In the conscious animal, somatotopy can be investigated in more detail. It has already been mentioned that the area responsive during movements of the upper limb is well defined, and that within this area, neurons associated with distal movements were found as frequently as those related to proximal movements. When the distribution of these two groups within the area was plotted, the former group tended to occupy an area located more rostrally than the latter (Fig. 6). There was some overlap between the areas (Brinkman and Porter, 1979). Similar results were obtained by Tanji and Kurata (1979, *this volume*) who also found that neurons related to distal movements tended to be found deeper on the medial surface than proximal ones, a reversal of the arrangement described by Woolsey et al. (1952) (Fig. 1).

In both studies of conscious animals and stimulation studies, the forelimb area is restricted to cytoarchitectonic area 6. However, with regard to the hindlimb area, comparison of the MII simiusculus with cytoarchitectonic boundaries suggests that the leg representation extends into area 4 (Fig. 1), although it is generally assumed that SMA is confined to area 6. Wise and Tanji (1980) have investigated this discrepancy by systematic mapping of the area 6/4 boundary, using as criteria the patterns of response to intracortical microstimulation and the relationship of neuronal activity to voluntary movements of the monkey, and taking into account cytoarchitectonic subdivisions in which the recordings were made. In each animal, the MII hindlimb area described by Woolsey et al. (1952) was found instead to provide the MI tail representation. The SMA leg area was situated 4 to 6 mm more anteriorly and the SMA/MI boundary was found to correspond approximately to the border of areas 6 and 4. The SMA forelimb area found in conscious monkeys is somewhat larger than the MII map shows, although still confined to area 6 (Brinkman and Porter, 1979). In contrast to the above studies, which imply that SMA (in particular the forelimb area) is larger than inferred from the classical simiusculus and that this map should be shifted somewhat rostrally so as to occupy area 6 without impinging upon area 4. The somatotopy of SMA involves a forelimb area rostral to a hind-limb area within the boundaries of area 6 and within the limb representation areas, distal and proximal limb representations are separated but there is some overlap (cf. Macpherson et al., 1982).

Bilaterality of SMA Neurons

Almost all investigators of neuronal activity and movement performance have allowed their animals use of the contralateral limb only, even in studies of SMA which seems to control movements on both sides of the body as evidenced by lesion studies (Travis, 1955). After a unilateral lesion of SMA, motor impairment was seen bilaterally, especially at proximal joints. We found that the majority of movement-related neurons exhibited modulation of their activity in association with task performance regardless of whether contralateral or ipsilateral limb was used. Also, neurons related to distal movements were encountered as frequently as ones related to proximal movements, a somewhat unexpected result, since control of distal movement is generally thought to be completely lateralized (e.g., Brinkman and Kuypers, 1973). Tanji and Kurata (1981, *this volume*) investigated a monkey trained to perform flexion/extension of the ankle, and recorded from SMA contralateral and ipsilateral to the moving limb. Only 9% of neurons in the ipsilateral cortex were responsive, which

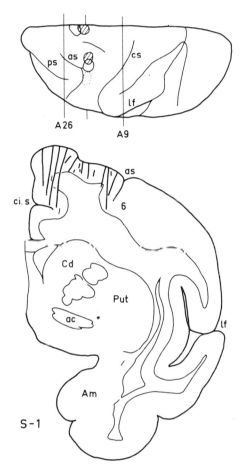

FIG 6. Below, electrode tracks in 1-mm section through area 6 of the brain of monkey S-1. The 4 most medial tracks are in the responsive area of SMA, the three most lateral ones in that of PM. Tracks in between did not yield neurons clearly related to movements of the upper limb. The level of the section is indicated on a drawing of the cerebral hemisphere involved of this animal, at the top of the figure, by the *interrupted line* in the middle. *Lines A 26* and *A 9* refer to the anteroposterior stereotaxic plane. The areas enclosed by *solid lines* show the somatotopy in SMA and PM as indicated by neuronal discharge with specific movements. *Hatched areas* are those where proximal neurons predominate; *open areas*, those with neurons mostly associated with distal movements. More than 90% of neurons in both SMA and PM in these areas were active regardless of whether contralateral or ipsilateral extremity was used. The area in PM enclosed by a stippled line is that found in monkey PM 2 (Fig. 10), where neurons associated with movement and influenced by visual events predominated. Neurons related to movement only in this area were almost all associated with contralateral distal movements only. *Abbreviations:* ac, anterior commissure; Am, amygdaloid complex; Cd, caudate nucleus; Put, putamen.

seems to contradict the findings of Brinkman and Porter (1979), it does not disprove a bilateral influence on motor control exerted by one SMA. Simple flexion/extension movements were not associated with bilateral changes in SMA activity in man, but complex distal movements were (Roland et al., 1980). It is this type of complex distal movement that was employed in the experiments of Brinkman and Porter (1979) in which SMA neurons were active with such movements, executed with either limb. The number of discharges associated with ipsilateral movement was often less than that associated with contralateral movement, and this was also the case for bilateral neurons in Tanji and Kurata's (1981) study. The pathways through which SMA exerts such bilateral control could be corticocortical or subcortical projections, or a combination of these. The SMA has a major projection to the ipsilateral primary motor area, but section of the PT does not significantly alter the MII motor map (Woolsey, 1975) which argues against involvement of contralateral primary motor area. Some of the bilateral influence of SMA might nevertheless be exerted through corticocortical interhemispheric connections. In a monkey in which corpus callosum had been transected, the proportion of neurons responsive during contralateral movement only was found to be increased even though two-thirds of the total sample continued to show modulation with movements of either extremity (Brinkman and Porter, 1979; Fig. 7), pointing to a role for subcortical connections in this bilaterality. Bilateral projections exist to the striatum (Kunzle, 1978), and lesions of the SMA projection area of striatum may result in motor impairment (Liles, 1975), but the contralateral corticostriate projection is a transcallosal one and should have been interrupted in a split-brain monkey. The bilateral projection to red nucleus (Kuypers and Lawrence, 1967; Kunzle, 1978; Hartman-von Monakow and Akert, 1979) would be a more promising candidate. Still, the most direct pathway through which SMA could influence movements would be through its connections with the spinal cord.

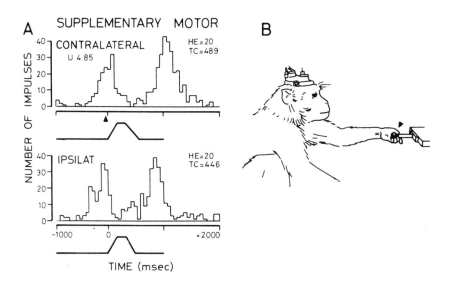

FIG 7. A: Activity of a neuron recorded in SMA of an animal with surgical transection of corpus callosum (Fig. 12) associated with finger flexion occurring with grasping of the lever (B) and grasping for food after the pull. Most SMA neurons were not affected in their bilaterality by callosal section.

After unilateral spinal injection of horseradish peroxidase (HRP), cells containing the tracer are found in the contralateral SMA (Coulter et al., 1975; Jones and Wise, 1977; Biber et al., 1978). These findings do not necessarily imply a purely contralateral projection, since labeled cells are likewise found only in the contralateral MI, and this area has a distinct bilateral spinal projection (Kuypers and Brinkman, 1970). Moreover, when SMA is stimulated, an evoked potential can be recorded from both the contralateral and ipsilateral cord at high cervical levels (Bertrand, 1956). Recent anatomical studies have demonstrated that the SMA corticospinal projection is also, in part, bilateral (Brinkman, *unpublished observation*). Thus, one SMA seems to be able to influence movements of both halves of the body, but the relative importance of its different efferent pathways in this bilateral control has not yet been established.

Comparison of SMA and MI in Recording Studies in Conscious Monkeys

(a) Relationship of Neuronal Activity to Movement

When monkeys were trained in a motor task which consisted of proximal (reaching, pulling a lever) and distal (grasping, holding and releasing a lever or food) movement sequences, many SMA neurons (93%) were found to be associated in their discharge with specific movements at a specific joint, such as elbow flexion or wrist extension (Brinkman and Porter, 1979). Neurons in MI behaving in a similar way in identical tasks have been described (Lemon et al., 1976; Lemon and Porter, 1976). Whereas MI neurons tended to show large changes in activity with very small changes in angle at a given joint (cf. Evarts and Fromm, 1978), SMA neurons as a group usually showed more prolonged modulation of discharge, such modulation preceding, and sometimes lasting for the duration of the associated movement. Whether this difference is significant still awaits analysis using a motor task requiring execution of very small, precise movements. A small group of SMA neurons found in the same experiment (7% of the sample) was not related to a specific movement but, rather, showed changes in discharge frequency throughout a movement sequence. Such neurons were not found in MI.

(b) Bilaterality

Of SMA neurons found using a complex trained movement sequence, 94% showed discharges that were associated with movements of both contralateral and ipsilateral extremities. In MI, neurons recorded in the hand representation area close to central sulcus (see Fig. 1) are almost exclusively related to contralateral movements only (Lemon et al., 1976). Bilateral MI cells have been described (Lemon et al., 1976; Matsunami and Hamada, 1981), but they were found only more rostrally in area 4. Moreover, they were mostly related to proximal movements only, which is in keeping with the differential origin of bilateral and contralateral spinal projections from MI (Kuypers and Brinkman, 1970). The SMA neurons display bilaterality for both proximal and distal movements.

(c) PTNs in SMA

Of the neurons tested, 5% sent their axons into the pyramidal tract (Brinkman and Porter, 1979), as compared with up to 20% of MI neurons (Lemon et al., 1976). The lower percentage in SMA could represent a sampling bias as many of the PTNs are located deeply in medial cortex, close to the cingulate sulcus, an area less extensively covered in the recording studies. Tracing of the corticospinal pathway from SMA has revealed that this projection is less dense than that from MI (Brinkman, *unpublished observation*) suggesting the difference in numbers found physiologically probably represents a realistic value. Functionally, SMA-PTNs differ from those of MI in that they show bilaterality. Such PTNs are extremely rare in MI (Lemon et al., 1976; Matsunami and Hamada, 1981).

(d) Peripheral Afferent Input into SMA

Only a small proportion of SMA neurons can be activated by afferent input from peripheral receptors. Percentages vary from 4% (Wise and Tanji, 1980) to 14% (Brinkman and Porter, 1979), and this result is the same, regardless of whether the animal is executing a movement, or is passive and relaxed. In contrast, about 80% of MI neurons can be influenced by such stimulation (Lemon et al., 1976; Lemon and Porter, 1976; Wise and Tanji, 1980). Responses elicited in the SMA hand area had latencies of more than 20 msec, originated from more than one joint, or from a wide area of body surface, and were difficult to reproduce faithfully with repeated stimulation. In MI, the shortest latency found was 6 msec; responses came from very restricted peripheral zones and were elicited consistently from trial to trial (Lemon et al., 1976). Similar results were obtained from the hindlimb areas (Wise and Tanji, 1980). From 4 to 10% of SMA neurons were responsive to peripheral inputs, whereas 86% of MI neurons were. Shortest latencies were 23 and 12 msec, respectively but, due to the small sample of SMA responsive neurons, the overall difference in latencies was not significant. Thus, SMA has a population of movement-related neurons that is quite distinct from that in MI.

Function of SMA in Motor Control

(a) Control of Posture

The lasting effect of a pure SMA lesion in monkeys, as described by Travis (1955) was hypertonia of proximal flexor muscles; bilateral lesions produced more pronounced deficits. It was postulated that SMA was bilaterally involved in control of muscle tone but others have not been able to reproduce these results (Coxe and Landau, 1965). Conflicting evidence also exists concerning SMA stimulation experiments when MI had been excised. One study reported that only crude proximal movements remained (Wiesendanger et al., 1973), another that no changes could be found (Woolsey, 1975). From lesion and stimulation experiments, only a weak case can be made for SMA as an area involved in the control of posture. Nevertheless, this idea has gained some following (Smith, 1979). In most recording studies of SMA, neurons related to maintenance of a joint position only or to postural adjustment of the body have been found rarely if at all (Brinkman and Porter, 1979; Tanji et al., 1980). Many movement-related neurons were found to be concerned with distal rather than proximal movements (Brinkman and Porter, 1979; Tanji and Kurata, 1979) and it seems now clear that SMA is not involved exclusively in control of posture or proximal parts of the limb and that its role is likely to be a more complex one which includes control over distal extremity muscles as well.

(b) Gating of Primary Motor Area Responses by SMA

Forced grasping has been a prominent if transient sign of SMA lesions in monkeys (Penfield and Welch, 1951; Travis, 1955; Bourbonnais et al., 1979; Smith et al., 1981), as well as in man (Erickson and Woolsey, 1951). This reflex has been thought of as depending on tight input-output coupling of MI responses (Rosen and Asanuma, 1972). A lesion of SMA would release the grasp reflex because some inhibitory action of SMA on MI would be removed. Virtually no experimental evidence exists to support this notion. Tanji et al.

(1980) have investigated whether SMA played a part in modifying motor cortex reflexes in conscious, behaving monkeys. Depending on the color of a light cue given a variable time before a disturbance occurred, the monkeys had to respond with a push or pull movement of a lever when the disturbance was imposed. It was assumed that changes in activity occurring during the period between light cue and lever disturbance were indications of the development of a preparatory motor set. In MI, many neurons showed such changes in activity depending on which instruction had been given, and reflex activity of MI neurons in response to the disturbance was also found to be modified (Evarts and Tanji, 1976; Tanji and Evarts, 1976). If SMA plays an important role in modulation of motor cortex responsiveness SMA neurons would be expected to show changes in activity during the preparatory period but no changes related to the movement itself. Ninety-four neurons showed differential responses to different instructions without visible movements or changes in EMG in arm or girdle muscles before the disturbance was imposed (Tanji et al., 1980). A puzzling aspect is the long latency of the change in response to onset of the light cue. The shortest latency was 140 msec, but latencies ranged to > 500 msec. These times are longer than the fastest reaction time that can be found in monkeys (Beck and Chambers, 1970). When the direction of the disturbance was the same in a number of consecutive trials instead of randomly distributed, the monkeys learned to anticipate in which direction the compensatory movement would have to be made, and the EMG of agonist muscle would increase at progressively shorter latencies; i.e., reaction time to the disturbance, normally about 120 msec, shortened. The increase in discharge did not move to a shorter latency after onset of the light cue, in parallel with the shift in EMG latency. Thus, it appears that there is in SMA a population of neurons that is not related to execution of a given movement or development of EMG activity directly, but instead may be concerned with some aspect of motor instruction using visual cues (Tanji and Kurata, *this volume*). Such neurons have been found by others (Sakai, 1978) and a small proportion of neurons influenced by some aspect of vision but not movement performance in SMA have been described (Brinkman and Porter, 1979).

(c) Preprogramming of Movements by SMA

The discharge characteristics of large numbers of SMA units are associated with particular movements, and many seemed to increase their discharge before onset of a particular movement, as detected on video records (Brinkman and Porter, 1979). For some cells, discharge modulation could occur before noticeable changes in EMG activity could be detected, as found by others (Smith, 1979; Tanji and Kurata, 1979, *this volume*). In this respect, SMA cells could have a direct role in specifying certain aspects of movement performance. This programming could influence lower levels of the central nervous system (CNS) either through MI (but not necessarily as gating), through subcortical structures or through direct connections with spinal cord.

Recently, in studies in man using measurement of regional cerebral blood flow (rCBF), the level of neuronal activity (Ingvar, 1979) yielded some insights on how SMA blood flow changes in normal humans during different motor tasks. Isometric contractions of hand muscles when compressing a stiff spring, or simple finger flexion movements resulted in increased rCBF in the contralateral primary motor and sensory areas only. When a complex motor task was performed, opposing the thumb against the fingers in a predetermined sequence, increases in rCBF were present in contralateral MI, but also in both SMAs. Finally, when the patients were asked not to execute this complex movement but instead just to think about performing it, rCBF increased in the SMAs only (Roland et al., 1980). Similar rCBF changes occurred in the face representation of SMA when, rather than counting out loud, subjects counted to themselves (Orgogozo and Larsen, 1979). Although the rCBF method does not tell about time relationship between changes in rCBF and particular events, the increase in rCBF in the SMAs in the absence of any peripheral motor events points to a role for SMA in the higher motor control. These findings argue against a role in the control of posture and postulate a role for SMA in motor programming and further support experiments in conscious monkeys.

Function of SMA as Suggested by Lesion Experiments

Lesion studies of SMA have focused attention on its role in the control of muscle tone. Lesions of SMA alone may (Travis, 1955) or may not (Coxe and Landau, 1965) affect muscle tone, but there seems to be good agreement that lesions of SMA together with damage to MI result in spastic paresis (Denny-Brown and Botterell, 1948; Travis, 1955). Animals developed marked hypertonia of both proximal and distal flexor muscles. In split-brain monkeys with combined SMA-MI lesions the spasticity was much more pronounced, with again suggests that the bilateral influence of SMA is exerted at least in part through the callosum (Brinkman and Kuypers, 1973). The neuronal mechanisms underlying control of muscle tone and spastic paresis are by no means well understood, and careful studies of SMA could help clarify some of these problems. Another recent development in SMA lesion studies is the deficit of distal extremity movement resembling apraxia, an inability to sequence movements properly, which is also bilateral in monkey (Brinkman, 1980) and man (Masdeu et al., 1978). These findings again indicate the importance of SMA in motor programming.

SMA: Summary and Conclusions

Recording experiments in conscious, behaving monkeys trained to perform motor tasks suggest that SMA is a restricted functional area of cerebral cortex located within cytoarchitectonic area 6 on the medial wall of the hemisphere, rostral to area 4 of the primary motor area. It is associated with movements of both the contralateral an ipsilateral extremities. Many neurons in SMA show modulation of their discharge in relation to particular movements, either proximal or distal ones. Such modulation often precedes the onset of the movement itself. In addition, there exists in SMA a group of neurons showing changes in their activity, which may be consistent with a function of modifying of primary motor area activity. These neurons are not associated with a particular movement per se. The SMA is not an area involved primarily in the control of posture. Despite its abundance of afferents from cortical sensory areas, SMA does not seem to act as a feedback center. Afferent input from peripheral receptors does not markedly influence SMA neurons in either the actively moving monkey or the passive, relaxed animal. (See Tanji and Kurata, *this volume*.) Findings in both monkeys and conscious humans point to a role for SMA in the programming of skilled learned movement sequences.

II. THE PREMOTOR AREA

Historical Background

The concept of a premotor (PM) area, located in area 6 rostral to area 4, was based on ablation studies showing spastic paresis rather than flaccid paresis such as was found after limited area 4 ablation (Fulton and Kennard, 1934; Hines, 1943). This concept of PM as found in monkeys and apes was questioned by Walshe (1935), who failed to find evidence in man to warrant separation of areas 4 and 6 on the basis of neurological symptoms. A separate unconfirmed PM area in monkeys in stimulation experiments (Woolsey et al., 1952) suggested that the motor map was not restricted to area 4, but extended rostrally into area 6 where the axial muscles and some of the proximal musculature were represented (Fig. 1). In lesion studies based on the MI and MII motor maps, spastic paresis was never found after a lesion of the lateral part of area 6 or of MI with part of area 6, but only after a combined lesion of MI and MII, including the supplementary motor area (Travis, 1955). As a compromise between these two conflicting sets of results, it has been proposed (Wiesendanger, 1982) that area 6 is not a functional unity but that its posterior portion forms part of the primary motor cortex representing axial and proximal muscles, whereas its anterior portion represents a true motor association area as originally conceived by Fulton (1935). This suggestion is derived in part from the movements obtained in awake monkeys with intracortical microstimulation (Kwan et al., 1978). In these experiments, movements at the elbow and shoulder joints were obtained most frequently in the border zone between areas 4 and 6 but often only with high currents 16–39 μA). Given the problems concerning activation of cortical neurons by microstimulation with repetitive shocks (Phillips and Porter, 1977, pp. 181–197) a distinction between areas 4 and 6 using microstimulation alone would be insufficient, and tend to favor expansion of the motor area rostrally beyond area 4. Histologically, the transition between areas 4 and 6 is not sharply defined. Nevertheless, the agranular cortex of the posterior bank of the arcuate sulcus and the zone of such cortex extending medially and laterally from this sulcus possess cytoarchitectonic features different from those of area 4, and it is this band of cortex that is considered as area 6 here (Fig. 1). From the above, the need for careful histologic analysis at the conclusion of chronic recording experiments, often a somewhat neglected item, is obvious. As a result of the stimulation and behavioral studies following lesion experiments, the possibility of a motor association area has been considered by few, yet the existence of such areas is taken as a matter of course in other systems. Only anatomical studies have indicated that the lateral part of area 6, like its medial part SMA, forms a separate functional area within the frontal cortex. There can be no doubt that the connections of area 6 from thalamus and other areas of cortex distinguish it from area 4.

Possible Functions of PM in Movement Control

The lateral part of area 6, called PM here, is an area upstream of the motor cortex of area 4 as it has a dense and topographically organized projection to area 4 (Pandya and Kuypers, 1969; Pandya and Vignolo, 1971; Matsamura and Kubota, 1979; Muakassa and Strick, 1979). It is also reciprocally connected with SMA (Pandya and Vignolo, 1971). Its subcortical projections likewise are directed to motor centers and match those found for SMA. Bilateral projections exist to the striatum (caudate nucleus, putamen, and claustrum) and red nucleus; ipsilateral connections with the ventrolateral thalamic complex of nuclei, in particular the (VA), the (VLc), and the nucleus X; and also with the centrum medianum and other parts of the intralaminar nuclei (Kunzle, 1978). As a consequence, PM also forms a station in the two important reentry circuits for area 4, the cortex-cerebellum-thalamus-cortex circuit and the cortex-basal ganglia-thalamus-cortex loop. The PM also projects to the ipsilateral pontine gray and to the reticular formation, to some extent bilaterally (Kunzle, 1978). Fibers from PM and SMA, although projecting to the same structures, do not share a common termination area within these structures, there is a clear topographic separation (Kunzle, 1978). Like SMA, PM has been shown to possess a direct projection to spinal cord (Jones and Wise, 1977). In the case of PM, this confirmed the original degeneration study of Hoff (1935). Interestingly, he found a bilateral corticospinal projection from area 6.

Thus, in its cortical and subcortical efferent projections, PM seems very similar to the other motor association area, SMA. However, in terms of its corticocortical afferents, PM differs from SMA. The PM area does not receive afferents from primary sensory cortical areas, and those from sensory association areas such as area 5 are derived from the more lateral parts of this area, and from area 7 (Chavis and Pandya, 1976). In addition, PM receives an input from visual cortical association areas, notably from the peristriate cortex which surrounds the primary visual cortex (Kuypers et al., 1965; Pandya and Kuypers, 1969). There is a precise topographical arrangement of these projections within PM, but this segregation does not seem to apply so much in the anteroposterior as in the mediolateral gradient, suggesting that area 6 is not divided into two convex strips, one close to area 4 and one further forward with different functions, as has been suggested (see above; Wiesendanger, 1982). The inputs from visual association areas distinguish PM from SMA, where inputs from somatosensory association cortex predominate. On the basis of its anatomical connections, a function can be postulated from PM in visuomotor integration, analyzing visual inputs and passing on instructions based on visual events to area 4 through which the motor apparatus is activated. Caution must be exercised because SMA has thus far not turned out to behave as the somatosensory integration center it was once thought to be (see above). Alternatively, PM could, like SMA, influence movement control directly through its subcortical connections. Long neglected in recording studies in conscious animals, PM has recently gained renewed acceptance as an area distinct from area 4.

Natural Activity of PM Neurons in Conscious Monkeys Performing a Learned Motor Task

Chronic studies of PM neurons in monkeys have used as motor tasks a self-paced, complex natural movement of upper limb involving reaching for and pulling of a horizontal lever and of manually collecting a food reward after the pull (Brinkman and Porter, 1979). In addition, a visually guided tracking task involving a wrist flexion-extension movement (Kubota and Hamada, 1978), a simple grasping task followed by a visually cued complex distal movement (Godschalk et al., 1980) or a complex visuomotor task have been used.

Identification of PM in the Conscious Animal

When successive electrode penetrations were made in the cortex of area 6 on the convexity of the cerebral hemisphere, starting from SMA, neurons discharging with movements of the upper limb could no longer be found more than 3 mm away from the midline. Only when the electrode reached the area located at the arcuate sulcus (Fig. 1) could such neurons again be found. Thus, in Fig. 6, such movement-associated neurons are found in the most medial and most lateral tracks only (Brinkman and Porter, 1979). Below the concavity of the arcuate sulcus they became much more difficult to find. Electrode penetrations a few millimeters caudal to the sulcus yielded neurons, which possessed the typical characteristics of MI neurons (cf. Lemon et al., 1976; Matsunami and Hamada, 1981), whereas neurons found anterior to it were easily recognized as frontal eye field neurons of area 8 (Fig. 1) due to their obvious relation to eye movements (cf. Bizzi, 1968). Therefore, there exists within lateral area 6, a region associated with upper limb movements, which is distinct from the motor areas of neighboring cortex.

Movement-Related Neurons in PM

Brinkman and Porter (1979) studied 37 PM neurons in 2 animals also used for SMA, using the same task. The monkeys were required to pull a horizontal lever with either hand for a food reward, and cells recorded were tested for sensory inputs from peripheral receptors while the animals remained passive and relaxed. Many of the cells were also tested for antidromic activation by stimulation of the PT (see Fig. 2). Thirty-six neurons showed modulation of their discharge during the movement task; 6 changed their discharge through the entire movement sequence, regardless of whether the contralateral or ipsilateral hand was used, and 30 showed changes in activity associated with specific forelimb movements. Of these 30, 27 also displayed modulation with movements of either hand. Nineteen (63%) were associated with distal extremity movements, 11 (37%) with proximal ones. The neurons associated with a particular movement tended to show changes in discharge patterns before the movements could be detected on the video records. Figure 8 shows such a neuron always associated with proximal movement, elbow

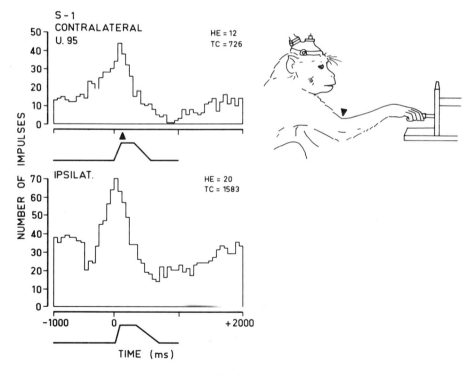

FIG. 8. Neuron recorded in PM always related in its activity with elbow flexion during the lever pull as is shown on the right. The discharge increased well before lever movement starts at time 0. The histogram obtained with contralateral movement shows the same pattern but is made up of fewer impulses because of a lower number of pulls.

flexion, which occurred when the monkey pulled the lever (figuring on the right). Figure 9 shows such a neuron always associated with a distal movement, extension of the wrist, which was seen when the animal took its hand off the lever. For this neuron, a slight difference in timing is evident when histograms for the contralateral and ipsilateral hand are compared and is due to a difference in execution speed of movement task between the normally preferred contralateral hand of the animal and its nonpreferred ipsilateral hand.

In these 2 monkeys, the recording area was just above where the arcuate sulcus often shows a spur pointing caudalwards (Fig. 6, top). In a third monkey, recordings were made in the area just lateral below this spur (stippled line in Fig. 6, and hatched in Fig. 10). In this region, two groups of neurons could be distinguished. One was similar to those found earlier in that the neurons were related to specific movements, but almost all cells were associated with contralateral movements only, with a preponderance associated with distal movements. Such cells made up about one-third of the more than 90 cells collected in this animal. The majority of neurons was related to movement as well, but in a more complex way, as these cells seemed strongly influenced by visual events, e.g., reorientation toward the lever after collection of food reward, or orientation to the site where the food reward was presented (Brinkman, *unpublished observation*). The importance of vision has been found for PM cells by other workers as well. Kubota and Hamada (1978) trained monkeys to track a light presented in one of 7 positions by aligning a second light through movement of a lever by flexion or extension of wrist. Of 88 recorded cells, 6 (7%) responded to the light cue only, with a latency of 60 msec; 25 (28%) were associated with the flexion or extension of wrist, and the changes preceded movement onset by 100 to 150 msec, whereas 41 (53%) were activated with both the flexion and extension movements. The rest of the neurons were depressed during the task. This experiment also points to different populations of neurons in PM, some being related to visual events or orientation and attention, others to movement. Support for this view comes from the study by Godschalk et al. (1980) in which monkeys were trained to squeeze a rubber bulb until a light cue appeared, after which the animals had

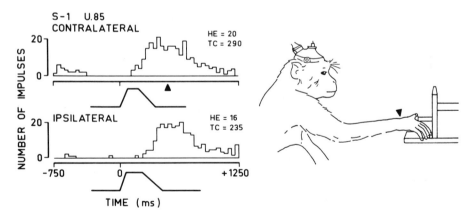

FIG. 9. Discharge pattern of a PM neuron always associated with a distal movement, extension of the wrist during lever release, as is shown on the right. Increase in activity occurred somewhat later for the ipsilateral hand because of a longer pulling time. The monkey normally preferred its contralateral hand.

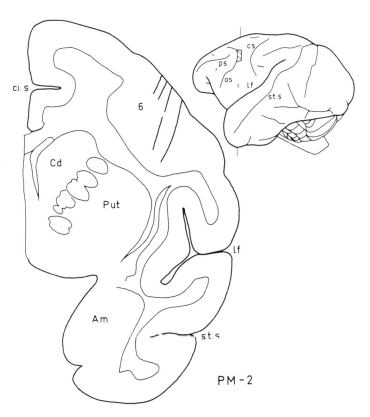

FIG. 10. Electrode tracks in 1-mm sections through the brain of monkey PM-2. **Right:** The recording area *(hatched)* and the level of the section. In this area, most neurons were related in their activity to movement and to visual events. Neurons associated with movement only were mostly related to contralateral distal movements.

to retrieve a food reward from a special board known to be a test for discrete distal movements which can only be executed satisfactorily using vision (Brinkman and Kuypers, 1973; Haaxma and Kuypers, 1975). Neurons recorded fell into 3 groups: one in which neuronal activity changed before movement occurred, one related to collection of the food reward, and one related to licking and chewing movements. In another study (Godschalk et al., 1981), a monkey pressed a switch (the hold phase) in order to visualize a food reward placed randomly in three different positions. After 1 sec of this visible phase, the animal gained access to the food (movement phase). In recordings made in the periarcuate area, cells were most active during a different phase, and cells more active with the movement phase occurred with more posterior electrode penetrations. In the posterior bank of arcuate sulcus neurons were often associated with the visible phase, or with the visible and movement phases. These neurons could transmit to the motor cortex information regarding the spatial orientation of an object and contribute to the guidance of a subsequent reaching movement.

In summary, there are at least two groups of neurons in PM at the concavity of arcuate sulcus: one group associated with visual events or orientation and another related to details of movement performance (Kubota and Hamada, 1978; Brinkman and Porter,

1979; Godschalk et al., 1980). A third group could be formed by neurons modulated in their activity throughout the movement performance (Kubota and Hamada, 1978; Brinkman and Porter, 1979).

PTNs in PM

Only Brinkman and Porter (1979) have tested PM neurons by antidromic stimulation of the pyramidal tract, probably because it was assumed that no corticospinal projection from PM exists despite the old anatomical studies (Hoff, 1935). Of 28 neurons tested, 9 (31%) were found to be PTNs with antidromic latencies from 1 to 3 msec. Like those in SMA, they were related to movements of both the contralateral and ipsilateral extremity. An example of such a PTN is shown in Fig. 10. Only the histogram for the contralateral hand is shown (top left). The discharge pattern for the ipsilateral one was similar. The neuron was always associated with extension of the fingers at the moment of lever release after the pull (Fig. 11, bottom left). Peak neuronal

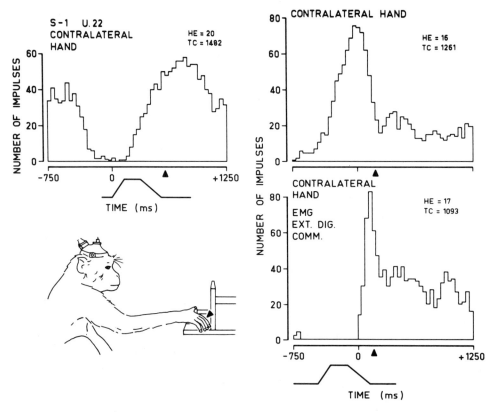

FIG. 11. Discharge pattern of a PTN in PM, shown for the contralateral hand only on the **top left**. The neuron behaved in a similar fashion when the ipsilateral hand was used and was associated with finger extension which occurred during the reach for the lever and after the pull during lever release, shown on the **bottom left**. The relation to lever release is shown more clearly in the **top right** histogram, where the reference point in time is not the start of the lever movement, but instead, a point in time midway along the return of the lever to its starting position (see position trace below histogram), i.e., a point closer in time to the moment of lever release. It is now much more evident that the neuron increases its activity before extensor movement onset as is also clear from the histogram at the **bottom right** obtained for the EMG activity of a finger extensor, m. extensor digitorum communis, during the same pulls. Peak neuronal activity leads the EMG by 120 msec.

activity leads the EMG activity by 120 msec (cf. Kubota and Hamada, 1978).

Peripheral Afferent Input into PM Neurons

On the basis of its afferent corticocortical connections, PM neurons should not be readily influenced by stimulation of peripheral receptors. This was the case in one study (Brinkman and Porter, 1979): Only 4 out of 28 neurons tested in 2 monkeys (14%) could be so driven in the passive relaxed animal. None of these neurons were encountered in over 60 neurons tested in the visual area of PM in a third animal (PM-2, Fig. 10). The peripheral inputs for responsive cells were not well defined, being derived, for example, from cutaneous stimulation of a large body surface, or from passive movements of more than one joint; and responses often could not be obtained from one trial to the next. In the actively moving animal, disturbances imposed randomly upon the lever movement never resulted in reflex-type changes in neuronal activity.

Somatotopy in PM

Anatomical experiments have shown a precise topographic relationship of the cortocortical connections of PM. The area at the concavity of arcuate sulcus projects to the forelimb representation area of MI (Matsamura and Kubota, 1979; Muakassa and Strick, 1979). Tracks in which neurons related to proximal movements predominated were more medial than those containing neurons associated with distal movements, but some overlap existed between areas (Fig. 6). Comparison of these areas with the one where neurons related to movements and visual events and to contralateral movements only were found, (Fig. 11) suggests that the latter area is lateral to the former two (Fig. 6, stippled outline). The PM may show a topographic arrangement with proximal movements represented more medially than distal ones, and visuomotor neurons lateral to both. These conclusions must remain tentative, however.

Bilaterality of PM Neurons

Many movement-related neurons in PM showed modulation regardless of whether the contralateral or ipsilateral limb was used in the task (Brinkman and Porter, 1979). As described above, some subcortical projections from PM are bilateral and could enable PM to steer movements of either limb. Alternatively, PM could do so by its connections through corpus callosum with PM, SMA, and MI in the opposite hemisphere (Pandya and Vignolo, 1971). In one monkey with corpus callosum transected (Fig. 12), the bilaterality for SMA neurons was not altered significantly, but more than two-thirds of the neurons recorded in PM were only active with contralateral movements and virtually silent when the ipsilateral limb was used (Fig. 13), indicating that PM seems to exert its bilateral influence on movement through the corpus callosum. Anatomical studies provide a basis for this, as area 6 receives a dense callosal projection almost throughout its extent on the convexity of the hemisphere, with the exception of a small part at the level of the spur of arcuate sulcus (Karol and Pandya, 1971) where, in intact animals, neurons tended to be related to contralateral movements only.

Comparison of PM, SMA and MI in Chronic Recording Studies

(a) Relationship of Neuronal Activity to Movement

There seem to exist within PM at least three functionally different groups of movement-related neurons: (1) A group of neurons modulated with specific movements, with changes of activity often preceding

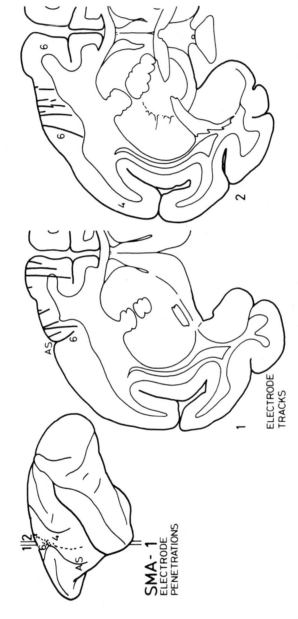

FIG. 12. Electrode tracks in SMA and PM as found in two 1-mm sections through the brain of a callosum-split monkey, SMA-4. The extent of callosal section is indicated in black in the figurine on the **bottom left**. *Hatched* area in this figurine is the responsive area of SMA. A neuron recorded in SMA in this animal is shown in Fig. 7, one recorded in PM in Fig. 13.

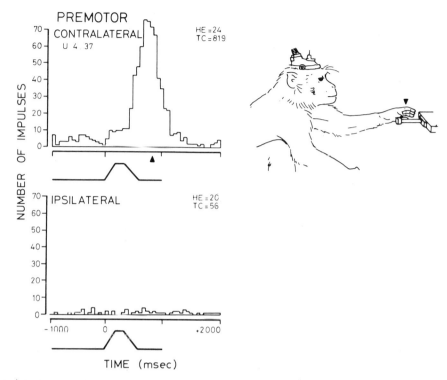

FIG. 13. Discharges of a neuron in PM in a monkey with callosal section. The neuron was related to extension of the fingers while reaching for the food as is shown on the right. Almost no impulses were associated with movements of the ipsilateral extremity. This was the case for a large proportion of PM neurons in this animal and this finding contrasts with the bilaterality of PM neurons found in intact animals in the same region. It suggests that PM exerts a bilateral influence on movement at least in part by way of the corpus callosum.

movement onset (Kubota and Hamada, 1978; Brinkman and Porter, 1979). A proportion of this group is active with movements of either limb (Brinkman and Porter, 1979). (2) A group related to movements and to visual events (Brinkman, *unpublished observation;* Kubota and Hamada, 1978; Godschalk et al., 1981). (3) A group that shows modulation throughout a movement sequence (Kubota and Hamada, 1978; Brinkman and Porter, 1979). There may be neurons related to visual events only (Kubota and Hamada, 1978; Godschalk et al., 1981). Of these groups, only the first one is similar to neurons found in SMA and MI where the cells discharge in relation to specific movements only, with changes in activity preceding movement onset. In SMA, very few cells can be influenced by visual events (14/271, 4%; see also; Sakai, 1978; Tanji and Kurata, *this volume*). In many studies in MI, such cells have not been reported.

(b) Bilaterality

Many PM neurons were active with movements of either limb. Such neurons were rare in MI close to the sulcus, although they could be found in larger numbers on the surface of the precentral gyrus (Matsunami and Hamada, 1981). These neurons were not very common even there, and were associated mainly with proximal movements. In PM, bilaterally related neurons were found in large numbers in a restricted area, and almost two-thirds of these were related to distal movements, a situation very much like that found in SMA. In some other aspects, these neurons resembled those in SMA (see d below).

(c) PTNs in PM

Of PM neurons tested with antidromic stimulation of the pyramidal tract 30% were PTNs (Brinkman and Porter, 1979). This figure is considerably higher than that found for SMA (5%) or even MI (e.g., 20%, Lemon, 1981). Taking into account the possibility of a sampling bias for the high PM figure, it must nevertheless be assumed that PM resembles MI rather than SMA. The figure also provides physiologic evidence for a projection from PM to spinal cord, long neglected in theories of motor control, but recently described again in anatomical studies (Jones and Wise, 1977). The PTNs in PM differed from those in MI, and resembled those in SMA, however, in that they were associated with bilateral movement performance and not with contralateral movements only.

(d) Afferent Input into PM

The PM neurons, like those in SMA, were seldom affected by natural stimulation of peripheral receptors in the passive, relaxed animal (Brinkman and Porter, 1979) or in the actively moving monkey (Brinkman, *unpublished observation*). Peripheral zones from which cells could be influenced were wide and driving of response as in SMA. These data contrast sharply with those in MI, where faithful driving from restricted peripheral zones is the rule for about 80% of neurons (Lemon, 1981; Lemon et al., 1976; Wise and Tanji, 1980; Wise and Tanji, 1981). Of the visual neurons in PM, none could be driven by somatosensory stimulation. In summary, PM seems to resemble SMA more than MI. In addition, PM differs from both SMA and MI in that it contains a large number of neurons that are related to aspects of vision and movements with responses that are unique to this area, as is a smaller number of neurons related to visual events or orientation responses only.

Function of PM in Motor Control

Lesion studies have been given no clear indication as to the function of PM in the control of movement. Spasticity observed after premotor ablations in older studies (e.g., Fulton and Kennard, 1934; Welch and Kennard, 1944) may have been due to inclusion of SMA in these lesions and involvement of MI since only lesions of MI in combination with SMA give rise to spastic paresis, whereas PM and MI combined do not result in flaccid paresis (Brinkman and Kuypers, 1973; Travis, 1955). Transient forced grasping is a feature of lesions of SMA (Travis, 1955; Smith et al., 1981) as well as of PM ablations (Goldberger, 1972). In this premotor syndrome distal movements are most affected, which argues against PM being a mere rostral extension of the primary motor area. In such a theory, forced grasping is supposed to be a manifestation of a release of brainstem postural control mechanisms after the cortical representation of proximal musculature has been removed (Bieber and Fulton, 1938).

Moreover, inasmuch as proximal muscles are involved in the forced grasping syndrome, these recover first despite the fact that their cortical representation area would have been removed (Goldberger, 1972). In the recording studies, neurons of PM have been found that were related to either proximal or distal extremity movements, and this seems to argue against a role for PM mainly in maintaining postural reflexes. This finding is more in keeping with the view of PM function as an association area for motor cortex, since complex rather than discrete movements were elicited by PM stimulation

(Hines, 1943), both of the proximal and distal extremity. In this concept, forced grasping would form only one expression of a more general motor deficit which resembles apraxia (Castaigne et al., 1972; Delacour et al., 1972; Deuel, 1977; Lhermitte et al., 1972). Such deficits occur after SMA lesions as well, again placing PM closer to SMA in relation to movement control. In addition, many PM cells showed changes in discharge well before movement implying a role for PM in the programming of movements.

However, PM seems to be unique among motor areas of the frontal lobe in the large number of its neurons related to both visual events and movements. Until careful studies have clarified the way in which visual information affects neuronal activity in relation to movement along the lines followed for the posterior parietal association cortex (Mountcastle et al., 1975; Sakata et al., 1980), the role of these visuomotor cells remains unclear. Studies in patients with premotor damage have suggested that such lesions may give rise to visuomotor ataxia, a difficulty in grasping objects with visual control (Derouesne, 1973; Jeannerod and Prablanc, *this volume*). Also, monkeys with area 6 ablations have great trouble steering the affected hand through a hole in a perspex plate to reach for food visible under it. The animals continued to try and reach straight for the food, hitting the plate (Moll and Kuypers, 1977). Taken together, the data from lesion and recording studies suggest that PM is involved in visuomotor integration.

Another possibility has come from studies of increased rCBF in the conscious human performing a number of motor tasks (Roland et al., 1980). When blindfolded patients were asked to direct their finger to another place in a gridlike maze, as instructed by the experimenter, rCBF was increased in the contralateral primary sensorimotor cortex, parietal area, SMA, and PM. When the ipsilateral hand was used, all areas on the same side except the sensorimotor cortex showed increased rCBF. It seems that association areas in cerebral cortex are activated regardless of which hand is used, and this is reflected also in the bilateral neurons found in SMA and PM in conscious monkey. When patients were asked to think about complex motor tasks, no changes were observed in PM but only in SMA. The authors concluded that PM, unlike SMA, is involved not so much in the preprogramming of a learned complex movement, but in motor learning. Such a role for PM has also been put forward in studies of slow pre-motion surface potential studies in monkeys learning a self-paced movement task (Hashimoto et al., 1979). With time and repetition, the potentials over PM which preceded movement by about 1 sec grew larger while those over MI diminished in size. Some cells in SMA have been found which seem to change their activity when a movement is learned (Tanji et al., 1980) and in man, SMA shows changes in rCBF in the learning situation as well. The PM lesions reportedly interfered with a monkey's retention of learned motor habits (Deuel, 1977). Tests aimed specifically at motor learning in chronic experiments, such as that of Gilbert and Thach (1977) on cerebellum, may help to elucidate the relative importance of PM in motor learning and/or preprogramming.

PM: Summary and Conclusions

In the conscious, behaving monkey trained to perform a motor task either self-paced or involving light cues, a restricted area responsive during performance of the task can be found in cytoarchitectonic area 6 at the level of the concavity of the arcuate sulcus. Within this area, different groups of neurons can be identified in relation to experimental variables. First a group related in its activity to specific aspects of movement performance. Second, a group associated with movements and visual events, and third, a group modulated in its activity throughout the execution of the task. A few neurons are related to visual cues only. In the first group, a subpopulation exists of neurons active with movements of either extremity; this group resembles similar neurons in SMA. Such neurons can be associated with proximal or distal movements suggesting PM is not an area involved mainly in the maintenance of body posture. The PM may function as a visuomotor integration center as evidenced by the large number of neurons influenced by visual aspects of the tasks. This is also suggested by findings in lesion studies in monkeys and man. Studies in conscious monkeys and man point to a role for PM in motor learning, when a new motor program is established or a previously learned one modified.

Motor Control Mechanisms in Health and Disease,
edited by J. E. Desmedt.
Raven Press, New York © 1983.

Functional Organization of the Supplementary Motor Area

*Jun Tanji and Kiyoshi Kurata

Department of Physiology, Hokkaido University School of Medicine, Sapporo, Japan

Horsley and Schaefer (1888) and the Vogts (1919) knew that stimulation of a portion of primate cerebral cortex rostral to the precentral leg area could produce movements of opposite limbs and face but Penfield and Welch (1951), however, were the first to distinguish a secondary or supplementary motor area (SMA), situated in the superior frontal gyrus on the medial wall of the hemisphere and in the upper bank of the cingulate sulcus, from the primary motor area in the precentral gyrus. The SMA lies within area 6 of Brodmann and in primates it corresponds largely to the mesial part of area 6aα of the Vogts (see Wiesendanger, 1973, 1981; Brinkman and Porter, 1979, *this volume*). After a brief review of previous key reports, this chapter describes recent studies using the technique of single-unit recordings in awake, behaving animals to describe functional organization of SMA.

Anatomical Connections

Among efferent paths originating from SMA, the existence of corticospinal neurons has been controversial (Bertland, 1956; DeVito and Smith, 1959). However, in recent studies with the retrograde tracer horseradish peroxidase (HRP) injected in the spinal cord, labeled neurons were found unequivocally in SMA (Biber et al., 1978; Murray and Coulter, 1981). In these reports, rostrocaudal arrangement of neurons projecting to cervical and lumbar spinal cord was demonstrated. The corticocortical path to the primary motor area (Jones and Powell, 1969, 1970c; Pandya and Vignolo, 1971) was confirmed by the tracer technique with additional evidence for a somatotopic relationship (Matsumura and Kubota, 1979; Muakkassa and Strick, 1979). Thus, the SMA has direct and indirect access to the spinal cord. The indirect output includes the pathway via the red nucleus (DeVito and Smith, 1959; Kuypers and Lawrence, 1967; Wiesendanger, 1973; Fromm, *this volume*), pontine nuclei (DeVito and Smith, 1959; Dhanarajan et al., 1977; Wiesendanger et al., 1979), basal ganglia (Jones et al., 1977; Kemp and Powell, 1970; Wiesendanger et al., 1973) and thalamus (DeVito and Smith, 1959; Wiesendanger et al., 1973). Major inputs to SMA are from other parts of the cortex, especially from the sensorimotor cortex. A prominent input from somatosensory area I and II and from area 5 (Jones and Powell, 1969; Jones et al., 1978; Vogt and Pandya, 1978) seems to provide the somatosensory input detectable in discharges of single neurons (Brinkman and Porter, 1979; Smith, 1979; Wise and Tanji, 1981). Connections from precentral motor cortex and from con-

---

*To whom correspondence should be addressed: Department of Physiology, Hokkaido University School of Medicine, West 7 North 15, Sapporo 060 Japan.

tralateral SMA are reciprocal (DeVito and Smith, 1959; Pandya and Vignolo, 1971). The connection from the lateral part of area 6 (Pandya and Vignolo, 1971) as well as from thalamus (Kalil, 1975; Kievit and Kuypers, 1977; Künzle, 1978) seems important for supposed role of SMA in higher-order motor control.

Electrical Stimulation Studies

Electrical stimulation of cortex, although highly artificial, provided basic knowledge about representation in cortical motor areas. The figurine map of precentral gyrus drawn by Woolsey and his colleagues (1958) is the outcome of applying the technique in an area where tight and oligosynaptic connections are focused to a limited portion of the spinal motor apparatus, giving rise to a useful topographic map. However, the figurine map for SMA, using the same technique (Woolsey et al., 1952), has been a subject of serious criticism (Wiesendanger et al., 1973), and surface stimulation seems inadequate for studying functional organization of SMA for two reasons: (1) Its output is not tightly linked to the motor apparatus as in the precentral gyrus; and (2) the effect of stimulation is inevitably contaminated by electrical currents spread into the precentral gyrus located just next to SMA. With combined techniques of intracortical microstimulation and surface positive and negative stimulation, Wiesendanger (1973) questioned the existence of topographic organization in SMA. Stimulation of SMA in man (Penfield and Welch, 1951; Bates, 1953; Talairach and Bancaud, 1966) elicits speech arrest, vocalization, adversive movements of eyes and head, and complex, synergistic movements of the proximal forelimb. Functional implication of these responses remains obscure.

Deficits Following SMA Lesions

After ablation of SMA, the effects are subtle and often transient. In man, unilateral large lesions produce little or no lasting deficits (Laplane et al., 1977; Penfield and Jasper, 1954). Erickson and Woolsey (1951) observed a transient grasp reflex. Penfield and Jasper (1954) also noted a tendency for forced grasping in their patients as well as a slowness of movements of the opposite limbs. When an object was placed in the patient's hand he grasped it reflexively and had to make a considerable effort to open his own hand again. The patient seemed to have lost his ability to inhibit the hand closure. Travis (1955) found that unilateral ablation of SMA in the monkey gave rise to a transient grasp reflex and a bilateral hypertonia, notably at the shoulders. Bilateral lesions resulted in increased flexor muscle tone and eventual development of contractures. These findings led her to postulate a role for SMA in the control of muscle tone and posture. Coxe and Landau (1975), however, did not observe persistent changes in muscle tone or posture. Thus, ablation studies have not been conclusive thus far and more sophisticated testing may reveal subtle changes in motor performance after an SMA lesion.

Fore- and Hindlimb Representation Areas in the SMA

Brinkman and Porter (1979) noticed a general tendency that neurons active in relation to fore- and hindlimb movements were located in rostral and caudal part of SMA, respectively. The observation agrees with Woolsey but not with Penfield and Welch (1951), who described the supplementary leg areas as extending 10 mm anteriorly from the precentral tail area Penfield and Welch, 1951, Fig. 5). These results point to a need for detailed survey of neuronal activity related to fore- and hindlimb movements in the entire extent of SMA, including the upper bank of cingulate sulcus. To study the topographical organization, it is essential to specify the movement as the one involving muscle activities in only a limited part of the body, since unrestrained forelimb movements, for example, are accompanied by supporting activity in axial body as well as hindlimb muscles. In our studies, therefore: (1) animal's limbs were encased in form-fitting casts and fixed at optimal joint angles for movements limited at appropriate joints with his body weight stably sustained on a chair; (2) extensive training was made over months until the pattern of muscle activities was stabilized; and (3) the electromyogram (EMG) was thoroughly monitored in numerous limb and axial muscles to make sure that muscle activities were actually limited to a particular portion of the limb.

Location of Neurons Related to Fore- and Hindlimb Movements

A monkey was trained to perform fore- and hindlimb movements alternatively. The forelimb movements, wrist extension and flexion, and the hindlimb movement, pedal-press movement by foot plantar flexion, were triggered by visual signals and temporally separated by several seconds. Seventy-four neurons were found to be active in association with wrist extension and/or flexion (upper row of Fig. 1). The discharge increase is evident prior to wrist extension, much less prior to wrist flexion and absent with the key press movement with the foot. Of these 74 neurons, 38 exhibited different activity depending on the direction of the wrist movement. None of the 74 neurons showed significant activity changes with movement of the foot. In contrast, 86 neurons were active with the foot movement (lower part of Fig. 1), but exhibited no significant activity changes with forelimb movements. Thus, no neuron exhibited positive relation to both the fore- and hindlimb movements.

The depth of each extracellularly recorded neuron was recorded with respect to the cortical surface, the large-celled layers of cortex, and the subcortical white matter. Recording sites of neurons were located using the data thus obtained and by histologic reconstruction of the lesion sites and electrode tracks. The recording sites were located in one of three possible regions: dorsal surface of the hemisphere, medial cortex dorsal to cingulate sulcus, and dorsal bank of cingulate sulcus. In Fig. 2, these regions of cortex are illustrated as if unfolded. Triangles indicate the recording sites of neurons related with the movement. At sites indicated with dots, neurons exhibit slight and nonsignificant changes or pos-

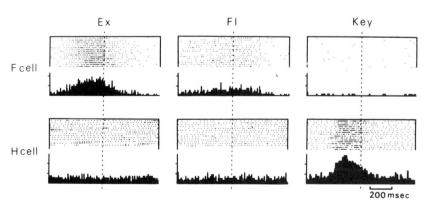

FIG. 1. Discharge of two SMA neurons whose activity changes prior to wrist extension (F cell, **top row**) and key-press movement by the hindlimb (H cell, **bottom row**). In raster displays, discharges before and after the onset of the wrist movements or the key-press movement (*interrupted lines*) are displayed as *dots*. In each periresponse histogram, discharges are summated in successive 8-msec periods.

sessed no relation to the task. They were found to be related to the animal's voluntary movements involving proximal muscles of forelimbs, hindlimbs, or axial body, during manipulative procedures attempted by experimenters. At sites indicated with horizontal bars, no movement-related activity in relation to the task or to any of the monkey's voluntary limb movements was detectable. The boundary between areas 4 and 6 was drawn approximately on the basis of transition to the typical cellular appearance of area 6 (cf. Wise and Tanji, 1981).

Three major findings emerged from Fig. 2: (1) Neurons related to forelimb as well as hindlimb motor task are distributed not only superficially, but in deeper portion of the medial face of the cortex and in the upper bank of the cingulate sulcus; (2) locations of these two classes of neurons are not intermingled but separated anteroposteriorly; and (3) neurons related to the hindlimb task are located in area 6, just anterior to hindlimb and tail areas of area 4 (confirmed by microstimulation, indicated as open circles, squares and diamonds in the figure). The third finding confirmed results by Wise and Tanji indicating that no discrepancy exists between physiological evidence for the location of hindlimb area of SMA and the posterior boundary of area 6

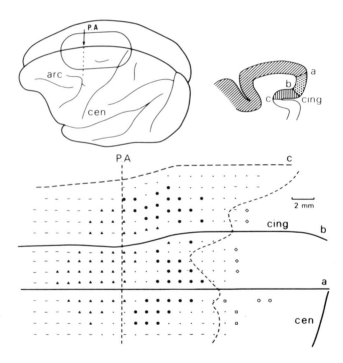

FIG. 2. Distribution of recording site of neurons related to the forelimb and hindlimb movements. **Top left:** Surface view of the rhesus monkey hemisphere used in the experiment showing the sulcal pattern and the area covered by the recording chamber. *P.A.:* Location of the anteroposterior reference line drawn where the posterior end of the arcuate sulcus comes to the surface. *Dashed line:* Location of the coronal section shown at top right. **Top right:** Three regions of cortex explored—dorsal and medial surface of the hemisphere and dorsal bank of the cingulate sulcus—are filled with the corresponding symbols of *diagonal lines, stipples,* and *vertical lines*. **Bottom:** Three portions of the cortex are displayed as if unfolded. *Triangles* and *filled circles:* Recording sites of neurons related to the fore- and hindlimb movements, respectively. *Diamonds, open circles,* and *squares:* Sites where microstimulation yielded muscle twitches in the tail, digits, and proximal hindlimb. Approximate boundary of areas 4 and 6 is drawn with an *interrupted line.*

determined cytoarchitectonically. Thus, the hindlimb part of SMA is to be placed within area 6, not extending into area 4 as reported by Woolsey et al. (1952).

Location of Neurons Related to Distal and Proximal Forelimb Movements

To study how neurons related to distal and proximal limb movements are distributed in the forelimb area of SMA, a paradigm of motor performance was designed where either distal or proximal limb musculatures are exclusively activated. Two monkeys were trained to sit in a chair and perform a motor task having two phases. One phase involved activity of digit muscles and the other called for activation of proximal forelimb muscles. The monkey's right forearm and upper arm were fixed to an L-shaped plastic cast so that the joint angles at the elbow and wrist were fixed at 90 and 180 degrees, respectively. The cast was attached to the sitting chair with a pivot in such a way as to be movable by inner or outer rotation at the shoulder joint. A key was attached to the distal end of the cast which could be pressed by flexion of 4 fingers, second to fifth digits. The monkey was required to shift the cast inwards or outwards to one of two correct holding zones of 4 degrees that were separated by 40 degrees. The correct holding was signaled by a green light. After the holding period of 2 to 4 sec, which varied unpredictably, a red light came on and instructed the monkey to press the key within 480 msec after the light onset. The monkey had to keep holding the cast in the correct zone during and after the key press. If the cast remained in the zone 2 sec after the key press, a reward was given and the animal had to shift the cast to the other zone. The EMG was recorded with silver wire electrodes chronically implanted in extensors and flexors of fingers, extensors and flexors of wrist and elbow joints, as well as in muscles in the shoulder, chest, girdle, and paravertebral region.

As a result of appropriate fixation of the forelimb and extensive training, dissociation of distal and proximal muscle activations in two phases of the motor task was achieved. Finger flexors, showing obvious activity increase prior to key press, exhibited no alterations of activity with shifting the cast (Fig. 3). In contrast, triceps, supra- and infraspinatus muscles appeared to be prime movers for the outward shift, and biceps and pectoralis muscles were active with the inward shift movement (middle and lower column of Fig. 3). None of their activity exhibited any alterations in association with the key press. Paravertebral muscles as well as many other muscles in the girdle changed their activity in various time relations with the motor task, but their activity changes were not at all times locked to either the shift movements or key press. Of 562 neurons recorded in the SMA of 2 monkeys, 225 neurons were found to change their discharge activity before starting the forelimb movements. Sixty of them appeared related to key press, 150 to the cast-shifting movement, and the remaining 15 neurons to both motor acts. Figure 4A shows an example of an SMA neuron, the activity of which increased prior to the key press but remained unchanged in association with inward or outward shift movements. The increase of discharge preceded the key press by 338 msec. Since the interval between the first increase of EMG activity in finger flexors and the key press was 204 msec, the neuronal activity preceded the muscle activity by 134 msec.

Figure 4B shows an SMA neuron whose activity changed in association with outward and inward shift movements with no alteration of activity with the key press. The activity increase started 234 msec and the decrease started 255 msec before the start of the shift movement detected by a potentiometer. The histogram in Fig. 5 shows distribution of SMA neurons related with the distal and proximal forelimb movements. Their rostrocaudal distribution was depicted against their depth from the sur-

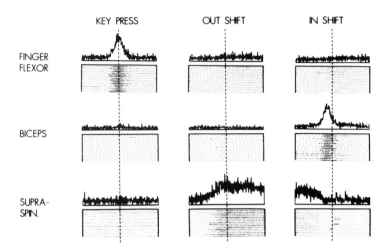

FIG. 3. EMG activity associated with distal and proximal forelimb movements. EMG is digitized with an A/D converter and displayed in rasters *(bottom)* and periresponse histograms *(top)* of each display. Finger flexor muscle is only active with the key press and biceps, and supraspinatus muscles are selectively active with outer and inner rotation movements at the shoulder joint. The duration of display period is 1,028 msec. *Dashed lines:* Time of occurrence of key press or onsets of shift movements.

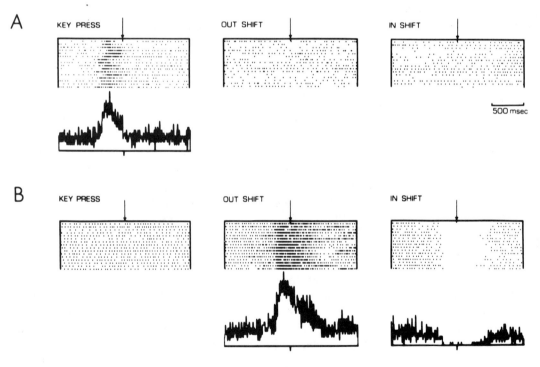

FIG. 4. Discharge of two SMA neurons whose activity changed prior to distal (**A**) and proximal (**B**) forelimb movements. **Top columns:** discharge before and after the onset of key press or cast-shifting movement *(arrows)* is displayed as *dots*. **Bottom columns:** histograms with bins representing summation of activity in successive 8-msec periods. Histogram; data obtained from 32 trials are summed which include the data obtained in 16 trials shown in the raster display above. (From Tanji and Kurata, 1979, with permission.)

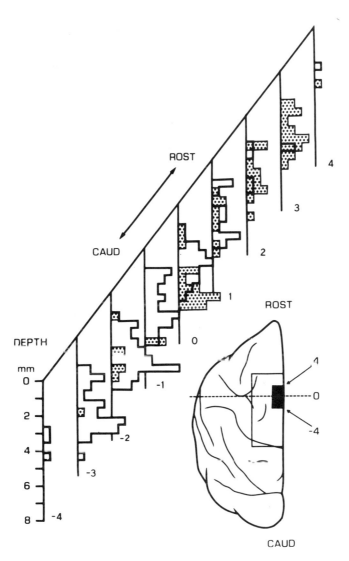

FIG. 5. Bottom: Area where points of microelectrode entry in all penetrations are located *(open rectangle)* and an area where neurons related to the forelimb movements are obtained *(shaded rectangle)*. Top: Distributions of neurons related to distal *(stippled bars)* and proximal *(open bars)* forelimb movements. A site corresponding to the caudal end of the arcuate sulcus is taken as a reference zero point in the rostrocaudal axis. Positive and negative numbers denote the distance in millimeters rostral and caudal from the reference point. Rostrocaudal distribution of neurons and their depth from the dorsal surface of the hemisphere is indicated in the histogram.

face, taking the posterior end of the arcuate sulcus as a reference zero line. Neurons related with distal forelimb movements (expressed as stippled bars) were located more rostrally and those related with proximal movements (open bars) were located more caudally, although with considerable overlap. In the overlapping region, neurons related with distal movements were located more deeply.

This study thus provided conclusive evidence that a class of neurons in SMA are active in association with movements performed exclusively by distal forelimb. The rostrocaudal arrangements of distal and proximal forelimb area agrees with Woolsey et al. (1952) and Brinkman and Porter (1979). However, the presence of neurons related to distal forelimb movements in deeper portion of the SMA, including the upper bank of the cingulate sulcus does not agree with the motor figurine chart drawn by Woolsey.

Comparison of Movement-Related Activities in the SMA and Precentral Motor Area

Because a class of SMA neurons started changing their activity prior to muscular contraction, just as precentral neurons (Evarts, 1967), one must know how neuronal movement-related activity differs in the two areas. Experiments were designed to compare quantitatively neuronal responses in the two areas using animals performing forelimb movements in response to sensory signals of three different modalities. Monkeys were trained to perform alternate flexions and extensions at the wrist in response to visual, auditory, or somatosensory signal, one of which was given to the animal in a random sequence. The visual signal was a red light placed in front of the animal. The auditory signal was a tone burst of 1 kHz, 30 dB above background noise level. The tactile signal was a vibratory stimulus to the monkey's hand. The vibration was generated by a servocontrolled direct current (DC) torque motor driven by a burst of square-wave oscillation at 500 Hz. The reward was given when the correct response was initiated within 280 msec after the onset of stimuli.

Difference of Activity

A representative example of neuronal activity in the precentral motor cortex is shown in Fig. 6. The neuron, identified as a pyramidal tract neuron (PTN) with a response latency of 0.9 msec to antidromic stimulus to the medullary pyramid, was active with wrist extension but not with flexion. In the left column of the figure, discharges are aligned at the onset of the signals calling for wrist extension. Neuronal response latency was calculated on the basis of the peristimulus histogram. The latencies to the visual, auditory, and tactile signals were 122, 100, and 40 msec, respectively. These values correspond well to those obtained in previous reports (Evarts, 1968; Evarts, 1973). In the right half of the figure, discharges are aligned at onset of wrist extension detected by the potentiometer. It is apparent that the discharge increase is similar, regardless of the modality of the sensory signal. The activity of the PTN resembled that of EMG activity recorded in the forearm, except for earlier onsets of activity changes.

An example of SMA neurons with marked movement-related activity is shown in Fig. 7. The activity of this SMA neuron is indistinguishable from that of precentral neurons except that the response latency to auditory signal is shorter (78 msec). More commonly, the magnitude of activity changes of SMA neurons was smaller (Fig. 8). In 124 out of 399 movement-related SMA neurons, the magnitude of activity is different depending on the modality of signals to which the animal responded (Fig. 8A). The neuron was most active in relation to visually triggered and less active with auditory-triggered wrist extension. It was almost inactive when the triggering signal was tactile.

Another feature of SMA neurons, different from motor cortex neurons, is the presence of response time locked to the visual or auditory stimulus. In the motor cortex, no neurons were found to be active with close time relation to visual or auditory signals, although the presence of precentral neurons driven by somatosensory input is well known. However, some SMA neurons such as the one in Fig. 8B exhibited discharges time locked with such sensory signals and started discharge 124 msec after onset of the visual signal. The temporal

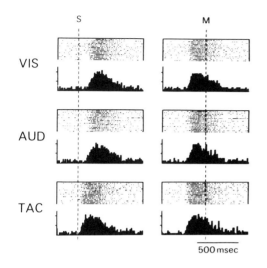

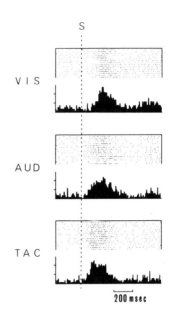

FIG. 6. Discharges of a PTN in the precentral motor cortex the activity of which increased with wrist extension. **Left**: Discharges are aligned at the onset of visual, auditory, and tactile signals. **Right**: Discharges are aligned at movement onsets. One step in the ordinate scale of the perievent histogram denotes 64 impulses/sec.

FIG. 7. Discharges of an SMA neuron having marked discharge increase in association with wrist extension.

locking of discharge onset to signal onset is evident in the top part of left column of the figure, whereas in the right column the interval between the discharge onset and the movement onset is variable. Sixty-eight SMA neurons exhibited discharges time locked to visual and/or auditory signals. A class of SMA neurons was activated by the tactile stimulus, but the magnitude of such somatosensory-evoked responses was smaller than observed in the motor cortex.

Quantitative Analysis

An attempt was made to compare the magnitude of movement-related activity in the two motor areas. For this purpose, the amount of increase or decrease of discharge frequency of individual neurons during 200 msec of premovement period above or below background level of discharge was calculated for each movement triggered by three different signals. In visually triggered

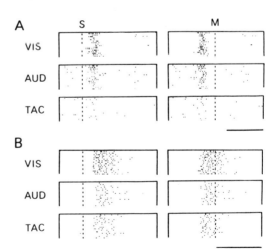

FIG. 8. **A**: Discharges of an SMA neuron, the activity of which is different depending on the modality of signals calling for wrist extension. **B**: Discharges of an SMA neuron exhibiting temporal locking of onset of discharge to signal onsets, especially with the visual signal. Discharges are aligned at the onset of signal appearance (**left**) and movement onset (**right**) in both **A** and **B**. *Horizontal bars:* 500 msec.

movement, mean magnitudes of SMA neurons and precentral neurons were 32 and 53 impulses/sec, respectively. The difference was found to be significant at $p < 0.00001$ by Mann-Whitney U test. Similar differences were obtained in auditory- and tactile-triggered movements. In the subsequent analysis, the latencies of neuronal responses to three signals were compared. Mean latencies of SMA neurons and precentral neurons is shown in Table 1. The latencies of SMA neurons to visual and auditory signals were significantly shorter than those of precentral neurons. For the tactile signal the reverse was true, although the difference was much smaller.

In summary, four major findings emerged from this series of experiments: (1) Magnitude of movement-related activity in SMA is smaller than in precentral motor cortex. (2) Response latency to visual and auditory signals are shorter in SMA than in motor cortex. Not much difference is found in response latency to tactile stimulus. (3) A considerable number of movement-related neurons in SMA to start their activity changes time-locked to the visual or auditory stimulus; and (4) Some neurons in SMA respond differently to three triggering signals, which is unusual in the motor cortex. Thus, SMA neurons are more closely linked with visual and auditory inputs but less intimately connected to somatosensory input, compared to precentral neurons. Judging from smaller magnitude of the movement-related activity, SMA seems to play a subsidiary role in motor control, as far as a simple motor task such as the one performed in the present response paradigm is concerned. In this sense, the term supplementary appears to be appropriate.

Role in Higher-Order Motor Control

In what aspects of motor performance does SMA possess more significant roles? Recent studies in man and monkeys suggest the role of SMA in generating a preparatory state for or even in programming forthcoming movements, rather than in actual execution of them. The readiness potential preceding initiation of movements by up to about one second, is known to have its maximum at the vertex (Kornhuber and Deeke, 1965). Deeke and Kornhuber (1978) found that, in patients with bilateral Parkinson's disease who failed to have the readiness potential in precentral region, the vertex maximum of the potential persisted. Judging from the location of the vertex electrode presumably on SMA, this suggests a role of SMA in the process of developing a preparatory state for impending movement. A newly developed technique of measuring regional blood flow of cortex (presumably reflecting the level of cortical activity) provided evidence that SMA was active when subjects were simulating a motor sequence internally, without executing it (Roland et al., 1980). Orgogozo and Larsen (1979) found that SMA exhibited higher blood flow during execution of complex movement sequences than during simple muscle contractions, which suggests a role of SMA in programming the motor sequence. These reports in man invite further studies to know what parameters of movements are programmed and in what way the preparatory process is achieved, in more precisely designed experiments. In our laboratory (Tanji et al., 1980), an attempt was made to observe how neuronal activity in SMA is altered with development of a preparatory state of primates for an impending move-

TABLE 1. *Latency of change in discharge frequency (msec)*

	Signal modality		
	Visual	Auditory	Tactile
SMA	128	108	68
Precentral	157	136	59
p^a	< 0.00001	< 0.00001	< 0.001

[a] By Mann-Whitney U test.

ment. The movement required the animal to respond properly to a sudden perturbing stimulus delivered to the forearm in one of two different directions. An instruction as to the direction of the monkey's movement (pushing or pulling the forelimb) was given several seconds before the stimulus. A number of SMA neurons exhibited instruction-induced changes of activity during the period intervening between the instruction and the perturbation-triggered movement. These particular neurons were not active in association with the movement per se. This substantiates the view that the SMA plays a part in modifying sensory-triggered motor responses known to be profoundly modified by establishment, in advance, of preparatory state of animals (Evarts and Tanji, 1974) or man (Hammond, 1956) to initiate intended movements.

Reciprocal Ia Inhibitory Pathway in Normal Man and in Patients with Motor Disorders

*Reisaku Tanaka

Department of Neurobiology, Tokyo Metropolitan Institute for Neurosciences, Tokyo, Japan

Reciprocal Ia inhibition is evoked by low-threshold afferents from the antagonist muscles, and it has the central latency of disynaptic linkage (Araki et al., 1960). It has been extensively investigated in acute animal experiments and is considered one of the basic mechanisms for reciprocal innervation of limb movement (Lloyd, 1946; Granit, 1970; Lundberg, 1970; Matthews, 1972; Hultborn, 1976). In man, reciprocal innervation is demonstrated in electromyograph (EMG) activities of muscles at voluntary flexion or extension of the limb. In Fig. 1A, electrical activities were observed in the soleus muscle but not in the pretibial muscles in voluntary ankle extension (plantar flexion), and vice versa at ankle flexion (dorsiflexion). Although very weak activities were recognized in the antagonist EMG records, they were due to potential spread from other muscles, since they became less discernible when implanted wire electrodes, rather than surface electrodes were used for EMG recording. Electrical inactivity of a muscle does not necessarily mean active inhibition of the homonymous α motoneurons (MNs) but may simply be a lack of excitatory drive or a facilitation too weak for discharge. More than 60 years ago Hoffmann (1918) had already shown the depression of the "Hoffmann-reflex" of the triceps surae muscle during voluntary dorsiflexion, which suggested that active inhibition underlies the antagonist's inactivity (Fig. 1B). The contribution of the Ia inhibitory pathway to the reciprocal organization of human motor activities is reviewed in this chapter.

Methodology

The method for revealing Ia inhibition is well established in animal experiments which allow direct access to the spinal cord and muscle nerves for stimulation, and intracellular recording from MNs. No such radical procedures are permitted in human experiments. Fortunately, the monosynaptic Hoffmann or H-reflex (Hoffmann, 1918, 1934) reflects the excitability of the homonymous MN pools (Fig. 1B). Therefore, Lloyd's technique (1946) for investigating the central connections of spinal reflexes could be applied with some technical modifications (Mizuno et al., 1971; Tanaka, 1974, 1980; Yanagisawa et al., 1976).

Ankle extensors (triceps surae muscle) and flexors (pretibial muscles) were used as a pair of antagonistic muscles because of the technical ease in evoking an H-reflex. The posterior tibial nerve was stimulated at the popliteal fossa, and the peroneal nerve at the caput fibulae level. Electrical pulses (0.5 to 1.0-msec duration) were given percutaneously for both conditioning and test

---

*Department of Neurobiology, Tokyo Metropolitan Institute for Neurosciences, 2-6 Masashidai, Fuchu-City, Tokyo, 183 Japan.

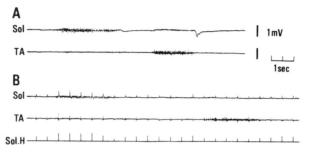

FIG. 1. A: EMG activities of ankle extensor (Sol) and flexor muscles (TA) at voluntary plantar flexion and movements. **B:** Isometric contraction. Changes in the soleus H-reflex size (Sol.H, full-wave rectified and integrated values) at ankle movements.

stimuli. Needles, which were insulated except for 1 to 2 mm of the tip, were also used on some occasions. The muscle site from which the H-reflex could be recorded at the lowest threshold was used throughout the experiment. In cases without an H-reflex in the pretibial muscles, the site of lowest threshold for the M-wave was chosen. The evoked EMGs were recorded from bipolar surface electrodes on the triceps and pretibial muscles. The reciprocal reflex effects were examined by applying, as a conditioning stimulus, a single pulse or train of pulses at a rate of 300/sec to the flexor or extensor nerve, and examining the change of amplitude of the antagonistic muscle H-reflex at variable intervals. Paired stimuli were applied every 3 sec, and more than 3 trials were recorded at every conditioning-test interval. The control H-reflex was recorded at frequent intervals to check its stability.

In assessing the results, special care was taken to ensure that the reciprocal effect was generated by low-threshold afferents from antagonist muscles, and that the central latency was disynaptic. The stimulus strength was controlled by monitoring M-wave threshold and was expressed in multiples of the threshold strength for the M-wave (x motor threshold: xMT). The threshold of the H-reflex, which is lower than that of the direct motor response (M-wave) in the triceps indicates that group Ia afferents have a lower threshold than α-efferents in this mode of stimulation. However, stimulation subliminal to the M-wave does not secure selective stimulation of group Ia afferents by itself. It often accompanies cutaneous sensation. In fact, the stimulated nerves are mixed, with afferents of cutaneous and articular origin. On the other hand, it is generally believed that group I afferents elicit no conscious sensation (cf. Matthews, 1972; McCloskey et al., 1980, *this volume*) in this respect, a separate experiment was made on 8 normal subjects in which the effect on sensory perception was tested with needle stimulation of the peroneal nerve (Yanagisawa and Tanaka, 1978). No sensation was aroused in any of the subjects when the intensity of stimulation was less than 1.0 xMT. At 1.0 xMT or slightly above, a sensation of local muscular contraction occurred accompanied by visible contraction. Paresthesia radiating to the toe could only be evoked at an intensity of 1.2 to 1.9 xMT. Measuring central latency is more difficult, since the conduction distance from the stimulated portion to the segmental spinal cord is much longer, and the conduction velocity (CV) of group I fibers is slower in man (fastest velocity, CV of 60 m/sec; Magladery and McDougal, 1950) than in cats (120 m/sec; Lloyd, 1943). Both factors result in less synchrony of the afferent discharges at entry of spinal cord.

We presumed that the conduction velocity of Ia fibres does not differ significantly between the tibial and the peroneal nerves as Lloyd himself claimed (1946; Lloyd and Chang, 1948). Then, the central latency can be estimated from adjustment of the minimal effective interval between conditioning and test shocks by conduction time for Ia volleys to reach the spinal cord from both sources. In short, the procedure established by Lloyd is applicable to human experiments.

Disynaptic Ia Inhibitory Pathway

Figure 2 shows an example from a normal subject, relaxed in a prone posture. The conditioning stimulation of the peroneal nerve evoked an inhibition of the triceps H-reflex (compare Fig. 2A and C). In this case, the minimum effective strength of the conditioning stimulus was 0.85 xMT and the maximum effect was attained at liminal strength. Because a train of 3 shocks at 3-msec intervals was used for conditioning, it is difficult to precisely estimate central latency. Since the effect was virtually abolished by withdrawal of the last shock (Fig. 2B), the last shock produced the inhibition, with the aid of temporal summation by the preceding two shocks. Time course of effects by double (squares) and triple (circles) conditioning shocks is illustrated in Fig. 2F. The abscissa is arranged to show the time of the third stimulus as 0 msec (which is 6.6 msec after the first shock). Notice, however, that the preceding two shocks had evoked an inhibition, which had a longer latency of 7.6 msec from the first shock and a slower time course. The inhibitory effect directly evoked by the third shock was obtained by subtraction of the former curve from the latter: starting just after 1 msec of the conditioning-test interval, reaching the maximum at 2 msec and subsiding within 6 msec. This time course is (except for the onset latency) analogous to the curve for Ia inhibition obtained from the cat spinal cord (Fig. 7 of Lloyd, 1946). This onset latency suggests a disynaptic linkage if the effect is indeed mediated by group Ia afferent. The test H-reflex is, of course, evoked by Ia volleys. In man, the motor pools of pretibial muscles are situated in the fourth and fifth lumbar segments, overlapping those of triceps in the fifth lumbar and 1st sacral segments (Sharrard, 1955). The electrode for stimulation of the peroneal nerve was usually located 5 to 7 cm distally from that for the tibial nerve. Since the CV of Ia afferents in tibial nerve in man is approximately 60 m/sec, the conduction time for Ia afferents in peroneal nerve to reach the motor pools should be about 1 msec longer than that of the tibial Ia volleys. Correction of the minimal effective interval of 1 msec by this delay shows that the central latency of the reciprocal inhibitory effect on the triceps surae MNs is about 0 msec. This matches the central latency of Ia inhibition in the cat hindlimb (Lloyd, 1946; Laporte and Lloyd, 1952; Araki et al., 1960). The same effect was evoked with needle electrode for conditioning stimulation without sensory perception. It can be concluded that this inhibition is evoked disynaptically via the Ia reciprocal inhibitory pathway.

As illustrated in Fig. 2D, the conditioning shock evoked not only Ia inhibition of the triceps H-reflex, but also a H-reflex in pretibial muscles. Reciprocal Ia inhibition of pretibial muscles was also investigated in the subject using this reflex (Fig. 3). As shown in Fig. 3B–E, the reciprocal inhibition appeared when the single tibial shock (arrows in Fig. 3 C–F) was given close to the third peroneal shock, which was linked with the test H-reflex (Fig. 3B). The brief time course of the effect was similar to Fig. 2F, except for its onset interval. The effect appeared even when the conditioning shock followed the test shock by 1 msec. However, with the calculation given above, its central latency could be estimated to be

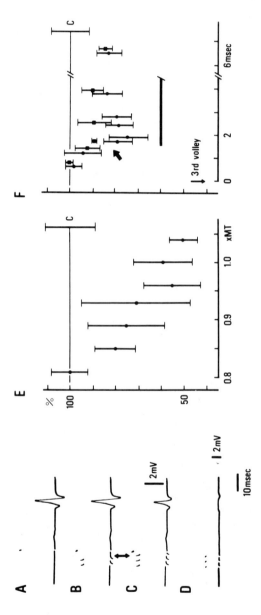

FIG. 2. Reciprocal Ia inhibition of triceps surae MNs. Normal subject at rest. A–C: Electromyographic records of the control H-reflex in the triceps surae muscle (A) and the effects of peroneal nerve stimulation on it with double (B), and triple shocks (C), 1.0 xMt, 333/sec). The third peroneal shock is set 2 msec prior to the test tibial shock. D: Response in the pretibial muscles by the triple peroneal volley. E: Effect of changing the strength of conditioning stimuli. *Abscissa* shows strength of the stimuli in xMT. The time relation of conditioning-test interval is the same as for C. F: Time course of the effects of double (*squares*) and triple peroneal stimuli (*circles*); 1.0 xMT, 300/sec). The *abscissa* shows the time interval between the third conditioning and test stimuli. The inhibition is obvious from 1.7 msec (*arrow* and *horizontal bar*). (From Tanaka, 1980, with permission.)

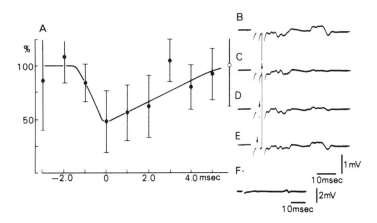

FIG. 3. Reciprocal Ia inhibition of the pretibial MNs. The same subject as in Fig. 2. A: Time course of the effect evoked by single tibial shocks (1.05 xMT). The *abscissa* shows time intervals between the tibial and the third peroneal shock, which evoked the test H-reflex. Negative values indicate that the tibial shock followed the peroneal shock and positive values the reverse. B–E: Specimen records of the effect from the pretibial muscles. *Vertical line* indicates the time point of third peroneal shock and *arrows* point the tibial shocks. F: Response from the triceps surae muscle by the tibial shock. (From Tanaka, 1974, with permission.)

about 0 msec again, indicating the same synaptic linkage. Therefore, the reciprocal Ia inhibitory pathway is constructed symmetrically in man between ankle extensors and flexors, as in the cat.

Another finding is that the H-reflex in a muscle, either ankle flexor or extensor, and Ia inhibition of its antagonists, tend to come together. It implies that there is a parallel between MNs activity and inhibitory interneurons, both of which are monosynaptically connected by Ia afferents from the same muscles.

Ia Inhibition in Normal Subjects

It is well known from animal experiments that interneurons mediating reciprocal Ia inhibition are under supraspinal control from pyramidal and extrapyramidal systems (Lundberg, 1970; Hultborn, 1976; Jankowska et al., 1976). As integrative center for reciprocal inhibition, the activity of the Ia inhibitory pathway is modulated in various physiological states.

(A) At rest

The subject in Figs. 2 and 3 is rather exceptional, because he showed a clear H-reflex in the pretibial muscles and disynaptic Ia inhibition of the triceps surae muscle at rest (Hohman and Goodgold, 1960; Tanaka, 1980). In most normal cases, peroneal nerve stimulation evoked neither (Hoffmann, 1934; Teasdall et al., 1951; Mizuno et al., 1970; Tanaka, 1974). Figure 4 shows the lack of Ia inhibition of the tricep H-reflex by peroneal stimulation in two other experiments. Strong single stimuli evoked longer latency inhibition, starting at about 7 msec, reaching a peak at 15 msec and subsiding to the control value at 40 msec (Fig. 4A, the inhibition at 80 msec was evoked by mechanical movement of the foot and is not considered here; Mizuno et al., 1971). The triple peroneal volleys with liminal M-wave intensity evoked a similar late depression of triceps H-reflex. A longer latency and time course would have been caused by weaker and repetitive stimuli, respectively. The fact that this inhibitory effect was manifest

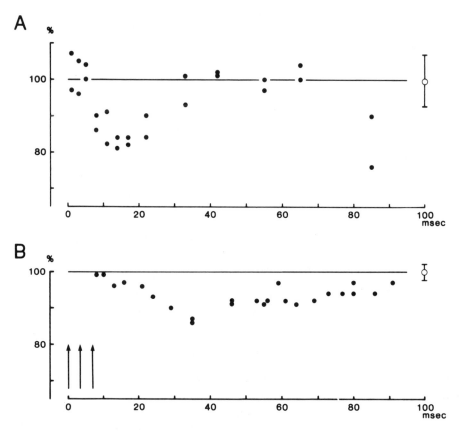

FIG. 4. Time course of effects of peroneal nerve stimulation on the triceps surae H-reflex in two normal subjects. **A:** Single conditioning stimulation with 1.34 xMT (near maximal for M-wave). (Modified from Mizuno et al., 1971, with permission.) **B:** Triple stimulation with 1.06 xMT (300/sec). *Arrows* indicate the stimuli. (Modified from Tanaka, 1974, with permission.)

at a subliminal stimulus for α-efferents and saturated at about 1.3 xMT, indicates the origin of group I afferents. The onset latency and slow onset of the effect conform well to presynaptic inhibition (Eccles et al., 1962). Low-threshold cutaneous and articular afferents were not involved since the same effect was evoked by peroneal stimulation with needle electrodes without somatic sensation (see Methods).

The lack of Ia inhibition to triceps and of pretibial H-reflex in many subjects indicates a low excitability of the flexor Ia interneurons and α-MNs. In the exceptional group with the pretibial H-reflex (which suggests a relatively higher excitability of the flexor system) tibial nerve stimulation, even with single pulses, easily evoked Ia inhibition of the flexor MNs (Fig. 3). Thus, the extensor Ia system has a higher excitability than the flexor Ia system.

(B) At voluntary contraction

As shown in Fig. 1B, the triceps surae MNs are inhibited during voluntary ankle dorsiflexion. Peroneal nerve stimulation with group I intensity generated the inhibition (Fig. 2) in subjects who showed no Ia inhibition at rest (Tanaka, 1974, 1980). Thus, the excitability of the Ia interneurons inhibiting the extensor MNs was increased during voluntary tonic flexor excitation. Since the agonist Ia afferents increase their discharges during voluntary contraction

(Vallbo, 1970, 1973), it is concluded that the Ia inhibitory pathway participates in reciprocal inhibition of the antagonists. The parallel facilitation of pretibial MNs and Ia inhibitory interneurons acting on extensor MNs was also observed just prior to the onset of ankle dorsiflexion from a resting state (Simoyama and Tanaka, 1974) and from the active plantar flexion (Tanaka, 1979 and *unpublished observation*). This supports the concept of α-γ linkage in reciprocal inhibition (Lundberg, 1970; see Prochazka and Hulliger, *this volume*). One disadvantage of our method is that it is impossible to test the excitability of Ia inhibitory interneurons without H-reflex in the antagonist muscles. In subjects in whom the pretibial H-reflex could not be evoked at active plantar flexion, we were unable to test the Ia inhibitory pathway to the flexor MNs.

Ia Inhibition in Central Motor Diseases

The reciprocal Ia inhibitory pathway can thus be controlled by the supraspinal motor centers in the same way reciprocal innervation is achieved. Therefore, it must be possible for the inhibitory pathway to be affected by the central disorders.

(A) Spastic hemiplegia

Lesions in the internal capsule and adjacent structures cause spastic hemiplegia. In lower limbs spasticity is prominent in the physiological extensor system, while paralysis is more severe in the flexor system. These clinical signs would seem compatible with reciprocal inhibition being still preserved in spastic patients. The electrophysiological study, however, indicated an exaggeration of the stretch reflex in flexor muscles by revealing the frequent occurrence of pretibial H-reflex at rest in patients (Hohman and Goodgold, 1960; Yanagisawa et al., 1976). Spasticity may also exist in flexor muscles even if it is not discovered in a clinical examination. Since the H-reflex is obtained from both triceps and pretibial muscles in spastic patients at rest, the excitability of the Ia inhibitory pathway to extensors and flexors was tested equally (Yanagisawa et al., 1976; Yanagisawa, 1980). The study revealed a marked Ia inhibition from tibial nerve to pretibial α-MNs, but no significant Ia inhibition from peroneal nerve to triceps α-MN. This pattern was rather similar as that in the normal subject at rest. Extensor spasticity can be reduced by a procaine block of the extensor nerve without significant loss of its motor power (Rushworth, 1960; Matthews, 1972). This effect was interpreted as caused by selective blocking of γ-efferents, which were supposed to be hyperactive in the spastic state. An alcohol block of motor points of triceps generated a similar but more persistent effect in the spastic muscle. Furthermore, it resulted in a definite increase of pretibial muscle power at voluntary dorsiflexion testing and increased EMG activities.

Existence of the pretibial H-reflex, the pattern of Ia inhibition between antagonists, and the effect of alcohol block led us to the following interpretation of mechanisms in capsular hemiplegia. A cerebrovascular lesion in the capsular and adjacent areas interrupts descending tracts from cerebral cortex and subcortical nuclei. This may cause a release from higher centers and a blockade of descending motor commands (Fig. 5). The release mechanism would act on both flexor and extensor systems, and also enhance the monosynaptic reflex activity in pretibial muscles. Its action would be much stronger in the extensor system (actually displaying spasticity) and this would, in turn, suppress the flexor system through the segmental Ia inhibitory pathway thereby masking flexor spasticity (Fig. 5A). Thus, the weakness of flexor muscles would be partly generated by tonic Ia inhibition from the extensors as well as by reduced driving power from the brain (Fig. 5B). In contrast, the extensor muscles are better off, since the increased excit-

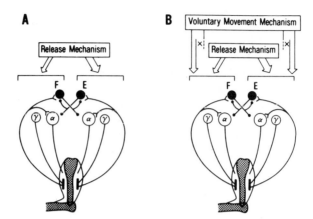

FIG. 5. Schematic illustration of the neural mechanism underlying the spastic hemiplegia at ankle joints. Neurons shown as *filled circles* indicate interneurons mediating the disynaptic Ia inhibition. F and E: Flexor and extensor systems, respectively. (From Yanagisawa et al., 1976, with permission.)

ability of their MNs may compensate a reduced central driving power to some extent. The segmental interaction on the descending motor commands is indicated by the effect of an alcohol block of nerves of spastic muscles. Flexor motor power would be improved by a therapy achieving disinhibition from Ia inhibitory interneurons (see Yanagisawa et al., 1976; Yanagisawa, 1980).

(B) Athetotic cerebral palsy

This condition is caused by a disorder in the basal ganglia without known spinal cord lesion and shows a defect in movements reciprocity (Carpenter, 1950; Yanagisawa and Goto, 1971; Mizuno et al., 1971). Special care was taken to select a category of adult patients with mild athetosis who were of normal intelligence, emotionally stable, and showed few or no signs of rigidity or involuntary movement when resting. They were able to maintain an active or passive posture as long as emotional stability was kept, which was essential for stability of experimental conditions. Peroneal stimulation evoked distinct Ia inhibition on the triceps surae H-reflex in 4 of 6 subjects (in the other 2 cases, confirmation was made impossible by fluctuations in test H-reflexes). The peroneal liminal stimulation did not evoke the pretibial H-reflex (cf. Tanaka, 1980).

A question arose from the above results: Why is an exaggerated Ia inhibition of triceps MNs present at rest in a disease characterized by disturbances in reciprocal innervation at voluntary movement? Yanagisawa (1980) proposed that the examined patients could have acquired this hyperactivity as a result of long, continuous attempts to somehow control their limbs. He based this hypothesis of a compensatory neural mechanism on the facts that the motor signs of athetosis in childhood usually showed much greater disturbances than grown-up patients, and that reciprocal Ia inhibition to triceps in normal patients is inactive at rest but active in voluntary dorsiflexion of the foot (Tanaka, 1974, 1980). This interesting hypothesis should be tested further by developmental studies.

Ia inhibition in spinal cord lesions was investigated and reported (Yanagisawa, 1980; Yanagisawa and Tanaka, 1978).

General Comments on Reciprocal Inhibition

The aim of present studies is to elucidate the activity of the reciprocal Ia inhibitory pathway. Naturally, criteria for demonstration of genuine Ia inhibition, i.e., control of stimulus intensity, etc., were fairly strict, and apparent discrepancies of others from the present study could be explained by differences in experimental conditions. For example, peroneal inhibitory effects on tonic unitary activity of triceps during voluntary plantar flexion (Ashby and LaBelle, 1977; Kudina, 1980) had a latency as late as 35 or 40 msec, and duration as long as 15 msec. This time course does not agree with that of classical reciprocal Ia inhibition but seems to conform to the later depression illustrated in Fig. 4A. A reciprocal inhibition was reported in active pretibial muscles by tibial nerve stimulation (Agarwal and Gottlieb, 1972; Ashby and Zilm, 1978). In this case, a slightly earlier latency suggests that the initial brief part could be caused by disynaptic Ia inhibition, since relatively high excitability of this pathway to flexors might not be fully suppressed by weak dorsiflexion, and could be activated by synchronous extensor group Ia volleys (Tanaka, 1974). The later and major part, however, seems to have different properties as mentioned above.

Although reciprocal innervation is one of the basic mechanisms in motor control (Basmajian, 1978), co-contraction may be important under certain circumstances (Tilney and Pike, 1925; see Humphrey and Reed, *this volume*). In fact, many types of movement also exist among repertoires of human motor ability, in which simultaneous activities of antagonistic muscles were proved by EMG recordings. Distal muscles, which have more complicated reciprocal relations and perform subtler and more skillful movements, are more likely to show this phenomenon (Long and Brown, 1964). In more proximal muscles of the ankle, knee, and elbow joints, co-contraction has been more frequently observed in conditions that require extraordinary muscle power or velocity, or during inexperienced motor performance. It is, however, less frequent after training and in skilled subjects (Basmajian, 1978; Patton and Mortensen, 1970; Tanaka, *unpublished observation*). It should also be noted that co-contraction of antagonistic muscles is sometimes a major clinical sign in some motor diseases of central origin. Admitting the existence of co-contraction of antagonistic muscles under certain physiological conditions, in our experiments, continuous EMG recordings indicated that ankle movement followed the rule of reciprocal innervation exactly (Fig. 1). With this information, we have examined the activity of the Ia inhibitory pathway, which was modulated according to the rule of reciprocal innervation.

Functional Organization of Recurrent Inhibition in Man: Changes Preceding and Accompanying Voluntary Movements

*,†E. Pierrot-Deseilligny, †R. Katz, and **H. Hultborn

†Service de Rééducation Neurologique, Hopital de la Salpêtrière, Paris, France; and **Neurofysiologisk Universitet Institut, Panum Institutet, Belgdamsveg 3 C, Copenhagen 2200, Denmark

Recurrent inhibition of motoneurons (MNs) first described by Renshaw (1941, 1946), is a well-known example of negative feedback in the CNS. Motor axons give off recurrent collaterals, which activate interneurons, now called Renshaw cells (RCs); these RCs, in turn, inhibit MNs. If there were no input other than from recurrent collaterals onto RCs, the larger the motor discharge, the larger the resulting recurrent inhibition of the MNs responsible for this discharge would be. However, it is no longer possible to consider that the recurrent pathway, like the MN after-hyperpolarization (AHP), only contributes to limit and stabilize the frequency of MN discharge. Indeed, the RC connections appear to be much more complex than those of a simple negative feedback from MNs to MNs. Renshaw cells have inhibitory projections not only to α-MNs, but also to γ-MNs (Ellaway, 1971), other RCs (Ryall, 1970), and Ia inhibitory interneurons (Hultborn et al., 1971a,b). Furthermore, RCs receive synaptic inputs from many sources (primary afferents and descending pathways) other than from recurrent collaterals (see Haase et al., 1975).

Changes in RC excitability during various voluntary contractions have been studied in man (Pierrot-Deseilligny et al., 1977, Hultborn and Pierrot-Deseilligny, 1979a) to know to what extent supraspinal controls can modify transmission in the recurrent pathway during natural movements. These experiments have led to the formulation of new assumptions concerning the functional role of recurrent inhibition (Hultborn and Pierrot-Deseilligny, 1979a; Hultborn et al., 1979a).

EVIDENCE FOR RECURRENT INHIBITION IN HUMAN SUBJECTS

The first step was the development of a method suitable for selective activation of RCs in man. The method used in animal experiments, i.e., the selective stimulation of motor axons, cannot be used in human subjects. Obviously, it is impossible, when dorsal roots are intact, to stimulate motor axons antidromically without inducing two additional factors: (1) a volley via Ia pathways whose facilitatory action on MNs may interfere with recurrent inhibition; (2) a muscle contraction that elicits a complex

---

*To whom correspondence should be addressed: Rééducation Neurologique, Hopital de la Salpêtrière, 47 boulevard de l'hopital, Paris 75651 Cedex 13 France.

afferent discharge. In order to overcome these difficulties, a method has been developed in which orthodromic activation of RCs in human subjects is elicited by a monosynaptic reflex discharge (Pierrot-Deseilligny and Bussel, 1975). Even small monosynaptic reflexes are sufficient to produce a sizable recurrent inhibition in cats (Hultborn et al., 1979b). Thus, we have activated RCs by conditioning monosynaptic discharges, and the resulting recurrent inhibition was assessed by a subsequent monosynaptic reflex. Both conditioning and test reflexes were soleus (Sol) H-reflexes, the same unipolar electrode delivering conditioning and test stimuli. It was essential to prevent excitatory postsynaptic potentials (EPSPs) created by the Ia conditioning volley from interfering with recurrent inhibition in the tested MNs. We have, therefore, developed a method in which only MNs that have already fired due to the conditioning volley have their excitability assessed by the test volley (Pierrot-Deseilligny et al., 1976). Indeed, in these MNs, the conditioning Ia EPSPs are eliminated by the spikes they have elicited (Coombs et al., 1955).

(1) The Test Reflex is Only Produced by MNs That Have Already Fired in the Conditioning Discharge

As shown below, this is due to a collision in the motor axons. Figure 1A (upper trace) shows the H1 conditioning reflex generated by the S1 conditioning stimulus. The second trace in Fig. 1A shows the maximum direct M response (M max) elicited by the test stimulus, which is supramaximal (SM) for α motor axons. If such a supramaximal stimulus is applied alone, the resulting direct and maximal motor response (M_{max}) is not followed by an H-reflex (Fig. 1A, second trace). This is because the antidromic motor volley collides with and eliminates the H-reflex evoked by the supramaximal stimulus (Hoffmann, 1922) at or close to the MN soma level (see MN "X" in Fig. 1C and D). Provided that the conditioning-test interval is adequate, the H1 reflex discharge (due to the afferent conditioning volley) collides with the antidromic motor volley caused by the test stimulus (cf. MN "Z" in Fig. 1C). This collision eliminates the antidromic volley so that the H' text reflex due to the SM test stimulus can now pass along the motor axons (cf. MN "Z" in Fig. 1D) and appear in the EMG (Fig. 1A, third trace). In other words, the collision of the H1 reflex with the antidromic motor volley following the supramaximal stimulation opens the way for an orthodromic reflex response to the same supramaximal test stimulus. Obviously, this collision also prevents the appearance of the H1 EMG response (Fig. 1A, third trace). Since H' can only pass along the motor axons in which previous collision has taken place, the H' test reflex response recorded in the EMG is only produced by MN that have already given rise to the H1 conditioning reflex (therefore H' can, at the very most, be equal to H1).

Note: Although the H' response is secondary to a stimulus supramaximal for α motor fibers, it is not contaminated by any F wave (Pierrot-Deseilligny et al., 1976). To obtain the H' response, the time interval between S1 and SM stimuli must be such that the conditioning reflex volley brought about by S1 does not reach the point of stimulation by the time SM is applied. It should also be pointed out that the conditioning stimulus must be adjusted so that there is no significant direct M wave in gastrocnemius muscles; otherwise, it is not possible to evoke the H' test reflex, because it is suppressed by additional recurrent inhibition created by the simultaneous antidromic motor volley.

(2) Inhibition of the Test Reflex After the Conditioning Reflex

Figure 2 shows the variations in the H' test reflex amplitude when the H1 conditioning reflex was increased by augmenting

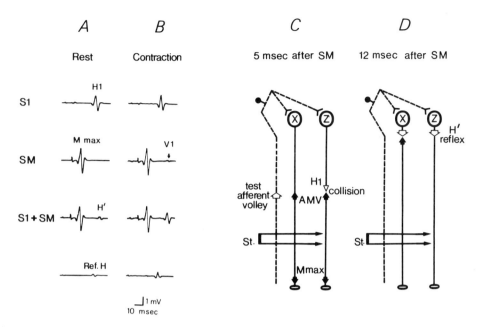

FIG. 1. Method to investigate the inhibition following a conditioning H-reflex. A and B. EMG responses in soleus following the same sequence of stimuli to the posterior tibial nerve at rest (A) and during a strong tonic voluntary contraction of the triceps surae (B). **Upper traces:** Isolated S1 conditioning stimulus evoking a maximum H1 conditioning reflex. The amplitude of the H1 conditioning reflex and the very small M1 response are the same at rest and during contraction. **Second traces:** Isolated supramaximal stimulation (SM) causing a maximal direct motor response (M_{max}); during contraction the V1 response appears after M_{max}. **Third traces:** Combined conditioning and test stimuli (10-msec conditioning-test interval); H1 is no longer present but a new response—the H' test reflex—appears. **Fourth traces:** Weak stimulus adjusted to produce a reference H-reflex, which has an amplitude identical to that of the H' test reflex at rest. The facilitation of the H' test reflex is larger than that of the reference H during contraction. C and D: Illustrations of different impulses (*arrows*) propagating along nerve fibers when using a 15-msec conditioning-test interval in a subject 1.65 m tall. Two MNs (X and Z) are activated by the Ia afferent volley elicited by the SM test stimulus. *Small white arrow* in C represents the H1 reflex discharge in axon Z. *Large white arrows* represent the Ia afferent test volley due to the test stimulus (C) and the following reflex discharge (D). *Black arrows* represent the orthodromic motor volley evoking M_{max} and the antidromic motor volley (AMV) due to stimulation of motor axons by the SM test stimulus (C). C: 5 msec after the SM test stimulus, as a result of this stimulation, impulses travel both orthodromically in Ia fibers and antidromically in motor axons. The H1 response, which runs along axon Z collides with and eliminates (collision) the antidromic motor volley. D: 12 msec after the SM test stimulus; a reflex response develops in both MNs, X and Z. This response is blocked in MN X but not in MN Z, because the antidromic impulse in MN Z was erased by the H1 response (as shown in C). (From Hultborn and Pierrot-Deseilligny, 1979a, with permission.)

the S1 conditioning stimulus strength. At low conditioning reflex amplitudes, the test reflex remained equal to the conditioning reflex, but as the H1 conditioning reflex continued to increase, there was a gradual decrease in the test reflex amplitude. This inhibition of the test reflex, which increases when the conditioning reflex discharge and/or the conditioning stimulus strength increase, can have three possible origins: (1) postspike AHP (Coombs et al., 1955), since the text reflex is only produced by MNs that have already fired in the conditioning reflex; (2) enhancement of the recurrent in-

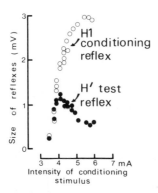

FIG. 2. Inhibition of the H' test reflex when the amplitude of the H1 conditioning reflex is increased via augmentation of the S1 conditioning stimulus (10 msec S1–SM interval). Amplitudes of H1 *(open circles)* and H' *(filled circles)* are plotted against intensity of S1 conditioning stimulus. Each symbol represents the mean of five measurements.

hibition caused by the increased conditioning reflex discharge, since it has been demonstrated in parallel animal experiments that there is a linear increase in recurrent inhibition with increasing conditioning monosynaptic reflexes (Hultborn et al., 1979b); (3) activation, when the conditioning stimulus strength is increased, of a larger number of fibers, like Ib fibers, inhibitory for Sol MNs.

(3) The Inhibition of the Test Reflex is Only Related to the Size of the Conditioning Reflex

Provided that the conditioning-test interval is more than 9 msec, the effect of the Ib fiber stimulation by the conditioning stimulus can be eliminated, since it has been demonstrated that in humans, the duration of Ib inhibition from triceps to Sol MNs is less than 10 msec (Pierrot-Deseilligny et al., 1979, 1981a). To eliminate the possibility of the conditioning stimulus activating other inhibitory afferent fibers, the size of the conditioning reflex was increased without modifying the conditioning stimulus strength. Thus, the number of fibers that could have possibly inhibited the test reflex was kept constant. It is emphasized that the inhibition of the test reflex with increasing conditioning reflexes (curve with open triangles in Fig. 3B) was the same whether the conditioning reflex is increased via an augmentation of the conditioning stimulus strength, or via a facilitation of the conditioning reflex (by a radial nerve stimulation of a soleus stretch). Provided that the conditioning-test interval was more than 9 msec, it was invariably found that the amount of inhibition of the test reflex was independent of the conditioning stimulus strength and depended only on the size of the conditioning reflex (Pierrot-Deseilligny and Bussel, 1975; Bussel and Pierrot-Deseilligny, 1977).

(4) Respective Roles of AHP and Recurrent Inhibition

Since the depression of the test reflex is not due to an activation of inhibitory fibers by the conditioning stimulus, it is caused by both AHP and recurrent inhibition. According to the size principle of Henneman et al. (1965a,b; Clamann et al., 1974; Henneman, 1981), increasing the conditioning reflex amplitude results in recruiting larger and larger MNs. As long as the conditioning reflex is of low amplitude, the test reflex is equal to it, indicating that the AHP has been overcome by the Ia test volley in all the smallest MNs. It is conceivable that the AHP can no longer be overcome by the Ia test volley in larger MNs, since such MNs are less sensitive to Ia inputs (Burke, 1968). This could explain why the test reflex does not follow the conditioning reflex at high amplitudes. However, if AHP was acting alone, there would be a fixed limit to the number of MNs available for the test reflex. Increasing the conditioning reflex should not decrease this number, and the variations in test reflex versus those in conditioning reflex should exhibit a plateau. Thus, the decreasing part of the curves in Fig. 3B,

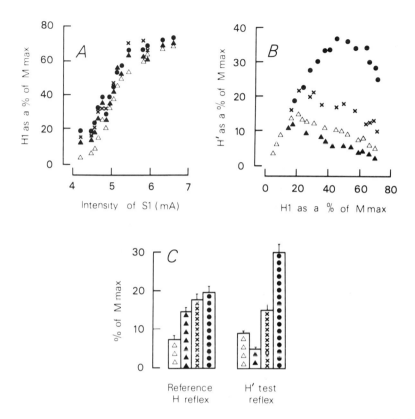

FIG. 3. Influence of soleus voluntary contractions of different force on recurrent inhibition brought about by the H1 conditioning discharge. *Open triangles:* Rest; *filled triangles:* Weak soleus contraction (10% of maximum); *crosses.* Medium contraction (40% of maximum); *filled circles:* Strong contractions (80% of maximum). A: Amplitude of the H1 conditioning reflex is plotted against the intensity of the S1 conditioning stimulus. B: Amplitude of the H' test reflex (expressed as a percentage of M_{max}) is plotted against that of H1. Each symbol in A and B represents the mean of five measurements. C: Histograms showing the size of reference H-reflex (**left** group of columns) and H' test reflex (**right** group of columns) at rest *(open triangles)* and during weak *(filled triangles)*, medium *(crosses)*, and strong *(filled circles)* soleus voluntary contractions. Each column represents the mean of 20 measurements; *vertical bars* at the top of the columns represent 1 SEM above the mean values. (From Hultborn and Pierrot-Deseilligny, 1979a, with permission.)

which implies a reduction in the number of MNs involved in the test reflex when the conditioning reflex is increased, can only be explained by recurrent inhibition elicited by the conditioning reflex.

METHOD OF ESTIMATING CHANGES IN RECURRENT INHIBITION DURING TRICEPS CONTRACTIONS

The amplitude of the H' test reflex can, therefore, be used to estimate the recurrent inhibition elicited by the H1 conditioning reflex. Indeed, if neither the Sol MN excitability nor the Sol MN AHP are modified, the size of the test reflex will depend only on the amount of recurrent inhibition caused by the H1 conditioning reflex. In such a situation, for a given amplitude of the conditioning discharge, changes in the size of the test reflex in relation to its size in control situation do reflect changes in recurrent inhibition elicited by H1. Although subjects stand without contracting the tested Sol, the test reflex was significantly smaller than in the sitting position. Since both Sol MN excitability and Sol MN AHP are similar in the two situations, recurrent inhibition was increased while standing (Pierrot-Deseilligny et al., 1977; Pierrot-Deseilligny and Morin, 1980). The situation is more complex during Sol voluntary contractions since both Sol MN excitability and Sol MN AHP are modified, which leads to more complicated experiments.

(1) Comparison of the H' Test Reflex with a Reference H-Reflex.
During Sol voluntary contractions, there is a facilitation of the Sol H-reflex (Paillard, 1955; Gottlieb et al., 1970, 1973). Hence, the size of the H' test reflex reflects not only possible variations in the depression elicited by the conditioning discharge, but also the increased excitability of Sol MNs. An appreciation of the latter factor can be obtained by simultaneous studies of variations in amplitude of a reference H-reflex (Fig. 1, fourth traces). The stimulus strength to tibial nerve was adjusted to obtain at rest a reference H-reflex of the same size as the H' test reflex, and the stability of stimulation conditions during contractions was verified by studying the constancy of the direct motor response (M1) caused by a conditioning stimulus strong enough to activate a few α motor axons (Fig. 1A and B; first traces). Figure 1A and B shows that the H' test reflex was significantly more facilitated than reference H during a strong tonic Sol contraction. It was verified by control experiments that, during contractions as at rest, polysynaptic effects evoked by the conditioning stimulation did not modify the test reflex (Hultborn and Pierrot-Deseilligny, 1979a). The only difference between the two reflexes is the depression elicited by the conditioning reflex, that the MNs fired in the test reflex undergoes; otherwise, the MNs responsible for both reference and test reflexes are subject to the same other influences (segmental and supraspinal) related to the voluntary contraction. If there were no changes in this depression, both reflexes should exhibit a similar amount of facilitation. A differential net effect on these two reflexes may therefore be ascribed to variations in the depression brought about by the conditioning H1 discharge.

In quantitative terms, this conclusion must be qualified. Indeed, if the MNs responsible for the two reflexes do receive the same influences related to the contraction, their sensitivity to these influences is not identical. It has been shown that the H' test reflex is much less facilitated than the reference H-reflex by a preceding stimulation of IA fibers from triceps (Hultborn and Pierrot-Deseilligny, 1979a). The weaker sensitivity of the MNs responsible for the test reflex is due to the AHP they are undergoing after the conditioning discharge. During AHP there is an increase in membrane conductance, which should result in a partial "shunting" of excitatory inputs (Coombs et al., 1955). Thus, no conclusion concerning changes in depression elicited by H1 can be drawn if, during contraction, the facilitation of the H' test reflex is less than that of reference H. Indeed, such a result could be explained by the weaker sensitivity to excitatory inputs related to voluntary contraction of the MNs fired in H'. In contrast, a facilitation of the H' test reflex larger than that of the reference H reflex indicates a reduction in depression elicited by the conditioning reflex, and the weaker sensitivity to excitatory inputs of the MNs fired in H' leads one to underestimate this reduction. A decrease in the test reflex in relation to rest, given the increase in the reference H reflex, cannot be explained by differences in sensitivity to excitatory inputs either, and indicates an increase in the depression following the conditioning reflex.

(2) Relative Roles of Changes in AHP and Recurrent Inhibition During Sol Voluntary Contractions.
Changes in depression following the conditioning reflex can reflect changes in recurrent inhibition and/or in AHP. Some MNs participating in the conditioning reflex may have discharged recently in the voluntary activity. Due to temporal summation of AHP (Ito and Oshima, 1962; Baldissera and Gustafsson, 1974), the AHP will be larger in these MNs following their iterated discharge in the H1 conditioning reflex. Parallel animal experiments have shown that the AHP summation would cause an additional MN depression comparable to that of the maximum autogenetic recurrent inhibition when a MN is activated in a voluntary movement 25 msec before it is fired by the conditioning reflex (Hultborn et al., 1979b). Hence, this phenomenon can contribute to a significant reduction in size of the test reflex during Sol contractions, and

the larger the number of MNs recruited by the contraction and/or the higher their discharge frequency, the greater this reduction of the H' test reflex. Under these conditions, a decrease in the test reflex, given the increase in the reference H-reflex, can reflect an increase in recurrent inhibition and/or in AHP, and supplementary experiments are necessary to show the respective roles of these two factors (see below). In contrast, a facilitation of the test reflex larger than that of reference H does indicate a decrease in recurrent inhibition following the conditioning reflex, and the temporal summation of AHP also contributes to the underestimation of this decrease.

(3) Contamination of the H' Test Reflex by the V1 Response During Triceps Surae Contractions.

As explained above, the test stimulus eliciting the H' response is supramaximal for motor axons. When such a stimulus is applied alone during strong contractions a so-called V1 response can be seen after the maximum M wave (Fig. 1B, second trace). This V1 response is, in fact, a small H-reflex which can appear in the EMG because of a collision between the natural orthodromic impulses generated by the voluntary effort and the antidromic motor volley due to the supramaximal stimulus (Upton et al., 1971). It is conceivable that at least part of the V1 response adds to the H' reflex, thus causing an overestimation of the H' test reflex during contraction. To avoid such an overestimation, the values of V1 were always subtracted from the corresponding H' test reflexes. Furthermore, experiments were only accepted when V1 was less than 1% of M_{max} during weak contractions or less than 5% of M_{max} during strong contractions. During phasic ramp contractions the experiments were only retained for further analysis when V1 was stable throughout the contraction.

(4) Experimental Procedure.

The voluntary soleus EMG activity and the output from a strain gauge fixed on the plate against the foot sole were displayed on an oscilloscope, so that the subject could perform the requested movements. In order to prevent the soleus contraction from producing an excessive plantar flexion, the foot was fastened to a rigid sole by a leather strap placed in front of the ankle joint. This contraction was not perfectly isometric. During the strongest soleus contractions 20° of supplementary plantar flexion occurred. Soleus contractions therefore produced a passive stretch of pretibial muscles. Since the amplitude of the H' test reflex depends on the amplitude of the conditioning discharge, changes in size of the test reflex, as between voluntary contractions and rest, only indicate changes in recurrent inhibition if the corresponding H1 conditioning reflex remains constant. For each type of contraction, the strength of conditioning stimulus was therefore adjusted so that the H1 conditioning reflex had the same size at rest as during contraction. The size of the conditioning H1 discharge was made large enough to elicit a sizable recurrent inhibition, i.e., it was always within the range in which the corresponding H' amplitude was decreasing (see above) at rest as well as during contractions (cf. Fig. 3B). Details of the experimental procedure have been given by Hultborn and Pierrot-Deseilligny (1979a).

CHANGES IN RECURRENT INHIBITION DURING TONIC SOLEUS VOLUNTARY CONTRACTIONS OF VARIOUS FORCES

Variations in the H' test reflex, seen while increasing the size of the conditioning reflex, via an augmentation of the conditioning stimulus (conditioning-test interval = 10 msec), were compared at rest (open triangles) and during three levels of tonic contraction. These contractions ranged from a weak force of about 10% of maximum (filled triangles) to a strong contraction about 80% (filled circles) including a medium force about 40% (crosses) (Fig. 3A). For small H-reflexes due to weak stimuli, the size of the H1 conditioning discharge was enhanced by tonic contraction when compared with rest. The development of the H' test reflex, although increasing H1 conditioning reflexes (Fig. 3B), was similar for all the curves, since increasing H1 beyond a certain value resulted in a progressive decrease in the H' test reflex. Nevertheless, for a given amplitude of H1, the size of H' decreased with the weakest voluntary force as compared with rest, whereas it gradually increased with stronger forces. A more precise analysis is illustrated in Fig. 3C. The size of H1 (60% of

M_{max} in this case) was identical both at rest and during contractions of different forces. Variations of the H' test reflex are compared with those of a reference H-reflex under different contraction forces. Figure 3C shows that the size of reference H increased with each level of contraction, even though the most important increase was seen between rest conditions and the weakest contraction (from 7.5 to 14.5% of M_{max}). By contrast, for the weakest contractions, the H' test reflex amplitude decreased from 9 to 5% of M_{max}, although the Sol MN excitability was increased due to the voluntary effort (as shown by the increase in reference H). This indicates a significant increase in the depression elicited by H1. With greater contraction forces there was no longer any inhibition of the H' test reflex, but, instead, a facilitation which largely increased along with contraction force. The test reflex eventually exceeded the reference H-reflex amplitude at the strongest contractions. Qualitatively similar results were obtained when using 20-msec conditioning-test time intervals.

Evidence for a Supraspinal RC Inhibition During Strong Tonic Contractions

As discussed above, a facilitation of the H' test reflex larger than that of reference H, indicates a decrease in the additional recurrent inhibition brought about by the conditioning reflex. Since this was observed despite both the lower sensitivity to excitatory inputs of MNs fired in H' and the temporal summation of AHP (two factors that contribute to reduce the facilitation of the test reflex, see above), the decrease in recurrent inhibition during strong contractions is likely to be strongly underestimated in the present experiments. Several arguments from human and animal experiments exclude the possibility that this reduction in additional recurrent inhibition elicited by H1 is caused by an occlusion in the recurrent pathway.

Note: Several factors could have caused the tonic MN firing due to the voluntary motor discharge to reduce the RC output caused by the conditioning reflex: occlusion at RC level (Eccles et al., 1954), mutual inhibition of RCs (Ryall, 1970), presynaptic transmitter depletion at the recurrent collateral terminals (Hultborn and Pierrot-Deseilligny, 1979b). In fact, parallel animal experiments have largely refuted this possibility (Hultborn and Pierrot-Deseilligny, 1979b). Even a strong tonic discharge in MNs (eliciting RC firing as high as 120 cycles/sec) failed to occlude a phasic burst of spikes due to phasic excitation; on the contrary, the RC response actually increased. Several arguments from experiments in normal man also disagree with occlusion as the cause of reduced recurrent inhibition during strong contractions (cf. Hultborn and Pierrot-Deseilligny, 1979a). Furthermore, in 2 patients with brain lesions, able to produce a voluntary MN discharge as large as that produced by normal subjects during the strongest contractions, such contractions caused either no facilitation or an inhibition of H'. Since it appears very unlikely that brain lesions are able to modify occlusion in the recurrent pathway, this strongly suggests that the H' facilitation observed in normal subjects is not due to such an occlusion.

Since occlusion cannot account for the reduction of additional recurrent inhibition elicited by H1 during strong tonic contractions, it must result from an inhibition of RCs. Renshaw cells are known to receive inhibitory influences from both primary afferents and descending tracts (see Haase et al., 1975). The inhibitory control acting on RCs during strong contractions is very likely supraspinal in origin (Hultborn and Pierrot-Deseilligny, 1979a), and Koehler et al. (1978) have shown that stimulation of the capsula interna efficiently inhibits RCs.

Evidence for a Supraspinal RC Facilitation During Weak Tonic Contractions

During the weakest contractions, there is a significant increase in the depression caused by the conditioning reflex. As seen above, this can be directly related to the voluntary motor discharge which produces both a temporal summation of the motoneuron AHP and a RC facilitation (via recurrent collaterals), resulting in an increase in recurrent inhibition elicited by H1. The possibility of a facilitation of RCs from a supraspinal source (see Haase et al., 1975) is supported by results obtained in spastic patients with brain lesions, able to perform Sol voluntary contractions producing the same amount of EMG activity as normal subjects. Representative data of the two populations are presented in Fig. 4. During such weak contractions all 7 patients examined exhibited an increase in the reference H-reflex, and in 6 of them there was also an increase in the test reflex. Even though this increase was smaller than that of reference H, it contrasts with the decrease in test reflex observed in all normal subjects. Thus, despite the same amount of voluntary MN firing, which should have produced in the two populations the same increase in the depression elicited by H1, this depression was larger in normals than in spastics.

Thus, during the weakest tonic contractions, there is normally (in addition to the RC facilitation via recurrent collaterals) another facilitation, likely supraspinal in origin, which is missing in spastic patients.

CHANGES IN RECURRENT INHIBITION PRECEDING AND ACCOMPANYING PHASIC SOLEUS CONTRACTIONS

Variations in Reference and Test Reflexes During Ramp Contractions

Figure 5 illustrates the results obtained while the subject performed a ramp contraction of 1 sec. Tension (represented by the continuous line) and Sol EMG increased for 1 sec and then remained constant during the following tonic contraction

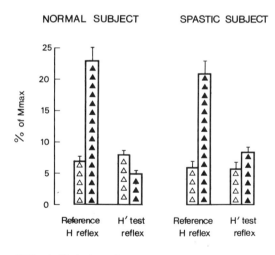

FIG. 4. Variations in reference and H' test reflexes during weak soleus voluntary contractions comparison of a normal subject and a spastic patient. Amplitudes of reflexes are expressed as a percentage of maximum M. *Open triangles:* Rest; *filled triangles:* Weak soleus contraction. Each column represents the mean of 20 measurements; *vertical bars* at the top of the columns represent 1 SEM above the mean values.

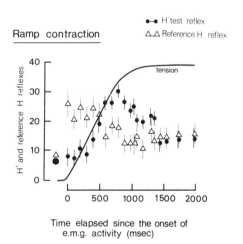

FIG. 5. Time courses of the variations in the test *(filled circles)* and reference *(open triangles)* H-reflexes during a ramp contraction lasting 1 sec. *Continuous line:* Tension. The amplitudes of the different responses (expressed as a percentage of M_{max} are plotted against time elapsed between the onset of EMG activity and H-reflex stimulation. Each symbol represents the mean of 10 measurements; *vertical bars* represent 1 SEM above and below the mean values. Left: *Large symbols* represent the rest values. (From Hultborn and Pierrot-Deseilligny, 1979a, with permission.)

(60% of the maximum force). As illustrated by the large symbols on the left of Fig. 5, the test and reference H-reflexes were equal at rest. Throughout the ramp phase, the time courses of the variations in the test (filled circles) and reference (open triangles) H-reflexes were almost the inverse. The H' test reflex continuously increased, while the reference H-reflex progressively decreased.

Note: It is well known (Paillard, 1955; Gottlieb et al., 1970) that an ordinary H-reflex, such as reference H, is most facilitated at onset of contraction. The following decrease in the H-reflex amplitude, while the EMG increases throughout the ramp, has been attributed to an increase in presynaptic inhibition exerted on Sol Ia fibers (Gottlieb et al., 1970). Special experiments were performed to determine whether presynaptic inhibition of the Ia fibers acts differentially on the maximum Ia afferent volley responsible for the test reflex and on the weak volley eliciting the reference H-reflex. The differential changes in the H' test and reference H-reflexes during contraction cannot be ascribed to presynaptic inhibition of Ia fibers (Hultborn and Pierrot-Deseilligny, 1979a).

Similar results were obtained with faster ramps (250, 500 msec). Whatever the ramp velocity, the time courses of the test and reference H-reflexes were reciprocal (Fig. 5), and at the end of the ramp, the H' test reflex largely overshot the reference H-reflex. This indicates a decrease in recurrent inhibition elicited by the test reflex. During the following tonic contraction the amplitude of the test reflex progressively decreased, so that both reference and test reflexes exhibited a similar amount of facilitation. Hence, recurrent inhibition elicited by H1 was smaller at the end of the ramp than during the following tonic contraction, although the force was the same. This is a general finding: For a given level of force reached during the ramp, recurrent inhibition is always smaller than when the same force is achieved during a tonic contraction.

Variations in Reference and Test Reflexes Prior to Ramp Contractions

Figure 6A shows the progressive facilitation of the Sol H-reflex during the 80 msec preceding a Sol ramp contraction (Pierrot-Deseilligny and Lacert, 1973).

Note: The maximum M wave, and to a lesser extent the monosynaptic reflex, are followed by a silent period which lasts for 50 to 100 msec (see Shahani and Young, 1973a). This silent period makes it impossible to appreciate precisely the time interval between the test (or reference) H-reflex and onset of the voluntary contraction during the 50 to 100 msec immediately preceding the movement. The subjects were therefore trained to perform simultaneous voluntary contractions in soleus of both sides. If onset of the 2 contractions differed more than 10 msec, the subjects were not retained for further experiments. For each movement, we measured the time interval between the reflex (test or reference) and the onset of the following voluntary contraction, as detected by the appearance of EMG activity in the contralateral Sol. Only test and reference H-reflexes preceding voluntary contraction by 30 to 60 msec were retained for further analysis. Results obtained during this period preceding Sol voluntary contractions were pooled and averaged. Figure 6B shows that the reference H-reflex exhibited a facilitation, whereas the amplitude of the test reflex was smaller than at rest. As seen above, this indicates an increase in the depression elicited by the conditioning reflex.

Evidence for a Progressive RC Inhibition Throughout Ramp Contractions

This increase in the depression elicited by H1 occurs prior to contraction, although there is no voluntary motor discharge, i.e., neither temporal summation of motoneuron AHP nor excitation of RCs via recurrent collaterals. This indicates, therefore, an RC facilitation from a source other than recurrent collaterals, probably supraspinal in origin. Conversely, at the end of the ramp contractions there is a decrease in recurrent inhibition, reflecting an RC inhibition.

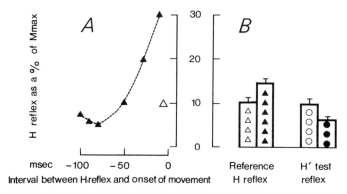

FIG. 6. Variations in the test *(circles)* and reference *(triangles)* H-reflexes prior to a soleus ramp contraction. The amplitude of H-reflex responses are expressed as a percentage of maximum M. **A**: Time course of variations in the reference H-reflex prior to a soleus ramp contraction. The amplitude of the reference H-reflex *(filled triangles)* is plotted against the time interval between the reflex and the onset of the movement (indicated by 0); Right: *open triangle* represents the value of the reflex at rest. **B**: Amplitudes of the reference *(filled triangles)* and test *(filled circles)* reflexes have been averaged during the 50 msec preceding the movement and are compared to their values at rest *(open symbols)*; each column represents the mean of 20 measurements; *vertical bars* at top of the columns represent 1 SEM above the mean values.

Note: The inverse time courses of the reference and test reflexes indicate that the depression elicited by the conditioning reflex progressively decreases throughout ramp contractions. This occurs despite the continuous increase in voluntary motor discharge (attested by increasing EMG activity). The resulting progressive enhancement of both the temporal summation of motoneuron AHP and excitation of RCs via recurrent collaterals should produce a continuous increase in the depression following H1. Again, the opposite finding can only be attributed to a decrease in supplementary recurrent inhibition caused by H1, and this decrease is likely to be strongly underestimated. During the strongest tonic contractions, the reduction of the supplementary recurrent inhibition elicited by H1 is not due to an occlusion in the recurrent pathway between the tonic voluntary firing and the phasic H1 discharge. The same evidence refutes occlusion as the cause of the decrease in recurrent inhibition at the end of ramps when EMG activity is smaller than during the strongest tonic contractions. Therefore, the reduction in recurrent inhibition at the end of ramp contractions reflects an inhibitory control, probably supraspinal in origin, acting on RCs.

Thus, from onset to end of ramp contractions there is a shift from RC facilitation to RC inhibition. Despite the increasing excitation of RCs via recurrent collaterals, the net result of the different influences converging onto RCs is an inhibition increasing throughout ramp contractions. For a given level of force reached during the ramp, the RC inhibition is more marked than when the same force is achieved during a tonic contraction.

Absence of Evidence for Changes in the Supraspinal Control of RCs Preceding and Accompanying Ballistic Contractions

In this type of contraction, the subjects were asked to perform the fastest possible contraction and then to relax immediately. The EMG exhibited a very early peak at 30 to 40 msec after onset and a rapid decline. Figure 7A shows the time course of variations in the test (filled circles) and reference (filled triangles) reflexes. The onset of EMG activity was used to trigger the stimulators, and the earliest test responses cannot appear until 40 msec later because of

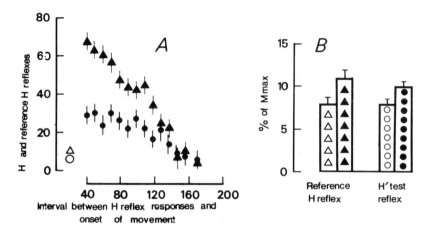

FIG. 7. Variations in the reference and test reflexes accompanying and preceding ballistic soleus contractions. A: Time courses of the variations in the test *(filled circles)* and reference *(filled triangles)* reflexes during a ballistic contraction; amplitudes of the different responses (expressed as a percentage of M_{max} are plotted against the time elapsed between the onset of EMG activity and the H-reflex; Left: *open symbols* represent the rest values. B: Amplitudes of the reference *(filled triangles)* and test *(filled circles)* reflexes have been averaged during the 50 msec preceding the ballistic contraction and are compared to their values at rest *(open symbols)*. Each symbol (A) or column (B) represents the mean of 10 (A) or 20 (B) measurements; *vertical bars* represent 1 SEM.

conduction time in the reflex pathway plus conditioning-test interval. Therefore, the Sol MN excitability was only assessed during the decreasing part of the EMG activity. At the peak EMG activity there was a considerable facilitation of the reference H-reflex, and thereafter its amplitude decreased rapidly and continuously. The test reflex was also facilitated at the beginning, although to a lesser extent than reference H, and thereafter decreased to reach the same values as the reference H-reflex. Figure 7B shows that in the 50 msec preceding the contraction both reflexes exhibited a small facilitation, the test reflex being slightly less facilitated than the reference H-reflex.

The temporal summation of MN AHP and excitation of RCs via recurrent collaterals tend to increase the depression elicited by H1 and therefore to cause the test reflex amplitude to be smaller than that of the reference H-reflex. Since these two factors decrease with voluntary motor discharge, this could explain why the test reflex is much smaller than reference H at the peak EMG and why the difference between the two reflexes diminishes as EMG declines. The lower sensitivity to excitatory inputs related to voluntary contraction of the MNs fired in H' (see above) can also account for the smaller amplitude of H'. Hence, during ballistic contractions, there is no evidence for changes in RC excitability from sources other than recurrent collaterals. Similarly, before ballistic contractions, in contrast with the period preceding ramp contractions, there is no evidence for changes in recurrent inhibition (Fig. 7B). Indeed, the slightly smaller facilitation of the H' test reflex than that of reference H can be explained by the lower sensitivity of MNs fired in H'.

FUNCTIONAL SIGNIFICANCE OF THE CONTROL OF RCs DURING VOLUNTARY MOVEMENTS

A general formulation on the role of recurrent inhibition during natural movements has to take account of the fact that RCs inhibit not only MNs but also Ia interneurons.

Recurrent Inhibition Serving as a Variable Gain Regulator for the Motor Output

It has been shown that during the weakest tonic voluntary contractions there is an RC facilitation, whereas there is an RC inhibition during the strongest contractions. These results agree with the hypothesis (Hultborn et al., 1979a) that recurrent inhibition could operate as a variable gain regulator for the motor output. Figure 8 is a tentative representation of the input–output relation for the MN pool. Muscular force depends on the number of recruited MNs and on the firing frequency of individual MNs. This motor output is simply indicated in Fig. 8A as the number of recruited MNs times firing frequency ($n \times f$). This output from MNs itself depends on the input they receive from supraspinal centers via descending pathways. Without recurrent inhibition, the input–output relation for the MN pool (continuous line in Fig. 8B) would be largely determined by the intrinsic properties of the MNs. For example, for a given synaptic excitation, the firing frequency of individual MNs depends predominantly on their AHP (Gustafsson, 1974). For simplicity's sake, this relation has been drawn linearly.

If the recurrent pathway is active, this will lower firing frequencies of individual MNs at a given input level and also postpone recruitment of most MNs to higher input levels, factors which add up to reduce the slope of the input–output relation. This reduction in slope will be maximum when RCs are facilitated (dotted line in Fig. 8) as observed during the weakest contractions. This gives a low input–output gain for the MN pool and allows the supraspinal centers playing on a considerable part of their working range to cause only small changes in muscle force. A facilitated recurrent pathway would thus give an improved resolution to control of motor output. By contrast, the RC inhibition, observed during strong contractions, would secure a high input–output gain for the MN pool (dashed line in Fig. 8), which favors a large tension output.

Recurrent Inhibition Serving as a Control of Reciprocal Ia Inhibition

For a given level of force reached during ramp contractions, RCs are much more inhibited than when the same force is achieved during a tonic contraction. This finding, which is difficult to interpret in

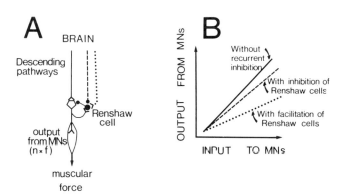

FIG. 8. A: Schematic diagram illustrating the output from MNs and the input to MNs; supraspinal control of RCs has also been represented. B: Schematic representation of the input–output relation for the MN pool in 3 cases: without recurrent inhibition *(continuous line)*, with facilitation of RCs *(dotted line)*, and with inhibition of RCs *(dashed line)*. (From Hultborn et al., 1979a, with permission.)

terms of regulation of motor output by projections of RCs onto MNs, may be explained by the projections of RCs onto Ia interneurons. Figure 9 shows the pathway of reciprocal Ia inhibition from Sol to tibialis anterior (TA), with the Ia interneurons from Sol projecting onto both TA MNs and "opposite" Ia interneurons excited by the TA muscle. During Sol voluntary contractions, this reciprocal Ia inhibition to TA is activated by both the Ia discharge from the contracting Sol (Vallbo, 1974) and excitatory inputs from descending pathways (Tanaka, 1980). However, the same Ia interneurons are inhibited by RC firing consecutive to voluntary motor discharge (Fig. 9). The efficacy of reciprocal Ia inhibition depends therefore on the level of

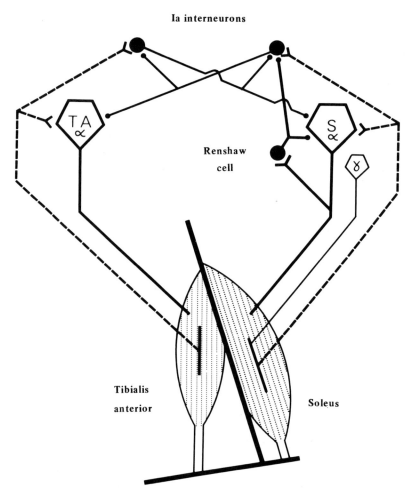

FIG. 9. Diagram summarizing the connections between Renshaw cells directed to soleus motoneurons and Ia inhibitory interneurons directed to motoneurons innervating the antagonist, tibialis anterior. The mutual inhibition between the Ia interneurons projecting to soleus MNs and those projecting to tibialis anterior MNs is also represented. *Large filled circles*, inhibitory interneurons; *Small filled circles*, inhibitory synapses; *interrupted lines*, Ia afferent fibers. S, soleus α-MN; TA, tibialis anterior α-MN; γ, soleus γ-MN. (From Hultborn and Pierrot-Deseilligny, 1979*a*, with permission.)

recurrent inhibition. It can be assumed that the supraspinal control of RCs regulates muscle tension and reciprocal Ia inhibition in parallel with regard to the requirements of the movement.

During Sol contractions, the supplementary plantar flexion produces a passive stretch of TA which causes a discharge in its Ia afferents. This Ia discharge is obviously larger during ramp contractions, which cause a static but also a dynamic stretch of the TA muscle. Such a Ia discharge excites both TA MNs and Ia interneurons inhibiting Sol MNs. This "undesirable" inhibition of Sol MNs may obviously be counteracted by the reciprocal Ia inhibition from Sol to TA, since Ia interneurons excited by the contracting Sol inhibit the opposite Ia interneurons excited by the TA (Fig. 9) (Hultborn et al., 1976). Nevertheless, it is necessary that this pathway of reciprocal Ia inhibition not be inhibited by RC discharge. The strong inhibition of RCs during ramp contractions could therefore allow reciprocal Ia inhibition to manifest itself. On the contrary, the weakest contractions are accompanied by a facilitation of RCs, which inhibits the reciprocal Ia inhibition. This prevents a deep depression of antagonistic MNs enabling them to fire and correct such accurately tuned movements at any moment.

ACKNOWLEDGMENTS

Our thanks are due to Anne Rigaudie for technical assistance and to David Macgregor for scrutinizing the English. This work was supported by grants from the Pierre and Marie Curie University, INSERM (816.021) and ETP (Tw 523).

Recurrent Inhibition of Motoneurons During the Silent Period in Man

*Jun Kimura

Division of Clinical Electrophysiology, Department of Neurology, College of Medicine, University of Iowa, Iowa City, Iowa 52242

Despite continued effort, action potentials of a voluntarily contracting muscle are transiently suppressed following a compound muscle action potential evoked by electrical stimulation of the nerve innervating that muscle (Hoffmann, 1919). Although designated as the silent period (SP), suppression of electrical activity during this interval is not absolute, since it can be interrupted by increasing voluntary muscle contraction. A burst of electrical activity then appears following the F-wave in the midst of an otherwise silent interval. This activity may be termed voluntary potential (VP), since it results from greater volitional effort breaking through the relative inhibition of the motoneuron (MN) pool. The SP is the result of several physiological mechanisms (Shahani and Young, 1973a): (1) collision of voluntary impulses along the course of the peripheral nerve with antidromic activity generated by the stimulus, (2) unloading of the spindle (Merton, 1951; Struppler et al., 1969, 1973), and (3) activation of the Golgi tendon organ (Granit, 1950). Recurrent inhibition by Renshaw cells (Renshaw, 1946) is also postulated to occur in the middle portion of the SP secondary to invasion of axon collaterals by antidromic impulses. Not all antidromic impulses generated by nerve stimulation reach the central MN pool during voluntary muscle contraction because some of them are extinguished by collision with orthodromic voluntary impulses in the same axon. Increased effort to contract the muscle obviously increases the probability of collision, because more axons carry voluntary impulses at the time of stimulation (Kimura, 1976, 1977). Stimulation of the nerve distally also enhances this probability as compared to proximal stimulation, because the collision occurs in proportion to the length of the nerve segment between the stimulus site and the cell body of the MN (Fig. 1). Thus, the greater the voluntary effort and the more distal the nerve stimulation, the less antidromic invasion of the MN pool. Therefore, recurrent inhibition of MNs, if present in man following antidromic activity, is predicted to be less effective with distal, rather than proximal nerve stimulation at a given level of muscle contraction. This hypothesis can be tested using the magnitude of VP, which breaks through the SP, as an index of MN excitability.

Method

The subject was seated with the hand pronated near the edge of the examining table on which a rigid force transducer was mounted in such a way that the lug was

---

*Division of Clinical Electrophysiology, Department of Neurology, College of Medicine, University Hospitals, Iowa City, Iowa 52242.

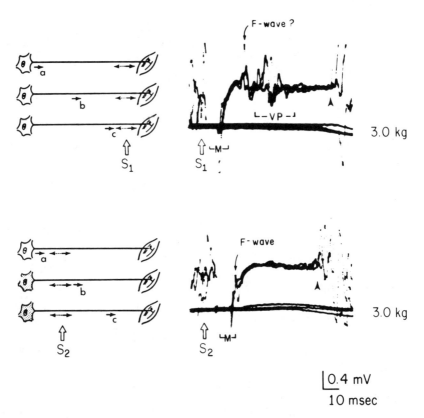

FIG. 1. Simultaneous recording of muscle force of 3.0 kg *(straight line)* and the SP from the voluntarily contracting, first dorsal interosseous muscle (3 trials superimposed). Site of stimulus: **Top,** wrist *(open arrow,* S2); **bottom:** axilla *(open arrow,* S2). The SP was broken by the VP with a stimulus at the wrist but not at the axilla, indicating greater inhibition of MNs with proximal than distal nerve stimulation. With distal stimulation, most antidromic activity is extinguished by collision with the voluntary impulses, a, b, and c, before reaching the MN pool. With proximal stimulation, antidromic activity, escaping collision, presumably invades recurrent axon collaterals, inhibiting the MNs *(shaded).*

against the radial aspect of the index finger. To abduct the finger effectively, the thumb was adducted against the edge of the table fixating the origin of the first dorsal interosseous muscle. A metal bar located between the index and middle fingers prevented a shift of the hand toward the force transducer despite the contraction of the adductor pollicis. The tension produced was displayed on an oscilloscope, which allowed the subject to adjust and maintain a steady voluntary muscle force at a given level. In our previous study, using a similar experimental design (Kimura, 1977), the maximal force recorded varied considerably not only among different subjects but also depending on the means by which the force transducer was pressed. If the middle phalanx was placed against the lug of the force transducer, the maximal tension produced was usually up to 2.0 to 3.0 kg; however, it was as high as 6.0 kg or more if the proximal phalanx was used because of increased mechanical advantage. When the proximal phalanx was used, however, it was often difficult to abduct the index finger selectively, especially at higher levels of effort, resulting in possible contribution of other intrinsic hand muscles to the total force. Therefore, the proximal interphalangeal joint was used, because it allowed reasonably selective abduction of the index finger and yet produced sufficient tension.

The SP was recorded from the voluntarily contracting first dorsal interosseous muscle. Surface electrodes were placed over the muscle and the first metacarpal bone. The stimulating electrodes were placed over the ulnar nerve at the wrist, elbow, and axilla. Using a stimulator coupled with a transformer, a stimulus of 0.1-msec duration and of supramaximal intensity was delivered. Regardless of the site of stimulation, a nearly identical compound muscle action potential was elicited, since all the motor axons were presumably excited. At least 3 trials were recorded for each site of nerve stimulation in each subject. The magnitude of VP was determined by estimating the area covered by the response on the photo (average amplitude × duration of response in mV msec).

Results

For the first part of the study, the effect of nerve stimulation at three different sites—wrist, elbow, and axilla—was compared while the subject was maintaining a steady force of 3.0 kg. Fifteen healthy subjects (9 males), 19 to 40 years of age, were tested bilaterally.

In a typical tracing (Fig. 1), the SP began immediately following the compound muscle action potential evoked by the stimulus and lasted approximately 100 msec before the voluntary muscle activity returned. This period of electrical silence was frequently broken by two separate muscle potentials. The first of the two, a well-synchronized potential of a stable latency, was tentatively equated with the F-wave (muscle response evoked by discharges of antidromically activated MNs), although the exact origin of this response in a voluntarily contracting muscle remains to be elucidated (McComas et al., 1970; Upton et al., 1971). The second response, designated VP in this study, was much more variable than the F-wave in latency, amplitude and duration.

The VP was present in all 30 muscles tested (right and left sides from 15 subjects) when the nerve was stimulated at the wrist. In contrast, the VP was absent in 5 with stimulation at the elbow, and in 9 with stimulation at the axilla. Furthermore, the size of VP (mean ± SD in 30 muscles) decreased progressively as the stimulus site was moved proximally from the wrist (52.8 ± 37.4 mV/msec) to the elbow (29.8 ± 26.8 mV/msec) and finally to the axilla (12.6 ± 14.2 mV/msec). In some muscles, the VP obtained with stimulation at the wrist was so large that it often overlapped with the F-wave and, less commonly, with the eventual return of muscle activity, practically abolishing the SP. This was rarely the case with stimulation at the elbow or axilla. Statistical analysis revealed that the size of VP was significantly smaller with progressively more proximal sites of nerve stimulation ($p < 0.01$).

In the second part of the study, the effect of increased force of voluntary contraction on the SP was evaluated with stimulation at the wrist and axilla (Fig. 2). Ten healthy subjects (5 males), 19 to 40 years of age, were tested bilaterally. The subject was instructed to abduct the index finger against a rigid force transducer with a muscle force ranging from 0.5 to 3.0 kg in increments of 0.5 kg. The results are summarized in Fig 3. When the nerve was stimulated at the wrist, the VP was 1.7 ± 3.4 mV/msec in size (mean ± SD of 19 responses from 10 subjects, right and left sides combined) at the muscle tension of 0.5 kg. The VP then increased for each increment of 0.5 kg, to 7.7 ± 10.5, 13.4 ± 11.6, 19.5 ± 16.3, and 29.7 ± 28.0 reaching 55.1 ± 39.2 mV/msec at 3.0 kg. The corresponding values with nerve stimulation at the axilla were 0.8 ±

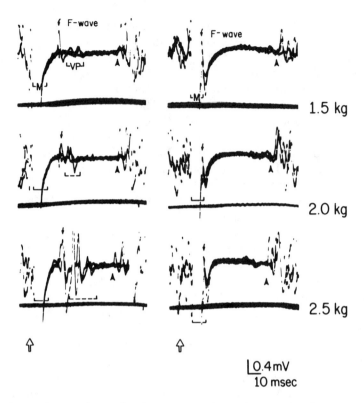

FIG. 2. Stimulation and recording as in Fig. 1 at muscle tension ranging from 1.5 to 2.5 kg (*open arrows* = stimulus). **Left:** wrist; **right:** axilla. Stimulating at the wrist, the VP was progressively greater in size at increasingly higher muscle force. With stimulation at the axilla, no VP was recorded at any level of muscle force, and SP duration was shortened as muscle force was increased.

2.8, 1.1 ± 2.6, 4.0 ± 6.1, 6.3 ± 8.7, and 10.8 ± 11.4 reaching 15.6 ± 15.7 mV/msec at 3.0 kg. Statistical analysis revealed that the size of VP was significantly larger with distal as opposed to proximal stimulation at muscle tensions of 1.0 kg and higher ($p < 0.01$). The difference in size of VP between the two stimulus sites was progressively greater at increasingly higher levels of muscle force (Fig. 3).

The SP measured from the stimulus artifact to the eventual resumption of muscle potential became progressively shorter with increased force of voluntary muscle contraction (Fig. 2). The exact relationship of the SP to muscle force and to the site of nerve stimulation was not systematically assessed in this study. It was apparent, however, that the duration of SP was dependent to a considerable extent on the size of VP in that a large VP was usually followed by prolonged electrical silence before the final recovery of voluntary activity.

Comments

The SP induced by electrical stimulation of the nerve (Merton, 1951; Shahani and Young, 1973a) or by unloading the muscle spindle (Struppler et al., 1969, 1973) has previously been studied in normal subjects (Higgins and Lieberman, 1968; McLellan, 1973) and in patients with neurological disorders. Few papers, however, have dealt with the relationship between the SP and the force produced by voluntarily contract-

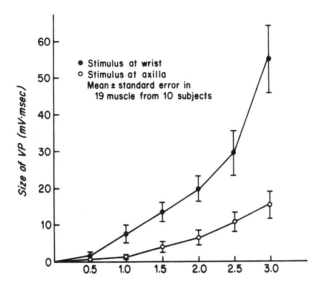

FIG. 3. Muscle tension (kg) *(abscissa)* and the size of VP breaking through the SP. For muscle forces of 1.0 kg and above, VP was significantly larger with stimulus at wrist than at axilla, indicating that MNs were more inhibited by proximal as opposed to distal nerve stimulation during voluntary muscle contraction. The difference in inhibitory effects of proximal versus distal stimulation became progressively larger as muscle force was increased.

ing human muscle. In particular, electrical potentials which break through the SP have rarely been systematically investigated, although their presence is well recognized. It is generally assumed that the duration of SP is independent of muscle tension (McLellan, 1973) and is dependent on the time course of mechanical change in the muscle (Merton, 1951).

Upton et al. (1971), studying the effect of voluntary contraction on the SP, recognized two separate potentials that appeared in the midst of the SP. They considered the first of the two to consist mainly of the H-reflex rather than the F-wave. Their conclusion was based on the assumption that during voluntary contraction the antidromic activity that presumably generates the F-wave could not reach the central MN pool because of collision with voluntary impulses. We have tentatively equated the same potential with the F-wave, as had been previously suggested (McComas et al., 1970), because at low levels of muscle contraction voluntary impulses are present in relatively few axons at any given moment, allowing substantial antidromic activity to reach the central MN pool (Kimura, 1977). With higher muscle tension, most antidromic activity will be extinguished by collision with the more frequent voluntary impulses and, therefore, the first potential, if elicited in this situation, must depend substantially on discharges of either reflexively or voluntarily activated MNs. In a number of our subjects, however, this potential was not clearly recorded during maximal or near-maximal voluntary contraction. The second potential breaking through the SP, designated VP in this study, was attributed by Upton et al. (1971) to either descending volitional or polysynaptic reflex inputs to MNs. Shahani and Young (1973a) also recognized voluntary potentials in the midst of the SP at 50 to 80 msec after the shock and occasionally between 30 and 50 msec, especially when weaker stimuli were used.

In the present study, we have documented, by using shocks of a supramaximal intensity, that the size of VP becomes progressively smaller as the site of nerve stimulation is moved proximally from wrist to elbow and finally to axilla. Since the magnitude of VP is an index of MN excitability during the SP, these findings indicate that MNs are more inhibited by proximal as opposed to distal nerve stimulation at the same voluntary muscle tension. Because antidromically directed impulses have greater probability of collision with voluntary impulses in proportion to the increased length of the nerve segment (Fig. 1), less antidromic activity reaches MN pool from distal compared to proximal stimulation during a given muscle contraction. Moreover, a greater number of motor axons are activated by proximal than distal stimulation of the same nerve. Thus, the greater the antidromic activity, the more the MN inhibition.

In the aforementioned discussion, we have not considered the possible influence on SP of ascending volleys via sensory fibers. To produce a well-defined SP by cutaneous nerve stimulation (Kranz et al., 1973), a much higher intensity than used in this study is ordinarily required. However, selective stimulation of cutaneous nerves at the wrist is known to cause inhibition at about 50 msec after the shock (Shahani and Young, 1973), and, therefore, a more proximal stimulus may inhibit MNs earlier, during the middle portion of the SP. Furthermore, group Ib afferent fibers from tendon organs are also excited, which could be a significant factor in the inhibition (Hufschmidt, 1960; Hufschmidt and Linke, 1976). If so, a proximally evoked volley is more likely to inhibit MNs because, with proximal stimulation, more afferent fibers are stimulated than with distal stimulation. Hence, a part of our findings could indeed be accounted for by assuming inhibitory effects of the ascending volley in sensory fibers. The results of the second part of our experiment, however, indicate that the antidromic activity plays a substantial, if not exclusive, role in the inhibition of MNs. With a given increase in voluntary muscle contraction, the increased number of orthodromic volleys will result in an increase in the probability of collision with antidromic impulses. This increased probability of collision is disproportionately large for antidromic impulses from distal points of stimulation compared to proximal points, because of the difference in length of the nerve segment between the stimulus sites and the central MN pool. Therefore, if antidromic invasion of MNs is the decisive factor, distal stimuli must be progressively less effective than proximal stimuli as muscle force is increased. Thus, the expected increase in VP with distal stimulation should become even more marked when voluntary muscle contraction is increased. We have indeed shown that the proximal stimulus is not only more effective than the distal stimulation at a given muscle tension, but also the discrepancy between the two stimuli increases with progressively greater muscle force (Fig. 3). This finding cannot be explained solely on the basis of sensory inhibition. Sensory volleys reaching the central MN pool are unaffected by voluntary motor impulses. Therefore, if the inhibition were sensory, the relative effectiveness of proximal versus distal stimulation would not vary exponentially with the degree of muscle contraction.

This study is consistent with the observations by Pierrot-Deseilligny and Bussel (1975), and Pierrot-Deseilligny and Morin (1980), who demonstrated recurrent inhibition following the passage of an orthodromic impulse via the H-reflex pathway. Our results further document the importance of inhibitory antidromic invasion of the MN pool, which has been shown to occur in cats (Ryall et al., 1972). Under our experimental design, the MN escapes antidromic activation only if the antidromic impulse is eliminated by collision with the orthodromic impulse in the same axon. Since the degree of inhibition is related closely to the amount of antidromic activity reaching the MN pool, it follows that the orthodromic passage of impulses is associated with less than full activation of recurrent inhibition. Alternatively, antidromic impulses of the neighboring motor axons may cause additional Renshaw inhibition after the passage of the orthodromic impulse. In either case, it is likely that in man, as in cats (Ryall et al., 1972), antidromic impulses produce Renshaw inhibition more effectively than orthodromic impulses. An additional inference from our findings is that the middle portion of the SP is at least in part caused by the Renshaw loop and that the VP tends to appear when recurrent inhibition is decreased because of reduced antidromic invasion of the MN pool. The presence of Renshaw inhibition does not preclude other mechanisms of the SP such as unloading of spindle (Merton, 1951; Struppler et al., 1969) and activation of Golgi tendon organ (Granit, 1950) associated with muscle contraction (Shahani and Young, 1973). Indeed, Hufschmidt (1960) and Hufschmidt and Linke (1976) have emphasized autogenic and orthodromic inhibition in SP.

ACKNOWLEDGMENTS

David Walker, M.S., E.E., provided electrical engineering assistance. Sheila Mennen, Deborah A. Gevock, and Cheri L. Turner gave technical assistance.

Segmental Versus Suprasegmental Contributions to Long-Latency Stretch Responses in Man

*Christina W. Y. Chan

School of Physical and Occupational Therapy, and Department of Physiology, McGill University, Montréal, Canada

During the past two decades, attention has focused on components of muscular response to stretch, which occur later than those normally associated with segmental reflexes. We owe it to the intriguing observation, first made by Hammond (1954, 1960; Hammond et al., 1956), that sudden muscle stretch in human subjects instructed to oppose the stimulus, produced early and late electromyographic (EMG) responses separated by a silent period of about 35 msec in the biceps. Although the early synchronous burst at about 15 to 20 msec probably corresponded to the tendon jerk, the late response presented puzzling features. Thus, it occurred before the earliest EMG activity attributable to a purely voluntary response, but it was dependent explicitly on the subject's mental set (Hammond, 1956). Important questions have since arisen concerning the physiological nature of these late responses: To what extent can they be accounted for in terms of known segmental mechanisms? Do they constitute response to long-loop reflexes extending up to higher centers in the CNS and back? To what extent is their behavior governed by input trajectories in accordance with servo reflex action? Do they represent elements of preformulated patterns of intended movement triggered by the arrival of afferent impulses?

HISTORICAL NOTE

There is a growing belief that impulses evoked in peripheral afferents—hitherto thought to subserve only segmental and/or intersegmental effects—may reach higher centers of the CNS before descending the cord to modulate the excitability of spinal motoneurons (MNs). The first experimental evidence for long-loop reflexes via spino-bulbo-spinal (SBS) pathways came from Shimamura and his colleagues. The SBS reflexes are well documented in animals (Gernandt and Shimamura, 1961; Shimamura and Livingston, 1963; Shimamura et al., 1964; Shimamura, 1973). They are mediated by impulses originating from cutaneous (Shimamura and Akert, 1965) and high-threshold (groups II and III) muscle afferents (Devanandan et al., 1969), which ascend the spinal cord via the spinoreticular and spino-cervico-reticular tracts (Shimamura et al., 1976). After a relay in the bulbar reticular formation, they descend the cord bilaterally to excite flexor MNs (Shimamura and Aoki, 1969) and presynaptically inhibit extensor MNs (Shimamura et al., 1967; Magherini et al., 1971).

The theory that impulses originating from

---

*School of Physical and Occupational Therapy and Department of Physiology, McGill University, Drummond Street 3654 Montréal, P.Q., H3G 1Y5 Canada.

low-threshold muscle afferents also traverse through long-loop pathways was put forward by Eccles in 1966. He surmised that the intercurrent facilitation of the test H-reflex following a conditioning H-reflex (Magladery, 1955; Paillard, 1955) or muscle stretch (Taborikova et al., 1966) probably represents long-loop effects mediated by group Ia muscle afferents, which relay in the cerebellum and brainstem before descending via the vestibulospinal tract to influence spinal MNs. Although his proposal was supported by Yap (1967), Taborikova et al. (1966), and Taborikova and Sax (1969), it was opposed by several lines of evidence (e.g., Gassel, 1970). In particular, Pompeiano and Barnes (1971) failed to observe any modulation of either reticular or vestibular neuronal discharge whereas group Ia fibers of the gastrocnemius-soleus muscles were activated by high-frequency sinusoidal vibration. Thus, Eccles' (1966) hypothesis that group Ia muscle afferents mediate long-loop reflexes through cerebellum and brainstem remains to be substantiated.

However, a growing body of research suggests that effects evoked by low-threshold muscle afferents probably extend higher in the central nervous system (CNS) than the brainstem. Phillips (1969) first advanced the concept of a transcortical stretch reflex. The afferent limb of this loop was seen to be mediated via group Ia fibers and the efferent limb by corticomotoneuronal (CM) projections. This proposal has sparked a large number of investigations of the extent to which long-latency components of the EMG responses to muscle stretch are attributable to such a transcortical pathway. In fact, considerable controversy has since been raised with regard to the supraspinal versus segmental origin of the long-latency stretch responses. The first objective of this chapter is to discuss the evidence both for and against a supraspinal participation in the generation of these late responses.

ORIGIN OF LONG-LATENCY STRETCH RESPONSES

Evidence for a Transcortical Contribution

In man, such evidence is necessarily less direct than in primate studies because of limitations inherent with human experiments. However, results from various sources are compatible with a supraspinal contribution to the long-latency stretch responses.

Latency Measurements

It was originally surmised that if components of the late response to stretch [termed here the functional stretch response, or (FSR) after Melvill Jones and Watt (1971a)] were mediated via a long-loop pathway, the time required for transmitting and processing the FSR within the CNS should increase in muscles with MNs situated progressively caudal to the brain. We tested this hypothesis by examining the EMG response elicited in a variety of limb muscles in 10 normal human subjects instructed to oppose sudden limb displacements (Chan et al., 1979c). As illustrated in Fig. 1, the time between onsets of tendon jerk [presumed monosynpatic stretch reflex (MSR)] and FSR increased with descending spinal level of muscle innervation. The mean ($\pm$SE) values (FSR $-$ MSR) were 32.1 $\pm$ 0.4 msec for biceps (C5–C6), 47.2 $\pm$ 0.8 msec for quadriceps (L2–L4), and 73.1 $\pm$ 1.5 msec for gastrocnemius (L5–S2). Insofar as these durations reflect the time required for central conduction (to be discussed later), these findings are generally compatible with the view that the FSR is the manifestation of a long-loop response.

A similar observation was noted by Marsden et al. (1973, 1976a) working on different muscles. Briefly, the late stretch response times exceeded the tendon jerk latencies by

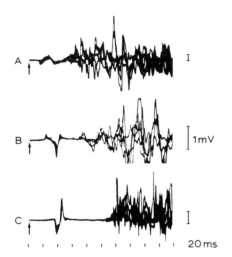

FIG. 1. EMG responses in (A), biceps; (B), quadriceps; and (C), gastrocnemius, obtained from the same subject opposing a suddenly applied and maintained stretching force. In each muscle, the stretch *(arrow)* produced a tendon jerk response, followed after a variable silent period, by a late asynchronous burst of activity termed the FSR. The time between onsets of tendon jerk and FSR latencies was seen to increase from biceps (C5–C6), through quadriceps (L2–L4) to gastrocnemius (L5–S2) and thus appeared sequentially related to the segmental level of muscular innervation. (From Chan et al., 1979c, with permission.)

5 msec for the jaw muscle (see Lund et al., *this volume*), by 22 msec for the thumb muscle and by 38 msec for the toe muscle, again conforming with the view that response durations are related to the distance of the spinal MNs from the brain. The added feature was that the measured conduction time to and from the cortex was found to be appropriate for mediating the long-latency response in the thumb muscle, which appeared at a latency of about 45 msec. For example, median nerve stimulation evoked a cortical response at a latency of just below 20 msec (cf. Desmedt and Cheron, 1981). Given a few milliseconds for the appearance of EMG in the thumb muscle, the conduction time from finger muscle to the cortex would be about 20 msec (Marsden et al., 1976a). Milner-Brown et al. (1975a) found that percentral stimulation in man evoked EMG discharge from finger muscles with a latency of about 20 msec. The conduction time to and from the brain (about 40 msec) is, therefore, suitable for driving the long-latency response in the thumb muscle.

However, there is, as yet, no proof for a causal relationship between the relevant afferent input and the cortical-evoked response and between the latter and the FSR. In addition, the validity of these measurements being indicative of central conduction time in part rests on the presumption that the afferent pathways generating the early- and late-stretch responses are similar. One could certainly argue that, although the early tendon-jerk response is due to the monosynaptic activation of the α-MNs by spindle primary afferents, the FSR might be attributed to the facilitatory influences of more slowly conducting spindle secondary afferents (Kirkwood and Sears, 1974, 1975, 1980; Stauffer et al., 1976). Alternately, Hultborn and Wigstrom (1980) have documented long-latency spinal activities elicited by Ia afferents through polysynaptic pathways. Such pathways are thought to be involved in the progressive recruitment of MNs during the tonic vibration reflex (TVR) (cf. Lance et al., 1983; Desmedt and Godaux, 1980; Desmedt, *this volume*). Another suggestion that the long-latency stretch response might be due to secondary afferents → dynamic γ-cells → nuclear bag fiber contraction → primary afferents → α-MNs → main muscle activity has been put forward by Appelberg et al. (1977). In both instances, the increase in the difference between tendon jerk and FSR latencies could be attributed to increase in distance of muscle from the cord. However, Marsden et al. (1976a) found that the difference between tendon jerk and FSR latencies in proximal shoulder muscles was equal to, or even greater than, that in distal finger muscles with similar segmental innervation. This finding therefore invalidates the above argument that FSR − MSR differences might be accountable in terms of differential peripheral conduction times in shoulder-finger projections.

Another presumption is that the muscles studied are under similar segmental influences. In this connection, it must be appreciated that the silent period separating the early and late responses is not solely due to conduction delay. For example, postexcitatory and segmental inhibitory influences such as (1) after-hyperpolarization of synchronously discharged MNs; (2) recurrent inhibition; (3) pause of spindle discharge; (4) postsynaptic inhibition from homonymous tendon organs; and (5) presynaptic inhibition of the Ia fibers from synergistic Ia afferents (Barnes and Pompeiano, 1970) may well contribute to the silent period (SP). Moreover, the finding that the silent period is consistently longer in gastrocnemius (an extensor) than biceps (a flexor) could be attributed to the effects of cutaneous fibers evoked by limb displacement. Thus, in accordance with flexor (Eccles and Lundberg, 1959) and SBS (Shimamura et al., 1967; Shimamura and Aoki, 1969) reflex action, these cutaneous inputs would likely be inhibitory to extensors and facilitatory to flexors.

In view of these uncertainties, the role of cutaneous and joint afferents in generating the MSR-SP-FSR sequence of response was examined, by comparing the gastrocnemius response to a standard dorsiflexing torque applied to the ankle with and without regional anesthesia of the ankle and foot (Chan et al., 1979b). Figure 2A shows five superimposed tendon jerk responses evoked by sharp taps to the Achilles tendon. The mean latency values as calculated from onset of stretch that triggered the oscilloscope, were similar to that obtained in

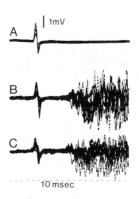

FIG. 2. EMG responses elicited in the gastrocnemius by applying A: Sharp taps to the Achilles tendon. B: A sudden and maintained dorsiflexing torque to the ankle. C: The same stimulus as in B during complete anesthesia of the ankle and foot. Note that, despite loss of cutaneous and joint sensations in the ankle and foot, there were no obvious changes in the response pattern and latencies. Five traces were superimposed in A and 10 in B and C. (From Chan et al., 1979b, with permission.)

response to a dorsiflexing torque during complete ankle block (Fig. 2C), thus providing good evidence that group Ia fibers from gastrocnemius were probably unaffected by the local anesthetic. Figure 2C illustrates that despite complete loss of cutaneous and joint sensations during ankle block, the pattern and temporal sequence of the EMG response to the dorsiflexing torque were similar to that of control response in the same subject (Fig. 2B). The mean values ($\pm$SE) of the FSR latencies (measured from onset of stretch) before, during, and after recovery from complete regional anesthesia of the ankle and foot for the 10 subjects studied were 99 $\pm$ 4.6, 98 $\pm$ 4.8, and 99 $\pm$ 4.1 msec, respectively and were indistinguishable from one another, as was their response magnitude. It should be noted that the late stretch responses in the long flexor of the big toe, infraspinatus and pectoralis major, were also unaffected by peripheral anesthesia (Marsden et al., 1977a). Hence, any contribution from cutaneous input to these muscles is probably negligible in the initiation of response. However, these findings cannot be generalized to all the muscles. Thus, Marsden et al. (1972, 1977a) showed that anesthesia of the thumb suppressed the late response in the long flexor of the thumb (cf. also Gandevia and McCloskey, 1977c). Consequently, different segmental mechanisms can be operating on different muscles. We therefore turned to clinical patients with suitable central lesions to gain further insight into the significance of supraspinal contributions.

Lesion Studies

On the basis that a late response mediated by a long-loop pathway would be abolished by a spinal transection, responses to stretch in spastic patients with a high probability of complete midthoracic transection were compared with those of normal subjects (Chan et al., 1977, 1979c). Figure 3 shows that no EMG response was evident in the affected muscles of the spastic subject (Fig 3B, lower trace) in the time bracket normally occupied by the FSR (Fig. 3A, lower trace), despite the fact that the duration of the early tendon jerk response was prolonged. Its absence therefore suggests that at least in man, the mechanisms responsible for hyper-reflexia do not by themselves generate the FSR. These findings point against a predominant segmental origin of the late response, in contrast to the results obtained in spinal cats (Ghez and Shinoda, 1978) and monkeys (Tracey et al., 1980b). It could be argued that segmental inhibition might have increased in spastic subjects. This seems unlikely in view of the findings that (1) the tendon jerk latency was comparable in normal and spastic subjects, thus the rise time to threshold would also be similar; (2) the duration of the tendon jerk response was significantly longer ($p < 0.001$) in spastic (mean = 25 msec) than normal subjects (mean = 13 msec, cf.

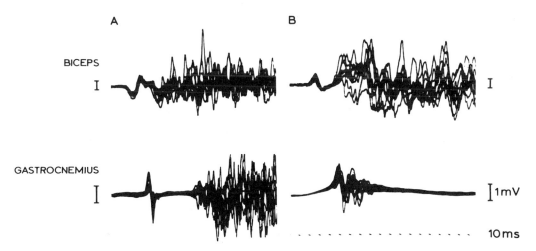

FIG. 3. EMG responses in biceps *(top traces)* and gastrocnemius *(bottom traces)* in a normal (A) and a spastic (B) paraplegic subject opposing a sudden and maintained stretching force. The biceps response evoked in the spastic subject (*top trace* in B), being above the level of transection, revealed a similar pattern to that of the normal subject (*top trace* in A). In contrast, the duration of the tendon jerk response was prolonged in the spastic gastrocnemius (*bottom trace* in B), but no EMG activity was evident in the time bracket normally occupied by the FSR in the normal subject, which in this instance (*bottom trace* in A) occurred at some 100 msec after the stimulus. (From Chan et al., 1979c, with permission.)

also Dietrichson, 1971, 1973); and (3) the facilitation of the test H-reflex in spastic patients was larger and occurred earlier than that of normal subjects (Zander-Olsen and Diamantopoulos, 1967). Furthermore, various workers (Delwaide, 1973; Ashby et al., 1974) have provided indirect evidence that presynaptic inhibition is decreased in spastic subjects.

We next turned to the question of cortical contribution to the FSR, by comparing responses on the affected side of spastic hemiplegic patients with those on their normal side (Chan et al., 1977, 1979c). Figure 4B shows the absence of the FSR on the affected side of these patients—in spite of prolonged tendon jerk responses (Fig. 4B). The pattern and latency of the EMG responses on the unaffected side (Fig 4A), however, were similar to those of the control subjects (Fig 3A). These findings are in accord with Marsden et al. (1977b), who reported that the late response in the thumb was either abolished or considerably reduced with lesions of the internal capsule and sensorimotor cortex. It should also be noted that the long-latency response in the upper limb muscle, but not the tendon jerk, was eliminated by a dorsal column lesion (Marsden et al., 1973, 1977c; Lee and Tatton, 1975). The dorsal column is known to transmit forelimb group Ia impulses to cortical area 3a in both cats (Oscarsson and Rosén, 1963) and primates (Phillips et al., 1971). Thus, the integrity of segmental pathways would seem unable to account for the long-latency stretch response if large afferent fibers to the brain are interrupted.

The absence of the long-latency stretch responses in spastic subjects could theoretically be due to decreased α- and/or γ-MN excitability resulting from the loss of tonic supraspinal facilitatory influences. However, this contention is not substantiated by the findings that α- and γ-MN excitability in spastic patients is either greater than or equal to that of normal subjects (Rushworth, 1960; Landau and Clarke, 1964; Dietrichson, 1971, 1973). It may also be argued that pathological lesions of the CNS in man are often not well localized. In this connection, it is important to note that the above

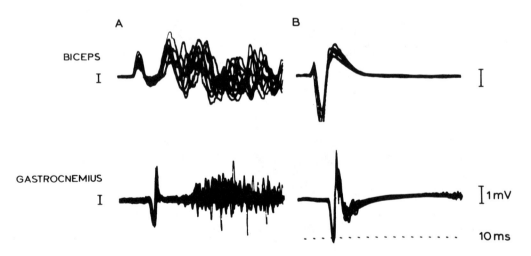

FIG. 4. EMG responses evoked in biceps *(top traces)* and gastrocnemius *(bottom traces)* on the (A) normal and (B) affected side of a spastic hemiplegic patient resisting a sudden stretching force. Note that despite a prolonged early discharge, no EMG burst could be observed in the spastic muscles (B) in the time gate normally occupied by the FSR in the unaffected muscles (A). (From Chan et al., 1979c, with permission.)

clinical findings are supported by *lesion experiments* in monkeys. In the upper limb studies of Tatton et al. (1975, 1978), three phases of EMG responses were evoked by mechanically induced perturbations, termed M1, M2, and M3 responses. Of particular importance is that lesion of the postcentral arm region involving areas 1, 2, 3b, most of 3a and some of 5, selectively abolished the M2 response (Tatton et al., 1975). These results thus suggested an important contribution from postcentral cortex to the M2 response. Certainly from an anatomical point of view, connections between corresponding pre- and postcentral areas have been well established in monkeys (Pandya and Kuypers, 1969; Jones, *this volume*). However, these findings have recently been challenged by Miller and Brooks (1981), who reported that lesion or cooling of sensorimotor cortex decreased the amplitude of M2 and M3 responses in some monkeys but did not abolish the responses in any animal.

On the other hand, evidence for a cortical contribution to long-latency stretch responses has emerged from another source. In patients exhibiting myoclonic jerks, electrical (or mechanical) stimulation of the median nerve elicited large somatosensory potentials with a mean latency of 24 msec, followed by an EMG wave representing the myoclonic jerk some 50 msec after the stimulus (Sutton and Mayer, 1974; Rosen et al., 1977; Halliday and Halliday, 1980). Animal experiments using experimentally induced epileptogenic foci have likewise implicated the motor cortex in the generation of myoclonic jerks. In monkeys, when such a focus was located in the precentral foot area, low-threshold percutaneous stimulation of the tibial nerve elicited a complex cortical response together with three distinctive EMG bursts in the soleus muscle. The second of these EMG waves occurred at such a latency (mean = 40 msec) as to permit its production by the evoked cortical spike some 10 msec earlier (Chauvel et al., 1978). Certainly, in lower mammals (rats, rabbits, and cats), lesion of the contralateral sensorimotor cortex has been shown to abolish the second component of the EMG responses associated with the myoclonic jerk (Angel and Lemon, 1975). Thus, the myoclonic jerk has been viewed as a pathologically exaggerated transcortical stretch reflex resulting from the epileptogenic focus.

Electrically Induced Reflexes

Another line of indirect evidence comes from electrical stimulation of muscle nerves in man. Upton et al. (1971) observed that electrical stimulation of the median nerve during isometric contraction produced a number of EMG peaks in the thumb muscle, termed M1, V1, and V2 responses. Both of the latter peaks were potentiated by voluntary efforts. These findings were confirmed by Milner-Brown et al. (1975b). Since the evoked somatosensory potential occurred some 20 msec after median nerve stimulation, they argued that there was time for a transcortical response to reach the finger muscles by 56 msec, the latency of their V2 wave. This interpretation was strengthened by the abolition, reduction, or delay of the V2 response with lesions of sensorimotor cortex, when no changes of V1 could be seen (Conrad and Aschoff, 1977).

Additional support for a transcortical response mediated by low-threshold muscle afferents comes from the study of H-reflex recovery curves in monkeys (Lachat et al., 1977; Hofflon et al., 1982). As in a late facilitation of the test H-reflex could be observed following a subthreshold conditioning H-stimulus, starting at a latency of about 40 msec and reaching a peak of 60 to 80 msec. This late facilitation was abolished after pyramidotomy, or when the motor cortex was cooled or lesioned. Furthermore, the time course of the late facilitation was in agreement with latencies of the cortical potentials evoked by the conditioning stimulus and of the motor responses elicited by electrical stimulation of the motor cortex.

Cortical Unit Recordings

From the above, it appears that several reports point to a cortical contribution to the FSR. What remains to be determined is whether such a contribution takes the form of a transcortical long-loop influence, or some other mechanism such as selective gain changes in segmental reflex pathways. The former view is favored by findings from single cortical unit recordings in behaving monkeys. Evidence first came from Evarts' (1973) work in which he simulated the human experiments of Hammond et al. (1956) by training monkeys to consciously oppose sudden displacement of a handle that they had learned to hold in a given position. In agreement with findings in man, EMG records showed a late response at a latency of 30 to 40 msec in elbow muscles whose lengths were changed by handle displacement. The intriguing observation was that excitation of pyramidal tract (PT) cells could be demonstrated to occur in time to contribute to the late EMG response (probably in M2 response of Tatton et al., 1975). This view was supported by Conrad et al. (1974, 1975). Having trained monkeys to move a handle between two targets by repetitive elbow movements, they injected torque pulses to perturb these movements. Again, many precentral cortical neurons (88%, Conrad, 1978) were found to exhibit an early response (latencies 20–40 msec) the timing of which was such that it could contribute to the M2 response in the elbow muscles. An added feature was that these early discharges often increased when the torque pulse opposed the elbow movement and decreased when the perturbation assisted the movement (Conrad et al., 1975). This latter finding was confirmed by Evarts and Tanji (1976). Furthermore, prior instruction was found to modify the early precentral discharge (Evarts and Tanji, 1974) in parallel with the long-latency response. For example, if the instruction was to pull, perturbation of the handle gave rise to increased early response in precentral neurons, which were active with pull movements. If the instruction was to push, perturbation evoked decreased response in the same neurons.

Thus, activity of precentral cortical neurons was appropriately timed, and covaried with direction of displacement as well as instruction in such a way as to make tenable their contribution to long-latency stretch response. However, proper timing and covariation suggest, but do not establish, a causal relationship between the two. Fetz et al. (1976) tackled this problem by means of postspike averages triggered from the action potentials of precentral (including some PT) cells that covaried with wrist movements, which the monkeys were trained to perform against a programmed load. For one quarter of these cells, the postspike averages revealed an increase in the mean EMG activity of wrist muscles at a latency of 4 to 12 msec. Most of these cells increased their discharges 20 to 60 msec prior to onset of the increased EMG activity and were driven by passive movements in a direction opposite to the active movement with which the cells covaried. These findings reinforce the view that not only is there a tight input-output coupling in movement-related precentral neurons, but also the discharges of these neurons occur at such a latency that they can and do evoke an increased EMG response in the related muscles. Taken together with the results of latency measurements from human studies, the data are consistent with the participation of a long-loop phasic contribution, rather than that of a tonic descending facilitation, in the generation of the late stretch response.

Role of Cerebellum

There is evidence that the cerebellum may be involved in the production of long-latency stretch responses, although not in the manner suggested by Eccles (1966). Thus, Marsden et al. (1978b) found that the late response of the long thumb flexor to stretch was reduced or delayed in patients with unilateral cerebellar lesions without signs of pyramidal or sensory deficit. A note should be added at this juncture with regard to the possible neurophysiological origin of the M3 response of Tatton et al. (1975). In primate experiments, many workers have also observed a second precentral response following arm perturbations in addition to the early precentral responses (Conrad et al., 1974, 1975; Evarts and Tanji, 1976). This second precentral component occurred at a peak latency as short as 50 to 60 msec, which would permit its contribution to the M3 response. It was shown to be dependent on the direction of movement, which the monkeys were instructed to perform, rather than the direction of limb perturbation, as was the case with the early precentral response (Evarts and Tanji, 1976). This second response is probably dependent on cerebellar input. Thus, cooling of the dentate (Meyer-Lohmann et al., 1975, 1977) and interpositus nuclei (Vilis et al., 1976) decreased the intensity and/or delayed the second precentral response together with deterioration of the M3 response, whereas the early precentral (and M2) responses were unaffected. In this context, evidence for a transcerebellar loop involving the interpositus nucleus with input from high-threshold muscle afferents for the second precentral response, has been demonstrated in cats. Thus, Murphy et al. (1974, 1975) observed that motor cortical neurons responded to low- (principally group Ia) and "high"-threshold (primarily group Ib and II) muscle afferents at a mean latency of 11 and 18 msec, respectively. These neurons were shown by intracortical stimulation to coincide with the efferent columns for contraction of the same muscles. Significantly, the second cortical response was reversibly blocked by cooling of the interpositus nucleus. These authors therefore suggested a transcerebellar pathway for high muscle afferent input involving the rostral spinocerebellar tract (shown to be excited principally by Ib afferents, Oscarsson and Uddenberg, 1965), to the interpositus nucleus, and from there through the contralateral ventrolateral (VL) nucleus of the thalamus to the motor cortex (Rispal-Padel and Latreille, 1974). The findings in primates that cooling of interpositus reduced the second precentral response might indicate a similar transcerebellar pathway in the generation of the M3 response. This view is particularly appealing in light of the report that VL thalamic neurons projecting to motor cortex (Strick, 1976a) discharged before onset of arm movement (Strick, 1976b, 1978).

However, it should be remembered that the cerebellar input to the second precentral response may be simply one of tonic facilitation; nor can the possibility that the second precentral response represents a transcerebellar feedback of the first response be excluded.

Evidence for a Segmental Contribution

From the above review, there are ample experimental findings compatible with a transcortical (and transcerebellar) contribution to the long-latency stretch responses. However, an alternative view is that the complexity of late EMG responses evoked by muscle stretch could simply be due to the interaction of excitatory and inhibitory influences at a spinal level. For example, the long-latency stretch responses might represent the return of activity due to ongoing Ia discharge as the aforementioned segmental postexcitatory and inhibitory influences subside. It is possible that additional excitatory input from homonymous spindle group II afferents—either directly on α-MNs (Kirkwood and Sears, 1974, 1975; Stauffer et al., 1976) or indirectly via their influence on dynamic γ-MNs (Appelberg et al., 1977)—might aid in overcoming this inhibition.

In this context, Newsom Davis and Sears (1970) first suggested that segmental mechanisms could account for the inhibitory-excitatory sequence of response observed in the intercostal muscles resisting a sudden increase in load (cf. Sears, 1973). By means of a simulated model of known spinal reflex mechanisms, Kearney (1978) showed that segmental reflexes were theoretically capable of generating the whole range of late responses thus far observed in the lower limb. His proposition concurred with the findings of Ghez and Shinoda (1978) that torque perturbations produced three discrete components of EMG responses in the triceps of both decerebrate and spinal cats, and that these resembled those observed in normal cats. Similar observations have also been reported in monkeys. Briefly, Tracey et al. (1980) demonstrated that stretching the extensor of either the fore- or hind-limb evoked early (M1) and late (M2) EMG responses in acutely spinalized monkeys. In other words, segmental mechanisms were capable of generating complex sequences of EMG responses similar to those noted in intact animals.

It is worth noting, however, that in the acutely spinalized monkeys, the amplitude of both M1 and M2 responses was reduced and their latency prolonged. Furthermore, the threshold velocity of the stretch necessary to evoke both responses was increased. Although these observations can be taken to reflect the generally depressed state of the CNS during spinal shock, the suppression of the M1 and M2 responses might originate from different sources. An analogy can be drawn from the tendon jerk and TVR, both of which manifest the monosynaptic influences of group Ia fibers (for TVR, cf. Brown et al., 1967b; Desmedt and Godaux, 1975, 1980), with additional polysynaptic effects being involved in the latter (e.g., Tsukahara and Ohye, 1964; Westbury, 1972; Homma and Kanda, 1973; Lance et al., 1973; Desmedt, *this volume*). Although the tendon jerk is well accepted as a segmental reflex, the TVR has been shown to be dependent upon supraspinal drive (Matthews, 1966; Gillies et al., 1971). However, both reflexes were suppressed in patients suffering from spinal shock (Ashby et al., 1974). In the same light, the reduction of the M2 response in acute spinal preparations might be taken as being consistent with a supraspinal contribution to the generation of the normal response in intact animals. A final note is that the stretch used in both the cat and primate experiments was slow, their fastest rise time to peak displacement being 80 msec (cf. Fig. 2, Tracey et al., 1980), compared with that used in our studies being some 20 msec (cf. Fig. 5). The well-known dynamic characteristic of the group Ia fiber response (cf. Matthews, 1972) during a prolonged stretch, might therefore itself contribute to the segmentation of the EMG response. In agreement with this view is the fact that fast ankle displacements were consistently used in our experiments, and we never observed an intermediate EMG response at 70 msec in the stretched calf muscle, which was reported by Gottlieb and Agarwal (1980) when "more prolonged stimuli" were used.

Hagbarth et al. (1980, 1981), using microelectrode recording of proprioceptive afferents in human subjects, showed that in response to sudden quick stretches, the discharges of group Ia fibers tended to be segmented, with two or three bursts of afferent inpulses occurring during stretching. This phenomenon has also been seen in cats (Tracey et al., 1980). Hagbarth et al. (1980) suggested that the segmentation of group Ia

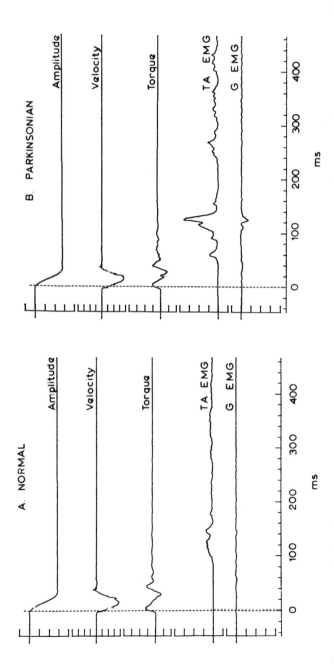

FIG. 5. Responses elicited by sudden servo-controlled plantar-flexing displacements of ankle position in (A) normal and (B) parkinsonian subjects instructed to relax. In this and all subsequent figures, each curve (or data point) is the ensemble average of 10 responses. Left: One division of the scale equals 0.125 for amplitude, 3.75 rad/sec for velocity, and 50 Nm for torque, with arbitrary units for the EMG scale. Plantar-flexing movements are denoted by downward deflections of the traces. Torque and EMG traces are assigned polarities commensurate with the direction of the movement they tend to produce. Note the much larger EMG responses (fourth traces) to the same stretching stimulus (amplitude = −0.34 rad; velocity = −14 rad/sec) in particular, the PSR at some 90 msec, in parkinsonian (B) vs normal (A) subjects. (From Chan et al., 1979a, with permission.)

discharges was due to some inherent mechanical properties within the limb (see Prochazka and Hulliger, *this volume*). However, on close examination of their records, these afferent bursts were often preceded by small irregularities in the position trace. Thus, they might not be entirely due to nonlinearities in muscular compliance as had been proposed but to minute vibrations associated with the stretching stimuli. Significantly, the segmentation of Ia discharges was not always followed by the segmentation of EMG responses. Thus, when flexor muscles were relaxed, generally only one EMG burst was present. In other words, any causal relationship between the two phenomena has yet to be established.

Another interesting dimension has been brought forward by Bawa and Tatton (1979). Using single motor unit (SMU) recordings of stretched wrist muscles in primates, they showed that 83% of the SMUs responded over an interval corresponding to only one of the three surface EMG peaks. It is, therefore, difficult to imagine that the M1, M2, and M3 responses merely represent either (1) synchronized oscillatory facilitation of the MNs as a result of segmental excitatory-inhibitory sequences; or (2) the consequence of segmented discharges from the same (Group Ia) afferent input. Rather, the separately responding MN subpopulations will be more in keeping with different reflex pathways being responsible for generating each of the three EMG bursts. This view is consistent with the report by Hendrie and Lee (1978) that although vibration suppressed the M1 (equivalent to the tendon jerk) response, it had negligible effects on the long-latency stretch responses in the human wrist muscles.

Conclusion: A Symbiotic Relationship

At first sight, it appears difficult to reconcile the ample evidence for a supraspinal contribution to the long-latency stretch responses with that suggesting sufficiency of segmental mechanisms. However, it would be a mistake to confine attention to one or the other view, when in all probability long-loop and segmental influences normally act symbiotically. Certainly, Vilis and Cooke (1976) have provided evidence that the segmental reflex mechanism could modify the magnitude of the M2 response. However, the contribution by each mechanism probably varies wtih a number of factors. Thus, the intention to resist might have shifted the emphasis to supraspinal mechanisms. On the other hand, more prolonged stretching stimuli might contribute to the grouping of EMG responses at a segmental level because of the dynamic characteristics of group Ia fiber response during stretch. Bearing in mind the process of encephalization on ascending the phylogenetic scale, it is also doubtful if the findings in cats and even monkeys are necessarily applicable to man. The results from human studies certainly indicated that the involvement of a supraspinal component is a necessary element in mediating long-latency stretch responses, particularly under the experimental paradigms thus far investigated.

PHYSIOLOGICAL NATURE OF LONG-LATENCY STRETCH RESPONSES

At one time these late responses were thought to be reflex in nature, because their latency was significantly shorter and less variable than that of a strictly voluntary response (Hammond et al., 1956; Melvill Jones and Watt, 1971*a*). However, it has since been shown that kinesthetically triggered reaction-time movements could be as short as 70 msec in elbow muscles (Evarts and Vaughn, 1978; Houk, 1978). Some interesting questions thus arise: To what extent are these late responses reflex in origin? If they are, is their output correlated to input parameters in a systematic manner, as would be expected of reflexes which are servo controlled? Alternately, to what extent do these responses manifest characteristics of reaction-time movements? In this connection, do they represent preprogrammed patterns of motor activity which, once triggered by afferent signals, are not modifiable by input trajectories?

Evidence for a Servo Reflex

Phillips (1969) originally proposed that the transcortical stretch reflex operated to compensate automatically for any change in muscle length by means of α-γ coactivation, termed by Matthews the servo-assistance mechanism of movement control (cf. Matthews, 1972, Chapter 10). Since then, some workers have presented evidence in favor of the long-latency stretch responses being servo controlled. Thus, Marsden et al. (1972, 1976b) showed that the late EMG activity of the long thumb flexor during voluntary tracking movement against an initially consistent resistance, increased after a halt or a stretch and decreased after a release. This behavior was interpreted as manifestations of automatic servo action. In addition, the gain of the servo was found to be proportional to initial load. Therefore, when the initial resistance to movement was increased, the late responses after a stretch, halt or release were augmented. Tatton and Bawa (1979) confirmed these observations with SMU recordings of primate wrist muscles. They showed that for the units discharging over the M2 or M3 (as well as M1) intervals, the probability of their firing increased with increasing magnitude of the step load or the initial displacement velocity. The response of these MNs is thus consistent with servo action. So is the behavior of many movement-related neurons at the level of motor cortex.

As previously mentioned, precentral cells (including PT neurons) exhibited an early response, which often increased when the perturbation opposed (i.e., mismatch between actual and intended movement) the related active movement and decreased when the disturbance assisted the movement (Conrad et al., 1974, 1975; Evarts and Tanji, 1976). In other words, a transcortical servo hypothesis predicts that the CM activity will change in opposite direction depending on whether the movement is active or passive. This was indeed found by Evarts and Fromm (1978) for finely controlled (1 to 2°) versus large ballistic (20°) movements. Thus, 26 of 29 PT cells which showed increased discharge for a given direction of fine, active movement, reduced their discharge for that direction of passive movement. Furthermore, in anesthetized baboons, Sakai and Preston (1978) showed that the activity of motor cortical (some identified PT) units, which were facilitated by ramp stretch of ankle extensor muscles, did reflect changes in both amplitude and velocity of muscle stretch, a finding that is again consistent with transcortical servo action.

Evidence for a Preprogrammed Response

However, there are also findings indicating that in some circumstances the late response may represent the triggered release of preprogrammed motor activity. Evarts and Granit (1976) demonstrated that when subjects were instructed to contract the biceps in response to a perturbation, a long-latency EMG response could be elicited regardless of whether the perturbation stretched or shortened the muscle. A response which assists rather than opposes changes in muscle length cannot be viewed in terms of a length servo. Crago et al. (1976) argued that the long-latency response to disturbance in elbow muscles displayed the characteristic features of reaction-time movements, in that the response latency became longer and more variable when choice was introduced in the task; for example, when the subjects did not know the direction of disturbance in advance. The short minimal latency and the automaticity suggested to these authors that it might be preprogrammed in nature (cf. Houk,

1978). However, Gottlieb and Agarwal (1980) reported no significant differences in the latency of the late response in the ankle muscles when choice was introduced in the task.

From the foregoing discussion, it becomes apparent that the nature of control of the long-latency stretch responses remains a matter of dispute. On one hand, there are ample findings in accord with these responses being servo driven, in which case they would be considered reflex in origin. On the other hand, there is also evidence that they are in the nature of reaction-time movements, which might contain the triggered release of preprogrammed motor elements. In this connection, patients afflicted with akinesia are known to have increased reaction times (Angel et al., 1970; Flowers, 1975), but their reflex latencies are not necessarily modified (DeJong and Melvill Jones 1971; Melvill Jones and DeJong, 1971). A key question thus arises. Are the latencies of late components of the stretch response in human ankle muscles increased or unaffected by parkinsonian akinesia? The findings should aid in the distinction between reflex and reaction-time components of the late responses to ankle displacement. The EMG responses evoked in tibialis anterior (TA) by suddenly applied servo-controlled plantar flexion of the ankle was compared between 9 parkinsonian and 9 age-matched normal subjects, with and without the subject's intention to oppose (Chan et al., 1979a).

If the subject was instructed to relax when ramps of different amplitudes and velocities were applied to plantar flex the ankle, a tendon jerk response was occasionally just observable in TA at 40 msec in 3 of 9 normal subjects; so was the presence of a small response at about 90 msec (Fig. 5A, fourth trace). This response, which we called polysynaptic stretch reflex (PSR) was found in all except one parkinsonian patients examined (Fig. 5B, fourth trace). Of particular interest is that the mean latency estimate of the PSR was indistinguishable between parkinsonians and normals, being 90 ± 3.9 and 93 ± 0.7 msec, respectively. This finding indicates that the PSR is not subject to parkinsonian akinesia, and hence suggests that it is in the nature of a reflex response. Indeed, commensurate with reflex behavior, which is proportional to input trajectories, the PSR area (determined by integrating the EMG burst over the time interval of its occurrence from 70 to 160 msec, these values having been verified by visual inspection) increased linearly with increasing magnitude of displacement velocity. This is illustrated in Fig. 6, which plots the PSR area as a function of displacement velocity. Each data point represents the averaged value of 10 observations, normalized with respect to the maximum value of the PSR area in each subject, with different symbols denoting the 4 patients studied. The linear regression line drawn through these points had a correlation coefficient (r) of -0.82. Note that displacement velocities have been assigned a minus sign, in keeping with the negative polarity being given to plantar-flexing movement (cf. Fig. 5 legend). That is why the r value is negative.

In contrast, the FSR evoked by subjects opposing the displacements exhibited a significantly longer latency in parkinsonians than age-matched normals, the respective mean values being 183.4 ± 20.3 and 118.1 ± 7.4 msec. In the same patients, the EMG response latency to a visual signal was similarly delayed (mean = 245.9 ± 32.5 msec, vs the normal mean = 179.8 ± 4.5 msec). In fact, as shown in Fig. 7, the increase in FSR latency bore a linear relationship ($r = 0.96$) to the increase in visual response latency.

The above findings have two important implications. First, the akinetic delay of FSR in parkinsonians argues against the FSR being of purely reflex origin. Second, the akinetic delay of reaction time in parkinsonians has been attributed to impair-

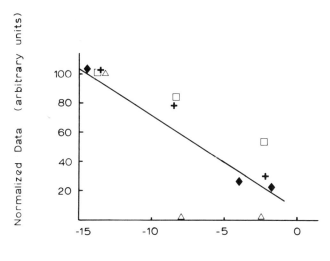

FIG. 6. Relationship between the PSR area of TA and displacement velocity. Response evoked by ankle plantar-flexion at an amplitude of −0.25 to −0.30 rad, with the subject instructed to relax. Data of the PSR area, normalized with respect to the maximum value of each of the 4 subjects studied, is denoted by the 4 different symbols. The *linear regression line* plotted had a correlation coefficient *(r)* of −0.82. (From Chan et al., 1979a, with permission.)

ment of the ability to generate preprogrammed reponse by Flowers (1975, 1976). The delay of the FSR thus suggests that at least its initial component might be preprogrammed in nature. Such a response is thought to be preformulated which, once triggered by the arrival of afferent signals, is released "as a unit, in which case it has

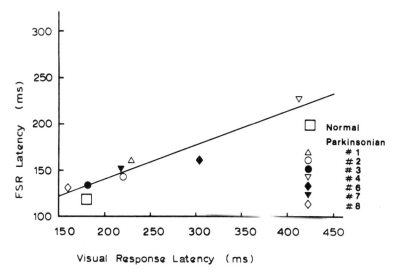

FIG. 7. Linear relationship ($r = 0.96$) between FSR latency and visual response latency in parkinsonian subjects. Note that 2 patients (● = no. 3 and ◇ = no. 8) exhibited latencies close to the mean value (□) of age-matched normals, showing that they probably did not suffer from akinesia like the rest of the patients. (From Chan et al., 1979a, with permission.)

to run its full course without the possibility of modifications" (Desmedt and Godaux, 1978a, p. 21). As Bizzi and Evarts (1971) described, a preprogrammed response is one whose pattern is stereotyped and relatively independent of stimulus characteristics.

We investigated this possibility by comparing the EMG responses elicited in the human ankle muscles by displacements having widely differing trajectories (Kearney and Chan, 1976; Chan and Kearney, 1977, 1982a). If the FSR were due to triggered release of a preprogrammed response, it should show no changes despite different input patterns. Alternatively, if the FSR were due to long-loop servo action, it should be dependent on the dynamics of applied stretch and, therefore, should vary according to the input profiles. We first mapped the FSR evoked in the ankle muscles of 8 to 12 normal subjects as a function of displacement amplitude and velocity. Subsequently, we compared the FSR evoked by sustained ramp displacements (lasting 500 msec) with those elicited by transient pulse displacements (lasting 60 msec) having widely different amplitude and velocity trajectories. As shown in Fig. 8A, the findings revealed a general lack of systematic relationship between the characteristics of the initial component of FSR and the stimulus (displacement) profiles, although occasional exceptions were noted. Neither the FSR latency evoked in gastrocnemius (G) nor its rise time (calculated by subtracting mean FSR latency measured at 15% of its maximum amplitude from that at 50%) showed any consistent velocity dependence. Figure 8B also demonstrates the absence of any systematic effect of displacement amplitude on the FSR rise time. Although there was a tendency for FSR latency to decrease with increasing displacement amplitude, the phenomenon was observed in less than half of the subjects studied.

In addition to the lack of a systematic correlation between input and output, the early phase of FSRs elicited by sustained ramp displacements in G (Fig. 9, fifth solid trace) and TA (Fig. 10, fourth solid trace) were found to be almost identical to those evoked by transient pulse displacements of widely different amplitudes and velocities (Fig. 9, fifth dotted trace; Fig. 10, fourth dotted trace). This was borne out quantitatively by the lack of a significant difference in latency or rise time of FSRs evoked by ramp and pulse displacements in the 8 subjects studied (Table 1).

Thus, none of the characteristics of the initial components of FSR systematically reflected any of the dynamics of applied muscle stretch—a behavior that is not consistent with the expected output of a length control servo. The similarity of FSRs evoked by muscle stretch of widely different trajectories, however, is strongly suggestive of the triggered release of a preprogrammed response. It is therefore concluded that at least the initial part of the FSR is governed by a preformulated pattern of intended movement, independent of specific limb displacement patterns. Of course, this does not rule out a possible contribution by a strictly voluntary component to the latter part of FSR, nor any reflex participation from the afferent fibers excited by the stimulus. In fact, the interaction between supraspinally triggered and segmental reflex components probably explains why exceptions to the general lack of correlation between input and output have been noted in individual subjects. For example, using multiple regression analysis, the TA FSR slope (determined by fitting a straight line to the first 35 msec of its occurrence) was found to correlate with displacement amplitude in two subjects and with velocity in another (cf. Chan and Kearney, 1982a).

SIGNIFICANCE OF SUPRASPINAL CONTRIBUTION TO LONG-LATENCY STRETCH RESPONSES

The finding of a substantial contribution by supraspinal structures to the generation

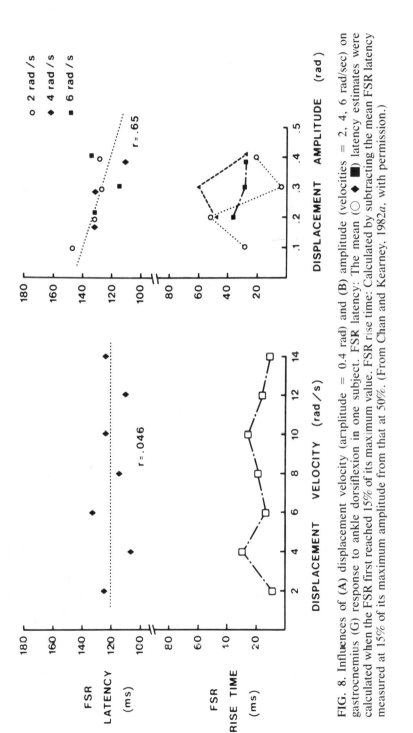

FIG. 8. Influences of (A) displacement velocity (amplitude = 0.4 rad) and (B) amplitude (velocities = 2, 4, 6 rad/sec) on gastrocnemius (G) response to ankle dorsiflexion in one subject. FSR latency: The mean (○ ◆ ■) latency estimates were calculated when the FSR first reached 15% of its maximum value. FSR rise time: Calculated by subtracting the mean FSR latency measured at 15% of its maximum amplitude from that at 50%. (From Chan and Kearney, 1982a, with permission.)

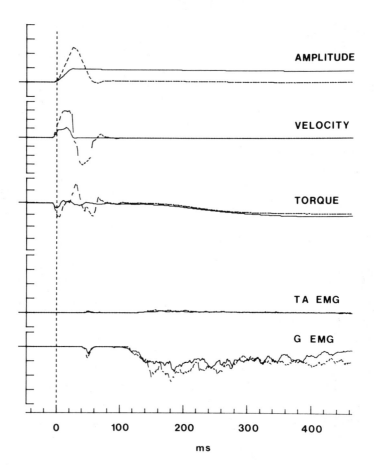

FIG. 9. Superimposition of the G responses to ramp *(solid traces)* and pulse *(dotted traces)* dorsiflexing displacements elicited in a subject instructed to oppose such stimuli. Traces and scales are the same as in Fig. 5. (From Chan and Kearney, 1982a, with permission.)

of FSR in ankle muscles of man has important implications for the general organization of human motor control.

(i) With the participation of supraspinal structures, a greater range of flexibility could be introduced into the myotatic response system than would be feasible in a purely segmental pathway. For example, in the control of lower limb movements, automatic muscle response to stretch might be widely adjustable in advance by setting the characteristics of supraspinal processing mechanisms according to expectation and behavioral set. In this context, Melvill Jones and Watt (1971a,b) showed that the EMG activity associated with landing from single steps or unexpected falls commenced well before the actual landing and was completed before time would permit participation of FSR. Indeed, at the time when FSR would have been expected to occur as a result of ankle dorsiflexion on landing, a relatively silent period was usually observed. This implies that the FSR plays no role in the control of landing from single steps or sudden falls and that during these movements, it must have been greatly reduced by supraspinal processing mechanisms. However, during rhythmic hopping and presumably running, an active FSR would be appropriately timed to contribute to the EMG associated with accelerating the individual

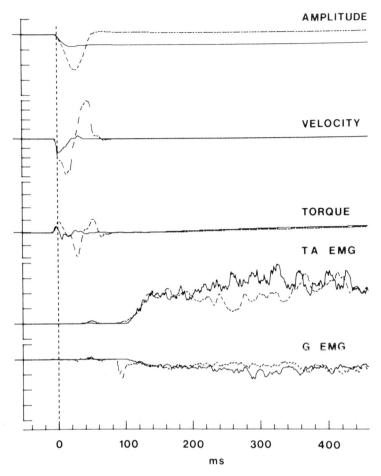

FIG. 10. Superimposition of the TA responses to ramp *(solid traces)* and pulse *(dotted traces)* plantar-flexing displacements elicited in a subject instructed to resist such stimuli. Traces and scales are the same as in Fig. 5. (From Chan and Kearney, 1982a, with permission.)

upwards into the next cycle of movement (Melvill Jones and Watt, 1971a).

(ii) Adaptive changes of FSR, depending on its usefulness in stabilizing stance in man, has also been demonstrated. By translating a servo-driven platform on which the subject stood in order to induce forward sway, Nashner (1976) found that in some (5 of 12) subjects, a FSR could be elicited in gastrocnemius as early as 120 msec after onset of forward sway, which operated under the circumstance to reduce postural sway. Interestingly, when the ankle was directly dorsiflexed by suddenly rotat-

TABLE 1. *Comparison of FSR latency and rise time between ramp and pulse responses*

	FSR latency (msec)			FSR rise time (msec)		
	Ramp	Pulse	p	Ramp	Pulse	p
G mean ± SE	126.9 ± 3.9	127.1 ± 4.7	$p > 0.9$	31.1 ± 6.9	21.4 ± 2.4	$0.3 > p > 0.2$
TA mean ± SE	133.1 ± 8.0	137.6 ± 9.9	$0.8 > p > 0.7$	26.5 ± 3.5	21.5 ± 2.3	$0.3 > p > 0.2$

ing the platform upward (in which case a FSR would have caused adverse enhancement of backward sway), the FSR was attenuated during the succeeding 3 to 5 trials, leading to reduced body sway. Furthermore, adaptability of FSR, although not its production, was lost in patients with clinically diagnosed cerebellar deficits. Thus, it seems that the flexibility, introduced into the myotatic response system by involvement of a supraspinal processing mechanisms, would have profound functional significance.

(iii) Another role of supraspinal participation could be to enable "learning" effects due to training. Milner-Brown et al. (1975b) showed that weight-lifters had more prominent late responses to electrical stimulation of muscle nerves (in particular, the V2 component) than normal subjects, but no significant difference was found in the earlier reflexes, which are spinal in origin. Mortimer and Webster *(this volume)* also observed an interesting potentiation of long-latency stretch responses in karate experts. It seems that supraspinal pathways mediating long-latency responses can be selectively potentiated (or presumably suppressed) as a result of a specific training regimen.

(iv) The finding of a substantial supraspinal component in the myotatic response system might well have a bearing on the organization and control of clonus. Clonus is usually observable in spastic, but only in exceptional circumstances in normal, subjects. It has been attributed to an oscillatory condition due to loading and unloading of spindle primary endings (Hagbarth et al., 1975; Wallin and Hagbarth, 1978), which are thought to be more excitable in spastic than normal subjects (e.g., Dietrichson, 1971, 1973; Szumski et al., 1974). In this context, the long-latency stretch response consists of a much more asynchronous activation of motor units than the monosynaptic stretch reflex (MSR). As pointed out by Wiesendanger et al. (1975), their temporally dispersed impulses may effectively counteract the inherent tendency of MSR to oscillate. This concept is supported by the model study of Stein and Oguztoreli (1978) in which simulated spinal reflexes were found to generate oscillations, although at a frequency (8–12 Hz) slightly higher than that reported by clonus (5 Hz). Increasing the gain of the supraspinal pathway quite effectively reduced this oscillation. An interesting idea therefore emerges. Is it possible that long-latency stretch responses exert a stabilizing influence tending to prevent the occurrence of clonus in normal subjects? If this hypothesis proves correct, it could have important clinical applications, for example, in the design of an electrical stimulation system for the control of clonus in patients.

(v) Possible disturbance in the supraspinal contribution to long-latency stretch responses might have to be considered when seeking a better understanding for the pathophysiology of some disease process(es). In this connection, Lee and Tatton (1975, 1978) have introduced an entirely new outlook concerning the origin of parkinsonian rigidity. On the basis of finding a much facilitated M2 component of upper-limb responses which was hardly modifiable by prior instruction in parkinsonians, they proposed that rigidity could be the manifestation of an increased gain in the long-loop pathway. Their theory was formulated on the grounds that the striatum projects mainly via globus pallidus and ventrolateral nucleus of thalamus to the precentral motor cortex (Kemp and Powell, 1971), and that parkinsonism is known to be associated with decreased dopamine content in basal ganglia (Hornykiewicz, 1966, 1972). If striatal disease led to an increased globus pallidus output (which is mainly facilitatory to PT neurons; Newton and Price, 1975), a situation might be created in which the gain of the transcortical reflex is constantly augmented. This hypothesis was subsequently supported by Mortimer and

Webster (1978), who established a significant correlation between the magnitude of M2 and the amount of rigidity, as measured separately by passive joint displacement in parkinsonians. In contrast to Tatton and Lee (1975), however, they found that the long-latency stretch response in parkinsonians is modifiable by prior instructions. In view of these upper limb observations, it is interesting to note our findings that, although the PSR in tibialis anterior was more regularly evoked and of lower threshold in parkinsonians than in normals (cf. Fig. 5), its facilitation was not found to be associated with any increase in the torque generated by the same displacement (Chan et al., 1979a).

CONCLUDING REMARKS

From the above review, it seems that the automatic response to compensate for limb displacements might very well be the manifestation of (1) reflex components that can be driven segmentally (the tendon jerk or MSR), or suprasegmentally [the PSR which disappeared in spastic patients with cord transection (Chan and Kearney, *unpublished observation*)]; and (2) a triggered component (the FSR), which is dependent on an intact cortex, interacting with segmental effects. One can imagine that in disease processes, these components can be differentially affected. Thus, in subjects with spasticity, while the tendon jerk is hyperexcitable, the FSR is abolished. In parkinsonian subjects, on the other hand, a facilitation of the PSR is manifested, while the FSR is significantly delayed.

When considering all of the related findings, the following concepts emerge. First, whereas we have shown that the FSR in ankle muscles is preprogrammed, other workers have provided evidence that the M2 component of the upper limb response is servo controlled (Marsden et al., 1976b; Tatton and Bawa, 1979). Second, Evarts and Fromm (1978) have demonstrated that the behavior of precentral cortical neurons controlling the same muscle depends on the task specified. Thus, precentral neurons that were recruited by ballistic movements showed the same response to either direction of active as well as passive movement. This behavior contrasted greatly with that of the PT cells activated by small, precisely controlled movements, which showed increased discharge for a given direction of active movement and decreased discharge for that direction of passive movement.

A key question thus arises. Do the differences in findings between our studies and those of others reveal real differences in control strategies adopted by the CNS, depending on the muscle involved and/or the task specified? Therefore, two issues are raised for future experiments: (1) Given the same muscle, would the long-latency stretch response be switched from one mode of operation to another, depending on the task strategy? (2) Given the same task, would the CNS switch from one mode of control to another, depending on the muscle (e.g., whether upper or lower limb, or distal versus proximal) studied?

ACKNOWLEDGMENTS

The author is deeply grateful to Dr. G. Melvill Jones for his constructive criticism of the manuscript. Special thanks are due to Dr. R.E. Kearney, co-author of many of the articles cited in this chapter.

Long-Latency Myotatic Reflexes in Man: Mechanisms, Functional Significance, and Changes in Patients with Parkinson's Disease or Hemiplegia

*Robert G. Lee, John T. Murphy, and William G. Tatton

Department of Clinical Neurosciences, University of Calgary, Calgary, Alberta; and Department of Physiology and Playfair Neuroscience Unit, University of Toronto, Ontario, Canada

In monkeys and in man sudden angular displacements of a joint result in a prolonged burst of electromyographic (EMG) activity from the muscle being stretched. This EMG response consists of a short-latency component (M1) and one or more long-latency components (M2, M3) (Tatton et al., 1975; Lee and Tatton, 1975; Tatton et al., 1978). The M1 component has a latency similar to that of a tendon jerk and is likely generated by spinal circuitry including monosynaptic and polysynaptic reflex pathways. However, the origins of the long-latency components of the stretch reflex remain uncertain and have become a matter of controversy. The timing of these late components led to suggestions that they could be mediated by long-loop transcortical reflex pathways (Phillips, 1969; Marsden et al., 1973). This view received support from observations that motor cortical neurons in monkeys responded to limb perturbations with delays, which would be appropriate for them to be participating in a transcortical feedback loop (Evarts, 1973). Furthermore, it was shown that lesions involving dorsal columns, postcentral cortex, or internal capsule reduce or completely abolish the long-latency reflexes (Tatton et al., 1975; Marsden et al., 1977a,b; Lee and Tatton, 1978).

Although the long-loop concept was favored by many workers in this field it was recognized that several alternative mechanisms could account for the delay between the onset of displacement and the appearance of late reflex activity in the EMG response (Tatton et al., 1978; Bawa and Tatton, 1979). For example, the M2 and M3 components might be mediated by slowly conducting polysynaptic pathways within the spinal cord (Hultborn and Wigstrom, 1980). They could also occur as a result of oscillations or repetitive firing of motoneurons (MNs) in response to reflex inputs. Finally, it is possible that continuous afferent activity occurring throughout the course of a prolonged displacement results in an EMG response with multiple components. Several investigators (Ghez and Shinoda, 1978; Tracey et al., 1980; Miller and Brooks, 1981) have shown that long-latency reflexes can still be obtained in cats and monkeys following transections of the brainstem or spinal cord. Whether these re-

---

*To whom correspondence should be addressed: Department of Clinical Neurosciences, University of Calgary, Faculty of Medicine, Health Sciences Center, 3330 Hospital Drive, N.W., Calgary, Alberta, Canada T2N 4N1.

sponses in lesioned animals are homologous to the M2-3 component in man and in intact monkeys trained to oppose displacements is still in question. The late component in the cat, which Ghez and Shinoda identified as M2, occurred with a variable latency of 25 to 40 msec, suggesting that it represents something different from the M2 response occurring with a latency of 30 msec in the monkey forearm and 55 msec in humans. Nevertheless, these observations suggest that there are mechanisms at the spinal level that are capable of generating long-latency responses to mechanical perturbations. The long-loop hypothesis has also been challenged by observations on the pattern of afferent discharge from muscle spindles during perturbations of an extremity in both humans and in monkeys (Hagbarth et al., 1980, 1981; Tracey et al., 1980). These investigators have shown that afferent input occurs throughout the displacement and in some cases the spindle discharges show a series of peaks with a configuration similar to the individual components of the EMG response.

It is obvious that our present understanding of the mechanisms underlying long-latency reflexes is incomplete, and there are still a number of unanswered questions. This chapter reviews some recent studies from our laboratories on the nature and timing of the afferent signals which generate the long-latency responses. Also discussed is the possible role that these late reflexes play in the control of normal movement. Finally, data obtained from human subjects with lesions affecting various components of the motor system is reviewed.

MECHANISMS UNDERLYING LONG-LATENCY REFLEXES

Although much attention has been focused on the issue of whether long-loop mechanisms are responsible for the late components of the stretch reflex, there are several other important questions: What type of afferent input generates the M2-3 component? Are the receptors and afferent pathways the same as those which mediate the M1 component? Do the same MNs generate the early and late components of the stretch reflex, or is there evidence for functional specialization within the MN pool? What effect do changes in the amplitude, velocity, or duration of displacement have on the long-latency component? Mechanical perturbations produced by a torque motor activate not only muscle-stretch receptors and Golgi tendon organs, but also cutaneous and joint receptors. The contribution of cutaneous and joint inputs to the long-latency EMG responses is relatively small, since blocking these inputs with local anesthetic does not consistently change the M2-3 component *(unpublished observation)*. Activation of spindle primary endings and Ia afferents probably accounts for most of the M1 component. However, it remains uncertain whether these same inputs are responsible for the long-latency components of the reflex.

Selective Effects of Vibration on Early and Late Components of the Stretch Reflex

Vibration is a potent inhibitor of monosynaptic reflexes and effectively suppresses tendon jerks and the H-reflex (cf. Lance et al., 1973; Desmedt, *this volume*). If the mechanisms that generate the late components of the stretch reflex are the same as those for the early component, then vibration should suppress both the M1 and M2-3 components of the EMG response to mechanical perturbations. To test this hypothesis, Hendrie and Lee (1978) studied the effects of vibration on EMG responses to angular displacements of the wrist in a group of 20 normal subjects. A torque motor was used to generate randomly timed displacements which stretched the wrist flexors. To accentuate the long-latency reflex activity, the subjects were instructed to oppose the displacements. During some

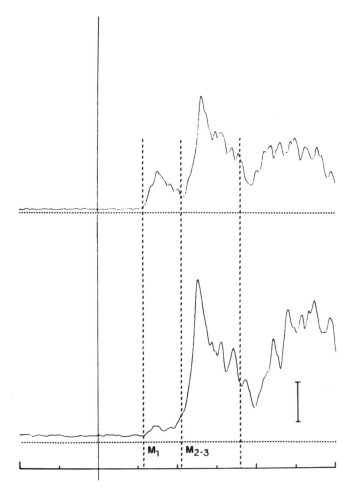

FIG. 1. Averaged rectified EMG responses from the wrist flexors of a normal human subject following angular displacement of the wrist joint in an extensor direction. (*Lower trace*, vibration; *upper trace*, control.) The subject was instructed to oppose the displacement. Solid vertical line indicates the time at which the torque motor was turned on to initiate the displacement. *Dotted lines* represent placement of cursors used with the computer display to integrate EMG activity over the M1 and M2-3 intervals. During vibration of the forearm muscles, there was marked suppression of M1 but no decrease in size of M2-3. Timing marks on baseline represent 25 msec. *Voltage calibration* = 100 μV.

runs a 100-Hz commercial vibrator was applied over the forearm muscles. Averaged EMG responses for 1 subject under control conditions and during vibration are show in Fig. 1. Although the M1 component is markedly suppressed during vibration, there is no reduction in the size of the M2-3 component. In fact, in this example, it is slightly accentuated during vibration. Similar results were obtained in all 20 normal subjects who participated in this study, and the results are summarized in Fig. 2.

The EMG activity was integrated over the M1 interval and over the M2-3 interval, and ratios were calculated to compare the amount of activity during vibration with that which was present under control conditions. The relative lack of effect of vibration on the long-latency components is illustrated by the fact that the ratios for the

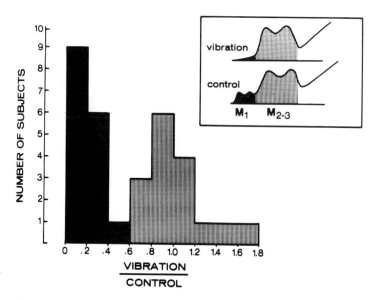

FIG. 2. Ratios of EMG during vibration to control EMG for 20 normal subjects. Ratios were calculated after integrating EMG activity over the M1 interval *(dark hatching)* and the M2-3 interval *(light hatching—see inset).* The suppression of M1 during vibration is indicated by ratios with values, all less than 0.6. For M2-3, the ratios are clustered around 1.0, indicating that vibration has no effect on this component. (From Hendrie and Lee, 1978, with permission.)

M2-3 component are clustered around a value of 1.0. It has recently been shown that vibration produces similar selective suppression of the early component of reflexes in the lower extremity elicited by torque perturbations at the ankle (Agarwal and Gottlieb, 1980).

Interpretation of these results is dependent on the mechanism by which vibration suppresses monosynaptic reflexes. Although a simple peripheral occlusion of Ia afferents has been proposed (Hagbarth, 1973) there is evidence that other factors such as presynaptic inhibition of Ia afferent terminals are involved (Gillies et al., 1969; see Desmedt, *this volume*). However, regardless of the exact mechanisms involved, these observations that M2-3 remains unchanged, while M1 is suppressed by vibration, suggest that the long-latency components are mediated by afferent pathways or central connections that are different from those which generate the monosynaptic reflex. Additional evidence to support this concept is provided in a paper by Stanley (1978), who investigated the early and late EMG responses from the intrinsic hand muscles following electrical stimulation of peripheral nerves. Conduction velocities were calculated for afferent pathways responsible for the V1 and V2 components of the EMG response, which have latencies comparable to M1 and M2. The V2 component was mediated by slower conducting afferent pathways than the V1. On the basis of this evidence, Stanley suggested that group II afferents might contribute to the V2 response.

Single Motor Unit Responses to Mechanical Perturbations

Further evidence that the mechanisms responsible for the early and late components of the stretch reflex are different is provided in a study by Bawa and Tatton (1979), who examined the discharge patterns of single motor units (SMUs) in the forearm muscles of monkeys following angular displacements at the wrist joint. Average response histograms were constructed and were correlated with the averaged EMG response recorded with surface electrodes. The majority of SMUs showed excitatory peaks in the histograms, which coincided in time with only one of the peaks in the averaged EMG response. Figure 3 shows an example of a motor unit from one of these monkeys, which fires in association with the M2 component of the surface EMG response. Of the single motor units examined in this study, 60% showed excitatory responses corresponding to a single peak in the averaged EMG response; 28% responded over the time intervals corresponding to two peaks in the EMG response. Thus, there is definite evidence in this study for functional specialization within the MN pool that participates in the EMG responses. These results do not argue in favor of either a spinal or supraspinal mechanism for the long-latency components of the stretch reflex, but they make it seem unlikely that the late components can be accounted for entirely by either prolonged discharges in a single population of afferent fibers or repetitive firing or oscillation of the same group of MNs.

Dependence of Long-Latency Reflexes on Duration of Displacement

The concept of long-loop reflexes has been based to a large extent on latency measurements. Onset of the M2-3 component occurs at a time that is close to what would be predicted for a response occur-

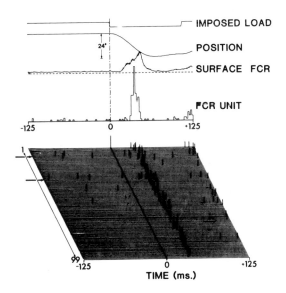

FIG. 3 Single motor unit responses to an imposed displacement of the monkey wrist. *Upper traces* show average records for the imposed step load [0.18 Newton-meters (Nm)], wrist position (downward deflections stretch the flexors), the rectified surface EMG recorded over flexor carpi radialis (FCR) and a response histogram (bin width 2.0 msec) for a SMU simultaneously recorded in FCR. The averages were constructed for 99 randomly imposed step loads. The response plane *below the average traces* shows the timing of the SMU responses on each of the 99 trials used to construct the histogram. Note that the SMU response peak is limited to the M2 interval of the average surface EMG. Also note that the SMU does show inconstant spikes in the response plane over the M1 interval in trials where the background firing levels are increased (as indicated by the *arrows* at the left side of the plane), suggesting that the SMU may be receiving weaker, subthreshold excitatory inputs during the M1 interval as discussed by Bawa and Tatton (1979).

ring as a result of afferent activity traveling over fast-conducting pathways to sensorimotor cortex and from there back to spinal cord and muscle. However, this argument requires an assumption that the effective stimulus occurs at, or very soon after, the onset of the displacement. The mechanical properties of the large torque motors used in human studies are such that displacements or stretch continue for 100 to 150 msec. These perturbations cannot be considered as discrete events comparable to electrical stimuli. Microelectrode recordings from peripheral nerves reveal that various stretch receptors are activated throughout the entire time course of a displacement (Burke et al., 1978a, 1980a; Hagbarth et al., 1981; Tracey et al., 1980; see Prochazka and Hulliger, *this volume*).

An important question is, what happens to the long-latency reflex if only very brief perturbations are applied to the extremity? This was investigated in a recent study on normal human subjects, using an adjustable mechanical stop to suddenly arrest movement of the handle at various times following the onset of the displacement (Lee and Tatton, 1982). With this technique it was possible to compare EMG responses to a variety of displacements of varying duration, but with identical velocity and other characteristics during the initial part of the displacement (Fig. 4). With very brief displacements, shown at the bottom of the figure, there is a well-developed M1 component but essentially no activity over the M2-3 interval. As the duration of displacement is increased beyond 40 msec, the long-latency components begin to appear and increase rapidly, reaching maximum values with displacements which continue beyond 50 to 60 msec. Increasing the duration of the displacement also results in an increase in amplitude of displacement, and it is possible that amplitude rather than duration is the important factor in determining whether or not EMG activity develops over the M2-3 interval. Figure 5 shows the results of a further study that was done to investigate this possibility. In this example, the amplitude of displacements was kept constant but the velocity, and, thus, the duration of displacement was varied. The EMG responses shown on the right of Fig. 5 reveal that brief, high-velocity displacements are accompanied by a well-defined M1 component, but no long-latency activity. Again, as the duration of displacement is progressively increased, long-latency activity begins to appear. The EMG response shown at the bottom of Fig. 5 reveals that a well-developed M2-3 component is present even when the velocity of displacement has been reduced to a level at which there is almost no M1 component. This could be interpreted as further evidence to suggest that M1 and M2-3 may be mediated by separate populations of afferent fibers.

To determine more precisely the relationship between duration of displacement and the long-latency EMG response, EMG activity was integrated over the M2-3 interval (55–100 msec) and plotted as a function of duration of displacement. This was done for 12 normal subjects; results from 4 of these are illustrated in Fig. 6. It is apparent that there is a critical minimum duration of displacement independent of velocity, which is required to generate the long-latency component of the stretch reflex. This value can be determined by measuring the point at which the curves in Fig. 6 begin to rise from baseline values. The mean value for this critical duration in the 12 subjects studied was 43.8 msec.

On preliminary inspection, these results could be interpreted as support for the concept that the long-latency components occur as a result of continued activation of afferent fibers over the course of the displacement. It could thus be argued that the M2-3 component is neither a long-loop response nor a long-latency response. If the effective stimulus occurs at some point after the onset of the displacement, then the latency may, in fact, be no longer than what

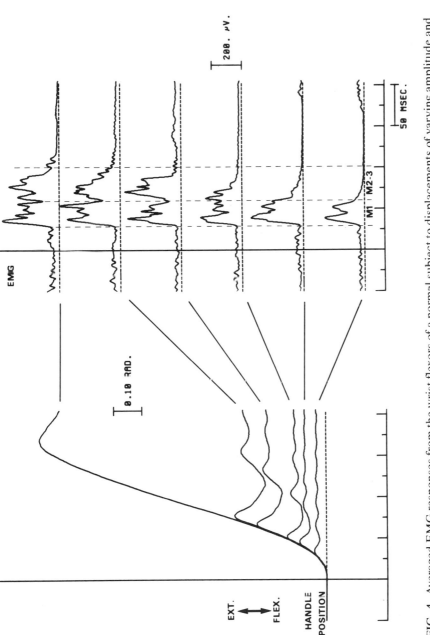

FIG. 4. Averaged EMG responses from the wrist flexors of a normal subject to displacements of varying amplitude and duration. An adjustable mechanical stop was used to suddenly arrest handle movement. Superimposed position traces on the left show that the mechanical stimulus was identical in each case up to the point at which the handle struck the stop. Note that the M2-3 component is absent with very brief displacements and appears only when the displacement is continued beyond a critical minimum duration. (From Lee and Tatton, 1981, with permission.)

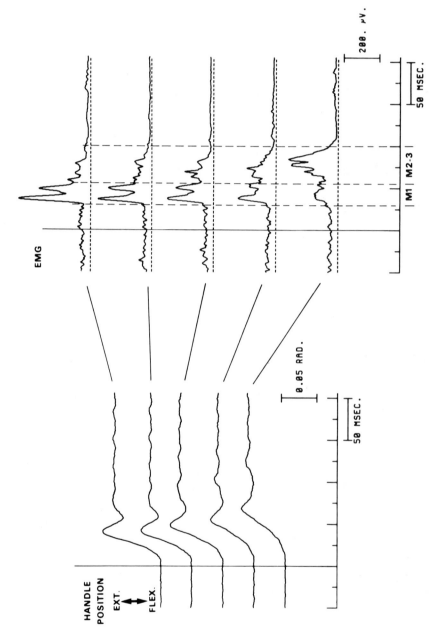

FIG. 5. Averaged EMG responses from the wrist flexors of a normal subject to a series of wrist displacements of constant amplitude but varying velocity and duration. Brief high-velocity displacements produce very little EMG response over the M2-3 interval (*top EMG trace*). Lower velocity displacements do produce an M2-3 response if they are continued beyond a critical minimum duration (*bottom EMG trace*). (From Lee and Tatton, 1981, with permission.)

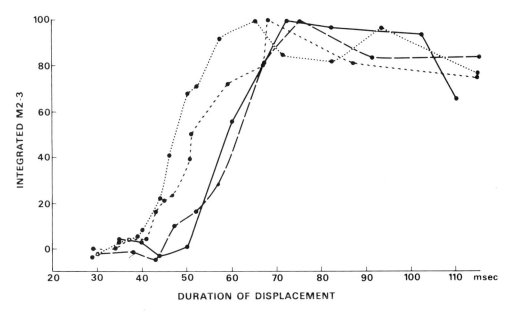

FIG. 6. Relationship between size of the M2-3 component and duration of displacement for 4 normal subjects. EMG activity was integrated over the 55- to 100-msec interval following onset of displacement. Values have been normalized with respect to the largest M2-3 response for each subject. Note that there is no EMG activity over the M2-3 interval until the duration of displacement is increased beyond 40 to 50 msec.

could be accounted for by spinal mechanisms. However, closer inspection of the data in Fig. 6 suggests that the explanation for the absence of an M2-3 component with brief displacements is not so simple. In the human forearm, the long-latency components begin approximately 55 msec after the onset of the displacement. If these components were entirely due to spinal reflexes, the afferent activity which gives rise to them should have been initiated at least 25 msec earlier (the shortest time for a spinal reflex in these muscles). If slower conducting afferent pathways were involved, the afferent activity should have been initiated even earlier. In this situation, sudden arrest of handle movement at any point beyond 30 msec after the onset of the displacement should not affect the long-latency components. However, as indicated above, the critical minimum duration of displacement is beyond 40 msec. The difference in time between the onset of M2-3 and this critical duration is not equivalent to the latency of a spinal reflex. In fact, it is equal to about half the spinal reflex latency, or the time required for afferent impulses to travel from the periphery to the spinal cord. This suggests that segmental afferent input interacts with other excitatory inputs on MNs to produce long-latency reflex activity.

Convergent Facilitatory Effects on MNs

These results suggest that it is not realistic to attribute all long-latency reflex activity to a single mechanism at either the peripheral or central level. The mechanical stimuli used to generate these responses activate numerous afferent pathways in addition to muscle stretch receptors. Some of these form direct synaptic connections with MNs at the spinal level. However, the sensory information associated with muscle stretching is also transmitted rapidly to the sensorimotor cortex, cerebellum, and other supraspinal centers. Much of the work on long-latency reflexes has been based on averaged surface EMG recordings. These include discharges of multiple motor units responding to a variety of excitatory inputs.

An alternative, and perhaps more accurate approach is to look on the long-latency reflex activity as the response of MNs to multiple excitatory inputs, some segmental and some suprasegmental. This concept of convergent facilitation is illustrated in Fig. 7. Three major excitatory influences on MNs are shown: (1) segmental inputs exciting MNs either directly or through one or more spinal interneurons; (2) excitation occurring as a result of feedback over long-loop supraspinal pathways; and (3) excitation from supraspinal centers associated with voluntary muscle activation. Segmental excitation could occur repetitively due to asynchronous activation of peripheral receptors during stimulus application. In most circumstances excitation from all 3 of these sources is required to generate long-latency reflex activity. The exact sites and mechanisms of interaction between these excitatory inputs are not yet fully understood. However, elimination of any one of them will reduce or abolish the late component of the stretch reflex. For example, the M2-3 component is either absent or very small during passive stretch, when there is no volitional intent to activate the muscle. Despite the reports of retained long-latency reflex responses in cat or monkeys following spinal cord or brainstem transection (Ghez and Shinoda, 1978; Tracey et al., 1980; Miller and Brooks, 1981), human studies indicate that lesions interrupting the proposed long-loop pathways on either the afferent or efferent side eliminate the late component of the reflex (see Chan, *this volume*).

Finally, as described above, removal of ongoing segmental excitation by sudden arrest of the displacement eliminates the long-latency reflex, even when the other mechanisms for excitation of MNs are functioning. It is relevant at this point to note that ongoing excitation at the segmental level is not always necessary to generate long-latency reflexes. Under certain conditions, electrical stimulation of peripheral nerves gives rise to long-latency EMG responses, which may be analogous to those obtained following angular displacements (Upton et

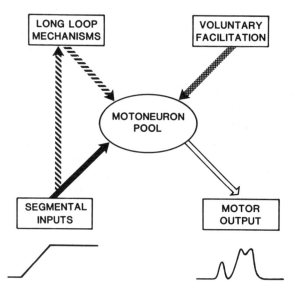

FIG. 7. Schematic diagram illustrating the various convergent excitatory inputs to the MN pool, which may be involved in the generation of long-latency reflexes.

al., 1971; Milner-Brown et al., 1975b; Conrad and Aschoff, 1977). Since these studies all involved stimulation of mixed peripheral nerves, it is possible that the late EMG response results from reafferent input associated with the muscle twitch. These long-latency responses to electrical stimulation are also dependent on MNs receiving simultaneous excitation from other sources, in this case voluntary muscle activation.

FUNCTIONAL SIGNIFICANCE OF LONG-LATENCY REFLEXES

Since the original observations by Hammond (1954, 1956) that long-latency stretch reflexes vary according to the volitional set of the subject, there has been considerable interest in the role that these late responses may play in controlling voluntary movement. Their exact function has remained an enigma, but a number of hypotheses have been proposed. One of the more popular has been the servo hypothesis, in which the late components of the reflex are regarded as the output of an error-feedback control system which restores the position of the limb following a perturbation (Hammond et al., 1956; Marsden et al., 1976a). Although there is an anatomical substrate for such an error-feedback system, physiological confirmation is still lacking. In recent experiments which attempted to assess the efficacy of this servo action directly, the long-latency components were consistently found to be inadequate to produce effective control of limb position (Allum, 1975; Bizzi et al., 1978; Kwan et al., 1979; Gottlieb and Agarwal, 1980). Further observations by Crago et al. (1976) of steady-state overcorrection, occasional inappropriate motor responses, and variable response latency all argue, on theoretical grounds, against a simple servo hypothesis. Crago et al. (1976) concluded that the long-latency components of the reflex represented a triggered response.

Allum (1975) has proposed a modified hypothesis known as a servo-information hypothesis. It states that the later components of the reflex response act as a pulse test signal to inform the central nervous system (CNS) of the current loading on the muscle; based on this information, CNS selects a motor response appropriate to the loading situation. For the long-latency component to act as an effective test signal, there must be sufficient delay between this test signal and later voluntary motor responses. This appears to be the case in certain experimental conditions (Kwan et al., 1979); however, more definitive observations are needed to lend support to this hypothesis.

Recent experiments point to a third hypothesis concerning the functional role of the long-latency stretch reflex. This is based on the concept of limb stiffness (Nichols and Houk, 1976). Stiffness was computed from the torque and joint position signals for a group of normal subjects responding to angular displacements of the elbow joint (Kwan et al., 1979). When the subjects were required to oppose the displacements, the long-latency EMG response was followed by a constant level of stiffness, which was maintained until the effective restorative movement occurred (Fig. 8). When no voluntary opposition was required, the late component of the EMG response was absent and the level of stiffness fell rapidly to preperturbation values. These observations suggest that the late components of the stretch reflex may be involved in controlling limb stiffness to produce a set of known initial conditions on which ensuing voluntary responses can be based. The basis for such control may be an internal model, within the CNS, of external conditions such as limb position, load, and moment of inertia.

In another set of experiments, reflex responsiveness was assessed at different phases of a cyclical movement. Repeated small, randomly timed torque pulses were applied to the arms of normal subjects while they carried out alternating flexion-exten-

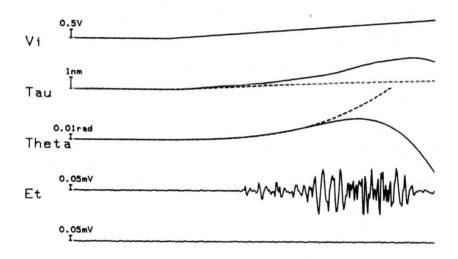
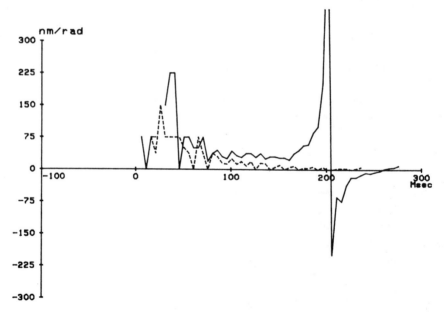

FIG. 8. Control of limb stiffness by long-latency reflex. *Top traces* show limb response to a ramp perturbation (Vi) in the flexion direction, with the resulting torque (Tau), and angular displacement (Theta). The upper triceps surface EMG, E_t was recorded in the oppose mode and the lower in the non-oppose mode. The *bottom graph* shows the computed stiffness of the limb system in Newton-meters per radian (Nm/rad) with respect to time after perturbation was presented at the time 0. *Solid lines*: Oppose mode records; *broken lines*: Non-oppose mode records. The long-latency component occurs at a latency of 75 msec, followed by a constant level of stiffness until the onset of the effective restorative movement at a latency of about 165 msec.

sion movements at the elbow joint. It was observed that there was an orderly modulation of reflex gain, which could be correlated with the different phases of the movement (Fig. 9). The timing of the gain modulation was consistent with the interpretation that reflex gain is appropriately adjusted for the generation of reflex tension to help counteract increases in dynamic load. Again, this interpretation assumes an inter-

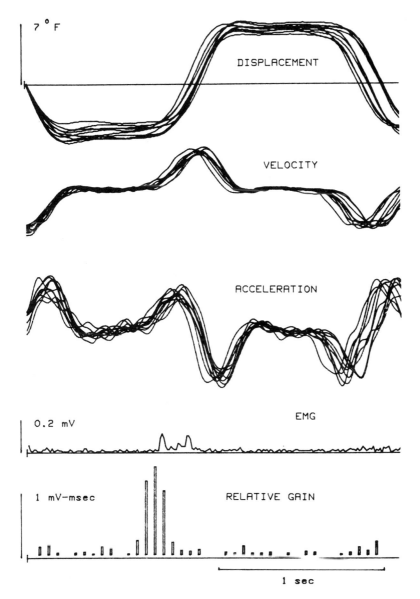

FIG. 9. Modulation of reflex gain during movement. *Top traces* show superimposed displacements of 10 movement cycles, followed by their respective velocities and accelerations, and the averaged integrated EMG record derived from surface electrodes above brachialis muscle. The *bottom histogram* shows the relative gain of the late reflex component (latency 64 msec) as determined by the variation in the responses to constant, test torque pulses applied at each 50-msec interval of the cyclic movement period. f: flexion.

nal model of external dynamic loading conditions. There are two common denominators in the interpretations of the above experiments. First, in both situations, the existence of an internal model of the external limb conditions and loading is required as a basis for predictive or feedforward control processes. Second, limb stiffness is controlled in both instances, although in different ways and for different uses in the two cases: presetting of initial conditions on the one hand (Kwan et al., 1979), and counteracting dynamic load increases on the other.

Three generalizations emerge from this brief survey of the functional implications of long-latency reflexes. First, the same reflex mechanism may have different uses in different movement contexts, and there is no *a priori* reason for these uses to be mutually exclusive. Second, in conjunction with an obligatory internal model, there is evidence that the long-latency reflex serves in feedforward control processes in the CNS. Finally, behavioral confirmation of the long-discussed servo control or error-feedback model is still lacking.

CLINICAL APPLICATIONS

Parkinson's Disease

If long-latency reflexes are involved in regulation of muscle tone (i.e., resistance to passive stretch), changes in the late components of the stretch reflex should occur in patients with clinical disorders resulting in increased muscle tone. Initial observations of patients with parkinsonian rigidity confirmed this prediction (Lee and Tatton, 1975, 1978; Tatton and Lee, 1975). It was shown that there was marked enlargement of the M2-3 component of the EMG response to displacements at the wrist joint. Furthermore, parkinsonian patients have a reduced ability to modulate long-latency reflex activity according to volitional intent. The M2-3 component is almost as large during passive stretch as when the subject is actively opposing the imposed displacement. Similar changes in long-latency reflexes in parkinsonian patients have been reported by others (Mortimer and Webster, 1978, *this volume;* Chan et al., 1979a). These observations led us to suggest that defective gain control in long-latency reflex pathways might be one of the factors responsible for increased muscle tone in Parkinson's disease (Lee and Tatton, 1975). It was proposed that under normal conditions, there is a gating mechanism which regulates feedback from muscle receptors to sensorimotor cortex.

Disruption of this mechanism could create a situation in which reflex gain was constantly set at high levels so that even small stimuli would generate a near-maximal reflex response. Figure 10 illustrates the relative sensitivity or gain of the M2-3 activity to varying velocities of imposed displacement in rigid parkinsonians as compared to normal subjects. Even for lower levels of background EMG activity, parkinsonians show markedly increased responses over the M2-3 interval for a given input velocity (compare the 87°/sec trace for the normal, background 7.9% of maximum to the 86°/sec trace for the parkinsonian, background 2.7% of maximum). Tatton and Bedingham (1981) have shown that graphs of integrated reflex EMG measured as a percentage of maximum voluntary EMG levels plotted against the logarithm of the velocity of the initial displacement offer a reproducible index of reflex gain. The logarithmic plots show that the parkinsonian M2-3 activity is highly sensitive to low input velocities as compared to the normal population, whereas the M1 velocity sensitivity is within the range of normals for a given level of background EMG activity.

Another abnormal feature that has been noted in the EMG responses of many parkinsonian patients is a prominent burst of activity during an interval that overlaps that of the M2-3 component in muscles which

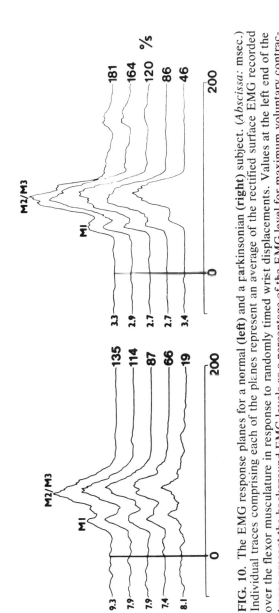

FIG. 10. The EMG response planes for a normal (**left**) and a parkinsonian (**right**) subject. (*Abscissa*: msec.) Individual traces comprising each of the planes represent an average of the rectified surface EMG recorded over the flexor musculature in response to randomly timed wrist displacements. Values at the left end of the traces present the background EMG levels as a percentage of the EMG level for maximum voluntary contractions. Values at the right of each trace present the average velocity over the initial 30 msec of the displacement evoking the responses. (Courtesy of William Bedingham.)

are shortened by the imposed displacement. This paradoxical shortening reaction, sometimes referred to as the Westphal phenomenon, has also been described in parkinsonian patients by Mortimer and Webster (1978). It may suggest disruption of mechanisms which normally act to produce reciprocal inhibition of agonist-antagonist muscle groups, although it can be seen in normal subjects or experimental animals (see Fig. 11). In some parkinsonian patients, the accentuated M2-3 component in the EMG response is followed by a series of oscillations occurring at frequencies in the 5- to 10-Hz range. We initially suggested that underdamped feedback from muscle receptors was capable of either resetting an existing parkinsonian tremor or initiating a new tremor at a more rapid rate. More recent studies show that resting 5-Hz parkinsonian tremor is relatively resistant to resetting by mechanical perturbations (Lee and Stein, 1981).

Several of these abnormalities noted in parkinsonian patients have been reproduced in monkeys rendered parkinsonian by the chronic administration of a phenothiazine (Tatton et al., 1979). As illustrated in Fig. 12, these animals also show marked accentuation of the long-latency EMG response in the muscle being stretched and a prominent paradoxical burst of EMG activity in the antagonist muscle being shortened. In addition, phenothiazine intoxication results in an abnormal response pattern from motor cortical neurons (MCNs) in these monkeys (Fig. 11). In intoxicated animals, all responding MCNs showed excitatory responses following imposed wrist displacements, whereas in normal monkeys a proportion of MCNs are inhibited by mechanical stimuli of this type. In addition, the time course of the excitatory peak in the averaged response histograms is prolonged in comparison to normal MCNs for a given duration, velocity, and magnitude of imposed displacement. Autocorrelograms of MCN activity frequently show oscillations occurring at the same frequency as parkinsonian tremor, whereas the firing in normals is always aperiodic when the displacements are randomly presented. At present, it is not possible to determine whether these abnormal motor cortical responses contribute to the generation of parkinsonianlike reflex activity or merely reflect altered reafference in the rigid animals.

Hemiplegia

Previous studies have shown a selective loss or reduction of long-latency reflexes in patients with focal lesions involving postcentral cortex or internal capsule (Lee and Tatton, 1978; Marsden et al., 1978b). However, it is uncertain what role, if any, long-latency reflexes play in the generation of spasticity associated with hemiplegia. In a recent study (Lee et al., 1979), EMG responses to wrist displacement were recorded from 21 patients with hemiplegia resulting from infarcts or intracerebral hemorrhages involving one cerebral hemisphere. This was not a homogeneous group of patients. Some had large lesions affecting cortex and subcortical white matter, whereas others had small, deeply situated lesions close to the internal capsule. Some patients had profound hemiplegia at the time of the examination; others had only mild weakness. In 10 patients spasticity in the hemiplegic arm was graded as moderate to severe; in the remainder, muscle tone was only slightly increased, or in some cases normal or even mildly hypotonic. As expected, there was considerable variation in the EMG responses, but it was possible to identify 3 distinct patterns (Fig. 13). All the hemiplegic patients showed accentuation of the M1 or short-latency component of the response. Seven patients showed long-latency components similar to the example in

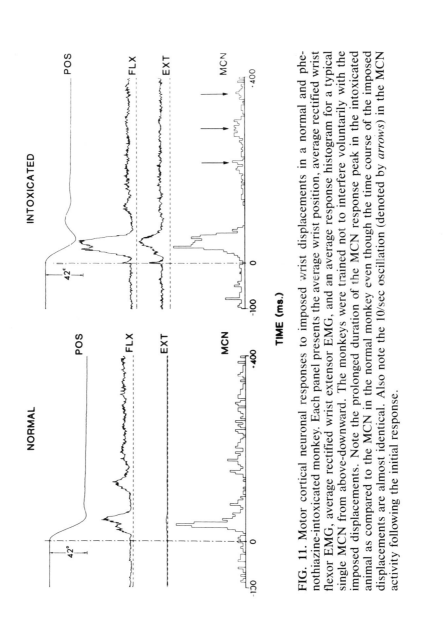

FIG. 11. Motor cortical neuronal responses to imposed wrist displacements in a normal and phenothiazine-intoxicated monkey. Each panel presents the average wrist position, average rectified wrist flexor EMG, average rectified wrist extensor EMG, and an average response histogram for a typical single MCN from above-downward. The monkeys were trained not to interfere voluntarily with the imposed displacements. Note the prolonged duration of the MCN response peak in the intoxicated animal as compared to the MCN in the normal monkey even though the time course of the imposed displacements are almost identical. Also note the 10/sec oscillation (denoted by *arrows*) in the MCN activity following the initial response.

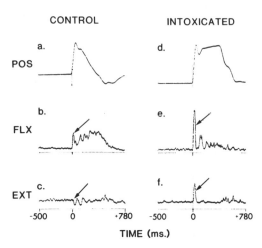

FIG. 12. Reflex and voluntary EMG responses to imposed wrist displacements in a monkey prior to and during chronic phenothiazine intoxication. The monkey was trained to compensate for the imposed displaced. **a** and **d**: Average wrist position, displacements stretching the flexor muscles are upward. **b** and **e**: Average rectified EMG for the wrist flexors. **c** and **f**: Average rectified EMG for the wrist extensors. *Arrows* indicate the reflex components of the EMG responses. Note the animal showed decreased EMG activity in the shortened muscle coincident with the excitatory reflex responses on the stretched muscle (many normal animals show a weak excitatory component coincidently with the M2-3 interval. see records in Fig. 11). In the intoxicated state, the records show a slowed return to the central wrist position, markedly increased M2-3 activity in the stretched muscle, the Westphal phenomenon, and oscillatory activity in the stretched muscle following the reflex responses in an identical manner to that found in rigid parkinsonians.

Fig. 13A. The M2-3 component was markedly enlarged, and its duration was prolonged in comparison to normal responses. A second pattern occurred in 4 of the 21 patients (Fig. 13B). In this group, there was no increase in reflex EMG activity following M1, and in some subjects the late component of the reflex was actually reduced. The third pattern (Fig. 13C) was present in 10 of the 21 subjects. Again, there was a very large M1 component, but during the 60- to 100-msec interval following the stimulus, when the long-latency components normally occur, the level of EMG activity was only slightly greater than that seen during the prestimulus control period. However, beginning at a latency of approximately 100 msec, there was a second large burst of activity, the initial part of which appeared very similar to the M1 component. This late activity occurred at the time where we normally see voluntary EMG responses if the subject is actively opposing the displacement. However, these recordings were all obtained during passive displacement, and in several cases the patients had clinically complete paralysis of the affected arm and were unable to make a voluntary response. We questioned whether this might be the initial burst of a repetitive discharge, possibly something analogous to clonus. However, analysis of responses over a longer time period than what is shown in Fig. 13 did not reveal any subsequent bursts of EMG activity. Furthermore, the interval between the two EMG bursts in these cases is approximately 50 msec. This would mean repetitive activity occurring at a rate of 20/sec, which is much faster than the usual frequency for clonus.

Sequential recordings were carried out in several patients who showed this peculiar double burst of EMG activity (Fig. 14). Recordings were obtained on days 10, 25, and 36 following a large left hemisphere infarct. Over this time, there was progressive shortening of the latency of the late second component. At the time of the initial recording, the patient had total paralysis and hypotonia of the right arm. By day 36 when the third recording was obtained, weak voluntary power had returned to the affected arm. Muscle tone had returned to normal, but clinically there was no evidence of increased tone despite the presence of hyperactive tendon jerks. Analysis of the clinical

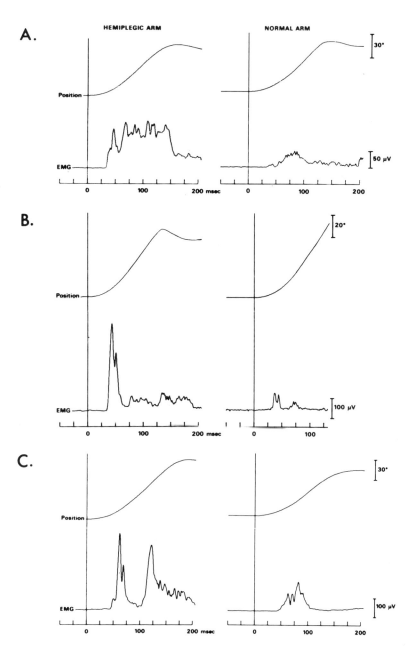

FIG. 13. The EMG responses in wrist displacements in hemiplegic patients. Handle position is shown at the *top* of each pair of traces with upward deflection representing extension at the wrist joint. The EMG activity was recorded from the wrist flexors. A, B, and C are recordings from three different patients and illustrate the 3 patterns of long-latency response that were found among 21 patients examined. The short-latency component (M1) was accentuated in the hemiplegic arm in all cases.

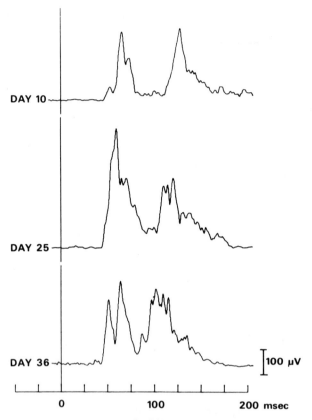

FIG. 14. Serial recordings of EMG responses to wrist displacements in a hemiplegic patient. Recordings were obtained 10, 25 and 36 days after an acute cerebral infarction. Note progressive shortening of latency of the delayed second component of the response.

and electrophysiologic data from all 21 patients did not reveal any consistent correlation between the presence or absence of clinical spasticity and the pattern of the late EMG responses. Therefore, in contrast to parkinsonism, it seems unlikely that long-latency reflexes play a major role in the pathogenesis of increased muscle tone following hemiplegia.

ACKNOWLEDGMENT

This research was supported by grants from the Medical Research Council of Canada.

Motor Control Mechanisms in Health and Disease,
edited by J. E. Desmedt.
Raven Press, New York © 1983.

Long-Latency Automatic Responses to Muscle Stretch in Man: Origin and Function

*C. D. Marsden, J. C. Rothwell, and B. L. Day

University Department of Neurology, Institute of Psychiatry and King's College Hospital Medical School, London, United Kingdom

HISTORY

Hammond (1956, 1960) is credited with suggesting that responses to muscle stretch might use pathways extending to the brain, but his experiments are widely misquoted. Hammond observed that when the human elbow was suddenly and forcibly extended, there occurred not only a small burst of electrical activity in biceps at around the spinal monosynaptic latency of the tendon jerk but also later events after about twice that delay. The subject had been instructed to resist the pull when it occurred, but if he was told to relax as soon as he felt the tug, these later events disappeared (Fig. 1). Hammond considered these responses to be automatic, for their latency was well below that of a normal voluntary reaction to a tap delivered to the string on which the forearm was pulling. Hammond was forced to conclude that stretch of human biceps elicited long-latency responses, which he believed to occur too fast to have been a simple voluntary response to the tug on the arm, but which were clearly voluntarily controlled. These considerations led him to suggest, that they might represent the impact of input to the brain, where voluntary effort could determine their occurrence.

As it turned out, Hammond's interpretation was incorrect, although the hypothesis that stemmed from it, namely the possible existence of long-loop stretch reflex mechanisms, may be valid. It transpires that voluntary reaction times to the known occurrence of a proprioceptive stimulus are remarkably short, very much shorter than had been appreciated on the basis of classical visual or cutaneous sensory reaction times. It is now known that what Hammond observed was a voluntary reaction to the stimulus (see Rothwell et al., 1980). However, Hammond's original records do contain true long-latency automatic responses, which were not abolished by voluntary intent. These events, which prefaced the subsequent voluntary burst of activity, were very inconspicuous, but, in retrospect, are quite evident (Fig. 1B).

Little was made of possible long-loop cerebral responses to muscle stretch until Phillips (1969) picked up the theme again in his Ferrier Lecture. Attracted by the dense cortico-motoneuron (MN) projection to the primate hand, and the recent discovery of a direct projection of spindle afferents to sensorimotor cortex, Phillips speculated that there might exist a transcortical stretch reflex mechanism. He suggested that evolution of the hand might have gradually led to the transplantation of certain reflex mechanisms normally resident in the spinal cord into the brain, where they would be under

---

*To whom correspondence should be addressed: Department of Neurology, Maudsley Institute of Psychiatry, DeCrespigny Park, Denmark Hill, London SE5 8AF, United Kingdom.

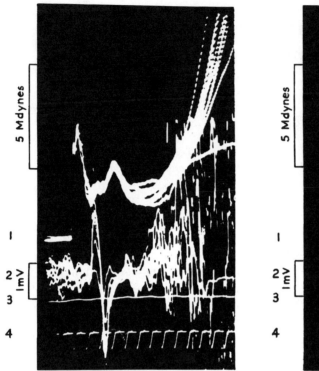
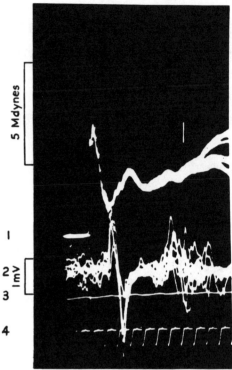

FIG. 1. Long-latency reflexes in biceps. The subject pulled on a wristlet using his elbow flexors. Without warning the wristlet was pulled away at a constant velocity. The tension in the tape (1) with its baseline (3), and the surface EMG from over biceps (2) were recorded. **Left:** A series of single trials in which the subject was instructed to resist the impending pull. **Right:** A series of single trials when instructed to let go. The time pulses (4) are at 10-msec intervals. Even when the subject attempts to let go on perceiving the stimulus, there is a burst of long-latency activity at 50 to 60 msec recorded from biceps. (From Hammond et al., 1956, with permission.)

greater and more direct control of the senses and of the will. At that stage, however, there was not direct evidence for the existence of any such mechanism.

A year earlier (1968), Marsden and colleagues had begun certain experiments to investigate the properties of the human stretch reflex in muscular control. In particular, we had set out to examine whether or not it could provide servo assistance to movement in the face of external perturbations, and if so, how powerful such assistance could be. We chose to investigate the top joint of the thumb, for this has the great advantage of being operated by only a pair of muscles, both of which lie in the forearm. By simple surface electrode recording it is possible to monitor the activity of flexor pollicis longus (FPL); needle electrodes are required for uncontaminated recording from the long extensor of the top joint of the thumb. Our first experiments involved introducing, at random, small perturbations during the course of visually controlled flexion movements of the top joint of the thumb. One of our earliest discoveries was that the latency of the muscles' responses to such perturbations was greatly in excess of that of the monosynaptic tendon jerk in FPL (Marsden et al., 1972, 1973). This clue led us to seek evidence for a possible long-loop transcortical stretch reflex pathway, along the lines proposed by Phillips (1969).

At about the same time, Evarts published

his pioneering papers on the firing characteristics of pyramidal tract neurons (PTNs) in the motor cortex in awake monkeys undertaking manual acts. Evarts (1973) showed that PTNs could respond very rapidly to a peripheral disturbance so as to alter their firing characteristics within a time interval appropriate to a fast transcortical reflex pathway. Our own calculations, based on the conduction of afferent information from the hand to the cortex, as indicated by somatosensory evoked potentials (SEPs), and from the cortex back to the forearm muscles, as taken from the results of direct cortical stimulation in man, led to the same conclusion (Marsden et al., 1973, 1976). There certainly appeared time for afferent information from the stretched thumb and its muscles to reach motor cortex, and to activate a descending cortico-MN volley such as to produce the long-latency automatic responses we had recorded in the FPL.

We set about providing other evidence to support this conclusion by examining the response to stretch of other muscles in the body at different distances from the brain and spinal cord. Comparison of the long-latency stretch reflex in jaw muscles, flexor pollicis longus, and flexor hallucis longus showed that their latency, compared to that of the monosynaptic spinal tendon jerk in the same muscles, was appropriate for such a transcortical stretch reflex mechanism (Marsden et al., 1973, 1976). Lee and Tatton (1975) independently came to a similar conclusion. They had studied the effect of sudden displacement of the wrist while the subject was instructed to hold a steady position, and had found three bursts of muscle activity in response to such a perturbation. They named these responses M1, M2, and M3, attributing M1 to spinal monosynaptic latency events, and M2 (and M3) to long-latency automatic responses, which they too thought might use the transcortical stretch reflex mechanism. Both our group, and Lee and Tatton, turned to human pathological material for further support of the transcortical long-loop hypothesis. Marsden et al. (1973) had noted that these long-latency automatic responses to stretch disappeared after a lesion of the dorsal columns in the high cervical region, and might be enhanced in certain types of myoclonus characterized by greatly exaggerated cortical SEPs. The effect of posterior column lesions was confirmed by Tatton et al. (1975) and by Marsden et al. (1977a) in a much larger series of patients. Both groups went on to show that lesions of the sensorimotor cortex itself could abolish long-latency automatic responses to muscle stretch (Marsden et al., 1977b, 1978a; Tatton et al., 1975; Lee and Tatton, 1978).

By the mid-1970s, the concept of long-latency automatic responses to muscle stretch in man utilizing long-loop transcortical stretch reflex pathways had become widely accepted; but there were those who still had their doubts (cf. Desmedt, 1978a). It was true that all the evidence that had been garnered in support of the concept was, necessarily, indirect. Then there were those who argued that these events were not reflex at all, because, according to Hammond's misinterpreted original experiment, they were under voluntary control. More recently, evidence has been provided that in certain circumstances response to muscle stretch consists of repetitive bursts of spindle firing. This has led to the suggestion that long-latency stretch reflexes are no more than a spinal response to a second burst of spindle activity (Hagbarth et al., 1981). Indeed, Lundberg (1975) has consistently upheld the view that there exists sufficiently complex machinery in the spinal cord to account for the long delay in those responses to thumb displacement (see Hultborn and Wigstrom, 1980).

Not only has the significance of long-latency automatic stretch responses been hotly debated, but their function has provoked spirited and challenging differences of opinion. There are those who hold that these events do provide servo assistance to movement, whereas others vigorously deny this. At the present moment, both the origin and the function of long-latency responses to muscle stretch are controversial.

This chapter reviews the status of long-latency stretch reflexes under the following headings: (1) Are such responses reflex? (2) Are such responses merely spinal events? (3) What is their function, if any? We will conclude that the long-latency responses to muscle stretch that occur in certain human muscles are truly automatic (reflex) responses, which cannot be explained on the basis of known spinal circuitry, and which do, in certain circumstances, exert powerful and important collaborative servo assistance to movement in man. We will argue that one of the main confusions surrounding this subject has been the unwarranted generalization of conclusions drawn from the study of one muscle group to apply to the whole body. We believe that the transcortical stretch reflex mechanism is a system evolved to its greatest extent for the control of the human hand but which is much less evident, if not rudimentary, in most other parts of the human body.

LONG-LATENCY STRETCH REFLEXES

Long-Latency Stretch Reflexes of Flexor Pollicis Longus

We had always held that the automatic responses to stretch of the human long thumb flexor occurring at twice tendon jerk latency were not significantly influenced by the subject's intent or set. Figure 2A illus-

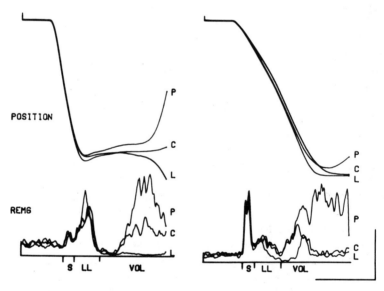

FIG. 2. Average positon (top) and rectified EMG responses (bottom) recorded from flexor pollicis longus *(left traces)* and biceps brachii *(right traces)* in the same individual. Sixteen stretches were given randomly to the muscle every 5 to 6.5 sec, interspersed with control trials with no stretch stimulus (not shown). In separate series of trials, the subject was instructed to maintain a constant effort (C), to pull/press hard on perceiving the stimulus (P), or to let go (L). All three sets of trials are superimposed and the spinal latency tendon jerk (S), the long-latency stretch reflex (LL), and the voluntary components (VOL) are labeled for each muscle on the EMG traces. In the FPL, the spinal component begins about 22 msec after the stretch and the long latency reflex after some 40 msec. The corresponding times for biceps are 17 msec and 38 msec. *Horizontal calibration:* 100 msec; *vertical calibration:* 3° or 150 μV.

trates the electromyographic (EMG) responses of FPL, recorded by surface electrodes over the muscle, in response to passive extension of the interphalangeal joint of the thumb. The subject held the thumb flexed 10° against a steady torque of 2 Newtons which suddenly increased to 6 Newtons, 50 msec after the start of the recording sweep. In one series of runs the subject was instructed to compensate for the disturbance and to restore the thumb to its original position; in a second series, he was instructed to flex his thumb as hard and as fast as possible as soon as he perceived the random stretch; in a third series of experiments, he was asked to relax his thumb completely the moment he felt the stretch.

Figure 2A indicates that stretch of the human long thumb flexor provokes a series of EMG responses. At tendon jerk latency (about 22 msec in this muscle), there is a small response. In fact, monosynaptic responses to stretches of this velocity can only be recorded in about 50% of normal human beings, and then are small (Marsden et al, 1978). Next, at about double tendon jerk latency, there occurs what we would term the automatic long-latency stretch reflex. This is more or less uninfluenced by voluntary intent (the slightly larger response is actually the control). There follows, at about 90 msec, in records in which the subject was asked carefully to reposition the thumb, a clear burst of activity, which could be vastly augmented or abolished at will. This latter response, however, was not a conscious effort by the subject to flex the thumb on perceiving the stimulus; it occurs without the subject having to think, yet obviously is easily modifiable at will. This activity blends into later events produced by the conscious action of the subject.

Thus, we see in Fig. 2A a series of EMG responses to stretch of the human muscle which lie on the continuum of most automatic to least automatic responses as suggested by Hughlings Jackson (1931). The spinal monosynaptic response (M1 of Lee and Tatton) would be Hughlings Jackson's most automatic event. The long-latency reflexes (M2 of Lee and Tatton) would qualify as automatic events. The first burst of voluntary activity, which occurs without thought but is subject to the will, would qualify as a less automatic response according to Hughlings Jackson. Whether or not this is the M3 as described by Lee and Tatton is open to question. It is not clear whether their M3 was as easily subject to alteration by intent as our event clearly is. We have never seen a response lying between the long-latency stretch reflex and our subsequent less automatic responses in FPL. Finally, in trials in which the subject was asked to pull as hard as possible, it is clear that the less automatic events grade into muscle activity produced by the subject consciously reacting to the stimulus, which would amount to least automatic events along the continuum discussed by Hughlings Jackson.

The spinal monosynaptic response undoubtedly is a reflex, but there are those who have claimed that the subsequent automatic response is not due to the erroneous belief that such automatic events are under voluntary control. However, as can be seen (Fig. 2A), the long-latency response of the long thumb flexor is more or less uninfluenced by the subject's intent, which produces an obvious effect only after about 110 msec. In any case, proponents of the view that these events were reflex could point to the common sneeze, which also can be suppressed at will on occasions. Another criticism has been that their size is not always constant; i.e., a constant input does not always generate the same output. This is undoubtedly true, and much will be made of it later, but the reasons for such variability deserve consideration.

In our earlier studies, one of the striking features of long-latency automatic stretch responses was that they were subject to the intensity of background contraction. The greater the force with which the subject was contracting flexor pollicis longus, the greater the size of the long-latency automatic responses to a given amplitude of displacement of the thumb. Taking the relationship of size of response proportional to force of contraction (termed gain control of the response) to its logical conclusion, we were able to show that during active relaxation the responses might not occur at all; i.e., at zero force there was zero gain (Marsden et al., 1976a). Such behavior certainly seemed very different from that of the monosynaptic tendon jerk. However, reappraisal shows that the tendon jerk itself is subject to gain control depending on the extent of background force of contraction, its size increasing the harder the muscle works (Fig. 3), so that this criterion of variability cannot be used decisively to separate reflex from other events. Our own position is that such an argument cannot be resolved to everybody's satisfaction. The events in question are more rather than less automatic along Hughlings Jackson's continuum, so we will choose to refer to them as reflexes.

Long-Latency Stretch Reflexes in Other Muscles

The automatic nature of long-latency stretch reflexes in the human thumb flexor seems incontrovertible. How then, did others come to entirely the opposite conclusion? One of the main reasons was that they studied other, usually more proximal, muscles, the behavior of which is not identical to that of the long thumb flexor. In Fig. 2B we show responses of human biceps to stretch in circumstances similar to those in Fig. 2A. The subject sat with his upper arm resting before him on a table, at shoulder height. The forearm was held vertical with the elbow at 90°, and a short chain placed around the semipronated wrist was attached to the spindle of torque motor acting to flex the elbow. The force in the motor

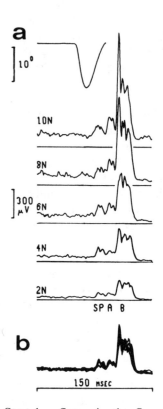

FIG. 3. Stretch reflexes in the flexor pollicis longus (FPL) of a single subject elicited over a range of background forces from 2 to 10 Newtons. **a:** *upper trace,* the first 100 msec of a typical record of the angular position of the thumb during a stretch; the same stretch was produced at all force levels. The 5 center traces show the rectified EMG with baseline obtained from FPL at each force level. The first 150 msec of the response are shown, with the stretch coming after 50 msec. **b:** The EMG responses have been scaled over the first 50 msec of the sweep so that their baselines become aligned. The superimposition of the resulting traces shows that each component of the stretch reflex was directly proportional to the background activity. Each trace is the average of 32 trials. The EMG traces are labeled to show the spinal, tendon jerk, component of the response (SP), and the long-latency reflex (A and B). The A and B components that make up the long-latency response of FPL have been described previously (Marsden et al., 1978). (From Adam and Marsden, 1978, unpublished observation.)

was increased from 5 to 15 Newtons, 50 msec after the start of recordings.

The main difference in the responses from biceps, compared with those from the long thumb flexor, is the presence in the former of a large spinal-latency component, which is absent or very small in the long thumb flexor. Although the peak velocity of stretch, in terms of angular rotation of the joint, is much less than that for the thumb, the difference in the distance between the point of muscle insertion and the center of rotation of the joint for, on the one hand biceps, and on the other hand FPL, makes the rate of linear muscle stretch comparable. It appears, threfore, that the spinal component of the biceps muscle's EMG response is much more sensitive to stretch than that of the long thumb flexor. This observation correlates with clinical experience of the distribution of tendon jerks. It is difficult to obtain a monosynaptic response to a tap with a tendon hammer to the pad of the thumb, but it is easy to obtain a biceps tendon jerk. In general, the tendon jerk response in biceps is stronger than that in triceps and, likewise, that in finger flexors is much more powerful than that in finger extensors.

Indeed, a general rule for the muscles of the arm would seem to be that the larger the direct corticospinal projections to a muscle, the smaller is its spinal response to muscle stretch, and the more disabling the effect of interruption of these descending pathways after, for example, a stroke. Thus, following a capsular hemiplegia, the predominant weakness is typically in the extensors of the elbows and fingers rather than in the flexors. Despite the pronounced early monosynaptic tendon jerk response in biceps, longer latency automatic events can be seen occurring later. Their timing corresponds, after due allowance for peripheral conduction delay, to that of similar long-latency automatic reflexes in the thumb flexor.

Most other workers who have studied the influence of voluntary intent on these long-latency automatic responses have concentrated on muscles around the elbow. Those who have found that voluntary set does affect these events have also usually used an experimental paradigm aimed at producing the fastest possible voluntary reaction time. Thus, the intertrial intervals have been short and fairly predictable, and, under these conditions, the voluntary reaction of the subject can be extremely fast so as to blend into the automatic stretch reflex responses. If the reaction time to the muscle stretch is lengthened by increasing the variability of the stimulus timing and the average interstimulus interval, then the effect on the size of the long-latency reflexes is proportionately smaller even for proximal muscles (Rothwell et al., 1980). This finding is illustrated in Fig. 4 for both biceps and flexor pollicis longus muscles. Stretches were applied in separate experiments to both muscles under two paradigms: (1) with the stretches every 3 to 3.15 sec with 50% of randomly interspersed control trials having no stretch stimulus; and (2) regular presentation of the stretch every 3 to 3.15 sec with no control trials. In both conditions the subject performed two sets of trials in which he was instructed either to pull (P) or to let go (L) on perceiving the stimulus. The effect of this prior instruction clearly is greater in (1) than in (2), since addition of control trials in (1) increased the subject's reaction time. In other words, if the intent of the subject to respond to the muscle stretch is separated from the time taken to react to the stimulus, then the effect on the size of the long-latency stretch reflex can be seen to depend on reaction time, rather than the intention of the subject to respond. Because of this, we have suggested that the effect of voluntary reaction time on the size of the long-latency stretch reflexes is caused simply by interaction between the reflex response and the first, very short latency, impulse of the voluntary motor command.

In conclusion, we believe that review of

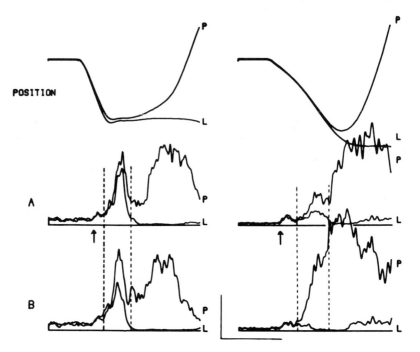

FIG. 4. Influence of voluntary intention on long-latency stretch reflex in flexor pollicis longus (**left**) and biceps brachii (**right**). *Upper records:* limb position; *lower records:* rectified surface EMG. Stretches applied in **A** and **B** as described in text. Each trace is the average of 16 trials, and runs in which the subject was instructed to pull (P) or to let go (L) have been superimposed. *Vertical arrow* indicates the start of the spinal component of the EMG response in FPL (22 msec) and biceps (17 msec). *Vertical dotted lines* approximately enclose the long-latency stretch reflex. Position traces are appropriate to condition B only. *Horizontal calibration:* 100 msec; *vertical calibration:* 150 μV or 5°.

the available evidence indicates that once true allowance is made for fast voluntary reaction time, long-latency automatic responses to muscle stretch are truly automatic, and are not easily influenced by the subject's intent or set. Therefore, they justify being called long-latency stretch reflexes.

ARE LONG-LATENCY STRETCH REFLEXES TRANSCORTICAL?

Evidence for and against transcortical reflexes is summarized in Tables 1 and 2, respectively. We will not review all the evidence provided in favor of the concept, and the reader is referred to the original articles for such data. We will, however, discuss the arguments raised against the notion.

A Spinal Mechanism for Long-Latency Responses to Stretch?

From his extensive knowledge of spinal cord circuitry, Lundberg (1975) has long believed that the extra delay of long-latency stretch reflexes can be explained without suggesting brain involvement. Lundberg's view apparently received support from the discovery by Ghez and Shinoda (1978) that similar long-latency events on stretch of forelimb muscles could be recorded after spinal transection in the cat. This observation was subsequently confirmed in the monkey (Tracey et al., 1980b). However, the anesthetized, spinalized cat or monkey is unable to make a voluntary contraction.

In contrast, long-latency automatic stretch reflexes can *only* be obtained regularly in man during voluntary contraction of

TABLE 1. *Evidence to support the concept that long-latency stretch reflexes are due to transcortical stretch reflex mechanisms*

Authors	Facts
Hammond et al., 1956; Marsden, Merton, and Morton, 1972 Lee and Tatton, 1975	Long-latency responses to muscle stretch occur at approximately twice tendon jerk latency.
Marsden, Merton, and Morton, 1976b	Differential timings of long-latency and tendon jerk responses in jaw, thumb, and big toe favor long-loop explanation.
Marsden, Merton, and Morton, 1972; 1976a	Long-latency responses can be seen without preceding tendon jerk component in thumb muscles, and in all muscles studied in a subject with no clinical tendon jerks.
Hendrie and Lee, 1978	Selective effect of vibration, which may abolish tendon jerk component, leaving long-latency response unchanged.
Evarts, 1973; Conrad et al., 1974, 1975; Evarts and Tanji, 1976	Motor cortex neurons fire at a latency midway between muscle stretch and long-latency stretch reflex. Many such neurons show opposite changes to active passive movements.
Dawson, 1946; Hallett, Chadwick, and Marsden, 1979	Greatly enlarged and cortical-evoked potentials may occur over somatosensory area midway between muscle stretch and large, long-latency myoclonic jerks.
Lucier, Ruegg, and Weisendanger, 1975	Demonstration of direct muscle spindle afferent projection to motor cortex.
Marsden et al., 1973, 1977a,b; Tatton et al., 1975	Lesions of dorsal columns, somatosensory, or motor cortex, and internal capsule abolish long-latency responses, leaving tendon jerks intact.

the appropriate muscle. When relaxed, such responses disappear (Marsden et al., 1976). Furthermore, lesions of the sensorimotor cortex or cortico-MN pathway, such as occur in a stroke, abolish these long-latency events (Marsden et al., 1977b), whereas in the monkey, cooling or ablation of the arm area of the sensory and caudal motor cortices do not (Miller and Brooks, 1981). Thus, the long-latency responses seen in animals under these conditions are not necessarily analogous to those seen in the intact human upper limb. However, recent experiments by Berardelli et al. (1981) indicate that the late EMG responses observed in the human leg may be more directly comparable. These authors were able to show that long-latency responses to ankle stretch in triceps surae appeared more likely to be due to spinal mechanisms than to involvement of any hypothetical transcortical pathway.

The conclusion is that the leg may be different from the arm, and that is exactly what Marsden et al (1976b) found in their original experiments. It took little or no training to produce obvious and typical long-latency stretch reflexes in flexor pollicis longus, but when we turned to undertake similar experiments in flexor hallucis longus, we immediately encountered difficulties. In fact, it took many hours of rigor-

TABLE 2. *Evidence against transcortical concept of long-latency stretch reflexes*

Authors	Facts
Ghez and Shinoda, 1978; Miller and Brooks, 1981; Tracey et al., 1980b	Demonstration of long-latency reflexes in spinal cats and monkeys, and also in monkeys with cortical lesions.
Hagbarth et al., 1981	Muscle stretch produces multiple bursts of spindle activity corresponding to the multiple EMG bursts. No long latency EMG responses seen without a second burst of spindle activity.

ous training of the long flexor of the big toe before appropriate long-latency events became apparent (Fig. 5). We believe, along with Berardelli et al. (1981) that what most people have studied in the leg is not the same as the long-latency stretch reflex in the human hand.

Stretch Reflexes in the Leg

The responses that we eventually recorded in the flexor hallucis longus appeared at a latency of about 75 to 90 msec. However, the most regular and familiar response found in calf muscles after sudden disturbance of the ankle joint is that described by Melvill Jones and Watt (1971), somewhat unfortunately, as the functional stretch reflex. This response only occurs when subjects actively oppose the disturbance and has a rather variable latency of ≥ 120 msec. In these respects, it resembles more the voluntary portion of the response which follows the stretch reflex in the upper limb. In fact, the voluntary response time to an Achilles tendon tap is often of a similar latency (Freedman et al., 1976). Like other voluntary responses, the functional stretch reflex is delayed in patients with Parkinson's disease and Chan et al. (1979) now suggest that this response should be regarded as a preprogrammed reaction released by the stimulus, rather than an automatic reflex response in the terms discussed above.

Nashner (1976) also has described a functional stretch reflex in the legs of standing subjects. Stretch of triceps surae was accomplished either by direct rotation of the ankle joint or by anteroposterior sway of the subject. Again, the muscle responses occurred after about 120 msec and only in 50% of the subjects. They tended to destabilize the body if produced by forcible dorsiflexion of the foot, but stabilized the body if stretch was produced by forward body sway. When the method of stretch was changed, there was appropriate facilitation or inhibition of the response over 2 to 4 consecutive trials. However, Nashner emphasizes that these responses in erect subjects could not be influenced by effort of will; thus, they may be different again from those originally described by Melvill Jones and Watt (1971).

Other authors have described a much smaller and more variable response, occurring 70 to 90 msec after sudden ankle perturbations, lying between the tendon jerk component and the functional stretch reflex. (The PSR of Chan et al., 1979, in tibialis anterior; the labile component following the tendon jerk of Gottlieb and Agarwal, 1979, 1980). However, these long-latency responses are much more difficult to obtain than those in the upper limb and may be the same as the responses recorded following very rapid ankle displacements by Berardelli et al. (1981). The latter authors found that these responses showed little similarity to the long-latency reflexes in the hand, since they were abolished by vibration and they never occurred without a preceding tendon jerk component. The leg does, indeed, appear different from the arm.

We conclude that virtually all long-latency responses hitherto described in leg muscles after mechanical displacement of the lower limb are not analogous to the long-latency transcortical stretch reflexes demonstrated in the human long thumb flexor. Indeed, we would emphasize again that it took us 3 to 6 months of practice to produce such analogous responses in our long toe flexors. No doubt a toe painter would be more adept!

Are Long-Latency Stretch Reflexes Due to Segmentation of Spindle Volley?

Implicit in the concept of a transcortical stretch reflex pathway is the belief that a single spindle volley can evoke both spinal and later transcortical responses. Until recently, this concept had not been ques-

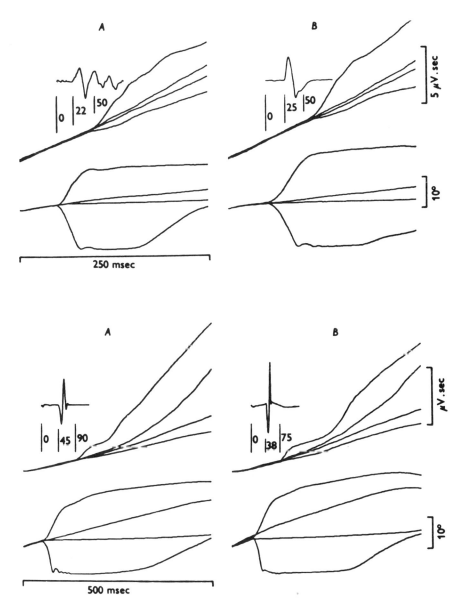

FIG. 5. "Tulip" responses recorded in two subjects (A and B) in the flexor pollicis longus *(top)* and flexor hallucis longus *(bottom)*. In each half of the figures traces are integrated EMG *(upper)* and limb position *(lower)*. Insert on the same time scale shows the latency of the tendon jerk in each muscle, produced by tapping the thumb pad or the Achilles tendon briskly with a tendon hammer. The subject tracked a moving spot on an oscilloscope screen before him by flexing his thumb or toe against the lever arm of a low-inertia torque motor. Control trials were interspersed at random with stretches (increase in motor torque, producing downward deflexion of the position record and upward deviation of the integrated EMG trace due to the stretch reflex), release (upward positional deviation, downward EMG deviation), and halts, in which the motion of the motor was halted by a servo-feedback device (halt in positional record, upward EMG deviation). The point of separation of the integrated EMG records at the start of the tulip indicates the latency of the stretch reflex response. Timing lines are labeled in milliseconds from the start of the disturbance. (From Marsden, Merton, and Morton, 1976*b*, with permission.)

tioned, but Hagbarth et al. (1981) have produced evidence from study of human wrist movements to show that stretch of a human muscle in certain circumstances can provoke not one, but a series of bursts of spindle discharge. The timing of these bursts, in the situation in which they have been recorded, is such as to be compatible with the notion that they give rise to repetitive EMG spinal monosynaptic volleys, the second of which corresponds to the long-latency stretch reflex.

Before jumping to the conclusion that these observations destroy the transcortical concept, it is worthwhile to carefully examine the circumstances of Hagbarth and colleagues (1981) experiments. The rates of stretch delivered were slow such that mechanical displacement was continuing up to, and even after, the onset of long-latency responses. Indeed, the authors relate the multiple bursts of spindle activity recorded in group Ia fibers to ripples in mechanical displacement. Even after rapid displacements lasting only about 10 to 15 msec, there were fairly large oscillations in the mechanical displacement records, oscillations which continued for about 20 to 30 msec and which in all probability were responsible for the segmented spindle volley. Adam and Marsden (*unpublished observation*) noted that multiple repetitive stretching of the long thumb flexor, produced by the stretching device bouncing off its mechanical backstop, can indeed produce extra-long-latency reflex EMG bursts. However, this extra activity was always superimposed on the long-latency events which followed a single stretch alone (Fig. 6).

Despite the apparent simple explanation for long-latency stretch reflexes proposed by Hagbarth et al. (1981), there is other evidence incompatible with this notion as far as the human hand is concerned. As indicated earlier, at least half of normal subjects have no spinal monosynaptic response in flexor pollicis longus. Events at longer latency are the first to appear in EMG rec-

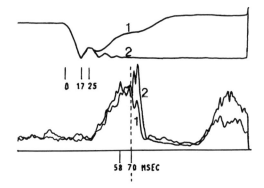

FIG. 6. Extra long-latency reflex EMG activity in flexor pollicis longus produced by a second muscle stretch caused by the motor bouncing off its backstop. Control (1) and double-stretch (2) conditions superimposed. Top: Motor position; bottom: surface-rectified EMG. Each trace is the average of 32 trials. The second stretch caused by the bounce begins 25 msec after the initial 10° stretch and produces an extra peak of EMG activity about 45 cm later, which is additional to the initial long-latency response. (From Adam and Marsden, *unpublished observations*.)

ords following stretch of that muscle. If the spindle response to such a stretch is segmented, as suggested by Hagbarth et al. (1981) the spinal cord can, therefore, only respond to the second spindle burst, which seems unlikely. Second, Hendrie and Lee (1978) have shown that vibration of stretched muscle, presumably by occluding Ia afferent discharge, abolishes the initial monosynaptic response to stretch, but not the later long-latency event. Third, such an explanation cannot account for the repeated observation, in both man and animals, that lesions confined to the sensory pathways in the posterior columns abolishes these long-latency events. Such lesions high in the cervical cord or brainstem (Marsden et al., 1977a) did not interfere in any way with spinal monosynaptic tendon jerks, yet abolished long-latency stretch reflexes (Fig. 7). Fourth, the rates of stretch employed to elicit long-latency stretch reflexes in the thumb were such that the imposed displacement could be complete well before the long-latency events began.

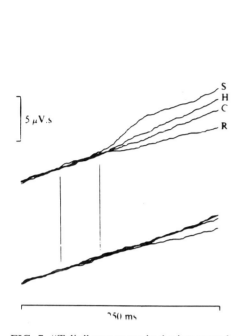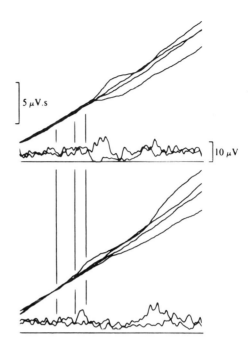

FIG. 7. "Tulip" responses in the integrated surface EMG from flexor pollicis longus in 2 patients with lesions of the sensory pathways. Experimental details as in Fig. 5; control trials (C) intermixed with stretch (S), release (R), and halts (H). Perturbations introduced 50 msec after the start of the sweep *(first timing bar)*. The patient on the left had a normal left hand *(upper traces)* and unilateral posterior column sensory loss in the right *(lower traces)*. The patient on the right exhibited spinothalamic sensory loss on the left *(upper traces)* and posterior column sensory loss on the right *(lower traces)*. In this patient, rectified superimposed EMG traces also are shown to stretch and release of the thumb. Timing bars are at the onset of the perturbations and 50 msec later in the left hand records, and 25 and 42 msec later in the right hand records. In both patients long-latency responses are present on the normal side 40 to 50 msec after the disturbance, yet are absent on the side with lesion of the posterior columns. In the patient on the right, a short-latency (25 msec) spinal response is present in the lower traces, with no sign of any later EMG events. (From Marsden et al., 1977*a*, with permission.)

Indeed, even extremely rapid pulse displacements of the thumb which are complete within 25 msec can produce long-latency reflexes (Fig. 8). Under these circumstances, it seems extremely unlikely that the spindles could produce repetitive afferent bursts, especially following the period of spindle unloading after the stretch. Finally, a direct test of the multiple spindle burst hypothesis can be made for the thumb using the H-reflexes recorded in flexor pollicis longus after stimulation of the median nerve at the elbow (Day et al., 1981). Even during voluntary activation of the long flexor, the period of inhibition in the H-reflex recovery curve following a half-maximal H-reflex conditioning shock, means that the monosynaptic reflex arc would find it difficult to respond to the second of two Ia afferent volleys spaced only about 25 to 35 msec apart (Fig. 9). In the case of the thumb, where the later, longer-latency responses are always much larger than any monosynaptic response, this period of inhibition seems particularly relevant.

What Hagbarth et al. (1981) have shown is that human muscle spindles, specifically group Ia afferent endings, are exquisitely sensitive to minor deviations in mechanical displacement. This means that late events

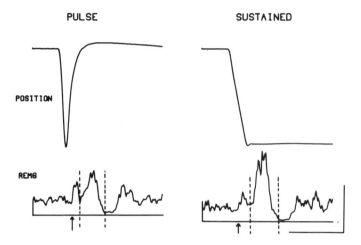

FIG. 8. Spinal and long-latency EMG responses from flexor pollicis longus in response to very rapid pulse stretches (**left**) or rather slower sustained stretches (**right**). Approximate duration of the long-latency EMG responses is indicated by the *vertical dotted lines*. Note that the earlier preceding spinal component *(vertical arrow)* is larger following the more rapid pulse stretch, whereas the long-latency response is larger after the sustained stretch. Traces are the average of 32 trials each in the same subject. *Horizontal calibration:* 100 msec; *vertical calibration:* 5° or 100 μV.

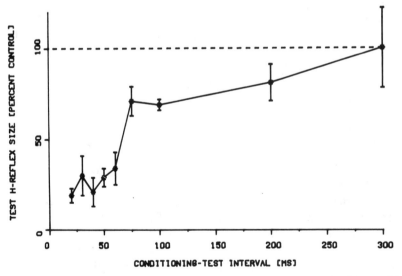

FIG. 9. Average (± 1 SE) H-reflex recovery curve in FPL of 4 normal subjects. Median nerve stimulated at the elbow to produce approximately half maximal H-reflex in the relaxed subject, with conditioning shocks followed at various intervals from 20 to 300 msec, by similar test H-reflex volleys. Size of the test reflex expressed as a percentage of the size of control H-reflexes alone. In each subject, 10 shocks at each interval were given, with 6-sec rest between each. It is difficult to limit H-reflexes, even partially, to the FPL, and these were the only 4 subjects with whom we succeeded. Nevertheless, recovery curves for the general body of superficial wrist flexors in the forearm are strikingly similar. Later phases of the recovery curve also may be apparent, lasting longer than the time scale illustrated here.

evoked by spindle discharge do not, of necessity, involve different pathways than earlier monosynaptic spinal reflexes. Latency by no means establishes the origin of such late events, as had been argued earlier. However, the findings of Hagbarth et al. do not disprove the existence of a transcortical stretch reflex mechanism, despite what some of the authors claim. The evidence accumulated to refute that suggestion has all been derived from the thumb or arm, but as indicated above, matters may be different as far as the leg is concerned. Here, if the conclusions of Berardelli et al. (1981) are to be accepted, long-latency muscle responses to ankle stretch may well represent fragmentation of spindle discharge. Of course, here lies one of the major differences between thumb and leg. The inertia of the ankle is so great that it is virtually impossible to produce mechanical displacement that does not continue throughout the period of generation of long latency events, which may, therefore, very easily represent the impact of mechanical ripples generating delayed responses.

Although we believe that the evidence detailed above refutes the arguments of Hagbarth et al. (1981) against long-latency transcortical reflexes in the forearm, we do not deny that events occurring in the monosynaptic arc may influence the size of reflexes traversing the long-loop mechanisms. Thus, it is clear in Fig. 8 that the size of the long-latency components is much larger following slower, prolonged stretch of the muscle than during the rapid pulse disturbance. Adam and Marsden (*unpublished observation*) suggested that maintained stretching provides continuous excitatory input to the α-MNs, which modulates the long-latency components produced by the initial muscle stretch. This concept also has been discussed by Vilis and Cooke (1980), and more recently by Lee and Tatton (1982). However, unlike the long-latency thumb responses, those seen in the wrist could not be produced unless the displacement lasted for longer than 40 to 50 msec.

An important, but unresolved, question is whether there really is any true analog in the leg of the long-latency stretch reflexes elicited in the human hand, which, on evidence provided above, we still believe to represent operation of a long-loop transcortical stretch reflex pathway. We suspect there may be, on the grounds of our experience in training the long flexor of the big toe. It is possible, although we have no evidence other than our own subjective experience, that the time it took to produce reliable responses in flexor hallucis longus may well have represented a period of training of such a transcortical mechanism. Certainly, they seemed not to be similar to the sort of events that can be averaged almost immediately from leg muscles in a subject simply asked to maintain a constant position. Without pressing the point too hard, we would speculate that the stretch reflex mechanisms of the feet in a thalidomide child with no hands might prove very different from those of a normal child, and may have remarkable similarity to those of the normal human hand.

Transcortical or Transcerebral?

Having mounted a defense for the concept that, with regard to the human hand, there may exist a long-loop transcortical stretch reflex pathway, is there evidence that this is a direct pathway to sensorimotor cortex and back? Studying the effects of various lesions in human patients certainly indicates that the sensorimotor cortex and cortico-MN pathway are likely to form the output pathway of such a long-loop system (Marsden et al., 1977*b*). The input, however, could pass directly to sensorimotor cortex through known projection of spindle afferents to that area, or indirectly via the cerebellum and nucleus interpositus (Allen and Tsukahara, 1978). It is unlikely that any reentrant system via basal ganglia is involved, for under normal circumstances, basal ganglia output neurons are not susceptible to proprioceptive stimuli (Iansek and Porter, 1980).

Evidence on timing favors the direct sensorimotor cortex route, for careful measurement gives little room to spare for any cerebellar diversion. Thus, the earliest volley of somatosensory information to reach the human cortex occurs at about 20 msec on stimulating the hand (Desmedt and Cheron, 1981), and the latency from cortical stimulation to forearm flexors is of the order of 18 msec. The onset of the long-latency automatic stretch reflexes can be about 40 msec, which leaves little time for anything other than cortical transfer. However, such latency arguments are not decisive. Turning to pathologic evidence, the picture of the effect of cerebellar lesions in man on long-latency stretch reflexes is not clear. Delay has been recorded in some patients, but most were diagnosed as having multiple sclerosis (Marsden et al., 1978), which might have introduced extra delay. Other patients with primary restricted cerebellar pathology had evidence of a reduction or distortion of long-latency stretch reflexes, but not of the opposite response to unloading, which evokes a silent period at the same long-latency. In animals, Tatton and Lee were unable to influence the earlier M2 response by cerebellar cooling, and thus concluded that this was likely to be a direct transcortical reflex. That could be the general conclusion, based on the evidence presently available, but involvement of the cerebellum in some way has not been excluded.

FUNCTIONS OF LONG-LATENCY STRETCH REFLEXES

No single unequivocal function for long-latency stretch reflexes has emerged. This fact is evidenced by the number of divergent proposals that have been put forward. For example, there are those who claim that its function is to linearize muscle properties (Houk, 1978; Murphy et al., 1979), damp out limb vibrations (Neilson and Neilson, 1978), help maintain a limb on an intended trajectory (Cooke, 1980), or to servo assist movements (Marsden et al., 1972, 1976a). Our own position has been that, from evidence available as far as the thumb is concerned, the long-latency stretch reflexes do play a significant part in automatic positional control of the digit. We use this term positional control in its broadest sense to encompass maintenance of a posture and accuracy of movement to a new position. We will now describe our own experiments dealing with: (1) the function of the long-latency stretch reflex in dealing with unexpected step changes in external force, in both holding and tracking tasks, and (2) the role of the long-latency stretch reflex when faced with unpredictable transient or dynamic force changes during movement.

Effect of Step Changes in Position on Long Thumb Flexor

We have analyzed the function of the long-latency stretch responses by studying the response to sudden extension of the thumb while the subject was asked either to maintain a constant position at about 10° of thumb flexion, or to execute controlled flexion of the top joint of the thumb through about 20° in 1 sec, with reference to a visual tracking task. In discussion of movement control, these two types of task are frequently classified as postural or dynamic, but we have never found the behavior of long-latency stretch reflexes in the thumb to differ between these two conditions. Accordingly, we will combine discussion of both types of task.

We introduced sudden static changes in motor torque such as to extend the thumb at random intervals during the task and have measured the effectiveness of the long-latency responses in compensating for this disturbance. For convenience, we have divided analysis of the results between the responses to large disturbances, that is, to those which are unequivocally perceived by the subject, and those to small disturbances, which lie around and even below the threshold for conscious perception.

Properties of the Long-Latency Responses Produced by Small Disturbances

Pushing these reactions toward the limit of their sensitivity, we find that changes in load of only ± 1% when the subject is tracking a moving target are rapidly and fully compensated (Marsden et al., 1979) (Fig. 10). Subjects were given a visual indication of thumb position by a line of an oscilloscope screen before them, and had to match their position with a second tracking line, which moved at a constant rate across the screen. The torque change lasted 200 msec, beginning 50 msec after the start of the recording sweep and the screen was blank throughout this period. Changes in EMG activity may be detected on integrated records some 50 msec after the change in load, indicating that the reflex mechanism is sensitive to changes in intended muscle length of only a few micrometers. Presumably, the high-sensitivity of muscle spindles to very small length changes (Matthews and Stein, 1969) is reflected here in the sensitivity of the servo

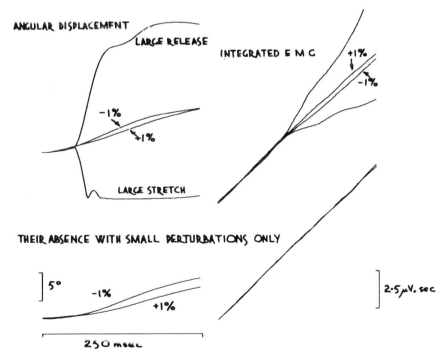

FIG. 10. Complete servo compensation following very small disturbances to thumb tracking movements. In control runs (not illustrated), the subject was tracked against a constant torque of 0.10 Nm, and in random trials this torque was changed 50 msec after the start of the recording sweep, by ± 1%. In the *upper* records, when larger disturbances also were included, complete positional correction is evident *(left traces)* by the end of the sweep, following the ± 1% torque changes. There are corresponding changes in the integrated EMG records **(right)** from surface electrodes over FPL. When small disturbances are given alone *(bottom traces),* there is no positional correction and no change in EMG. (From Marsden, Merton, and Morton, *unpublished observations;* see Marsden et al., 1979.)

mechanism. For disturbances of this size, servo compensation is indeed effective. Since perturbations of ± 1% are an order of magnitude below the threshold of conscious perception, the reflex nature of these responses are beyond doubt. However, such extreme sensitivity appears only when the small disturbances are intermixed with much larger changes in torque, a feature for which we have as yet no explanation. When large disturbances are not present, there is no EMG change in response to the ± 1% perturbations and no positional servo compensation (Fig. 10).

Properties of the Long-Latency Responses Produced by Large Disturbances

When larger, long-lasting increases in torque are applied to the thumb while the subject performs a task against an initially small load, long-latency stretch reflexes are evoked which initially are approximately proportional to the size of the disturbance. When the size of stretch is increased above a certain point, the response saturates, usually well before the disturbance has become maximal (Marsden et al., 1976a; see Fig. 2). This is also illustrated in Fig. 11, in which a subject was asked to make accurate tracking movements every 5 to 6 sec by flexing his thumb, which was disturbed at irregular intervals 50 msec after the start of the recording sweep, by sudden increases in the torque load to one of 7 new values. Subjects were given no specific instructions on how to respond to the torque change, and most preferred to continue tracking after vision was restored and the torque increment removed. The size of the EMG response to the stretch was measured from integrated records as the percentage increase over basal levels of activity calculated for the duration of the reflex burst.

As the size of the torque disturbance was increased, the distance the thumb was forced to move passively also increased in proportion up to the maximal torque increment of 0.25 Nm. However, the EMG traces show that the long-latency reflex began to saturate much earlier, with torque increments of only about 0.20 Nm. This saturation point occurred when the size of the EMG burst was still smaller than the maximum burst size seen when the subject was asked to make fast voluntary maximal thumb flexion (bottom trace). These results are shown graphically in Fig. 12, where means (± 1 SEM) from 6 normal subjects are depicted. The relationship between the size of the torque disturbance and maximal thumb extension was approximately linear (Fig. 12A), whereas that between torque and size of EMG response was saturated (Fig. 12B). In Fig. 13 the degree of positional correction produced by the long-latency stretch reflex has been measured from the thumb position records and is plotted against the size of the initial disturbance. The solid line denotes the line of identity, showing where points should lie if the stretch reflex produced complete positional compensation. The dotted line is the line of 50% effectiveness. It can be seen that for torque steps producing a maximal disturbance of about 20°, the long-latency stretch reflex provided approximately 50% correction for the initial displacement. Saturation occurred for larger disturbances.

Two main features are clear from these results. First, automatic long-latency responses to muscle stretch of the thumb flexor are unable to compensate for very large disturbances (over 30% of maximal thumb extension), since saturation of the reflex mechanism occurs. However, in the midrange of disturbances (from 5 to 20° thumb extension), such long-latency stretch reflexes may produce a moderately effective and quite sizeable contribution to control of thumb position. Second, these long-latency events do not behave like any classical positional servo mechanism, since they are pulsatile outputs from the nervous system which frequently are terminated despite large remaining positional errors (see Fig. 11).

Despite the fact that the reflex output sat-

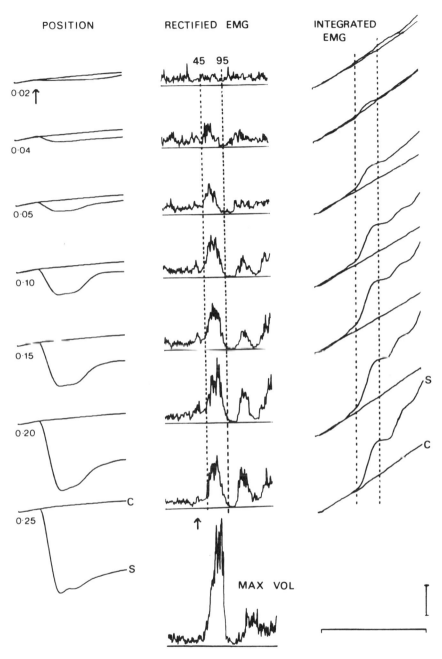

FIG. 11. Average responses of flexor pollicis longus in one subject to increasing sizes of stretch while tracking against a background torque of 0.1 Nm. Traces are left, thumb position with size of added torque given below; **middle**, rectified EMG with baseline; **right**, integrated EMG. Appropriate control trials (c) are superimposed on the responses to stretch (s) in the position and integrated EMG records. Lowest rectified EMG trace is the average muscle activity of the same subject during 16 maximum fast voluntary flexions of the thumb (max vol) made in his own time. Other traces are the average of 24 (3 × 8) trials. *Dotted vertical lines* in the middle row of records delineate the approximate duration of the reflex from 45 to 95 msec, and the *vertical arrow* indicates the onset of the small, spinal-latency component of the EMG response, which can be observed in some records. Stretch starts after 50 msec, as indicated by the *arrow* below the top record in the left row of records. *Calibration marker:* 10°, 50 µV, or 1.5 µV sec. (From Marsden et al., 1981, with permission.)

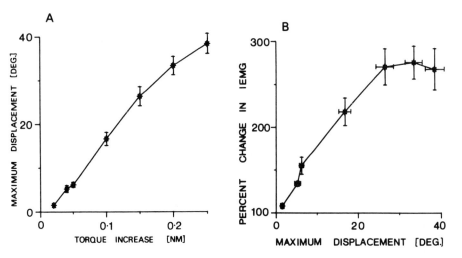

FIG. 12. A: Relationship between size of torque step (Nm) added to the background torque of 0.1 Nm, and the maximum displacement of the thumb (degrees). B: Relationship between maximum thumb displacement and the increase in EMG as a percentage of control levels of activity during the period of long-latency response between about 45 and 90 msec (depending on the characteristics of each individual) after the stretch, measured from integrated EMG records. (From Marsden et al., 1981, with permission.)

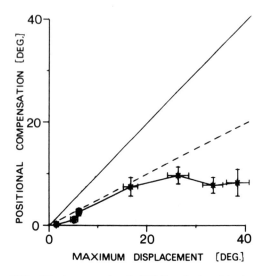

FIG. 13. Average (± 1 SEM) relationship between maximum thumb displacement and the amount of positional compensation achieved by the long-latency stretch reflex during thumb tracking tasks. *Solid line:* Line of identity which indicates complete positional compensation to the displacement produced by the different stretches. *Dotted line:* 50% positional compensation. (From Marsden et al., 1981, with permission.)

urates and is pulsatile, it cannot be dismissed as ineffective in the thumb as it has been in the muscles of the neck (Bizzi et al., 1976) or shoulder (Allum, 1975). We tentatively ascribe this difference to the specialized function of the muscles of the hand as opposed to the more proximal muscles of the trunk. We have already commented on the proposition that evolutionary specialization of hand function in primates and man may have led to parallel development of the sensitivity and power of long-latency transcortical stretch reflex mechanism. We would also point out that if shoulder muscles were to provide 50% reflex compensation against forces producing arm disturbances of 30° in a free-standing subject, equally large changes in posture would have to occur for balance to be maintained.

Reliability of Reflex Compensation

In the experiments described so far, the results of stretch reflex experiments have been averaged so as to clarify the usual

EMG and positonal responses. But the very variability in response from trial to trial, which makes averaging necessary, must be of crucial importance to the manner in which the central nervous system (CNS) controls movement. If such reflexes are unreliable, then one might even expect them to interfere with, rather than to assist, normal movement.

In fact, variability of the reflex response from trial to trial may be very considerable. A plot of 6 consecutive responses of the same subject to the same torque perturbation introduced at random intervals during a holding task is shown in Fig. 14A on a time scale encompassing later voluntary (VOL) phases of correction. The long-latency EMG response of FPL is difficult to measure in single trials, because its duration often is not clear. However, the positional

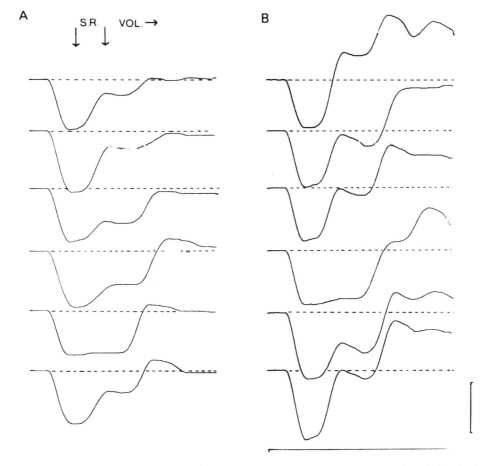

FIG. 14. Sequences of position records from the same subject after the same size of thumb disturbance (0.1 Nm rising to 0.16 Nm, 50 msec after the start of the records). A: Subject instructed to compensate for the disturbance; B: Subject instructed to produce a voluntary response of constant size irrespective of the preceding reflex correction. Visual feedback was blanked for the duration of the recording sweep. The positional correction achieved by the long-latency stretch reflex (SR) is highly variable in both sequences. A: Later voluntary response (VOL) produces a postional correction inversely related to the size of the preceding automatic response, and compensation is accurate (*dotted line*). B: Voluntary response produces an approximately constant positional change, despite variation in the size of the earlier correction. *Horizontal calibration:* 500 msec; *vertical calibration:* 8°. (From Marsden, Rothwell, and Traub, 1981, with permission.)

correction that it achieves in each run is quite evident, ranging in this subject from almost complete compensation (top) to virtual absence (next to bottom). We repeated these experiments in 10 different normal subjects, expressing the size of the individual long-latency responses as the percentage of positional correction that they achieved after the disturbance. The results are shown in the scatter diagram in Fig. 15, the heavy horizontal bars indicating the mean values for each subject. A surprising range of variation in the size of the response, after repeated presentations of the same disturbance poses considerable problems for any theory of stretch reflex function. For example, it would be useless to have a positional compensation device that corrected for disturbances by a different amount each time. The CNS would have to wait for one reaction time after each perturbation to see where the thumb had ended up before proceeding with the next movement. The theoretical advantage of fast automatic correction by reflex events would be lost, for no time would be gained over and above that involved in reacting to the initial disturbance alone. Alternatively, if such reflex were just a system of producing a linear springlike behavior in the muscle (Houk, 1978), the error in the CNS's estimation of the spring constant would be extremely high.

The problem with this type of approach to analysis of stretch reflex function is that the response and its functions are considered in isolation. In reality, the automatic stretch reflex is part of a complete sequence of muscular reactions to a disturbance. Many theories of stretch reflex function implicitly acknowledge this, whether it be to maintain preset muscle stiffness (Murphy et al., 1979), to linearize the mechanical behavior of the muscle (Nichols and Houk, 1976), or even to act as a pulse test signal to the limb (Allum, 1975). All these theories assign to the automatic stretch reflex some functional role in determining the size of the

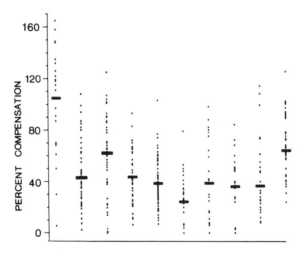

FIG. 15. Range of variation in the amount of positional compensation provided by the long-latency stretch reflex in individual trials in 10 subjects. Each point represents the percentage of positional correction provided by the automatic stretch response, and *heavy horizontal bars* indicate the mean value for each subject. The thumb was flexed to about 5° against a standing torque of 0.1 Nm, which, at random intervals (4.5 ± 0.75 sec), was increased to 0.16 Nm for 1 sec. Subjects were instructed to return the thumb to its original position as rapidly as possible. Visual feedback was blanked. (From Marsden et al., 1981, with permission.)

subsequent important muscular reaction to the disturbance. These later events themselves produce effective compensation.

Interaction Between Stretch Reflex and Subsequent Responses to Perturbations

We devised a simple experiment to investigate the interaction of the long-latency reflex response to extension of the thumb with the subsequent more automatic or voluntary reaction. Subjects were instructed to hold their thumb at a constant position, with reference to a visual indication on an oscilloscope screen before them, against a steady torque supplied by the motor. Every 5 to 6 sec at random, the torque suddenly was doubled for a period of 1 to 1.5 sec, and the subjects were instructed to compensate for this disturbance as rapidly as possible. This usually was achieved well within 500 msec during which time the screen display was blank, and data collected by a PDP 12 computer.

Three typical consecutive individual responses from the same subject is shown in Fig. 16. The thumb position was restored in two discrete phases, produced by two corresponding bursts of EMG activity. The first phase was the long-latency stretch reflex, which as described above, was extremely variable in size; the second phase was the voluntary response. The remarkable feature of these records is that despite the variation in size of the early reflex response, the subject restored his thumb to the starting position on every occasion within approximately one reaction time. In effect, thumb position was restored by a sequential series of different routes, rather as suggested originally for complex movements by Bernstein (1967).

This accurate compensation, despite the variable reflex response, was achieved by

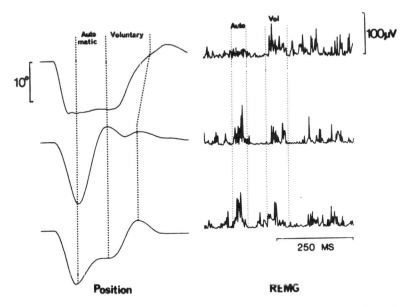

FIG. 16. Three consecutive responses from the same subject given the same size of disturbance (0.1Nm rising to 0.2 Nm, 50 msec after the start of the sweep), when instructed to compensate for the disturbance and restore initial thumb position. Visual feedback was removed for the duration of the sweep. Left: Thumb position records show the full range in size of automatic stretch reflex and later voluntary phases of the response *(dotted lines)*. Right: Corresponding phases in the surface-rectified EMG from flexor pollicis longus are more difficult to observe in single records, and are approximately indicated by the *dotted lines*.

accurate grading of the subsequent voluntary burst appropriate to the size of the earlier reflex event. In fact, there was a close inverse relationship between the size of the long-latency reflex effect and that of the subsequent voluntary compensation. This is shown graphically in one subject in Fig. 17A, where the size of the reflex and voluntary positional correction has been measured and plotted for each single trial. Similar high correlation coefficients have been recorded in a series of 10 normal subjects.

The simplest explanation for this phenomenon is that the automatic and voluntary volleys share a common neuronal pool, which is made refractory by an amount proportional to size of the preceding automatic volley. The size of the refractory pool adjusts the magnitude of a later voluntary response of constant size. The interval between the automatic and voluntary bursts of EMG activity is of the order of 50 msec, which is too long for any simple neuronal refractory period but could be caused, for example, by recurrent inhibitory effects from Renshaw cells. However, in a second series of experiments, we showed that this reciprocal relationship between reflex and voluntary responses was not a necessary consequence of the sequence of muscle activation, as expected from the above explanation. It turned out that the size of the voluntary response was entirely dependent on the subject's intent. If so instructed, the subject could produce a voluntary burst of the same size at the same latency regardless of the size of the variable preceding reflex volley.

To help subjects perform this task, they were given only a visual display of the size of their voluntary positional output, with no indication of stretch size or reflex correction. They were then told to use the torque disturbance to thumb as a signal to move to a constant distance, rather than to compensate for the original perturbation as in the

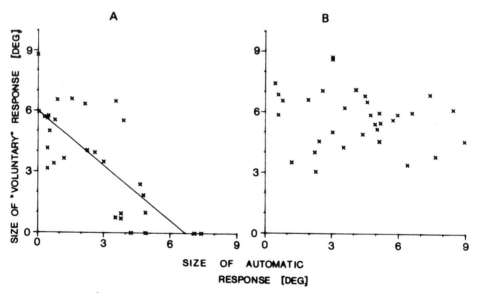

FIG. 17. Relationship in single trials between the size of the automatic long-latency stretch reflex and later voluntary positional compensation after thumb disturbances of the same size. A: Subject was instructed to compensate for the disturbance and restore the initial thumb position. Slope of the regression line = -1.1, correlation coefficient = -0.70 ($N = 28$; $p < 0.0001$). B: Subject was instructed to produce a constant size of voluntary output, approximately the same as the average voluntary output in A. Correlation coefficient = 0.01 ($N = 38$; $p > 0.05$). (From Marsden, Rothwell, and Traub, 1981, with permission.)

previous set of trials. A sequence of runs from the same subject as illustrated in Fig. 14A are shown again for this experiment in Fig. 14B, with the relationship between the size of the reflex and voluntary events shown in Fig. 17B. The usual inverse relationship between reflex and voluntary positional correction was abolished completely by this change in strategy, in all 10 subjects. Thus, the inverse relationship between reflex and voluntary responses was only operational when required by the task. In different circumstances, the two responses could become completely independent, such as to give a greater degree of flexibility to the motor command.

How then does the subject, in normal circumstances, adjust the voluntary volley to be appropriate to an unpredictable earlier reflex event when trying to calculate the muscle activation required to overcome an unexpected perturbation? It is conceivable that he monitors the mechanical consequence of the earlier reflex response and then makes a judgment as to the size of the subsequent volley required to restore the original position of the thumb. This seems unlikely for a number of reasons. First, the interval between the two EMG bursts of activity is usually less than 50 msec which is much shorter than any known reaction time to a peripheral stimulus, even to a kinesthetic stimulus to the same digit. Earlier work has shown that it is possible to respond voluntarily to a limb disturbance within about 90 msec, but the interval between voluntary and reflex events seen here seems too short to fall into any of even the fastest voluntary reactions. In any case, the subject is quite unaware of any voluntary perceptual judgment in deciding on the size of the second burst of activity, which makes the final complete correction for the thumb displacement.

Some form of automatic correction could be made by the brain subconsciously, once it knew the effect of the first reflex response to the thumb displacement. However, as far as we know the signals that could provide such information to the brain must be either those of the maximal velocity of reflex compensation, or the final positional correction achieved by the reflex volley. The latter causes an EMG burst of muscle activity beginning some 40 msec after the displacement but the mechanical effect of this EMG activity is not felt until some time later. The peak velocity of the reflex compensation usually occurred about 60 to 70 msec after the displacement, and the maximal positional correction was not achieved until even later, around 80 to 100 msec after the displacement. This would leave intervals of only about 50 msec from peak velocity of reflex compensation to onset of second voluntary burst of activity, or 20 msec from final positional correction achieved by the reflex response to the onset of the voluntary burst. Both latencies would be remarkably short for the operation of any mechanism involving transmission of information from the periphery to the brain to allow the latter to form some judgment, albeit subconscious, as to the size of the next voluntary volley required to complete accurate correction for the original thumb displacement. However, it is just possible that this delay could be encompassed within the operation of some very fast conducting transcortical reflex pathway. Alternatively, we can envision some form of complex spinal cord machinery achieving the correct balance between the reflex and voluntary corrections.

Whatever the mechanism producing these results, the experiments show clearly that there can be interaction between the long-latency stretch reflex and later muscular events which produce rapid and accurate responses following externally applied disturbances to movement. The effect of the subject's conscious intent is not only to change the size of each component of the response, but also to influence interactions between them according to the movement strategy used.

Role of Long-Latency Stretch Reflexes in Positional Accuracy to Dynamic Force Changes

The results discussed above relate to the behavior of the stationary or moving limb attempting to maintain positional accuracy despite unpredictable step changes in force being applied externally to the digit. Thus, we have discussed the function of the long-latency stretch reflex in positional limb control when faced with a steady, external force change. We will now consider the role of the long-latency stretch reflex in compensating for an unexpected transient, that is a pulse change in torque, or dynamic force changes applied to the moving limb during transit from one position to another. Bizzi et al. (1976), working on head or arm (Polit and Bizzi, 1979) movement in normal and deafferented monkeys have concluded that the stretch reflex has no role in maintaining positional accuracy in the face of transient torque changes applied to the moving limb. The main experimental findings were that transient changes in torque applied to the moving part during or before movement to a new final position, both in normal and deafferented animals, did not affect final positional accuracy of the arm. (Bizzi and Abend, *this volume*). To account for this behavior, Bizzi proposed an open-loop hypothesis of movement control, suggesting that the moving part be regarded as a spring whose parameters of spring stiffness and rest length are set by the CNS. The proposal was that stiffness and rest length are independently set and defined by adjusting agonist and antagonist muscle activities. In this way, transient force changes introduced unpredictably to the moving limb would have no effect on final position accuracy and, furthermore, final limb position would be independent of initial limb position. Kelso and Holt (1980) interpreted their experiments on movement of the human finger as compatible with Bizzi's hypothesis. However, the sphygmomanometer cuff that was used to functionally deafferent the hand inevitably left the muscle receptors intact, since the finger flexor and extensor muscles are located in the forearm. Therefore, we would question their conclusion.

Stimulated by Bizzi's hypothesis, we have examined the role of the long-latency stretch reflex during unpredictable dynamic force changes applied to the thumb moving accurately to a new learned target location. The method used was to train subjects to flex rapidly the top joint of the thumb against a constant initial load through 20° from a fixed starting positon to a new learned target position. The load was provided by a DC torque motor whose output shaft was linked to an accurate potentiometer, and whose positional output was used to generate negative velocity feedback drive to the torque motor. The initial load consisting of the inertia of the system and a small constant torque (0.02 Nm) acted to extend the thumb. The moment the thumb began to flex, the negative velocity feedback circuit increased motor torque proportional to the velocity of movement, thus simulating an increase in viscous friction encountered during thumb movement. This viscous friction acted as a force to oppose dynamic changes of thumb position and was proportional to lever velocity. Once the subject had learned to move his thumb accurately from the rest to the final target position, he began a series of trials in which occasionally and unpredictably (approximately 1 run in 5) the viscous friction of the load was changed prior to the lever being moved. The control value of viscous friction was set at 0.0042 Nm/rad/sec; the test values were 2.0, 1.3, and 0.7 times the control value. The results are shown in Fig. 18.

In the normal subject, an increase in viscous friction altered the trajectory of the thumb by slowing the rate of movement, but the subject effortlessly compensated for this change by producing a burst of automatic electromyographic activity in the long thumb flexor after an interval of 80 to 120 msec from the start of movement of the lever.

We interpret this change as evidence of operation of the long-latency stretch reflex mechanism, the slightly longer latency being due to the time taken for the change in viscosity to deviate the thumb from its intended path by an amount sufficient to evoke the automatic response. Increasing viscosity would result eventually in halting the movement, as studied by Marsden et al. (1976a), which evoked a compensatory muscle response with exactly the same latency as the response of the long thumb flexor to direct stretch or release. We believe that the increase in viscosity encountered by the thumb led to a mismatch in the rate of interfusal muscle fiber shortening relative to that of extrafusal muscle fiber contraction, resulting in an increased firing of muscle spindle afferent fibers sufficient to produce an automatic long-latency stretch reflex output.

Despite the increased viscosity slowing the trajectory of movement, the automatic compensation produced by the long-latency stretch reflex was such as to restore final thumb position back to its intended target with a considerable degree of accuracy. Indeed, there was a fair degree of correspondence between the size of the long-latency stretch reflex burst and the extent of slowing of the trajectory of movement introduced by the viscosity change, bigger changes in viscosity evoking greater reflex responses than smaller changes, with accuracy remaining unimpaired in both situations.

The next stage in the experiment was to study the effect of anesthetizing the thumb, by ring block with local anesthetic, on the capacity of the individual to compensate for changes in viscous friction during thumb flexion movement. Previously, we have shown (Marsden et al., 1972, 1977, 1979) that anesthetizing the thumb reduces the size of long-latency stretch reflexes in many subjects, at least when done in the naive individual. When the experiments were repeated in the same subjects after thumb anesthesia, the long-latency response to increased viscosity encountered during thumb flexion was abolished, and positional accuracy suffered proportionately (Fig. 18). Thus, under anesthetic the thumb tended to undershoot the intended end position when viscous friction was unexpectedly increased, and the amount of the final position error introduced by such a change of load after anesthesia was highly correlated with the size of the long-latency EMG response lost as a result of the anesthesia. Similarly, when viscous friction was reduced during thumb flexion, the thumb overshot its target point, because activity of the long thumb flexor no longer silenced when the trajectory of movement accelerated faster than expected.

These data suggest that the long-latency stretch reflex plays an important role in maintaining positional accuracy in the face of unpredictable dynamic force changes. This statement is clearly incompatible with Bizzi's "open-loop spring" hypothesis for movement. In our experiments, the final load against which the thumb operated once it had achieved its final position was identical to that encountered by the thumb before it started to move the lever. If that positional change had been achieved solely by altering the spring constants of agonist and antagonists, final position would not have been changed in any way by the introduction of an alteration in viscous friction during movement, whether or not the thumb was anesthetized. The fact that anesthesia, which abolished an obvious response to the viscous friction change, led to a severe positional error indicates that the thumb, at least, is not moved in this fashion. This important point can be illustrated by examining the behavior of a simple spring trying to return to its rest length against the load used in the present experiment. It always succeeds, regardless of the value of viscous

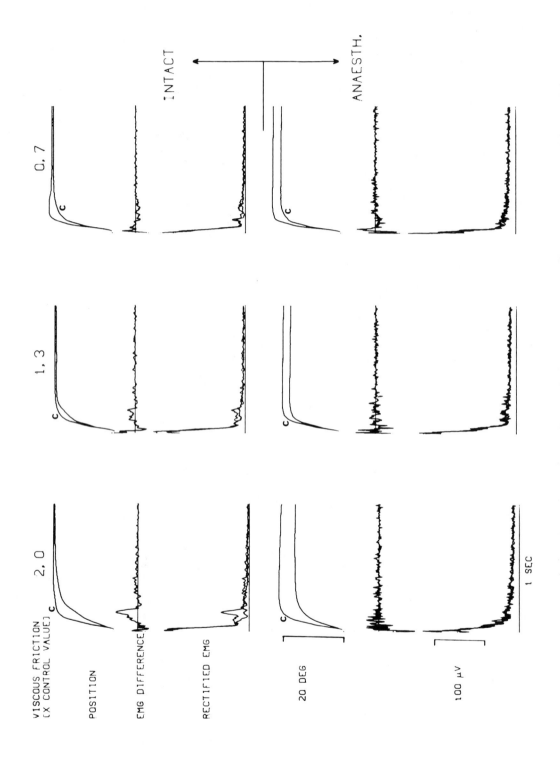

friction of the load that it is working against (Fig. 19). Clearly the human thumb, at least, is not like a simple spring. Our results in these experiments were verified by observations performed on a single male patient with a complete deafferented hand caused by a severe but restricted peripheral sensory neuropathy. As expected, introduction of an unpredictable increase in viscous friction of the load in this patient during thumb flexion did not evoke a muscle response, and, consequently, the thumb constantly undershot its intended end position (Fig. 20).

The final implication of the results of these experiments is that movement of the thumb against this sort of background resistance and through this sort of trajectory, can be achieved without very accurate knowledge of the dynamic characteristics of a load. In our experiments on normal subjects, increasing viscous friction by a factor of 2 was not sufficient to create any consistent positional errors. The long-latency stretch reflex was quite capable of accurately compensating for such dynamic changes in load to achieve a highly accurate final position. When the viscous friction was increased by a factor of 5, positional accuracy was still maintained, but the long-latency stretch reflex was accompanied by later muscle responses (Fig. 20). The implication is that the subject attempting to move a load to a specified end position as

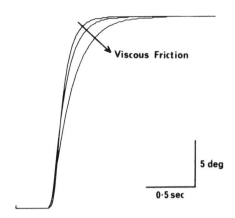

FIG. 19. Trajectory of a load moved by an initially stretched spring released approximately 300 msec after start of the sweep. The final equilibrium position is independent of the magnitude of the viscous friction of the load.

rapidly as possible can underestimate the dynamic behavior of the load, and yet still maintain positional accuracy when aided by an effective long-latency stretch reflex.

CONCLUSIONS

We have argued that the long-latency stretch reflexes in different animals and even in different muscles of the human body, may utilize different anatomical pathways. Multiple repetitive spindle bursts may produce long-latency peaks of excitation via the monosynaptic spinal reflex arc or, conversely, a single spindle volley could produce long-latency effects by traversing a long-loop transcortical route. We suggest that the dominance of the latter pathway is dependent on the density of the direct corticospinal connections to the muscles, which, in turn, is reflected in the degree of flexibility of control required of the limb. The very large cortico-MN projections to the human hand lie at the extreme of this spectrum, and it is here that the long-latency stretch reflex is most likely to use the transcortical pathway. In contrast, spinal mechanisms may play a greater role in pro-

◄─────────────────────

FIG. 18. The effect of unpredictably altering the viscous friction of the load on thumb trajectory and motor outflow to the long thumb flexor. The 6 conditions shown are 3 sizes of change in viscous friction ($\times$ 2.0, $\times$ 1.3, and $\times$ 0.7) from the control value set at 4.2×10^{-3} Nm/rad/sec, both with the thumb intact (**top**) and with the thumb anesthetized (**bottom**). Each panel shows from the *top*, thumb position with an upward deflection denoting thumb flexion (control run is denoted by c), EMG difference (test run minus control run), and rectified EMG. The averages of 32 control and test runs are shown superimposed in each panel.

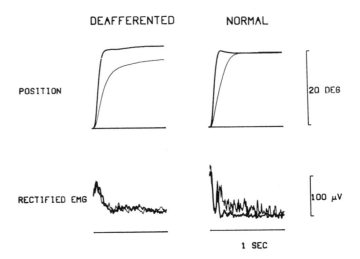

FIG. 20. Comparison of a subject deafferented below the elbow with a normal subject when faced with an unpredictable fivefold increase in viscous friction of a load. The subject attempted to flex the thumb against the load so as to attain a designated end position with the eyes shut. The *top trace* shows thumb position (upward deflection denotes flexion movement) and the *bottom trace* rectified EMG. *Thick lines* indicate traces from the control runs, and each trace represents the average of 32 runs.

ducing the long-latency effects seen in the muscles of the leg and trunk. The degree to which either mechanism operates, therefore depends on the muscles under investigation and explains the apparent contradictory results reported by workers in this field. Thus, we believe that for the human hand, the long-loop reflexes obtained in response to muscle stretch are mediated by a strong transcortical pathway which has supplanted spinal stretch reflex mechanisms.

Together with the debate over the anatomical pathway mediating long-latency stretch reflexes, is the concurrent discussion regarding their role in movement control. We have confined our studies on the functional significance of the long-latency responses to the muscles controlling the human hand, where they are large and very easily obtained. We have demonstrated that the long-latency stretch reflex plays a part in positional servo assistance but not according to classical control principles. Thus, although a small disturbance to a stationary or moving limb at or around the threshold of conscious perception is perfectly compensated by the reflex, larger disturbances are not. In these conditions, the responses are pulsatile, variable, and inadequate, in a manner that would appear to be designed to interfere with movement control rather than to assist it.

The potential unpredictability of the reflex size renders it useless if the CNS needs to monitor its effect before subsequent less automatic compensatory responses can be dispatched. However, we find that for large disturbances, these later less automatic events interact at short latency with the preceding long-latency stretch reflex to achieve the intended limb position quickly and accurately. The type of compensation produced depends on the size of the initial disturbance. Small disturbances are corrected solely by the reflex response. As the disturbance increases in size, the long-latency stretch reflex saturates and is assisted by later, less automatic responses which are graded according to the individual's prior intention to respond to such events.

In conclusion, we view the control of

movement by centrally generated commands in the following way. Since the loads against which a limb acts are seldom totally predictable, afferent feedback is required to adjust the motor program to suit the external conditions. When prediction of the external conditions is approximately correct, mechanisms involving the long-latency stretch reflex machinery automatically tune the motor program to match the actual loading conditions on the limb. When prediction of the load is grossly inaccurate, stretch reflex mechanisms can no longer adequately adjust the program to cope with its task, and conscious intervention from the individual is necessary if the original movement intention is to be fulfilled. Such a control system trades off automaticity against flexibility in the face of a largely unpredictable world.

Motor Control Mechanisms in Health and Disease,
edited by J. E. Desmedt.
Raven Press, New York © 1983.

Dissociated Changes of Short- and Long-Latency Myotatic Responses Prior to a Brisk Voluntary Movement in Normals, in Karate Experts, and in Parkinsonian Patients

*James A. Mortimer and David D. Webster

Geriatric Research, Veterans Administration Medical Center, and Department of Neurology, University of Minnesota Medical School, Minneapolis, Minnesota 55417

The question of whether the gain of the myotatic response can be controlled independently of homonymous α-motoneuron (MN) discharge has important implications for posture and movement control. Evidence on this point is contradictory, with some investigators claiming independent control of stretch reflex gain (Dufresne et al., 1980; MacKay et al., 1981; Soechting et al., 1981) and others (Crago et al., 1976; Houk, 1979) maintaining that apparent changes in myotatic gain reflect superimposed reaction-time responses. On initial consideration, the idea that the gain of myotatic pathways may not be independently controlled seems contradictory to reported changes in amplitude of the H-reflex or tendon jerk, approximately 100 msec before initiation of voluntary α-MN discharge (Kots, 1969; Coquery and Coulmance, 1971; Gottlieb and Agarwal, 1973). However, in these studies, reflexes were evoked in electrically silent muscles and subthreshold changes in MN excitability prior to the voluntary discharge may have occurred without being detected. The adaptive nature of long-latency (> 50 msec) myotatic responses appears to have greater support (Hammond, 1960; Thomas et al., 1977; Marsden et al., 1978b; Mortimer and Webster, 1979; Chan et al., 1979a; Kwan et al., 1979; Soechting et al., 1981). However, most studies failed to control the baseline EMG discharge, which is known to influence the amplitudes of both short- and long-latency myotatic responses (Marsden et al., 1976b; Mortimer and Webster, 1978). These studies therefore suffered from the same technical limitations as those demonstrating changes in short-latency myotatic responses preceding movement.

This chapter will first consider the evidence for and against task-dependent changes in short- and long-latency myotatic responses in normal subjects, and then will examine adaptive changes in myotatic reflex gain during transition from posture to movement in normal subjects, highly trained karate instructors, and in patients with idiopathic Parkinson's disease.

---

*To whom correspondence should be addressed: Geriatric Research, Educational, and Clinical Center (11G), Veterans Administration Medical Center, 54th Street and 48th Avenue South, Minneapolis, MN 55417.

CHANGES IN THE GAIN OF SHORT- AND LONG-LATENCY MYOTATIC PATHWAYS DURING MOVEMENTS AND POSTURE-MOVEMENT TRANSITIONS: NORMAL SUBJECTS

Changes in Myotatic Responses During Movement

Amplitude of myotatic responses evoked during movements is potentially influenced by several factors including (1) magnitude and time course of the perturbation; (2) sensitivity of muscle stretch receptors, which varies with the time course and rate of shortening or stretch of the muscle and with the intrafusal tension produced by dynamic and static fusimotor activity; (3) level of facilitation or discharge in the MN pool; and (4) segmental and suprasegmental influences on gain at various synaptic relays. When the muscle is actively contracting and the limb is in motion, all of these factors may be changing simultaneously, making interpretation difficult. Thus, apparent increases or reductions in myotatic gain during movements may be due to centrally controlled changes in gain, to changes in the facilitation of the MN pool, to alteration in compliance of muscles leading to different mechanical consequences for the same torque perturbation, or to changes in dynamic sensitivity of muscle-stretch receptors.

Dufresne et al. (1980) studied myotatic reflex gain in normal subjects during sinusoidal eye-arm tracking and found that the times of changes in reflex gain did not coincide with the times of alterations in the rectified EMG discharge. They also showed that alterations in myotatic gain were evident during force tracking under nearly isometric conditions. The largest change in amplitude during the tracking cycle occurred for the myotatic response between 50 and 100 msec following the torque perturbation. Soechting et al. (1981) studied the changes in myotatic reflex gain as subjects tracked a ramp, and he found that the peak amplitude of the myotatic response was attained prior to the peak of the intentional electromyographic (EMG) discharge associated with the movement.

By direct recording of muscle spindle afferents in monkeys trained to make hold-ramp-hold movements, Schieber and Thach (1980) demonstrated that fusimotor activity can be dissociated from the discharge of homonymous α-MNs. Despite large variations in the time course of EMG activity, the pattern of spindle afferent discharge was basically identical, with firing frequency maximal at the beginning of the movement. Other evidence of α-γ decoupling suggests that reflex gain may be controlled independently from α-MN discharge during locomotion (Prochazka et al., 1976; Loeb and Duysens, 1979; Prochazka and Hulliger, *this volume*).

Changes in Myotatic Responses During Posture-Movement Transitions

Changes during the transition from posture to a rapid ballistic movement have been studied by Hallett et al. (1981), who examined long-latency responses to stretch of the long thumb flexor in the interval between the signal to move and actual commencement of thumb flexion. In some subjects, increases in long-latency myotatic responses were observed, but these were accompanied by increases in baseline EMG discharge in anticipation of movement. Soechting et al. (1981), using pseudorandom torque perturbations, studied changes in myotatic response (defined as EMG occurring up to 100 msec after the load change) as subjects initiated ballistic flexion movements of the forearm from an initial posture. They found that coincident with the intentional EMG discharge, the myotatic response of the agonist muscle decreased in amplitude. Since baseline levels of discharge were not well controlled in these experiments, it is not possible to judge whether the changes in myotatic gain prior to movement were the result of alterations in MN excitability or of changes in gain at other relays. However, changes due to shortening or stretch of muscle can be excluded, because the limb was not in motion during the period over which the change in gain was demonstrated. In the experiments of Hallett and Soechting, comparisons between changes in short- and long-latency pathways could not easily be made because of the small size or absence of responses at the latency of the tendon jerk. To address this issue and to control for background levels of EMG activity, we studied changes in myotatic responses of the biceps muscle as subjects prepared to make rapid ballistic flexions or extensions of the forearm. All movements were initiated from a fixed posture (90° of elbow flexion) maintained against a constant preload. Biceps and triceps baseline EMG activity was controlled in view of Evarts and Vaughn's finding (1978) that, even with a constant preload, the background discharge of the stretched muscle may vary due to different degrees of co-contraction in agonist and antagonist muscles.

Methods

Methods have been reported elsewhere (Mortimer et al., 1981). Briefly, torque pulses were delivered to the subject's right forearm, which was strapped into a rigid, lightweight arm support that could be rotated about the elbow in a horizontal plane. Surface EMGs were recorded from biceps and lateral triceps, amplified, high-pass filtered (30 Hz to 10 Hz) and full-wave rectified. Amplified signals for torque, angular position, and biceps and triceps rectified EMG were sampled at intervals of 1.6 msec and averaged on a MED-80 computer. To reduce the contribution of brachioradialis muscle to the flexion torque, the forearm was maintained in a supine position by a strap attached to the arm support.

Subjects maintained a 90° elbow angle against a steady, clockwise 2-Nm preload on biceps and were instructed to respond to a series of tones by flexing or extending their forearm as rapidly as possible. To encourage maximum velocity movements, the ballistic momentum was arrested by a foam pillow (for the flexion movements) or a karate bag (for extension movements). No trial was initiated until subjects held their arm within a 1° window and demonstrated electrical silence of triceps for 5 sec. Torque pulses of 2-Nm amplitude and 500-msec duration loading biceps were presented at 8 different times to sample the interval from just before tone onset to the intentional EMG response. For each interval 16 or 20 trials were given. To ascertain the response to the tone alone, no torque pulse was given on the same number of trials. Trials were presented in random order and sorted to yield average responses for each time interval.

Figure 1 shows the average response on 20 trials to the tone presented alone (I) and to the tone plus a torque pulse delayed 75 msec following tone onset (II). The average biceps response to the torque pulse was computed by subtracting trace II from trace I. In thise case, subtraction results in a discrete response lasting from about 25 to 100 msec following torque pulse onset. However, in many cases presentation of the torque pulse prior to or shortly after the onset of the tone produced not only this early reflexlike response but also a prolonged discharge, suggesting anticipatory responses (see Hallett et al., 1981). To quantify the EMG responses, we calculated motor response indices. These normalized measures of integrated EMG activity were obtained for 3 time intervals (25 to 50 msec, 50 to 75 msec, and 75 to 100 msec) after torque perturbation (Mortimer and Webster, 1978; Mortimer et al., 1981). These intervals correspond approximately to the M1, M2, and M3 intervals (Lee and Tatton, 1975, 1978; Mortimer and Webster, 1978).

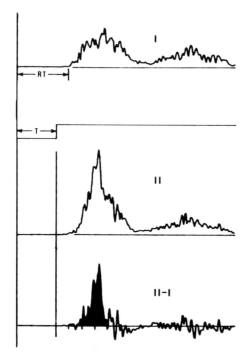

FIG. 1. Computation of EMG responses to torque pulses. The average response to the torque pulse on 20 trials *(bottom trace)*, obtained by subtracting the average response to the tone presented alone *(top trace)* from the average response to the tone plus a torque pulse delivered 75 msec after tone onset *(third trace from top)*. Torque pulse is shown schematically in *second trace*. Tone onset corresponds with *left vertical axis*. Time calibration: T = 75 msec. Subject D4. (Adapted from Mortimer et al., 1981, with permission.)

Increases in Myotatic Responses Before Ballistic Movements

Figure 2 shows data from a normal subject where the relative magnitudes of 3 motor response indices (A: 25–50 msec; B: 50–75 msec; C: 75–100 msec) are plotted against the time between tone onset and the midpoint of the interval over which the motor response was calculated. The average response to tone alone is plotted in the bottom trace. Motor response indices for longer latency responses (B and C) changed before onset of voluntary EMG responses to tone alone *(arrow* in bottom trace), while the magnitude of the short-latency motor response (A) did not change until after the onset of intentional EMG activity. For all 3 indices, the peak values exceeded those during maintenance of posture (points to the left of time 0). The early increase in myotatic responses was followed by a decline during the movement itself (cf. Fig. 3) in all normal subjects tested ($N = 18$). In no case did an increase in the short-latency myotatic response (A) precede the initiation of voluntary EMG activity, whereas increases in the size of long-latency responses (B and C) occurred as early as 50 msec after tone onset.

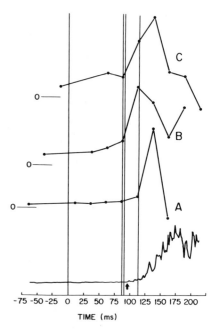

FIG. 2. Motor response indices (A, B, and C) plotted against time relative to tone onset. Indices are normalized to percentage of maximum value attained. Peak values shown are 100%. *Bottom trace* shows the average response to the tone presented alone. Points to the left of time 0 show size of motor response indices when the subject is engaged in the postural task. Computed times of change of motor response indices are indicated by *vertical lines*, with B changing first, followed by C and then A. Subject D6. (From Mortimer et al., 1981, with permission.)

Suppression of Myotatic Responses During Ballistic Movements

Decreases of stretch responses of agonist muscles during fast movements have been described (Desmedt and Godaux, 1978a, 1979; Cooke, 1980; Gottlieb and Agarwal, 1980; Soechting et al., 1981). Although there is a temptation to explain these decreases by central regulation of reflex gain (Desmedt and Godaux, 1978a; Gottlieb and Agarwal, 1980b), other explanations must be considered, including unloading of muscle spindle receptors and autogenic inhibition from increased Golgi tendon organ discharge during muscular contraction (Prochazka and Wand, 1980a). Studies of muscle spindle afferents during voluntary

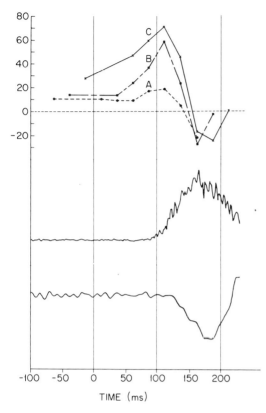

FIG. 3. Temporal comparison of changes in motor response indices (A, B, and C), EMG response to the tone presented alone *(middle trace)* and acceleration of the limb in the tone-only condition *(lower trace)*. Downward deflection of acceleration trace corresponds to forearm flexion. *Scales:* Motor response indices: Actual values; EMG response: arbitrary scale; angular acceleration: 5,000°/sec². Subject D7. (From Mortimer et al., 1981, with permission.)

Figure 4 summarizes differences between the times of change of A, B, and C, determined by computing the time at which the magnitudes of motor response indices exceeded 20% of the difference between their maximum and baseline values (see vertical lines in Fig. 2). With few exceptions, the times of change of C (t_C) preceded those of B (t_B), and those of B preceded those of A (t_A): i.e., the longest latency responses (75–100 msec) were modified first, followed by those of medium latency (50 to 75 msec).

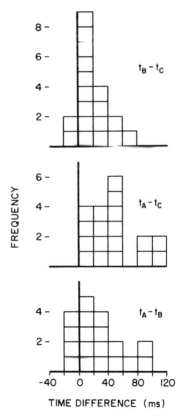

FIG. 4. Frequency histograms of differences between computed times of change of motor response indices (t_A, t_B, t_C). (From Mortimer et al., 1981, with permission.)

contractions demonstrate that both primary and secondary endings fall silent above certain velocities of muscle shortening (Burke et al., 1978b; Prochazka et al., 1979). The maximum rate of shortening of the biceps muscle in our normal subjects was between 2 and 4 resting lengths per second. Velocities of 0.2 resting lengths per second, which produce suppression of spindle discharge in unobstructed cat muscles (Prochazka et al., 1979), were attained within 10 msec after movement initiation. This may have contributed to the early decrease in both short- and long-latency myotatic responses. Suppression of the myotatic response in soleus muscle of man has been shown even if actual shortening of the muscle is prevented mechanically during contraction (Gottlieb and Agarwal, 1980). However, the fact that the decrease in myotatic response was smaller when shortening was obstructed suggests that unloading of muscles spindles contributes to the reduction of myotatic response during shortening. Prochazka and Wand (1980) have suggested that Golgi tendon organ inhibition may be the dominant influence on homonymous MNs during voluntary isotonic movements involving muscles shortening at velocities greater than 0.2 resting lengths per second. In our subjects during maximum limb acceleration, the calculated response was frequently negative in sign (Fig. 3). This might be explained by an unmasking of Golgi tendon organ inhibition when spindle receptors are presumably unloaded. In several subjects, increases in amplitude of short- and long-latency myotatic responses were seen as the acceleration of the limb began to decrease (Fig. 3), consistent with a reloading of muscle spindle receptors or by a decrease in Golgi tendon organ inhibition as the contractile force declined.

Specificity of Changes in Myotatic Gain Preceding Movement

When subjects are instructed to respond to imposed torques by resisting or by initiating movements in the opposite direction, the magnitude of long-latency myotatic responses is increased in comparison to situations in which subjects assist or move in the same direction as the torque input (Mortimer and Webster, 1978). To ascertain the time course of changes in the myotatic response of the intended antagonist of a voluntary movement, subjects were asked to respond to a tone by rapid extension of the forearm as torque pulses stretching the biceps were given at different time delays relative to the tone signal. Results of experiments in 2 subjects are shown in Fig. 5, where a and c illustrate the time course of A, B, and C, when the task was to flex the arm rapidly (biceps agonist), and b and d, the time course when the task was to extend the arm rapidly (biceps antagonist). In place of the increase in gain, which occurred when biceps was the intended agonist, a decrease of the myotatic response occurred when biceps acted as the antagonist in the intended movement. The time course appears to be different from that seen when the stretched muscle acted as agonist, however, with the decrease in gain occurring at a longer latency. These results provide support for the specificity of increases in long-latency responses prior to movement and their dependence on the intent of the subject.

In summary, our experiments suggest that adaptive changes in long-latency responses occur in normal subjects in preparation for voluntary movements. These changes are controlled independently from both α-MN discharge and gain of short-latency myotatic pathways.

CHANGES IN MYOTATIC RESPONSES DURING POSTURE-MOVEMENT TRANSITIONS: PARKINSON'S DISEASE PATIENTS AND KARATE EXPERTS

Changes in the myotatic response during the transition from posture to movement were assessed in 16 patients with idiopathic Parkinson's disease. The same paradigm

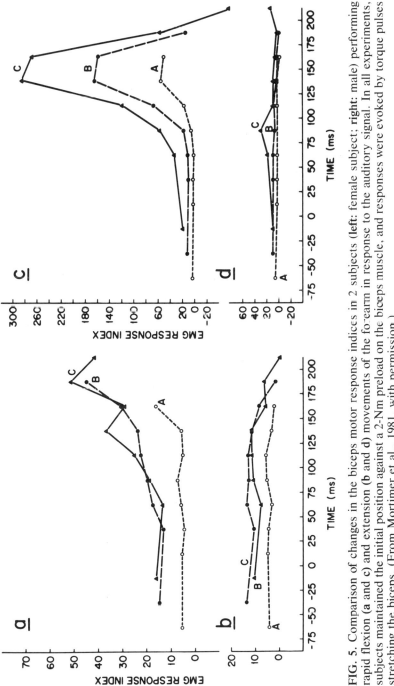

FIG. 5. Comparison of changes in the biceps motor response indices in 2 subjects (left: female subject; right: male) performing rapid flexion (a and c) and extension (b and d) movements of the forearm in response to the auditory signal. In all experiments, subjects maintained the initial position against a 2-Nm preload on the biceps muscle, and responses were evoked by torque pulses stretching the biceps. (From Mortimer et al., 1981, with permission.)

was used and patients were encouraged to make maximum velocity flexion movements of the forearm in response to tone cues. As in normal subjects, no change was seen in the short-latency (A) motor response until after the intentional motor volley reached the MNs. However, in contrast to normal subjects, only 9 of the 16 Parkinson's disease patients exhibited the normal pattern of an initial increase in long-latency myotatic responses prior to the intentional EMG discharge. An example of a normal response in a Parkinson's disease patient is shown in Fig. 6A, where a combined motor response index representing activity from 50 to 100 msec after onset of torque pulse (ordinate) is plotted against the time interval from tone onset. Despite a longer median reaction time, the temporal pattern of change is similar to that seen in a 57-year-old man with normal motor function (Fig. 6B). Four patients showed no clear change in long-latency myotatic responses preceding onset of the intentional EMG; whereas 3 showed the opposite response, i.e., a decrease of the long-latency response prior to the voluntary EMG response to the tone (Fig. 6C). Although the 3 groups of patients did not differ systematically on quantitative measures of bradykinesia (Mortimer and Webster, 1978), those patients with abnormal changes or no changes in myotatic response gain tended to have greater rigidity than those with a normal increase in gain before movement. This was particularly true of the 3 patients in whom a decrease in gain preceded the intentional motor response. One possible implication is that the abnormally high gain of long-latency myotatic pathways in Parkinson's disease patients with rigidity (Lee and Tatton, 1975, 1978; Tatton and Lee, 1975; Mortimer and Webster, 1978, 1979) may interfere with the initiation of intentional movements, and, therefore, the optimum strategy may be to decrease this gain prior to initiation of fast movements.

Among the 20 normal subjects studied, 3 persons had attained a black belt in karate and 2 had achieved lesser status. The change in long-latency myotatic responses in one such subject is shown in Fig. 6D, plotted with a compressed vertical scale. Subject with this type of training tended to have much larger increases in long-latency response magnitudes prior to maximum velocity movements in comparison with the other normal subjects.

FUNCTIONAL INTERPRETATION

Findings presented in this chapter lend support to a growing body of evidence suggesting that different pathways are involved in mediating short- and long-latency myotatic responses (see Lee et al., *this volume*). Although multiple peaks in the EMG response to torque perturbations can be seen in spinal animals (Ghez and Shinoda, 1978; Tracey et al., 1980b; Miller and Brooks, 1981) and can be observed in the discharge of primary muscle spindle afferents following a torque perturbation (Hagbarth et al., 1980, 1981; Prochazka and Hulliger, *this volume*), this does not necessarily eliminate the possibility that long-latency myotatic responses may involve supraspinal pathways. It was, therefore, pertinent to look for experimental conditions in which these long-latency responses would be modified independently of the short-latency myotatic responses.

The present findings demonstrate that the changes in the magnitudes of the short- and of the long-latency myotatic responses can be dissociated during the period that precedes a voluntary movement, suggesting that the amplitude of these responses must be influenced by different pathways. The alteration in both intermediate-latency (B,M2) and long-latency (C,M3) myotatic responses before the onset of intentional motor discharge implies that the gains of long-latency myotatic responses are modifiable on the basis of the subject's intention to use the

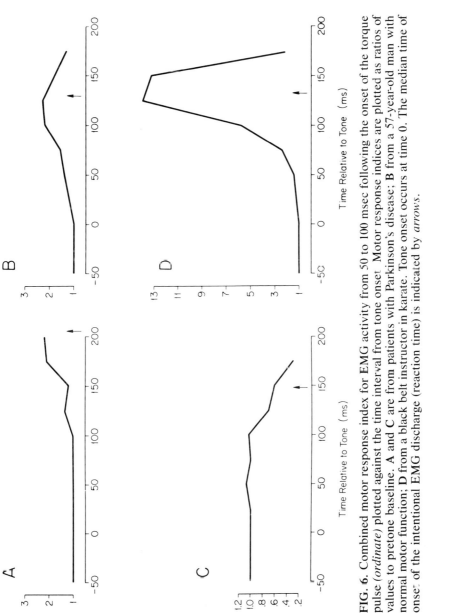

FIG. 6. Combined motor response index for EMG activity from 50 to 100 msec following the onset of the torque pulse (*ordinate*) plotted against the time interval from tone onset. Motor response indices are plotted as ratios of values to pretone baseline. **A** and **C** are from patients with Parkinson's disease; **B** from a 57-year-old man with normal motor function; **D** from a black belt instructor in karate. Tone onset occurs at time 0. The median time of onset of the intentional EMG discharge (reaction time) is indicated by *arrows*.

muscle in subsequent movement. However, in agreement with the findings of Houk and co-workers (Crago et al., 1976; Houk, 1979), our experiments provide no evidence that the short-latency myotatic responses are modified to meet task requirements. Indeed, the gain of the short-latency reflex seems to depend principally on the level of discharge or facilitation of the MN pool, changing only after the intentional motor volley reaches the MNs. Previous studies showing very early changes in the gain of the H-reflex and tendon jerk before movement have lacked sufficient control over the preexisting level of MN activity (Houk, 1979).

A mode of control that permits the subject to modify the mechanical impedance of the neuromuscular apparatus to match requirements of task and external (and internal) load would seem beneficial. It is conceivable that this control is provided by alterations in the gain of long-latency myotatic pathways. Long-latency responses have many properties that are not characteristic of reflexes or servo systems (Crago et al., 1976; Evarts and Vaughn, 1978; Houk, 1979). However, the presence of a variable-gain myotatic response at a latency longer than the tendon jerk and shorter than the intentional motor response to auditory stimuli is clear. There is reason to believe that these long-latency myotatic responses are influenced by converging inputs from segmental and suprasegmental pathways (Lee et al., *this volume*) and are, therefore, subject to central control.

In considering the possible functional significance of changes in long-latency myotatic responses during or preceding voluntary movement, the question is raised of whether the pathways mediating these responses are being used in a feedforward (predictive) or a feedback (nonpredictive) mode. Evidence for involvement of long-latency myotatic responses in compensating automatically for sudden load perturbations indicates that they provide only a fraction of the torque required to return the limb to its intended trajectory (Allum, 1975; Bizzi et al., 1978; Kwan et al., 1979). However, sudden changes in load are rarely encountered under natural conditions, and these studies give little indication of the load compensation that would be provided when encountering an unexpected load with slowly changing temporal characteristics. The low-pass filter properties of muscle assure that brief transient changes in EMG activity, as would result from synchronized discharge of MNs following sudden muscle stretch, will produce relatively little force.

Most changes in load encountered in normal life are quite predictable, and, therefore, the neuromuscular system dynamics can be preset according to an internal model (see Lee et al., *this volume*). The fact that long-latency myotatic responses change prior to initiation of voluntary EMG discharge suggests that they may prepare the neuromuscular system for predictive alterations in load that occur because of internal or external load changes resulting from the movement itself. In our experiments, the transition from posture to maximum velocity movement, required the forearm flexors or extensors to overcome quickly the inertia of the forearm and torque motor (which was equal to about 10% of forearm inertia). The increase in gain of long-latency myotatic pathways prior to movement may help to overcome this load. Indirect support for this mechanism is provided by comparison of data from patients with Parkinson's disease, karate experts, and other normal subjects. Figure 7 shows intentional EMG discharge and forearm angular velocity for 2 normal subjects without karate training (A,B), one karate black belt (C) and one bradykinetic Parkinson's disease patient (D) as they made maximum velocity flexion movements. Although there was a great diversity in the pattern of discharge among normal subjects, all showed an appreciable initial burst of activity that was instrumental in overcoming inertia. In contrast, the

Parkinson's disease patient (D) lacked this initial burst, which was replaced by a gradual onset of EMG activity. Despite the fact that the slow normal subject (B) and the Parkinson's disease patient had similar minimum reaction times (131 msec and 162 msec, respectively), the accelerations differed markedly (68.2 rad/sec^2 versus 25.0 rad/sec). Along similar lines, Evarts et al. (1981) have recently described Parkinson's disease patients with normal reaction times, but slow onset of EMG characteristic of bradykinesia. In the skilled karate subject (C), a very high level of acceleration was attained with a large initial burst of EMG activity, which reached a peak soon after its initiation. Time to peak EMG and peak acceleration were highly correlated as shown in Fig. 8.

The possibility that myotatic response pathways may be involved in increasing the acceleration of ballistic movements is supported by evidence obtained during selective fusimotor blockade in human subjects. Smith et al. (1972) found that fusimotor blockade resulted in a marked decrease in acceleration of rapidly initiated arm movements, such as those used in dart throwing. In our normal subjects, peak acceleration was significantly correlated with the maximum increase in gain of long-latency pathways prior to movement initiation (Fig. 9). However, early facilitation of MNs through a feedforward input mediated by long-latency myotatic pathways may play a more important role in supporting α-MN discharge when movements are carried out at moderate speed or against larger inertial loads. In these situations, there would be more time available for fusimotor-induced discharge of muscle spindle receptors to influence the acceleration of the limb. The large increase in gain seen in our experiments during maximal velocity movements carried out against only a small load may reflect an extreme condition in which gains are adjusted to a maximum value, which would assure the most rapid movement possible against any load that might be encountered.

CONCLUSIONS

The magnitude of myotatic responses observed during movements is influenced by a number of factors, including the mechanical characteristics of the perturbation, the time-varying sensitivity of muscle spindle receptors, the level of facilitation of the MN pool, and segmental and suprasegmental influences on gain at various synaptic relays. When the muscle is actively contracting and the limb is in motion, all of these factors may be changing simultaneously, making interpretations of changes in the magnitude of the myotatic response difficult. During transitions from posture to movement, the time course of change of myotatic responses can be assessed prior to the onset of the movement, allowing a more unique interpretation to be placed on the experimental findings. Previous studies failed to control the baseline level of MN discharge. Our experiments (controlled for this factor) suggest that increases of long-latency components of the myotatic response precede the earliest change in intentional EMG, whereas the first change in the short-latency myotatic response follows the increase in voluntary EMG activity. The absence of an increase in long-latency myotatic response in the antagonist before intended movements suggests that the initial increase in agonist myotatic gain serves a specific function.

When subjects carried out ballistic flexion movements of the forearm, the initial increase in the agonist myotatic response was followed by a suppression which may be due to unloading of the muscle spindle receptors or to an increase in autogenic Golgi tendon organ inhibition. Parkinson's disease patients exhibited a variety of patterns of change in the magnitude of long-latency myotatic responses preceding movements. Although the most common

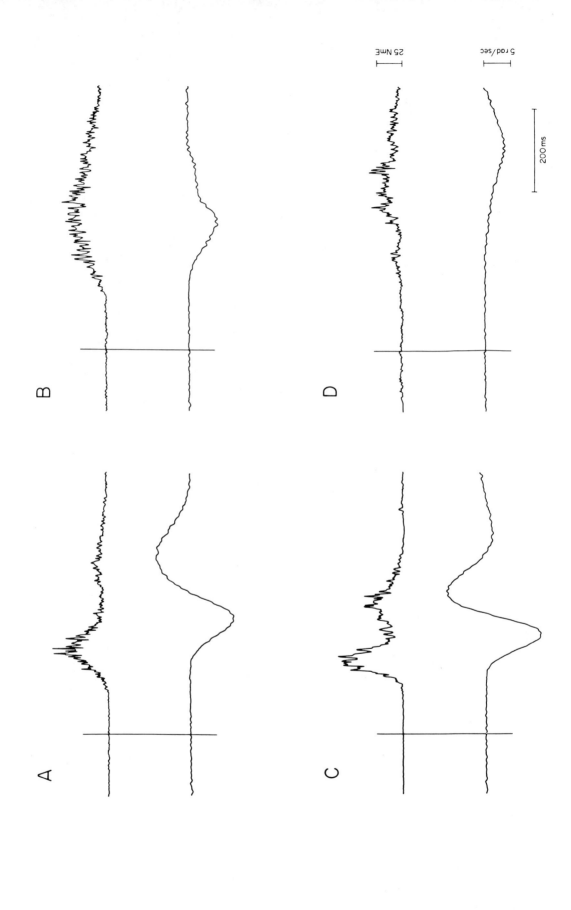

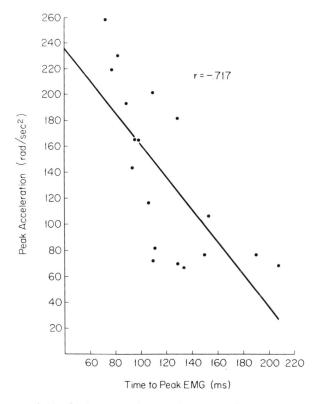

FIG. 8. Time to peak EMG plotted against peak acceleration for the 18 normal subjects.

pattern was similar to that seen in normal subjects, a substantial minority of patients showed either no change in gain preceding the onset of the intentional EMG or a decline. The latter patients tended to be more rigid, suggesting that the abnormally high gain of long-latency myotatic pathways in Parkinson's disease patients with rigidity may be voluntarily suppressed prior to onset of the intentional motor activity. The possibility that the initial increase in myotatic gain preceding movement assists in overcoming the inertial load of the forearm was suggested by comparison of data from Parkinson's disease patients, karate experts, and other normal subjects instructed to make maximum velocity movements. Karate experts had briefer rise times in the initial agonist burst, greater limb acceleration, and larger increases in the gain of long-latency myotatic pathways preceding movement than other normal subjects.

Our findings provide evidence for independent control of the gains of short- and

---

FIG. 7. Intentional EMG discharge and forearm angular velocity in response to tone. A and B are from normal subjects without karate training; C from an individual with a black belt in karate; D from a bradykinetic patient with Parkinson's disease. *Vertical lines* denote times of tone onset. Upward deflections of angular velocity correspond to forearm extension. Averages of 20 trials are shown. Calibration of EMG responses is in Newton-meter equivalents (Nm E): 1 Nm E is the level of motor discharge necessary to maintain a constant position of 90° of elbow flexion against a 1-Nm load when the antagonist is silent. Same calibrations apply to A–D.

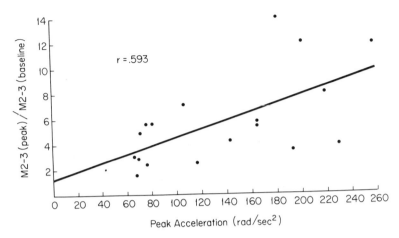

FIG. 9. Ratio of the peak value of the combined long-latency motor response index prior to movement onset to its baseline (pretone) value plotted against peak acceleration for the 18 normal subjects.

long-latency myotatic responses, demonstrating an adaptive capability of the latter, but not of the former. The precise role that regulation of long-latency myotatic gain may serve requires further investigation, but present evidence favors compensation for slow, predictable changes in load rather than regulation of movement trajectories in the presence of unexpected load perturbations.

ACKNOWLEDGMENTS

This research was supported by the Veterans Administration. Dr. Peter Eisenberg gave helpful comments on the manuscript.

Muscular Contractions Elicited by Passive Shortening

*Ronald W. Angel

Department of Neurology, Veterans Administration Medical Center, Palo Alto, California 94304

More than a century ago, Westphal (1880) described "a kind of paradoxical muscle contraction" that occurs with passive shortening of muscle, now called the shortening reaction (SR). The paradoxical side of SR is best revealed by contrasting it with a more familiar response. The stretch reflex is elicited by an elongation of the muscle, the SR by passive shortening. The neuronal mechanism of the stretch reflex has been identified and studied exhaustively, but the mechanism of SR is virtually unknown. The stretch reflex is exaggerated in the upper motor neuron syndrome, whereas the SR is increased in extrapyramidal syndromes. Although the stretch reflex is familiar to all clinicians, the SR is relatively obscure, being neglected in most neurology textbooks. Some authors have suggested that the SR plays a role in the pathophysiology of parkinsonism, athetosis, and dystonia (Rondot and Metral, 1973). Enlarging on that suggestion, the present chapter will propose that muscle tone is normally regulated by the SR mechanism, in health as well as in disease. Although it was first described in patients with neurological disease (Andrews et al., 1972, 1973; Foerster, 1921; Rondot and Scherrer, 1966; Westphal, 1880), the SR is easily demonstrated in normal subjects. If the SR is viewed as a normal phenomenon rather than a sign of disease, then a number of questions can be raised. What is the neuronal mechanism of the SR? What is its biological function or purpose? Why is the response exaggerated in some diseases and reduced in others?

THE SR FAMILY

It can be hypothesized that Westphal's phenomenon belongs to a family, which includes all of the muscular contractions that occur with passive shortening. Another member of the "SR family" has been reported by Crago et al. (1976), who found notably asymmetric electromyographic (EMG) responses to symmetric disturbances in normal biceps. Stretch of the biceps produced a marked increase in EMG, whereas decreased force allowed the muscle to shorten and produced either an EMG decrease of smaller magnitude or an actual increase. The EMG increase during shortening was of functional interest, because it represents a reflex action assisting length change rather than opposing it.

Another candidate for membership in this family is the second contraction of the agonist muscle during ballistic movement (Wachholder and Altenburger, 1926; Terzuolo et al., 1973; Hallett et al., 1975c). When

*Department of Neurology, Veterans Administration Medical Center, 3801 Miranda Avenue, Palo Alto, CA 94304.

a limb is moved rapidly from one position to another, the agonist muscle commonly shows two bursts of EMG activity separated by a period of electrical silence during which the antagonist muscle becomes active. This sequence has been called the triphasic pattern. Since it occurs while the limb is moving passively (by inertia), the second agonist contraction deserves the term paradoxical. When it begins, the limb has already been accelerated by the force of the initial contraction. One could argue that it counteracts an excessive contraction of the antagonist muscles, but that seems unlikely. A typical two-burst pattern has been recorded from triceps following paralysis of all the elbow flexors by blockade of their muscle nerves. During the second half of a ballistic movement, the limb is moving on the momentum acquired during the first half. The origin and insertion of the agonist are brought together passively, and a second contraction may be required to restore optimal stiffness. Since the SR has a dynamic, as well as a static, component (Andrews et al., 1972, 1973), it is sensitive to the rate of passive shortening, being larger when the rate is increased. The second burst during a ballistic movement of the posterior deltoid muscle during rapid abductions of the arm was indeed found to be rate-sensitive. On random trials, the movement activated a torque motor, which caused an unexpected increase of velocity. Amplitudes of the initial (A) and second burst (B) were measured for each trial. During movements that were accelerated by the motor, the ratio B/A was significantly increased (Figs. 1 and 2). During voluntary movements at different speeds, the size of B was correlated with peak velocity (Fig. 3). Hence, the second burst is rate sensitive.

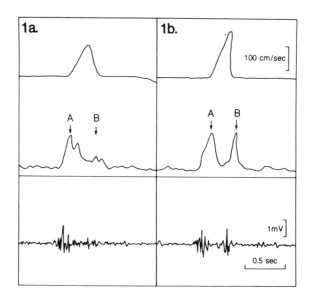

FIG. 1. a: Voluntary abduction of the right upper limb, normal subject. *Top line:* Velocity of hand; *second line:* Rectified and filtered EMG of posterior deltoid muscle; *bottom line:* Unfiltered EMG, same muscle. As the trace begins, the upper limb is extended directly forward. The deflection of the velocity curve indicates voluntary abduction of the limb, moving the hand 30 cm to the subject's right. The EMG shows two peaks of activity. b: Voluntary abduction of same limb, with artificial acceleration. Onset of voluntary movement triggers a torque motor, which accelerates the movement. The initial peak **A**, at acceleration of the limb, is approximately the same size as in **1a**. Peak **B** is increased.

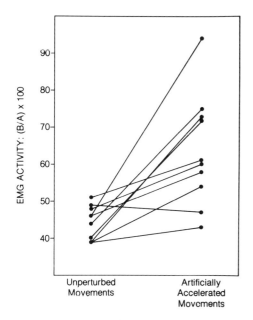

FIG. 2. Effect of artificial acceleration on the second EMG burst in posterior deltoid muscle. Left: Mean B/A ratios obtained during 25 ballistic abductions of the arm without perturbation. Right: Mean ratios during 25 movements artificially accelerated by the torque motor.

Another candidate for membership in the SR family is the burst of EMG activity that occurs after the silent period when a muscle is unloaded during a sustained isometric contraction (Alston et al., 1967). This burst after unloading (BAU) occurs while the limb is moving passively (by inertia) and is reduced in limbs affected by the upper motor neuron syndrome, as might be expected if BAU is a form of SR. Since SR is known to be exaggerated in muscles affected by parkinsonian rigidity, one might expect the BAU to be increased in the rigid muscles. The BAU was compared in 9 patients with Parkinson's disease and in 6 control subjects. In each case, the unloading reflex was elicited 40 times and averaged. The size of BAU was expressed as a percentage of the baseline EMG just before unloading. In parkinsonism, the mean BAU was 198 ± 54 (SD). In the control group, it was 199 ± 153 (SD). The difference was not statistically significant. This negative result would suggest that the BAU is not, in fact, a special case of the SR. Instead, one might conclude that the two phenomena are mediated by different neuronal mechanisms.

MEASUREMENT OF SR IN TRICEPS BRACHII

The subject is seated with the right arm abducted to 90° and supported against gravity. The elbow is flexed to 90°, with the hand extending straightforward to grasp a recording handle coupled to a torque motor. When the motor is energized, it exerts a force of approximately 20 Newtons, pulling the handle to the subject's right and thus extending the forearm. Passive movement is arrested by a mechanical stop, so that the distance traveled by the hand is 20 cm. Each subject is instructed as follows: (1) keep a firm grip on the handle; (2) relax the arm and shoulder muscles; (3) do not resist the pull of the motor; and (4) do not assist the movement that is caused by the motor. Surface EMGs of the right biceps and triceps are full-wave rectified and filtered. Signals representing the velocity of the hand and the EMGs are averaged in blocks of 8 and stored on magnetic tape to be later displayed by an X-Y plotter. In comparing SRs obtained from different subjects, it is necessary to normalize EMG signals. To provide a standard EMG for triceps, the subject holds the elbow at 90°, contracting the muscle against a force of 20 Newtons at the hand, the resulting EMG being processed as described above. The same procedure is repeated during contraction of the biceps against a 20-Newton force, providing a standard for that muscle. For each block of 8 SR tests, the maximal deflection of the triceps EMG curve is located, and its distance above the zero baseline is measured in millimeters. The deflection above baseline during the steady 20-Newton contraction, also summed over 8 sweeps of the signal averager, is used as the standard de-

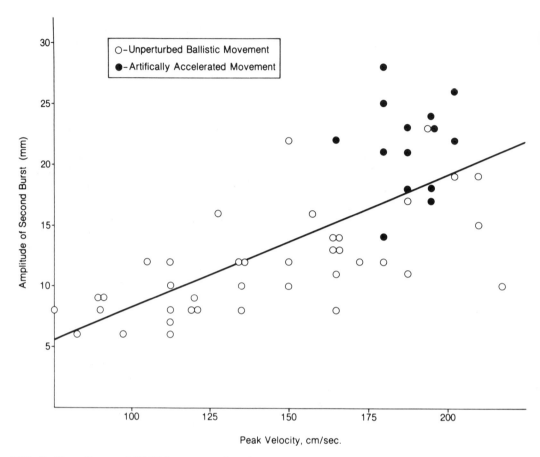

FIG. 3. Size of second EMG burst as a function of velocity. The arm was abducted voluntarily at various speeds, some of the movements being accelerated by the torque motor. Each point represents the size of the second burst on one trial. There is a significant correlation between velocity and size of the second burst ($r = 0.71$).

flection. The size of the SR in triceps is defined by the formula:

SR = maximal EMG deflection (mm) ÷ standard deflection (mm) × 100.

The duration of movement is the interval between rise of the velocity curve and return to zero velocity; the distance covered by the handle is 20 cm. Hence, the mean velocity is equal to 20 cm divided by the duration.

To determine the effects of reinforcement, some of the passive extensions are produced during active contraction of the elbow flexor muscles. On these reinforced trials, an initial force is applied to the handle by means of a coil spring. In order to hold the elbow at 90°, the subject has to counteract the spring by contracting elbow flexors. On some trials the initial force (F) is 10 Newtons, and on others it is 20 Newtons. These conditions are designated as F10 and F20, respectively. When no spring is attached to the handle, the condition is designated as F0. Under conditions F10 and F20, the subject is instructed to maintain the starting position until he senses the pull of the motor, at which time he should relax the flexor muscles, allowing the motor to extend the forearm passively. Passive extension is produced in blocks of 8 trials.

The conditions are presented in a counterbalanced order that is designed to obviate any effects of practice or fatigue: F0, F10, F20, F20, F10, F0.

THE SR IN PARKINSONISM

Shortening reactions were compared in parkinsonian patients with those in normal controls. A control group consisted of 8 men age 34 to 77 years (mean: 58 years); they had normal strength, tone, coordination, reflexes, and sensation in the right upper limb. The parkinsonian group consisted of 8 men from 55 to 84 years (mean: 65.5 years), with moderate to severe clinical signs, and receiving treatment with various drug regimens that were not interrupted for this study.

All of the records obtained under condition F0 had several features in common. Although the subjects were supposed to relax the arm prior to the passive movement, a certain amount of EMG activity was often present in biceps and triceps before the external force was applied. A sudden rise of the velocity curve marked the onset of passive extension, after which the biceps usually showed a transient EMG increase (stretch reflex). This was followed by a peak in the triceps, which was the electrical sign of the SR (Fig. 4). Under condition F0 (without reinforcement), the SR is relatively small and the passive movement is short. Figure 5 shows that reinforcement increases size of SR and duration of the passive movement.

Records from a parkinsonian patient (Figs. 6 and 7) are strikingly different. Un-

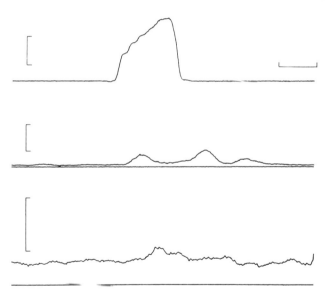

FIG. 4. Shortening reaction in a normal subject, condition F0. From top downward, (1) velocity of hand, (2) rectified biceps EMG, (3) biceps baseline, (4) rectified EMG, and (5) triceps baseline. Extension of elbow causes upward deflection of velocity trace. As trace begins, elbow is flexed to 90°. Although subject was instructed to relax the arm, each muscle shows some EMG activity. At time indicated by deflection of velocity curve, forearm is abducted passively by an external force of 20 Newtons, which subject has been instructed not to resist or assist. During passive abduction, biceps shows an increase of activity (stretch reflex), and triceps shows a relatively small increase (SR). Calibrations for Figs. 4–9: Velocity, 100 cm/sec; biceps and triceps EMG, 50% of standard; time, 100 msec. Each line represents the computed sum of 8 trials.

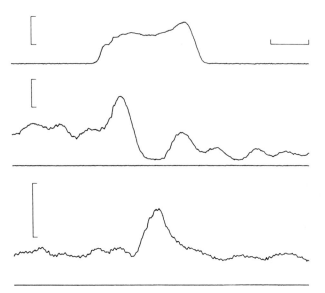

FIG. 5. Shortening reaction in same subject, condition F20. As trace begins, elbow is at 90°, and flexor muscles are resisting an external force of 20 Newtons. An additional force of 20 Newtons causes passive abduction of forearm. A prominent stretch reflex in biceps is followed by a relatively large SR in triceps, showing effect of increased reinforcement. Mean velocity of movement is less than it was under condition F10.

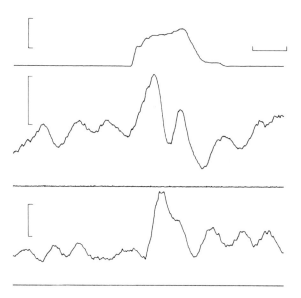

FIG. 6. Shortening reaction in patient with Parkinsonism, condition F0. Initial conditions are the same as in Fig. 4. Both muscles show considerable baseline EMG. Passive abduction, shown by *upward deflection of top line*, is followed by stretch reflex in biceps and large SR in triceps. As compared with Fig. 1, the duration of passive movement is longer.

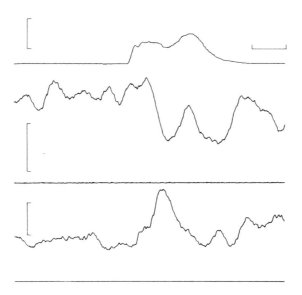

FIG. 7. Shortening reaction in same patient, condition F20. Initial conditions are the same as in Fig. 5. Shortening reaction is again very prominent but not larger than in Fig. 4.

der condition F0, the SR is larger than normal, and its size does not increase on the trials with reinforcement. The velocity of passive movement is smaller than normal under all three conditions. Figure 8 shows that, as expected, the size of SR was significantly larger in parkinsonism patients and that reinforcement had no effect on the size of SR. This unexpected observation suggests that parkinsonism interferes with the supranuclear factors associated with preparatory set, or it could indicate that SR is saturated in parkinsonism.

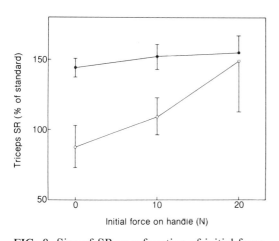

FIG. 8. Size of SR as a function of initial force exerted by flexor muscles. *Abscissas:* Initial force on handle. *Ordinates:* Size of triceps SR, expressed as percentage of standard. *Open circles:* Means for normal subjects. *Closed circles:* Means for Parkinsonian patients. *Vertical lines:* Standard error.

THE SR IN CEREBELLAR ATAXIA

If the SR is exaggerated in patients with extrapyramidal disease and reduced in those with pyramidal tract lesions, it is pertinent to ask whether the SR is affected by lesions of the cerebellum. Accordingly, the test described above was applied to a patient with unilateral ataxia, caused by a CT-scan-proven lesion of the right cerebellar hemisphere. The clinically normal left limb served as a control. In the ataxic arm, the SR was significantly larger than in the opposite limb, and it did not show the normal increase in reinforcement (Figs. 9 and 10).

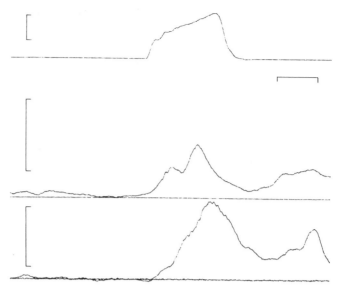

FIG. 9. Shortening reaction from ataxic arm of cerebellar patient. Initial conditions are the same as in Fig. 4.

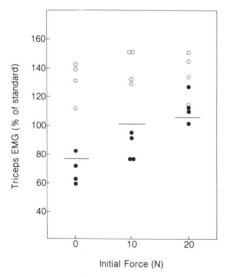

FIG. 10. Size of SR in ataxic and normal limbs of patients with right-sided cerebellar lesion. *Abscissas:* Initial force exerted by flexor muscles. *Ordinates:* Amplitude of triceps EMG response, expressed as percentage of EMG during standard contraction. Each point is the mean of 8 measurements. *Open circles:* Right arm. *Solid circles:* Left arm. Each *horizontal line* indicates the mean of the corresponding means obtained from 5 control subjects. In the patient's left (unaffected) arm, as in controls, the SR tends to increase with the initial force exerted by flexor muscles (reinforcement). In the right (ataxic) arm, the SR is greater than in the unaffected arm, and it does not increase significantly with reinforcement.

The SR was also measured in a patient with bilateral ataxia due to cerebellar disease and compared with the SRs from a group of 5 normal controls. In the control group, the size of the SR was 75% of standard ± (SD) under condition F0. In the ataxic patient, the corresponding value was 203%. At each level of initial force, SR was significantly larger in the ataxic limb than in control limb ($p < 0.01$), and did not show the normal increase with reinforcement.

MECHANISM AND SIGNIFICANCE OF THE SR

Attempts to define the receptors and the central pathways responsible for the SR have not given definitive answers. Andrews et al. (1972, 1973) found that the dynamic SR was reduced or abolished by limb ischemia, whereas the static SR was unchanged. Procaine infiltration reduced or abolished the dynamic and static stretch reflexes of the muscle infiltrated. The SR disappeared at first, but soon returned. Their data suggested that the dynamic SR depends on large afferent fibers, and the static SR depends on smaller fibers from the antagonist. Andrews and co-workers concluded that the distribution of the SR in flexor and extensor muscles is not readily explained by the known properties of group I and group II reflex effects.

Although the mechanism and functional significance of the SR have not been determined, measurement of the SR could have clinical as well as theoretical value. For example, Andrews and Burke (1972) have found that decrease in amplitude of the triceps SR is correlated with clinical movement in parkinsonism. The technique described above could be used to measure such correlations, monitor the effects of therapy, and facilitate research on movement disorders.

ACKNOWLEDGMENTS

This research was supported by the Veterans Administration. Dr. Peter Eisenberg gave helpful comments on the manuscript.

Motor Control Mechanisms in Health and Disease,
edited by J. E. Desmedt.
Raven Press, New York © 1983.

Neuronal Organization of the Premotor System Controlling Horizontal Conjugate Eye Movements and Vestibular Nystagmus

*Hiroshi Shimazu

Department of Neurophysiology, Institute of Brain Research, University of Tokyo School of Medicine, Tokyo, Japan

Characteristic patterns of neuronal discharges related to slow and fast eye movements have been found in a wide area of the brainstem in the behaving animal. Besides ocular motoneurons (MNs), a keen interest has been taken in neurons located in the vestibular nuclei (Duensing and Schaefer, 1958, Luschei and Fuchs, 1972; Miles, 1974; Fuchs and Kimm, 1975; Keller and Daniels, 1975; Keller and Kamath, 1975; Waespe et al., 1977; McCrea et al., 1980), the reticular formation (Duensing and Schaefer, 1957; Sparks and Travis, 1971; Cohen and Henn, 1972a,b; Luschei and Fuchs, 1972; Keller, 1974), the prepositus hypoglossi nucleus (Baker et al., 1976), and interneurons in the abducens nucleus (Delgado-Garcia et al., 1977). Eye movement-related neurons are classified as tonic, burst, burst–tonic, or pause neurons according to their discharge pattern associated with slow or fast eye movements. If proper coupling of several types of brainstem neuron is hypothesized in a wiring model, it could theoretically generate characteristic patterns of eye movement. It seems now crucial to know what types of neuron in what structures actually make direct connection with each other and what kinds of functionally identified neuron are immediate premotor neurons connecting directly with MNs to generate eye movements.

The oculomotor function consists of slow and fast eye movements. Slow eye movements comprise pursuit movements and slow phases of vestibular and optokinetic nystagmus, and fast eye movements include saccades and quick phases of nystagmus. Unit activities recorded in various structures of the brainstem exhibit, in general, firing patterns in a similar fashion for both visual- and vestibular-induced eye movements. Since vertibular nystagmus can easily be reproduced, even in acute experiments, either by electrical stimulation of the vestibular nerve or by rotation of the turntable on which the animal is mounted, we have been investigating MN and premotor neuronal activities related to vestibular nystagmus in the cat under carefully controlled local anesthesia. Neural connection between single neurons and/or the target sites of their axonal projection were examined on each neuron whose discharge characteristics were functionally identified in association with nystagmus.

Membrane potential changes in abducens

---

*Department of Neurophysiology, Institute of Brain Research, University of Tokyo School of Medicine, 7-3-1 Hongo, Bunkyo-ku, Tokyo, Japan.

MNs during nystagmus were investigated to reveal excitatory and inhibitory inputs composing slow and quick phases of MN activity (Maeda et al., 1972). Figure 1 represents schematically component postsynaptic potentials (PSPs) in abducens MNs on both sides which act as the agonist and the antagonist during horizontal conjugate eye movements. In the agonist, the membrane potential gradually progressed in the depolarizing direction during the slow phase (Fig. 1B), causing a gradual increase in firing rate (Fig. 1A). This depolarization was found to be caused by a summation of excitatory PSPs (EPSPs). In the antagonist, the membrane potential shifted slowly in the hyperpolarizing direction with a mirror image of the membrane potential of the agonistic MN (Fig. 1D). This hyperpolarizing deflection was caused not only by reduction of EPSPs but also by increase in the summation of inhibitory PSPs (IPSPs). During the quick phase directed to the opposite side, the former antagonist became the agonist in which a burst of spikes was induced (Fig. 1C). The burst was caused by a steep depolarization of the MN, which consisted of sudden production of EPSPs and a decrease in preexisting IPSPs (disinhibition) (Fig. 1D). In the antagonist, the spike activity was rapidly suppressed at the time of steep hyperpolarizing deflection of the membrane potential (Fig. 1A). The hyperpolarization was found to consist of sudden production of IPSPs in addition to a decrease in preexisting EPSPs (disfacilitation) (Fig. 1B). Similar results were obtained by intracellular recording from trochlear MNs in the oblique oculomotor system as well (Baker and Berthoz, 1974). The steep hyperpolarization in the antagonist and the steep depolarization in the agonist were suggested to occur synchronously by observing field potential changes in the abducens nuclei at the beginning of both saccades and quick phases of vestibular and optokinetic nystagmus in the behaving cat (Nakao et al., 1977). The succeeding sections describe functional properties of premotor neurons in the vestibular nuclei and reticular formation, which may explain the component PSPs in abducens MNs associated with slow and quick phases of vestibular nystagmus.

PREMOTOR NEURONS IN THE MEDIAL VESTIBULAR NUCLEUS

Since there are various kinds of neurons in the vestibular nuclei in terms of their input from labyrinthine receptors and the destination of their axonal projection, we

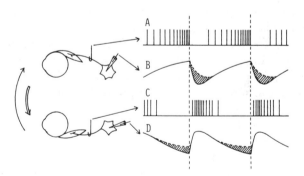

FIG. 1. Schematic representation of abducens motor activity during slow and quick phases of vestibular nystagmus. Long and short *arrows* in front of the eyes (**left**) indicate the direction of slow and fast eye movements, respectively. A and C: Impulse activity of abducens MNs on both sides. B and D: Intracellularly recorded membrane potentials in abducens MNs on both sides. *Hatched area* represents the contribution of IPSPs. *Vertical broken lines* indicate the onset of quick phases.

identified each vestibular neuron as receiving monosynaptic input from the ipsilateral horizontal canal and projecting to either the contralateral or the ipsilateral abducens nucleus. Spikes of single neurons were recorded extracellularly in the medial vestibular nucleus (MVN) and were identified as horizontal type I neurons by their excitatory response to ipsilateral horizontal rotation of the turntable and inhibitory response to contralateral rotation (Duensing and Schaefer, 1958; Shimazu and Precht, 1965). Secondary neurons were identified by their monosynaptic activation following stimulation of the ipsilateral vestibular nerve (Fig. 2B) (Precht and Shimazu, 1965).

Excitatory Vestibular Neurons

The secondary vestibular neurons projecting to the contralateral abducens nucleus have been shown to excite abducens MNs (Baker et al., 1969; Highstein, 1973; Hikosaka et al., 1980). We have used three techniques to reveal that the axon of individual neurons recorded in the MVN projects to and terminates in the contralateral abducens nucleus.

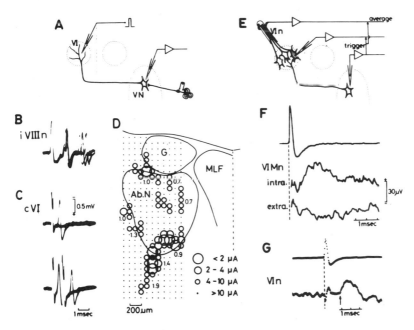

FIG. 2. Identification of excitatory premotor type I MVN neurons. A: Experimental arrangement for B–D. B: Monosynaptic activation of a type I neuron after stimulation of the ipsilateral vestibular nerve. C: Antidromic activation of the neuron after single- (top) and double- (bottom) shock stimulation of the contralateral abducens nucleus. D: Distribution of effective sites in and around the contralateral abducens nucleus for antidromic activation of a single type I MVN neuron. Thresholds for antidromic activation are shown by the size of circles in each stimulus site. Numerals indicate antidromic latencies in milliseconds (msec) from different effective sites. E: Schematic representation of experimental arrangements for F and G. F: Spike-triggered average of the membrane potential of a contralateral abducens MN (middle) and the extracellular field potential (bottom) after spikes of a single type I MVN neuron (top) (400 sweeps). G: Averaged potential of the contralateral abducens nerve (bottom) triggered by spikes of a single type I MVN neuron (top) (6065 sweeps). Arrow indicates the onset of increase in spike density. VI or Ab.N.: abducens nucleus, VIn: abducens nerve, VIIIn: Vestibular nerve, G: genu facialis, MLF: medial longitudinal fasciculus, VN: vestibular nuclei.

(1) During extracellular recording of spike activity of an identified secondary type I MVN neuron, systematic tracks with the stimulating microelectrode were made in and near the contralateral abducens nucleus (Fig. 2A) to find effective sites for antidromic activation of the MVN neuron (Fig. 2C). Figure 2D exemplifies the distribution of low-threshold effective sites for a single vestibular neuron. Antidromic activation was obtained from scattered, isolated spots within and near the abducens nucleus with varying stimulus thresholds of less than 10 µA. The patchlike distribution of effective sites, separated by ineffective spots, indicates multiple axonal branching within the nucleus (Nakao et al., 1981). Furthermore, the latencies of antidromic activation from different areas in the nucleus were widely divergent. The shift in latencies from nearby effective spots rules out the spread of stimulus current to a single axon passing through the nucleus or to neighboring structures such as the medial longitudinal fasciculus (MLF). To confirm the above view, the recently developed technique of intraaxonal staining of single neurons with horseradish peroxidase (HRP) (Snow et al., 1976; Jankowska et al., 1976; Cullheim and Kellerth, 1976; Kitai et al., 1976) was applied to electrophysiologically identified secondary type I vestibular neurons. The axon projecting to the contralateral abducens nucleus gave off a number of local collateral branches and distributed terminals in a three-dimensional area in and around the nucleus (Ishizuka et al., 1980; McCrea et al., 1980). The morphophysiological consistency justifies the interpretation that the multiple effective foci for antidromic activation of a single MVN neuron indicate its axonal termination in the contralateral abducens nucleus.

(2) More direct evidence for monosynaptic connection of MVN neurons with abducens MNs was provided by the technique using spike-triggered average (STA) of PSPs (see recent review in Kirkwood and Sears, 1980). During extracellular recording of spike activity of an identified secondary type I MVN neuron, intracellular recording was simultaneously made from a contralateral abducens MN. The membrane potential in the MN was triggered from spontaneous or glutamate-induced repetitive spikes of the single MVN neuron and averaged over a 10-msec sweep duration after spikes (Fig. 2E). The effectiveness of this technique has been well demonstrated in previous studies (Mendell and Henneman, 1971; Jankowska and Roberts, 1972; Watt et al., 1976; Rapoport et al., 1977; Hikosaka et al., 1978). When the averaged intracellular potential revealed an EPSP (Fig. 2F), it was interpreted as a unitary EPSP which was produced by synaptic action of the single MVN neuron. By comparing the latency of the unitary EPSP measured from the foot of the triggering spike (0.9 msec) with the latency of antidromic activation of the MVN neuron on stimulation of the contralateral abducens nucleus (0.6 msec), the unitary EPSP was concluded to be induced monosynaptically. This provides direct evidence indicating that the MVN neuron makes a direct excitatory connection with abducens MNs (Hikosaka et al., 1980b).

(3) When the abducens nerve was spontaneously activated, a technique of correlating the firing of a single MVN neuron with abducens MN activity could be employed. As shown schematically in Fig. 2E, spontaneous discharges of the whole contralateral abducens nerve were monopolarly recorded at its cut end in the orbit with a DC amplifier, and were averaged over a 10-msec sweep duration before and after the triggering MVN neuron spikes. When the averaged potential trace exhibits a positive deflection, it is interpreted as an increase in the density of spikes in the whole nerve (Hikosaka et al., 1978). Since the probability of spike generation in MNs may increase even during small unitary EPSPs, and a single axon diverges into a number of terminal branches connecting with a num-

ber of MNs (Ishizuka et al., 1980), the total effects of the single axon would be reflected in an increase in spike density of the whole nerve. In fact, the averaged potential in Fig. 2G exhibited a transient positive deflection with the latency of 1.4 msec measured from the foot of the triggering spike. The latency corresponded to the sum of the latency of the monosynaptic unitary EPSP of MNs and the conduction time of impulses from the abducens nucleus to the orbit (Nakao et al., 1981). This indicates that the MVN neuron, whose spikes were used as triggering spikes, makes monosynaptic excitatory connection with MNs.

Almost all (34/35) type I MVN neurons, which were identified as projecting to the contralateral abducens nucleus by means of the above described methods, exhibited periodic activity closely related to nystagmic rhythm (Nakao et al., 1981). The occurrence of their spikes was consistently in phase with contralateral abducens nerve discharges, and tonic activity during the slow phase was abruptly suppressed when the contralateral abducens nerve was silenced at the quick phase (Fig. 3B). The mean time course of discharge frequency during the slow phase was represented by averaging successive discharge frequencies of identified MVN neurons, which were examined with similar nystagmic cycles (Fig. 3C). There was a slight tendency for the discharge frequency to increase toward the end of the slow phase. It is, therefore, concluded that tonic activity of excitatory MVN neurons contributes to the slowly increasing EPSP of contralateral abducens MNs during the slow phase and that its abrupt cessation is responsible, at least in part, for the disfacilitation of MNs at the quick phase (see Fig. 1B).

When the direction of nystagmus was reversed, these MVN neurons exhibited a train of spikes during the quick excitatory phase of the contralateral abducens nerve

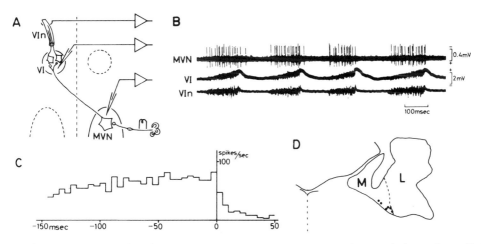

FIG. 3. Discharge pattern of excitatory premotor type I MVN neurons in association with vestibular nystagmus. A: Experimental arrangements. B: Spike activity of a single MVN neuron (top), field potential in the contralateral abducens nucleus (middle), and discharges of the contralateral abducens nerve (bottom). In this and succeeding figures, abducens field potentials were simultaneously recorded during nystagmus as an indicator of membrane potential changes in abducens MNs at the beginning of the quick phases. (From Maeda et al., 1972, with permission.) C: Averaged time course of successive discharge frequency in excitatory premotor type I MVN neurons. In this and Figs. 5 and 7, the time at zero in the diagram indicates the onset of the quick phase revealed by field potential changes in the abducens nucleus. D: Location of identified MVN neurons. M: medial vestibular nucleus. L: lateral vestibular nucleus.

or showed no activity. These characteristics were similar to those of axons of secondary type I vestibular neurons recorded in the contralateral abducens nucleus (monosynaptic V_c axons) in a previous study (Hikosaka et al., 1977), and their activity at the quick phase contributes to a fraction of EPSP of contralateral MNs (Fig. 1D).

Recording sites of these MVN neurons were marked by electrophoretic ejection of dye from the recording microelectrode. They were located in the rostral and ventrolateral part of the MVN, and some were on the border between the MVN and the lateral vestibular nucleus (LVN) (Fig. 3D). These sites were within the area of neurons projecting to the contralateral abducens nucleus examined in anatomical studies (Gacek, 1971; Maciewicz et al., 1977).

Inhibitory Vestibular Neurons

The secondary vestibular neurons projecting to the ipsilateral abducens nucleus have been shown to inhibit abducens MNs (Baker et al., 1969; Highstein, 1973; Hikosaka et al., 1980b). The three techniques used for identification of excitatory vestibular neurons were used to detect axonal projection of individual type I MVN neurons to the ipsilateral abducens nucleus.

(1) Systematic tracking for antidromic stimulation was made to identify secondary type I MVN neurons, which projected to the ipsilateral abducens nucleus. Before tracking on the ipsilateral side, the contralateral abducens nucleus was first stimulated with strong currents to confirm no antidromic activation, excluding the possibility that these MVN neurons were not passing through the ipsilateral nucleus and continuing contralaterally. Figure 4D exemplifies the threshold distribution of effective spots for antidromic activation of a single MVN neuron. The effective spots were found in the ventromedial part of the ipsilateral abducens nucleus and were separated by ineffective sites, suggesting the ex-

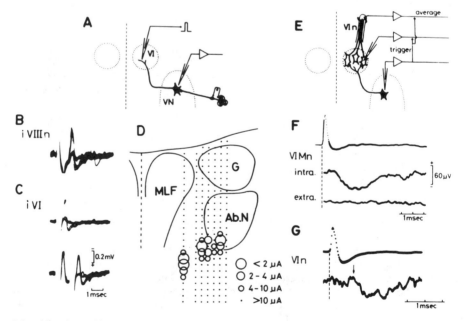

FIG. 4. Identification of inhibitory premotor type I MVN neurons. Same arrangement as in Fig. 2, but showing axonal projection to the ipsilateral abducens nucleus. Averaged unitary IPSP in F was recorded with 300 sweeps. *Arrow* in G indicates the onset of decrease in spike density of the whole abducens nerve (average of 600 sweeps).

istence of local axonal branching in this area. In general, the extent of distribution of effective spots for antidromic activation in the abducens nucleus was narrower in ipsilaterally projecting neurons than in contralaterally projecting neurons. This is in agreement with morphological findings obtained by intraaxonal staining with HRP, indicating that ipsilaterally projecting MVN neurons had the tendency toward localized distribution of terminal branches in the abducens nucleus (Ishizuka et al., 1980).

(2) The technique of SPA of PSPs was used to detect monosynaptic inhibitory connection of a single MVN neuron with ipsilateral abducens MNs. Figure 4F shows the averaged membrane potential of an abducens MN triggered from spikes of a single type I MVN neuron, revealing a unitary IPSP with a monosynaptic latency after triggering spikes.

(3) Spike-triggered average of whole abducens nerve activity could also be used for identification of inhibitory MVN neurons. As in the case of excitatory neurons, this technique was available only when the abducens nerve was spontaneously active. In Fig. 4G, the averaged potential trace of the ipsilateral abducens nerve triggered from spikes of a single type I MVN neuron exhibited a transient negative deflection, indicating an inhibitory effect on the MNs. The latency of inhibition was 1.4 msec, which was estimated to be in a monosynaptic range, indicating that the axon of the MVN neuron made monosynaptic inhibitory connection with ipsilateral abducens MNs.

All (12/12) inhibitory type I MVN neurons identified as projecting to the ipsilateral abducens nucleus exhibited a clear nystagmus-related firing pattern; i.e., tonic firing during the slow silent phase of the ipsilateral abducens nerve and suppression of firing during the quick excitatory phase (Fig. 5B) (Nakao et al., 1981). The averaged time course of inhibitory neuron discharges during the slow phase (Fig. 5C) showed a slightly steeper increase in frequency than that of excitatory neurons shown in Fig. 3C. The maximum frequency at the end of the slow phase was close to 100/sec, similar to that of excitatory neurons. These results indicate that tonic activity of inhibitory type I neurons contributes to the slow summation of IPSPs in ipsilateral abducens MNs during the slow eye movement directed to the

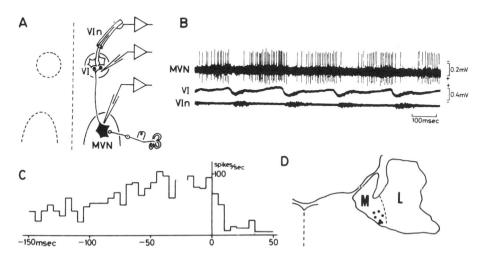

FIG. 5. Discharge pattern of inhibitory premotor type I MVN neurons in association with vestibular nystagmus. Same arrangement as in Fig. 3. Note that slow-phase discharges in the MVN neuron are in phase with silent period of the ipsilateral abducens nerve.

contralateral side and that the sudden cessation of their activity causes disinhibition of MNs at the quick phase (see Fig. 1D). Like excitatory MVN neurons, inhibitory type I neurons exhibited a train of spikes at the quick inhibitory phase of ipsilateral abducens MNs, thus contributing to a fraction of IPSP in MNs (see Fig. 1B). Inhibitory type I MVN neurons were also located in the rostral and ventrolateral part of the nucleus and intermingled with excitatory type I neurons (Fig. 5D).

In summary, these findings provide direct evidence for the view that both excitatory and inhibitory secondary type I MVN neurons terminating in the abducens nuclei participate in generation of both the slow and quick phases of nystagmic rhythm (Maeda et al., 1971, 1972; Hikosaka et al., 1977). In chronic experiments with alert monkeys, spike activity of a class of neurons in the vestibular nuclei has been found to pause during saccadic eye movements or quick phases of nystagmus (Miles, 1974; Fuchs and Kimm, 1975; Keller and Daniels, 1975; Waespe, 1977). This class of neurons in the monkey vestibular nuclei may correspond to secondary vestibular neurons projecting to the ocular motor nuclei and participating in the generation of nystagmus.

ORIGIN OF SPIKE SUPPRESSION OF SECONDARY TYPE I MVN NEURONS DURING THE QUICK PHASE

The question arises, is spike suppression of secondary MVN neurons at the quick phase caused by active inhibition or merely by disfacilitation and, if active inhibition contributes to it, where are inhibitory interneurons located? To answer this question, an intracellular study was performed on secondary MVN neurons, which showed nystagmus-related activity (Hikosaka et al., 1980*b*). In Fig. 6A, the membrane potential proceeded slightly in the depolarizing direction with increasing frequency of spikes during the slow phase. At the time of suppression of contralateral abducens nerve activity, the membrane potential shifted steeply in the hyperpolarizing direction, and pikes were suppressed abruptly. In order to examine the possible contribution of IPSPs to the steep hyperpolarizing deflection of the membrane potential, Cl^- ions were injected into the cell through the recording microelectrode to reverse the commissural IPSP into a depolarizing direction (Fig. 6B) (Mano et al., 1968). Under this condition, the gradual depolarizing shift during the slow phase was followed by an additional, steep depolarization at the beginning of the quick phase (Fig. 6C), instead of the steep hyperpolarizing shift consistently found in the control record. The potential changes were almost entirely attributable to the transmembrane potential as revealed by comparison with extracellular field potentials recorded during nystagmus (Fig. 6D). Thus, the synaptic mechanism, having produced a hyperpolarization at the quick phase in the control, induced a depolarization when the IPSP was inverted to a depolarizing direction. These results indicate that an abrupt production of IPSPs underlies the suppression of spike activity in secondary type I MVN neurons during quick phases.

The location and dynamic properties of inhibitory interneurons responsible for the above described nystagmus-related periodic inhibition of secondary type I neurons were investigated (Nakao et al., 1981). As a candidate for these inhibitory interneurons, type II vestibular neurons were selected for study, since they have been suggested to be inhibitory interneurons acting on type I neurons in the same nucleus (Fig. 7A) (Shimazu and Precht, 1966). Type II denotes the response in which spike discharges increase in frequency during contralateral horizontal angular acceleration and decrease during ipsilateral angular acceleration (Duensing and Schaefer, 1958). Among these neurons, those that were activated by stimulation of the contralateral

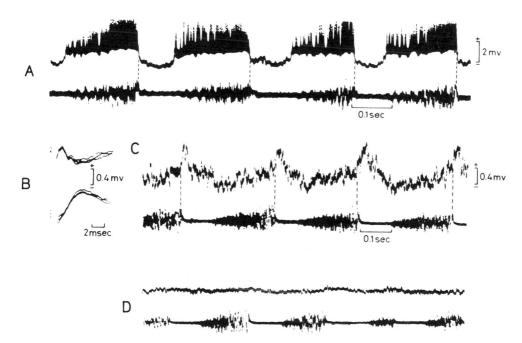

FIG. 6. A: Membrane potential changes and spike activity of a secondary MVN neuron (top) in association with nystagmic modulation of contralateral abducens nerve activity (bottom). Extremely high frequency of discharges is probably caused by injurious effects of penetration. B: Commissural IPSP induced by stimulation of the contralateral vestibular nerve (top) and its reversal to depolarization by injection of Cl⁻ ions into the cell (bottom). C: Nystagmus-related changes of the membrane potential after reversal of the IPSP. *Vertical broken lines* in A and C indicate the onset of the quick phases. D: Extracellular field potential as a control for C. (From Hikosaka et al., 1980b, with permission.)

vestibular nerve with short latencies (2.5–3.5 msec) were further selected for study (Fig. 7B). This activation was shown to be mediated through the commissural fibers interconnecting the bilateral vestibular nuclei (Shimazu and Precht, 1966). Most (19/21) of the type II neurons had a nystagmus-related rhythm, showing an abrupt increase in discharge frequency in quick excitatory phase of the ipsilateral abducens nerve (Fig. 7D and E). The onset of the burst was coincident with suppression of spikes of ipsilateral type I neurons (see Fig. 5B) and the duration of the burst was approximately the same as that of suppression of type I neurons. When the direction of nystagmus was reversed, the type II neurons exhibited tonic activity with gradually increasing frequency during the slow excitatory phase of the ipsilateral abducens nerve (Fig. 7F and G). This tonic activity of type II neurons was in phase with that of contralateral type I neurons (see Fig. 3B), which is in agreement with the view that type II neurons are activated from contralateral type I neurons through the commissural pathway. These nystagmus-related type II neurons were also located in the ventrolateral part of the rostral MVN and intermingled with the excitatory and inhibitory premotor type I neurons (Fig. 7C).

The inverse phase relation of firing between type I and type II vestibular neurons during nystagmic cycle is consistent with the hypothesis that the type II neurons are contributing to inhibition of type I neurons at the quick phase. This was more directly demonstrated by STA of the membrane po-

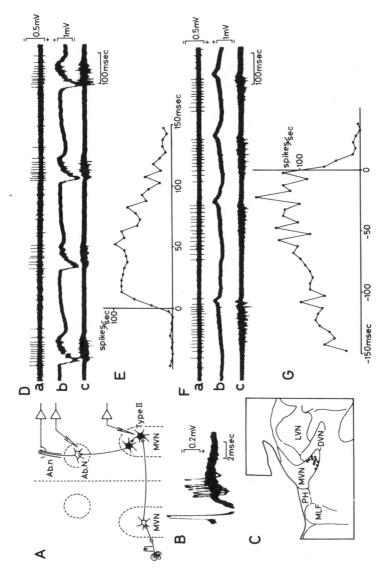

FIG. 7. Discharge pattern of type II MVN neurons in association with vestibular nystagmus. A: Experimental arrangements. B: Transcommissural activation of a type II neuron after stimulation of the contralateral vestibular nerve. C: Locations of nystagmus-related type II neurons in the MVN. D: Spike activity of the type II neuron shown in B (a), field potential in the ipsilateral abducens nucleus (b), and discharges of the ipsilateral abducens nerve (c) during nystagmus with quick phases directed to the ipsilateral side. E: Averaged time course of successive discharge frequency during quick phases. F: Same as in D, but the direction of nystagmus was reversed. G: Same as in E, but measured during slow phases directed to the ipsilateral side. (From Nakao et al., 1981, with permission.)

tential of a type I neuron after spikes of a single type II neuron (Nakao et al., 1981). The averaged potential revealed a unitary IPSP with a monosynaptic latency (Fig. 8), indicating that the type II neurons activated through the commissural pathway make direct inhibitory connections with secondary type I neurons. It is, therefore, reasonable to conclude that the type II vestibular neurons mediating commissural inhibition play a role in the origin of the nystagmic modulation of activity of type I neurons projecting to the abducens nuclei and thereby contribute to the generation of nystagmic rhythm of MNs. The next question is, where does excitatory input causing spike burst of type II neurons come from? This problem will be discussed after describing functional properties and axonal projections of burst neurons in the reticular formation.

PREMOTOR BURST NEURONS IN THE RETICULAR FORMATION

A group of neurons in the pontine reticular formation has been reported to exhibit a high-frequency burst of spikes prior to fast eye movements, and they have been called burst neurons or bursters (Sparks and Travis, 1971; Luschei and Fuchs, 1972; Cohen and Henn, 1972a,b; Keller, 1974). On the basis of a close functional correlation of activity of burst neurons with eye movement, these neurons were presumed to project to the ocular motor nuclei either directly or through a few synapses and provide an excitatory motor command signal for fast eye movements. However, there was little evidence for the destination of the axon of these burst neurons. There could be nonpremotor burst neurons which project to the sites other than the ocular motor nuclei. In fact, some burst neurons in the paramedian part of the pontine tegmentum were found to send their axon to the cerebellar flocculus (Nakao et al., 1980b). Therefore, to understand the premotor neuronal organization in the brainstem for generation of fast eye movements, it was thought to be essential to study whether or not there are burst neurons directly con-

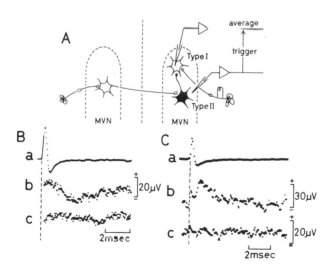

FIG. 8. Direct inhibitory synaptic connection of type II MVN neurons with type I MVN neurons. A: Schematic representation of experimental arrangements. B: Averaged unitary IPSP of a type I neuron (b) and averaged extracellular field potential (c) triggered from spikes of a single type II neuron (a) (600 sweeps). C: Same arrangement of records as in B, but the unitary IPSP was recorded with another pair of single type II and I MVN neurons, when the commissural IPSPs induced by contralateral vestibular nerve stimulation were inverted into a depolarizing potential by electrophoretic ejection of Cl$^-$ ions into the cell (1,200 sweeps). (From Nakao et al., 1981, with permission.)

nected with MNs, whether they are excitatory or inhibitory in nature, and where they are located. These questions have been investigated in a series of experiments in our laboratory (Hikosaka et al., 1977; Hikosaka and Kawakami, 1977; Hikosaka et al., 1978; Igusa et al., 1980; Sasaki and Shimazu, 1981a).

Inhibitory Burst Neurons

Recording of axonal spikes within the abducens nucleus revealed the existence of a class of axons which exhibited high-frequency burst activity at the quick inhibitory phase of MNs in the same nucleus (Hikosaka et al., 1977). Their intraburst frequency was 300 to 800/sec and the beginning of their spike burst was coincident with the onset of IPSPs in MNs at the quick phase. Assuming that these axons terminate on abducens MNs, the IPSPs produced in MNs at the quick inhibitory phase (Fig. 1B) were suggested to be mainly attributed to synaptic action of these burst axons. The origin of cells of these axons was explored in the brainstem and was found to be located predominantly in the dorsomedial part of the pontomedullary reticular formation caudal to the contralateral abducens nucleus (Hikosaka and Kawakami, 1977). The burst neurons recorded in these regions were antidromically activated from the contralateral abducens nucleus. Systematic tracking for antidromic stimulation in and near the abducens nucleus showed that the low-threshold effective spots for antidromic activation of a single burst neuron were separated by ineffective sites in the nucleus. The patchlike distribution of effective sites suggested multiple axonal branching within the nucleus. This has recently been demonstrated in a morphological study using intraaxonal staining with HRP (Yoshida et al., 1981). These neurons exhibited a burst of spikes when the activity of the contralateral abducens nerve was suppressed during the quick phases directed to the ipsilateral side.

Their firing characteristics corresponded to those of unidirectional medium-lead burst neurons found in the pontine reticular formation of the alert monkey (Luschei and Fuchs, 1972; Keller, 1974). Electrical stimulation applied to the region where these burst neurons were densely located induced monosynaptic IPSPs in contralateral abducens MNs. It was, therefore, suggested that these neurons were inhibitory in nature and they were termed inhibitory burst neurons (IBNs).

To demonstrate that the IBNs are actually inhibitory premotor neurons directly connected with abducens MNs, the SPA technique was used (Hikosaka et al., 1978). Inhibitory burst neurons were first identified by the occurrence of burst activity during vestibular nystagmus induced by horizontal rotation (Fig. 9A), and then their low-frequency, spontaneous discharges under non-nystagmic conditions (Fig. 9B) or spikes induced by glutamate ions ejected from the recording microelectrode were used as triggering spikes. Under the non-nystagmic condition, the cross correlation between two simultaneously recorded IBNs was examined, and no correlation was found in any pair in the analysis interval of 10 msec. This indicates that the averaged PSPs described below were unitary events evoked via the axon terminals of single IBNs.

While recording spikes of an IBN, localized stimulation was applied to the contralateral abducens nucleus to confirm antidromic activation from the nucleus (Fig. 10D). The membrane potential recorded intracellularly from a contralateral abducens MN was triggered from the IBN spikes (Fig. 10E) and averaged over a 10-msec sweep duration after the spikes. The averaged potential revealed a unitary IPSP with a monosynaptic latency measured from the foot of the triggering spikes (Fig. 10F). Several MNs were impaled during recording from a single IBN. Seventy-six pairs of IBNs and abducens MNs, consisting of 29 IBNs and 76 MNs, were investigated. Of

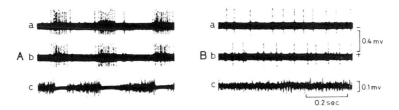

FIG. 9. Simultaneous recording of unit spikes of two IBNs. A: During nystagmus. B: Under nonnystagmic condition. A and B: Spikes of a pair of IBNs on the right side. C: Abducens nerve discharges on the left side. (From Hikosaka et al., 1978, with permission.)

the 76 pairs, 45 (60%) revealed unitary IPSPs, suggesting a widely divergent connection of single IBNs with contralateral abducens MNs. Locations of IBNs that were identified as immediate premotor neurons are shown in Fig. 10B.

In summary, burst activity of IBNs participates in active inhibition of contralateral abducens MNs during fast eye movements directed to the ipsilateral side. The major part, especially the initial phase, of IPSPs produced in antagonistic abducens MNs (Fig. 1B) is attributed to the summation of unitary IPSPs caused by the burst activity of IBNs.

Excitatory Burst Neurons

The results described above are put in order and schematically represented in Fig. 11. Briefly, tonic activity of left abducens MNs (L.ABMn) during the slow phase is

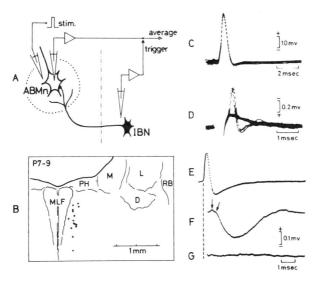

FIG. 10. Identification of premotor IBNs. A: Schematic representation of experimental arrangements. B: Location of IBNs (dots), which were identified as making monosynaptic inhibitory connections with contralateral abducens MNs. The IBNs located in a rostrocaudal extent from P7 to P9 are shown by projecting on a frontal plane. PH: Prepositus hypoglossi nucleus, D: descending vestibular nucleus, RB: restiform body. C: Intracellular spike potential antidromically induced in an abducens MN. D: Extracellular spikes of an IBN activated antidromically from the contralateral abducens nucleus. E–G: Averaged unitary IPSP of the contralateral abducens MN (F) and averaged extracellular field potential (G) triggered from spikes of the IBN (E) (360 sweeps). In F, *first arrow* indicates the arrival of the presynaptic impulse and *second arrow* the onset of IPSP. (From Hikosaka et al., 1978, with permission.)

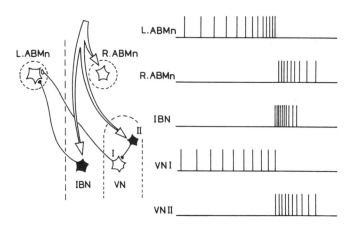

FIG. 11. Nystagmus-related activities of abducens MNs on both sides, the right IBN and types I and II vestibular neurons (VN I and VN II). Left scheme illustrates identified connections among these neurons. Neurons of *open* and *filled* shapes represent excitatory and inhibitory neurons, respectively. *Trifurcate arrow* indicates a hypothesized common excitatory burst input which was later confirmed.

maintained, at least in part, by tonic excitatory input from right type I vestibular neurons (VN I) and is inhibited at the quick phase by a spike burst of right IBN. The tonic activity of type I vestibular neurons induced by horizontal canal input is also inhibited at the quick phase by burst activity of type II vestibular neurons (VN II). The burst activity of right abducens MNs (R.ABMn) is attributed to production of steep EPSPs in the MNs. The origin of these EPSPs, however, was not fully elucidated except for a partial contribution of excitatory input from left type I vestibular neurons (not illustrated in Fig. 11). The coincidence of burst activities of abducens MNs, IBNs, and type II vestibular neurons on the same side appears to be essential for well coordinated activity of bilateral abducens MNs during fast, conjugate eye movements. The question then arises of how the synchronism of burst activities in the different cell groups can be assured? The simplest hypothesis would be the existence of a common source of excitatory input to these three kinds of neuron through the axon collaterals of single burst neurons.

To examine the above possibility, a wide area of the pontine reticular formation was explored. Two classes of burst neurons were found in the dorsomedial part of the pontine reticular formation immediately rostral to the abducens nucleus (Fig. 12C and D) and were identified as excitatory in nature, thus being called excitatory burst neurons (EBNs) (Igusa et al., 1980; Sasaki and Shimazu, 1981a). A class of EBN exhibited high-frequency spikes at the beginning of the quick phases (Fig. 12A). The maximum intraburst frequency was 300/sec in this neuron and attained 400/sec in other neurons. When the direction of nystagmus was reversed, the EBNs were silent or fired sporadically without any relation to nystagmic cycle. The onset of burst activity preceded ipsilateral abducens nerve discharges by 5–10 msec, corresponding to firing characteristics of unidirectional medium-lead burst neurons. Thus, their firing patterns were essentially similar to those of caudally located IBNs, except that intraburst frequency of IBNs was approximately twice as high as that of rostrally located EBNs. The other class of EBNs was differentiated from the above by their characteristic onset time of spike activity related to the quick phases. They exhibited irregular low-frequency activity preceding rapid activation of the ip-

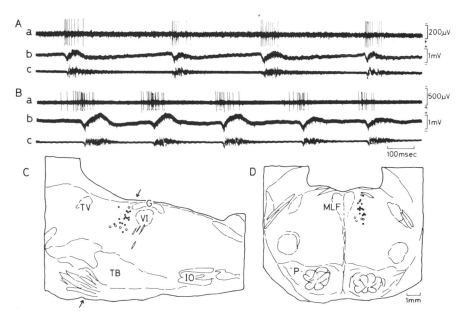

FIG. 12. Discharge patterns and locations of EBNs. A: Medium lead EBN. B: Long-lead EBN. In A and B, a: Unit spikes of EBN, b: field potentials in the ipsilateral abducens nucleus, c: ipsilateral abducens nerve activity. C and D: Locations of EBNs identified by antidromic activation from the ipsilateral abducens nucleus or by STA of unitary field potential. Marked spots for recording sites are projected to the parasagittal plane (C) and the transverse plane (D) indicated by *oblique arrows* in C. *Filled* and *open circles* represent medium-lead and long-lead EBNs, respectively. IO: Inferior olive, P: pontine nucleus, TB: trapezoid body, TV: ventral tegmental nucleus, VI: abducens nucleus. (From Sasaki and Shimazu, 1981*a,* with permission.)

silateral abducens nerve by 20 to 80 msec and displayed a burst of spikes (150–400/sec) immediately before abducens nerve activity (Fig. 12B). This neuron class resembles unidirectional long-lead burst neurons in the monkey pontine reticular formation in terms of their firing characteristics (Luschei and Fuchs, 1972; Keller, 1974).

In marked contrast to IBNs, the rostrally located EBNs were antidromically activated from the ipsilateral abducens nucleus. In order to investigate the extent of descending axonal projection of single EBNs, microelectrode tracking was made in a wide area caudal to the level of the abducens nucleus for antidromic activation of single EBNs. Figure 13 exemplifies the distribution of effective sites for antidromic activation of a single long-lead EBN with stimulus currents of less than 20 μA. A collision block test was employed to confirm the antidromic nature (Fig. 13A). The effective spots were found not only in the ipsilateral abducens nucleus (Fig. 13C) but also further caudally along the zone ventrolateral to the ipsilateral MLF (Fig. 13B–E). In the frontal plane 2 and 3 in Fig. 13B, the effective region in the dorsomedial reticular formation corresponded well to the area of IBN (Fig. 13D and E). The effective sites were found further laterally in the ventral part of the rostral MVN as well (Fig. 13E). It was noted that the low-threshold spots were separated by relatively high-threshold sites in the region of IBN as well as in the ventral area of the MVN. These results suggest that the axon of the single EBN extends caudally and that collaterals emerging from the stem axon give off a number of ramifications in the ipsilateral IBN region and the MVN. Antidromic latencies were not always shorter at the proximal sites

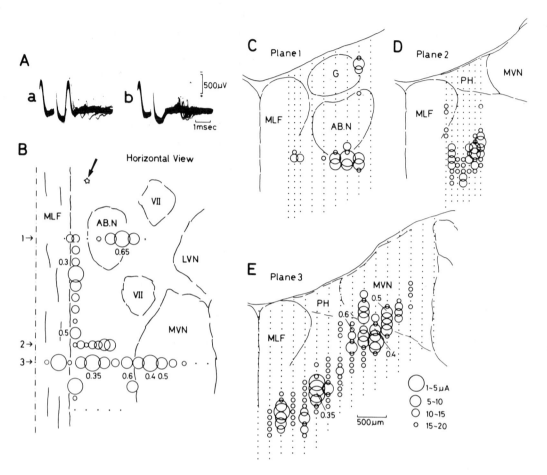

FIG. 13. Antidromic activation of a single long-lead EBN from the ipsilateral abducens nucleus, IBN area, and MVN. **A:** An example of collision block test. The sweeps were triggered by rotation-induced spikes, which were followed by antidromic stimulation. Stimulation in **b** failed to evoke spikes in contrast to full activation with slightly delayed stimulation in **a**. **B:** Horizontal view of the brainstem, showing effective sites for antidromic activation of the EBN the location of which is indicated by *oblique arrow*. **C–E:** Transverse view of three planes indicated by *horizontal arrows* 1, 2, and 3 in **B**. The diameter of each circle indicates the threshold current as shown in the *bottom-right*. *Dots without circles* indicate threshold currents > 20 μA. The diameter of each circle in **B** represents the lowest threshold value in each track of stimulating microelectrode. Numerals in **B** and **E** indicate latencies in milliseconds of antidromic response of the EBN from respective spots. AB.N: Abducens nucleus, VII: facial nerve. (From Sasaki and Shimazu, 1981a, with permission.)

than the distal sites relative to the location of the EBN cell. There was a shift in latencies from nearby effective spots. Therefore, the distribution of effective spots in Fig. 13 cannot be attributed to the simple course of a single passing axon, but indicates that a single EBN axon issues collateral branches and may terminate in the IBN region and the ventral part of the MVN. Some medium-lead EBNs were also antidromically activated from these regions. None of EBNs were activated antidromically from the spinal cord at the C_2 level.

To investigate possible target neurons with which EBN axons might make direct connection, the area immediately rostral to the abducens nucleus (EBN area) was stimulated with a tungsten microelectrode (Fig. 14A). Abducens MNs, IBNs, and types I and II MVN neurons were selected for studying effects of microstimulation of the EBN area. The effects on abducens MNs were investigated with intracellular recording (Fig. 14C). Stimulation of the ipsilateral EBN area at a relatively weak intensity (20 μA) induced EPSPs in a MN with the latency of 1.0 msec (0.8–1.2 msec in other MNs), indicating a monosynaptic excitatory connection with MNs in consideration of antidromic latencies of EBNs (0.4–1.3 msec) after microstimulation of the abducens nucleus. With double-shock stimulation at short intervals, the monosynaptic EPSP in response to the second shock was

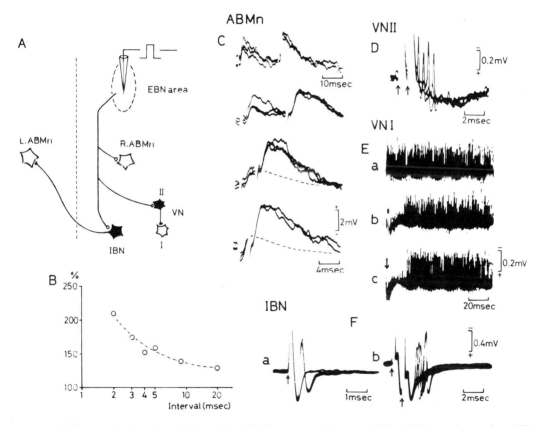

FIG. 14. Effects of microstimulation of the EBN area on abducens MNs, IBNs, and type I and II MVN neurons. A: Relevant neural connections. B: Facilitation of second EPSPs in an abducens MN with double shocks to the EBN area at varied intervals. C: Specimen records of EPSPs in the same abducens MN. Intervals between two shocks were varied. Time scale of 4 msec applies to the second to fourth traces. *Dotted lines* are drawn on the basis of averaged EPSPs induced by a single shock. D: Spikes of a type II vestibular neuron (4 superimposed traces) induced by double shocks *(arrows)* to the EBN area. Single shock stimulation induced no spikes. E: Inhibition of spikes of a type I MVN neuron (50 superimposed traces). a: Spontaneous spikes without stimulation; b: response to triple shocks (mark) to the EBN area; c: response to single shock *(arrow)* to the contralateral vestibular nerve. F: Response of an IBN. a: Antidromic response to microstimulation *(arrow)* of the contralateral abducens nucleus; b: spikes induced by double shocks *(arrows)* to the EBN area. Single-shock stimulation induced no spikes. Intensity of stimulus currents at the EBN area was 20 μA (0.1-msec duration) in each case. (From Sasaki and Shimazu, 1981a, with permission.)

markedly facilitated, the facilitation lasting for approximately 20 msec (Fig. 14B). The IBNs, which were identified by their discharge pattern in association with nystagmus and antidromic activation from the contralateral abducens nucleus (Fig. 13Fa), were transsynaptically activated by stimulation of the ipsilateral EBN area. In Fig. 13Fb, single-shock stimulation did not induce spikes, whereas double shocks with a short interval evoked spikes, probably due to the frequency potentiation effect described above. The latencies of induced spikes ranged from 1.3 to 2.2 msec from the effective shock. This suggests the existence of monosynaptic excitatory connection with IBNs from the EBN area, taking into account that the antidromic latencies of EBNs from the IBN area were 0.3 to 1.0 msec. Horizontal type II MVN neurons, which were identified by horizontal rotation and activation after contralateral vestibular nerve stimulation, were also activated transsynaptically by double-shock stimulation of the EBN area (Fig. 14D). The latencies of evoked spikes were 1.1 to 2.5 msec from the effective second shock, suggesting a monosynaptic excitatory connection with type II neurons. In contrast, spike activity of horizontal type I MVN neurons was suppressed for approximately 20 msec after stimulation (2 to 3 shocks at intervals of 1 to 2 msec) of the EBN area (Fig. 14Eb). Suppression of type I spike activity can well be explained by above-mentioned activation of type II neurons, because the latter exerts an inhibitory action on the former (Fig. 8). As expected, the duration of suppression of type I spikes was similar to that of commissural inhibition through type II neurons (Fig. 14Ec).

The results described above suggest that EBNs are a common source of excitatory input to ipsilateral abducens MNs, IBNs, and type II MVN neurons. However, the possibility could not be excluded that direct stimulation of the EBN area would activate passing fibers of some excitatory neurons located elsewhere, which might terminate on the above-mentioned target neurons. To confirm that individual EBNs in themselves make monosynaptic excitatory connection with abducens MNs, the technique of spike-triggered average was used. The results in Fig. 15 were obtained with a medium-lead EBN. It was antidromically activated with low-threshold currents applied to the restricted sites along a track through the center of the abducens nucleus (Fig. 15A and C), its antidromic latency being 0.4 msec (Fig. 15B). This indicates that the EBN sent its axon to or through the ipsilateral abducens nucleus. To differentiate between axons that terminate within the nucleus and those that merely pass through the nucleus, STA of the field potential was used with spontaneous spikes of the EBN recorded under non-nystagmic condition or glutamate-induced spikes (Fig. 15D). Within the effective sites for antidromic activation, the STA of the field potential revealed an early negative or positive-negative spike followed by a later, slow-negative wave. No appreciable field was obtained more dorsally or ventrally along this track. Judging from the latency after triggering spikes, the initial spike field in the averaged record was attributable to the arrival of impulses coming from the burst neuron. The interval between the initial negative spike and the onset of the late negativity (first and second arrows in Fig. 15D) was 0.3 to 0.4 msec, which corresponded to a single synaptic delay time. The late, slow-negative field potential was therefore attributed to postsynaptic currents induced by synaptic action of the single EBN. This indicates that the axon terminated within the nucleus and was not merely traversing through the nucleus. The negative polarity of the late unitary field potential suggests that the synaptic action of the single EBN is excitatory in nature, in agreement with the findings obtained in the spinal cord (Watt et al.,

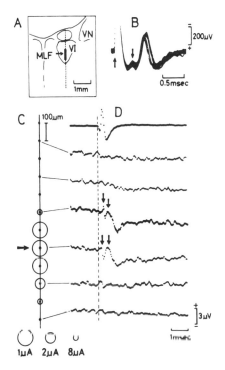

FIG. 15. Identification of premotor EBN. A: Microelectrode track *(dotted line)* for antidromic stimulation through the abducens nucleus. *Thick line* indicates effective sties of stimulation with < 10 μA for antidromic activation of a single medium-lead EBN. B: Antidromic spikes induced by stimulation at the site indicated by a *horizontal arrow* in A or C. *Upward* and *downward arrows* indicate stimulus artifact and the time of spike initiation, respectively. *Spikes* show an initial small positivity (see also **top** record in D). C: Thresholds for the antidromic activation measured at 100-μm step along the electrode track shown by a *thick line* in A. The diameter of each circle indicates the threshold current as shown below the diagram. *Dots without circles* indicate ineffective sites with < 10 μA. D: Averaged unitary field potentials triggered from spikes of the EBN shown in top (2,700 sweeps). Records were taken at 200-μm step shown in C. Voltage calibration applies to field recordings (Igusa et al., 1980, with permission).

1976) and trigeminal motor nuclei (Taylor et al., 1978). Monosynaptic excitatory connection of EBNs with ipsilateral abducens MNs has been further confirmed by STA of abducens nerve discharges, showing facilitation of MN activity with a monosynaptic latency after spikes of a single EBN (Sasaki and Shimazu, 1981b). Similar results were obtained with long-lead EBNs. They were antidromically activated from the ipsilateral abducens nucleus and STA of the field potential revealed a monosynaptic unitary negative field, indicating that long-lead EBNs also terminate in the abducens nucleus.

Since the onset of burst activity of medium-lead EBNs was coincident with steep depolarization of ipsilateral abducens MNs at the quick phase (Igusa et al., 1980), a part of steep EPSPs in agonistic MNs (Fig. 1D) is attributed to the burst input from medium-lead EBNs. The functional role of long-lead EBN input in generation of rapid activation of MNs is interesting. Despite their low-frequency prelusive activity, abducens MNs are completely suppressed before the quick excitatory phase, probably because of coexistence of an intense IPSP arising from inhibitory type I MVN neurons during the late period of the slow phase (Fig. 1D). The prelusive activity of long-lead EBNs may provide abducens MNs with a background excitatory input which effectively causes rapid activation of MNs at the time of disinhibition due to suppression of tonic activity of inhibitory MVN neurons.

It has been hypothesized that medium-lead burst neurons receive an inhibitory influence from pause neurons which exhibit tonic discharges except during fast eye movements (Keller, 1974). A recent report has stated that pause neurons are antidromically activated from the IBN area and that stimulation of the pause neuron area produces monosynaptic inhibition of IBNs (Nakao et al., 1980), supporting the view that burst activity of IBNs may, in part, be caused by disinhibition due to suppression of pause neurons during the quick phase. The excitatory input from EBNs plus the disinhibition may be effective for producing much higher intraburst frequency of IBNs than that of EBNs, similar to the mechanism of bursting of abducens MNs as described above.

The question arises as to the source of excitatory input to EBNs. As far as rotation-induced quick phases of nystagmus are concerned, it should originally arise from the ipsilateral horizontal canal. In a recent study, EBNs were found to be activated in a burst fashion with the latency of 5 to 10 msec after stimulation of the ipsilateral vestibular nerve (Sasaki and Shimazu, 1981b), suggesting that the pathway from the vestibular nerve to EBNs is polysynaptic. There is also visual input to EBNs. Stimulation of the ipsilateral optic nerve or the contralateral superior colliculus induced spikes in a burst fashion of functionally identified EBNs (Sasaki and Shimazu, 1981b). Relevant to this finding is the study of Raybourn and Keller (1977), showing that train pulse stimulation of the contralateral superior colliculus in the alert monkey induced burst discharges outlasting the stimuli in pontine neurons, which burst in association with naturally occurring fast eye movement. There are also several works revealing that stimulation of the superior colliculus activates neurons in the vicinity of the contralateral abducens nucleus (Peterson et al., 1974; Precht et al., 1974; Grantyn and Grantyn, 1976). Anatomical evidence for the tectal projection to the contralateral pontine reticular formation has been provided by Kawamura et al. (1974). The projection sites appear to include the EBN area described in the present paper. Thus, visual and vestibular inputs are likely to converge on EBNs. This convergence may provide a neural basis for a common feature of both visual- and vestibular-induced fast eye movements (Cohen and Henn, 1972a; Ron et al., 1972; Nakao et al., 1977; Raphan and Cohen, 1978), by operating the brainstem neuronal machinery controlled by EBNs.

Finally, the location of EBNs must be considered. Anatomical studies have reported a projection to the abducens nucleus from the ipsilateral pontine reticular formation apparently extending more rostral and ventral to the EBN area (Büttner-Ennever and Henn, 1976; Graybiel, 1977). Physiologically, stimulation of the paramedian pontine reticular formation (PPRF) at the border between the nucleus reticularis pontis oralis (NPRO) and pontis caudalis (NRPC) (Brodal, 1957) induced monosynaptic EPSPs in ipsilateral abducens MNs (Highstein et al., 1976). The site of stimulation appears to be more rostral to the EBN area. The ventral-most part of the EBN area may be overlapping with the dorsocaudal portion of the abducens-related PPRF delineated by Graybiel (1977), but many EBNS are located further dorsally outside the PPRF. A class of neuron in the dorsocaudal portion of the PPRF was identified as projecting to the spinal cord as well as the ipsilateral abducens nucleus (Grantyn et al., 1980). These reticulospinal neurons having collateral projection to the abducens nucleus did not, however, exhibit burst activity related to the quick phases of vestibular nystagmus (Sasaki and Shimazu, 1981a). Furthermore, we have observed that many unidentified neurons being scattered in the wide region of the PPRF more ventral and rostral to the EBN area were antidromically activated by stimulation of the ipsilateral abducens nucleus in agreement with above-mentioned anatomical and physiological studies, but they were not bursting at the quick phases of nystagmus. Further study will be required to clarify the functional role of these PPRF neurons.

PREMOTOR INTERNEURONS IN THE ABDUCENS NUCLEUS

Conjugate horizontal eye movements are caused by coordinated antagonistic actions of the lateral and medial rectus muscles. Vestibular nucleus neurons projecting to the oculomotor complex control medial rectus MNs in a reciprocal manner opposite to those for abducens MNs, i.e., ipsilateral excitation and contralateral inhibition. Although these vestibular neurons may function during nystagmus in a similar fashion to those projecting to the abducens nuclei, synaptic inputs to medial rectur MNs in association with vestibular nystagmus are quite different from those to abducens MNs. In particular, there appears to be little contribution of IPSPs at the initial phase of laterally directed fast eye movement, and disfacilitation is primarily responsible for the suppression of MN activity (Furuya et al., 1979; Nakao and Sasaki, 1980).

It has recently been suggested that interneurons in the abducens nucleus are important as a prime modulator of medial rectus MN activity. Anatomical studies have revealed the projection of a class of interneurons within the abducens nucleus to the contralateral oculomotor complex through the ascending MLF (Graybiel and Hartwieg, 1974; Gacek, 1977; Bienfang, 1978; Steiger and Büttner-Ennever, 1978). Physiologically, interneurons in the abducens nucleus exhibited a burst-tonic type of discharge similar to that of abducens MNs during conjugate horizontal eye movements (Del Gado-Garcia et al., 1977). These interneurons were presumed to be excitatory in nature on the basis of the finding that, after acute or chronic partial isolation of the abducens nucleus from adjacent structures, stimulation of the nucleus produced monosynaptic EPSPs in contralateral medial rectus MNs (Highstein and Baker, 1978).

Since some neurons in the abducens nucleus have been found to send their axon to regions other than the oculomotor nucleus such as certain areas in the cerebellum (Kotchabhakdi and Walberg, 1977), it was investigated whether individual interneurons in the abducens nucleus exhibiting MN-like spike activity do make monosynaptic excitatory connection with contralateral medial rectus MNs (Nakao and Sasaki, 1980). The monosynaptic excitatory connection of a single interneuron was ascertained by the technique of spike-triggered average of the membrane potential of contralateral medial rectus MNs, revealing a unitary EPSP with a monosynaptic latency after spikes of the interneuron. All of the thus identified interneurons exhibited spike activity similar to that of MNs in the same nucleus during both slow and quick phases, probably due to nystagmus-related common input to these two kinds of neurons. Contralateral medial rectus MNs therefore receive a common motor signal similar to the activity of abducens MNs through direct excitatory connection from abducens interneurons. The onset of spike burst of interneurons at the quick phase was slightly earlier than that of MNs, presumably because of lower threshold for spike initiation in the former neurons. This compensates the possible delay of medial rectus activity caused by conduction of impulses along the MLF and synaptic transmission in the oculomotor nucleus, resulting in synchronous burst of abducens and medial rectus MNs. These mechanisms obviously contribute to well-coordinated activity of the lateral rectus muscle of one eye and the medial rectus muscle of the other during conjugate horizontal eye movements and explain well the genesis of internuclear ophthalmoplegia following a lesion of the ascending MLF.

MODE OF OPERATION OF PREMOTOR MECHANISMS UNDERLYING GENERATION OF HORIZONTAL NYSTAGMUS

Discharge characteristics of various kinds of premotor neuron have been described in earlier sections in conjunction with PSP changes in abducens MNs during nystagmus. Each premotor neuron has been identified as excitatory or inhibitory in nature, and synaptic connections among these neurons have been ascertained. The findings are synthesized and schematically represented in Fig. 16. The diagram shows experimentally verified connections of identified excitatory and inhibitory premotor neurons and their discharge patterns during one cycle of nystagmus induced by activation of the right horizontal canal. The excitatory and inhibitory secondary type I vestibular neurons on the right side exhibit tonic firing with gradually increasing frequency, acting on abducens MNs and interneurons on each side to induce vestibulo-ocular reflex directed to the left side. The slow phase discharges of excitatory type I vestibular neurons reach the ipsilateral EBN area, probably through polysynaptic pathways (broken line). When the slowly increasing excitatory effects attain to the threshold of EBNs, they are converted into burst activity either due to membrane properties of EBN itself or caused by local neuronal circuits in the EBN area. The burst activity of efferent fibers of EBNs excites ipsilateral abducens MNs and IBNs in a burst fashion, the latter exerting inhibitory

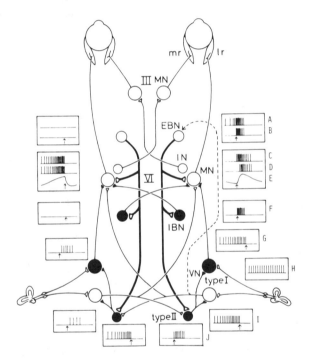

FIG. 16. Premotor neuronal organization of horizontal conjugate eye movements. Synaptic connections among neurons are verified as described in the text. *Open* and *filled circles* indicate excitatory and inhibitory neurons, respectively. Axons of EBNs are drawn with *thick lines* to make the scheme easy to see. VN: Vestibular nucleus, VI MN: abducens MN, VI IN: abducens interneuron, III MN: medial rectus MN, mr and lr: medial and lateral rectus muscles. Four kinds of synaptic input to abducens MN are likely to apply to abducens interneuron as well, but the latter inputs are taken off to avoid complexity of the scheme. *Broken line* represents polysynaptic pathway, in which locations of mediating interneurons have not yet been examined. Each neuron is identified by nystagmus-related activity. Discharge patterns shown in rectangles represent responses in one nystagmic cycle of each neuron on both sides to increased input from the right horizontal canal. *Arrow* in each rectangle indicates the onset of quick phase determined by membrane potential change in abducens MN. A: Long-lead EBN; B: Medium-lead EBN; C: abducens interneuron; D: abducens MN; E: membrane potential of abducens MN; F: IBN; G: inhibitory type I vestibular neuron; H: primary afferent from the horizontal canal, I: excitatory type I vestibular neuron; J: type II vestibular neuron. The *left rectangles* correspond to A–J on the *right*.

effects on contralateral abducens MNs to induce antagonistic inhibition. Medial rectus MNs on each side receive reverse effects due to crossed innervation from interneurons in the abducens nuclei, thus causing conjugate fast eye movements directed to the right side.

Descending collateral effects of EBNs reach the ipsilateral MVN and produce a spike burst in type II vestibular neurons, which, in turn, cut off the tonic activity of type I neurons. The cessation of tonic input from type I neurons to EBNs may be effective for discontinuation of burst activity of EBNs. On the other hand, slow-phase discharges of right type I vestibular neurons cause tonic activity of left type II neurons. When right type I neurons cease to fire at the quick phase, the activity of left type II neurons also ceases probably due to cessation of excitatory input from right excitatory type I neurons and, in addition, arrival of inhibitory input from right IBNs (Hikosaka et al., 1980) (not illustrated in Fig. 16). The suppression of tonic activity in left type II neurons causes disinhibition of left type I neurons, producing a train of spikes which intensify activation of right type II neurons through the commissural pathway. In fact, it has been confirmed that at least a part of commissural fibers conveys the same nystagmus-related signal as that of premotor fibers driving abducens MNs (Nakao et al., 1981).

The diagram in Fig. 16 could be simplified by removing MNs and neuronal connections directly related to projection to MNs. Figure 17 depicts a thereby obtained diagram of vestibulo-reticular connection which appears to be relevant to generation of nystagmic rhythm in vestibular and reticular premotor neurons. The essential mode of operation of this circuit has been described above. Interestingly, this circuit contains two chains, both of which appear to be effective to generate a periodic rhythm (Harmon, 1964). First, the circle consisting of type I vestibular neuron,

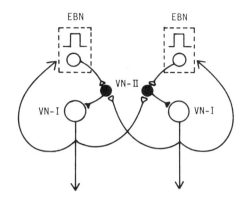

FIG. 17. Simplified diagram of vestibuloreticular connection explaining the mode of operation of premotor mechanisms underlying generation of vestibular nystagmus. VN-I and VN-II: type I and II vestibular neurons. The connection of VN-I with EBN *(arrow)* appears to be polysynaptic, but it is drawn as direct projection for simplification. *Broken rectangle* on each side represents the EBN area, in which burst activity shown by pulse is created and output EBN projects to ipsilateral type II vestibular neuron.

EBN, type II vestibular neuron, and back to type I vestibular neuron functions as a delayed recurrent inhibitory circuit. This could by itself produce a circling rhythm when input from other sources such as primary vestibular afferents tonically activate the type I vestibular neuron. Second, the reciprocal inhibitory connection between type I vestibular neurons on both sides through type II vestibular neurons could, by itself, produce an alternating rhythm, provided that type I neurons are tonically activated from other sources and have a mechanism causing a decrease in frequency of spikes (Brown, 1911; Reiss, 1962). The mechanism of frequency decrease could be either intrinsic within the cell or extrinsic from the outside source. Intracellular recordings from type I vestibular neurons during nystagmus showed that there was no intrinsic adaptation mechanism in these neurons before the moment of production of steep IPSPs at the quick phase. This suggests that the periodic suppression of spike activity is caused by extrinsic inhibitory synaptic input generated through more or less remote pathways (Hikosaka et al., 1980). This presumed remote pathway appears to correspond to the above-described delayed recurrent inhibitory circuit. The two fundamental circuits are united by convergence of inputs from ipsilateral EBNs and contralateral type I vestibular neurons onto a single type II neuron. Compared with separate projection of the two inputs to different neurons, their convergence onto the same type II neuron seems important for functional coupling of the two systems, because the summation of two excitatory actions would be effective to far exceed the threshold of the type II neuron and produce its intense activity enough to cut off impulse transmission from the vestibular nerve to the secondary type I neuron.

Finally, it comes into question whether the discharge characteristics of premotor neurons in the MVN are sufficient to explain quantitatively the postsynaptic activity of abducens MNs during the slow phase. The increase in firing rate or premotor secondary MVN neurons during the slow phase seems to be too slight on the average to create the postsynaptic activity of abducens MNs. This view agrees with the considerable phase differences between frequency responses of secondary vestibular neurons and ocular MNs. Therefore, the slow-phase discharges of abducens MNs may not be entirely attributed to premotor vestibular neurons. There might be other classes of immediate premotor neurons related to to slow phases of MN activity. A class of axon spikes recorded within the abducens nucleus is activated disynaptically from the contralateral vestibular nerve (disynaptic V_c axons) and shows firing characteristics similar to those of abducens MNs during the slow phase (Hikosaka et al., 1977). Thus far, however, we have failed to locate the origin of this class of premotor neurons in the vestibular nuclei. Some, or the majority, of disynaptic V_c axons might be the axons of interneurons in the abducens nucleus, but no evidence for their synaptic connection with abducens MNs has been obtained (Nakao and Sasaki, 1980). Anatomical studies revealed efferent projections of neurons in the prepositus hypoglossi nucleus to the abducens nuclei (Maciewicz et al., 1977; McCrea et al., 1979). Physiologically, some prepositus hypoglossi and nearby reticular neurons were suggested to send axonal branches to the abducens nucleus, but the branches within the nucleus appeared to be sparse (Hikosaka and Igusa, 1980). Thus far, no evidence has been found for their significant functional connection with abducens MNs.

Dynamic Characteristics of Vestibular and Visual Control of Rapid Postural Adjustments

*,**M. Lacour, †P. P. Vidal, and **C. Xerri

**Laboratoire de Psychophysiologie, Université de Provence, Centre Saint Jérome, Marseille; and †Laboratoire de Physiologie du Travail CNRS, Paris, France*

This chapter reviews some findings concerning the dynamics of vestibular and visual control of rapid postural reactions and, more specifically, the control of postural reactions to fall. A first series of results demonstrates the role of vestibular afferences in triggering these reactions. A second set of results point to visual-vestibular interactions and to early directional influence of visual motion cues in rapid postural control during linear motion.

VESTIBULAR CONTROL OF RAPID POSTURAL ADJUSTMENTS

The vestibular system is involved in many compensatory reflexes. The vestibulo-ocular reflex stabilizes the eye in space when the head is rotated and results from stimulation of semicircular canals. The vestibulo-spinal reflexes participate in postural control and equilibrium function and mainly involve stimulation of the otoliths.

Dynamics of the Otolith System

The otolith system is composed of the utricular and saccular maculae disposed at almost right angles to one another. The utricular maculae lie approximatively in the horizontal plane when the head is tilted 20 to 25° forward. The saccular maculae are in the vertical place when the head is in its normal position, at about 45° relatively to the sagittal plane. The sensory cells in each macula exhibit 50 to 100 stereocilia and only one eccentrically located kinocilium. These sensory cells may be considered as linear accelerometers. The adequate stimulus consists of a shearing force, due to the component of acceleration directed in the plane of the macula, which bends the sensory hairs. The microstructure of maculae shows a morphological polarization of hair cells which is opposite to either side of the striola (Spoendlin, 1966). The inertial receptors of the maculae are, therefore, able to detect linear accelerations of the head in the fore-aft (X axis), lateral (Y axis), and vertical (Z axis) directions. They also respond to positions by detecting the component of gravity. Afferent fibers synapse either at the bottom of hair cells (type II cell) or by enveloping a single hair cell (type I cell). The first-order vestibular neurons are located in the Scarpa's ganglion. They send their axons (vestibular nerves) to the vestibular nuclei. The rostroventral area of the Deiters' nucleus is heavily fed by both utricular and saccular primary afferents. The

---

*To whom correspondence should be addressed: Laboratoire de Psychophysiologie, Université de Provence Centre de Saint Jérome, Marseille 13397 Cedex 13 France.

rostral part of the descending nucleus and the medial nucleus receive also some otolith afferents, whereas the Y group is only in relation with saccular afferents (See Fig. 2A).

Primary afferents exhibit a spontaneous discharge around 60 spikes/sec in the monkey (Fernandez et al., 1972). Most units have a tonic behavior (regular firing units), whereas some behave more phasically (irregular firing units). The utricular afferents respond to static tilts of the head in the frontal (pitch) and sagittal (roll) planes (Fig. 1A). 75% of units show a maximal firing rate for a 90° pitch up or a 90° roll ipsilateral. The remaining units present an opposite response pattern. The discharge is roughly proportional to the component of gravity and follows a sigmoid function with a linear relationship in the -1 to 1 g range. The saccular afferents show a resting discharge which is either maximal or minimal when the head is in the horizontal plane. Many units increase their firing rate for a 180° tilt in the frontal or sagittal planes. One can note that these static positions induce a hair-cell deflection similar to that produced by a downward vertical acceleration. The response of primary afferents to dynamic stimuli has been studied in the monkey (Fernandez and Goldberg, 1976). The firing rate of regular units parallels the applied force for both prolonged stimulation and rapid force transitions. The response exhibits a constant gain from 0- to 2-Hz stimulus frequencies, with a phase lag relative to acceleration not exceeding 10°. In contrast, the irregular units adapt to maintained stimulation. They show a velocity sensitivity for rapid force transitions and an enhanced gain when increasing stimulus frequency (Fig. 1B).

Rather different responses occur in second-order vestibular neurons to both static and dynamic stimuli. According to Duensing and Schaefer (1959), the second-order otolith units are of the α (⅔) or β type (⅓). The α-neurons increase their resting rate for ipsilateral tilts and decrease their discharge for contralateral tilts. The β-neurons show an opposite response pattern (Fig. 1C) γ- and δ-neurons, probably third-order ones, increase and decrease their firing discharge, respectively, whatever the head-tilt direction.

Adrian (1943) first recognized that second-order otolith units activated by a static head-tilt were also activated by a linear horizontal acceleration of the head in the opposite direction. Melvill-Jones and Milsum (1969) described the transfer function of otolith units during sinusoidal linear horizontal acceleration in the cat. They found a phase lag relative to head acceleration, which increases with stimulus frequencies in the range 0, 1 to 3 Hz. The response is in phase with acceleration at very low frequencies and with position at frequencies

FIG. 1. Responses of otolith afferents and second-order vestibular neurons to static tilts and to linear accelerations. **A**: Responses of utricular afferents (1–4) and saccular afferents (5) in the squirrel monkey to roll and pitch static tilts. The discharge rate (spikes/sec) is shown as a function of the tilt in degrees. (Based on Fernandez and Goldberg, 1976, and modified from Goldberg, 1979; with permission.) **B**: Responses of otolith afferents to 5-sec (**a**) and 0.5-sec (**b**) centrifugal force transitions in regularly discharging afferents (2) and irregularly discharging afferents (1) in the squirrel monkey. (*Ordinates:* time, sec.) (Based on Fernandez and Goldberg, 1976b, and modified from Goldberg, 1979, with permission.) **C**: Responses of lateral vestibular nucleus neurons to static tilts toward the ipsilateral and contralateral sides relative to the head horizontal position (reference: 0°) in the cat. Responses are of α-, β-, and γ-types, according to Duensing and Schaeffer (1959). (From Peterson, 1970, with permission.) **D**: Responses of second-order vestibular neurons to sinusoidal linear acceleration in the cat. The phase of the response is expressed relatively to acceleration (in degrees) and as a function of the stimulus period in seconds. (From Melvill-Jones and Milsum, 1969, with permission.)

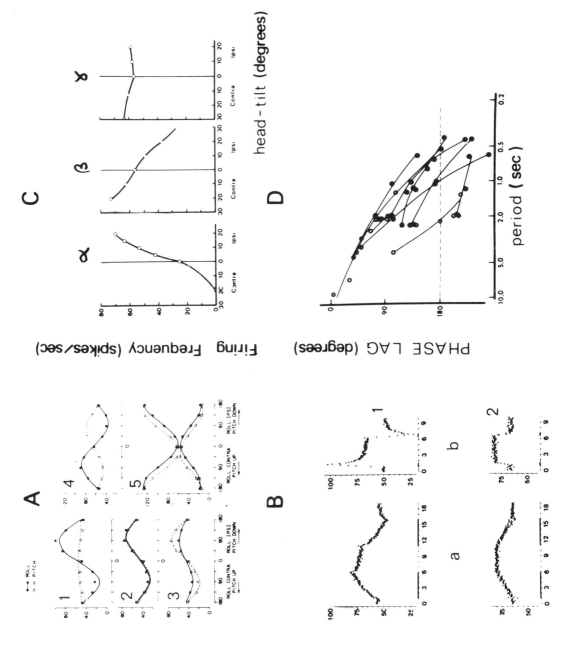

up to 1 to 2 Hz (Fig. 1D). Similar results were obtained by Lannou et al. (1980) in rat. Thus, vestibular nuclei are not a simple relay for otolith afferents, but a site where sensory information is integrated.

This processed information reaches the spinal neurons via vestibulospinal pathways (Fig. 2B). The lateral vestibulospinal tract originates from Deiters' nucleus. It connects monosynaptically and polysynaptically with spinal motoneurons (MNs) at all spinal levels. These ipsilateral connections facilitate α- and γ extensor MNs. The medial vestibulospinal tract originates from the medial, descending and lateral vestibular nuclei. It reaches the cervical and high thoracic spinal levels only and conveys both facilitatory and inhibitory influences on neck muscles on both sides. A third vestibulospinal tract (the caudal) originating from the medial and descending nuclei sends fibers at all spinal levels (Peterson and Coulter, 1977). Its functional properties remain unknown. The vestibulospinal reflexes can also be mediated by the reticulospinal pathways, important projections from vestibular nuclei into the pontine reticular formation having been described.

Postural Reactions to Fall

deKleyn and Magnus (1921) first described in the cat a reflex due to downward vertical acceleration ("Sprungbereitschaft" response). These postural reactions to fall were absent in bilateral labyrinthectomized cats. Rademaker (1935) confirmed these results. Thus, responses to fall had a vestibu-

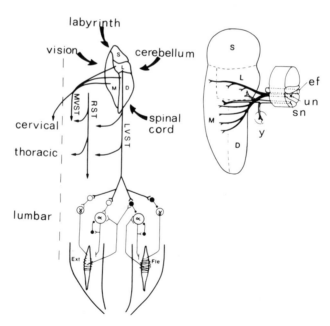

FIG. 2. Anatomical and physiological organization of the vestibulospinal pathways. **Right**: Projection of the otolith afferents onto the vestibular nuclei (From Gacek, 1975, with permission). ef: Efferent fibers; un: utricular nerve; sn: saccular nerve; y: y group; S,L,M and D: superior, lateral, medial and descending vestibular nuclei. **Left**: Origin and termination of the vestibulospinal tracks. The reciprocal influences on extensor and flexor muscles by means of the α- and γ-MNs are shown at the lumbar spinal cord level. *Filled* and *open circles* represent inhibitory and excitatory spinal interneurons, respectively. LVST: Lateral vestibulospinal tract; MVST: medial vestibulospinal tract; RST: reticulospinal tracts. Note that the vestibular nuclei receive both labyrinthine, visual, cerebellar, and spinal inputs. (From Lacour, 1981, with permission.)

lar and very likely otolith origin because they were absent in mice with hereditary absence of otoliths (Lyon, 1951). In recent works on the mechanisms involved in controlling rapid postural adjustments, most authors used the electromyographic (EMG) method and free-fall test, but we also report findings concerning muscle responses to voluntary step, jumping, or running. One can distinguish between motor activities developing before ground contact (early responses), and those occurring after landing.

Muscle Responses Before Landing

Melvill-Jones and Watt (1971a) found a gastrocnemius EMG activity in man landing from a 25-cm jump to the ground. This substantial muscle response advanced the moment of foot contact on the platform by a considerable amount (100 msec) and was completed 100 msec after ground contact. In repetitive hopping movements, burst activities start well before landing (84 msec) but last longer after ground contact. From these results, Melvill-Jones and Watt (1971a) proposed that body deceleration resulted from a preprogrammed motor sequence. When human subjects are suddenly and unexpectedly dropped from 2.5 to 20.3 cm above ground, EMG activities in both gastrocnemius and tibialis anterior muscles begin 75 msec after onset of fall. This latency was independent of the height of fall. Falls of 2.8 cm (duration 7.5-msec) or 5.1 cm, which do not allow the EMG activity to build up led to a very uncomfortable landing, while landing from higher heights was relatively smooth (Melvill-Jones and Watt, 1971b). Therefore, this early EMG activity would be an otolith-originating reflex involved in a preprogrammed control of landing. The fact that the response developed in both flexor and extensor leg muscles, with a relatively long latency, suggests mediation, not by direct vestibulospinal pathways but, rather, by vestibuloreticulospinal ones.

Watt (1976) compared EMGs from three pairs of muscles (splenius, gastrocnemius, and tibialis anterior) in cats suddenly dropping from 45 to 50 cm above ground. Watt compared normal to both canal-plugged and bilabyrinthectomized cats. He showed that muscle responses to fall occurred in normal cats as two peaks. The early muscle activity (latency: 55 msec in gastrocnemius) was totally suppressed after labyrinthectomy but remained in canal-plugged preparations. The later EMG activity began about 70 to 100 msec after release and depended on the height of fall. It was roughly centered on the time of ground contact. These two EMG patterns were not modified by blindfolding the cats. All cats failed to land properly when the early response was lacking, the second muscle response being unable to correctly decelerate the cat. Watt distinguished between an otolith-originating reflex assisting landing (early response) and a nonlabyrinth-originating reflex (later response), which likely depends on visual, tactile, and proprioceptive cues.

Greenwood and Hopkins (1974, 1976, 1980) in man and Lacour et al. (1978a) in monkey pointed to globally similar results. The postural reactions to fall developed as two peaks (Fig. 3A). The early one (81.6 msec in man and 32 msec in monkey's soleus), was reduced in patients with absent labyrinthine function and virtually suppressed in labyrinthectomized monkeys. This early EMG activity was observed in muscles throughout the body, and its latency was independent of the height of fall. Moreover, the energy of the early responses decreased in normal monkeys and man as the magnitude of acceleration reduced, in the range 0.2 to $0.9g$ (Fig. 3B). The second peak was centered around the time of contact. It rapidly disappeared in our experimental conditions where the monkey's fall was mechanically braked, so that the animal did not have to make the postural adjustments necessary for landing (Fig. 3A). Different interpretations were proposed for the initial peak. Greenwood and Hopkins (1976) indi-

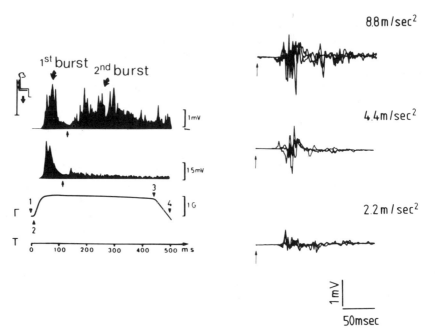

FIG. 3. Motor reactions during free-fall in the monkey. **Left:** Typical detected and integrated EMGs from the soleus muscle during free-fall (acceleration peaks: 8.8 m/sec^2; average 40 trials for 4 baboons). From *top* to *bottom:* EMG activity recorded during falls 0 to 10, falls 30 to 40, downward acceleration of the head and time scale in msec from moment of release. *Arrows* indicate initial trigger (1), beginning of downward acceleration of the head (2), beginning of the deceleration (3), and virtual moment of ground contact (4). Note that the second burst of EMG activity disappears with repetition of falls. (From Lacour et al., 1978a, with permission.) **Right:** Three superimposed raw EMGs from splenius muscle for three consecutive falls at 8.8 m/sec^2, 4.4 m/sec^2 and 2.2 m/sec^2, with normal vision. One can observe a decrease in the early EMG activities (first burst) and a longer latency as acceleration decreases. (From Lacour et al., 1981a, with permission.)

cated that the early peak occurred even during free-fall from 120 cm, when it appeared superfluous to landing, and that it was strongly reduced when the subject himself released the fall. They concluded that it was a startle reaction, triggered by the synchronous stimulation of otoliths due to the sudden change of 980 cm/sec^2 in vertical linear acceleration. In fact, it is well known that a loud sound elicits in man a startle reaction, characterized by EMG activities developing in various muscles with latency close to that reported for the initial peak. Paillard (1955) showed a H-reflex facilitation starting around 80 msec after a brief and strong sound, with a maximal increase 100 to 150 msec after the sound. However, we believe that these similarities in latencies are purely coincidental. In addition, repetition of such a sound stimulation rapidly leads to habituation, i.e., disappearance of the effects. We never observed such habituation in monkeys. Furthermore, the latency of the first burst of EMG was very short (32 msec) compared to the latency (60 to 80 msec) in a startle reaction to a brief, strong and unexpected sound. These arguments strongly suggest that the first peak cannot be a startle reaction to release, but represents a functional response that automatically and reflexively controls landing. In other experiments in which human subjects were suddenly pushed forward, Bussel et al. (1978) found EMG activity in soleus starting 56 msec after postural perturbation. This EMG was reduced in labyrinthectomized subjects and was reversed with reversal of push direction (backward), occurring then in the antagonist flexor muscles (tibialis anterior). This early muscle response is, therefore, functionally involved in counteracting body displacement, which also militates against the startle-reaction interpretation.

The modulation of T and H spinal reflexes during fall was studied by Greenwood and Hopkins (1977) in man and by Lacour et al. (1978a) in monkeys. Facilitation of the H-reflex started just prior to onset of soleus EMG and lasted until the end of the fall. The T-reflex modulation exhibited a similar time course except in the first 50 msec where T-reflexes were depressed (Fig. 4). This was consistent with the hypothesis of muscle spindles unloading at release of fall. Matthews and Whiteside (1960) pointed to strong depressions of both H- and T-reflexes from 50 to 100 msec after release in subjects seated in a drop chair, but their data were artifactual (see Greenwood and Hopkins, 1977, Lacour et al., 1978a). Our results in monkeys suggest that the early EMG activity to fall was due to direct activation of α-MNs and not by means of the γ-loop. We cannot exclude a α-γ coactivation, but the γ activation, if any, remained unable to overcome the unloading.

Recent data (Prochazka et al., 1976; Lewis et al., 1979b) showed no major modification in activity of plantaris spindle primary afferents as well as of ankle extensor spindle secondary afferents in the conscious cat during free fall.

We found a H-reflex facilitation corresponding to an increase in spinal MN excitability starting at about 16 to 18 msec after onset of downward acceleration (Lacour et al., 1978a). This delay does not take into account the unknown delay in the otoliths which, however, have a very low threshold ($< 0.005g$). Therefore, the early motor reactions to fall may involve direct vestibulospinal pathways, or rapid vestibuloreticulospinal ones since muscular responses were observed in both flexor and extensor muscles. Fiorica et al. (1962) showed strongly increased discharge at Scarpa's ganglion in the unanaesthetized cat suddenly dropped from 10 m, but did not identify latencies nor nature of recorded units (semicircular, utricular, or saccular units). However, the early muscle responses to fall appear very likely due to predominant activation of saccular afferents. In man, Watt and Zucker (1980) found that altered gravity obtained by rotating the gravity vector by 90° relative to the head (or in parabolic flight conditions in aircraft) strongly reduced the otolith-spinal response. When suspended horizontally in the supine condition, the EMG activity was decreased about 30% when compared to vertical testings.

Muscle Responses After Landing

The problem is to determine the relative contribution to decelerating the body of EMG activities prior to landing or after ground contact. In other words, are the former sufficient alone or is there a significant contribution of short-latency segmental monosynaptic response or [stretch reflex (SR)] and long-latency reflexes such as the so-called functional stretch reflex (FSR). The SR latency is around 38 msec in man when elicited by a sharp tap to the Achilles tendon. The FSR develops well after the SR, with a latency of approximately 120 msec to a sudden forceful dorsiflexion of the foot. The fastest voluntary plantar flexion never produces EMG activity before 200 msec (see Paillard, 1955; Melvill-Jones and Watt, 1971a).

Melvill-Jones and Watt (1971a,b) reported no useful contribution to landing events from SR as well as FSR in different situations such as landing from free-fall, single step, or jump. The FSR would occur too late to play a role in body deceleration, and active inhibition was present at the time

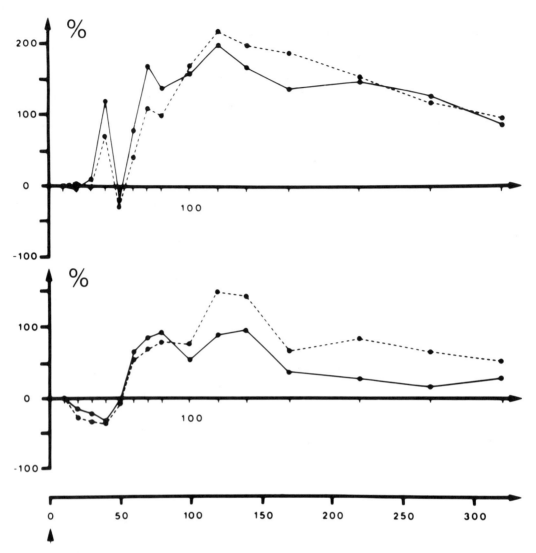

FIG. 4. Modulation of monosynaptic spinal reflexes during free-fall in the monkey. Mean results (6 baboons, 54 samples per point) concerning the H-reflex *(upper graphs)* and the T-reflex *(lower graphs)* recorded on the left *(dashed line)* and right *(solid line)* soleus muscles. The peak of acceleration is 8.8 m/sec^2. Ordinate: Amplitude of reflexes during falls as a percentage of control reflexes. *Abscissa:* Time between release *(arrow)* and moment when H- or T-stimulus is given (msec), increased by 20 msec representing the latency of these reflexes in baboon. (From Lacour et al., 1978a, with permission.)

when the FSR would have been supposed to develop. Their suggestion was that landing was reflexively and functionally controlled by preprogrammed activities triggered by otolith afferents. In fact, they only tested landing from heights up to 20 cm and could not observe the second EMG burst, which occurs only when falls are of sufficient height. This second burst, roughly centered on the moment of foot contact, represents the voluntary control of landing and necessitates vestibular and visual cues, as well as knowledge of the fall height and learning of the experimental conditions.

Others, however, found significant modulations of EMG after landing, at times consistent with SR and FSR. During a 15-cm downward step in man, Greenwood and Hopkins (1976, 1980) showed a peak of EMG of about 60 msec after landing. This peak was absent during expected landings with unexpectedly lengthened steps, implying it was of reflex origin and dependent on mechanical events of landing. Similar conclusions were drawn by Prochazka et al. (1977), Prochazka (1980) and Lewis et al. (1979b) in the cat. Landing control implies rapid reflex responses in hindlimb muscles. By recording primary and secondary muscle spindle afferents from ankle extensors, they showed a rapid burst of firing due to muscle stretch. Dietz and Noth (1978) studied the relative role of preinnervation and stretch-induced EMG of triceps brachii muscles in man when self-initiating a fall forward onto a platform. They found EMG beginning 130 to 200 msec prior to the impact, the latency of which was independent of the fall height under visually controlled falls. This burst was followed by a second one, which peaked 18 to 30 msec after arms contact on the platform. This latency is similar to that of monosynaptic response (15 msec) to sudden unexpected stretch of biceps muscle in man (Hammond et al., 1956). Another burst occurred about 100 to 150 msec after impact, a latency compatible with a segmental or long-loop reflex. They agreed, therefore, on the presence of both pre-programmed muscle activity and reflex responses, likely being of proprioceptive origin. Dietz et al. (1979) also showed a burst of EMG developing 35 to 45 msec after ground contact in gastrocnemius muscle in man while running. This burst, which has also a latency consistent with a SR response, was strongly reduced after Ia afferent block (ischemia).

Conclusions

Four types of muscular responses can be observed in subjects submitted to vertical linear motion. Two occur before landing and two others after ground contact. Prelanding activities prior are composed of an early EMG response, which does not habituate with repetition of falls, but the amplitude of which depends on the amplitude of acceleration. This first peak disappears after labyrinthectomy but remains in canal-plugged preparations. It is reduced when the subject himself triggers the fall. It would constitute an automatically triggered landing preparation capable of decelerating the body by developing tension in all concerned muscles. This initial response is predominantly due to otolith stimulation resulting from sudden transition from 1 to 0 g and would be functionally efficient especially when the height of fall is short and when the fall is unexpected. The early response is followed by a second EMG peak, when the fall is of sufficient height (about 20 cm). The timing of this second response is related to timing of landing; its amplitude is function of fall acceleration and its latency depends on knowledge of fall height. It is likely that this second response represents a pre-programmed response concerned with the voluntary control of landing. It necessitates both vestibular and visual cues for adequate timing of coordinated muscle responses. Learning phenomena interact in the building up of this response. The muscular activities developing after ground contact appear to depend on experimental conditions. It is difficult to compare the EMG activities in such different situations when body motion is or is not expected, or whether or not it results from voluntary maneuvers. We believe that the short-latency (SR) as well as long-latency (FSR) responses constitute regulatory mechanisms that assist landing. These mechanisms involving proprioceptive afferents elicited by the mechanical fall events fulfill an important function in situations such as walking and running on a suddenly rough ground, downward stepping with unexpected, lengthened or shortened steps, landing from unexpected free-fall etc.

VISUAL CONTROL OF RAPID POSTURAL ADJUSTMENTS

Role of Vision in Postural Control

All experiments reported above did not show great modifications when postural perturbations were produced either in total darkness or with blindfolded subjects. This was particularly evident for postural reactions to fall, and many authors took these results as an argument against the role of vision in rapid postural reactions. In fact, the absence of modifications with the removal of visual cues supported the general statement that vision intervenes only in the low-frequency range from 0 to 0.2 Hz (see Berthoz et al., 1979). Talbott (1974) showed that low-frequency postural oscillations were reduced by 50% in man and dogs when the subject's eyes were open instead of closed. The dynamic effect of vision on low-frequency perturbation of posture was also evidenced in man or dogs standing in front of a visual stimulus in sinusoidal linear or circular movement. The postural instability was maximum in the 0.1 to 0.3 Hz range of stimulus frequencies (Lestienne et al., 1977; Mauritz et al., 1977). In addition to the slower dynamics of the visual system as compared to other sensory systems (such as the vestibular), the visual system was found to intervene in postural control with greater latency. For example, the visually induced changes in EMG leg muscle activity in man was 0.5 sec (Lestienne et al., 1977). Changes in the subjective vertical occurred after 1 to 2 sec (Dichgans et al., 1972) and circular or linear vection phenomenon developed after approximately 1 sec (Dichgans and Brand, 1974; Berthoz et al., 1975). However, Anderson et al. (1972) and Maeda et al. (1977) showed physiological evidence in favor of fast-acting visual-spinal pathways. One hypothesis was that darkness or blindfolding did not test the role of vision but, rather, the successful capacity of the subject to compensate for an expected lack of visual cues. This assumption led to the examination of other paradigms capable of producing an unexpected and unusual lack of visual cues. That was done by Nashner and Berthoz (1978) in man and by our group in monkeys (Lacour et al., 1978b; Vidal et al., 1979) by means of the visual stabilization method.

Early and Directional Influences of Visual Cues During Rapid Postural Perturbations

We shall first present some findings concerning the EMG modifications in monkey falling in total darkness. In contrast with the results of Melvill-Jones and Watt (1971a) and Greenwood and Hopkins (1976, 1980) in man and of Watt (1976) in cat, we found in monkey a slight reduction and longer latency of the early EMG (Fig. 5). This discrepancy remains unexplained. However, it may be suggested that total darkness induces a change in alertness or in the gain of reflex pathways. One can also expect that the monkey modifies his motor strategy because of the expected lack of visual cues. Such a motor strategy change was described in man when submitted to identical postural perturbations in different environments (Vidal et al., 1981).

The visual stabilization method consists of suppressing visual cues about body motion selectively and only at the very precise moment when postural perturbation occurs. In this situation, the static visual input relative to position remains. In man, Nashner and Berthoz (1978) showed that once the visual environment of a man standing on a mobile cart has been stabilized, a striking increase in body pitch and decrease in the early soleus EMG response were observed (Fig. 6B) when the carriage was moved backwards. The decrease in EMG activity occurred within the first 100 to 150 msec. Quantitative evaluation gave about 5% mean difference in maximum body pitch between normal (N) and closed-eye conditions, whereas about 40%

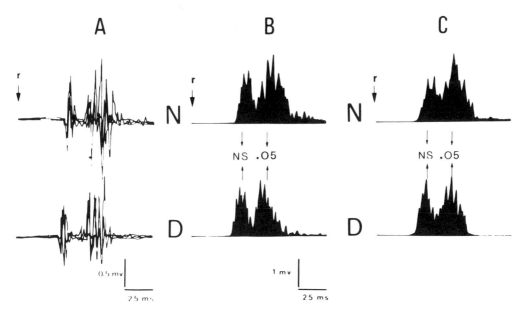

FIG. 5. Early EMG activities during free-fall in two conditions of visual surroundings and influence of repetition. The soleus muscle responses are shown when falls occur with normal vision (N) or in total darkness (D). A: Three superimposed raw EMGs. B: Mean value of integrated EMG during the first three days of experimentation. C: Mean value of integrated EMG one week later. Note the absence of habituation and the slight decrease in muscle energy when totally eliminating the visual cues. Only the second subcomponent of the first burst is slightly reduced ($p < 0.05$). Data from 2 monkeys, 90 trials in each condition in B and C. (From Vidal et al., 1979, with permission.)

difference was seen between N and visual stabilization (SV) conditions (Vidal et al., 1978). The authors concluded, therefore, that vision intervenes at an early stage in postural control.

Similar results were obtained by our group in monkeys (Lacour et al., 1978; Vidal et al., 1979). While falling, the monkey's head was covered by a box, the inside of which was lit and covered by black and white checkerboard paper. In this manner, the full visual surroundings were fixed with respect to the monkey's head. In this SV condition the early motor responses to fall were strongly reduced (Fig. 6A). In the N condition, the early response (latency 30 msec) was composed of two components. The first subcomponent (30 to 60 msec after release) was rarely modified in the SV condition, while the second (60 to 100 msec) exhibited in most cases a powerful reduction essentially in axial and distal extensor muscles, with a maximum decrease of about 30% on average in neck muscles.

To determine if the effects of visual stabilization resulted from a presetting of the gain of reflex pathways in conjunction with expected lack of visual cues, another SV method (Fig 7A) was developed. A V-shaped stationary retroprojection screen was placed in front of the monkey. A random black and white pattern was projected onto the screen. The speed of this pattern on the screen was commanded by a servo-system, the input of which consisted of the integral of the acceleration of fall, i.e., a velocity signal (see Lacour et al., 1981a). When the speed of the film was equal, at any moment, to that of the falling monkey, the animal experienced total exclusion of visual motion cues. This method provided an unexpected visual stabilization. Trials were also made under stroboscopic light with a strobe frequency of 4.5 Hz. Results

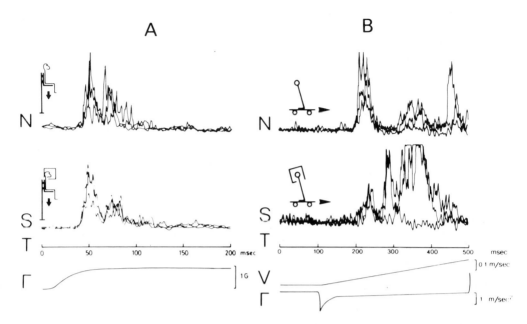

FIG. 6. Influence of visual stabilization on muscle responses to rapid postural perturbations. **A:** Three superimposed selected raw EMGs from soleus muscle in the falling baboon (acceleration peak: 8.8 m/sec^2) in two conditions of free-fall: With normal (N) and stabilized (S) vision. T: Time scale, msec, Γ: recording of downward acceleration of the head. Note the strong motor depression occurring in the interval 60 to 100 msec after onset of fall. **B:** Three superimposed selected raw EMGs from gastrocnemius muscle in a man standing erect on a moveable carriage rapidly moving backward at random times. The subject had either normal (N) or stabilized (S) vision. T: Time scale, msec, V: velocity of the carriage; Γ: acceleration of the carriage. Note the strong decrease in the first burst of EMG activity 100 msec after onset of postural perturbation in the S condition. (From Vidal et al., 1978, with permission.)

in both conditions again showed a strong decrease in the early motor responses to fall (Fig. 7B). The reduction was particularly accentuated during falls at low acceleration (2.2 m/sec^2) when compared to free-falls (8.8 m/sec^2), and specifically in the splenius and soleus muscles. We concluded, therefore, that decrease in EMG did not result from a presetting of the gain of reflex pathways but from the selective elimination of the visual motion cues per se. This conclusion is in close agreement with that of Amblard and Cremieux (1976), who found a postural destabilization induced by stroboscopic light in standing subjects; this implied that a velocity visual information is required to stabilize posture efficiently.

Elimination of visual motion cues in the SV condition can induce an intersensory conflict between visual and vestibular cues, the otoliths detecting a downward acceleration while the visual cues indicating no motion. A sensory conflict theory was proposed to explain the self-body motion perception felt by subjects at rest when submitted to a visual stimulus in linear (linear vection) or circular (circular vection) movement. We thus developed a new paradigm for changing the relative speed of the visual scene and the direction of the image motion during fall (see Fig. 7A and Lacour, 1981; Lacour et al., 1981a,b). The enhanced vision (EV condition) corresponded to a film motion directed upward, in the opposite direction of fall, with speeds equalling 0.6, 0.75, 1, 1.5, and 3 times the speed of

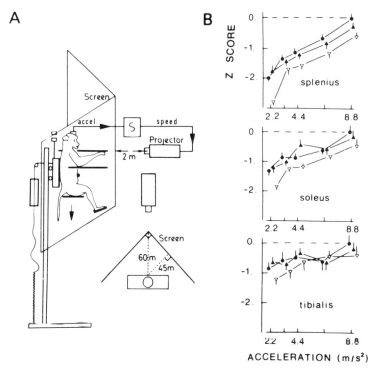

FIG. 7. Energy of the early EMG responses to fall in the monkey as a function of amplitude of acceleration and of visual conditions. A: Experimental set-up. A projector, the input of which consisted of the integral of vertical acceleration, and the output of which commanded the film motion, projected a black and white random pattern onto a V-shaped screen placed in front of the baboon. The EMG activities were recorded during different accelerations of fall (8.8, 6.6, 4.4, 3.3, and 2.2 m/sec²) and in 6 visual surrounds. (1) Visual scene remained stationary on the screen (normal vision, NV); (2) falls occurred in total darkness (D); (3) the visual scene moved downward at a velocity equal to that of the falling monkey (visual stabilization, SV); (4) the visual scene moved downward at a velocity slower than that of the monkey (reduced visual input, RV); (5) the visual scene moved upward at variable speed (enhanced visual input, EV); (6) falls occurring during stroboscopic light at 4.5 Hz. B: Mean values recorded from splenius, soleus and tibialis anterior muscles in 3 normal baboons during falls in normal vision *(dots)*, in total darkness *(filled triangles)*, and with visual stabilization *(open triangles)*. Ordinate: Energy of the response expressed as the Z score relative to the energy recorded at 8.8 m/sec² in normal vision (values normalized to zero). *Abscissa:* Amplitude of fall acceleration. Each point is represented with its 1% confidence interval. (From Lacour et al, 1981a, with permission.)

the falling monkey. In contrast, the reduced vision (RV condition) corresponded to a film motion directed downward.

Since visual motion cues predominantly intervened during slow falls, different accelerations of fall from 8.8 m/sec² to 2.2 m/sec² were tested to favor either the vestibular or visual cues, respectively. If the visually induced modulations were direction-specific and depended on the relative speed of the visual scene, this would be a major argument in favor of an early and specific influence of vision in postural control.

Figure 8 shows that during rapid falls (8.8 m/sec² max. acceleration) the EMG responses were reduced in the RV condition, with a maximum decrease when all motion cues were suppressed (SV condition). The energy of muscle responses did not differ from control values in the EV condition. During slow falls (2.2 m/sec² max. acceleration) a greater decrease was observed in the

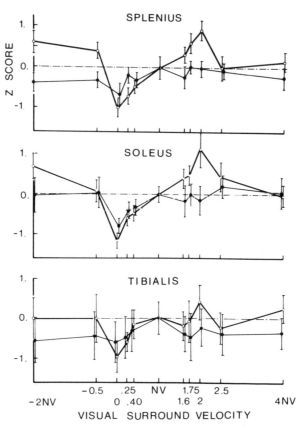

FIG. 8. Early directional influence of visual motion cues on postural control in the falling monkey. Mean values recorded in 3 normal baboons from splenius, soleus, and tibialis anterior muscles during slow falls (2.2 m/sec^2: *circles*) and rapid falls (8.8 m/sec^2: *dots*). *Ordinate:* Energy of the response expressed as the Z score relative to the energy recorded in normal vision (0, *dashed line*). *Abscissa:* Relative velocity of the visual surround. Values between 0 (SV condition) and NV correspond to a reduced visual input (RV condition) values above NV to a reinforced visual input (EV condition). Note the roughly linear increase of energy from 0- to 2-NV occurring during slow falls, which points to visually induced modulations depending on the relative velocity of the visual scene. (From Lacour et al., 1981a, with permission.)

RV condition. In contrast to rapid falls, the muscle response energy was strongly accentuated when the visual input was reinforced, with a maximum effect for a velocity twice the normal velocity of the visual surround. Thus, the visually induced modulation of muscle responses to fall was direction specific, at least for slow falls. The energy of the responses depended on the relative speed of the visual surround and was roughly proportional to it in a certain range. These results provide a further argument against the startle reaction theory. They also confirm the observations of Thoden et al. (1977), who have shown a direction-specific modulation of monosynaptic reflexes in cat limbs depending on the direction of a visual scene rotation, and which is opposite for optokinetic and vestibular stimuli.

This early directional influence of visual motion cues on postural control in the falling monkey was also observed in another set of experiments performed in unilateral

(UN) and bilateral (BN) vestibular neurectomized monkeys (Lacour and Xerri, 1980; Lacour et al., 1981a). It is well known that visual cues are involved in compensation of postural deficits after labyrinthectomy (Kolb, 1955; Courjon et al., 1977; Putkonen et al., 1977). It was then interesting to analyze the visually induced modulations of muscular responses during the time course of recovery when manipulating the visual surround. Figure 9 illustrates those results in UN and BN baboons. One can see that the partial recovery of the early muscle responses in BN baboons is mainly due to visual motion cues (Fig. 9A). The EMG recorded during falls with visual stabilization are almost totally suppressed whatever the postoperative time. Visual motion cues from peripheral retina represent the most relevant sensory input capable of substituting for exclusion of labyrinthine afferents and of eliciting the most appropriate muscle responses. In the UN baboons, visual motion cues intervene at early stages (the first two postoperative weeks) of the recovery process, fulfilling a transient substitution function in vestibular compensation (Fig. 9B). The motor depressions recorded in the

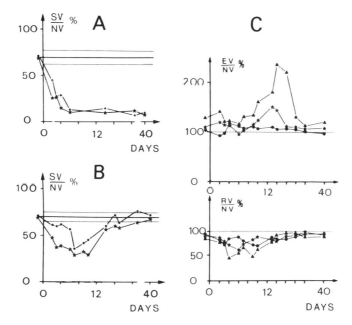

FIG. 9. Role of the visual motion cues in compensation of postural reactions to fall in the vestibular neurectomized monkey. Results from the splenius muscle. A: Mean results in 3 bilateral vestibular nerectomized baboons during falls at 8.8 m/sec^2 with normal (NV) or stabilized vision (SV). Ordinate: Mean evolution ratio SV/NV in left (solid circles) and right (stars) muscles as a function of the postoperative time in days (abscissa) and in percent of the control SV/NV ratio (15 values per point). B: Mean results in 5 unilateral vestibular neurectomized baboons during rapid falls (8.8 m/sec^2) in the NV and SV conditions. Results from the ipsi (dots) and contral (stars) splenius muscles. Same presentation as in (A), 25 values per point. The thick and thin horizontal lines represent the mean value of SV/NV before surgery and its 1% confidence interval, respectively (N = 125 samples). (From Lacour and Xerri, 1980, with permission.) C: Mean values from the contralateral splenius muscle in 3 unilateral vestibular neurectomized baboons during falls at 8.8 m/sec^2 (dots), 4.4 m/sec^2 (stars) and 2.2 m/sec^2 (triangles) in the EV (relative velocity of the visual scene = 1.6 NV) and the RV (relative velocity of the visual scene = 0.4 NV) conditions. Ordinate: Value of the EV/NV (upper graphs) and RV/NV (lower graphs) ratios expressed in percent of control. Abscissa: Postoperative time in days. (From Lacour et al., 1981a, with permission.)

SV condition were strongly accentuated when compared to normal monkeys. Furthermore, the modulations were markedly accentuated in both the EV and RV conditions and were greater for slow falls than for rapid ones (Fig. 9C). The visually induced modulation of the EMGs also depended on localization of the muscles relative to the side of the lesion, suggesting that visual-vestibular interactions may be mediated by vestibulospinal pathways.

In summary, stabilization of the visual surround with respect to the monkey's head produces a strong motor depression. This result supports the suggestion of Nashner and Berthoz (1978) that vision intervenes at an early stage in postural control. The dramatically enhanced effect of visual stabilization in labyrinthectomized monkeys confirms this view. In addition, the visual modulations are a function of the relative speed of the visual surround, direction-specific, and are dependent on the amplitude of acceleration. Our results in hemilabyrinthectomized monkeys and in man (Wicke, *personal communication*) confirm the present findings. Rapid visual influences and vestibular cues interact in the elaboration of rapid postural adjustments to fall.

Functional Role of Visual-Vestibular Interactions

Visual-vestibular interactions were described in the vestibular nuclei in goldfish (Dichgans et al., 1973; Allum et al., 1976), rabbits (Kubo et al., 1979), cats, and monkeys (Henn et al., 1974; Keller and Daniels, 1975; Waespe and Henn, 1977a,b). In all these studies, the authors recorded units activated by semicircular canal afferents. The main conclusion was that visual input improved the poor performance of the vestibular system for low-frequency stimulation. The greatest motor depressions recorded in falling monkeys during lower velocities of fall and visual stabilization support this general statement that visual input would dominate the response at low accelerations but vestibular input at high accelerations. Recordings from canal-dependent units in vestibular nuclei in alert monkey during conflicting visual-vestibular stimulation showed a reduction of the neuronal discharge (Waespe and Henn, 1978). This experimental paradigm is equivalent to our visual stabilization condition, and the observed decrease could be the neural correlate of the motor depressions.

However, the fall mainly activates the otolith system, and there is a lack of data concerning visual-vestibular interactions in otolith-dependent units. However, visual modulation of otolith-dependent units was recently shown by Daunton and Thomsen (1979) in cat vestibular nuclei. Of the units that responded to a linear acceleration, 76% were also activated by a large visual stimulus moving linearly in the opposite direction. If this was confirmed, the problem would be to know the functional role of rapid visual influences in postural control during linear motion. We know that the otolith system indeed provides good directional information, as shown in primary afferents (Fernandez and Goldberg, 1976) and second-order vestibular neurons (Daunton and Melvill-Jones, 1973), and that it has a very low threshold. It is also well known, however, that the vestibular system does not detect head motion when performed at constant velocity and that it is unable to distinguish between a linear acceleration directed in a particular direction and a deceleration in the opposite direction. That otolith afferent information is not without ambiguity is also evidenced when considering the peripheral mechanism by which linear acceleration is transduced. The statoconial mass can be similarly deflected by linear accelerations and by gravity. Other sensory information is necessary to clarify these ambiguities, of which visual input is one. It may be suggested that visual motion cues contribute to distinguish between a translational movement relative to the inertial space and a static position relative to the gravitational-vertical. It may be why Melvill-Jones and Young (1978) found that human subjects remained confused about direction of movement during low vertical accelerations in the absence of vision.

Visual motion cues provide complementary and indispensable information for body motion perception and, therefore, for landing preparation. Visual influences and vestibular cues are highly interdependent in building up the adequate muscle responses to rapid postural adjustments. This conception, which considers that sensory information arising from different sensory channels is, in fact, complementary gives a functional significance to the multisensory convergences found in various central nervous system (CNS) structures. It also suggests that sensory substitution may constitute a key mechanism in the compensation of sensory defects such as labyrinthectomy (see Bach-y-Rita, *this volume*).

Sites of Visual-Vestibular Interactions

Many CNS structures receive visual, vestibular, and somatic inputs. We have only considered the vestibular nuclei, because they have been extensively studied in recent years. These nuclei may receive visual inputs via the accessory optic nuclei (Simpson et al., 1979), the inferior olive and the flocculus. However, Keller and Precht (1979) found that in cats the visually induced modulation in vestibular neurons discharge still persisted after cerebellectomy. Other structures, such as the subparafascicular area (Barmack et al., 1978) and the tegmenti pontis reticular nucleus (Cazin et al., 1980; Precht and Strata, 1980) may relay the visual inputs. The exact pathway remains unknown, but Azzena et al. (1978) reported an activation in the vestibular nuclei with a mean latency of 28 msec by photic stimulation, suggesting the existence of fast pathways from the retina to vestibular neurons. However, the visual effects induced in vestibular nuclei by rotating visual scenes showed a slow dynamics. They are mainly observed for low-frequency stimulation (< 1 Hz), for low velocities ($< 60°/sec$), and for low accelerations ($< 5°/sec^2$). In these ranges, the vestibular neurons code the velocity of the visual scene motion. Furthermore, the visually induced modulations seem to have a long time constant (2.4 sec; Waespe and Henn, 1977a). It is apparently not the case in the parietal cortex (area 5) of the monkey, which receives both visual and vestibular inputs (Buttner and Buettner, 1978). Visual-vestibular interactions were also demonstrated in brainstem structures, which could be of functional importance for rapid postural control. For example, in contrast to the slow onset of visual effect of vestibular neurons and to the relatively long latencies of linear and circular vections, intense neuronal activity was induced by motion of large, high spatial frequency visual surrounds in superior colliculus (Bisti et al., 1974), pontine nuclei (Collewijn, 1975) with latencies as fast as 40 to 60 msec according to the species. These latencies are compatible with our own results on the early directional influence of visual motion cues on the postural control in the falling monkey.

ACKNOWLEDGMENTS

This work was supported by ERA 272 CNRS, and CRL 806007 INSERM.

Motor Control Mechanisms in Health and Disease,
edited by J. E. Desmedt.
Raven Press, New York © 1983.

Analysis of Movement Control in Man Using the Movable Platform

*Lewis M. Nashner

Neurological Sciences Institute, The Good Samaritan Hospital and Medical Center, Portland, Oregon 97209

A challenge faced by students of human movement is to use only noninvasive measurements to explore a highly complex control process. The surface electromyographic (EMG) activity of muscles, the resultant forces exerted on support surfaces, and the motions of body parts can be documented, but precise knowledge of incoming sensory signals and of resultant activities within spinal and brain motor centers is denied. Despite these limitations, current ideas about the organization of motor actions owe much to observations of human motor behaviors. For example, ideas about multilevel hierarchical organization were conceived during the 19th century based in part on the observation that infantile reflexes evolved during maturation into coordinated adult behaviors but then could be unmasked in adults suffering central nervous system (CNS) lesions.

A hierarchically organized motor system is one in which complex behaviors are synthesized from a limited repertoire of elementary units of action (e.g., Gallistel, 1980). The elementary units of action are simple behaviors organized within lower levels of the system, whereas higher levels within the system are required to coordinate these simple behaviors within the broader context of a complex motor act. The human posture and equilibrium studies described in this chapter are based on this hierarchical concept. An experimental approach has been developed which first identifies several units of action common to different posture and equilibrium control tasks and then defines organizational principles with which these units of action are coordinated during the execution of more complex motor actions. In this chapter, the term automatic behavior describes elementary units of postural action during which the relations between sensory inputs and resulting temporal and spatial patterns of muscle EMG activity are relatively stereotyped. One class of higher-level coordinating action has also been described, in which the spatial and temporal coordination of automatic behaviors is altered as a consequence of changes in the environmental conditions. Actions fitting this description have been termed adaptive motor behaviors.

Movable Platform Techniques

Studies of human posture and equilibrium control have provided experimental evidence which supports the concept of hierarchical organization. They have revealed that much of this task is accomplished by

---
*Neurological Science Institute, Good Samaritan Hospital, 1120 Northwest Twentieth Avenue, Portland, Oregon 97209.

motor actions which, by virtue of their latencies (short with respect to the most rapid voluntary actions) and their organizational constancy with respect to stimuli, are automatic in nature. Construction of complex behaviors from these automatic actions has been explored by exposing subjects to altered conditions: 1. Standing subjects were exposed to moving support and visual surfaces; 2. external objects in the subject's grasp moved unexpectedly; 3. subjects were provided different configurations of external support while performing stance tasks. Identification of individual automatic actions and definition of principles controlling their central coordination using the above paradigms have been aided by studying the posture and equilibrium controls of young children at chronologically different stages of motor development and by studying patients with clinically well-defined motor dysfunctions. In addition to helping in the identification of elementary units of action, these latter studies have also provided a means by which hierarchical concepts can be applied to understanding the functional basis for normal and abnormal motor development.

All platform experiments have focused on the sensory and motor processes which maintain the center of body mass in the correct vertical orientation with respect to the center of foot support as a subject stands freely, walks, and performs free-standing voluntary arm movements. The organization of motor responses has been studied by examining the patterns of leg muscle EMG which stabilize the anterioposterior (AP) sway motions of the center of body mass. These sway motions frequently occur during free stance as a consequence of the inherently unstable properties of the upright human body or of support-surface perturbations. During locomotion, unexpected perturbations in orientation arise from similar sources; however, the relation between the center of body mass and the center of foot support is now a dynamic one that is concerned with both the instantaneous stability of the walking subject and his continued forward travel. In contrast, the subject preparing to exert force on a handheld manipulandum while standing unsupported has advanced knowledge of the forces that will potentially destabilize the body. In all of the above instances, the forces acting on the body would lead to instability, were it not for the exertion of appropriate compensatory forces by the feet on the support surface. How compensatory motor actions are initiated following unexpected AP displacements of the body center of mass or in anticipation of voluntary AP disturbances and how these actions are coordinated among a large group of functionally related muscles of the legs, trunk, and arms are major questions that which have been addressed by this volume.

Unexpected alterations in the sensory conditions have been used to explore the hierarchical organization of sensory inputs. Specifically, the contributions of individual sensory inputs to equilibrium have been quantified by measuring the equilibrium adjustments of standing subjects deprived of vision and/or support surface inputs. To learn more about the adaptive processes that modify the context-dependent strategy for the relative weighting of support surface, vestibular, and visual inputs, responses to the perturbation of individual inputs were remeasured at intervals immediately following unexpected changes in sensory conditions. Progressive changes in the stimulus-response properties of the system following unexpected changes in sensory conditions were used to quantify the relative weighting of support surface, visual, and vestibular inputs.

The platform system illustrated in Fig. 1 is a key component of the human experimental program, because its movable support surfaces and visual enclosure can alter two critical variables of the posture and equilibrium control system. A separate surface of the platform supports each foot and

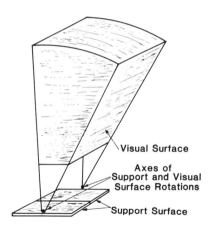

FIG. 1. The movable platform system: support surfaces and the visual enclosure.

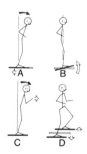

FIG. 2. Kinds of surface movements used to perturb the posture of a subject. A: Backward (or forward) displacement of both support surfaces *(open arrow)* causes AP sway in the opposite direction *(filled arrow)* centered principally about the ankle joints. B: Toes up (or toes down) rotations of both support surfaces *(open arrow)* rotate the ankle joints, but now this ankle motion is uncorrelated with AP sway. C: Forward (or backward) displacement of a handheld manipulandum *(open arrow)* causes AP sway *(filled arrow)* in the same direction as the handle pull. D: Upward displacement of one support surface and simultaneous downward displacement of the other *(open arrow)* cause the subject to sway laterally to the side of the lowering leg.

is independently movable in two linear and one rotational degree of freedom *(df)*; the longitudinal axis, the vertical axis, and a rotational axis co-linear with the ankle joint. The enclosure, which completely surrounds the subject's field of view, can independently sway about a rotational axis that is also co-linear with the ankle joints. While standing on the platform the subject may also grasp a manipulandum that is either fixed to resist voluntary pushes and pulls or can be unexpectedly displaced forward to disturb the subject's equilibrium. The contact surfaces of the supports and manipulandum are instrumented with strain gauges which quantify the reaction forces. A potentiometer belt system measures the displacements of the body center of mass with respect to the center of foot support.

Several different waveforms of brief, support-surface movements have been used to disturb the subject's AP equilibrium. Unexpected, forward or backward displacement of the two support surfaces (Fig. 2A) induces a predictable AP sway rotation principally about the ankle joints and in a direction opposite that of the platform movement (Nashner, 1977). Sway rotations of the center of body mass about the ankles can be induced in two other ways: unexpectedly rotating the support surfaces (Fig. 2B) or unexpectedly displacing a hand-held manipulandum (Fig. 2C). Although each of the above three perturbations involve a very different combination of sensory inputs, each requires a similar compensatory action: contraction of the calf and thigh muscles appropriate to resist forward or backward AP sway displacements of the body center of mass. The above three perturbations have been used to identify an automatic behavior which is common to all AP compensatory actions.

Unexpected brief upward displacement of one support surface and simultaneous downward displacement of the other produces a very different motion (Fig. 2D); the body sways laterally to the side of the lowering leg (Nashner, et al., 1979). This condition requires a very different compensatory action than that described above for AP sway perturbations; the lowering leg is actively extended and the elevating leg actively flexed to maintain the appropriate distribution of lateral forces. Reciprocal vertical and forward-backward surface displacements have been interposed in sequences in order to determine how rapidly the system can reorganize from one pattern of automatic action to another. Unexpectedly alternating the stimulus waveform between horizontal and reciprocal vertical displacements and by superimposing these two stimulus waveforms has provided additional information about the organization of automatic actions.

Although support-surface and visual input information is always perceptually correct, the sense of orientation derived from the support surface and visual inputs can be altered by moving these surfaces continuously. Although the position of the ankle joints with respect to the support surface provides useful vertical information under fixed surface conditions, this input can be reduced to below threshold levels by a procedure termed support surface stabilization (see Fig. 3A). Under this unusual condition, the platform support surfaces rotate in direct proportion to changes in the AP orientation of the center of body mass, so that the relative orientation of subject and support surface is fixed (Nashner, 1971). An analogous procedure can be used to approximately eliminate the normal sway-related movements of the subject with respect to his visual surround (see Fig. 3B). This technique is termed visual stabilization (Nashner and Berthoz, 1978). Under this unusual condition, the visual enclosure sways in direct proportion to the changes in AP orientation, so that relative orientation of subject and his surround remains fixed.

Equilibrium Maintained Principally by Automatic Actions Mediated by Local Surface Inputs

Dominance of support surface inputs in the control of upright equilibrium during both free stance and locomotion was dramatized by the clinical observation that patients with different degrees of vestibular dysfunction could stand and walk when deprived of vision by eye closure. However, these patients required an absolutely rigid and flat support surface (Martin, 1965).

Equilibrium actions of normals and vestibular deficit patients have been quantified by imposing a variety of brief support-surface displacements and measuring the resultant EMG activities of leg muscles and compensatory forces exerted by the feet on the support surface. In normal subjects, EMG responses correlated with significant changes in the forces exerted on the support surface occurred at latencies of 95 to 110 msec following the imposition of surface perturbations. Forward and backward surface displacements excited the ankle muscles stretched by the resulting sway rotation about the ankle joints and, after an addition 10- to 15-msec delay, the proximal thigh muscles on the same dorsal or ventral aspect of the legs (examples in Fig. 4) (Nashner, 1977). When patients with moderate to severe, but well-compensated vestibular deficits were exposed to identical perturbations, the organization of leg muscle EMG adjustments was temporally and structurally the same as those of normal

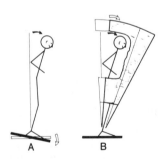

FIG. 3. Continuous platform surface movements that alter the sensory inputs to equilibrium control. **A:** Rotating the support surface *(open arrow)* in direct proportion to the AP rotations of the body center of mass *(filled arrow)* stabilizes the support surface with respect to the subject and eliminates orientation inputs from this surface. **B:** Rotating the visual enclosure *(open arrow)* in direct proportion to the AP sway rotations of the body center of mass *(filled arrow)* stabilizes the visual surface and eliminates orientation inputs from this surface.

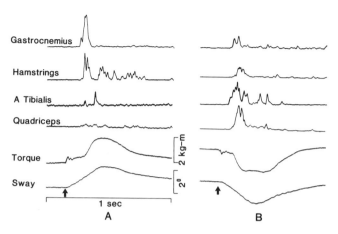

FIG. 4. EMG, ankle torque, and AP sway responses (average of 5 consecutive trials) of a subject to forward (A) and backward sway (B) perturbations. Within 100 msec of the forward sway perturbations *(filled arrow),* gastrocnemius, and hamstrings (additional 10–20-msec delay) are activated in fixed proportion. Tibialis anterior and quadriceps muscles are activated in a similar temporal and structural pattern during backward sway perturbations.

subjects (Nashner et al., *in press*). Despite significant trial-to-trial variations in the amplitude of responses, the ratios of contractile activity and the temporal relations between distal and proximal synergists were relatively fixed in both normal and vestibular patients.

Following reciprocal, vertical displacements, a very different pattern of muscle activity occurred at the same 95- to 110-msec latencies. The ankle muscles shortened (rather than stretched) by the resulting ankle rotation were first excited, followed in 10 to 20 msec by the thigh muscles on the opposite (rather than the same) dorsal or ventral aspect of the leg (Nashner et al., 1979). Again, despite significant variations in the amplitude of individual actions, the ratios of contractile activity and the relative timing between distal and proximal synergists were relatively fixed.

The EMG adjustments and resulting active muscular force associated with AP and reciprocal vertical perturbations were large enough to maintain the subject's equilibrium. The force and AP sway traces of Fig. 4 show that the compensatory EMG activity was associated with a three- to fivefold increase in the forces resisting sway. Similarly, EMG activities in following reciprocal, vertical displacements of the two support surfaces were correlated with the redistribution of vertical forces appropriate to maintain lateral stability. Linkages between distal and proximal muscles observed during postural adjustments are also consistent with the known dynamic interactions among ankle, knee, and hip joint motions. During AP sway, linkages between distal and proximal muscles on the same dorsal or ventral aspect of the legs are useful in coordinating the mechanically coupled oscillation of ankle and hip joints. During lateral sway, proportionate activation of muscles on the opposite dorsal or ventral aspect of the legs coordinate the flexion and extension movements of ankle and knee joints.

Several observations have established that the organizational structure of the above two equilibrium actions are automatic in nature. Subjects were exposed to displacements over a range of velocities to show that response amplitudes are graded to the support surface stimulus (Nashner and Cordo, 1981). Displacement waveforms were unexpectedly changed from horizontal to vertical to show that response latencies do not vary, and directional or organizational errors do not occur with choice;

these latter two criteria are commonly used to distinguish voluntary reaction time movements from reflex actions (Nashner and Woollacott, 1979). When subjects performed reaction time, voluntary sway movements at the ankle joints, triggered by the support-surface perturbations described above, they were unable to modify voluntarily the latency or organizational patterns of the automatic component of the equilibrium action (Nashner and Cordo, 1981).

Under normal (fixed) support-surface conditions, the ankle joints are the critical axis for sensing shifts in body orientation (Nashner, 1977). Although sway includes the knee and hip joint motions, these tend to be coordinated during stance to minimize changes in the position of the center of body mass (e.g., Gurfinkel et al., 1971). Subjects were exposed unexpectedly to changes in the rotational orientation of the support surfaces to test whether or not the ankle joint proprioceptive inputs alone can mediate the automatic adjustments. Under this stimulus condition, the ankle joints were rotated in place at approximately the same rate as during AP sway perturbations, although visual and vestibular inputs were not immediately affected. Despite the lack of correlated vestibular and visual inputs, support surface rotations elicited EMG adjustments that were temporally and structurally the same as those elicited by horizontal displacements (Nashner, 1977). The ankle joint muscles stretched by the rotation were activated at 95- to 110-msec latencies, followed by the proportionate activation of the thigh muscles on the same dorsal or ventral aspect of the legs.

Automatic Actions Provide Stability and Maintain Stepping Rhythm During Locomotion

Equilibrium in the AP plane is considerably more difficult to maintain during locomotion, because of the requirement for a dynamic balance between motions of the center of body mass and the center of support. During the single-support phase of the step cycle, the inherent instability of the body is exploited as a propulsive mechanism; under the influence of gravitational forces, the center of body mass sways forward relative to the support leg, but it is subsequently arrested by placing the other leg forward of the center of body mass. The motions of the body relative to the support during the single-support phase (prior to placement of the other foot) therefore resemble the AP sway motions studied during free stance. Presumably, the walking subject must continuously adjust the rate of forward sway relative to the timing and placement of the other foot in order to maintain the rhythm and metrics of the stepping pattern.

The compensatory mechanisms that continuously adjust the rate of forward progression of the body during the single support phase of the step cycle were recently examined by staggering the two movable platform surfaces at one-step intervals within a walkway 4 m long (Nashner, 1980). As subjects walked back and forth, steps centered on one of the platform surfaces were occasionally perturbed by horizontally or rotationally displacing the surface. The mechanical consequences of these surface displacements were similar to those occurring in the stance condition; forward or backward displacement altered both the AP orientation of the center of body mass relative to the support foot and the ankle joint of that foot, while rotational displacement changed the angle of the ankle joint but not the AP orientation.

The EMG signals in ankle muscles attributable to platform movements were 2 to 5 times larger than that which was characteristic of the unperturbed steps, even though the ankle rotational motions caused by the support surface perturbations were approximately one half the rate of the normal step related ankle motion (Nashner,

1980). Equilibrium adjustments occurred at 100-msec latencies in the gastrocnemius (onset of EMG activity), when this muscle was stretched faster than normal due to backward surface displacement (Fig. 5A), or in the tibialis anterior, when this muscle was shortened slower than normal due to forward surface displacement (Fig. 5B). Following both backward and forward displacements of the surface, force changes related to the EMG activity were sufficient to appropriately alter the rate of forward progression of the body. Gastrocnemius activation slowed the rate of forward progression to help realign the center of body mass with an abnormally backward placed support foot, while tibialis anterior activation hastened the rate of progression to realign the body with an abnormally forward placed foot. Differences in EMG activity and mechanics of motion between perturbed trajectories (heavy lines) and adjacent unperturbed trajectories (fine lines) are emphasized by vertical grids in Fig. 5. Unexpected rotations of the support surface were also imposed during walking to demonstrate that ankle position is the critical input mediating these rapid automatic adjustments. Following surface rotations, subjects responded by resisting errors in ankle trajectory, which in these instances, resulted in the inappropriate adjustments in the rate of forward progression.

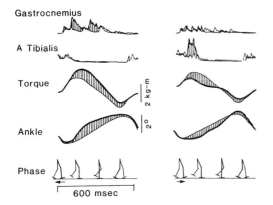

FIG. 5. EMG, ankle torque, and ankle angle recordings (average of 4 consecutive trials) during perturbed (heavy lines) and unperturbed steps (fine lines). Left: Backward displacements of the platform beginning immediately following heel-strike (arrow) dorsiflex the ankle abnormally fast, and the body center of mass is displaced abnormally ahead of the center of foot support. Differences between unperturbed (normal) and perturbed (abnormal) steps are emphasized by vertical grid lines. Activation of the gastrocnemius muscle at abnormally high levels beginning approximately 100 msec after perturbations is associated with a larger than normal ankle torque. These responses bring the body back into approximately normal alignment by the end of the step cycle. Right: Response organization following forward displacements of the platform (arrow) is similar. Abnormally slow dorsiflexion of the ankle is compensated by abnormally high activation of tibialis anterior muscle beginning at 100-msec latencies.

Automatic Actions Recruited in Association with Voluntary Movements to Provide Postural Support

Few everyday voluntary movements are insulated from the requirements of postural stabilization. For example, a freely standing subject's equilibrium is upset whenever he grasps and then exerts force on an external object. Several research groups previously reported that the destabilizing effect of such voluntary movements is anticipated by the activation of postural muscles providing support for the movement (e.g., Pal'tsev and El'ner, 1967; Marsden et al., 1978c, this volume). Movable platform studies have shown that automatic postural actions are an integral part of rapid voluntary arm movements.

In one study, freely standing subjects executed reaction time ankle joint movements in response to tone stimuli (Nashner and Cordo, 1981). In some instances, tones were given while the subject was in a state of equilibirum whereas in others, support-surface displacements which disturbed the subject's equilibrium coincided in time with the tones. Under stable conditions, reaction time latencies of calf muscles varied over a wide range, but some were below 100 msec and therefore were as rapid as automatic EMG adjustments. However, when required to perform the identical reaction time movement in a state of disequilibrium, compensatory automatic adjustments (95- to 110-msec latencies) always occurred sequentially before the instruction movement, which were now delayed until a minimum of 140 msec.

When freely standing subjects initiate voluntary arm movements against an external object, the appropriate calf musculature is activated first. The sequence of action then radiates upward to thigh muscles, and last to the focal musculature of the arm (Cordo and Nashner, 1982). Furthermore, the organizational characteristics of the stabilizing components of these movements are similar to the automatic adjustments which stabilize AP sway induced by unexpected support surface perturbations. However, stabilizing components are absent from the movement, and the focal musculature is activated earlier when an external support eliminates the need for active postural stabilization. In terms of hierarchical principles, then, automatic and voluntary actions are temporally and structurally coordinated to produce a complex motor act which includes both postural and focal components. However, the coordination of postural and focal components is flexible, depending on the stability requirements of the act.

In the second study, subjects performed tone-triggered reaction time arm pulls against a manipulandum while standing freely or with trunkal support (Cordo and Nashner, in press). During unsupported, tone-triggered, voluntary pulls, the temporal and spatial pattern of activation of ankle (gastrocnemius) calf and thigh (hamstrings) muscles was similar to that produced by unexpected forward AP sway perturbations. Ratios of EMG activity level were approximately the same in both instances, and ankle muscles were activated 10 to 20 msec in advance of the synergist muscles of the thigh and 30 to 50 msec in advance of the arm muscles executing the pull (Fig. 6A).

In a third study, unexpected displacements of the manipulandum were used to demonstrate that the linkages between leg and arm muscles are mediated, at least in part, by fast-acting pathways rather than by

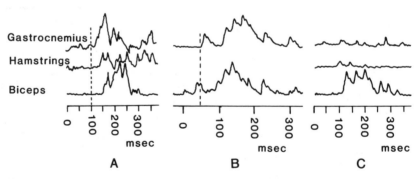

FIG. 6. EMG responses of leg and arm muscles during the performance of standing arm movements. A: An unsupported subject was instructed to pull the handle as rapidly as possible following a tone (beginning at 100 msec). Subject responds by first activating gastrocnemius, then after a 10- to 20-msec delay, the hamstrings, and, finally, the biceps of the arm after an additional 50-msec delay. The temporal and structural organization of leg muscle activity is therfore very similar to that observed during forward AP sway perturbations (see Fig. 4A). B: The handle is unexpectedly displaced forward. The gastrocnemius muscle is activated within 55 to 70 msec of the onset of the pull. C: Subject responds to tone as in part A, except now he leans against a chest support. Activity of leg muscles is suppressed, and activation of biceps of the arm is significantly faster.

higher centers. Although the trunk of the unsupported subject did not begin to move forward until 75 to 100 msec after the unexpected displacement, the ankle (gastrocnemius) muscles appropriate to resist the unexpected pulls by the handle were activated at latencies of 55 to 75 msec (Fig. 6B). In contrast, ankle and thigh muscles of supported subjects were inactive during both reaction time arm pulls and unexpected manipulandum displacements (Fig. 6C).

Automatic Actions Reordered to Produce Different Condition-Specific Strategies of Equilibrium Control

A well-know clinical observation is that the instability of some patients can be dramatically reduced by allowing them to touch a fixed surface lightly with a single finger. Presumably, performance is improved by a reorganization of the strategy for utilization of sensory information, since a single outheld finger does not in itself provide significant mechanical stability. The ability to reorganize strategy for control is a fundamental attribute of the hierarchically organized system, in which a rich behavioral repertoire can be achieved by coordinating elementary units of action in different temporal and spatial patterns. This same organizational attribute may explain how subjects alter their strategy of equilibrium control when asked to stand with different configurations of mechanical support or under altered sensory conditions.

One example of the reorganization of automatic actions has already been introduced in the immediately preceding section. This section described the abolition of associated postural adjustments in subjects provided trunkal support while pulling on a handheld manipulandum. Apparently, eliminating an associated postural adjustment in the supported condition enabled subjects to move their arms more quickly in response to tones. This adaptive change removed one component of the movement which, under unsupported stance conditions, always preceded the initiation of action by the arm. Another example of the reorganization of equilibrium actions was described in the same study; subjects were allowed to grip the fixed manipulandum while being exposed to forward and backward displacements of the platform support surface (Cordo and Nashner, 1982). While appropriate calf and thigh muscles were activated in fixed temporal sequence beginning at 95- to 110-msec latencies under free-stance conditions, these automatic leg actions were abolished, and the now supportive arm muscles were activated within 60 to 100 msec of the onset of support surface motions. Thus, the associated postural component of a voluntary movement can be eliminated when stabilization is not required, and the focus of postural activity can be placed on whichever member provides the most efficient support.

How subjects adapt their strategy of equilibrium control under altered sensory conditions is a complex question, which has historically been a major interest to physiologists and psychologists. A major contribution to this area of study has been the efference copy principle advanced by von Holst and Mittelstadt (1950). The principle of efference copy explains how a system might distinguish sensations produced by self-motions from those produced by motions of external objects relative to the individual. The same kinds of sensory conflicts are also possible during free stance. Support surface and visual inputs can be generated either by sway motions or by motions of support and visual surfaces relative to the subject. However, because the spontaneous sway motions of the inherently unstable body are not necessarily generated by voluntary actions, efference copy alone cannot be used to resolve these sensory conflicts. Hence, a major focus of platform studies has been to describe the process that enables subjects to distinguish sensations produced by sway motions from those generated by motions of support and visual surfaces.

Forces generated by automatic actions mediated by support-surface inputs are sufficiently strong under normal (fixed) support-surface conditions to compensate AP sway within 1 to 2 sec. If inappropriately recruited by a conflicting support-surface input, however, these same automatic actions are equally potent in upsetting the equilibrium of the subject. An earlier platform study was designed to discover how the posture control system adapts to conflicting support-surface inputs when subjects were exposed unexpectedly to changes in the slope of the surface (Nashner, 1976). Subjects adapted to ankle rotational inputs not correlated with AP sway by attenuating the automatic EMG adjustments within 3 to 5 trials. Following this adaptation, EMG adjustments to AP sway perturbations were more delayed (≥ 175) and were now related to motions of the body with respect to vertical but not to ankle motions. Presumably, these more delayed actions were mediated by vestibular and visual inputs. A similar adaptive process was found to modify the weighting of visual inputs (Nashner and Berthoz, 1978). Immediately following the elimination of visual feedback (stabilization of the visual surround), EMG adjustments were weaker than normal and subjects' sway excursions were abnormally large. However, after platform perturbations were reimposed 6 times under this unusual condition, the strength of automatic EMG adjustments increased to near-normal strength. Presumably, the weighting of a visual inputs, which, under this condition provided conflicting information, was reduced.

Although the above studies of normal adult subjects have demonstrated that the weighting of inappropriate inputs can be decreased under altered sensory conditions, these studies alone did not reveal the organizational characteristics of this adaptive process. Specifically, adult studies did not describe the process by which conflicting inputs are identified. A missing element in the above adaptive control studies was an experiment in which vestibular inputs were altered. Unfortunately, experiments of this type could not be performed noninvasively on normal subjects. Therefore, some of the most useful information about adaptive reorganization of sensation has been obtained in the several recent studies of patients with vestibular dysfunction and of young children with incomplete motor development.

Vestibular Inputs Provide an Inertial-Gravitational Reference Against Which Unexpected Conflicts in Sensory Conditions are Quickly Resolved

The patient with a motor deficit limited to clinically well-documented vestibular impairment is an ideal subject to study the adaptive process, because patients in this population who are not dizzy can stand and walk normally under most circumstances and yet are highly unstable under others (e.g., Martin, 1965; Begbie, 1967). In a recently completed platform study (Nashner et al., 1982), a protocol was developed to examine the adaptive capabilities of a group of 12 patients whose vestibular deficits were clinically well quantified and ranged from complete to very mild impairment. The performance of these patients was assessed as they stood freely under 6 different conditions: (1) normal (fixed support and visual surfaces); (2) normal with eyes closed; (3) fixed-support and stabilized visual conditions; (4) stabilized support-surface conditions with normal vision; (5) stabilized support-surface conditions with eyes closed; and (6) stabilized support- and visual-surface conditions. When imposed in the above numerical order for intervals of 50 sec each, these tests exposed the patient to sensory conflicts of increasing complexity. During stance under altered conditions, performance ratios were computed by full-wave rectifying and numerically integrating each 50-sec sway record (after removal of DC bias), and then comparing the integrated values to those resulting if sway oscillations had been at maximum amplitudes possible for the feet-together stance. Thus, small ratios were consistent with stable posture, while ratios approached unity as the subject swayed near the limits of stability.

The performance of patients and normals was stable and approximately the same under the first three conditions, those in which support surface inputs were normal. Performance ratios ranged between 0.1 and 0.3. This observation reinforces the conclusion that automatic EMG adjustments mediated by support-surface inputs are functional in vestibular deficit patients and are the most potent component of equilibrium control. In contrast, the performance of patients and normals varied considerably under the three conditions in which support-surface inputs were eliminated by stabilization of this surface. The single most significant observation was that all but two patients (one with complete and one with severe disruption) could stand with ankle joints stabilized and eyes closed, whereas most of the remaining patients became more unstable or lost balance on subsequently reopening their eyes within a stabilized visual environment (Fig. 7). In contrast, normals performed equally well under these last two conditions. Apparently, patients with subtle vestibular deficits were able to perform this task without vision but were unable to suppress the influence of a perceptually correct but orientationally inappropriate visual input. In all but one instance, patients whose vestibular deficits were sufficiently severe to prevent adaptive suppression of inappropriate visual inputs were equally unable to adaptively attenuate inappropriate support-surface inputs when exposed to unexpected surface rotations.

The above observations have led to the hypothesis that the inertial-gravitational input provided by the vestibular system is a

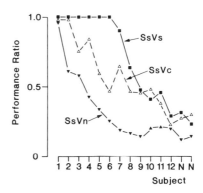

FIG. 7. Quantification of the amount of sway of 12 vestibular deficit patients and 2 age-matched normals (Ns) as each stood for two 50-sec intervals under three different conditions, stabilized support surface with normal vision (S_SV_N), stabilized support surface with eyes closed (S_SV_C), and stabilized support and visual surfaces (S_SV_S). Performance ratios of 1.0 indicate instability, whereas ratios near zero indicate stable stance. Note that both patients and normals swayed more on eye closure. However, those with vestibular deficits were completely destabilized or nearly so when subsequently reopening their eyes within a stabilized surround, whereas normals and several patients in remission at the time of testing were not further destabilized by this condition.

vertical reference against which conflicts in the two externally referenced sensory systems (support surface and visual inputs) are quickly resolved. When confronted with conflicting orientation information, the vestibular reference is, in the short run, always presumed correct. The normal subject does not orient to simultaneously stabilized support and visual surfaces, even though these two inputs are perceptually in agreement with one another. This conceptualization helps clarify the functional disability of the patient with subtle vestibular abnormality, who usually performs quite normally and, yet, reports intermittent instances of disorientation. To serve as a vertical reference, the vestibular system must not only be accurate, it must also be a highly reliable motion transducer. Because many patients with subtle vestibular deficits may experience variability as well as loss of transductive function, their greatest losses are probably in adaptive control functions.

Development of Automatic and Adaptive Equilibrium Controls

The most convincing experimental evidence that spinal circuits can independently generate automatic postural and locomotor behaviors has been that derived from the spinal cat preparations (e.g., Grillner, 1975). When suspended above a moving treadmill and provided tonic activation of spinal circuits, this preparation walks, trots, and even gallops in a coordinated manner. If the stepping trajectory of a limb is perturbed, reflex adjustments are coordinated with the stepping movement (e.g., Forssberg et al., 1977; Duysens and Pearson, 1976). However, this preparation is incapable of unsupported locomotion. The presence of reflex stepping behaviors in the newborn human suggests that similar automatic postural and locomotor behaviors can also be generated by the human spinal cord prior to the full development of central pathways (Forssberg and Wallberg, 1980). A study of posture and equilibrium controls in young children, beginning with those just learning to stand freely and including others up to the age of 10 years, has been recently completed (Forssberg and Nashner, 1982). The principal aim of this study was to exploit the developmental process as a means of distinguishing between the automatic and the adaptive behaviors of the system. The experimental protocol for evaluating adaptive controls of 18 children was identical to that already described for vestibular deficit patients; each child stood for 50-sec intervals under 6 different conditions in which sensory conflicts were increased and then was exposed to support surface rotations. In addition, the spatial and temporal organization of automatic EMG adjustments was also measured during support surface perturbations and during arm pulls on a manipulandum.

The automatic postural actions are organizationally intact in the youngest children able to stand and walk unsupported. Despite significantly greater variability in the latency and amplitude of automatic EMG adjustments elicited by support-surface perturbations in children below the ages of 6 to 8 years, the distal to proximal timing and the fixed ratios of muscular actions were adultlike in all including the youngest, ages 1½-years-old. In addition, the sequencing of action beginning at the ankle muscles and then radiating upward to thigh and then arm was evident in several 1½- to 2½-year-old children tested in the arm-pull paradigm. Although the literature suggests that vestibular function is fully developed in children of preschool-age (Eviatar and Eviatar, 1978), children below the critical age of approximately 7.5 years were unable to quickly resolve conflicts among sensory inputs. Specifically, children below the critical age did not attenuate automatic EMG adjustments elicited by rotations of the support surface. Furthermore, the performance of these younger children resembled that of the patients with subtle vestibular deficit; they were able to stand with ankles stablized and eyes open or closed but then lost balance when subsequently opening their eyes within a stabilized surface and visual environment. Young children performed successfully without vision but not with a conflicting visual input, whereas the older children and normal adults performed equally well under these two conditions. Thus, study of children has demonstrated an instance in which an automatic action is well developed, and yet a higher level action necessary to appropriately coordinate it is relatively underdeveloped. The slow kinetics of maturation of the sensory mechanisms involved appears rather remarkable. There is little evidence about rates of maturation of central sensory pathways in man, but Desmedt et al. (1976,1980) documented that the central lemniscal conduction of somatosensory input from fingers also takes about 7 years from birth to reach adult values in normal man.

Concluding Remarks

The sensory and motor actions of human subjects standing and walking on a movable platform surface have been broken down into a number of automatic units of action and a set of organizational principles for their central coordination. Although the hierarchical structuring of motor actions is not a conceptually new approach to movement control, platform studies are providing a depth of understanding heretofore achieved only with reduced animal preparations. Specifically, identification of automatic components of human postural and locomotor actions is helping to bridge some of the gaps between previously only qualitative human motor studies and physiologically detailed descriptions of postural and locomotor control processes in animal preparations.

Human platform studies have added another dimension of analysis which has yet not been adequately addressed in the reduced animal preparation; namely, the observation of central adaptive and coordinating actions. The performing human is perhaps the best experimental model through which to develop a deeper conceptual understanding of adaptive mechanisms, as the human is motorically one of the more highly adaptable creatures and can be easily incorporated into complex experimental paradigms. Specifically, platform studies have shown that subjects quickly alter the weighting of support surface, vestibular, and visual inputs using a fixed strategy for reorganization following change in environmental conditions. Furthermore, subjects eliminate associated postural components of voluntary actions or refocus these postural actions on different muscle groups, depending on the configuration of postual support.

Hierarchical models of human posture and equilibrium control derived from platform studies have also been conceptually useful in thinking about the functional basis for normal motor development and abnormal movement control, at a time when the clinical disciplines tend to be dominated by anatomical and pharmacological rather than physiological models. In the two examples included in this chapter, performance was impaired by the lack of a critical sensory input to the process of sensory adaptation (vestibular patients) or by a developmental immaturity within this central process (children below the ages of 7 to 8 years). In a previous platform study, patients with cerebellar dyssynergia showed breakdowns in the timing and metrics of automatic postural actions (Nahsner and Grimm, 1978). In a related study, Traub et al. (1980) have suggested that some of the motor impairment of the Parkinson's disease patient may be due to the loss of coordination between voluntary motor activity and the associated actions which provide postural support. The results of such studies may prove useful not only to the neurologist as a complement to sophisticated anatomical and pharmacological measures, but equally or more so to the physiatrist responsible for the retraining of motor-impaired patients.

Motor Control Mechanisms in Health and Disease,
edited by J. E. Desmedt.
Raven Press, New York © 1983.

Tonic Labyrinthine Reflex Control of Limb Posture: Reexamination of the Classical Concept

Christina W. Y. Chan

School of Physical and Occupational Therapy, McGill University, Montréal, Québec, Canada

Vestibulospinal limb reflexes may be divided into static and dynamic stabilizing influences, of which only the former derive from the otolith organs (cf. Wilson and Melvill Jones, 1979). Thus, Money and Scott (1962) found that normal cats, in the absence of vision, retained posture and balance with both slow and rapid tilting of the supporting surface. Cats in which the canals had been inactivated could adjust to slow but not rapid tilts. Subsequent removal of otolith function led to a consistent failure to retain balance during both slow and rapid tilting. In this connection, tonic labyrinthine reflexes were first described by Magnus and de Kleijn (1912; also Magnus, 1926) as acting in the same direction on all four limbs. Extensor tone of all four limbs was maximal when the animal was supine with the snout 45° above the horizontal plane, and minimal when the animal was 180° out of phase with the snout 45° below the horizontal plane. These observations have since been challenged by various investigators. Questions have also been raised about whether findings from quadripeds could be extrapolated to the bipedal mode of postural control in man. The objective of this chapter is therefore to reexamine Magnus' classical concepts in light of recent experimental findings in both animals and humans. To provide a more comprehensive picture, a brief overview of the transducing process and of the pathways through which otolithic influences could be exerted on the limb MNs in the spinal cord, will first be presented. For more detailed information, the readers are referred to reviews by Goldberg (1979), Wilson and Peterson (1981) and Wilson and Melvill Jones (1979).

INFORMATION TRANSDUCED BY OTOLITHIC RECEPTORS

The peripheral vestibular organs comprise two main elements, the semicircular canals and the otolith organs. The canals transduce specifically dynamic rotational movements of the head relative to space, whereas the otolith organs respond to *static* displacements of the head relative to gravity, as well as dynamic linear acceleration stimuli. Consequently, it is the otolith organs that provide the primary vestibular response activating tonic vestibulospinal reflexes participating in postural control. The mechanical response of the otolith organs has been the subject of much speculation. However, recent work (Lindeman, 1973; Smith and Tanaka, 1975; Hudspeth and Corey, 1977) has demonstrated that elements in the end organ act as miniature linear accelerometers, detecting both the magnitude and direction of the imposed resultant field of linear acceleration. Dynamic characteristics of response have been elucidated in an extended study of discharge patterns in identified primary afferent otolithic vestibular nerve fibers in monkeys (Fernandez and Goldberg, 1976a,b,c; cf. review by Goldberg, 1979). Broadly speaking, they can be divided into two main types: (1) regular units which display little change in gain or phase over a wide range of frequencies of natural stimulation from DC to 2 Hz, and whose response is predominantly tonic; and (2) irregular units showing a continuous increase of gain over the same range and whose response is more phasic. Presumably it is the regular components of primary afferent activity that are mainly responsible for tonic vestibulospinal influences. The sensitivity of the otolith response is emphasized by the fact that in man the threshold for preception of linear acceleration is about 0.005 g (Melvill Jones and Young, 1978).

It is clear that otolith-spinal influences extend to motoneuronal (MN) pools in the lower cord. Thus, Watt (1976) has demonstrated that the early electromyographic (EMG) response evoked in the ankle muscles of the cat during sudden falls is an otolith-dependent reflex (cf. p. 631). At an electrophysiological level, single shocks to the vestibular nerve of decerebrated cats (Gernandt et al., 1957) evoked reflex discharges in lumbar ventral roots, and stimulation of the vestibular nerve caused primary afferent depolarization of group I and cutaneous afferents to lumbar cord, indicating the presence of presynaptic inhibition (Cook et al., 1969). The question therefore arises as to what pathways are available for transmitting vestibular information to the cervical and lumbosacral cord for the control of fore- and hindlimb muscles.

PATHWAYS MEDIATING VESTIBULOSPINAL INFLUENCES TO THE SPINAL CORD

Anatomical and physiological studies have revealed that information from the vestibular and related nuclei is relayed to the spinal cord by three main tracts: the medial vestibulospinal, the lateral vestibulospinal, and the reticulospinal tracts (cf. Wilson and Melvill Jones, 1979).

The Medial Vestibulospinal Tract

This tract, which originates in the medial (Nyberg-Hansen, 1964), descending (Wilson et al., 1967), and Deiters' nuclei (Rapoport et al., 1977), and descends the cord bilaterally (Nyberg-Hansen, 1964), has few fibers reaching the lumbar level. These fibers are thought to control neck and other axial musculature (cf. Pompeiano, 1975; Wilson, 1979), and are therefore not directly relevant to the present issue of limb control.

The Lateral Vestibulospinal Tract

This tract, which originates mainly in Deiters' nucleus (Brodal et al., 1962) with some in the descending nucleus (Rapoport et al., 1977), on the other hand, has the majority of its fibers extending as far as the ipsilateral lumbosacral cord. Fifty percent of these neurons, the axons of which branch to the cervical enlargement, also project to the lumbosacral region (Abzug et al., 1974). Thus, activity conducted through these fibers can contribute to reflex coordination between fore- and hindlimb muscles. Lateral vestibulospinal tract neurons can often be activated at short latency, sometimes monosynaptically, by electrical stimulation of the ipsilateral semicircular nerve (Peterson et al., 1980). More importantly, these neurons can be facilitated by natural labyrinthine input. Thus, lateral tilt (or side-down rotation of the head) excites ipsilateral Deiters' neurons (Boyle and Pompeiano, 1979), as well as identified lateral vestibulospinal tract neurons projecting ipsilaterally (Orlovsky and Pavlova, 1972) and bilaterally to the lumbar level (Peterson, 1970).

What is the nature of these descending influences on spinal cord mechanisms? Stimulation of Deiters' nucleus, which receives predominantly otolithic inputs (cf. Wilson and Peterson, 1981) produces widespread di- and polysnaptic excitation of both fore- and hindlimb extensor α-MNs, with monosynaptic excitation of some hindlimb extensor MNs (Lund and Pompeiano, 1968; Wilson and Yoshida, 1969; Grillner et al., 1970). In contrast, influences on limb flexor α-MNs are mainly inhibitory. The interneurons in the polysynaptic pathways are known to be also involved in other reflexes such as the Ia reciprocal inhibitory pathway (Hultborn and Udo, 1972; Hultborn et al., 1976) and the crossed extensor reflex (Bruggencate and Lundberg, 1974). These findings suggested that vestibulospinal effects will interact with somatosensory inputs at a spinal cord level. For example, vestibulospinal influences may be augmented or attenuated depending on the bias of the interneurons.

Broadly speaking, γ-MNs appear to be affected in much the same way as α-Mns. Thus, electrical stimulation of Deiters' nucleus mainly facilitates limb extensor γ-MNs (Grillner et al., 1969) and inhibits limb flexor γ-MNs (Kato and Tanji, 1971). However, exceptions have been noted. Thus, Poppele (1967) observed that γ- and α-gastrocnemius MNs can be activated independently by natural vestibular inputs, at least in decerebrate cats with lesion of the contralateral labyrinth.

The Reticulospinal Tract

This tract, which arises from the medial pontine and medullary reticular formation (Nyberg-Hansen, 1965), has its bulk reaching the lumbar cord bilaterally (Peterson et al., 1975b). Reticulospinal neurons are excited by electrical stimulation of the vestibular or semicircular nerve on either side (Peterson et al., 1975a, 1980), possibly as a result of extensive connections between brainstem reticular neurons and vestibular nuclei (Peterson and Abzug, 1975). Reticular units also respond to natural excitation of otolithic receptors (Spyer et al., 1974). Thus, stimulation of reticulospinal fibers descending in the medial longitudinal fasciculus evokes monosynaptic excitatory (Grillner and Lund, 1968) as well as disynaptic excitatory and inhibitory influences on lumbar MNs, which are generally reciprocal to those of the lateral vestibulospinal tract (Wilson and Yoshida, 1969; Grillner et al., 1971). For example, gastrocnemius MNs are monosynaptically excited by lateral vestibulospinal neurons and inhibited disynaptically by reticulospinal neurons.

In summary, it is clear that there are extensive pathways by which vestibular, in particular otolithic, information can reach as far as the lower cord. These pathways can exert their influences either (1) directly by monosynaptically changing the excitability of α- and γ-MNs; (2) indirectly, by changing the bias of interneurons; or (3) via presynaptic inhibitory action on certain primary (notably group Ia) afferents.

TONIC LABYRINTHINE REFLEX CONTROL OF LIMB POSTURE

Animal Studies

As mentioned in the introduction, Magnus and de Kleijn (1912; Magnus, 1926) studied tonic labyrinthine reflexes evoked by whole head-body rotation in decerebrate cats whose necks were either denervated (by cutting C_1–C_3) or immobilized with a cast. They found that the extensor tone of all four limbs was facilitated when the head was supine and decreased when the head was prone. In other words, tonic labyrinthine reflexes exerted symmetric effects on all four limbs. These classical observations have since been challenged by Roberts (1968, 1973, 1979). Using decerebrate cat preparations as well as examples of characteristic animal postures, he showed that in the absence of neck reflexes, as for example when the animal's head is normal with respect to the body (denoted by the middle row in Fig. 1), tonic labyrinthine reflexes actually influence the fore- and hindlimbs in opposite rather than the same manner. Thus, upward tilt (or dorsiflexion) of the head with respect to gravity (Fig. 1, left middle panel), as when a dog is taking off for a jump, causes extension of both hindlimbs and flexion of both forelimbs. Downward tilt (or ventroflexion) of the head, as when a dog is standing face down on an incline (Fig. 1, right middle panel), elicits flexion of both hindlimbs and extension of both forelimbs. These behavioral observations have been supported by experimental findings (Lindsay et al., 1976). In this study, ingenious devices were used to fix the decerebrate cat with independent support for head, neck, and trunk so that labyrinthine and neck reflexes could be separately elicited. More specifically, tonic labyrinthine reflexes were evoked by moving the head with the neck fixed at the axis vertebra and C_1–C_2 (which together with C_3 mediate the tonic neck reflexes, cf. McCouch et al., 1951) denervated. Nose-down rotation (or ventroflexion) of the head produced a sustained shortening of the forelimb extensor, as revealed by monitoring the length of the medial triceps (Fig. 2A), whereas nose-up rotation (or dorsiflexion) of the head resulted in a sustained relaxation (Fig. 2B). These findings are, therefore, in direct contrast to those reported by Magnus (1926) that upward tilt of the head produced increased extension of all four limbs.

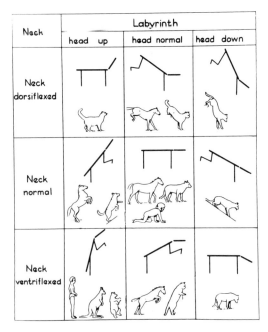

FIG. 1. Combined influences of tonic labyrinthine and neck reflexes on the limbs as proposed by Roberts (1968). The *middle horizontal row* shows tonic labyrinthine reflexes with the neck normal, whereas the *middle vertical column* illustrates tonic neck reflexes with the head normal. The interaction between these two sets of reflexes is demonstrated by the panels in the four corners. More specifically, the **left top** and **right bottom** panels show the antagonistic organization of labyrinthine and neck reflexes when they are evoked simultaneously by the same direction of head and neck displacement, with the result that the limb posture remains unmodified. The **right top** and **left bottom** panels illustrate how they can reinforce each other when they are elicited by *opposite* direction of head and neck displacements. (From Roberts, 1968, with permission.)

Another deviation from Magnus' description noted by Lindsay et al. (1976) is the presence of asymmetric tonic labyrinthine reflexes due to lateral head tilting. Briefly, in the absence of neck reflexes, side-down rotation of the head caused shortening of the ipsilateral forelimb extensor (Fig. 3, left top panel) and side-up rotation of the head evoked the opposite response (right top panel). Interestingly, similar tonic labyrinthine influences could be superimposed on a preexisting tonic vibration reflex (TVR) elicited in medial triceps, in that side-down rotation of the head facilitated the TVR and side-up rotation diminished it (Rosenberg et al., 1980). This finding is intriguing, since electrical stimulation of the vestibular nerve gives rise to primary afferent depolarization of Ia afferents (Cook et al.,

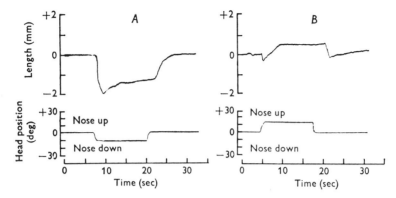

FIG. 2. Isotonic length changes evoked in the medial triceps by A, nose-down (or ventroflexion) and B, nose-up (or dorsiflexion) rotation of the head in a decerebrate cat, with C_1 and C_2 denervated and the axis vertebra clamped. (From Lindsay et al., 1976, with permission.)

1969), which are known to be predominantly involved in the TVR (Brown et al., 1967b). It therefore implied that the combined facilitatory influences from otolithic receptors, acting either directly or (more likely) indirectly on α- and γ-extensor MNs, could overpower their presynaptic inhibitory action on Ia terminals. The general pattern of tonic labyrinthine reflexes is seen by Roberts (1979) as tendency to act on antigravity limb muscles to oppose head displacement relative to space.

However, the above observations are at variance with recordings from lumbar MNs in decerebrate cats by Ehrhardt and Wagner (1970). As illustrated in Fig. 4B, with neck innervation blocked, facilitation of lumbar extensor MNs was often seen regardless of

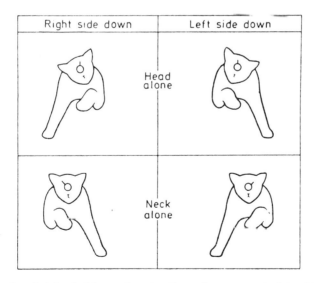

FIG. 3. Asymmetrical tonic labyrinthine and neck reflexes in response to lateral tilts in a decerebrate cat with C_1 and C_2 cut. Upper figures: action of asymmetric tonic labyrinthine reflexes with the axis vertebra clamped. Lower figures: antagonistic action of the asymmetrical tonic neck reflexes by rotating the axis vertebra (where the tilt of the spinous process is denoted by a line on a circle) with the head fixed. (From Roberts, 1975, with permission.)

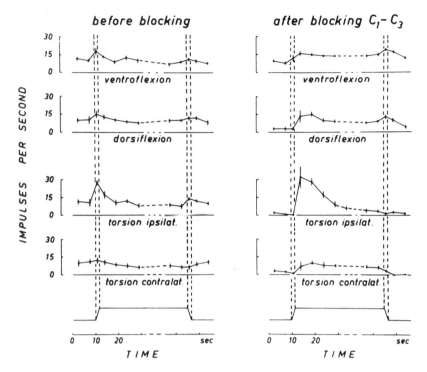

FIG. 4. Discharge of an extensor (gastrocnemius) MN in response to head displacements (fifth trace) plotted against time in a decerebrate cat. Mean and standard deviation of discharge frequency computed for every 5-sec period are shown. **Left:** In the presence of the neck reflexes, a transient facilitatory influence during the head movements and a slight inhibition during the position periods could be observed. **Right:** After blocking C_1–C_3 bilaterally, head movements produced no effect, but strong tonic labyrinthine reflexes, suppressed before by the neck reflexes, could be seen during the position periods. Note that the MN was facilitated regardless of the direction of head movement, but was most marked during ipsilateral torsion. (From Ehrhardt and Wagner, 1970, with permission.)

the direction of head movement, but was graded for the different maintained positions. First, in keeping with Roberts' (1979) scheme, most lumbar extensor MNs were mildly facilitated during dorsiflexion of the head and maximally facilitated during ipsilateral torsion (or side-down rotation). However, inhibition of lumbar extensor units was not usually found with ventroflexion, or contralateral torsion (i.e., side-up rotation; cf. also Poppele, 1967 for γ-MNs, whose discharge was also increased), as expected.

An intriguing point of conflict has also been raised by Soechting et al. (1977), who subjected decerebrate cats to sinusoidal rotations in the vertical plane at frequencies below 0.2 Hz when otolithic (versus vertical semicircular canal) contribution would be dominant. The EMG of forelimb extensor (triceps) was maximal when the ipsilateral side was tilted up, in direct contrast to that found under static conditions by Lindsay et al. (1976). Their data were more confounding given the fact that many otolithic receptors in monkeys (Fernandez et al., 1972), as well as most Deiters' neurons (Boyle and Pompeiano, 1979) and identified vestibulospinal tract neurons (Orlovsky and Pavlova, 1972) in cats (at least in decerebrate preparations with intact cerebellum), were excited during ipsilateral side-down rotation of the head. Two explanations are possible. First, there are some otolithic receptors whose polarization vectors are such that they can be excited by ipsilateral side-up tilt. Second, the difference between static and dynamic labyrinthine effects may reflect a distinction between the responses to changes in the orientation of the gravity vector and transient linear acceleration (cf. Anderson et al., 1979).

The above results with natural labyrinthine stimulation under dynamic conditions are certainly at variance with those obtained by electrical stimulation of the vestibular nerve. In particular, Kim and Partridge (1969) found that a sinusoidally modulated train of electrical stimuli to the utricular nerve generally increased the tension produced by mechanical stretch of triceps surae in decerebrate cats, although the opposite effect was noted in 3 of 14 cats. Bearing in mind that electrical stimulation is relatively artificial, this study showed that utricular influence generally facilitates hindlimb extensor MNs, in agreement with Ehrhardt and Wagner (1970) using natural labyrinthine stimulation.

Lesion Studies

Lesion studies confirmed the importance of labyrinthine input in the control of posture. Briefly, unilateral labyrinthectomy resulted in typical postural deficits in various vertebrates (cf. Schaefer and Meyer, 1974). These consisted of tilting of head and body as well as gait deviation toward the side of lesion, an extension of limbs on the intact side and a flexion on the operated side. Furthermore, the tonic labyrinthine reflexes described by Lindsay et al. (1976) were dependent on an intact cerebellum. Thus, in acutely cerebellectomized cats with neck reflexes eliminated, nose-up tilt (dorsiflexion) of the head produced shortening instead of lengthening of medial triceps, and side-down rotation of the head in either direction resulted in a simultaneous shortening of *both* triceps (Lindsay and Rosenberg, 1977; Dutia et al., 1981).

There are extensive interconnections between (1) the anterior (and to a lesser extent the posterior) cerebellar lobe(s) and the Deiters' nucleus; (2) the flocculus (and nodulus) and the other three vestibular nuclei; as well as (3) the fastigial nucleus and all vestibular nuclei (cf. Wilson and Peterson, 1981). Briefly, activity of vestibulospinal tract cells has been inhibited by Purkinje cells of the anterior and posterior cerebellar vermis (Akaike et al., 1973). This inhibition can be brought about by various inputs—including neck (Berthoz and Llinas, 1974) and vestibular (Precht et al., 1977)—which converge upon vermal Purkinje cells. In this connection, low-frequency (0.015 to 0.15 Hz) sinusoidal rotation of the whole animal (otolithic input) and of the axis vertebra with the head stationary (neck input) influenced vermal Purkinje cells of the anterior cerebellar lobe in the opposite manner (Denoth et al., 1979). More specifically, the majority of these cells was excited by ipsilateral side-down rotation of the neck, and inhibited by side-down rotation of the whole animal. It is, therefore, not surprising that the cerebellum is required for the production of normal tonic labyrinthine reflexes to head tilt. Certainly, Boyle and Pompeiano (1979) observed a reciprocal pattern of response in Deiters' neurons to low-frequency sinusoidal rotation of the whole animal (i.e., excitation during ipsilateral and inhibition during contralateral tilt) in decerebrate cats with intact cerebellum. This was in contrast to the lack of reciprocality in their response (i.e., increase in firing rate during both directions of tilt) often found in decerebellate animals (Peterson, 1970).

Human Studies

From the above, it appears that static changes in head position relative to gravity evoke certain facilitatory changes in limb extensors, which are probably graded under cerebellar control for the different head positions. From a functional point of view, otolith-spinal reflexes are generally seen to counteract the effect of head displacement relative to space on antigravity limb muscle, although exceptions have been noted. A question is therefore raised concerning the extent to which these findings from quadripeds could be generalized to man, given his or her bipedal mode of postural control.

The study of tonic labyrinthine influence on limb muscle tonus in human subjects has been rather limited. Recent experiments examined the effects of caloric stimulation of the labyrinth in normal man by measuring the Achilles tendon and H reflexes in soleus (Delwaide and Juprelle, 1977), as well as the TVR in lower limb muscles (Delwaide et al., 1976). Interestingly, the monosynaptic reflexes and the TVR, which contains polysynaptic components (Lance et al., 1973; Desmedt, *this volume*) in extensor muscles were all facilitated from the beginning of irrigation to the end of nystagmus, whereas the TVR in flexor muscles were not affected. However, proper interpretation is difficult, for the use of a cutaneous stimulus might have evoked nonspecific reticular activation, which would be impossible to distinguish from the combined descending vestibulo- and reticulospinal influence expected from labyrinthine stimulation. This is particularly relevant, since the reflexes were facilitated to a similar extent whether warm or cold water was used, or whether the same caloric stimulation was applied *ipsi*laterally or *contra*laterally—maneuvers known to produce nystagmus of opposite directions. Stejkal (1979) examined the effect of tilting the whole body in the sagittal plane in spastic patients. He found no correlation between the amount of activity at rest in spastic muscles and the position of the head in space. It is difficult to generalize the findings in patients with upper motor neuron lesions to normal subjects. Furthermore, visual inputs, which are known to modify the early EMG response to sudden falls (Vidal et al., 1979), were not eliminated in these experiments.

We have examined how tonic labyrinthine reflexes, evoked by changes in static position of the head and body relative to gravity, modulate the excitability of the soleus MN pool in man (Chan and Kearney, 1982b). Subjects were blindfolded and fixed to a circular bed in a supine position by means of body and hand straps, as well as partial casts at the legs. Given the well-documented effects of tonic neck reflexes which are known to oppose tonic labyrinthine reflexes (Magnus, 1926, Roberts, 1968, 1979; Ehrhardt and Wagner, 1970; Abrahams, 1972; Lindsay et al., 1976), extreme care was taken to ensure a constant angle between the head and body by means of a neck cast to keep the neck input constant. Changes in static orientation of the head and body relative to gravity was applied by rotating the bed in random steps of 15° or 30° over a range of 30° (head supine or nose-up) to 150° (head prone or nose-down). The excitability of the soleus MN pool was measured by means of the H-reflex technique with the subject at the test position, as well as at the control position (60°) both before and after each change in head orientation.

Figure 5 is a graph of the H-reflex amplitude, normalized with respect to the mean of the pre- and post-test control H-reflexes, and plotted against the angle of head tilt in degrees for 11 of 13 subjects examined. Generally, the H-reflex amplitude was minimal when the head was near the vertical position at 90°, and progressively facilitated when the subject was tilted either forward to about 150° (nose-down) or backward to about 30° (nose-up). Although there were differences in the relative magnitude of the change and the exact location of the minimum of the curves among the 11 subjects, these findings were very consistent in a given subject, as denoted by the similarity of the results obtained in subjects 11, 12, and 13 tested on separate occasions.

Figure 6 illustrates that, when the observed H-reflex data for subject #5 (solid line) were fitted with a sine wave (dotted line), 92% of the variance was accounted for

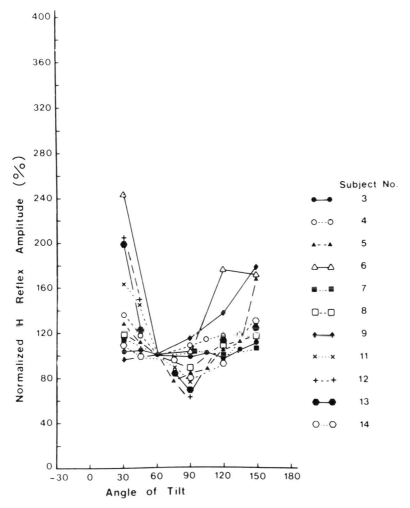

FIG. 5. Plot of normalized H-reflex amplitude as a function of the angle of head tilt (degrees) in 11 subjects (indicated by symbols and curves). Each data point represents the mean of 25 to 75 responses. (From Chan and Kearney, 1982b, with permission.)

(% VAF)-defined as the percentage of the variance of the observed H-reflex amplitude which was accounted for by the results of the equation (cf. legend of Fig. 6). Good fits, indicated by 80 to 100% VAF, and with minimum H-reflex excitability close to the vertical (15°) were actually observed in 7 of 10 subjects for whom sufficient data points were obtained. Thus, the soleus MN pool excitability in man is modulated as a function of changes in static orientation of the head and body relative to gravity. The H-reflex amplitude was generally minimal when the head and body were near vertical, and maximal when the subject was tilted to about 60° forward or backward.

One may argue that these influences are exerted on the ankle extensor by cutaneous receptors which are known to contribute to the control of posture. However, we had tried to minimize at least the effect of rapidly adapting cutaneous receptors by allowing one minute to elapse after each change in head orientation before recording. Under this condition, we found no significant influence on the H-reflex amplitude by changing the cutaneous input with the subject supine. It seems likely, therefore,

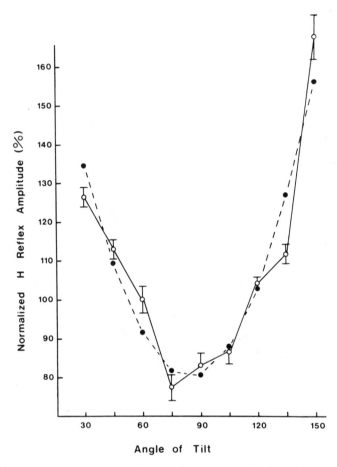

FIG. 6. Observed *(solid line)* and predicted *(dotted line)* normalized H reflex amplitude in subject N5, showing an extremely good fit with 92% variance accounted for. The observed data points are the mean of 25 responses ± SE and are fitted with a sine wave *(dotted line)*, using the following equation: H predicted = $K_1 + K_2 \sin(\theta + K_3)$, where K_1 = tonic level of MN excitability (%); K_2 = gain of vestibulospinal input (%); θ = angle of tilt of bed (degree); K_3 = angle between bed and direction of maximum sensitivity of otolith receptor (degree). (From Chan and Kearney, 1982b, with permission.)

that these influences are mediated by otolithic receptors, some of which have been found in primate experiments to respond in a similar fashion (Fernandez and Goldberg, 1976a; Goldberg, 1979).

Our findings in normal man concurred with those obtained from unit recordings in decerebrate cats. Briefly, Ehrhardt and Wagner (1970) observed that lumbar extensor MNs were usually facilitated regardless of the direction of head movement. Certainly electrical stimulation of the utricular nerve has often been found to increase the tension produced by mechanical stretch of the ankle extensor muscles in decerebrate cats (Kim and Partridge, 1969).

On the other hand, our results are at variance with those reported by Magnus (1926) and Roberts (1968, 1979; Lindsay et al., 1976) in decerebrate cats, but they appear appropriate for postural control in man. To elaborate: standing is a position of equilibrium for man, there is no cause for increased excitability of the lower limb extensor. However, any offset from this position of equilibrium, e.g., standing face down on an incline, the subject is more prone to fall, hence the need for increased excitability of the lower limb extensor arises. Evidence for a *dynamic* otolith-dependent limb reflex during sudden downward acceleration in man has been demonstrated. Melvill Jones and Watt (1971) first observed a short-latency (75 msec) EMG response in gastrocnemius during sudden falls. The amplitude of the early component was explicitly dependent on the magnitude of the linear acceleration (Greenwood and Hopkins, 1976a,b, 1980). Subsequently, the early component of similar EMG responses evoked during sudden falls in cats was found to be abolished by total labyrinthectomy, but not by inactivation of the canals alone (Watt, 1976). It therefore seemed reasonable to conclude that the early response to sudden falls in the human gastrocnemius is an otolith-dependent reflex. Our results on static otolith-spinal effects due to head rotation in the fore-aft direction thus corroborate with their dynamic effects due to downward linear acceleration, although different populations of otolithic receptors (e.g., regular versus irregular units, receptors with different polarization vectors) are probably involved in the 2 situations.

INTERACTION BETWEEN TONIC LABYRINTHINE AND NECK REFLEXES

A review of the tonic labyrinthine reflexes would not be complete without some comments as to how they may interact with tonic neck reflexes, since any static changes in head position in real life are likely to elicit the two reflexes simultaneously. In this connection, Magnus and de Kleijn (1912) and Magnus (1926) first showed that in labyrinthectomized animals, dorsiflexion of the head causes extension of both forelimbs and flexion of both hindlimbs. Ventroflexion of the head evokes the opposite response. Furthermore, rotation or lateral flexion of the head on the neck produces extension of both fore- and hindlimbs on the side towards which the jaw is turned. These observations have since been confirmed by Roberts (1968, 1973, 1979) and Thoden and Wenzel (1979). Briefly, in decerebrate cats, in which neck and labyrinthine reflexes could be elicited independently or in combination, right sidedown rotation of the neck (i.e., rotating the body with the head fixed) produces relaxation of the right medial triceps, and right side-up rotation of the neck evokes the opposite response (Lindsay et al., 1976; cf. Fig. 3, bottom panels). Moreover, neck effects on limb muscles, evoked by lateral flexion or dorsiflexion of the body on the stationary head, are reciprocally organized (Thoden and Wenzel, 1979). Thus, ipsilateral flexion of the body facilitates both ipsilateral fore- and hindlimb extensor monosynaptic reflexes (MSRs) while inhibiting ipsilateral limb flexor MSRs, with reversed connections on the contralateral limb muscles.

The neck reflexes described above act on limb muscles in such a way as to oppose the effects of tonic labyrinthine reflexes (Roberts, 1968, 1973, 1979; cf. Fig. 1, left top and right bottom panels). Thus, when the two reflexes are simultaneously evoked by moving the head on the body with intact neck innervation, no net change in the length of medial triceps could be observed (Lindsay et al., 1976). It is noteworthy that the antagonistic organization of neck and labyrinthine effects, evoked either as a result of electrical (Abrahams, 1972) or natural (Ehrhardt and Wagner, 1970) stimulation, has also been observed in recordings taken from lumbosacral ventral roots (Abrahams, 1972) as well as single lumbar MNS (Ehrhardt and Wagner, 1970; cf. Fig. 4A in contrast to Fig. 4B). This interaction is thought to permit free movement of the head without affecting the stability of the trunk (Lindsay et al., 1976; Roberts, 1979).

In this connection, it should also be noted that postural deficits following unilateral lesion of C_1–C_3 dorsal roots (Manzoni et al., 1979a,b; however, Cohen, 1961) are essentially opposite to those produced by a unilateral labyrinthine lesion (cf. Schaefer and Meyer, 1974).

An important and quite different kind of interaction between labyrinthine and neck reflexes has been demonstrated by Nashner and Wolfson (1974) in man. Briefly, galvanic stimulation of the vestibular system in a standing man elicited opposite responses (excitation or inhibition) in the gastrocnemius depending on whether the head was turned to the left or the right. Since this reversal of vestibulospinal reflex is exactly what was required in these circumstances, these findings amounted to a demonstration of short-term adaptive interactions, which must imply relatively complex information processing in the central nervous system.

Patterns and Mechanisms of Postural Instability in Patients with Cerebellar Lesions

*,**Johannes Dichgans and †Karl-Heinz Mauritz

**Department of Neurology, University of Tübingen, Tübingen, Federal Republic of Germany; and †Department of Neurology, University of Freiburg, Freiburg, Federal Republic of Germany*

The functional subunits of the cerebellum are rather well understood, but a wide gap still exists between our knowledge of microprocesses at a neuronal level and our understanding of their functional significance in everyday motor activities. This is also true for posture control in a bipedal stance. Neurophysiological examinations of cerebellar patients with well-defined lesions were performed to explore these problems (Silfverskiöld, 1977; Gurfinkel and Elner, 1973; Dichgans et al., 1976; Nashner and Grimm, 1978; Mauritz et al., 1979; 1981). Characteristic patterns of disturbances now make it possible to distinguish three kinds of postural ataxia in cerebellar patients (Mauritz et al., 1979). Consequently, postural cerebellar ataxia should not be treated as a single entity, since different pathophysiological mechanisms underlie the three syndromes of postural ataxia.

QUANTITATIVE ANALYSIS OF POSTURAL INSTABILITY

Force-moment-measuring platforms proved useful for quantification of body sway and stance ataxia. Computer processing of the displacement of the center of force results in a parametric documentation of sway area and sway path per unit time, mean amplitude of sway, and histograms of sway direction as well as sway position (Dichgans et al., 1976; Mauritz et al., 1979; Hufschmidt et al., 1980). These parameters and Fourier analysis of the anteroposterior and lateral sway components (Bensel and Dzendolet, 1968; Nashner, 1980) allow for clinical follow-up studies and for the differentiation of different kinds of postural ataxia in cerebellar lesions.

To study the underlying pathophysiological mechanisms, these measurements were supplemented by EMG recordings of leg and trunk muscles and by goniometer recordings of ankle angle, the angle of hip displacement, etc. (Nashner, 1971, 1980; Dichgans et al., 1976; Gurfinkel et al., 1976).

Special techniques such as suddenly displacing the supporting platform (Nashner, 1976, 1977; Gurfinkel et al., 1976; Lestienne et al., 1977; Mauritz et al., 1979; Allum and Büdingen, 1979), electrical stimulation of the tibial nerves in standing subjects (Mauritz et al., 1981), balancing on seesaws (Dietz et al., 1980; Mauritz et al., 1980), ischemic blockage of group I afferents (Mauritz and Dietz, 1980), displacement of the visual surroundings (Dichgans et al., 1972, 1975, 1976; Lee and Lishman, 1975; Lestienne et al., 1977; Mauritz et al.,

---

*To whom correspondence should be addressed: Neurologischeklinik, Eberhard-Karls-Universität Tübingen, 18, Liebermeisterstrasse, Tübingen 7400 Federal Republic of Germany.

1977), or vestibular stimulation (Nashner and Grimm, 1978; Nashner, 1972) make it possible to test the function of the three main control loops involved in stabilization of posture.

THREE FORMS OF STANCE ATAXIA IN CEREBELLAR PATIENTS

Clinical observations differentiate among several cerebellar syndromes (see Dow and Moruzzi, 1958), and ablation experiments in animals show specific disturbances with lesions of the paleocerebellum, neocerebellum, and vestibulocerebellum (Carrea and Mettler, 1947; Sprague and Chambers, 1953). A similar grouping is appropriate for patients with cerebellar postural ataxia (Mauritz et al., 1979).

Late Cortical Atrophy

Patients with late cortical cerebellar atrophy have lesions predominantly in the upper vermis and paravermis of the anterior lobe, that is, in the paleocerebellum (Victor, 1975) which is where the afferents from the legs project (Oscarsson, 1965). The disease mainly damages Purkinje cells, occurs mostly in alcoholics, and probably results from nutritional deficiencies (Victor et al., 1959; Victor, 1975; Mancall, 1975). The susceptibility of this paleocerebellar region to degenerative processes has been explained by a difference in enzyme pattern, with low pyruvate dehydrogenase levels, especially in the anterior cerebellar vermis (Reynolds and Blass, 1976; Blass and Gibson, 1978; Kark et al., 1978).

Mean amplitude of sway and length of path are greatly increased without exception in these patients. Characteristic of this paleocerebellar syndrome is a postural tremor around 3 Hz in the anteroposterior direction (Silfverskiöld, 1969, 1977; Dichgans et al., 1976; Mauritz et al., 1979, 1980) (Fig. 1A). The 3-Hz sway component can easily be differentiated from the tremor of Parkinson's disease (Fig. 1B). The typical 3-Hz oscillation can be evoked in incipient cases, before it is detected clinically, either by a sudden mechanical destabilization or by electrical stimulation of tibial nerves.

Follow-up studies in these patients over 5 years showed a decay in the dominant anterior–posterior sway frequency component with progression of the disease. This finding parallels the gradual decrease in the tremor frequency observed with progressive cooling of the dentate nucleus after limb perturbations in monkeys (Vilis and Hore, 1980). Less characteristic and frequently smaller in amplitude is a 0.5 to 1 Hz frequency peak mainly in the spectrum of the lateral sway component (Fig. 1A). This low-frequency peak is also seen with other kinds of postural disequilibrium including normals who have been visually destabilized, patients with tabes dorsalis (Dichgans et al., 1976), and patients with afferent disturbances (Mauritz and Dietz, 1980).

Patients with late cerebellar atrophy, although they have severe disturbances in stance and gait, almost never fall. This may be because of their intact intersegmental movements between head, trunk, and legs. The importance of these synergies (Bernstein, 1967; Gurfinkel et al., 1971; Nashner, 1977) for postural stability may be demonstrated in patients with vestibulocerebellar lesions in whom they are absent. In contrast, patients with late cerebellar atrophy show exaggerated intersegmental reaction patterns. Since trunk and leg movements are roughly 180° out of phase, a backward movement of the legs (that is increase of the ankle angle) is compensated for by a forward inclination of the trunk.

Visual stabilization of posture is frequently observed in patients with cortical cerebellar atrophy. Its amount does not correlate with the general instability of posture. Nor does *its absence correlate* with deficit of visual pursuit, optokinetic nystagmus, or inability to suppress vestibuloocu-

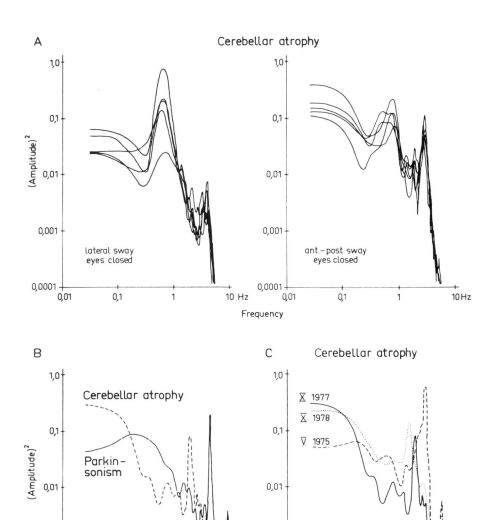

FIG. 1. A: Fourier power spectra of the lateral (**left**) and anterior–posterior body sway (**right**) in a patient with late cortical cerebellar atrophy. The five spectra represent measurements on five consecutive days. Note the high reproducibility of the 3-Hz peak in the anterior–posterior direction and a 0.7-Hz peak mainly in the lateral component. B: Comparison of Fourier power spectra of a patient with parkinsonism *(solid line)* with that of a patient with late cortical cerebellar atrophy *(dotted line)*. The prominent tremor frequency is clearly separate in the two syndromes. C: The dominant tremor frequency shifts to lower frequencies with progressive cerebellar atrophy as seen in one typical patient repeatedly studied over 3½ years.

lar reflexes by visual fixation (Mauritz et al., 1979). Again, this is in contrast to the patients with vestibulocerebellar lesions, who are always more insecure or even unable to keep balance with their eyes closed.

In summary, the characteristic features of these patients that distinguish them from neocerebellar and vestibulocerebellar patients are the 3-Hz anteroposterior sway component and the increase in intersegmental reflex movements of stabilization. This is most easily evidenced by Fourier spectra (Fig. 1A) or visualized by histograms of sway direction or sway position (Fig. 2). The 3-Hz component can be mechanically evoked in patients with the incipient disease and can be synchronized in overt cases.

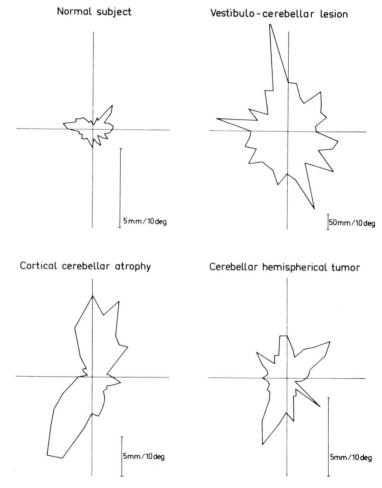

FIG. 2. Comparison of position histograms (excentricities summed within every 10° polar bin, sampling interval 30 msec, integration over 1 min) of a typical normal, a patient with a vestibulocerebellar medulloblastoma, a patient with late cortical cerebellar atrophy, and a patient with a cerebellar hemispherical lesion. The patient with the hemispherical lesion has only a slightly greater instability than the other two patients. The patient with the anterior lobe atrophy has a predominant (high-frequency, low-amplitude) anterior–posterior instability which would be even more impressive in the direction histogram (not shown). The patient with the vestibulocerebellar lesion exhibits omnidirectional sway of an extremely large amplitude. His low-frequency, high-amplitude instability is more pronounced in the position histogram presented here. Note the different scaling which was necessary for technical reasons.

Neocerebellar Lesions

Lesions of the neocerebellum (cerebellar hemispheres), if they disturb stance at all, cause only slight postural instability. No characteristic feature, neither directional preference nor specific frequency peaks within their sway spectrum, can be observed in these patients (Fig. 2). Sway parameters are within the 2 SD range of normals, and there is no significant difference from normals in these parameters, even when the eyes are closed. A 3-Hz tremor can never be provoked either by suddenly tilting the platform or by electrical stimulation of tibial nerves. The cerebellar hemispheres interact with the cerebral cortex and are mainly concerned with the temporospatial organization of goal-directed movements of the limbs and with speech but not with body posture.

Vestibulocerebellar Lesions

Lesions of the vestibulocerebellum (posterior vermis including flocculus and nodulus) cause an extreme instability of stance without preferred axis or frequency (Fig. 2). The average amplitude of sway is unusually large. Sway, on average, is definitely slower than in patients with spinocerebellar atrophy. In contrast to cases with anterior lobe lesions, these patients, mostly suffering from a medulloblastoma, are characterized by the absence of intersegmental movements and a lack of a set value determining the upright (Mauritz et al., 1979). Thus, they may even fall when sitting. Neocerebellar functions, such as pointing, may be intact.

NORMAL PREPARATORY POSTURAL ACTIVITY WITH GROSS VOLUNTARY MOVEMENTS IN CEREBELLAR PATIENTS

When normal subjects are instructed to rise as quickly as possible from the normal standing position to the toe tips and to hold that position for a short time, there is invariably a preactivation of the (antagonistic) anterior tibial muscle before the (agonistic) gastrocnemius muscle discharges (Gurfinkel et al., 1981). The difference in latency between these two antagonistic muscles depends on the initial position of the upright body and amounts to 100 msec or more. From a teleological viewpoint, this "postural preactivation" of the antagonist is necessary to shift the center of gravity forwards and to prevent a fall backwards. This innervation pattern is preprogrammed, involuntary, and, once fired off, not modifiable by a rapid tilt of the supporting platform.

Patients with paleocerebellar and neocerebellar syndromes present the same postural preactivation (Fig. 3). The patients with vestibulocerebellar lesions examined so far were unable to perform the task.

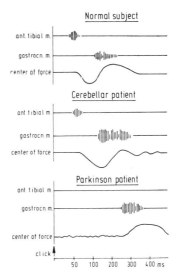

FIG. 3. Anticipatory postural activation of anterior tibial muscle (Gurfinkel et al., 1981). Subjects were instructed to rise quickly to tiptoes after a click-signal. A preceding tibial discharge is observed both in normals and in cerebellar patients (with anterior vermis and hemispherical lesions) but not in parkinsonian patients. Therefore, the backward displacement of the center of force is also absent in parkinsonian patients. Note the increased reaction time in parkinsonian patients.

They generally were too unstable. In patients with Parkinson's disease, this anticipatory postural component of a voluntary movement was absent (Fig. 3). Traub et al. (1980) reported a similar finding: postural reflexes in distant muscle groups elicited by a mechanical disturbance were preserved in cerebellar patients but disturbed in patients with Parkinson's disease. Intersegmental mechanisms of postural stabilization (head versus trunk, chest versus hip, hip versus legs) that normally compensate for the displacement of their adjacent segment are also deficient in patients with Parkinson's disease.

ALTERED REFLEX RESPONSES AS THE POSSIBLE CAUSE OF POSTURAL TREMOR IN SPINOCEREBELLAR ATROPHY

Recent evidence in normals suggests that myotatic reflexes at 40 to 50 msec latencies are seen only with a very sudden angular displacement of the ankle joint and, if occurring, are of little functional significance (Nashner, 1976; L. M. Nashner, *this volume*; Gurfinkel et al., 1976; Burke and Eklund, 1977; Allum et al., 1979). Functional stretch reflexes occurring at latencies between 100 and 120 msec in normals are the first to be mechanically effective. They are probably mediated by suprasegmental structures (Phillips, 1969; Evarts and Tanji, 1974; Evarts and Fromm, 1978; Marsden et al., 1978a,b; cf. Desmedt, 1978a). These issues are discussed in detail elsewhere in this volume. Responses occurring later than 180 msec after body displacement may be vestibular in nature (Nashner, 1976; Nashner and Grimm, 1978).

Reflex responses are stereotypically organized into a pattern of fixed synergisms of leg muscles as a preprogrammed subroutine (Gelfand et al., 1971) controlled by movement inputs from the entire leg. They are also selected according to the environmental conditions and their behavioral significance to the subject. According to Nashner (1980),

> the adjustments are too complexly interwoven into an organizational structure involving many leg muscles to be characterized simply as the response of each individual muscle to its own stretch input.

It may then be investigated whether the spinocerebellum, which is intensely supplied by proprioceptive afferents from the legs (Oscarsson, 1965), is in the long loop of the functional stretch reflex and/or plays any role in organizing motor patterns for postural stabilization.

We tested more than 30 patients with late cortical cerebellar atrophy of the anterior lobe using a platform that could be rapidly tilted in pitch about an axis aligned to the ankle joint or by applying a current to the tibial nerve to directly stimulate afferent fibers from muscle spindles. Stimulus efficiency was directly monitored by recording the H reflex while the subject was standing on the platform. Surface electrodes recorded EMG activity over the gastrocnemius, soleus, and tibial muscles. The EMG was full-wave rectified and filtered. Displacement of the center of foot pressure was recorded by strain gauges connected to the four corners of the platform. Also recorded were the angular displacements of ankle angle, hip, and head. Recordings were frequently averaged to eliminate noise. Electrical stimuli to the tibial nerve were applied in the popliteal fossa. Stimulus intensity was usually adjusted to 50% of that necessary to evoke a maximal H-reflex response (and rarely an M response). Thus, mainly Ia afferents were excited, and subsequent mechanical effects reflected reflex responses only.

Electrical stimulation was preferred over mechanical tilting of the platform since the latter displaces the body, possibly excites reflex responses, and, by the displacement of the ankle joint, involves viscoelastic forces of the stretched muscle and inertial reaction forces of the body. The methods used are described by Mauritz et al. (1981).

In contrast to Nashner and Grimm (1978) who, in a study of posture in cerebellar atrophy patients, found little alteration of functional stretch responses, mainly in terms of a deficit in adaptation, we observed an increased gain and delayed appearance of postural long-loop reflexes. These and increased intersegmental reflexes were shown to most probably cause the 3-Hz tremor observed in these patients.

Short-Latency Segmental Stretch Reflex Responses

Electrical stimulation in normals and patients with anterior lobe atrophy (usually without demyelinizing neuropathy) similarly elicits the H reflex after a normal latency of about 30 msec. This response exerts a forward torque on the platform starting 45 to 50 msec after the stimulus. While the center of force is still displaced forward, the ankle joint is extended, and the tibial muscle is stretched. This stretch starts 80 msec (for a maximal M response, it is 50 msec) after the stimulus. Stretch of the tibial muscle after another 30 to 40 msec (total latency 150 to 120 msec) results in a short-latency response of this muscle termed tib_1 (Fig. 4). Because of the reaction force, a more tonic backward sway of the center of foot pressure follows. This is automatically compensated for by an intersegmental flexing forward of the trunk (and head), thereby stabilizing posture (Fig. 5). The amplitude of tib_1 frequently increases with eye closure.

The tib_1 response, in principle, as suggested by Taborikova (1973) and Allum and Büdingen (1979), could also be a long-loop response to tibial nerve stimulation. We argue that this is not the case on the basis of the following evidence. (1) The amplitude of tib_1 depends on the rate and amplitude of the forward displacement of foot pressure. (2) If body displacement and tibialis stretch are reduced by loading of the trunk with 30

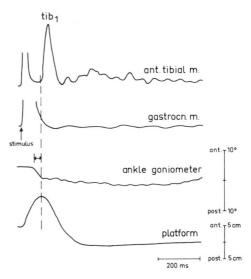

FIG. 4. Averaged stimulus responses ($N = 16$) after electrical stimulation of both tibial nerves. Rectified recordings of the surface EMG of anterior tibial and gastrocnemius muscles, the ankle angle, and the displacement of the center of force. The initial gap in the recording includes the stimulus artifact and, in gastrocnemius, the H reflex and direct response of the muscle to stimulation. The latter two responses cause an ankle angle displacement, which is followed by a discharge of the stretched anterior tibial muscle within 40 msec *(dotted line)*.

kg, an identical stimulation of Ia fibers causes a smaller tib_1 potential. (3) With supramaximal stimuli, when M responses occur exclusively, and H responses are occluded by antidromic collision, tib_1 latencies are shortened by an amount roughly equal to the time interval between M and H responses (24–35 msec). If one considers the mechanical delays caused by body inertia and friction and takes into account a threshold for exciting the muscle receptors, the 120-msec latency seems to be compatible with the assumption of a spinal stretch reflex causing tib_1.

The assumption that Ia afferents are involved is supported by an experiment with ischemic blockade of Ia afferents (Dietz, 1978) but preserved motor innervation. This occurs 20 to 25 min after cuff insufflation, when H reflexes are minimized and tib_1 is equally reduced, although, with higher stimulus intensities, M responses are preserved.

The possibility that late EMG responses are flexor reflexes triggered by the painful stimulation of cutaneous receptors was excluded by displacing the stimulating electrodes a few centimeters in the popliteal fossa or to several places on the leg and foot. Even when stimulus strength was increased until it was almost intolerable, no motor reactions in tibialis anterior were monitored by EMG. The fact that tib_1 is unchanged in gain and loop time in cerebellar patients compared to normals shows that tib_1 does not depend on intact cerebellar function.

Long-Latency Suprasegmental Responses

Long-latency suprasegmental reflex responses, in contrast to short-latency responses (tib_1), show marked differences

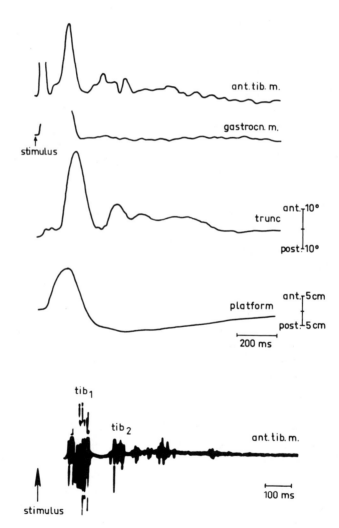

FIG. 5. Typical stimulus response in a normal subject (average of rectified EMG responses, $N = 16$). Note the large tib_1 and the smaller tib_2 discharge in the anterior tibial muscle. This is also seen in the original recording (**below**).

between normals and patients with cerebellar anterior lobe atrophy. Whereas normals may or—more frequently—may not show a second discharge in tibialis anterior (tib_2), this is regularly the case in patients with anterior lobe disease. Normals exhibit an interpeak (tib_1–tib_2) interval of 90 to 120 msec, whereas in patients with anterior lobe syndrome, this interval is increased by an average of 80 msec (compare Figs. 5, 6).

We suggest that tib_2 is a long-loop response to stretch of the tibial muscle. This assumption is based on its latency, which seems appropriate when one considers the rather low velocity of stretch (Nashner, 1976; Chan et al., 1979). The delay of tib_2 increases with progression of the disease and reached 350 msec in a single case. Whereas a third discharge is very exceptional in normals, this and later discharges occur in these patients. They parallel the postural tremor regularly evoked or synchronized by this stimulation. The tremor has the same frequency as spontaneous oscillations occurring in clusters in advanced cases. The increased delay in interpeak latency of long-loop reflexes coincides with the decrease in tremor frequency (Fig. 1C). Both long-loop reflexes and tremor are inhibited by visual stabilization of posture with the eyes open (Fig. 6).

Long-latency responses in triceps surae are inconsistent in normals, occurring simultaneously with tib_1 or 30 to 60 msec later. In patients with anterior lobe atrophy, an alternating discharge of the tibial and gastrocnemius muscles is present which is not seen in normals. Alternation of dis-

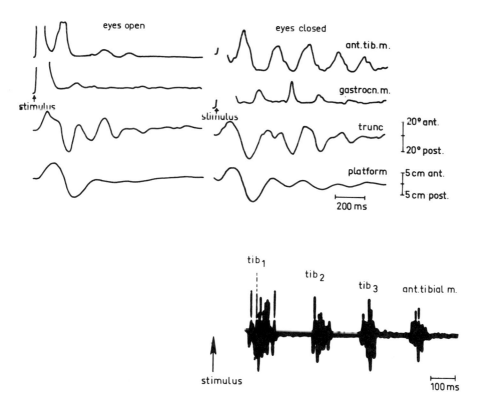

FIG. 6. Rhythmical discharges in the anterior tibial muscle after electrical stimulation of the tibial nerve are evoked only with subject's eyes closed in a patient with incipient anterior lobe atrophy. Correspondingly, a damped oscillation is seen in the platform and trunk (angle) recording. **Bottom:** Original recording of poststimulus tibialis activity with subject's eyes closed.

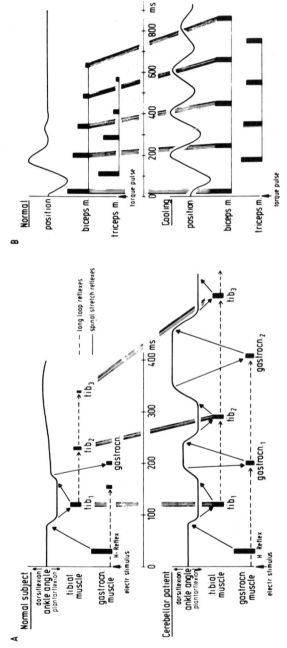

FIG. 7. Schematic presentation of poststimulus responses and of postural tremor mechanisms in patients with late cerebellar cortical atrophy (A) and limb tremor in a primate induced by cerebellar nuclear cooling (B). A: Upper half: Normal subject. Lower half: Patient. Electrical stimulation of the tibial nerve evokes the H reflex, plantar flexing the ankle and exerting a stretch to the anterior tibial muscle. This stretch activates Ia afferents and elicits a segmental stretch reflex (solid arrow) in the tibial muscle (tib_1). Simultaneously, the stimulated Ia afferents are responsible for a long-loop response (dotted arrow) in the gastrocnemius muscle (gastrocn). Tib_1 dorsiflexes the ankle, stretches the gastrocnemius, and elicits a segmental stretch reflex. This stretch reflex is not synchronous with the long-loop reflex in gastrocnemius. Tib_1 is followed by repetitive long-loop reflexes (tib_2, tib_3, . . .). Whereas in the normal, the segmental response and the long-loop reflex arrive at different times in the gastrocnemius, they coincide in the patient because of a delay in long-loop reflex time. Tib_1 is also followed by a delayed long-loop reflex (tib_2) which is synchronous with, and enhanced by, the segmental reflex evoked in the tibial muscle by reflex contraction of the gastrocnemius ($gastrocn_1$), thereby inducing postural tremor. B: In the upper part, the biceps is rapidly stretched, and a segmental discharge of the stretched muscle and an alternating activation of the antagonistic triceps are recorded. After cooling, the latency of the first (segmental) discharge in the stretched muscle is similar to that in the control. This parallels the observation of normal segmental reflex latencies in cerebellar patients. The latency of the later discharges, however, is delayed, as indicated by the shadowed stripes. (7B is modified after Vilis and Hore, 1980, with premission.)

charges in flexors and extensors constitutes and sustains tremor. So far, we have not been able to similarly synchronize postural tremor in Parkinson patients using this stimulation technique. The tremor seems to be the consequence of reflex facilitation increasing the gain of long-latency responses and their loop time. This is not the case with segmental reflex responses. The tremor also results from temporal coincidence of segmental stretch reflexes with exaggerated and delayed suprasegmental responses adding to each other. This is consistent with the findings of Vilis and Cooke (1976) who observed large M_2 responses only if motor activity over short and long pathways coincided in time.

The schematic drawing in Fig. 7A illustrates the hypothetical mechanisms of postural tremor in our patients. Stretch of the gastrocnemius or electrical stimulation elicits a monosynaptic reflex which in turn elicits a stretch response in the antagonist, and so on. Stretch responses in agonist and antagonist then coincide with the long-loop response following a previous stretch of the same muscle. In addition, it seems possible (but so far has not adequately been tested) that the hypotonia in some cases also contributes to the tremor in that the mechanical damping of possible oscillations by viscoelastic forces is decreased.

The pathways responsible for tib_1, which is not delayed with anterior lobe lesions, cannot be responsible for tib_2. The latter obviously depends on intact cerebellar function. The former is independent and segmental. A transcortical loop for these reflexes must be assumed in man (Lee and Tatton, 1975, 1978; Marsden et al., 1978; Chan et al., 1979; C. D. Marsden, J. C. Rothwell, and B. L. Day, *this volume*). That the loop depends on intact cerebellar function is further evidenced by the delayed long-loop reflexes of the arm in patients with cerebellar lesions (Marsden et al., 1978b) and by the data of Vilis and Hore (1980) who found normal short-latency but delayed long-latency responses (of the antagonist) in animals with reversible cooling of the cerebellar nuclei (Fig. 7B). Tremor frequency decreased with progressive cooling, just as it did in our patients with progression of the disease (Fig. 1C).

Vilis and Hore (1980) assume that the reflex is delayed by lack of a predictive input from cerebellum to precentral motor cortex from which the intended cortical response of the antagonist (Evarts and Tanji, 1976) originates. Rather than being preprogrammed by the cerebellum on the basis of prior stretch information from the agonist, the reflex now corresponds to a suprasegmental response of the antagonist to its own stretch. The reflex in their paradigm normally occurs together with the end of the long-latency response in the agonist and breaks its action, terminating the movement. Cooling delays the antagonist response and thereby causes an initial overshoot of the agonist movement and an oscillatory rebound. According to our own experiments, this explanation does not hold for tib_1, since tib_1 is not delayed in patients with anterior lobe atrophy. The explanation of Vilis and Hore, however, may be valid for the subsequent EMG bursts. Our model clearly does not advocate abnormal fixed patterns of rapid postural responses (Nashner, 1977) as the basis for tremor but rather the "falling back" into having to use the reflex response of each individual muscle to its own stretch input after a lesion damaging predictive capabilities of spinocerebellum.

Our understanding of postural disorders in cerebellar patients is still incomplete, and the challenge to clinical neurophysiologists is demanding. The methods developed for analysis already help to differentiate groups of patients and are promising.

Motor Control Mechanisms in Health and Disease,
edited by J. E. Desmedt.
Raven Press, New York © 1983.

Rapid Postural Reactions to Mechanical Displacement of the Hand in Man

**C. D. Marsden, *,†P. A. Merton, and ‡H. B. Morton

**The National Hospital, Queen Square, London; †The Physiological Laboratory, Cambridge; and ‡University Department of Neurology, Institute of Psychiatry and King's College Hospital Medical School, London, England

A decade ago detailed evidence was finally obtained that human muscles possess servo-like control mechanisms of the type that had been under investigation for many years before (Merton, 1953; Marsden et al., 1971b, 1972). Such mechanisms provide a measure of rapid, automatic compensation for unexpected changes in load and also for fatigue in muscles performing voluntary movements. In these early experiments, which were on the thumb, the subject was seated with the forearm supported and the proximal phalanx of the thumb in a clamp. The question arose as to what would happen in a standing subject with the arm free. Servo responses in the thumb (or in the periphery in general) would be likely to be ineffectual unless the proximal joints were braced to back up their efforts. And in the longer term, they might be negated unless the trunk was steadied so that its swaying was controlled.

When these distant muscles were looked at, responses of the kind needed to steady the limb and trunk could readily be found and these electrical reactions have a strikingly short latency, 55 to 110 msec in different muscles. Several experiments to elucidate the mechanism of these rapid reactions are described, leading to the conclusion that they constitute a distinct, and apparently new, class of motor reaction. Preliminary accounts were given by Marsden et al (1976c, 1978a). Several of the observations are described in the paper of Marsden et al. (1981b), which is complementary to the present chapter. The first investigation of these postural reactions in neurological patients was made by Traub et al. (1980).

METHODS

The methods were those described in detail in earlier papers (Marsden et al., 1976a,b, 1981b) with some additions. The subject made repeated voluntary movements, about one every 5 sec, at a standard speed imposed by a tracking task. He pulled on a wire looped over his thumb (or his fingers or wrist). The wire was tensioned by an electric motor. Initially, the tension was always 300 g (2.9 N). In the middle of a movement, the current in the motor could be altered to increase the force in the wire (to 570 g or 5.6 N), to decrease it (to 120 g or 1.2 N), or to halt the movement. Such perturbations occurred at random and were interspersed with control trials in which the force opposing movement re-

---
*To whom correspondence should be addressed: Physiological Laboratory, University of Cambridge, Downing Street, Cambridge CB2 3EG United Kingdom.

mained constant. All of the forces are modest for a manual task. The linear movement of the wire was recorded by means of a potentiometer on the motor spindle. Several trials of each kind were averaged, usually eight or 16. Muscle activity was recorded as the rectified and integrated EMG from surface leads over the belly of the relevant muscle. The recording sweep always lasted for 250 msec, with perturbations imposed 50 msec after the start of the sweep. For measuring displacement of the elbow, knee, shoulder, or trunk, a sensitive photoelectric method was used. A 6-V electric bulb was stuck to the part of the body in question. Its light was focused onto a differential photocell (United Detector Technology, Inc., type PIN-SC/25). The output was amplified with a condenser coupling and clamped until the beginning of each sweep, so that the records started from the same level. The subject was given a direct-coupled meter on the photocell preamplifier, and he could move slightly to keep the preamplifier near the middle of the range.

In some experiments, a second low-inertia electric motor was used, acting on a different part of the body from the first motor. It also had a drum and a flexible wire. The calibration of the displacement potentiometer on the motor spindle was adjusted to be the same as that on the first motor. The initial force in the second wire was always 150 g (1.5 N); it fell to zero in a release trial and rose to 300 g (2.9 N) in a stretch or "pull" trial (as it is often called here). When the second motor was in use, experimental runs had six modes instead of the usual four. The six modes were presented in the order called random cyclic by Marsden et al. (1976a); i.e. successive groups of 12, each containing two of each mode in use, were randomized, with a fresh random order for each successive group of 12. Other motors were used for special purposes as described in **RESULTS**. The subjects were two of the authors, P.A.M. and C.D.M.

RESULTS

The first experiment that showed the participation of distant muscles during thumb movements was done with a shoulder muscle, pectoralis major. The arrangements are illustrated in Fig. 1. The subject's task was to pull on a wire tensioned by an electric motor by flexing his thumb. Records were

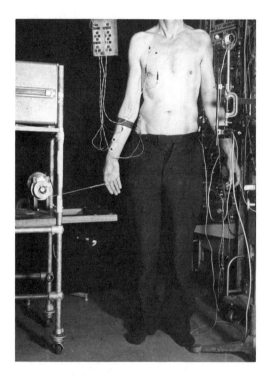

FIG. 1. Arrangements for recording simultaneous "tulips" from flexor pollicis longus, pectoralis major, and TFL. Pairs of electrodes (blackened for photography) are shown on the three muscles. Note that the trousers were below their normal level in order to uncover TFL (used for Fig. 4). To the left is seen the motor and the wire running from the drum to the pad of the thumb. To the right, the photocell for detecting body sway, in a black box clamped to a stand, looked at a pea bulb clipped to the top of the trousers. Above the photocell, at shoulder height, is seen the small meter used by the subject to keep the photocell amplifier in its sensitive range. In an actual experiment, the pea bulb and photocell were on the same side as the recording electrodes.

taken simultaneously over the long flexor of the thumb and over the pectoralis major. Pectoralis is a muscle that contracts to prevent the arm being pulled away from the subject's trunk when the thumb pulls on the wire. Repeated thumb movements were made, with randomly applied perturbations in midmovement, as described in **METHODS**.

The resulting averaged records of integrated EMG give what we have called "tulip" records of the responses to the three standard perturbations, "halt," "release," and "stretch." The records in Fig. 2 are taken from the experiment of Marsden et al. (1976b, Fig. 10). The tulip in the long flexor (Fig. 2B) is similar to that seen with the thumb clamped and the limb supported, when effectively only the long flexor as prime mover is in action. With the arm hanging free, a similar tulip is seen in pectoralis too (Fig. 2A), revealing that when the thumb contracts harder, so does pectoralis, and when it contracts less, pectoralis also slackens. What is more, the latency of the tulip in pectoralis (about 40 msec) is, if anything, shorter than that in the thumb muscle. With the arm stationary and only the thumb moving, the responses in pectoralis can be regarded as postural adjustments to steady the arm, and we refer to them as giving a "postural tulip." But the line between posture and movement is indistinct; it makes no difference to the records if the subject attempts to keep his thumb stationary relative to the hand and pulls on the wire by adducting his shoulder, and this was actually what was done in the particular experiment illustrated in Fig. 2.

It was originally thought that the pectoralis tulip was of the same nature as the thumb tulip and was caused by responses of similar mechanism, based on the stretch reflex, which came into play when, under the influence of perturbations applied to the thumb, pectoralis was stretched or was allowed to shorten. To account for the responses when the movements were made by the thumb with a stationary arm, it was thought necessary to suppose that the sensitivity or "gain" of the stretch reflex in pectoralis was raised (Marsden et al., 1976c). This view was abandoned after we began to record the movements of the elbow; when our standard perturbations are applied to the pad of the thumb, the elbow does not move significantly for 45 msec, by which time responses in pectoralis have already started (illustrated in Marsden et al., 1981b). So movements of the upper arm apparently do not cause the pectoralis tulip, which must therefore be caused by afferent signals arising in the forearm or hand from perturbations to the thumb. Such heteroge-

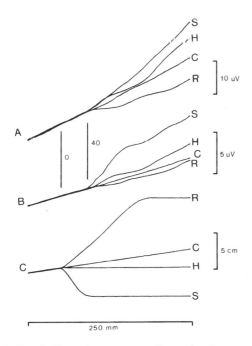

FIG. 2. Experiment recording simultaneous tulip responses in the EMG from A: Pectoralis major and B: flexor pollicis longus C: the corresponding records of linear displacement of the thumb. The four modes are labeled: C, Control; H, halt; R, release; S, stretch. The *timing line* is at 40 msec after the time of imposition of perturbations. Each trace is the average of 48 (5 May 1973.)

netically driven postural responses we refer to as "driven" tulips to distinguish them from ordinary autogenetic reflex tulips in prime-moving muscles (caused by receptors in those muscles themselves), which we refer to as "reflex" tulip. Tulips that are presumably "reflex" in this sense were seen in pectoralis when perturbations were applied to the elbow (Marsden et al., 1976b).

To clinch the demonstration of the "driven" nature of the postural tulip in pectoralis, a second motor was called into play with which perturbations could be applied to the wrist concurrently with those to the thumb. The second motor was placed on the opposite side of the subject to the thumb motor (it is not shown in Fig. 1). The subject held his arm slightly forwards, so that the wire from the second motor could run in front of his abdomen to be looped around the wrist. The two motors thus pulled in opposite directions, one on the thumb and the other on the wrist. The subject tracked, as before, by flexing his thumb, the intention being that the wrist should remain stationary. In this way, it was insured that pectoralis could fulfill a postural role; in the earlier paper, the shoulder was doing the tracking. In the experiment of Fig. 3, four traces are plotted, but they are not the usual four conditions (control, halt, stretch, and release). Instead, in two conditions (stretch and release of the thumb), the wrist motor caused the wrist to move in the opposite direction to the thumb and, in another pair, to move in the same direction. So the wrist motor was used either to reverse the effect of the thumb on the movement of the upper arm or to augment it. The remaining two conditions were the usual control and halt of the thumb, with constant current in the wrist motor.

To get sizeable arc movements from the wrist perturbations, the latter were applied before the thumb perturbations but, for clarity in the electrical records, not so early that they caused much response in pectoralis before the driven thumb tulip began.

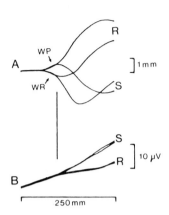

FIG. 3. Postural responses in pectoralis major from perturbations applied to the thumb to show their indifference to movements imposed on the arm during the latent period. A second motor (not shown in Fig. 1) applied pulls and releases to the wrist (opposite in direction to those to the thumb) mixed in the same experimental run with thumb perturbations, as described in the text. Pulls and releases were applied to the wrist 35 msec after the start of the recording sweep; each might be combined with either a stretch or a release applied by the other motor to the thumb at the usual time of 50 msec. A: Displacement records from the elbow, made photoelectrically. A pull on the wrist moves the trace up, whereas a stretch to the thumb moves it down. The two traces with stretches to the thumb are labelled S, and the two with releases to the thumb R; in each case, one trace is associated with a wrist pull (WP) and the other with a wrist release (WR). B: The corresponding four elements of the postural tulip in pectoralis major superimposed, showing that the two thumb stretch responses (S) and release responses (R) are apparently indifferent to the direction of movement of the elbow in the early stages. The *timing line* is at 45 msec after perturbations to the thumb. Each trace is the average of 32. Halt and control traces not plotted. (5 July 1979.)

By trial, a time 35 msec from the start of the sweep (i.e., 15 msec before the thumb perturbations) was found satisfactory. With this timing, the latent period of wrist responses was expiring just as the driven tulip started, some 45 msec after the thumb perturbations. The wrist perturbations lasted 75 msec. The results are plotted in Fig. 3A and B, omitting the control and halt records.

The responses in pectoralis to thumb stretch and release are similar irrespective of whether the wrist motor assists the thumb motor or opposes it in moving the upper arm, despite striking differences in the displacement records of the upper arm. This experiment thus establishes that the tulip in pectoralis is "driven" in the sense that it is indifferent to the initial direction of movement of the upper arm (which moves pectoralis) and must, therefore, depend on some source of afferent input other than pectoralis. The argument is that moving the wrist so as to cause an earlier and larger movement of the upper arm than that caused by thumb perturbations alone (as seen in Fig. 2B) does not augment the responses in pectoralis; neither does moving the upper arm in the opposite direction diminish them. Thus, in these circumstances, the "driven" tulip is dominant over the "reflex" tulip in pectoralis.

After we had demonstrated postural responses in the shoulder, attention turned to muscles stabilizing the trunk against sway. The first muscle looked at was tensor fasciae latae (TFL). In the experimental set-up shown in Fig. 1, contraction of TFL on the side of the arm in use has the effect of moving the pelvis laterally away from the arm. An experiment was conducted in exactly the same way as that in Fig. 2 but with an extra recording channel over TFL as shown in Fig. 1. Body stance was adjusted so that there was initially a moderate amount of activity in TFL. The TFL then gave a "postural tulip" at a latency of about 85 msec, as shown in Fig. 4A. In this tulip, the responses are in the same sense as those in flexor pollicis longus; i.e., a stretch perturbation causes an increase in activity, and a release causes a diminution. Since contraction of TFL tends to oppose movement towards the motor, these responses will tend to oppose body sway caused by perturbations applied to the thumb. These postural tulips in TFL invert, as they ought to, when the direction of pull on the thumb is reversed, thus providing a useful check on the above interpretation. To demonstrate this, the subject, after an ordinary run (Fig. 4E), turned through 180°. He took the pull of the wire with the same (right) thumb, the pendant arm in this case passing across his abdomen. When he again adjusted his stance for moderate activity in the right TFL, an inverted tulip was immediately and effortlessly obtained in this muscle (Fig. 4F). This is correct, for in this posture a pull by the wire causes the body to sway in such a direction as to diminish the length of TFL, so that a stretch reflex in flexor pollicis longus should be accompanied, as it is, by a release response in TFL.

The latency of tulips in TFL is 85 msec or less, perhaps just too long to be guaranteed to be involuntary on the evidence of latency alone (see Marsden et al., 1976a, Figs. 11, 12). But we have little doubt that they are true automatic responses, for they appear without specific effort on the part of the subject and even while his attention is elsewhere. Indeed, the subject is quite unaware, unless he attends specially, that he is doing anything to steady his pelvis against the perturbations. If his pelvis is steadied for him by the experimenter, the tulip in TFL disappears (Fig. 4G) without the subject being aware of any change in his behavior. In the light of what follows, it would seem that steadying the trunk acts by removing the possibility of sway rather than by preventing early sway which might reflexly cause the tulip in TFL.

An experiment was also done that showed that sensation in the feet was not important in this situation. Pneumatic cuffs occluded the circulation at both ankles for more than an hour, rendering both feet deeply anesthetic but having no detectable effect on postural tulips in TFL.

Postural responses similar to those in TFL were also found in the long flexor of the great toe, in the ankle extensors (triceps surae), in the thigh muscles (hamstrings and knee extensors), and in the muscles of the

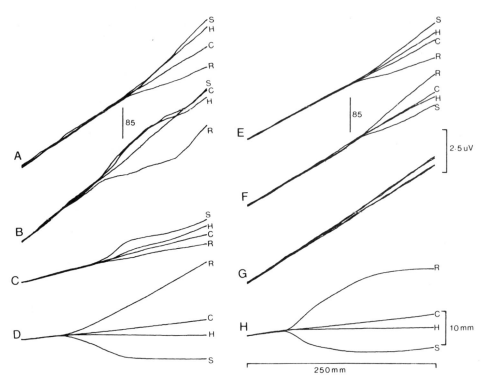

FIG. 4. Tulip responses in the integrated averaged EMG from **A**: TFL, **B**: pectoralis major, and **C**: flexor pollicis longus during adduction of the right arm. The load and perturbations were applied by a wire around the pad of the thumb as shown in Fig. 1. Simultaneous tulips were recorded in the three muscles. **D**: Records of the linear displacement of the thumb taken at the same time. The timing line is at 85 msec after the time of imposition of perturbations. Each trace is the average of 32. (31 December 1975.) **E–G**: Records from TFL in another experiment. **E**: A postural tulip obtained exactly as in A. **F**: Subject turned through 180°, and the movement of the arm was one of abduction (to the right across the abdomen): the responses in TFL were inverted, showing a decrease during stretch applied to the thumb and an increase during a release trial, as indicated by the labels. **G**: The subject turned back to the position of Fig. 1, but his trunk was steadied by a seated experimenter, who pulled on a rope around the subject's hips so as to press the trunk against a support; the responses vanished. The record of displacement of the thumb given in H was similar for E, F, and G. Each record in E is the average of 64, in F, of 32, and in G, of 24. The displacement calibration applies to both displacement records, and the calibration of integrated voltage to all the electrical records. (16 March 1976.)

back (erector spinae). In all of these experiments, the task was the same as before (pulling on the wire with the wire around the thumb or sometimes around the wrist), and the subject was unsupported, either standing or kneeling.

Evidence supporting the view that these distant postural tulips were "driven" in the sense defined above was obtained, in the case of TFL and flexor hallucis longus, by recording the displacement of the pelvis. It was found (Marsden et al., 1981b,) that for most of the latent period of the postural tulip, movement caused by the perturbations applied to the thumb was not more than a few micrometers. So it was very unlikely that the responses could be caused by movement eliciting a stretch reflex (or its inverse) in the postural muscle in question.

Even firmer evidence of their driven nature was desired for postural tulips in the ankle extensors (triceps surae), because this muscle group has been fixed on for investigation of postural responses in patients (Traub et al., 1980). In an experiment of this kind, the subject stands freely without support, facing the motor. With his elbow flexed to bring the hand into the midline, he pulls the wire straight towards him with his thumb. To show that the postural tulips obtained in triceps surae are driven and not caused by sway of the body towards the motor (or, in release responses, away from it), the second motor was used to apply perturbations to the hips, either so as to oppose any sway caused by the perturbations to the hand or so as to augment it, on the same principle as the experiment described above for pectoralis. A double cord was tied tightly around the subject at the level of the greater trochanter and connected to the second motor in front by a flexible wire. The line of pull was more or less the same for the two wires. Fore-and-aft movement of the knee, which gives an index of stretch or release of the ankle extensors with the subject standing, was recorded photoelectrically with the pea bulb taped to the skin over the head of the tibia.

In the illustration of this six-condition experiment (Fig. 5E–H), the records from the control and halt trials are not plotted. Perturbations were applied to the hips at start of sweep. It is seen that the responses in

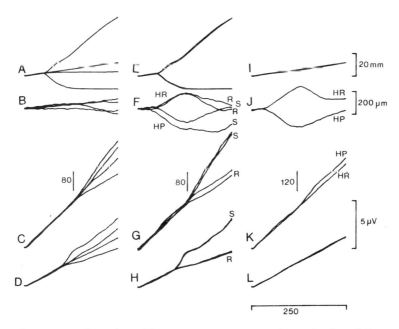

FIG. 5. Postural responses from the ankle extensors, gastrocnemius and soleus (triceps surae). The pull of the wire was taken on the left thumb. In each column, the records from above downwards are: Displacement of left thumb, displacement of right knee (photoelectric recording), integrated EMG of right ankle extensors, integrated EMG of left flexor pollicis longus. A–D: Simultaneously recorded results with the three standard thumb perturbations and controls. The postural tulip in the ankle extensors (C) has a latency of 80 msec after the thumb perturbations; movements of the knee (B) in the latent period are slight and irregular. Each trace is the average of 32. E–H: Experiment with two motors in which pulls and releases to the hips (HP and HR) were combined with either stretches or releases to the thumb (see text). F: Movement of the knee towards the motor causes the trace to move downwards. Each trace is the average of 16. I–L: Records with perturbations applied to the hips only (HP and HR), as described in the text. K: Timing line is 120 msec from the start of sweep (not, as usual, from 50 msec at the time of thumb perturbations). Each trace is the average of 32. (9 July 1979.)

ankle extensors are effectively the same whether the knee moves in the same direction as the thumb or in the opposite direction during either stretch or release of the thumb. This is taken to show that the responses, because they are indifferent to the direction of knee movement, are not reflex responses to the movement of the leg but are driven from the thumb. This conclusion tallies with the evidence of displacement records taken during a standard tulip (Fig. 5B) which show only very small and uncertain movements of the knee.

Figure 5I,J shows the effect of the hip perturbations, pull, and release, on their own. These records are taken from a six-way experiment, of which the first four conditions were the standard thumb conditions (control, halt, release, and stretch), the records from which are those in Fig. 5A–D, and the last two (giving Fig. 5I,J) were the two hip perturbations alone. The movements of the knee in Fig. 5J only differ by $^{1}/_{10}$ mm at their largest but nevertheless are associated with a small divergence in the integrated electrical records with a latency of about 120 msec from the start of the sweep, which can be regarded as a small reflex postural tulip (Fig. 5K). This divergence, although small, is believed to be genuine, for it is seen in the individual runs of which the illustrated records are the grand average. The low threshold of these responses is, perhaps, surprising, for it represents rotations of only about a minute of arc at the ankle, but we have met with other similar instances of great sensitivity (Marsden, et al., 1981b). They emphasize the need for care in identifying postural tulips as driven and explain why we did what might have appeared unnecessarily elaborate experiments to establish this point.

In all of the above experiments, the subject was standing or kneeling without other support. In another series of experiments, he knelt and steadied himself by holding onto the table that carried the motor with his right hand, as shown in Fig. 6. He pulled

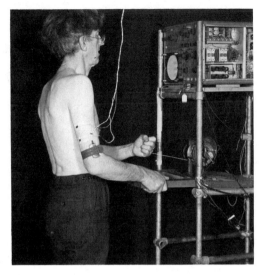

FIG. 6. Arrangements for recording postural tulips in the opposite (right) triceps and biceps during flexion movements of the left elbow. The load and perturbations were applied by a wire around the left wrist. The subject knelt, steadying himself by holding the wooden table on which the motor was mounted with his right hand. Electrodes over triceps and the grounding plate are seen.

on the wire in the usual manner with the wire looped round his wrist (or, in certain experiments, with the hand supinated and the wire round the pad of the thumb or around the tips of the middle three fingers). Brisk postural tulips were then found in biceps and triceps of the right arm at a latency of some 55 msec. The tulip in triceps is illustrated in Fig. 7B. The responses were so large and regular that they could be heard clearly on the loudspeaker in the single trials. In release trials, the muscle was not infrequently completely silenced (examples of this are in Fig. 10A,B). As might be expected, the responses are of such a nature as to oppose body sway caused by perturbations to the left arm; e.g., a stretch in the wire, which would tend to cause the body to sway forwards, is accompanied by an increase of activity in triceps, which would extend the elbow and push the trunk backwards.

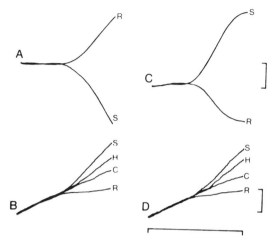

FIG. 7. Crossed triceps tulips with reversed shoulder movement to show that they are "driven." For the simultaneous records A and B, the arrangements were as shown in Fig. 6. Records of shoulder displacement are given in A (only stretch and release plotted), and the triceps tulip in B. C and D: Subject pulled on the wire with his left arm everted to point directly away from the trunk (see text). Because of twisting of the trunk about its vertical axis, the movements (C) of the right shoulder were reversed compared with A, but the triceps tulip (D) was unchanged. Each trace is the average of eight. (19 February 1977.) Horizontal calibration, 250 ms; upper vertical calibration, 100 μm; lower vertical calibration, 2 μV (for B) and 3 μV (for D).

The briskness and short latency of this postural tulip in the opposite (crossed) triceps naturally suggest that again it is a drive tulip and not caused directly by body sway. To establish its driven nature, the first step was to record the movements of the right shoulder (above the active triceps). Figure 7A gives the displacement records taken during stretch and release trials in the same run as for Fig. 7B. During the latent period of the tulip, there is a virtual absence of shoulder movement. This makes it very likely, of course, that the triceps tulip is driven, but better evidence was desired.

In the original set-up for Fig. 7, shown in Fig. 6, the motor was directly in front of the subject, and the direction of the wire was more or less towards the midline of the trunk. The subject's right hand held the edge of the motor table 12 cm to the right of the midline and of the wire. This was changed. A board was clamped across the table; the subject moved 75 cm to the right and gripped the board with his right hand in the same attitude as before. The line of the wire passed to the left of the subject's trunk. His left forearm was everted to a right angle. He pulled on the wire loop with the back of his wrist, moving as in a backhand stroke at tennis. With this arrangement, the first effect of a perturbation is to twist the trunk about its vertical axis; e.g., in a stretch trial, the left wrist is pulled forward, but the right shoulder initially moves backwards; the converse occurs with a release. The postural tulip in right triceps (Fig. 7D) was found to be unchanged (in both subjects) even though the record of displacement of the shoulder (Fig 7C) showed that it was now moving, at the start (and until near the end of the sweep), in a direction opposite to that required to elicit the tulip through a stretch reflex in triceps; e.g., in a stretch trial, triceps contracted at normal latency although the shoulder was moving backwards and, therefore, taking the stretch off triceps. Hence, the response is seemingly driven and not reflex.

In another series of experiments pointing to a similar conclusion, the right hand gripped a handle that was attached to an eccentric on a second electric motor. The motor could be caused to rotate 180° between stops in such a way as to pull or push the right hand through roughly 1 mm either at the same time as, or just before perturbations were applied to the opposite wrist. Pulling the wrist forward will relax, and pushing it back will stretch, the right triceps. But causing the right triceps to relax during a stretch trial or to be stretched during a release trial did not appear to alter the postural tulips driven by perturbations applied to the left wrist. Finally, the stops were removed and the motor set to revolve continuously as fast as it would, unsynchronized to the perturbations. This caused a considerable shaking of the whole arm, but a postural tulip elicited at the same time was remarkably unaffected, presumably because it was driven rather than reflex from the shaken arm. Figure 8 shows this for a frequency of 20/sec and an amplitude of oscillation, measured at the elbow, of ¾ mm. By reducing the amplitude to ¼ mm, the frequency could be pushed up to 56/sec; the answer obtained was the same.

These experiments with the second motor also guard against a possible source of error; for example, in a stretch trial, the reaction from the increased tension in the wire will pull the motor, the table on which it is mounted, and the whole frame towards the subject. Any such movement would push on the subject's right hand holding the table and might be responsible for a stretch response in the right triceps. However, this factor cannot be of significance, for the above experiments show that much larger movements of the hand towards or away from the subject produced by the second motor are without effect. Another experiment that controls for this error is given below in the arrangement for two thumbs.

The briskness of these postural tulips in triceps suggested that they might be elicited by much smaller perturbations than those used as standard (+270 g for stretch and −150 g for release on an initial force of 300 g). With smaller perturbations, responses could still be elicited down to stretches and releases of only +7.5 g and −7.5 g on the usual initial 300 g, i.e., of only ±2.5% change in load (Fig. 9). (An increase of load of +2.5%, although referred to as a stretch, only retards the tracking movement and does not actually reverse it.) It is perhaps surprising that such very small perturbations, which cannot offer much threat to bodily stability, should cause distant muscles to contract or relax in order to counteract any effect they might have. It also seems very improbable that the direct mechanical effect of the ±2.5% perturbations could spread rapidly to distant muscles—a consideration that again favors a driven mechanism for the distant responses that occur.

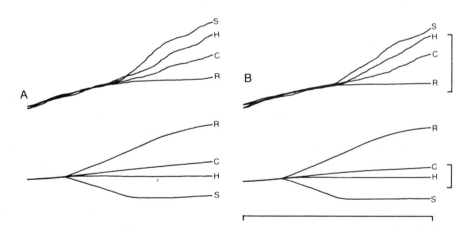

FIG. 8. Lack of effect of vibration of the supporting hand on the postural tulips in crossed triceps, with arrangements similar to Fig. 6. For this experiment, the right hand held a handle that for run A was stationary and for run B, immediately following, was driven by an electric motor to oscillate backwards and forwards. The vibration was at about 20/sec and was not locked to the recording sweeps. The amplitude of oscillation was ¾ mm, as measured by a mark at the elbow moving against a scale. Each trace is the average of eight. (28 September 1977.) Horizontal calibration, 250 mm; upper vertical calibration, 5 μV; lower vertical calibration, 10 mm.

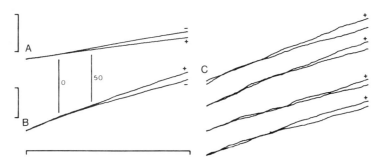

FIG. 9. Postural responses resulting from minor perturbations. The arrangements were as illustrated in Fig. 6. A: Displacement records for the left wrist. B: Stretch and release responses in right triceps for +2.5% and −2.5% load changes (± 7.5 g) applied at the usual time of 50 msec to the left wrist. The control records are not plotted. A and B: each trace is the average of 32 trials made up of four experimental runs of eight trials for each mode. C: EMG records for the four component runs are plotted separately to show that the responses are regularly present. (18 August 1977.) Horizontal calibration, 250 mm; upper vertical calibration, 10 mm; lower vertical calibration, 2.5 µV.

In these last experiments, the subject reported that he was unaware of the smaller perturbations. On this subject (P.A.M.), some forced-choice threshold determinations were done to put this impression on a quantitative basis. To start with, trials with perturbations of +5% or −5% on 300 g initial force were presented in random cyclic order without control trials or other perturbations. The subject pulled on the wire with the tips of his fingers. In two runs, each of 32 trials, he made 78% and 88% correct choices. These percentages show that a substantial fraction of the ±5% perturbations were detected correctly, although at a level well below confident recognition. Next, six similar runs were made with ±2.5% perturbations, which gave 47, 65, 47, 53, 55, and 59% correct choices (54% overall). This is close to pure guessing, which gives an expectation of 50% correct. Random cyclic was then changed to fully random presentation, and three longer runs of 88 trials were done with ±2.5% perturbations. Correct choices were 49, 66, and 63% (59% overall). This is all that was done on this question. The indication is that for ±2.5%, performance was little better than chance and well below conventional threshold criteria for a forced-choice situation. Thus, if the subject has any conscious perception of ±2.5% perturbations, it will be of a very uncertain kind. In any case, he cannot, of course, be using conscious information, because the latency of response is too short. Responses to ±2.5% perturbations were seen with the pull taken both by the wire around the wrist or by the pads of the fingers. The fingers did not give better postural responses to small perturbations than the wrist. As to where perturbations are detected by the postural mechanisms, skin and forearm muscles were ruled out as inessential by an experiment, with the pull taken at the wrist, in which the whole lower arm was anesthetized by a cuff just below the elbow; the responses to ordinary perturbations were unaltered (Fig. 10A,B). We got no further with the question than that.

As regards the right arm, with which the subject steadies himself, it was no surprise to find that if the subject let go of the table top, the responses in triceps vanished. The triceps, apparently, does not take part in postural responses unless its actions can assist in stabilizing the body, irrespective of the presence or absence of a background contraction in triceps. Whereas excellent crossed tulips in triceps were obtained with the fixed handle, there was no trace of response with an identical floating handle slung from above. In another experiment,

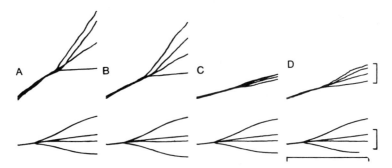

FIG. 10. Effect of peripheral anesthesia on crossed triceps tulips. A is a control tulip obtained exactly as in Fig. 6 except that the right hand gripped a handle instead of the table. B was taken 55 min after occluding the circulation with a cuff below the left elbow (on the arm that pulled on the wire). C shows the effect of anesthetizing the right hand (holding the handle) by a cuff around the wrist; before anesthesia the responses were as in B. D is the tulip obtained during recovery, immediately after deflating the cuff on the right wrist. Each trace is the average of 16 (22 August 1977.) Horizontal calibration, 250 mm. Upper vertical calibration, 2 μV. Lower vertical calibration, 10 mm.

the right wrist was slung by a cord from above. With the hand clear, there were no responses in triceps, but as soon as the fingertips touched the frame, involuntary postural responses in triceps could be heard in the loudspeaker.

At this time (1977), crossed tulips in subject P.A.M. were greatly attenuated when the right hand was anesthetized by a cuff around the wrist for the usual period of about 80 min. This is shown in Fig. 10C. So that the subject could grip the handle properly with an anesthetic hand, his fingers were carefully bandaged to the handle. In this experiment, a recording run was started at the moment when the cuff was finally deflated. About half-way through the run, sensation returned to the fingers, and it was at this time that large responses began to be heard in the loudspeaker. The record of the whole run is shown in Fig. 10 D. A second experiment gave similar results. There is no reason to attribute the results to poor bandaging of the fingers, but these results appear to be exceptional. Traub et al. (1980) found that anesthesia of this kind does not depress crossed tulips in every subject. In light of this, P.A.M. was retested in 1979 in an identical experiment to that of Fig. 10C,D, and he now showed no depression of crossed triceps tulips. Habituation to the effects of anesthesia has already been described for reflex tulips in flexor pollicis longus (Marsden et al., 1977a). An interesting observation is that when the right arm failed as a support, even though the subject was unaware that it had done so, it lost its postural responses. Thus, the decision as to whether a muscle can usefully engage in postural activities must be made at a subconscious level.

The foregoing examples have reiterated that the latency of heterogenetically driven postural tulips is not a great deal longer than that of ordinary autogenetic reflex tulips in the same muscles. In order to get a better estimate of the extra time needed to produce a postural as compared with an ordinary tulip, the same muscle, flexor pollicis longus, was used both as the prime mover and as a postural muscle. The subject knelt in front of the motor and pulled on the wire by flexing his left thumb. His left hand was supinated in front of his abdomen, and the wire was hitched around the pad of the thumb. The pad of the other thumb (and no other part of that limb) pressed lightly on the frame at head height to support the subject as he leaned slightly forwards. The right forearm was roughly vertical, and the upper arm was horizontal, directed forwards. To avoid fatigue at the shoulder, the weight of the right arm was taken above the elbow by a long sling from above. Tulips were obtained in simultaneous recordings over the two long thumb flexors. Then, the hands were exchanged as a precaution against an effect of cerebral dominance or other asymmetry; the left thumb rested on the left-hand side of the frame, and the wire was pulled on by the right hand. Tulips were again recorded simultaneously from both muscles.

The records (Fig. 11) show that it does not matter much which way the experiment is done. The accurate latencies on the timing lines were from measuring at high gain the records of single runs of eight trials averaged. The reflex and driven tulips of the left thumb both measure 4 msec earlier than those of the right thumb, but it is not known if this is significant. The subject P.A.M. is left handed. The conduction time from motor cortex to the electrodes over flexor pollicis longus, by the method of Merton and Morton (1980a) appeared to be the same in the two hands.

The postural tulip is very like the direct one and is 22 msec later in the same muscle in both cases. The postural tulip is presumably driven, for mechanical transmission of perturbations from one thumb to the other in the early part of the latent period must be excessively small. The estimate that it takes some 20 msec longer to produce a crossed driven tulip than a direct reflex tulip rests on the assumption that the afferent signals responsible for the crossed driven tulip also arise in the prime-moving long flexor. It was argued above that this was the case for the driven tulip in pectoralis when perturbations were applied to the thumb, as they are here. The upper arm did not begin to move for 45 msec—too late to drive the pectoralis tulip. Similar considerations probably apply here, but if they do not, alternative signals to drive the crossed tulip, arising in the upper arm and shoulder, would not be likely (bearing in mind inevitable mechanical delays in the arm) to arrive at the CNS more

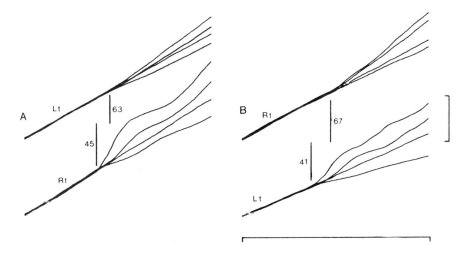

FIG. 11. Experiment using one long thumb flexor as a postural muscle and the other, simultaneously, as a prime mover (see text). The postural tulips are above, and the direct tulips in the prime mover below. A: left thumb was used for posture and the right as prime mover. B: hands were the other way around. (The runs for B were, in fact, done before those for A). Each trace is the average of 64 trials. (2 October 1977.)

than a few milliseconds before those from the thumb flexor. Hence, our estimate of 20 msec would be little upset.

DISCUSSION

Postural responses in muscles distant from the prime mover were deliberately sought in the first place, but it was by no means expected that they would turn out to be so similar in appearance to the tulips in the prime mover or that they would occur at a latency so little longer. Whether the two types of tulip share some common mechanism, presumably in the brain, possibly in the basal ganglia or cerebellum, remains to be seen. One thing seems clear: that postural tulips are not caused by length changes in the postural muscles themselves; that is to say then, they are not based on the autogenic stretch reflex. This emerged as a new and significant point, and much of the chapter has been devoted to somewhat complicated experiments designed to establish it as unequivocally as possible.

If postural tulips are not based on local reflexes in the postural muscles, they must be caused by afferent signals from elsewhere in the body, presumably from the limb to which perturbations are applied. This heterogenetically excited type of response we have called driven; ordinary autogenetic tulips, based on the stretch reflex in the muscles in which they occur, we refer to as reflex. There is no reason why both kinds of tulip should not be caused by afferent signals from the same receptors, for example, muscle spindles. The proposed mechanism for driven tulips would be akin to the mechanism of the crossed extensor reflex of the cat when the limb in which the crossed extension occurs is deafferented by dorsal root section.

One further conclusion about the mechanism of driven tulips seems fairly safe to draw: they cannot be initiated by the γ route, by γ activation of the spindle in the postural muscles setting into operation a follow-up length servo to cause the muscle to contract (or vice versa for relaxation). This follows because such a mode of contraction would involve extra conduction time to and from the muscle in addition to the conduction time for ordinary motor impulses. Driven tulips, as we have seen, only get an extra 20 msec or so, and that is not long enough for the two extra journeys to and from, for example, flexor pollicis longus (Fig. 11).

Presumably, postural tulips are driven for the sake of speed. With perturbations applied to the thumb or wrist, driven tulips start before the body begins to sway appreciably. They anticipate (to used the word in its strict sense) disturbances of posture and will tend to prevent them from ever developing. If we had to wait for body sway to set up ordinary stretch reflex action, everything would be many milliseconds later, and the reactions correspondingly less effective.

Postural tulips are not only rapid but unexpectedly sensitive. Perturbations of only 7.5 g to the thumb or wrist cause distant responses. Measurement of the effective mass of the forearm and calculation shows that the accelerations or decelerations produced by such perturbations are not more than 0.6% g (where g is the acceleration caused by gravity), that being an upper limit. This is the acceleration of a ball bearing on a smooth plane inclined at only 0.3° to the horizontal. The body must perform quite a feat in sensory discrimination to detect such disturbances, particularly as it has necessarily to be done in the first 20 msec or so of the latent period of the response. Perhaps this is one of the reasons that muscles have such a rich sensory innervation. It should also be mentioned that in order to get responses to small perturbations, it may be necessary to mix them in with large perturbations in the same experimental run, as was also found in the case of ordinary reflex tulips (Marsden et al., 1979, 1981b). Why this should be so is not yet understood.

Postural tulips in triceps disappear if the hand ceases to grip a firm object, and those in tensor fasciae latae disappear if the trunk is otherwise steadied. These observations are taken to show that muscles only exhibit postural responses if, in fact, they can, at the same time, perform a useful postural function. We have argued from the experiments that recognition that circumstances are appropriate for a postural response is made subconsciously. Other experiments on this topic are reported elsewhere (Marsden et al., 1981b). It is also shown there that if the nontracking hand, instead of holding a tabletop, engages in a task equivalent to holding a cup of tea, the responses in triceps reverse (the tulip turns upside down), which is what they have to do if the tea is to be prevented from spilling when the other hand encounters perturbations. Whether the muscle is to produce a "postural" or a "tea-cup" tulip is also apparently decided at a subconscious level. These various complexities of behavior are one reason for our belief that the phenomena described have a cerebral or cerebellar mechanism.

The Jendrassik Maneuver: Quantitative Analysis of Reflex Reinforcement by Remote Voluntary Muscle Contraction

*P. J. Delwaide and P. Toulouse

Department of Neurology and Clinical Neurophysiology, University of Liège, Liège, Belgium

The Jendrassik maneuver (JM) (Jendrassik, 1885) is well known to clinicians who use it to increase the amplitude of a weak tendon reflex or to determine whether a spinal lesion is partial or total. It consists of an effort to separate clenched interlocked hands during or just before percussion of the tendon. The JM augments the amplitude of proprioceptive monosynaptic reflexes and also that of exteroceptive reflexes (Hugon, 1973a; Delwaide and Young, 1973; Delwaide et al., 1975).

The mechanism of the JM remains open to clinical discussion. Microneurographic techniques have recently been applied. On the one hand, Burg et al. (1973, 1974), supporting an earlier hypothesis (Sommer, 1940; Paillard, 1955; Buller, 1957), observed an increased discharge from Ia afferents of the conditioned muscle which is presumed to reflect a γ-MN discharge; it was therefore postulated that the facilitatory effects were caused by fusimotor activity. On the other hand, Hagbarth et al. (1975a) and Burke et al. (1980a) failed to find an increase in IA activity and attributed the facilitation to influences working directly on α-MNs (Landau and Clare, 1964; Gassel and Diamantopoulos, 1964b). Thus, putative explanations of the JM facilitation remain controversial and require further elucidation.

Other problems have received little attention. It is not known to what extent the JM could be used as a paraclinical test of motor function. There are few data in the literature comparing quantitative results from normal subjects and neurological patients (Bishop et al., 1968). Furthermore, the functional significance of intersegmental facilitation has only rarely been discussed (Gottlieb and Agarwal, 1973). However, if muscular contraction in upper limbs is able to modify the excitability of reflex arcs in a lower limb, inquiry should be made as to its effects, beneficial or otherwise, on postural mechanisms and voluntary contraction.

The work reported here first examines some methodological aspects of the JM in order to make the conditioning effect reproducible and capable of quantitative expression both in normal subjects and in neurological patients. Factors influencing the intensity of facilitation and its mechanisms are subsequently examined; finally, the functional role that might be attributed to remote muscular contraction is considered.

---

*To whom correspondence should be addressed: Department of Neurology and Clinical Neurophysiology, Institut de Médecine, University of Liège, Hopital de Bavière, boulevard de la Constitution, Liège 4000 Belgium.

METHODOLOGICAL CONSIDERATIONS

One of the reasons the JM has been so little explored in clinical neurophysiology lies in the difficulties of standardizing the conditioning contraction and analyzing its parameters. Many muscles are, in fact, involved in the classical JM, and the force they exert is difficult to measure. We have attempted to simplify the JM and have limited the voluntary conditioning procedure to an isometric contraction of wrist extensors in the forearm on the same side as the lower limb to be tested. The forearm is immobilized in a horizontal position with the palm downwards; a strain gauge and an accelerometer are attached to the back of the hand. Cutaneous electrodes measure the electromyographic (EMG) activity of the wrist extensors while other electrodes confirm that contraction is limited to these muscles. The subject is ordered to extend the wrist on hearing an auditory signal. The isometric contraction is performed as quickly as possible, so that the force plateau is reached in 200 msec, and the force is maintained at a constant level for about 3 sec. This is controlled by having the subject watch a storage oscilloscope which records the strain gauge output.

We have called this conditioning maneuver the selective contraction (SC) to distinguish it from the classical JM. Even though simplified, it is fairly complex, being ballistic at the start and subsequently maintained. In some experiments, we have limited the conditioning contraction to the ballistic component, the subject performing an unopposed (isotonic) extension of the wrist. In others, we have studied a ramp movement (Delwaide and Toulouse, 1981).

As a preliminary, we investigated whether the low-intensity auditory stimulus was itself capable of modifying the amplitude of the reflexes under test and found this not to be the case. In that respect, our results differ from those of Rossignol and Melvill Jones (1976).

The first stage of the definitive investigation consisted of an intense SC compared with a classical JM performed by pulling apart the conjoined clenched fingers. Figure 1A illustrates two examples of such a comparison and shows that there is no difference between the facilitations elicited by the two types of conditioning. In 10 subjects, we found no significant differences. It is not therefore necessary to activate many muscle groups in order to obtain maximal facilitation of the reflex. These results lead to the conclusion that the conditioning brought about by the classical JM is clearly supramaximal and that SC is sufficient to obtain quantitatively similar effects. Therefore, because of the advantages accruing from control of the parameters, selective contraction should be not only retained but preferred.

Another methodological problem is raised by the manner in which results are expressed. Sometimes, as in Fig. 1A, the facilitation is expressed as an absolute value (Sommer, 1940; Clarke, 1967; Ott and Gassel, 1969). Although this is a legitimate estimate, it does not permit comparison between subjects or in the same subject at different times. It is, of course, known that these values are affected by the recording conditions, such as electrode position, skin resistance, etc. Expressing results as percentage of reference values (Sommer, 1940; Struppler and Preuss, 1959; Kawamura and Watanabe, 1975) can be objected to since this does not relate amplitude change to the baseline value, and comparisons of results between different subjects or different myotatic arcs make no sense. The most meaningful way of expressing results seems to be to relate the increase in amplitude of the reflex to the maximal value of the motor response recorded from the muscle. We thus propose to make use of this relationship by analogy with the known advantages of the relationship H_{max}/M_{max} (Angel and Hoffmann, 1963; Pierrot-Deseilligny and Lacert, 1973). Table 1 gives examples of

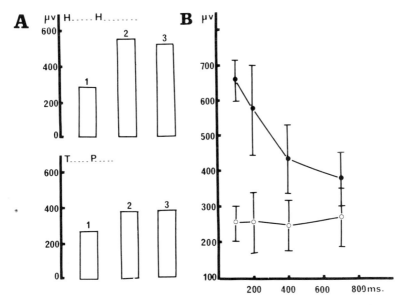

FIG. 1. A: Comparison of the facilitation of control responses *(column 1)*, following a maximal JM *(column 2)* and an SC of the wrist extensors *(column 3)* in two subjects (HH and TP). B: Time course of the test soleus reflex *(empty circles)* at various time intervals after the onset of the auditory command stimulus; the *filled circles* represent the mean of 30 to 50 measurements of the reflex conditioned by SC of the ipsilateral wrist extensors. *Vertical lines:* standard deviations.

this form of expression. In expressing the facilitation as a fraction of M_{max}, an idea is gained of the number of MNs recruited by the conditioning maneuver (Delwaide and Toulouse, 1981).

By the use of visual monitoring of the contraction of forearm muscles, it is possible to repeat more or less identical conditioning procedures. Figure 1B illustrates the results obtained by repeating the SC as exactly as possible and shows mean values obtained in a series of 30 reflex measurements taken 100, 200, 400, and 700 msec

TABLE 1. *Facilitation of soleus tendon reflex as measured in 10 subjects*

Time delay (msec)	N	Control reflex (%)			$M_{max}(\%)$		
		JK	Wrist ext	Δ	JK	Wrist ext	Δ
300	10	156	156	0	12	11,3	0,7[a]
750	10	131	129	2[a]	6,1	6	0,1[a]

[a] Not significant.

after the onset of the auditory signal. The standard deviations are relatively small, especially at 100 msec, and are of an order of magnitude normally acceptable in clinical neurophysiology. In other words, reproducible results can be obtained at fixed intervals following the onset of the command signal.

In order to be certain that the increase in amplitude of a reflex is specifically caused by the contraction of a remote muscle, it must be ascertained that the lower limb muscle under examination is not itself displaying EMG activity (Mark, 1963). This precaution has been taken by many (Clarke, 1967; Kawamura and Watanabe, 1975; Hagbarth et al., 1975a) but not all workers. Control recordings should be taken regularly from the muscle to be conditioned, either by skin electrodes or, better, by intramuscular needles (see A. Prochazka and M. Hulliger, *this volume*). When SC rather than the classical JM is used, EMG activity is only rarely observed in the muscle under test (one case in 20), perhaps because the axial muscles are only minimally involved.

FACTORS INFLUENCING FACILITATION

The temporal relationship between the conditioning contraction and the reflex facilitation will be considered first; it has previously been investigated by Paillard (1955), Gottlieb and Agarwal (1973), and Kawamura and Watanabe (1975). Under the experimental conditions used, the following sequence of events was observed (Fig. 2): EMG activity in wrist extensors begins, on average, 182 msec after onset of the auditory signal. Deflection of the accelerometer tracing follows this almost immediately, whereas the tension recorded by the strain gauge remains unchanged until some 60 to 80 msec after onset of EMG changes. As shown in Fig. 2, facilitation of the soleus tendon reflex begins about 100 msec after the start of the auditory stimulus—that is to say, before the onset of EMG activity in the conditioning muscle. This initial facilitation is, however, not very marked and does not last more than 100 msec (phase I). Some 25 to 35 msec later, an explosive facilitation supervenes (phase II). This second phase lasts about 350 msec and terminates about 550 msec after the start of the auditory stimulus. The curve is asymmetrical, the ascending phase being steeper than the descending. From about 600 msec onwards, a third phase of moderate facilitation can be discerned, which lasts until the end of the contraction (phase III).

When the isometric conditioning contraction is replaced by a ballistic movement executed by the same muscles, the first two phases are similar, but the facilitation then terminates at about 550 msec. It should be emphasized that this is of longer duration than the conditioning contraction, which lasts 200 msec at most. When, on the other hand, the soleus tendon reflex is conditioned by a ramp movement, facilitation of the second phase is very evidently less

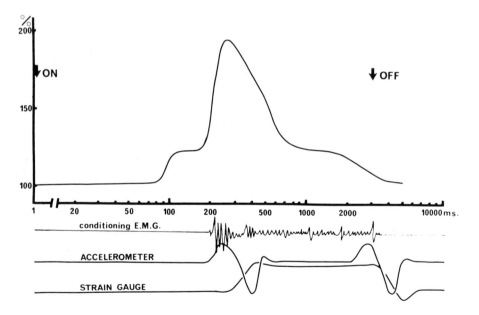

FIG. 2. Changes with time of amplitude of the soleus tendon reflex and various parameters of the conditioning contraction. By the use of a logarithmic scale, three phases of facilitation can be demonstrated.

marked but continues to increase in the third phase after 600 msec (Delwaide and Toulouse, 1981). This type of contraction is therefore able to influence differentially the second and third phases of facilitation in the conditioned reflex.

Although the importance of temporal relationships between the conditioning contraction and the test reflex has been emphasized by earlier authors, a time course in three phases has not been previously described. Contrary to our findings, Buller and Dornhorst (1957) considered the facilitation to precede EMG activity and to persist unchanged until the end of the conditioning contraction. Our results indicate that facilitation should necessarily be described in terms of the type of conditioning contraction and of the time interval separating the beginning of the command to condition and the test reflex itself (Delwaide and Toulouse, 1981).

The type of test reflex is also a determining factor (Toulouse and Delwaide, 1981). Like the tendon jerk, the Hoffmann reflex with a value of $H_{max}/2$ is significantly increased by SC at an interval of 300 msec (peak of the second phase), using Student's test. This increase is sometimes less marked than that of a tendon reflex of the same amplitude. However, there is no statistically significant difference when groups of subjects are considered. At an interval of 750 msec (during the third phase), facilitation of the H reflex is no longer significant in relation to control values.

These results could explain why it has been maintained both that the H reflex is facilitated (Struppler and Preuss, 1959; Landau and Clare, 1964; Gassel and Diamantopoulos, 1964b) and, contrarily, that it is unaffected (Sommer, 1940; Paillard, 1955; Buller and Dornhorst, 1957), the difference probably depending on the time interval between conditioning and test.

The degree of facilitation depends on the initial amplitude of the test reflex and, particularly in the case of the Hoffmann reflex, on its position in the recruitment curve. Expressed in relation to M_{max}, the threshold H reflex is increased less than the $H_{max}/2$ reflex, and the facilitation of H_{max} is less than that of $H_{max}/2$. A low-amplitude tendon reflex is facilitated less than a high-amplitude reflex. It should be noted that in normal subjects, the highest amplitude tendon reflex habitually corresponds to the value $H_{max}/2$. This difference in effect according to the amplitude of the test reflex can be observed in both the second and third phases.

The intensity of facilitation is also a function of the force exerted by the conditioning contraction (Bishop et al., 1968). By monitoring it with the strain gauge, it is easy to grade the conditioning contraction in a repeatable fashion and to use values corresponding to maximal force (F_{max}) or fractions thereof ($F_{max} \times 3/4$; $F_{max}/2$; $F_{max}/4$). As shown in Fig. 3, the facilitation elicited by F_{max} is the most intense; both the peak and the plateau are higher. This facilitation is similar to that elicited by a maximal JM. The curves corresponding to lesser degrees of force are successively lower, showing a reduction of facilitation in both the second and third phases. As shown on the right side of Fig. 3, the peak of facilitation is more sensitive to an increase in force of the conditioning contraction than is the plateau, lending emphasis to the notion of considering the different phases separately. The graph further suggests that, within limits, the degree of facilitation is directly proportional to the force of the conditioning contraction (Delwaide and Toulouse, 1981). Thus, the facilitation can be graded and does not obey an all-or-none law.

The intensity of facilitation depends on the site of both the conditioning contraction and the test reflex. Figure 4A shows that a contraction of equal force in either hand muscles or deltoid does not elicit quantitatively equal effects on the same tendon reflex, the facilitation being greater when the intrinsic muscles of the hand are con-

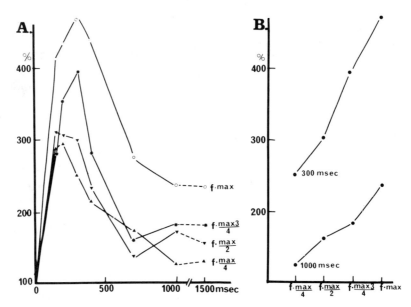

FIG. 3. **A:** Facilitation of the soleus tendon reflex following different force intensities of the conditioning contraction of wrist extensors. **B:** Facilitation of a reflex as a function of the force developed, respectively, during the peak of facilitation and during the later plateau. It can be seen that the facilitation is influenced to a greater extent at the peak than during the plateau.

tracted. If instead of exerting the same force, the conditioning muscles develop half their maximal force, the facilitations elicited are similar. We have also observed that contraction of wrist extensors elicits a greater degree of facilitation than a contraction of equal force in wrist flexors. Some muscle groups facilitate distant monosynaptic reflexes more intensely than others. Muscles most effective in this respect are those that are most influenced by the pyramidal motor system. When the conditioning contraction is effected by contralateral wrist extensors, the facilitation is equal in duration and intensity to that produced by ipsilateral muscles. We have not found any variation related to handedness of subjects (Toulouse and Delwaide, 1981).

Figure 4B illustrates the effects of the same conditioning contraction on different tendon reflexes in the ipsilateral lower limb. The percent facilitation of the quadriceps tendon reflex is greater than that of the soleus tendon reflex; the short biceps femoris tendon reflex holds an intermediate position. This difference also appears when the results are expressed in relation to M_{max}, so it is not dependent on the lesser amplitude of the control reflex in quadriceps (Toulouse and Delwaide, 1981).

Isometric contraction of the muscles of the anterior compartment of the leg (rapid onset, then maintenance for more than 2 sec) also facilitates the biceps brachii and the masseteric tendon reflexes. Facilitation plotted against time shows three phases similar to those in Fig. 2. Remote muscle contraction thus increases tendon reflexes in the caudorostral as well as in the rostrocaudal direction (Toulouse and Delwaide, 1980; Delwaide and Toulouse, 1981). However, the amount of facilitation is often less marked after the contraction of muscles of the leg.

MECHANISMS OF THE CONDITIONING

The definition of the three distinct phases shown in Fig. 2 depends on particular ex-

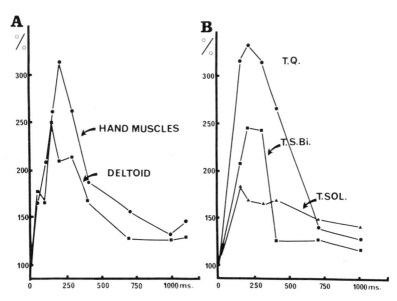

FIG. 4. A: Conditioning of the soleus tendon reflex by the intrinsic hand muscles and the deltoid, respectively. The force developed by the two muscles is the same, but the facilitation elicited by the intrinsic hand muscles is greater. **B:** Following conditioning contraction of the ipsilateral wrist extensors, the tendon reflex of quadriceps is facilitated to a greater extent that that of the short biceps femoris and to an even greater extent than that of soleus.

perimental conditions. For example, the difference between phases II and III is accentuated by more intense conditioning contractions. Moreover, if the conditioning contraction is ballistic, there is no phase III. A conditioning tonic vibration reflex evokes a moderate and stable remote facilitation similar to phase III. On the other hand, in patients with radial palsy, attempts at maximal voluntary contraction on the paralyzed side bring about only a small facilitation which begins at the same latency as phase I. These findings indicate that it is possible to manipulate each phase of facilitation separately, and thus, separable mechanisms for each are suggested.

Phase I seems to be related to voluntary command, as it can be seen in patients with radial palsy but is absent during a tonic vibration reflex of the conditioning muscle or after its passive lengthening. Phase II could be related to proprioceptive afferent impulses set off by muscle contraction transmitted by group Ia (and perhaps group II) afferent fibers. In fact, phase II is only clearly seen in experimental situations known to make the muscle spindles discharge. In contrast, it does not appear in cases of radial palsy. An interesting parallelism can be drawn between procedures that increase the facilitation and the results of microneurography in man in the same experimental conditions. The mechanism of phase III is more hypothetical and could involve the activity of IA afferents also coming from the conditioning contraction but, at that delay, as a tonic and lower frequency discharge.

FUNCTIONAL ROLE

The facilitation of tendon reflexes in the lower limb by remote muscular contraction is not negligible under resting conditions. There exists a potential recruitment of 15% of the motor units in quadriceps and a 11% in soleus when the tendon reflex is used as the test. This facilitation lasts several hundred milliseconds even for a very short conditioning contraction. An α-MN facilitation is shown by the facilitation of electrically elicited reflexes, although in this case, the effect is of lesser degree and shorter duration. As illustrated in Fig. 4, the facilitation exerted on reflex arcs in the lower limb does not differentiate between flexor and extensor function. It can be proposed that the JM puts the myotatic arcs in a state of alertness, in readiness for postural adjustments. This role can help maintain the upright posture and insure a better fixation of lower limbs. The greater degree of facilitation of quadriceps fits with this interpretation. From this point of view, it should be noted that the facilitation is exerted very rapidly and is reinforced by the ballistic character of the conditioning movement, which is particularly prone to disequilibrate the axis of the body. Facilitation elicited by voluntary contraction of upper limb muscles thus appears as a relatively simple mechanism for intersegmental coordination.

When the lower limbs are themselves the site of a voluntary contraction, an undifferentiated facilitation of antagonist MNs might upset the balance of reciprocal actions between various myotatic arcs and counteract motor programs. To test this, we have examined the influence of a remote contraction on the voluntary EMG activity of soleus. This activity was first rectified and then summed by an averaging computer. In subjects in whom the facilitation of the tendon reflex is intense, SC of wrist extensors did not measurably intensify the EMG activity. Even feeble contraction was not reinforced (Fig. 5).

Thus, if the Jendrassik effect persists, it is of little functional importance when the MNs of the lower limb are voluntarily activated. The mechanisms of intersegmental coordination do not come into play, either because they normally follow the same pathway as motor commands and are occluded or because the interneurons mediating the Jendrassik effect are inhibited under resting conditions.

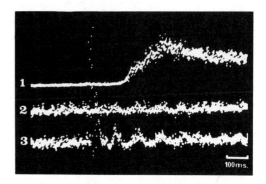

FIG. 5. Rectified and summed EMG. Note stimulus artifact. *Upper trace:* Recording of wrist extensors. *Middle trace:* Low-amplitude EMG of soleus, unmodified by ipsilateral wrist extensor contraction. *Lower trace:* The same EMG can nevertheless be modified by stimulation of sensory branches of the radial nerve applied at the time of the stimulus artifact.

CONCLUSIONS

Although known for a long time, the JM brings about effects that have been little investigated. By taking some methodological precautions, it is possible to define certain characteristics of intersegmental facilitation that are reproducible in a single subject but which vary in intensity from person to person. Following a selective isometric contraction of the forearm muscles lasting 3 sec (designated here as the selective contraction, SC), the reflex amplitude exhibits complex changes which vary with time following the command to contract and with the type of conditioning contraction. Three phases can be distinguished in the facilitation. The facilitatory effect is graduable, and its intensity is related to certain parameters of the conditioning contraction, notably, its force and locus. The type of reflex used as a test, its amplitude, and its proximal or distal location in the lower limb also affect the intensity of the facilitation.

The different phases of facilitation, particularly the peak and the final plateau, can be differentially manipulated by suitable experiments, favoring the notion of separate mechanisms in their genesis. The peak of facilitation is likely related to the stimulation of Ia afferents originating from the conditioning muscles either by fusimotor excitation in voluntary contraction of by maneuvers that stimulate the spindle receptors mechanically.

A systematic study of remote facilitation by voluntary contraction makes it possible to envisage clinical applications, particularly in spasticity and rigidity, and to think of the phenomenon in quantitative rather than qualitative terms. Finally, it must be emphasized that although remote muscular contractions represent the most classical mode of intersegmental facilitation of monosynaptic reflexes, it is not the only way, since it has been shown that the position of the upper limb can bring about a pattern of facilitation and inhibition in motor nuclei of the lower limb (Delwaide et al., 1977).

Motor Control Mechanisms in Health and Disease,
edited by J. E. Desmedt.
Raven Press, New York © 1983.

Mechanisms of Vibration-Induced Inhibition or Potentiation: Tonic Vibration Reflex and Vibration Paradox in Man

*John E. Desmedt

Brain Research Unit, University of Brussels Faculty of Medicine, B-1000 Brussels, Belgium

Vibration of muscle or tendon can markedly influence motor control in several ways. The vibration effects associated with mechanical tools under certain working conditions may have detrimental effects on motor activities. On the other hand, clinical vibrators can have therapeutic uses in the rehabilitation of patients with certain central motor disorders (cf. Hagbarth, 1973). The afferent fibers from the muscle spindle are extremely sensitive to vibration, which thus provides a convenient noninvasive method for investigating myotatic mechanisms in man. In the cat, vibration was shown to rather selectively activate the primary Ia afferents from the muscle spindles (Bianconi and Vandermeulen, 1963; Brown et al., 1967; Matthews, 1973; A. Prochazka and M. Hulliger, *this volume*). In primary afferents responding to vibration, the action potentials are closely synchronized to the vibration cycle, whether they discharge one-to-one or at another mean frequency (Burke et al., 1980c).

The reflex effects of vibration raise a number of questions. Vibration activates the motoneurons (MN) in the tonic vibration reflex (TVR) (DeGail et al., 1966; Lance et al., 1966, 1973; Eklund and Hagbarth, 1966; Hagbarth, 1973) but simultaneously inhibits the phasic myotatic reflexes involving the MNs of the same pool (DeGail et al., 1966; Lance et al., 1973). These concurrent opposite effects have been dramatically illustrated for the normal quadriceps muscle by the pioneering studies by DeGail et al. (1966) (Fig. 1). The knee jerk elicited every 5 sec is clearly depressed throughout the period of vibration of the patellar tendon, whether a TVR of the same quadriceps muscle is present (Fig. 1B) or absent (Fig. 1A) because of unidentified experimental conditions. The contrast between the concomitant slow tonic response known as TVR and the inhibition of the phasic myotatic reflexes involving the same MN pool is currently designated as the vibration paradox, and this has been discussed extensively (Gillies et al., 1969; Delwaide, 1971, 1973; Hagbarth, 1973; Lance et al., 1973; Ashby et al., 1974, 1980; Dindar and Verrier, 1975; Burke et al., 1980d; Desmedt and Godaux, 1977c, 1978c).

Another problem related to vibration has been raised by the unexpected finding that although the phasic reflexes are thus inhibited both in lower limb and in upper limb (Deschuytere et al., 1976), the masseteric myotatic reflex is instead clearly potentiated by chin vibration (Godaux and Desmedt, 1975a). Finally, there is also an in-

---

*Brain Research Unit, University of Brussels, 115, Boulevard de Waterloo, B-1000 Brussels, Belgium.

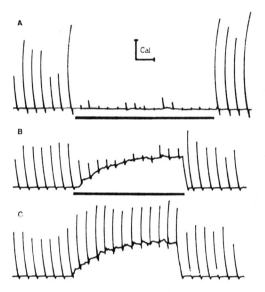

FIG. 1. The vibration paradox. Knee jerks are elicited every 5 sec and recorded with a myograph in a normal subject. A, vibration of the quadriceps muscle (*black bar*) suppresses the tendon jerk and elicits no tonic contraction (TVR) in this example. B, other example in which quadriceps vibration inhibits the knee jerks and elicits a TVR. C, voluntary contraction of quadriceps without vibration in same subject as B does not suppress knee jerks. Calibration: vertical 0.4 kg for A, 0.6 kg for B and C; Horizontal 10 sec. (From DeGail et al., 1966, with permission.)

triguing contrast between the slow build-up and rather long latency of the TVR and the fact that the MN firing can be rather tightly coupled to the vibration cycles in certain muscles and/or under certain conditions in man (Desmedt and Godaux, 1975; Hagbarth et al., 1976). The present chapter reviews these issues and emphasizes data in man that are more pertinent to the clinical uses of vibration in patients. Studies in experimental animals, and especially in the decerebrate cat (in which a powerful reticulospinal facilitation potentiates the vibration responses), helped identify certain features of the vibration effects (Matthews, 1966; Homma et al., 1972; Kanda, 1972; Hultborn and Wigstrom, 1980), but they involve conditions that are distinct from those prevailing in intact man.

EFFECTS OF VIBRATION PARAMETERS

The spindle afferents discharge throughout the vibration period with abrupt onset and offset in marked contrast to the reflex response (TVR) which presents a gradual onset (Fig. 1B). The slow development of the human TVR suggests involvement of polysynaptic pathways (Hultborn and Wigstrom, 1980) in conjunction with synaptic facilitation through the repetitive spindle input. The gradual recruitment of TVR is recorded as a progressive increase in gross EMG activity over seconds and in the progressive recruitment of single motor units at various latencies after vibration onset. The motor unit in Fig. 2 only starts firing about 440 msec after vibration onset, and the latency of the first spike varies for different motor units (see below). As a rule, the EMG waves of the TVR measured a few seconds after onset of vibration are larger when either amplitude or frequency of vibration is increased. When single motor units are recorded in masseter, an increase of either frequency or amplitude of vibration recruits more units and increases the discharge rate of units already recruited (Desmedt and Godaux, 1975). Within limits, a higher vibration frequency can compensate for a decreased vibration amplitude for maintaining discharge rate of a given unit.

The parametric sets are remarkably different for the vibration-induced inhibition of monosynaptic reflexes in human soleus. The Hoffmann or H reflex was elicited by an electric pulse to posterior tibial nerve every 5 sec under standard conditions (cf. Hugon, 1973a; Hugon et al., 1973). The intensity of the stimulus of 1.0 msec duration was adjusted to elicit a near-maximal H reflex with only a small preceding direct muscle M response (Fig. 3A). During steady vibration at 80 Hz, the H reflex inhibition increased for larger amplitudes of the applied vibration (Fig. 3B–D), although it decreased for higher vibration frequencies from 65 to 200 Hz (Fig. 3E–G). Thus, the vibration-induced

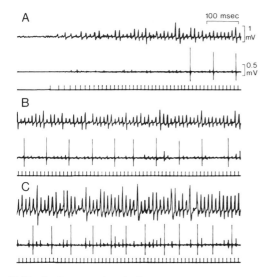

FIG. 2. Progressive buildup of masseter TVR in man. Continuous recording on camera of 3 cathode ray oscilloscope traces. (a) EMG of masseter picked up with a concentric needle electrode; (b) discharge of a single motor unit recorded with a tungsten (5 μm tip) extracellular microelectrode inserted into the masseter; and (c) timing of each vibration cycle. A, onset of vibration up to the 46th cycle. EMG waves synchronized with vibration cycles are small at first, and buildup progressively. They maintain their one-to-one relation to the vibration cycles throughout. B, 64th to the 121st vibration cycle of the same uninterrupted series. C, 574th to the 632nd vibration cycle of the series. The single motor unit discharges at intervals which are somewhat variable but remain multiples of the vibration period. Vibration frequency, 84 Hz; vibration amplitude, 2.5 mm.

inhibition of monosynaptic reflexes in soleus consistently increased with vibration amplitude (Fig. 3I), but decreased with vibration frequency.

FEATURES OF THE VIBRATION PARADOX

The so-called vibration paradox in limb muscles is a prominent feature of steady vibration of the Achilles tendon of normal man whereby two opposed effects are simultaneously elicited in soleus muscle: tonic contraction of TVR and inhibition of phasic reflexes such as the H reflex (Fig. 4) or Achilles tendon reflex (see Fig. 8A) (cf. De Gail et al., 1966; Lance et al., 1973; Ashby et al., 1980).

The vibration-induced inhibition involves presynaptic inhibition (Schmidt, 1971) of the spindle afferents (Gillies et al., 1969; Delwaide, 1973). It is accompanied by primary afferent depolarization and is abolished by picrotoxin. It is not abolished below a spinal transection. By contrast, the TVR is abolished by a spinal transection (DeGail et al., 1966) which eliminates an essential facilitatory influence from vestibulospinal and reticulospinal pathways (Gillies et al., 1971) on polysynaptic TVR pathways in spinal cord (Hultborn and Wigstrom, 1980). There is no evidence that the TVR pathways would actually include supraspinal levels, because the TVR abolished by spinal transection in cat can be temporarily restored during the posttetanic potentiation elicited by faradization of heteronymous synergist muscle spindle afferents (Kanda, 1972) or by potentiation of monoaminergic synapses that mediate the supraspinal facilitation (cf. Lundberg, 1975) through intravenous injection of L-DOPA (Goodwin et al., 1973).

VIBRATION-INDUCED INHIBITION IN PATIENTS WITH MOTOR DISORDERS

The TVR is abolished, but vibration-induced inhibition is potentiated during the spinal shock following spinal lesions in man (Fig. 4B, D) (Ashby et al., 1974). This persists for 6 months or more after the lesion even though the tendon jerks may become normal or exaggerated in a few weeks (Diamantopoulos and Zander Olsen, 1967; Ashby and Verrier, 1975). One year or more after a complete spinal transection, the inhibition is no longer enhanced (Fig. 4C), and it may indeed become less than normal (Burke and Ashby, 1972) although this is variable (Ashby and Verrier, 1975). Patients with incomplete spinal lesions show similar but less profound alterations (Ashby et al., 1980). After a cerebral lesion producing

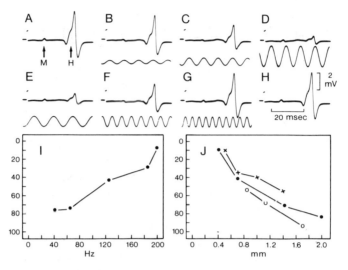

FIG. 3. Parameters of vibration-induced inhibition of H-reflex of soleus in man. A to H, belly-tendon electrogram of soleus. The lower trace presents an analog signal of the steady vibration when this is applied to the Achilles tendon. The single electric pulse to tibial nerve in popliteal fossa (see artifact 3 ms after start of the oscilloscope sweeps) has the same intensity throughout and elicits near-maximum H-reflexes with only a small preceding direct muscle M response (*arrows* in A). A, H control H-reflexes without vibration. B, C, D steady vibration at 80 Hz with an amplitude of 0.4 (B), 0.7 (C), and 2.0 mm (D), respectively. E, F, G steady vibration of 1.0 mm and frequencies of 65 (E), 125 (F), and 200 Hz (G), respectively. I, J diagrams showing vibration-induced inhibition of H-reflex in percent of control (*ordinate*) as a function of vibration frequency at constant 1 mm amplitude (*abscissa*, in I), or of vibration amplitude (*abscissa*, in J) at 3 chosen vibration frequencies. These frequencies in J are 140 Hz (*crosses*), 80 Hz (*dots*) or 40 Hz (*circles*). (From Desmedt and Godaux, 1978c, with permission.)

flaccid hemiplegia, vibration-induced inhibition is increased in the paralyzed limbs. However, after several months, as the pyramidal spasticity subsequently develops, the vibration-induced inhibition of the phasic proprioceptive reflexes decreases markedly (Delwaide, 1969, 1973; Ashby and Verrier, 1976; Ashby et al., 1980), whereas the TVR is potentiated and presents a more abrupt onset in spastic muscles (Hagbarth, 1973). Such dissociations between TVR and vibration inhibition in the course of hemiplegia suggest that descending supraspinal pathways exert distinct influences on the respective spinal circuits involved. No consistent alterations in vibratory inhibition are observed following cerebellar lesions giving rise to ataxia in man (Ashby et al., 1980).

MECHANISMS INVOLVED IN THE VIBRATION PARADOX

For the vibration paradox to be resolved, the distribution of the two concurrent effects among MNs of the soleus pool must be directly analyzed (Desmedt and Godaux, 1978c). A first point is that the vibration paradox cannot be dismissed as resulting from trivial occlusion of the phasic reflex afferent volley by vibration-induced afferent discharge ("busy line" effect considered by Hagbarth, 1973). Some degree of occlusion may well be a contributing factor in tendon reflexes involving spindle activation by a tendon tap, but this is rather unlikely for the monosynaptic H reflex that is directly elicited through electrical stimulation of Ia af-

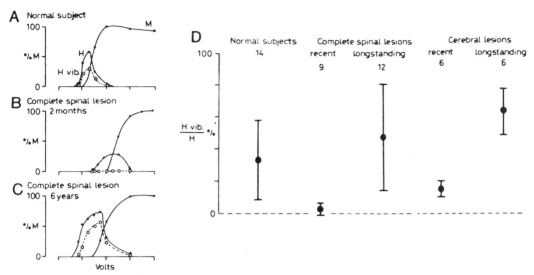

FIG. 4. Alterations of vibration-induced inhibition of the soleus muscle H-reflex in patients with motor disorders. A, recruitment curve of H-reflex without (*solid line*) or with vibration (*broken line*) and of direct M response in normal adult. B, recruitment curve in a patients 2 months after complete spinal transection (increased vibration-induced inhibition). C, same six years after a complete spinal transection. D, means and standard deviations of percentage inhibition of H-reflex by Achilles tendon vibration (maximum H-reflex during vibration divided by maximum H-reflex without vibration) in 14 normals, in 9 recent and 12 longstanding (over a year) complete spinal cord lesions, and in 6 recent and 6 longstanding pyramidal lesion producing hemiplegia. (From Ashley et al., 1980, with permission.)

ferents (Hugon, 1973a). The afferent volley of the H reflex indeed does not appear to be reduced by tendon vibration (Lance et al., 1973). Moreover, the vibration-induced inhibition is not fully developed until about 100 msec after vibration onset and sometimes outlasts the vibration by several seconds, in marked contrast to the spindle afferent discharge which has an abrupt onset at the start of vibration and is confined to the vibration period (Burke et al., 1980c). The fact that picrotoxin, a drug that antagonizes GABA-mediated presynaptic inhibition (Eccles, 1969; Schmidt, 1971), eliminates vibration-induced inhibition in cats is also incompatible with a "busy line" effect (Gillies et al., 1969). In any case, the delayed onset of the inhibition and its persistence after vibration offset fit in with the slow kinetics of presynaptic inhibition (Eccles et al., 1962).

Experiments were carried out in man to estimate the recruitment threshold and the susceptibility to vibration-induced inhibition of single soleus motor units. H reflexes were elicited by 1.0 msec electric stimuli to the posterior tibial nerve at intervals of 5 sec (Hugon, 1973a; Hugon et al., 1973). Figure 5A illustrates a standard recruitment curve for the H reflex (latency 30–35 msec) as well as the direct M response (latency 5–10 msec) corresponding to orthodromic excitation of soleus motor axons by the tibial nerve shock. The progressive decrease of the H reflex with stronger stimuli results from occlusion by the antidromic motor axon volley (cf. Hoffmann, 1922; Magladery et al., 1950, 1951; Hugon, 1973a). Highly selective tungsten microelectrodes with a 5-μm tip provided stable reproducible records of single soleus motor units, even when the muscle produced strong H-

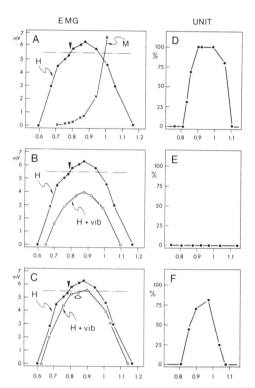

FIG. 5. Effect of Achilles tendon vibrations on H-reflex recruitment curve and on monosynaptic reflex threshold of a single soleus motor unit. **A,** recruitment of H-reflex (*dots*) and of direct muscle M response (*crosses*) when increasing intensity of electric pulse to tibial nerve. The peak amplitude of belly-tendon soleus electrogram (mV, *ordinate*) is plotted against shock intensity (*abscissa:* units relative tibial nerve shock that elicits a half-maximum direct M response). The horizontal interrupted line corresponds to the monosynaptic reflex threshold of a single motor unit simultaneously recorded in soleus (50% probability of firing). The diagram in D presents for each intensity of tibial nerve shock the percentage of trials (10 trials at each intensity) for which that unit fired one spike. **B, C** recruitment curve of soleus H-reflex before (*dots*) and during (*circles*) vibration of Achilles tendon at 80 Hz, with an amplitude of 0.8 (B) or 0.4 mm (C). In B, H-reflex fails to reach the threshold of the single motor unit which fails to fire (**E**). In C, H-reflex less inhibited by smaller vibration; the single motor unit is activated but discloses a reduced probability of firing (**F**). (From Desmedt and Godaux, 1978c, with permission.)

reflex contractions. The H-reflex threshold of each single soleus motor unit was estimated, on the basis of 10 trials, relative to the peak voltage of the corresponding belly tendon H reflex (arrowhead in Fig. 5A). When stimuli of increasing strength were used, any given soleus motor unit first fired intermittently as its threshold was reached and then consistently discharged one spike at higher intensities, eliciting larger H reflexes. To provide a consistent scaling for data from different experiments, the percent probability of firing of each unit (with or without tendon vibration) was plotted relative to the shock intensity eliciting a half-maximum direct M response (Fig. 5D), the latter being taken as a consistent feature of the soleus recruitment curve (Desmedt and Godaux, 1978c).

Single soleus motor units with reflex thresholds corresponding to a wide range of H-reflex amplitudes (0.2 to 4.7 mV in belly tendon recordings) were tested under five different vibration conditions. With an appropriate vibration, the H reflex is considerably depressed and fails to reach the size at which the simultaneously recorded unit had been recruited in the absence of vibration (Fig. 5B); the unit actually fails to discharge in the presence of such a vibration (Fig. 5E). With a smaller vibration that only slightly reduces the H reflex (Fig. 5C), the same unit can discharge, but its probability of firing is smaller at all shock intensities (Figs. 5, 7). Throughout, the direct M response was not affected by vibration. Such data make it possible to relate the units' monosynaptic threshold to the amplitude of the simultaneously recorded belly tendon H reflex. In the presence of vibration, the tibial nerve shock must be increased to augment the H reflex up to its previous size in order to fire the motor unit being examined. This means that each soleus motor unit is recruited by the monosynaptic volley when about the same fraction of the soleus MN pool is made to fire, irrespective of the presence or absence of vibration-induced inhibition.

In agreement with this finding, the low-threshold motor units recruited for small H reflexes indeed require stronger vibration in order to be silenced. Figure 6 illustrates two soleus motor units with different reflex thresholds that are recorded simultaneously along with the belly tendon H reflex. When tendon vibrations of increasing amplitude are applied, the higher-threshold unit (2) is silenced before the lower-threshold unit (1) (Fig. 6 F–H). Thus, the units are de-recruited in reverse order. The last-recruited soleus MNs are the most susceptible to being silenced by steady vibration. Thus, each soleus unit ceases to respond to the monosynaptic excitation when the H reflex is reduced by vibration to roughly the size at which the same unit had first been recruited in the absence of vibration.

In other words, the vibration inhibition works in reverse on the same rank order of the soleus MNs. This makes sense of the vibration paradox by showing that for any slight change of a given parametric set, either the next lowest threshold (still silent) motor unit will be recruited or the next highest threshold (already firing) motor unit will be silenced. On the other hand, the motor units are recruited in TVR in the same order as in graded phasic reflexes or in voluntary contractions (Desmedt and Godaux, 1980). Thus, the vibration-induced steady spindle afferent input recruits motor units in the usual order but concomitantly elicits a certain level of presynaptic inhibition onto the Ia afferents whereby the further recruitment of MNs through the polysynaptic pathways of TVR is more and more restricted or limited as the phasic monosynaptic reflexes are reduced below their control value in the absence of vibration (Fig. 7B,C) (Desmedt and Godaux, 1977c, 1978c, 1980).

The recruitment order of motor units in phasic proprioceptive reflexes is not modified by vibration inhibition, and any given soleus unit starts firing when the H reflex reaches an adequate amplitude regardless of whether vibration is present or absent. This implies that the vibration-induced presynaptic inhibition involves spindle afferents fairly homogeneously. In the presence of vibration inhibition, only low-threshold MNs can thus be recruited either in TVR (which is indeed correspondingly rather small in soleus) or in monosynaptic reflexes. The so-called vibration paradox observed for the global muscle responses no longer appears "paradoxical" when considered in terms of the effects impinging on single MNs of the soleus pool (Fig. 7C).

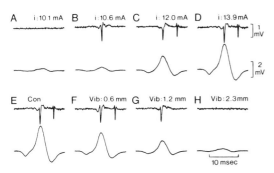

FIG. 6. Recruitment and de-recruitment order of 2 single motor units of soleus in man (**upper panel**). **Lower panel**: simultaneous belly-tendon recording of H-reflex. A–D, recruitment of motor units and H-reflex when the tibial nerve shock is increased from 10.1 to 13.9 mA. E same conditions as in D. F–H inhibition of H-reflex to the 13.9 mA stimulus by steady vibration at 80 Hz applied to Achilles tendon. The vibration amplitude is 0.6 (F), 1.2 (G), and 2.3 mm (H), respectively. Unit number 2 is silent in G when unit number 1 still fires a spike. Unit number 1 is in turn silent in H. Thus, the 2 motor units are de-recruited in reverse order for increasing tendon vibration. (From Desmedt and Godaux, 1978c, with permission.)

VIBRATION-INDUCED POTENTIATION OF THE MASSETER REFLEX

Although phasic proprioceptive reflexes are inhibited by tendon vibration in the limb muscles (Fig. 8A), they are potentiated in

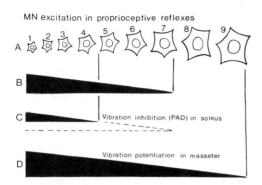

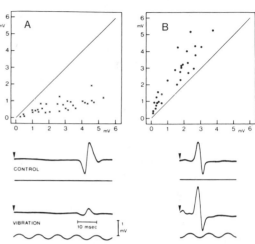

FIG. 7. Cartoon sketching the steady vibration ability to recruit MN in a pool. These MNs are recruited in a fixed order from left to right in proprioceptive phasic or tonic reflexes, and also in voluntary contractions (Henneman's size principle). The actual amount of resulting synaptic activation (indicated by vertical width of the black triangles in **B** to **D**) is larger for MNs first recruited, and it decreases from left to right. **B**, synaptic activation by a standard proprioceptive volley recruits MNs 1 to 7. **C**, in the presence of steady vibration of soleus, presynaptic inhibition of spindle afferents reduces the amount of synaptic activation proportionally and the standard volley only recruits MNs 1 to 4. **D**, by contrast in masseter, steady vibration elicits no presynaptic inhibition of spindle afferents and therefore potentiates the synaptic activation proportionally whereby the standard volley now can recruit MNs 1 to 9.

FIG. 8. Contrasting effects of steady vibration on different tendon reflexes in normal man. Pooled results on 12 adult subjects. The diagrams present the peak voltage of the belly-tendon reflex response of soleus (**A**) or masseter (**B**) elicited by percussion of the Achilles tendon or the mandible, respectively. *Abscissa:* control. *Ordinate:* same reflex during steady vibration at 100 Hz. Each symbol represents the mean voltage of 10 consecutive reflexes. **Bottom panel,** oscillogram of reflexes without **(upper)** or with **(lower)** vibration. (From Godaux and Desmedt, 1975b, with permission.)

jaw-closing muscles (Fig. 8B). The mean increase in masseter reflex by vibration of chin is about 30% in normal man, and it is recorded for a wide range of masseter reflex amplitudes from 0.1 to 4 mV. The potentiation appears within the first few seconds after vibration onset. It persists for about 5 sec when vibration is turned off. The effect is also recorded for the masseter H reflex elicited through electrical stimulation of the masseter nerve (Godaux and Desmedt, 1975b).

POSSIBLE MECHANISMS UNDERLYING VIBRATION-INDUCED POTENTIATION OR INHIBITION

The vibration can activate Golgi tendon organ afferents (Burke et al., 1980c) which are known to elicit disynaptic inhibition in the homonymous MN pool. However, there seems to be no reason why Golgi afferents would not be activated by vibration in jaw closer muscles.

There is experimental evidence that vibration inhibition of the soleus H reflex depends primarily on activation of homonymous spindle afferents. Reflex inhibition is also recorded during experimental stretches of the Achilles tendon, whereby the steady mechanical stimulation is restricted to the triceps spindle afferents, and this effect occludes with that elicited by Achilles tendon vibration (Delwaide, 1971, 1973).

The question of other muscles of the limb being involved by soleus tendon vibration must nevertheless be considered. The H reflex of soleus can be depressed when vibration is applied anywhere in the lower limb (Rushworth and Young, 1966), but the maximum effect is recorded for vibration of Achilles tendon (Delwaide, 1973). On the other hand, the H-reflex inhibition by Achilles tendon vibration persists when antagonistic flexor muscles have been paralyzed as a result of local anesthesia or of traumatic lesion of peroneal nerve (Dindar and Verrier, 1975; Ashby et al., 1980). This problem has been further examined in two macaque monkeys in whom all nerves of the right lower limb (crural, obturator, peroneus, short biceps, semitendinosus, semimembranosus) had been surgically transected except for the nerve branches innervating the triceps surae. Under these conditions, the Achilles tendon reflex was still markedly inhibited by tendon vibration, even on the second day after surgery (Desmedt and Godaux, 1980). The data emphasize that homonymous spindle afferents represent the main route mediating vibration-induced inhibitory effects on soleus phasic reflexes (Gillies et al., 1969; Delwaide, 1973). Thus, vibration-induced inhibition in soleus primarily involves autogenic presynaptic inhibition of spindle afferents.

A strikingly different situation prevails for jaw closer muscles, since chin vibration not only fails to inhibit but actually potentiates masseter reflexes, and it has been proposed that there is actually no presynaptic inhibition in the proprioceptive reflex circuits of the jaw elevators (Desmedt and Godaux, 1978c, 1980). It would indeed appear that if these circuits did include any mechanism (interneurons and axoaxonic synapses) for homonymous presynaptic inhibition of spindle afferents, this should show up as a vibration-induced depression of the phasic masseter reflexes. Experimental data on the cat suggest that spindle afferents from the masseter in cat do not exert any presynaptic inhibitory effect on spindle afferents involved in reflex jaw closing, whereas they elicit PAD in lingual afferent fibers and presynaptically inhibit the jaw-opening reflex evoked by stimulation of the lingual nerve (Nakamura and Wu, 1970; Nakamura et al., 1973a; Nakamura, 1980). If so, the chin vibration will not gate or reduce the phasic afferent volley of masseter H or tendon reflexes but will facilitate these reflexes by adding homonymous spindle afferent excitation to the phasic volley (Fig. 8B). In the presence of vibration, a given proprioceptive input will thus recruit a larger fraction of the masseter or temporal MN pool (Fig. 7D) but a smaller fraction of the soleus or limb muscle MN pool (Fig. 7C). As a matter of fact, the TVR itself is definitely stronger in jaw elevator muscles than in limb muscles, as indeed expected if the barrage of spindle afferent impulses elicited by steady vibration undergoes no presynaptic inhibition in jaw elevators and can thus create unopposedly a stronger excitation in the MN pool.

A side issue in this discussion is whether chin vibration effects in masseter may not suffer any interference from possible reciprocal effects emanating from jaw opener muscles that are rather closely located. Although the jaw closer muscles are richly endowed with muscle spindles, anatomical studies have emphasized the absence of muscle spindles in jaw opener muscles (Szentagothai, 1949). As could be expected from the lack of spindles, strong vibration applied below the chin does not elicit any TVR in the jaw-opening digastric, even when it is put under slight tension by having the jaw closed and head extended (Desmedt and Godaux, 1980). It is indeed unlikely that chin vibration could elicit any reciprocal postsynaptic inhibition of jaw-opener origin onto the masseter and temporal MNs. In cat experiments, the supramaximal stimulation of the digastric nerve does not evoke any inhibitory postsynaptic potential in masseter MNs (Kidokoro et al., 1968a; Nakamura, 1980). As a matter of fact, the jaw openers appear to have rather little significance for the reflex displacements of the jaw in man. For example, the exteroceptive reflex opening of the mouth that is elicited by stimulation of lip mucosa or gums involves essentially an inhibition of jaw closer muscles, and the concomitant reflex contraction of digastric muscles is small and highly susceptible to habituation (Desmedt and Godaux, 1976).

MOTOR UNIT RECRUITMENT AND DISCHARGE PATTERNS IN JAW CLOSER AND LIMB MUSCLES

The responses to muscle vibration provide a useful technique for titrating tonic stretch reflexes in man or patients with motor disorders (cf. Lance et al., 1973; Hagbarth, 1973). In discussions of these problems, it must be emphasized that the features of the human TVR differ from those in decerebrate cat, where strong reticulospinal facilitation results in abnormally enhanced TVR responses of abrupt onset (Matthews, 1966; Homma et al., 1972; Kanda, 1972). Such studies of tonic reflexes to vibration in cat are no doubt valuable; however, the human TVR must be investigated in its own right to elucidate current issues.

In normal man, the TVR elicited by vibrators placed on the skin overlying muscle or tendon presents a progressive force buildup and motor unit recruitment over several seconds (Figs. 2, 9A). The force of TVR is a modest fraction of the maximum tetanic or voluntary force of the muscle. These features of the human TVR, namely its slow onset and decline and its marked susceptibility to volition or postural influences and to anesthetic drugs, suggest that polysynaptic spinal pathways activated by the steady spindle afferent input of vibration are the main mechanism of the human TVR (DeGail et al., 1966; Lance et al., 1973; Homma et al., 1972; Desmedt and Godaux, 1975, 1980; Hagbarth et al., 1976; Burke and Schiller, 1976). In other words, progressive facilitation and recruitment in polysynaptic proprioceptive pathways (Hultborn and Wigstrom, 1980) gradually increase the depolarizing pressure in MN pool, eventually leading to the growing tonic contraction of TVR.

This statement does not imply that the monosynaptic excitation by Ia spindle af-

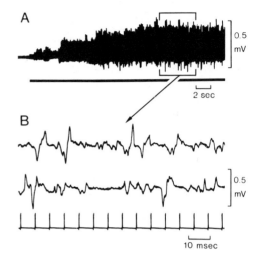

FIG. 9. TVR in human brachial biceps. Belly-tendon electrograms during vibration at 133 Hz with 1.8 mm amplitude. A, oscillogram at slow speed to show progressive buildup of EMG response; the onset of vibration is indicated by thick line under the trace. B, two samples of EMG response recorded with a concentric needle electrode at faster speed from same muscle when TVR is fully established. The vibration cycles are indicated by the lower trace. The EMG discharge is asynchronous and not closely related to vibration cycles.

ferents would play no part in TVR but suggests that monosynaptic Ia EPSPs only influence TVR discharges when the polysynaptically induced slow rising depolarization has brought the membrane potential of homonymous MNs close to firing threshold (Homma et al., 1972; Desmedt and Godaux, 1975). The analysis of motor unit firing in TVR points to a difference between limb and jaw closer muscles in the patterning of the MN discharges (Desmedt and Godaux, 1975). The EMG recorded with a concentric electrode is typically asynchronous in limb muscles such as the brachial biceps (Fig. 9B) or soleus, the motor units discharging irregularly in the range of 6 to 12/sec (Lance et al., 1973). By contrast, in masseter or temporalis, the EMG presents a pattern of synchronized waves at vibration frequency, even during the initial TVR buildup (Figs. 2, 10A). When analyzed on a fast oscilloscope sweep synchronized to vibration, the

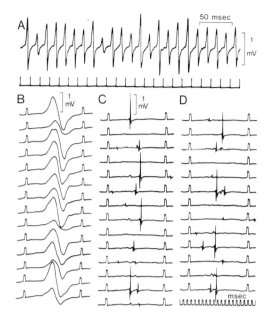

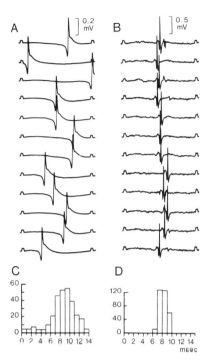

FIG. 10. TVR in human masseter. A, B, EMG recorded with concentric needle electrode during steady vibration of lower jaw. A, continuous record of synchronized waves with a one-to-one relation to vibration cycles (indicated by pulses of lower trace). B, each vibration cycle is displayed on transverse sweeps in a raster; the timing pulses correspond to an arbitrary, but constant, point in two adjacent vibration cycles (photoelectric device mounted on shaft of vibration-generating motor). The EMG waves exhibit constant latency but their amplitude varies at random in successive cycles. C, D, similar raster display of single motor units simultaneously recorded in masseter: they fire independently of one another at variable intervals which are multiples of the vibration period.

FIG. 11. Single motor unit discharges during TVR in soleus (A, C) and masseter (B, D). Steady vibration of Achilles tendon (A, C) or lower jaw (B, D) at 71 Hz and 1.8 mm amplitude. The timing pulses in each oscilloscope sweep correspond to a constant point in two adjacent vibration cycles. Only the sweeps with actual firing of the single motor unit are displayed in this figure. The time range of firing (jitter) extends to the full vibration period for soleus, but is restricted to a small fraction thereof for masseter. C and D, histograms of the firing jitter for the two motor units.

EMG waves disclose a one-to-one relationship to vibration cycles in spite of marked irregular fluctuations in amplitude (Fig. 10B). The raster display has been produced from the FM tape by using a digital counter set to trigger for the next sweep chosen for photography on film.

The EMG waves recorded by the concentric electrode are made up of many motor unit potentials. Simultaneous recording from the masseter with a microelectrode discloses single mV spikes (Fig. 10C,D). Another masseter motor unit is illustrated in Fig. 11B in a similar display in which only the sweeps with actual unit firing are shown. The phase locking of motor units in masseter or temporal TVR is expressed by the rather small jitter (or time range over which a given unit fires) of roughly 2 msec; the standard deviation about the modal latency is about 0.4 msec (Table 1). The jitter tends to increase for vibration rates lower than about 60 Hz, presumably as a result of the decreasing rate of stretching of spindles for slower vibrations (Desmedt and Godaux, 1975, p. 436).

In human limb muscles, single motor units discharge throughout the vibration cycle (Fig. 11A), and the jitter is virtually

TABLE 1. *Latency and jitter features of proprioceptive responses in soleus and masseter muscles*

	Onset latency (msec)	SD of latency (msec)	Jitter (msec)
Achilles tendon reflex	36	0.5	2
H reflex in soleus	32	0.2	1
TVR in soleus (80 Hz vibration)	—	1.8–2.4	8–15
Masseter reflex (jaw jerk)	7.5	0.43	2
H reflex in masseter	5.4	0.22	1.2
TVR in masseter (80 Hz vibration)		0.4	1.5–3.8

equal to the vibration period (Table 1). In soleus, the standard deviation about the modal latency is 1.8 to 2.4 msec, much larger than for phasic soleus proprioceptive reflexes or for masseter TVR (Table 1). The unique features of masseter or temporal TVR have been confirmed by Hagbarth et al. (1976) who ascribed the synchronization to shortness of reflex arcs, which would allow for less dispersed Ia volleys. However, proximity to spinal cord provides no explanation because vibration of dorsal intercostal muscles at a distance of only 10 to 12 cm from the cord elicits a TVR with quite asynchronous EMG (Desmedt and Godaux, 1980), as is found in limb muscles.

Another hypothesis could be that the vibration of the lower jaw would insure a better mechanical coupling to the spindles of jaw closer muscles than can be achieved in limb or body. Any direct comparison of actual mechanical efficiency of vibrations in soleus or masseter would be difficult. However, the latency of firing of single motor units after vibration onset provides a useful estimate; for example, a given unit can be recruited after a fraction of a second for a strong vibration, but it only starts firing after many seconds for smaller vibrations. In the masseter, the jitter of units is always below 4 msec, even when the onset latency is made to exceed 10 sec by using a small vibration amplitude of 0.1 to 0.25 mm (Fig. 12). In soleus, the jitter varies from 8 to 14 msec in different units and does not seem to exhibit systematic variations for vibration amplitudes that change onset latency over a wide range (Fig. 12). Thus, the different mechanical efficiency of vibrations, as reflected by time taken to reach MN threshold, fails to blur or remove the marked difference in jitter between the masseter and soleus TVRs. Therefore, differences in the central mechanisms involved in the TVR must be responsible for the obviously bet-

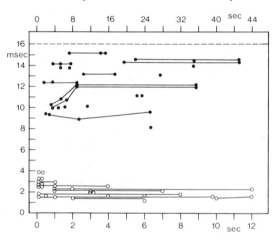

FIG. 12. Relation between jitter of single motor units during steady vibration and latency of firing after vibration onset in human masseter and soleus. All vibrations are at 62 Hz, but their amplitudes changed in different tests. *Ordinate:* jitter of motor units in ms. The horizontal interrupted line at 16 ms represents the maximum possible value of jitter which corresponds to a full vibration period. *Abscissa:* latency of firing of each unit after vibration onset in seconds. Abscissa scale is different for soleus (*dots*) and masseter (*circles*). Data for the same motor unit have been joined by a line and correspond to different tests in which the vibration amplitude had been changed whereby the latency was modified. The latency jitter of masseter motor units range from 1.5 to 3.8 ms, while the jitter for soleus motor units extends from 8 to 15 ms. These values are not much affected by onset latencies.

ter efficiency of monosynaptic Ia EPSPs in actually locking the times of firing of jaw closer MNs to the vibration cycle. The following items deserve consideration.

The lack of presynaptic inhibition onto spindle afferents from jaw closer muscles, as revealed by the vibration-induced potentiation of their phasic proprioceptive reflexes (Fig. 8), must facilitate homonymous spindle excitation of MNs. Even with small vibration amplitude and rather modest spindle input, the size of the Ia monosynaptic EPSPs is thus better preserved throughout and beyond the phase of slow build-up of polysynaptic Ia depolarization of jaw closer MNs. By contrast, in limb muscles, the strong homonymous presynaptic inhibition of spindle afferents rapidly reduces the size of Ia monosynaptic EPSPs after onset of vibration (Figs. 5,6,7C,8A). As a result, the depressed monosynaptic Ia EPSPs elicited by each vibration cycle (cf. Kanda, 1972) are unable to impose any tight locking of the MN discharge in limb TVR (Fig. 11A,C).

The spindle afferents from jaw closer muscles are located in the trigeminal mesencephalic nucleus (Szentagothai, 1949; Kawamura et al., 1958; Jerge, 1963; Smith et al., 1969; Taylor, 1976). Cell clustering with nexus formation has been described in cat, as has physiological evidence for electrotonic coupling with mutual excitatory effects (Baker and Llinas, 1971). The electrotonic coupling only influences the timing of homonymous proprioceptive excitation by a fraction of a millisecond during vibration. This effect could be significant for the monosynaptic Ia EPSPs in trigeminal MNs but not for the slowly rising polysynaptic Ia depolarization.

In conclusion, the gradual tonic buildup of TVR depends on progressive facilitation in polysynaptic pathways of spindle afferents (Hultborn and Wigstrom, 1980) whereby a slowly rising depolarization is elicited in homonymous MNs. In jaw closer muscles, the vibration oscillations have, in addition, a synchronizing effect through superposition of sizable monosynaptic Ia EPSPs on the slow MN depolarization of polysynaptic origin. As a result, the jaw closer TVR has a unique EMG pattern of synchronized waves (synchronized motor units) with a one-to-one relationship to the vibration cycles. The single motor units fire independently of one another (at intervals that are different multiples of the vibration period), but they are each tightly phase-locked to the vibration cycle and occur in a time slot of about 2 msec (Table 1) (Desmedt and Godaux, 1975). By contrast, in limb muscle TVR, the EMG is desynchronized (Figs. 10,12A,C) with a smaller relative contribution of monosynaptic Ia EPSPs.

The so-called vibration paradox is a clinically relevant feature of limb muscles (Figs. 1,4,6) but not in jaw closer muscles where phasic reflexes are potentiated rather than depressed by vibration (Fig. 8). To understand this vibration paradox in limb muscles, it must be realized that MNs that fail to discharge in TVR or in phasic reflexes are of larger size and higher threshold than the MNs contributing to the recorded TVR. The two concomitant phenomena making up the vibration paradox in limb muscles, namely, vibration-induced inhibition of phasic reflexes and vibration-induced excitation of homonymous MNs that results in TVR, are influenced differentially in patients with spinal cord lesion or with pyramidal hemiplegia and change markedly in the course of recovery after such lesions.

Motor Dyscontrol as a Hazard in Massive Body Vibration in Man

Gabriel M. Gauthier, Jean Pierre Roll, *Maurice Hugon, and Bernard Martin

Laboratoire de Psychophysiologie, Université de Provence, Centre Saint Jérome, Marseilles, France

Modern vehicles generate vibrations which are transmitted to pilots and passengers through solid contacts such as the seat (arm rest, back, and bottom cushions), command levers, and floor. The helicopter is an example in which these factors combine to cause vibrations of higher magnitude than are present in other forms of transportation. The generated frequency spectrum covers a 1 to 30 Hz bandwidth with local and intermittent high-power peaks. Though local resonances may persist, a generalized high-frequency filtering occurs through skeletal and muscular masses (Hornick, 1973). High-intensity vibrations induce mechanical and physiological effects that alter human performance at postural and movement control levels (Berthoz, 1971; Hornick, 1973; Malcolm and Melvill Jones, 1973; Jex and Magdaleno, 1978). Most of these alterations are significant during vibration but may persist minutes afterwards. Vibration signals enter the neurological network through sensory receptors of cutaneous, muscular, articular (Hagbarth and Eklund, 1966; Talbot et al., 1968; Millar, 1973; Burke et al., 1976a,b; Roll et al., 1980a), visceral (Tyler and Bard, 1949), and vestibular (Young and Jacob, 1968; Young and Oman, 1969) origins. The major effects may be classified as sensory, perceptual, and motor. The net result is inevitably a decrease of sensory–motor performance.

We studied the effects of whole body vibration (WBV) on (1) spinal reflexes; (2) tracking performance in tasks requiring position, velocity, and force controls; (3) postural equilibrium and postural adjustments.

SENSORY EFFECTS

The numerous mechanoreceptors in skin, tendons, joints, and muscles are all activated by local vibration. Superficial (touch) and deep (pressure) cutaneous receptors respond to high-frequency vibration up to 250 Hz (Talbot et al., 1968). The density of receptors in some skin areas, their low threshold, and their high firing frequency (Johansson and Vallbo, 1979; Johansson et al., 1980) suggest that vibration applied to the skin per se may result in massive afferent volleys which are likely to alter exteroceptive perception.

Microneuronography in man allows detection of these effects and correlation with the perception experienced (Hulliger et al.,

---

*To whom correspondence should be addressed: Laboratoire de Psychophysiologie, Université de Provence, Centre Saint Jérome, rue Henri Poincaré, Marseilles 13397 Cedex 13 France.

1979; Jarviletho, 1977; Johansson and Vallbo, 1980). The tendon receptors (Golgi tendon organs) are also sensitive to vibration but do not follow the stimulus beyond 120 Hz and generally fire at subharmonic frequencies (Burke et al., 1976a,b). The extreme sensitivity of muscle primary endings to very small (20 to 25 μm) and rapid muscle stretch has been demonstrated in the cat (Echlin and Fessard, 1938; Bessou and Laporte, 1962; Bianconi and Vandermeulen, 1963; Brown et al., 1967; Matthews and Stein, 1969). The activation of secondary endings requires greater displacement amplitudes (Grusser and Thiele, 1968; Bessou and Laporte, 1962). Vibration applied to a tendon in the absence of muscle contraction can be presumed to be a specific stimulus for primary endings in man (Burke et al., 1976a,b). Primary endings follow vibration frequencies up to 220 Hz. Direct recordings from nerve fibers have also shown that primary ending discharge frequency rarely exceeds 50 Hz during normal movements (Vallbo, 1981; Hagbarth, 1981; Roll and Vedel, 1981). Consequently, the vibration-induced afferent impulses in the low-frequency range (10 to 50 Hz) may constitute a major invasion of the sensory channel and mask out information related to the ongoing motor activity. Limited data are available on activation of articular receptors by vibration (Millar, 1973). The vestibular organ receptors may be excited by low-frequency vibration. In fact, the perceptual threshold levels have been shown to be $0.0035°/sec^2$ for angular acceleration (Clark, 1964) and $3 cm/sec^2$ for both vertical and horizontal accelerations (Walsh, 1962, 1964). For example, a 10-Hz vertical vibration as low as 10 μm in amplitude is perceived and can affect performance.

PERCEPTUAL EFFECTS

Muscle vibration may cause perceptual phenomena in man (Goodwin et al., 1972; Eklund, 1972; Craske, 1977; Lackner and Levine, 1979; Roll and Vedel, 1981): alterations of sense of position and occurrence of kinesthetic illusions resulting from vibrations of muscle tendons. Goodwin et al. (1972) described an illusory motion of the forearm of a blindfolded seated subject when a 100-Hz vibration was applied to biceps or triceps of the immobilized elbow. When two identical vibrations were applied to the Achilles tendons of a blindfolded man immobilized in a standing position, a sensation of forward body movement was experienced (Bonnet et al., 1977; Lackner and Levine, 1979).

MOTOR EFFECTS

Tonic Vibration Reflex

In man, vibration evokes a tonic contraction of the vibrated muscle and a relaxation of its antagonists (Homma et al., 1971, 1972; Roll et al., 1972, 1973; Desmedt and Godaux, 1975, 1978c). These characteristics depend on central factors (Hagbarth and Eklund, 1968; Delwaide, 1971, 1973; Gillies and Burke, 1971). Thus, the tonic vibration reflex (TVR) results from activation of monosynaptic and polysynaptic pathways under the influence of the spindle primary afferents from the vibrated muscle (J. E. Desmedt, *this volume*).

The TVR is a self-limiting phenomenon because the contraction of the extrafusal muscle reduces the efficacy of the vibration in the absence of consistent fusimotor drive. Furthermore, the Golgi tendon organs can induce autogenic inhibition; finally, the input evoked by vibration might limit this input through presynaptic inhibition (Gillies et al., 1969; Barnes and Pompeiano, 1970; Desmedt and Godaux, 1978c). The TVR is facilitated by light homonymous isometric voluntary activity which both facilitates the MNs and increases the dynamic sensitivity of the spindles (α–γ linkage).

Exteroceptive TVR

Local vibration applied to the volar aspect of the fingers causes a strong exteroceptive (tangoceptive) finger flexion reflex in man (Eklund et al., 1978). The flexion needs a low resting level of tonic activity to develop; it is involuntary, brisk, and outlasts the cessation of vibration. Activities in extensors are depressed by such a vibration. The authors conclude that a reflex assistance to the central command, using short-loop spinal mechanisms or a long supraspinal loop and central controls is involved. Such skin effects of vibration are relevant to manipulation of vibrating tools.

The Antagonist Vibration Response

A vibration applied to the tendon of the biceps (elbow immobilized) of a blindfolded subject may induce a motor response in the triceps muscle. This phenomenon has been named antagonist vibration response (AVR) (Roll et al., 1980). The subject experiences a sensation of extension of his forearm. When vision is allowed, and the subject is convinced of the immobility of his forearm, the AVR ceases instantly, and a TVR develops in the vibrated muscle. The switching of motor effects produced by vibration is evidence that the encoding of afferents depends largely on central settings conditioned by supraspinal information.

Tendon Vibration and Inhibition of the Monosynaptic Reflexes

Vibration depresses the monosynaptic reflexes in lower limb (Hagbarth, 1973; Lance et al., 1966, 1973) and upper limb (Deschuytere et al., 1976) but not in the masseter muscle (Desmedt and Godaux, 1975). It is generally agreed that presynaptic inhibition of Ia afferents causes the depression of the H reflex by tendon vibration. The premotoneuronal mechanism does not prevent polysynaptic excitation of the MNs. Thus, the "vibration paradox" introduced by Desmedt and Godaux (1978c), may be resolved. Still, a part of the monosynaptic reflex inhibition might be caused by a spurious activation of the antagonist muscle giving rise to a reciprocal inhibition (Ashby et al., 1980).

To study the effects of vibration on human performance, we examined the effects of vibration on (1) monosynaptic spinal reflexes; (2) sensory motor tasks—visual motor tracking, force and movement controls; and (3) posture and equilibrium.

METHODS

Vibration was applied to human subjects in two different situations. In whole-body vibration (WBV), the subject was seated in a cushioned helicopter pilot chair attached to a large platform bolted to the head of a powerful vertical hydraulic jack. The jack was driven at 18 Hz by a pump servo-controlled in amplitude. The accelerations used ranged from 0.1 to $0.5g$ in the vertical direction at platform level. Lower values were measured at knee, hip, chest, and forehead levels. An arm manipulandum was also mounted on the platform to be used in visuomanual tracking tasks and movement perception. The rotating foot rest used in foot tracking tasks could be immobilized in a given angular position to permit the application of torques in isometric conditions. Strain gauges provided a measure of the force developed. For body part vibration, in one experiment type, the feet rested on the vibrating platform while the subject was seated in a separate, nonvibrating chair (leg vibration, LV). In another situation, the trunk and the head were vibrated while the feet rested on a nonvibrating stand (head–trunk vibration, HTV). In a third situation, the head alone was submitted to vibration (head vibration, HV).

Spinal Reflexes in Lower Limbs

The Hoffmann (H) reflex was elicited in subjects exposed to various vibration conditions. Short cathodic stimuli (1 msec) to the sciatic nerve stimulated mainly the group I afferents. The Ia afferents caused a monosynaptic activation of the soleus MNs. The muscular activity was recorded by surface electrodes (Hugon, 1973a) and computer averaged over 20 stimulations. The tendon reflex (T) was elicited in soleus by mechanical taps delivered to Achilles tendon by an electromagnetic hammer (Ling Dynamic System type 201). The tonic vibration response was elicited in soleus by the hammer used to induce the T reflex. Monosynaptic reflexes and TVR were studied during whole-body vibration, leg vibration, and head–trunk vibration (Fig. 1).

In contrast to what is generally observed with local vibration, whole-body vibration did not elicit any tonic motor activity (TVR or AVR) or illusory sensations of movement. The soleus and tibialis anterior remained silent during the vibration period. In all subjects, WBV or LV induced an immediate, deep, and long-lasting inhibition of H and T reflexes. The depression of reflexes under WBV was significantly stronger than that under HTV. After cessation of the stimulus, recovery of H and T reflexes fol-

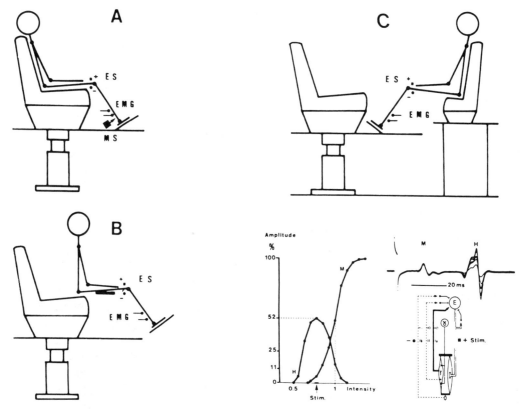

FIG. 1. Experimental conditions: **A:** Whole body vibration was applied to seated Ss by a vertical hydraulic jack. Epicutaneous electrical stimulation (ES) was delivered to the posterior tibial nerve to evoke H-reflex monitored with EMG electrodes placed on the soleus muscle. Mechanical stimulation (MS) of the Achilles tendon was also used to induce T-reflex and TVR. **B** and **C:** Similar electrical stimulation was applied during head and trunk vibration, and during leg-only vibration. The diagram represents the neuromuscular circuit activated by electrical stimulation to evoke M-response and H-reflex. The record shows typical responses (10 runs). Based on these curves, an appropriate stimulation amplitude (*arrow*) was determined to satisfy experimental requirements. (Modified from Roll et al., 1980a).

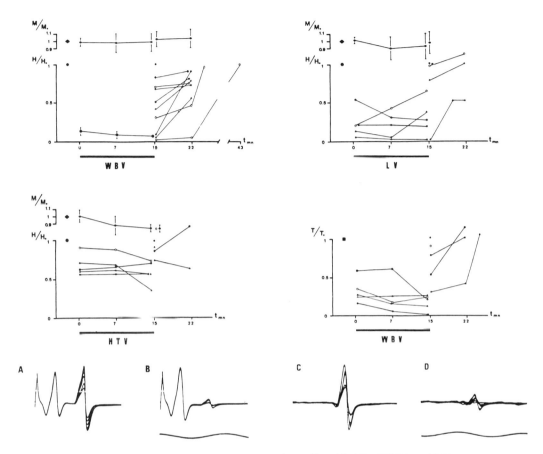

FIG. 2. Effects of vibration on H-reflex (1,2,3) and T-reflex (4). The WBV and LV graphs show a markedly depressed H-reflex. The inhibition persisted throughout the 15-min vibration period and generally continued beyond the recording period. Records (5 sweeps) are shown to illustrate the decreased H-reflex under WBV (**B**) compared to control level (**A**). The effects produced by HTV were much weaker suggesting that the vibration acts upon various leg receptors rather than on the vestibular organs. The upper curves of each graph show the evolution of the M-response as a function of time and demonstrates that vibration did not alter electrically induced muscular contractions. WBV was seen to significantly depress T-reflex as illustrated by its change as a function of time for five Ss and by superimposed individual EMG responses recorded before (**C**) and during vibration (**D**). The reflex inhibition persisted after stimulus offset. (Modified from Roll et al., 1980b).

lowed a time course that depended on the individuals (Fig. 2). The TVR induced in soleus by repetitive mechanical taps was readily inhibited during WBV or LV (Fig. 3, top). After the cessation of the vibration, the TVR remained at a low amplitude for a few seconds and then returned to control level. Figure 3 (bottom) showed the mean results obtained in five subjects.

Discussion

A moderate vibration applied to the legs of a seated resting subject deeply depresses monosynaptic reflexes and TVR. The TVR recovers after a few seconds, but the monosynaptic reflex depression lasts for minutes after stimulus cessation. The leg vibration does not by itself induce TVR or illusory movements. Comparison of HTV graph

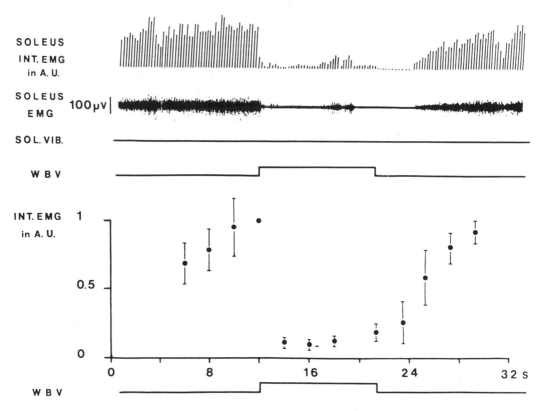

FIG. 3. Effects of vibration on tonic vibration response. WBV produced a profound decrease of TVR induced by high-frequency taps applied to the Achilles tendon. Direct and integrated EMG records from one S are shown on the top to illustrate the observed effects. The lower curves drawn on the same time scale as that of the records show the results from five Ss. In order to facilitate comparison between individual curves, the data were normalized for each S, with respect to the integrated EMG value obtained just prior to the beginning of the vibration period. A unit value was attributed to that data point. (From Roll et al., 1980b, with permission.)

with the WBV and LV graphs (Fig. 2) indicates that the vibration used is only effective through local somesthetic afferents. Published data suggest that the absence of inhibitory effects in HTV and HV situations may be related to the fact that the stimulus frequency (18 Hz) is too high to activate the vestibular system.

The neck proprioceptors activated by a similar vibration do not affect the monosynaptic reflexes of a seated resting subject. In our experiment, the observed alterations of leg monosynaptic reflexes are probably caused by the leg vibration itself. It is likely that in the leg muscles at rest, a vibration indirectly applied to the tendons results in the activation of a limited number of primary endings responding at a subharmonic of the stimulus frequency (Hagbarth et al., 1976; Burke et al., 1976a,b; Roll and Vedel, 1981). The absence of illusory motion may be attributed to the weakness of the vibration-induced afferents or to the fact that the activation affects symmetrically antagonistic muscles. An excitation of the superficial rather than deep plantar surface receptors is also probable with the vibration used in our experiments (Talbot et al., 1968). However, the relative influence of cutaneous and muscular receptors cannot be easily determined without microneurographic recordings.

Central inhibitory effects also deserve consideration. The deep prolonged inhibition of H and T monosynaptic reflexes induced by WBV and LV contrasts with the weaker inhibition observed when vibration is applied locally to the tendon of the tested muscle (Homma, 1973; Lance et al., 1966, 1973; Van Boxtel, 1979). Moreover, this inhibition persists throughout the 15-min vibration periods and outlasts the stimulus for minutes, depending on the subject. These results are different from those observed when vibration is applied locally on one muscle. This could be related to presynaptic inhibition evoked by vibration in the tested reflex arc and to a postsynaptic inhibition simultaneously elicited by the stimulus in antagonist muscles.

The peripheral afferents stimulated by vibration also affect perceptive and motor central activities. The presynaptic inhibition that reduces the reflex efficiency could affect all sensory modalities. The mechanisms involved are spinal and last only a few hundred milliseconds. However, these mechanisms are potentiated by higher mechanisms in bulbar, caudal pontine, and cerebellar structures. According to Thoden et al. (1971) and Magherini et al. (1973), the pontobulbar mechanism may be activated by cutaneous and high-threshold (group III) muscular afferents but not by low-threshold (groups I and II) afferents. Repeated stimulation of the skin can depress the H reflex in man, possibly through the spinobulbospinal system just considered.

Another possible route may be considered. Various types of somesthetic information may reach cortical levels and return to spinal levels via the lateral cortical spinal pathway and be responsible for primary depolarization (Andersen et al., 1964) and/or interneuron inhibition (Fetz, 1968). This loop, which does not involve reticular structures, might produce a long-lasting depression of monosynaptic reflexes in a subject exposed to vibration. Thus, a cortical control would be topically exerted over the sensory input from a given activated region.

The monosynaptic reflex inhibition may be seen as the product of a vibration-induced decrease of activity in afferents at presynaptic levels: H and T reflexes and TVR would be reduced by both presynaptic inhibition and by a decrease of activity polysynaptic assistance loop. From a functional point of view, the monosynaptic reflex depression may involve protective mechanisms that suppress some part of the afferents induced by vibration. The cortical control over interneuron activity may be discriminative, since Lundberg et al. (1977) suggested that low-threshold cutaneous and group II muscular afferents might converge with Ib afferents onto inhibitory interneurons, thereby inducing postsynaptic inhibition at the MN level. This effect may be facilitated by rubro- and corticospinal pathways so that a supraspinal control may be exerted over the cutaneous polysynaptic excitation of the MNs. This would constitute another way of "protecting" the local MNs against vibratory stimulus "aggression."

Sensory–Motor Tasks

The subjects were vibrated in a seated position at a frequency of 18 Hz and an acceleration of 0.1 g at platform level. The effects of whole-body vibration were derived from the comparison of the performance of the subjects obtained prior to, during, and after short periods of vibration. Whole-body vibration applied to a seated subject altered sensory–motor performance. Subjects were not aware of the alteration unless visual feedback of performance was provided. They described difficulties in continuing or resuming a given task, especially at stimulus onset and offset. They equally reported a loss in the definition of kinesthetic feedback information and sense of effort during vibration and for some time after stimulus cessation.

Effects of Vibration on Forearm Positioning

Blindfolded subjects were trained to execute forearm rotation from a fixed position to a defined target position. When performance was stabilized, WBV was applied to the platform. All subjects tended to position their forearms beyond the target (past pointing). The systematic overshooting of the position target was paralleled by an increase of data dispersion from 8% to 10% before vibration to 20% to 30% under vibration. The averaged performance was still altered after vibration arrest. Large past pointings and significant data spread resulted from both flexion and extension. Larger angular displacements showed similar results, although to a much lesser extent.

Effects of Vibration on Forearm and Foot Tracking of Visual Targets

The forearm rotation around the elbow was detected by a potentiometer whose output signal drove a spot displayed on an oscilloscope next to the target spot. The performance was analyzed in terms of position and velocity errors for flexion and extension separately. Figure 4 shows time recordings of foot (top) and forearm (bottom) tracking before, during, and after WBV. Tracking alterations caused by vibration affect both position and velocity controls. The histograms of Fig. 4 quantify the results in 10 runs of 10 cycles per subject in

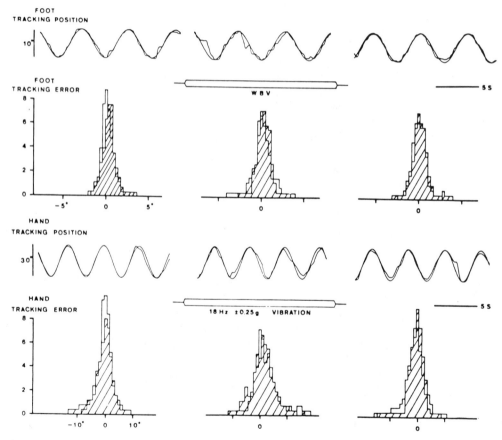

FIG. 4. Forearm and foot tracking of visual target. Foot **(upper)** and forearm **(lower)** tracking average (10 Ss) performance is described in terms of tracking error before (*left*), under and after (*right*) WBV. Time recordings of target and S's tracking response are also shown as examples in the 3 tested conditions. Vibration altered tracking precision evidenced by broadenings of tracking error histograms. (From Gauthier et al., 1981, with permission.)

each situation. The angular error during flexion (hollow area) is of similar magnitude to that during extension (hatched area). All histograms centered near zero, which means that subjects led as well as lagged the target. Vibration significantly increased position error as seen in foot and, more so, in arm tracking by a widening of the histogram and larger maximum error amplitude. Velocity amplitude was also altered by vibration. After vibration, tracking with normal control resumed rapidly.

Effects of Vibration on Foot Control of Static and Dynamic Torques

In tests of reproduction of predefined torques, vibration applied to the platform supporting both subject and pedal definitely altered performance. Both the torque maximum amplitude and trial-to-trial dispersion increased by as much as 20% for low torques (12 Nm). In addition, torque stability as attempted during the 2-sec torque holding was totally disrupted. Following stimulus cessation, the subjects tended to produce even larger amplitude torques. The graph of Fig. 5 represents mean performance of the 10 subjects as a function of the reference torque in the 10 to 25 Nm range before, during, and after vibration.

In tests of rapid torque alternation, when the subject, involved in a task consisting of modulating a torque around a given value, was suddenly vibrated, the oscillation lost its previbration regularity, and its amplitude increased. At stimulus cessation, the subject experienced severe difficulty in continuing the task. The effect is characterized by decrease of the torque-modulating wave frequency and amplitude and, occasionally, a decrease of mean torque level. The EMG shows an increase of modulation amplitude during vibration.

Discussion

The WBV affects sensory–motor performance by altering position, velocity, and force controls. We mentioned that vibration activates cutaneous, articular, and muscular receptors. Among these, muscle spindles and Golgi tendon organs are most sensitive to the stimulus. These receptors convey to the CNS information for motor program control. Thus, it is not surprising that alterations result from vibration applied massively to a subject during a sensory–motor task (Levison, 1978; Roll et al., 1976).

The vestibulospinal afferents were not significantly affected by the 18-Hz vibration (Martin et al., 1980). Still lower-frequency resonance of the head may excite the labyrinthine receptors and reach the vestibuloocular system. Indeed, the vestibuloocular reflex may undergo some modifications and produce a response up to frequencies larger than the usually cut-off frequency of 8 Hz (Keller, 1978; Young and Oman, 1969).

Whole-body vibration applied to a resting seated subject deeply decreased spinal reflexes. In active situations, WVB does decrease spinal MN excitability but to a much lesser extent than in a passive state where it drops almost to zero. Thus, the decrease of spinal MN excitability because of WBV is responsible for some sensory–motor con-

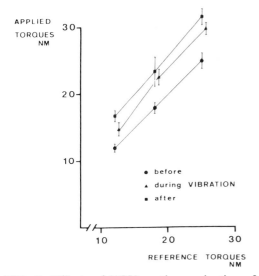

FIG. 5. Effects of WBV on the production of defined torque. The effects outlast the duration of the stimulus. (From Gauthier et al., 1981, with permission.)

trol alterations, perhaps through defects in the servo assistance of central command by spinal systems. Such defects might be compensated for by central mechanisms whose activity outlasts the duration of vibration.

The subjects were not directly aware of the alterations of their performance unless visual feedback was provided. Vibration may affect sense of position and sense of effort (see D. I. McCloskey, S. Gandevia, E. K. Potter, and J. G. Colebatch, *this volume*). The kinesthetic information may be greatly altered. Indeed, WBV induces afferents in all vibrated muscles, including agonistic and antagonistic muscular groups involved in the movement. With WBV, the large vibration-induced volleys occurring in both muscular groups partly mask the movement-induced afferents. It follows that position and velocity control signals normally used for accurate positioning and tracking become partly inefficient. The subject does not seem to be aware of this impairment and executes the programmed movement in order to reach the expected feedback sensation. Overshooting occurs in the pointing task. Overshoot and high-velocity movement result in tracking tasks.

Vibration applied to the whole body or to the legs resulted in kinesthetic perceptive defects whose poststimulus duration was much shorter than that of monosynaptic reflex inhibition. This observation raises a problem as to a possible correlation between these two effects. Desmedt and Robertson's (1977) work (see also Hillyard et al., 1979) suggests that corticofugal controls exerted over the relays of ascending pathways do not participate in selective attention. The attenuation–selection processes handling the "filtering" of the information conveyed to cortex by sensory pathways take place at the cortical level. According to the model of Hillyard et al., the filtering would be mediated by the discrimination activities feeding back onto the inputs. With constant-amplitude, long-duration vibration, a stable predictive filtering could be exerted. On vibration cessation, the predictive filtering becomes excessive with respect to sudden decrease of the afferents.

Other factors may contribute to the deterioration of performance, such as the vibration-induced TVR which may be dramatic in tasks requiring weak force development. In fact, the vibration-induced contraction adds to voluntary command executed by the subject, and excessive force is developed.

POSTURE AND EQUILIBRIUM

Seated subjects were vibrated for periods of 30 min at 18 Hz (vertical acceleration, 0.25 or 0.5 g). Postural signals were recorded during 15-sec periods of quiet standing, eyes closed and arms hanging along the body. The effects of vibration on postural equilibrium were derived from the comparison of position and velocity histograms of the postural forces recorded before and immediately after stimulus cessation. The effects of vibration were also studied on quiescent subjects or subjects performing defined postural adjustments during stimulus application. For that purpose, the subjects stood on the posturograph mounted directly on the vibrating platform. Similar findings were derived from WBV and LV situations.

Immediately after vibration, the subject standing with eyes closed on the posturograph reported a feeling of perfect stability though postural movements were paradoxically very large. The sensation of "postural comfort" vanished after 30 to 45 sec. Posturograms showed that anteroposterior and lateral position and velocity signals were markedly larger after vibration as compared to previbration readings. Figure 6 illustrates a control posturogram and one obtained immediately after a 30-min vibration of the leg.

When the vibration was applied to the head alone or to the head and trunk, exclud-

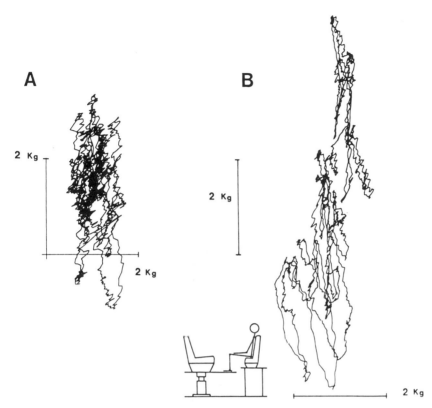

FIG. 6. Pre- and post-vibration posturograms. The alterations induced by vibrations were evaluated by 30 s X-Y recordings of lateral (horizontal deflection) and anteroposterior (vertical) components of postural forces. Before vibration (A), the overall extent of postural force variations was of the order of 1 Kg and 4 Kg in the lateral and sagittal directions, respectively. After 30 min of vibration (B) applied to the legs (situation illustrated by drawing), there was more than a two-fold increase in the overall lateral and anteroposterior force component excursions (From Martin et al., 1980 with permission.)

ing the legs, no significant differences were observed between pre- and postvibration measures (Fig. 7). The alteration of postural equilibrium was evaluated as the mean deviation from resting previbration position (in kg) for the 10 subjects. The data suggest that the main site of action of vibration is the legs.

The posturograph signals were also monitored during vibration of standing subjects (Fig. 8). Immediately after stimulus application, the subject's mean posture shifted forward, and oscillation amplitude increased by two- to threefold. Subjects were not aware of the postural shift and increase of oscillation amplitude but felt perfectly stable. A series of tests designed to evaluate the effects of vibration on voluntary control of postural adjustments showed that the subject's performance was considerably altered. Postural sways around the ankle joint executed with eyes closed were reproduced under vibration with a 25 to 50% amplitude increase.

Discussion

Vibration applied to the whole body of a subject results in alterations of postural equilibrium that persist for minutes after stimulus cessation. The observation that HTV and HV did not induce postvibrational

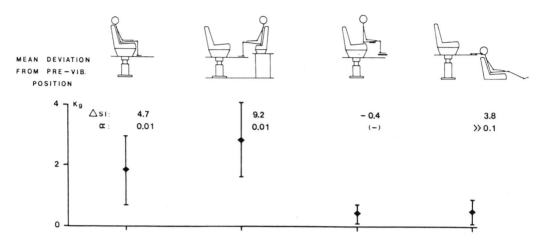

FIG. 7. Mean postural equilibrium alterations in the situations illustrated by the drawings. For each S, a mean deviation value was determined as the average over six trials of the difference between mean resting position before vibration and mean resting position after vibration. The graph represents the average in absolute value (diamond) and confidence interval ($z = 0.05$) of the mean deviation values obtained from the 10 tested Ss. Likewise, SI is the average of individual differences between pre- and post-vibration semi-interquartiles and α is the corresponding probability threshold. The drawings picture the experimental situation. As can be readily seen, WBV and LV induced on average, a significant shift of resting position and an increase of oscillation amplitude (ΔSI) while situations HTV and HV had no effect. This suggests that the main site of action of vibration is located in the legs. (From Martin et al., 1980, with permission.)

effects suggests that vestibular organs and vestibulospinal systems are probably only moderately affected by vibration. The kinesthetic deficits observed during vibration outlast the stimulus by only a few seconds.

Assuming that the main postural program is not modified by the vibration, the observed changes may be related to one or a combination of three possibilities: (1) a sensorimotor reflex effect evoked by the vibration and developing independently of the ongoing postural program; (2) a servo-assistance defect resulting from alteration of the spinal reflex systems; and (3) an alteration of the afferents normally involved in the feedback regulation of the central program.

Leg vibration has been shown to result in an increase of both soleus tonic activity and amplitude of repeated torques; the subject was not aware of these changes. In addition, the TVR and monosynaptic reflexes were depressed. The motor commands from central origin are controlled by cutaneous and muscular afferents. Monosynaptic reflex and TVR depression observed under vibration suggests that these commands would result in postural activity changes comparable to those recorded with respect to arm positioning.

An alteration of movement perception may add its effects to the decrease of efficiency of the servo-assistance mechanisms. The detection threshold of errors around a set point would be considerably increased, and an increase in the amplitude of corrective movements would be observed. The signal/noise ratio would be somehow reduced.

CONCLUSION

Low-frequency (18 Hz) vibration applied massively to a human subject alters static and dynamic characteristics of sensory–motor performance. The vibration applied to body elements excites a large number of

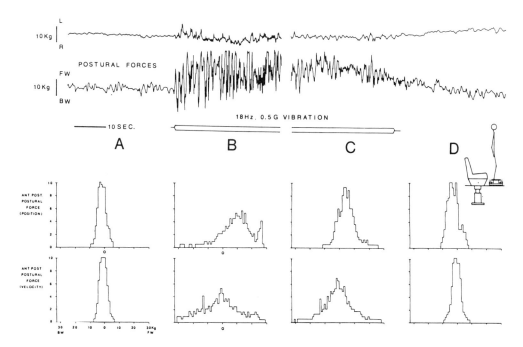

FIG. 8. Per-vibration effects in standing man. Stabilometer signals were recorded during vibration applied to a standing S (drawing). Immediately after stimulus application (**B**), the equilibrium oscillation amplitude increased markedly in both directions. After 3 min of vibration (**C**), posture was still altered compared to pre- (**A**) and post- (**D**) vibration periods. The lower histograms, derived from anteroposterior force and its rate of change, were used to characterize the observed effects. Per-vibration histograms were significantly broader, demonstrating an increase of both oscillation amplitude and velocity. (From Martin et al., 1980, with permission.)

muscular, cutaneous, and articular receptors. These signals constitute a neuronal noise which alters motor programming and execution. The sensitivity of signal detection systems decreases, and regulation amplitude around set points increases. Futhermore, the duration of the alterations beyond cessation of vibration suggests that the central neural machinery attenuates the afferent signals at spinal and central levels. Vibration thus may constitute a noxious stimulus both uncomfortable and dangerous for the subject. Though some habituation to long-duration vibration might eventually take place, this ability is limited.

Particular care should be taken in vibrating vehicles to prevent vibration from reaching muscular masses, especially those involved in motor tasks requiring position, velocity, and force controls.

ACKNOWLEDGMENT

This work has been supported by SNIAS and CNRS ERA 272 grants.

Reflex Control of Bipedal Gait in Man

*E. Pierrot-Deseilligny, C. Bergego, and L. Mazieres

Service de Rééducation Neurologique, Hôpital de la Salpêtrière, 75651 Paris Cedex 13, France

That spinal reflexes can regulate centrally programmed movements has been assumed for a long time (Sherrington, 1910). During fictive locomotion (Severin et al., 1967) or in the freely moving cat (Prochazka et al., 1976; M. Prochazka and M. Hulliger, *this volume*), significant discharges in both Ia and Ib fibers have been recorded. The question arises then how, and to what extent, proprioceptive group I discharges provide reflex actions that may assist locomotion. The distribution pattern of excitation and inhibition from group I muscular afferents, which has been extensively investigated in cat spinal cord (Lloyd, 1946; Laporte and Lloyd, 1952; Eccles et al., 1957; Eccles and Lundberg, 1958, 1959; Engberg, 1964), shows that some connections, but not all, cross a joint. A particular role for such connections has been postulated for cat locomotion. For example, the presence of Ia connections from knee muscles to hip extensors has been correlated with the fact that in cat during stepping, movements in hip and knee joints do not occur in phase. So it has been suggested that these Ia projections may assist hip extension during locomotion (Engberg and Lundberg, 1969; Lundberg, 1969). Lundberg (1969) also suggested that Ib inhibition may operate to regulate the yield of knee, ankle, and toe extensors during the stance phase of cat locomotion: the strong Ib inhibitory interconnections between these yielding muscles may help to explain the remarkable similarity in angular movements at knee, ankle, and toe joints. Furthermore, a long-lasting yield in these extensors, which is of great importance since it contributes to the soft spring characteristic of feline gait, may be assisted by facilitation of Ib reflex pathways from low-threshold cutaneous afferents (Lundberg et al., 1977).

In man, it has recently been demonstrated that Ia reflex pathways are not vestigial since they play a role in locomotion. Indeed, Dietz et al. (1979) have shown that during running the gastrocnemius–soleus (GS) spinal stretch reflex significantly contributes to the muscle force exerted by extensors of the leg during ground contact. If group I reflex pathways assist locomotion, their organization in man's lower limb can be expected to differ from that of the cat hindlimb, since the requirements of human heel bipedal gait differ from those of feline digitigrad and quadripedal locomotion. In particular, the quadriceps (Q) contraction is of paramount importance during the stance phase of human gait, since, at the moment of knee yield, this contraction supports all the body weight in a biped. It is therefore significant that in man, contrary to cat, strong Ia projections from GS to Q and depression of Ib inhibition of Q MNs by cutaneous stimulation of the foot sole have been found (Pierrot-Deseilligny et al.,

---

*To whom correspondence should be addressed: Service de Rééducation Neurologique, Hôpital de la Salpêtrière, 47, boulevard de l' Hôpital, 75651 Paris Cedex 13, France.

1981a,b). Indeed, both operations may be expected to assist Q contraction during bipedal locomotion.

In investigating the function of a given spinal pathway in man, where the methods must obviously be indirect, it is essential to insure that the pathway is selectively explored. Hence, this chapter first describes experiments that demonstrate that the methods used do allow us to explore group I reflex pathways. A detailed description of the pattern of group I fiber projections from ankle muscles is then given. The pattern of control of Ib reflex pathways to various MN pools by cutaneous afferents from different skin regions is also presented in detail. Finally, the timing of knee and ankle angular movements and of different muscle EMG activity at these joints during human locomotion is considered in conjunction with the organization of group I reflex pathways. This organization in man fits several requirements of bipedal gait.

PATTERN OF GROUP I FIBER PROJECTIONS IN MAN'S LOWER LIMB

In the spinal cat, Laporte and Lloyd (1952) explored the pattern of projections of Ia and Ib fibers from different muscular nerves. Similarly, we have used the size of the H reflex to assess the effects of conditioning stimuli applied to different muscular nerves. The excitability of four motoneuron (MN) pools was so explored: quadriceps (Q), soleus (Sol), short head of the biceps femoris (Bi), and tibialis anterior (TA). Conditioning stimuli were single pulses of 0.5 msec duration given through bipolar surface electrodes to different nerves: the inferior branch of the soleus nerve (Inf Sol), the gastrocnemius medialis nerve (GM), the common peroneal nerve (CPN), and the femoral nerve (Fem). The strength of the conditioning stimuli was controlled by monitoring the direct muscle response. Since it is not possible to record incoming volleys in man, stimulus strength is expressed in multiples of the threshold strength of M wave ($\times$ motor threshold, MT). To avoid any recurrent inhibition, the conditioning stimulus strength was kept lower than $1 \times$ MT. Details of the experimental procedure are given elsewhere (Pierrot-Deseilligny et al., 1981a,b).

Projections of Group I Fibres from the Inf Sol Nerve

Time course of the variations in Sol and Q H reflexes.

Figure 1 shows the time course of variations in the H reflex of Sol (left part) and Q (right part) when preceded by an Inf Sol nerve stimulus, the strength of which was $0.8 \times$ MT. In both cases, biphasic variations were observed, with an early facilitation and a subsequent inhibition. The amount of facilitation and inhibition was similar in both cases. Such a quantitative comparison is possible, since the size of the unconditioned test reflex was identical in both cases (about 15% of maximum M).

The conditioning stimulus stimulated not only the Inf Sol nerve but also the skin and sometimes the sural nerve, which is anatomically close. To insure that the variations in the test reflex did not result from these cutaneous stimulations, experiments were performed with a selective cutaneous stimulation of either skin or sural nerve by moving the electrode delivering the conditioning stimulus 3 cm lower and more lateral, to a site at which this stimulation did not involve the Inf Sol nerve, and also by selectively stimulating the sural nerve at the ankle. Such a pure cutaneous stimulation does not modify the test reflexes (Pierrot-Deseilligny et al., 1979, 1981a,b). The biphasic variations in Sol and Q H reflexes shown in Fig. 1 can therefore be considered effects of Inf Sol nerve stimulation. When conditioning stimuli were applied to other nerves (GM, CPN, Fem), it was also checked that variations in the different test reflexes did not result from stimulation of either the skin underneath the conditioning

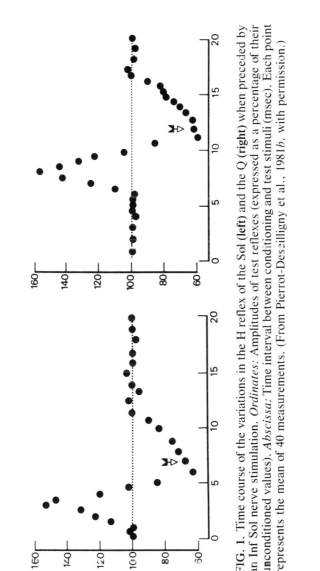

FIG. 1. Time course of the variations in the H reflex of the Sol (**left**) and the Q (**right**) when preceded by an Inf Sol nerve stimulation. *Ordinates*: Amplitudes of test reflexes (expressed as a percentage of their unconditioned values). *Abscissa*: Time interval between conditioning and test stimuli (msec). Each point represents the mean of 40 measurements. (From Pierrot-Deseilligny et al., 1981b, with permission.)

electrodes or cutaneous fibers contained in the nerve (CPN and Fem).

When the conduction velocity (CV) of the fibers responsible for the early facilitation and the distance between conditioning and testing electrodes were compared, it was found that the latency of this facilitation is compatible with a monosynaptic linkage (Pierrot-Deseilligny et al., 1981b). It was also shown that the time courses of the early facilitation and of the subsequent inhibition overlap each other (see below) and that the real latency of the inhibition was only 1 msec longer than that of the facilitation. This suggests that the early facilitation is Ia in origin and that the subsequent inhibition is Ib in origin. The evidence is, however, indirect, as it always is with human experiments. In investigating the function of a certain neuronal pathway, we must take care that we are really dealing with just that pathway. In other words, this implies that several congruent results from different experiments are necessary.

Evidence for Ia and Ib effects.

In order to separate the thresholds of the two subsequent and opposite effects, experiments were performed in which the conditioning stimulus strength was changed, while the conditioning–testing interval was kept constant. This interval (arrows in Fig. 1 and figures within brackets in Fig. 2) was chosen when subsequent inhibition was near maximum. Figure 2 shows that increasing conditioning stimulus strength for a given interval also resulted in biphasic variations, thus indicating that the time courses of the two effects overlap each other. The earliest effect, i.e., facilitation, appeared with the lowest threshold and thus involves the thickest fibers. According to animal data, this strongly suggests that such facilitation is Ia in origin. That the threshold

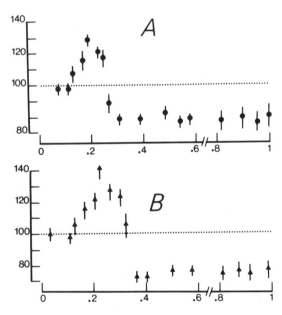

FIG. 2. Effect of changing the strength of conditioning stimuli applied to the Inf Sol nerve. A: Variations in Sol H reflex (7 msec). B: Variations in Q H reflex (12 msec). *Ordinate:* Amplitudes of test reflexes (expressed as a percentage of their unconditioned values). *Abscissa:* Conditioning stimulus strength (expressed in multiples of the threshold strength of the M wave, MT). The conditioning–test interval is indicated in each figure. Each symbol represents the mean of 30 measurements. The *vertical bars* = ± 1 SEM. (From Pierrot-Deseilligny et al., 1981b, with permission.)

of Ia fibers was extremely low (0.15 × MT) can be related to the fact that the Inf Sol nerve was stimulated very distally. In such a distal location, α motor fibers are known to branch on their way to the muscle (Eccles and Sherrington, 1930) (hence their smaller diameter and higher threshold), whereas there is almost no branching of the Ia fibers (Barker, 1974).

Figure 2 shows that increasing conditioning stimulation strength first increased the initial facilitation, reaching a peak at 0.22 × MT. On further increases in conditioning stimulation, the facilitation declined and was replaced by an inhibition of both Sol and Q H reflexes. The stimulation strength at which the facilitation began to decline (0.22 × MT) can be considered the threshold for inhibition. This inhibition has a slightly higher electrical threshold than that of Ia facilitation, which indicates that the fibers responsible for it are slightly thinner. According to classical animal data, this suggests that such inhibition is Ib in origin. It has been shown recently in cat that Ia afferents also contribute to these "so-called Ib effects" (Fetz et al., 1979; Jankowska et al., 1978), although to a much lesser extent than Ib afferents. Jankowska et al. (1978) have shown convergence from Ia and Ib afferents onto common inhibitory interneurons acting on MNs, and Ib transmission to MNs may therefore be facilitated by simultaneous Ia activity. It is likely that the inhibition from group I afferents in Figs. 1 and 2 is evoked mainly from Ib afferents, but although it is described below as Ib inhibition, a contribution from Ia afferents cannot be excluded.

To further confirm that the two successive and opposite effects were, respectively, Ia and Ib in origin, a progressive ischemic blockade of group I fibers was performed, since ischemia is known to affect the thickest fibers first (Magladery et al., 1950). Ischemia was produced by a sphygmomanometer cuff inflated to 300 mm Hg around the upper part of the leg just below the electrode eliciting the Sol test reflex but above the electrode delivering the conditioning stimulus (Fig. 3); the efficacy of the ischemic blockade of Ia fibers was controlled by recording the Achilles tendon jerk. The shaded area in Fig. 3 indicates the period (17 to 21 min after onset of ischemia) during which the Achilles tendon jerk began to decrease and eventually disappeared, thus indicating a blockade of Ia

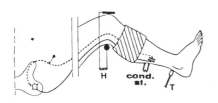

Conditioning: inferior soleus nerve

Test : soleus

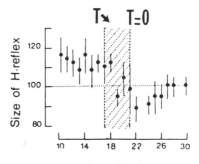

Duration of ischaemia (min)

FIG. 3. Effect of ischemia of the leg on soleus H-reflex facilitation caused by stimulation of the Inf Sol nerve (3.5 msec conditioning–test interval; 0.8 × MT stimulus strength). Left: The experimental design. Right: *Ordinate,* Size of the soleus H reflex (expressed as a percentage of its unconditioned value). *Abscissa:* Duration of ischemia (min). Each symbol represents the mean of five measurements. *Vertical bars:* 1 SEM. *Shaded area* indicates the period during which the Achilles tendon jerk began to decrease and finally disappeared. (From Pierrot-Deseilligny et al., 1981*b*, with permission.)

fibers from the GS. In this experiment, the conditioning–testing interval was kept constant at 3.5 msec, an interval at which, before ischemia, facilitation still overshot inhibition. This facilitation began to decrease at the same moment as the Achilles tendon jerk (Fig. 3), thus indicating that the facilitation was caused by stimulation of Ia fibers. For 5 min after the disappearance of the tendon jerk, there was even a reversal of the effect, since facilitation was replaced by an inhibition and then the test reflex returned to its control value. This is consistent with a progressive blockade of Ia fibers, exposing Ib inhibition, followed by a blockade of Ib fibers.

Since Ia connections from Sol to Q do not exist in the cat (Eccles et al., 1957), a final experiment explored whether facilitation of Q MNs after the Inf Sol nerve stimulation is of Ia origin. Figure 4 shows a comparison between the effects of a tibial nerve electrical stimulation and those of an Achilles tendon tap on the Q test reflex. Both the mechanical and electrical stimuli were very weak: the electrical stimulation was 0.4 times the threshold of the Sol H reflex, and the mechanical stimulation was 0.33 times the threshold of the Achilles tendon reflex. In both cases, there was a huge facilitation of the Q H reflex. The facilitation evoked by the tendon tap occurred 5.5 mscc later, because the conditioning stimulation was applied more distally than the electrical stimulation. It can be assumed from animal data (Lundberg and Winsbury, 1960) that at rest such a very weak tendon tap activates predominantly primary spindle endings. The Ia origin of this facilitation is also demonstrated by its disappearance during an ischemic blockade of Ia fibers from the triceps surae. There was no inhibition following facilitation in the case of the tendon tap, thus confirming that the inhibition elicited by the electrical stimulation is Ib in origin.

Group I projections from Inf Sol to MNs supplying flexor muscles.

Thus, in man, there is an autogenetic Ia excitation and Ib inhibition from Sol to Sol, and the pattern of group I fiber projections from Sol to Q is exactly the same. An in-

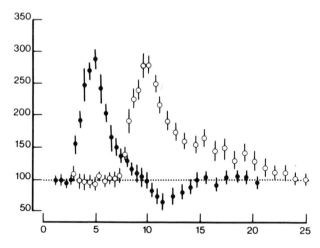

FIG. 4. Comparison of variation in the Q H reflex when conditioned by an electrical stimulus applied to the posterior tibial nerve *(filled circles)* or by a tendon tap applied to the Achilles tendon *(open circles)*. *Ordinate:* Amplitude of the Q H reflex (expressed as a percentage of its unconditioned value). *Abscissa:* Time interval between conditioning and test stimuli (msec). Each symbol represents the mean of 30 measurements. *Vertical bars:* 1 SEM. (From Pierrot-Deseilligny et al., 1981*b*, with permission.)

verse pattern of projections was found when the conditioning and test muscles were antagonistic, as in Fig. 5, where variations in the monosynaptic reflex of the short head of the biceps femoris (knee flexor) were studied after a conditioning stimulation applied to inferior branch of the nerve supplying Sol (extensor muscle). The time course also exhibited biphasic variations, but in this case, there was an early inhibition and a subsequent facilitation. Given the distance between conditioning and testing electrodes, the latency of early inhibition is compatible with a disynaptic linkage. Increasing the conditioning stimulus strength also resulted in biphasic variations, the inhibition having the lowest threshold. It is of importance that the threshold of this inhibition ($0.15 \times MT$) was exactly the same as that of the Ia facilitation of Sol and Q MNs, whereas the threshold of the facilitation ($0.22 \times MT$) was the same as that of the Ib inhibition of Sol and Q MNs. Early inhibition, which has the lowest threshold, can thus be attributed to stimulation of Ia fibers, and subsequent facilitation, which has a slightly higher threshold, to that of Ib fibers. So, in accordance with animal data, the pattern of projections onto antagonistic muscles is an Ia inhibition and an Ib facilitation. In only two subjects was it possible to record an H reflex in TA. As described by Tanaka (1974, 1980), in these two subjects, there was weak reciprocal Ia inhibition from Sol to TA and no evidence for subsequent Ib facilitation (Pierrot-Deseilligny et al., 1981b).

Pattern of Projections of Group I Fibers from Ankle Muscles

Similar experiments were performed with conditioning stimuli applied to either the GM or the CPN nerves. In almost all cases, this resulted in significant biphasic variations. Figure 6 shows the time course of variations in Q H reflex: there was an early Ia facilitation and a subsequent Ib inhibition after GM nerve stimulation and an inverse pattern from pretibial flexors.

Table 1 shows the pattern of projections of Ia fibers from ankle muscles. The Ia effects are excitatory for autogenetic and synergistic MN pools and inhibitory for antagonistic MN pools. This agrees with classical data in spinal cats (Lloyd, 1946; Laporte and Lloyd, 1952; Eccles et al., 1957b). After GM nerve stimulation, Ia facilitation of both Bi and Q MNs, which supply antagonistic muscles, seems to contradict this general pattern of projections. The GM muscle is, however, bifunctional: as a knee flexor, it is a synergist of the biceps femoris; as an ankle extensor, it can also be considered as a synergist of the quadriceps muscle, the knee extensor.

In fact, when this pattern of projections of Ia fibers is quantified, there are extremely large quantitative differences between men and cats. In contrast to the spinal cat, Ia effects in man onto some close synergistic (GM to Sol) or direct antagonistic (TA to Sol) muscles are weak or even absent. On the other hand, the Ia projections from ankle muscles to Q, which are either absent in cat (Eccles et al., 1957b) or negligible when they exist (E. Jankowska, R. Mackel, and D. Mc Crea, *personal communication*), are potent in man. There are significant differences between cat and baboon with respect to Ia projections from the different heads of triceps surae onto the MNs supplying the other heads (T. Hongo, A. Lundberg, C. G. Phillips, *personal communication*). When man is compared to baboon, the discrepancy between Ia projections from triceps surae onto various MN pools is much more marked. This different organization in cats, baboons, and humans suggests that the widespread Ia connections in man are not vestigial but have been subject to phylogenetic adaptation. The potency of Ia connections from ankle to knee muscles fits the requirements of heel bipedal gait.

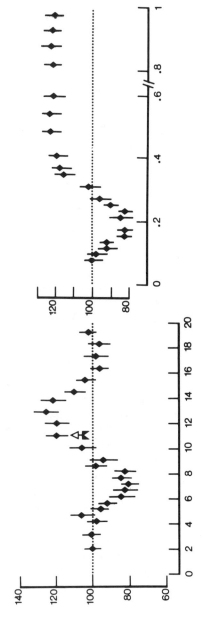

FIG. 5. Variations in the Bi H reflex when preceded by a conditioning stimulation applied to the Inf Sol nerve. *Ordinate*: Size of reflex (expressed as percentage of its unconditioned value). **Left**: *Abscissa*, Time course of the variations (msec). **Right**: *Abscissa*, Effect of changing the strength of the conditioning stimulus (in multiples of the threshold strength of the M wave, MT). Explored conditioning–test interval is 11 msec. Each symbol represents the mean of 30 measurements. *Vertical bars*: 1 SEM. (From Pierrot-Deseilligny et al., 1981*b*, with permission.)

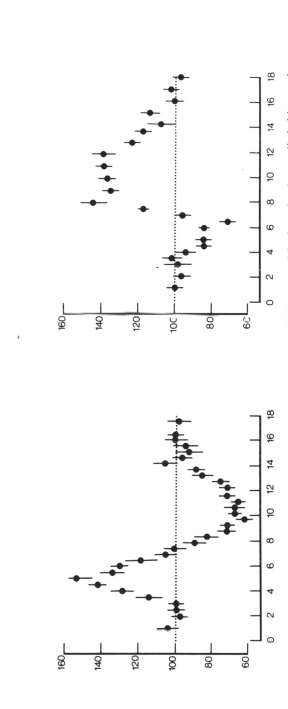

FIG. 6. Time course in the variations of the Q H reflex when preceded by a conditioning stimulus applied either to the GM nerve (**left**) or to the common peroneal nerve (**right**). *Abscissae:* Time interval between conditioning and test stimuli (msec). *Ordinates:* Amplitude of the test reflexes. Each symbol represents the mean of 40 measurements. *Vertical bars:* 1 SEM. (From Pierrot-Deseilligny et al., 1981b, with permission.)

TABLE 1. *Pattern of projections of Ia fibers from ankle muscles in man*[a]

MN pools	Nerves stimulated		
	Inf sol	GM	CPN
Sol	+++	NS	−
TA	−	−	++
Q	+++	+++	−−
Bi	−	+++	+

[a] The conditioning stimulation was applied to Ia fibers contained in three different nerves (Inf sol, inferior soleus nerve; GM, gastrocnemius medialis nerve; CPN, common peroneal nerve). The effects of these conditioning stimuli were tested by the H reflex in four MN pools (Sol, soleus; TA, tibialis anterior; Q, quadriceps; Bi, short head of biceps femoris). Type [excitatory (+) or inhibitory (−)] and strength (number of symbols) of Ia effects are indicated (NS, no sizable effect).

CUTANEOUS CONTROL OF Ib REFLEX PATHWAYS TO MOTONEURONS

The action of a cutaneous stimulation on Ib effects constitutes the second striking difference between men and cats. Lundberg et al. (1977) have shown that in spinal cats, stimulation of low-threshold cutaneous afferents from ipsilateral limb facilitates transmission of synaptic action from Ib afferents to MNs. In contrast, we have found that such cutaneous stimulation depresses Ib reflex pathways to MNs.

Depression of Ib Inhibition from GM to Q MNs

The curve with open circles in Fig. 7 shows early Ia facilitation and subsequent Ib inhibition of the Q H reflex when it is preceded by GM nerve stimulation. Care was taken to prevent electrical diffusion of the GM nerve stimulus to sural nerve fibers supplying the skin of the lateral part of the foot or any other cutaneous fibers supplying the foot sole. Sural nerve stimulation was applied 7 msec before GM nerve stimulation, below the lateral malleolus so that the subject felt a tactile sensation clearly irradiating to the lateral part of the foot.

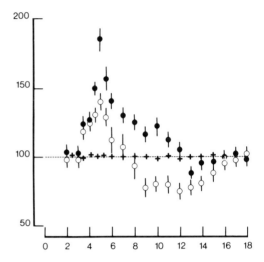

FIG. 7. Effects of sural nerve stimulation on variations in the Q H reflex induced by stimulation of group I fibers from GM. *Ordinate:* Size of the test reflex (expressed as a percentage of its unconditioned value). *Abscissa:* Time interval between GM nerve stimulation and the stimulus eliciting the test reflex (msec). *Open circles,* control conditions (Q H reflex is conditioned by only GM nerve stimulation); *filled circles,* Q H reflex is conditioned by both sural and GM nerve stimulations (sural nerve stimulation preceding GM nerve stimulation by 7 msec); *crosses,* effects of sural nerve stimulation when applied alone. Each symbol represents the mean of 40 measurements. *Vertical bars:* 1 SEM. (From Pierrot-Deseilligny et al., 1981a, with permission.)

This sural nerve stimulation caused the inhibition to disappear, thus exposing a time course (filled circles) identical to that observed after a pure Ia facilitation (cf. Pierrot-Deseilligny et al., 1981b).

Since cutaneous stimulation caused no variation in the test reflex when applied alone (cf. crosses), it can be concluded that cutaneous afferents depress Ib inhibition from GM to Q MNs. The only alternative hypothesis would be a facilitation of Ia excitatory pathways. A careful examination of the effects of sural nerve stimulation onto the Q H reflex early facilitation eliminated this hypothesis. It has been shown (Pierrot-Deseilligny et al., 1981b) that for the earliest conditioning–test intervals (between 3.5 and 4.1 msec), Ia facilitation from GM was as yet uncontaminated by subsequent Ib inhibition, which started to manifest itself at a 4.2-msec interval. Figure 7 clearly shows that as long as Ia facilitation is pure (3.5- and 4-msec conditioning–test intervals), sural nerve stimulation does not modify it. Sural nerve stimulation only starts to modify the time course of group I effects at the 4.2-msec interval which corresponds to the onset of Ib effects (Pierrot-Deseilligny et al., 1981a,b). This strongly suggests that cutaneous stimulation acts through a depression of Ib effects.

Effect of varying cutaneous stimulus strength.

In the following, cutaneous stimulus strength is expressed in multiples of the perceptual threshold (PT). The effects of varying the strength of sural nerve stimulation were studied when the interval between sural and GM nerve stimulations was 7 msec and that between GM nerve and Q H-reflex stimulation was 9 msec. It was found (Pierrot-Deseilligny et al., 1981a) that the depression of Ib inhibition by a preceding sural nerve stimulation appears at the perceptual threshold (1 × PT); increasing cutaneous stimulus strength up to 2 × PT increases the depression of Ib inhibition, thereby unmasking the Ia facilitation, but further increases in cutaneous stimulation strength up to the pain threshold do not further increase depression of Ib inhibition of the Q H reflex. In the experiment illustrated in Fig. 7 and in all experiments described in the following, the cutaneous stimulus strength was 2 × PT, perceived by the subject as a very slight tactile sensation.

Time course of the cutaneous depression of Ib effects.

The interval between Ib and test reflex stimulations was kept constant at 9 msec, an interval at which Ib inhibition of Q MNs was maximum (Fig. 8; cf. Fig. 7). The big open circle on the right and the dotted line represent the value taken by the test reflex when conditioned only by Ib stimulation. Figure 8 shows that in spite of the distance between sural and GM nerve stimulation (sural nerve stimulation was applied 37 cm more distally than GM nerve stimulation), cutaneous stimulation started to decrease Ib inhibition of Q MNs for an interval as short as 5.5 msec. This depression was maximum at 7 msec, when it was strong enough to expose the facilitation caused by stimulation of GM Ia fibers. Further increases in the interval between sural and GM nerve stimulations resulted in a decline of this depression, to be replaced by a weak reinforcement maximal at 14 msec. This reinforcement was not observed in all subjects and, when present, always remained weak. The Ib inhibition returned to its control value (i.e., Ib inhibition of the Q H reflex without cutaneous stimulation) at 17 to 18 msec.

An attempt was made to estimate the central latency of this cutaneous depression of Ib effects. The earliest interval at which cutaneous stimulation starts to depress Ib effects corresponds to the action of the fastest cutaneous fibers on the latest Ib effects, i.e., those evoked by the slowest Ib fibers that are recruited by the conditioning stimulation. The CV for the fastest cutaneous fibers contained in the sural nerve measured between the ankle and the poplitea fossa was found to be as fast as 65 m/sec, which agrees with Hallin and Torebjork (1973). In this subject, cutaneous impulses required 15.4 msec to travel the 100 cm separating the sural stimulation site from the spinal cord. Since there is a linear relationship between diameter and CV in peripheral axons, and since electrical excitability decreases with CV, the slowest Ib fiber recruited by the GM nerve stimulation almost certainly has a CV close to that of the fastest GM motor fibers, i.e., 53 m/sec (Yap, 1967); indeed, the conditioning stimulation applied to the GM nerve was just below the

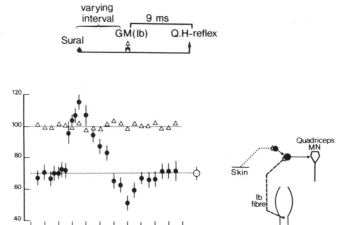

FIG. 8. Time course of the cutaneous depression of Ib effects from GM onto Q MNs. **Left:** *Ordinate:* Size of the Q H reflex (expressed as a percentage of its unconditioned value). *Abscissa:* Time interval between sural and GM nerve stimulations (msec). *Filled circles,* Q H reflex conditioned by both sural and GM nerve stimulations; *large open circle and dotted line,* Q H reflex conditioned only by GM nerve stimulation; *open triangles,* effects of sural nerve stimulation when applied alone. Each symbol represents the mean of 40 measurements. *Vertical bars:* 1 SEM. **Right:** Wiring diagram of the presumed linkage. (From Pierrot-Deseilligny et al., 1981a, with permission.)

motor threshold (0.95 × MT). The slowest Ib impulses from GM thus traveled 63 cm with a CV estimated at 53 m/sec and should therefore have reached the spinal cord 11.9 msec after Ib stimulation. If one takes into account only CVs and distances between stimulation sites and spinal cord, sural and Ib volleys should arrive simultaneously at the spinal cord given an interval between stimuli of 3.5 msec, representing the difference (15.4 − 11.9) between the arrival at the spinal level of cutaneous and Ib impulses. The difference between this figure and the earliest interval at which sural stimulation depressed Ib effects, 2 msec (5.5 − 3.5), may be taken as the central latency of inhibitory action from sural to Ib effects.

Spinal pathway.

At this brief latency and for so short a duration, depression cannot be exerted presynaptically onto Ib fibers (see Eccles et al., 1962). Since Q MN excitability is not modified by sural nerve stimulation applied alone, it can be assumed that depression is caused by a postsynaptic inhibition of Ib interneurons. In the low spinal cat, Lundberg et al. (1977) presented evidence from MN recording for a convergence of monosynaptic Ib excitation and disynaptic cutaneous excitation (from low-threshold cutaneous afferents) onto Ib interneurons. In the same preparation, low-threshold cutaneous afferents also produced trisynaptic IPSPs in interneurons that may transmit Ib effects to MNs (E. Jankowska, R. Mackel, and D. Mc Crea, *personal communication*). The very short central latency of the cutaneous depression of Ib effects found in man suggests an oligosynaptic linkage (diagram in Fig. 8).

Control of Ib Reflex Pathways by Cutaneous Afferents

Effect of ipsilateral sural nerve stimulation on different Ib reflex pathways.

It has been shown (Pierrot-Deseilligny et al., 1981a) that Ib inhibition from Inf Sol to

Q MNs is similarly depressed by ipsilateral sural nerve stimulation. The distance between sural and Inf Sol nerve stimulation sites was smaller than the distance between sural and GM nerve stimulation sites (19 cm instead of 37 cm) so that depression of Ib inhibition from Inf Sol appears earlier (2.3 msec instead of 5.5 msec). It is of prime importance to note in Fig. 9 that sural nerve stimulation also strongly depressed Ib facilitation from CPN to Q MNs. It has also been found that sural nerve stimulation depresses autogenetic Ib inhibition from Q to Q (Pierrot-Deseilligny et al., 1981a). So all the Ib pathways to Q MNs, homonymous or heteronymous, inhibitory or excitatory, are depressed by sural nerve stimulation. Similarly, all Ib pathways to Bi MNs, either inhibitory from GM and TA or excitatory from Sol, are depressed by cutaneous stimulation (Pierrot-Deseilligny et al., 1981a). By contrast sural nerve stimulation does not modify Ib pathways to Sol MNs: neither Ib inhibition from GM onto Sol nor autogenic Ib inhibition from Sol to Sol is depressed by cutaneous stimulation (Pierrot-Deseilligny et al., 1981a).

Effects of changing the cutaneous stimulation site.

Cutaneous stimulation, which was always perceived as a slight tactile sensation, was applied either to cutaneous nerves or directly to the skin at different points on the ipsilateral lower limb. Figure 10, which deals with Ib inhibition from GM to Q, shows that the efficacy of cutaneous stimulation in depressing Ib effects depends very much on the stimulation site. In this case, cutaneous stimulation was applied directly to the skin, and care was taken to prevent this stimulation from radiating. Columns with black symbols (Fig. 10) show that Ib inhibition was strongly depressed if cutaneous stimulation was applied to the skin of the foot sole and that no significant differences were found between anterior and posterior parts of the sole. By contrast, no changes in Ib inhibition from GM to Q MNs were found when cutaneous stimulation was applied either to the dorsum of the foot, or to the leg or thigh. Stimulation applied to the dorsum of the foot depressed Ib pathways to Q MNs only if such stimulation encroached on the DP or SP nerves and gave a sensation irradiating up to the tip of some toes. It should be pointed out that whatever the site of cutaneous stimulation applied to the ipsilateral lower limb, such stimulation never elicited any depression of Ib effects from GM or Inf Sol nerves to Sol MNs (Pierrot-Deseilligny et al., 1981a).

On the other hand, all Ib pathways to MNs supplying muscles operating at the knee (Q and Bi) were depressed by cutaneous stimulation provided that the resulting cutaneous sensation was perceived in a skin area (foot sole, tip of the toes) in contact with the ground during locomotion. Furthermore, stimulation of low-threshold cutaneous afferents from the contralateral

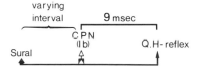

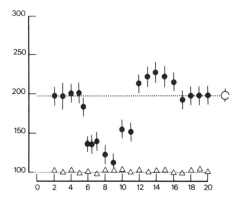

FIG. 9. Time course of the cutaneous depression of Ib effects from CPN onto Q MNs. See caption to Fig. 8, but CPN used instead of GM. (From Pierrot-Deseilligny et al., 1981a, with permission.)

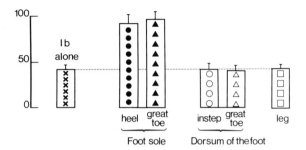

FIG. 10. Effect of cutaneous stimulation of various sites on Ib inhibition from GM to Q. *Ordinate:* size of reflex (expressed as percentage of its unconditioned value). *Column with crosses* and *dotted line* represent the Ib inhibition of the Q H reflex when GM nerve stimulation is applied alone. Other *columns* represent the values taken by the Q H reflex when preceded by both cutaneous and GM nerve stimulation. On the basis of the distance between electrode sites and the spinal cord, the time interval between cutaneous and GM nerve stimulations was chosen so that cutaneous depression of Ib effects, if any, would be maximum.

foot sole produces exactly reverse effects, i.e., facilitation of transmission of synaptic action from Ib afferents to MNs supplying knee muscles (Bergego et al., 1981). Such facilitation of Ib reflex pathways is obtained by stimulating contralateral cutaneous afferents from the foot sole only. When cutaneous afferents from the soles of both feet are stimulated together, contralateral facilitation is completely masked by ipsilateral depression of Ib effects, this depression being as strong as when the skin of the ipsilateral foot sole is stimulated along (Bergego et al., 1981).

Evidence for tonic depression of Ib effects from the skin of the ipsilateral foot sole.

In the experiments just described, cutaneous stimuli were phasic and electrically applied. In order to see to what extent natural tonic stimuli could modify Ib effects, a complete anesthetic blockade with 1% lidocaine of cutaneous afferents contained in both plantar and sural nerves was performed. This suppressed the tonic cutaneous discharge caused by the permanent contact of the foot sole with the rigid device. The curve with open circles in Fig. 11 shows the time course of early Ia inhibition and subsequent Ib facilitation of Q MNs in control conditions when preceded by a CPN stimulation. The curve with black circles shows that the anesthetic blockade of cutaneous afferents from the ipsilateral foot sole produced a considerable reinforcement of Ib facilitation, which lasted longer than in control conditions and almost completely masked early Ia inhibition from CPN. This indicates that in control conditions, Ib effects were tonically depressed by cutaneous afferents from the foot sole. It is difficult to know the pathway involved, but one can assume that tonic cutaneous depression of Ib effects operates through the same pathway as that used by the phasic cutaneous discharges above.

FUNCTIONAL SIGNIFICANCE

In man's lower limb, the organization of group I reflex pathways differs from that of the cat hindlimb in two essential aspects: strong Ia projections from ankle to knee muscles and depression of Ib reflex pathways to MNs supplying knee muscles by cutaneous afferents from the ipsilateral foot sole. This section considers whether these group I reflex discrepancies between man and cat correspond to some of the requirements of heel bipedal gait.

In cat during walking, there is no need for Ia connections from GS to Q, since the Ia

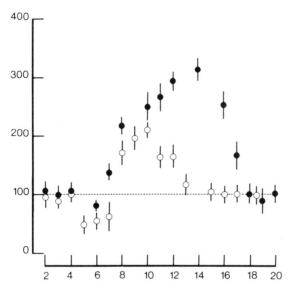

FIG. 11. Effect of anesthetic blockade of cutaneous nerves from the foot on Ib effects from CPN onto Q MNs. *Ordinate:* Size of the Q H reflex (expressed as a percentage of its unconditioned value). *Abscissa:* Time interval between the CPN stimulation and the test stimulus (msec). *Open circles,* control conditions (without blockade); *filled circles,* with anesthetic blockade of the sural and plantar nerves. Each symbol represents the mean of 30 measurements. *Vertical bars:* 1 SEM. (From Pierrot-Deseilligny et al., 1981a, with permission.)

discharges from these two extensors are similar; indeed, both muscles exhibit a similar EMG activity and are simultaneously and similarly stretched, since the movements of ankle and knee joints are remarkably similar (Engberg and Lundberg, 1969). By contrast, the time relationships between the movements of ankle and knee joints and those between GS and Q (vasti) EMG activities are much more complex during human locomotion. Figure 12 schematically represents such time relationships insofar as they can be drawn from congruent data originating from many authors (Eberhard et al., 1968; Sutherland, 1966; Brandell, 1977; Pedotti, 1977). The first striking difference is that in man, the movements of ankle and knee joints are almost completely out of phase during stance.

Role of group I projections from GS onto Q MNs.

In man, during normal walking, the GS muscle only contracts during the stance phase, where it exhibits a lengthening contraction (Fig. 12); GS contraction progressively increases while there is a continuous ankle dorsiflexion (Eberhard et al., 1968; Herman et al., 1976). According to animal data, this lengthening contraction most probably produces a strong discharge in both Ia and Ib fibers from GS. As described above, these Ia and Ib fibers have similar projections onto both Q and GS MNs. Nevertheless, cutaneous stimulation of the foot sole during stance can be expected to inhibit transmission in Ib reflex pathways to Q but not to GS MNs (cf. Fig. 13). Because of this cutaneous depression to Q MNs, the net result of the Ia and Ib discharges from GS will have very different effects on Q and GS MNs in conjunction with the different role played by these two extensors during the stance phase of human walking.

In early stance, there is a slight flexion of the knee (yield of knee) and the Q contraction (Fig. 12) reaches its peak when this yield is at its maximum (Brandell, 1977). It

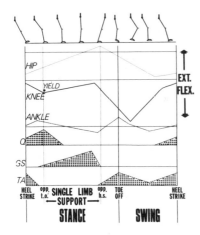

FIG. 12. Time relationships between the movements of hip, knee, and ankle joints and the EMG activity in Q, GS, and TA muscles during human gait.

should be pointed out that this moment also corresponds to the opposite toe off, i.e., to the beginning of the period of single limb support (Fig. 12). At that moment, the quadriceps contraction supports all the body weight. Thus, both Ia excitatory discharge from GS and removal of Ib inhibition (by cutaneous stimulation) could be crucial in giving a high safety factor to this quadriceps contraction. The role of the triceps surae is different. Indeed, triceps tension resists and restrains the ankle dorsiflexion produced by the resultant of the extrinsic forces (Sutherland et al., 1980), but it is imperative that triceps tension be overcome by this resultant if the body is to be brought forward. That the Ib inhibition of GS MNs is not depressed by cutaneous afferents could thus be required to counteract the autogenetic excitatory Ia discharge, thereby limiting the GS contraction force.

Figure 14 deals with the knee extension occurring after the yield of the knee. During this phase, there is a change in the extrinsic knee torque from flexion to extension (cf. Fig. 14). The termination of EMG activity in Q closely parallels the transition from extrinsic flexion to extrinsic extension torque (Sutherland et al., 1980). The phasic activity of Q during this phase thus appears to compensate exactly for the extrinsic flexion torque which will be more pronounced when there is some force counteracting progression of the subject. This is the case, for example, when walking uphill (Fig. 14). In such a situation, the increased extrinsic knee flexion torque is still compensated for by the Q contraction, which is increased

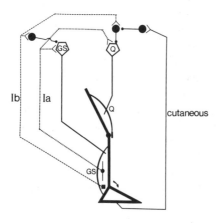

FIG. 13. Ia and Ib fiber projections from GS onto GS and Q MNs. Inhibition of Ib pathways to Q MNs by cutaneous afferents from the foot sole is also represented.

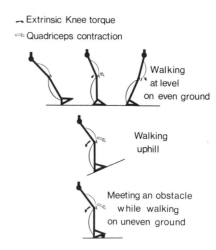

FIG. 14. Extrinsic knee torque (*black arrows*) and Q contraction (*white arrows*) during stance phase of human gait. The size of *arrows* is roughly proportional to the strength of torque and contraction.

and prolonged during the entire extension (Brandell, 1977).

Both GS contraction and initial ankle dorsiflexion are also increased, which most probably increases the Ia discharge from GS. Such an increased Ia excitatory discharge from GS, uncounteracted by Ib opposite effects (cutaneous depression), could be important in assisting increased quadriceps contraction. A similar mechanism with an automatic reflex compensation could become even more important in the case of a sudden and unexpected increase in the extrinsic knee flexion torque. Such a sudden increase occurs, for example, when the subject meets an obstacle in lifting his forefoot when walking on uneven ground (Fig. 14). Under these conditions, there is also an increased and prolonged Q contraction (C. Morin and E. Pierrot-Deseilligny, *unpublished data*), and Ia discharge from GS also increases, since the initial ankle dorsiflexion is increased (Fig. 14). That the increased Ia discharge from GS would trigger the increase in Q contraction is of considerable interest, since it would allow an automatic and very rapid compensation for increased load. Indeed, the ankle is the first joint whose angle is modified by uneveness of the ground, and the monosynaptic Ia connections from GS to Q favor a rapid Q contraction in response to such increased stretch of the GS muscle.

During the swing phase of normal walking, Ib pathways from GS to Q are no longer depressed by cutaneous stimulation of the foot sole. Such Ib pathways are not used during this period, since there is no GS contraction and thus most probably no Ib discharge from GS. The question arises then concerning the functional role of strong Ib inhibitory projections from GS to Q. They could be used in running, since during the swing phase of running there is a GS contraction (Dietz et al., 1979). The resulting Ib discharge, inhibitory for Q MNs, could limit knee extension at the end of swing phase: indeed, in running, in contrast to walking, the knee is slightly flexed at touchdown (Matsushita et al., 1974). It is also possible that these Ib pathways are used in motor activities other than locomotion. This assumption is supported by the fact that during voluntary contractions there is a complete reversal of cutaneous effects since Ib reflex pathways are facilitated by an ipsilateral cutaneous stimulation (Pierrot-Deseilligny et al., 1982). The Ib pathways from GS, for example, are facilitated by a cutaneous stimulation applied only to the anterior part of the foot sole, i.e., the skin field that may come into contact with an obstacle when this muscle is contracting. This cutaneous facilitation of Ib inhibitory pathways could serve, as proposed by Lundberg et al. (1977), as a mechanism helping to curtail an exploratory movement when the moving limb meets an obstacle.

Role of group I projections from TA onto MNs supplying knee muscles.

Unlike the cat, most of the knee flexion in man takes place at the end of the stance phase (cf. Fig. 12). Figure 12 shows that during this preswing phase, the TA contraction progressively increases while there is a plantar flexion of the ankle. This lengthening contraction of the TA most probably results in a strong Ia discharge, facilitatory for Bi MNs and inhibitory for Q MNs. These Ia effects, uncounteracted by opposite Ib effects because of cutaneous stimulation of the toes until the toe-off, could play an important role in the contemporaneous knee flexion. In particular, Ia inhibition of Q MNs could prevent a stiff gait by inhibiting the Q stretch reflex that would be otherwise triggered by the rapid knee flexion.

During the end of the swing phase, the TA muscle also exhibits a lengthening contraction, but there is a quadriceps contraction with knee extension (Fig. 12). At this moment, Ib excitatory pathways from TA onto Q are operative since they are no longer inhibited by cutaneous stimulation of ipsilateral foot sole but are facilitated by cutaneous stimulation of the contralateral foot sole. This shows the usefulness of nondepressed Ib connections from ankle to knee muscles: Ib excitatory effects from TA are desirable to counteract the opposing Ia inhibitory effects which otherwise could hinder knee extension.

The above interpretation considers the functional role of group I reflex pathways in the light of requirements of human locomotion and, though speculative, could be supported by the fact that cutaneous stimulation of the two foot soles produces complementary effects (ipsilateral depression and contralateral facilitation) on transmission of Ib effects to MNs supplying muscles operating at the knee. Such a cutaneous control of Ib reflex pathways can indeed be used in locomotion in which both foot soles are alternatively stimulated by ground contact. Thus, it is assumed that the ipsilateral cutaneous depression of Ib reflex pathways operates in association with Ia connections from ankle to knee muscles to provide Ia reflex actions which then assist heel bipedal gait during its stance phase. Foot contact with the ground could play the role of a switch in the transmission of Ib effects: during the swing phase, Ib effects are no longer depressed but are facilitated (contact of the contralateral foot sole with the ground) and counteract Ia discharges from ankle muscles, thus neutralizing the Ia reflex actions which are no longer required.

ACKNOWLEDGMENTS

The authors are grateful to Hans Hultborn for valuable comments. Our thanks are also due to Anne Rigaudie for technical assistance and to David Macgregor for scrutinizing the English. This work was supported by grants from the Pierre and Marie Curie University, INSERM (816.021) and ETP (TW 523).

Pathophysiological Aspects of Human Locomotion

*B. Conrad, R. Benecke, J. Carnehl, J. Höhne, and H. M. Meinck

Department of Clinical Neurophysiology, University of Göttingen, 3400 Göttingen, Federal Republic of Germany

Disturbances of gait represent one of the most frequent problems in neurological diagnosis and therapy. It is well known that clinical features of an altered gait can be related to distinct neurological syndromes and objective parameters of gait would be useful.

There have been numerous studies on the physiological mechanisms of walking in normal subjects (see Eberhart et al., 1968; Murray, 1967; Herman et al., 1973; E. Pierrot-Deseilligny, C. Bergego, and L. Mazieres, *this volume*) but only few on patients with neurological diseases (Peat et al., 1976; Murray et al., 1978; Chong et al., 1978; Knutsson and Richards, 1979; Dietz et al., 1981; E. Knutsson, *this volume*). Systematic quantitative analyses of disturbed gait patterns, which are a prerequisite for defining types of gait disorders, have seldom been carried out.

LOCOMOTION AS A RESULT OF INTEGRATIVE SENSORIMOTOR ACTIONS

Human locomotion is the result of many complex interactions (Grillner, 1975): (1) the subject must execute the actual locomotor movements of each limb involved according to a rather stereotyped "efferent" plan; (2) in order to accomplish purposeful locomotion, the subject must continuously adapt his movements to expected and unexpected external conditions using afferent information (Thomson, 1980; Roberts, 1976) and additional systems that anticipate the efferent plan and compare the plan to what actually occurs; (3) he must maintain equilibrium while walking by stabilizing his center of gravity within narrow limits (postural adjustment). One of the essential features of gait is that each of the individual elements of this integrative process can be regarded as influencing and modifying the others while in turn being influenced and modified by them. From the complexity of motor coordination in gait behavior, numerous sources of disturbances can arise (Eberhart, 1976).

In studying pathophysiological aspects of human locomotion in neurological diseases, it is important to differentiate between one-sided disturbances in asymmetric or hemisyndromes and symmetric disturbances or parasyndromes which affect both legs. Patients with unilateral disabilities may exhibit different types of movement disturbances, and the unaffected limb does not exhibit the normal pattern of movement because compensating mechanisms occur; on

---

*To whom correspondence should be addressed: Department of Clinical Neurophysiology, University of Göttingen Medical Faculty, 40 Robert-Kochstrasse, 3400 Göttingen, Federal Republic of Germany.

the other hand, the disabled limb may show a disturbed pattern different from that seen in bilateral disabilities. In order to reduce the complexity of the pathophysiological factors involved, we are here concerned exclusively with symmetric gait disorders, especially with paraspastic gait.

METHODOLOGICAL PREREQUISITES FOR GAIT ANALYSIS IN NEUROLOGICAL DISEASES

A kinesiological analysis should not be only confirmative (Winter, 1976) but should give specific clues in guiding the therapy program for a patient. A detailed analysis of gait requires recordings of a great number of parameters. For practical purposes, one must restrict collections of data to essential points: stride dimensions and temporal aspects of walking (cadence, stance, swing times, step length); joint motions; and EMG activity of the main leg muscles. Since neurological patients often cannot be investigated for more than a limited period of time, restricting the parameters as well as reducing the data by means of computerized analysis which shortens the turnaround time of the kinesiological examination is necessary.

From a kinesiological point of view, there is no essential difference between locomotion on a treadmill and locomotion on the ground (Visser and Berntsen, 1968; van Ingen Schenau, 1980). There may be differences with respect to afferent and efferent integrative systems involved, such as differences in visual information which aid in maintaining equilibrium and walking speed (in locomotion on ground the surroundings move in relation to subject, whereas this is not the case in treadmill locomotion). In addition, there is more stress in a treadmill situation, since speed of locomotion must be kept constant within narrow limits. However, the treadmill offers the advantage of being better able to adjust to walking speeds and lengths of stride, especially when averaging procedures are used; this allows better standardization of gait patterns and reduces handicaps caused by trailing cables.

Microswitch shoes make it easy to measure different temporal dimensions of gait by means of heel and toe contacts with the ground. Joint movements of ankle, knee, and hip are most easily recorded by goniometers. The recording of EMG activity with bipolar surface electrodes ought to include at least four representative leg muscles: tibialis anterior (pretibial flexors), medial gastrocnemius (triceps surae), biceps femoris (hamstring group), and rectus femoris (quadriceps). Gait analysis ought to be performed at a standardized speed of 2 to 3 km/hr, since the majority of patients cannot exceed this velocity. Walking speeds below 1 km/hr are inappropriate because at walking speeds this low large interindividual gait variations appear even in normal subjects. To obtain representative results for any given patient, gait analysis ought to be performed by averaging at least 8 to 10 consecutive step cycles after a steady state is reached.

PARASPASTIC GAIT AS AN EXAMPLE OF "EFFERENT" GAIT DISTURBANCE

The paraspastic state of lower extremities results, in the majority of cases, from lesions in the spinal cord. Regarding the pathophysiological mechanisms involved, one would initially suspect a lesion primarily affecting the descending systems, thus representing a disorder of efferent movement programs or of descending control of spinal locomotor mechanisms or both (see R. J. Grimm, *this volume*). The following questions are important. (1) Are there any characteristics specific to paraspastic gait disorders? (2) How consistently can they be observed? (3) How can they be quantified? (4) Are they dependent on the integrity of afferent systems?

An attempt to answer these questions

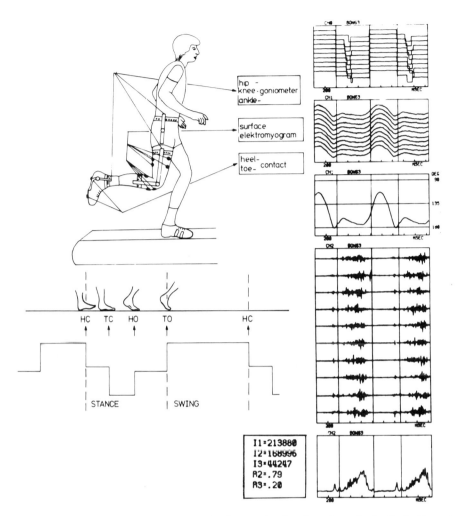

FIG. 1. Experimental arrangement and quantitative evaluation in gait analysis. **Left:** Experimental arrangement. **Bottom:** Foot contacts with corresponding digital signal configuration. *Broken vertical lines* indicate the onsets of the stance and swing phases (HC, heel contact; HO, heel off; TO, toe off). **Right:** Averaging of gait parameters (foot contact, knee joint movements, EMG) as derived from 10 consecutive step cycles. I1, integral of the rectified and averaged EMG activity of the entire cycle; I2, integral of the stance phase; I3, integral of the swing phase; R2, percentage of activity in the stance phase and the swing phase, respectively.

was made using the following method (Fig. 1). Measurements were performed on a treadmill at velocities of 2 and 3 km/hr. Shoes equipped with microswitches indicated the times at which heel or toes or both were in contact with ground. The ankle, knee, and hip joint excursions were recorded with goniometers. Records of EMGs were made using bipolar surface electrodes fixed over m. tibialis anterior, m. gastrocnemius, m. rectus femoris, and m. biceps femoris. The EMGs were fed through amplifiers with a frequency response down 3 dB at 60 and 700 Hz. The EMG data, joint movements, and foot contact times were stored on a magnetic tape. The full-wave-rectified EMG and the goniometer signals consisting of 10 consecutive step cycles were averaged by means of a PDP-11-40 computer. The time reference for triggering the averaging procedure was the strike of the heel. The PDP-11-40 calculated the mean duration of stance and swing phases. In addition, it calculated the EMG integrals of the muscles for the whole step cycle as well as separately for stance and swing phases.

Figure 2 shows an example of knee joint movements and gastrocnemius EMG in a normal subject compared to those in a paraspastic patient. In normal subjects, the typical knee rotation pattern during the swing phase is characterized by large rapid excursions into both directions of flexion (providing early foot–floor clearance) and extension (projecting the extremity forward for

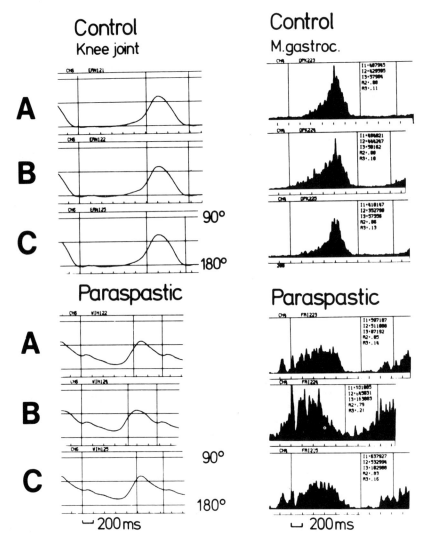

FIG. 2. Comparison of normal and pathological knee joint movements and EMG of the gastrocnemius muscle, 2 km/hr walking speed. A: Walking on a treadmill with support by holding on to a railing. B: Walking blindfolded. C: Blindfolded with support. For explanation of the integral values, see legend of Fig. 1.

the next step). At low walking speeds, knee joint movements during the stance phase (Fig. 2, top left) do not yet show the "flexion–extension wave" that is observed at higher walking speeds or in running (see also Fig. 1). Patients with paraspastic gait disorders show striking deviations from this pattern (Fig. 2, bottom left). The overall angle range of knee rotations is clearly narrowed, with a relatively more pronounced decrease in the extension phase than in the flexion phase. Walking cycles are shorter because of a decrease in length of stride and an increase in frequency of steps. The strike of the heel occurs while the knee is still in a considerably flexed position. Thereafter, the continued extension of the knee joint is necessarily performed during the stance phase. Quality and quantity of pathological gait patterns depend on the amount of sensory information available (kinesthetic afferents from the upper extremities and vision) to serve to stabilize the elements of gait (Fig. 2).

It is well known that in normal subjects (Fig. 2, top right), the gastrocnemius muscle is only active during the stance phase, gradually increasing in activity with progressive passive dorsiflexion of the ankle joint and ceasing abruptly when knee flexion sets in. In patients with paraspastic gait disturbances (Fig. 2, bottom right), the EMG patterns of the gastrocnemius are highly disintegrated. The recruitment period is broadened, with the onset of EMG activity appearing by the end of the swing phase. There is a flattening of the activity shape (the peak of EMG activity at the end of the recruitment period observed in normal subjects is lacking). Both amplitude and shape of EMG activity of the gastrocnemius show a striking dependence on the amount of sensory information that assists in gait performance, as was demonstrated above for the knee joint movement. Synchronized EMG peaks occur immediately after the heel strikes ground, supporting the assumption that exaggerated monosynaptic stretch reflexes are involved. The characteristics of altered EMG activity (Fig. 2) were basically similar for all muscles investigated (Fig. 3).

In order to answer the question of whether the deviations from normal gait patterns described are specific for paraspastic gait, we compared the paraspastic pattern to gait instability induced by alterations of various afferent systems.

GAIT PATTERN AND AFFERENT INFORMATION

There is little doubt that the precision of locomotor performance depends on feedback from visual, vestibular, and somatosensory channels. The greater the external demands, the more necessary a large spectrum of sensory clues becomes. It is, however, an open question to what extent, and how specifically, the different afferent systems contribute to the elaboration of complex sequences of motor commands in gait. In order to elucidate the role of the different sensory signals in gait performance, gait patterns were analyzed with and without participation of individual sensory channels.

The simplest model of reduced sensory information is walking blindfolded. Figure 4 (left column, D) illustrates how the activity pattern of the gastrocnemius is changed when a normal subject walks the treadmill without visual information: there is a persistent EMG activity throughout the step cycle, and gradually increasing EMG activity during the stance period (left column, A) can hardly be identified. The length of step is also distinctly reduced, but the "flatfoot" period of simultaneous ground contact of heel and toes (not illustrated) and the ratio of double support period to stride period are increased.

Anesthesia of the soles of both feet by means of a bilateral infiltration block of both plantar nerves does not affect the nor-

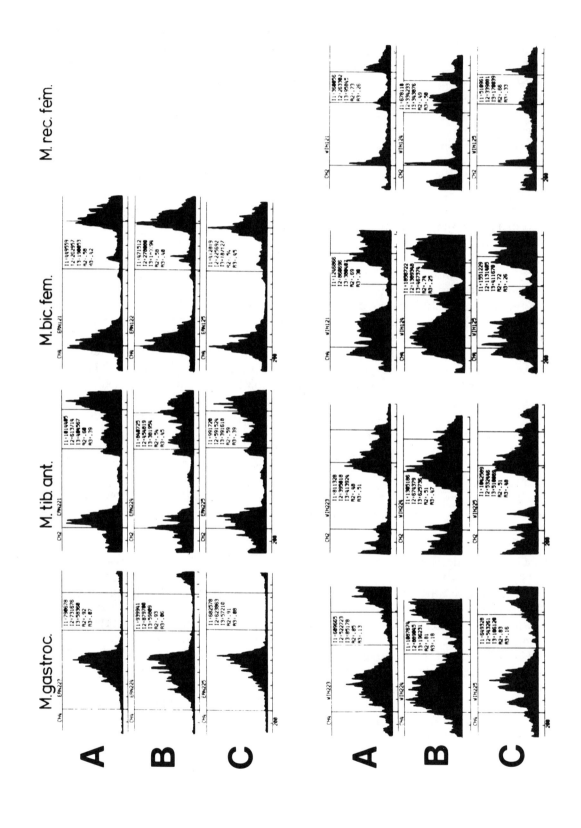

mal recruitment of the gastrocnemius (Fig. 4A–D) and merely diminishes the total EMG integrals, suggesting a nonspecific facilitatory effect of cutaneous afferents from the sole of the foot on the gastrocnemius MN pool.

Other models for studying the effect of reduced sensory information on motor gait performance can be taken from neurological patients with isolated sensory deficits. Figure 5 shows the altered pattern of gastrocnemius activity in a case of spinal ataxia resulting from tabes dorsalis with almost complete loss of deep sensation and tendon reflexes. It is easily recognized that length of step and EMG during stance phase deviate from the pattern of a normal subject mainly with respect to the condition of walking freely (Fig. 5D). Slight kinesthetic information, e.g., touching a railing with a fingertip (Fig. 5B, C), normalized the pattern.

It was of interest to study locomotor pattern in lesions of the vestibular system. Since the vestibular system plays an important role in the regulation of posture, deviation from normal locomotor pattern may be caused mainly by a disturbed interrelationship of movement and posture. Figure 6 illustrates various gait parameters (foot contact, knee joint movement, EMG of gastrocnemius and tibialis anterior) of a young male patient 4 weeks after his recovery from acute bacterial meningitis. The disease had resulted in total and irreversible lesions of both eighth cranial nerves. With kinesthetic support (Fig. 6A), all gait parameters were normal and independent of the presence of visual information (Fig. 6C). When the patient was walking freely, however (Fig. 6B), all gait parameters were distinctly changed: there was a small but consistent decrease in the length of step, affecting the swing phase more than the stance phase; the period of "flat-foot" contact (heel and toe) was remarkably increased; the ratio of double-support to stride period was increased (not illustrated); the strike of the heel occurred at a more flexed position of the knee; and the muscle activity patterns of gastrsocnemius and tibialis anterior during the stance phase were highly disintegrated.

PROTECTIVE GAIT MECHANISMS

Pathologic gait patterns caused by either a predominantly efferent or a predominantly afferent type of sensorimotor disturbance show a number of common features. They all exhibit a shortening of the length of step, a prolonged contact phase of both heel and toe ("flat-foot" phase), an increase in the ratio of double-support period to stride period, and a striking increase in total EMG activity, as well as a broadening of the recruitment range and a disintegration of the activity profile in all the muscles investigated. These features are mainly observed when no additional kinesthetic support is available.

It might be argued that these common characteristics of altered gait patterns might be a stereotyped pathological sign for various sensorimotor deficits. However, this is not necessarily the case. Similar deviations from normal patterns also occur in healthy subjects in situations when—in reality or in their imagination—stability is threatened, e.g., in the expectation of walking on slippery ground. Whenever there is a real or imagined threat, normal subjects and patients rely more strongly on locomotion strategies with prolonged bipedal contact and monopedal "flat-foot" contact and with a reduction in walking speed. Apparently,

FIG. 3. Average EMG activity of four leg muscles from 10 consecutive step cycles in a healthy subject (above) and a patient with a paraspastic syndrome (below). A: Walking with kinesthetic support (holding on to a railing). B: Walking without support. C: Walking blindfolded with kinesthetic support. In the healthy subject, detectable EMG activity in the rectus femoris failed to appear.

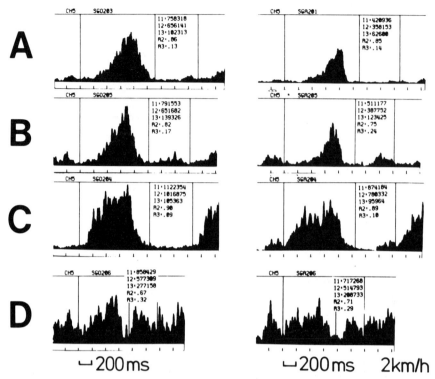

FIG. 4. Changes in normal average EMG activity while walking blindfolded **(left)** and after anesthesia of the soles of the feet **(right). Left** before, **middle** after anesthesia of the soles of the feet. **A:** Walking with kinesthetic support of both arms. **B:** Walking with kinesthetic support of the right arm. **C:** Walking without kinesthetic support. **D:** Walking blindfolded without kinesthetic support.

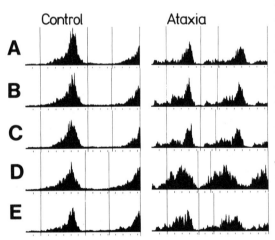

FIG. 5. Average EMG activity of the gastrocnemius muscle in a healthy subject **(left)** and a patient with spinal ataxia (tabes dorsalis; **right). A:** Walking with kinesthetic support from both arms. **B, C:** Walking with kinesthetic support from the right **(B)** or left arm **(C). D:** Walking freely without kinesthetic support. **E:** Walking blindfolded with kinesthetic support. *Time scale:* 200 msec.

the more complex a locomotive demand, the lower the threshold for changing over to using protective gait mechanisms. The features of this protective strategy, executed automatically, demonstrate a general tendency of the CNS to stabilize balance at the expense of walking speed (Herman et al., 1976). These gait characteristics, particularly the prolonged "flat-foot" phase and double-support period, are better suited for preventing the danger of unexpected disturbance of balance. The marked increase in EMG activity in each muscle that occurs while walking without additional kinesthetic information corresponds to a stiffening of the joints. The observation of a strong increase in the integrals of EMG activity in all muscles while walking freely, as compared to supported walking, confirms that a greater expenditure of energy per step takes place in the unsupported walking state (Ralston, 1976).

In light of these considerations, all of the

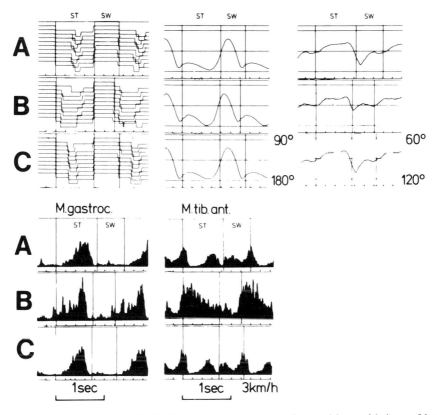

FIG. 6. Mechanical (above) and EMG (below) gait parameters in a subject with loss of both labyrinthine functions. Top: Foot contact (*left column*, 10 consecutive steps), average knee (*middle column*), and ankle joints (*right column*). Bottom: Average EMG activity of the gastrocnemius muscle (*left column*). Vertical lines indicate beginning and end of stance (ST) and swing (SW) phases.

above mechanisms are apparently nonspecific. They merely represent stabilizing and thus protective actions. The relatively low walking speed alone in the experiments demonstrated, which is in itself combined with reactions of a more postural nature, cannot be the prime reason for the appearance of a protective gait pattern, since only minor effects on the quality of gait performance can be observed in normal subjects within the applied range of speed. Therefore, with regard to pathological gait in neurological patients, it must be considered that specific gait disturbances caused by lesions of certain pathways are widely superimposed by nonspecific protective gait mechanisms occurring in almost all gait disorders. These protective gait strategies strongly depend on psychological elements such as emotional stress, anxiety, and consciousness of gait impairment.

In order to uncover specific changes of altered efferent motor aspects, for example, in paraspastic gait, presumptive or real dangers for the subject have to be reduced as much as possible (e.g., by supplying additional support in stabilizing balance). The less a gait pattern is positively influenced by additional kinesthetic information via the upper extremities, the fewer nonspecific protective factors actually contribute to the alterations of gait pattern. From these considerations, it becomes obvious that the proportion of specific pathological changes of gait primarily caused by afferent disturbances (e.g., spinal ataxia) cannot easily be differentiated from nonspecific gait mechanisms participating in this gait disorder.

For the above reasons, analyzing rhythmic step-like leg movements by completely eliminating the problem of body support may be preferable. This can be achieved by analyzing locomotion-like movements during bicycling. This method has an additional advantage, since it can also be employed in patients unable to walk (Benecke et al., 1980; R. Benecke, B. Conrad, H. M. Meinck, and J. Höhne, *this volume*).

Motor Control Mechanisms in Health and Disease,
edited by J. E. Desmedt.
Raven Press, New York © 1983.

Anatomical and Functional Organization of Reflexes Involving the Trigeminal System in Man: Jaw Reflex, Blink Reflex, Corneal Reflex, and Exteroceptive Suppression

*B. W. Ongerboer de Visser

Department of Clinical Neurophysiology, Municipal Hospitals Slotervaart, Louweswegg, Amsterdam 1066EC, The Netherlands

Electrophysiological methods may provide information concerning reflexes passing through the trigeminal system such as the jaw reflex, blink reflex, corneal reflex, and the reflexogenic exteroceptive suppression (or exteroceptive silent period) in jaw-closing muscles that cannot be obtained by direct observation. Correct interpretation of abnormal findings in relation to the anatomical situation may help to determine and localize disorders.

The trigeminal nerve (fifth cranial nerve) can be divided into two parts: a sensory root conveying superficial sensation from the skin and the subjacent mucous membranes of the ipsilateral side of the face and a motor root (Fig. 1). The cell bodies of the afferent fibers running through the sensory root are located in the trigeminal ganglion (semilunar or Gasserian ganglion). The peripheral processes of these ganglion cells are distributed via the three large branches of the nerve, i.e., ophthalmic (first branch), maxillary (second branch), and mandibular (third branch). The central processes come together in the middle of the brachium pontis. The ophthalmic division runs into the cavernous sinus and leaves the cranial cavity via the superior orbital fissure. It supplies the skin of the forehead (supraorbital nerve) as far as the vertex, the upper eyelid, the skin of the anterior half of the nose, the conjunctiva, cornea, and iris, and the mucous membranes of the frontal sinus and upper nose. The maxillary division leaves the cranium via the foramen rotundum and supplies the skin of the posterior half of the nose, the lower eyelid, the upper cheek (infraorbital nerve), the upper anterotemporal region, the mucous membranes of the lower nose, the upper jaw, the oral part of the palate, and the upper teeth. The mandibular division leaves the cranium via the foramen ovale and supplies the skin of the ipsilateral half of the chin (mental nerve), the posterior part of the cheek, the lower jaw, part of the tongue, the floor of the mouth, and the lower teeth.

On reaching the medial portion of the branchium pontis, the trigeminal sensory root bifurcates into ascending and descending divisions (Figs. 1 and 2). The ascending division terminates in the principal (main) sensory nucleus and is concerned with touch. The descending root (trigeminal spinal tract) continues through the lateral por-

*Department of Clinical Neurophysiology, Municipal Hospitals Slotervaart, Louweswegs, Amsterdam 1066EC, The Netherlands.

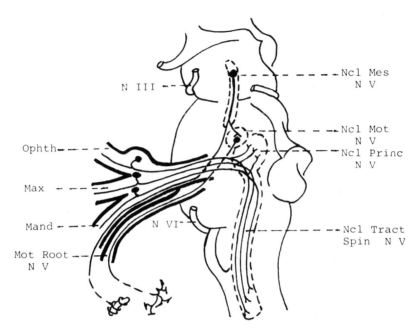

FIG. 1. Trigeminal sensory and motor pathways projected on the lateral surface of the brainstem. It is suggested that, for the jaw reflex, afferent fibers from muscle spindles travel in the mandibular division of the sensory root, and efferent fibers from the trigeminal motor nucleus in the motor root. Abbreviations used: Ncl Mes NV, trigeminal mesencephalic nucleus; Ncl Mot NV, trigeminal motor nucleus; Ncl Princ NV, trigeminal principal sensory nucleus; Ophth, trigeminal ophthalmic division; Max, trigeminal maxillary division; Mand, trigeminal mandibular division; Mot Root NV, trigeminal motor root; NIII, oculomotor nerve; NV, trigeminal nerve; NVI, abducens nerve.

tion of the tegmentum of the pons and into the lateral part of the medulla oblongata. Fibers in this root terminate on cells of the closely adjacent nucleus of the descending root (spinal trigeminal nucleus). These fibers relay impulses of pain and temperature and possibly also touch.

From experimental findings in animals, it was conjectured that in man proprioceptive fibers from spindles in the jaw-closing muscles pass through the motor root (Willems, 1911; Allen, 1919; Corbin and Harrison, 1940; Szentagothai, 1948; McIntyre, 1951; McIntyre and Robinson, 1959; Smith et al., 1968). Entering into the pons, they course adjacent to the motor and principal sensory nuclei and ascend in a bundle (trigeminal mesencephalic root) to their cell bodies, which are unipolar and located in the mesencephalic nucleus of the fifth cranial nerve (Figs. 1 and 2). Branches of these cells descend through the mesencephalic root to the trigeminal motor nucleus. Motor fibers leave the motor center in the dorsolateral pons. They pass first under the sensory fibers and then under the Gasserian ganglion, join the mandibular division of the fifth nerve, and leave the base of the skull through the foramen ovale. The trigeminal motor fibers supply the masseter, pterygoid, temporal, and mylohyoid muscles, the tensors tympani and palati, and the anterior belly of the digastric muscle.

The blink and corneal reflexes are mediated through the trigeminal and facial nerve, whereas the jaw reflex and silent period reflex in jaw-closing muscles induced by exteroceptive stimuli to the orofacial regions are mediated only by the trigeminal nerve.

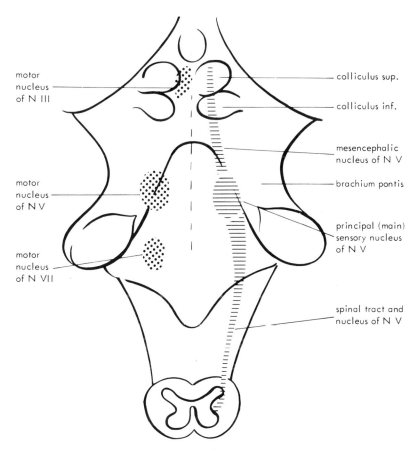

FIG. 2. Sensory trigeminal nuclei on the right side and the motor nuclei of the third (oculomotor nerve), fifth (trigeminal nerve), and seventh (facial nerve) cranial nerves on the left side projected on the dorsal surface of the brainstem.

THE JAW REFLEX

The jaw reflex can be induced by a tap of the tendon hammer on the chin, depressing the lower jaw. The response is a bilateral contraction of the jaw elevators. The reflex was first reported in 1886 by de Watteville and has also been called jaw jerk, masseter reflex, temporomasseter reflex, and mandibular reflex.

The jaw reflex can be recorded electrically by using a tendon hammer that has a microswitch to trigger the sweep of a dual-trace persistence oscilloscope at the moment of the tap (Kugelberg, 1952; McIntyre and Robinson, 1959; Goodwill, 1968; Kimura et al., 1970; Ongerboer de Visser and Goor, 1974; Godaux and Desmedt, 1975a). The right and left masseter responses are recorded simultaneously with needle or surface electrodes (Fig. 3). The reflex can be evoked bilaterally in normal subjects up to the age of about 70 years but can be absent over that age. The wide range of the latency times, 6.4 to 9.2 msec, means that comparison between different subjects is of little practical value. However, comparison between latency times on both sides in a single subject appears to be of great value, because there is no difference between the latency times in normal subjects, or the difference is very small and

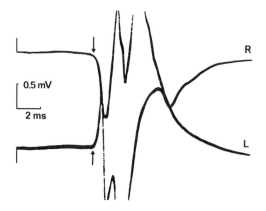

FIG. 3. Normal jaw reflex response in the right and left masseter muscle. R, right; L, left in this and subsequent figures.

should not exceed 0.5 msec (Ongerboer de Visser and Goor, 1974; Ferguson, 1978).

The current view of the jaw reflex pathway in man is largely based on work by McIntyre and Robinson (1959). They believed that afferent fibers from muscle spindles and efferent fibers share a common pathway in the motor root of the trigeminal nerve, as in the cat. However, other authors have recently reported jaw reflex abnormalities in patients with peripheral lesions of the trigeminal nerve who showed no clinical or EMG evidence of masseter denervation (Goor and Ongerboer de Visser, 1976; Ferguson, 1978). Further, we recently studied prospectively jaw-reflex latencies in a group of 16 patients, who underwent surgery of the sensory trigeminal root as treatment for trigeminal neuralgia (B. W. Ongerboer de Visser, *this volume;* 1982). In all of them, masseter muscles were functioning normally both clinically and electromyographically before and after surgery. Postoperation reflex responses disappeared in nine and were delayed in seven patients ipsilaterally to the lesion. These data suggest that the proprioceptive fibers of the jaw reflex pass mainly, if not entirely, through the sensory division of the fifth cranial nerve, probably the mandibular, and not the motor root (Fig. 1). From the observation that in peripheral lesions of the mandibular branch an absent reflex can be found, it may be deduced that afferent fibers for the human jaw reflex do not cross the midline in the brainstem.

With respect to the central pathways, disturbed jaw reflex lateral to clinically demonstrated midbrain lesions have been reported (Hufschmidt and Spuler, 1962; Ongerboer de Visser and Goor, 1976a), and recently, abnormal reflex responses were found on the side of midbrain lesions verified by postmortem findings *(vide infra)*. This means that as in animals, the jaw jerk in man seems to run centrally through the midbrain, probably through the mesencephalic nucleus of the trigeminal nerve. From the total length of the afferent and efferent limb pathways, the diameter of the largest fibers in the trigeminal motor root, and the mean latency time of the reflex, it can be postulated that the jaw jerk in man is mediated by a reflex arc of only two neurons (McIntyre and Robinson, 1959).

THE BLINK REFLEX

Reflex blinking, i.e., bilateral contraction of the orbicularis oculi muscle evoked by a tap on one side of the forehead, was first described by Overend in 1896. Rushworth (1962b) reviewed the first reports on facial reflexes elicited from many areas of the face. In the more recent literature, the blink reflex is also called the orbicularis oculi reflex, but from the clinical view the original term seems more appropriate.

The first EMG examination of the blink reflex was undertaken by Kugelberg in 1952. The substitution of supramaximal electrical stimulation of the supraorbital branch of the trigeminal nerve for the mechanical tap led to standardization of the technique. The reflex responses are recorded simultaneously with needle or surface electrodes from the lower eyelids. The blink reflex is shown by EMG analysis to consist of two discharges in the orbicularis

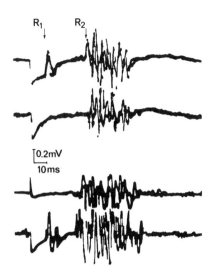

FIG. 4. Normal blink reflex. R_1 is the early ipsilateral response, and R_2 is the late bilateral response. *Upper pair of traces:* reflex responses after electrical stimulation of the right supraorbital nerve. *Lower pair of traces:* reflex responses after electrical stimulation of the left supraorbital nerve.

oculi muscles (Fig. 4). The first discharge is an early and brief response, R1, and occurs ipsilateral to the side of stimulation with a latency of approximately 10.6 msec. The second discharge is a more prolonged response, R2, and occurs both ipsilateral and contralateral to the side of stimulation with a latency of 21 to 43 msec (Kimura et al., 1969). The common afferent limb of the blink reflex is formed by the sensory root of the trigeminal nerve, and the facial nerve is the common efferent limb, as is shown by the general finding that peripheral damage to their fibers produces abnormal reflex responses. The difference of the two responses is attributed to differences in their central neural pathways.

Most authors assume on the basis of findings in the cat and the clinical and electrophysiological features seen in patients with pontine lesions caused by strokes or multiple sclerosis that the early blink reflex is conducted through the pons and is relayed through an oligosynaptic arc including one or more interneurons (Tokunaga et al., 1958; Kimura, 1970, 1971, 1973, 1975; Namerow and Etamadi, 1970; Lindquist and Martenson, 1970; Hiraoka and Shimamura, 1977; Trontelj and Trontelj, 1978).

Some authors concluded from electrophysiological studies that the early reflex is cutaneous and nociceptive (Lindquist and Martenson, 1970; Shahani and Young, 1973a; Penders and Delwaide, 1973). However, others (Kugelberg, 1952; Rushworth, 1962b; Gandiglio and Fra, 1967; Bender, 1968; Ferrari and Messina, 1968; Brown and Rushworth, 1973; Moldaver, 1973; Messina, 1975) presume that the reflex is myotatic and proprioceptive in nature, although muscle spindles have not been found in the facial muscles of man or several other mammalian species. If the early blink reflex is myotatic and transmitted via the proprioceptive mesencephalic nucleus, it may also be affected by lesions involving the mesencephalon. A study of the myotatic jaw reflex, which runs through the mesencephalic trigeminal nucleus, with the early blink reflex in the same patient may clarify this possibility. Recently, we have demonstrated in three patients absent jaw reflexes ipsilateral to unilateral midbrain lesions although the early blink reflexes were normal (Fig. 5). Postmortem findings revealed a tumorous lesion (Fig. 6) in two and a vascular lesion (Fig. 7) in the other patient. The three lesions also involved other structures in one half of the midbrain, the mesencephalic tract, and the nucleus of the trigeminal nerve, but the trigeminal motor nucleus and fibers in the pons were not affected (Figs. 6 and 7). These findings make it very likely that the early blink reflex does not pass through the mesencephalic nucleus and is therefore not myotatic in nature. However, the exact pathway that leads from the supraorbital branch of the sensory trigeminal root to the ipsilateral facial nucleus in the pons is unknown.

The late blink reflex is considered to be nociceptive. From findings in patients who

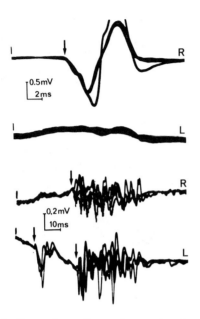

FIG. 5. Absent jaw reflex and normal early blink reflex on the right side. The jaw reflex on the left side and the late blink reflex are normal. The postmortem finding is shown in Fig. 6.

have undergone a tractotomy of the trigeminal nerve (Sjoqvist operation; Kugelberg, 1952) and in patients with Wallenberg's lateral medullary syndrome (Kimura and Lyon, 1972; Ongerboer de Visser and Kuypers, 1978) with involvement of the spinal trigeminal tract and its nucleus, it may be concluded that fibers of the first-order neurons that are responsible for the late blink reflex pass through the spinal tract of the trigeminal nerve in the lateral medullary region and terminate in the area of the ipsilateral spinal trigeminal nucleus (Fig. 8). The latter authors postulated that from this nucleus the late reflex is conducted through polysynaptic medullary pathways running both ipsilaterally and contralaterally to the stimulated side before making connections with the ipsilateral and contralateral facial nucleus, respectively. From their anatomical examination, it is suggested that these trigeminofacial connections pass through the lateral reticular formation of the lower brainstem lying medial to the spinal trigeminal nucleus (Fig. 8).

With respect to the late reflex pathways from the spinal fifth nerve complex to the facial nuclei, it is of interest to note that according to Holstege and Kuypers (1977) and Holstege et al. (1977), the bulbar lateral reticular formation in the cat contains a lateral and a medial propriobulbar fiber system; the former distributes fibers to the bulbar motor nuclei mainly ipsilaterally, whereas the latter tends to distribute fibers to these nuclei bilaterally. The medial propriobulbar system also contains fibers arising from neurons at the level of the obex and passing to the intermediate facial subnucleus on both sides. This subnucleus is thought to innervate the orbicularis oculi muscle (Courville, 1966). If these connections also exist in man, they could support the assumption that ipsilateral and contralateral trigeminofacial connections subserving the late reflex pass through the bulbar lateral reticular formation.

THE CORNEAL REFLEX

The corneal reflex can be evoked clinically by a light mechanical stimulus, such as a wisp of cotton wool, on the cornea. The response is a bilateral blinking following contraction of the orbiculares oculi muscles. The reflex is considered to be nociceptive and serves to protect eyes from injury. Electrophysiologically, the reflex can be evoked by touching the cornea with a 2-mm-diameter metal sphere that is connected to an electronic trigger circuit. As soon as contact between subject and sphere is established, the touch-sensitive circuit delivers a pulse which triggers the sweep of a dual trace persistence oscilloscope (Ongerboer de Visser et al., 1977). The reflex responses are recorded simultaneously from the lower eyelids with surface electrodes (Fig. 9). The actual lid movement seen with the recorded corneal reflex is as-

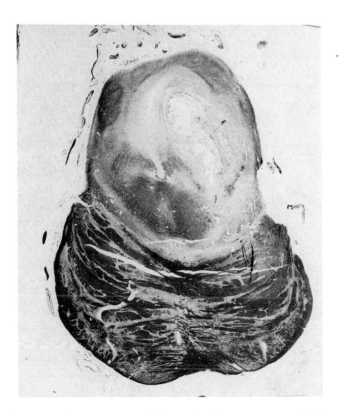

FIG. 6. Cross-section through the mesencephalon, showing a tumorous lesion on the right side comprising the mesencephalic tract and its nucleus (see also Fig. 5). Klüver, ×3.

sociated with a bilateral response, R2, and is not preceded by an early reflex response as in the blink reflex (Kugelberg, 1952; Rushworth, 1962b; Magladery and Teasdall, 1961; Tatcher and Van Allen, 1971; Ongerboer de Visser et al., 1977). The latency times range widely from 36 to 64 msec in normal subjects, which means that interindividual comparison is of little practical value. However, comparison of response latencies in a single subject is of great value, because normally there is a high degree of constancy of the differences between the latency times.

A network of free nerve endings within the cornea forms the receptive part of the corneal reflex. Afferent impulses are conducted along small unmyelinated and myelinated fibers in the long ciliary branch of the ophthalmic division of the trigeminal nerve to the pons (Windle, 1926; Tower, 1940; Lele and Weddell, 1959). From the pons, they descend into the medulla oblongata along fibers of the spinal trigeminal tract lying in the lateral medullary region and terminate on cells of the ipsilateral spinal nucleus of the trigeminal nerve (Windle, 1926; Sjoquist, 1938; Rowbotham, 1939; Smyth, 1939; Moffie, 1971). From recent electrophysiological findings (Ongerboer de Visser and Moffie, 1979, 1980) in patients with lateral medullary infarctions (Wallenberg syndrome), it is concluded that the corneal reflex is conducted along medullary polysynaptic pathways running both ipsilaterally and contralaterally from the stimulated cornea before connecting, respectively, with the ipsilateral and con-

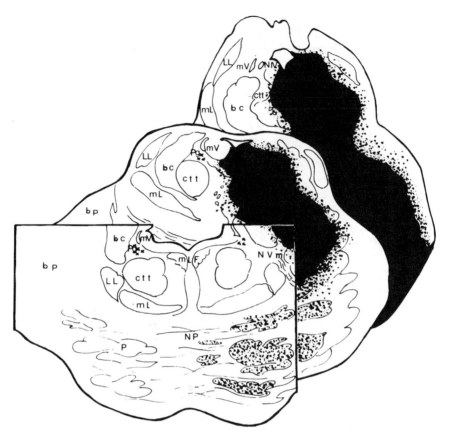

FIG. 7. Location and extent of the infarction in the midbrain and rostral portion of the pons investigated in detail. *Fine dots* indicate the area of fiber degeneration. Abbreviations used: bc, brachium conjunctivum; bp, brachium pontis; ctt, central tegmental root; ll, lemniscus lateralis; mlf, medial longitudinal fasciculus; ml, medial lemniscus; mV, ramus mesencephalicus; NP, nuclei pontis; P, pyramidal tract; Pg, nucleus pigmentosus (cocus coerulus); NVm, nervus masticatorius; NIV, nervus trochlearis.

tralateral facial nucleus. From anatomical findings, it is suggested that the ascending pathways from the spinal trigeminal complex to the facial nuclei are located in the lateral reticular formation of the lower brainstem (Fig. 10).

These data lead to the idea that the corneal reflex and the late blink reflex utilize similar trigeminofacial connections passing through the medulla oblongata. However, there are important differences between the two reflexes. The corneal reflex shows a high degree of constancy of intraindividual latency times and no facilitation or habituation (Magladery and Teasdall, 1961; Ongerboer de Visser et al., 1977), which is in contrast to the late blink reflex which clearly shows these phenomena (Penders and Delwaide, 1973; Desmedt and Godaux, 1976; Boelhouwer and Brunia, 1977; Ferguson et al., 1978). Further, afferent impulses for the late blink reflex pass to a large extent along myelinated fibers of all diameters and to a lesser extent along small unmyelinated fibers of the supraorbital branch of the ophthalmic division of the trigeminal nerve (Windle, 1926). Afferent fibers in the ciliary branch of the ophthalmic trigeminal division from the cornea, on the other hand, are for the most part small in diameter and slow

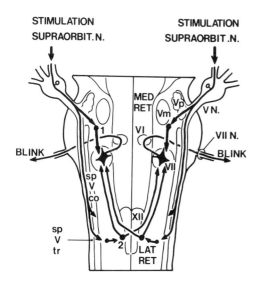

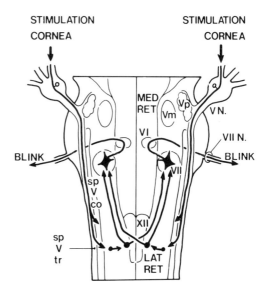

FIG. 8. Presumed location of the bulbar interneurons subserving the two responses of the blink reflex: (1) the ipsilateral early response (R1), and (2) the bilateral late response (R2). Abbreviations used. VN, trigeminal nerve; Vm, trigeminal motor nucleus; Vp, trigeminal principal sensory nucleus; spVtr, spinal trigeminal tract; spVco, spinal trigeminal complex; VI, abducens nucleus; VII, facial nucleus; VIIN, facial nerve; XII, hypoglossal nucleus; MED RET, medial reticular formation; LAT RET, lateral reticular formation.

FIG. 10. Presumed location of the bulbar interneurons subserving the two responses of the corneal reflex. For key to abbreviations, see legend to Fig. 8.

in conduction (Tower, 1940; Lele and Weddell, 1959). In the spinal descending trigeminal root, they comprise a small fiber bundle and a narrow fiber spectrum in comparison with afferent fibers of the late blink reflex. Because of these anatomical and physiological differences, it can be expected that lesions of the trigeminal nerve or the lateral medullary region will affect the corneal reflex more often than the late blink reflex.

EXTEROCEPTIVE SUPPRESSION IN MASSETER MUSCLES

The EMG silent period (SP) refers to a transitory relative or absolute decrease of EMG activity evoked in the midst of an otherwise sustained contraction (Shahani and Young, 1973b).

The SP in human masseter muscles was first described by Hoffmann and Tonnies (1948) as the inhibitory component of the tongue–jaw reflex seen after electrical stimulation of the tongue. Later authors described a silent period occurring in masse-

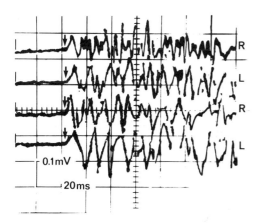

FIG. 9. Normal corneal reflexes. *Upper pair of traces:* stimulation applied to the right cornea. *Lower pair of traces:* stimulation applied to the left cornea.

ter muscles after a tap on the chin, an electrical stimulus to the masseter muscle belly, the gums, or the mucosa inside the mouth, or an acoustic stimulus (Struppler et al., 1960; Huffschmidt and Spuler, 1962; Shahani and Young, 1973b; Meier-Ewert et al., 1974; Godaux and Desmedt, 1975b; Desmedt and Godaux, 1976).

We studied the SP reflex in masseter muscles during maximal contraction after square-wave electrical stimulation lasting 0.2 msec at 100 to 200 V to the mental or infraorbital nerve (Ongerboer de Visser and Goor, 1976b). The SPs were recorded simultaneously from both masseter muscles with needle or surface electrodes (Fig. 11). The subject maintains maximal clenching of the jaws, which is monitored by both the subject and the observer through the loudspeaker of the EMG and by the observer via the interference pattern on the oscilloscope. The duration of the SP in normal persons after mental nerve stimulation ranges widely from 15 to 98 msec (mean duration 52.9 msec); that after infraorbital nerve stimulation ranges from 13 to 94 msec (mean duration 51.4 msec). Some of the SPs show one or more interruptions caused by a recurrence of voluntary activity (Fig. 11b). The latency time of the first masseter SP ranges from 10 to 14 msec (mean duration 11.6 msec). The difference between the right and left latency times does not exceed 2.0 msec. The intraindividual constancy of the latency time and the small difference between the latency times on the two sides make recording of the SP reflex valuable for objective assessment of trigeminal nerve lesions. From findings in normal subjects and in patients with trigeminal nerve or brainstem lesions, it is concluded that the SP reflex evoked with the above-described technique is cutaneous in origin. Therefore, the reflex is called the cutaneous SP reflex in masseter muscles. Another designation that has been proposed for such effects is "exteroceptive suppression" (Godaux and Desmedt, 1975b; Desmedt and Godaux, 1976), and it seems important to stress the exteroceptive origin of the phenomenon, since it must be clearly distinguished from the classical "silent period" of proprioceptive origin.

Anatomical data were derived from patients with the following trigeminal nerve or brainstem lesions. A bilaterally absent SP reflex was found in patients with a selective trigeminal sensory root interruption when the analgesic side was stimulated. The reflex was bilaterally normal when the unaf-

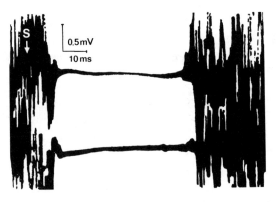 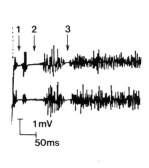

FIG. 11. Left: Example of a normal cutaneous silent period reflex in the right (upper trace) and left (lower trace) masseter muscle after stimulation of the right mental nerve. No breakthrough activity is seen. Right: Example of a normal cutaneous silent period reflex in the right (upper trace) and left (lower trace) masseter muscle after stimulation of the right mental nerve. The reflex consisted of three SPs (1, 2, 3) and of two recurrences of voluntary activity.

fected side was stimulated (Fig. 12). Further, in patients with a selective trigeminal motor root interruption, a normal SP reflex could be obtained on the normal side when the mental nerve on that side or on the contralateral paretic side was stimulated (Fig. 13). These findings indicate that the SP reflex is conducted through the trigeminal sensory root and crosses the midline.

In patients with unilateral midbrain lesions in whom the jaw reflex was absent on that side and in patients with lateral medullary lesions in whom late blink and corneal reflexes were absent bilaterally when the affected side was stimulated, the cutaneous SP reflex appeared to be normal after stimulation on both the affected and unaffected side. However, the cutaneous SP reflex was abnormal in patients with unilateral pontine lesions and also in a patient with a midline lesion located in the pons. In the former patients, the SP reflex could only be evoked on the normal side (Fig. 14), and in the latter patient, the SP reflex appeared only on

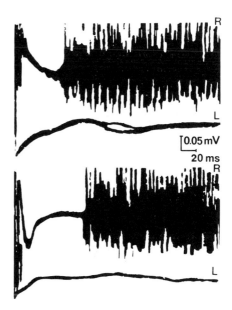

FIG. 13. Cutaneous silent period in masseter muscles in a patient with functional loss of the trigeminal motor root on the left side. *Upper pair of traces:* stimulation of the right mental nerve. *Lower pair of traces:* stimulation of the left mental nerve.

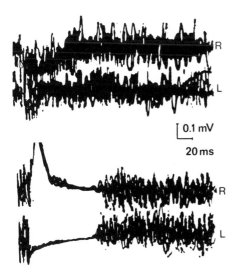

FIG. 12. Cutaneous silent period in masseter muscles in a patient with functional loss of the trigeminal sensory root on the right side. *Upper pair of traces:* stimulation of the right mental nerve. There is no inhibition. *Lower pair of traces:* stimulation of the left mental nerve.

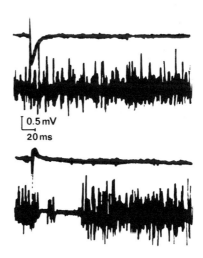

FIG. 14. Cutaneous silent period in the left masseter muscle in a patient with a tumor in the right half of the pons and showing a functional loss of both the trigeminal sensory and motor root on the right side. *Upper pair of traces:* stimulation of the right mental nerve. *Lower pair of traces:* stimulation of the left mental nerve.

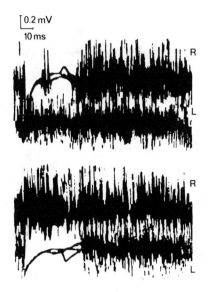

FIG. 15. Cutaneous silent period in masseter muscles in a patient with a midpontine tumor. No crossing over takes place. *Upper pair of traces:* stimulation of the right mental nerve. *Lower pair of traces:* stimulation of the left mental nerve.

the stimulated side, but no SP was seen in the contralateral masseter muscle (Fig. 15). These findings suggest strongly that the cutaneous SP reflex in masseter muscles is relayed through the pons and crosses the midline in this region.

With respect to the cutaneous SP reflex and its pontine pathways from the trigeminal sensory root to the trigeminal motor nuclei, it is of interest to note that the findings in man are consistent with the findings in the cat made by Kidokoro et al. (1968). After electrical stimulation of trigeminal exteroceptive fibers, they invariably found inhibition in trigeminal motoneurons. Furthermore, their findings suggest that the SP reflex is conducted along the region dorsomedial to the trigeminal motor nucleus. From this nucleus, fibers are distributed ipsilaterally and contralaterally to the motor nuclei of the fifth nerve. If these connections also exist in man, they would support the assumption that the arc of the cutaneous SP reflex in man is restricted to the pons.

GENERAL REMARKS

The abovementioned reflexes utilize different pathways, and so it may be expected that a combined electrophysiological examination can provide valuable supplementary clinical information (Yates and Brown, 1981). It must be kept in mind that recording is indispensable because some of the reflexes such as inhibitory phenomena and the early blink reflex cannot be evaluated by clinical means only. This also holds true for other factors such as differences between the jaw reflexes in the two sides, the two components of the late blink reflex and the corneal reflex, and the latencies of all reflexes.

Human Jaw Reflexes

*J. P. Lund, Y. Lamarre, G. Lavigne, and G. Duquet

Faculty of Dental Medicine, Department of Physiology, and Center for Research in Neurological Sciences, University of Montreal, Montreal, Quebec H3C 3J7, Canada

This chapter reviews observations and results on the human jaw reflexes with the view of assisting pathophysiological studies of clinical patients with disorders of the orofacial neuromuscular system. Further data can be found in reviews by Matthews (1975), Dubner et al. (1978), and Hannam (1979).

LOADING JAW-CLOSING MUSCLES

The jaw-jerk or masseteric reflex is readily elicited by stretch of the jaw-closing muscles (Hugelin and Bonvallet, 1956; McIntyre, 1951). Although it is the trigeminal analog of monosynaptic myotatic reflexes in other parts of the body, the short conduction distances produce responses with a latency as short as 7 to 12 msec following taps to the chin of normal subjects (McIntyre and Robinson, 1959; Hufschmidt and Spuler, 1962; Goodwill, 1968; Hannam, 1972; Matthews, 1976).

In our studies and those of Cooker et al. (1980), the jaw-jerk reflex was produced by applying pressure to the lower incisor teeth with a lever driven by a torque motor (Lamarre and Lund, 1975) while the upper teeth bit onto a fixed bar. The time between the start of the torque pulse, measured by a strain gauge attached to lever arm, and the first deflection of the reflex response was slightly less than most latencies measured in experiments in which the chin was tapped with a clinical hammer, probably because the tissues covering the chin are less rigidly coupled to the mandible than are the teeth. When the subjects maintained the jaw in a steady position, the latency ranged from 6 to 8 msec (mean 7.5 msec), but it was sometimes as short as 5 msec when the torque pulses were applied as the jaw was closing. The segmental EMG response was followed shortly by an increase in the velocity of jaw closure proportional to its amplitude. This is illustrated in Fig. 1, where both the EMG amplitude and the mechanical response are increased by the Jendrassik maneuver. Cooker et al. (1980) also reported that the masseteric reflex response to step loading came after about 8 msec and produced "a very substantial force."

If the subjects were instructed to oppose the action of the torque pulse, some of them were able to increase the amplitude of the segmental response (Lund et al., 1978). The longer-latency activity that these subjects also generated began at approximately the same latency as voluntary jaw closure in response to electrical stimulation of the lip. Neither we nor Cooker et al. (1980) could find evidence of long-loop responses (cf. Marsden et al., 1976a) to loading in the jaw muscles (Lamarre and Lund, 1975).

---

*To whom correspondence should be addressed: Centre de Recherches en Science Neurologique, Université de Montréal, C.P. 6128, Succursale A. Montreal, P.Q. H3C 3J7, Canada.

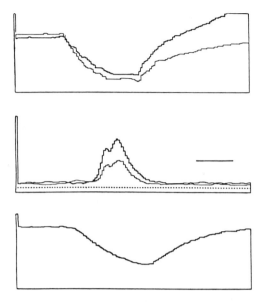

FIG. 1. Response to loading while the subject attempted to maintain his jaw at a fixed height before *(fine lines)* and during the Jendrassic maneuver. *Upper traces,* averaged displacement (closing upwards); *middle traces,* rectified, integrated, and averaged masseter EMG; *lower trace,* averaged force applied by torque motor with bars held together. First bin represents 100 counts, and the total number of bins is 127. *Horizontal calibration:* 10 msec; 100 counts = 0.25 mm, 550 µV, 1.25 kg. Number of averaged trials, 8. Inertial load, 1 kg. *Dashed line* indicates OV when the instrument base line is less. (From Lamarre and Lund, 1975, with permission.)

Marsden et al. (1976b) examined the stretch reflex responses of a number of muscles including the masseter and temporalis to test mechanisms of servo action. They tapped the chin with a reflex hammer, and the latency and form of the jaw-jerk reflex response were compared with those of the response to loading torques applied by an apparatus similar to our own. The latencies of the jaw-jerk reflex of two subjects tested were 7 msec and 8 msec, whereas the latencies of the loading responses during slow tracking were calculated to be 12 msec and 13 msec, respectively. The difference in response latencies was taken as evidence that the stretch response is normally mediated via a long servo loop. However, it is equally plausible that the difference in latency is caused by the difference in the two procedures. Taps to the chin are usually brisk and would be expected to rapidly excite spindle afferents synchronously, but, if the rate of application of force by the lever arm was initially slow, or if muscle activity is low, the start of the segmental reflex could have been significantly delayed.

The amplitude of the short-latency masseteric response to loading is dependent on the strength of the stimulus and on the state of the muscle. During isometric contractions, the response to a given stimulus is proportional to the preexisting level of EMG activity (Fig. 2), suggesting that it is proportional to the excitability of masseteric motoneurons (MN). Since the same relationship holds true for closing movements at velocities up to about 75 mm/sec measured between the incisors (Fig. 3), the rate of intrafusal muscle fiber shortening brought about by fusimotor neurons must be about equal to the rate of extrafusal contraction to maintain the responsiveness of the muscle spindle afferents. We know from animal experiments that γ MNs and Ia afferents are active during such movements (Lund et al., 1979; Taylor and Cody, 1974; Goodwin and Luschei, 1975). However, when the muscle shortens more rapidly, muscle spindles are unloaded, and their afferents fall silent (Taylor and Cody, 1974; Goodwin and Luschei, 1974). The amplitude of the human masseteric loading response also falls at higher velocities (Fig. 3). Eventually, when the subjects close the jaw as fast as they can, the EMG is transiently decreased about 10 to 20 msec after the start of the torque pulse (Fig. 4), presumably because inhibitory effects from Golgi tendon organs and periodontal receptors remain while the spindle afferents are insensitive.

When a stretch is applied to the contracting muscles under either isometric or low-

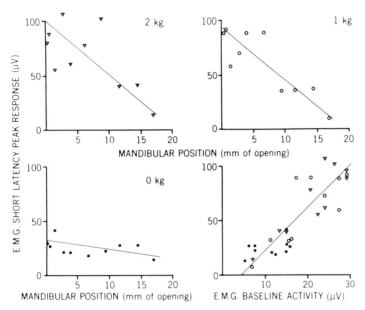

FIG. 2. Graphs were drawn from calculations made on averaged records like those in Fig. 1. A torque load of 600 mg was applied while the subject tried to maintain a steady position. Base-line EMG activity is the average activity in the 10 msec preceding the torque pulse. The peak response is the maximum amplitude of the reflex response in the 10 to 20 msec after loading began minus the base-line activity. Sixteen trials were averaged per series. The relationships between the reflex response and jaw position (measured between the incisors) are shown under the three inertial load conditions (0, 1, and 2 kg). The series of trials at different muscle lengths and inertial loads were randomly assigned. In the final graph, all peak responses are replotted against EMG base-line activity. All regression lines are significant at the 0.5 level.

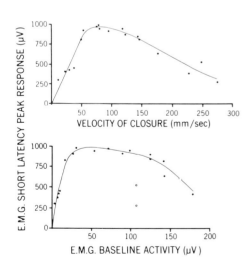

FIG. 3 Similar graphs to those of Fig. 2 for 19 trials of closure at constant velocity with no inertial load. Lines were drawn by hand.

velocity isotonic conditions, the amplitude depends on the rate of lengthening (Fig. 5A). It is probable that the relationship to the velocity of stretch is governed by the mechanical properties of the muscle spin-

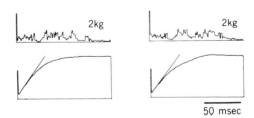

FIG. 4. Examples of transient inhibition of the masseteric EMG without a preceding jaw-jerk reflex from two subjects who tried to close the jaws as fast as possible against an inertial load of 2 kg. The point of loading occurs where the movement trace deviated from the projected trajectory (drawn by hand). Thirty-two trials were averaged; 100 counts = 730 μV and 7.5 mm.

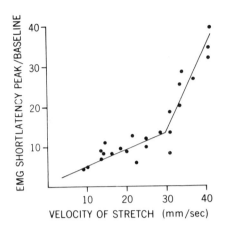 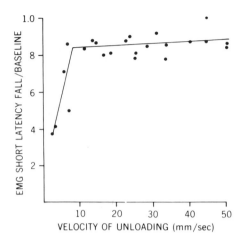

FIG. 5. Effect of varying the velocity of lengthening (**left**) and shortening (**right**) the masseter muscle by the application of ramp torques on the loading and unloading reflex responses of one subject. Subjects attempted to maintain a constant jaw position. The average velocity was calculated from the averaged displacement trace of a series of 16 trials.

dles, because both primary and secondary endings are less sensitive to a given increment in stretch velocity at the bottom of their response curve than at higher velocities (Matthews and Stein, 1969).

We proposed that the jaw jerk reflex provides some active compensation for unexpected increases in resistance during jaw closure (Lamarre and Lund, 1975), but it is unlikely that an automatic response to loading is fundamental to mastication because monkeys seem to be able to chew quite well if the input from muscle spindle afferents from jaw muscles is cut by making bilateral lesions of the trigeminal mesencephalic nucleus (Goodwin and Luschei, 1974). However, Cooker et al. (1980) made a detailed analysis of the stiffness of the human mandibular system by analyzing the mechanical responses to sinusoidal displacements. They were able to estimate that the myotatic jaw reflex contributes about 50% of the total stiffness at low displacement frequencies (1–8 Hz). The head moves up and down during walking and running in an approximately sinusoidal oscillation within this frequency range, and the mandible hanging beneath it is subjected, therefore, to cyclical changes in gravitational load. It is prob-

able that load-induced variations in spindle input may be of great importance in helping to maintain a stable mandibular position relative to the skull during locomotion.

Hufschmidt and Spuler (1962) made the interesting observation that the jaw-jerk reflex response to a chin tap was absent if the molar teeth were clenched together. Goldberg (1972) found that if subjects bit on a wooden stick on one side, there was no response on that side but a very large response on the other (Fig. 6). He wrote that two explanations were possible for these results: (1) the jaw-closing muscle spindles may become insensitive to the stimulus during clenching, or (2) the periodontal afferents excited by pressure on the teeth could reduce the effectiveness of the Ia afferent volley. The second hypothesis is supported by the observation that local anesthesia of the teeth of subjects biting on the levers of a loading apparatus increases the size of the jaw-jerk reflex (Fig. 5; Lamarre and Lund, 1975).

Although it is generally accepted that reciprocal inhibition in response to antagonist muscle stretch does not occur in the jaw muscles (Dubner et al., 1978, p. 260), Matthews (1975) found that tonic activity in the

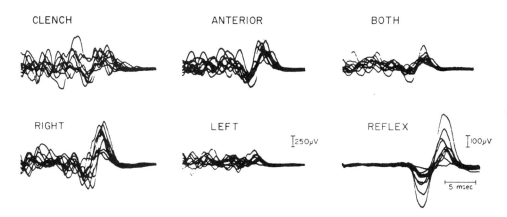

FIG. 6. Electrical activity recorded from the left masseter muscle after chin tap. Each sweep is triggered by contact of the reflex hammer with the chin. The biting position is indicated over each record, which consists of the photographic superimposition of 10 sweeps. In reflex, the jaw muscles were relaxed prior to the tap. The time base is the same for all records; the voltage calibration on the left is the same for all positions except reflex. All records are from the same subject. (From Goldberg, 1972, with permission.)

digastric muscle was briefly inhibited during and after the jaw-jerk reflex.

THE MASSETERIC H REFLEX

Although the examination of the H reflex and the M response of limb muscles has become a routine procedure, the comparable events in jaw muscles have been rarely examined in man. This is undoubtedly because of the inaccessibility of the motor branches of the trigeminal nerves, which obliges one to insert needle electrodes through the masseter muscle to stimulate the masseteric nerve as it emerges from under the skull (Godaux and Desmedt, 1975a). It has been shown in the monkey that it is possible with this technique to stimulate spindle afferents and α and γ MNs (Lund et al., 1979).

Godaux and Desmedt (1975a) recorded both M responses and H reflexes with onset latencies of 1.3 to 1.9 msec and 4.6 to 5.8 msec, respectively. A small F wave was sometimes present when supramaximal stimuli were used. The changes in stimulus current altered the H and M responses of the masseter in the same way as in the classic lower limb recruitment curve. The threshold of the H response was less than that of the M response because the Ia spindle afferents have a lower stimulus threshold than the motor axons. Increasing stimulus intensity increased the amplitude of the H response to a maximum at about the threshold of the M response. More intense stimulation progressively increased the M response at the expense of the H response. However, vibration had opposite effects on the amplitudes of the massateric and soleus H responses: the masseteric H reflex increased during vibration applied to the chin, and a similar potentiation was recorded for the masseteric jaw jerk. These potentiations contrast with the inhibition of limb myotatic or H reflexes that is elicited by muscle vibration (cf. Desmedt and Godaux, 1980; J. E. Desmedt, *this volume*).

Despite the inaccessibility of the masseteric nerve, a transcutaneous electrical stimulator, the Myo-monitor®, has been produced that is supposed "to accomplish transcutaneous electrical neural stimulations through the fifth and seventh nerves to relax the masticatory and facial muscles and to precisely control their contraction"

(Martinis et al., 1980). De Boever and McCall (1972) and Bessette and Quinlivan (1973) attempted to verify these claims. Both groups found that stimulation of the skin excited the masseter muscle beneath the electrodes but not the temporalis or digastric muscles on the same side, although these are also innervated by branches of the mandibular nerve. Bessette and Quinlivan (1973) reported that the facial muscles contracted at a lower threshold than the masseter. The facial nerve travels forward on the surface of the masseter muscle and was probably directly excited by the stimulus. However, EMG records from the masseter muscle showed that suprathreshold stimulation excited the muscle fibers after a delay of less than 2 msec and that there was no evidence of an H reflex.

Jankelson et al. (1975) attempted to see whether contraction caused by the Myo-monitor® was neurally mediated. Their approach was very indirect: threshold intensity–duration curves were constructed, and the chronaxie calculated. They claimed that their results were consistent with neurally mediated stimulation because their curve resembled an intensity–duration curve of "neurally mediated stimulation" redrawn from Lenman and Ritchie (1970) and not a "direct muscle stimulation" curve from the same source (Figs. 4 and 5 of Jankelson et al., 1975). The "direct muscle curve" seems to be the same as that of a muscle denervated for 100 days, whereas the "neurally mediated stimulation" curve appears to have been produced by a stimulator having an output impedance of 250 Ω (Fig. 5.4 of Lenman and Ritchie, 1970). Since the human masseter muscles stimulated by Jankelson et al. (1975) were probably normally innervated, and the Myo-monitor® had an output impedance of nearly 3KΩ, the first curve would have been shifted to the left, and the second far to the right. They may even overlap. In addition, the manufacturer's choice of an invariant stimulus duration of 2 msec assures that muscle fibers will be readily stimulated: the duration of 2 msec is close to the chronaxie of skeletal muscle fibers and much higher than the chronaxie of α motor axons (cf. Desmedt, 1950). For selective motor nerve stimulation, the electric stimuli should have a duration of only 0.1 or 0.2 msec.

Since in all probability, this apparatus excites the subadjacent muscle fibers directly, the "myocentric occlusal position," which is claimed to be "the vertical and horizontal position of occlusion most comparable with individual musculature of the patient" (Myo-monitor® manual, Myo-tronics Research Inc., quoted by Azarbal, 1977), would be expected to vary with the orientation of the stimulated fibers. The superficial fibers of the masseter pull the mandible upwards and forwards, and it has been repeatedly confirmed that the "myocentric occlusal position" is anterior to the position of maximum intercuspidation of the teeth (Remein and Ash, 1974; Azarbal, 1977). The finding that the position can be displaced laterally probably results from an imbalance in the effect of stimulation on the two sides.

LOADING THE JAW-OPENING MUSCLES

Jaw-opening muscle responses to stretch are definitely not a mirror image of those in jaw-closing muscles. There is no evidence that a monosynaptic myotatic reflex exists in suprahyoid muscles; indeed, they contain few, if any, muscle spindles (Cooper, 1960). However, stretching these muscles by unloading the contracting jaw-closing muscles or by pushing the mandible upwards does produce a response, but at much longer latency than would be expected for a monosynaptic reflex (Hannam et al., 1968; Lamarre and Lund, 1975). This period of excitation begins 24 to 34 msec after loading and coincides with the second phase of inhibition in the unloaded jaw-closing muscles (Fig. 7).

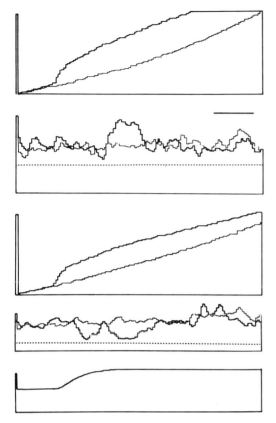

FIG. 7. Response of the digastric muscle to loading and of the masseter muscle to unloading during two series of jaw-closing movements. The same load was applied during the two series of trials. Control records, in which no loads were applied, are traced with *finer lines*. Calibration: 20 msec; 100 counts = 2 mm, 140 μV (A), 35 μV (B), 1.25 kg. Sixteen trials per series. Inertial load, 1 kg. (From Lamarre and Lund, 1975, with permission.)

UNLOADING THE JAW-CLOSING MUSCLES

The responses of the masseter and temporalis muscles to unloading during a voluntary contraction were first described by Hannam et al. (1968). In their experiment, the subjects bit on a Perspex tube until it fractured. The resulting upwards acceleration of the jaw was followed by a long-lasting reduction in EMG activity, the first phase of which begins 6.5 to 11 msec after unloading, and, if the unloading occurs during isometric conditions or slow closure, this is followed by a second (Fig. 7) and perhaps a third period of depression. The duration and degree of reduction of isometric EMG activity are dependent on the rate of unloading (Fig. 5B), duration of the stimulus (Fig. 9), and EMG base-line activity (Fig. 8).

The unloading reflex is probably important in both chewing and biting. When one is biting on brittle objects such as nut shells or hard biscuits, activity in the jaw-closing muscles gradually builds up until the object fractures. Miles and Wilkinson (1982) have shown that under these circumstances there is also a cocontraction of the digastric muscles. When the object between the teeth breaks, the lower jaw first closes very rapidly, but a combination of the unloading reflex and digastric stiffness end the movement. During chewing of brittle food such as carrots or biscuits, there are transient pauses in the jaw closer muscle EMG burst when the teeth break through the food (Ahlgren, 1969).

It is probable that, as in the limb muscles, the earliest part of the unloading response during isometric biting or slow closure is caused by a removal of spindle afferent input. However, as discussed above, muscle spindles are probably silent during very rapid jaw closure, but, nevertheless, a strong unloading response still occurs (Fig. 10). Under these circumstances, it is likely that the fall in EMG activity is primarily a result of inhibition caused by other groups of afferents rather than of a withdrawal of spindle input. The direct evidence comes from Goodwin and Luschei (1974): lesions of the trigeminal mesencephalic nucleus to remove the spindle afferents do not abolish unloading responses.

TONIC VIBRATION REFLEX

Ito (1974) showed that vibration causes a gradually rising tonic contraction of the

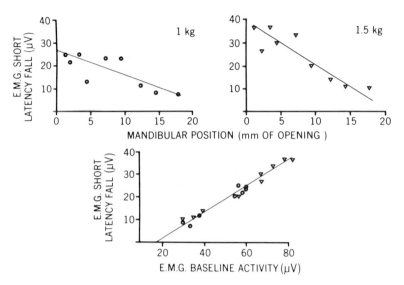

FIG. 8. Relationship among the amplitude of the unloading reflex response, maintained jaw position, and masseteric EMG base-line activity. Inertial loads of 1 and 1.5 kg were used. Sixteen trials were averaged. See Fig. 2 for details.

human masseter similar to that in limb muscles: as in the limbs, it is accompanied by a gradual inactivation of the antagonists (Hagbarth et al., 1976). Hagbarth et al. (1976) also found that the masseter tonic vibration reflex (TVR) could be voluntarily suppressed if the subjects were given visual feedback from a force transducer.

Work on limb muscles suggests that the TVR depends to a large extent on poly-

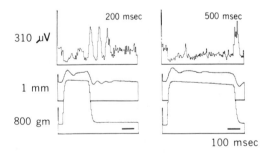

FIG. 9. Effect of increasing the duration of unloading while trying to maintain a fixed position. Masseter EMG activity remains below base line throughout the period of unloading. Note the successive oscillations and reflex responses caused by reapplication of the inertial load. Sixty-four trials per series. Inertial load 2 kg.

synaptic projections from primary spindle endings (Homma et al., 1972; Hultborn and Wigstrom, 1980). Polysynaptic pathways activated through Ia afferents during chin vibration also provide the main excitatory drive for the masseter TVR which has a slow progressive onset, as does that in limb muscles (Desmedt and Godaux, 1975, 1980; Hagbarth et al., 1976). However, a feature of the masseter TVR that is not present in limb muscle TVR is that the firing of motor units is closely locked to the vibration cycle, even during small vibrations requiring a long excitatory build-up through polysynaptic pathways (Fig. 11) (Desmedt and Godaux, 1975, 1980; J. E. Desmedt, *this volume*).

There is another difference in the effects of vibration on limb and jaw muscles. The amplitude of tendon or H reflexes in the leg or arm is reduced during the TVR (DeGail et al., 1966), but the masseteric reflex is potentiated (Godaux and Desmedt, 1975a). Desmedt and Godaux (1980) propose that phase locking of motor units and jaw-jerk reflex potentiation result from a greater "efficiency of monosynaptic Ia EPSPs in

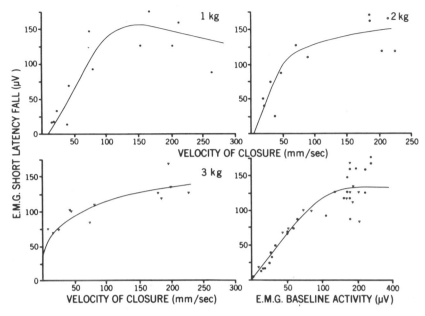

FIG. 10. Relationship between the unloading response movement velocity and masseter EMG baseline activity. See Fig. 2 for details. Sixteen trials per series. Inertial loads 1, 2, and 3 kg.

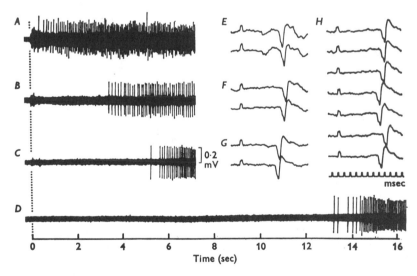

FIG. 11. The effect of vibration amplitudes on the recruitment of a well-isolated masseter unit. The vibration frequency is 85/sec throughout. A: 2 to 6 mm vibration amplitude. B: 1 to 6 mm. C: 0 to 6 mm. D: 0.25 mm. The *vertical dotted line* indicates the onset of vibration. The larger vibration amplitudes in A and B resulted in some mechanical artifacts which increased the thickness of the base line. E: First two spikes of the test in A displayed on a fast sweep triggered by the vibration cycle (26th and 43rd vibration cycles after vibration onset). F: First two spikes of the test in B (290th and 302nd cycles after onset). G: First two spikes of the test in D (105th and 1,065th cycles). H: First seven spikes of the test in C (465th, 500th, 509th, 517th, 534th, and 542nd cycles after onset). (From Desmedt and Godaux, 1975, with permission.)

masseter MNs than in spinal MNs." They suggest that, in contrast to findings in the spinal cord, there is no presynaptic inhibition of jaw closer spindle afferents during the TVR. In addition, electrotonic coupling between the afferents within the trigeminal mesencephalic nucleus (Baker and Llinas, 1971) may increase the synchronization (see J. E. Desmedt, *this volume*).

There are some additional factors that could account for the peculiar features of the masseteric TVR. Spindle secondary endings are now known to be stimulated during the TVR (Burke et al., 1976, 1980b). Furthermore, Appenteng et al. (1978) used spike-triggered averaging techniques to show that secondary spindle afferents of the cat jaw-closing muscles make monosynaptic excitatory connections with jaw-closing MNs that are about 70% of the strength of the monosynaptic Ia input. However, this connection is not exclusive to jaw muscles, because spindle secondaries also make monosynaptic connections on spinal MNs (Kirkwood and Sears, 1974, 1980; Stauffer et al., 1976). It is possible that the properties of the jaw fusimotor system differ from that of the limbs. We know that the sensitivity to stretch of spindle afferents in limb nerves falls during the TVR (Burke et al., 1976), but comparable data for the jaw closing muscles are lacking. However, Sessle (1977) has published some evidence that masseter fusimotor MNs receive positive feedback from spindle afferents in cat. Hair afferents, which can be activated by vibration of the skin (Burke et al., 1976), also cause increased firing of trigeminal fusimotor neurons (Appenteng et al., 1980). If these mechanisms exist in man, vibration could increase fusimotor output, increasing the rhythmical monosynaptic volley and also the response to stretch.

Desmedt and Godaux (1980) were unable to produce a TVR in the digastric muscle by vibration under the chin with the jaw closed and the neck extended, but Hellsing (1977) reported that a TVR in the digastric, coupled with inhibition of the jaw-closing muscles, could be evoked if the subjects bit on a strain gauge; local anesthesia of the skin under the vibrator did not abolish the response. The possibility that the periodontal receptors around the teeth biting on the strain gauge were responsible for exciting the digastric MNs and inhibiting the jaw-closing MNs was rejected by Hellsing because electric stimulation of the inferior dental nerve at the mental foramen at 1 to 40 Hz did not mimic the muscle responses to vibration at 130 Hz. This evidence is not conclusive, because the mental nerve does not innervate the teeth, and one cannot be sure that the intramandibular branch was stimulated. Indeed, his observation that there was an immediate activation of the masseter muscle and rebound closure when the subject dropped the strain gauge during vibration, thereby unloading the teeth, is good evidence that the periodontal pressoreceptors do participate in the digastric response. The described phenomenon may not be a true TVR, and the anatomical evidence for lack of spindle in these muscles must be considered (cf. J. E. Desmedt, *this volume*).

THE JAW-OPENING REFLEX AND EXTEROCEPTIVE SUPPRESSION OF JAW-CLOSING MUSCLE ACTIVITY

Humans do not appear to give the full jaw-opening reflex response to mechanical or electrical stimulation of facial or oral tissues that is described in animal experiments. In animals, the digastric and lateral pterygoid muscles are excited (Kamamura et al., 1968) through a disynaptic pathway (Sumino, 1971), and two periods of inhibition of jaw-closing MNs occur (Kidokoro et al., 1968).

Two periods of inhibition of the jaw-closing muscles of comparable latency and duration occur in man, but the digastric or lateral pterygoid muscles have not been shown to be excited at short latency (Hoffmann and Tonnies, 1948; Beaudreau et al., 1969; Yemm, 1972a,b; Matthews, 1975; Gillings and Klineberg, 1975). Instead, tonic digastric activity is inhibited about 10 msec after stimulation of the palate (Matthews, 1975). However, Meier-Ewert et al. (1974) found that loud sounds cause a similar pattern of suppression of jaw closer muscle activity, and, in a small fraction of their subjects (3/20), the digastric was activated at short latency. Desmedt and Godaux (1976) were able to record an excitatory response in the digastric muscle following electrical stimulation of the skin or mucosa of the lips. They imply that this digastric activity is equivalent to that recorded in animals and concluded that it had not previously been observed because rapid habituation occurs at relatively low rates of repetition (0.2–1 Hz). However, the latency of the response recorded by Desmedt and Godaux varied from 60 to 75 msec. It seems to occur during the second phase of suppression of jaw-closing muscle activity in man, whereas the digastric jaw opening reflex response in animals is disynaptic and coincides with the onset of the earliest suppression of the jaw closing muscle activity.

Many people have described the responses of the jaw-closing muscles as "silent periods" or "exteroceptive silent periods," but we support the suggestion of Godaux and Desmedt (1975b) that it is better to use the term "exteroceptive suppression" to avoid confusion with the silent period that follows the tendon jerk or direct electrical stimulation of muscle.

Adequate Stimulus for Exteroceptive Suppression

It appears that stimulation of adequate intensity anywhere within the mouth (Hoffman and Tonnies, 1948; Yemm, 1972a) or on the facial skin of the maxillary and mandibular trigeminal divisions (Yu et al., 1973; Godaux and Desmedt, 1975b) will suppress jaw-closing muscle EMG activity, but even painful stimulation of the adjoining ophthalmic division or cervical nerve territories is reported to be ineffective (Yu et al., 1973; Godaux and Desmedt, 1975b), although Ongerboer de Visser and Goor (1976) found that percutaneous stimulation of the supraorbital nerve caused suppression of masseter activity in four of 15 subjects.

What is the nature of the stimulus at the "adequate intensity" required to produce exteroceptive suppression? It has been supposed that nociceptive afferents are responsible for exteroceptive suppression, and the best evidence for this was provided by Godaux and Desmedt (1975b). They electrically stimulated the upper and lower lips and tongue and compared the threshold and form of the EMG response to the subjects' reports of their sensory experiences. They found that the threshold stimulus for exteroceptive suppression always caused slight pain. Similarly, liminal tooth pulp stimuli were reported always to be painful by Bratzlavsky et al. (1976), and adequate electrical stimulation of the oral mucosa was described as "uncomfortable" by Yemm's 1972a) patients.

However, two more recent studies of the effects of electrical tooth pulp stimulation contradict the aforementioned hypothesis. Although Fung et al. (1978) interpreted exteroceptive suppression of masseteric activity as a nociceptive response, they also reported that it sometimes occurs when the stimulus is too weak to cause pain. McGrath et al. (1981) carried out an experiment that was similar in design to that of Godaux and Desmedt (1975b) but which gave a different result. They found that the threshold for suppression of masseter muscle activity was approximately the same as each subject's sensory detection threshold and that both were far below the pain threshold (Fig. 12). On a few occasions, the suppression threshold was even less than the sensory detection threshold. Other experiments in which the lips, face, or oral

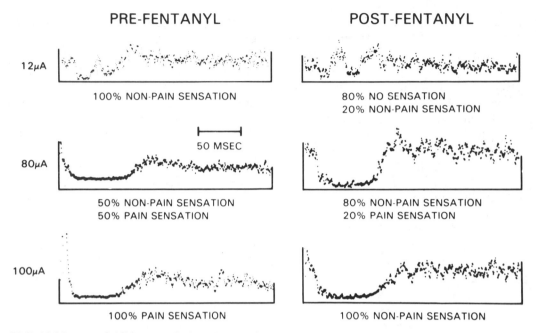

FIG. 12 Masseter inhibitory periods and sensations produced by electrical tooth pulp stimulation at intensities of 12 μA, 80 μA, and 100 μA, before (left) and after (right) the administration of fentanyl (1.7 μg kg i.v.). Postfentanyl times were 7, 11, and 15 min, respectively, for the upper, middle, and lower traces shown on the right. The percentage of the 30 pulses that was not detected, or was detected as nonpain sensations or pain sensations, is indicated below each trace. (From McGrath et al., 1981, with permission.)

mucosa were tapped or stimulated electrically also led to the conclusion that the threshold of the reflex response is below the pain threshold (Yu et al., 1973; Bratzlavsky, 1975; Matthews, 1975). In addition, a similar pattern of muscular responses can be produced by loud but nonpainful sounds (Meier-Ewert et al., 1974).

Furthermore, one of the findings of Godaux and Desmedt (1975b) appears to be in line with the view that exteroceptive suppression can indeed be elicited by nonpainful stimuli. They stimulated afferent fibers in the mucosa of the lower lip and also as they emerged from the inferior alveolar canal at the base of the ramus of the mandible. From the reduction in the latency of onset of the suppression that occurred when the nerves were stimulated nearer to the skull, they calculated that the conduction velocity of afferent fibers responsible for the suppression was roughly 40 msec. This would place them in the Aβ group which contains only mechanoreceptors. The fastest conducting nociceptors innervating the face (high-threshold Aδ mechanoreceptors) conduct at less than 25 msec, and the other classes of nociceptors are even slower (Price and Dubner, 1977).

Response Pattern

Two periods of exteroceptive suppression occur (Fig. 12), which have been called ES_1 and ES_2 by Godaux and Desmedt (1975b). They occur simultaneously in the masseter and temporalis muscles on both sides (Yemm, 1972a; Matthews, 1975; Godaux and Desmedt, 1975b; Bratzlavsky et al., 1976; Meier-Ewert et al., 1974; Ongerboer de Visser and Goor, 1976). The ES_1 component is generally reported to have a latency between 10 and 17 msec and to last from 10 to 25 msec. When the second ES_2 component is observed, its latency is more variable, and values from 30 to 60 msec have been reported (Matthews, 1975; Godaux and Desmedt, 1975b; McGrath et al., 1981). In cats, the two periods of depression have been shown to coincide with postsynaptic hyperpolarization of jaw-closing MNs (Kidokoro et al., 1968; Nakamura, 1980).

It has generally been found that the depth of the suppression is proportional to the intensity of stimulation, and this is best illustrated in the papers by Godaux and Desmedt (1975b) and McGrath et al. (1981). In addition, the majority report that the ES_1 response has a lower stimulus threshold than ES_2 (Bratzlavsky, 1975; Yemm, 1972a; Yu et al., 1973). Bratzlavsky and Yu et al. concluded that ES_1 is a response to the stimulation of nonnoxious, presumably tactile, afferent fibers and that the stimulation of nociceptive afferents is necessary for ES_2. However, there are some discrepancies in their results that are not in accordance with such a simple hypothesis. According to Yu et al. (1973), innocuous stimulation of mechanoreceptors in the upper lip led to inhibition of masseter muscle activity beginning at 15 to 20 msec and lasting 8 to 18 msec, but the inhibition that followed a noxious stimulus to the same region only began 40 to 50 msec after the stimulus. Since electrical stimulation of nociceptors must also excite the lower-threshold mechanoreceptor afferents, why did these not produce an inhibition beginning at 15 to 20 msec? Conversely, why did a noxious tooth pulp stimulus only cause an ES_1 response (latency 10–15 msec, duration 10–30 msec) in the reports by Bratzlavsky et al. (1976)?

McGrath et al. (1981) concluded that no pattern of response (ES_1, ES_1 + ES_2, or a merging of the two) (Fig. 12) was unequivocally associated with pain. However, ES_1 sometimes occurred alone at low stimulus intensities, and this pattern was never associated with pain. Furthermore, the reflex response was unaltered by fentanyl, which abolished the pain. In contradiction to this, the stimulus thresholds of ES_1 and ES_2 were found to be identical and above the pain threshold by Godaux and Desmedt (1975b) and Desmedt and Godaux (1976). Furthermore, Desmedt and Godaux (1976) suggest that ES_1 is sometimes recorded alone because it is less sensitive to habituation than the ES_2 response. In some subjects, ES_2 is completely abolished after several shocks at 1 Hz, which is the frequency of stimulation commonly used by the other groups (Desmedt and Godaux, 1976).

The two periods of suppression are separated by a short period of activity which has generally been considered to represent a brief reappearance of the voluntary EMG interference pattern (Schenck and Lauck-Koehler, 1954; McLellan, 1973; Meier-Ewert et al., 1974; Ongerboer de Visser and Goor, 1976). However, this brief burst can be of much greater amplitude than the averaged EMG activity preceding the stimulus (Fig. 12, 12 μA, postfentanyl), suggesting that MN excitability is transiently increased by a postinhibitory rebound or by additional excitation. There are two possible sources of excitation: the exteroceptive stimulus itself or an increase in jaw-closing muscle spindle activity.

The latter explanation was proposed by Yemm (1972b), who showed that even though the digastric is not activated by electrical stimulation of the palate, the jaw opens slightly (about 0.3 mm) 8 to 12 msec after the start of ES_1. He also found that the waveform in the period of activity preceding ES_2 was synchronized to the stimulus (Yemm, 1972a) and suggested that this is a consequence of stimulation of the masseter muscle spindles by the opening movement, giving a jaw-jerk reflex. If this is so, ES_2 could be a classic silent period following the stretch reflex, as Yemm suggested. However, it has been pointed out by Godaux and Desmedt (1975b) that, as the stimulus strength increases, ES_1 and ES_2 become stronger and tend to fuse, wiping out the brief excitation (Fig. 12). The ES_2, therefore, cannot be a consequence of a prior stretch reflex which, even if it occurred, would be strongly suppressed by the exteroceptive stimulus (Desmedt and Godaux, 1976). It is probable that the waveform in the brief excitatory response is synchronized to the stimulus because it is directly caused by the palatal stimulus. Similar stimuli have been shown in paralyzed animals to produce a brief facilitation of jaw-closing MNs between the two periods of inhibition (Kidokoro et al., 1968; Sumino, 1971).

If the exteroceptive response is caused by tapping the teeth together (Ahlgren, 1969; Hannam et al., 1969; Matthews, 1975), hitting a tooth with an instrument (Goldberg, 1971; Sessle and Schmitt, 1972; Hannam et al., 1970), or electrically stimulating the gingivae over the root of the tooth (Goldberg, 1971), a transient increase in masseter muscle activity (latency approximately 7 msec) sometimes occurs. Goldberg (1971) and Sessle and Schmitt (1972) claim that periodontal pressoreceptors cause this response, because it is at least partially abolished by local anesthesia of the tooth and gums (Fig. 13). If this is so, the reflex is probably a vestige of the powerful monosynaptic reflex present in sharks, which snap their jaws together if the teeth are pressed upon (Roberts and Witkovsky, 1975). However, Hannam et al., (1970) and Matthews (1975) believe that this is really a jaw-jerk reflex caused by the transmission of vibration to the muscle spindles of the jaw-closing muscles.

Clinical Significance

Exteroceptive suppression and the jaw-jerk reflex can be used to assess the extent of trigeminal nerve damage. In patients with a pure unilateral sensory root lesion, stimulation of the ipsilateral facial skin causes no exteroceptive suppression on either

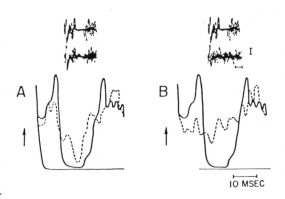

FIG. 13. Electrical stimulation of the gingiva overlying the labial surface of the root of the upper left central incisor and its effect on the left masseter muscle before and after local anesthesia of the stimulus area. The three computer tracings are shown in A and B. The sweep was triggered by the stimulus pulse and is indicated by the *arrows*. The *solid lines* in A and B that descend from the decay of the artifact are the averaged response to 20 shocks delivered to a subject biting on a stick placed between the left first molars; the dashed lines are the averaged response to 20 pulses beginning 4 min (A) and 7 min (B) after injection of the anesthetic. Above A and B are photographic superimpositions of 10 oscilloscope EMG records before (*upper records*) and after anesthesia (A, *lower record*, 4 min after; B, *lower record*, 7 min after). The shock artifact appears after the beginning of the sweep in the oscilloscope tracings. Stimulus frequency, 0.5 Hz. (From Goldberg, 1971, with permission.)

side, but the reflex is present on both sides when the contralateral face is stimulated (Hufschmidt and Spuler, 1962; Ongerboer de Visser and Goor, 1976). The jaw-jerk reflex can be normal if the lesion is restricted to the sensory root (McIntyre and Robinson, 1959; Hufschmidt and Spuler, 1962; Ongerboer de Visser and Goor, 1976) because muscle spindle primary and secondary afferent cell bodies lie in the trigeminal mesencephalic nucleus, and their axons enter the brainstem through the motor root (Cajal, 1909; Szentagothai, 1948; McIntyre, 1951; McIntyre and Robinson, 1959). However, in many cases of retro-Gasserian neurotomy, the jaw-jerk reflex is absent

(Hufschmidt and Spuler, 1962; Ongerboer de Visser and Goor, 1976), suggesting that the motor root had been damaged during surgery. A pure motor root lesion causes wasting of the masseter muscle, denervation potentials, and a loss of the jaw-jerk reflex on the affected side. Sensation is normal, and stimulation of the skin on the affected side will evoke an exteroceptive silent period contralaterally (Ongerboer de Visser and Goor, 1976).

Ongerboer de Visser and Goor (1976) report on an interesting group of six patients who apparently had a unilateral lesion restricted to spindle afferent fibers. There was an ipsilateral loss of the jaw-jerk reflex, although voluntary control of the masseter muscle and exteroceptive suppression were normal (see B. W. Ongerboer de Visser, *this volume*). The jaw-jerk reflex and exteroceptive suppression are reported to be useful in the evaluation of the extent of brainstem lesions.

THE SILENT PERIOD

If the jaw-jerk reflex is evoked while the subject voluntarily contracts the jaw-closing muscles, it is immediately followed by a silent period (Hufschmidt and Spuler, 1962), and the term is also used for the period of inactivity in a contracting muscle that follows stimulation of the motor nerve or direct stimulation of the muscle fibers. Fulton and Pi-Suner (1928) attributed the silent period to the withdrawal of excitation caused by unloading of the muscle spindle receptors during the tendon jerk. In addition, the refractoriness of MNs following their synchronous activation and inhibition from various sources (Renshaw cells, Golgi tendon organs, skin receptors excited by the tap, periodontal and joint receptors) has been implicated (Agarwal and Gottlieb, 1972; Matthews, 1972; Matthews, 1975). The axons of the trigeminal MNs do not appear to have recurrent collaterals (Lorente de No, 1933), and there is no evidence that recurrent Renshaw inhibition occurs when cat masseteric MNs are antidromically stimulated (Kidokoro et al., 1968).

Because of the unique partition of trigeminal primary afferent somata between the Gasserian ganglion and trigeminal mesencephalic nucleus, it has been possible to assess the effect that removing all tactile, tendon organ, temporomandibular joint, and many periodontal receptors has on the silent period. The fact that the silent period still follows the jaw-jerk reflex after retro-Gasserian neurotomy is strong evidence for the fundamental role of spindle afferents (Hufschmidt and Spuler, 1972), but the other afferents traveling in the sensory root could increase the intensity and the duration of inhibition.

In recent years, there have been many reports that the duration of the masseteric silent period is longer than normal in patients suffering from pain in jaw muscles or the temporomandibular joint [temporomandibular joint pain dysfunction (TMJ) syndrome or myofascial pain dysfunction (MPD) syndrome]. In the first of these studies by Bessette et al. (1971), the jaw-jerk reflex was evoked by a tap on the chin while the subjects clenched their teeth together as hard as possible in centric occlusion (the position in which the contact area between the upper and lower dental arches is greatest). The duration of the silent period of normal subjects ranged from 20 to 30 msec (Bessette et al., 1971, Table I), but patients with TMJ syndrome had longer silent periods, ranging from 23 to 152 msec. The patients were then given an acrylic occlusal splint which they wore for 3 weeks. One week later, all were asymptomatic, and the silent period duration was now within the normal range. Many subsequent studies have confirmed these findings (e.g., Bessette et al., 1974; McNamara, 1976; Bailey et al., 1977b; Bessette and Shatkin, 1979; Skiba and Laskin, 1981), and the diagnostic value of these measurements in the detection of myofascial pain and for the evaluation of treatment has often been asserted.

However, we remain skeptical that there is really any fundamental difference between the reflex responses of TMJ or myofascial pain patients and normal subjects because of what we believe are fundamental problems in experimental design and the measurement of the silent period in the studies cited above. For instance, we believe that the following criticisms detract from the initial study. (1) Only a single measurement of the silent period duration appears to have been made for each masseter muscle (Bessette et al., 1971, Table I). Therefore, we cannot tell how much within-subject variation exists. (2) The chin was tapped by hand, and the stimulus was therefore variable, but its intensity was not recorded. (3) Biting force was not recorded. (4) Measurements from records were not made by independent observers. (5) The errors in the method of measurement were never discussed or quantified, and we find it hard to make exact measurements on their published records.

In defense of the decision not to standardize or monitor the input parameters and biting force, some studies have been carried out that purport to show that variations in stimulus intensity (Bailey et al., 1977a) and biting force (Bessette et al., 1973) are insignificant. However, there is contradictory evidence that all five factors previously listed are important. Silent period duration has been found to depend on stimulus order (Layne and Rugh, 1981) and biting force (McNamara et al., 1977; Bernstein et al., 1981). In addition, Bailey et al. (1977a) did find that the silent period duration increased with the force of the chin tap, and it was perhaps only their small sample size that prevented the achievement of statistical significance.

We have recently completed a study (G. Lavigne, R. Frysinger, and J. P. Lund, in press—unpublished data) to assess the error in one method of measurement. The apparatus that we used was basically the same as that illustrated by Lamarre and Lund (1975, Fig. 1). One subject was instructed to maintain a fixed jaw position against a load of 3 kg, and loading torque pulses of constant amplitude were applied at random intervals. Six photographs of the masseteric jaw-jerk reflex and following silent period were chosen for each jaw position (18 records in all). Copies were made, and 30 people were asked to measure the duration of the silent period. The rating method of McCall et al. (1978) was employed, which defined the silent period as "beginning at the peak of the last significant spike preceding the inhibition and ending at the first significant spike that was part of the ongoing activity." There were significant differences between observer groups (technicians and dental students), individual observers, repeated measurements of the same record, and between trials. In two examples of our recordings shown in Fig. 14, the duration of the silent period measured by our observers ranged from 22.0 to 77.5 msec for A and 32.0 to 64.0 msec for B.

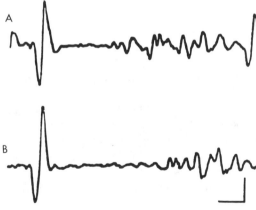

FIG. 14. Two photographs of surface EMGs recorded on an oscilloscope from the right masseter muscle of a subject biting on two bars placed between the incisor teeth. The jaw-closing muscles were stretched by a torque pulse applied to the axis of the lower bar at the start of each sweep. Vertical calibration, 85 μV; horizontal calibration, 10 msec.

Our results suggest that measurements made by one observer cannot be directly compared with those made by a colleague who used the same method of analysis, and it is even more unjustified to compare between laboratories or clinics. The differences between observers are so large that if an arbitrary limit is used to distinguish a "normal" silent period from an "abnormal" one, as has been advocated by McNamara (1977) and Bessette et al. (1979), then the same subject could be placed in different categories by different observers. If he were found to be "abnormal" and had deflective occlusal contacts but no pain, he could be advised to undergo "prophylactic" treatment (McNamara, 1977).

Until well-controlled studies have been carried out that prove definitely that the silent period of myofascial pain patients is longer than normal when all other variables have been standardized, we believe that it is wrong to classify patients by this criterion. Even if there is a real difference, it will not provide justification for modifying the occlusion of a symptomless patient just because his masseteric silent period exceeds some arbitrary level.

SUMMARY

Although the jaw reflexes are analogous in many ways to the corresponding limb reflexes, important differences do exist. The myotatic reflex appears to contribute more stiffness to the jaw-closing muscles than to limb muscles. The jaw tends to swing up and down during locomotion, and, to maintain a stable position in relationship to the skull, it is necessary that the muscles be made stiff by tonic contraction and/or through a powerful servoreflex. The short conduction pathway and rapid contraction of jaw muscles allow reflex effects to act with little phase lag and to provide efficient compensation. If limb muscle reflexes were equally powerful, their effects could be of more nuisance than help in overcoming expected loads because they occur so late. Perhaps the lack of Renshaw cell inhibition of trigeminal MNs and the potentiation of the jaw jerk reflex by chin vibration are features designed to maintain the strength of the myotatic reflex during locomotion. The jaw-opening reflex (including exteroceptive suppression of jaw-closer muscle activity) is bilaterally symmetrical rather than bilaterally reciprocal, as are the analogous spinal flexor withdrawal reflexes. Bilateral braking is necessary to stop closure, because the mandible crosses the midline, whereas withdrawal of a limb often needs to be compensated for by extension of the other to maintain balance.

It has recently been shown in animals that limb and jaw reflex responses are highly context dependent: the size and direction of limb reflexes depend on the phase of locomotion (Forssberg et al., 1977), and the gain of the jaw-opening reflex is increased during the closing phase of mastication (Lund et al., 1981).

ACKNOWLEDGMENTS

Our research is supported by a group grant from the Canadian Medical Research Council which also provided a Fellowship to G. Lavigne. Part of this review was written at the Section of Physiology, Institute of Stomatognathic Science, Tokyo Medical Dental University when J. P. Lund was a guest of Dr. Yoshio Nakamura and the Japanese Society for the Promotion of Science.

Motor Control Mechanisms in Health and Disease,
edited by J. E. Desmedt.
Raven Press, New York © 1983.

Comparative Study of Corneal and Blink Reflex Latencies in Patients with Segmental or with Cerebral Lesions

*B. W. Ongerboer de Visser

Department of Clinical Neurophysiology, Municipal Hospital Slotervaart, Amsterdam, The Netherlands

The corneal and blink reflexes serve to protect the eyes from damage and are considered to be nociceptive. The corneal reflex can be evoked by light mechanical stimulation of the cornea, e.g., with a wisp of cotton wool. The reflex response is bilateral blinking.

From a simple receptive network of free nerve endings within the cornea, afferent impulses pass through fibers in the long ciliary branch of the ophthalmic (first) trigeminal division (Windle, 1926; Tower, 1940; Lele and Weddell, 1959). Entering the pons, they descend through the spinal trigeminal tract in the lateral medullary region to the area of the ipsilateral spinal nucleus of the fifth cranial nerve (Sjoqvist, 1938; Rowbotham, 1939; Smyth, 1939; Moffie, 1971). From this nucleus, multiple intramedullary synaptic pathways run both ipsilaterally and contralaterally to the stimulated side before connecting, respectively, with the ipsilateral and contralateral facial nuclei. The ascending multisynaptic pathways are located in the lateral reticular formation, lying medial to the trigeminal complex (Ongerboer de Visser and Moffie, 1979;

Ongerboer de Visser, 1980; B. W. Ongerboer de Visser, *this volume*).

The blink reflex can be elicited clinically by mechanical stimulation, such as a tap with a tendon hammer, on a wide area of the face, especially around the eyes (Overend, 1896; Rushworth, 1962b). The response is bilateral blinking. Kugelberg's (1952) use of electrical stimulation of the supraorbital nerve led to standardization of the technique. The EMG of the blink reflex is composed of an early oligosynaptic ipsilateral reflex response and a late multisynaptic bilateral reflex response. Only the second component is associated with clinically visible blinking.

Afferent impulses for the blink reflex pass through the supraorbital nerve of the first trigeminal division to the pons. There are indications that the early blink reflex is transmitted through the pons and relayed through an oligosynaptic arc (Tokunaga et al., 1958; Kimura, 1970; Namerow and Etamadi, 1970; Lindqvist and Martenson, 1970; Hiraoka and Shimamura, 1977; Trontelj and Trontelj, 1978). As in the corneal reflex, impulses for the late blink reflex are conducted centrally through the spinal trigeminal tract to the area of the spinal trigeminal nucleus (Kugelberg, 1952; Rushworth, 1962b; Kimura and Lyon, 1972;

---

*Department of Clinical Neurophysiology, Municipal Hospitals Slotervaart, Louwesmeg 6, Amsterdam 1066 EC, The Netherlands.

Ongerboer de Visser and Moffie, 1978). The uncrossed and crossed multisynaptic trigeminofacial connections in the late blink reflex are also located in the lateral reticular formation of the lower brainstem (Ongerboer de Visser and Kuypers, 1978; B. W. Ongerboer de Visser, *this volume*).

Although the corneal and blink reflexes pass through the same trigeminoreticular structures, there are important differences. Afferent fibers from the cornea are small, unmyelinated and myelinated, and conduct relatively slowly. The fiber spectrum of the supraorbital nerve is broad. The large fibers are of high conduction velocity. The afferent fibers from the cornea form a small bundle compared with the large fiber bundle from the supraorbital nerve (Windle, 1926; Tower, 1940; Lele and Weddell, 1959). On electrophysiological examination, the corneal reflex is not preceded by an early ipsilateral response as in the blink reflex (Kugelberg, 1952; Magladery and Teasdall, 1961; Tatcher and Van Allen, 1971; Ongerboer de Visser et al., 1977). When the corneal reflex is evoked by touching the cornea with a small sphere, the reflex has a high degree of constancy of intraindividual latency times and no facilitation or habituation (Magladery and Teasdall, 1961; Ongerboer de Visser et al., 1977). This is in contrast to the late blink reflex, which shows variable latencies because of facilitation and habituation (Penders and Delwaide, 1973; Boelhouwer and Brunia, 1977; Ferguson et al., 1978).

For a long time, it has been a common clinical experience that in cerebral lesions both the direct and consensual corneal reflexes may be reduced or absent following stimulation of the cornea opposite to the lesion (Wolff, 1913; Claude, 1922; Gans, 1934; Purves-Stuart, 1937). Clinical studies on corneal reflex changes in patients with cerebral lesions are reported by Oliver (1952), Richwien (1966), and more recently by Ross (1972) who postulated from findings in one patient with a glioma in the parietal lobe that this site might influence the contralateral corneal reflex. However, Magladery and Teasdall (1961) concluded from electrophysiological studies that latency times of the corneal reflex were only slightly, if at all, increased by cerebral lesions. These contradictions are also reported for the late blink reflex. Some authors (Deken et al., 1976) have not observed changes in latency, whereas others (Kimura, 1974; Fisher et al., 1979) reported that in some patients with cerebrovascular lesions, the late blink reflex opposite to the lesion may be changed.

In the present study, corneal and blink reflexes were examined electrophysiologically in patients who had a cerebral lesion, a medullary lesion, or a lesion of the trigeminal nerve outside the brainstem in an attempt to find an answer to the following questions. First, which site in the contralateral cerebral hemisphere has an effect on the two reflexes? Second, are the reflex changes that might be seen in suprasegmental lesions similar to those seen in segmental lesions involving the trigeminal system? Third, would it be possible to indicate by electrophysiological means to which brainstem structures a possible suprasegmental effect is directed? Fourth, are the corneal and blink reflex results in the three groups of lesions different from each other?

MATERIAL

The study was carried out in 18 patients with unilateral lesions of the trigeminal root (age range 26–72 years, mean 52 years), in 18 patients with Wallenberg's lateral medullary syndrome (Currier, 1969) (age range 45–81 years, mean 63 years) and in 51 patients with unilateral cerebral lesions (age range 36–74 years, mean 61 years).

Patients with Trigeminal Nerve Lesions

Seven of the 18 with trigeminal root lesions had undergone surgery for trigeminal neuralgia: Gasserian ganglion coagulation in five and retrogasserian root section in two patients. Eleven patients were operated on for a tumor in the cerebellopontine angle. In the former cases, there was ipsilateral numbness and hypalgesia in the third trigeminal division in one case and in the second and third trigeminal divisions in six cases. The corneal reflex was clinically normal in four and decreased in the other three cases. The 11 patients who had undergone surgery for a tumor in the cerebellopontine angle showed pareses of the jaw-closing muscles ipsilateral to the lesion. In three of them, a hemifacial hypalgesia and in the others a hemifacial analgesia were found. Pareses of facial muscles were excluded clinically and myographically in all 18 patients.

Patients with Wallenberg's Syndrome

The 18 patients showing clinical signs and symptoms of Wallenberg's syndrome were divided into three groups according to the severity of their illness. All were examined electrophysiologically within the first week after the onset of the symptoms. They were in an alert state during the examination.

Group 1 consisted of six patients showing the typical syndrome and having the most severe degree of vertigo, imbalance, nausea, vomiting, dysphonia, and hiccups. Autopsy was performed on three of these patients. In the first, the infarction involved, besides other structures, mainly the spinal fifth nerve complex and its tract and the lateral reticular formation, whereas the medial reticular formation was largely unaffected. In the other two patients, the infarctions included, besides other structures, the spinal fifth nerve complex and its tract and both the bulbar lateral and medial reticular formation.

Group 2 consisted of nine patients. Clinical signs and symptoms were less severe than in group 1. Autopsy in one of them confirmed the clinical diagnosis: an infarction was found in the dorsolateral region of the medulla oblongata including the descending spinal tract of the trigeminal nerve and sparing the lateral reticular formation.

Group 3 consisted of three patients in whom vertigo, nausea, and vomiting lasted only 1 day. Pain sensation was slightly diminished in the ipsilateral half of the face and the contralateral half of the trunk. Cerebellar signs were minimal, but Horner's syndrome was present on the side of the lesion.

Patients with Cerebral Lesions

Fifty-one patients were included in the study. In one of them, the lesion was restricted to the cerebral peduncle (Ongerboer de Visser, 1981). Each patient was examined neurologically immediately before the electrophysiological testing. The lesion sites were analyzed by scintigraphy or CT scan and, in some cases, by angiography. Autopsy was performed in 10 patients, including the patient with a lesion of the cerebral peduncle (which appeared to be an infarction) (Table 1).

In patient 16, who suffered from a tumor in the upper one-third of the right postcentral gyrus, there was hypalgesia in the left leg as well as loss of postural sensibility and tactile discrimination. Patients 18 through 20 had tumorous lesions in the right temporal lobe causing uncinate fits. In two of them, an upper quadrantic anopia was found. All of the other patients were suffering from vascular lesions in one hemisphere. Patients 13 through 15 and 17 had an infarction in the territory of the anterior cerebral artery. In patient 17, the lesion was restricted to the upper one-third of the right postcentral gyrus. The remaining patients had lesions situated in the territory of the middle cerebral artery: an intracerebral hematoma in cases 24 and 29 (Fig. 1) and cerebral ischemia in the other cases. The patients 5 to 12 and 51 showed a pure motor hemiplegia without sensory deficits (Moffie

TABLE 1. *Sites of lesions in 51 patients with cerebral lesions*

Diagram of lesion (Capsula interna)	Patient no.	Hemi-paresis/plegia	Hemi-hyp/an-algesia	Aphasia	Epilepsy	Corneal reflex abnormal (Patient no.)	Late blink reflex abnormal (Patient no.)
	1–4	R	—	Expr	—	—	—
Capsula Interna	5–12	R	—	—	—	—	—
	13–15	R leg	R leg	—	—	—	—
	16,17	—	L leg	—	—	—	—
	18,20	—	—	—	Temporal lobe	—	—
	21–23	—	—	Rec	—	—	—
	24–26	—	R	Rec	—	R (3)	R (2)
	27,28	L face	L face	—	—	L (2)	L (1)
	29–33	—	L	—	—	L (5)	L (4)
	34–40	L	L	—	—	L (7)	L (5)
	41–50	R	R	Expr/rec	—	R (10)	R (7)

Abbreviations used: R = right, L = left, Expr = expressive, Rec = receptive.

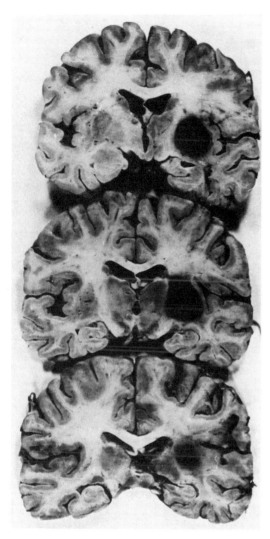

FIG. 1. Intracerebral hematoma on the right side in a patient with left-sided hemianalgesia and loss of corneal and late blink reflexes. The lesion interrupted the ascending fibers to the parietal lobe, the presumed descending projection from the parietal lobe to the lateral reticular formation, and the nucleus of the spinal trigeminal tract in the lateral medullary region.

et al., 1979). A CT scan revealed a lesion situated in the posterior part of the posterior limb of the internal capsule (Fig. 2), interrupting the pyramidal motor fibers. Patients 27 and 28 showed weakness of the angle of the mouth and hemifacial hypalgesia corresponding to a lesion in the lower one-third of the pre- and postcentral gyri. Patients 24 to 26 and 29 to 33 showed symptoms corresponding to the lower two-thirds of the postcentral gyrus. Patients 21 to 23 had ischemic lesions in the posterior part of the left parietal lobe (Fig. 3), causing disturbance of speech on its receptive side, alexia, and agraphia.

METHODS

The methods used to record corneal and blink reflexes and their parameters have been reported elsewhere (Ongerboer de Visser et al., 1977; Ongerboer de Visser and Goor, 1974). The corneal reflex can be elicited electrophysiologically by touching the cornea with a 2-mm-diameter metal sphere which is connected to an electronic trigger circuit. As soon as contact between the cornea and the sphere is established, the touch-sensitive circuit delivers a pulse which triggers the sweep of a dual-trace persistence oscilloscope. During application of the metal sphere to the cornea, the upper eyelid is held up by the examiner's finger. The reflex responses in both orbicularis oculi muscles are recorded with surface electrodes placed over the lower eyelids. The bandpass extended from 30 Hz to 1.6 kHz.

The wide range of latency (34–64 msec) means that an interindividual comparison is of no practical value. However, comparison in one subject of latency simultaneously recorded on the two sides is of great value. A difference between latencies of the corresponding direct and consensual responses elicited by right-sided and left-sided corneal stimuli of 10 msec or more is regarded as abnormal and points to an afferent delay (trigeminal nerve lesion). The same holds for bilateral absence of the responses to stimulation on the affected side, pointing to an afferent block (trigeminal nerve lesion). A latency difference of 8 msec or more

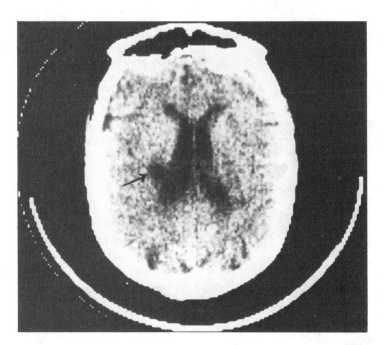

FIG. 2. Computed tomography in a patient with "pure motor hemiplegia" and normal corneal and blink reflexes. The lesion (arrow) is situated in the posterior part of the posterior limb of the internal capsule.

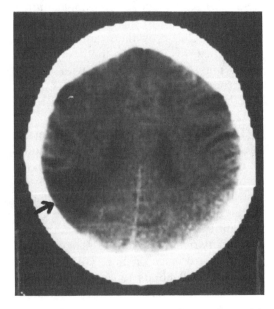

FIG. 3. Computed tomography in a patient with receptive aphasia, alexia, and agraphia. The corneal and blink reflexes were normal. The lesion (arrow) is situated in the posterior part of the left parietal lobe.

between the direct and simultaneously recorded consensual response indicates an efferent delay (facial nerve lesion). This also holds for a consistent unilateral absence of both the direct and consensual reflex response after stimulation of the affected and unaffected sides, respectively, pointing to an efferent block (facial nerve lesion).

Blink reflexes were obtained by supramaximal square-wave electrical stimuli on the supraorbital nerve. The reflex responses were recorded simultaneously from the inferior half of the right and left orbicularis oculi muscle by coaxial needle or surface electrodes. A difference in latency time between the right and left side exceeding 1.5 msec for the early reflex and 8.0 msec for the late reflex was considered abnormal. The amplitudes of the reflex responses were only considered significant when they showed a decrease of more than 50% with respect to the normal side.

RESULTS

Patients with Trigeminal Nerve Lesions

One of the seven cases who had undergone surgery for trigeminal neuralgia showed normal corneal and blink reflexes. In three, only the corneal reflex was abnormal electrophysiologically, showing an afferent delay (range 10–24 msec). In the other three, the corneal and early blink reflexes were abnormal, showing, respectively, an afferent delay of the reflex responses in three (range 12–24 msec) and a delayed early response (Fig. 4) in two cases (5 and 9 msec) and an absent early response (Fig. 5) in one case. In these three patients, the late blink reflexes fell within the normal range.

In all of the 11 patients who had undergone surgery for a tumor in the cerebellopontine angle, the corneal reflex and the two components of the blink reflex were abnormal. The corneal reflex showed an afferent delay (range 16–28 msec) in eight and an afferent block in three patients. The early blink reflex was delayed (range 2–7

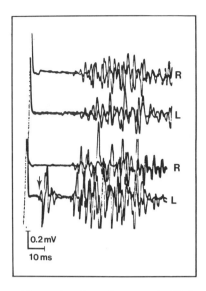

FIG. 5. Example of an absent early blink reflex in trigeminal nerve lesions. The lesion is on the right side. *Upper pair of traces:* absent early blink reflex with normal late blink reflex responses after stimulation of the right supraorbital nerve. *Lower pair of traces:* normal blink reflex after stimulation of the left supraorbital nerve. Two responses are superimposed.

msec) in seven and absent in four patients. Nine patients had an afferent delay (range 10–18 msec) of the late blink reflex (Fig. 6). They included two patients with an absent early reflex, seven with a delayed early reflex, eight with an afferent delay of the corneal reflex, and one with an afferent block of the corneal reflex. The remaining two patients had an afferent block of the blink (Fig. 7) and corneal reflex.

Patients with the Wallenberg Syndrome

The latency times of the early blink reflex fell within the normal range in all 18 patients. In the corneal and late blink reflexes, five types of abnormalities (types A–E) could be differentiated.

Type A (Figs. 8 and 9) consisted of an absence of the reflex responses on both sides when the stimulation was applied to the side of the lesion. A stimulus to the normal side elicited a normal reflex response

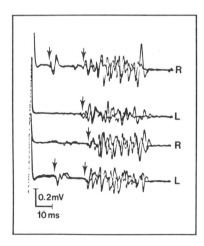

FIG. 4. Example of a delayed early blink reflex in trigeminal nerve lesions. The lesion is on the left side. *Upper pair of traces:* normal blink reflex after stimulation of the right supraorbital nerve. *Lower pair of traces:* delayed (5 msec) early blink reflex with normal late blink reflex responses after stimulation of the left supraorbital nerve. Two responses are superimposed.

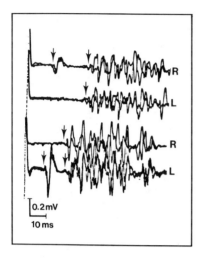

FIG. 6. Example of an afferent delay of the early and late blink reflex responses in trigeminal nerve lesions. The lesion is on the right side. *Upper pair of traces:* delayed early (2.5 msec) and late (11 msec) blink reflex after stimulation of the right supraorbital nerve. *Lower pair of traces:* normal blink reflex after stimulation of the right supraorbital nerve. Two responses are superimposed.

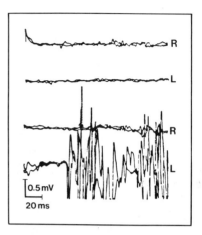

FIG. 8. Example of type A abnormality of the corneal reflex in lateral medullary lesions. The lesion is on the right side and involves, among other structures, the spinal fifth complex, its tract, and the lateral reticular formation. *Upper pair of traces:* afferent block of the corneal reflex after stimulation of the right cornea. *Lower pair of traces:* normal corneal reflex response on the left (the intact side) and an absence of the reflex response on the right side (the side of the lesion) after stimulation of the left cornea. Two responses are superimposed. R, right; L, left in this and subsequent figures.

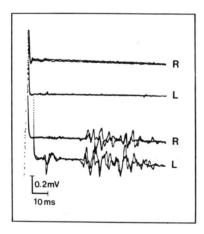

FIG. 7. Example of an afferent block of the blink reflex in trigeminal nerve lesions. The lesion is on the right side. *Upper pair of traces:* afferent block of the blink reflex after stimulation of the right side. *Lower pair of traces:* normal blink reflex after stimulation of the left supraorbital nerve. Two responses are superimposed.

on that side but no response on the contralateral (affected) side.

Type B (Figs. 10 and 11) consisted of an absence of the reflex responses on both sides when the stimulation was applied to the side of the lesion as in the type A abnormality. However, a stimulus to the normal side elicited a normal reflex response on that side, whereas the contralateral reflex response showed a latency time that was markedly longer than the latency time of the normal reflex response.

Type C (Figs. 12 and 13) consisted of a delay of the reflex responses on both sides (the latency difference between the two responses fell within the normal range) when the stimulation was applied to the side of the lesion. However, a stimulus to the normal side elicited a normal reflex response on that side, whereas the contralateral reflex response showed a latency time that

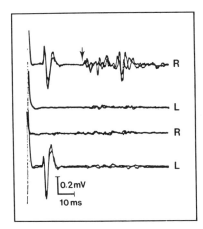

FIG. 9. Example of type A abnormality of the late blink reflex in lateral medullary lesions. The lesion is on the left side. *Upper pair of traces:* normal early and late blink reflex responses on the right (the intact side) and an absence of the late reflex response on the left (the side of the lesion) after stimulation of the right supraorbital nerve. Two responses are superimposed. *Lower pair of traces:* afferent block of the late blink reflex with a normal early blink reflex after stimulation of the left supraorbital nerve.

was markedly longer than the latency of the normal reflex response.

Type D (Figs. 14 and 15) consisted of an absence of the reflex response on both sides when the stimulation was applied to the side of the lesion. The reflex responses were normal bilaterally when the normal side was stimulated.

Type E (Figs. 16 and 17) consisted of a bilateral delay of the reflex responses following stimulation on the affected side. The reflex responses were normal bilaterally when the normal side was stimulated.

In some cases, a marked decrease of the amplitudes of the reflex responses was seen. The decrease was present in the two reflex responses following stimulation on the side of the lesion (Fig. 18) and in the responses following stimulation on the side of the lesion as well as in the response ipsilateral to the lesion when the normal side was stimulated (Fig. 19).

The relationship between the corneal and

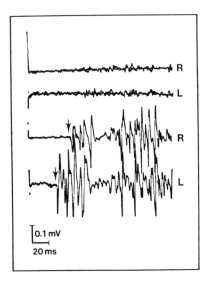

FIG. 10. Example of type B abnormality of the corneal reflex in lateral medullary lesions. The lesion is on the right side. *Upper pair of traces:* afferent block of the corneal reflex after stimulation of the right cornea. Two responses are superimposed. *Lower pair of traces:* normal corneal reflex response on the left (intact side) and a delay (26 msec) of the reflex response on the right (the side of the lesion) after stimulation of the left cornea.

late blink reflex abnormalities was studied (Table 2). Type A abnormality of the corneal reflex was observed in three cases of whom two showed type A and one type B abnormality of the late blink reflex response. Type B abnormality of the corneal reflex was seen in two cases who showed type C abnormality of the late blink reflex. Type C abnormality of the corneal reflex was seen in two cases of whom one had type C and the other type D abnormality of the late blink reflex. Type D abnormality of the corneal reflex was found in five cases of whom three showed type D and two type E abnormality of the late blink reflex. Type E abnormality of the corneal reflex was observed in six cases of whom three showed type E abnormality of the late blink reflex. In the other three cases, the late blink reflex was normal.

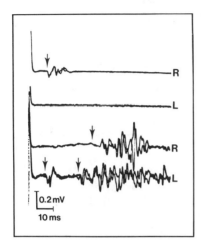

FIG. 11. Example of type B abnormality of the late blink reflex in lateral medullary lesions. The lesion is on the right side. *Upper pair of traces:* afferent block of the late blink reflex with a normal early blink reflex response after stimulation of the right supraorbital nerve. *Lower pair of traces:* normal early and late blink reflex responses on the left (the intact side) and a delay (10 msec) of the late blink reflex response on the right (the side of the lesion) after stimulation of the left supraorbital nerve.

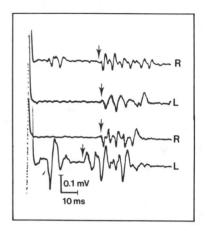

FIG. 13 Example of type C abnormality of the late blink reflex in lateral medullary lesions. The lesion is on the right side. *Upper pair of traces:* afferent delay (10 msec) of the late blink reflex with a normal early blink reflex after stimulation of the right supraorbital nerve. *Lower pair of traces:* normal early and late blink reflex responses on the right (the intact side) and a delay (9 msec) of the late blink reflex response on the right (the side of the lesion) after stimulation of the left supraorbital nerve.

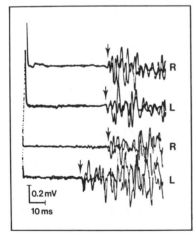

FIG. 12. Example of type C abnormality of the corneal reflex in lateral medullary lesion. The lesion is on the right side. *Upper pair of traces:* afferent delay (15 msec) of the corneal reflex after stimulation of the right cornea. *Lower pair of traces:* normal corneal reflex on the left (the intact side) and a delay (16 msec) of the reflex response on the right (the side of the lesion) after stimulation of the left cornea. Two responses are superimposed.

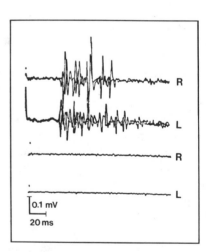

FIG. 14 Example of type D abnormality of the corneal reflex in lateral medullary lesions. The lesion is on the left side. *Upper pair of traces:* normal corneal reflex after stimulation of the right cornea. *Lower pair of traces:* afferent block of the corneal reflex after stimulation of the left cornea ipsilateral to the lesion. Two responses are superimposed.

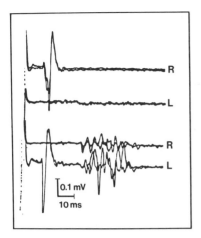

FIG. 15. Example of type D abnormality of the late blink reflex in lateral medullary lesions. The lesion is on the right side. *Upper pair of traces:* afferent block of the late blink reflex with a normal early blink reflex response after stimulation of the right supraorbital nerve. *Lower pair of traces:* normal blink reflex after stimulation of the left supraorbital nerve. Two responses are superimposed.

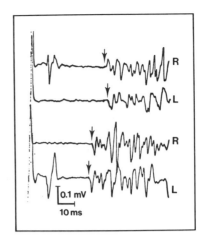

FIG. 17. Example of type E abnormality of the late blink reflex in lateral medullary lesions. The lesion is on the right side. *Upper pair of traces:* afferent delay (9 msec) of the late blink reflex with a normal early blink reflex response after stimulation of the right supraorbital nerve. *Lower pair of traces:* normal blink reflex after stimulation of the left supraorbital nerve.

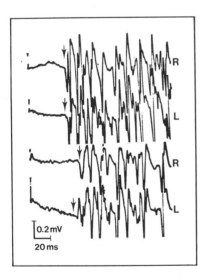

FIG. 16. Example of type E abnormality of the corneal reflex in lateral medullary lesions. The lesion is on the right side. *Upper pair of traces:* normal corneal reflex after stimulation of the right cornea. *Lower pair of traces:* bilateral delay (10 msec) of the corneal reflex after stimulation of the right cornea.

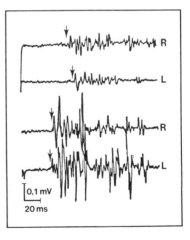

FIG. 18. Example of type F abnormality of the corneal reflex in lesions of the lower postcentral region (Table 1). The lesion is in the left hemisphere. *Upper pair of traces:* afferent delay (on the right 22 msec; on the left 28 msec) of the corneal reflex after stimulation of the right cornea. The amplitudes of reflex responses are markedly decreased. *Lower pair of traces:* normal corneal reflex after stimulation of the left cornea.

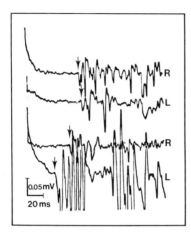

FIG. 19. Example of type C abnormality of the corneal reflex in lesions of the lower postcentral region (Table 1). The lesion is in the left hemisphere. *Upper pair of traces:* afferent delay (on the right 32 msec; on the left 30 msec) of the corneal reflex after stimulation of the right cornea. The amplitudes of the bilateral reflex responses are markedly decreased with respect to the left reflex response after stimulation of the left cornea. *Lower pair of traces:* normal reflex response on the left (the intact side) and a delay (16 msec) of the corneal reflex response on the right (the side opposite to the hemispheral lesion) after stimulation of the left cornea. The amplitudes on the right are markedly decreased.

Patients with Cerebral Lesions

In the early blink reflex, the absolute values of the latency times as well as the differences between the latency times on the right and left side fell within the normal ranges in all 50 patients. The corneal reflex was normal in 23 patients (numbers 1–23) and abnormal in 28 patients (numbers 24–51). The late blink reflex was normal in 31 patients (numbers 1–23, 26, 28, 33, 39, 40, 48–50) and abnormal in 20 patients (numbers 24, 25, 27, 29–32, 34–38, 41–47, 51).

Five types of abnormalities were found in both the corneal and the late blink reflex responses. They were similar to those observed in the patients with Wallenberg syndrome. When the corneal reflex abnormalities were related to the late blink reflex abnormalities, the following observations were made (Table 3).

Type A abnormality (Fig. 8) of the corneal reflex was found in six patients, of whom three showed type A abnormality of the late blink reflex, two showed type B, and one showed type C. Noteworthy is that type A abnormality of the corneal and late blink reflexes was seen in the patient with the infarction of the cerebral peduncle (Ongerboer de Visser, 1981). Type B abnormality (Fig. 10) of the corneal reflex was seen in four patients, of whom one showed type B, two showed type C, and one showed type D abnormality of the late blink reflex. Type C abnormality (Fig. 12) of the corneal reflex was found in three patients, of whom one had type C and two type D abnormality of the late blink reflex. Type D abnormality (Fig. 14) of the corneal reflex was found in five patients, of whom one

TABLE 2. *The electrophysiological types of abnormality in the 18 patients with Wallenberg's lateral medullary syndrome*

Type of abnormality	Corneal reflex (N)	Late blink reflex (N)
A	3	2
B	2	1
C	2	3
D	5	4
E	6	5
Normal	—	3

TABLE 3. *The electrophysiological types of abnormality in the 27 patients with parietal lobe lesions or lesions in which the parietal lobe was involved*

Type of abnormality	Corneal reflex (N)	Late blink reflex (N)
A	5	2
B	4	3
C	3	3
D	5	4
E	10	7
Normal	—	8

showed type D and four type E abnormality of the late blink reflex. Type E abnormality (Fig. 16) of the corneal reflex was seen in 10 patients, of whom three showed type E abnormality of the late blink reflex. The remaining seven patients showed normal late blink reflexes.

As in the patients with Wallenberg's syndrome, in some cases there was, besides the abnormal latency times, a marked decrease of amplitudes of the corneal and late blink reflex responses. The decrease in amplitude was present in both of the reflex responses following stimulation on the side contralateral to the cerebral lesion (Fig. 18) or stimulation on the side contralateral to the cerebral lesion as well as in the reflex response opposite to the cerebral lesion, when the normal side (ipsilateral to the cerebral lesion) was stimulated (Fig. 19).

DISCUSSION

A diminished corneal reflex may be an early sign of a lesion of the trigeminal nerve and may occur before any cutaneous anesthesia can be detected (Walton, 1977). In the present series, seven patients with lesions of the fifth cranial nerve had no sensory disturbances in the ophthalmic trigeminal division. Corneal reflexes were diminished clinically in three and were normal in four patients. On electrophysiological examination, the corneal reflex was also normal in one of the latter four patients, but in three, the latencies of both the direct and the consensual reflexes were significantly delayed (afferent delay) when compared with those following stimulation on the normal side. This indicates that the corneal reflex can be abnormal even though it seems normal on clinical observation. With respect to the blink reflex, the early component was abnormal in three of the seven patients, and the late component was normal in all seven patients. Because the early blink reflex runs unilaterally, differentiation between lesions of the fifth and those of the seventh cranial nerves is impossible on the basis of an isolated abnormal early blink reflex. Therefore, recording the corneal reflex is a better way of detecting intracranial lesions of the ophthalmic division of the trigeminal nerve than recording the blink reflex.

The 11 patients who had operations in the cerebellopontine angle had clinical hemifacial sensory disturbances with abnormal corneal reflexes. In all of these patients, abnormal electrophysiological findings correlated well with the sensory disturbances. In some patients, recording corneal reflex appeared helpful because spontaneous contraction of eyelids interfered with the clinical examination. Furthermore, repeated examination at follow-up in eight of the 11 patients demonstrated the usefulness of the objective assessment. With clinical improvement, reflexes can return (one case), or differences between latencies can become smaller (one case) or normal (one case). On the other hand, reflexes can disappear (two cases) or latency differences become larger (three cases) with clinical regression.

With regard to the fact that the corneal reflex was abnormal more often than the blink reflex, the following factors may play a role. Afferent fibers for the corneal reflex are fewer, and a lesion may damage a great part or all of afferents for the corneal reflex although enough fibers are left for the blink reflex. Another interesting feature is the finding of an isolated abnormal early blink reflex, which has also been reported by others (Kimura, 1970; Goor and Ongerboer de Visser, 1976; Ferguson, 1978). This might be explained by an alternative pathology or related to the ranges in latency differences between sides, which are much smaller for the early than for the late blink reflex.

In all 18 patients with Wallenberg's syndrome, the early blink reflexes showed normal latency times. This indicates that the supraorbital nerve and the second branch of the facial nerve were intact and that the pontine trigeminofacial connections were functioning normally (Namerow and Etamadi, 1970; Kimura, 1975; Goor and Ongerboer de Visser, 1976; B. W. Ongerboer de Visser, *this volume*).

In the type D (Figs. 14 and 15) and E (Figs. 16 and 17) abnormalities, an afferent block and afferent delay following stimulation on the affected side and normal reflex responses following stimulation on the normal side suggest that the lesion can be expected in the descending spinal trigeminal tract with or without its nucleus (Figs. 20 and 21). Because the interneuronal network in the lateral reticular formation is spared, the bilateral reflex responses are normal at stimulation on the unaffected side. In the type A (Figs. 8 and 9), B (Figs. 10 and 11), and C (Figs. 12 and 13) abnormalities showing an afferent block or delay of the bilateral responses following stimulation on the affected side and an abolished or delayed reflex response on the affected side with a normal reflex response on the normal side when this side is stimulated, the lesion involves not only the spinal trigeminal tract but also the adjoining lateral reticular formation (Figs. 20 and 21). This lesion interrupts the trigeminal access route to MNs of orbicularis oculi muscles on both sides and also the crossed ascending fibers that are distributed from the corresponding neurons on the intact side to MNs of the orbicularis oculi muscle on the affected side.

In three patients of group 1 with the most severe symptoms and type A abnormality, anatomical examination disclosed extensive lesions in the lateral medullary region. As a result of this finding, it is reasonable to presume that when type A and probably also B or C abnormality occur in Wallenberg's syndrome, the lesion can be expected to extend into the lateral reticular formation.

In some patients with Wallenberg's syndrome, both the clinically diminished or absent corneal reflex responses and the facial sensory disturbances correlated well with the delayed (type E, Figs. 16 and 17) or

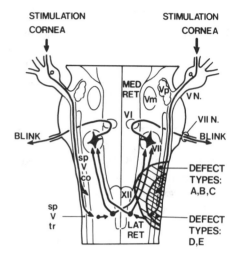

FIG. 20. Presumed location of the bulbar interneurons subserving the two responses of the corneal reflex. A lesion involving the *double-shaded area* on the right would only interfere with the bilateral response elicited by stimulation of the cornea on the affected side. A larger lesion involving the *lightly shaded area* as well would also interfere with the bilateral response on the affected side when the reflex is elicited by stimulation of the cornea on the intact (left) side. VN, trigeminal nerve; Vm, trigeminal motor nucleus; Vp, trigeminal principal sensory nucleus; SpVtr, spinal trigeminal tract; SpVco, spinal trigeminal complex; VI, abducens nucleus; VII, facial nucleus; VII N, facial nerve; XII, hypoglossal nucleus; MED RET, medial reticular formation; LAT RET, lateral reticular formation.

abolished (type D, Figs. 14 and 15) reflex responses. However, in others, there was no relationship between clinical and electrophysiological observations. The abnormal reflex responses on the side of the lesion in type A, B, and C abnormalities (Figs. 8–13) following stimulation on the normal side were not seen on clinical examination, where the reflexes were considered to be normal bilaterally. Thus, recording the corneal and blink reflex can reveal clinically undetectable abnormalities.

The corneal reflex was disturbed in 18 and the late blink reflex in 14 patients. This finding indicates that the corneal reflex is more susceptible to ischemic lesions than

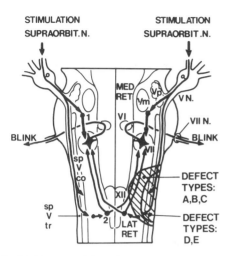

FIG. 21. Presumed location of the bulbar interneurons subserving the two components of the blink reflex: (1) interneurons subserving the ipsilateral early component; (2) interneurons subserving the bilateral late component. A lesion involving the *lightly shaded area* (right) would only interfere with the bilateral late component elicited by stimulation of the nerve on the affected side. A larger lesion also involving the *double-shaded area* would also interfere with the late component on the affected side elicited by stimulation of the nerve on the intact (left) side. For key to abbreviations, see legend to Fig. 20.

the late blink reflex. This might be explained by accepting that, similar to the afferent fibers, the number of interneurons subserving the corneal reflex is smaller than the number along which the late blink reflex is mediated. A lesion in the lateral medullary region might then destroy a great part or all of the interneurons subserving the corneal reflex, whereas enough interneurons are left over for the late blink reflex.

In the 51 patients with cerebral lesions, the absolute values of the latency of the early blink reflex and the differences between the right and left sides all fell within the normal ranges. These findings are in agreement with those of Kimura (1974) and Fisher et al. (1979). The latter authors found in some of their patients with a stroke, delayed early blink reflexes within the first week after the stroke. After that time, these changes resolved. In the present series, blink reflexes were studied after at least 1 week following a stroke. Further, all patients were in an alert state during the test. This is important because a diminished consciousness may prolong latency and decrease the response amplitudes of the late blink reflex regardless of which side is stimulated. In comatose patients, late reflexes may even disappear (Lyon et al., 1972).

The corneal reflex appeared to be abnormal in 28 patients (Tables 1 and 3; patients number 24–51), among whom the late blink reflex was abnormal in 20. Eight of the 28 patients (Table 1; patients number 24–26, 29–33) showed sensory disturbances in the face, the upper limb, and trunk contralateral to the lesion and no motor deficits. They had a lesion in the lower two-thirds of the postcentral region. In two patients (Table 1; patients number 27 and 28), definite sensory disturbances were restricted to the face. They had lesions in which the lower one-third of the postcentral region was involved. In the remaining 17 patients, not only was the postcentral region involved but also the precentral and occipitotemporal regions. Thus, mainly the lower third of the postcentral region exerts an influence on the corneal and late blink reflex. The absent corneal and late blink reflexes after a lesion of the cerebral peduncle indicate that this descending influence is mediated by fibers that run through the peduncle.

Both the corneal and late blink reflexes were normal in the first 23 patients. They had the lesions in the internal capsule, the frontal and temporal lobes, as well as the posterior and superior areas of the parietal lobe. With respect to postcentral influences on corneal and late blink reflexes, several authors (Jones and Wise, 1977; Brodal, 1978) noted that fibers from the lower part of the postcentral gyrus project to the spinal trigeminal complex. If these projections are involved in the corneal and late blink reflexes, their lesion might explain the loss of these reflexes.

Another striking feature is the long latencies that can occur in both segmental and suprasegmental lesions. A delay of some milliseconds can be explained by a decreased conduction velocity in demyelinated fibers, but a delay up to 28 msec is difficult to explain by demyelinization alone. In lateral medullary lesions in which the lateral reticular formation is involved (type C abnormality; Figs. 12 and 13) latency delays up to 28 msec were found. This indicates that loss of interneurons in the lateral reticular formation contributes the greatest part of delay. The long delays (up to 28 msec) in trigeminal nerve lesions may be explained by assuming that there is a relationship between the number of afferent fibers in the trigeminal nerve and the number of interneurons in the lateral reticular formation that can be reflexly activated. Loss of afferent fibers may then result in a decrease of interneurons that can be activated, leading to a reduced conduction through the interneuronal network. In lesions of the caudal postcentral region, similar delays (up to 32 msec) as in trigeminal nerve lesions were found. It is thus attractive to postulate that the crossed excitatory influence on the spinal trigeminal nucleus and the lateral reticular formation is related to the number of postcentral cells that can be excited and to the number of descending fibers along which excitatory impulses are conducted. Loss of parietal cells or descending fibers may then result in reduced excitability of the spinal trigeminal nucleus and the lateral reticular formation. This may lead to slowing of conduction through these structures.

The observation that corneal and late blink reflexes were unaffected in pure motor hemiplegia caused by a lesion in the posterior part of the posterior limb of the internal capsule where the precentral fibers converge (Moffie et al., 1979) may be explained by assuming that fibers from the postcentral region were not affected by the lesion. However, loss of late blink reflexes in some patients with pure motor hemiplegia is reported by Fisher et al. (1979b). In these patients, the lesions might interrupt postcentral fibers descending to the cerebral peduncle.

In conclusion, from the present study, it can be deduced that recording corneal and blink reflexes provides objective information that cannot be obtained by clinical observation. Caution must be used in interpreting the corneal and late blink reflex changes produced by lesions at different anatomical levels.

Clinical Uses of the Electrically Elicited Blink Reflex

Jun Kimura

Division of Clinical Electrophysiology, Department of Neurology, College of Medicine, University of Iowa, Iowa City, Iowa 52242

The blink reflex is a contraction of the orbicularis oculi muscle caused by reflexly activated motoneurons (MN) of the facial nerve (Kugelberg, 1952; Rushworth, 1962a,b; Ferrari and Messina, 1968; Kimura et al., 1969; Penders and Delwaide, 1969, 1973; Shahani and Young, 1972, 1973; Ongerboer de Visser and Goor, 1974; Desmedt and Godaux, 1976). Stimulation of the supraorbital nerve elicits the reflex response which consists of an early R_1 and a late R_2 reflex. Whereas R_1 is evoked only on the side of stimulation as a pontine reflex (Kimura, 1970; Shahani and Young, 1972), R_2 is recorded bilaterally and presumably relayed through a more complex route including the pons and lateral medulla (Kimura and Lyon, 1972; Ongerboer de Visser and Kuypers, 1978). Stimulation of the infraorbital or mental nerve also elicits R_2 in the orbicularis oculi bilaterally and, less consistently, R_1 on the side of stimulus (Gandiglio and Fra, 1967).

Of the two components, R_1 is more stable with repeated trials and is, therefore, better suited for assessing conduction of the trigeminal and facial nerves. Analysis of R_2, however, is essential in determining whether the afferent or efferent arc of the reflex is primarily involved (Kimura and Lyon, 1972; Goor and Ongerboer de Visser, 1976). With a lesion of the trigeminal nerve, R_2 is slowed or diminished bilaterally when the affected side of the face is stimulated (afferent delay). With a lesion of the facial nerve, R_2 is abnormal on the affected side regardless of the side of stimulation (efferent delay).

This chapter reviews a 10-year experience with the blink reflex as a diagnostic test and discusses its clinical value and limitations in patients with trigeminal and facial neuropathy, acoustic neuroma, Guillain–Barré syndrome, Charcot–Marie–Tooth disease, diabetic polyneuropathy, multiple sclerosis with pontine lesions, Wallenberg syndrome, and other disorders causing facial hypesthesia.

METHOD

Subjects lay supine on a bed in a warm room with eyes gently closed. Surface electrodes were used for stimulation and recording (Kimura et al., 1969). The recording electrode was placed on the upper or lower aspect of the orbicularis oculi muscle laterally, with a reference electrode on the lateral surface of the nose and a ground electrode around the arm. Using a constant-current unit, the supraorbital, infraorbital, or mental nerve was stimulated with the cathode placed over the respective foramen on one side, and the reflex responses were recorded on both sides simultaneously. Shocks were of such intensity that reflex responses were just maximum and nearly stable with repeated trials. Shocks of the same intensity were delivered to each side in each subject to compare reflexes elicited by right- and left-sided stimulation. If R_1 was unstable or not excitable, shocks of higher intensities or paired shocks at 5 msec were used to facilitate the response. The shock artifact was large when the supra- or infraorbital nerve was stimulated, since the active recording electrode (G_1) was located only 2 to 3 cm away from the stimulating electrode.

A specially designed amplifier with short blocking time (1.0 msec) and low noise (0.5 μV RMS at bandwidth of 2,000 Hz) was developed to overcome the problem of stimulus artifact (Walker and Kimura, 1978). Briefly, the input and interstage coupling capacitors are eliminated so that the feed-forward path from electrodes to output is strictly a DC amplifier. The amplifier thus recovers virtually immediately, because it contains no significant capacitors to store charge. An integrator feedback loop subtracts an offset signal from the stage-1 input to establish a low-frequency cutoff and to eliminate DC electrode offset. To achieve fast recovery, a circuit detects if the stage-1 output exceeds limits near saturation in either direction and, via a control circuit, opens the feedback loop. Frequency response used was 20 to 2,000 Hz for sensory nerve and 20 to 32,000 Hz for muscle action potentials.

All responses were recorded on an FM tape recorder or photographed on Polaroid® film. The reflex latency of R_1 was measured from the stimulus artifact to the initial EMG deflection. With paired shocks, R_1 latency was measured from the stimulus artifact of the second shock of the pair. For each subject, at least eight responses were measured, and the minimal latency was determined.

The conduction of the facial nerve was also routinely tested by stimulating the nerve with the cathode placed just anterior to the mastoid process. The compound action potential recorded from the ipsilateral orbicularis oculi muscle was designated the direct response. To compare the conduction through the distal facial nerve to that of the reflex arc that includes the trigeminal nerve and the proximal segment of the facial nerve, the latency ratio of R_1 to the direct response (R/D ratio) was calculated.

NORMAL VALUES IN ADULTS AND INFANTS

Table 1 shows the normal latency range of the direct response, R_1, the R/D ratio, and R_2 elicited by stimulation of the supraorbital nerve in 83 healthy subjects 7 to 86 years of age (average age, 37) and in 30 full-term

TABLE 1. *Normal subjects and patients with bilateral diseases (mean ± SD)*

	Number averaged	Direct response (msec)	R_1 (msec)	R/D ratio	Ipsilateral R_2 (msec)	Contralateral R_2 (msec)
Normal adults	166	2.9 ± 0.4	10.5 ± 0.8	3.6 ± 0.5	30.5 ± 3.4	30.5 ± 4.4
Normal neonates	60	3.3 ± 0.4	12.1 ± 1.0	3.7 ± 0.4	35.9 ± 2.5	Often absent
Diabetic polyneuropathy	172	3.4 ± 0.6	11.4 ± 1.2	3.4 ± 0.5	33.7 ± 4.6	34.8 ± 5.3
Guillain–Barré syndrome	108	4.5 ± 3.4	15.1 ± 6.9	3.8 ± 1.0	38.2 ± 8.9	38.3 ± 8.7
Charcot–Marie–Tooth disease	92	5.7 ± 3.1	15.7 ± 4.1	3.1 ± 1.0	38.5 ± 6.7	38.6 ± 6.5
Multiple sclerosis	126	2.9 ± 0.5	12.3 ± 2.7	4.3 ± 0.9	35.8 ± 8.4	35.7 ± 8.0

neonates (Kimura, 1975; Kimura et al., 1977). Of the two components, R_1 was recorded in all but three infants. Its latency in neonates (mean ± SD; 12.1 ± 1.0 msec) was significantly greater than that in adults (10.6 ± 0.8 msec) despite a considerably shorter length of reflex arc. Unlike the consistent response in adults, R_2 was difficult to elicit in infants (Hopf et al., 1965; Clay and Ramseyer, 1976) and was recorded in only 20 of 30 neonates, mostly on the side ipsilateral to the stimulus (Kimura et al., 1977).

In 51 other healthy subjects 12 to 77 years of age (average age, 40), the infraorbital and mental nerves were tested in addition to the supraorbital nerve on each side of the face. The reflex response was recorded from the orbicularis oculi muscle on both sides regardless of sites of stimulation. With stimulation of the supraorbital nerve, both R_1 and R_2 were regularly elicited in all subjects tested. When the infraorbital nerve was stimulated, R_2 was present in all, but R_1 was inconsistent; on the other hand, stimulation of the mental nerve elicited R_2 inconsistently and R_1 only rarely (Table 2). The latencies of R_1 and R_2 were similar for the supraorbital and infraorbital nerves; R_2 elicited by stimulation of the mental nerve was considerably greater in latency.

Responses from the right and left sides were pooled together to determine the normal range, since there was no significant difference between them. The mean latency plus three standard deviations was taken as the upper limit of normal. Based on the first group of 83 healthy subjects, the direct response was considered delayed if it exceeded 4.1 msec, and R_1 delayed if it exceeded 13.0 msec. Additionally, difference in latency of the direct response between the two sides in one subject had to be less than 0.6 msec and that of R_1 less than 1.2 msec. The R/D ratio was considered abnormal if it fell outside the range of 2.6 to 4.6, two standard deviations above and below the mean in normals.

The normal range of R_2 was determined on the basis of the second group of 51 normal subjects in whom all three divisions of the trigeminal nerve were tested. With stimulation of the supraorbital nerve, the upper limit of normal for latency of R_2 was 40 msec on the side of stimulus and 41 msec on the contralateral side. Two types of com-

TABLE 2. *Normal subjects and patients with facial hypesthesia*

Response	Nerve stimulated	Normal subjects (N = 51)				Patients with unilateral disease		Patients with bilateral disease (N = 6): right and left sides combined (msec)
		Number of subjects with unequivocal response			Right and left sides combined (msec)	Stimulation on normal side (msec)	Stimulation on affected side (msec)	
		Both sides	One side	Neither side				
Direct response	Facial	51	0	0	3.0 ± 0.4			
R_1	Supraorbital	51	0	0	10.1 ± 0.9	10.4 ± 0.5	12.8 ± 2.2	11.1 ± 3.0
	Infraorbital	17	6	28	10.3 ± 1.2	11.2 ± 0.6		
	Mental	1	4	36	10.6 ± 3.4			
Ipsilateral R_2	Supraorbital	51	0	0	28.8 ± 3.6	30.7 ± 2.8	34.5 ± 6.9	33.9 ± 4.8
	Infraorbital	51	0	0	27.9 ± 4.2	30.5 ± 4.5	37.2 ± 10.8	
	Mental	43	0	8	33.3 ± 5.6			
Contralateral R_1	Supraorbital	51	0	0	29.9 ± 3.4	32.6 ± 4.0	35.6 ± 6.2	34.1 ± 4.4
	Infraorbital	51	0	0	29.9 ± 4.0	32.6 ± 4.5	40.2 ± 12.8	
	Mental	43	1	7	34.9 ± 5.2			

parison were possible. First, a latency difference between the ipsilateral and the contralateral R_2 simultaneously evoked by stimulation on one side should not exceed 5 msec. Secondly, a latency difference between R_2 evoked by right-sided stimulation and the corresponding R_2 evoked by left-sided stimulation should be less than 7 msec. With stimulation of the infraorbital and mental nerves, the upper limits were 41 msec and 50 msec, respectively, on the side of stimulus and 42 msec and 51 msec, respectively, on the contralateral side.

TRIGEMINAL NEURALGIA AND PARATRIGEMINAL SYNDROME

The blink reflex may be used to test conduction of the trigeminal nerve which constitutes the afferent arc of the reflex pathways (Kimura et al., 1970; Ongerboer de Visser and Goor, 1974). The present series (Table 3) consisted of 96 patients with trigeminal neuralgia and 17 patients with paratrigeminal syndrome. The diagnosis of the latter group included meningioma of sphenoidal ridge or petrous apex (four), herpetic neuropathy of the trigeminal nerve (three), tumors including metastasis (three), intracavernous aneurysm (one), Tolosa–Hunt syndrome (one), and undetermined etiology (five).

In patients with trigeminal neuralgia, R_1 was abnormal in seven and normal in the remaining 89. In three of the seven patients with delayed or absent R_1, however, alcohol block or nerve evulsion had been performed prior to the test. Thus, only four of 93 patients with "idiopathic" trigeminal neuralgia showed a block or CV slowing attributable to the irritative process (Fig. 1). In contrast, 10 of 17 patients with demonstrable cause for facial pain showed a significant delay of R_1 on the affected side (Fig. 2). In patients with delayed R_1 as a result of lesions of the trigeminal nerve, R_2 was often, though not always, delayed bilaterally when the affected side was stimulated, indicating an involvement of the afferent arc of the blink reflex. The R/D ratios were usually increased.

BELL'S PALSY

The direct response elicited by stimulation of the facial nerve helps to detect Wallerian degeneration but fails to reveal the function of the proximal nerve segment, which is primarily affected in Bell's palsy. In contrast, the blink reflex tests conduction of the entire length of the facial nerve including the involved interosseous portion (Rushworth, 1962; Kimura et al., 1969, 1976; Penders and Delwaide, 1971, 1973; Penders and Boniver, 1972; Schenck and Manz, 1973).

The blink reflex was studied in 144 patients who had acute unilateral facial weak-

TABLE 3. *Patients with unilateral diseases (mean ± SD)*

	Stimulus on	Number averaged	Direct response (msec)	R_1 (msec)	R/D ratio	Ipsilateral R_2 (msec)	Contra-lateral R_2 (msec)
Trigeminal neuralgia	Affected side	89	2.9 ± 0.4	10.6 ± 1.0	3.7 ± 0.6	30.4 ± 4.4	31.6 ± 4.5
	Normal side	89	2.9 ± 0.5	10.5 ± 0.9	3.7 ± 0.6	30.5 ± 4.2	31.1 ± 4.7
Paratrigeminal syndrome	Affected side	17	3.1 ± 0.5	11.9 ± 1.8	3.9 ± 1.0	36.0 ± 5.5	37.2 ± 5.7
	Normal side	17	3.2 ± 0.6	10.3 ± 1.1	3.4 ± 0.6	33.7 ± 3.5	34.8 ± 4.1
Bell's palsy	Affected side	100	2.9 ± 0.6	12.8 ± 1.6	4.4 ± 0.9	33.9 ± 4.9	30.5 ± 4.9
	Normal side	100	2.8 ± 0.4	10.2 ± 1.0	3.7 ± 0.6	30.5 ± 4.3	34.0 ± 5.4
Acoustic neuroma	Affected side	26	3.2 ± 0.7	14.0 ± 2.7	4.6 ± 1.7	38.2 ± 8.2	36.6 ± 8.2
	Normal side	26	2.9 ± 0.4	10.9 ± 0.9	3.8 ± 0.5	33.1 ± 3.5	35.3 ± 4.5
Wallenberg syndrome	Affected side	23	3.2 ± 0.6	10.9 ± 0.7	3.6 ± 0.6	40.7 ± 8.6	38.4 ± 7.1
	Normal side	23	3.2 ± 0.4	10.7 ± 0.5	3.4 ± 0.4	34.0 ± 5.7	35.1 ± 5.8

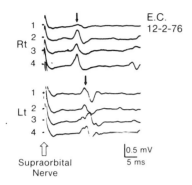

FIG. 1. Delay of R_1 in a 68-year-old woman with typical clinical symptoms of "idiopathic" trigeminal neuralgia on the left. Four tracings were recorded on each side to show consistency. Subsequent studies on several occasions over a 2-year period showed no essential change. No demonstrable lesion has so far been demonstrated despite extensive search.

ness of peripheral type without any demonstrable cause for neuropathy. Abnormalities of blink reflex were detected in all patients with Bell's palsy during the first week, although they were not necessarily apparent at the onset. In two patients not included in this study, the blink reflex remained normal throughout the entire course of minimal unilateral facial weakness lasting 1 to 2 days, perhaps representing an unusually mild form of Bell's palsy. Alterations of the blink reflex were always efferent in type in that R_1 and R_2 were absent or delayed on the affected side of face regardless of the side of stimulation.

In 100 of 127 patients tested serially, the previously absent R_1 or R_2 returned, and the direct response remained relatively normal. This indicates recovery of conduction across the involved segment without substantial distal degeneration (Fig. 3). These patients generally showed a good clinical recovery within a few months after onset. The latency of R_1, however, was considerably prolonged initially, suggesting demyelination of the involved segment. The R/D ratios were much greater than in normals, indicating conduction abnormalities in the proximal segment of the facial nerve. The latency of R_1 decreased during the second month and returned to a normal level during the third or fourth months. In the remaining

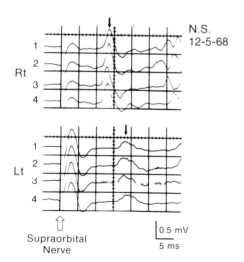

FIG. 2. Delay of R_1 in a 46-year-old woman with paratrigeminal syndrome caused by petrous ridge meningioma on the left. Four tracings were recorded on each side.

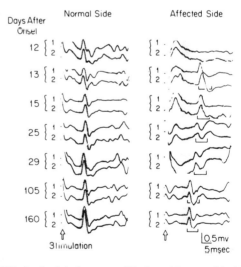

FIG. 3. Serial changes of R_1 in a 16-year-old girl with Bell's palsy on the right. Two consecutive tracings were recorded on each side to show consistency of R_1 on a given day. On the affected side, R_1 was delayed when it first appeared on the 13th day of onset but recovered progressively thereafter. *Shaded areas* indicate normal range (mean ± SD in 83 subjects).

27 patients, direct response was markedly diminished without return of the reflex response during the first 2 weeks, indicating distal degeneration (Kimura et al., 1975).

ACOUSTIC NEUROMA

Because of its strategic location, a cerebellopontine angle tumor frequently compresses the trigeminal and facial nerves as well as brainstem structures, and the blink reflex is of unique diagnostic value in localizing acoustic neuromas (Bender et al., 1969; Lyon and Van Allen, 1972; Kimura and Lyon, 1973; Eisen and Danon, 1974). In all 33 patients included in this review, the diagnosis was confirmed surgically. The direct response was absent in seven, including five tested only after surgical sacrifice of the facial nerve during operative removal of the tumor. In these patients, neither R_1 nor R_2 was elicitable on the side of the degenerated facial nerve. Of the remaining 26 patients with intact direct response, R_1 was absent on the affected side in five, delayed in 17, and normal in four. In six of the 22 patients with abnormal R_1, R_2 was delayed bilaterally when the affected side of the face was stimulated, suggesting a compression of the trigeminal nerve (afferent delay). In six others, R_2 was altered on the affected side whether the right or left side was stimulated, suggesting involvement of the facial nerve (efferent delay). In seven patients, abnormalities of R_2 were of mixed pattern, i.e., delay of R_2 bilaterally after stimulation of the affected side but with greater delay ipsilaterally and a delay of contralateral R_2 after stimulation of the normal side. These findings were taken to indicate a combined lesion of the trigeminal and facial nerves or involvement of the brainstem. In the remaining three, R_2 was normal despite delayed R_1 on the affected side.

POLYNEUROPATHY

There have been many clinical descriptions of facial or trigeminal nerve involvement in various polyneuropathies, but the cranial nerves have rarely been tested electrophysiologically. Because slowed CV is characteristic of demyelination (McDonald and Sears, 1970), studies of the blink reflex are particularly useful in a demyelinating type of polyneuropathy (Kimura, 1971a). Of 186 patients included in the present series (Table 1), 54 had Guillain–Barré syndrome (GBS), 46 Charcot–Marie–Tooth disease (CMT), and 86 diabetic polyneuropathy.

The right and left sides were considered together since there was no difference between them. Of the 108, 92, and 172 sides tested in GBS, CMT, and diabetic polyneuropathy, respectively, R_1 was absent in nine, seven, and one, delayed in 49, 57, and 17 (Fig. 4), and normal in the remainder. Similarly, the direct response was absent in six, seven, and two, delayed in 44, 45, and 20, and normal in the rest. Excluding absent responses, average latency of R_1 was significantly greater in each of the three neuropathies than in normals. The direct response was also slowed in the neuropathies when compared to the normal value. The R/D ratio was slightly increased in GBS, moderately decreased in CMT, and slightly decreased in diabetics. In the polyneuropathies, abnormalities of R_2 were bilateral and could not be categorized into an afferent or efferent type. Although latencies of R_2 were commonly within the normal range when analyzed individually, the average value was significantly greater in the neuropathies than in the control (Table 1).

MULTIPLE SCLEROSIS

It is now well known that alterations of electrically elicited blink reflex are not always caused by disorders of the peripheral reflex arcs but may be attributable to lesions affecting central reflex pathways. In multiple sclerosis (MS), the incidence of blink reflex abnormality varies a great deal depending on the selection of patients. In general, the longer the history of clinical symptoms, the higher the rate of abnormal-

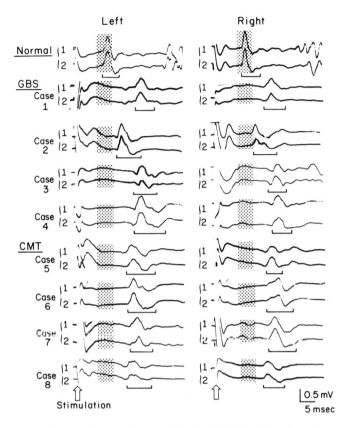

FIG. 4. Bilateral delay of R_1 in four patients with the Guillain–Barré syndrome (GBS) and four patients with Charcot–Marie–Tooth disease (CMT). Two tracings were recorded on each side in each subject. *Top tracings* are from a normal control, with *shaded areas* indicating the normal range.

ity. The incidence of abnormality is considerably lower in our current series than in earlier studies since patients are now referred for electrophysiological testing much earlier than before. Of various lesions affecting the brainstem, the delay of R_1 is most conspicuous in MS, as might be expected from the nature of the pathology (Kimura, 1970; Namerow and Etemadi, 1970; Lyon and Van Allen, 1972; Kimura, 1975; Lowitzsch et al., 1976; Paty et al., 1979). To determine the effect of central demyelination on the blink reflex, 63 patients with unequivocal clinical signs of extensive pontine involvement were selected from a list provided by the local MS society. Most, but not all, had clinical evidence of anatomical dissemination or historical evidence of relapse.

Conduction of the facial nerve as measured by latencies of the direct response was normal in all. In contrast, R_1 was abnormal on one or both sides in 33 of 38 patients with definite diagnosis, 12 of 14 patients with probable diagnosis, and four of 11 with possible diagnosis (Fig. 5). The incidence of abnormal R_1 rose with increasing duration of illness in each category. Considering the right and left sides together and excluding three absent responses, the average latency of R_1 in 123 responses from the 63 patients was substantially greater than in normals, but the degree of slowing was less when compared to GBS and CMT. Since the direct response was normal in MS, the R/D ratio was significantly increased. In Fig. 6, the latency distribution of the direct response and R_1 in MS is compared to

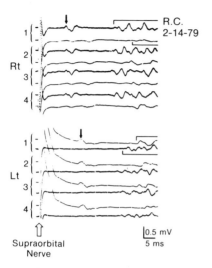

FIG. 5. Efferent delay of the blink reflex in a 35-year-old woman with multiple sclerosis and mild facial and abducens paresis on the left. Four consecutive stimuli were delivered on each side. Reflex responses were simultaneously recorded from the ipsilateral (*upper tracing* in each frame) and the contralateral (*lower tracing* in each frame) orbicularis oculi. With right-sided stimulation, R_1 is normal, and R_2 is delayed contralaterally, whereas with left-sided stimulation, R_1 is slowed, and R_2 is delayed ipsilaterally. This finding suggests a lesion involving the efferent arc of the reflex on the left.

those in normal subjects and the two demyelinating types of polyneuropathy.

WALLENBERG SYNDROME

Because of inherent latency variability, R_2 is not as dependable as R_1 in determining nerve conduction along the reflex arc. However, alteration of R_2, either in latency or magnitude, is commonly associated with lesions affecting the lateral medulla, as demonstrated in patients with the Wallenberg syndrome (Kimura and Lyon, 1972; Ongerboer de Visser and Kuypers, 1978). Of 23 patients with this syndrome included in the present analysis, five were tested with stimulation of the infraorbital and mental nerves as well as supraorbital nerve.

As shown in Table 3, the average latency of R_1 was normal in this syndrome, although, analyzed individually, the values were sometimes slightly greater on the affected side than on the normal side. When the hypesthetic side of the face was stimulated, both ipsilateral and contralateral R_2 were absent in seven and significantly delayed in 10 (Fig. 7). In the remaining six patients, stimulation on the affected side elicited ipsilateral and contralateral R_2 of essentially normal latency, although they were minimal in size when compared to those elicited by stimulation on the other side of the face. With stimulation on the normal side, R_2 was normal bilaterally in 20 of 23 patients. In the remaining three patients, R_2 was normal on the side ipsilateral to the stimulus but absent on the affected side (Ongerboer de Visser and Kuypers, 1978). Although there were some individual variations, the findings were essentially the same whether the supraorbital, infraorbital, or mental nerve was tested. Stimulation of the mental nerve, however, often failed to elicit R_2 regardless of the side stimulated. This was considered a nonspecific finding in view of the inconsistency of this reflex in normal subjects.

FACIAL HYPESTHESIA

An afferent delay of R_2 is by no means specific to the Wallenberg syndrome, since similar alteration may be seen in some patients with contralateral hemispheral lesion (Messina and Quattrone, 1973; Kimura, 1974; Dehen et al., 1976; Fisher et al., 1979). An afferent type of impairment of R_2 with cerebral disease is seen commonly, though not exclusively, in those with a sensory disturbance in the face. Thus, it is tempting to postulate that R_2 may be affected by any disease process associated with sensory deficit regardless of the site of the responsible lesion.

In the present study, this hypothesis was tested in 25 patients who presented with subjective complaints of facial hypesthesia

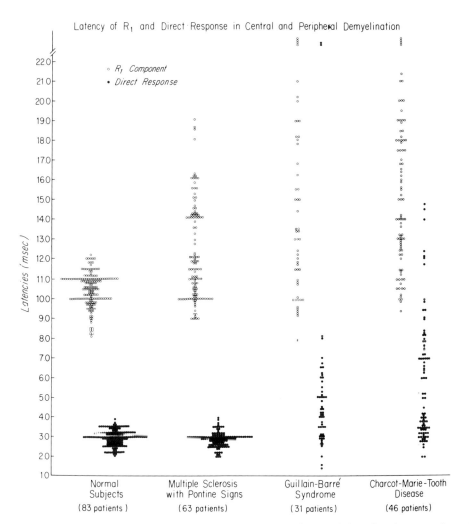

FIG. 6. Latency distribution of the direct response and R_1 of the blink reflex in normals and in patients with central or peripheral demyelination of the reflex pathways. The direct response is slowed in Charcot–Marie–Tooth disease and to a slightly lesser extent in the Guillain-Barré syndrome but remains normal in multiple sclerosis. Delay of R_1 is comparable between the two polyneuropathies but slightly smaller in multiple sclerosis.

(Kimura and Yamada, 1980). There were six patients with bilateral sensory deficit, all on the basis of isolated trigeminal neuropathy, and 19 patients with unilateral disease of either the trigeminal nerve (10) or brainstem nine. Of the 25 patients, eight were tested for all three divisions of the trigeminal nerve, five for both supraorbital and infraorbital nerves, and 12 for supraorbital nerve only.

In studies of the supraorbital nerve in the six patients with trigeminal neuropathy, R_1 was slowed or absent bilaterally in three and normal in the others. Similarly, R_2 was delayed or difficult to elicit regardless of the side of stimulation in four and normal in the remainder. When the 19 patients with unilateral hypesthesia of the face were tested, R_1 was absent on the affected side in six, delayed in seven (Fig. 8a), and normal in the others. When the affected side was stimulated, R_2 was bilaterally absent or small in

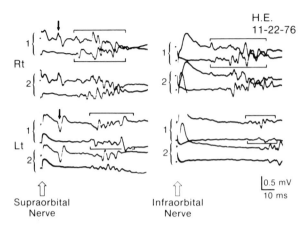

FIG. 7. Alteration of R_2 in a 63-year-old man with the Wallenberg syndrome on the left. Two consecutive stimuli were delivered on each side. Reflex responses were simultaneously recorded from the ipsilateral (*upper tracing* in each frame) and the contralateral (*lower tracing* in each frame) orbicularis oculi. With stimulation of either the supraorbital or infraorbital nerve, R_2 was normal bilaterally on right-sided stimulation but significantly delayed bilaterally on left-sided stimulation. An afferent delay of R_2, when seen in conjunction with a normal R_1 as in this case, suggests a lateral medullary lesion, although it is not specific for it.

four, delayed in seven (Fig. 8b), and normal in the rest. The response was generally smaller the more complete the sensory loss of the stimulated area. Thus, R_2 was usually absent when the anesthetic part of the face was stimulated in contrast to a normal R_2 elicited by stimulation of the corresponding part of the face on the normal side. When sensory change was equivocal, the difference in R_2 elicited by right- and left-sided stimulation was also inconsistent and difficult to determine because of variability from one trial to the next. In these cases, the right and left sides were alternately stimulated every 5 sec or so for several shocks. This procedure was helpful in excluding an apparent asymmetry caused by random variations that tended to shift from one side to the other.

When sensory change was limited to one division of the trigeminal nerve, R_2 was selectively diminished only when the affected dermatome was stimulated. Otherwise, the findings were essentially the same whether the supraorbital or infraorbital nerve was tested. The average value of R_2 elicited with stimulation of the supraorbital or infraorbital nerve on the affected side was significantly greater in latency, shorter in duration, and smaller in amplitude than the corresponding R_2 elicited by stimulation on the normal side of the face (Table 2). Stimulation of the mental nerve was not suited for this purpose because of its inherent inconsistency even on the normal side.

COMMENTS

Because it is noninvasive and easily available, the electrically elicited blink reflex has found wide clinical applicability (Table 4). Of the two components, R_1 is more reliable than R_2 for nerve conduction studies: R_1 is most conspicuously delayed in diseases associated with extensive demyelination either centrally (Kimura, 1970, 1973, 1975; Namerow and Etemadi, 1970; Namerow, 1973; Lowitzsch et al., 1976) or peripherally (Kimura, 1971a). It has also been shown that R_1 is slowed in a more localized lesion of the trigeminal nerve (Kimura et

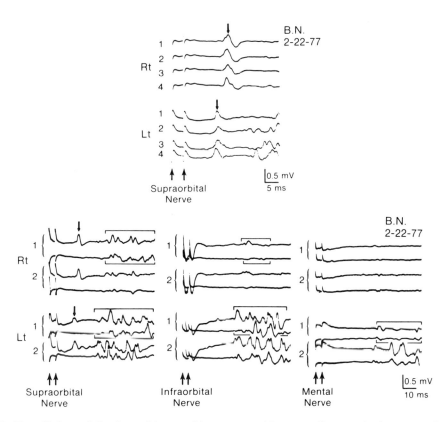

FIG. 8. Top: Delay of R_1 in a 54-year-old woman with parasellar meningioma causing facial hypesthesia on the right. Four tracings were recorded on each side. Paired stimuli were used after single shocks failed to elicit consistent responses on either side. **Bottom:** Alteration of R_2 in the same patient as shown in **top** figure. Two consecutive paired stimuli were delivered on each side. Reflex responses were simultaneously recorded from the ipsilateral (*upper tracing* in each frame) and the contralateral (*lower tracing* in each frame) orbicularis oculi. Regardless of the nerve stimulated, R_2 was diminished or nearly absent bilaterally on right-sided stimulation and normal bilaterally on left-sided stimulation.

al., 1970; Ongerboer de Visser and Goor, 1974), facial nerve (Rushworth, 1962; Kimura et al., 1969; Penders and Delwaide, 1971, 1973; Penders and Boniver, 1972; Schenck and Manz, 1973; Kimura et al., 1976) or both (Bender et al., 1969; Lyon and Van Allen, 1972; Kimura and Lyon, 1973; Eisen and Danon, 1974).

Alteration of R_1, however, does not necessarily indicate a pathological process of the reflex arc itself, since indirect effects of lesions outside the primary reflex pathway may also contribute (Kimura, 1975). For example, reversible block of R_1 may occur in comatose patients when the brainstem is severely suppressed with acute supratentorial lesions or massive drug intoxication (Lyon et al., 1972). A significant change in latency of R_1 was described in acute hemispheric strokes by Fisher et al. (1979) who used mechanical rather than electrical stimulation. These changes were transient, recovering almost completely within a few days. In contrast, the latency of R_1 usually remains normal even during acute stages of hemispheric disease, provided an adequate stimulus is given to compensate for a possible reduction in excitability of the reflex

TABLE 4. *Direct response and R_1 and R_2 of the blink reflex*

Disorders	Direct response	R_1	R_2
Paratrigeminal syndrome	Normal	Abnormal on the affected side (59%)	Abnormal on both sides when affected side stimulated (afferent abnormality)
Bell's palsy	Normal unless distal segment degenerated	Abnormal on the affected side (99%)	Abnormal on the affected side regardless of side of stimulus (efferent abnormality)
Acoustic neuroma	Normal unless distal segment degenerated	Abnormal on the affected side (85%)	Afferent and/or efferent abnormality
Guillain–Barré syndrome	Abnormal on either side (52%)	Abnormal on either side (61%)	Normal or abnormal on both sides
Charcot–Marie–Tooth disease	Abnormal on either side (57%)	Abnormal on either side (72%)	Normal or abnormal on both sides
Multiple sclerosis	Normal	Abnormal with pontine lesion	Normal or complex abnormality, not necessarily efferent or afferent
Wallenberg syndrome	Normal	Normal or borderline	Afferent abnormality
Facial hypesthesia	Normal	Abnormal with lesions of the trigeminal nerve or pons	Afferent abnormality
Comatose state	Normal	Abnormal with pontine lesion, otherwise reduced excitability in acute stage	Absent on both sides regardless of side of stimulus

pathway (Kimura, 1974; Dehen et al., 1976).

The excitability of the blink reflex may be reduced in patients with supranuclear lesions. Single electric shocks may then activate R_1 only partially. When this is suspected, paired stimuli with a 5-msec interval may be helpful in eliciting a much larger R_1 with a considerably shorter latency (Kimura, 1971b, 1974, 1975). If R_1 is fully activated, then its latency generally indicates the conduction characteristics of the reflex arc itself rather than an excitability effect of a remote lesion. In particular, a delay of R_1 in some MS patients by as much as several milliseconds is difficult to explain on the basis of altered excitability alone.

In central demyelination of MS, the average latency of R_1 is only 12.3 ± 2.7 msec (mean $\pm$ SD) as compared to 15.1 ± 6.9 msec in GBS and 15.7 ± 4.1 msec in CMT. These findings presumably reflect that, unlike the polyneuropathies, demyelination in MS detectable by this method is restricted to a short segment in the pons. It is interesting in this regard that the average delay of R_1 in Bell's palsy (12.8 ± 1.6 msec) is very close to that found in MS as expected from a focal involvement of the facial nerve. In most patients with trigeminal neuralgia, the reflex latency is usually normal, indicating that the irritative process responsible for facial pain in this disease affects nerve CV relatively little. This may be at least in part because the first division of the trigeminal nerve is relatively spared in this condition. In contrast, the paratrigeminal syndrome is frequently associated with a significant increase in blink reflex latencies.

The degree of slowing in the direct response and R_1 is much the same in CMT and GBS (Fig. 6). The decreased R/D ratio found in the former suggests that slowing of facial CV is more distal than proximal. In the latter, the average R/D ratio is slightly increased, indicating more proximal involvement of the facial nerve provided, of course, that the trigeminal nerve is relatively intact. The R/D ratio is also significantly increased in MS and paratrigeminal syndrome as expected from normal CV of the facial nerve in these entities. The ratio is also increased in Bell's palsy unless the facial nerve is degenerated distally.

Unlike R_1, R_2 is not a reliable test of nerve conduction, since it varies considerably from one trial to the next. Analysis of R_2, however, is important in differentiating a lesion of the afferent from that of the efferent reflex arc when R_1 is delayed. Moreover, R_2 is altered not only by lesions affecting the reflex pathways directly, as is likely in the Wallenberg syndrome (Kimura and Lyon, 1972; Ongerboer de Visser and Kuypers, 1978), but also by lesions indirectly influencing the excitability of the polysynaptic connections. Thus, R_2 is absent or, if present, markedly diminished or delayed in any comatose state regardless of the site of the responsible lesions (Messina and Micalizzi, 1970; Lyon et al., 1972). It is also significantly affected by a hemispheric lesion showing either afferent or efferent pattern of delay, perhaps depending on the site of involvement (Messina and Quattrone, 1973; Kimura, 1974; Dehen et al., 1976; Fisher et al., 1979), and by the state of arousal (Shahani, 1968; Ferrari and Messina, 1968; Kimura and Harada, 1972, 1976; Desmedt and Godaux, 1976; Boelhouwer and Brunia, 1977).

Unlike R_1, R_2 is consistently elicited not only by stimulation of supraorbital nerve but also by shocks applied to the dermatomes of the infraorbital nerve. As shown in the present study, R_2 is significantly smaller when a hypesthetic area of the face is stimulated as compared to the corresponding area on the normal side. This finding, however, does not localize the lesion precisely, since R_2 is also affected by a supratentorial lesion indirectly reducing excitability of the polysynaptic pathways. Whereas sensory deficits of the face are often associated with alteration of R_2, the reverse is not always true. Indeed, a similar reduction of R_2 is sometimes seen in cases of pure motor hemiplegia showing facial weakness without apparent sensory change (Kimura, 1974; Dehen et al., 1976; Fisher et al., 1979). In these cases, minor sensory deficits may not be detected clinically. It is more likely, however, that certain supratentorial lesions outside the somatosensory pathways can cause inhibition or reduced facilitation of the reflex pathways.

CONCLUSION

The blink reflex has only recently been used in clinical tests. It is valuable in differentiating the paratrigeminal syndrome from trigeminal neuralgia, since the reflex latencies are usually affected in the former but not in the latter. In Bell's palsy, the return of R_1 and R_2 offers a reasonable assurance that remaining axons will survive without undergoing further deterioration. Even in this group of patients, however, R_1 is delayed during the first 4 weeks, suggesting demyelination rather than physiological or functional block of the involved nerve segment. A marked delay in latency of R_1 and the direct response offers conclusive evidence that the facial nerve is severely involved in some polyneuropathies. This is expected in the Guillain–Barré syndrome, which is characteristically associated with facial weakness, but not in Charcot–Marie–Tooth disease, in which weakness of the facial muscle, if any, is clinically very subtle.

Conduction through a demyelinated segment in the central reflex arc can also be measured objectively by the blink reflex. Delay of R_1 in multiple sclerosis and alteration of R_2 in the Wallenberg syndrome suggest that in the right clinical context, the test may be used to evaluate lesions in the brainstem that may or may not be clinically manifest. The test may also provide an objective means to assess sensory disturbances in the face, although it should be kept in mind that various other factors can influence the excitability of the blink reflex significantly.

ACKNOWLEDGMENTS

The author wishes to thank Dr. David Walker, MSEE, for engineering advice and Ms. Sheila Mennen, Deborah A. Gevock, and Cheri L. Turner for technical assistance.

Motor Control Mechanisms in Health and Disease,
edited by J. E. Desmedt.
Raven Press, New York © 1983.

Cutaneomuscular (Flexor) Reflex Organization in Normal Man and in Patients with Motor Disorders

*H. M. Meinck, R. Benecke, S. Küster, and B. Conrad

Department of Clinical Neurophysiology, University of Göttingen, 3400 Göttingen, Federal Republic of Germany

The human flexor reflex is understood to be a protective sensorimotor mechanism, securing withdrawal from an offensive stimulus; the more complex functions of the flexor reflex system as a "lower motor center" (see Lundberg, 1979) are far from being established in human neurophysiology. Electromyographically, flexor reflexes consist of complex patterns of excitation and inhibition in both flexors and extensors, representing an interdependent system of cutaneomuscular organization finely adjusted to the site and intensity of stimulus (Hagbarth, 1960; Kugelberg et al., 1960; see Young, 1973). Animal experiments have shown that the spinal polysynaptic pathways of this reflex are controlled from several supraspinal structures (see Lundberg, 1975). In man as well, adequate cutaneomuscular organization and adjustment depend, in all likelihood, on differential supraspinal control of reflex paths to flexors and to extensors; damage to suprasegmental structures causes disturbance of cutaneomuscular reflex mechanisms.

This chapter deals with the organization and disturbances of reflexes to cutaneous stimuli in normal subjects and in patients. Since the term "flexor reflex" does not cover the entire spectrum of these reflex effects, the more general term "cutaneomuscular reflex" is preferred.

METHODOLOGICAL CONSIDERATIONS

Three neuronal test systems of increasing complexity may be used for investigating cutaneomuscular reflex effects, with, in each case, the effector motoneuron (MN) pool forming the final common pathway (Fig. 1). Evidently, the simplest method (Fig. 1A) is to record stimulus-induced EMG activity from a primarily inactive muscle. However, with this method, inhibitory and subliminal excitatory reflex effects remain undetectable. This problem is overcome by introducing a test activity, either a monosynaptic (preferably H) reflex (Fig. 1B) or a tonic voluntary EMG activity (Fig. 1C), both being conditioned by a cutaneous stimulus. The possibility of investigating cutaneomuscular reflex effects by means of vibration-induced test activity has occasionally been tried (Schenck and Lauck-Koehler, 1950; Godaux and Desmedt, 1975b). However, the introduction of a test activity is accompanied by new problems

---

*To whom correspondence should be addressed: Abteilung für Klinische Neurophysiologie, Neues Klinikum, University of Göttingen, Robert-Kochstrasse 40, Göttingen 3400, Federal Republic of Germany.

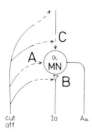

FIG. 1. Different neuronal test systems for investigating cutaneomuscular reflex effects. Cut. aff., cutaneous afferents; Ia, Ia afferents; Aα, Aα afferents.

which are partly the result of the increasing complexity of the neuronal test system itself: a conditioning cutaneous volley might influence a monosynaptic test reflex pre- or postsynaptically and possibly also via the γ system. A tonic voluntary EMG activity may additionally be influenced at suprasegmental generating or relaying structures. Moreover, in resting and active states, a different setting of interneurons and MNs may lead to different gating of certain reflex effects (cf. Meinck and Piesiur-Strehlow, 1981).

The advantages of using H reflexes as a test parameter are: (1) the relative simplicity of the neuronal test system and our more extensive knowledge about it; (2) the applicability even to plegic muscles; and (3) the high sensitivity of H reflexes to inhibitory and facilitatory influences, particularly if the test reflex is elicited through barely suprathreshold stimuli (Meinck, 1980). However, the main drawbacks of H-reflex testing are its applicability to only a limited number of muscles and the extensive time involved, which makes the method wearisome for some patients.

On the other hand, an investigation of stimulus-induced changes in tonic voluntary EMG activity is possible in any muscle and only requires a short time but is less sensitive than H-reflex testing (Granit and Job, 1952; Gassel and Ott, 1970). For cutaneomuscular reflex investigation, the "EMG modulation curve" technique (Gassel and Ott, 1970) seems to be widely accepted. Nowadays, superimposition of single EMG sweeps is considered inadequate. Summation of the raw EMG seems inadequate because it favors stimulus-locked, short-latency effects and neglects dispersed, long-latency effects, and it has poor resolution capacity for inhibitory reflex phases, particularly if they are weak (see Fig. 2; cf. Gassel and Ott, 1970). In contrast, computer summation of the rectified EMG allows qualitative and quantitative evaluation of both inhibitory and excitatory reflex effects. The application of poststimulus time histograms and the "CUSUM" technique (Ellaway, 1978) is limited to single motor unit studies.

Quantitative analysis of cutaneomuscular reflex effects by means of tonic EMG testing is subject to several problems, e.g., the influence of the degree of basic activity. With enhancement of basic EMG activity, the amplitudes of inhibitory phases remain

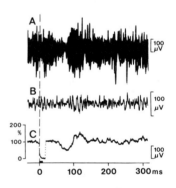

FIG. 2. Comparison of different methods for displaying cutaneomuscular reflex effects. Record from the abductor pollicis brevis muscle after stimulation of digit V. Stimulus train (400/sec, train duration 10 msec) delivered via ring electrodes at 0 msec (vertical broken line). The records show essentially identical signals: A, raw EMG, 32 sweeps superimposed; B, same sweeps, summated; C, same sweeps, rectified before summation. Left scale in C is based on the distance between the zero level of the computer (indicated by short-circuiting the input of the rectifier to the ground for about 20 msec) and prestimulus activity level, which was taken as 100% (cf. Godaux and Desmedt, 1975b).

fairly constant, but those of excitatory phases decrease (H.M. Meinck, *unpublished data*). If an excitatory phase is primarily weak, it might be completely occluded by a strong tonic activity (cf. Caccia et al., 1973). In order to keep basic activity constant and weak, a continuous feedback of the degree of tonic EMG activity should be presented to the subject via an EMG integrator display (Godaux and Desmedt, 1975*b*). Motor units of different recruitment thresholds may be unequally susceptible to inhibition and facilitation (Garnett and Stephens, 1980; Desmedt and Godaux, 1978*c*). Moreover, the synaptic connectivities of a given cutaneous input to different motor units of one pool are inhomogeneously organized (Garnett and Stephens, 1980; K. Kanda and J.E. Desmedt, *this volume*).

The stimulus site and the receptive field should be carefully checked: Hagbarth (1960) and Caccia et al. (1973) have shown that the stimulus site influences the type of response in a given muscle (Figs. 3 and 4). Stimuli can be applied either to the skin or to cutaneous or mixed nerves. To obtain comparable data from all subjects including patients with central sensory loss, we prefer mixed-nerve stimulation at distal sites, which permits an objective determination of stimulus strength relative to motor threshold and renders recordings from sensory nerve unnecessary. In our experience, reflex effects evoked by muscle afferent stimulation can be neglected in this experimental situation (Meinck et al., 1981*c*).

The appropriate stimulus for eliciting cutaneomuscular reflexes is a 10- to 20-msec train of impulses. Single pulses and shorter trains often fail to evoke consistent effects because of lack of temporal facilitation (cf. Kranz et al., 1973 versus Caccia et al., 1973; Conrad and Aschoff, 1977; Garnett and Stephens, 1980; Willer et al., 1978). The efficiency of such a stimulus train depends not only on the number of pulses applied but also on the duration of, and interval between, individual pulses as well as on the

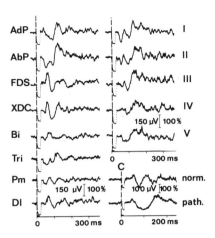

FIG. 3. Reflex effects of finger nerve stimulation in hand and arm muscles. Stimulus parameters identical to those in Fig. 2; the signals are recorded as in Fig. 2C. A: Cutaneomuscular organization in hand and arm muscles after index finger stimulation in a normal subject, records being obtained from the mm. adductor pollicis (AdP), abductor pollicis brevis (AbP), flexor digitorum superficialis (FDS), extensor digitorum communis (XDC), biceps (Bi), triceps (Tri), pectoralis medius (Pm), and deltoideus lateralis (Dl). B: Reflex effects induced in the AbP muscle by stimulation of digits (digs.) I to V. C: Reflex effects in a patient suffering from central monoparesis of the right arm; comparison of normal (norm.) and affected (path.) sides.

stimulus intensity (Tørring et al., 1981; Willer et al., 1978). Adequate (mechanical or thermal) stimuli may be used for studying special problems.

In our experience the stimulus recurrence frequency is not of prime importance in a given range of between 0.1 and 1/sec in investigating normal subjects (see Desmedt and Godaux, 1976). In patients with CNS lesions distinct reflex alterations are observed with varying recurrence frequencies (Figs. 5 and 6). Stimulus regularity or irregularity influences the amplitude but neither the shape nor the time course of reflex effects (Caccia et al., 1973). With regularly repeated stimuli, the normal reflex response in inactive muscles begins to habituate after 10 to 50 repetitions (Dimitrijevic and Nathan, 1970). However, if the muscle

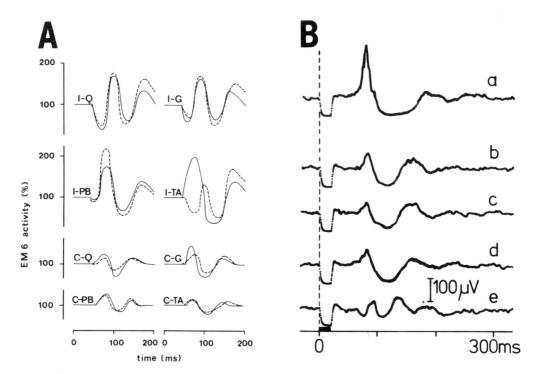

FIG. 4. Cutaneomuscular reflex organization in the leg. Labeling of panels A and B as in Fig. 3. A: Reflex effects in various ipsi- (I) and contralateral (C) leg muscles after tibial *(unbroken line)* and peroneal *(broken line)* nerve stimulation. Average reflex patterns drawn to an identical scale. Q, quadriceps (vastus lateralis); PB, posterior biceps (long head); G, medial gastrocnemius; TA, tibialis anterior. B: Receptive field of the TA reflex pattern. Stimuli were applied with fixed intensities to the medial plantar nerve (a), the lateral foot sole margin (b), the lateral (c) and medial (d) foot sole close to the heel, and to the deep peroneal nerve at the retinaculum flexorum (e). (From Meinck et al., 1981c, with permission.)

is tonically activated, rhythmic stimulus repetition results in a successive stabilization of the reflex response, possibly because of a different preset of interneurons or MNs or both (Brune, 1955; Caccia et al.,

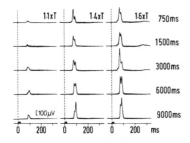

FIG. 5. Influence of increasing both stimulus intensity (**left** to **right**) and stimulus interval (**top** to **bottom**) on the normal TA reflex pattern in the inactive state. Medial plantar nerve stimulation; $1 \times T = A\alpha$ threshold determined in the flexor hallucis brevis muscle.

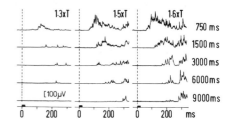

FIG. 6. Influence of increasing both stimulus intensity and interval on the TA reflex response in a hemiparetic subject. Experimental arrangement and composition of the figure as in Fig. 5.

1973; Meinck et al., 1981; Desmedt and Godaux, 1976).

The status of the subject is of prime importance: anxiety and apprehension as well as being cold or having a filled bladder or bowel may exaggerate cutaneomuscular reflexes with accentuation of late components (cf. Hagbarth and Finer, 1963; Young, 1973). This phenomenon is possibly related to sensitization of such reflexes by an unexpected additional—independent and distant—stimulus (Hagbarth and Finer, 1963; Dimitrijevic and Nathan, 1971).

CUTANEOMUSCULAR REFLEXES OF THE HAND AND ARM

The few reports dealing with cutaneomuscular reflexes in hand and arm muscles present a relatively uniform picture (Caccia et al., 1973; Garnett and Stephens, 1980). Digital nerve stimulation evokes alternating excitation and inhibition in hand muscles beginning with latencies at about 40 msec (Fig. 3). Afferents responsible for threshold elicitation of both inhibition and excitation have been assumed to belong to fast-conducting mechanoceptor (group II) fibers (Caccia et al., 1973; Garnett and Stephens, 1980). If the stimulus is successively moved from the thumb to the little finger (Fig. 3B), the reflex response of a given muscle gradually changes. However, there is no reciprocal activity in the antagonistic forearm and hand muscles, nor is there any difference between dorsal and ventral finger skin stimulation (Caccia et al., 1973; Garnett and Stephens, 1980; see below).

Single motor unit recordings have furnished insight into the intrinsic organization of the reflex pathways. Of the first dorsal interosseous motor units, only those with low recruitment thresholds, slow contraction times, and low twitch tensions showed a distinct early excitation (Garnett and Stephens, 1980). Lesions of the sensorimotor cortex or the descending tracts result in a selective suppression of the second excitatory reflex phase (Conrad and Aschoff, 1977; Jenner and Stephens, 1979; Fig. 3C), suggesting a long-loop mediation of this particular reflex component via the sensorimotor cortex. Selective suppression of one reflex phase seems to be a minor alteration compared to cutaneomuscular reflex disintegration in leg muscles after rostral lesions (see below). Such differences of supraspinal influence on reflex mechanisms of hand versus leg are not yet clear.

CUTANEOMUSCULAR REFLEXES OF THE LEG

Research in cutaneomuscular reflexes of the leg originated from studies of the Babinski sign and related phenomena (Kugelberg et al., 1960; Landau and Clare, 1959; Grimby, 1963, 1965; Nathan and Smith, 1955). Parallel to the progress of animal experiments on spinal organization and supraspinal control of flexor reflexes (see Lundberg, 1979), interest focused on human flexor reflexes in a stricter sense (for review, see Desmedt, 1973a). Systematic application of the EMG modulation curve technique revealed oscillating sequences of excitation and inhibition with onset latencies often under 50 msec (Meinck et al., 1981). Reflex patterns in the antagonistic muscles showed reciprocity during initial reflex phases. Late reflex activity, however, exhibited a homonymous course in the antagonists (Fig. 4A). Reflex patterns evoked from different nerves in various muscles exhibited common characteristics such as similar (mainly low-threshold) afferents from the skin for all reflex phases as well as habituation of reflexes in inactive muscles and dishabituation in active muscles (see Desmedt and Godaux, 1976). Moreover, in synergistic and antagonistic muscle pairs, the time courses of reflex sequences as well as the amplitudes of individual reflex phases revealed correlation, supporting the impression of an interdependently coordinated reflex system (cf. Young, 1973). Such a reflex coordination might result either from similar characteristics of neuronal connections between cutaneous afferents and different motor nuclei or from a conjoint operation of neuron circuits in the individual pathways (Meinck et al., 1981).

In patients suffering from suprasegmental lesions, the following changes of human cutaneomuscular reflexes in the leg were described: breakdown of cutaneomuscular organization (Bathien and Bourdarias, 1972; Dimitrijevic and Nathan, 1968; Grimby, 1965b; Kugelberg et al., 1960), suppression of early reflex components (Fisher et al., 1979; Horstink and Notermans, 1979; Shahani and Young, 1973c), and dishabituation of the reflex activity (Dimitrijevic and Nathan, 1971). Correlation with clinical data revealed that reflex alterations occurred in a high percentage of hemiparetic patients but did not correlate systematically to clinical severity or duration of the disease or to particular clinical symptoms such as presence of a Babinski sign (Fisher et al., 1979).

A greater clinical value than that suggested by these authors was derived by systematically investigating the influence of facilitatory mechanisms in the undisturbed and disturbed reflex pathway. It should briefly be remembered that the normal reflex pattern in the slightly activated tibialis anterior (TA) muscle on plantar nerve stimulation consists of two bursts separated by inhibition (Fig. 4Ba). If the muscle is inactive, the earlier burst can be easily observed, but the later burst is effected only by fairly high stimulus intensities (cf. Meinck, et al., 1981; Shahani and Young, 1971). In normal subjects, enhancement of excitatory and inhibitory reflex phases was observed with increasing stimulus strengths; the recurrence frequency of the stimulus (intervals varied between 0.75 and 9 sec), however, proved to exert no systematic influence either on shape and time course or on amplitudes of the reflex response (Fig. 5). This is true for both inactive and active conditions. However, the inactive state is preferred for testing to allow a more appropriate basis of comparison with paretic patients.

In patients with suprasegmental lesions (Fig. 6), reflex activity at low stimulus intensity ($1.3 \times T$) occurs only at faster repetition rates. However, under this condition, the reflex burst is desynchronized, and its latency in comparison to normal values is delayed (70 ± 15 msec; cf. Fisher et al., 1979). Increasing the stimulus intensity results in an unstructured enhancement of the reflex response, its onset latency shifting successively to shorter and, in some cases (see Fig. 6), eventually to normal values. If, at high stimulus intensity the stimulus interval is increased, onset latency again becomes delayed, and overall reflex activity distinctly decreases. This type of pattern disintegration, as well as rapid fatigue and delay of the reflex response, were observed in all 60 patients suffering from acute or chronic upper motoneuron diseases. No clear correlation was found between reflex alterations and anatomic levels of the lesions. There was, however, a correlation between the degree of reflex alterations and clinical severity of disease, but only when the symptoms were mild. Moreover, such reflex alterations were also found in many cases in whom a suprasegmental lesion would have been expected or was equivocal (e.g., beginning MS or ALS) but could not definitely be confirmed by neurological examination. However, no such changes were observed in patients suffering from extrapyramidal (cf. Delwaide et al., 1974) or cerebellar disorders.

Single motor unit studies in hemiparetic patients revealed a dependence of firing latency on stimulus intensity and recurrence frequency on the MN level. The firing latency of the TA motor unit in Fig. 7 shortened successively from about 250 msec (top sweep $0.9 \times T$, right) to the normal value of

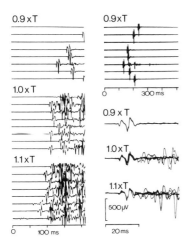
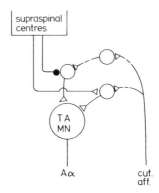

FIG. 7. Influence of increasing stimulus strength on the reflex response of a single motor unit of the primarily inactive TA muscle in a hemiparetic subject. Medial plantar nerve stimulation, $1 \times T = A\alpha$ threshold. Stimulus intervals fixed at 1.5 sec. Each panel in the **left column** and the **top right panel** represents eight consecutive reflex responses (from **top** to **bottom**). **Top left** and **right panels** show identical sweeps, with the sweep speed being reduced in the top right panel. The first motor unit to respond is identical in almost all sweeps. Motor unit identification at different stimulus intensities in the **bottom right panel** (five spike-triggered sweeps superimposed).

FIG. 8. Tentative wiring diagram to explain differential supraspinal gating of reflex pathways in normal and diseased subjects. Further explanations in the text.

about 70 msec (1.1 × T, sweeps 6 and 7) with increasing stimulus strength. Thus, shortening the onset latency of reflex response with increasing size and recurrence frequency of an incoming volley is at least in part caused by an increasingly rapid transmission to one particular motor unit. The interneuronal and synaptic mechanisms that cause such a distinct shortening of firing latencies in reflex pathways deprived of supraspinal control are obscure.

Alterations of the TA reflex pattern on plantar nerve stimulation might be explained by assuming different reflex paths under differential supraspinal control, a concept proposed by Lundberg (1979). Figure 8 shows a simple hypothetical model drawn on the basis of latencies and types of reflex effects in normal and diseased subjects. In normal subjects, activity in cutaneous afferents from the sole of the foot evokes a sequence of excitation–inhibition–excitation in the TA MN pool. The two excitatory reflex phases are symbolized by two pathways of differing lengths. Intercalated inhibition is not represented in the scheme as it might be explained, at least in part, as a silent period secondary to the initial burst. Under normal conditions, the two pathways are closely controlled from supraspinal structures: the short-latency excitatory route is facilitated, whereas the long-latency route is inhibited.

Loss of supraspinal control causes disfacilitation of the short-latency reflex route and disinhibition of the long-latency route. An increase in spinal cord input (e.g., by increase in stimulus intensity or recurrence frequency or both) in such cases might compensate partly for the lack of supraspinal facilitation in the short-latency excitatory route and might also cause excessive activity in the disinhibited long-latency route. Lack of intercalated inhibition might be secondary to a suppression of the early burst. Such an explanation is no more than a working hypothesis, and other explanations, at least for the disintegration of the reflex pattern and enhancement of its later parts, might be found in denervation supersensitivity of MNs and in enlargement of the dorsal horn neurons' receptive fields, respectively, as a result of chronic suprasegmental lesions (see Brenowitz and Pubols, 1981).

INTERLIMB CUTANEOMUSCULAR REFLEXES

Reflex patterns evoked in hand and leg muscles by cutaneous afferent stimulation have latencies short enough to suggest segmental transmission of initial reflex phases (Garnett and Stephens, 1980; Meinck et al., 1981c). However, the question remains as to whether longer loops contribute to later parts of cutaneomuscular reflex responses: transcortical (Conrad and Aschoff, 1977; Jenner and Stephens, 1979) and spinobulbospinal (SBS) reflex loops (Gassel and Ott, 1973; Meier-Ewert et al., 1972, 1973; Shimamura et al., 1964) seem to be the most likely. Reflex effects subjected to both transcortical and SBS pathways (elicited through muscle afferent or cutaneous stimulation) occur in distant muscles (Gassel and Ott, 1973; Marsden et al., 1978b; Meier-Ewert et al., 1972). Therefore, far-field cutaneomuscular reflex effects were reinvestigated and compared to "local" reflex responses.

After brachial nerve stimulation, we found nonreciprocal inhibition (60 msec) followed by excitation (80 msec) in various tonically active lumbosacral extensors and flexors (Meinck, 1976; Piesiur-Strehlow and Meinck, 1980; Meinck and Piesiur-Strehlow, 1981). H reflexes in extensors revealed the excitation but not the inhibition (Fig. 9); yet, apart from this, both methods led to similar results: low-threshold skin afferents from the arm and hand were capable of evoking reflex inhibition as well as excitation. The receptive field covered the entire surface of the body rostral to the buttocks in-

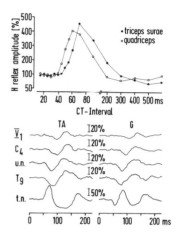

FIG. 9. Descending interlimb reflexes. **Top panel:** Triceps surae and quadriceps H-reflex recovery curves after single painful shocks delivered to the ipsilateral ulnar nerve at the elbow. Each spot represents the ratio of average amplitudes of 10 conditioned to 10 unconditioned H-reflexes, which were elicited in random sequence. *Abscissa:* interval between conditioning and test pulses (CT-Interval). **Bottom panel:** EMG modulation curves of the tibialis anterior (TA) and gastrocnemius (G) muscles after a triple shock (pulse interval: 3 msec) to the supraorbital nerve, the paraspinal skin at C_4 and T_9 levels, the ulnar nerve at the elbow (u.n.), and the tibial nerve at the ankle (t.n.). Signals are displayed as in Fig. 2C. Note different amplitude scales.

cluding the face (Fig. 9). Since percutaneous stimulation of dorsal roots C_4 and T_9 (Belke, 1948) induced the same reflex effects, with inset latencies shorter with T_9 than with C_4 stimulation, a directly descending (i.e., intraspinal) transmission rather than transmission via a supraspinal loop was assumed. Several characteristics of these reflex effects (e.g., nonreciprocity of reflex patterns in antagonistic muscles, extensive receptive field, distinct facilitatory convergence from both sides) have been described in animal experiments (e.g., Illert et al., 1978; Koehler et al., 1978; Schomburg et al., 1978) and may thus also suggest a relationship to directly descending intraspinal, most probably propriospinal, pathways. Moreover, Faganel (1977;

Faganel, *personal communication*) reports facilitation of leg tendon reflexes to noxious stimulation below the level of spinal transection, which also supports the assumption of a directly descending intraspinal pathway. This type of reflex system, with polytopic somatosensory convergence onto its neurons, may be involved in the integration of various descending and segmental signals to MNs (Illert et al., 1978). It possibly also plays a role in reinforcing recruitment of lumbosacral MNs during the step cycle in response to a rostral stimulus (cf. Schomburg et al., 1977).

Ascending reflex effects from leg afferents onto arm muscles (Kearney and Chan, 1979) consist of low-amplitude oscillations of inhibition and excitation with onset latencies of about 70 msec. The pathophysiology of both ascending and descending interlimb reflexes requires further investigation.

OPEN QUESTIONS AND PROBLEMS

Among the various questions raised, three are essential. (1) Which pathways and structures are involved in the transmission and control of cutaneomuscular reflexes? (2) How can the differences of cutaneomuscular reflex organization in arm and hand versus leg and foot muscles be explained? (3) Is there any clinical relevance in such reflex investigations?

For the first question, only indirect evidence has been obtained in patients with circumscribed CNS lesions or by analogy with animal experiments (e.g., Lundberg, 1979). The present discussion concerning the transcortical reflex hypothesis illustrates the difficulty in drawing conclusions on this basis. The occurrence of similar reflex effects after both cutaneous and muscle afferent stimulation (Conrad and Aschoff, 1977; Garnett and Stephens, 1980) and the observation of gating effects from the skin on reflexes to muscle afferent stimulation (Marsden et al., 1978a) might be explained not only by cortical but also by brainstem and spinal mechanisms such as convergence of different afferent channels onto the same interneurons and build-up or withdrawal of presynaptic inhibition (cf. Lundberg, 1979). Concerning the specific value of reflex latencies, it must be remembered that the isolated human spinal cord is capable of generating cutaneomuscular reflex responses with latencies of nearly 400 msec (Shahani and Young, 1971, 1973c).

Obvious differences between the cutaneomuscular reflex organization in hand and leg are difficult to interpret. The surprising observation that dorsal and volar finger nerve stimulation evoke identical reflex effects which, moreover, exhibit a nonreciprocal course in antagonists (Caccia et al., 1973; Garnett and Stephens, 1980) raises doubts concerning the functional importance of cutaneomuscular reflex mechanisms for the sensorimotor apparatus of the hand (but see K. Kanda and J.E. Desmedt, *this volume*). By analogy with the well-known influence of the pyramidal tract on flexor reflex transmission in cats (see Lundberg, 1979), a detailed cortical control of cutaneomuscular reflexes of the hand might be inferred from widespread pyramidal representation of the latter (see Kuypers, 1973). However, reflex disintegration caused by suprasegmental lesions appears less distinct in hand than in leg muscles. One might speculate on whether phylogenetically older mechanisms such as cutaneomuscular reflexes may be functionally less important in the highly corticalized neuronal organization of the hand than in the less corticalized leg.

For the application of reflex investigations to pathophysiological questions, a more extensive comparison of reflex alterations and clinical findings is necessary in order to better correlate clinical syndromes with types of reflex alterations. A special clinical value of such investigation lies in the development of neurophysiological techniques for detection of subclinical disorders and for the quantitative control of therapeutic efforts (see P.J. Delwaide, J. Schoenen, and L. Burton, *this volume*). The results reported here suggest that reflex analysis can serve as a useful tool in neurological diagnosis.

ACKNOWLEDGMENTS

The authors are indebted to Mrs. Annegret Ende and Mrs. Karin Thäter for technical assistance and to Mr. Peter Wenig for design of electronic equipment. This work was supported by the Deutsche Forschungsgemeinschaft (SFB 33).

Exteroceptive Influences on Lower Limb Motoneurons in Man: Spinal and Supraspinal Contributions

*,**P. J. Delwaide and †P. Crenna

**Department of Neurology, Institut de Médecine, University of Liège, Liège, Belgium; and
†Department of Human Physiology, University of Milan, Milan, Italy*

The interaction between afferent inputs of proprioceptive and cutaneous origin in motoneuron (MN) pools is important in motor organization. Questions may be raised about the possible predominance of one or the other of these influences, about whether the resultant effects are facilitatory or inhibitory, about their intensity and timing, about their topographical distribution, and, above all, on the way in which such interactions are integrated into mechanisms of reflex control and voluntary contraction. The problem is, of course, not new. It has been studied in animals, particularly by Lloyd (1943) and Lundberg (1969), and in man, several investigations have been made on the influence of cutaneous afferents on MN excitability (Kugelberg, 1948; Hagbarth, 1960; Hugon, 1973b; Gassel and Ott, 1970, 1973b; Shahani and Young, 1971, 1973; Pierrot-Deseilligny et al., 1973; Piesiur-Strehlow and Meinck, 1980; Meinck et al., 1981). Different approaches have included the recovery curve of a monosynaptic reflex after stimulation of a sensory nerve, elicitation of flexion reflexes, and the study of changes in voluntary EMGs following stimulation of sensory nerves.

Despite this, previous studies have only partly answered the questions raised. Some authors only studied one MN nucleus, either flexor or extensor (Gassel and Ott, 1970; Shahani and Young, 1971); others only stimulated a single nerve trunk (Hugon, 1973b) or only investigated events occurring at brief latency after stimulation. Furthermore, a sensory nerve contains fibers of different diameters mediating different sensory modalities, and graded stimulation intensities have rarely been studied.

In addition, there has in recent years been great interest in the control of voluntary contractions by long-loop reflex pathways relaying through supraspinal structures both in animals and in man (Phillips, 1969; Evarts and Tanji, 1974; Marsden et al., 1976a,b, 1978a,b; Lee and Tatton, 1975, 1978; Conrad, 1978; cf. C. D. Marsden, J. C. Rothwell, and B. L. Day, *this volume*). The long-loop circuits appear to be involved in supplying additional responses to short-latency segmental reflexes, and they must exert some control on the spinal mechanisms (cf. Tracey et al., 1980; Miller and Brooks, 1981; Hagbarth et al., 1981; Gydikov et al., 1981; R. G. Lee, J. T. Murphy, and W. G. Tatton, *this volume*; C. W. Y. Chan, *this vol-*

---

*To whom correspondence should be addressed: Department of Neurology and Clinical Neurophysiology, Institut de Médecine, Hopital de Bavière, Université de Liège, Boulevard de la Constitution, Liège, 4000 Belgium.

ume). It may be asked whether reflex supraspinal regulatory mechanisms also intervene in the case of exteroceptive cutaneous stimulation. Results that could support such a hypothesis have been reported in animals (Shimamura and Akert, 1965; Angel and Lemon, 1975) and in normal man (Garnett and Stephens, 1980; Jenner and Stephens, 1979). In some clinical instances, such as cortical (Sutton and Mayer, 1974; Sutton, 1975; Halliday and Halliday, 1980) and reticular reflex myoclonus (Hallett et al., 1977; Young and Shahani, 1979), the possible role of the long-loop reflex has been proposed. For that reason, the possible long-loop control of spinal circuits concerned with the flexion reflex deserved reinvestigation.

EXCITABILITY OF SOLEUS AND TIBIALIS ANTERIOR MN POOLS ON STIMULATION OF THE IPSILATERAL SURAL NERVE

A simultaneous study was made of the excitability curves of the motor nuclei of soleus and tibialis anterior. For soleus, the Hoffmann (H) reflex was used; for tibialis anterior, as often as possible, use was made of the tibialis anterior H reflex obtained by weak stimulation of the external popliteal nerve at the head of the fibula. In those subjects in whom the tibialis anterior H reflex could not be elicited, the F response to supramaximal stimulation of the nerve in the same place was used. Conditioning stimulation was carried out by cutaneous electrodes applied over the sural nerve at the lateral malleolus using two 1-msec shocks separated by 2.5 msec. The perception threshold was established, and subsequent stimulation was performed at various multiples of the sensory threshold. The type of sensation (tactile, painful) was reported by the subjects. The conditioning and conditioned stimuli were applied to the same limb.

Figure 1A–C, illustrates some significant

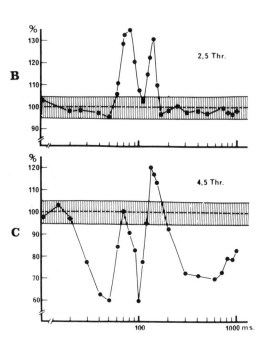

FIG. 1. Recovery curves of the soleus H reflex ($H_{max}/2$) after stimulation of the ipsilateral sural nerve at different intensities. **A**: At threshold, **B**: At 2.5 × threshold intensity. **C**: At 4.5 × threshold intensity. *Abscissa* (semi-log): delays (msec) between the stimulation of the sural nerve and that of the posterior tibial nerve. *Ordinate*: the amplitude of the conditioned responses expressed as a percentage of reference values. The limits of standard deviations appear on both sides of the control values.

results. The novelty of our experimental protocol consists in having studied many latencies by increasing the test interval by 10-msec steps for the first 150 msec and then by longer intervals, sometimes as much as 100 msec. Low-intensity ipsilateral stimulation (Fig. 1A) close to threshold slightly modulates the soleus H-reflex amplitude; increased values are seen for a delay of 70 msec to 150 msec. However, even at low intensity, this facilitation consists of two successive components. More intense stimulation (Fig. 1B) at 2.5 times threshold produces tactile sensation and results in two obvious phases of facilitation. The first, called F(a), begins at 60 msec, peaks at 80 msec, and disappears at 100 msec. The second, called F(b), starts at 120 msec, reaches its peak at 150 msec, and fades away after 200 msec. The latencies of F(a) and F(b) correspond approximately to the facilitations already seen at threshold stimulation (Fig. 1A). An even more intense stimulus (4.5 × T) applied to the sural nerve (Fig. 1C) elicits a painful sensation and causes reflex inhibition after 20 msec. This inhibition is still present after 1,000 msec. However, it is not uniform but is interrupted by a period of disinhibition at the same latency as F(a) and perhaps even by a slight facilitation corresponding to F(b).

The first problem raised by these results was whether F(a) and F(b) represented two distinct episodes of facilitation or whether they were the first and last parts of a single facilitatory process interrupted by facilitation of the antagonist tibialis anterior MN pool which occurs at 100 msec. To resolve this, recovery curves of tibialis anterior and soleus motor nuclei were studied (Fig. 2). The test is represented by the respective H reflexes of tibialis anterior and soleus after ipsilateral sural nerve stimulation. Similar results have been obtained using the F response of tibialis anterior. It is striking to note the parallelism of the excitability curves of the two motor nuclei and the existence of two peaks corresponding to F(a) and F(b). In particular, there is no facilitation of tibialis anterior at latencies corresponding to the reduced facilitation of soleus. The F(a) peak of tibialis anterior facilitation is superimposed on a background facilitation well evident at a latency as short as 20 msec. When, instead of a medium stimulus as in Fig. 2, a painful stimulus (4.5 × T) is used, this early facilitation increases in amplitude; it begins after about 15 to 20 msec and is followed by two peaks at latencies corresponding to F(a) and F(b);

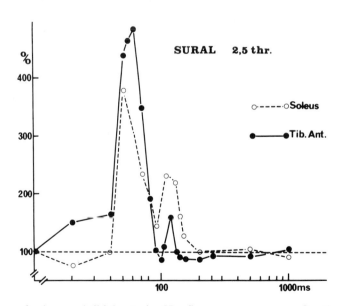

FIG. 2. Comparison of soleus and tibial anterior H reflex recovery curves after the same conditioning (ipsilateral sural nerve stimulation at 2.5 × threshold intensity). Two curves are drawn from the same experimental session in one subject. Coordinates as in Fig. 1.

between them, an inhibitory phase appears, which is more marked as the stimulus strength is increased further.

Table 1 documents for seven subjects the similarities in time course of the facilitations for soleus and tibialis contralateral responses and the fact that tibialis is not facilitated (at about 100 msec) when the soleus facilitation is reduced. As a result, it became evident that the excitability curves of these two antagonistic motor nuclei following ipsilateral sural nerve stimulation showed at low intensity, two well-defined peaks of facilitation, F(a) and F(b), at the same latencies for both. This phenomenon does not obey the rules of reciprocal innervation. At a higher intensity of stimulation, in addition to the F(a) and F(b) facilitations, there is a reciprocal effect consisting of short-latency facilitation of tibialis anterior MNs and inhibition of soleus MNs. These reciprocal effects become apparent after about 20 msec. The reciprocal inhibition and facilitation are not quite unexpected. It is known that painful stimulation facilitates flexor and inhibits extensor MNs through the flexor reflex afferents (FRA) (cf. Lundberg, 1975). The finding that the facilitation occurred simultaneously in MNs of antagonistic function was surprising.

Double facilitation has been reported by Gassel and Ott (1970) who attributed the two peaks to recruitment of two different types of afferents. But Fig. 1A shows that the two peaks are both elicited by even a low-intensity stimulus. Thus, the second peak cannot be attributed to afferent impulses transmitted by a more slowly conducting group of fibers (group III, for example). These results led us to investigate whether the two peaks of facilitation were a metameric phenomenon limited to a few segments of the spinal cord or could be evoked by distant cutaneous stimuli.

Stimulation of Other Sensory Nerves

The contralateral sural nerve was stimulated (Fig. 3), and tests were again carried out on the soleus and tibialis anterior MN pools. Low-intensity stimulation causes the appearance in the soleus nucleus of facilitation at the same latencies as F(a) and F(b) (Fig. 3A). Stimulation of medium intensity ($2.5 \times T$) makes the two peaks appear clearly (Fig. 3B). Painful high-intensity stimulation ($4.5 \times T$) elicits a small facilitation of short latency and long duration (Fig. 3C) in soleus MNs as well as two peaks of facilitation at latencies corresponding to F(a) and F(b). The amplitudes of these two peaks of facilitation elicited by contralateral stimulation are equal to, or even greater than, those evoked by ipsilateral stimulation. Inspection of tibialis anterior excitability curves shows that at low stimulus intensity, there is a discrete facilitation which can be divided into two phases; they are

TABLE 1. *Latencies of F(a), F(b), and intermediate point in tibialis anterior and soleus after contralateral sural conditioning*

	Tibialis anterior			H Soleus		
	F(a)	interm.	F(b)	F(a)	interm.	F(b)
M... Christine	90	120	190	80	100	130
G... Pascal	80	110	140	90	120	140
B... Dany	70	140	160	90	120	160
H... Henriette	90	120	150	70	120	170
G... Anne	60	90	150	60	90	110
C... Paolo	60	80	120	70	90	130
D... Henri	50	100	140	80	100	140
Mean	71	108	150	77	106	140

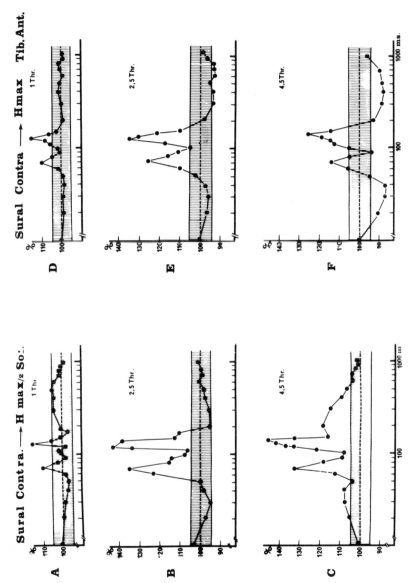

FIG. 3. Recovery curves of H reflex in soleus (left) and tibialis anterior (right) after stimulation of the contralateral sural nerve at different intensities. A and D at threshold; B and E at 2.5 × threshold intensity; C and F at 4.5 × threshold intensity. Coordinates as in Fig. 1

more evident when medium (2.4 × T) stimulus intensity is employed. Following painful stimulation (4.5 × T), they supervene on a background of inhibition which begins at 20 msec.

Other nerves, such as the lateral femoral cutaneous, the internal saphenous, the digital nerves of the hand, and even the ophthalmic branch of the trigeminal, were stimulated. Stimulation of the latter, even at very painful intensities, elicits only the two peaks of facilitation simultaneously in soleus and tibialis anterior; no short-latency facilitation or inhibition is seen in either soleus or tibialis motor nucleus. The peak of facilitation are similar whether the trigeminal nerve branch stimulated is ipsi- or contralateral to the lower limb tested. Identical results were obtained from stimulation of nerves in the upper limb and trunk. Thus, for the antagonistic muscles tibialis anterior and soleus, stimulation of a large variety of purely sensory nerves consistently elicits two peaks of facilitation. In addition, when nerves entering the spinal cord in or near the S_1 segment are stimulated, whether ipsi- or contralateral to the test reflex, reciprocal facilitation or inhibition is also seen. This phenomenon is not peculiar to the calf muscles. Two similar synchronous facilitation peaks can be seen in the quadriceps and in the short head of biceps in the thigh following sural or femoral cutaneous nerve stimulation with excitability being tested by the tendon reflexes (P.J. Delwaide and M. Gadea, *unpublished data*).

Do Exteroceptive Afferents Activate Long Loops?

Because the two phases F(a) and F(b) of facilitation are elicited by stimuli distant from the motor nuclei concerned, the mechanism is not related to circuits involving one or a small number of spinal segments. The possibility of facilitation transmitted through a long loop should therefore be considered. Latencies, even for the first peak, are compatible with such a hypothesis. In experiments undertaken for a different purpose, Marsden et al., (1976b) found that the somatosensory evoked potential to foot stimulation had a latency of 34 msec and that the M_2 response of flexor hallucis longus appeared after 75 msec. The F(a) peak in tibialis anterior appears after a mean latency of 71 msec (Fig. 2). However, this analogy is indirect, so we set out to measure the sequence of appearance of peaks of facilitation as a function of site of stimulation. In the same experiment, we stimulated the ophthalmic branch of the trigeminal and the posterior branch of the iliohypogastric nerve. The latter was stimulated 7.5 cm from its entry into the cord, which corresponds to the conduction distance from the ophthalmic branch of the trigeminal nerve to the brainstem (Kugelberg, 1952).

The soleus H reflex was conditioned by stimulation of the two nerves. The beginning and the maximum of the first peak [F(a)] were measured every 5 msec after conditioning stimulation. Figure 4 shows that trigeminal stimulation evokes a facilitation about 8 msec earlier than that elicited from the iliohypogastric nerve. This result suggests that the soleus motor nucleus is not activated directly by afferents entering the spinal cord close by but rather through a supraspinal relay for the first part of this facilitatory influence. When account is taken of the heights of the subjects, the difference in the latencies of the first peaks following stimulation of the two nerves is compatible with a conduction velocity in ascending fibers of the order of 55 to 60 m/sec.

From this, it appears that nonsegmental diffuse phenomena have to be taken into account to explain, at least in part, these facilitations, which occur simultaneously in a large number of pools; they might involve a supraspinal generator activated by sensory stimulation distant from the motor nucleus involved. A direct segmental circuit

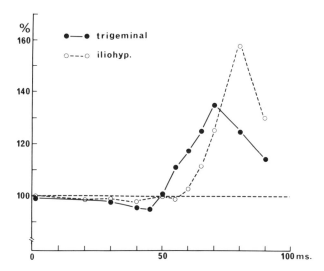

FIG. 4. Amplitude changes of the soleus H reflex after stimulation of the ophthalmic branch of the trigeminal nerve *(filled circles)* and the posterior branch of the iliohypogastric nerve *(empty circles)*. *Abscissa:* time (msec) between the sensory nerve stimulation and the evocation of the H reflex. *Ordinate:* amplitude of the conditioned reflex expressed as a percentage of the control values.

between the afferents and the motor nuclei seems unlikely to be the only mechanism involved.

Is the Supraspinal Circuit Specific?

Other experimental situations are known, in fact, in which stimuli, for example, acoustic, give rise to facilitations at latencies of the same order as in the startle reaction. We therefore sought to determine whether the peaks of facilitation may be regarded as a special case of the startle reaction. If such were the case, descending impulses would use the same pathway as that activated by an acoustic stimulus. To investigate this, we tested the excitability of soleus following acoustic stimulation or stimulation of fifth nerve (Fig. 5A) or both (Fig. 5C). In these experiments, the electrical stimulation intensity was high. After intense acoustic stimulation at an irregular frequency, the soleus H reflex (H-Sol) exhibits marked but nonuniform facilitation. Several peaks can be distinguished, of which the earliest is at 80 msec (Fig. 5B); marked habituation occurs when the stimulus is repeated. When the two stimuli are delivered together (Fig. 5C), the facilitation of H-Sol is no larger than when each is given individually. Thus, cutaneous and acoustic stimulations evoke nonadditive facilitations.

Why is the facilitation biphasic? Since both peaks occur following stimulation at threshold, it seems unlikely that they would be elicited by separate sets of afferents. On the other hand, it is possible that they arise from differential central processing of input or from separate circuits within the central nervous system. The difference in the jitter of the peak latencies, low for F(a) and (± 5 msec in the same experience) and higher for F(b) (± 20 msec), could argue in favor of such an interpretation. Again, F(b) exhibits a more marked habituation, even when low stimulation rates (0.1/sec) are employed. The distinction between the two peaks is in line with effects of sleep: if the subject falls asleep during the experiment, the amplitude of the second [F(b)] peak is dramatically reduced, whereas that of the first remains unchanged (Fig. 6). Finally, in a single experiment, administration of 10 mg of di-

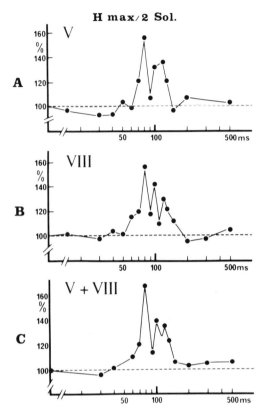

FIG. 5. Amplitude changes of the soleus H reflex ($H_{max}/2$) after stimulation of the ophthalmic branch of the trigeminal nerve (A), acoustic stimulation (B), and simultaneous stimulation of both (C). Coordinates as in Fig. 1.

azepam did not change the amplitude of the first peak but reduced that of the second. We conclude that because both peaks occur for stimulation at threshold, there is a single ascending spinal pathway excited by large-diameter primary afferents which, at a supraspinal level, activates the two distinct circuits responsible, respectively, for the F(a) and F(b) facilitations.

Investigation of facilitations following sensory stimulation does not require cooperation from the subject and could be applied both to normal volunteers and to patients with neurological lesions in a more standardized way than the study of proprioceptive long loops (Lee and Tatton, 1975, 1978; Mortimer and Webster, 1978; Marsden et al., 1978a,b; C.D. Marsden, J.C. Rothwell, and B.L. Day, *this volume*). The effects are not contaminated by the elicitation of monosynaptic reflexes (no M_1 response).

Functional Significance

Beyond any eventual practical application, our results raise two questions with regard to functional significance.

Relationship to flexion reflex.

Sural stimulation occasionally causes the appearance, in healthy subjects, of two bursts of EMG activity in ipsilateral flexor muscles such as the short head of biceps and tibialis anterior (Hugon, 1973b; Bathien and Bourdarias, 1972). These bursts of EMG activity have been considered spinal flexion reflexes and named RA_{II} and RA_{III} because they are attributed, respectively, to afferents of groups II and III. These responses may be compared with the activities described by Shahani and Young (1971) in tibialis anterior following stimulation of the sole of the foot. The latencies of these responses coincide with the F(a) and F(b) peaks of facilitation. As the same peaks can be elicited by sensory stimuli applied at a distance, the purely spinal nature of the EMG responses may be questioned. Indeed, sural nerve stimulation of medium intensity occasionally causes the appearance of bursts of EMG activity in soleus, both in the normal subject (Martinelli et al., 1979) and in the parkinsonian patient (Delwaide et al., 1974). Furthermore, in spastic patients, stimulation of the sural nerve does not elicit two distinct bursts of activity; the EMG response occurs with a longer latency and has a longer duration than in control subjects (Delwaide, 1971; Shahani and Young, 1971). Actually, the few reports of multicomponent flexor reflex responses in patients with complete spinal cord division show quite abnormal features (Shahani and Young, 1971; Faganel, 1973)

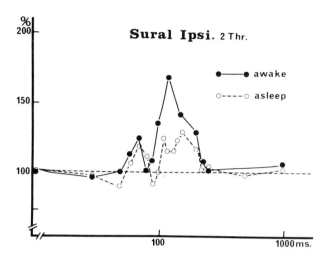

FIG. 6. Modifications of the H reflex recovery curves (ipsilateral sural nerve stimulation) in a subject when awake *(filled circles)* or asleep *(empty circles)*. Same experimental session. Coordinates as in Fig. 1.

or deal with investigations performed late in time, after the stage of spinal shock (Hugon, 1973b), when neuronal plasticity phenomena occurring in the transected cord could have determined compensatory reorganization of the synaptic system (cf. P. Bach-y-Rita, *this volume*). From this, it may be suggested that the burst activity observed in flexor muscles does not simply represent the activation of polysynaptic spinal activity but is caused by peaks of facilitation superimposed on moderate facilitation occurring in the motor nucleus of tibialis anterior following ipsilateral painful stimulation.

In the pyramidal syndrome, the more progressive and long-lasting facilitation could induce anterior tibial MNs to discharge. In this way, it is possible to interpret the late onset of the response and the shortening of the latency when the intensity of stimulation is increased. In any event, further investigations are required before the flexion reflex in spastic patients can be equated with phasic facilitations seen in healthy subjects. These latter could represent the long-loop activity discussed above and not solely that of polysynaptic spinal circuits. Nevertheless, this interpretation by no means excludes the operation of a simultaneous segmental reflex component.

Relationship between excitability curves and EMG activity changes.

Facilitations in phase in antagonistic muscles were obtained at rest and need not occur similarly during voluntary contraction. Sural nerve stimulation at low intensity is able to reinforce EMG activity in both muscles, contracted separately, whose activity is maintained constant by voluntary contraction; at least two higher amplitude bursts are seen, one early (50 msec onset latency) and one later (160 msec) (Fig. 7A). If the stimulation intensity is increased (Fig. 7B), the first burst of activity in tibialis anterior increases and occurs earlier, whereas that in soleus is reduced in amplitude and has longer latency. On average, the latency of the first reinforced burst is 36 msec in tibialis anterior and 65 msec in soleus. Thus, a relative alternation appears as stimulation intensity increases; such a pattern is in conformity with classical notions of reciprocal innervation.

Therefore, the intensity of conditioning stimulation is important in bringing about the alternation of EMG facilitation. On the

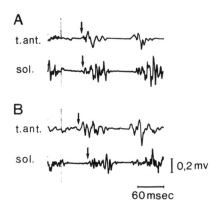

FIG. 7. Modulation of EMG activity of two antagonist muscles after low-intensity (2.5 × threshold in A) and high-intensity (4.5 × threshold in B) stimulation of the ipsilateral sural nerve. In A and B, the upper row is the EMG of tibialis anterior, the lower that of soleus. Low-intensity stimulation (A) brings about simultaneous reinforcements (see *arrows*). With high-intensity stimulation, the reinforcements are no longer in phase.

other hand, even intense stimulation elicits simultaneous facilitation in the motor nuclei at rest, as illustrated by the excitability curves. Voluntary contraction could activate interneurons that facilitate afferent transmission towards the flexor motor nucleus; conversely, they could inhibit and slow down that for the extensor MN pool. This kind of action is similar to that attributed by Tanaka (1974, 1980) to Ia interneurons that transmit inhibition from Ia afferents of the antagonist muscle to the extensor motor nucleus when this is activated by voluntary contraction.

However, before such a hypothesis is adopted, the possibility must be considered that the progressively earlier burst of EMG activity in tibialis anterior—and, reciprocally, progressively later in soleus—with increasing intensity of stimulation may be explained by the reciprocal phenomena of facilitation and inhibition shown in Figs. 1 and 2. The more intense the conditioning stimulation, the more marked is the facilitation of the tibialis anterior motor nucleus. Thence, supervening on this background of facilitation, the very beginning of the F(a) phase may bring about the early and synchronous discharge of an increased number of MNs resulting in the burst discharge. On the other hand, the inhibition of the soleus motor nucleus would be responsible for the fact that only the summit of F(a)—necessarily later—could bring about supplementary facilitation of the motor nucleus.

CONCLUSIONS

Stimulation of sensory nerves in healthy subjects brings about complex modifications in the excitability of motor nuclei that depend on the intensity of stimulation and the topographical relationships between the nerves stimulated and the MN pools tested. Phenomena in contiguous zones between the spinal terminations of the stimulated nerve and the motor nuclei consist of reciprocal facilitation and inhibition: ipsilateral painful stimulation facilitates the tibialis anterior motor nucleus and inhibits that of soleus; contralateral stimulation facilitates the flexor motor nucleus and has little effect on the soleus nucleus. These facilitations and inhibitions begin after 20 to 30 msec, develop progressively up to 200 to 400 msec, and decline progressively after 1,000 msec. They undoubtedly depend on spinal mechanisms. Subsequently, whatever sensory nerves are stimulated or motor nuclei are tested, even if they are several dozen centimeters apart, two successive intense but relatively brief facilitations can be seen, F(a) and F(b). When the motor nucleus tested is close to the conditioning stimulus, these facilitations are superimposed on a background of facilitation or inhibition as described above. These peaks of facilitation occur more or less simultaneously, even in motor nuclei of antagonistic function situated in the same segment. On the other hand, there is a rostrocaudal progression of these facilitations which suggests that they depend on long loops activated by exteroceptive afferents.

The F(a) and F(b) facilitations present striking analogies with the startle reaction, and the hypothesis is put forward that, at least in part, they could be equivalent following brief exteroceptive stimuli. If such is the case, one may speculate about the nature of the EMG discharges elicited in flexors following sensory nerve stimulation and considered as the equivalent in man of the flexion reflex. The peaks of facilitation could explain the reinforcement by bursts of EMG activity following voluntary contraction. When stimulation is moderate in intensity, these bursts occur in phase in antagonist muscles. If the stimulation is painful, the bursts occur with disorganization in time.

Nociceptive Flexion Reflexes as a Tool for Pain Research in Man

*Jean Claude Willer

Laboratorie de Physiologie, Faculté de Médecine Saint Antoine, 75571 Paris Cedex 12, France

In spite of progress made in understanding pain transmission and modulation, the search for a method of measuring pain in humans remains difficult. This may be partly explained by the polymorphic nature of pain sensation and by the absence of limits in pain appreciation. Numerous methods have been described for measuring pain in man, using thermal (Hardy et al., 1948; Mor and Carmon, 1975), mechanical (Russel and Tate, 1975), chemical (Keele, 1962; Lim, 1968), or electrical (Hill et al., 1952; Notermans, 1966) cutaneous stimulations. However, all of these methods found their limits rapidly, since they were based on the subjective sensations reported by the subjects. This chapter presents an objective method for measuring pain in man using nociceptive flexion reflexes. The method is based on the existence of a strong correlation between a pricking pain and a long-latency reflex (RA-III reflex; Hugon, 1973b) recorded from a flexor muscle of the lower limb and elicited by electrical stimulation of the ipsilateral sural nerve or of the skin in the distal receptive field of this nerve (Willer, 1977).

PAIN MECHANISMS

Pain Transmission

It is now largely accepted that nociceptive input is conducted to the spinal cord via small-caliber myelinated fibers (Aδ fibers) and unmyelinated fibers (C fibers). Activation of Aδ fibers is known to give a feeling of a short-lasting well-localized pricking pain, whereas C fibers are responsible for a longer-lasting poorly localized sensation of burning pain (Burgess and Perl, 1973; Collins et al., 1960, 1966; Heinbecker et al., 1933; Hallin and Torebjork, 1973; Van Hees and Gybels, 1972). However, it has been recently shown that during some specific experimental conditions, the larger myelinated cutaneous afferent fibers (Aα and Aβ fibers) are also able to convey noxious information (Boureau et al., 1978; Willer et al., 1978).

At the spinal level, dorsal horn neurons located in laminae I, II, and V have been shown to be the best candidates for relaying noxious messages (Wall, 1967; Christensen and Perl, 1970; Perl, 1971; Besson et al., 1973; Menetrey et al., 1977; Le Bars et al., 1975, 1976a, b). However, layer VI, VII, and VIII neurons also seem to play a non-negligible role in pain transmission (Albe-Fessard et al., 1974; Levante and Albe-Fessard, 1972).

---
*Laboratoire de Physiologie, Faculté de Médecine Saint Antoine, 27 rue de Chaligny, 75571 Paris, Cedex 12 France.

From spinal cord to brain, the spinothalamic tract (STT) is one of the classical pathways for ascending pain volleys (Cassinari and Pagni, 1969; Noordenbos and Wall, 1976). It originates from some laminae I, II, and V cells and runs in the anterolateral quadrant (ALQ) until reaching thalamic nuclei (Trevino et al., 1973).

The spinoreticular pathway, also running in the contralateral ALQ, has recently been shown to be an alternative route for pain messages (Wall and Dubner, 1972). Originating from numerous laminae (I, V, VI, VII, VIII), this tract ends in the medial brainstem reticular formation and in a part of a central pathway associated with aversive motivational aspects of pain, including the medial thalamus, hypothalamus, and limbic structures (Casey, 1966). The spinocervical tract (SCT) also seems to be involved in pain transmission, since most SCT neurons receive C fiber inputs (Brown et al., 1973), and bilateral lesions of the cat dorsolateral funiculus including SCT are required to block pain (Kennard, 1954).

Although the lemniscal system is not generally concerned with transmission of noxious messages, a dorsal column (DC) pathway involved in this function is suggested by several data: (1) DC lesions at the thoracic level decrease the forelimb force exerted to escape a noxious stimulus applied to the hindlimb in monkeys (Viereck et al., 1971); (2) there is an overlapping between cells of origin of SCT and DC tracts in cervical and lumbar enlargements where cells projecting to DC are particularly concentrated in laminae V and VI (Rustioni and Kaufman, 1977); (3) some C fiber inputs to the nonprimary afferent axons in the DC have been described, and some DC neurons have been shown to respond maximally with noxious stimulations (Dart and Gordon, 1973). In supraspinal structures, it is more difficult to characterize the different pathways and structures involved in pain messages. The difficulties result from the multiplicity of ascending paths which are not necessarily corresponding from one species to another, and from the fact that somatic messages converge on central structures and are mixed with other sensory inputs (Albe-Fessard and Besson, 1973).

Control of Pain Input

At the spinal level, which is what is considered here, noxious messages are strongly modulated by powerful segmental and suprasegmental inhibitory influences.

Segmental control.

According to the "gate control theory of pain" of Melzack and Wall (1965), stimulation of the skin evokes nerve impulses that are transmitted to three spinal cord systems: (1) the cells of substantia gelatinosa (SG), (2) the DC fibers, and (3) the first central transmission cells (target cells, T cells) in the dorsal horn. According to Melzack and Wall, three facts are important: (1) the background activity that precedes the stimulus, (2) the stimulus-evoked activity, and (3) the relative balance of activity in large-versus small-diameter fibers. The gate control role played by SG consists of (at least) a presynaptic inhibition or the release of this inhibition of the afferent inputs before they influence the T cells, with opposite effects of the inputs of large and small caliber fibers. The final effect will be that an input in large fibers will induce a presynaptic inhibition, but an input in small fibers will decrease or stop this inhibition. According to this theory, the output of T cells is crucial for deciding whether pain will result, and this decision is a function of the number and rate of firing of active fibers and of the balance of large and small fiber activity in the afferents.

In a recent paper, Wall (1978) reexamined some features of the gate control theory of pain:

The mechanism by which the control is achieved remains completely unknown. Presynaptic inhibition as a phenomenon isolated from postsynaptic inhibition is in doubt. The role of SG in any function is unknown. That a gate control exists is no longer open to doubt, but its functional role and its detailed mechanism remain open for speculation and for experiments.

However, even if this theory is still argued and criticized (Nathan, 1976), it has been productive of fruitful investigations in pain research and therapy.

Endogenous opiate brainstem control.

A profound and long-lasting analgesia is produced by electrical stimulation in the midbrain periaqueductal gray (PAG), the dorsal raphe, and the nucleus raphe magnus (NRM) (Reynolds, 1969; Oliveras et al., 1974, 1977; Basbaum et al., 1976). This analgesia has been compared to that obtained with large doses of morphine and was, like the morphine analgesia, completely reversed by the specific opiate antagonist naloxone (Mayer and Liebeskind, 1974; Mayer, 1975; Oliveras et al., 1977). Moreover, the distribution of endogenous opiates (enkephalins, endorphins) and that of opiate receptors overlap in numerous brain areas, particularly in the midbrain PAG (Simantov et al., 1976; Goldstein, 1976; Hughes, 1975; Terenius, 1978; Simon and Hiller, 1978). This brainstem analgesic system acts via activation of a descending serotoninergic bulbospinal system which inhibits the activity of dorsal horn neurons involved in pain transmission, especially the lamina V cells (Fields et al., 1977; Guilbaud et al., 1973).

Direct spinal effect of opiates on pain transmission.

In spinal animals, morphine and derivatives produce an important naloxone-reversible depressive effect on lamina V cells' nociceptive activity without significantly affecting the transmission of the nonnoxious messages (Besson et al., 1973; Le Bars et al., 1975, 1976a,b). This direct effect of morphine at the spinal level can be partly explained by the following findings. There is a very dense concentration of both opiate receptors and enkephalin in layers I and II of the spinal dorsal horn in different mammal species (LaMotte et al., 1976; Pert et al., 1976). Substance P (SP), an 11-amino-acid peptide synthesized in the body of the C fibers, is a putative neurotransmitter for pain messages (Henry, 1977) and is also highly concentrated in the presynaptic endings of C primary afferents at the level of laminae I and II (Takahashi and Otsuka, 1975). Duggan et al. (1976) offered an explanation of the mechanism of the spinal opiate inhibition of the layer V cell activity. When recording from these neurons (Fig. 1A), they found that iontophoresis of morphine or of met-enkephalin was without any effect when directly injected in lamina V, although a strong depressive naloxone-reversible action was observed when these drugs were applied in substantia gelatinosa (lamina II). Furthermore, Jessel and Iversen (1977) have shown that the release of SP in the spinal trigeminal nucleus (equivalent of spinal dorsal horn) was stereospecifically inhibited by morphine and met-enkephalin in a naloxone-reversible effect. All of these data led Jessel and Iversen to propose an integrative hypothesis (Fig. 1B): the nociceptive input conveyed in C fibers induces a release of SP in laminae I and II which strongly excites the dendritic arborization of lamina V cells which run up to superficial layers (I, II). This SP release is blocked by morphine and met-enkephalin when these drugs are bound by the opiate receptors localized in the presynaptic endings.

FLEXION REFLEXES AND PAIN SENSATIONS IN MAN

Ipsilateral flexion reflexes of the lower limb have been extensively studied with electrophysiological methods in man (Kugelberg et al., 1960; Dimitrijevic and Nathan, 1967; Shahani and Young, 1971, 1973; Bathien and Bourdarias, 1972;

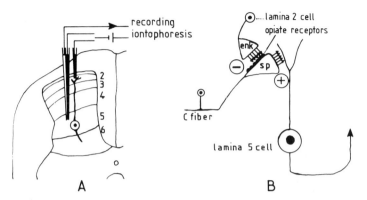

FIG. 1. A: Experimental set-up showing that morphine acts in the substantia gelatinosa (SG). Administered in SG with a micropipette, morphine was found to reduce the nociceptive responses of the layer 5 neurons, recorded with another micropipette. In contrast, morphine was without effect when applied directly on lamina 5 (Duggan et al., 1976). **B:** According to Jessell and Iversen (1977), messages in primary afferent C fibers induce a release of substance P (SP) which exerts an excitatory ($\oplus$) action on the dendritic endings of lamina 5 cells. This SP release is blocked ($\ominus$) when met-enkephalin (enk) binds on the presynaptic opiate receptors.

Hugon, 1973b). In the following, nociceptive flexion reflexes of the lower limb are presented as a useful tool for pain research in humans, since the pain threshold and that of a nociceptive reflex are well correlated when either the sural nerve or the skin of its distal receptive field is stimulated. The skin is innervated by cutaneous nerves which run with blood vessels through the subcutaneous tissues. These nerves branch and penetrate the dermis, forming a double plexus lying parallel to the skin surface. The first description of these structures was put forward by Ruffini (1905) who described a deep dermal plexus from which fibers branched at irregular intervals towards a second, more superficial dermoepidermal plexus. According to Winkelman (1960), the fibers of the deeper plexus are of larger diameter (Aα, Aβ) than those of the more superficial plexus (Aδ, C). It is thought that the fibers of this superficial dermoepidermal plexus, terminating in complex open loops or around the superficial capillaries, act here as adequate pain receptors (Weddell, 1941, 1945; Weddell and Sinclair, 1953; Lim, 1968).

Methodology for Nociceptive Reflexes

Experiments were performed on 127 healthy volunteers (men and women) and on patients affected with different neurological disorders related to pain sensations. During the sessions, they were comfortably seated in an armchair so as to obtain good muscular relaxation.

The technical details for stimulation and recording have been extensively described elsewhere (Willer, 1977). In brief, the sural nerve was stimulated percutaneously with surface electrodes or transcutaneously with a pair of steel needle electrodes insulated to the tip and placed through the skin on the sural nerve at its retromalleolar site. Electrical stimulation of the skin in the distal receptive field of the sural nerve was achieved through surface electrodes placed on the scratched and degreased surface of the IV and V toes. The stimulus consisted either of a volley of 10 to 12 rectangular pulses (1 msec duration each at 100 to 300 Hz) or a single rectangular shock (0.1 to 1 msec) delivered by a constant-current stimulator at intervals of 5 to 8 sec. The stimulus intensity was monitored using a passive probe placed in the stimulating circuit.

Muscle reflex activity was recorded from the biceps femoris, capitis brevis muscle (Bi), since it previously had been shown that Bi was the earliest reflexly generated activity in the lower limb of normal man (Bathien and Bourdarias, 1972; Hugon, 1973b). For this, a pair of surface electrodes was placed on the scratched and degreased skin above the desired muscle. The temperature of the skin near the stimulating and recording sites was monitored with thermocouples and maintained at a stable level of 34 ± 1°C during the experiment. A stimulation eliciting 80 to 90% of reflex responses was chosen as threshold. The measurement of sensations was performed using a classical rating scale (see Willer, 1977). The thresholds of both nociceptive reflexes and pain sensations were determined using the staircase limits method (Sidowski, 1966) with four series of increasing and decreasing stimulations. Each session lasted between 45 and 70 min, and the subjects showed stable values of thresholds at successive tests repeated at 48- to 72-hr intervals.

Stimulation of sural nerve.

As fully described previously by Hugon (1973b), when a train of shocks is used, electrical stimulation of sural nerve elicits two types of reflex responses in Bi (Fig. 2A). The first (RA-II) is of short latency (50–70 msec) and of low threshold (5 ± 0.6 mA). This kind of stimulation is never painful and is reported as a tactile sensation comparable to that produced by rhythmic percussion of the skin with a reflex testing hammer. The sensation is very well localized at the stimulated locus and is accompanied by a projected tingling sensation in the receptive field of the sural nerve to the fifth toe. The second response (RA-III) is of longer latency (80–150 msec) and of higher threshold (10 ± 1 mA). Such a stimulus provoked a liminal pain described as a sharp pinprick localized at the point of stimulation. There was no significant difference in the results when surface or needle electrodes were used for stimulation.

Stimulation of the Skin

When a 10-mA stimulation was applied to the skin in the distal receptive field of the sural nerve, an RA-III response was always recorded without the RA-II (Fig. 2B). This response, elicited by a 10-mA stimulus, was of large amplitude and long duration, corresponding clearly to a suprathreshold response. Simultaneously, the subjects experienced an intense drilling pain lasting 15 to 30 sec after the stimulus. In case of skin

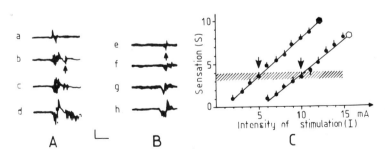

FIG. 2. A and B: polysynaptic reflex responses recorded in Bi at various intensities. A: Stimulation to sural nerve (a,b,c,d: 6,10,12,14 mA). B: Stimulation of the skin (e,f,g,h: 5,7,9,11 mA). C: Regression curves of sensation (S) versus Intensity (I) of stimulation of the skin (●), S = 0.9I − 0.9, and of sural nerve (○), S = 0.7I − 0.4. *Hatched area* shows the pain threshold. *Arrows* indicate the threshold of RA-III in each situation. Note the close relationship between the pain and RA-III thresholds (Willer, 1977).

stimulation, the thresholds of both RA-III and pain were of 5 ± 0.5 mA.

These results show a relationship (Fig. 2C) between the threshold of RA-III and that of a pain sensation, whether stimulation is applied to the sural nerve (10 mA) or to the skin (5 mA). The discrepancy between these two values, and also the fact that RA-II was never observed in case of skin stimulation, suggested an inhibitory influence from the RA-II response on the RA-III. In terms of afferent volleys, this suggested that large-diameter cutaneous fibers (Aα, Aβ) should exert an inhibitory action on the smaller cutaneous fibers (Aδ in this case), since Hugon (1973b) stated that the RA-II reflex resulted from activation of the larger fibers, whereas the Aδ fibers are involved in the genesis of the RA-III response. In order to test this hypothesis, an ischemic block of the larger diameter fibers of the sural nerve was achieved using a pneumatic tourniquet placed at the ankle between the stimulating and recording electrodes. This was performed during test stimulations of sural nerve at RA-III threshold (10 mA). Ischemia resulted in a parallel decrease in RA-II and increase in RA-III responses with time. After 15 min of ischemia, RA-II was abolished whereas RA-III was clearly suprathreshold. By the same time, the sensation that was at pain threshold before ischemia became stronger and stronger and was felt as a long-lasting burning and drilling pain at the 15th minute of ischemia. At this time, the thresholds of both RA-III reflex and pain sensation decreased from 10 to 5 to 6 mA.

These results, showing in part that activation of larger diameter fibers is able to inhibit nociceptive responses elicited by smaller diameter impulses, are in good agreement with those of Hugon (1973b) who noted that a tactile stimulus preceding a painful stimulation (both to the sural nerve) reduced the RA-III reflex activity. They are also consistent with the gate control theory of pain (Melzack and Wall, 1965).

The Afferent Volleys Involved in RA-III Reflex and in Pain

It is generally accepted that activation of Aδ fibers in man elicits a pricking pain (Collins et al., 1966; Burgess and Perl, 1973), whereas the larger diameter fibers would not be concerned with pain (Heinbecker et al., 1933). We have reexamined the role played by the larger diameter cutaneous fibers in transmission of nociceptive input (Boureau et al., 1978; Willer et al., 1978). We have studied afferent volleys in sural nerve and compared them to the RA-III reflex and to the sensations elicited by electrical stimulation of the sural nerve. The first point explored was the relationship between the RA-III reflex and the intensity of stimulation as a function of duration and number of shocks. For a single shock, the RA-III reflex and pain sensations are evoked at gradually lower intensities as the shock duration increases (Fig. 3). Similarly, for shocks of the same duration, the liminal intensity for evoking the RA-III reflex decreases as the number of shocks in a repetitive train (100 Hz) increases (Fig. 3).

These data raise the question of which fibers conduct the afferent volley responsible for RA-III response in cases of a single shock and of a repetitive stimulation. Hence, we recorded simultaneously the sural sensory action potentials (Buchthal and Rosenfalck, 1966) and the RA-III reflex (Fig. 4), using single shocks of various durations or trains. When a 0.5-msec duration was used (single shock), the first wave to appear in the neurogram was characteristic of large-diameter, fast-conducting, and low-threshold fibers (Aα fibers: 55–60 msec). At the maximal Aα amplitude, the sensation reported was never painful. No accompanying reflex activity was ever observed in Bi (Fig. 5A). Increasing slightly the intensity of stimulation resulted in an enlargement of

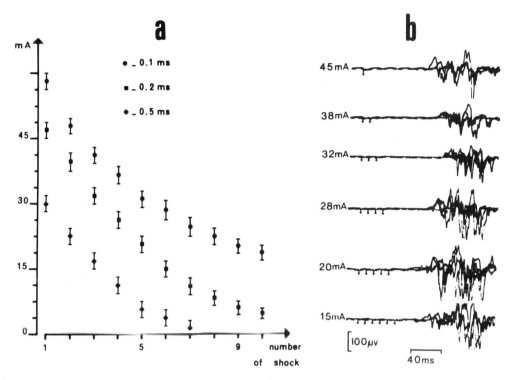

FIG. 3. A: Relationships among intensity, duration, and repetition (100 Hz) of the shocks that are necessary to evoke a nociceptive reflex and a pain sensation. **B:** RA-III reflexes elicited by various intensities of stimulation according to the number of the shocks (0.2 msec duration each). (From Willer et al., 1979, with permission.)

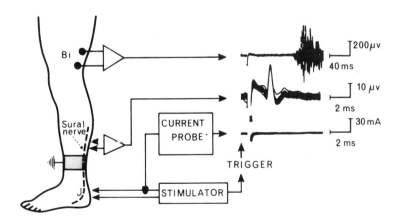

FIG. 4. Experimental set-up for recording simultaneously sural nerve afferent volleys and reflex activities from the biceps femoris muscle.

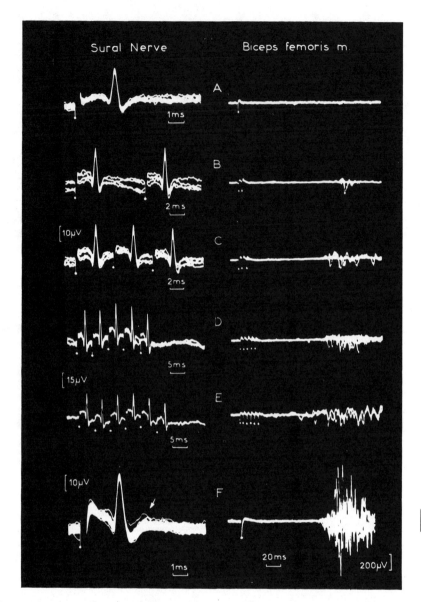

FIG. 5. Comparison of sensory nerve potentials in the sural nerve (left column) and reflex responses in the short biceps femoris muscle (right column) in normal man. A: Single electrical stimulus of 0.5 msec duration that elicits an Aα wave in the sural neurogram but no reflex response. B–E: Effects of repetitive stimulation (100 Hz) on the neurogram and on the reflex activity. The reflex RA-III appears with two shocks and is progressively facilitated as the number of shocks increases. F: Stimulation of sural nerve eliciting both a maximal Aα wave and an Aδ component (at *arrow*) on the neurogram and a large RA-III in the flexor muscle. (From Willer et al., 1978, with permission.)

the Aα wave through recruitment of Aβ fibers, according to Collins et al. (1960). Reflex activity still did not appear, nor was any pain sensation reported. When the stimulus was increased to higher intensities (45–50 mA), a small delayed response appeared on the neurogram, corresponding to higher threshold and slower conducting cutaneous fibers (Aδ fibers: 17–28 msec).

A RA-III response was always observed in Bi, and a pricking pain was always reported by the subjects. However, lower intensities are sufficient to evoke both pain and RA-III reflex provided repetitive stimulations are used. As can be seen from Fig. 5A, a single shock eliciting an inframaximal Aα wave on the neurogram is not followed by the reflex activity in Bi. On repetitive stimulations (100 Hz), a RA-III reflex appears in the muscle and increases with the number of shocks (Fig. 5B–E). Simultaneously, subjects experienced pricking and vibrating pains which often persisted after the stimulation. These results indicate that repetitive Aα waves are able to generate both RA-III reflexes and pain sensations, since there was no evidence that repetition recruited responses in the Aδ group (Buchthal and Rosenfalck, 1966).

In order to demonstrate more unequivocally that repetition of Aδ spikes is capable of mediating pain and RA-III reflex, a block of the finest cutaneous fibers of sural nerve was achieved with lidocaine. Before the block, a stimulation intensity (using a single shock) was chosen in order to elicit an Aδ wave on the suralogram, a RA-III reflex in Bi, and a corresponding pain sensation. As seen from Fig. 6, the Aδ component disappeared about 10 min after the block, as did the RA-III and pain responses. At this moment, a repeated shock eliciting only Aα waves again evoked the RA-III reflex and pain sensation. Furthermore, 20 min after the lidocaine injection, some Aα fibers were also blocked (Fig. 6). Then, a repeti-

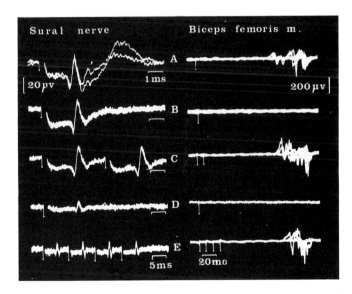

FIG. 6. Effects of lidocaine infiltration on both sural volleys and Bi activity. A: Control recordings. The sural nerve is stimulated to elicit an Aδ wave and a large RA-III reflex. B: Ten minutes after infiltration. Note the disappearance of both the Aδ and RA-III whereas the Aα wave remains unchanged. In that case, a repeated shock elicits two Aα waves and a single RA-III (C). D: Twenty minutes after infiltration. The amplitude of the Aα wave has decreased, showing that some of these fibers have also become blocked. A repetitive stimulation again elicits a RA-III reflex accompanied by its pain sensation. (From Willer et al., 1978, with permission.)

tive stimulation again elicited RA-III and pain responses. So, it appears clearly that the afferent volleys generating the RA-III reflex and the painful pinprick depend mainly on the parameters of stimulation used: the Aδ fibers are effectively involved in the case of a single shock, whereas the larger diameter fibers (Aα, Aβ) can evoke these nociceptive phenomena when they are repetitively and adequately excited (Willer et al., 1978).

Supraspinal Influences on RA-III Reflex and on Pain Sensation

Since it is well known that specific psychological situations (attention, anxiety) are able to modify the perception of pain (Hill et al., 1952) and spinal excitability (Hugelin, 1972), we studied their effects simultaneously on the thresholds of both pain sensation and RA-III reflex (Willer et al., 1979). During increased attention induced by a mental task, inhibition of both pain sensation and RA-III reflex was observed (Fig. 7). In contrast, during anxiety or stress generated by anticipation of a strong pain, a facilitation of these effects was noted (Fig. 7). These data are in good agreement with those of the literature (Bathien, 1971; Hill et al., 1952; Willer, 1975). A mechanism of reticular activation can be involved during attention, whereas limbic structures would be involved in the mechanisms of anxiety according to Routtenberg's hypothesis (1968), since "arousal system I" (corresponding to reticular activation) produces inhibition of nociceptive reflexes and of pain sensations, whereas the effects of "arousal system II" (corresponding to activation of limbic structures) have completely opposite effects on the same parameters.

In contrast with this parallel modulation of supraspinal descending influences on both pain sensation and motor nociceptive activity, we have also shown the possibility of a dissociation between these two parameters when a heterotopic as well as a homotopic pain is used as a conditioning test (Willer et al., 1979). The heterotopic pain was achieved with a nociceptive stimulus applied to the contralateral ulnar nerve to avoid intersegmental influence between sural and ulnar afferent volleys. In these conditions, when a nociceptive (10 ± 0.7mA) ulnar stimulus was given simultaneously with the 10-mA sural one, the pricking pain initially described to the sural electrodes decreased to become significantly tactile, although no significant change occurred in the RA-III response (Fig. 7). The homotopic pain was elicited with a very strong and noxious stimulation (one shock of 70 mA) applied to the sural nerve itself. Immediately after this kind of stimulus, both pain and RA-III reflex were enhanced when a 10-mA stimulation was given to the sural nerve. However, a dissociation between the recovery curves of RA-III and pain was observed: 10 to 12 sec after the very noxious test stimulation, the pain sensation returned to its control values (threshold pain), although the RA-III reflex was still facilitated (about +80%) and only recovered its initial values 28 to 30 sec later (Fig. 7). These data suggest that supraspinal descending influences can act differently on spinal dorsal horn neurons in case of noxious ascending

FIG. 7. Different supraspinal modulations on RA-III and on pain sensation. **Upper left:** Effects of a mental task on a suprathreshold (14 mA) pain and reflex. Note the decrease in the reflex amplitude and the sensation becoming tactile during the test. **Lower left:** Effects of stress. Note the increase in pain and in the reflex activity during the stress. **Upper right:** Effects of an heterotopic pain. Note the dissociation between the pain sensation (decreased) and the reflex (unchanged). **Lower right:** Effects of an intense homotopic pain. Note the initial and parallel increase in both pain and reflex and the dissociation between the time course of the recovery curves of sensation and reflex. *Arrows* indicate the onset of a nonsignificant difference between the control and the poststimulation values. (From Willer et al., 1979, with permission.)

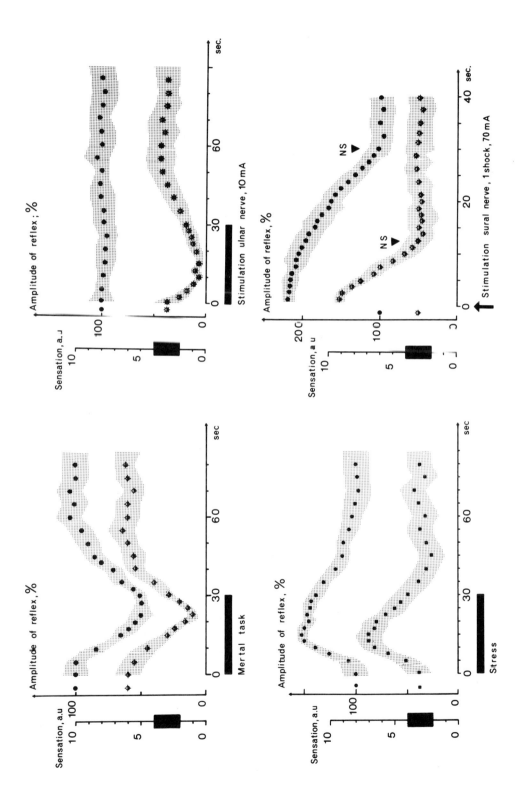

messages and in case of spinal motor nociceptive activity (Willer et al., 1979).

Pharmacological Studies

In previous papers (Willer and Bathien, 1975, 1977; Willer et al., 1976), we have shown the possibility of modifying the pain sensation and the RA-III reflexes with either algogenic chemical agents, local anesthetics, or analgesic drugs. However, the most interesting findings (reported here) concern the effects of morphine on the spinal cord in spinal human patients.

It is now well known from animal studies that morphine exerts its depressive effects on spinal noxious message through two different mechanisms: (1) indirectly by reinforcing the descending inhibitory systems from brainstem origin (Satoh and Takagi, 1971; Takagi et al., 1955; Vigouret et al., 1973) and (2) directly at the spinal level by inhibiting nociceptive transmission (Bodo and Brooks, 1937; Koll et al., 1963; Le Bars et al., 1975; McClane and Martin 1967; Wikler and Frank, 1944). We have reconsidered the direct spinal depressive action of morphine on nociceptive flexion reflexes compared to the monosynaptic H reflex in chronic spinal patients for this study (Willer and Bussel, 1980). When usual therapeutic doses of morphine chlorhydrate (0.2 mg/kg) were given i.v., there was a rapid depression of nociceptive reflexes (by 70%) while the H reflex remained unchanged (Fig. 8). Moreover, there was a very significant dose–effect relationship in these effects, since 0.3 and 0.35 mg/kg morphine induced, respectively, a 90% and 96% depression of the nociceptive reflexes. This morphine-in-

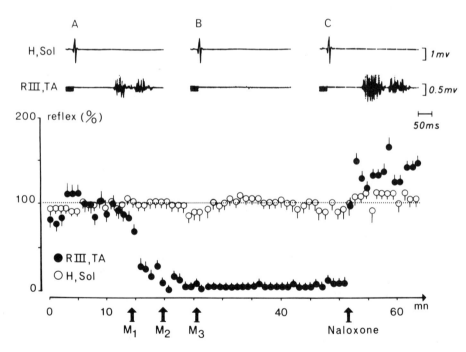

FIG. 8. Effects of morphine on spinal reflexes in spinal man. **Upper:** Monosynaptic (H-sol) and polysynaptic (RA-III, TA) reflexes during control (**A**) and injection of 0.35 mg/kg morphine (**B**) and of 0.02 mg/kg naloxone (**C**). **Lower:** Effects of morphine and naloxone versus time on H-sol and RA-III and TA. The results are expressed in percent of reflex amplitude (100% is control). M_1, 0.2 mg/kg; M_2 and M_3, respectively, additional doses of 0.1 and 0.05 mg/kg morphine. (From Willer and Bussel, 1982, with permission.)

duced depression was specific, since the specific narcotic antagonist naloxone (0.02 mg/kg i.v.) totally reversed these effects in the first minute following injection (Fig. 8). It must be noted that naloxone alone did not produce any change in these reflexes. Therefore, it appears clear from this study in chronic spinal man that morphine, and probably opiates in general, play an important role in the modulation of nociceptive messages directly at the spinal level, since usual therapeutic doses of the drug are able to exert a major depression on nociceptive reflexes without affecting monosynaptic reflexes.

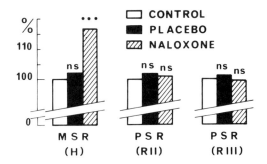

FIG. 9. Effects of naloxone and placebo on spinal reflexes in normal awake man. MSR, monosynaptic reflex (H reflex); PSR, polysynaptic reflex (R-II is tactile, whereas R-III is nociceptive); ns, not significant; ***$p < 0.001$. (From Boureau et al., 1978, with permission.)

Testing the Endogenous "Morphine-like" Systems

The use of the specific opiate antagonist naloxone made possible numerous studies on pain research, especially on neuropeptide (endorphins and enkephalins) regulation and function.

Findings in Normal Man

Since naloxone, when used alone, induced facilitation of nociceptive reactions in mice, rats, and dogs (Jacob et al., 1974; Jacob and Michaud, 1976) and enhanced spinal reflexes in spinal cats (Goldfarb and Hu, 1976), we have examined this latter possibility in normal and spinal man. For this purpose, the effects of naloxone (0.8 mg, 3 cc i.v.) compared with a placebo (saline, 3 cc i.v.) were studied on spinal reflexes: monosynaptic (H reflex), polysynaptic tactile (RA-II), and nociceptive (RA-III) reflexes were tested in a double-blind investigation in normal and spinal man (Boureau et al., 1978; Willer and Bussel, 1980). In normal and spinal man, naloxone did not produce any significant change in RA-II or RA-III responses, although an increase (+20%) was found in the H reflex of normal subjects (Fig. 9). These findings are consistent with others that have shown that naloxone (used alone) had no effect on respiration, heart rate, blood pressure, and vigilance (Evans et al., 1974; Aronski et al., 1975; Finck et al., 1977; Willer et al., 1979b). Therefore, it appears that in normal pain-free man, the endogenous "morphine-like" systems are not operative, i.e., the opiate receptors are not permanently activated by the endogenous narcotics, and, consequently, there is no permanent activation of the serotoninergic bulbospinal descending pathways that are known to be involved in the pain-suppressive system emanating from the ventral gray matter and which have inhibitory effects on transmission of nociceptive flexor afferents in the spinal cord (Akil and Mayer, 1972).

In contrast with these findings, the endogenous analgesic systems seem to be tonically active in patients with congenital insensitivity to pain and activated by some peripheral maneuvers such as transcutaneous nerve stimulation (TNS) or electroacupuncture (EA).

Congenital Insensitivity to Pain

This syndrome remains without any satisfactory physiopathological explanation and must be precisely defined according to clinical criteria (Thrush, 1973). Patients affected with congenital insensitivity to pain have a lack, from birth, of sensation and

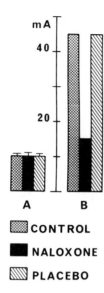

FIG. 10. Diagram comparing the results obtained in normal subjects (A) and in a patient with congenital insensitivity to pain (B) following injections of naloxone or of placebo. (From Dehen et al., 1977, with permission.)

reaction to pain in the whole body. Other sensations are preserved: e.g., they can easily discriminate between hot and cold. The tendon and cutaneous reflexes are normal. The peripheral sensory nerves and central nervous system are histologically normal, and postmortem examination has so far failed to disclose any underlying abnormality (Baxter and Olszewski, 1960).

These characteristics make it possible to distinguish this syndrome from (1) familial central autonomic dysfunction (Riley et al., 1949), an autosomal recessive condition characterized by subtotal or total loss of unmyelinated fibers in the peripheral somatic and autonomic nervous system (Aguayo et al., 1971), (2) congenital insensitivity to pain with anhydrosis, in which the Lissauer tract and small neurons in the dorsal root ganglia are absent (Swanson et al., 1965), and (3) some hereditary or acquired neuropathies in which pain sensations are partly or totally lost (Appenzeller et al., 1972; Comings and Amromin, 1974; Denny-Brown, 1951; Johnson and Spalding, 1964).

We had the opportunity to explore three cases of congenital insensitivity to pain (Dehen et al., 1977, 1978; Willer et al., 1978). We found that the threshold of the RA-III reflex from Bi was spontaneously stable at 45 to 50 mA, i.e., about 350% high-

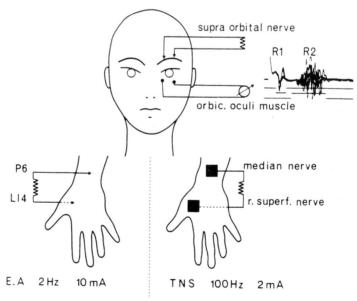

FIG. 11. Experimental procedure for the blink reflex, electroacupuncture (EA) and transcutaneous nerve stimulation (TNS). For clarity of the figure, segmental conditioning TNS sites are not shown.

er than the control values (10 mA) found in normal subjects in similar experimental conditions. However, naloxone injections (0.8 to 2 mg, i.v.) caused a large (70%) and rapid fall in the threshold of RA-III from 45 to 13 to 14 mA within 5 to 6 min following injections. At 17 to 20 min after injection, the threshold rose towards its initial values. By the same time, patients did not complain of pain but reported only discomfort and impressions of warmth. In constrast, double-blind injections of saline as placebo did not produce any change in the RA-III threshold which remained stable at its initial values (Fig. 10).

In one patient, a powerful morphine-like drug (fentanyl citrate) was given to test opiate receptor saturation. We found that fentanyl brought about a rapid and transient increase (35 to 40%) in the RA-III threshold. The lowering in the RA-III threshold (abnormally high) by naloxone can be interpreted as evidence for partial suppression of a morphine-like tonic inhibition exerted on neurons transmitting nociceptive flexor afferent input. This effect could be the

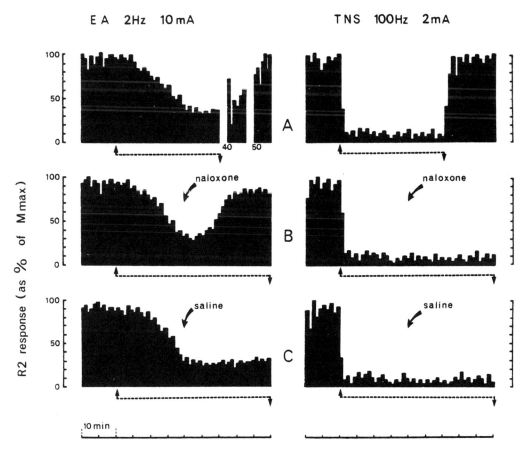

FIG. 12. Effects of EA and TNS on the nociceptive component (R2) of the blink reflex in the same subject. R2 responses are expressed as percentage of the maximal direct motor response (M_{max}) response = 100%) from the orbicularis oculi muscle, elicited by stimulation of the ipsilateral facial nerve. **Left**: the progressive and moderate naloxone sensitive depression in the R2 response obtained with EA. **Right**: the rapid and major depression in the R2 response which is not modified by naloxone.

same as that exerted by the inhibitory raphe spinal system activated by morphine and derivatives. In this case, this descending system appears to be permanently hyperactive, and neurons believed to be spinothalamic thus subject to abnormal (supraphysiological) inhibition. In any case, according to our observations, congenital insensitivity to pain would result from a permanent hyperactivity of a morphine-like analgesic system. This system seemed to be unsatured, since it still reacted to injection of a powerful narcotic.

However, the nature of the endogenous opioid involved in this pathology remains unknown. In one of our patients, the levels of beta-endorphin and dynorphin were estimated in CSF and blood, independently by Dr. Girard in Paris, and by Dr. Goldstein in Palo Alto, and found to be within the normal range. Therefore, the pathophysiology of this condition cannot be related to the presence of abnormally elevated levels of these endogenous opiates in the patient. Endogenous opiates, like morphine when administered in repetitive injections, present the phenomenon of pharmacological tolerance and it is conceivable that congenital insensitivity to pain would be related to some disturbance of opiate receptors rather than to an abnormally increased release of an endogenous opioid.

Electro-Acupuncture and Transcutaneous Nerve Stimulation Analgesia

Two different methods of peripheral nerve stimulation are in use for clinical pain relief. This first method, electro-acupuncture, has been found by some authors to

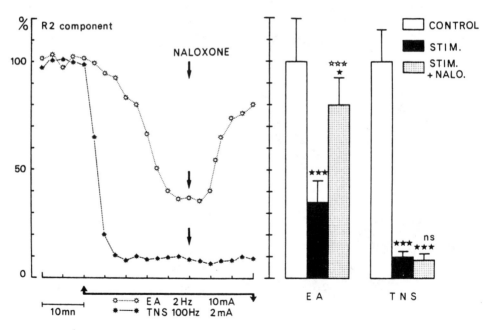

FIG. 13. Global results. **Left**: the temporal aspect of EA- and TNS-induced depression in the R2 response. *Arrows* indicate naloxone injection. **Right**: histograms showing the respective maximal depressive effect of EA and TNS upon R2 response before and after naloxone injection. Black stars indicate the significance of the variations between controls and other conditions while open stars and ns concern the significance between the pre- and post-naloxone injection situations. As in Fig. 2, the R2 response is expressed as percentage of M_{max}.

produce a good pain relief and to elevate the subjective pain threshold (Andersson and Holmgren, 1976; Chapman et al., 1976). It involves low-frequency noxious stimuli delivered through needle electrodes inserted into muscles innervated either from the same spinal segment or by distinct spinal segments (heterosegmental stimulation). The mechanism of electro-acupuncture seems to involve the release of endogenous opiates since it can be reversed by naloxone (Mayer et al., 1977; Pomeranz and Chiv, 1976). We found that electro-acupuncture increased the threshold of the RA-III nociceptive reflexes of the biceps femoris muscle (Boureau et al., 1978).

In the other method, high-frequency noxious stimulation is delivered through surface electrodes over cutaneous nerves belonging to the same spinal segment as the painful region. This method of transcutaneous nerve stimulation has been shown by most authors to act via segmental non-opiate interactions that probably involve a gate mechanism (Willer et al., 1982). However, the basic mechanisms by which either of these two methods produce analgesia are unclear and remain a matter for discussion.

New data have been obtained by analyzing how each method affects the nociceptive component R2 of the human blink reflex (Boureau et al., 1978; Willer et al., 1982). This reflex provides a genuine test

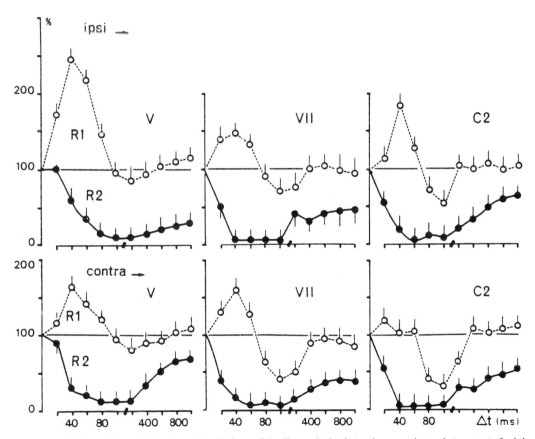

FIG. 14. Effects of conditioning stimulation of ipsilateral (ipsi) and contralateral (contra) facial afferents (supraorbital (V) and facial (VII) nerves and the facial receptive field of C2 cervical root (C2)) on the R2 component. R2 responses are expressed here as percentage of control responses (100% corresponding to the mean of 50 control responses).

since it is a nociceptive reflex in the Sherrington sense and actually protects the eye through eyelid closure. Moreover we have been using electro-acupuncture and transcutaneous nerve stimulation routinely for the treatment of facial pains. The method is summarized in Fig. 11. Electro-acupuncture at 2 Hz induced a progressive, naloxone-reversible depression of the R2 component of the blink reflex (Fig. 12, left side). The effect persisted for 8 to 17 minutes after stimulation. In contrast, transcutaneous nerve stimulation induced a rapid and powerful depression of the R2 nociceptive reflex which was not influenced by naloxone (Fig. 12, right side; Fig. 13) and this effect did not persist after cessation of stimulation. These data concur with the view that the R2 reflex depression by electro-acupuncture involves the activation of an endogenous opioid system. By contrast, the R2 depression by transcutaneous nerve stimulation presents distinct properties and rather involves another mechanism such as a local spinal inhibitory circuit.

To assess this hypothesis, the two components (R1 and R2) of the blink reflex were conditioned by stimulation of different nerves of the facial region (ipsilateral or contralateral supraorbital nerves, facial nerves, Cervical 2 dermatome). In all cases the conditioning stimulus produced a facilitation of the R1 component in parallel with a major depression of the simultaneously recorded R2 nociceptive component (Fig. 14). This R2 depression was very similar to the one observed with transcutaneous nerve stimulation. One could argue that afterhyperpolarization of facial MNs having fired in the R1 early blink reflex response could play a role in the effect. However, this can be excluded because the depression of the R2 nociceptive blink response was still

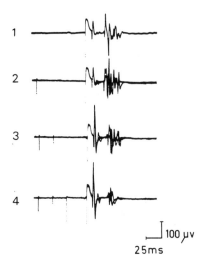

FIG. 16. Effect of infraliminar conditioning stimulations on the blink reflex responses, showing a convergence between all these afferents. 1: control response. 2: infraliminar conditioning stimulation of the ipsilateral cutaneous branch of the radial nerve. This does not modify clearly the control response. 3 and 4: summation of infraliminar conditioning stimuli of radial (as in 2) plus contralateral C2 area (in 3) plus contralateral supraorbital nerve (in 4). Note the progressive increase in R1 parallel to a decrease in R2 response with the number of conditioning stimuli.

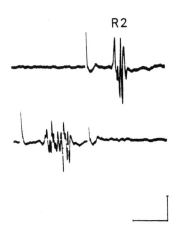

FIG. 15. This figure shows that the depression of R2 observed in Fig. 14 is not due to an increase in R1. An R2 response is elicited alone by stimulating the contralateral supraorbital nerve *(upper trace)*. This response is abolished *(lower trace)* by a conditioning stimulation of the contralateral facial nerve. Calibrations: vertical = 100 μV; horizontal = 40 msec.

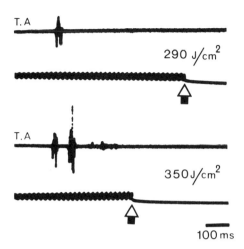

FIG. 17. Typical reflex responses recorded in tibialis anterior (TA) muscle and onset of pain sensation (at *arrows*) elicited by a threshold (290 J/cm²) and a supra-threshold (350 J/cm²) noxious laser radiant heat stimulation applied on the skin receptive field of the sural nerve. (From Willer et al., 1979, with permission).

elicited (Fig. 15) when the test stimulus evoking R2 was delivered to the contralateral supraorbital nerve: under such conditions, the R2 response is not preceded by a R1 response since the latter only occurs ipsilaterally to the evoking stimulus.

Electrophysiological as well as anatomical studies have shown that many polysynaptic pathways (segmental and heterosegmental) converge upon the trigeminal nucleus (Willer et al., 1982). This convergence is documented, for example, by the observation that addition of several infraliminar conditioning afferent stimulations elicited an increase of R1 that paralleled a depression of R2 (Fig. 16). This suggests that converging inhibitory mechanisms interacting at a pool of interneurons (probably the spinal trigeminal nucleus) could indeed be involved in the depression of the nociceptive R2 blink reflex by transcutaneous stimulation.

CONCLUSIONS

Nociceptive flexion reflexes constitute a practical tool for pain research in man when they are studied with adequate methods. In this connection, it must be noted that nociceptive flexion reflexes elicited by noxious laser radiant heat (Fig. 17) is an interesting technique that would be worthy of development, since thermal stimuli are more specific and natural than electrical ones, especially for pain research (Willer et al., 1979).

Motor Control Mechanisms in Health and Disease,
edited by J. E. Desmedt.
Raven Press, New York © 1983.

Deep Sea Diving: Human Performance and Motor Control Under Hyperbaric Conditions with Inert Gas

*,**Maurice Hugon, †Laurent Fagni, and ‡Kunihiro Seki

**Laboratoire de Psychophysiologie, Université de Provence, Centre de Saint Jerome, Marseille, France; †Goupement d'Intérêt Scientifique de Physiologie Hyperbare, Marseille, France; and ‡J.A.M.S.T.E.C., Yokosuka 237, Japan

Close to a quarter of the world's petroleum comes from the continental shelf. Drilling and transport facilities lie 100 to 300 m beneath the surface, sometimes even deeper. Underwater operations are essentially automated. However, numerous detailed operations require the direct intervention of divers, acting as intelligent, highly adaptable operators. The diver utilizes scuba gear as light and flexible equipment, but, as a costly side effect, he is subject to the environmental hydrostatic pressure [10 m of seawater (MSW) = 1 Bar]. The pressure effects on open cavities (lungs, respiratory tract, and middle ear) are neutralized by the diver by his use of a mixture of gases at the appropriate environmental pressure. The convenient mixture is supplied by scuba tanks or by an auxiliary source aboard the accompanying boat. The breathing mixture flows through a valve system according to the diver's needs.

In actual practice, deep dives for extended periods of time can be described as follows (Hugon, 1981): Caissons able to support high internal pressures are installed on board the accompanying boat. The divers enter the caissons with their equipment at atmospheric pressure. By injection of the appropriate breathing mixture, the pressure in the caissons is raised to the hydrostatic pressure of the intended underwater work site. Once this condition is reached, the divers move to a separable compartment of the caisson, the diving bell, and seal themselves inside. The diving bell is then lowered to the work site. Once bell–seawater pressure equilibrium has been established, the divers can open the door in the floor of the bell and swim out to work. Dry resting periods are possible in the bell, and nights are spent under pressure after the bell returns aboard the companion boat. The boat equipment furnishes the breathing mixture and power (light and heating) to the bell and the immersed divers. A complex multichannel communication system produces televised images from the work site, orders, calls, and mechanical transfer of tools and materials. A sanitary system and various security systems are also included.

Severe environmental conditions confront the diver: cold (a few degrees above 0°C), effects of buoyant force, scarcity of geocentric vertical references, and vision problems arising from the use of a mask with artificial lighting. However, his main problem is to cope with physical effects of

---

*To whom correspondence should be addressed: Laboratoire de Psychophysiologie, Université de Provence, Centre de Saint Jérome, Rue Henri Poincaré, Marseille 13397, Cedex 4, France.

pressure per se as well as the effects that result from dissolution of compressed gas in body fluids and structures. Moreover, the high density of the compressed breathing mixture considerably increases the work load of ventilation and tends to reduce the speed of diffusion of gases in the pulmonary alveoli. Additionally, the respiratory rhythm is modified. These difficulties are partly palliated by use of a light gas, helium, which presents practically no biological problems at depths less than 800 MSW. Nitrogen is too heavy and narcotic to be used beyond 50 to 60 MSW. The partial pressure of oxygen is normally maintained above 200 mbar to facilitate the pulmonary transfer of oxygen and the oxygenation of tissues. However, to avoid toxic effects, the oxygen partial pressure must be less than 500 mbar. Compressed air (dense, narcotic, and toxic) cannot be utilized on long dives at depths greater than 30 MSW.

The direct or indirect effects of pressure can be obtained in a dry caisson, where the subjects are submitted to a breathing mixture pressure equal to the hydrostatic pressure of the work site ("simulated" dive).

HUMAN PERFORMANCE UNDER HIGH PRESSURES

To improve the quality of diving procedures that maintain the safety and fitness of the diver, the following parameters must be determined: type of gas and partial pressure of oxygen, rates of compression for any given depth, duration of stay at constant pressure, and decompression procedures.

The usual characteristics of the dive (Sagittaire III) described by Fig. 1 enable sub-

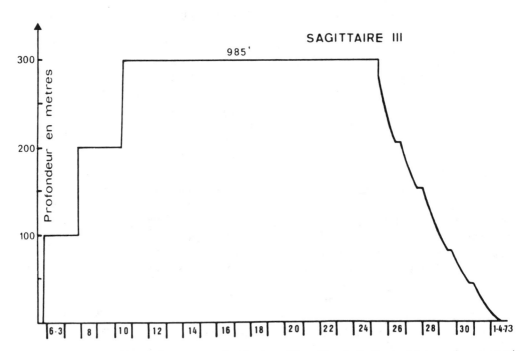

FIG. 1. Sagittaire-III dive (helium–oxygen). *Abscissa:* Time (days). *Ordinate:* Meters of sea water in dry caisson (10 MSW = 1 bar). P_iO_2 = 300 mbar, P_iN_2 = 0. T = 32° C. Rates of compression: 84 MSW/hr between 0 and 200 m and 60 MSW/hr between 200 and 300 MSW. Stays at constant pressure: 48 hr. Sojourn at depth: 15 days. Decompression 7 days. (From Seki, 1976, with permission.)

jects to be remarkably efficient at 300 MSW (31 atm) throughout the entire duration of the stay. Tests of the subject's state include body temperature, respiratory and cardiac activity and brain activity. Psychometric tests can be close to the actual tasks performed under water; however, conservation of psychophysiological capacities can coexist with severe physiological defects, and a particular psychological effort could conceal a real deficiency during the test period. Evaluation of the diver's condition requires a critical inventory of his physiological and psychological capacities in order to permit suggestions for improvements in diving procedures.

Studies of fatigue, sensorimotor activities, perception, and mental performance have been undertaken in Philadelphia (Lambertsen et al., 1978), Durham (Bennett et al., 1974, 1980), Alverstoke (Hempleman 1980), and Marseille–Toulon (Lemaire, 1979; Hugon et al., 1979; Hugon and Seki, 1979; Seki and Hugon, 1976).

Fatigue Under High Pressure

Fatigue described as a "painful sensation with difficulty of action" (Littré, 1976), is, by definition, subjective in the presence of objective physiological changes. With Kogi's method, Seki, and Hugon (1976) conducted a study using three separate questionnaires given to 16 divers in dry caissons simulating six dives to from 400 to 610 MSW with a He–O$_2$ atmosphere. The dives were accompanied by marked lowering of alertness and by bodily discomfort. The compression phases are the most wearing, in proportion to the rate of the compression. Rests at constant pressure provoke a marked improvement in the subject's state, according to the divers, which deteriorates again if the compression recommences. The practice of short "excursions" for a few hours to deeper levels (455 MSW) after a daily rest at constant pressure (390 MSW) seems to be especially wearing (Janus IIIa). There is a persistent feeling of fatigue for a few days after returning to sea level. In conclusion, the overall state of fatigue, rated on a scale of 1 to 9 by the subjects, is less with decreasing compression rate and reduced by constant-pressure stages.

Critics point out that the reports of the subjects can be biased by errors, illusions, interest, etc. Moreover, the "fatigue sensation" can be caused not only by the pressure but also by anxiety over the risky situation, by insomnia, or simply by boredom in an experiment that lasts 2 weeks to a month. But, taking into account all data relative to deep diving, Seki's conclusions appear rather well supported: compression and decompression are major causes of hyperbaric fatigue. Rests at constant pressure are accompanied by an improvement in subject condition as long as the pressure is less than that at 500 to 600 MSW. For greater values, the condition of the diver can degenerate.

In Seki's data, bodily symptoms of discomfort emphasize joint and muscle pains which develop during compression, saturation, and decompression (Fig. 2). The compression bends are attributed to the difficulty of saturating poorly vascularized tissues with helium (causing osmotic dysbarism?), and decompression bends are attributed to local gas embolism. These discomforts are reduced when nitrogen is added to the helium–oxygen mixture (for example, 5% partial pressure of nitrogen), and adaptation to bottom conditions is faster. Nitrogen could help to compensate for pressure effects on cell membrane. It is also possible that nitrogen, acting as a narcotic, simply reduces the subjective manifestations of high-pressure stress without attacking the causes. It could thus produce a masking effect, hiding the real modifications.

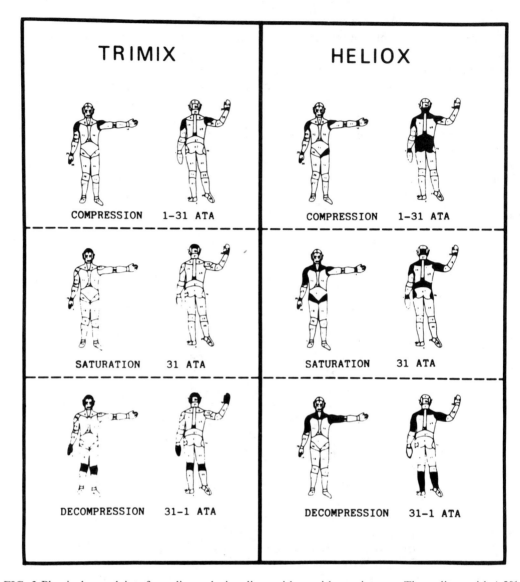

FIG. 2 Physical complaints from divers during dives with or without nitrogen. Three dives with 4.5% nitrogen (**left column**, pooled data from five divers) or one dive with helium–oxygen (**right column**, two divers). Mean rate of compression in all cases was 75 MSW/hr. Note: (1) bends develop during compression; (2) the subjective situation improves at constant pressure (300 MSW); (3) decompression produces new bends; (4) bends are significantly reduced by the use of nitrogen (compare left and right columns).

Sensorimotor Abilities

The subject is asked to maintain a sustained force, to execute precise phasic movements, to manipulate small items with recognition and specific placement, and so forth. The performance depends on the quality of sensory coding, muscle coordination, and central processing (attention, perception, problem solving, decision) which improve through learning.

Local static work capacities are well conserved during two dives at 300 and 487 MSW. The ratio of the integrated EMG to the static force displayed does not vary with depth. Dynamic work is probably not altered at 500 MSW (C. Lemaire, *unpublished data;* Lambertsen et al., 1978). Neuromuscular deficiency is apparently not a limiting factor in performance at these depths.

Manual or digital dexterity and tapping speed are practically unchanged at 200 MSW, even after rapid compression (100–200 MSW/hr). At 350 MSW, even after slow compression (a few meters per hour), there is usually some observable degeneration of these abilities. Performance improves during the stay at bottom and during decompression. The aftereffects are negligible and of short duration. Performance can be lowered by as much as 20 to 30% in, for instance, simple visual reaction time. However, as a general rule, deficiencies under experimental conditions do not appreciably affect divers' efficiency at work (Lambertsen et al., 1978).

Maintenance of balance

The maintenance of the standing posture depends on a deliberate motor program assisted by proprioception and cutaneous, graviceptive, and visual inputs. This program uses mesencephalic and spinal control circuits in conjunction with local myotatic regulations (Fig. 3). At depth, posture oscillations registered on a balance platform are significantly increased (Berghage et al., 1975; Lambertsen et al., 1978). Equilibrium on a narrow rail is also affected. During the dive of Entex V (450 m, helium–oxygen–nitrogen; Cerb-Gismer, and Gis, 1981), these sensitive tests revealed an alteration in the state of divers who were otherwise in perfect clinical condition. To date, the physiology of this defect has not been established.

Perception

This term includes the sensory coding and CNS processing that lead to recognition of a change in the field of stimulus. Curiously, this type of detection–discrimination task has not been the object of many studies. Seki (1976) and Seki and Hugon (1976) observed that the visual critical fusion frequency (CFF) decreases during dives with helium–oxygen (Fig. 4). A greater variation of the delay between two signal lights is necessary to create a noticeable separation (Fig. 5). Recognition of the temporal separation of two electrical stimuli applied to the thumb and little finger of the same hand suffers from a similar degeneration. These effects could be great for certain divers. In the depths studied (300 MSW and 450 MSW), such changes are reversible within a few hours after the end of compression. Changes are attributed not to a variation in decision-making processes but to losses in perception of sensory messages (Blanc-Garin et al., 1979).

Mental Activities

The tests used present the subjects with binary- or multiple-choice questions under time pressure or arithmetical operations (count, add, etc.) or logical operations (ordering of items, classification, grouping, etc.) or language tests or memory tests (visual, verbal, etc.). A certain instability of alertness or slips of attention (presence of microsleeps) (Hempleman, 1980), have been reported, but as a whole, the authors regard performance deficiencies in mental tests at depths as not attributable to defects in alertness and/or attention, even when the divers have slept rather poorly.

The essential results can thus be summed up as follows (Fig. 6). At a depth greater than 200 to 300 m, mental alacrity is reduced without a definite loss of capacity for solving complex problems. Nevertheless, the short-term memory (tested by retention of a series of letters or numbers) is altered. Learning capacities may be decreased at depths. High compression rates increase these losses. Rests at constant pressure improve performance. The addition of nitro-

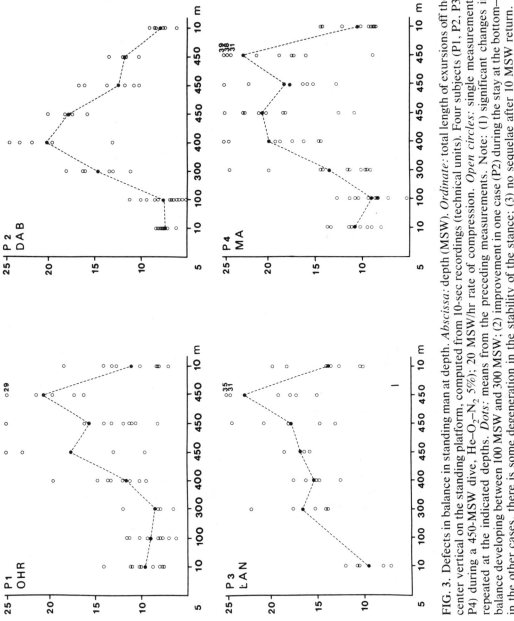

FIG. 3. Defects in balance in standing man at depth. *Abscissa*: depth (MSW). *Ordinate*: total length of excursions off the center vertical on the standing platform, computed from 10-sec recordings (technical units). Four subjects (P1, P2, P3, P4) during a 450-MSW dive, He–O$_2$–N$_2$ 5%; 20 MSW/hr rate of compression. *Open circles*: single measurements repeated at the indicated depths. *Dots*: means from the preceding measurements. Note: (1) significant changes in balance developing between 100 MSW and 300 MSW; (2) improvement in one case (P2) during the stay at the bottom—in the other cases, there is some degeneration in the stability of the stance; (3) no sequelae after 10 MSW return.

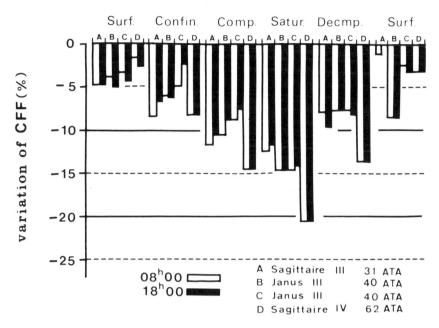

FIG. 4. Critical flicker frequency (CFF) degenerates with depth. CFF is checked by the use of a foveal periodic illumination at decreasing frequency (light/dark ratio, 1). Measurements regularly undertaken at 8 a.m. *(clear bars)* and 6 p.m. *(black bars)* during four dives (He–O$_2$). The *zero line* is for the CFF as checked by the method of constant stimulus (cf. Fig. 5). Strict criterion changes are expressed by (CFF$_2$ − CFF$_1$) × 100, with CFF$_1$ the value at sea level. Four dives; 12 divers. From left to right: Surf., CFF at surface level as checked by the method of limits (strict criterion); Confin., CFF in heliooxygen at 10 MSW during a stay for habituation to the new situation in the caisson; Comp., CFF during compression; Satur., CFF at depth, at the beginning of the stay; Decmp., CFF during decompression. The overall picture is consistent with a reversible defect in CFF.

gen to the breathing mixture (a small percentage of the total pressure) improves underwater performance and recovery.

Conclusion

Despite numerous methodological ambiguities, the convergence of the results is striking. At high pressures, helium–oxygen mixtures induce deficiencies in a variety of psychomotor or mental performances. These deficiencies are not evident above 300 MSW reached through slow compression or 200 MSW with rapid compression (several dozen meters per hour). Rests at constant pressure start some regression of the changes. With nitrogen-containing mixtures, fatigue is reduced, performance improved, and recovery accelerated. Dives to less than 300 MSW leave the subjects in full possession of their psychomotor and mental capacities. Deeper than 300 MSW, even in a dry caisson, the deficiencies raise questions about physiological mechanisms. Is the cause of trouble the hydrostatic force, or is helium a biologically active molecule? Why are compression periods especially distressing, and how do constant-pressure periods induce at least a partial recovery of capacities? Which neurological structures are affected by hyperbaric stress?

HIGH-PRESSURE NEUROPHYSIOLOGY

High-pressure atmosphere creates neurophysiological as well as behavioral defects which can be investigated with technical modifications of clinical neurophysi-

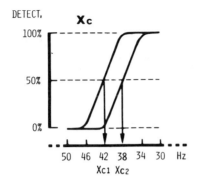 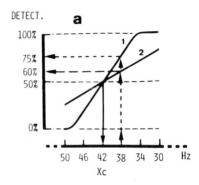

FIG. 5. A method to ascertain changes in perception (constant-stimulus technique for critical flicker frequency). The S are presented short series of flickering lights at given constant frequencies (Light/darkness duration ratio 1). S are asked to tell their perception (fusion, flickering, or possible flickering) for 64 presentations for each frequency in pseudorandom order. **Left hand graph:** *Abscissa,* flicker frequencies; *ordinate,* number (relative to 64 × 100) of positive responses for flickering. The *curve* on the right of **left graph** expresses the statistic of responses (one subject) that only include clear positive answers for flickering. The curve on the left expresses a statistic that includes the two answers (flickering and possible flickering) for detection of flicker. Xc is the critical frequency for decision at 0.5 probability. In a given experiment, Xc depends on the attitude of the S facing the risk to make a decision. Xc also depends on the experimental method, using strict criterion (Xc = 38) or loose criterion (Xc = 42). The slopes of the two curves are similar. Index **a** is the ratio of a change of probability for detection to the correlated change in frequencies. This index is an expression of the perceptual sensitivity to changes in frequency. **Right hand graph:** The graph illustrates the case for a loss in perceptual sensitivity (curve 2: **a** is low relative to curve 1). With depth, **a** usually lessens, an indication of perceptive defect. Xc may increase or decrease depending on the subject without any correlation with depth. (From Hugon et al., 1979, and Seki, 1976, with permission.)

ology and animal physiology methods (Brauer, 1975; Macdonald, 1975, 1980; Hygon, 1981).

Comparative Study of Pressure and Helium Effects

It has been known since the work of Regnard (1891; in Macdonald, 1975) that hydraulic compression of crustaceans provokes, in the absence of compressed gas, motor excitation and convulsions that worsen progressively between 1 and 50 atm. At 200 atm, paralysis occurs, which reverses with decompression but does not reverse when pressure has increased beyond 500 atm. When the rate of compression is low enough (25 MSW/hr, the animals retain their normal mobility until 100 atm. Similar data have been obtained from tadpoles, fish, or liquid-breathing mice (Kylstra, 1967; in Macdonald, 1975) whose lungs are filled with fluorocarbon saturated in oxygen. These animals slowly breathe liquid and extract enough oxygen to survive at rest, at 20° C. Hydraulic compression of such preparations causes tremor and spasms between 50 and 80 atm and ends with paralysis of the diaphragm around 150 atm. Thus, pressure per se creates nervous disorders.

Even though the hydrostatic pressure developed with compressed gas has effects that are comparable to those caused by hydraulic compression, it is nonetheless probable that the gas could also act as a chemical. This is clearly the case for N_2 which is narcotizing when compressed (excitation and euphoria at 30 MSW; slowing of mental function, reduction of memory and dexterity at 40 MSW; deep narcosis at 100 MSW). Helium, which has long been con-

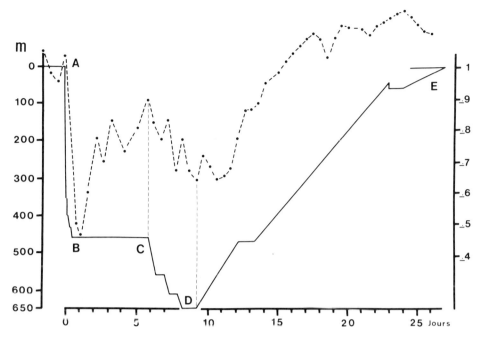

FIG. 6. Mental performance with depth (Atlantis II, Duke University). *Abscissa:* time (days). *Ordinate: left axis,* depth (MSW); *right axis,* performance relative to surface efficacy (mental calculus). Helium–oxygen plus nitrogen (10%). Compression rate: slowing from 108 MSW/hr to 15 MSW/hr to 460 MSW. Compression induces a large defect in mental performance. Stay at 460 MSW results in large recovery. New compression (very slow) induces new (moderate) decrease in performance. Decompression demonstrates the reversibility of the defects (overperformance at the end of decompression and after return to sea level is caused by some overlearning after so many repetitions of the test). The overall picture demonstrates degeneration of mental capacities with depth and improvement by adaptation. (Modified from Bennett and Coggin, 1981, by Hugon, 1981b, with permission.)

sidered an excitant and was then believed to be biologically inert, now appears to be able to counteract the effects of pressure above 1,000 MSW (Brauer et al., 1980).

Cell Physiology Under Pressure

Pressure physiology of neuron and synapses has been studied in crustacea, molluscs, batrachians, and rats. Complex molecules and supramolecular structures can undergo modifications under high pressure. Changes in proteins (Fish et al., 1979) and in lipoprotein association (Galla and Trudell, 1980) in phospholipid bilayers (Miller, 1972; Miller et al., 1973) are said to develop under pressure. As a general rule, any system susceptible to any change in volume will be pressure sensitive.

Apparently such changes can develop at rather low pressures (20 atm) in biochemical structures such as myosin, hemoglobin, or lipoprotein membrane matrix (Friess et al., 1975). Passive and active ionic membrane transport can be changed, usually with depolarization (Macdonald, 1975). Myelinated nerve fibers tend to produce repetitive activity (frog: Kendig and Erickson, 1981). In contrast, sinoatrial activity in mouse heart is slowed under pressure (Ornhagen, 1979). Electrically induced neuromuscular twitch increases under high pressure (rat diaphragm) because of slowing of the muscle action potential and/or modification of excitation–contraction coupling. Synthesis, release, and postsynaptic uptake of neural transmitters appear to be modified (Martin et al., 1972; Parmentier et al., 1981). Such defects can be attributed to some impairment in mutual "stereorecognition" between transmitter and receptor sites (Lehmann, 1978), but the defect could also occur in the development of the activity of the "second messenger," because of some change in enzyme activity. In general, such synaptic changes occur only beyond 600 to 1,000 MSW. However, Henderson et al. (1977) have demonstrated some slowing in postsynaptic response in squid at 35 atm.

Cellular disorders can noticeably decrease during phases of constant pressure. This result (Wann et al., 1980), if general, is of prime importance and significance. Such an improvement in cell activity can be interpreted as resulting from homeostatic cell regulation which results in "adaptation." Hugon (1981), and Hugon and Fagni (1981) tentatively extrapolate to the whole organism the role of cell regulation.

Anesthetics and nitrogen oppose the excitant effect of pressure, and raising the pressure can revive animals from a mild anesthesia. A current hypothesis suggests that anesthesia increases the fluidity of the phospholipid membrane matrix. Pressure could have an opposite effect (Miller, 1972; Seeman, 1972). Other works place emphasis on the possible alteration by anesthetics (and pressure) of the proteins involved in synaptic transmission. These studies induce the development of a "high-pressure pharmacology" different from "normal-pressure pharmacology" because molecules and reactions can be changed by pressure (Fish et al., 1979; Walsh, 1980).

Clinical Neurophysiology Under High Pressure

The Hoffmann (H) reflex and polysynaptic reflexes can be elicited in man or monkey as adaptations from sea-level procedures (Hugon, 1973a,b; 1974; Bonnet et al., 1973).

To test the H reflex, the subject (man or monkey) is seated on a chair in a relaxed state. (The experimental limb is restrained by a plaster cast.) The sciatic nerve is stimulated at the knee by skin (or chronic) electrodes (1 msec duration). Records from soleus show direct motor (M) and reflex (H) responses (interstimulation delay, 3 sec). A strain-gauge device records the isometric force of M and H responses. The recovery cycle of monosynaptic excitability is studied by applying two equal sciatic stimuli (S_1 and S_2) eliciting H reflexes of about 25% of the maximal M response (in man) or 10% (in monkey). The ratio of the second reflex to the first one (H_2/H_1 ratio) measures spinal excitability for each interval between the S_2 and S_1 shocks (Fig. 7; cf. Figs. 10–12). (Hugon, 1973a).

The stimulation of the sural nerve (one to three shocks, 0.1 msec duration in monkey) elicits a polyphasic polysynaptic reflex recorded in tibialis anterior as a typical flexor muscle. The sciatic or sural stimulus evokes cortical potentials which are recorded at the best S-I contralateral and ipsilateral location (focal chronic macroelectrode in skull of monkey). Twenty somesthetic evoked potentials (SEP) are averaged in awake monkeys.

Reflexes and evoked potentials are studied at sea level, during compression at various rates, and under various absolute pressures with He–O_2 or He–O_2–N_2 mixtures. The temperature is set at 30 to 33° C in He, and humidity is 30–50%. A system for gas mixture maintenance eliminates carbon dioxide and various pollutants of biological or technical origin and maintains oxygen at a fixed value.

The so-called "divers" are totally separated from the experimenter (Fig. 8), lack any direct assistance, and perform the required experimental operations under pressure. Signals (stimulus, reflexes, evoked

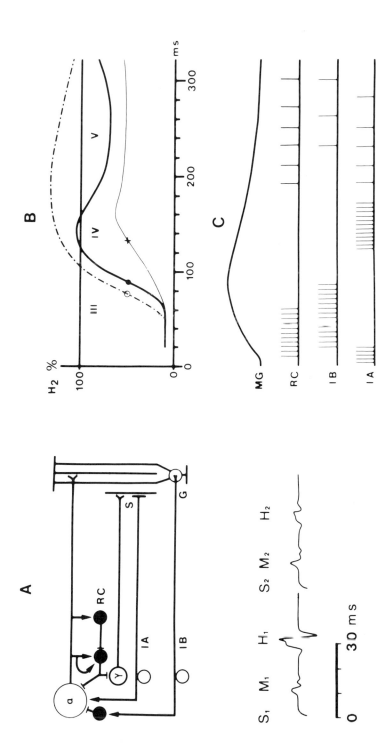

FIG. 7. Monosynaptic reflex circuits and associated recovery cycle of excitability. A; α, alpha MN and motor axon to muscle; RC, Renshaw cells and recurrent circuits; γ, γ-fusimotor neurons (and axons to spindle, S); IA, primary afferent fiber from spindle; IB, Golgi afferent fiber from the tendon organ (G); S_1, sciatic stimulus for Hoffmann reflex; M_1, direct motor response; H_1, Hoffmann reflex (latencies for man); S_2, M_2, H_2, same after a S_1–S_2 delay. The peak-to-peak F_2/H_1 ratio expresses the reduction (or facilitation) in the capacity of the spinal cord to produce the second reflex (H_2) after the inital (S_1, M_1, H_1) complex event (S_2 is made equal to S_1). If $M_2 = M_1$, S_2 is said to have the same efficacy as S_1 for motor recruitment; the conclusion is then extrapolated to sensory recruitment. Any change in the H reflex is caused by spinal changes provoked by the (S_1, M_1, H_1) complex event, not by changes in stimulus efficacy. B: H recovery curves (in monkey). Ordinates, $H_2/H_1 \times 100$; Abscissas, S_1–S_2 delay (msec); III, early inhibition phase; IV, late facilitation phase (or "rebound" of excitability); V, late inhibition. Normal recovery cycle (thick line). Interrupted line (open circle) is for depth (600 MSW) after fast compression (100 MSW/hr). Thin line (star) is for slow compression (600 MSW at 20 MSW/hr). C: Proprioceptive events caused by the H_1 reflex contraction (schematic). MG, H mechanogram (scleus, ankle angle 90°, isometric). Recurrent activites and proprioceptive events combine to contribute to the different phases of the recovery through postsynaptic and possibly presynaptic effects.

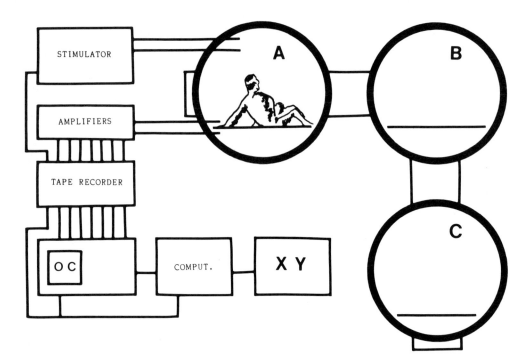

FIG. 8. Dry caissons and experimental setting for neurophysiology. A, B, caissons for daily activity, hygiene facilities, and rest; C, wet pot for swimming exercises. The experimenter has no direct access to divers except through pressure lock normally used for meals and other items.

potentials, current EMG, EEG, EKG activities) are conducted through the walls of the caisson and are processed in the usual ways. Experiments are of long duration (Fig. 1) because of slow compression (in man especially, 10–30 m/hr), and very slow decompression (231 hr from 610 MSW of depth to sea level); the sojourn at the bottom can range from a few hours to 2 or 3 weeks. A high-pressure experiment is very costly in time, facilities, qualified crew, and gas, and consequently experiments are relatively rare. Data from higher organisms under pressure accumulate slowly from year to year.

High-pressure neurological syndrome.

High-pressure neurological syndrome (HPNS) is the combination of neurological changes elicited in man by compression and pressure: tremor, myoclonus, convulsions without EEG correlates (type 1 convulsions: Brauer et al., 1979), then convulsions with EEG correlate paroxysms (type II). The thresholds for tremor and convulsion are respectively, approximately 200 to 300 MSW and 600 to 800 MSW for monkey, 400 and 1,000 MSW for mice; 700 and 1,300 MSW for fish (Hunter and Bennett, 1974; Bennett, 1975; Brauer, 1975; Brauer et al., 1980). Beyond these upper limits, irreversible changes develop (phase III of Brauer: Rostain et al., 1981). The most neurologically evolved organisms are most sensitive to pressure; similarly, they show the greatest sensitivity to a high rate of compression (Brauer, 1975; Brauer et al., 1977). Interindividual variability is high. Of course, human dives are limited to low rates of compression (less than 30 MSW/hr) and limited final pressures (less than 700 MSW). The EEG slowing and paroxystic elements in EEG reported by Rostain are other components of HPNS. No clear correlation between HPNS and mental or perceptual defects has been documented to date.

Autonomic disorders (vertigo, nausea) are normally absent after low compression rates but can be severe under fast compression in man. Bradycardia as seen by Ornhagen (1979) in an *in vitro* sinus node is disputable in whole animal because of homeostatic controls. The progressive addition of nitrogen during compression reduces the motor components of the HPNS (Hunter and Bennett, 1974; Brauer et al., 1974; Bennett et al., 1974; Rostain et al., 1977; Gardette and Rostain, 1981; Bennett et al., 1980), but the EEG does not improve (Rostain, 1980).

The HPNS is also sensitive to general anesthetics and to various antiepileptic drugs used experimentally or clinically, which may in turn lead to practical procedures for preventing or postponing HPNS development.

Reflexes under high pressure.

The latency, shape, and duration of M, H, and tendon jerk (T) EMG responses are not modified with pressure in man (Roll et al., 1978, 610 MSW; Harris, 1979, 420 MSW) or in monkey (Bonnet et al., 1973, 980 MSW). For instance, the mean latency value for the H reflex in a man was 32 ± 1 msec (SD) at sea level and at 610 MSW. Such a constant latency suggests that the velocity of afferent and efferent impulses does not change with depth. The recording of EMG motor unit response elicited in adductor pollicis in man by nerve stimulation at the arm (Hugon and Lemaire, 1975; 300 MSW) showed no change in latency, no indication of repetitive fiber activity, and no change in the recovery cycle of the motor fiber (Fig. 9). Such results indicate a physiological integrity of the components of the sensorimotor pathways.

Roll et al. (1978), Harris (1979), Harris and Bennett (1981), and unpublished results from our group demonstrate a clear but moderate facilitation of the H reflex under pressure (400 MSW and deeper). The T reflex is facilitated. The variability of the H and T reflexes increases with depth. Polysynaptic reflexes are also facilitated and develop a lasting afterdischarge (Bonnet et al., 1973). Roll et al. (1973) made reference

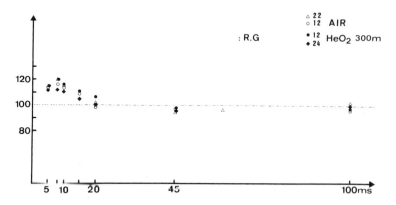

FIG. 9. Recovery cycle of excitability for single motor fiber (median nerve in man). The activity of a single motor unit is recorded in the adductor pollicis after a juxtathreshold single stimulus (V_o) at the median nerve. A second stimulus is applied Δt msec afterwards and set at the V_t voltage needed for a second muscle response of the motor unit. *Abscissa*, interstimulus delay (t); *ordinate*, V_o/V_t ratio × 100. One subject studied in normal air at sea level, then at 300 MSW in helium–oxygen at different times in a day. There is no change with depth in the classic phases of hyper- and hypoexcitability of the motor axon. Same results for three other divers during the same dive. (From Hugon and Lemaire, 1975, with permission.)

to some defect in corticospinal pre- or postsynaptic controls with release of the segmental reflexes. Four days after thoracic total transection of the cord, myotatic reflexes dramatically increase; myoclonia develops with episodes of tremor at 800 to 900 MSW (Fagni et al., 1981) but not at sea level before or after the dive. The polysynaptic reflexes are also facilitated with depth.

Myoclonus under pressure.

In intact monkey, cat, rat, mouse, and man (as seen by Brauer in a personal dive), a high rate of compression (100 MSW/hr or more) induces local myoclonus, first rare and of short duration, then frequent and long lasting. Such large myoclonus develops bilaterally in axial, then proximodistal and distal, muscles from a common central source as judged by the synchrony. Myoclonus in neck muscles only precedes by 10 to 20 msec the myoclonus in limb flexors and extensors. The EEG correlates are not at first apparent but then develop as sustained and repetitive paroxysms (see type I and II convulsions of Brauer). The picture varies with individuals and diving procedures. Myoclonus also develops in the spinal cat, locally and erratically caudal to the transection and bilaterally and in synchrony in the rostral cord *(unpublished data).*

Recovery cycle of H reflex.

Studies were restricted to phases III (early inhibition), IV (late facilitation), and V (late inhibition) of the cycle (Fig. 7). Beyond 300 MSW, published (Bonnet et al., 1973; Roll et al., 1978; Harris, 1979; Hugon et al., 1980; Harris and Bennett, 1981) and personal results suggest that the rate of compression has a striking influence on the H recovery cycle. In cases of slow compression (less than 40–60 MSW/hr, Fig. 10), the general trend is towards a reduction of the late facilitation with an increase in late inhibition. High compression rates (Fig. 11), in contrast, accentuate the late facilitation when pressure increases beyond 600 to 700 MSW (monkey experiments). The duration of the early inhibition is reduced. The early first part of phase III (inhibition) is generally but not always preserved with depth (see Bonnet et al., 1973). An enduring facilitation of the H reflex develops instead of the usual late inhibition. Modifications reverse on return to sea level.

The sciatic stimulus liminal for M responses and supraliminal for H responses is also supraliminal for Ib fibers, which have central inhibitory effects on homonymous MNs. The H_1 response consists of a twitch that causes the tendon organs to fire and the spindles to pause, whereas during the muscle relaxation, spindles fire and tendon organs tend to pause. Such alternation in afferent volleys and central excitatory–inhibitory effects causes phase IV in the cycle (Bouaziz et al., 1975). As a side effect, this input is supposed to induce a sustained presynaptic inhibition, a possible cause for late inhibition, since bicuculline reduces this late inhibition in monkey, and oxyaminoacetic acid increases it (M. Lucciano and M. Hugon, *unpublished data*).

Curtis and Ryall (1964) and Ryall (1970) have shown that the recurrent Renshaw interneurons present an early discharge (cholinergic nicotinic) and late long-lasting discharge (muscarinic, see Fig. 12). Atropine (3 mg i.v.) suppresses the late inhibition without significantly changing the early inhibition. From the blocking effect of atropine on muscarinic synapses, we conclude that late Renshaw cell activity plays a decisive part in the development of late inhibition. The use of dihydro-β-erythroidin effectively suppresses the major part of the early inhibition (except the very beginning of the phase which is caused by Ib afferents: Bouaziz and Hugon, 1973 in monkey). Thus, the role of the early nicotinic recurrent inhibition in phase III is clearly demonstrated.

From all of this, we can tentatively propose the following interpretations of recov-

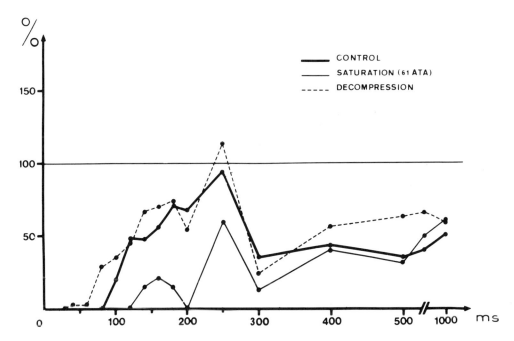

FIG. 10. Hoffmann reflex recovery curve in man with depth at low rate of compression. (Compare with Fig. 7.) *Abscissa*, interstimulus delay (msec); *ordinate*, $H_2/H_1 \times 100$. H_1, half H_{max}, angle at 90°, isometric restraint. *Thick line*, h recovery cycle at sea level with phases III, IV, and V. The hump between 120 and 200 MSW is believed to be caused by Ib inhibition evoked by the H_1 contraction. *Thin line*, same subject and technique at 610 MSW after slow compression (helium–oxygen). The general subsidence of the curve could be caused by presynaptic inhibition elicited by the Ib then Ia afferents in response to the H_1 contraction and relaxation. *Dashed line*, recovery of normal cycle at the end of the decompression. (From Roll et al., 1978, with permission.)

ery cycle variations under pressure. Slow compression results in increase in presynaptic inhibition which decreases phases IV and V of the cycle. Such an effect results from an increase of supraspinal control on spinal input which restrain spinal reflex hyperexcitation. Fast compression would have an atropinic antimuscarinic effect on recurrent inhibition. The late inhibition is suppressed; meanwhile, the intermediate phase IV (the rebound of excitability) is facilitated. In most of the cases, these two effects develop without any modification of the early inhibition, thus suggesting preserved nicotinic action of acetylcholine. Finally, the supposed blocking effect of pressure on muscarinic activation of Renshaw cells would result in a sustained disinhibition of the MNs which would cause myotatic and H-reflex facilitations with depth.

It is interesting that the initial effect of pressure would develop at the synaptic level; a cholinergic muscarinic defect would be the result of pressure modification in production, release, or binding of the transmitter or its subsequent cell effect. These effects appear quite reversible with decompression as a result of cell regulation or supraspinal control on Renshaw cell activity (Pierrot Deseilligny and Morin, 1979, 1980).

Finally, the abnormal facilitation of the monosynaptic reflexes by galvanic labyrinthine stimulation (Lacour et al., 1978) could be related to spinal changes in excitability. When studying eye performance with

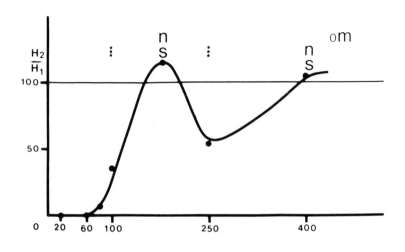
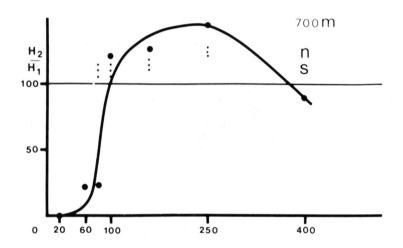
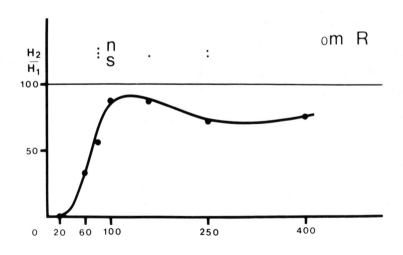

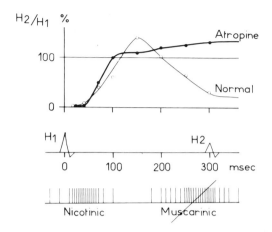

FIG. 12. Hoffmann reflex recovery curve with atropine at sea level. (Compare with Fig. 7.) Recovery cycles, normal and after 3 mg/kg atropine (monkey). *Middle line,* H reflexes. *Low trace,* recurrent activity caused by the H$_1$ response (early and late Renshaw cell activity) (M. Hugon and L. Fagni, *unpublished data*).

depth, Gauthier (1976) found no nystagmus or abnormal labyrinthine activities. Similarly, we consider the defect in balance at depth (Fig. 3) as being caused by spinal (or supraspinal) changes in motor control rather than by labyrinthine or vestibular defects.

High-pressure tremor.

Present in normal subject at sea level under steady contraction, fine tremor is present at depth with a similar frequency of 10 ± 2 Hz but with higher intensity (man, monkey, cat, and other mammals). It is also present in muscles caudal to a spinal transection in cat. Such a tremor demonstrates that the spinal system has the capacity to produce oscillations in the myotatic loop.

When the dorsal roots for leg muscles are cut, rare episodes of weak activity can be seen in soleus under pressure at 800 MSW (helium–oxygen). Such episodes are often divided into unequal bursts at 20 to 25 bursts/sec in spinal or normal preparations (M. Hugon and D. Garcin, *unpublished data*). These data suggest a local excitatory effect of pressure on the spinal cord.

We tentatively conclude with the following model, using concepts from Mori and Ishida (1978), Elble and Randall (1970), and from Desmedt's volume on Tremor mechanisms (1978b) along with our own results (Fagni et al., 1981): pressure reduces the muscarinic recurrent Renshaw activity at the spinal level, which makes the MNs hyperexcitable; spinal discharges develop as short myoclonus or sustained activites. The nicotinic recurrent inhibition, the afterhyperpolarization of the MNs, and the postinhibitory rebound of excitability combine to make the ventral root activity repetitive in the absence of late recurrent inhibition. Such a repetitive bursting induces muscle responses, and the afferents from them impinge on the preceding mechanisms at delays and for durations that depend on the contraction–relaxation process (see Fig. 7). Recurrent circuit and myotatic loop resonate at lower frequencies, which depend on the muscle length and load.

Central supraspinal mechanisms do not play any basic role in such a tremor under pressure, but in normal subjects, they control spindle sensitivity and MN excitability when active contraction is ordered. Motoneuron facilitation can be obtained by reduction of recurrent inhibition (Pierrot-Deseilligny and Morin, 1980) or reduction of presynaptic inhibition (which, conversely, can be increased if the subject is asked to

FIG. 11. Hoffmann reflex recovery curve in monkey with depth at high rate of compression. *Abscissa,* interstimulus delay; *ordinate,* H$_2$/H$_1$ × 100; n.s., nonsignificant difference. Phase IV is largely increased, begins earlier, and is long lasting. 0 m, at sea level in normal air; 700 m, under 700 MSW pressure in helium–oxygen; O m R, after return to sea level. The large phase IV facilitation has disappeared during decompression. After compression (Figs. 7 and 12), we suggest there is a blocking effect of pressure on muscarinic recurrent inhibition.

rest for H testing; see Fig. 7). Such controls make the spinal tremorogenic mechanisms active during tonic contraction but do not impose the tremor frequency.

The cerebellum, which is well known to participate in low frequency tremor (3–4 Hz) and in harmaline tremor (10–12 Hz), does not appear to play any role in high-pressure tremor (see Fig. 13), nor does it appear to have any decisive role in HPNS paroxysms.

Thus, high-pressure tremor appears to depend mainly on some modification of excitability in the spinal cord in combination with supraspinal normal control for movement. Abnormal excitability in spinal cord apparently depends on changes in recurrent inhibition and/or changes in presynaptic inhibition. High pressure is the initial cause for such modifications.

Cortical somatic evoked potentials in monkey.

Somatic evoked potential (SEP) recruitment with depth (Fig. 14) is elicited by posterior tibial or sural nerve stimulation and can be recorded by focal electrodes implanted in the skull over the sensorimotor area S-I of the leg and foot. After rapid compression (100 MSW/hr, for instance), the afferent volley elicited by the Hoffmann stimulus set for a just noticeable M response evokes a cortical SEP that is strongly facilitated with respect to the SEP at sea level (Fig. 14A). The SEP elicited by sural stimulus is facilitated in a similar way (Fig. 14B).

Early waves are specifically facilitated. Above 400 MSW, after a Hoffmann stimulus set for half-maximum M responses, an early unusual (P20) ipsilateral wave develops, homologous to the usual P20 contralateral. A late P200 bilateral wave parallelly increases. The latencies of the early waves do not significantly change with depth. Any SEP modification observed with depth reverses with return to sea level.

After electrical stimulation of the optic chiasm in the rat, Kaufmann et al. (1980) observed a facilitation of the cortical poten-

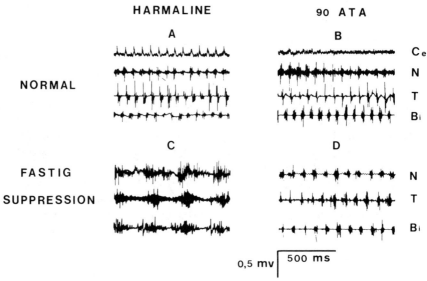

FIG. 13. Tremor in cat: harmaline versus pressure. Cerebellum is not involved. *Upper row,* activity in cerebellum (Ce) and EMG in neck muscles (N), Triceps (T), and upper arm biceps (Bi) with harmaline or under pressure (890 MSW). *Low row,* after bilateral fastigial electrolysis, EMG with harmaline and under pressure. The fast tremor (10 ± 2 Hz) is still present at depth where the fast harmaline tremor has disappeared. We conclude that the role of the fastigium (and olivocorticofastigiospinal circuit) is unimportant in pressure tremor (L. Fagni, *unpublished data*).

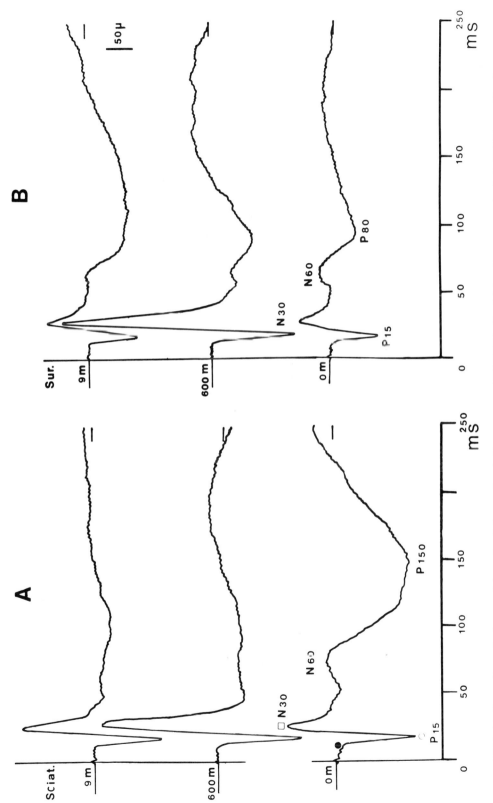

FIG. 14. Somatic evoked potentials at depth in monkey helium–oxygen 600 MSW after high-rate compression). **A**: SEP to popliteal nerve stimulus at threshold for M response. 9 m, contralateral SEP in helium–oxygen under 9 MSW before dive; 600 m, same at 600 MSW; 0 m, same after return to sea level. Monofocal recordings (S-I area) from skull-implanted electrode; 20 responses averaged. **B**: SEP to stimulus at depth. Note the facilitation of early waves at depth.

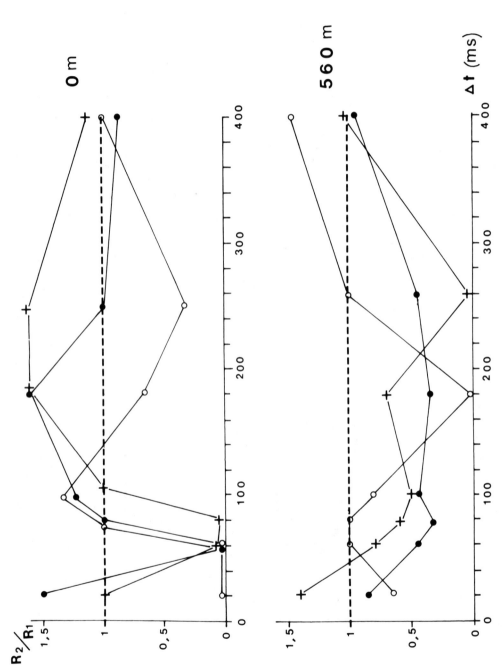

FIG. 15. Recovery cycle of somatic cortical evoked potentials after sciatic double stimulus. *Abscissa*, interstimulus delay (Hoffmann stimulus at M threshold; H reflex about 10% M_{max}); *ordinate*, peak-to-peak ratio of the P35 (*circles*), N60 (*dots*), and P120 (*crosses*) waves of the second SEP relative to the same measurements from the first SEP alone (mean value for 20 responses from focal skull-implanted electrode on the contralateral S-I area for limb projection). Monkey at 0 MSW and arrival at 560 MSW (helium–oxygen). Fast compression (mean value 100 MSW/hr). The SEP recovery curves undergo strong modifications at depth (L. Fagni, *unpublished data*).

tials but not those at the geniculate level. They concluded that there is a cortical hyperexcitability. Phenytoin prevents neither the facilitation of the evoked potentials nor the progression of seizures in rat.

Recovery cycles in SEPs.

After rapid compression, the recovery cycle of the SEP (Fig. 15) elicited by double stimulation in monkey undergoes strong modifications (Fagni, 1979). The early inhibition of the SEP elicited by the Hoffmann stimulus (at M threshold) becomes weak and long lasting with depth (100–300 msec) instead of being deep and short-lived (20–80 msec). The only conclusion at this stage is that pressure undoubtedly induces changes in the dynamics of the SEP.

SEPs do not undergo marked facilitation with pressure after slow compression in monkey (helium–oxygen, 980 MSW). In man, the visually evoked potentials (VEP) undergo moderate reduction (200 MSW) and facilitation (500 MSW) and then return to normal value during the stay at depth. Details depend on the subject. The morphology and latencies of the VEPs do not change appreciably. On the whole, after slow compression, the changes in cortical evoked responses are rather moderate. We suppose that there is an active slow adaptation of the organism which reduces pressure modifications during long stays at depth and slow changes in pressure. Of course, such adaptation does not overcome very high pressure strain. Over 700 to 800 MSW pressure, defects can continue increasing towards critical accidents (Rostain et al., 1981).

Clinical neurophysiology with pressure provides us with considerable data that enlarge the initital HPNS clinical concept: monosynaptic and polysynaptic reflexes are modified at depth as well as cortical evoked potentials. Dynamic processes (recovery cycles of excitability) are especially affected.

GENERAL CONCLUSION

Recent decades have witnessed an exploding development of high-pressure studies in living subjects. The industrial need for oil has acted as a permanent and general stimulus for deep diving. Progress in research has improved safety and efficiency of work at depth. Such research is also of practical interest if we have to make a reliable selection of those subjects who will be highly adaptive to pressure. Finally, high-pressure research has now developed sufficiently to ask its own basic physiological questions in relation to biophysics (MacDonald, 1980), anesthesiology, and biology (Brauer, 1975).

ACKNOWLEDGMENTS

The universities Aix–Marseille I and II, CNRS, DRET, and CNEXO have supported the studies by ERA CNRS 272 and GIS Physiologie Hyperbare at Marseille. We are indebted to CEH-COMEX (Marseille) and Cerb–Gismer (Toulon) for sustained collaboration and to L. Agate and F. Joubaud for illustration and typing.

Pathophysiology of Dystonias

*J. C. Rothwell, J. A. Obeso, B. L. Day, and C. D. Marsden

University Department of Neurology, Institute of Psychiatry, and King's College Hospital Medical School, London SE5 8AF, United Kingdom

Torsion dystonia is the term applied to describe a motor syndrome characterized by the presence of abnormal movements and postures produced by prolonged spasms of muscle contraction that distort the body into typical postures. *Athetosis* refers to distal dystonic movements in the limbs (Marsden and Harrison, 1974), but to avoid confusion, torsion dystonia will be used here to describe both proximal and distal dystonic movements.

Idiopathic torsion dystonia may take a number of forms. In childhood, the disease usually is progressive, particularly if it starts in the legs, and may spread to involve all four limbs (generalized dystonia). When the illness commences in adult life, it usually does not affect the legs and often remains confined to its site of onset (focal dystonia). Examples of focal dystonia include blepharospasm and oromandibular dystonia, spasmodic torticollis or retrocollis, truncal dystonia, and dystonic writers' cramp. Not infrequently in adults, the illness may start in the neck and spread to involve one or both arms or vice versa (segmental dystonia).

The excessive cocontraction of antagonist muscles that characterizes dystonia was first noted by Oppenheim (1908) using simple palpation of the muscle bellies. This was confirmed with a mechanical recording technique by Wilson (1928) and later, with EMG recording, by Hoefer and Putnam (1940) and Herz (1944). Foerster (1921) also noted the presence of remote muscle contraction ("overflow"), another characteristic of dystonia, which was aggravated particularly during fine manipulative movements. Relatively little has been added to this picture in the last 40 years.

A paradoxical contraction of passively shortened muscle (Westphal phenomenon) has been noted by Rondot and Scherrer (1966) and Yanagisawa and Goto (1971; see R. W. Angel, *this volume*), but the significance of this finding in the production of dystonic symptoms is obscure. The same feature is also present in spasticity and in Parkinson's disease (Bathien et al., 1981).

In view of the lack of knowledge of the pathophysiological mechanisms underlying dystonia, treatment is difficult and mainly empirical (Marsden, 1981). Conventional pathological studies of a few patients with hereditary or idiopathic torsion dystonia (Zeeman and Dyken, 1968) also have failed to reveal any convincing abnormality. In this chapter, we present our data on the pathophysiology of dystonia collated from the study of 35 patients (Table 1) all of whom were diagnosed according to the criteria of Marsden and Harrison (1974).

METHODS

In all patients, we made routine EMG recordings from surface electrodes over up

---

*To whom correspondence should be addressed: University Department of Neurology, Institute of Psychiatry, Denmark Hill, London SE5 8AF, United Kingdom.

TABLE 1. *Clinical classification of 35 patients with dystonia*

Type	N	Mean age ± 2 SD
Generalized	19	32 ± 4.5
Segmental	9	29 ± 5.2
Focal		
Arm	8	40 ± 5.7
Blepharospasm/ oromandibular	3	67 ± 7.8
Torticollis	4	39 ± 4.5

to eight different muscles involved in the dystonic spasms.

"Ballistic" flexion movements of the thumb were made through a 15° angle against a small opposing torque of 0.04 Nm supplied by a low-inertia torque motor which pressed against the thumb pad (see Hallett and Marsden, 1979). In two patients, we recorded the activity of the antagonist, extensor pollicis longus, during the movements with 0.003-inch platinum–iridium wire electrodes inserted into the muscles. "Ballistic" elbow extensions were made without any opposing force. Subjects were seated comfortably with the shoulder abducted to 90° and elbow flexed to 80°. The forearm rested on a light aluminum arm pivoted coaxially with the elbow joint, which had a vertical handle mounted at the end that subjects could grasp. Movements were made in the patients' own time through an angle of about 20°.

Long-latency stretch reflexes were elicited in the flexor pollicis longus and triceps brachii according to the method of Marsden, et al. (1976). The subject maintained a constant initial position against a force of 2.5 N (thumb) or 8 N (elbow) with reference to a visual display before him. Every 4 to 5 sec at random, this force was increased by a factor of three for 200 msec so as to stretch the active muscle. Averages of 24 responses were made in each subject. The size of the stretch reflexes was measured from integrated EMG records as the percentage increase in activity during the period of the reflex compared to the activity in the same time interval during control trials. The duration of the long-latency response was measured from the rectifed EMG records

Reciprocal inhibition in the forearm was studied by eliciting test H reflexes every 5 sec in the relaxed forearm flexor muscles with low-intensity electrical stimulation of the median nerve at the elbow (see Day et al., 1981). In alternate trials, a conditioning stumulus at motor threshold was applied to the radial nerve (which supplies the extensor muscles) in the spiral groove at various times before or after the test shock was given. By convention, we referred the timing of the test shock (median nerve) to the conditioning shock (radial nerve). Thus, if the median nerve stimulus was given before the radial nerve stimulus, the interval was assigned a negative value (see Fig. 9A). Averages of 10 control and 10 conditioned H reflexes were studied at each interval.

Peristimulus time histograms (PSTH) of regularly firing motor units in the forearm flexor muscles were made using 0.003-inch platinum–iridium wire electrodes inserted into the muscle. Subjects were instructed to maintain a steady unit contraction at about 10 Hz with the aid of an audio monitor, and electrical stimuli were given at random intervals to the radial nerve every 3 to 5 sec. Histograms with bin widths of 1 msec were constructed from 100 msec before to 400 msec after the stimulus after 400 or 500 stimuli had been given.

RESULTS

EMG Patterns in Dystonia

At rest.

There was no involuntary muscle activity in many of the patients classified as having focal or segmental dystonia if they were completely relaxed in the laboratory environment. Even those patients with spasmodic torticollis showed a considerable diminution in the number of spasms when relaxed, with their head and neck well supported. However, involuntary movements persisted at rest in all 12 patients with generalized dystonia. We have classified the involuntary muscle contractions in these 12 patients into three groups according to the length of the EMG burst:

1. Continuous periods of activity lasting 30 sec and terminated by short periods of silence (Fig. 1A; $N = 5$).
2. Repetitive, sometimes rhythmical, spasms lasting 1 to 2 sec each and separated by equal periods of relative EMG silence (Fig. 1B; $N = 3$). Herz (1944) termed this "myorhythmia."
3. Rapid, irregular, and brief jerks lasting only some 100 msec and resembling those seen in myoclonus (Fig. 1C; $N = 4$).

Voluntary movement.

During voluntary movement, we have confirmed that prolonged, excessive, cocontracting activity is present in antagonist muscle groups (Fig. 2). In the patients with generalized dystonia, this replaced the pattern of activity seen at rest in the same muscles. Nevertheless, not all voluntary movements were accompanied by muscular cocontraction. In all of our patients, we observed alternating activity during rapid flexion/extension movements of the wrist and elbow (Fig. 3 left), even in patients whose movements were slow. The typical cocontraction with overflow of muscle activity appeared preferentially during more delicate and precise maneuvers (Fig. 3 right). Also, in six patients (four with focal arm dystonia, one with blepharospasm, and one with torticollis), we saw an alternating postural tremor in the outstretched arms with a frequency of 6 to 10 Hz.

Myoclonic Dystonia

We noted above that patients with focal or segmental dystonia usually showed no dystonic muscle spasms at rest. However, in nine of the 24 patients with focal or segmental dystonia whom we studied, we observed small jerks in the affected limb which appeared clinically to be similar to those of myoclonus. In all but one of these cases, the jerks were aggravated considerably by voluntary action and could be superimposed on dystonic spasms. We term this myoclonic dystonia. A summary of the results of EMG recording in these patients is given in Table 2. The muscle jerks were usually of long duration (up to 500 msec), occurring irregularly but synchronously in the muscles concerned. Somatosensory evoked potentials (SEPs) were normal, and back-averaging of the cortical activity failed to reveal any time-locked EEG event preceding the individual muscle jerks in any of the cases that we studied.

"Ballistic" Elbow and Thumb Movements

Fast limb movements in man are characterised by a bi- or triphasic pattern of activation in agonist and antagonist muscles. This may be recorded even in patients with complete limb deafferentation (Hallett et al., 1976), indicating its independence of afferent feedback. Fast flexion of the terminal phalanx of the thumb and fast extension of the elbow were studied in 13 patients with dystonia. In each patient, we examined the reciprocal activation of agonist/antagonist muscles and measured the duration of the first agonist burst of EMG activity. There was a wide variation in the pattern of ballistic arm movements in those with arm dystonia, ranging from normal to unrecognizable (Fig. 4). The data are summarized in Table 3. At the elbow, one of the four patients we examined with segmental dystonia affecting the shoulder and arm and all of the patients with generalized dystonia had prolonged or unrecognizable agonist EMG bursts (Fig. 4C). Even when the antagonist burst was not accompanied by reciprocal silence of the agonist, we often saw an initial silence in the same muscle preced-

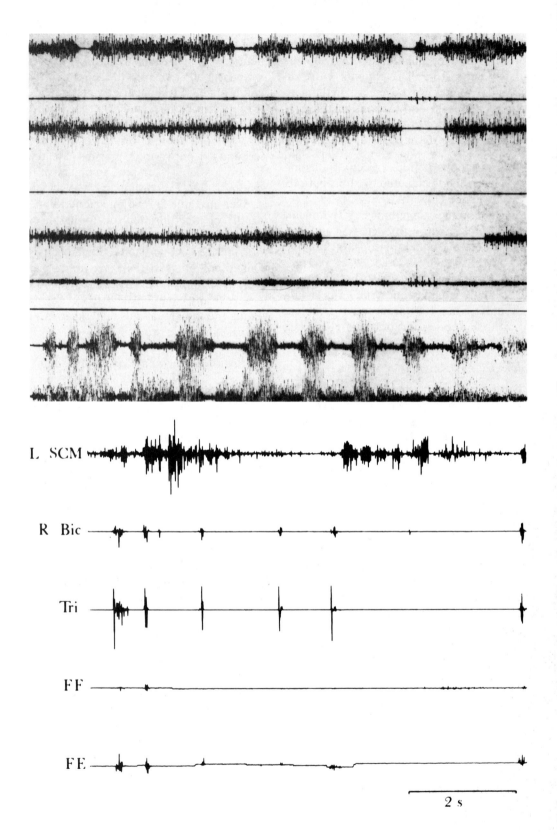

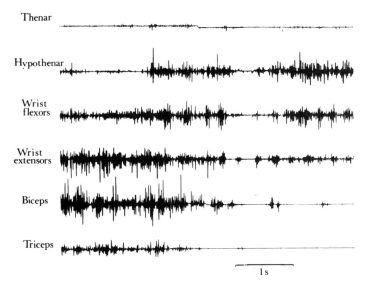

FIG. 2. Cocontraction of antagonist muscles (wrist flexor/extensor, biceps/triceps) in patient AR with segmental dystonia affecting the right arm, shoulder, and neck, during an attempt to pick up a small object from a table.

ing the first signs of agonist activation (Fig. 4B). In three of the six patients with focal dystonia of the hand and forearm, the duration of the initial agonist burst during fast thumb flexions was within the normal range, and in two of the patients in whom we inserted wire electrodes into the antagonist, extensor pollicis longus, reciprocal muscle activation was observed. Three patients showed abnormally prolonged agonist EMG bursts.

Long-Latency Stretch Reflexes

We studied the stretch reflex in the affected flexor pollicis longus of 16 patients and in the affected triceps brachii of six patients with dystonia. The amplitude and duration of the stretch reflex are shown in Table 4 together with those of a group of age-matched normal individuals. There was no change in the size and duration of the reflex in either proximal or distal muscles. However, in five patients, there was overflow of the reflex response to distant muscles not normally activated in the task, reminiscent of the overflow observed during voluntary muscle activation (Fig. 5).

Using a 250-msec collection period, we never observed any paradoxical shortening reactions in antagonist muscles. These occur at a much longer latency than the reflex responses studied here (Bathien et al., 1981).

Reciprocal Inhibition

The cocontraction of antagonist muscles in dystonia is most easily explained as a disorder of the normal reciprocal inhibitory mechanism. In normal subjects, reciprocal inhibition is mediated by two different

FIG. 1. Three types of involuntary muscle activity in different individuals with generalized dystonia. **Top:** Continuous periods of EMG activity lasting from 2 to 30 sec. **Middle:** Repetitive, rhythmical, and cocontracting spasms lasting only 1 to 2 sec each ("myorhythmia"). **Bottom:** Rapid, irregular, and brief jerks lasting only some 100 msec. Muscles are left sternocleiodmastoid (LSCM), right biceps (Bic), triceps (Tri), forearm flexors (FF), and forearm extensors (FE).

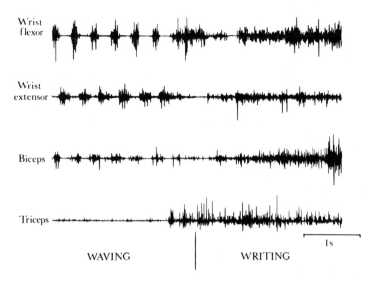

FIG. 3. Normal reciprocal activation of wrist flexors and extensors during waving **(left)** and resumption of typical cocontraction when the subject stops to pick up a pen and write his name **(right).** Same patient as in Fig. 2.

mechanisms (Day et al., 1981), both of which probably utilize the same spinal Ia inhibitory interneuron. These are termed peripheral and central inhibitions (Fig. 6).

To investigate the peripheral mechanism of Ia inhibition in the arms of dystonic and normal subjects, we gave a single conditioning volley at motor threshold to the radial nerve. This produced inhibition of the flexor motor neuron (MN) pool, revealed by eliciting test H reflexes in the relaxed forearm flexors at various times before and after the conditioning volley. From this we could construct the time course of the effect of the afferent radial nerve volley (Day et al., 1981). Our timing convention is shown diagrammatically in Fig. 7 (top). The timing of the median nerve test shock is given relative to the radial nerve conditioning shock along a time line at the bottom of the figure. Representative average control and test H reflexes are superimposed above the line.

TABLE 2. *The EMG and EEG characteristics of patients with myoclonic dystonia*

Patient	SSEP following digital nerve stimulation	Muscles involved[a]	Length of EMG bursts (msec)
WO	Asymmetrical: larger over contralateral scalp	FCU	100–150
GI	N.D.	FF, Bic	100–150
AR	Normal	FF, Bic, Tri	<500
GL	Normal	Trap, SCM (action only)	500–1,000
YH	N.D.	Trap, Delt, Tri (action only)	300–500
FU	Normal	FF, FE (action only)	100
RI	N.D.	FF, FE, Bic, Tri	300–500
SP	N.D.	FE, Bic, Delt (action only)	<1,000

[a] FCU, flexor carpi ulnaris; FF, forearm flexors; FE, forearm extensors; Bic, biceps; Tri, triceps; Delt, deltoid; SCM, sternocleidomastoid.

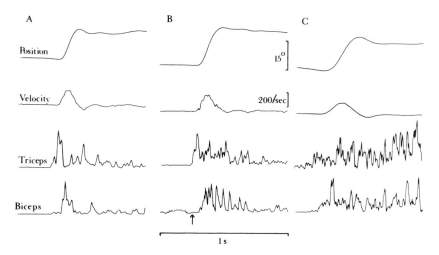

FIG. 4. Single "ballistic" extension movements of the elbow through 20° in three different subjects. A: Patient YH with segmental dystonia of the arm and shoulder. A normal, alternating pattern of activity is seen in triceps and biceps. B: Patient AR with segmental dystonia of the arm, shoulder, and neck. The initial agonist burst is prolonged, and there is no reciprocal silence of triceps when the antagonist is activated. Note, however, the small silence in the biceps trace *(arrow)* preceding triceps activation. C: Patient AL with generalized dystonia. Only the order of muscle activity (triceps followed by biceps) is recognizable.

TABLE 3. *Duration of the first agonist burst (Ag1) during rapid thumb flexions or elbow extensions*[a]

Joint	Patient/group	Ag1 (msec)	Antagonist activation[b]
Thumb	Focal hand/arm dystonia		
	WO	115	
	MA	80	
	PR	>150	
	SH	150	Reciprocal
	ST	90	
	MI	90	Reciprocal
Thumb	Segmental dystonia (arm/shoulder)		
	AR	>200	
	LO	100	
	RT	85	
Elbow	Segmental dystonia (arm/shoulder)		
	RI	90	Reciprocal
	LO	150	Reciprocal
	YH	110	Nonreciprocal
	AR	200–400	No reciprocal agonist silence
Elbow	Generalized dystonia		
	AL	>200	Nonreciprocal
	CO	>200	Nonreciprocal
	LA	150	Rarely reciprocal

[a] Normal ranges are 50 to 90 msec (FPL) and 60 to 110 msec (triceps). Agonist activity is defined as reciprocal if it begins after the agonist and is accompanied by EMG silence in the agonist.

[b] Activity of the antagonist was studied in only two patients during thumb movements, since this required insertion of wire electrodes into the extensor pollicis longus.

TABLE 4. *Comparison of the size and duration of the long-latency stretch reflex in flexor pollicis longus (FPL) and triceps in a group of neurologically normal individuals and patients with segmental or focal arm dystonia*

		FPL		Triceps	
Group	Age (years)	Size	Duration (msec)	Size	Duration (msec)
Normals ($N = 11$)	46.5 ± 4.7	375 ± 28	53 ± 1.2	562 ± 55	53 ± 1.9
Dystonics ($N = 16$)	40 ± 5.1	403 ± 76	51 ± 2.5	388 ± 78	50 ± 3.2

[a] The size of the reflex (± SE) is given as a percentage increase over control levels of activity over the same time period.

In normal individuals, the time course of depression of the flexor H reflex showed three distinct phases (Fig. 7, middle and bottom traces; continuous lines). The timing of the first (Fig. 7B) was compatible with Ia disynaptic inhibition from extensor muscles. It was abrupt in onset and short in duration. It was followed by two much longer phases of inhibition lasting up to 1 sec (Fig. 7 bottom), which were separated by a return to normal control values after about 50 msec.

The average recovery curve from the forearm flexors of the affected limb in a group of eight patients with focal or segmental dystonia is shown as the dotted lines in Fig. 7 (middle and bottom traces). Patients with generalized symptoms could not

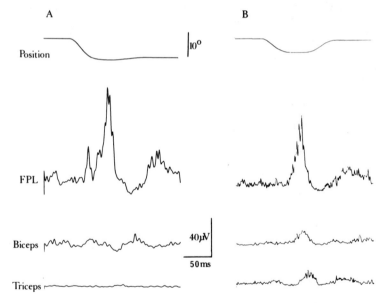

FIG. 5. Stretch reflexes in the flexor pollicis longus (FPL) of a normal subject **(left)** and patient AR **(right)**. This patient was one of the five who showed "overflow" of the reflex response to distant muscles (biceps and triceps). In the normal subject, the stretch reflex in FPL consists of a small initial spinal latency response, starting about 22 msec after the disturbance *(top trace)*, and a larger, long-latency response with a latency of about 45 msec. No responses are seen in biceps and triceps. In AR, no spinal component of the stretch reflex in FPL can be seen, but the longer-latency response "overflows" into biceps and triceps.

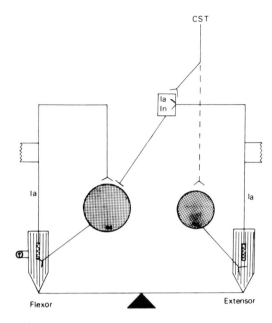

FIG. 6. Diagrammatic summary of the central and peripheral components of reciprocal inhibition. Both the direct corticospinal tract (CST) projection to the extensor muscles of the wrist (central inhibition), and the Ia afferent fibers (peripheral inhibition) send branches to the Ia inhibitory interneurons (IaIn) in the spinal cord. These then project monosynaptically to inhibit the flexor MN pool. The excitability of the α MNs was tested by eliciting H reflexes in flexor muscles at different times after a conditioning shock had been given to the extensor afferent fibers in the radial nerve.

be studied because of their inability to relax completely. In dystonia, the second period of inhibition from 10 to 60 msec was much less pronounced than that in normal subjects, although it was striking that the early, short-duration (disynaptic) inhibition of the flexor muscles was completely normal in the same patients.

The Ia disynaptic inhibition also could be seen quite clearly as changes in the firing probability of single flexor motor units after randomly timed radial nerve shocks. Single motor units were recorded in five normal individuals and in three patients with arm dystonia with intramuscular wire electrodes while they maintained a constant small contraction of the flexor muscles of the forearm to produce a fairly regular firing frequency of 10 Hz. Single electrical shocks were then applied at random intervals to the radial nerve, and a peristimulus time histogram of the flexor motor unit firing frequency was constructed over 200 to 400 stimuli. In both the normal subjects and those with arm dystonia, the short-latency disynaptic inhibition appeared as a distinct decrease in firing probability 20 to 25 msec after the radial nerve stimulation (Fig. 8A,B). However, the later phases of inhibition that were seen in the H-reflex curve were not apparent in the PSTH of either normal or dystonics; indeed, excitatory events appeared to dominate at intermediate latency. We discuss this in more detail below.

DISCUSSION

In their EMG studies of dystonia, Hoefer and Putnam (1940), Herz (1944), and Yanagisawa and Goto (1971) all observed a tonic, nonreciprocal pattern of activity in agonist and antagonist muscles during any voluntary or postural contraction. This was broken up into various different patterns by a regular grouping of action potentials. Stretch reflexes were not enhanced, and frequently, a paradoxical shortening (Westphal) reaction was observed. We have confirmed all of these observations in a group of patients with a wide range of physical disabilities. In addition, we also have characterized a group of patients who, besides showing typical dystonic contractions during voluntary movement, also had superimposed myoclonic muscle jerking. The duration of these bursts was fairly long in comparison with most forms of myoclonus, and, in the absence of any preceding time-locked EEG activity, we suggest that they may be produced by discharge of a subcortical focus.

We also noted a wide range of abnormalities in the EMG pattern accompanying

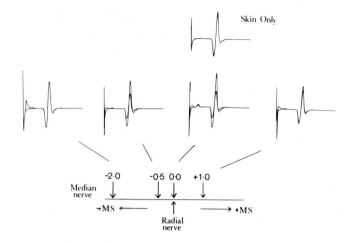

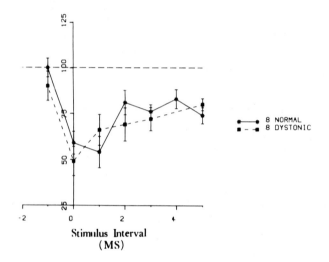

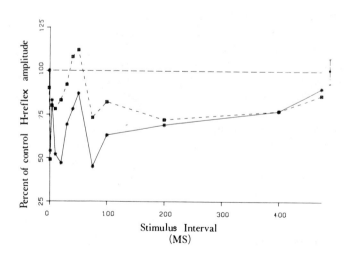

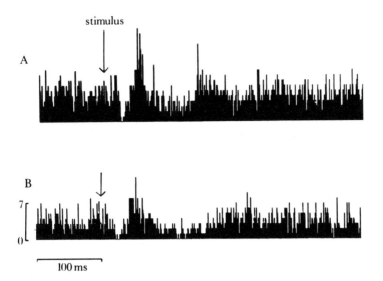

FIG. 8. Peristimulus time histograms recorded from the flexor carpi ulnaris of a normal subject (A) and a patient with focal dystonia of the hand and forearm (B). The bin width was 1 msec. The subjects attempted to maintain a single motor unit in the flexor muscle firing at about 10 Hz while small, sub-motor-threshold electric shocks were given to the radial nerve. The PSTH of the firing probability of the motor unit was constructed from 100 msec before to 400 msec after the shock (arrow). The disynaptic Ia inhibition is evident as a reduction in firing probability some 22 msec after the stimulus. No clear long-duration second phase of inhibition is seen following this; the main event is a large increase in firing probability peaking at about 70 msec after the stimulus.

rapid self-paced ("ballistic") movements. Although the general trend of our results indicated that more severely affected dystonic patients showed larger agonist bursts and a lack of reciprocal activation, we frequently saw patients in whom normal patterns of activity occurred in muscles that were normally involved in dystonic spasms. Thus, changes in the bi- or triphasic EMG pattern in "ballistic" movements cannot be regarded as a diagnostic feature of dystonia.

The major result of our physiological investigations was the finding that the time course of reciprocal inhibition from forearm extensors to forearm flexor muscles was abnormal in patients with dystonia when evaluated using the H reflex technique. In nor-

FIG. 7. **Top:** Summary of the timing convention used. The occurrence of the median nerve test shock was expressed relative to that of the radial nerve conditioning shock (*time line* at bottom of figure). Above the time line are representative average control and test H reflexes from the forearm flexors of a normal subject, superimposed at four different time intervals. Maximum inhibition is seen at −0.5 and 0.0 msec. If the conditioning radial nerve shock is applied to the surrounding skin alone, then no effect is seen on the flexor H reflex. **Middle:** Time course of the first phase of flexor H reflex inhibition in eight normal (*continuous line*) and eight dystonic (*dotted line*) subjects. Average values of the size of the test H reflex, expressed as percent of control response, are plotted ±1 SE for time intervals from −1 to 5 msec. The curves from the two groups are the same. **Bottom:** Plot of the total time course of flexor H reflex inhibition from −1 to 500 msec. The first phase plotted in B is hardly visible on this time scale. Normal individuals show two further phases of inhibition separated by a return towards control levels at 50 msec. In the group of dystonic subjects, the second phase of inhibition is much reduced. For the sake of clarity, standard errors have not been plotted on all points, but a representative error bar is shown on the right hand edge of the figure.

mal individuals, this inhibition is a complex affair, having three phases which may last for up to 1 sec, and is very similar to the reciprocal inhibition seen in the muscles of the calf (Mizuno et al., 1971). We have argued before that the first phase of this inhibition is mediated by the disynaptic Ia inhibitory pathway (Day et al., 1981). This phase was normal in our dystonic patients whether evaluated using the H-reflex technique or by the PSTH. The main change in dystonia was a reduction in the amount of inhibition seen in the second phase of the H-reflex curve from 10 to 60 msec. In normal subjects, this phase of inhibition was as strong as the disynaptic phase but lasted considerably longer and therefore presumably had a much stronger effect on spinal cord excitability.

We did not observe this phase of inhibition in the PSTH of normal or dystonic subjects. Even taking into account the fact that the PSTH may not represent the true time course of postsynaptic events in the MN but may be the differential of the summed IPSPs and EPSPs at the neuronal membrane (Knox, 1974; Knox and Poppele, 1977), we saw little evidence of this second phase of inhibition in either normal or dystonic patients. We suggest, therefore, that it may be produced by presynaptic inhibition of Ia afferent fibers from the flexor muscle (see Eccles et al., 1962). Hence, it cannot be detected by the PSTH technique, which relies on direct corticospinal tract activation of the MN pool. Reduction in the amount of this inhibition in dystonia could be, at least in part, responsible for the muscular cocontraction during voluntary movement. However, whether this is produced by changes in the activation of spinal interneurons by extensor muscle afferents or by changes in the central biasing of the presynaptic terminals, we cannot say. In fact, it is interesting to note that diazepam, the drug most widely accepted as having some therapeutic action in dystonia (see Marsden, 1980), is believed to potentiate presynaptic inhibition in the spinal cord (Delwaide, 1973; P. J. Delwaide, J. Schoenen, and L. Burton, *this volume*). Alternatively, the reduction in inhibition seen in dystonic patients may be produced by increased facilitation from 10 to 60 msec after the stimulus, although in this case it is not clear which pathways would be involved.

Two other physiological defects have been noted in patients with dystonia: an enhanced shortening reaction (Westphal phenomenon) and an increase in the H reflex recovery rate following a conditioning H reflex in the calf. We do not know what relationship these findings have to the defect in reciprocal inhibition; however, it is evident that an abnormality in peripheral reflex mechanisms cannot be the whole cause of dystonic symptoms in these patients. The most striking feature of all the individuals we studied was that not all movements were equally affected. Many of the most severely incapacitated patients could make rapid flexion/extension movements of the elbow or wrist with normal activation of agonist and antagonist muscles. Likewise, many individuals with focal or segmental dystonia could produce a relatively normal "ballistic" pattern of muscle activity even in muscles usually affected by dystonic spasms; some patients, despite showing changes in their reciprocal inhibition, had normal alternating agonist/antagonist muscle activation and even showed initial silencing of the active antagonist before a foot movement (Fig. 4B).

Thus, the abnormality producing dystonic spasms in these patients cannot lie solely in the common motor outflow pathways, since some categories of movement are unaffected. A fault must also lie in the production of the motor command as well. In the extreme case, dystonia may only be present during one type of movement, as, for example, in dystonic writers' cramp, and clinically, it is quite common to en-

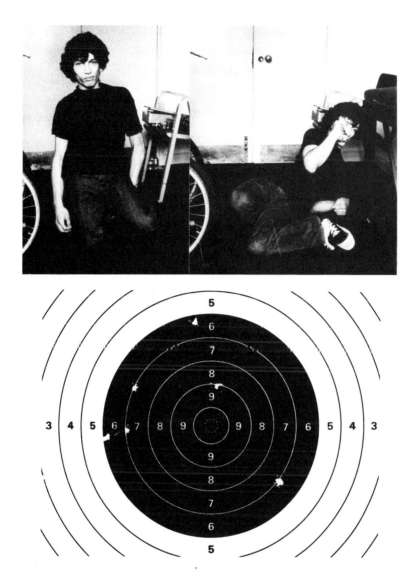

FIG. 9. Top: Photographs of a young man with generalized dystonia. **Bottom:** His performance on the local pistol range using his right hand.

counter patients who are unable to walk forwards, but can run, climb stairs, or walk backwards relatively easily. Indeed, one young man whom we studied could not stand or walk unassisted, yet could drive his own car and shoot with remarkable accuracy on the local pistol range (Fig. 9).

Thus, we have no complete pathophysiological explanation of all categories of dystonic muscle activity. Nevertheless, it is encouraging to have found one more physiological abnormality in a group of patients with a disease once thought to be of hysterical origin.

Clinical Neurophysiology of Muscle Jerks: Myoclonus, Chorea, and Tics

*C. D. Marsden, J. A. Obeso, and J. C. Rothwell

University Department of Neurology, Institute of Psychiatry, and King's College Hospital Medical School, London SE5 8AF, United Kingdom

Sudden jerks of the limbs or whole body occur in a variety of neurological conditions characterized by movement disorders such as myoclonus, chorea, and tics. In this chapter, we describe and compare the physiological characteristics of these abnormal movements.

MYOCLONUS

The pathophysiology of myoclonus has been the subject of several reviews in recent years (Halliday, 1967, 1975; Marsden, 1980; Marsden et al., 1982), each of which has used a different terminology to classify the various types of myoclonic jerking.

Halliday (1967) divided myoclonus into three major categories: pyramidal, extrapyramidal, and segmental. This classification, which had strong anatomic–functional implications, was modified later by Halliday (1975, 1980) when he discarded the extrapyramidal type because it seemed to represent fragments of other movement disorders rather than a distinct entity. Nevertheless, it is as well to bear this early classification in mind, for it highlights what is still a basic gap in our knowledge of the pathophysiological mechanisms of myoclonus.

In this chapter, we follow the classification of Marsden et al. (1982) (see Table 1), which also classifies myoclonus into three main pathological types: cortical myoclonus, which, like Halliday's pyramidal myoclonus, results from abnormal activity of a localized cortical focus, probably in the motor cortex; spinal myoclonus (Halliday's segmental myoclonus), arising from overactivity of spinal cord mechanisms; and subcortical myoclonus, under which may be grouped all those forms of myoclonic jerking whose physiological origin lies between the cerebral cortex and the spinal cord. Animal studies have pointed to the reticular formation (Zuckerman and Glaser, 1972) as a likely site for the origin of some of these subcortical jerks, and this has been confirmed physiologically in a few human patients. Nevertheless, the extent and details of subcortical myoclonus still remain in a twilight zone of our pathophysiological knowledge.

Cortical myoclonus has been subdivided into epilepsia partialis continua, spontaneous cortical myoclonus, and cortical reflex myoclonus, three categories which lie along a continuum describing the sensitivity of a motor cortical focus to discharge. In epilepsia partialis continua, repetitive spontaneous bursting of a small

---

*To whom correspondence should be addressed: University Department of Neurology, Institute of Psychiatry, Denmark Hill, London SE5 8AF, United Kingdom.

TABLE 1. *Classification of myoclonus*

Cortical myoclonus
  Epilepsia partialis continua
  Spontaneous cortical myoclonus
  Cortical reflex myoclonus

Subcortical myoclonus
  Spontaneous reticular myoclonus
  Reticular reflex myoclonus
  Ballistic movement overflow myoclonus
  Oscillatory myoclonus
  Cortico–subcortical myoclonus

Spinal myoclonus

From Marsden et al. (1982), with permission.

group of (motor) cortical cells produces repeated, rhythmic firing of pyramidal cells and regular, repetitive myoclonic jerking. In spontaneous cortical myoclonus, the focus in the motor cortex is not so excitable and discharges intermittently but spontaneously to produce occasional jerking which is neither repetitive nor rhythmic. In cortical reflex myoclonus, the focal group of cells may only be provoked to discharge on receipt of a volley of afferent input from the periphery (see Marsden, 1980, for a discussion of the relationship between epilepsy and cortical myoclonus). Cortico–subcortical myoclonus has also been added to describe those forms of myoclonus that may take origin from a subcortical focus but produce their effects via the motor cortex and pyramidal tract or vice versa.

Although this classification is useful as a working hypothesis on the origin of myoclonus, it must be recognized that investigation of new patients not uncommonly reveals unsuspected physiological features that are difficult to explain in the light of our present knowledge. Indeed, one of the most puzzling features encountered in the study of myoclonic jerking is the enormously wide range of pathologies that may result in a remarkably similar clinical picture of muscle jerking.

This chapter presents our approach to the physiological characterization of myoclonus in man. This is based on four main sources of information: clinical findings, EMG recordings, EEG–EMG correlation, and somatosensory evoked potentials (SEPs). Careful analysis of information from all of these sources provides details on which the classification outlined above is based.

Clinical Data

Myoclonus may be focal, multifocal, or generalized. The first term refers to a jerk restricted and limited to one part of the body; multifocal jerks occur asynchronously in different muscle groups, whereas generalized myoclonus produces the well-known massive synchronous jerks involving trunk and limb muscles. Myoclonus may arise spontaneously in the central nervous system (spontaneous myoclonus) or may depend on sensory afferent input (reflex myoclonus) or voluntary muscle activation (action myoclonus) to trigger the abnormal contractions.

In general, cortical myoclonus is focal and affects distal muscles, especially those of the arm. When it is stimulus sensitive ("cortical reflex myoclonus"), jerks are evoked only in the limb that has been stimulated, reflecting the somatotopic organization of afferent input and motor output in the motor cortex (Hallett et al., 1979). Subcortical myoclonus usually is generalized and may affect both proximal and distal muscles. It may be spontaneous, reflex, as in "reticular reflex myoclonus" (Hallett et al., 1977b), or triggered by certain movements, as in "ballistic overflow myoclonus" (Hallet et al., 1977a).

EMG Correlates of Myoclonus

Simple surface EMG recording of several pairs of antagonist muscles provides valuable information distinguishing the different types of myoclonus. The main features to consider are the length of each EMG burst, the pattern of contraction of antagonist muscles, and the order of muscle activation.

Duration.

Most of the commonly occurring types of myoclonus have muscle bursts lasting between 10 and 50 msec (Fig. 1). A few more unusual categories, such as ballistic overflow myoclonus or postanoxic oscillatory myoclonus, present jerks lasting between 50 and 120 msec (Hallett et al., 1977a) (see Fig. 2). Long-lasting jerks (> 120 msec) have been described in spinal myoclonus and a few varieties of subcortical myoclonus.

Innervation pattern.

In many patients with cortical myoclonus, only one major group of synergistic muscles may be activated (see Fig. 1). Most other forms of myoclonus show a simultaneous, cocontracting activation of agonist and antagonist muscles. To date, there are only two exceptions in which alternating activity of antagonist muscles has been recorded: (1) in ballistic overflow myoclonus (Hallett et al., 1977a), where jerks were triggered by fast (ballistic) movements of a limb, and the EMG showed the typical alternating activity that characterized a normal ballistic movement in man, and (2) transient, sudden onset jerks, with alternating EMG activity resembling a burst of tremor, which were the main factors distinguishing postanoxic oscillatory myoclonus (Fig. 2).

Activation order.

In a few patients, it is possible to distinguish accurately the timing of contraction of different muscles. This has been particularly well documented in patients with cortical and reticular reflex myoclonus (Hallett et al., 1977b, 1979; Leigh et al., 1980). In cortical myoclonus, the activation

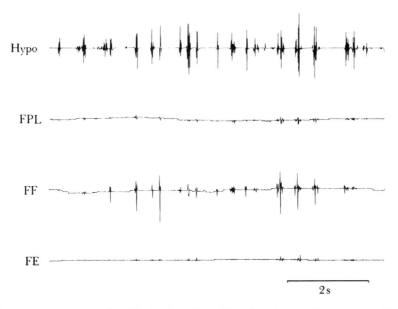

FIG. 1. Spontaneous muscle jerking in the relaxed hand and arm of a patient with spontaneous cortical myoclonus. The jerks consist of short-duration bursts of EMG activity (50–100 msec) and occur synchronously, at irregular intervals, in the forearm flexor (FF) and hypothenar (Hypo) muscles. The flexor pollicis longus (FPL) and forearm extensor muscles (FE) are less affected. (Patient J.B.)

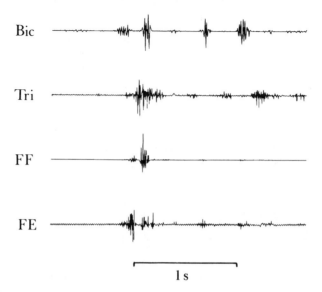

FIG. 2. The EMG pattern of postanoxic oscillatory myoclonus. Each clinically observed jerk actually consists of fairly long-duration bursts of alternating EMG activity in the biceps (Bic) and triceps (Tri), and forearm flexor (FF) and extensor (FE) muscles, lasting for up to 2 sec. No abnormal muscle activity is seen before or after the jerk. In this record, the patient was about to raise her arms to shoulder level when the jerks began. (Patient D.M.)

order follows a rostrocaudal direction (Fig. 3), whereas in the reticular type, the excitatory volley propagates up the brainstem and down the cord (Fig. 4).

EMG–EEG Correlation

Spontaneous EEG spikes preceding EMG discharges can be observed in a number of patients with myoclonic epilepsy (Grinker et al., 1938; Van Bogaert et al., 1950; Kugelberg and Widen, 1954). In the majority of cases, however, that relationship is not apparent by simple visual inspection of conventional EEG records. Recently, using small computers, it has become possible to collect and average EEG data that precede an input EMG trigger point, so that EEG activity occurring before a jerk can be collected. Random brain activity is canceled out using the averaging technique, and only that cortical activity that is time-locked to the muscle jerk is seen on the records. In this way, it is possible to demonstrate EEG activity time-locked to the jerks which may have not been suspected from the routine EEG–EMG studies (Shibasaki et al., 1975, 1978, 1981; Hallett et al., 1979).

Patients with cortical myoclonus usually show a large, well-defined brain wave preceding the jerks by 20 to 25 msec for the upper limbs (Fig. 5) and by 30 to 35 msec for the lower limbs. Bipolar recording of these spontaneous time-locked EEG potentials allows precise definition of the brain area responsible for the abnormal discharge (Fig. 6). This focus almost invariably is located in the motor cortex, corresponding to the C3 or C4 electrode positions.

It is important to note that even using these techniques there are a number of patients in whom no preceding wave may be evident using scalp recording but in whom depth recording from the cortex may show large focal spikes (see, for example, Thomas et al., 1977; Wieser et al., 1978). This is presumably because the orientation

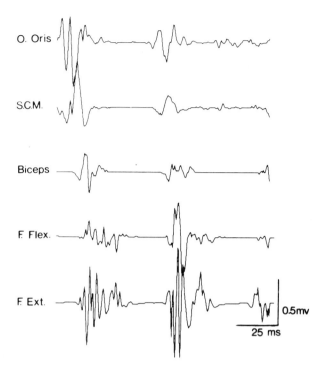

FIG. 3. Order of muscle activation in the jerks of a patient with cortical reflex myoclonus. The two jerks illustrated were induced by voluntary tonic activation of facial and right arm muscles. The cranial nerve nuclei are activated in a descending order. O.Oris, orbicularis oris; S.C.M., sternocleidomastoid. (From Hallett et al., 1979, with permission.)

of the dipole in the focus is so anteroposterior (e.g., in the bank of the central sulcus itself) that it is not detected on the surface of the scalp.

Cortical activity also may be seen in the EEG record in patients with some forms of subcortical myoclonus, and the presence of these waves may lead to erroneous characterization of the jerks. However, with the back-averaging technique, it may be seen that this activity is not time-locked to the muscle jerking, and in many cases, it may actually follow the jerk rather than precede it (Hallett et al., 1977b).

A bilateral brain wave may precede the myoclonus in a few patients (Fig. 7), sometimes by intervals as long as 50 to 100 msec (Dawson, 1947; Watson and Denny Brown, 1955; Shibasaki et al., 1981). This is obviously much slower than would be expected from a direct corticospinal activation of the muscles (see C. D. Marsden, P. A. Merton, and H. B. Morton, *this volume*), and it is therefore not clear whether this type of myoclonus is indeed arising in the cortex and being conducted very slowly down some extrapyramidal pathway or whether some form of cortico–subcortical interaction is taking place.

Another common source of error arises when multifocal jerks are recorded. If there are muscles that contract regularly some milliseconds before the muscle used to trigger the recording sweep, it is quite possible to pick up EEG activity preceding the analyzed jerk that is simply afferent activity reaching the cortex from the contraction of these distant muscle groups. The only way to overcome this difficulty is to take care to average muscle activity from all those mus-

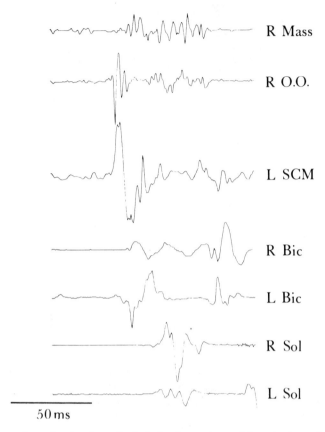

FIG. 4. Order of muscle activation in a single jerk of a patient with reticular reflex myoclonus. The jerk illustrated was precipitated by a light tap to the right shoulder while the patient lay supine on a couch. The cranial nerve nuclei are activated in ascending order. Mass, masseter; O.O., orbicularis oculi; S.C.M., sternocleidomastoid; Bic, biceps brachii, Sol, soleus. (Patient A.M.)

cles that are jerking in order to detect which, if any, show regular activation preceding the brain wave.

Somatosensory Evoked Potentials

Somatosensory EPs to electrical nerve stimulation (usually of the median nerve at the fingers or wrist) are commonly employed in the evaluation of patients with myoclonus. Giant SEPs restricted to the hemisphere contralateral to the jerking limb may be recorded in cortical myoclonus. In patients with cortical reflex myoclonus, the peripheral stimulus often evokes a large SEP (P-26/N-35) which usually is followed by a jerk some 20 to 25 msec later in the same arm (Fig. 8). Halliday (1967, 1975, 1980) has pointed out that in several instances that patients with focal jerks may only have enhanced SEPs while actually jerking. Although changes in the excitability level of the cortical neurons may be responsible in part for this fact, we have observed that SEPs may be larger when the stimulus is given randomly (Fig. 9A) than when the stimulus is intercalated between the jerks (Fig. 9B). Presumably this is because of artifactual summation of the spontaneous and electrically evoked potentials during the jerking.

The SEPs are normal in subcortical and spinal myoclonus. The origin of the SEPs and spontaneous EEG activity time-locked

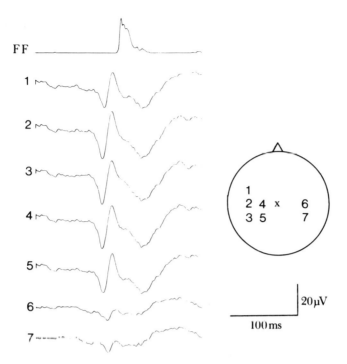

FIG. 5. Brain activity preceding spontaneous muscle jerks recorded with scalp electrodes in a patient with spontaneous cortical myoclonus. The EEG records are monopolar, referred to a linked mastoid reference. Electrode 2 is at the C3 position. The sweeps were triggered by the jerks in the right forearm flexor (FF) muscles and are the average of 64 trials each. The EMG trace was rectified prior to averaging. Note the absence of any ipsilateral brain activity. (Patient J.B.)

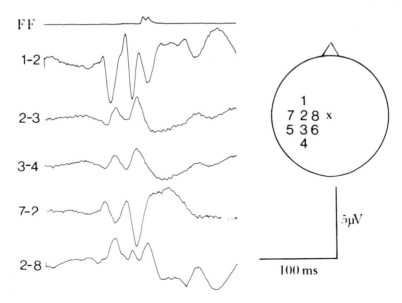

FIG. 6. Bipolar EEG montage from the patient illustrated in Fig. 5. Sixty-four spontaneous jerks in the right forearm flexors were back-averaged as before. Polarity reversal of the preceding brain waves is seen about electrode 2.

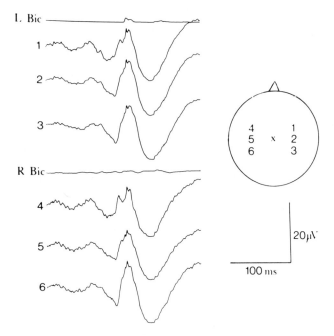

FIG. 7. A bilateral brain wave preceding action-triggered jerks of the right biceps brachii (Bic) in a patient with subcortical myoclonus. The patient was attempting to hold his arm out horizontally in front of his shoulder. The sweeps were back-averaged from the jerks occurring on the right side. Jerks also occurred in the left biceps muscle, but these were not time-locked to those occurring on the right side and therefore could not have been responsible for the bilateral wave. (Patient L.O.)

to the jerks in cases of cortical myoclonus seems similar from bipolar records. Indeed, the primary cortical event (N20) to somatosensory stimulation is usually of normal amplitude (see also Young and Shahani, 1979), and it is only the later P26/N35 that is increased in size. It appears in these cases that the initial afferent volley reaches the cortex as usual and only later causes discharge of an abnormally active focus in the motor cortex.

CHOREA

A physiological evaluation of chorea was first accomplished by Wilson (1928) who, by means of mechanical recording techniques, beautifully demonstrated that all different combinations of agonist/antagonist patterns of contraction can be seen in chorea. A few years later, Hoefer and Putman (1940) and Herz (1944) confirmed Wilson's observations by EMG recording. Choreic movements were characterized both clinically and electromyographically by an unpredictable pattern of muscle activity. Recently, Hallett and Kaufman (1981) studied a patient with Sydenham's chorea and noted an abnormal EMG pattern during ballistic movements.

The following discussion on the pathophysiology of chorea is based mainly on our own extensive studies with five patients with typical Huntington's disease and one case of benign hereditary chorea.

EMG Pattern

The irregular, flitting character of choreic movements is seen during EMG recording as a continuous change in the activation order of each muscle and of the length of each EMG burst. It is therefore impossible to define a single pattern of EMG activity in

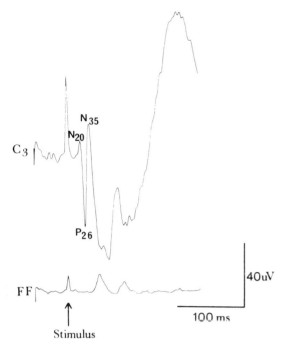

FIG. 8. Enlarged somatosensory evoked potentials recorded from the C3 electrode (referred to linked mastoids) after electrical stimulation of the digital nerves of the right hand in a patient with cortical reflex myoclonus. The P26/N35 component is particularly large, whereas the N20 is within the normal range. Note the double reflex response seen in the rectified EMG records from the forearm flexor (FF) muscles, starting some 45 msec after the stimulus. This is seen clinically as a small, brief jerk of the fingers. The patient was relaxed and seated comfortably. (Patient E.A.)

chorea. Indeed, many of the abrupt and brief bursts of muscle activity in chorea are reminiscent of myoclonus (Fig. 10 top), whereas the long-lasting contractions (200–300 msec) that affect mainly the more proximal muscles, are in many respects similar to the muscle spasms of torsion dystonia (J. C. Rothwell, J. A. Obeso, B. L. Day, and C. D. Marsden, *this volume*) (Fig. 10 bottom). Even repetitive EMG bursts occurring for a few seconds in a single muscle can be present (Hoefer and Putnam, 1940). However, none of these patterns follows an organized sequence. It is the continuous change from one muscle to another and from one type of movement within an individual muscle to another that really characterizes chorea.

The Hung-Up Tendon Jerk

In Huntington's and Sydenham's chorea, tapping the knee elicits a late muscle contraction following the classical spinal stretch reflex known as the "hung-up tendon jerk." Clinically, this response resembles a spontaneous choreic movement; hence its great pathophysiological interest. The EMG recording reveals a recruiting pattern of muscle activity starting about 80 msec after the tap and reaching its maximum amplitude some 200 msec later (Fig. 11). The long latency of onset of this response suggests the involvement of a long-loop pathway, but there is not yet any positive indication as to the actual mechanism.

Voluntary Movements

The results of this section are limited to one patient with Huntington's chorea who presented florid choreic movements with little mental deterioration. He was thus quite capable of understanding the nature of the different motor tasks we asked him to perform. Rapid alternating movements of the elbow and wrist were executed without any difficulty and showed normal reciprocal muscle activation (Fig. 12). Fast movements (ballistic) of the elbow to an end position also showed a normal pattern of activation of the antagonist (Fig. 13), but the duration of the agonist and antagonist bursts was abnormally long, confirming the recent data of Hallett and Kaufman (1981) in their case of Sydenham's chorea.

EEG Correlates

We have found it impossible to attempt to record any EEG activity time-locked to the involuntary movements of chorea. As ex-

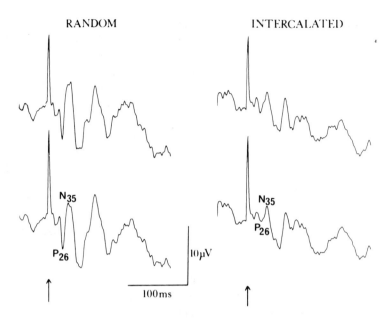

FIG. 9. The influence of stimulus timing on the size of the evoked cortical potential in a patient with regular spontaneous jerking of the right flexor carpi ulnaris. On the **left**, electrical stimuli *(arrow)* were given to the digital nerves of the right hand at random intervals between 1 and 2 sec. On the **right**, the stimuli were given 50 msec after the end of each jerk. In the latter case, the size of the P26/N35 component was much reduced in size. Electrical stimulation had no effect on the muscle jerking. Each record in the average of 128 trials. The electrodes were positioned over C3 **(top)** and 2 cm anterior **(bottom)**. (Patient M.W.)

plained above, the irregular and asynchronous pattern of activity, which continually moves from muscle to muscle, makes it difficult to be certain that any brain activity that we may see using the back-averaging technique is not caused by afferent activity provoked by distant, unrecorded muscle jerks. In fact, we could not even record any EEG activity time-locked to the large "hung-up" tendon jerk in the patient illustrated in Fig. 11.

TICS

In contrast to the attempted pathophysiological classification of myoclonus, the classification of tics is purely clinical and is based on age of onset, symptomatology, and evolution. For practical reasons, we have concentrated our attention on the tics of Gilles de la Tourette's syndrome. This is a condition with onset in early life (2–15 years) in which multiple muscle and verbal tics persist chronically into adult life. Tics appear mostly in the upper half of the body and characteristically affect the face, neck, and shoulder. The cause of Gilles de la Tourette's syndrom is unknown, and there is still argument as to whether it should be considered primarily as a psychological or neurological disorder.

The tics of Gilles de la Tourette's syndrome may be divided clinically in two categories: simple tics, which are brief muscle jerks occurring irregularly and apparently at random in different muscle groups and which the patient is usually able to suppress by an effort of will for a short period of time, and complex tics, which take the form

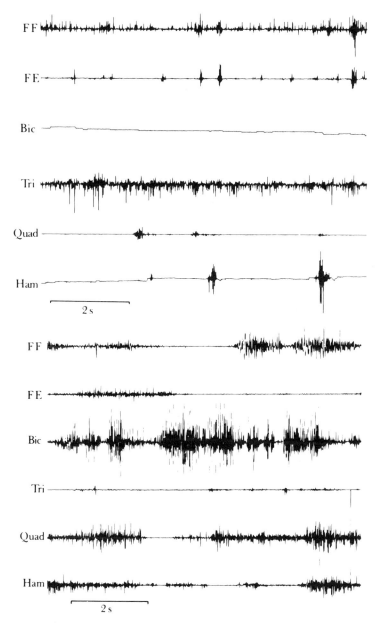

FIG. 10. Examples of EMG pattern seen in chorea. Top: Brief bursts of activity resembling myoclonus, seen in the forearm extensors (FE), quadriceps (Quad), and hamstring (Ham) muscles. This activity occurred spontaneously while the patient was apparently at rest. B: Long-lasting contractions similar to those of dystonia occurring spontaneously in a different patient.

of mannerisms more than sudden muscle jerks, such as wiping, twiddling the ear, smacking or jumping, accompanied by vocalization.

EMG Correlates of Tics

Simple tics consist of short bursts of muscle activity lasting up to 200 msec but most

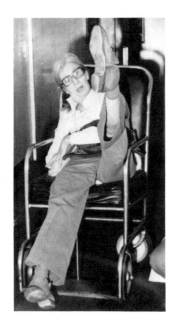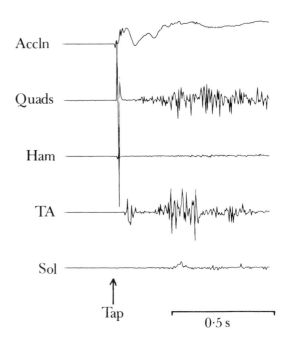

FIG. 11. A single "hung-up" tendon jerk in a patient with Huntington's chorea. On the **right** are the EMG records taken from a different jerk. The top trace is from an accelerometer attached to the front of the tibia, and the other records are from quadriceps (Quads), hamstrings (Ham), tibialis anterior (TA), and soleus (Sol). A slow build-up of muscle activity begins in Quads and TA some 100 msec after the initial tendon tap *(arrow)* and the usual monosynaptic response. It is this that raises the leg into the remarkable posture shown in the picture on the **left**.

commonly less than 100 msec, involving agonist and antagonist muscles (Fig. 14). Simple tics can be recorded asynchronously in different muscle groups.

Complex tics may consist of many different EMG patterns. Burst length may be considerably prolonged, and cocontraction may be present (Fig. 15B,C). Normal reciprocal activation of antagonist muscles may also be seen during some fast complex tics (Fig. 15A), which gives the appearance of a normal ballistic movement. The frequency of tics recorded electromyographically is quite irregular, and any one individual may exhibit a number of different combinations over a short period of time. Indeed, simple and complex tics can occur simultaneously or independently of each other.

EEG–EMG Correlation in Simple Tics

Routine EEG recording has revealed abnormalities in about 45 to 65% of patients

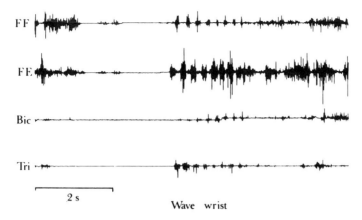

FIG. 12. Rapid alternating movements of the wrist in a subject with Huntington's chorea. Normal alternating activity is seen in the forearm flexor and extensor muscles (FF, FE) and in biceps and triceps (Bic, Tri). Spontaneous choreic movements begin when the patient stops waving.

with Gilles de la Tourette's syndrome (Shapiro et al., 1978). In most cases, these findings were considered to be mild or insignificant, but the incidence of such abnormalities was higher in patients with Gilles de la Tourette's syndrome than in the general population. However, there was no evidence of paroxysmal activity time-locked to the tics in the one case who had simultaneous EMG–EEG monitoring (Sweet et al., 1973).

For the last 2 years, we have studied the EEG activity preceding simple tics using the back-averaging technique in six patients with typical Gilles de la Tourette's syndrome (Obeso et al., 1981). Somatosensory evoked potentials following median nerve stimulation were normal in all these patients, so we extended the analysis time of the EEG activity preceding the tics to 1 sec instead of the more common 250 msec before and after the myoclonic jerks.

It is well known that a slow negative EEG wave can be recorded bilaterally beginning some 500 to 800 msec before a voluntary movement (Kornhuber and Deecke, 1965).

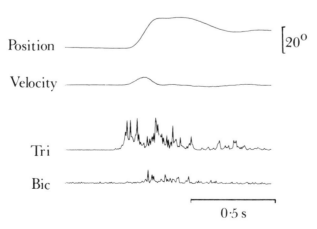

FIG. 13. A single rapid "ballistic" extension movement of the elbow in the same patient as in Fig. 12. The duration of the initial agonist burst in triceps brachii (Tri) is prolonged (150 msec compared to the normal range of 60–110 msec). Nevertheless, the triphasic pattern of activation is still quite clear.

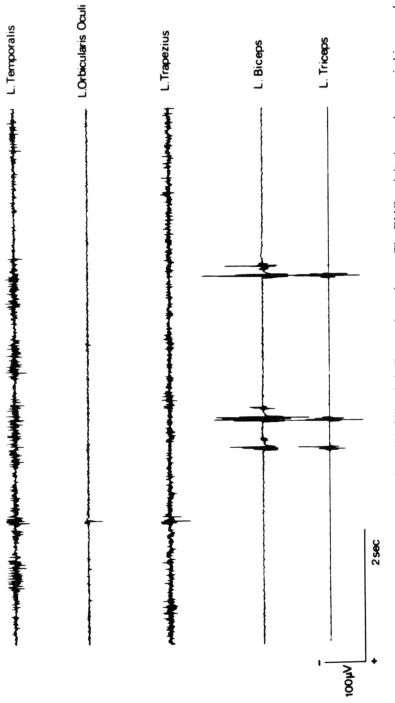

FIG. 14. Simple tics of the left arm in a patient with Gilles de la Tourette's syndrome. The EMG activity is synchronous in biceps and triceps and consists of short bursts of muscle activity lasting about 100 msec. (Patient A.S.)

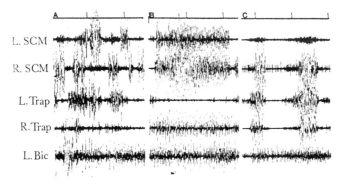

FIG. 15. Examples of complex tics in three different patients with Gilles de la Tourette's syndrome. The muscle activity may be either reciprocally organized (**A**) or not (**B,C**). The duration of the EMG bursts is also quite variable. The *timing bars* on the top are at 1 sec intervals.

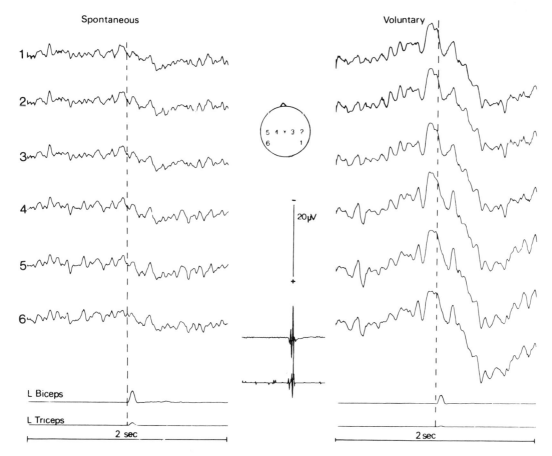

FIG. 16. The EEG activity preceding spontaneous and voluntary jerks in a patient with simple tics of the left arm (same patient as in Fig. 14) (**inset**). The brain activity was recorded monopolarly, referred to linked mastoid electrodes, using amplifiers with a very slow low-frequency time constant (0.16 Hz). Each trace is the average of 100 trials and was triggered by the occurrence of a jerk in the left biceps (see rectified records on two bottom traces). A single jerk is illustrated in the EMG records in the **middle insert**. When the patient mimicked his tics voluntarily, an obvious bilateral premovement potential preceded the EMG burst by some 800 msec. This was completely absent during involuntary tics. The *dotted line* refers to the onset of activity in biceps. (From Obeso et al., 1981, with permission.)

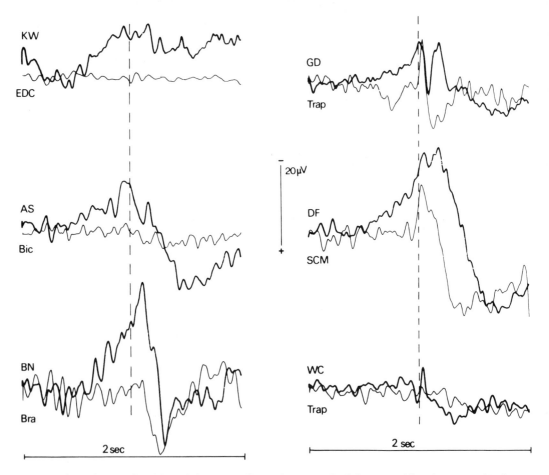

FIG. 17. Superimposed EEG activity preceding voluntary *(thick lines)* and involuntary *(thin lines)* tics in six patients with Gilles de la Tourette's syndrome. The timing of the jerk is indicated by the *vertical dotted line,* and the muscle studied is labeled beneath each trace. Records are the average of between 30 and 128 trials each. EDC, extensor digitorum communis; Bic, biceps; Bra, brachioradialis; Trap, trapezius; SCM, sternocleidomastoid. (From Obeso et al., 1981, with permission.)

Such a *Bereitschaftspotential,* or premovement potential, probably represents the activity of widespread cortical areas preparing for movement. The premovement potential is terminated by a sharper negative wave starting some 70 msec before the movement and restricted to the contralateral motor cortex, called the motor potential, and this is followed by a large positive wave, the postmovement potential, recorded bilaterally. If the tics of Gilles de la Tourette's syndrome are voluntary movements following a psychological conflict or are triggered by unpleasant sensory feelings (Bliss, 1980), a premovement potential should be evident prefacing the tics.

The method and results of this study in our six patients have been explained in detail (Obeso et al., 1981). In a typical experiment, the subject sat comfortably in a chair and was instructed to relax, allowing his tics to take place. The EEG recording from seven scalp electrodes and one EMG electrode from a tic-affected muscle were collected in the computer. Between 100 and 200 individual tics were collected before a final average was made in each patient. The subjects were asked to inform the experimenter about any voluntary movement occurring during this period.

In Fig. 16, a representative record from one of the patients studied is shown. It is evident that no premovement potential occurred in any of the channels in relation to the biceps jerk that characterized his tic. A similar finding was obtained in the other five patients. In the second part of the experiment, each subject was asked to mimic voluntarily the movement that characterized his tic. After some practice, all patients found it very easy to imitate their tics, and the EEG events were studied as before. A premovement potential was obviously present in five of the six cases in these circumstances (Fig. 17), and one could be demonstrated using a cumulative sum (CUSUM) technique in the sixth.

CONCLUSIONS

It is apparent that our knowledge of the origins of sudden muscle jerking is at present rudimentary. In the case of myoclonus, we have begun to build up a sketchy picture of the physiological mechanisms underlying the jerks. However, tics can be classified only by the negative finding that they are not true voluntary movements, and in chorea, our electrophysiological data can do little more than supplement clinical observation. Clearly, a good deal more work is necessary before we can form a clear picture of the pathophysiology of sudden muscle jerks.

Visuomotor Control of Leg Tracking in Patients with Parkinson's Disease or Chorea

*,**Nobuo Yanagisawa, **Sadakazu Fujimoto, and †Reisaku Tanaka

**Department of Medicine (Neurology), Shinshu University School of Medicine, Matsumoto, †Department of Neurobiology, Tokyo Metropolitan Institute for Neurosciences, Tokyo 183, Japan

Reaction time and pursuit tracking by arms have been studied in parkinsonian patients to elucidate motor disturbances and therapeutic effects (Schneider, 1968; Britz et al., 1970; Cassell et al., 1973; Flowers, 1976; Evarts et al., 1981). In the present chapter, we report on visuomotor tracking by the lower limbs which is interesting in view of the clinical disturbances seen in parkinsonian patients ("frozen gait"). We also studied patients with chorea who show rapid irregular involuntary movements which offer a marked contrast to the anomalies in Parkinson's disease. EMG was recorded with surface electrodes bipolarly from the tibialis anterior, the triceps surae as well as from the thigh flexors and extensors.

MATERIALS AND METHODS

Ten patients with Parkinson's disease (37–69 years of age, median 53), five patients with chorea (27–66 years, median 48), and six normal subjects (22–65 years, median 36) were studied (Fig. 1). Each subject sat on a reclining chair, and the lower extremity was fixed at thigh and foot with straps. The axis of rotation of the footplate was adjusted to that of ankle joint, and torque around ankle was recorded in isometric contraction of leg muscles. A cathode ray oscilloscope was set before the subject's eyes at a distance of 1 m, and a horizontal zone 1 cm wide (target, T) was presented on the screen; T could be moved upward or downward by a function generator (Iwatsu FG-330). A horizontal line (control cursor, F) was also displayed which moved in proportion to the torque on footplate.

Upward deflection indicated ankle dorsiflexion, and downward plantar flexion. The subject was asked to keep the control cursor within the target zone. Two modes of target movements were tested: (1) one-way step shift for 5 sec for rapid tracking and (2) one cycle of sine wave at 0.1 Hz for slow tracking. In some sessions, a vocal warning signal (WS) was given which preceded the target movement by an unpredictable interval between 1.5 and 4.5 sec; no WS was given in other sessions.

Records were made on an eight-channel inkwriter with high linearity (Nihon Khoden Recticorder WI-640G) and on a four-channel magnetic tape data recorder (Sony DFR-4515). Twelve trials were made on each task. Records of responses with starts earlier than stimulus (anticipation) and with obvious failure to notice stimulus were discarded. In step mode experiments, reac-

---

*To whom correspondence should be addressed: Department of Medicine (Neurology), Shinshu University School of Medicine, Asahi-Cho, Matsumoto 390, Japan.

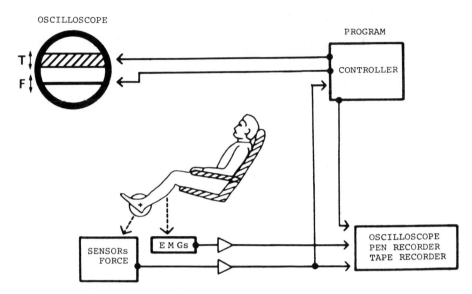

FIG. 1. Block diagram of experimental set-up.

tion time of movement and initial movement time were measured. The reaction time was the duration from target jump to onset of movement, and the initial movement time was that from onset of movement to the time the shifted target zone was reached. In sine-wave mode experiments, number of errors and total duration (time of error) outside of the target zone were measured, and their average values per trial were calculated.

RESULTS AND COMMENTS

Quick Tracking Movement in Parkinsonism

Figure 2 shows records of step mode tracking movements without warning signal (WS) in normal and parkinsonian subjects. Quick and accurate catch-up and hold movements in normal subjects are best indicated in the error trace (T–F, Fig. 2A). Abrupt appearance and silence of agonist EMG activities corresponding to the force trajectory of the step-tracking movement are also clear in the EMG of tibialis anterior (TA).

On the other hand, the reaction of the parkinsonian subject was generally much retarded in EMG and force generation (Fig. 2B).

Reaction time measurements (Fig. 3; Table 1) indicated that the value in leg muscles was comparable to that of arm in both normal and parkinsonian subjects (Cassell et al., 1973; Heilman et al., 1976) and that reaction time was prolonged in parkinsonism. Furthermore, the delay was more marked in cases with more severe disability (Fig. 3). Initial movement time was also larger in parkinsonians than in normal subjects (Table 1). Parkinsonians showed marked variability in both reaction time and initial movement time in every trial, especially in cases with greater impairment. Delay of movement initiation and retarded performance on fast voluntary contraction of leg muscles may be essential expressions of bradykinesia in parkinsonism.

Figure 3 also indicated a marked improvement of reaction time in the presence of WS in most parkinsonian patients. Many patients achieved reaction time values com-

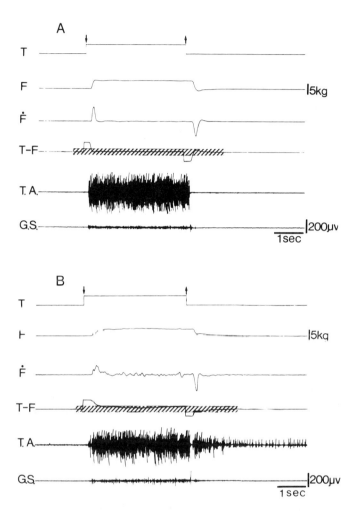

FIG. 2. Specimen records of step tracking movement without a warning signal. A, normal; B, Parkinsonian. T, target; F, torque by contraction of leg muscles; ankle dorsiflexion produces upward shift; Ḟ, differential of F; T–F, T minus F; T.A. and G.S., EMGs from pretibial and triceps surae muscles. *Hatched zone* on T–F is width of target zone.

parable to those of normals. In contrast, the WS effect was less marked in normal subjects. Initial movement time was not reduced by WS, in contrast to the case of reaction time (not shown). Shortening of reaction time by WS is attributed to attention or readiness of response (Teichner, 1954). The significant effect of WS on reaction time may suggest the existence of psychological factors as well as motor control disorders in parkinsonism.

Slow Tracking Movement in Parkinsonism

Figure 4 shows specimen records of slow sine wave mode tracking movements. Although the maneuver was much more difficult than that in step tracking because of the need for control of continuously changing muscle force, tracking was remarkably well performed in the normal subject (Fig. 4A). Parkinsonian subjects often missed the target zone and could not catch up quickly

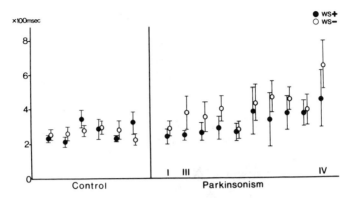

FIG. 3. Reaction time with and without warning signal (WS) for step movements in parkinsonian and normal subjects. Roman numerals for parkinsonian patients indicate severity of each case by stages according to Hoehn and Yahr (1967). I is a patient of slightest impairment, III moderate, and IV severer motor impairment.

(Fig. 4B). Thus, their number of errors missing the target zone was only slightly larger than for normals, but the mean time of error was significantly greater in parkinsonians (Fig. 5; Table 1). This is an expression of the characteristic bradykinesia of Parkinson's disease.

Tracking Movements in Chorea

The main characteristics of chorea were revealed in slow tracking movements rather than step mode tracking. Reaction time and initial movement time were less impaired in step mode tracking in chorea (Table 1).

The specimen record of Fig. 6 indicates frequent misses of the target in smooth pursuit movement. Despite the increased number of errors, time of error was only slightly affected (Table 1). Abrupt, irregular and brief cessations of EMG activities were observed in the phase of increasing contraction, particularly in pretibial muscles. It was not associated with active involuntary contraction in the antagonists. In chorea, the most prominent disorder is an irregular, phasic, and spontaneous movement of extremities and face, as also revealed by phasic EMG activities in muscles (Thiebaut and Isch, 1958; Yanagisawa, et al., 1975, 1976). Disturbance of voluntary movement is also observed in chorea, which has been commonly considered to be caused by incoordination resulting from involuntary contraction of antagonistic muscles. In contrast, the present observation indicates that in chorea, involuntary cessation of active

TABLE 1. *Summary of reaction time, initial movement time, number and time of errors in parkinsonism, chorea and normal subjects*

	Normal (N = 6)	Parkinsonism (N = 10)	Chorea (N = 5)
Reaction time (msec)			
WS (+)	276 ± 54	326 ± 77	277 ± 102
WS (−)	273 ± 28	413 ± 106	332 ± 80
Initial movement time			
[WS (+)] (msec)	187 ± 96	639 ± 460	350 ± 160
Number of errors	1.83 ± 1.27	3.21 ± 0.57	5.27 ± 2.43
Time of error (sec)	0.59 ± 0.42	1.52 ± 0.73	0.71 ± 0.44

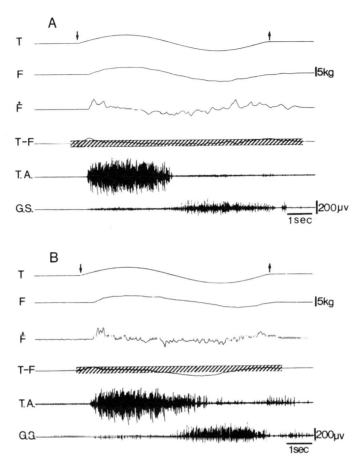

FIG. 4. Examples of records of smooth tracking in normal (**A**) and parkinsonian (**B**) subjects. Recording was similar to Fig. 2.

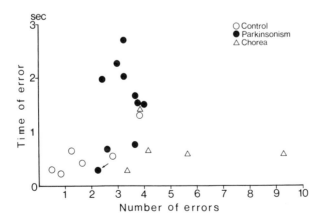

FIG. 5. Number of errors *(abscissa)* and mean time outside of target zone (time of error) *(ordinate)* in smooth tracking in parkinsonism, chorea, and normal subjects. The *arrow* indicates the parkinsonian patient with the slightest impairment (see Fig. 3).

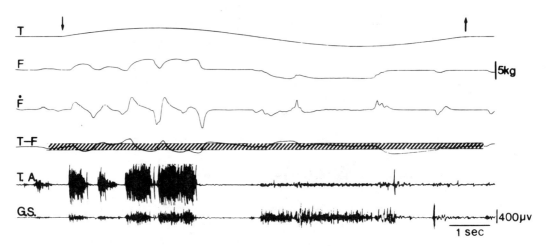

FIG. 6. Smooth tracking by a choreic patient. Recording was similar to Fig. 2.

contraction is also an aspect of this central motor disorder. Similar irregular brief inactivations of muscles without abnormal excitation of antagonists were observed in steady contraction of leg muscles in Huntington's disease, Sydenham's chorea, and hemiballism (Yanagisawa et al., 1975, 1976). Phasic involuntary inactivation, as well as excitation, was improved by adequate therapy, e.g., by corticosteroid administration for the case of Fig. 6 who suffered from systemic lupus erythematosus and chorea.

Motor Control Mechanisms in Health and Disease,
edited by J. E. Desmedt.
Raven Press, New York © 1983.

Slow Visuomotor Tracking in Normal Man and in Patients with Cerebellar Ataxia

**Hirokuni Beppu, **Minami Suda, and *Reisaku Tanaka

**Department of Neurology, Tokyo Metropolitan Neurological Hospital, and *Department of Neurobiology, Tokyo Metropolitan Institute for Neurosciences, Tokyo 183, Japan

Recent studies of cerebellar function in voluntary movements (Brooks, 1979; Flowers, 1976; Hallett et al., 1975b; Mano and Yamamoto, 1980; Meyer-Lohmann et al., 1977; Thach, 1978; Vilis and Hore, 1980) have been primarily concerned with ballistic or rapid movements, thus revealing the importance of the cerebellum in open-loop control of movement. They have shown that the cerebellum plays a critical role in initiation of the movement and in accuracy of performance, which are affected by cerebellar disorders. On the other hand, we are often required in daily life to make slow and accurate movements. Some investigators have stated that slow movements are less affected by cerebellar dysfunctions (Kornhuber 1974; Hallett et al., 1975b; Deecke and Kornhuber, 1977; Brooks, 1979). We therefore decided to study slow voluntary movements in cerebellar disorders using ramp visuomotor tracking in elbow flexion–extension movements.

METHODS

The investigation was performed on nine healthy adult volunteers (ages 23–42 years; six males and three females) and 12 patients with cerebellar disorders (ages 26–65 years; 10 males and two females; 10 spinocerebellar degeneration and two Wallenberg's syndromes). For display (Fig. 1), a TV screen was divided into upper and lower halves, in each of which a vertical strip was displayed. The upper strip (target, T) was moved to the right or left from the screen center by a ramp voltage. The lower strip (control cursor, D) moved in proportion to the angular displacement of the handle (potentiometer coupled to the handle's axis). The handle was adapted for elbow flexion–extension movements in a horizontal plane. The subject's task was to pursue the target by controlling the handle with elbow flexion. The initial angle of the elbow joint was 90°. The range of the target was 30°, and the velocity was any of 7.5°, 15°, and 30°/sec in terms of the handle's angular displacement.

In order to keep the subject's alertness stable, an acoustic warning signal (1 kHz, 0.8 sec) was given an unpredictable period between 1.3 and 3.8 sec before the ramp. The subject was usually informed of the direction and velocity of the target movement in advance. Twelve trials were repeated in each session with direction and velocity of target movement kept constant. The intervals between trials were between 14 and 17 sec. A typical experiment consisted of six sessions, two for each target velocity.

---

*To whom correspondence should be addressed: Department of Neurobiology, Tokyo Metropolitan Institute for Neurosciences, 2-6 Musashidai, Fuchu-City, Tokyo, 183, Japan.

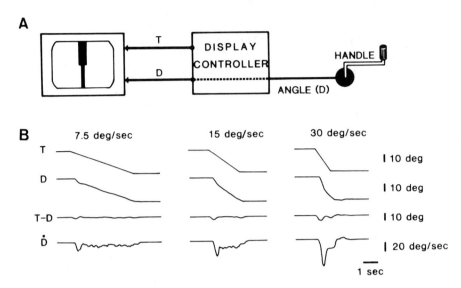

FIG. 1. A: Schematic illustration of the test system for visuomotor elbow tracking movement. **B:** Specimen records of the tracking flexion movement in a normal subject (Y.F., female, 42 years). T, D, T–D, and Ḋ show target trajectory, handle displacement, error, and handle velocity, respectively. The tested target velocity in proportion to the handle displacement is given for each column. The traces were redrawn from the original record.

RESULTS AND COMMENTS

An example of visuomotor elbow tracking movements for a target moving with a constant velocity with a normal subject is shown in the lower part of Fig. 1. Its characteristics are well demonstrated in the traces of error (T–D) and handle velocity (Ḋ) during the slowest tracking trial (left column). At the initial stage, the handle moved quickly soon after the onset of the target movement (reaction time), catching up with target. Later, the trajectory of handle displacement matched that of target very well, the handle's velocity being almost equal to the target's until a few hundred milliseconds prior to target arrest, when it tended to lag behind, anticipating the terminal arrest. The control cursor in the display was kept within the width of the moving target strip during task except at onset (see MATCH trace in Fig. 5, left column). Thus, we classified the whole process into three phases: initial catch-up phase, middle pursuit phase, and terminal phase. The normal subject achieved this pattern after a few trials and subsequently maintained it.

The above pattern remained qualitatively the same when the target velocity increased to 15° and 30°/sec (middle and right columns, respectively). The initial transient increase of handle velocity, later referred to as the initial peak velocity (IPV), augmented its peak size in accord with the target's velocity increase. The handle's velocity during pursuit also increased, but the duration of the pursuit phase was reduced because of shorter movement duration. In many trials at the fastest target velocity (30°/sec), the duration of the pursuit phase was greatly shortened, and the initial catch-up phase was immediately followed by anticipatory deceleration of the terminal phase.

On the other hand, the tracking pattern in patients with cerebellar disorders was quite different. The initiation of the tracking movement (reaction time) was generally delayed and more fluctuating, the initial catch-up reaction was less accurate, and the pursuit phase showed a conspicuous saccadic pattern (see Fig. 3B). The control strip was largely dislocated from the target strip (see Fig. 5, right column). Furthermore, the tracking pattern was not stable and varied among trials within a session.

Initial Catch-Up Phase

Quicker catch-up reactions would result from shorter reaction times and larger IPV sizes. The latter, however, must not be so large as to surpass the preceding target, sacrificing accuracy of the performance. Figure 2A is a scatter diagram in which IPV sizes were plotted against reaction times of corresponding trials for different target velocities. The reaction time ranged between 170 and 300 msec, and trials with faster target velocities tended to show shorter reaction times and smaller variations. In this case, the mean and standard deviation ($m \pm SD$) of reaction time at target velocities of 7.5°, 15°, and 30°/sec were 234 ± 43, 219 ± 27, and 189 ± 19 msec, respectively. Although there were individual differences in reaction time values, this tendency was observed in other normal subjects and was interpreted as showing that detection of slower target movements requires longer time. The amount by which reaction time shortened with increasing target velocity was variable in different subjects.

Figure 2A also implies that the IPV size is a function of the target velocity, increasing significantly at faster target velocities. The IPVs (±SD) at target velocities of 7.5°, 15°, and 30°/sec were 15.3 ± 2.1, 32.6 ± 5.2, and 59.9 ± 5.8°/sec, respectively. Although these IPV sizes happened to be approximately twice the respective target velocities tested, this is not the rule. The absolute sizes as well as the increasing ratios were variable in each subject.

Thus, the target velocity seemed to be a basic parameter for generating adequate sizes of IPVs. This, however, does not necessarily indicate that the subject detects the target velocity itself and adjusts the IPV size to it. The positional signal of the preceding target, or the amount of error between moving target and handle, is also important for an accurate catch-up (Nagaoka and Tanaka, 1980).

Since the present task was a learned vol-

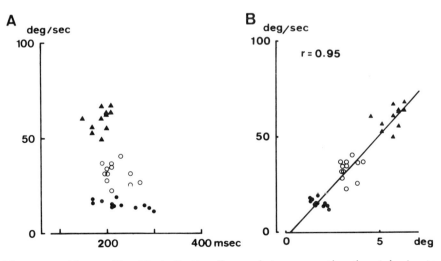

FIG. 2. The same subject as Fig. 1B. **A**: Scatter diagram between reaction time *(abscissa)* and IPV *(ordinate)*. *Filled circles, open circles,* and *filled triangles* show the results from the sessions with target velocities of 7.5°, 15°, and 30°/sec, respectively. **B**: Relationship between the initial error *(abscissa)* and IPV. The regression line and correlation coefficient *(r)* are shown.

untary reaction, and the target movement was predictable, the subject could have foreseen the error distance to be covered and preprogrammed his response just prior to initiation, resulting in a precise catch-up reaction. The reaction time that is characteristic for the subject must be considered for estimation of the error. When IPV amplitudes were plotted against the initial error, the graph showed a linear relationship (Fig. 2B).

The regression line calculated from the plots was steep (slope = 9.7) and converged towards the original point (Y-intercept = -2.4). The correlation coefficient value was very high ($r = 0.95$). Individual differences in slope, Y-intercept, and r were small between normal subjects. They ranged between 4.2 and 10.4, -5.5 and 6.6, and 0.87 and 0.95, respectively. This may be because of the simplicity of the tested task. In other experiments, the three or four target velocities were tested in random sequences so that the subject could not foresee the target velocity of the next trial (M. Nagaoka and R. Tanaka, *unpublished data*). Here again, a linearity between IPV and initial error was revealed. Therefore, this linear relationship represents an adaptiveness of the subject's catch-up reaction to a moving target.

On the other hand, this linearity deteriorated significantly in the cerebellar ataxia patients (Figs. 3, 4). First, the reaction time was greatly delayed and variable with larger initial errors (Fig. 3A; see also Holmes, 1939; Flowers, 1975. The IPV could not attain proper size and was generally smaller than expected for quick catch-up. Thus, its regression line showed a smaller slope, a deviation of Y-intercept from the origin, and a decrease in the correlation coefficient. The slope, Y-intercept, and r in the patients ranged between 0.2 and 6.7, -3.5 and 33.9, and 0.28 and 0.87, respectively. The values of slope and correlation coefficients matched the clinical severity well. Figure 4 shows the recovery time course of a patient suffering from Wallenberg's syndrome. The slope of the regression lines increased, and the correlation coefficients were higher. The patient recovered 1 year later when the function returned to normal range.

Nagaoka and Tanaka (1980) suggested

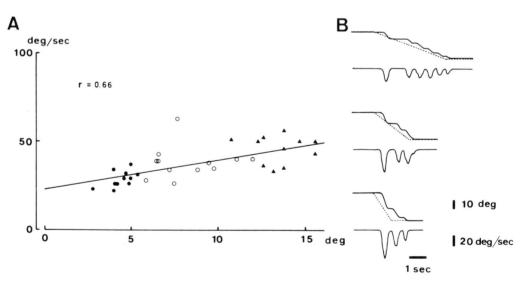

FIG. 3. A patient with spinocerebellar degeneration (K.N., female, 27 years). A: Relationship between the initial error and IPV. Same illustration as Fig. 2B. B: Specimen records of her tracking performance.

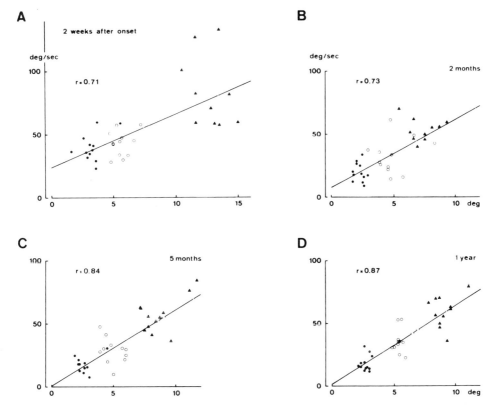

FIG. 4. A patient with Wallenberg's syndrome (T.I., male, 35 years). Recovery of the relationship between the initial error and IPV. The times of examination in relation to the onset of the disease are given in the figures from A to D.

that the IPV is centrally programmed. The major reasons were its brief duration, corresponding EMG burst in agonist muscles (see below), relatively stereotyped shape, and independence from somatic sensory information. The loss of adaptive IPV reaction in cerebellar ataxia supports the above idea.

The present finding that cerebellar patients cannot control limb velocity effectively may be related to a monkey experiment by Mano and Yamamoto (1980). They showed that some Purkinje cells that are tightly related to wrist movement showed discharge patterns of simple spikes conforming very well to that of velocity and suggested that velocity is the major information coded by Purkinje cells in the cerebellar hemisphere.

Pursuit Phase

As shown in Fig. 3B, one characteristic of the pursuit phase in cerebellar ataxia was the saccadic pattern in which the patient alternated catch-up movement and arrest and could not keep the velocity of the handle constant (as normal subjects did to achieve smooth pursuit pattern). In other words, he repeated the catch-up reactions. This pattern was observed in most patients who suffered from various cerebellar disorders, spinocerebellar degeneration (SCD), and Wallenberg's syndrome. The undulation of the velocity traces was not regular, fluctuating between 1 and 3 Hz; it varied its shape in every trial and tended to increase frequency at faster target velocities. These properties are quite different from those of the pursuit phase of patients with lesions in the superior peduncle (Holmes, 1939). The handle movement of these patients showed very regular oscillations which were only slightly affected by change in target velocity, going forwards and backwards instead of forwards and stopping. Accordingly, alternation of EMG activities was observed in the antagonistic muscles.

The inability of the cerebellar patient to keep a constant velocity during the pursuit phase does not necessarily indicate his inability under all conditions. Most patients could make tonic contractions of the tested muscles when resisted. Furthermore, the pursuit pattern tended to become smoother when the visual cue of handle displacement was erased during pursuit. The irregular undulation may be caused by a mechanism involving visual information and its processing for corrective reaction (see Holmes, 1939). Smoother movement by elimination of the visual cue does not necessarily mean that the tracking performance becomes more accurate. The handle trajectory often deviated from that of the target. In contrast, the same maneuver did not affect the performance of normal subjects (Nagaoka and Tanaka, 1981). Thus, the cerebellar system is essential in accurate control of slow arm movement as well as rapid movement.

The EMG Activities

The EMG activities of the agonist elbow flexor (m. biceps brachii and brachialis) and the antagonist extensor muscles (m. triceps brachii) were recorded during the tracking task by implanted wire electrodes. In normal subjects (left column of Fig. 5), the initial catch-up movement was represented by a burst activity in the agonist muscle which was followed by a brief silent period and succeeding tonic activities of the pursuit phase. After movement arrest, the EMG gradually decreased but was maintained to some extent throughout the period during which the handle was held in the flexed position (see integrated EMG in Fig. 5). Reciprocal innervation was maintained throughout the slow target tracking as revealed by the silence of the triceps EMG.

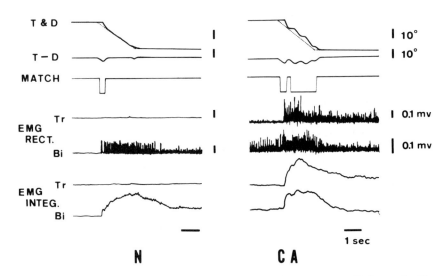

FIG. 5. Comparison of the tracking pattern between a normal (**left column**, H.B., male 42 years) and a cerebellar patient (**right**, T.F., male, 65 years). MATCH: the high level indicates that the control cursor was kept inside the width of the target strip; the low level, out. The EMG activities from the agonist (Bi) and antagonist (Tr) are also shown as rectified (RECT) and integrated (INTEG).

However, during the fastest target movement in which the catch-up movement was much larger and ballistic-like, the silent period just following the initial EMG burst in the agonist was associated with a burst of activity in the antagonist extensors, a pattern that has been reported in some ballistic movements (Wachholder and Altenburger, 1926; Wachholder, 1928; Hallett et al., 1975b; Desmedt, 1981; C. Ghez, D. Vicario, J. H. Martin, and H. Yumiya, *this volume*). The antagonistic EMG was silent at slow target velocities when movement braking was not necessary.

In the cerebellar patients, the EMG patterns were quite variable, reflecting the ever-fluctuating tracking movements in every trial (see Grimm and Nashner, 1978) (Fig. 5, right column). There were often background discharges during the waiting period. The initial catch-up phase did not show the brief EMG burst-and-silence pattern typical of the control subjects but rather a gradual increase which directly transformed to a tonic discharge. In about half of the patients, significant EMG activities were observed simultaneously in antagonist muscles. Such cocontractions were never observed in normals. Antagonistic cocontraction could not be related to clinical severity or to the specific types of cerebellar disorders. Cocontraction of the antagonistic muscles has been reported in patients with cerebellar lesions (Terzuolo and Viviani, 1974) and in cerebellectomized monkeys (Soechting et al., 1976). The reason the antagonistic cocontraction occurs in some of the cerebellar patients and not in others is not known. It could be the result of damage to certain substructures of the cerebellar system or a kind of compensatory reaction by the patient's endeavor to compensate for cerebellar signs, dysmetria, intention tremor, etc. Whatever its basis, the antagonistic cocontraction modified the tracking pattern of the cerebellar patients.

Motor Unit Control in Movement Disorders

*Jack H. Petajan

Neurology Department, Utah University Medical College, Salt Lake City, Utah 84112

In normal subjects, it is possible to reliably characterize single motor unit (SMU) firing characteristics using audiovisual feedback (AFB) of the SMU to establish control (Petajan and Philip, 1969). Subjects can adjust frequency from onset upwards until a second unit is recruited, the recruitment frequency being defined as that frequency existing just prior to recruitment. Following recruitment of a second motor unit, increasing the level of effort generally results in recruitment of additional motor units but little increase in frequency of the primary unit. Thus, the frequency change from onset to recruitment essentially constitutes the primary range of firing for the motor unit under study. Under circumstances of slowly increasing effort, the motor units investigated are the low-threshold generally tonic motor units that initiate muscle contraction.

Another feature of control is the subject's ability to sustain stable firing for a period of 1 min, generally near the onset level. Normal subjects are able to sustain firing for a period of 1 min without lapses (the modal value for lapses in firing is 0). A lapse is defined as an interspike interval over three standard deviations longer than the mean onset interval. Normal subjects can sustain firing for a period of 1 min without recruitment of a second motor unit (Petajan et al.,

1979). For normal subjects to attain the level of SMU control described above, it is essential to provide AFB of the SMU to the subject. Actually, audio feedback is sufficient for such control, but visual feedback is helpful in preventing boredom. Contractile force of the muscle is so small during the activation of SMUs that physiological feedback including tactile, proprioceptive, and visual (observation of the moving part) modes is not adequate for control. Careful observation of finger position using a grid as reference can permit subjects to control single unit firing for approximately 1 min, after which visual attention lags. A rare subject can also do this with proprioceptive feedback alone, but oscillations in effort and recruitment of more than two units in an oscillatory manner commonly occur (Petajan, 1981). In the evaluation of movement disorders, one can determine whether abnormalities of gross movement seen on neurological examination are also detectable in SMU control and whether any abnormality of SMU control is correctable by AFB.

In patients with movement disorders, four levels of SMU control have been investigated: (1) presence of spontaneous motor unit firing; (2) assessment of voluntary control of SMUs using AFB; (3) determination of the response of spontaneously firing motor units and those at rest to muscle stretch (reflex activation); (4) evaluation of SMU control in response to tactile, visual, or proprioceptive feedback and cutaneous electrical stimuli.

---

*Department of Neurology, Utah University Medical College, 50 N. Medical Drive, Salt Lake City, Utah 84112.

METHODS

Potentials from SMUs were recorded by a 26-gauge Teflon®-coated monopolar needle electrode. For Parkinson's disease and Huntington's chorea, attention was focused on the first dorsal interosseous (FDI) and biceps brachii (B). For patients with spastic paraparesis, tibialis anterior (TA), soleus (Sol), and gastrocnemius (G) muscles were investigated. Insertion of the needle electrode nearly parallel to the skin and mechanical support of the wire leads to stabilized electrode position during tests. Appropriately amplified SMU potentials were displayed on an oscilloscope and played back through a loudspeaker so that the subject could use AFB to establish control of firing. Depending on the disorder investigated, the subject was seated comfortably in a chair or was supine facing the electromyograph with limbs resting on a padded support to minimize tactile feedback. For example, the FDI of the dominant hand was investigated with the hand resting on its ulnar aspect. Indifferent and ground electrodes were attached in a manner to eliminate tactile feedback from the hand. All subjects were questioned concerning the presence of discomfort or a sense of tension in the skin so that electrode position could be adjusted to minimize this source of feedback. A gentle abduction of the first finger was sufficient to activate single units.

Subjects were allowed 10 to 15 min of "play time" to establish awareness of level of effort required for activation and control of the SMU. On occasion, tactile feedback was supplied to indicate to the subject the general direction of limb movement required for primary activation of motor units in the muscle. This was a tactic applied frequently in muscles of the lower extremities in patients with spastic paraparesis. All data were recorded on a Hewlett Packard 3955 FM tape recorder for subsequent display. Average frequency of motor unit firing was determined by use of a pulse counter (Atec), but stable 1-sec epochs of firing were measured by hand using a 100 msec/cm sweep speed. Two 5-mm silver disk electrodes were placed on the extensor surface of the forearm and attached to a Grass S48 stimulator. Square-wave pulses 0.05 to 0.1 msec in duration and perceived as a light touch stimulus were delivered according to the method described below.

The following tests were performed. (1) The subject's ability to produce minimal effort just sufficient to activate an SMU and his ability to sustain continuous firing for 1 min were determined. (2) On verbal command of "stop" and "start," subject ability to turn the unit on and off was assessed. Gradually increasing effort required for recruitment of a second unit was applied. Frequency adjustment tests were performed three or more times in order to determine the reliability of performance. Onset and recruitment frequencies were determined for the SMU. The onset frequency is the lowest frequency that still permits sustained firing without lapses greater than 500 msec. For FDI, 500 msec is approximately three standard deviations longer than the mean onset or lower limiting interspike interval. Recruitment frequency is the frequency that exists just prior to the recruitment of a second motor unit. (3) Subjects were instructed to interrupt firing as rapidly as possible in response to a "click" stimulus delivered to the forearm and heard over the loudspeaker. The "click" stimulus was delivered randomly within 5 sec following a "ready" signal delivered verbally by the tester seated behind the subject. The intensity of the stimulus was adjusted so that only a light "touch" was experienced by the subject. The stimulus was also recorded on the tape recorder for measurement at a later time. A second instruction to restart SMU firing as rapidly and smoothly as possible was also given. Subjects were allowed a practice period during which they responded to "click" stimuli. When they felt comfortable with their ability to perform, the test commenced. During the course of several minutes, attempts were made to deliver a minimum of 20 "click" stimuli. (4) Some subjects were supplied tactile feedback during tests of SMU control when activation was not possible with AFB. (5) All subjects underwent a complete neurological examination and were classified with respect to the movement disorder to be studied.

The following classes of patients were evaluated: (1) Parkinson's disease—the well known features of tremor, rigidity and bradykinesia constituted the criteria for diagnosis. Patients were minimally to moderately involved and not on medication (L-DOPA) at the time of the study. (2) Criteria for patients with Huntington's disease (HD) included the presence of chorea, adventitious movements, hypotonia, characteristic postures, and dementia. Offspring of proven Huntington's disease parents were investigated and considered to be at risk. Minor signs of HD recorded in these offspring included frequent adventitious movements, inability to sit still, abnormal postures assumed when sitting, rapid voluntary movements, exaggerated muscle activity during the performance of simple motor tasks, hypotonia, poor ability to concentrate, and early signs of dementia. (3) Patients with spastic paraparesis resulting from multiple sclerosis. Spastic paraparesis was defined as muscular weakness associated with increased resistance to passive flexion and extension of the lower extremity and hyperactive stretch reflexes. (4) Patients with truncal and distal ataxia and severe action tremor resulting from spinocerebellar tract degeneration. These patients had a positive family history of ataxia with onset in childhood with marked progression and disability occurring in adolescence.

RESULTS

Normal Subjects

Motor unit frequency control results were obtained from normal subjects (Petajan and Philip, 1969; Petajan, 1981). Five female and five male subjects in each of three age groups—12 to 20, 20 to 30, 30 to 40 years—were investigated for ability to sustain SMU firing without lapses or recruitment for 1 min. The modal value for lapses greater than 500 msec was 1/min, with no subject exceeding 2/min. Findings were identical for involuntary recruitment of a second motor unit. The majority of subjects sustained firing without lapses or recruitment (Petajan et al., 1979). The inhibitory response to cutaneous stimulation was performed on eight control subjects (Petajan et al., 1979). Subjects fired a mean of 1.7 motor unit potentials (ranged from 0.7 to 2.5) following 20 presentations of the "click" stimulus before motor unit firing could be inhibited. Thirty percent (range 5%–60%) of motor unit reactivations were ballistic, defined as a firing rate greater than 60 Hz and/or recruitment of additional units. The lapse period following the click was six (range 2.8–11.3) times longer than the prestimulus interspike interval.

Parkinson's Disease

Twenty-three patients with Parkinson's disease were investigated prior to the initiation of treatment with L-DOPA. Patients had bilateral and moderate to severe involvement. Motor units in biceps brachii (B) and the first dorsal interosseous (FDI) were investigated. Half of the patients (11) had no tremor at rest but could not produce a minimal contraction to activate an SMU. When activation did occur, a burst of firing was produced after a prolonged latency, and patients were unable to sustain or adjust firing rate (Fig. 1). Only one patient was able to initiate firing and increase effort sufficiently for recruitment of a second unit. This patient was minimally involved in clinical tests. The remaining 12 patients had tremor either continuously or with increased effort. When this occurred, the primary unit fired at tremor frequency. With increased effort, additional units firing at or about tremor frequency were recruited. In

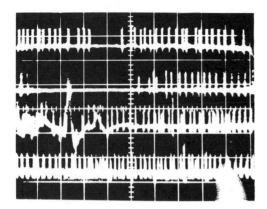

FIG. 1. Patient with Parkinson syndrome attempting to maintain the discharge of a single motor unit (SMU) in the first dorsal interosseous muscle (FDI) with the use of auditory feedback (AFB). The impersistence phenomenon is seen as long pauses and inability to restart firing of SMU rapidly. Horizontal calibration between *grid lines*, 500 msec.

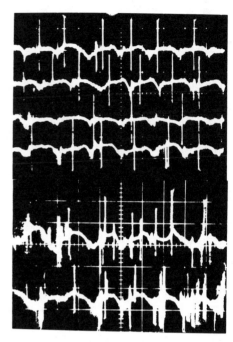

FIG. 2. Patient with Parkinson syndrome. As he increases effort to activate SMU in biceps brachii, more motor units are recruited into tremor bursts (two bottom traces). Horizontal calibration, 100 msec.

some instances, the unit was activated normally but developed tremor as effort was increased. Recruitment of motor units increased in proportion to effort, as occurs normally (Fig. 2). The order of recruitment of motor units within bursts of tremor was found to remain constant. On subsequent replay of recordings of tremor, using sweep speeds fast enough to allow discrimination of individual SMU, the intervals between units within a tremor burst varied considerably. Thus, the motor control of SMU imitated in most respects the deficiencies of gross motor behavior. These included increased latency to activation (bradykinesia) and impersistence of firing. Single units fired at tremor frequency with more units recruited into the tremor burst as effort increased. This was associated with normal awareness of increased level of effort. The low-threshold motor units initiating contraction came under improved control with L-DOPA therapy.

Nine elderly control subjects (66–84 years old) manifested episodic firing of SMUs with lapses in firing sometimes as long as 1 sec. Although inattentiveness could not be excluded, such lapses were not voluntarily preventable. Lapses required insistent reactivation. However, subjects were able to reactivate SMUs with little difficulty, so that lapses were generally of short duration (approximately 500 msec). In this respect, elderly control subjects differed from patients with Parkinson's disease who had great difficulty in SMU reactivation and bradykinesia. The SMU frequency characteristics and control of the elderly control group were otherwise normal.

Huntington's Disease

Thirty-seven patients were classified into three groups representing increasing ability to perform tests of motor control (Table 1) (Petajan et al., 1979):

TABLE 1. *Classification of Huntington's disease patients*

	Total patients	Group 1: poor control	Group 2: fair control	Group 3: good control
At risk for HD	20	7	4	9
Possible clinical HD	4	4		
Likely clinical HD	3	2		1
HD by clinical signs (proved by Varian scan, CT scan, or pneumoencephalogram)	10	9	1	

Group 1 consisted of those patients who were unable to achieve SMU control and had gross chorea. It is possible that they were unable to comprehend the task.

Group 2 comprised patients for whom SMU control was possible but firing could not be sustained. Lapses in firing greater than 500 msec were present in abnormally high numbers. Response to on and off commands was poor; there were delays in response or failure to respond. Small ballistic activations of motor units termed "microchorea" were seen (Fig. 3). Single units or clusters of two or more units firing in very brief high-frequency bursts greater than 60 Hz were seen, which exceeded the occasional gross contractions seen in control subjects.

Group 3 included those patients in whom SMUs were isolated easily. The number of

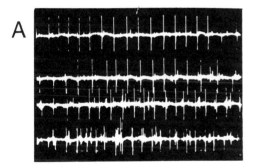
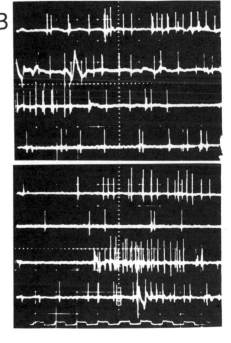

FIG. 3. Effort to sustain firing of SMU in the first dorsal interosseous muscle is shown for a normal subject (**A**) and for a patient at risk to develop Huntington's disease (**B**). No clinical signs of the disease were present. Note numerous lapses and bursts of recruitment, i.e., "microchorea." Horizontal calibration, 200 msec. Vertical calibration steps, 100 μV.

lapses exceeding 500 msec per minute did not differ from normal control subjects. No microchorea was present. Frequency control and response to verbal commands were normal. This group was divided into two subgroups: 3A (normal click response) and 3B (abnormal click response). Subjects in 3A fired a normal number of units following the stimulus; lapse time was normal, and motor unit reactivation was performed smoothly. Subjects in 3B fired an excessive number of SMUs following the "click" stimulus. Lapses were prolonged, and reactivations were generally ballistic (Fig. 4). Motor unit control data with respect to clinical classifications are shown in Table 1.

With one exception, all patients with the clinical diagnosis of HD had group 1 or poor control of SMUs; one subject had fair control. For the 20 subjects at risk without clinical signs of HD, seven had poor and four fair control, and nine manifested good control. For nine of these subjects with an abnormally high number of lapses per minute, a mean value of 16 and a range of 3 to 32 was obtained. Lapses were too numerous to count in two subjects and normal (0–2) in the remaining nine subjects. In three subjects with an excessive number of motor unit potentials following the "click" stimulus, a mean of 3.3 potentials (range 3–4.4) was obtained. When subjects with fair and poor control but not exhibiting clinical signs of HD were combined, a 55% incidence of abnormal control was obtained, which is close to the expected incidence of HD in subjects at risk. When responses to the "click" test were abnormal, other aspects of motor unit control were also abnormal, including an excessive number of lapses per minute or microchorea. The "click" test did not seem to increase the sensitivity of the motor control battery in detecting abnormality. The "click" test could not be performed on subjects with excessive recruitment and microchorea. Onset and recruitment frequencies were increased in HD patients, but the increase was not statistically significant.

In summary, HD patients and those at risk, when able to accomplish SMU control, had difficulty in initiating and sustaining motor unit firing. A special feature of the disorder was excessive recruitment manifested as microchorea and gross chorea. These subjects had considerable difficulty in producing small muscle contractions and frequently complained about this excessive effort.

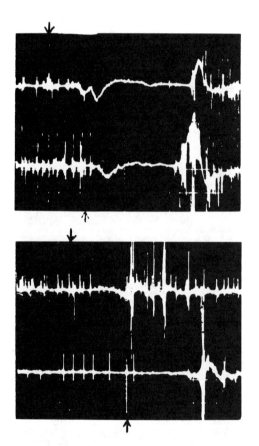

FIG. 4. Pattern of responses to click stimulus in patient at risk to develop Huntington's disease. Delay in interruption of SMU firing to click (arrow), and ballistic-type reactivations of SMU firing in the FDI muscle. This type of response occurs in 55% of patients at risk for HD. Horizontal calibration, 200 msec.

Spastic Paraparesis

Six patients, two males and four females with spastic paraparesis resulting from multiple sclerosis (Petajan, 1979), were studied. Two of the six patients were able to activate SMUs in TA but not in Sol or G muscles. Long latencies to activation, grossly impaired frequency modulation, and impersistence were characteristic of their control. Three patients could not activate units in TA without the assistance of tactile feedback or AFB, but with tactile feedback and AFB, control was ultimately established in this muscle utilizing AFB alone. Without feedback, motor unit firing could not be sustained at any level. In none of the subjects was SMU control possible in Sol or G. Awareness of the level of effort was extremely high during attempts to control SMU firing. In one patient, no voluntary control of SMUs was possible in any muscle despite application of feedback. In response to phasic or slow stretch, no motor units could be activated in tibialis anterior in any of the six subjects. Motor units were most easily activated in Sol during both phasic and slow (5 sec to maximal) stretch during manual dorsiflexion of the foot. The recruitment order of units within bursts of clonus was maintained, but intervals between potentials varied considerably (Fig. 5). With slow stretch, very high levels of tension were required for activation of SMUs. In four of six patients, SMUs firing at slow, highly regular rates were found in the Sol muscles.

Patients with Familial Ataxia

A female aged 44 years and a male aged 15 years experiencing severe familial truncal and limb ataxia diagnosed as Friedreich's ataxia were studied. Both had severe loss of position and vibratory sense. A mild peripheral neuropathy was present in the woman, but response to pinprick was preserved in feet. In this patient, despite severe ataxia of the lower extremities when attempting to move the limbs through space toward a target, the foot could be held steadily in any position of dorsiflexion. The SMU control in TA using AFB was normal. Discontinuation of AFB resulted in an abrupt increase in recruitment to three to five units, at which point she reported her

CLONUS BURSTS-SOLEUS.

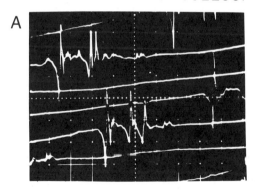
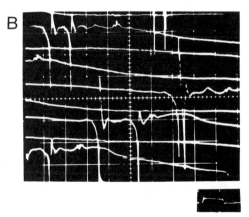

FIG. 5. SMU recording in clonus bursts of soleus muscle. A: Orderly recruitment. B: Varying intervals between motor unit potentials. Horizontal calibration, 10 msec. Vertical calibration step, 200 μV.

level of effort to be equal to that needed for SMU control (Fig. 6). Light touch applied to the dorsum of the foot in the absence of AFB resulted in stable control of one or two motor units. By adjustment of the pressure of tactile stimulation, it was possible to achieve SMU control.

No evidence of peripheral neuropathy was present in the second patient, who had a severe loss of position and vibratory sense. Recording from biceps brachii and triceps muscles during performance of finger-to-nose testing revealed a severe slow, irregular alternating tremor, but utilizing AFB to establish control, the patient was able to adjust the frequency of firing of SMUs. The learning period was somewhat prolonged, but normal SMU control was eventually established. During sustained firing of an SMU, AFB was discontinued, and an abrupt cessation of firing occurred. Restoration of AFB resulted in reactivation of the unit. This phenomenon was uniformly repeatable. Tactile stimulation improved motor unit control, and in some instances it was sufficient to permit control of a single unit for periods of approximately 1 min.

DISCUSSION

The SMU behavior is abnormal in Parkinson's disease, Huntington's disease, and spastic paraparesis. Anomalies of frequency are qualitatively identical to the characteristic gross movement abnormality seen in each condition. The abnormality does not depend on the action of proprioceptive, tactile, or visual feedback systems. In addition, these disorders have in common an increased latency to activation of SMU, impersistence in firing, and impaired frequency control. In simpler terms, in each disorder the gross inability to produce small movements constituted the major "negative" signs of these disorders.

Also, there are distinguishing features of the three disorders at the level of SMU. In Parkinson's disease, a regular tremor and recruitment of motor units into the tremor burst with increased effort are an essential feature. Recruitment order does not vary from one tremor burst to the next. Huntington's disease discloses excessive and accelerated recruitment accompanied by microchorea and gross chorea. Patients were aware that their muscular effort was excessive as they attempted to produce small muscle contractions. In spastic paraparesis, a distinguishing feature is the tonic firing of single units, especially noticeable in the Sol muscle. If voluntary effort was possible, it was seen in TA but not Sol or G. (These muscles can often be activated by means of the Jendrassik maneuver.)

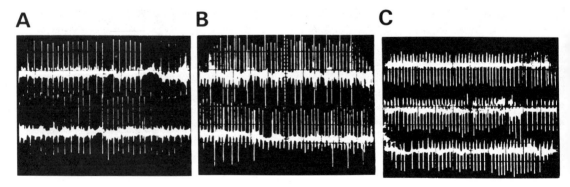

FIG. 6. Patient with familial spinocerebellar ataxia. A: SMU firing in tibialis anterior muscle with audio feedback. B: AFB discontinued: recruitment of SMUs increases, and variations of SMU firing rate are observed. C: Tactile feedback (manual support of leg) was sufficient to restore SMU control. Horizontal calibration, 500 msec.

Awareness of the level of effort was extremely high when the patient was able to activate an SMU. This suggests that the awareness of effort is not associated with peripheral input from actual muscle contraction or limb displacement and that the motor neurons (MN) activated may have high thresholds. When motor units were activated by slow passive stretch in Sol or G, very high levels of stretch were required. Low-threshold small MNs were apparently not accessible to activation in these muscles. In the two patients with familial ataxia, AFB of SMU potentials was successful in correcting motor unit control. These patients had no awareness of their level of muscular effort and were totally dependent on tactile or visual feedback to establish control.

Frequency control can be considered an intensive characteristic of all MNs. Included in this parameter is the initiation of firing as well as the adjustment of MN firing to various levels. Frequency modulation is impaired in Parkinson's and Huntington's diseases and spastic paraparesis and would seem to represent a fundamental property of MN behavior on which more complex behaviors depend. In other words, inability to initiate activity can be considered qualitatively the same as inability to adjust the level of activity. Absent frequency modulation results in a failure of the MNs to adapt to the demands required for movement.

In Parkinson's disease, recruitment occurs in proportion to effort, and the awareness of effort for small contractions is grossly normal, whereas recruitment is excessive in Huntington's disease, and there is awareness of this excessive level of activity. Therefore, the patient with Huntington's disease has a recruitment abnormality superimposed on abnormal frequency control. Spatial recruitment of motor units for performance of a given motor act must depend on a control system different from that mediating frequency control. Such a function would logically depend on a topographic representation for parts or movement in the central nervous system. In patients with spastic paresis, both frequency and recruitment abnormalities exist accompanied by an excessive awareness of effort when very small movements are performed. Inability to select motor units for activation or to adjust their frequency implies involvement of both cortical and striatonigral (basal ganglia initiator) systems in this disorder of motor control. Interactions between basal ganglia and thalamus may play a role in mediating the awareness of level of effort.

Finally, the MN output system in patients with familial ataxia is normal when physiological or normal feedback modes can be bypassed by the use of AFB. Thus, intensive (frequency modulation) and extensive (recruitment of MNs) properties of SMU control are normal. In these patients, conscious awareness of the level of effort is entirely absent, most likely associated with the degeneration of the spinocerebellar tract. Recent reports correlating the results of sensory examination with the location of lesions in the spinal cord (Ross et al., 1979; Nathan and Smith, 1973) suggest that the dorsal columns may not mediate position sense. The dorsal spinocerebellar tract, possibly in combination with other afferent systems, may mediate this complex sensory modality (Nathan and Smith, 1973). The disorder of movement results from the absence of an essential comparator of motor unit output (producing the actual movement) and the desired pattern of movement. The term "sensory ataxia" refers to this defect. Supplying such patients with feedback of movement by tactile stimulation or muscle activity might improve their motor control substantially.

Motor Control Mechanisms in Health and Disease,
edited by J. E. Desmedt.
Raven Press, New York © 1983.

Analysis of Abnormal Voluntary and Involuntary Movements with Surface Electromyography

*Mark Hallett

Section of Neurology, Department of Medicine, Brigham and Women's Hospital, Boston, Massachusetts 02115

Simple methods using surface EMG electrodes can be helpful in the differential diagnosis of diseases of the central nervous system. The usefulness of these techniques has already been demonstrated in the analysis of tremor (Desmedt, 1978b). The EMG patterns of simple voluntary movements and involuntary movements can be used similarly to classify different pathophysiological states. Even more can be learned by analysis of EMG responses to different stimuli such as muscle stretch, electrical shock, or vibration, but these methods lie beyond the scope of this chapter.

VOLUNTARY MOVEMENT

Different types of clumsiness of voluntary movement can be studied by asking patients to make a monophasic ballistic movement at a single joint while recording from agonist and antagonist muscles. Functionally, "ballistic" means that the patient is trying to make the movement as rapidly as possible. It is important to have well-defined starting and stopping positions, but the precise angular distance of the movement does not seem critical. Although it is useful to have a record of joint angle in addition to the EMG information, it is not necessary. Flexion of the elbow and top joint of the thumb have been well studied, but there is no reason why other movements could not be used.

The initial part of EMG pattern of a normal ballistic movement is triphasic, characterized by a burst in the agonist (Ag 1), followed by a burst in the antagonist (An 1), followed by return of activity in the agonist (Ag 2) (Wachholder and Altenburger, 1926; Angel, 1974; Hallett et al., 1975a; Hallett and Marsden, 1981). There is ordinarily a pause in agonist activity at the time of An 1. The Ag 1 and An 1 bursts range in duration from about 50 to 100 msec, and this seems independent of initial angle of the joint, distance moved, or force against which the limb moves (Hallett and Marsden, 1979). Variations of force needed to make different movements are accomplished by varying the amount of EMG activity within this fixed time interval. This can be done by varying the number of motor units recruited and their firing frequency (Desmedt and Godaux, 1977a,b, 1978b).

Derangements of this normal pattern in various pathophysiological states are summarized in Table 1. Patients with lesions of

---

*Section of Neurology, Brigham and Women's Hospital, 75 Francis Street, Boston, Massachusetts 02115.

TABLE 1. *Derangements of the ballistic movement pattern*

Disorder	EMG pattern
Pyramidal lesion	Slightly prolonged Ag 1 and/or An 1 (Decreased maximal interference pattern in routine EMG)
Loss of α motor neurons	Prolongation of Ag 1 and/or An 1 (Denervation in routine EMG)
Cerebellar ataxia	Prolonged Ag 1 and/or An 1
Parkinson's disease	Abnormal patterning, multiple bursts
Athetosis	Excessive activity in the antagonist blocking the movement or causing the limb to go the wrong way

the pyramidal tract, and most of these have been patients with the "upper motor neuron" form of amyotrophic lateral sclerosis (ALS), show mild prolongation of Ag 1 and/or Añ 1 (Fig. 1) (Hallett, 1979). This would

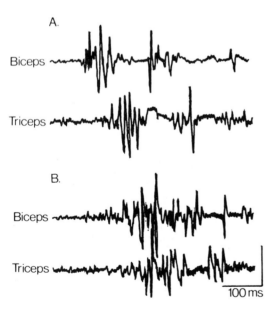

FIG. 1. Examples of the EMG pattern underlying rapid flexion movements in two patients with the "upper motor neuron" form of ALS. Patients were attempting 20° rapid flexion movements of the elbow. A shows a normal triphasic pattern. B shows a pattern with prolongation of Ag 1. (From Hallett, 1979, with permission.)

seem to be because motor unit recruitment is reduced and additional time is thus needed to generate sufficient force to make the movement. Patients with loss of anterior horn cells from ALS show moderate prolongation of the initial bursts of the triphasic pattern (Hallett, 1979). This would seem to be because of loss of power of the muscle, again requiring additional time to generate appropriate force.

Movements of patients with lesions of the cerebellum or cerebellar pathways are also characterized by preservation of the triphasic pattern with prolongation of Ag 1 and/or An 1 (Fig. 2). This has been seen in rapid movements of the elbow (Hallett et al., 1975b) and thumb (Marsden et al., 1977a; Hallett and Marsden, 1981). There is no known deficit in motor unit recruitment in cerebellar disease, nor are movements in general slow, and these facts separate these patients from those with pyramidal tract or anterior horn cell lesions. Prolongation of the initial bursts in the cerebellar patient would seem to predispose him to make dysmetric movements. For example, prolongation of Ag 1, which produces the accelerative force for the movement, would predispose to hypermetria.

Movements made by patients with Parkinson's disease are often slower than normal despite their attempts to move rapidly. At first glance, the EMG pattern appears normal, with alternate agonist and antagonist bursts of normal duration (Hallett et al., 1977). With a closer look, it can be appreciated that the movement is not completed with a single agonist–antagonist–agonist cycle as it should be normally. Instead, there can be multiple cycles of agonist–antagonist activity, and this abnormality is more frequent with longer movements (Fig. 3) (Hallett and Khoshbin, 1980). The derangement of bradykinesia appears to limit the amount of EMG activity in each burst.

Patients with athetosis have marked difficulty making voluntary movements. De-

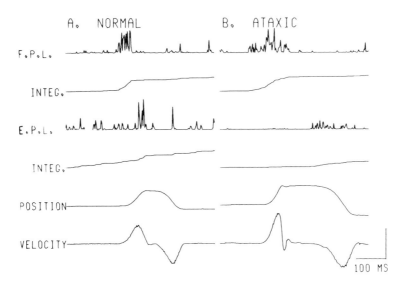

FIG. 2. The EMG during rapid thumb movements for the normal and ataxic thumbs of a woman with unilateral cerebellar ataxia. In A and B the patient was trying to make a 10° movement from an initial angle of 20° against a torque of 0.08 Nm. The traces are, from above downward: full-wave-rectified and then rectified and continuously integrated EMG from FPL; full-wave rectified and then rectified and continuously integrated EMG from EPL; position; and velocity. Ag 1 and An 1 have longer durations in the ataxic performance. (From Hallett and Marsden, 1981, with permission.)

tailed analysis of agonist–antagonist relationships shows that the antagonist muscle is often inappropriately active. It may fire before the agonist, causing a movement in the wrong direction, and it may fire simultaneously with the agonist, apparently limiting the ability of the agonist to move the limb. The duration of the EMG activity may be markedly prolonged during this synchronous activation of the antagonist muscles.

INVOLUNTARY MOVEMENTS

There are basically three types of EMG patterns that underlie involuntary movements, and each one resembles a normal physiological phenomenon. To recognize these patterns, it is necessary to record surface EMGs from a pair of antagonist muscles. One type, which can be called "ballistic," looks like a normal voluntary rapid movement and has the triphasic pattern described in the first section. Another type, which can be called "reflex," has the appearance of a normal monosynaptic tendon jerk or long-latency stretch reflex. The EMG bursts are often synchronous in the antagonist muscles and have a duration of about 10 to 30 msec. The third type can be called "tonic" and appears similar to what is seen with a normal voluntary tonic ramp movement. The EMG bursts last 150 msec or longer and occur synchronously or asynchronously in the antagonist muscles. The patterns seen in different types of involuntary movements are summarized in Table 2.

Myoclonus

Myoclonus means a rapid muscle jerk, and it is quite clear that the term encompasses a variety of completely different physiological phenomena. Different forms of myoclonus exhibit each of the three EMG patterns. We have previously reported the physiological details of two types of myoclonus with the reflex pattern (Fig. 4A). Re-

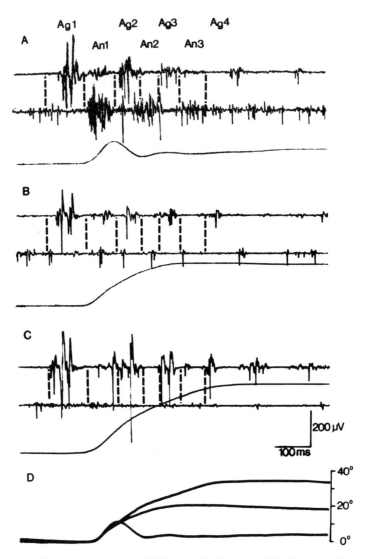

FIG. 3. Attempted ballistic movements of 10° (A), 20° (B), and 40° (C) by a 68-year-old man with Parkinson's disease. In A, B, and C, the traces are, from above downwards, biceps EMG, triceps EMG, and position of the elbow. D shows the three position traces superimposed. The parts of the figure were aligned so that the movements all began at the same time from the beginning of the traces. The *dashed vertical lines* indicate the correspondence of the timing of EMG bursts in the different movements. Note the additional cycles of bursts: the third and fourth agonist bursts are labeled Ag 3 and Ag 4, and the second and third antagonist bursts are labeled An 2 and An 3. In D, the initial part of the position traces are similar, indicating similar velocities of the different movements. In **A**, the subject did not sustain the attempted 10° position but returned quickly to near the starting point. (From Hallett and Khoshbin, 1980, with permission.)

ticular reflex myoclonus (Hallett, et al., 1977) is characterized by generalized muscle jerks, more proximal than distal and more flexor than extensor. The jerks can be produced by intention or somatosensory stimulation. The cranial nerve musculature is activated in an order as if the signal that produced the myoclonus traveled up the

TABLE 2. *EMG appearance in different types of involuntary movements*

Disorder	EMG pattern			Comment
	"Reflex"	"Ballistic"	"Tonic"	
Myoclonus	X			In reticular reflex and cortical reflex types; Creutzfeldt–Jakob
		X		Ballistic movement overflow myoclonus
			X	Dystonic myoclonus; SSPE myoclonus
Dystonia			X	Seen in dystonia, athetosis, and dyskinesias
Tic		X		In a stereotyped set of muscles; not fully involuntary
Chorea (Huntington's type)	X	X	X	Any type of movement in any muscle at any time

brainstem. The EEG shows generalized spike discharges which are associated with the myoclonic jerks but not time-locked to them (Fig. 5A). Cortical reflex myoclonus (Hallett et al., 1979) has the same EMG pattern but is characterized by jerks that are focal or multifocal rather than generalized and tend to be more distal than proximal. They can also be produced by intention or somatosensory stimulation. Cranial nerve musculature is activated in downward order. The EEG shows nothing apparent at first, but if the EEG is averaged time-locked to the myoclonic jerk, then a focal negative transient can be appreciated preceding the myoclonic jerk (Fig. 5B). The jerks in Creutzfeldt–Jakob disease also show the reflex pattern but have a different

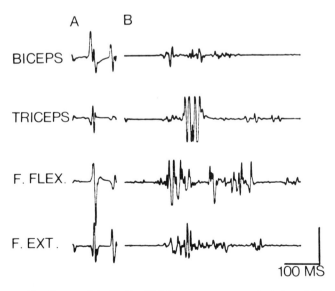

FIG. 4. Comparison of reflex and ballistic EMG appearance underlying different types of myoclonus. A is from a patient with reticular reflex myoclonus, and B is from a patient with ballistic movement overflow myoclonus. Calibration: 100 msec, 1 mV for A; 0.5 mV for B. B, biceps; T, triceps; FF, finger flexors; FE, finger extensors. (From Chadwick et al., 1977, with permission.)

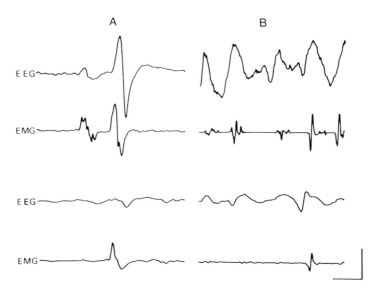

FIG. 5. Comparison of the EEG–EMG correlation in reticular reflex myoclonus (A) and cortical reflex myoclonus (B) in spontaneous jerks. The top half of each part shows a single sweep, whereas the bottom half shows an average of 84 sweeps for A and 32 sweeps for B. Data collection is triggered by an EMG event occurring three-quarters of the way along the sweep. In A, the EEG is from vertex referred to left ear (positive is downward), and the EMG from right sternocleidomastoid. In the single sweep, a large spike in the EEG is associated with the EMG burst that triggered the collection. A second myoclonic jerk, preceding the one that triggered the collection, is also associated with a cerebral event. The spike does not appear in the average, suggesting that its time relation to the EMG activity is not fixed. In B, the EEG is from a point 2 cm behind a point 7 cm down on a line from vertex to the right ear referred to the Fz electrode, and the EMG is from the left finger flexor. In the single sweep, each EMG burst is preceded by a positive wave in the EEG. This positive wave does remain after averaging, suggesting that its time relationship to the EMG is fixed. In A, the voltage calibration indicates 0.5 mV for the EEG and EMG; in B, the voltage calibration indicates 0.05 mV for the EEG and 1 mV for the EMG. (From Chadwick et al., 1977, with permission.)

EEG correlate (Shibasaki et al., 1981). There are undoubtedly other physiological types as well.

There are a variety of types of essential myoclonus, but one subtype has the ballistic pattern and is called ballistic movement overflow myoclonus (Hallett et al., 1977) (Fig. 4B). This disorder appears to be autosomal dominant and is not associated with epilepsy or mental retardation. The myoclonus represents the overflow of excessive movement into unnecessary muscles when a discrete voluntary movement is attempted.

Myoclonus can occur in patients with different forms of dystonia. This "dystonic myoclonus" can be characterized by EMG bursts lasting 200 msec or more (Fig. 6). Tonic EMG bursts are also ordinarily seen in SSPE.

It is useful to make this distinction among the physiological forms of myoclonus, since the reflex varieties may be treated with serotonin agonists or clonazepam (Chadwick et al., 1977), and the ballistic variety might be treated with propanolol (Ferro and Calhau, 1977); there is no good approach to therapy of the tonic variety.

Dystonias

The dystonias exhibit patterns of the tonic type (Yanagisawa and Goto, 1971). The EMG activity is present involuntarily

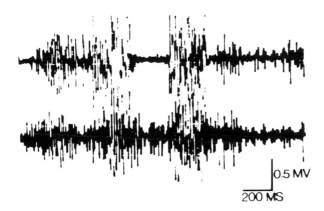

FIG. 6. The EMG appearance underlying dystonic myoclonus from a young boy with dystonia musculorum deformans. Top trace is from biceps; and the lower trace is from triceps. (From M. Hallett, B. T. Shahani, and R. R. Young, *unpublished data*.)

at rest but can be brought out further by attempted voluntary movement. The axial muscles tend to be involved, and the dystonia often looks like an abnormal posture. Rest or action tremors can be superimposed on this tonic pattern.

Athetosis

Athetosis has long been considered to be physiologically similar to dystonia, and tonic EMG patterns can be seen (Hoefer and Putnam, 1940; Herz, 1944; Yanagisawa and Goto, 1971). Limb muscles may be involved more than axial muscles. Our recent observations illustrate that the "involuntary movements" are almost entirely brought out by attempted voluntary movement, even of a distant body part. One charac-

teristic of athetosis is that there tends to be an alternation between two abnormal postures (Denny-Brown, 1968). Rest or action tremors can be superimposed (Hoefer and Putnam, 1940; Andrews et al., 1973; Neilson, 1974).

Dyskinesias

Dyskinesias, such as tardive dyskinesias produced by long-term dopamine blockade or L-DOPA dyskinesias produced by excessive dopamine stimulation, also show a tonic pattern (Fig. 7).

Sydenham's Chorea

The EMG pattern in Sydenham's chorea has been held to be similar to that of athe-

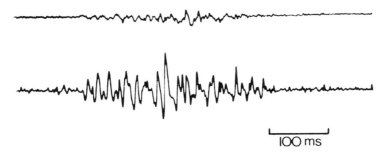

FIG. 7. The EMG of an L-DOPA dyskinesia in an elderly woman with Parkinson's disease. The top trace is from biceps, and the lower trace is from triceps.

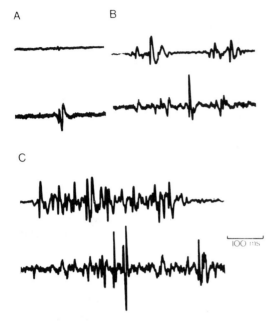

FIG. 8. The EMGs of three involuntary movements in a woman with senile chorea. Top trace is from biceps, and bottom trace is from triceps. A shows a reflex pattern; B shows a ballistic pattern; and C shows a tonic pattern.

tosis (Hoefer and Putnam, 1940; Herz, 1944). In any one involuntary movement, there is a tendency for one muscle of an antagonist pair to be more active than the other, and in different movements, the EMG burst length is quite variable (Herz, 1944). Rondot et al. (1978) have noted that antagonist muscles tend to be activated asynchronously, and we have recently confirmed this observation (Hallett and Kaufman, 1981).

Tic

Tic is a quick movement in a stereotyped set of muscles that often shows the ballistic pattern (Hallett and Marsden, 1979). These movements are not fully involuntary in that they can often be suppressed by the patient at the expense of mounting emotional tension. The emotional tension appears to be dissipated by the tic.

Chorea

Chorea, as seen in Huntington's chorea (Hallett et al., 1977) and senile chorea (Fig. 8), is characterized by any type of EMG pattern in any muscle at any time. Reflex, ballistic, and tonic patterns can be recognized in the same muscle pair. Tonic patterns have a tendency to show synchronous activity in antagonist muscles (Rondot et al., 1978).

CONCLUSION

These rather simple observations with surface EMGs of activity at rest and during a stereotyped rapid movement can be extremely useful in classifying and thinking about disorders of movement.

Motor Control in Man After Partial or Complete Spinal Cord Injury

*M. R. Dimitrijevic, J. Faganel, D. Lehmkuhl, and A. Sherwood

Texas Institute for Rehabilitation and Department of Rehabilitation, Baylor College of Medicine, Houston, Texas

The spinal motor activity above and below a spinal lesion presents different motor behavior patterns. The muscles may not be under volitional control below the level of the lesion. Yet, reflex, automatic, postural, and positioning regulatory mechanisms within the body and limbs are disturbed so that uncontrolled and noncoordinated abnormal movements or spasms still occur. This leads to several questions: Which of these abnormal movements or spasms are generated and controlled completely by spinal cord reflex mechanisms? How much brain influence is present in the paralyzed muscles of spinal cord injury patients? Are supraspinal mechanisms necessary to provide and maintain spinal reflex motor activities?

In this study, we examined phasic and tonic stretch reflexes in a group of 66 spinal cord injury patients with chronic lesions defined as complete on the basis of clinical signs. The effectiveness of reinforcement maneuvers in eliciting coactivation of gross reflex activity in paralyzed muscles has been studied, as well as tendon jerk responses conditioned by noxious cutaneous stimuli in an attempt to test the propriospinal interneuron system and the residual brain influences in patients with paralyzed limbs.

Riddoch (1917) compared the functions of the completely divided spinal cord in man with incomplete, substantial lesions and concluded that "paraplegia in flexion" related to complete transection of the spinal cord, whereas "paraplegia in extension" was the result of severe but incomplete lesions. He stressed, however, that no manifestations could indicate unequivocally the anatomical division of the spinal cord. Years later, Kuhn (1950) came to similar conclusions, stating that the terms "mass flexion" and "mass extension" may be of value in characterizing the patterns of flexor and extensor activity elicitable below the transection of the cord (Pollock et al., 1951). Pollock noticed that long after injury, the reflex characteristics of the distal segments of the spinal cord change to such a degree that it becomes impossible to differentiate between complete and incomplete transections.

Guttman's (1952) studies of traumatic paraplegics surviving from World War II led him to suggest that early management of the position of paralyzed limbs influenced the development of reflex synergies in later stages in both complete and incomplete lesions. On the other hand, the new knowledge of spinal reflex activity gained from

---

*To whom correspondence should be addressed. Texas Institute for Rehabilitation and Research and Department of Rehabilitation, Baylor College of Medicine, Texas Medical Center, Houston, Texas 77025.

experimental work in animals suggested that some characteristics of spinal reflex motor activities could not be explained unless we assumed the presence of clinically unrecognized, partially preserved, residual descending brain influences on spinal cord activities (Lundberg, 1975). At the same time, improvements in instrumentation enabled clinical neurophysiological methods to relate changes in tone, position, posture, and movement to motor unit activity and to study segmental motor control and its dependence on subclinical residual supraspinal control in paralyzed spinal cord injury patients.

METHODS

Observations were made with the patient sitting in a slightly reclining position, fully supported in a comfortable chair or prone on an examining table. The EMG recordings were made with Beckman recessed silver–silver chloride surface electrodes placed approximately 5 cm apart over the belly of the muscles examined. As many as 12 channels of EMGs were simultaneously recorded. The EMG signals were amplified with a bandwidth of 10 to 2,000 Hz and displayed on a Brush eight-channel recorder, a 16-channel Siemens Mingograph recorder, or a storage oscilloscope. Joint movements were monitored by placing an electrogoniometer over the center of the joint rotation. A strain gauge fixed at one end and attached to a strap around the ankle of the patient was used to monitor the force of knee extension. Two types of tendon tap hammers were used. The first was an automated, floor-mounted electrodynamic device that delivered a constant force (Simons and Lamonte, 1971). This hammer had a strain gauge attached to the head to measure the force of each tap. The second type was a small, hand-held electrodynamic hammer used for study of the conditioning effects of noxious stimuli (Trontelj et al., 1968). The noxious stimulus consisted of a 2-kHz, 20-msec train of 0.2-msec electrical pulses. This was applied to various dermatomes through Copeland–Davies electrodes applied approximately 2 cm apart so as to penetrate the epidermis. As a source of vibration, a modified Foredom Model 974 universal motor with a cam-driven head was used. This device provided a variable frequency of vibration from 0 to 128 Hz (normally set on the maximum and a constant head displacement of 3 mm.

Knee jerks were elicited by a floor-mounted hammer at a 0.5 Hz rate with the patient in a standard sitting position. The average of 20 peak-to-peak EMG responses recorded from the quadriceps femoris muscle was used to estimate responsiveness of the stretch reflex loop (Simons and Dimitrijevic, 1972). Ankle and biceps femoris jerks were elicited with a manual reflex hammer.

For the pendulum test, the subjects sat on the edge of the examining table with the leg arranged so that it could swing freely after being released from a horizontal position (Boczko and Mumenthaler, 1958). Knee joint movements were monitored by an electrogoniometer.

Ankle clonus was extensively used in this study. Clonus was usually elicited by a brisk jerk of the foot while the subject was prone on a comfortable examining table. In a few cases, the foot was allowed to hang free over the end of the table so that the weight of the foot served as the restoring force when clonus was initiated by other maneuvers. Patellar clonus was also similarly assessed by briskly jerking the patella of the supine patient.

Assessment of the vibratory reflex was done with the subject in a supine position on a padded examination table by applying the vibrator to the skin overlying the tendon of the muscle to be stimulated.

Reinforcement maneuvers were used both as a test of the ability to activate spasms suprasegmentally and as a means of modifying segmental reflexes such as clonus and the vibratory reflex. These maneuvers consisted of strong sustained activation of muscle groups involved in the following motor acts: breath holding, neck flexion against resistance applied manually to the forehead, Jendrassik, right-hand grip, left-hand grip, and clench of the jaw. Typically, the experimenter requested the subject to initiate the maneuver and simultaneously activated a switch denoting the time of onset. Each maneuver was attempted at least three times with a 3-sec rest after the cessation of any response. In patients with cervical lesions, the Jendrassik maneuver was accomplished by activating the biceps or shoulder elevators bilaterally.

Conditioning Tendon Jerks

Thirty taps at a frequency of 1/sec divided into three sets of 10 taps each were used to assess the magnitude of the tendon jerk responses to various conditioning stimuli. The first set was used to obtain a baseline response without conditioning. During the second set of tendon taps, conditioning noxious stimuli of slightly subthreshold intensity were delivered 50 to 250 msec prior to the tendon taps. A third set was finally applied to assess the aftereffects of the preceding set of stimuli. The conditioning stimulus was applied to the plantar surface of the foot (lateral border of the midfoot) as well as to the S_3, T_5, and T_{10} dermatomes. Whenever the relationship between the pattern of withdrawal reflex responses and the tendon reflex responses was studied, a suprathreshold intensity of noxious stimulus was used. In each case, the tendon jerk responses having an appropriate latency were full-wave rectified, integrated, and averaged. The effects of the conditioning stimulation were then estimated by calculating the percentage change of the mean value of the set with conditioning from the values obtained from the base-line set. Statistical significance for the results was derived from a t-test.

All recordings were made from patients with severe chronic lesions of the cervical and thoracic spinal cord. Those with complications such as decubitus ulcers and bladder infections were not included. All were adults, the majority being male. They were 18 or more months post-injury. The detailed tendon tap and pendulum test data were collected from a group of 16 patients from the Spinal Cord Injury Unit of the Houston VA Medical Center and 38 patients with complete motor paralysis from the Institute for Rehabilitation and Research, Houston. Studies of the propriospinal interneuron system were conducted in a group of 12 patients. The rest of the data on vibration reflexes, clonus, and suprasegmental activation of spinal reflexes were collected from the 38 patients with complete motor paralysis.

STRETCH REFLEX STUDIES

Tendon Taps

In a group of 38 patients, the excitability of stretch reflexes was examined bilaterally in quadriceps femoris and triceps surae muscles. The EMG responses were categorized as absent, low amplitude (below 0.4 mV), or brisk (Table 1). There is wide variability in the responses. In the group of 16 patients in whom only patellar reflexes were studied, the tendon tap responses also varied from an exaggerated clinical response and large-amplitude EMG signals to clinically weak or trace movements accompanied by marginally detectable EMG signals. In repeated measurements in five

TABLE 1. *Electromyographic responses to tendon tap in 38 patients with severe chronic spinal cord injury and without any other neurological disorders*

Tendon tapped	Absent	Low	Brisk
Left Achilles	10	16	12
Right Achilles	10	14	14
Left patellar	6	14	18
Right patellar	4	16	18

cases, the mean value did not change more than 40%. A comparison of the mean of the magnitudes of the left and right patellar tendon tap responses in 12 of the 16 patients examined revealed that the response pairs were highly correlated ($r = 0.96$).

Pendulum Test

A pendulum test of the quadriceps femoris was recorded from the left or right side in the 16 spastic patients with chronic cervical or high thoracic spinal cord lesions. The patterns of joint movement can be divided into three groups as shown in Fig. 1A. Sinusoidal oscillation of the knee joint angle was observed in nine of the 16 patients. In three patients, the goniometric traces revealed irregular, restricted joint movement, requiring a much longer time to come to the resting position. In four other patients, the pendulum test records are characterized by a combination of both patterns described above.

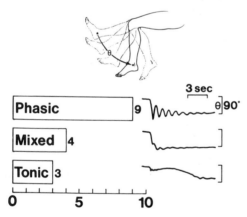

FIG. 1. Pendulum test results for 16 chronic spinal cord injury patients recorded from one paralyzed quadriceps muscle in each patient. The curves on the right are typical goniometric test results when the leg is allowed to swing freely downward from the horizontal position, as shown in the diagram above. Reflex responses to slow sustained stretch of the quadriceps during free fall of the leg were classified as: phasic, tonic, or mixed.

Clonus

Records of the 38 chronic, spinal-cord-injured patients referred for evaluation of spasticity were examined for the presence of clonus (Table 2). From 152 examined muscle groups producing extension of lower limbs (quadriceps, femoris and triceps surae) unsustained clonus was present in 60, and sustained clonus in six muscle groups. We tested whether the Jendrassik maneuver could modify clonus in patients with sustained and unsustained clonus of the ankle. In the patients with unsustained clonus, the Jendrassik maneuver did not modify the clonus (Fig. 2). On the other hand, in the patients with sustained clonus, the maneuver increased the amplitude of the clonus or even initiated it (Fig. 2d). In the paralyzed and spastic muscle groups of patients with sustained ankle clonus who were able volitionally to start and stop their clonus, the effort to start resembled a modified Jendrassik maneuver, whereas suppression of clonus was accomplished by a sustained effort to relax the limb.

Vibration Reflex

With strong vibration (3 mm, 120 Hz) applied to the tendon of a paralyzed muscle, a characteristic response consists of an early phasic contraction and a later tonic component which typically diminishes in size over a few seconds or fractions of a second. In

TABLE 2. *Number of muscle groups exhibiting sustained and unsustained clonus in a sample of 38 chronic paralyzed spinal-cord-injured patients*

Muscle	None	Less than 1 sec	More than 1 sec	Sustained
R. quadriceps	20	11	6	1
R. triceps	24	5	7	2
L. quadriceps	21	7	10	0
L. triceps	21	3	11	3

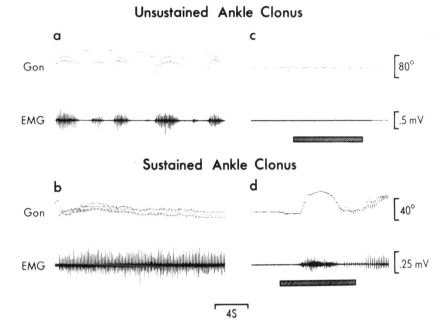

FIG. 2. Unsustained and sustained clonus. Monitoring each of two completely paralyzed subjects with an ankle electrogoniometer and surface EMG electrodes over the soleus muscle, typical patterns of unsustained (a) and sustained (b) ankle clonus can be seen, resulting from manually induced clonus. The effect of Jendrassik maneuvers when tension on the muscle was removed can be seen in (c) and (d). The reinforcement maneuver did not produce any detectable activation of the muscle in the patient with unsustained ankle clonus (c), although a marked effect was produced in the patient with sustained ankle clonus (d).

the triceps surae with absent volitional motor control of 38 chronic spinal cord injury patients, no response was elicited in 10 patients (Fig. 3). In 25 patients, responses elicited were classified as (1) short phasic, (2) unsustained tonic and phasic, (3) unsustained tonic, augmented by Jendrassik, or (4) persistent tonic (Fig. 3). Reinforcement through the Jendrassik maneuver selectively augmented a declining response—dishabituation (Fig. 3). Vibration-induced spasms were not common, and when the vibration induced a gross withdrawal response, the recordings were disregarded. The data of Fig. 3 were collected under conditions that prevented coactivation of another segmental input by passive movement or unrecognized skin stimulation.

Suprasegmental Activation of Spinal Reflexes

Chronic spinal-cord-injured patients are able to induce spasms with various maneuvers ranging from mental effort to movement of nonparalyzed extremities. When motor unit activity was recorded with surface electrodes from quadriceps, hamstrings, tibialis anterior, and triceps surae bilaterally in the 38 spinal cord injury patients, 15 demonstrated coactivation of gross reflex activity in paralyzed muscles during reinforcement maneuvers. Breath holding and neck flexion were equally potent in initiating muscle activity below the lesion, with the Jendrassik maneuver nearly as effective (Table 3).

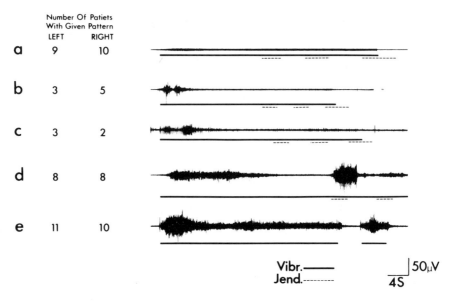

FIG. 3. Vibratory response patterns of triceps surae muscles in a group of paralyzed spinal cord injury patients. Examples of each of the five categories used in classifying the response to vibration are as shown: a, no response; b, short phasic response; c, unsustained tonic and phasic response; d, unsustained tonic response which could be augmented with Jendrassik maneuver; e, persistent tonic response present in repetitive episode of vibration.

Since movement of the nonparalyzed portion of the body can trigger reflex activity within the paralyzed portion through peripheral mechanisms, particularly in severely spastic patients, we applied several criteria: (1) consistent and repetitive activation with a stereotyped pattern of recruitment of motor units and muscle groups; (2) presence of motor unit activity in the distal foot muscles with absence of activity in abdominal, paraspinal, and proximal leg muscles; and (3) evidence for recruitment of motor units and muscles scattered around paralyzed limbs and without any biomechanical relation.

Conditioning the Phasic Stretch Reflex with Noxious Stimulation

The plurisegmental interneuron system was tested by conditioning tendon jerks with noxious cutaneous stimulation. When a noxious stimulus is applied regularly to the plantar surface of the foot with sufficient strength to activate the tibialis anterior muscle (withdrawal reflex) a typical pattern of increased excitability (sensitization) followed by a progressive habituation of plantar withdrawal to complete extinction of response occurs (Fig. 4). When the patellar reflex of the ipsilateral leg is elicited simultaneously with the plantar withdrawal reflex, the amplitude of the patellar reflex is diminished (Fig. 4). After the first 20 patellar responses, simultaneous noxious stimulation of the plantar surface of the foot was

TABLE 3. *Effectiveness of reinforcement maneuvers in eliciting coactivation of gross reflex activity in paralyzed muscles of severe chronic spinal cord injury patients*

Maneuver	Number responding
Clench jaw	1
Breath holding	9
Neck flexion	9
Jendrassik	6
Right hand grip	4
Left hand grip	3

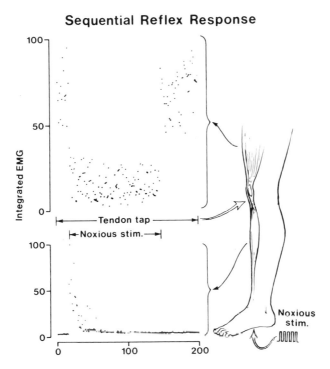

FIG. 4. Conditioning effect of plantar withdrawal reflex on patellar reflex. The top trace shows the suppressing effect of noxious plantar stimulation on the patellar tendon taps, with the corresponding withdrawal responses in the tibialis anterior shown in the bottom trace. *Abscissa:* time (msec).

introduced. Fluctuations in the amplitude of individual patellar reflex responses continued to occur, but at a much lower level. The suppressing effect on the patellar reflex was lost on cessation of the noxious stimulation. The noxious stimulus suppresses patellar reflex amplitude each time it is applied, regardless of its effectiveness in producing a withdrawal reflex response.

In 12 patients with chronic paralysis caused by severe injury to the cervical or upper thoracic segments of the spinal cord, Achilles, hamstring, and patellar jerks were elicited from muscles ipsilateral and contralateral to the conditioning stimulation applied 50 to 250 msec before the application of test taps. The plantar stimulation suppressed the Achilles and patellar jerks and facilitated the hamstring jerk (Fig. 5). This was a constant finding in eight of the 12 patients tested. In the same eight patients, the predominant contralateral effect was facilitation of the tendon jerk responses tested. Exceptions to these findings of selective suppression of extensor muscles and facilitation of tendon jerks of flexor muscles on the ipsilateral side as well as nonselective facilitation of tendon jerks of flexor and extensor muscles on the contralateral side have been found in four of the 12 patients. One had facilitation of the ipsilateral ankle jerk and suppression of the ipsilateral knee jerk, two had facilitation of both ankle and knee jerk ipsilaterally, and one had suppression of the contralateral Achilles jerk. Other than these exceptions, the only interpatient differences were in the degree of facilitation and suppression.

The ascending effect of cutaneous noxious stimuli applied over the S_3 dermatome was tested using tendon jerks with reflex arcs in S_1 and L_2 spinal segments of the same 12 patients (Fig. 6). The typical response was bilateral suppression of Achilles

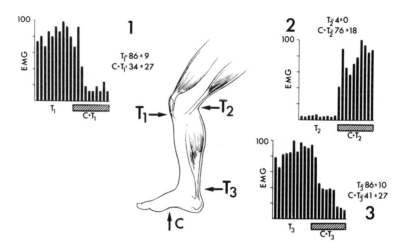

FIG. 5. Effects of plantar stimulation on patellar (1), hamstrings (2), and Achilles (3) tendon tap responses in a patient with chronic paralysis as a result of cervical spinal cord injury. The *bar* height represents the full-wave-rectified, integrated EMG from the indicated muscle group in response to the tap, the first 10 without, and the last 10 with, conditioning noxious stimulation to the plantar surface of the foot.

and patellar jerks. This pattern was found in nine of the 12 patients, and the suppression effect was more pronounced with the patellar jerk. Exceptions were found in three patients. In one, stimulation of the S_3 dermatome resulted in bilateral facilitation of the ankle jerk, and the second one had bilateral facilitation of the knee jerk. In only one patient did the effect of the ascending influence of the conditioning stimulus fail to be symmetrical bilaterally. In this case, there was contralateral facilitation and ipsilateral suppression of the ankle jerk. All three exceptional cases had cervical lesions ranging from C_4 to C_7.

When the cutaneous stimuli were applied to dermatomes cranial to the segments tested by the tendon taps (Fig. 7), the effects of the stimuli were facilitatory. When the noxious stimuli were applied to the T_5, T_{10}, and L_1 dermatomes, there was an immediate, bilateral facilitatory effect on both knee and ankle jerks. Exceptions to this pattern of response were seen in two of the 12 patients where cutaneous stimulation bilaterally suppressed the tendon taps.

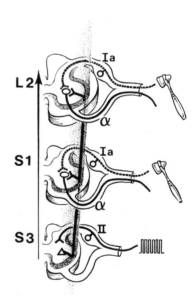

FIG. 6. Ascending influence of noxious cutaneous stimuli on Achilles and patellar tendon jerks. Diagrammatically illustrated is the relationship between the conditioning cutaneous stimulus and the spinal segments tested by tendon taps. The typical response was bilateral suppression of Achilles and patellar tendon jerks.

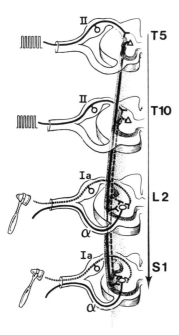

FIG. 7. Descending influence of noxious cutaneous stimuli on Achilles and patellar tendon jerks. Application of noxious stimuli to dermatomes cranial to the segments tested by tendon taps produced an immediate bilateral facilitatory effect on both knee and ankle jerks.

Since the descending facilitatory influence on the Achilles jerk was immediate, the minimum conditioning-to-test-stimulus interval was measured in one C_6-paralyzed spinal cord injury patient. The delay between conditioning cutaneous stimulus and ipsilateral Achilles tendon tap to produce the optimal tendon tap response was determined by delaying the test stimulus in 5-msec increments until a definite conditioning effect was obtained. A minimum delay of 15 msec at T_{10} and of 30 msec at T_5 were found to produce the facilitation. This yielded a conduction time of about 15 msec for five segments or about 3 msec per segment. The distance between the T_3 and T_8 spinous processes (corresponding roughly to spinal segments T_5 and T_{10}, respectively) was 20 cm, which gives a calculated value of 13 m/sec for the conduction velocity of the caudally spreading influence of the conditioning noxious stimulus.

DISCUSSION

Walshe (1914) hypothesized that there is a double system of motor innervation in spastic paraplegia and thought that "paraplegia in flexion" was under "pure spinal motor control" and that "paraplegia in extension" was the end result of spinal and residual brainstem segmental influence. Fulton (1930) observed that extensor hypertonus in the cat depends on transmission of postural reflexes along intact vestibulospinal and possibly ventroreticulospinal pathways to the cord centers below the level of the lesion, since only destruction of these tracts results in a preponderance of the flexor reflex of "paraplegia in flexion." These animal results should not be compared to the immediate and late effects in man of closed spinal cord trauma which varies greatly in the degree and distribution of actual structural damage to the cord. Persons with clinical signs of complete motor paralysis and loss of sensation below the lesions do not always demonstrate a complete interruption of all spinal tracts at postmortem examination. Moreover, features of spinal reflex organization, such as reciprocal inhibition or the local-sign cutaneomuscular reflex, were absent in EMG studies of patients with spastic paralysis (Dimitrijevic and Nathan, 1967a).

Previous studies of the tendon jerks revealed an afterdischarge that followed well-synchronized tendon jerk EMG responses and had a distinct phasic or tonic pattern (Dimitrijevic and Nathan, 1967b). When low-amplitude background motor unit activity was present in the antagonist muscle during muscle tap, there was evidence of reciprocal inhibition. This was consistent for the tendon jerk EMG and not for the later afterdischarge. Furthermore, the amount of afterdischarge was proportional to the ongoing afferent inflow in the spinal cord below the level of the lesion. If this input is diminished, the afterdischarge decreases and certain new features of spinal reflexes reappear (e.g., cyclical, alternating, rhythmical bursts of motor unit activity between antagonistic muscles, resembling clonus or tremor).

Afterdischarge can be understood as a segmental extension of afferent inflow to the interneuronal pool which has the capability of distributing activity along the spinal cord. When spinal cord excitability is increased, then the afterdischarge will radiate to the other muscle groups of ipsi- and contralateral limbs with phasic or tonic features. The presence of tendon jerk afterdischarge phenomena of widespread distribution within spastic muscles suggested the possible role of plurisegmental interneuron pathways in the irradiation of excitation within the spinal cord below the level of the injury. Excessive activity of the MNs that overrides segmental reciprocal organization and selective local response of cutaneomuscular reflexes could result from disinhibition of the spinal interneuron system that activates flexor and extensor muscles simultaneously, even after a very weak stimulus. Furthermore, since the amount of motor unit activity depends on the intensity of stimulation, the presence or absence of motor unit activity at the time of stimulation, and the phase of movement of the limb if it is moving at the time of stimulation, the spinal interneuron system of the isolated spinal cord deprived of brain influence must be able to develop certain functional organization (Dimitrijevic and Nathan, 1968). "Spontaneous activity" in EMG records from muscles of spastic paraplegics revealed significantly diminished muscle hypertonia and excessive reflex responses to all forms of stimulation after reduction of afferent inflow to the spinal cord (Dimitrijevic and Nathan, 1967a).

Thus, the propriospinal system is capable of augmenting the input by recruiting additional motor cells within the spinal cord. Moreover, latency measurements of the flexor withdrawal reflex in the tibialis anterior muscle have shown the same latency in paralyzed spastic spinal cord injury patients as in healthy adults (Dimitrijevic, 1973). The spinal cord interneuronal system can indeed substitute for absent supraspinal excitation of cutaneomuscular reflexes but cannot substitute for supraspinal inhibitory control to generate a well-defined local withdrawal response. However, repetitive stimulation at the same site will evoke withdrawal reflexes that decline progressively and eventually disappear. Such habituation of withdrawal and stretch reflexes occurs in spastic and paralyzed spinal cord injury patients but does not occur if stimulation is made random either in time or intensity (Dimitrijevic and Nathan, 1970, 1971, 1973; Dimitrijevic et al., 1972).

The propriospinal interneuron system functions in irradiating excitation and becomes an important network for integrating sensorimotor activities in the spinal cord deprived of supraspinal influences (Nathan and Smith, 1959). The main feature of this propriospinal interneuron system in patients with a transected spinal cord is the large variation in the amplitude and pattern of responses. If there is no interference from other distant segments, the reflex response reflects the strength and pattern of stimulation. However, a maximal intensity of stimulation produces a rigid pattern of activation of the muscle groups as the simple segmental organization of plantar withdrawal reflex overrides the plurisegmental propriospinal system (which provided plasticity of response and mediated excitation to the distant sacral, thoracic, and cervical segments).

This study disclosed a large variability of response amplitude for different reflexes in the same patient or for the same reflexes in different patients, even though reflex responses appeared within a normal latency. This probably relates to the presence of "patches" of anterior and lateral long extrapyramidal descending pathways. In the evaluation of any large group of spinal cord injury patients, there are those with paralysis and severe spasticity and those with paralysis and a mild degree of spasticity, even though both groups have similar degrees and levels of lesion (Pollock et al., 1951).

In spastic paraplegics, there were muscles that responded with an "idiosyncratic" stereotyped reflex response regardless of the nature and kind of stimulation, whereas other muscle groups in the same patient exhibited a wide variation in reflex patterns. Therefore, segmental reflex circuits are not the only mechanisms controlling the activity of muscles in paralyzed spinal cord patients. A supplementary mechanism may involve clinically unrecognized residual bulbospinal facilitation of reflex responses in paralyzed muscle groups. The assumption that residual bulbospinal pathways facilitate reflexes at different segmental levels can be compared with clinical findings of "patchy sensation" or "sacral sparing" of sensation in patients whose clinical picture otherwise indicates a complete spinal cord injury. Thus, a moderate survival of portions of ascending or descending pathways is not only a logical assumption but is also a clinical finding in patients with gross paralysis and severe sensory disturbances. This thesis is supported by evidence of a consistently wide variation of amplitudes of tendon jerks among 54 spastic and paralyzed patients with similar neurological deficits.

The finding of sustained clonus in only six of the 154 extensor muscles examined in the 38 patients with clinically complete spinal cord injury suggests that a decreased threshold for the phasic stretch reflex is not a sufficient condition for clonus. Clonus apparently requires a facilitatory mechanism of the MNs participating in the exaggerated phasic stretch reflex. Such a tonic facilitation can be provided by clinically unrecognized residual bulbospinal influence. The same assumption can be applied to explain the effectiveness of the Jendrassik maneuver in patients with sustained clonus but not in patients with unsustained clonus. Thus, sustained clonus can be taken as a clinical sign of the presence of "residual mechanisms of decerebrate rigidity" (Dimitrijevic et al., 1978).

Vibration of paralyzed and moderately spastic muscle produces a short phasic response followed by a sustained, slowly declining tonic component which depends on the degree of preservation of residual bulbospinal influence. The rate of habituation also reflects the degree of deterioration of bulbospinal pathways to a particular spinal segment (Dimitrijevic et al., 1977). In this study, we recorded different patterns of vibratory reflex response together with fast habituation rate and presence of dishabituation during the Jendrassik maneuver. Our finding that 15 of the 38 paralyzed spinal cord injury patients demonstrated suprasegmental activation of spinal reflexes is another bit of evidence for the preservation of brain influence to trigger reflex spasms, probably by activation of the propriospinal interneuron system. Partial survival of brain influence on the propriospinal interneuron system capable of activating muscles that cannot be activated voluntarily suggests (1) the existence of a dual supraspinal motor control system for regulating the contraction of skeletal muscles and (2) that the brain controls the spinal interneuron system to provide involuntary coordination of movement and posture in advance of, or together with, volitional control.

This dual supraspinal innervation of segmental MN pools and interneuron systems makes it possible for different degrees of impairment of the functions mediated by the two supraspinal motor systems to occur in spinal cord injury. Thus, in a population of patients who have sustained traumatic injury to the spinal cord, one would expect to find (1) patients with impaired but partially present volitional and postural motor control, (2) patients with absent volitional but partially present postural motor control, (3) patients with absent postural motor control but partially present volitional motor control, and (4) patients with complete absence of volitional and postural motor control but with segmental reflexes and modest irradiation abilities through the propriospinal interneuron system. In other words, the absence of volitional movement below a lesion does not always indicate a complete interruption of supraspinal fibers. Paralyzed muscles with easily elicitable sustained clonus or long-lasting tonic withdrawal reflexes suggest partial preservation of continuity of bulbospinal pathways through the injured portion of the spinal cord. By contrast, paralyzed muscles that have brisk tendon and withdrawal reflexes, localized responses that do not radiate to other muscles, no afterdischarge, and quickly declining amplitude of the response during rhythmical activation (unsustained or absent clonus, quickly declining withdrawal reflex response to repetitive activation) suggest a spinal cord deprived of residual brain influence.

The white matter component of the propriospinal interneuron system is very simple and diffuse and has widespread segmental input to the spinal MNs of different segments (Nathan and Smith, 1959). These spinal associative tracts mediate information caudally and cranially and potentiate MN excitability. The gray matter interneuron system, on the other hand, is more complex and has a kind of nucleus organization for flexor reflex and for withdrawal reflex movement. The presence of residual brainstem influence on nucleus-like structured interneurons and "noncommitted" interneurons can contribute to stereotyped reflex responses. Segments with absent residual brainstem influence can still be influenced by the white matter interneuron system conveying facilitation whenever sensory impulses enter another segment of the spinal cord; the gray matter interneuron system can excite MNs of the same segment when the combined stimulus exceeds the threshold. Diffuse irradiation of subclinical facilitation within the spinal cord by the white matter interneuron system is illustrated in Fig. 7 in which a subthreshold noxious cutaneous stimulus (eliciting no withdrawal reflex) modified the segmental stretch reflex in remote spinal segments. When brainstem residual influence is present, segmental reflex responses are even more variable; they irradiate easily to other segments; they are often sustained and exhibit afterdischarge; and they do not habituate.

SUMMARY

The essential features of motor control in spinal man can be understood in terms of segmental reflexes interacting with, and controlled by, the influence of distant segments and even by the brainstem. Thus, overall motor control in patients with spinal cord lesions can be classified according to structure as: (1) simple segmental stretch and withdrawal reflexes; (2) plurisegmental gross reflex movement of paralyzed muscles; or (3) propriospinal processes with partial brain influence (i.e., severe spasticity and traces of position and postural control). Because of the variable nature of the injuries, there may be exceptions to this rule. However, the basic mechanisms can be understood by studying stretch reflex responses to various stimuli. The segmental reflexes are under a powerful influence of the propriospinal interneuron system which conducts impulses up and down the spinal cord. Finally, the apparently "isolated" spinal cord in which clinical signs indicate complete motor paralysis and lack of sensation below the lesion is not always isolated from supraspinal control of involuntary motor activity. In a significant proportion of the clinically complete spinal injuries we studied, it was possible to demonstrate the presence of preserved bulbospinal influences on spinal reflex responses.

The Use of Monosynaptic Reflex Responses in Man for Assessing the Different Types of Peripheral Neuropathies

*P. Guiheneuc

*Laboratoire de Physiologie, Université of Nantes, Nantes 44035, France

Even careful clinical examination may fail to provide precise information about the severity, localization (proximal or distal), and nature (degeneration of fibers or sheaths) of nerve impairment. Needle EMG and determination of nerve conduction velocities (CV) are very useful but can only be performed in muscles or in distal segments of peripheral nerves, and proximal zones are not adequately investigated. Biopsies can also be taken from the distal end of a sensory nerve, but the pathological changes in proximal segments of nerves are at present relatively unknown. Several methods have been tested to disclose possible impairment of proximal nerve trunks including study of cerebral or spinal evoked potentials (see Desmedt, 1982) and analysis of the H or F responses. The present chapter surveys results obtained with reflex methods in the study of peripheral neuropathies.

The value of clinical tendon reflexes has long been known, but electrophysiological analysis of monosynaptic reflex responses is relatively recent. Malcolm (1951) measured amplitude and latency variations of knee and ankle jerks in patients suffering from nerve root compression. The H response elicited in soleus muscle by electrical stimulation of afferent Ia fibers led Paillard (1955) to develop a noninvasive method to explore the monosynaptic reflex arc in humans. By evoking such a reflex in hand and foot muscles of diabetic and uremic patients, Liberson (1963) showed that there is a marked reduction of CV in afferent proprioceptive fibers. In 1965, latency of the soleus H response was measured by Mawdsley and Mayer in alcoholic patients. The same year, Visser pointed out that this latency may increase following an acute L_5-S_1 disk prolapse.

We proposed a formula (H index) that corrects the latency of the H response as a function of reflex arc length, making it possible to measure the CV of the afferent and efferent pathways of the H reflex (Guiheneuc and Ginet, 1971; Guiheneuc et al., 1972; Guiheneuc and Ginet, 1973). Using this formula, and taking advantage of recent methodological standardization (Hugon, 1973a), we showed striking differences in both the type and course of proximal lesions in well-defined peripheral neuropathies (Bathien and Guiheneuc, 1974; Guiheneuc and Bathien, 1976). Simultaneously, evaluation of the H response in rectus femoris muscle (Guiheneuc and Ginet, 1974) allowed a functional study of mechanically injured lumbosacral roots to be car-

---

*Laboratoire de Physiologie, Université de Nantes, 1 rue Gaston Veil, Nantes 44035, France.

ried out (Guiheneuc et al., 1973; Descuns et al., 1973). One important advance in the use of reflex methods in peripheral neuropathies has been the possibility of relating modifications of proximal CV to amplitude changes in T and H responses (Guiheneuc and Bathien, 1976; Guiheneuc et al., 1980b) as well as the formulation of electrophysiological criteria to demonstrate the dying-back phenomenon in man (Guiheneuc et al., 1980a).

Pathological variations of the latency of the H response have been described in neuropathies of various origins: mechanical injuries (Visser, 1965; Dreschler et al., 1966; Bouquet and Cioffi, 1969; Deschuytere and Rosselle, 1973; Notermans and Wingerhoets, 1974; Braddom and Johnson, 1974), metabolic disorders or nutritional deficiencies (Liberson, 1963; Mawdsley and Mayer, 1965; Blackstock et al., 1972; Wager and Buerger, 1974; Panayiotopoulos and Lagos, 1980; Troni, 1981b), inflammatory or infectious diseases (Sebille, 1980), or lesions caused by toxic agents or drugs (Tobin and Sandler, 1966; McLeod and Penny, 1969; Casey et al., 1973; Sebille, 1978). Since the methodological report at the EMG Congress in Brussels (Hugon et al., 1973), the H reflex has been routinely used to study the peripheral nervous system.

METHODOLOGY

A monosynaptic reflex response can be obtained in many muscles of the upper and lower limbs in Man. However, polyneuropathies predominate in the longest nerve fibers of the body, and responses recorded in muscles of lower limbs have proven most interesting. In addition, the H responses evoked in soleus and rectus femoris muscles are the largest, and they may be compared to the T responses mechanically evoked in the same muscles. Special mention is also made of reflex responses evoked in foot muscles.

Recording of T and H monosynaptic reflex responses of soleus was carried out according to known and well-established methods (e.g., Hugon et al., 1973). Only some particular technical points will be mentioned here.

The subject is comfortably seated in a raised chair with knees flexed at 120° and ankles at 90° and is instructed to remain relaxed (Fig. 1). Such a sitting posture greatly facilitates appropriate relaxation and allows T responses to be easily elicited in quadriceps and soleus at optimal rest length. Flexion of the knee permits the stimulating electrode to be conveniently placed over the tibial nerve in the popliteal fossa. Room temperature for the experiment is kept between 20° and 25°C. The soleus T response is evoked by percussion of the Achilles tendon with an electromagnetic hammer constructed from a Bruel and

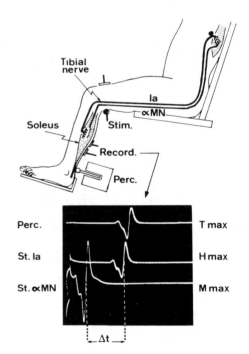

FIG. 1. T and H responses evoked in soleus muscle. Figure shows the position of the subject, the placement of the percussion device and of stimulating and recording electrodes, typical T_{max}, H_{max}, and M_{max} responses, and the time interval measured to calculate proximal CV.

Kjaer 4.809 vibrator. The triggering pulse (3 to 4 msec) and power amplifier are adjusted to obtain maximal speed and force of percussion. The vibrator is fixed on a frame attached to the chair and provides accurate percussion of the right or left Achilles and patellar tendons, whatever the length of the subjects lower limbs.

Electrical shocks are delivered by an electronic stimulator (Racia Neuro Binaire). The constant-current output can be set up to 50 mA, a power just sufficient for maximal stimulation of the tibial nerve in the popliteal fossa in some obese or edematous subjects. The soleus responses, recorded with stainless steel electrodes (Alvar Diamant), are passed through differential amplifiers (3 Hz–20 kHz, 3 db) to an oscilloscope (Tektronix R 7 613) and a magnetic tape recorder (Philips Analog 7).

For soleus reflexes, the surface recording electrodes are placed on the calf in line with the Achilles tendon, with the proximal midway between the two extremities of the fibula and the other 3 cm away distally. The H-reflex response is evoked by electrical stimulation of the popliteal branch of the sciatic nerve in the popliteal fossa. Square-wave shocks of 1.5-msec duration are delivered at a frequency of 0.3 Hz. In each recording session, the maximal amplitude of the direct motor response (M_{max}) is first determined; subsequently, the stimulus intensity is progressively decreased to evoke a maximal reflex response (H_{max}). In some subjects, complete recruitment curves of H and M responses are established. The M_{max} and H_{max} values are based on mean amplitude of three successive responses at a constant stimulus intensity. The soleus T response is evoked by maximal percussion of the Achilles tendon at a frequency of 0.3Hz with the electromagnetic hammer perpendicular to the tendon at the level of the lateral malleolus. The mean amplitude of the five best out of 20 consecutive responses is taken as the T_{max} response.

Quadriceps H reflexes are evoked in the rectus femoris muscle by monopolar stimulation of the femoral nerve just below the inguinal ligament and outside the femoral vessels. The proximal surface recording electrode is placed on the inferior part of the rectus femoris muscle belly, and the distal one 4 cm away on the tendon (Guiheneuc and Ginet, 1974) (Fig. 2).

H-reflex responses in distal muscles can be obtained in the upper and lower limbs. However, those evoked in foot muscles (flexor hallucis brevis, extensor digitorum brevis) have proven most useful for peripheral neuropathies. By stimulating the mixed nerve at two points, it is possible to measure CV in distal segments of afferent fibers (see Fig. 6). However, even in normal subjects, it is often difficult to ascertain that the reflex response evoked in these muscles is a true H response (caused by Ia fiber stimulation) and not an F response possibly resulting from antidromically activated motoneurons (MN). An H response may be

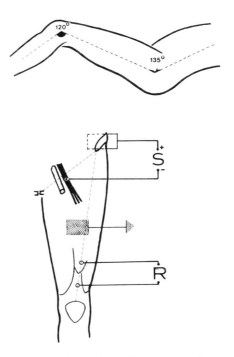

FIG. 2. Experimental conditions to evoke an H reflex in rectus femoris muscle. (From Guiheneuc and Ginet, 1974, with permission.)

assumed only if the shape of the reflex response is stable and similar for the two stimulation sites and if an increase of stimulus intensity so as to evoke an M_{max} response provokes a clear decrease or a complete disappearance of the reflex response. For the reflex response evoked in the flexor hallucis brevis muscle, such characteristics are present in only about 30% of normal subjects (Guiheneuc, 1974) and are lacking in most patients suffering from a peripheral neuropathy (Panayiotopoulos and Lagos, 1980). An H reflex can be obtained in several muscles of the upper limbs. The response evoked in the flexor carpi radialis is found in 90% of normal subjects (Deschuytere et al., 1976; J. Deschuytere, C. De Keyser, M. Deschuttere, and N. Rosselle, *this volume;* Jabre and Stalberg, 1980; Jabre, 1981).

Quantitative Methods and Data Handling

The amplitude of M, H, and T responses is measured from peak-to-peak voltages. The T/H and H/M ratios are calculated from the mean T_{max}, H_{max}, and M_{max} responses measured as described above. The recruitment curves of the H and M responses are recorded on three-dimensional graphs: the intensity of the stimulus *(I)*, as a multiple of the threshold intensity (I_0) of the reflex response, is on the abscissa; the amplitude of the H and M responses is indicated on the ordinate as a percentage of the maximal amplitude of the direct motor response (M_{max}); the latency of the responses is on the Z axis (see Fig. 11).

The time interval separating the M and H responses is measured between the points of sign reversal of the two waves (maximal response traces) (Fig. 1). Measurement of this time interval between H and M responses offers three advantages over determination of H response latency alone: (1) it corresponds to conduction time over two trajectories (afferent and efferent) of equal length (distance between popliteal fossa and spinal reflex center); (2) the neuromuscular synaptic transmission time is eliminated from the measurement; (3) the times are measured from the base line crossings of the H and M responses, which show least variations in latency when the stimulus intensity is changed (Paillard, 1955). This interval corresponds to the time that the reflex impulse takes to complete the arc: popliteal fossa–spinal reflex level–popliteal fossa. Although the length of this arc cannot be measured directly, studies on cadavers have shown that it can be considered a relatively constant fraction (80%) of the subject's height. Moreover, there is a strict correlation between the time separating the M and H responses and the height of a normal adult subject (Fig. 3).

A large number of animal studies, as well as the experiments of Magladery et al. (1951) in man, indicate that the spinal synaptic delay in the Hoffmann reflex arc is less than 1 msec. Thus, it is possible to calculate an average CV (afferent and efferent) over the proximal segments of the sciatic nerve:

$$\text{Proximal CV} = \frac{\text{Height in mm} \times 0.8}{\Delta t(M \to H) - 1 \text{ msec}}$$

Parabolic transformation of the preceding formula allows an H index to be calculated, offering the advantage of a useful expansion of the measuring scale into the zone of the usual variations in proximal CV (Fig. 4) (between 60 and 35 msec, the relationship between proximal CV and the H index is almost linear):

$$\text{H index} = \frac{\text{Height (cm)}^{-2}}{\Delta t (M \to H) \text{ (msec)}} \times 2$$

With this formula, the conduction value of the reflex pathway is expressed in arbitrary units. The coefficients have been chosen so that normal value of the H index is close to 100.

Linear correlations have been carried out by the usual methods; the 95% tolerance ellipses and hyperbolas have been com-

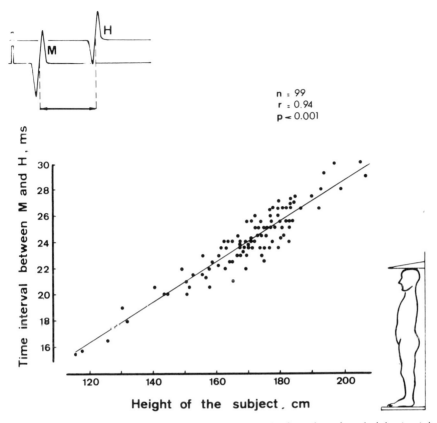

FIG. 3. Time interval between M and H responses (msec) plotted against height (cm) in normal subjects.

puted. For the determination of nonlinear correlations, the dispersion of points on the graph has been divided into zones on the abscissa, the mean being calculated according to X and Y values for the points in each zone. The increment between two zones is defined by the step that permits recovery of about 95% of the points in each zone (see Fig. 7).

NORMAL VALUES

Normal values for the H reflex were obtained from 68 adult subjects, 19 to 54 years old (mean 32 ± 6 years) (Table 1). None of these subjects had ever complained of neuralgia, root pain, or lumbago, and all had tendon reflexes normal in threshold and amplitude in all four limbs. The motor CV in the popliteal and/or tibial nerve(s) was normal in all subjects. In addition, we explored 105 normal infants from 1 day to 5 years of age, born at term and free of neurological disorders (Vecchierini and Guiheneuc, 1979, 1981), and 36 patients aged between 60 and 85 years, hospitalized for benign diseases, and whose clinical neurological examination was normal.

H/M Ratio in Soleus Muscle

In normal subjects, the H response often has a threshold lower than that of the M response and a maximal amplitude roughly half of the M_{max} voltage (H/M ratio = 0.54 ± 0.10 in adults). When stimulus intensity is increased to evoke an M_{max} response, the H response disappears.

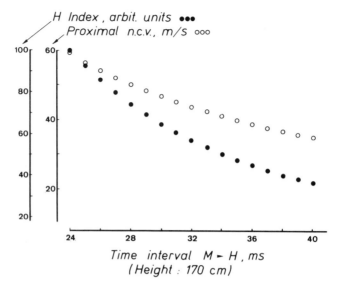

FIG. 4. Proximal CV (m/sec) and H index (arbitrary units) for similar values of the time interval between M and H responses (msec). Note the scale magnification obtained with the H index formula (100 to 36) compared to the change in proximal CV (60 to 35) for a similar increase of the time interval between M and H responses (24 to 40 msec). Values calculated for a subject 170 cm high.

In children (Fig. 5), the H/M ratio diminishes from birth (0.65 ± 0.10) to 1 year of age (0.40 ± 0.13) and then slightly increases up to 3 years, reaching nearly the same values as in adults. In subjects over 50 years of age, the mean value of the H/M ratio decreases, but the dispersion of individual values is enhanced (H/M = 0.32 ± 0.15 between 70 and 80 years).

This H_{max}/M_{max} ratio affords an estimate of the population of soleus MNs that is recruited by electrical stimulation of the afferent pathway of the monosynaptic reflex. Measurement of this ratio, and not of the absolute amplitude of H_{max}, avoids errors related to interindividual variations in the volume of the conductor between recording electrodes and origin of potentials (site of activation of the soleus in the end-plate zone) (Hugon, 1973a). However, the relatively large range of results in normal adults should be pointed out (0.31 to 0.81).

The T_{max} Response and T/H Ratio

The mean T_{max} value in normal adults is smaller than the corresponding H response (T/H ratio 0.76 ± 0.13) and M response (T/M ratio 0.41 ± 0.14). This T_{max}/M_{max} ratio makes it possible to estimate the percentage of MNs of soleus recruited by a maximal stimulus of the spindles, whereas the H_{max}/M_{max} ratio relates to the excitation of Ia fibers in the popliteal fossa. Consequently, the T_{max}/H_{max} ratio provides an indication of the functional value of the primary afferents (spindles plus extremities of the Ia axons) between the soleus muscle and the

TABLE 1. *Evaluation of normal H-reflex parameters*

	N	H/M	T/H	H latency (msec)	$\Delta t(M \to H)$	H index
Soleus	68	0.54 ± 0.10	0.76 ± 0.13	30.2 ± 1.1	24 ± 1.0	100.3 ± 7.1
Rectus femoris	18	0.36 ± 0.06		14.5 ± 0.95	9.9 ± 0.9	

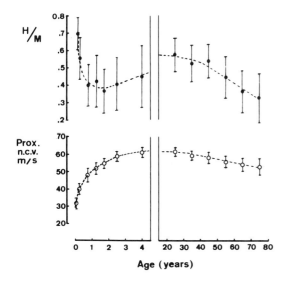

FIG. 5. Normal values (±1 SD) of H/M ratio and proximal CV in children from birth to 4 years of age (left part of the graph) and adults (right part). Values in children reproduced from Vecchierini and Guiheneuc (1979).

popliteal fossa. The fact that T response in normal subjects is smaller than the H response may result from an inadequate or insufficiently strong stimulation of annulospiral endings by the hammer. It may also be caused by an adaptational behavior of the fusimotor activity of γ-MNs during repeated spindle activation at a constant frequency.

Proximal CV and H Index

In normal adults, the proximal CV is 59.2 ± 2.6 m/sec, which corresponds to an H index of 100.3 ± 7.1 (arbitrary units). In children, the proximal CV progressively increases from 31.1 ± 3 m/sec at birth to 47.4 ± 5.3 m/sec at 9 months and 54.7 ± 3.4 m/sec at 18 months. Adult values are obtained at around 3 years of age. In adults and aged patients, the decrease of proximal CV we have noted is very slow, being equal to about 2 m/sec for each 10 years of age after the age of 30 (Fig. 5).

In normal subjects, the proximal CV appears to be constantly higher than the distal motor CV of the same nerve trunk. But it is important to point out that the CV measured in the two cases does not refer to the same structures: the distal motor CV is related to the impulse propagation velocity in α MNs of the tibial nerve. On the other hand, the time interval between the H and M responses corresponds to a combined afferent and efferent conduction time. The changes in proximal CV may involve the Ia fibers, the intramedullary circuit, or the MNs of the reflex arc. But, since synaptic delay is very short, changes of proximal CV and H index are generally caused by lesions of afferent and/or efferent fibers.

Finally, it may be asked if the height of the subject is closely correlated to the length of the reflex arc. First, height is the most easily measured body dimension. Second, laws of human growth indicate that there is a good relationship between the increase in length of a limb and the increase in height. Third, a biometric length measurement becomes increasingly reliable as it relates to a greater number of different inline body segments. But it must be kept in mind that these proximal CV and H index formulas cannot be used in subjects whose relative body segment growth has been uneven.

Normal Findings in the Quadriceps Muscle

In 18 normal subjects (17 to 33 years old), the H/M ratio in the rectus femoris muscle is lower than that in the soleus muscle (0.36 ± 0.16). But it should be noted that passive lengthening of the quadriceps, produced by a modification of the knee or hip joint angles, induces a sharp reduction of the H response. The value of the H/M ratio indicated here corresponds to the above-described experimental conditions (Fig. 2). The mean latency of the H response is 14.5 ± 0.9 msec and the time interval between M and H responses is 7.5 to 12.5 msec (mean 9.9 msec) (Guiheneuc and Ginet, 1974).

Afferent Proprioceptive CV

A clear H response was evoked in the flexor hallucis brevis muscle in 15 out of 49 healthy young men (17 to 33 years of age). In other subjects, only an F response was elicited. The amplitude of the H response was small (H/M ratio 0.10 ± 0.07) (Fig. 6).

Afferent CVs measured in these 15 subjects according to the Liberson method were between 39 and 65 m/sec (mean 49.2 ± 7.2 m/sec) (Guiheneuc, 1974). This value is rather similar to those found by Liberson (1963) and Wager and Buerger (1974). A distal sensory CV clearly related to Ia fibers is not easily measured even in normal subjects.

RESULTS IN POLYNEUROPATHIES

The results obtained in patients with polyneuropathies may vary considerably from one subject to another according to the cause, duration, severity, and localization of nerve lesions. However, systematic exploration of a large number of patients over the past 10 years has enabled us to distinguish response patterns observed in homogeneous populations of subjects presenting the same neuropathy.

The following method of presenting the results has proven the most effective: the amplitude of H responses (as expressed by the H/M ratio) and of T responses (T/H ratio) for each subject examined is plotted against the nerve CV (proximal CV or H index) of each subject (Fig. 7). From the points corresponding to individual values, a successive mean curve is calculated (see Methods), permitting relative changes to be described in the two parameters studied, amplitude versus CV. When there is a worsening of the polyneuropathy, the amplitude of responses decreases, and nerve CV diminishes; thus, the mean curves tend to shift from the upper right-hand part of the graph (normal values) toward the lower left-hand part. To simplify the presentation, here, only the successive mean curves appear on the graphs along with an indication of the standard deviation for each pair of parameters. But the method of calculation

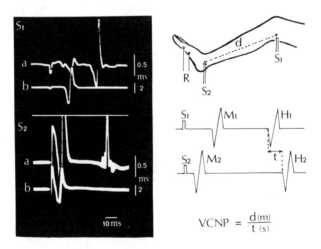

FIG. 6. Method to calculate afferent proprioceptive fiber CV (VCNP) according to Liberson (1963). A reflex response (a) is evoked in the flexor hallucis brevis muscle by stimulating the tibial nerve first at the popliteal fossa (S_1), then at the ankle (S_2). Afferent CV is obtained by dividing the distance separating the two stimulation points (*d*) by the difference of latencies of the two reflex responses (*t*). In b, recording of the maximal M motor responses. Note the difference in threshold and shape of the reflex responses, a not uncommon finding even in normal subjects.

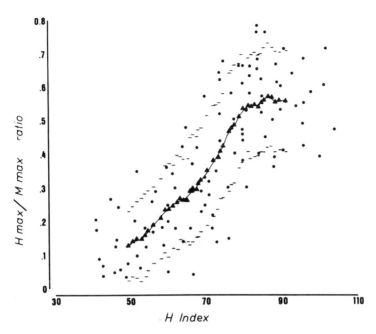

FIG. 7. Study of relationship between the amplitude of H reflex response (H/M ratio, *ordinate*) as a function of the proximal CV (H index, *abscissa*) in 103 uremic patients. From individual values *(points)*, successive means *(triangles)* and standard deviation *(dashes)* are calculated.

for all curves was the same as that used in Fig. 7.

Different Patterns of Reflex Data

Type I.

One hundred three patients aged 18 to 63 years (mean 38 ± 12) with terminal chronic renal insufficiency were examined. Their glomerular clearance was equal to or less than 10 ml/min, and all underwent systematic laboratory examination whether or not they had clinical signs of polyneuropathy. Electrophysiological investigation was undertaken prior to any hemodialysis. Recordings from patients suspected on clinical grounds of having some other cause of peripheral neuropathy, such as alcoholism, diabetes, amyloid disease, radicular lesions, or previous treatment with isoniazid, nitrofurantoin, immunodepressors, etc. were excluded. The results reported here thus concern only a homogeneous population of severe nondialyzed uremic patients (Fig. 8).

For these patients, the most notable fact was the significant decrease in proximal nerve conduction velocity (mean H index 71 ± 13.5). For certain patients, H index values were as low as 50 or 60 (corresponding to a proximal CV of 42 to 50 m/sec). The H/M ratio was significantly lower in comparison with normal values (0.38 ± 0.17) (Fig. 9). In fact, the H/M ratio seems to decrease only slightly when the H index is above 75 to 80. Below these values, the H-response amplitude diminishes more rapidly, and there is a simultaneous change in proximal CV such that the curve of means (H/M plotted against the H index seems to have two sections with different slopes. The study of recruitment curves (see Fig. 11) shows that in uremic subjects the threshold of the H response is most frequently below that of the direct motor response. In the majority of cases, the maximal amplitude of the H and M responses is obtained by stimulus intensities close to those found necessary in normal subjects.

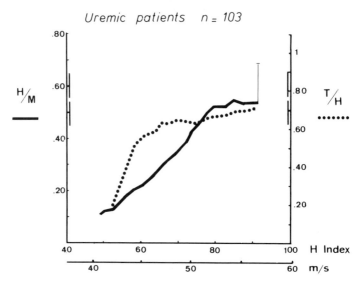

FIG. 8. Mean values of H/M ratio *(black solid line)* and T/H ratio *(dotted line)* plotted against proximal nerve CV *(abscissa)* in uremic patients. Curves are obtained according to the method described in Fig. 7. *Vertical bars* are indications of the SDs for the two parameters.

In lower limbs, the recordings are comparable on the two sides.

For uremic patients, the T/H ratio remained normal during a long period of the course of the polyneuropathy, indicating a parallel decrease in the amplitudes of the tendon and Hoffmann reflexes. When the polyneuropathy became more severe (H index below 60–65), the T/H ratio diminished rapidly.

Apart from chronic renal insufficiency, the same kinds of results were observed in other metabolic diseases (multiple myeloma, macroglobulinemia), in certain collagenoses (periarteritis nodosa, polyarthritis rheumatica), and in certain congenital neuropathies (metachromatic leukodystrophies. Dejerine–Sottas disease; a family afflicted with a hypertrophic form of Charcot–Marie–Tooth disease). The results were relatively similar in certain radiculopathies (see below).

Type II.

One hundred twenty-five chronic alcoholics aged 28 to 68 years (mean 46 ± 8) were examined. Interview revealed a long history of alcohol intoxication, in which it was almost impossible to date the beginning exactly. The patients were hospitalized either for a complication of hepatic cirrhosis (hemorrhage, jaundice, ascites, or edema) or for trophic complaints. Laboratory investigation of hepatic metabolism (BSP clearance, prothrombin level, serum protein analysis) always showed distinct abnormality; cirrhosis was verified by liver biopsy. All patients in whom the neuropathy could have been caused by something other than chronic alcoholism were excluded. Also excluded were all patients showing disturbance in glucose metabolism: fasting blood glucose equal to or above 1.2 g/liter glucose tolerance curve of diabetic type (glycemia above 1.8 g/liter in a sample and above 1 g/liter 1 hr after absorption). The results, therefore, concern only 125 nondiabetic alcoholics (Fig. 10).

For these patients, the H response was always sharply reduced (mean value of the H/M ratio 0.12 ± 0.11). For 21 subjects, it proved impossible to obtain a soleus H re-

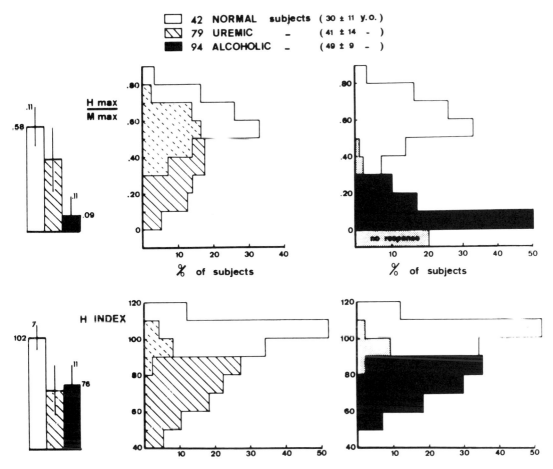

FIG. 9. Means (±1 SD) and distribution of H/M ratio and H index values in 42 normal *(white)*, 72 uremic *(cross-hatched)*, and 94 chronic alcoholic *(black)* subjects. (From Guiheneuc and Bathien, 1976, with permission.)

flex in either leg. Proximal nerve CV was generally moderately decreased, an H index below 65 being rare.

Figure 11, which compares recruitment curves for H and M responses in a normal and an alcoholic subject, shows a considerable reduction in amplitude and a flattening of the reflex response recruitment curve for the alcoholic as well as a threshold for the H response equal to that for the direct motor response. The latency of the H response is moderately increased (Fig. 11). The amplitude of the T response was diminished, but less than that of the H response, so that a T/H ratio greater than 1, and sometimes of 1.5 or 2, could often be noted. When the neuropathy was more severe, the T and H responses disappeared.

Similar results were observed during hereditary polyneuropathies (Thevenard's disease or type-I HSN) and during prolonged treatments with isoniazid and nitrofurantoin.

Type III.

Soleus T and H responses were studied in 27 patients (15 to 51 years of age) suffering from Hodgkin's disease and receiving 1.4

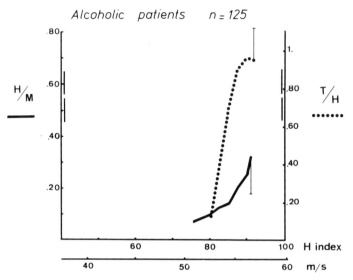

FIG. 10. Mean values of H/M ratio *(black solid line)* and T/H ratio *(dotted line)* plotted against proximal CV in alcoholic patients.

mg/m² of vincristine i.v. on days 1 and 8 during 3 consecutive months. The same protocol was repeated for certain patients as of the beginning of the fifth month of treatment. Each patient was subjected to an electrophysiological examination before and after each week of vincristine administration (Guiheneuc et al., 1980a). In these patients, a very rapid decrease in the T response was observed, with disappearance of the response once the dose of vincristine reached or exceeded 6 to 8 mg/m² (Fig. 12). At the same time, the CV over the reflex arc remained normal, the mean of the H index still being close to 90 when the T response disappeared. Likewise, the H/M ratio remained normal at the beginning and even increased over several days following each administration of vincristine (Fig. 13). When the neuropathy subsequently worsened, as a result of increasing doses of vincristine, the H response diminished without any serious alteration ever occurring in CV over the reflex arc.

Intermediary type responses.

A pattern between two of the types described above was frequently observed. In Fig. 14, the amplitude of the H/M ratio is indicated as a function of the proximal nerve CV in chronic alcoholics. The curve represented by a solid line corresponds to results in patients presenting no clinical or biological signs of diabetes (population of type II). The curve represented by a dotted line was obtained from results relative to alcoholics having an elevated glycemia when fasting or a perturbed curve of hyperglycemia. The results in this latter group seem to be intermediary between those of the "pure" alcoholics (type II) and those of the uremics (type I). Changes in soleus monosynaptic responses in other polyneuropathies are also found of the "intermediary" type between types I and II: primary amyloidosis (Portuguese families) and polyneuropathies during perhexiline maleate treatments.

Other patterns of results.

During certain cases of distal arteriolitis, responses were similar to those of type III, but after extended follow-up of these patients, it could not be determined that there was progression of the distal axonal lesions toward the proximal segments of nerve

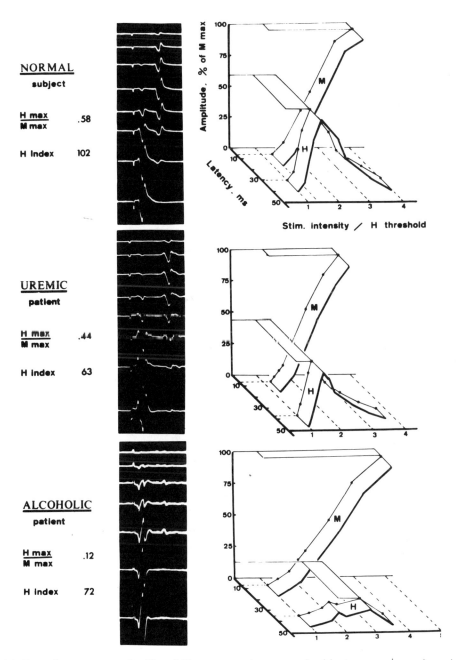

FIG. 11. Recruitment curves for H and M responses in a normal subject, a uremic, and an alcoholic patient. (From Guiheneuc and Bathien, 1976, with permission.)

trunks, the H/M ratio having remained normal or modified slightly *(unpublished data)*.

In diabetics, the results observed corresponded, according to the patients, to one or another of the types described above. It should be added that we obviously did not study all possible varieties of polyneuropathy. Only those etiologies are mentioned here for which we could observe a sufficient number of cases to provide a worth-

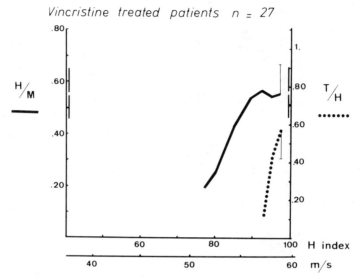

FIG. 12. Mean values of H/M ratio *(black solid line)* and T/H ratio *(dotted line)* in vincristine-treated patients.

while assessment of the form of the soleus T and H responses. Other types of reflex responses are possible during polyneuritis; for example, neuropathies occuring during leprosy, if they entail a moderate increase in soleus H-response latency, are accompanied by signs of local compression and axonal degeneration and do not correspond to any of the types described above (Sebille, 1980).

Significance of Results

The amplitude and latency of soleus T and H responses may thus be very different according to the type and severity of the polyneuropathy under study (Fig. 15).

Diminution of the H index.

A diminution of the H index indicates a reduction in the CV over the arc popliteal

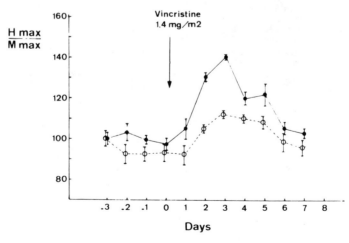

FIG. 13. Variations in the H/M ratio in percent of initial values in two patients from 3 days before to 7 days after a first vincristine injection of 1.4 mg/m^2. (From Guiheneuc et al., 1980, with permission.)

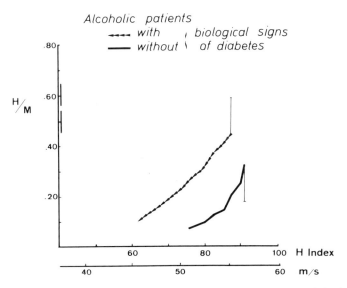

FIG. 14. Mean values of H/M ratio plotted against proximal CV in alcoholic diabetic patients *(arrow line)* and in alcoholics without biological sign of diabetes *(solid line)*. The *solid line* is the same as in Fig. 10.

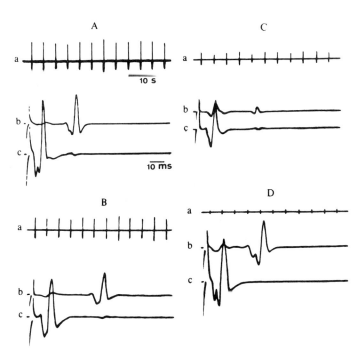

FIG. 15. Typical records of T_{max} (a), H_{max} (b), and M_{max} (c) responses evoked in the soleus muscle in a normal subject (**A**), a patient with a chronic renal insufficiency (**B**), an alcoholic patient (**C**), and a patient treated with vincristine (**D**). Note in **B** the lengthening of the H-response latency, the T/H and H/M ratios remaining almost normal; in **C** the low H-response amplitude with a T/H ratio above normal; in **D**, the T-response decrease, the H response being normal.

fossa–spinal reflex level–popliteal fossa, i.e., on the proximal segments of sciatic nerve fibers. The change in CV value may involve the afferent fibers, intramedullary circuit, or efferent fibers of the reflex. An increase in the time course over the reflex arc is the dominant characteristic of peripheral neuropathy in chronic renal insufficiencies and has been observed by other authors (Panayiotopoulos and Lagos, 1980). In fact an experimental segmental demyelination is accompanied by a sharp reduction in CVs (Kaeser and Lambert, 1962; McDonald, 1963; Lehmann and Ule, 1964; Mayer and Denny-Brown, 1964; Hall, 1967; Morgan-Hughes, 1968; etc.). Moreover, many authors have described extensive lesions of the Schwann cells and myelin sheaths in uremic patients (Asbury et al., 1963; Forno and Alston, 1967; Dinn and Crane, 1970; Kornfeld and Appenzeller, 1970; etc.). Thus, a sharp decrease in the H index is caused by early proximal demyelinating lesions of the peripheral nerves. The more moderate reduction in proximal CVs in alcoholics as well as the absence of change in the H index at the beginning of vincristine treatment indicate mild or inapparent functional alterations of the sheaths. However, a reduction in CV may appear before there is any clearly discernible lesion of the periaxonal sheaths (Sharma and Thomas, 1974). It is thus possible that a metabolic disorder of the neuron itself or a membrane defect contributes to the decrease in CV (Nielsen, 1973; Behse and Buchthal, 1978; Rasminski, 1978).

Proximal versus distal CV changes.

The proximal CV measured in a uremic or an alcoholic subject was plotted against the tibial nerve motor CV obtained in the same subject. Thus, CV in proximal and distal segments of the same nerve trunk could be compared. In both uremic and alcoholic patients there was a linear relationship between the two values, and the difference in slope of the two regression lines was not statistically significant (Fig. 16). These results are consistent with lesions diffuse along the whole length of peripheral nerves (proximal and distal parts). Relationships between proximal CV (obtained with the H reflex) and distal motor CV have not been

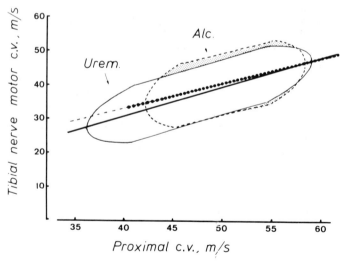

FIG. 16. Relationship between proximal nerve CV *(abscissa)* and tibial nerve motor CV *(ordinate)* in uremic *(clear area)* and alcoholic *(shaded area)* subjects. For each group of patients, the regression line and the area limited by the 95% tolerance *ellipses* and *hyperbolas* were obtained from individual values.

tested in other polyneuropathies except for some radicular lesions (see below). Some studies are concerned with findings obtained using the F wave in distal muscles (Kimura, 1978; J. Kimura, *this volume*).

Decrease of the H/M ratio.

The H_{max}/M_{max} ratio makes it possible to estimate the percentage of soleus MNs that can be recruited by Ia fibers in the popliteal fossa. A decrease of this ratio indicates that a portion of the MN pool remaining active (unimpaired) can no longer be affected by stimulation of the afferent pathway. If, in peripheral neuropathies at the onset, spinal synaptic transmissions are intact (central controls and cutaneous polysynaptic reflexes evoke normal responses), decrease of the H/M ratio would imply that there are proximal lesions (between the popliteal fossa and the α MNs) of the primary afferent fibers.

In alcoholic patients, the H/M ratio, plotted against proximal CV, shows an early sharp decrease. A 10% diminution of proximal CV corresponds to an 80% decrease of the H/M ratio from its normal value. In chronic alcoholics, distal axonal lesions have been detected by electrophysiological and histological examination on nerve biopsy (Walsh and McLeod, 1970; McLeod et al., 1973; Behse and Buchthal, 1978; Tackmann et al., 1977; Said and Landrieu, 1978). The decrease in the H response indicates that the main part of the MN pool can no longer be recruited by afferent fibers. Thus, the diminution of the H/M ratio is related to important early proximal lesions of Ia fibers, so it would seem more appropriate to describe this type of neuropathy as a diffuse axonal degeneration and not as a uniquely distal axonal lesion (Fig. 17). The change of slope in the lower part of the H/M curve in alcoholics occurs because, in many patients, when the polyneuropathy becomes more severe, no clear H response is found.

In uremic patients, the H/M ratio decreases only slightly when the H index is higher than 75 or 80. Below these values, H response amplitude decreases more rapidly, so the curve (solid line) appears to be made up of two segments with different slopes. We suggest that the early phase of this polyneuropathy involves pure or almost pure demyelinating lesions. The slope of the upper section of this curve would thus represent the physiological events related to demyelination, so that a 10% decrease of

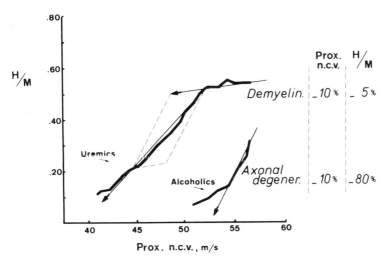

FIG. 17. Possible contribution of axonal degeneration and of demyelination to the changes of H/M ratio plotted against proximal CV in alcoholic and uremic patients.

proximal CV corresponds to a very slight decrease (about 5%) of reflex discharge. When the lesions worsen, the slope changes. A 10% decrease of proximal CV is now accompanied by a 30% decrease of the H/M ratio. We postulate that axonal degeneration of Ia fibers, together with demyelination, is responsible for this more rapid decrease in amplitude of the H response.

In alcoholic patients with diabetes, the curve of means, plotted from changing H/M values as a function of changes in the H index, is of the "intermediary" type, being located between the results of uremics and those of nondiabetic alcoholics. It may be assumed that these subjects develop a mixed polyneuritis associating axonal degeneration and demyelinating lesions. It is thus necessary in studying peripheral neuropathies in alcoholics to recognize cases involving a disorder in glucide metabolism. Other biological abnormalities (metabolism of certain vitamins, synthesis of immunoglobulins) may be relevant to correct interpretation of the results in these patients. These "intermediary" results were regularly observed in several families of patients afflicted with primary amyloidosis (Portuguese type) and in subjects treated with perhexiline maleate. It would thus seem likely that, as in the case of diabetic alcoholics, demyelination and axonal degeneration lesions coexist during these neuropathies (see also Sebille, 1978).

Increase of the H/M ratio.

An increase of the H/M ratio indicates a monosynaptic reflex hyperexcitability present distally from the fusimotor controls, since the H response biases neuromuscular spindles. It may be caused by a lowering of the threshold of the α MNs (either spinal or supraspinal or secondary to a decrease in inhibitory peripheral influence) or by an increase in afferent impulses (an increase in the discharge of the Ia fibers or diminished presynaptic inhibition). The cause of increase in H/M ratio during the 48 to 72 hr following the first administrations of vincristine is unknown. It may be linked to an increase in discharge of myelinated afferent fibers undergoing deterioration, a phenomenon that has been described subsequent to mechanical (Wall et al., 1974; Howe et al., 1976), ischemic (Torebjork et al., 1979), or dystrophic (Rasminsky, 1976) lesions. However, the possibility of a specific effect of the drug on α MN excitability cannot be excluded (Albuquerque et al., 1972).

Decrease of the T/H ratio.

The very rapid decrease in the T_{max}/H_{max} ratio during vincristine treatment is in very close correlation with the dose administered. It cannot be a result of hypoexcitability of the γ MNs since it is concomitant with an increase in monosynaptic reflex excitability (increase of H/M after vincristine injections). Lesions of the muscle spindle fibers also seem improbable if the drug doses administered to our patients are taken into account. Finally, the decrease in the T/H ratio with normal H/M ratio at onset of the polyneuropathy points to serious lesions of the distal Ia fibers (the proximal segments remaining unimpaired) and indicates that the neuropathy induced by vincristine has the characteristics of a quite typical retrograde axonal degeneration (Guiheneuc et al., 1980a).

Increase of the T/H ratio.

An increase in the T/H ratio may mean that there is a reflex hyperexcitability concerning solely or primarily the activity of the γ MNs. Such an increase is observed in chronic alcoholics when the polyneuropathy is not too advanced. At the the same time, the T and H responses of the quadriceps sometimes have an amplitude appreciably above normal, whereas the soleus H response is already sharply reduced. It is thus possible that in the phase of rapid aggravation of axonal lesions, a reflex hyperexcitability (α and γ) may be observed;

however, in the case of the quadriceps, it does not seem chiefly to involve the γ loop.

It seems more likely that the increase in the T/H ratio occurs because neuronal degeneration first affects afferent fibers of large diameter. Those that remain functional thus have a diameter and a threshold comparable to those of α MNs. Their stimulation cannot evoke an H response, because the reflex impulse collides with the recurrent discharge on simultaneously excited MNs. On the other hand, percussion of the Achilles tendon can provoke discharge of the Ia fibers that have not undergone degeneration, producing a T response which, although diminished, may be higher than the H response for the same subject.

RESULTS IN PATIENTS WITH ROOT LESIONS

Nerve Root Compression by Intervertebral Disk Prolapse

The H reflexes evoked in soleus and rectus femoris muscles were studied in 91 cases of lumbosacral disk prolapse confirmed by radiography and surgery: two L_3–L_4 disk hernias, 48 cases of L_4–L_5 hernia, and 41 cases of L_5–S_1 hernia (Guiheneuc et al., 1973; Descuns et al., 1973). Reflexes were studied in both lower limbs. The results obtained for the limb suspected of having radicular lesions were compared with those of the healthy limb or, in case of bilateral lesions, with normal values (Fig. 18). The amplitude of the H response was always greatly reduced when a disk compressed nerve fibers of the reflex arc studied. There was complete abolition of the rectus femoris H response in both cases of L_3–L_4 hernia and a very great decrease in the soleus H/M ratio in the 41 cases of L_5–S_1 disk hernia (mean 0.06 ± 0.02). In the 48 cases of L_4–L_5 disk prolapse, there was a moderate decrease in right anterior (0.19 ± 0.08) and soleus (0.31 ± 0.14) H/M ratios, but with considerable dispersion in the results, the H responses for the two muscles being in some cases normal and in others almost abolished (Table 2).

In the 41 cases of L_5–S_1 disk hernia, there was no correlation between the decrease in the amplitude of the reflex response or in the H_{max}/M_{max} ratio and the EMG signs at the level of the triceps surae. These signs could at rest be reduced to a few fasciculations, whereas the H response was abolished. Conversely, signs of severe denervation could be found on the EMG when the H_{max}/M_{max} ratio was above 0.10. Proximal nerve CV was generally only slightly modified by an L_5–S_1 disk prolapse (H index 88 ± 11.3, an insignificant varia-

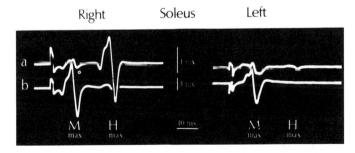

FIG. 18. H_{max} (a) and M_{max} (b) responses evoked in right and left soleus muscles in a patient suffering from S_1 left root compression as a result of an L_5–S_1 disk prolapse. The H/M ratio is normal in the right leg soleus (0.62) and decreased in the left side (0.07). Calibrations: 1 mV, 10 msec.

TABLE 2. *Effects of ipsilateral disk prolapse on H reflex*

	L_3-L_4	L_4-L_5	L_5-S_1
Number of patients	2	48	41
Soleus			
H/M	0.62	0.31 ± 0.14	0.06 ± 0.02
H index	99	92 ± 8.5	88.2 ± 11.3
Rectus femoris H/M	0	0.19 ± 0.08	0.28 ± 0.11

tion from normal values) or L_4-L_5 hernia. The complete abolition of the right anterior H response in both L_3-L_4 hernia cases prevented evaluation of proximal CV at that level. Thus, most disk hernias have little effect on nerve CV in reflex arcs, an opinion confirmed by others (Visser, 1965, 1973; Dreschler et al., 1966; Bouquet and Cioffi, 1969). However, certain disk hernias whose course extends over months or years and which present important signs of denervation in the region of the compressed root may be accompanied by and appreciable decrease in the H index, a reduction in the M_{max} response, and a paradoxical increase in the H/M ratio (Descuns et al., 1973).

Carcinomatous Radiculopathy

The H-reflex response evoked in soleus muscle was studied in 36 patients referred because of pain in the sciatic nerve region which before or after the electrophysiological examination was related to a carcinomatous radiculopathy. The results were compared with those of 41 subjects explored for an L_5-S_1 disk hernia confirmed by surgery (Fig. 19). The amplitudes of the H response and of the H/M ratio were reduced similarly in both groups of patients. In the case of mechanical compression of the first sacral root by an acute L_5-S_1 disk prolapse, proximal CV was only slightly decreased (55 ± 4.1 m/sec), whereas distal motor CV was almost normal. When pain along the sciatic nerve was caused by a carcinomatous radiculopathy (lymphopathies; lung, breast, kidney, or intestinal cancers, proximal CV was sharply reduced (50

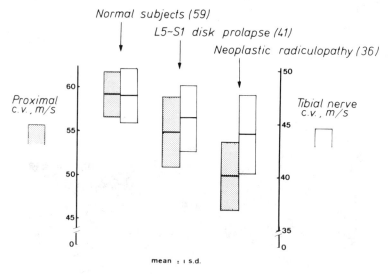

FIG. 19. Mean values (±1 SD) of proximal CV *(shaded columns)* and tibial motor CV *(white columns)* in normal subjects (left), patients suffering from an S_1 root compression as a result of a disk prolapse (middle), and patients having a carcinomatous radiculopathy (right). Note the striking decrease of proximal CV in this third group.

± 3.8 m/sec), often much more than the distal CV. On several occasions, such a discrepancy between proximal and distal CV led us to suggest complementary clinical investigations which allowed a previously undetected malignant disease to be found in some patients.

Infectious or Inflammatory Radiculopathies

Fifteen subjects with a Guillain–Barré syndrome were examined. For 13 of them, no H-reflex response could be evoked in soleus of either leg. In two subjects, it was possible to study an H response, for which, respectively, the amplitude of the H/M ratio was 0.05 and 0.08 and the H index 55 and 61. Motor nerve CV of the external popliteal sciatic nerve was also reduced in both cases (33 and 36 m/sec), and distal latency greatly increased (8.5 and 6.1 msec). However, the small number of cases in which it was possible to evoke an H response limits the relevance of this type of exploration. In subjects presenting a Guillain–Barré syndrome, the proximal segments of MNs can be studied by using the F wave (Kimura, 1978; J. Kimura, *this volume*).

RESULTS IN DRUG THERAPY TRIALS

The study of monosynaptic reflex responses can be useful in testing the effectiveness of drugs with potential beneficial influence. A new drug, isopropylamino-2-pyridine phosphate (IAPP, Nerfactor®, Laboratoires Ipsen), was administered according to a double-blind protocol to a group of 15 subjects afflicted with a Hodgkin's disease and receiving 2.8 mg/m^2 of vincristine at the beginning of each month for 3 months. At the end of the experiment, it was determined that eight subjects had received the active drug and seven a placebo of identical appearance. The results showed that the proximal nerve CV and the H/M response were not significantly different for the two groups. However, there was a clear difference in the T/H response (Fig. 20). Vincristine, it is to be recalled, induces a very rapid decrease in this ratio (see above). For patients treated

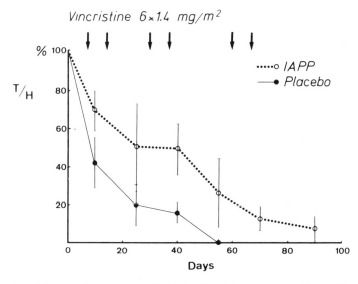

FIG. 20. Changes in T/H ratio (in percent of initial values) in patients suffering from a Hodgkin's disease treated by vincristine (six injections of 1.4 mg/m^2) and receiving orally either 1,500 mg/day of IAPP *(dotted line)* or a placebo *(solid line)*. The T/H ratio is significantly less impaired in IAPP-treated patients.

with IAPP, the mean of the T/H ratio between the 50th and 60th days was 0.26 ± 0.18, whereas for patients receiving the placebo it was 0.6 ± 0.8. These results suggest a protective effect of IAPP against distal axonal degeneration induced by vincristine *(unpublished data)*.

ADVANTAGES AND LIMITATIONS OF THE EXPLORATIONS OF T AND H REFLEXES IN PERIPHERAL NEUROPATHIES

Studies carried out in different laboratories on modifications of monosynaptic reflexes in peripheral neruopathies are still sparse. However, their many advantages should lead to more frequent use of these methods: they may easily be carried out in good technical conditions; the stimulations are not painful; recording of potentials is performed with surface electrodes; and most of the useful measures (amplitude and H index ratios) are simple to take.

The results already obtained indicate that these methods provide information not accessible by other EMG techniques and include CV determinations of proximal nerve trunks and the functional state of proprioceptive afferent pathways. These data have already led to differentiation of several types of polyneuropathies (types I, II, and III presented above) or of possible differences of responses in the same population (alcoholics). They have made it possible to accurately determine in man criteria for the appearance and course of retrograde axonal degeneration (the dying-back phenomenon during vincristine treatment). They offer the possibility of early suspicion of the carcinomatous origin of a sciatic pain and of assessing a radicular compression. Their application leads to the conclusion that not all neurotoxic drugs produce the same type

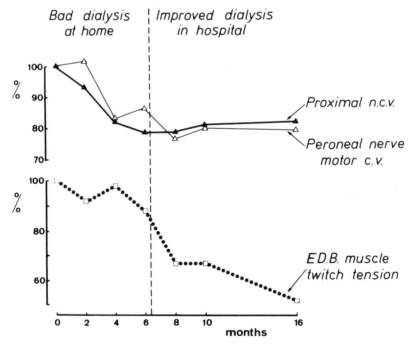

FIG. 21. Changes (in percent of initial values) of proximal and distal CV (upper part of the graph) and of the twitch tension of the extensor digitorum brevis muscle (lower part) in a patient undergoing hemodialysis. See text.

of lesion (for example, isoniazid, perhexiline maleate, and vincristine lead to three different kinds of results). They can also be a valuable tool during therapeutic trials affecting the peripheral nervous system.

Finally, it is difficult to determine a correspondence between the monosynaptic responses (amplitudes and latencies) and motor performance. We have noted a frequent absence of correlation between the modifications of proximal CV and the motor performances evaluated by measurement of muscle force or clinical testing. Figure 21 is an example of such a discrepancy. A patient with chronic renal insufficiency underwent hemodialysis, first at home for 6 months, where conditions of treatment were inadequate, then in our hospital, where dialysis was strikingly improved. The upper curves indicate changes in proximal CV and distal motor CV of the peroneal nerve. In the lower part of the graph, the amplitude of the mechanical response of the extensor digitorum brevis muscle is plotted versus time. The peroneal nerve was maximally stimulated at the ankle with a shock 1 msec in duration, and the twitch response of the muscle was measured using a special device including a strain gauge applied on the toes. During the 6 months of poor dialysis at home, a considerable decrease in proximal and distal CV occurred. Simultaneously, the muscle twitch force remained almost normal. With therapeutic improvement, the CVs no longer decreased, but muscle twitch amplitude further decreased. The use of complementary methods chosen according to the patinet to be examined can lead to a useful study of lesions. In this respect, the H reflex and associated techniques should be more frequently used to explore proximal nerve fibers.

Motor Control Mechanisms in Health and Disease,
edited by J. E. Desmedt.
Raven Press, New York © 1983.

H Reflexes in Muscles of the Lower and Upper Limbs in Man: Identification and Clinical Significance

*J. Deschuytere, C. DeKeyser, M. Deschuttere, and N. Rosselle

Center of Electromyography and Electrodiagnosis, Kliniek Heilige Familie, Antwerp, Belgium; and Department of Electromyography and Clinical Neurophysiology, University of Louvain, Louvain, Belgium

This chapter summarizes our studies on H reflexes elicited in the lower and upper limb muscles in man and considers mainly the standardized techniques and criteria. The H reflexes represent a prominent diagnostic aid when latencies can be accurately determined. The usual nerve conduction velocity (CV) measures provide information about peripheral motor fibers and sensory fibers, but monosynaptic reflexes may yield information about conduction in more proximal segments of peripheral nerve and about lesions of plexuses and roots (see P. Guiheneuc, *this volume*). Techniques of facilitation are sometimes required to bring out reliable H reflexes in some muscles. Moreover vibration-induced inhibitory effects on the homonymous α motoneurons (MN) can be used (cf. Hagbarth, 1973; Delwaide, 1973; Deschuytere et al., 1976; Desmedt and Godaux, 1978c, 1980). When visualized by superimposed oscilloscope recordings, H reflexes are more reliable than F waves whose latencies may vary in one subject when examined at different times (Fra and Brignolio, 1968; Fisher et al., 1979). When the F wave is derived from the hypothenar or thenar eminence, complications may result from either nerve entrapment (carpal tunnel, ulnar nerve at the wrist) or an aberrant course of motor fibers in the ulnar and median nerve, with or without an anastomosis in the forearm. This was one of the reasons why our studies of H reflexes in the flexor carpi ulnaris, flexor carpi radialis, or palmaris longus muscles were undertaken. In addition to motor fiber involvement, H reflexes may reveal Ia fiber involvement, and their elicitation is less inconvenient for the patient than F responses.

H REFLEXES IN THE LOWER LIMB

Electrically induced reflex activity in man was described by Hoffmann (1918). Animal experiments helped elucidate the monosynaptic nature of late motor responses to electrical stimulation of a mixed peripheral nerve. Renshaw (1940) showed central delays of about 0.5 msec in the cat, compatible with a single synaptic delay. Lloyd (1943) demonstrated that the afferent limb of the monosynaptic reflex was related to group I sensory fibers of muscular origin and that these afferents of lowest electrical threshold subserved the monosynaptic reflex which was identified as the two-neuron

---

* To whom correspondence should be addressed: Kliniek Heilige Familie, 89, Provinciale Steenweg, 2621-Schelle, Antwerp, Belgium.

pathway of the reflex response to brief stretch (Lloyd, 1943).

In a classic EMG study of the spinal cord reflex activity in man, Magladery and McDougal (1950) identified the indirect motor responses recorded from calf muscles as monosynaptic reflexes and, in honor of Hoffmann, designated them as H reflexes. Arbitrarily, these authors called "F reflexes" all secondary motor responses in the small muscles of the hand and the foot and also the indirect motor responses derived from the anterolateral muscles of the leg. They postulated that "F reflexes" were produced through a reflex arc of which the afferent fibers are more slowly conducting than motor fibers. They speculated, however, about the possibility that afferent impulses were delayed in multineuronal relays within the spinal cord. Subsequently, Magladery et al. (1951) described true monosynaptic reflex activity in the calf muscles, hamstrings, and quadriceps muscles following a tendon tap or a submaximal electrical stimulation of the mixed nerve supplying the muscle. The afferent fibers of the monosynaptic arc were identified as group I fibers with a CV that was about 10% greater than that of the efferent α axons.

Later investigations challenged the original interpretation of the F waves. Dawson and Merton (1956) proposed that the F waves in man, as in the animal, consisted of recurrent discharges from MNs that are antidromically activated. Mayer and Feldman (1967) demonstrated F waves in a patient with intradural section of the posterior C_7, C_8, and T_1 roots. When records are made from the hypothenar muscles following supramaximal electrical stimulation of the ulnar nerve at wrist or elbow, the latencies of the M wave and the F wave for both sites of stimulation were constant. In a normal subject, however, it was found that the F responses were more complex, with a later part at greater latency than the usual F wave. They concluded that the F wave consisted of an initial antidromic motor discharge and, in some cases, a later response to other slower conducting afferent fibers with polysynaptic connections. With careful testing, the H response could be distinguished from the F wave.

There have been some reports of genuine H reflexes in small muscles of hand and foot (Johns et al., 1957; Angel and Alston, 1964), and Hagbarth (1962) used posttetanic potentiation to facilitate monosynaptic reflexes in other muscles than the calf muscles. Liberson (1962) described H reflexes in other muscles in normal adults by using averaging techniques. Deschuytere and Rosselle (1971) demonstrated genuine H reflexes in the anterolateral muscles of the leg in normal adults. The reason special techniques of facilitation may be successful is the presence of a sufficient number of functionally intact low-threshold Ia fibers in the normal adult with a presumed higher CV than the motor fibers in the same segment of the peripheral nerve.

This was substantiated in newborns and infants up to 1 year in whom true H reflexes were reported in the small muscles of hand and foot (Thomas and Lambert, 1960; Mayer and Mosser, 1969, 1973) and also by the findings of Teasdall et al. (1952) that in neurological disorders (for instance, lower brainstem and spinal cord lesions), H reflexes may appear and replace F waves in the anterolateral muscles of the leg and in small muscles of hand and foot. These data have been repeatedly confirmed (Hofmann and Goodgold, 1961; Mayer and Mawdsley, 1965; Thorne, 1965; Mayer and Feldman, 1967). In infants, the descending pathways are not functionally myelinated, whereas in the abovementioned neurological disorders, the descending paths are interrupted. Therefore, supraspinal inhibition must influence electrically induced reflexes to some extent. Supraspinal inhibition is also active for the H reflexes in the calf muscles, since these responses can be enhanced in patients with upper MN damage (Angel and Hofmann, 1963). Thus, it must be possible to evoke true H reflexes in muscles other than calf muscles, quadriceps, and part of the hamstrings (i.e., physiological extensors) provided supraspinal inhibition is counteracted by facilitation and a sufficient number of Ia fibers are stimulated in a mixed peripheral nerve. Furthermore, convergence may play a role for synergic muscles (Deschuytere and Rosselle, 1971).

The question of whether true monosynaptic responses occur in the extensor digitorum longus muscle arose from our interest in root compression syndromes at the lumbosacral level (Deschuytere and Rosselle, 1970, 1973). The abovementioned muscle is representative of the fifth lumbar roots, and the medial gastrocnemius muscle for the first sacral roots (Knutsson, 1961). As in entrapment neuropathies, when spinal roots are compressed, the CV may be reduced in the afferent and/or efferent fibers, which produces an increase in latency of the corresponding monosynaptic reflex (Liberson, 1963). Prolonged latencies of H reflexes in medial gastrocnemius muscle were observed in chronic compression of the first sacral roots when the affected side was compared to the normal side. With identification of genuine H reflexes in the extensor digitorum longus muscle, the diagnostic use of monosynaptic reflexes in fifth lumbar and first sacral root compression was made possible (Deschuytere and Rosselle, 1971, 1973).

Our studies on the L_2–L_3–L_4 roots *(unpublished data)* are incomplete because the number of patients with root compression syndromes at these levels was insufficient. For electrical stimulation of the femoral nerve in the inguinal region, Teflon®-coated needle electrodes had to be used to stimulate quasiselectively the nerve fibers supplying either the vastus medialis or lateralis. There is an overlapping of nerve fibers from at least two lumbar roots for each of these muscles.

In our clinical studies, the tibial and the common peroneal nerves (at times the deep peroneal nerve selectively) were stimulated with needle electrodes or with fork electrodes (cathode proximal, and cathode–anode distance of 20 mm), respectively, in the popliteal fossa and at the level of the capitulum fibulae. Detection was performed by coaxial needle electrodes in both the medial gastrocnemius and extensor digitorum longus muscle. The coaxial needle was placed to pick up an M wave of short duration and stable latency. Distances from electrical stimulation to detection sites were measured and found to be similar on the ipsi- and the contralateral sides. The intensity of electrical stimulation was subliminal, liminal, or supraliminal but always inframaximal for the M wave when indirect motor responses were elicited. With such an intensity, the F wave was never observed in many thousands of trials. A rectangular current pulse of 2.0 msec duration is most effective for eliciting H reflexes. This is not unphysiological, as the afferent impulses in Ia fibers evoked by tendon tap may last for 2 msec (Lloyd, 1943).

It seems likely that apart from spatial summation, temporal summation may play an important role in the development of reflex activity (Deschuytere and Rosselle, 1971). It was reported for the gastrocnemius–soleus muscle in man (Paillard, 1955) that the 1.0-msec electrical stimulus duration favored the H reflex relative to the direct motor response (in comparison with a 0.1-msec duration), and that the strength of the electrical stimulus could be diminished when the duration of the shock was long (Mayer and Mawdsley, 1965). A tentative explanation of temporal summation is that the electrical stimulus duration of 1.0 or 2.0 msec applied to the peripheral nerve first has excitatory effects (depolarization) for some nerve fibers when the current is opened (cathode) and then has some smaller excitatory effects when the current is turned off (anode). The impulses in Ia fibers caused by the second depolarization may reach α MNs which are already at a higher level of excitability following the first stream of impulses in homonymous and heteronymous Ia fibers. A temporal summation has also been postulated for a tendon tap by Gassel and Diamantopoulos (1966).

Apart from the facilitation obtained during a remote contraction, i.e., the Jendrassik maneuver, we found that slight active contraction (Hoffmann, 1918; Liberson, 1962) or slight passive stretch (Angel and Hofmann, 1963) could facilitate the reflex. Posttetanic potentiation (Hagbarth, 1962) and postvibratory potentiation (Hagbarth and Eklund, 1966) can enhance electrically induced reflex activity, but these procedures are not convenient for clinical work. With paired stimuli of equal strength, autogenetic effects of a contraction of a flexor (the extensor digitorum longus muscle is a physiological flexor) on homonymous α MN excitability may provide facilitatory actions in the third and fourth phases, respectively, by group Ia and group II afferents, as reported in animal experiments. When one records from the extensor digitorum longus muscle, the most effective facilitation for eliciting late motor responses resulted from electrical stimulation with paired stimuli, even in the presence of a well-developed direct motor response (Deschuytere, and Rosselle, 1971). Olsen and Diamantopoulos (1967) had reported that the recovery curves of calf muscles with pairs of maximal H reflexes were identical whether the preceding direct motor response was large or small and concluded that antidromic invasion of spinal MNs should be of short duration.

We started our experiments with paired stimuli of equal strength to the common peroneal nerve so that the M wave evoked by the first stimulus could produce facilitation during the third phase, when there is a discharge in Ia afferents caused by stretch of the extensor digitorum longus muscle in the relaxation phase after about 170 to 250 msec. That was a time-consuming procedure (Deschuytere and Rosselle, 1971), and the next step was a repetition of the stimulation at about 4 to 5/sec using phase III after each preceding M wave and displacing continuously the stimulating electrodes in order to excite effectively a sufficient number of Ia fibers. This technique was more appropriate for clinical practice, and on superimposed recordings, late motor responses could be seen to have different configurations but stable latency when the first deflection was controlled. In many instances, the Jendrassik maneuver was combined, and the subject made a contralateral handgrip which caused the indirect motor response to suddenly appear or present greater amplitude.

When detecting with coaxial needle electrodes, one can easily distinguish a true late motor response from the concomitant contractions of the synergic muscles. In 1971, Deschuytere and Rosselle were able to derive in almost every normal adult an indirect motor response in the extensor digitorum longus muscle showing all the characteristics of the classical H wave. The latencies of the indirect motor responses derived from the extensor digitorum longus muscle varied from 27 to 33 msec in 50 normal adults from 24 to 64 years of age (Deschuytere and Rosselle, 1971). The latencies of the H reflexes recorded from triceps sural varied from 25 to 30 msec. On the average, there was a difference of 3 msec between the latencies of H reflexes of the above-cited muscles in any given subject, and this correlated with the distances measured by calipers from the site of stimulation to the spinous process of the 12th thoracic vertebra and back to the point of recording. For estimation of CV, one should take into account the delays at the monosynaptic relay and neuromuscular junction.

Aging is a hard-to-define factor with respect to CVs of Ia fibers and motor fibers, as there are concepts of chronological age and physiological age. Indeed, in subjects more than 70 years old, we have found greater latencies than those indicated above. We are reluctant to introduce formulas to predict the latencies of H reflexes in relation to age, and the height of the subject is noted; it is questionable if leg length measurements, as proposed by Braddom and Johnson (1974), are valid for evaluating the actual total pathway.

Caliper measurements are more accurate, and it is dubious if the formula of the H index based on the total height of the subject (Guiheneuc and Ginet, 1973; P. Guiheneuc, *this volume*), is appropriate. In our experience, measurements of latencies from stimulus to the first deflection of the H-reflex (in many trials) represent valid data. The latencies of the M responses have also to be considered when investigating the H reflex latencies on the ipsi- and contralateral side.

In chronic root compression syndromes, differences of H-reflex latencies between the affected and the normal side are of great clinical value (Deschuytere and Rosselle, 1973), even in the absence of other EMG findings (Fig. 1). Differences of 1 msec are indicative, and differences of 2 msec are quite significant. Variations of the H-reflex amplitude may reflect changes in excitability of the α MN pool: a decrease of amplitude may indicate a reduced number of motor and/or Ia fibers in the reflex arc. When detected with needle electrodes, the H-reflex represents only the activity of a small bundle of muscle fibers. Variations of H-reflex duration on an affected side may

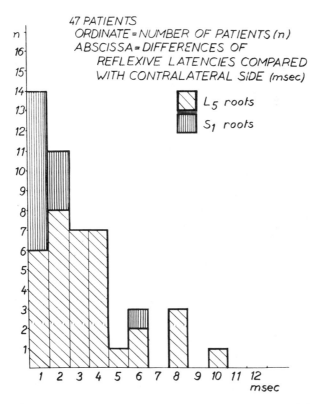

FIG. 1. Data on H reflexes in triceps surae (for S_1 root) and in extensor digitorum longus (for L_5 root) in 47 patients with chronic compression of L_5 or S_1 spinal root. *Abscissa*, difference in latencies of H reflex between the affected and the contralateral control sides; *ordinate*, number of patients.

be caused by a slowing (segmental demyelination) in afferent and/or efferent nerve fibers or by motor fiber regeneration. H-reflex latency studies in the lower limb may also contribute to the diagnosis of generalized peripheral neuropathies, especially alcoholic polyneuropathy (Blackstock et al., 1972; P. Guiheneuc, *this volume*) and Guillain–Barré syndrome (GBS) (Deschuytere and Rosselle, 1975; Deschuytere et al., 1976, 1981). Extremely prolonged latencies of H-reflexes have been recorded in patients with GBS. Follow-up of a patient with relapsing GBS showed that enhancements of H-reflex latencies promptly reflected exacerbations of the disease, in particular for the most affected limb.

H REFLEXES IN THE UPPER LIMB

On the basis of anatomical and phylogenetic considerations, it was assumed that the superficial forearm flexors were comparable to the gastrocnemius–soleus muscles and actually are physiological extensors and antigravity muscles. Elicitation and identification of late motor responses were rather easy in the flexor carpi radialis (FCR) and palmaris longus (PL) muscles (Deschuytere and Rosselle, 1974). Evidence for considering the late motor responses in the FCR and PL muscles as genuine H reflexes was provided by data on tendon vibration, recovery curves, spread of phasic stretch reflexes, and posttetanic potentiation (Deschuytere et al., 1974, 1976). Recruitment curves of the H and M waves from the FCR muscle were useful for the study of tendon vibration effects. There was a shift to the left of the M-wave curve as compared to those from soleus (Hugon, 1973*a*). Variations of the M wave curve in one subject are demonstrated in Fig. 2 with an invariant H-wave curve. Slight displacements of the stimulating electrodes may provoke these variations, and the shift to the left of the M-wave curve may be more accentuated.

The recruitment curves indicated that the maximal H wave was reached with an intensity that evoked an M response of about the same amplitude. The control of a quasiinvariant motor response of such an amplitude was used to maintain stable experimental conditions when vibration was applied to the tendon. During vibration of the FCR tendon (30 subjects), the H wave disappeared almost immediately in all subjects (Fig. 3). A complete abolition of the H wave was not surprising when detection was performed with a coaxial needle electrode, as earlier experiments with such an electrode inserted in the medial gastrocnemius muscle furnished identical data. When vibration was interrupted, the H wave reappeared and in most subjects reached the initial voltage 1 sec after cessation of vibration. Thereafter, a rebound effect was usually observed with a maximum after 3 sec; then it declined gradually and subsided 4 to 6 sec after its maximum.

Recovery curves of the H reflexes in FCR were studied in 15 normal subjects (Deschuytere et al., 1976) (Fig. 4) and were found strikingly similar to those in triceps surae. Furthermore, reinforcement maneuvers such as an isometric or isotonic contraction of the ipsilateral quadriceps muscle (Burg et al., 1974) without any concomitant activation of the forearm flexors enhanced H reflexes in the FCR and PL muscles.

The median nerve was stimulated in the cubital fossa, and coaxial needle electrodes were recorded from either FCR or PL muscles. The same distances for electrical stimulation and needle detection were chosen for both muscles on both sides (on average, 80 mm). The latencies of H reflexes in both muscles varied from 15 to 17 msec and were related to height and age of subject and in good correlation with caliper measurements. When in the normal individual the ipsi- and heterolateral side were compared, the differences in latencies were always less than 1.0 msec, and most were less than 0.5 msec. Latency measurements of H reflexes

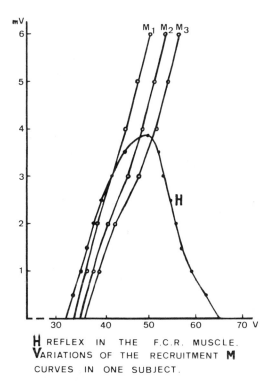

FIG. 2. Recruitment curves of M and H responses in flexor carpi radialis (FCR) in a normal subject. Slight displacement of the stimulating electrodes may result in slight changes in the M response curve while the H-reflex curve is stable. *Abscissa,* stimulus strength (volts); *ordinate,* response amplitude (mV).

of these wrist flexors represent a diagnostic tool for chronic compression of proximal segments of median nerve, medial cords (fasciculus ulnaris), middle trunks (truncus intermedius), and C_7 roots when the ipsi- and contralateral side were compared. In the absence of lesions of these peripheral structures, tendon vibration was helpful in the investigation of spastic hyperreflexia (Delwaide, 1973; Deschuytere et al., 1976; J. E. Desmedt, *this volume*) in the upper limb.

At the same time, we started to study H reflexes in the FCU muscle to explore lesions of the C_8 and T_1 ventral and dorsal roots, lower trunks (truncus caudalis), medial cords (fasciculus ulnaris), and proximal segment of ulnar nerves (Deschuytere et al., 1981). In contrast with FCR and PL, it was rather difficult in most subjects to evoke H reflexes in FCU without facilitation. This was a surprising finding, as the FCU muscle is a wrist flexor comparable to the FCR and PL muscles. Certainly there is a different motor supply; however, whereas the tibial nerve innervates physiological extensors of the leg and plantar and interossei muscles of the foot, the median and the ulnar nerves innervate the homologous muscles of forearm and hand. The phylogenetic relationship between the median and ulnar nerves is further illustrated by the fact that in some individuals (and this has been described as a familial trait), motor and sensory fibers may have an aberrant course in the trunks or cords, resulting in an anomalous innervation of the skin and small muscles of the hand.

Many of these subjects have an anastomosis in the forearm between median

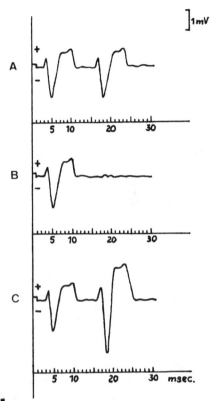

H REFLEX IN THE F.C.R. MUSCLE.

A - CONTROL M- AND H- WAVES.

B - VIBRATION APPLIED TO THE TENDON.

C - M- AND H- WAVES AFTER THE CESSATION OF VIBRATION. REBOUND EFFECT.

FIG. 3. The effects of tendon vibration on the M and H responses in FCR in a normal subject. **A:** Control before vibration. **B:** During vibration. **C:** Rebound effect on H reflex after cessation of vibration.

When the ulnar nerve is stimulated, and H reflexes are recorded from FCU, it is clear that there is very little convergence, because the muscles of the hand cannot be considered as synergic muscles *stricto sensu,* and impulses in Ia collaterals from flexor digitorum profundus may have only a small effect on the excitability of the FCU MN pool. Thus, the amount of synaptic convergence may be important for eliciting H reflexes.

In relation to our facilitation techniques, the recovery curves of H reflexes derived from FCR (Deschuytere et al., 1976) (Fig. 4) present their third phase after about 160 msec to 200 msec and this facilitation phase comes earlier than that of the recovery curves of triceps surae because the conduction distances to the spinal cord are smaller for upper than for lower limb. It appeared that the repetition rate of 5 to 6/sec was optimal for evoking H reflexes in FCU.

The latencies of H reflexes in FCU ranged from 16 msec to 18 msec, whereas latencies in FCR and PL ranged from 15 msec to 17 msec, although the same distances for electrical stimulation and EMG detection were chosen for all muscles. The latencies for the FCU were greater than those for FCR, perhaps because motor and Ia fibers traveling in the T_1 and C_8 roots have a longer pathway than those in the C_7 roots.

A cogent argument to consider the late responses of FCU as genuine H reflexes was furnished by the vibration of the tendon. There was always a complete abolition or a great reduction of response during vibration when detection was performed with coaxial needle electrodes. The rebound effect seen after the vibratory stimulation was comparable to that seen in the FCR and PL muscles (Deschuytere et al., 1976). In some individuals, complete abolition in FCU during vibration in many trials excludes contamination with an F component. Finally, a tendon tap to FCU resulted in reflex responses with latencies comparable to the latencies of the FCU H reflex.

and ulnar nerves (Marinacci, 1968). Since there was a need for facilitation of FCU H reflex in many subjects, the lack of convergence from Ia collaterals of synergic muscles (Eccles et al., 1957) onto the α MNs of FCU may be an explanation. It seems likely that, when the median nerve is stimulated to evoke H reflexes in FCR and PL, there may exist a prominent synaptic convergence (Deschuytere et al., 1976).

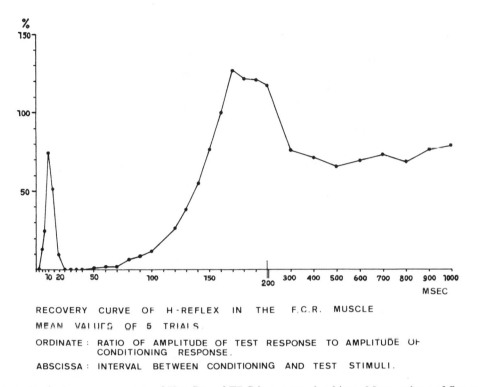

RECOVERY CURVE OF H-REFLEX IN THE F.C.R. MUSCLE
MEAN VALUES OF 5 TRIALS
ORDINATE: RATIO OF AMPLITUDE OF TEST RESPONSE TO AMPLITUDE OF CONDITIONING RESPONSE.
ABSCISSA: INTERVAL BETWEEN CONDITIONING AND TEST STIMULI.

FIG. 4. Typical recovery curve of H reflex of FRC in a normal subject. Mean values of five trials for each point. *Abscissa,* interval between conditioning and test stimuli (msec); *ordinate,* relative amplitude of second H reflex (percent of the control first H reflex).

The clinical significance has to be stressed. When the affected side was compared with the normal side in patients, it was found for the three wrist flexors that differences of H-reflex latencies of 1.0 msec were indicative and differences of 2.0 msec were significant for the diagnosis of chronic compression syndromes, the FCU muscle being of special value for the thoracic outlet syndrome. Obviously, these investigations also represent a diagnostic aid for patients with generalized peripheral neuropathies.

CONCLUDING REMARKS

Most commonly, CV measurements are performed on the distal segments of peripheral nerves for the diagnosis of peripheral nerve entrapment and generalized peripheral neuropathies. Studies of H reflex latencies may reveal lesions in the proximal segments of peripheral nerves, plexuses, and roots. These investigations do not yet appear to be appreciated for their true value.

It is surprising that H reflexes are still rarely used for early diagnosis of the Guillain–Barré syndrome. Guillain et al. (1916), in their original publication, pointed to the alterations of the tendon reflexes and stressed the importance of graphic recording of tendon reflexes to reveal the considerable delay of reflex response. As mentioned earlier, we have recorded extremely prolonged latencies of H reflexes in GBS in lower and upper limb even at early stages when the CVs of motor and sensory fibers in peripheral nerves, routinely examined, were normal or borderline, when routine EMG was inconclusive, and when cerebrospinal fluid examinations were normal. Based on our experience with H reflexes,

GBS may be excluded in the absence of significantly augmented H reflex latencies in clinically affected extremity.

In other generalized peripheral neuropathies, the increase of H reflex latencies is less spectacular. Nevertheless, the data obtained may be helpful to distinguish the predominant axonal type (dying back) from the segmental demyelinating type of involvement. Surprisingly, there are more reports using the F wave for this purpose than studies using H reflexes (Kimura and Butzer, 1975; Lachman et al., 1975, 1976, 1980).

Without denying the validity of F waves in chronic compression syndromes (Fisher et al., 1979) and peripheral neuropathies, we find H reflexes to be more reliable. A CV evaluation based on F waves derived from hypothenar or thenar muscles may give misleading data because of nerve entrapment. This also applies for the muscles of the foot, either plantar muscles (tarsal tunnel syndrome) or muscles of the dorsum pedis, which may be damaged by shoe wearing (Rosselle and Stevens, 1973). Therefore F-wave latencies in distal muscles can be misleading, and this also applies to H reflex latency studies in these muscles. Certainly, our studies on H reflexes in leg or forearm muscles are more important in diagnosis of chronic root compression syndromes, which are frequently seen. Although several publications on S_1 root compression recommend H reflex latency measurements, we are not aware of any study from other authors concerning H reflexes of the extensor digitorum longus muscle, which is representative of L_5 roots and has proved a valuable diagnostic aid in chronically compressed L_5 roots; in many instances it was the only objective finding (Deschuytere and Rosselle, 1973). L_5 root compression syndromes are more frequent than S_1 compression syndromes. If an S_1 root is compressed, we often find it together with a compression of the L_5 roots. Most of the herniated, prolapsed, or ruptured disks are found at the L_4–L_5 level, usually affecting L_5 roots and less frequently S_1 roots. Ruptured L_5–S_1 disks which usually result in an S_1 root compression, are less common and even rare. Another observation is that an antelisthesis of one lumbar vertebra may result in traction on the cauda equina and consequently in chronic compression of the L_5 and S_1 roots on both sides.

Beyond controversy was a demonstration of H reflexes in the forearm FCR and PL muscles, and different authors have put it to use (Garcia et al., 1979; Jabre and Stalberg, 1980). Again, we were primarily concerned with measurements of latencies but also with a study of tendon vibration with regard to hyperreflexia. Classic H-latency measurements were useful in different compression syndromes. Moreover, it was found that FCR was representative of the C_7 roots. The C_6–C_7 intervertebral disk is most susceptible to injuries and, if herniated or ruptured, may be responsible for chronic compression of the C_7 roots. H reflexes elicited in FCU after electrical stimulation of the ulnar nerve can test additional neurological structures and help in the diagnosis of thoracic outlet syndrome.

F-Wave Determination in Nerve Conduction Studies

Jun Kimura

Division of Clinical Electrophysiology, Department of Neurology, College of Medicine, University of Iowa, Iowa City, Iowa 52242

Conventional nerve conduction studies usually deal with the distal parts of peripheral nerves. For evaluation of polyneuropathies, the proximal nerve segments should also be assessed. This can be achieved for sensory nerve fibers by recording cerebral somatosensory evoked potentials (SEP) (Desmedt et al., 1966; Desmedt and Noel, 1973; Desmedt, 1983). For motor nerve fibers, reflex responses (see P. Guiheneuc, *this volume*) or F-wave responses can be used.

The F wave (Magladery and McDougal, 1950) is a late muscle potential which results at least in part from backfiring of antidromically activated motoneurons (MN) after stimulation of the nerve (Dawson and Merton, 1956); McLeod and Wray, 1966; Miglietta, 1973; Trontelj, 1973; Schiller and Stalberg, 1978). The latency of the F wave thus includes the conduction time of motor impulse to and from the spinal cord through the proximal segment of the nerve. The possibility of assessing motor nerve conduction in the segment proximal to the site of stimulation was first explored in patients with Charcot–Marie–Tooth disease (Kimura, 1974) and later documented for various other neurological disorders (Conrad et al., 1975; Kimura and Butzer, 1975; Kimura et al., 1975; Panayiotopoulos and Scarpalezos, 1977; Kimura, 1978). This chapter reviews 7 years' experience with F wave in nerve conduction studies and discusses its clinical value and limitations.

F-WAVE DETERMINATION

To evoke an F wave, a supramaximal stimulus may be applied practically at any point along the course of the nerve. The most distal stimulation at the wrist or ankle may be preferred to determine the motor conduction time to and from the spinal cord from the stimulus site. This is advantageous for detection of a conduction delay in a diffusely affected nerve which should be proportional to the length of the pathway covered by the volley. This accumulated delay sometimes allows detection of a relatively mild slowing which may escape conventional motor nerve conduction studies. However, increased latency of the F wave elicited at the most distal stimulation will not distinguish distal versus proximal conduction abnormalities. For this purpose, the nerve may be stimulated at a more proximal site, and the latencies of F wave and M response compared.

For clinical evaluation, the median, ulnar, peroneal, or tibial nerve is stimulated at two or three points along the nerve, and potentials are recorded with surface electrodes from the appropriate muscles (Kimura, 1974; Kimura et al., 1975). The latencies of M response and F wave are measured from the stimulus artifact to the beginning of the evoked potential. The F-wave latency is relatively consistent in some individuals but in others varies by a few milliseconds from one stimulus to the next. In our laboratory, therefore, at least 16 trials are displayed on a storage scope, automatically shifting successive sweeps vertically, and the shortest latency is determined. If the F wave is difficult to elicit, slight voluntary contractions often enhance the evoked potential. The study is considered inadequate unless more than eight F waves are clearly identified among these trials.

The F wave represents the muscle potential evoked by discharges of antidromically activated MNs. Thus, the F wave first travels in the centripetal direction toward the spinal cord before it is conducted distally to activate the muscle. With more proximal stimulating electrodes, the F wave moves closer to the M response because the latency of M response increases, whereas that of F wave decreases. The F wave is clearly separated from the M response with stimulation at wrist, elbow, ankle, or knee (Figs. 1–3). However, it is usually buried in the terminal portion of the M response when a stimulus is delivered at the axilla. In this situation, the F wave can be separated from the M response if an additional stimulus is delivered at the wrist (Kimura, 1974, 1976). With this technique, the orthodromic impulse from the axilla and the antidromic impulse from the wrist are extinguished by collision, leaving the M response evoked from the wrist and the F wave from the axilla intact. Since these two remaining evoked muscle potentials are clearly separate, it is now possible to measure the latency of the F wave elicited by the axilla stimulation.

The decrease in latency of the F wave is nearly the same as the increase in latency of the M response when the stimulating electrodes are moved from the wrist to the elbow and then to the axilla. This observation, in turn, allows the calculation of the F wave latency from axilla on the assumption that it must equal the sum of the latencies of the F wave and M response elicited by a distal stimulation minus the latency of the M response evoked by an axillary stimulation (Kimura and Butzer, 1975). Hence, $F(A) = F(W) + M(W) - M(A)$, where $F(A)$ and $F(W)$ are latencies of the F waves from the axilla and wrist, respectively, and $M(A)$ and $M(W)$ are latencies of the corresponding M responses from the same stimulus points. This equation has been found satisfactory in assessing conduction in the most proximal segment of the median and ulnar nerves.

For clinical evaluation of the F wave, the median, ulnar, peroneal, or tibial nerve may be stimulated at two or three points, but this is not always possible because of the time involved. To make clinical studies simple and brief, our routine procedures include study of the F wave only with stimulation at the wrist and elbow for the median and ulnar nerves and at the ankle and knee for the peroneal and tibial nerves. When deemed necessary, the latency of the F wave from any proximal sites of stimulation may be calculated by the equation described above as long as the latency of the corresponding M response is available.

CENTRAL LATENCY

The central latency is defined as the time required for passage of an impulse to and from the spinal cord. It is determined as $F - M$, where F and M are latencies of the F wave and M response, respectively (Fig. 4). The conduction time in the proximal segment from the stimulus site to the cord is calculated as $(F - M - 1)/2$, assuming an estimated delay of 1.0 msec for the turnaround time at the cell body. It is also postulated that the F wave with the shortest latency travels in the fastest-conducting motor fibers and, therefore, is directly comparable to the M response. No information is available for the exact central delay in

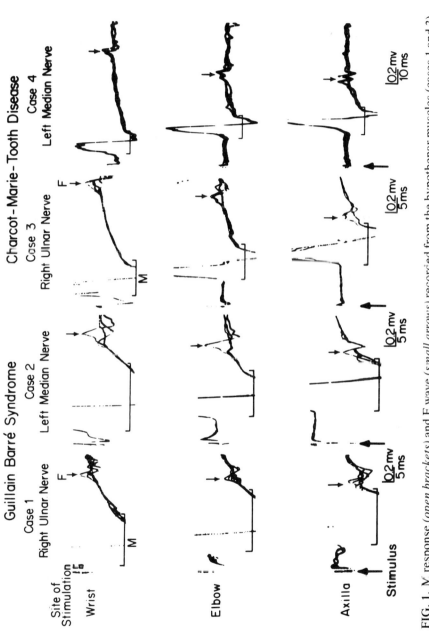

FIG. 1. M response (*open brackets*) and F wave (*small arrows*) recorded from the hypothenar muscles (cases 1 and 3) and thenar muscles (cases 2 and 4) in patients with polyneuropathies. Three consecutive trials are superimposed for each tracing. A slower sweep speed (10 msec per division instead of 5 msec per division) was necessary to record the substantially delayed F wave in Case 4. Because of slowing in nerve conduction, the M response and the F wave were distinctly separate even with stimulation at the axilla. The tracings for cases 1 and 2 were reproduced with permission from a previous publication (Kimura, 1978).

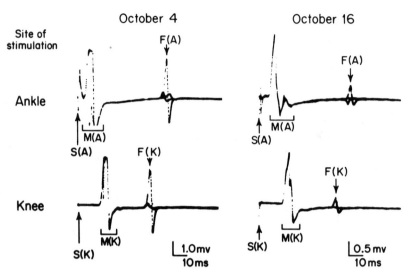

FIG. 2. M response *(open brackets)* and F wave *(small arrows)* recorded from the extensor digitorum brevis on two different occasions in a patient with mild Guillain–Barré syndrome. On October 4, shortly after onset, the F wave was slightly delayed in latency but of large amplitude. On October 16, the F wave was even greater in latency and much lower in amplitude.

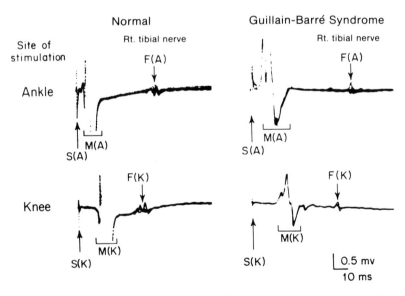

FIG. 3. M response *(open brackets)* and F wave *(small arrows)* recorded from the abductor hallucis in two subjects. Compared to the control, the F wave in the patient was increased in latency. The M response was normal in latency although reduced in amplitude.

MNs in man (Trontelj, 1973). Animal data indicate that this is very close to 1.0 msec (Renshaw, 1941; Lloyd, 1943). Furthermore, the absolute refractory period of the fastest human motor fibers is also about 1.0 msec or slightly less (Kimura, 1976). Should recurrent discharge occur earlier, therefore, the impulse is not likely to be propagated distally because of refractoriness of the axon near the cell body. For the F wave of minimal latency, therefore, the turnaround time of 1.0 msec seems a reasonable estimate.

A given F wave is only a portion of the M response, representing either fast or slow conduction fibers. A few milliseconds difference between the earliest and latest F waves is probably attributable to the difference between fastest and slowest motor fibers. The uncertainty that the F wave may not represent the fastest fibers can be circumvented by recording many F waves at each stimulus site so that the minimal F-wave latency can be selected (Kimura, 1974; Conrad et al., 1975; Kimura et al., 1975). In some diseased nerves, however, surviving fast fibers contributing to the M response do not produce F waves. In less extreme instances, an F wave may be elicited only in slow-conducting fibers although faster-conducting fibers still contribute to the M response, thus resulting in a false impression of proximal slowing when, in fact, fast-conduction fibers are blocked proximally.

The error must be small if the M response is not substantially reduced in amplitude and the F wave is easily elicitable, since in these nerves the increase in latency of the M response is nearly the same as the decrease in latency of the F wave when the stimulus is moved proximally (Mayer and Feldman, 1967; Kimura, 1974; Kimura et al., 1975). In doubtful cases, it is feasible to determine whether the sum of F latency and M latency is similar at distal and proximal stimulus sites. If so, one can be reasonably certain that motor fibers with the same conduction characteristics contribute to the F wave and the fastest components of the M response. This in turn provides a rationale for direct latency comparison of these two muscle potentials.

F-WAVE CONDUCTION VELOCITY AND F RATIO

For assessment of the F-wave conduction velocity (FWCV) analogous to the motor nerve conduction velocity (MNCV), the approximate length of the median and ulnar nerves from axilla to spinal cord is obtained by measuring the surface distance from the stimulus point at the axilla to the C_7 spinous process via the midclavicular point (Kimura, 1974). Likewise, the surface distance from the stimulus site at the knee to the T_{12} spinous process is determined by way of the greater trochanter of the femur for the peroneal and tibial nerves (Kimura et al., 1975). The FWCV in the segment to and from the spinal cord is calculated as $(D \times 2)/(F - M - 1)$, where D is the distance from the stimulus site to the cord and $(F - M - 1)/2$ is the time required to cover the length D.

Estimating the accurate length of the proximal segment may be subject to error, especially if a standardized method is not used. To adjust for different lengths of the nerve, the patient's height or arm and leg length may be used if a normogram is avail-

FIG. 4. (Overleaf). Latencies of F wave and M response for median (A), ulnar (B), peroneal (C), and tibial nerves (D), in normal subjects, patients with the Guillain–Barré syndrome (GBS), and patients with Charcot–Marie–Tooth disease (CMT). Only those nerves for which both F wave and M response were elicited were included. Sites of nerve stimulation are indicated in the key. The difference in latency between F wave and M response *(triangles)* is the time interval required for passage of the impulses to and from the spinal cord.

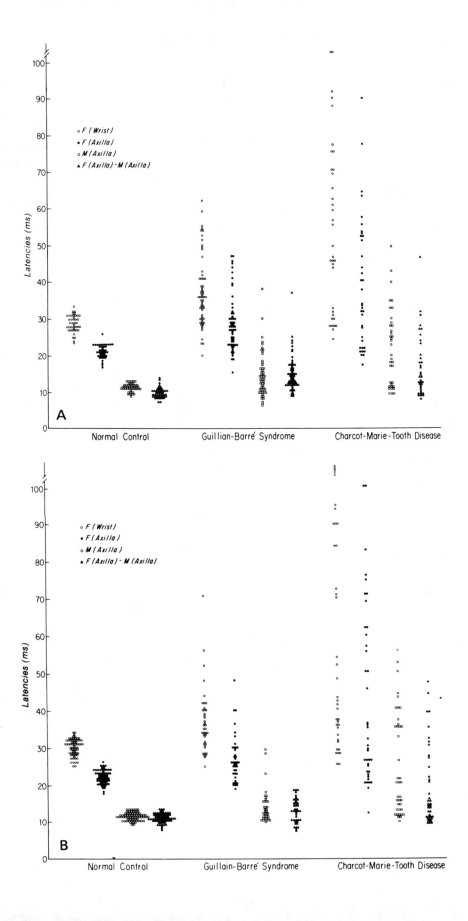

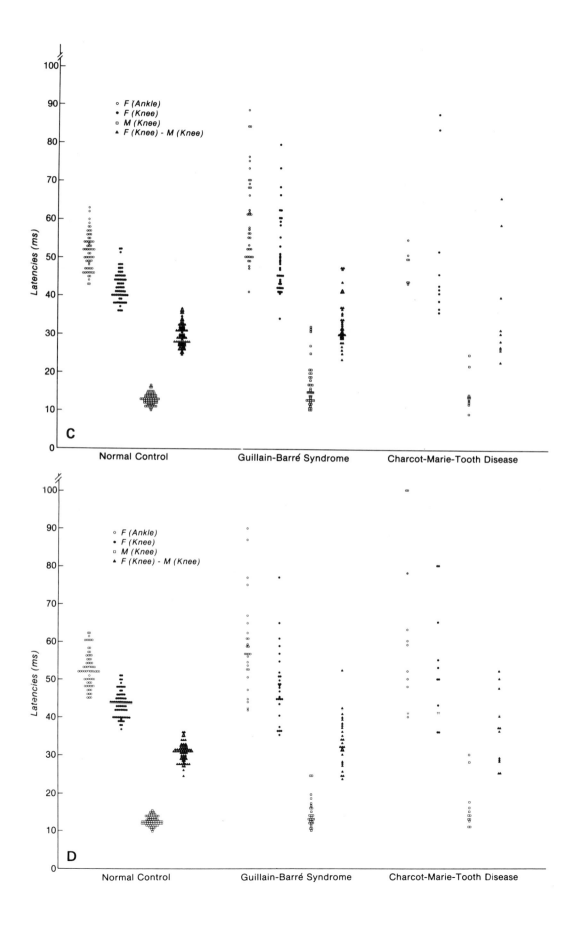

able, but the actual length of the nerve segment under study is more directly related to the F-wave latency under consideration. In the case of the lower extremities, the accuracy is based on observation in five cadavers in which surface determination and actual length of the nerve were compared (Kimura et al., 1975). No such data are available at present for the upper extremities. Absolute F-wave latencies may also be used provided that the patient's extremities are not unusually short or long. However, latency determination is in general more meaningful if one side is compared to the other in the same subject or if one nerve is compared to another in the same extremity.

To circumvent the controversy of surface distance determination, which may be difficult to reproduce from one laboratory to another, Eisen et al. (1977) proposed the latency ratio between F-wave and M response to assess proximal versus distal nerve compression syndromes. Using a similar ratio, $(F - M - 1)/2M$, it is also possible to compare the conduction time from cord to stimulus site and that of the remaining distal nerve segment to the muscle (Kimura, 1978). The time required for the impulse to travel from stimulus site to cord is expressed as $(F - M - 1)/2$, whereas that to muscle is represented by M. An estimated delay of 1.0 msec at the cell body is excluded from the proximal latency as is done for calculation of FWCV. Designated the F ratio, it provides a simple means to compare conduction of the proximal and distal segments. Unlike the FWCV, the F ratio is independent of the total length of the nerve but is based on the assumption that the proportion between the proximal and distal segment is the same among different subjects.

In man, the FWCV in the central segment is normally about the same as the MNCV in the proximal segment and slightly faster than that in the distal segment (Kimura, 1974; Conrad et al., 1975; Kimura and Butzer, 1975; Kimura et al., 1975). The F ratio is close to unity with stimulation at the elbow and knee, indicating that the time from the cord to the site of stimulation is approximately the same as that from the stimulus site to the muscle.

F WAVES IN NORMAL SUBJECTS AND PATIENTS WITH POLYNEUROPATHY

Table 1 summarizes the normal values of the F-wave latency, FWCV, and F ratio determined in 33 healthy subjects 22 to 58 years of age (average age 35) for median and ulnar nerve (Kimura, 1974) and in another group of 33 subjects, 21 to 47 years of age (average age 26), for peroneal and tibial nerves (Kimura et al., 1975). Two standard deviations above and below the mean were taken as the normal limits. Thus, the FWCV in the cord-to-axilla segment of the median and ulnar nerves was considered slow if it was less than 50 m/sec. The MNCV was considered slow if it was less than 50 m/sec in the axilla-to-elbow segment and 46 m/sec across the elbow or in the more distal segments. In the peroneal and tibial nerves, the lower limit of normal was 45 m/sec for the FWCV in the cord-to-knee segment and 40 m/sec for the MNCV in the knee-to-ankle segment. The F ratio was considered abnormal if it fell outside the range of 0.86 to 1.22, 1.18 to 1.62, 0.93 to 1.29, and 0.97 to 1.37 for median, ulnar, peroneal, and tibial nerves with stimulation at elbow, below elbow, knee, and above knee, respectively (Kimura, 1978).

The F-wave was tested in 121 patients with polyneuropathies, 72 with the Guillain-Barré syndrome (GBS), and 49 with Charcot–Marie–Tooth disease (CMT). The median, ulnar, peroneal and tibial nerves were studied in various combinations unilaterally in some, bilaterally in others (Figs. 1–4). Altogether, 321 and 250 peripheral nerves were available for analysis of GBS and CMT, respectively. Tables 2 to 5 summarize the latency and conduction velocity

TABLE 1. Normal values (mean ± SD)

Number of nerves tested	Site of stimulation	M latency (msec)	F latency (msec)	F ratio (F − M − 1)/2M	F ratio (R) / F ratio (L)	MNCV between two stimulus sites (m/sec)	FWCV from cord to stimulus site (m/sec)
66 Median nerves[a]	Wrist	3.5 ± 0.5	29.1 ± 2.3				59.2 ± 3.9
	Elbow	7.8 ± 0.8	24.8 ± 2.0	1.04 ± 0.09	1.01 ± 0.07	56.0 ± 5.0	62.2 ± 5.2
	Axilla	11.3 ± 1.0	21.7 ± 2.8			63.3 ± 6.0	64.3 ± 6.4
66 Ulnar nerves[b]	Wrist	2.9 ± 0.5	30.5 ± 2.0				56.7 ± 2.9
	Below elbow	6.7 ± 0.7	26.0 ± 2.0	1.40 ± 0.11	0.99 ± 0.09	55.9 ± 5.1	58.2 ± 2.9
	Above elbow	9.2 ± 0.9	23.5 ± 2.0			56.9 ± 4.6	61.1 ± 5.4
	Axilla	11.2 ± 1.0	21.9 ± 1.9			61.3 ± 6.8	63.0 ± 5.9
66 Peroneal nerves	Ankle	4.5 ± 0.9	51.3 ± 4.7				53.3 ± 3.7
	Knee	12.9 ± 1.4	42.7 ± 4.0	1.11 ± 0.09	1.02 ± 0.09	49.4 ± 3.8	56.3 ± 4.9
66 Tibial nerves	Ankle	4.1 ± 0.6	52.3 ± 4.3				51.3 ± 2.9
	Knee	12.8 ± 1.3	43.5 ± 3.4	1.17 ± 0.10	1.00 ± 0.10	46.8 ± 3.4	54.4 ± 3.6

[a] F wave was elicited by axillary stimulation in 42 of 66 nerves.
[b] Middle segment across elbow was tested in 34 of 66 nerves.

of the M response and the F-wave and the F ratio in GBS and CMT.

Guillain–Barré Syndrome

Of 102 median, 71 ulnar, 90 peroneal, and 58 tibial nerves tested in GBS, both M responses and F wave were recorded in 58 median, 40 ulnar, 39 peroneal, and 29 tibial nerves. Only these nerves were included in the analysis (Table 2). The latencies of M response and F wave as well as central latency calculated as $F - M$ were much longer (Fig. 4), and MNCV and FWCV slower, in GBS than in the normal subjects. In many patients, however, the F ratio remained normal, indicating that the nerves were affected to the same degree above and below the stimulus sites at the elbow and knee. Analysis of the F ratio showed that a majority of the nerves tested were slowed equally in the segments above and below the stimulus site, whereas the remaining nerves were predominantly slowed in either proximal or distal segments, but with equal preference between the two (Table 3).

Although the F ratio remained normal in a majority of patients, as discussed, it was not necessarily associated with uniformly slow conduction along the entire length of the peripheral nerve (Table 4). Slowing in FWCV of the cord-to-axilla segment was more frequent than that in MNCV of the elbow-to-wrist segment for both median ($p < 0.05$) and ulnar nerves ($p < 0.01$). Indeed, the FWCV in the central segment was slow when the MNCV in the more distal segments was normal or borderline in 16 median and 15 ulnar nerves, whereas the reverse was true in only seven median and three ulnar nerves. In the lower extremities, there was also significant difference in the incidence of abnormality between the proximal and distal segments for tibial ($p < 0.01$) but not peroneal nerve. In 10 peroneal and nine tibial nerves, the FWCV of the proximal segment was slow when the MNCV of the distal segment was normal, although the reverse was the case in four peroneal and one tibial nerve.

For the median nerve, conduction delay in the terminal segment from the wrist to the muscle was considerably greater than expected by comparison with the slowing of MNCV in the elbow-to-wrist segment. In calculating the F ratio, therefore, a marked increase in terminal latency sometimes more than compensated for a prominent slowing of nerve CV of the central segment. For the ulnar nerve, the F ratio with stimulation below elbow was often increased, indicating prominent slowing in the proximal segment which included the common site of compression at the elbow as well as the radicular portion of the nerve. For the peroneal and tibial nerves, reduced F ratios were often associated with an increase in terminal latency which was out of proportion to the degree of slowing of MNCV in the knee-to-ankle segment.

Charcot–Marie–Tooth Disease

Of 60 median, 60 ulnar, 51 peroneal, and 29 tibial nerves tested in 49 patients with CMT, the F wave was present in 36, 31, 10, and 12, respectively. As in GBS, the average latencies of the M response and F wave as well as central latencies calculated as $F - M$ were much greater (Fig. 4) and MNCV and FWCV slower in CMT than in normals (Table 5). When the incidence of abnormality was analyzed in individual nerves, the FWCV was slowed in 25, 25, 6, and 5 and normal in the remainder. In the upper extremities of these patients, the MNCV of distal segments was in general slightly slower than the FWCV of the proximal segment, although the difference was significant only for the median nerve ($p < 0.05$). The F ratio was decreased in both median and ulnar nerves, but the average value was significantly different from the normal only in the former. In the lower extremities there was no significant difference between MNCV distally and FWCV

TABLE 2. *Guillain–Barré Syndrome (mean ± SD)*

Number of nerves tested	Site of stimulation	M latency (msec)	F latency (msec)	F ratio $(F - M - 1)/2M$	$\dfrac{F \text{ ratio (R)}}{F \text{ ratio (L)}}$	MNCV between two stimulus sites (m/sec)	FWCV from cord to stimulus site (m/sec)
58 Median nerves	Wrist	5.8 ± 3.1	38.1 ± 12.7				48.6 ± 11.1
	Elbow	11.2 ± 4.8	32.6 ± 9.9	1.07 ± 0.35	1.02 ± 0.14	48.2 ± 12.1	49.1 ± 11.4
	Axilla	14.5 ± 5.7	29.4 ± 9.5			55.5 ± 14.1	47.5 ± 14.5
40 Ulnar nerves	Wrist	4.0 ± 2.0	36.8 ± 8.6				48.1 ± 9.7
	Below elbow	8.3 ± 2.5	32.1 ± 7.1	1.42 ± 0.29	1.02 ± 0.19	52.2 ± 10.7	47.4 ± 9.6
	Above elbow	11.2 ± 3.5	29.7 ± 8.7			47.7 ± 12.0	47.4 ± 10.7
	Axilla	13.7 ± 4.8	27.2 ± 6.2			56.8 ± 14.9	48.0 ± 12.3
39 Peroneal nerves	Ankle	7.6 ± 4.8	59.9 ± 11.5				42.5 ± 8.7
	Knee	16.9 ± 5.8	50.6 ± 10.3	1.06 ± 0.23	1.01 ± 0.17	43.0 ± 8.2	43.9 ± 11.8
29 Tibial nerves	Ankle	5.6 ± 2.3	56.4 ± 10.6				42.7 ± 8.8
	Knee	14.6 ± 3.8	47.9 ± 9.4	1.12 ± 0.20	1.02 ± 0.18	43.3 ± 9.0	43.8 ± 9.9

TABLE 3. F ratio with stimulation
at the elbow or knee in the
Guillain–Barré syndrome

Nerve	Number tested	Normal	Decreased	Increased
Median	58	27	17	14
Ulnar	40	23	4	13
Peroneal	39	20	13	6
Tibial	29	20	6	3
All nerves combined	166	90	40	36

proximally. The F ratio was normal for both peroneal and tibial nerves.

Diabetic Neuropathy and Carpal Tunnel Syndrome

The F ratio (Fig. 5) was determined in 102 diabetics with diffuse symmetrical polyneuropathy (DPN), 44 patients with the carpal tunnel syndrome (CTS), and 74 age-matched control patients. The F ratio was slightly but significantly smaller in DPN than in the control patients for median (mean ± SD: 093 ± 0.14 versus 1.05 ± 0.09), peroneal (1.02 ± 0.19 versus 1.08 ± 0.12), and tibial nerves (1.09 ± 0.19 versus 1.17 ± 0.13) but not for ulnar nerve (1.35 ± 0.20 versus 1.41 ± 0.12). The slope of the regression line through the origin was signifcantly different between DPN and the controls in all but the ulnar nerve ($p < 0.01$). These findings suggest that motor conduction abnormalities in diabetics are more intense in the distal than proximal segments.

As expected, the F ratio of the median nerve was reduced in CTS (0.90 ± 0.12). The value was much the same as in DPN and significantly smaller than that in the control patients ($p < 0.001$). The slope of the regression line was similar to that of DPN but significantly different from that of the control group ($p < 0.01$).

COMMENTS

A major limitation in clinical uses of the F wave is its inherent latency variability from one trial to the next. Determination of the shortest latency after several trials largely, although perhaps not completely, circumvents this uncertainty. When the F wave is difficult to elicit, it can be potentiated with a slight voluntary contraction. Recording as many as 100 F waves at each stimulus site may be useful in special studies, but it is not suited to the purpose of routine clinical evaluation. Calculation of central latency, FWCV, and F ratio may be subjected to additional errors because some of the assumptions required in the equations may not be entirely correct. These factors, however, do not alter the outcome of a study so long as the same standard equations are used to establish the normal range and to evaluate the patient group.

In GBS, conduction abnormalities are usually diffuse, involving any segment of the peripheral nerve. There is a tendency toward involvement of the most proximal, possibly radicular, portion of the nerve and the most distal or terminal segment, whereas the main nerve trunk is relatively less affected. Thus, the FWCV may be abnormal even when the MNCV is normal, suggesting that the pathological process in

TABLE 4. *The MNCV in the most distal segment versus FWCV in the most proximal segment in the Guillain–Barré syndrome*

Nerve	Number tested	MNCV in elbow-to-wrist or knee-to-ankle segment		FWCV in the cord-to-axilla or cord-to-knee segment	
		Normal	Slow	Normal	Slow
Median	58	35	23	26	32
Ulnar	40	30	10	18	22
Peroneal	39	27	12	21	18
Tibial	29	22	7	14	15
All nerves combined	166	114	52	79	87

TABLE 5. *Charcot–Marie–Tooth disease (mean ± SD)*

Number of nerves tested	Site of stimulation	M latency (msec)	F latency (msec)	F ratio $(F - M - 1)/2M$	$\dfrac{F \text{ ratio (R)}}{F \text{ ratio (L)}}$	MNCV between two stimulus sites (m/sec)	FWCV from cord to stimulus site (m/sec)
36 Median nerves	Wrist	6.4 ± 3.0	55.6 ± 26.1	0.94 ± 0.21	0.97 ± 0.17		33.7 ± 14.6
	Elbow	15.6 ± 7.8	46.1 ± 21.4			30.4 ± 14.6	36.4 ± 14.9
	Axilla	22.2 ± 10.6	39.3 ± 17.8			38.9 ± 20.2	38.4 ± 16.8
31 Ulnar nerves	Wrist	5.2 ± 2.9	55.5 ± 35.1	1.35 ± 0.29	0.99 ± 0.14		39.2 ± 18.7
	Below elbow	13.1 ± 7.9	48.2 ± 29.8			38.0 ± 18.3	40.2 ± 19.0
	Above elbow	18.0 ± 10.6	40.7 ± 27.2			36.6 ± 19.3	42.3 ± 20.8
	Axilla	21.3 ± 14.0	37.3 ± 23.6			42.5 ± 22.1	43.7 ± 18.9
10 Peroneal nerves	Ankle	5.6 ± 1.3	52.8 ± 10.6	1.12 ± 0.25	1.01 ± 0.14		47.2 ± 6.9
	Knee	15.0 ± 4.8	50.8 ± 19.1			40.7 ± 15.2	41.6 ± 6.8
12 Tibial nerves	Ankle	5.4 ± 1.4	62.8 ± 21.3	1.19 ± 0.19	0.88 ± 0.20		42.9 ± 14.2
	Knee	16.2 ± 6.3	52.5 ± 15.3			40.3 ± 14.9	43.9 ± 12.3

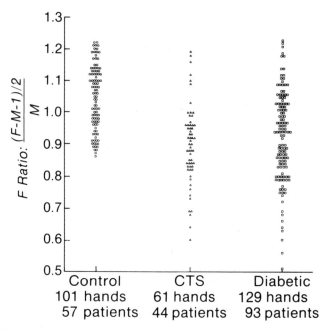

FIG. 5. F ratio of the median nerve in the control group, carpal tunnel syndrome (CTS), and diabetic polyneruopathy (DPN). Significantly reduced ratios in both disease groups indicate disproportionate slowing of motor conduction distally.

these nerves predominantly involves the central nerve segment (Ashby et al., 1969). The MNCV may be normal in 15% to 20% of cases in GBS tested within the first few days of onset (Humphrey, 1974; Eisen and Humphreys, 1974), possibly because the lesion is too proximal to detect using the ordinary techniques.

In CMT, both M response and F wave are often difficult to elicit in the lower extremities. The F wave in the upper extremities is recorded with ease, albeit markedly slowed. The lower extremities are most severely diseased in CMT. When the nerve is affected mildly, the motor conduction in the distal segment may be slowed with a normal conduction in the central segment. In advanced cases, both segments may be equally slow. It has been suggested that CMT may be subdivided into the hypertrophic type associated with marked slowing and the neuronal type with relatively normal CV (Dyck et al., 1975). The dichotomy is supported by a bimodal distribution of motor nerve CVs (Thomas et al., 1974), although there are also a number of intermediate values in latency of the M response and F wave as shown in the present series (Fig. 4).

Clinical neuropathic symptoms in DPN rather frequently occur as distal abnormalities. Distally predominant symptoms, however, do not necessarily indicate a distal pathological process, since probabilistic models can produce distal sensory deficit on the basis of randomly distributed axonal dysfunction (Waxman, 1980). The average value and distribution of the F ratio, however, indicate the CV abnormalities in DPN are indeed more common distally than proximally, although the entire length of the nerve is significantly slowed. In CTS, the F ratio of the median nerve is reduced as expected from localized distal slowing across the common site of compression.

Unlike the case in polyneuropathies, the

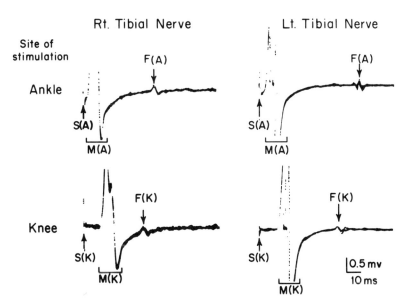

FIG. 6. M response *(open brackets)* and F wave *(small arrows)* recorded from the abductor hallucis in a patient with sacral plexus lesion on the left. The F wave was increased in latency on the affected side relative to the normal side.

F wave is often normal in clinically unequivocal cases of brachial or lumbosacral plexopathy. Thus, normal F-wave latencies do not preclude the presence of lesions. Conversely, an unequivocal delay of the F wave, which may be the only abnormality in plexopathy, is a reliable sign of proximal lesions (Fig. 6). The F wave is usually normal in mild cases of radiculopathy, at least in early stages, and thoracic outlet syndrome with predominantly vascular symptoms. Thus, in these conditions, the F-wave determination seems less helpful as a tool of early diagnosis than might have been expected on theoretical grounds. Using the latency ratio between F-wave and M response, Eisen et al. (1977) have shown clinical usefulness of F wave determination in patients with presumed root injuries. The F wave is also altered in the neuronal type of thoracic outlet syndrome (Wulff and Gilliatt, 1979). Since right–left difference of F wave latency, central latency, and F ratio are very small intraindividually, this comparison may provide a reliable means of assessing unilateral radicular or plexus lesions.

For overall electrophysiological studies of the peripheral nerve in a neuropathy, it is important to study the entire length of the nerve. F-wave conduction velocity and F ratio provide a simple means of comparing nerve conduction of the proximal segment to that of the remaining distal segment.

ACKNOWLEDGMENTS

The author wishes to thank Mr. D. David Walker, MSEE, for engineering advice, and Ms. Sheila Mennen, Deborah A. Gevock, and Cheri L. Turner for technical assistance.

Central Actions of Neurotropic Drugs Assessed by Reflex Studies in Man

*P. J. Delwaide, J. Schoenen, and L. Burton

Section of Neurology and Clinical Neurophysiology, Department of Internal Medecine, University of Liège, Liège, Belgium

Lloyd (1943) and Eccles (1961; Eccles et al., 1954) initiated remarkable progress in the physiological analysis of spinal reflex pathways. Shortly thereafter, and despite the difficulties inherent in human experimentation, Magladery et al., (1952) and Paillard (1955), continuing the pioneering studies of Paul Hoffmann (1918), made quantitative studies of monosynaptic reflexes in intact man. Kugelberg (1948, 1962), Hagbarth (1952), Hagbarth and Finer (1963), and Hugon (1973b) showed that these techniques could be extended to exteroceptive reflexes. Later, functions such as presynaptic inhibition (Delwaide, 1971, 1973; Burke and Ashby, 1972), recurrent Renshaw inhibition (Pierrot-Deseilligny et al., 1976; E. Pierrot-Deseilligny, R. Katz, and H. Hultborn, *this volume;* Bussel and Pierrot-Deseilligny, 1977; Pierrot-Deseilligny and Morin, 1980), and reciprocal inhibition elicited by Ia fibers (Tanaka, 1974, 1980; Yanagisawa et al., 1976; Yanagisawa, 1980) were demonstrated in man. Various reflexological techniques developed in man have provided a wealth of valuable findings in physiopathological research.

At the same time, a number of neurotransmitters involved in reflexes, particularly of inhibitory type, have been identified in animal studies (see Krnjevic, 1980). In fact, synaptic junctions are vulnerable points of reflex circuits, and spinally active drugs are most likely to act at this level to modify spinal reflexes (Ryall, 1979). On this basis, reflex tests have been used in animals for screening muscle relaxants and other centrally active drugs. In man, knowledge of spinal neurotransmission is fragmentary and must still rely largely on extrapolation of data from animal experiments. This raises many problems for drug evaluation in clinical pharmacology, with the notable exception of muscle relaxants. However, it is now possible to activate and monitor well-defined spinal circuits in man although knowledge of the nature of some of the neurotransmitters involved is presumptive. Furthermore, the pharmaceutical industry is developing a large series of drugs that can interfere with neurotransmission. It is thus tempting to see—within the limits of strict ethical considerations—if it is possible by pharmacological means to manipulate neurotransmission in human reflex pathways and to quantify the results. Such a study would have the following objectives:

1. To try to validate in man the pharmacological data derived from animal studies. The human spinal cord presents morphological (Schoenen, 1980),

---

*To whom correspondence should be addressed: Department of Neurology, Institut de Médecine, Hôpital de Bavière, Boulevard de la Constitution, Liège, 4000, Belgium.

as well as functional, differences from that of cat or other experimental animals, and its pharmacology might also disclose some particularities.
2. To establish in man the mechanism of synaptic action of various drugs developed by the pharmaceutical industry. In order to achieve this, it would first be necessary to establish the type of reflex response that is modified. It may prove possible to differentiate agents by a battery of electrophysiological tests and to select the one that is most effectively modified. Provided the neurotransmitter involved in this peculiar response is known, it could be concluded that the drug interferes with its action.
3. To study central nervous system pharmacokinetics for individual drugs in man.

Such pharmacological studies can be carried out on normal volunteers, but to date, they have chiefly been performed on patients suffering from neurological disorders in whom modifications are more marked. Early quantitative studies of reflexes have already shown them to be easily modified by drugs (e.g., succinylcholine, Brune et al., 1960; chlorpromazine, Stern and Ward, 1962; diazepam, Brunia, 1973; Stern et al., 1968).

NEUROTRANSMISSION IN SPINAL CIRCUITS

In the animal, the spinal distribution of peripheral or descending afferents and the cells of origin of several ascending pathways are fairly well known. Little information, however, exists regarding interneuronal connections (Ralston, 1980). In man, the terminal distribution of descending pathways and the origin of ascending fibers are patchily known (see Nathan and Smith, 1955; Kuypers, 1973; Schoenen, 1964; Schoenen, 1980). Projections of the corticospinal tract have been carefully studied and are not identical to those in other mammals. Furthermore, the human cord differs in its dendritic organization of many neuronal groups (Schoenen, 1980), which suggests differences in intrinsic spinal connections among species. Histoenzymologic techniques suggest that certain biochemical mechanisms related to neurotransmission may be peculiar to the human cord (Schoenen, 1973).

These considerations underline the danger of extrapolating to man data derived from animal experimentation. So far as neurotransmission is concerned, this is risky because the point of impact of many neurotransmitters or neuromodulators remains a matter of speculation, even in animals. Furthermore, certain pharmacological manipulations of spinal reflex activity in man lead to different results from those obtained in animals. For example, ketamine depresses the Hoffmann reflex in rats but facilitates it in man (Bastron et al., 1972), and ethanol reinforces presynaptic inhibition in animals (Davidoff, 1973) but not in man (Ashby et al., 1977). This chapter will not review findings on neurotransmitters (see Krnjevic, 1974, 1980), but will examine some reflex circuits that can be studied in man by the techniques of clinical neurophysiology and be pharmacologically manipulated.

The spinal gray matter may be divided into two main compartments: a "motor" compartment (laminae V to IX), encompassing the intermediate gray and the ventral horn, and a "sensory" compartment (laminae I to IV), corresponding to the dorsal horn. Now that various spinal neurons have been identified in man, immunocytochemical studies can identify certain neurotransmitters involved (Fig. 1).

"Motor" Compartment (Laminae V to IX)

In the ventral horn, recurrent inhibition by the Renshaw (1941) circuit involves recurrent collaterals of MNs, which are cholinergic (Eccles et al., 1954), and the Renshaw interneurons, which are glycinergic,

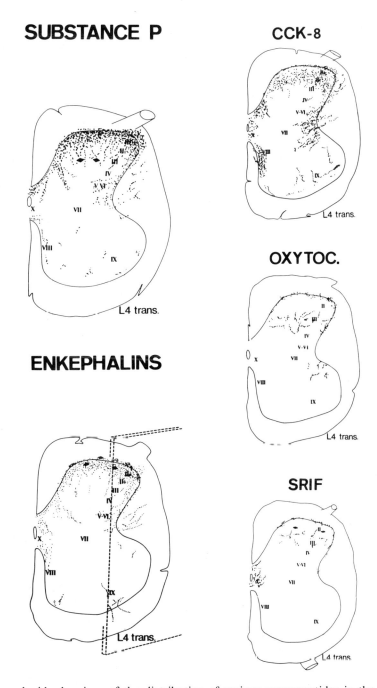

FIG. 1. Camera lucida drawings of the distribution of various neuropeptides in the human spinal cord (on a transverse section at the L_4 level; from Schoenen et al., 1982, with permission.) Compare with Table 1. Note dense projection of substance P-positive fibers in laminae I and external lamina II (IIe); cell bodies *(black lozenges)* are located in lamina III. Enkephalin-positive fibers are finer, more regularly beaded; positive cell bodies are found in laminae I and II. Cholecystokinin (CCK-8)-positive fibers are widely distributed in the dorsal horn as well as in the intermediate gray and the ventral horn. Oxytocin (OXYTOC.)-positive fibers are mainly concentrated in laminae I and III. Somatostatin (SRIF)-positive cell bodies are located in lamina II.

like other ventral horn inhibitory neurons (Boehme et al. 1976) such as the interneurons of the Ia reciprocal inhibition. In man, the Renshaw system can be studied physiologically (Pierrot-Deseilligny and Morin, 1980), but it has not yet been investigated pharmacologically. Recent work has brought indirect proof that glycinergic receptors are present on human MNs (Hayashi et al., 1981) as in animals.

Presynaptic inhibition acting on Ia spindle afferents (Eccles, 1964), whose reduction brings about the tendon hyperreflexia in the pyramidal syndrome (Delwaide, 1971, 1973), may be mediated by a GABAergic interneuron (Eccles et al., 1962; Smith et al., 1976). In man, the nature of the neurotransmitter in presynaptic inhibition remains a matter of speculation. GABA is a likely candidate, since diazepam, known to reinforce GABAergic action, augments presynaptic inhibition in man (Delwaide, 1977; Ashby and White, 1973), although other antispastics such as baclofen do not have this property.

A large number of descending fibers terminate in the ventral horn, and several neurotransmitters or neuromodulators have been identified within these terminals: serotonin and norepinephrine in bulbo- and pontospinal fibers in both animals and man (Nobin and Bjorklund, 1973); acetylcholine in reticulospinal endings in animals (McGeer et al., 1974); vasopressin, oxytocin, and TRH in terminals of hypothalamic origin in animals (Hokfelt et al., 1980) and in man. Glutamate is said to be the transmitter in corticospinal axons, but this amino acid, as well as aspartate, has also been implicated in transmission from primary spinal afferents (Rizzoli, 1968; Duggan, 1974; Krnjevic, 1980).

"Sensory" Compartment (Laminae I to IV)

In animals, neurotransmitters and neuromodulators concerned with pain-generated impulses have caught most attention (for reviews, see Basbaum and Fields, 1978; Willis and Coggeshall, 1978; Abdelmoumene, 1979; Cervero and Iggo, 1980). Although some links remain obscure, it is presumed that substance P (SP) is concerned in the transmission of nociceptive input carried by fine-caliber afferent fibers. The liberation of SP may be inhibited by an enkephalinergic interneuron in substantia gelatinosa (Jessell and Iversen, 1977). Other cutaneous afferents liberate somatostatin, glutamate, angiotensin, or cholecystokinin. Their actions may be inhibited by GABA. Neurons containing neurotensin have also been identified in the substantia gelatinosa (Hokfelt et al., 1980). Finally, the transmission of sensory impulses in the dorsal horn is controlled by pathways descending from brainstem (liberating enkephalins or 5-HT) (Basbaum and Fields, 1978) and cortex.

In man, terminals containing SP have been suggested by biochemical assay (Cuello et al., 1976) and identified by immunocytochemistry in laminae I and II (Hokfelt et al., 1980). Neurons containing SP have recently been demonstrated in lamina III. This peptide can also be detected in CSF, and its concentration diminishes when peripheral afferents degenerate, as in certain neuropathies (Nutt et al., 1980). As in the animal, enkephalin-positive afferents are numerous within laminae I and II of the human cord. Moreover, certain cell types of lamina II contain enkephalins. Opioid control of pain mechanisms similar to that identified in the animal cord is suggested in man by the changes in nociceptive reflex threshold induced by morphine or naloxone, even when the cord is transected (Willer and Bussel, 1980; J. C. Willer, *this volume*). Also, electroacupuncture in pain patients appears to augment CSF enkephalins (Sjölund et al., 1977). Cholecystokinin is also present in many dorsal horn afferents. Table 1 is a summary of the differential distribution of a various neuropeptides in the human spinal gray matter.

TABLE 1. *Differential distribution of various neuropeptides in the human spinal cord*[a]

Peptide	Fibers	Neurons
Substance P	+++: I, II ++: III, IV, V–VI lat. +: IX	III (radiate cells)
Enkephalins	+++: I, II, V–VI, IML ++: III, V–VI med., VIII (dors.), IMM +: IX	II (stellate cells) I
Vasopressin	+: II, V–VI med.	—
Oxytocin	++: I, III, IV +: IX d-1	—
Somatostatin	+: I, IIi, III, X, IML, IMM	II (islet cells)
Cholecystokinin	+++: I, V–VI lat., VII lat., VIII ++: V–VI med., X +: IX	—

[a] Data from Schoenen et al. (1982).

Table 2 shows the principal substances currently known to be involved in spinal neurotransmission. It still contains many question marks and may be profoundly changed as a result of progress in our knowledge.

SELECTION OF ELECTROPHYSIOLOGICAL TESTS

In man, techniques do not yet exist to selectively test each of the putative neurotransmitters. Before reviewing current possibilities, it is useful to define, from a theoretical point of view, the properties that an ideal reflex test should possess (Table 3):

1. It should be compatible with the requirements of medical ethics. Overtly painful stimuli or maneuvers that might be prejudicial to the subject (such as extended anesthesia or prolonged ischemia) could not be permitted. By the same token, the administration of drugs that are dangerous or could be upsetting should be avoided.
2. It should be simple and rapidly carried out. In practice, it is necessary to repeat the test several times during an experimental session, and technical complexities would prevent collection of adequate data.
3. Measurements should be stable under control conditions, and one should be able to attribute unambiguous significance to a variation in experimental values.
4. The active collaboration of the subject in the test should be reduced. This requirement derives in part from point 3 above but is particularly important for neurological patients, since their motivation may be inconstant, and they may suffer from various neurological deficits. However, patients have to be cooperative enough to remain immobile for 2 or 3 hr of testing.
5. The test should, however, be sensitive. Within the constraints of therapeutic dosage, a substance should manifest its effects in a measurable and graduable fashion.
6. Finally, the test should be specific for a particular mode of facilitation or inhibition and be capable of correlation with the activity of a defined neurotransmitter.

These requirements, which may sometimes appear contradictory (for example, stability and sensitivity), can be satisfied at the present time by only a limited number of tests, although some fulfill one criterion

TABLE 2. Currently known neurotransmitters

Transmitter	Fibers	Neurons	Site of action	Action	Agonists	Antagonists
Acetylcholine						
Nicotinic	Ventral horn	Motoneurons, recurrent collaterals	Postsynaptic Renshaw cells	Excitation	Physostigmine Choline Lecithin Methacholine Nicotine	Curare-like agents
Muscarinic	Diffuse	? Interneurons ? Supraspinal (reticular)	Postsynaptic ? Interneurons ? Motoneurons	? Excitation	Same as above	Atropine, Scopolamine
Monoamines						
Norepinephrine	Diffuse	Supraspinal (A_1, A_2, A_6)	? Interneurons	Inhibition	MAO inhibitors Nomifensine Nortryptiline Desipramine Clonidine	Reserpine Phenoxybenzamine Phentolamine β Blockers ? Bretylium
Epinephrine	Intermedio-lateral nucleus	Supraspinal (C_1, C_2)	IML	Inhibition	Same as above	Same as above
Serotonin (5-HT)	Laminae I–II	Supraspinal (raphe nuclei B_1, B_2)	Presynaptic: primary afferents	Inhibition	5 HTP Ipronizide ? Clonazepam Zimelidine Imipramine Amytriptiline Fenfluramine	Reserpine Methysergide Cyproheptadine R41-468
Dopamine	Ventral horn IML nucleus Ventral horn?	Supraspinal (A_{11}, ? A_{12})	Postsynaptic: many neurons Postsynaptic	Inhibition	L-DOPA Bromocriptine Pergolide Lisuride	Butyrophenones (haloperidol)
Amino acids						
GABA	Mainly laminae I to III; also laminae IV, VI, VII, X	Interneurons	Presynaptic: primary afferents (Ia, C, Aδ fibers) Postsynaptic: dorsal horn neurons, motoneurons	Inhibition depolarization Hyperpolarization	Valproic acid Progabide γ-Vinyl-GABA γ-Acetylenic-GABA (GAG) Muscimol L-Glutamate Benzodiazepines Barbiturates (?)	Hydrazides ? Tetanus toxin
Glycine	Ventral horn	Interneurons: Renshaw cells; Ia interneurons (?)	Postsynaptic: motoneurons	Inhibition: hyperpolarization	Glycine Benzodiazepines (?)	Thebaine Strychnine (Tetanus toxin)

Substance	Location	Cells	Action site	Effect	Agonist	Antagonist
Aspartate	Ventral horn	Interneurons	Postsynaptic: motoneurons	Excitation	L-Aspartate (?)	? Mg^{2+} ?Baclofen α-Aminoadipate
Glutamate	Diffuse	Ganglion cells (primary afferents ?) Corticospinal afferents (?)	Postsynaptic: interneurons, motoneurons	Excitation	L-Glutamate (?)	? Baclofen
Taurine	?	?	Postsynaptic: motoneurons	Inhibition	Taurine	?
Serine Proline	?	?	Postsynaptic	Inhibition	?	?
Peptides						
Substance P	Laminae I → II (III, IV, V–VI lat., IX)	Ganglion cells (primary afferents) Interneurons: lam. III radiate cells Supraspinal afferents	Postsynaptic: dorsal horn neurons ? Interneurons	Excitation Transmission or modulation ?	—	? Baclofen
Enkephalins	Laminae I, II, V–VI lat., IML (III, V–VI med., VIII, IX, IMM)	Interneurons: lam. II stellate cells, lam. I Supraspinal off	Presynaptic: primary afferents (C, Aδ ...) Preganglionic symp. neurons	Inhibition	Morphine analogs	Naloxone
Somatostatin	Laminae I, IIi, III, X, IML, IMM	Ganglion cells: Primary afferents (C, Aδ?) Interneurons: lam. II islet cells	Postsynaptic: dorsal horn neurons Preganglionic symp. neurons	Excitation	—	—
Angiotensin II	Dorsal horn Ventral horn IML nucleus	Ganglion cells: primary afferents Supraspinal	?	Excitation	—	? Captopril
Neurotensin	Laminae II–III	Interneurons	?	?	—	—
CCK-8	Laminae I, V–VI lat., VII lat., VIII (V–VI med.; IX, X)	Ganglion cells: primary afferents ? Supraspinal ? Interneurons IMM (rat)	?	Dopamine interaction	Caerulein	—
Vasopressin Oxytocin	I, II, III, IV	Hypothalamic	?	?	Lys-Vasopressin Oxytocin	—
TRH	Ventral horn	Hypothalamic	?	?	TRH	—

better than others. Table 3 indicates the value of various tests according to these requirements.

With respect to stability, reproducible quantitative results can be obtained by methods that are becoming more strict and well codified. As shown in Figs. 2 and 3, this is notably the case for values of H_{max}/M_{max} and even F_{max}/M_{max} ratios, for vibratory inhibition, for the RA-II and RA-III reflexes of the short head of biceps, and even for excitability curves.

The sensitivity of a test can be increased by modifying the experimental conditions. Thus, vibratory inhibition usually becomes more marked if the value $H_{max}/2$ is taken rather than H_{max}. However, under these conditions, stability is compromised. Sensitivity seems to be higher in neurologic patients. Diazepam, for example, in an i.m. dose of 10 mg, does not reinforce vibratory inhibition in normal volunteers but does so in spastic patients. Similarly, the blink reflex is not affected by L-DOPA in normal subjects, but it is in parkinsonian patients.

The specificity of a test is at present the most delicate point. It depends very much on physiological and pharmacological knowledge of spinal circuits. In principle, a reflex response elicited by stable stimulation can be modified by activity exerted at a synapse between primary afferent and MN, at interneuronal synapses, or at synapses between descending fibers and the same MN or interneuron. In order to eliminate these last possibilities, it may be useful to carry out control measurements on patients with stabilized spinal transections. This requirement is but rarely encountered in published investigations. It can be appreciated that the purely spinal localization of the reflex circuits and the limited number of synapses involved are factors enhancing selectivity of the test. Following conditioning of a reflex response, the first milliseconds after the conditioning stimulus are of particular interest because the delay is too short to allow the intervention of long loops. To refine the specificity, different checkings are necessary, and complementary tests prove useful. If, for example, a drug is tested by a flexion reflex, any change in the reflex amplitude cannot be attributed to an action exerted in the posterior horn unless the precaution is taken of testing MN excitability at the same time, (by the H reflex or the F response). If the latter is changed, no conclusion can be reached. This example illustrates the fact that there cannot be a unique experimental protocol involving only a single test but that methods must be adapted to the drug tested at its presumed point of action, and controls must be multiplied to achieve selectivity.

The influence of the state of consciousness is an important factor in any study of pharmacological effects on reflex activity. It is known that in the normal adult, reflexes are depressed during sleep. Hodes and Dement (1964) showed that the amplitude of the H reflex decreases by 20 to 30% during the passage from wakefulness to slow-wave

TABLE 3. *Evaluation of the various electrophysiological tests*

	H_{max}/M_{max}	T_{max}/M_{max}	F	Vibratory inhibition	Renshaw	Recovery curve
Ethically admissible	+	+	+	+	+	+
Simple and rapid	++	+	++	++	−	+
Limited participation	+	+	+	+	+	+
Stable in controls	++	+	+	++	+	−
Sensitive	+	+	+	++	+	++
Correlated to a patho-physiologic mechanism	−	−	−	++	++	−
Total score	7	5	6	10	6	5

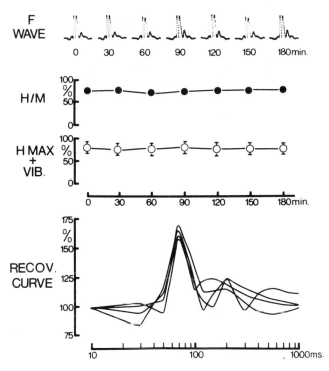

FIG. 2. Stability of various electrophysiological values over 3 hr; recordings are made in spastic patients. F wave: each oscillogram represents 20 rectified and summated responses obtained in tibialis anterior after supramaximal stimulation of the peroneal nerve. H/M: the value is higher than 50% and remains remarkably stable with an SD too small to be represented. H_{max} + vib: the value is obtained as H_{max} amplitide during vibration divided by H_{max} amplitude (control) × 100 and is higher than 50%. The recovery curves (bottom) obtained by conditioning the H reflex by stimulation of the tibial nerve at the ankle are repeated each 30 min and, at least at 80-msec delay, indicate similar values.

sleep. On the other hand, H-reflex amplitude remains appreciable during anesthesia and does not disappear until anesthesia is very deep (de Jong et al., 1967). The polysynaptic blink reflex is similarly modified (Shahani, 1968), its first component disappearing during slow-wave sleep.

Although dosages causing sleep are not used in pharmacological research, there is nevertheless a frequent risk of altering the state of alertness which should be monitored by a psychophysiological test. The most effective guarantee, however, is provided by control values of either monosynaptic or polysynaptic reflexes repeated during the experiments. Even when there is a change in the level of consciousness, some substances may specifically alter some reflexes (for example, the monosynaptic reflexes) more than others, which would suggest that the observed effects did not result simply from reduced alertness as a nonspecific cause.

ELECTROPHYSIOLOGICAL TESTS AND PHARMACOLOGICAL CORRELATIONS

In studies combining reflexes and pharmacology, two types of problems can, theoretically, be distinguished. (1) It must be discovered which electrophysiological test (or battery of tests) is sensitive to the action of the drug under consideration. Insofar as the test has precise physiological meaning,

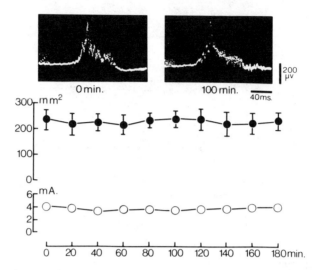

FIG. 3 Stability of polysynaptic responses recorded in control subjects over 180 min. **Upper row:** The rectified and summated (20 sweeps) responses recorded in the biceps femoris after identical painful stimulation of the sural nerve (RA-III). The response recorded after 100 min is very similar to control response. **Middle row:** The RA-III responses have been measured after enlargement. The values are indicated with the SD. **Lower row:** Threshold value of the sural nerve stimulation that elicits an RA-III response.

the mode of action of the drug can be deduced. However, one cannot say in advance which test will be influenced. Examples of unexpected results are reported below. Also, in a first phase, which may be called screening, it is necessary to investigate how a substance acts in a series of complementary explorations, which can be classified under three heads: excitability of the myotatic arc, interneuronal transmission, and posterior horn mechanisms. (2) Once the sensitive test has been defined, the kinetics of the product in the CNS beyond the blood–brain barrier can be investigated. The delay after which the test changes is easily measured; the intensity of the effects can be related to the dose given. The time after which no further changes are seen can also be measured.

Excitability of Myotatic Arc

This depends on many factors, segmental and suprasegmental. In clinical neurophysiology, some of the control mechanisms can be both demonstrated and quantitatively studied. Although studies of the excitability of the myotatic reflex arc have been carried out on normal subjects, they have been most extensively performed on spastic patients.

Measurements of the excitability of the MN pool.

The F response can test MN excitability (Schiller and Stalberg, 1978) but has never been used in pharmacological studies. This response has the particular interest of being related entirely to MNs, and its elicitation does not depend on reflex activity. However, its lack of stability makes it necessary to average several dozen summed responses. We have studied its variations under the influence of several substances (lysine-vasopressin, progabide, tizanidine). These variations are frequently parallel to those of the H_{max}/M_{max} ratio (see below. However, under the influence of 7.5 U of lysine-vasopressin, a reduction of the amplitude of the F response was seen for approximately 30 min in some normal subjects although the H_{max}/M_{max} ratio remained stable. In spastic patients, amplitude reduction was observed in all subjects. It is as yet too soon to assess the value of this test.

The Hoffmann reflex amplitude depends on the number of MNs recruited by the reflexogenous volley. If the excitability of the motor nucleus is increased, the same reflexogenous volley causes more MNs to discharge. The amplitude of the H response, depending on a single synapse, reflects the excitability of the motor nucleus so long as the presynaptic inhibition exerted on I_a fibers remains constant. But when the latter is susceptible to modification, it is of interest to perform complementary studies of the F response. The amplitude of the H reflex also depends on a number of experimental factors such as the resistance between the electrodes, the position of the electrodes, etc. (see Hugon, 1973a; Hugon et al., 1973). For this reason, Angel and Hofmann (1963) proposed relating the maximal amplitude of the H reflex (H_{max}) to the maximal amplitude of the motor response recorded under the same conditions (M_{max}). This ratio H_{max}/M_{max} expresses the proportion of MNs in the motor nucleus that are reflexly activated. This ratio is easily obtained and is stable under control conditions. Although overall excitability does not allow conclusions to be drawn about which synapses are influenced, the H_{max}/M_{max} ratio is sensitive and reliable.

In normal subjects, this ratio (or H reflex amplitude) is diminished by haloperidol (10 mg i.m.) (Paillard et al., 1961), nicotine (0.8–1.0 mg i.v.) and pentobarbital (50–100 mg i.v.) (Gassel, 1973), reserpine (10 mg i.v.), and chlorpromazine (75 mg i.v.) (Stern et al., 1968). As mentioned above, lysine-vasopressin does not change the H_{max}/M_{max} ratio, and neither does thymoxamine (Phillips et al., 1973). Certain points may be disputed. Brunia (1973) reported a reduction in H-reflex amplitude after an i.v. injection of diazepam, 10 mg., but we have failed to observe a reduction in the H_{max}/M_{max} ratio following an intramuscular injection of 15 mg. It is interesting to note that increases in the amplitude of the H reflex have been reported, for instance, after administration of meprobamate (800 mg i.m.) or nefopam, an analog, or orphenadrine (Gassel, 1973). Changes in the H_{max}/M_{max} ratio are useful to consider in man, as contradictory results have been reported in animal experiments according to the species studied. Thus, in the rat, reserpine increases the amplitude of the monosynaptic reflex (Grossman et al., 1975), whereas the opposite is observed in the cat (Geber and Dupelj, 1977).

Particular attention has been paid to the effect of muscle relaxant drugs on H_{max}/M_{max} ratio in cases of spasticity (Angel and Hofmann, 1963; Dietrichson, 1971, 1973; Delwaide, 1971; Garcia-Mullin and Mayer, 1972). The ratio is increased by spasticity, and it would appear entirely reasonable to assume that myorelaxants would reduce it. Matthews (1965) noted that chlorproethazine reduced the amplitude of the H reflex in spastic subjects. In a study on diazepam, the same author (1966) concluded that spasticity could be reduced without change in the H_{max}/M_{max} ratio. We have looked again at this problem in 16 patients (Fig. 4). Measurements were made every 30 min after the i.m. administration of 10 mg diazepam; 15 out of 16 patients showed a reduction in the ratio; after 90 min, the reduction averaged 13%.

In general, muscle relaxants with a central action modify the ratio, although to different extents. With baclofen (60 to 90 mg/day by mouth), Castaigne et al. (1973) found a reduction of 20%, and McLellan (1973) of 10%; however, Pedersen et al. (1974) found a reduction in only five out of 17 patients. In our own series of 13 patients, who each received 20 mg i.m., a reduction of 11% was

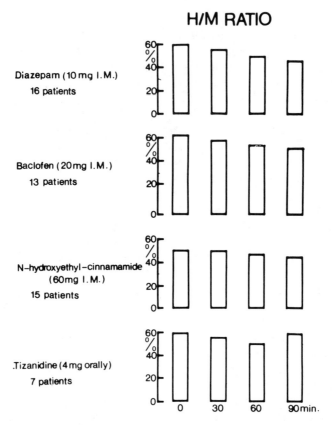

FIG. 4. Evolution of the H/M ratios under the influence of four myorelaxant drugs.

observed 90 min after injection. N-Hydroxyethylcinnamamide (60 mg i.m.) had no effect (−4%); tizanidine (4 mg orally) reduced the ratio for 60 min but this latter has returned to normal values after 90 min.

The influence of other drugs on this ratio has also been reported. Thymoxamine, an α blocker, reduces the amplitude by 74% in spastic patients according to Mai (1978), whereas propranolol, a β blocker, has no effect on the ratio, although it distinctly reduces clonus (Mai and Pedersen, 1976; Pedersen and Mai, 1978). Substances implicated in the physiopathology of spasticity, such as glycine (Barbeau, 1974; Stern and Bokonjic, 1964; Boehme et al., 1976; Hall et al., 1979), have apparently not been studied reflexologically.

As can be seen, the amplitude of the H reflex and the H_{max}/M_{max} ratio can be influenced by drugs both in normal subjects and in spastic patients. In the latter, there is a poor correlation between diminution of the ratio and clinical improvement. Although as a test it is both stable and sensitive, the H_{max}/M_{max} ratio represents a global measure that alone does not provide enough information on the neurochemical mechanisms that may be changed by a drug.

Comparison between H and tendon reflexes.

Following Paillard (1955), emphasis was laid on comparison between the H reflex and the tendon reflex (T) or, better, between the H_{max}/M_{max} and T_{max}/M_{max} ratios. The amplitude of the tendon reflex is a measure both of the excitability of the motor nucleus and of fusimotor tone. The difference between changes in the H and T reflexes might, therefore, yield information on alterations in the γ system. This has been widely used as a method, for example, by Buller and Dornhorst (1957). However, microneurographic studies (see D. Burke, *this volume*) and findings underlying the concept of α–γ linkage do not confirm this simple view and call for caution when one is considering differences in changes between H and T responses (Delwaide et al., 1980).

In healthy volunteers, the tendon reflex is depressed more than the H reflex by haloperidol (Paillard et al., 1961), thymoxamine (Phillips et al., 1973), and to a lesser extent by chlorpromazine (Brunia, 1973). Diazepam (Brunia, 1973), nicotine, and pentobarbital (Gassel, 1973) reduce the amplitude of the H and T responses in parallel. In spastics, we have shown that the T_{max}/M_{max} ratio is reduced by 10% 90 min after diazepam and by 9% after baclofen. Castaigne et al. (1973) also observed after baclofen a reduction of this ratio that was slightly less marked than that of H_{max}/M_{max}. N-Hydroxyethylcinnamamide does not influence the T response, but tizanidine reduces it by 5%. Matthews (1965), on the other hand, observed that chlorproethazine reduced the T response to a greater extent than the H.

Vibratory inhibition.

A vibratory stimulus applied to the Achilles tendon clearly reduces the amplitudes of the Hoffmann and soleus tendon reflexes (Hagbarth and Eklund, 1966; Lance et al., 1966; Delwaide, 1971). As a result of animal experiments by Gillies et al. (1969) and analyses carried out in man (Delwaide, 1971, 1973; Dindar and Verrier, 1975; Ashby and Labelle, 1977; Ashby and Zilm, 1978; Desmedt and Godaux, 1978c), this inhibition has been attributed to an autogenous presynaptic inhibitory mechanism exerted by Ia fibers coming from the soleus on other Ia fibers of the same origin. In this inhibition, inputs from skin or antagonist muscles play only a minor role (Delwaide, 1971; Ashby et al., 1980; J. E. Desmedt, *this volume*). Thus, vibratory inhibition has a good physiological basis in an interneuron intercalated between terminals of Ia fibers. This interneuron is undoubtedly also open to facilitatory suprasegmental influences. GABA is the presumed transmitter of the interneuron involved in presynaptic inhibition. Under constant experimental conditions, the intensity of vibratory inhibition remains stable and can be quantitatively expressed by the formula $H(vibrated)_{max}/H(control)_{max} \times 100$. When this value is measured several times in a single subject, similar results are obtained, indicating that the test is reproducible. There are marked differences among individuals. Increasing age reduces the amount of vibratory inhibition (Delwaide, 1973).

Vibratory inhibition of the tendon reflex could be used, but results are less consistent, partly because the tonic vibratory reflex (TVR) modifies the tension of the Achilles tendon so that percussion does not always produce identical T reflex responses. For this reason, vibratory inhibition of the H reflex is a better measure. In spasticity, vibratory inhibition is reduced or even absent (Delwaide, 1971, 1973; Burke and Ashby, 1972; Ashby and Verrier, 1976), but it may be increased immediately following spinal transection or stroke (Ashby and Verrier, 1976; see J. E. Desmedt, *this volume*). In Parkinson's disease or cerebellar syndrome, however, vibratory inhibition is of normal amplitude. Measurement of vibratory inhibition turns out to be a rather sensitive test in, for example, multiple sclerosis of moderate severity, probably because vibratory inhibition tests the spinal cord "in action," whereas other methods test it "at rest."

In normal subjects, it seems difficult to potentiate vibratory inhibition by drugs. In our experience, diazepam (15 mg i.m.) does not cause any change. Similarly, Ashby et al. (1977) did not observe any effect of alcohol, which is known to reinforce presynaptic inhibition in amphibia.

In contrast, in spastic subjects (Fig. 5), diazepam, 10 mg i.m., distinctly reinforces vibratory inhibition, bringing values closer to normal. This effect is much clearer than changes in the H_{max}/M_{max} ratio (Delwaide, 1971, 1973). In a recent series of 16 spastic patients, mean index of inhibition was 69% before and 49% 30 min after the injection of 10 mg i.m. diazepam. This has been confirmed by Verrier et al. (1975) who, however, did not observe the effect in spastic patients with spinal transection of long duration, as if changes in vibratory inhibition also involved a supraspinal action of the drug. This interpretation may be queried because it has been shown that diazepam exerts its relaxant effect at spinal level (Cook and Nathan, 1967).

Other drugs may also reinforce vibratory inhibition in spasticity; these include many benzodiazepines such as chlordiazepoxide, oxazepam, and R06-9098. Tizanidine (4 mg by mouth) in seven patients caused the mean value of the index to fall from 74.1% to 66% in 30 min (Fig. 5). Progabide, a GABAergic substance, occasionally reinforced vibratory inhibition, but its efficacy was not great, at least at the dosage used (6 × 300 mg). Thymoxamine increased vibratory inhibition by 8% (Mai, 1978). In general, the degree of reinforcement of vibratory inhibition by the various drugs correlates well with their muscle relaxant activity. However, not all muscle relaxants modify vibratory inhibition.

Baclofen, for example, does not have such an effect, at least when given intramuscularly (Delwaide et al., 1980) or by mouth (Ashby and White, 1973) (Fig. 5). In some cases, there is even a slight reduction of vibratory inhibition (Castaigne et al.,

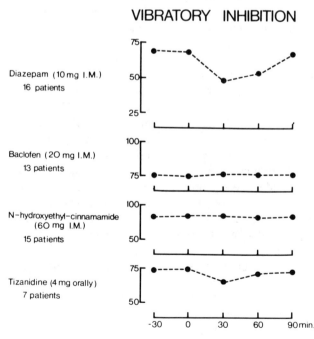

FIG. 5. Evolution of the vibratory inhibition index: [H_{max} (vibration)/H_{max} (control)] × 100 in spastic patients under the influence of four myorelaxant drugs. The drugs are administered at time 0.

1973; Delwaide et al., 1980). In some patients, however, following intravenous injection, Pedersen et al. (1974) found reinforcement of vibratory inhibition. Hydroxyethylcinnamamide also has no effect on vibratory inhibition (Delwaide et al., 1975), nor does lysine vasopressin.

Thus, vibratory inhibition, which seems to reflect presynaptic inhibition, may be reinforced in spastic subjects by some drugs. Insofar as the benzodiazepines are concerned, the findings are confirmed by animal pharmacology (Davidoff, 1978; Costa and Guidetti, 1979). It has also been shown in the cat that tizanidine reinforces presynaptic inhibition, but this is not the predominant action (D. M. Coward, *personal communication*). It has also been suggested that thymoxamine acts on the descending control of the interneuron responsible for presynaptic inhibition. The lack of effect in normal subjects may mean that vibration activates presynaptic inhibition maximally, so that a drug cannot further increase it. Finally, the vibratory inhibition test makes it possible to divide muscle relaxants into two categories: one group, represented by diazepam and tizanidine, reinforces vibratory inhibition and it also reinforces presynaptic inhibition. Another group, represented by baclofen and N-hydroxyethylcinnamamide, does not reinforce vibration inhibition and does not appear to affect presynaptic inhibition, in agreement with recent pharmacological findings (Koella, 1980).

More specific tests.

Other tests allow quantification of mechanisms that control the excitability of the myotatic arc. Pierrot-Deseilligny et al. (1976; E. Pierrot-Deseilligny, R. Katz, and H. Hultborn, *this volume*), described a method for studying the Renshaw interneuron in intact man. This interneuron is activated by a cholinergic recurrent collateral of the MN axon. Its inhibitory action is exerted in animals, and probably also in man, by glycine. This test has not yet been used in pharmacological studies.

The Ia interneuron (probably glycinergic) can be studied by estimating the H-reflex inhibition following stimulation of the lateral popliteal nerve during voluntary contraction of tibialis (Tanaka, 1974, 1980; R. Tanaka, *this volume*). This method has been applied to spastic patients (Yanagisawa et al., 1976; Yanagisawa, 1980; R. Tanaka, *this volume*) and discloses a reduced efficiency of the Ia interneuron. The test has apparently not yet been used in drug evaluations.

Interneuronal Transmission

The tests described above concern interneurons involved in the control of the myotatic reflex. We now consider the recovery curves of monosynaptic reflexes, particularly at latencies above 50 msec, and the nonnociceptive exteroceptive reflexes.

Recovery curve of H reflex with paired stimuli.

The five phases of this curve are now classical (Hoffmann, 1924; Magladery et al., 1952; Teasdall et al., 1951; Paillard, 1955; Diamantopoulos and Zander Olsen, 1967; Tabarikova and Sax, 1969). The curve is altered both in spasticity (Teasdall et al., 1952; Matsuoka et al., 1966; Zander Olsen and Diamantopoulos, 1967; Yap, 1967) and parkinsonism (Zander Olsen and Diamantopoulos, 1967; Yap, 1967; Takamori, 1967; Krassoievitch and Tissot, 1971; Sax et al., 1977). The fourth phase of intercurrent facilitation could reflect the perturbations brought about by the two disorders and be a sensitive criterion of drug action (Gassel, 1973). The recovery curve has therefore become a classical test. However, Paillard (1955) stressed the importance of methodology (values of conditioning and conditioned responses, angle of flexion of ankle, etc.) and showed that changes are often the result of variations in techniques. This has been confirmed (Delwaide, 1971), and there is no standardized methodology at the present time. The mechanisms reponsible for changes in amplitude are ill understood except perhaps during the first 80 msec. Both segmental and suprasegmental factors are implicated. The physiological correlate of changes in phase IV is not known. The test does not possess pharmacological specificity but is considered to be sensitive for documenting pharmacological effects.

In normal subjects, Gassel (1973) demonstrated a reduction in phase IV after nicotine and pentobarbital. In spastic subjects, Bergamini et al. (1972) showed a reduction in phase IV 10 and, particularly, 30 days after administration of baclofen. We have observed a marked inhibition of phase V after diazepam (Delwaide, 1971), and Mai (1978) reported a reduction in phase IV after thymoxamine. In parkinsonian subjects, the duration and amplitude of phase IV are reduced by treatment with L-DOPA (Krassoievitch and Tissot, 1971; Herbison, 1973; Sax et al., 1977).

In patients with tardive dyskinesias, Crayton et al. (1977) have shown that facilitation of phase IV is reduced by apomorphine and amphetamines. They suggest (Crayton and Tamminga, 1978) that phase IV is under dopaminergic control and propose, as a result of comparing prolactin levels and recovery curves, that the latter could be used as a test of dopaminergic transmission. However, this hypothesis does not rest on firm foundations and derives from argument by analogy.

Recovery curve after stimulation of tibial nerve.

Stimulation of the tibial nerve at the level of the internal malleolus by a brief train of shocks alters the amplitude of the soleus H reflex for more than a second. In normals, an inhibition develops progressively to a maximum at 80 to 100 msec, after which it diminishes progressively and disappears after 1,500 msec (Delwaide, 1971). Although these changes are elicited by stimulation of a mixed nerve, fibers of cutaneous origin play the main role in excitability changes (Martinelli et al., 1979). In spastic patients, a different curve is seen, with a marked facilitation between 60 and 100 msec (Delwaide et al., 1980). In parkinsonian patients, there is a reduction of the early inhibition (Martinelli and Montagna, 1979). The changes, although marked, still require physiopathological elucidation, since segmental and suprasegmental mechanisms are no doubt involved. This recovery curve presents certain advantages as a test, since it does not involve antidromic stimulation, and the changes in spasticity mainly involve the abnormal facilitation. The normal curve is not changed by diazepam or phenobarbital.

In spasticity, diazepam does not induce any change (Fig. 6), although it does alter the H-reflex response to paired stimuli (see above). Tizanidine is also without effect. On the other hand, baclofen clearly modifies this response (Fig. 6): it reduces the facilitation and increases the subsequent inhibition. Similar changes are seen with hydroxyethylcinnamamide.

These findings complete the catalog of opposite effects obtained with two groups of muscle relaxants. On one hand, diazepam and tizanidine reinforce vibratory inhibition but have no effect on recovery curves; by contrast, baclofen and 4-hydroxyethylcinnamamide have no effect on vibratory inhibition but change recovery curves towards normality. It may be proposed that baclofen depresses the activity of the neurons responsible for the facilitatory phase and that the reduction of facilitation could be caused by a cumulative effect occurring at each synapse. This concurs with pharmacological data (Henry, 1980), although other effects, notably those of GABA agonists, are possible by an action at receptors insensitive to bicuculline (Bowery et al., 1980).

RECOVERY CURVES

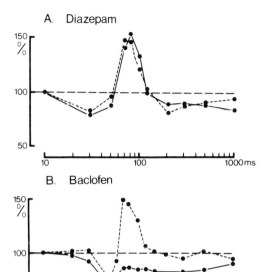

FIG. 6. Influence of diazepam (10 mg i.m.) and baclofen (20 mg i.m.) on the recovery curves of the soleus H reflex after stimulation of the tibial nerve at the ankle. In *abcissa* (log scale), delay (msec) between the conditioning and the H reflex; in *ordinate,* variations of H-reflex amplitude from control values (100%). The interrupted lines indicate the curve before any drug in two spastic patients (**A** and **B**). The *continuous lines* show how the curves are influenced by (A) diazepam and (B) baclofen, respectively. Compare the changes in **B** with the stability of similar curves shown in Fig. 2.

Exteroceptive reflexes.

Stimulation at nonnoxious intensity of some purely sensory nerves can evoke EMG responses in particular muscles. Two responses should be considered in relation to pharmacological studies. (1) Nonpainful stimulation of the supraorbital branch of the trigeminal nerve causes blinking (Kugelberg, 1952; Rushworth, 1962; Shahani and Young, 1968, 1973b; Penders and Delwaide, 1973; see B. W. Ongerboer de Visser, *this volume*). This reflex has two components. (2) Noxious stimulation of the sole of the foot elicits two bursts of EMG activity in tibialis anterior (Shahani and Young, 1971, 1973a).

All of these responses are easy to obtain but are more difficult to quantify or to maintain stable. Their specificity is low in relation to localization of pharmacological action except when the interneuron chain is short, as in the blink reflex. Before significance can be given in amplitude of polysynaptic reflexes, it is necessary to be certain that MN excitability remains stable under the influence of drugs. Although the amplitude of the polysynaptic reflex is not a good quantitative indicator, habituation thereof can be used to quantify drug effects. Habituation is defined as the progressive reduction in reflex response amplitude when repetitive stimulation is carried out above a critical frequency (see Desmedt and Godaux, 1976). In the blink reflex of normal subjects under the influence of diazepam, the first component was unchanged, but the second was reduced (Kimura, 1973; J. Kimura, *this volume.*)

This test has not hitherto been employed in spastic subjects but mainly in parkinsonian patients. Penders and Delwaide (1971) defined the habituation index as the inverse of the fastest stimulation frequency at which the fifth stimulus of a series still evokes a response R_2 with an amplitude at least 20% of that of the control value. Under normal conditions, the mean index is 5.3 sec, with a range of 3.2 to 8.0 sec. These values were found to be relatively stable over successive test sequences.

In parkinsonian patients, the mean habituation index is 1.6 sec. Following treatment for several weeks with L-DOPA and amantadine, it rises to 2.3 sec, which is still below the values in normals, even though clinical improvement is evident and clinical signs practically absent. These results have been confirmed by Messina et al. (1972). Conversely, anticholinergics do not change the habituation index. No study has yet been carried out on effects of dopaminergic agonists. Changes in habituation under L-DOPA therapy have also been described in lower limb exteroceptive reflexes in parkinsonian patients (Delwaide et al., 1974).

Few data exist for the lower limb flexor reflexes in spasticity. Baclofen depresses them (Shahani and Young, 1973c).

Spinal Cord Dorsal Horn Mechanisms

Neurons in the dorsal horn of the spinal gray matter play a minor role in the modulation and transmission of nociceptive inputs. A number of neurotransmitters have been implicated in animals in dorsal horn function, in both primary afferents and descending pathways (see J. C. Willer, *this volume*). Techniques for the exploration of the human dorsal horn are as yet poorly developed, although studies of lumbar evoked potential are promising. At present, the only available technique is the flexor reflex elicited by painful stimulation of the ipsilateral sural nerve (Hugon, 1973b). A high-intensity single shock elicits a polyphasic response in the short head of biceps femoris with a latency varying from 80 to 150 msec (RA-III). The amplitude of response could be used, but it is more convenient to measure the threshold intensity that elicits the RA-III response. The technique is simple, not very painful, and gives consistent results (Fig. 3). However, it is neither very sensitive nor very specific when account is taken of the different synaptic influences capable of modifying the amplitude of the reflex. The technique was chiefly used by Bathien and Willer (1971; Willer and Bathien, 1977) who showed that the threshold of the RA-III response was raised by the oral administration of 1.0 g of acetylsalicylic acid. Pentazocine (50 mg) first lowers, then raises the threshold; intradermal histamine reduces it, and glafenine raises it provided it is previously lowered by a chronically painful process.

Willer and Bussel (1980) studied patients with spinal transection, examining the amplitude of the reflex elicited in the tibialis anterior by stimulation of the sural nerve. Morphine strongly depresses the nociceptive reflex without modifying the monosynaptic reflex, which indicates that the excitability of MNs was not affected and that depression of the reflex was related to a purely spinal action of morphine, probably in the dorsal horn. Reduction of the polysynaptic reflex is completely reversed by naloxone (J. C. Willer, *this volume*).

Willer and Bathien (1977) have also recommended comparison of thresholds of the nociceptive reflex in short biceps following stimulation, respectively, of sural nerve and of the skin it innervates. The perception threshold of pain and the threshold of the nociceptive reflex are closely correlated. The threshold of the reflex was lower for cutaneous stimulation than for stimulation of the nerve by 50%. The relationship between the two thresholds remains stable from one experiment to another in the same subject. The cutaneous threshold is more easily modifiable than the nerve threshold by drugs, whether applied locally or systemically. Following application of an algogenic substance (naphthalene) to skin, the cutaneous threshold is clearly lowered, although the nerve threshold is little changed. On the other hand, a local anesthetic (lidocaine) raises the cutaneous threshold but has no effect on the nerve threshold. Acetylsalicylic acid given intravenously raises the cutaneous threshold only. Pethidine (200 mg i.m.) affects both thresholds: the cutaneous is raised, but the nerve, paradoxically, is lowered.

We have used the nerve threshold to investigate effects on dorsal horn of ceruletide (CRL), a drug similar to cholecystokinin which has been demonstrated in human dorsal horn neurons (Schoenen et al., 1982). Ceruletide might possess analgesic properties at low doses (50 ng/kg i.m.) since we have observed a slight lowering of nociceptive reflex threshold, although the monosynaptic reflex was unchanged.

VALUE IN CLINICAL PHARMACOLOGY

Some drugs may electively modify electrophysiological tests in conjunction with their mode of pharmacological action. The changes brought about by a drug can be quantified. Once screening has been carried out and a sensitive test recognized, it can be used as a specific method for determining the CNS actions of a drug:

1. The delay of a drug effect may be determined. This is not always related to the plasma concentration because it may be influenced by speed of penetration through the blood–brain barrier. For example, we have noted that after injection of baclofen, 10 mg, an effect can be observed 15 min after i.m. administration. With the help of electrophysiological methods, it is not necessary to carry out lumbar punctures in order to study CSF and find out when a drug has penetrated the blood–brain barrier (Knutsson et al., 1973).
2. The duration of action of a single drug dose can be determined, which is practically useful in fixing the intervals at which medication should be given.
3. A dose–effect relationship can be demonstrated. For example, if 20 mg of diazepam, instead of 10 mg, be given to spastic patients, vibratory inhibition is clearly increased. Similarly, pethidine, 0.2 mg/kg, reduces the tibialis anterior reflex response by 65% following stimulation of the sural nerve, whereas 0.35 mg/kg reduces it by 95% (Willer and Bussel, 1980).
4. Comparison can be made of substances belonging to the same chemical group. Thus, we have been able to compare the effects of diazepam and another benzodiazepine (R06-9098) in equivalent dosage on vibratory inhibition in spastic patients (Fig. 7). In this test, R06-9098 was found to be less active than diazepam in reinforcing vibratory inhibition just as clinically it

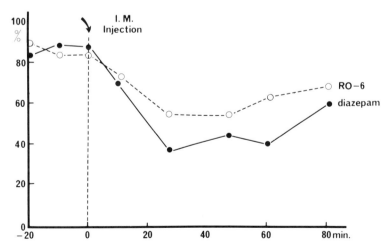

FIG. 7 Evolution of the vibratory inhibition index measured in the same spastic patient after administration of the same dose of diazepam and RO-6, respectively. After this latter product, the reinforcement of the vibratory inhibition is less marked than after diazepam.

has been shown to have a less marked muscle relaxant effect.

Such findings are important during the development phase of a new drug. It is difficult to derive the same data by other methods with which quantification is rarely possible. Repeated lumbar punctures are difficult to justify on ethical grounds to study pharmacokinetics beyond the blood–brain barrier. To provide similar information, electroencephalographic examination requires elaborate data reduction treatment and multivariant statistical analysis (Fink, 1978) and may provide less useful data. Evoked potentials (Saletu, 1977) yield interesting information but may have to be complemented by motor control studies to reveal specific pharmacological modifications at the synaptic level.

CONCLUSIONS

The association between electrophysiological exploration of the spinal cord and neuropharmacology is recent and has not yet been fully exploited in spite of the unique advantages outlined above. The appropriate tests for quantifying the spinal action of a drug must be determined for the drug under study. Most often, a given drug may only show an effect when electrophysiological tests are used in conjunction with a pathophysiological disturbance in patients.

ACKNOWLEDGMENTS

This work was supported by the grants 120.054 of INSERM (France) and 3.4565.81 of the FRSM (Belgium).

Motor Control Mechanisms in Health and Disease,
edited by J. E. Desmedt.
Raven Press, New York © 1983.

Quantification of the Effects of Muscle Relaxant Drugs in Man by Tonic Stretch Reflex

*Manuel Meyer and Csaba Adorjáni

Department of Neurology, University Hospital, CH-8091 Zürich, Switzerland

Increased resistance during slow passive movement of a limb appears "friction-like" in the spastic patient and "plastic" in the rigid patient. The resistance is mainly a result of the tonic stretch reflex (TSR). The TSR contributes to motor disability as well as to other dysfunctions such as loss of strength, release of flexor reflexes, and clonus in the spastic patient and, together with bradykinesis, to disturbed postural fixation in the parkinsonian patient. A decrease in TSR has been considered in therapeutic attempts to improve patients with pathologically raised muscle tone. Hence, it was logical to develop methods for quantitative analysis of TSR to improve the diagnosis and to permit a detailed and reliable assessment of therapeutic procedures.

Qualitative and quantitative analyses of the TSR can indeed serve to differentiate disorders with increased muscle tone. For elucidation of the pathophysiology of motor disorders, TSR analysis may be supplemented with additional methods such as evaluation of presynaptic inhibition by vibration-induced inhibition of H reflex, and intraneural recording of multiunit signals may be required. The prime advantage of TSR measurement is the detection of small changes in muscle tone and in actions of muscle-relaxing drugs.

THE MECHANOMYOGRAPHIC APPROACH TO TSR

The aim of this approach is to stretch the muscle with a passive movement whose amplitude and timing can be controlled while minimizing gravitational forces. A precise measurement is made of the resistance arising in muscles during and after such a movement. McKinley and Berkwitz (1928) developed a machine that induced a forearm motion with the help of weights and calculated torque exerted by flexor and extensor muscles on the elbow joint. More refined methods were achieved with electromotors, e.g., by Schaltenbrand (1937) who forced the forearm or the lower leg into a series of linear step-like flexion and extension movements. With this method complemented by two spring dynamometers, it was possible to also measure the effect of the degree of muscle lengthening on the resistance induced by the movement. Spiegel et al. (1956) applied a similar technique to measure resistance in parkinsonian patients using large-amplitude linear movements, and Boshes et al. (1960) and Brumlik and Boshes (1961) investigated passive sinusoidal elbow movements in parkinsonians. The

---

*To whom correspondence should be addressed: Department of Neurology, University Hospital, 100 Rämistrasse, Ch-8091 Zürich, Switzerland.

muscular resistance was measured as torque using strain gauges, and rotation of elbow joint was recorded by a potentiometer.

A similar method was used by Nashold (1966) with application of linear motions. Webster (1966), with his impressive mechanomyographic machine, also produced sinusoidal elbow movements with variable angle velocities. Here, the torque–position data were plotted as hysteresis loops. In addition, torque was integrated over the whole range of angular displacement. This "torque integral" was supposed to represent the work done by the patient's limb in response to passive motion. By varying stretch velocities, the total work over 10 cycles was taken as an adequate sampling of spasticity or rigidity and was called "spasticity index" in spastic patients and "rigidity" in parkinsonian patients. Knutsson et al. (1971) and Norton et al. (1972) analyzed hysteresis loops. The latter derived three values for each loop: total loop area, area of left half, and area of right half of loop. Leavitt and Beasley (1964) used two force transducers to measure manually induced passive movement of the knee joint. Herman et al. (1967) developed the "rotational-joint apparatus," which induced well-defined linear movements of varying amplitude and velocity about the ankle joint and measured length–tension relationships.

The majority of these "machines" (not all of them can be mentioned here) were only used sporadically on a few patients. Thus, the optimism of initial clinical investigators has often soon evaporated. Systematic clinical mechanomyographic research was pursued by only few investigators, e.g., Schaltenbrand (1958) and Herman et al. (1973). The main reason for the slow progress may be that in central motor disorders, disturbances of antagonistic interaction may complicate the picture: coactivation, disturbed antagonistic inhibition in extrapyramidal diseases, or shortening reactions of spinal origin. These disturbances may prevent measurements from reflecting the effective increase in TSR. In addition, severely disabled patients cannot be exposed to long experimental procedures. All of these variables may lead to a distortion of results.

THE ELECTROMYOGRAPHIC APPROACH

There is little doubt that the EMG or the combined mechanomyographic–EMG investigation of TSR has produced the most important results and clinical application. The EMG of slowly stretched and shortened muscles represents the reflex pattern. By combining EMG with mechanographic measurements, forces resulting from physical properties of joint and viscoelastic plastic properties of muscles can be distinguished from the reflex components of the resistance to stretch. Through integration of the EMG with surface electrodes (Foley, 1961) or wire electrodes (Herman, 1970) or by use of frequency analysis of groups of motor units (Meyer and Adorjani, 1980), reflex activity can be satisfactorily quantified. Simultaneous control of passive movement is necessary for analysis of the reflex pattern (Rushworth, 1960; Shimazu et al., 1962; Burke et al., 1970, 1971; Ashby and Burke, 1971; Andrews et al., 1972; Meyer and Adorjani, 1980).

CHARACTERISTICS OF PASSIVE TSR IN SPASTICITY AND RIGIDITY

In normal muscle slow passive movements generate no reflex activity. The resistance to passive stretch results from the mechanical properties of the joint and the viscoelastic–plastic properties of the muscles involved. Very rapid movements may produce a "phasic" reflex in normal muscle analogous to the monosynaptic tendon jerk. In addition, facilitating procedures such as the Jendrassik maneuver can induce TSR activity (Mark, 1963; Andrews et al.,

1973b). In spastic and rigid conditions, the slow passive muscle stretch produces a resistance caused by reflex activity with patients at rest (Fig. 1).

Linear and Sinusoidal Movements

The passive movements usually are induced using linear and sinusoidal displacements. These displacements should have a relatively large amplitude (min. 10°) and a relatively low frequency (not more than 2/sec), so that rheologic components of resistance and reflex transmission times are negligible. Linear movements were used by Foley (1961), Rushworth (1960), Rondot et al. (1958), Shimazu et al. (1962), Herman (1970), Burke et al. (1970), and Andrews et al. (1972). whereas sinusoidal movements were used by Esslen (1968), Burke et al. (1971a), Herman (1973), Meyer and Adorjáni (1980). In linear ramp-like movements, the reflex response during the moving phase (called "dynamic TSR") is distinguished from the one during sustained stretch ("static TSR"). Correspondingly, an EMG response occurring during shortening of a muscle is referred to as "dynamic shortening" and "static shortening reaction."

During sinusoidal movement, there is no sustained stretch, and the muscle moves from a lengthening into a shortening process. The peak activity of the reflex EMG corresponds to induced sinusoidal movement. If the degree of lengthening of muscle has a facilitating effect on reflex activity, which always happens during a static TSR response, then the peak of reflex EMG is near the point of maximum stretch of the sinusoidal movement. Hence, during exclusive static activity of TSR, the EMG peak would theoretically be "in phase" with the sinusoidal cycle (Fig. 2). In the case of exclusive velocity sensitivity of TSR, the EMG peak would theoretically be at the point of maximum velocity of sinusoidal muscle stretch, that is, 90° in advance of maximal lengthening of the muscle (Fig. 3). If there are both velocity sensitivity and an inhibiting influence of the muscle lengthening, the phase lead of TSR can be greater than 90°, as for spastic quadriceps muscle (Burke et al., 1970).

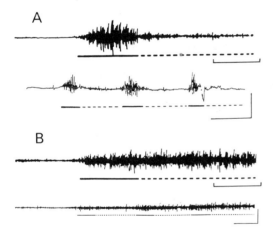

FIG 1. The TSR in spasticity and rigidity. *Solid line:* increasing stretch. *Broken line:* constant stretch. **A:** Reflex EMG of biceps brachii in hemiparetic spasticity. No discharge of motor units in resting muscle. Exaggerated "phasic" discharges at the initial stretching, with rapid reduction in spite of continued stretching. By an additional stretch, the discharges are again evoked but then rapidly decay. The number of reflex discharges depends on the velocity of stretching. **B:** Reflex EMG of biceps brachii in parkinsonian rigidity. Tonic discharges in the resting muscle. At the initial stretching, reflex discharges appear and continue with little tendency for adaptation. By stretching the muscle with smaller amplitudes and maintaining it in a steady position, increasing reflex discharges are evoked, which continue. The number of reflex discharges depends mostly on the extent of stretching. Calibration: 1 sec, 0.5 mV. (From Shimazu et al., 1962, with permission.)

The TSR in Spasticity

In patients with spasticity of spinal or cerebral origin, there is an obvious velocity sensitivity. The "dynamic" reflex reaction is dependent on degree of spasticity as well as on velocity of stretch. The reflex response is linearly related to velocity. Burke et al. (1970) used the line of best fit of integrated reflex EMG versus angular velocity

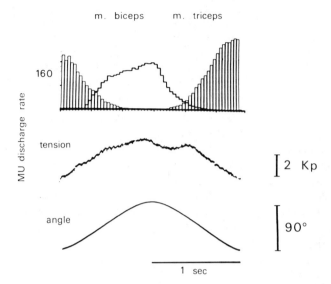

FIG. 2. The TSR in a patient with parkinsonian rigidity in biceps *(curve)* and triceps brachii *(bars)*. Position dependence of the peak of the reflex EMG; no phase lead. **Top:** Mean discharge rate of a group of motor units during 20 periods of sinusoidal movement. **Center:** Tension curve represents the resistance of the limb against the movement, calibrated in kiloponds. **Bottom:** Record of the output of the electronic goniometer representing the change in the elbow angle with time, which is proportional to the lengthening and shortening of the muscle. Maximum angular velocity is ca. 80°/sec. The trigger signal for computer analysis is given at maximal flexion. (Modified from Meyer and Adorjáni, 1980, with permission.)

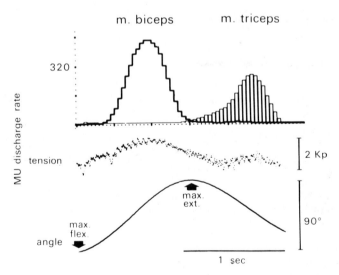

FIG. 3. The TSR in a patient with spastic hemiparesis in biceps *(curve)* and triceps brachii *(bars)*. Velocity dependence of the peak of the reflex EMG, phase lead of 55°. Maximum angular velocity 80°/sec. **Top:** Mean discharge rate of a group of motor units during 20 periods of sinusoidal movement. **Center:** Tension curve represents the resistance of the limb against the movement, calibrated in kiloponds. **Bottom:** Record of the output of the electronic goniometer representing the change in the elbow angle with time, which is proportional to the lengthening and shortening of the muscle. Maximum angular velocity is ca. 80°/sec. The trigger signal for computer analysis is given at maximal flexion. (Modified from Meyer and Adorjáni, 1980, with permission.)

(AV) of the manually induced linear movements to assess the degree of spasticity. In addition, Burke et al. (1971a) investigated the phase relationships of rectified reflex EMG to sinusoidal movement in quadriceps and hamstrings, and Ashby et al. (1971) repeated this for biceps and triceps.

In our studies, we used a motor-driven device that forced the arms into uniform repeated sinusoidal movements with a total excursion of 90°. The EMG of biceps and triceps was detected using wire electrodes. Twenty cycles with stepwise increases in angular velocity between 25° and 270°/sec, corresponding to 0.07 to 0.7 cycle/sec, were recorded and averaged. The EMG data were processed as poststimulus time histograms, and the resistance of the arms as tension curves. This method provided an analysis of the discharge rate of groups of motor units (6–8) or single motor units, selected by window techniques, in relation to sinusoidal movement. Similarly, a clear relationship was found between peak discharge rate of groups of motor units (Fig. 3) or of single motor units (Fig. 4) and maximum angular velocity of sinusoidal movement.

As in previous studies (e.g., Ashby et al., 1972), a correlation was found between clinical assessment of spasticity and EMG results. The slope of the reflex EMG (peak

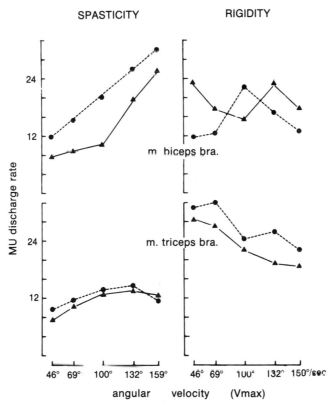

FIG. 4. Relationship of the discharge activity of a single motor unit to the maximum angular velocity in a patient with spastic hemiparesis **(left side)** and a patient with parkinsonian rigidity **(right side)**. In spastic muscles, the activity increases with increasing angular velocity, in triceps only up to a certain velocity. The Jendrassik maneuver *(dotted lines)* provokes a further increase in the discharge rate and improves the linear relationship with angular velocity. In the rigid muscle **(right side)**, no clear-cut relationship between the discharge rate and the maximal angular velocity is seen. In this motor unit, there is no functionally interpretable effect of the Jendrassik maneuver. (From Meyer and Adorjáni, 1980, with permission.)

discharge rate) versus angular velocity regression line, expressed as the angle between regression line and X axis, was the best parameter (Burke et al., 1970). The slope increased with increasing degree of spasticity (Fig. 5, upper panel), and the linear correlation coefficient was close to 1.

We evaluated the disability of nine patients with spinal spasticity and hemiparesis using clinical methods. The clinical investigation involved measurement of hand grip with a dynamometer and testing simple motor skills such as speed and endurance of the hands by operating a type-

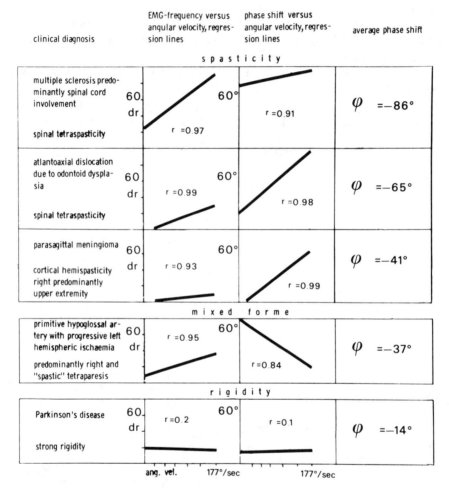

FIG. 5. Typical TSR results in five patients. **Top panel:** Three patients with varying degrees of spasticity, decreasing in severity from top to bottom. **Middle panel:** Mixed spastic–rigid increase in muscle tone. **Bottom:** Parkinsonian rigidity. The slopes of the regression lines of the relationship of peak multiunit discharge versus angular velocity, expressed as angle between curve and X axis, increase with the degree of spasticity. The linear correlation coefficient is close to 1 in spasticity, and there is no apparent correlation in rigidity. The correlation of the regression phase lead of the peak multiunit discharge versus angular velocity is close to 1 in the strongly velocity-sensitive muscles. The slope of this regression line flattens out with increasing spasticity, and the phase lead, even at low velocities, approaches 90°. The average phase lead increases with the degree of spasticity. In rigidity, the average phase lead is small. In rigidospastic states, the results are intermediate. (Modified from Meyer and Adorjáni, 1980, with permission.)

writer key with index finger, handling a safety pin, dialing a telephone number, counting the time to cover a distance of 20 m, or climbing stairs. A patient with maximum disability had a score of 20. As Fig. 15 (upper panel) indicates, there was a correlation between the clinically assessed disability score and the slope of the regression line of reflex EMG versus AV, even though the clinical scale did not include an evaluation of muscle tone and tendon jerks. The "threshold velocity" below which no reflex activity can be expected can be calculated from the Y intercept of the regression line. This threshold velocity decreases with increasing severity of spasticity (Fig. 5). This negative correlation with degree of spasticity has not nearly as good a fit as the positive correlation of the reflex EMG–AV regression line.

The value of the static TRS or "position sensitivity" in spasticity is still controversial. Using linear movements, Rondot et al. (1958) and Ashby and Burke (1971a) investigated upper limps, and Herman (1970) and Dietrichson (1971, 1973), triceps surae muscle. They found a decreasing reflex activity on sustained stretch. In quadriceps and hamstrings, Burke et al. (1971) found practically no such activity. From Herman's precise linear investigation (1970) and from phase relationships found in sinusoidal stimulation, it can be concluded that with slow stretch velocities a certain position sensitivity exists which becomes increasingly velocity sensitive as angular velocity increases (Fig. 6). In strongly velocity-sensitive muscles, the increase in phase lead with increasing AV was again shown to be a regression line with correlation coefficients close to 1. Since phase lead is limited to 90°, the slope of the regression line of phase lead versus AV flattens out with increasing spasticity, and the phase lead at low velocities approaches 90° (Fig. 5, upper panel, middle column). The average phase lead, i.e., average of phase shifts for the series of maximum AVs, increased with the degree of

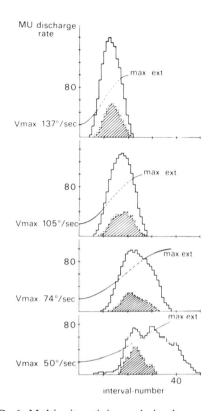

FIG. 6. Multiunit activity and simultaneous activity of a single motor unit in the triceps muscle of a spastic patient. At low angular velocity, there is a bimodal discharge pattern with a more static and a more dynamic component. With increasing angular velocity, a transition from the static to the dynamic pattern can be observed. (From Meyer and Adorjáni, 1980, with permission.)

spasticity (Fig. 5, upper panel, right column), and in the clinically tested patients, a linear correlation was found (Fig. 7).

The effect of muscle lengthening on the TSR was investigated by Burke at al. (1970) on quadriceps and later on other muscles (Burke and Lance, 1973). The effect of increasing muscle stretch on the quadriceps stretch reflex is inhibitory (clasp knife phenomenon) but facilitatory in all other spastic muscles investigated.

The TSR in Rigidity

The reflex pattern in rigid muscle tone, e.g., in parkinsonian patients, was de-

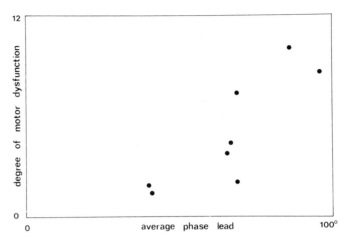

FIG. 7. Relationship of the clinically assessed disability (maximum disability score, 20) to the phase lead of the reflex EMG (in degrees) during sinusoidal passive movement (see text). Correlation coefficient = 0.84.

scribed with linear muscle stretch by Rondot (1958), Denny-Brown (1960), Rushworth (1960), Shimazu et al. (1962), Dietrichson (1971, 1973), and Andrews et al. (1972); the latter authors also used manually applied sinusoidal stimulation. We investigated parkinsonian patients, and pure rigidity was easily differentiated from pure or mixed spastic states. In mildly rigid muscles, a certain velocity sensitivity was seen by Andrews et al. (1972). Because velocity sensitivity is expressed as a slope of a regression line, and because the slope is influenced by threshold velocity too, these authors calculated a so-called "dynamic sensitivity" as the ratio between the increase in reflex EMG at a high angular velocity (400°/sec) and at a low angular velocity (100°/sec). The dynamic sensitivity was negatively correlated to severity of parkinsonian rigidity: the smaller the rigidity, the larger the dynamic sensitivity (Fig. 8). The typical pattern of a single motor unit in a rigid parkinsonian biceps and triceps is represented in Fig. 4. There is no clear relationship between discharge rate and maximum angular velocity. In this example, the triceps shows a slight decrease in activity with increasing angular velocity. In rigidity, there is little or no correlation seen in the TSR–AV regression line (Fig. 5, bottom).

The hallmark of rigid muscle is strong dominance of the static component of TSR or position sensitivity. This means maximum reflex activity at greatest stretch and, in addition, reflex activity at sustained stretch. In contrast, these features are poorly represented in "pure" spastic muscles. Andrews et al. (1972) found that static TSR was more pronounced in flexors than in extensors. They proposed that static TSR was positively correlated with severity of rigidity.

The relationship of "static" to "dynamic" reflex activity, i.e., of "velocity sensitivity" to "position sensitivity," is reflected in the phase relationship during sinusoidal stretching. If reflex activity, as in severe parkinsonian rigidity, reaches the highest point at greatest stretch, the peak EMG is located "in phase" with maximum extension of the muscle or angular position of joint (Fig. 2), and there would be no phase lead. However, this extreme condition was never observed. The low velocity sensitivity induces a small phase lead, usually less than 25°. Other important features of rigidity are the shortening reaction and the

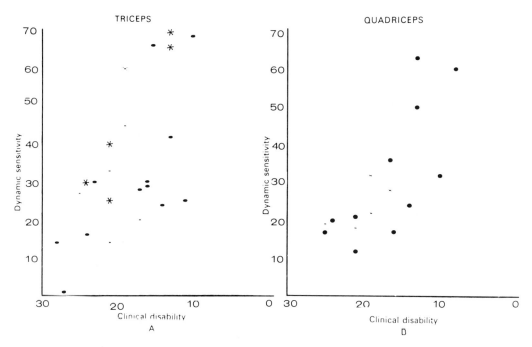

FIG. 8. Inverse correlation of clinical disability with the "dynamic sensitivity" of the stretch reflex (see text) in parkinsonian patients. A: Triceps brachii. Without resting tremor (ovals), with resting tremor (asterisks), thalamotomized without resting tremor (crosses). B: Quadriceps. Thalamotomized (crosses), unoperated (circles). (From Andrews et al., 1972, with permission.)

disturbed antagonistic inhibition (see R. W. Angel, *this volume*). Probably both express the same pathophysiological mechanism, which must have its origin in supraspinal structures.

The rigid muscle also reacts to passive shortening with discharge of motor units (dynamic shortening reaction. The motor unit discharge during maintained shortening is called the static shortening reaction (Andrews et al., 1972). The traditional clinical presentation of rigidity is based on the static TSR and the static shortening reaction. Also, in patients with spinal transection, a kind of shortening reaction is sometimes seen which may be caused by release of flexor reflexes. During sinusoidal stimulation, the shortening reaction may demonstrate either a static (position-dependent) or a dynamic (velocity-dependent) tendency. Sometimes unmodulated EMG activity occurs, which cannot be related to the sinusoidal lengthening or shortening of the muscle (Fig. 14).

EFFECTS OF MUSCLE RELAXANTS ON TSR

The use of TSR for demonstration of muscle-relaxing effects of therapeutic procedures was, until the end of the 1960s, restricted to mechanomyographic tests, i.e., reduction in tension or torque or overall EMG assessments. Vazuka (1958) tried to estimate the effect of methocarbamol, meprobamate, and zoxazolamine on TSR during sustained stretch with surface EMG and found that the changes of the EMG output by various drugs correlated well with the clinical relaxation. Struppler (1960) tested meladrazine with dynamic TSR and found a clear decrease. Kurtzke and Gylfe (1962) determined the resistance of knee and elbow joints to manually induced steady pull on wrist or ankle. The mean change in pounds from "predrug conditions" represented a decrease of 30% in spasticity. Webster (1964), using integrated torque, demonstrated an immediate plunge of the "spasticity index" after an intravenous injection of ethoxybutamoxane. Matthews (1965) and Matthews et al. (1972) investigated the effects of phenothiazines, chlorproethazine and dimethothiazine, on TSR with surface and needle electrodes in flexor and extensor muscles of upper and lower limbs. In almost all cases, a reduction of dynamic stretch activity was obtained. The improvement varied from complete abolition of all reflex activity to moderate reduction in response to slow stretch. Levine et al. (1969) studied the effect of oral diazepam by evoking TSR in quadriceps of spastic patients by gravity stimulation (suddenly dropping the leg rest) and quantifying EMG by integration. In the majority of cases, they found a decrease of EMG of 30 to 50% or more.

All of these procedures did not provide much more than confirmation of the clinical evaluation. This situation changed with a more systematic exploration of TSR by Herman (1970) and Herman et al. (1972, 1973) (Fig. 9) and by the EMG methodology of Lance's group who not only quantified accurately the degree of muscle relaxation but also specified which feature of the TSR was reacting.

Muscle Relaxants in Spasticity

The prime effect of muscle relaxants on spasticity is on the velocity sensitivity (Fig. 10). To achieve reliable results under comparable test situations, it is desirable always to examine TSR at the same amplitude of joint rotation and to apply a sufficient variety of angular velocities in order to calcu-

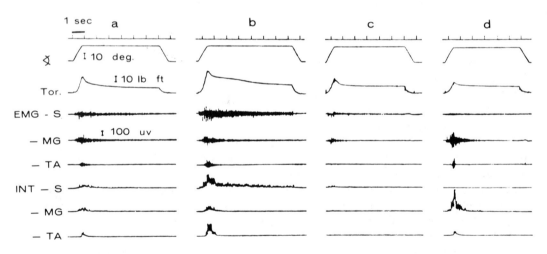

FIG. 9. Relationship of the reflex EMG to torque of the triceps surae (S, MG) and the tibialis anterior muscles (TA) during ramp stretches of the ankle joint (plantar flexion) of 30°/sec in a paraplegic patient. a, control; b, placebo; c, diazepam (10 mg/day); d, dantrolene sodium (150 mg/day). Int, integrated EMG. The shortening reaction or cocontraction of the tibilias anterior antagonizes muscle tension. Torque reduction in c is related to reduction in the amount and duration of the dynamic and static TSR under the influence of diazepam, and shortening reaction is abolished. In record d, despite a clearly increased dynamic TSR in the medial gastrocnemius muscle, the torque response under therapy with dantrolene sodium is diminished. This discrepancy, the combination of increased reflex activity with a reduction of torque, indicates that the drug acts mainly on the contractile mechanisms of muscles. (From Herman et al., 1972, with permission.)

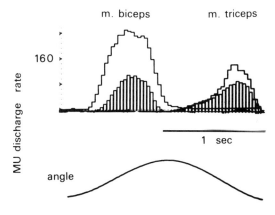

FIG. 10. Effect of an intravenously injected benzodiazepine (20 mg motrazepam) on the TSR in spastic biceps and triceps muscles of a hemiparetic patient. The size of the mainly velocity-dependent reflex EMG is markedly reduced 30 min after application *(bars)*. The phase shift of the peak EMG activity shows a slight reduction of the phase lead for biceps and triceps muscles. Total excursion of the forearm, 90°. Maximum angular velocity of the sinusoidal movement, 80°/sec. (From Meyer and Adorjáni, 1980, with permission.)

late the regression line of reflex EMG versus AV. To compare results in a trial, the slope of regression line and threshold velocity may be used (Jones et al., 1970; Burke et al., 1971b; Ashby et al., 1972). In this way, varying changes in dynamic stretch response can be recorded which could never be graded so accurately in clinical tests. This was clearly demonstrated by Jones et al. (1970) and is also shown in Fig. 11 for treatment with baclofen, 60 mg/day, and diazepam and baclofen, 75 mg/day, compared with a placebo.

The phase relationship of the TSR during sinusoidal movements is, with the exception of knee extensors (Burke et al., 1971b), qualitatively unaltered during antispastic medication. As Fig. 11 indicates, the decrease in reflex size is combined with a slight decrease in phase lead. The reduction in velocity sensitivity during antispastic therapy cannot be detected in a linear reduction of slope of the regression line of phase lead versus AV because the maximum phase lead obtainable is restricted to 90°, and the slope of the regression line may increase if the phase lead at low velocities is markedly reduced. Hence, it is the reduction of average phase lead, calculated from stimulation with a series of different velocities, that reflects the effects of antispastic therapy.

Antiparkinsonian Drugs in Rigidity

Mechanomyographic measurements are not suitable to measure drug effects on rigidity, as shown by Webster (1966) who measured the "integrated torque" in patients with Parkinson's disease. He found it necessary to test patients while they performed self-activating tests in order to prevent drowsiness during which resistance to stretch and "rigidity values" fell to almost zero. In contrast, rigidity was precipitated by an alerting reaction. Normal controls tended to show negative resistance values, indicating that they had been unconsciously driving the turntable. In trials with antiparkinsonian drugs, trihexyphenidyl and chlorphenoxamine, "resting rigidity" became more negative. The integrated positive torque did not correlate well with the disability of the patients. It was the negative integrated torque that correlated positively with drug effects and negatively with disability.

Antiparkinson drugs can reduce the dynamic and static components of rigid TSR. Figure 12 shows the effect of biperiden on rigid biceps and triceps after 30 min. The reflex EMG is greatly reduced, and the peak of multiunit activity in biceps shows a slight increase of the small phase lead. The static (position-sensitive) component of the triceps TSR is virtually removed. Andrews and Burke (1973) examined the effects of L-DOPA therapy of 3 months' duration in parkinsonian patients. They found a correlation of clinical improvement with a reduction in the dynamic component of TSR

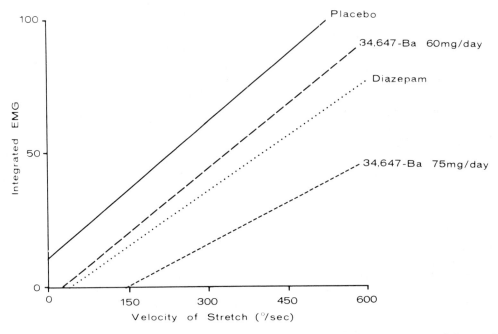

FIG. 11. Effects of muscle relaxants on the regression line of reflex EMG versus AV of the quadriceps muscle of a tetraparetic patient (injury to the lower cervical spinal cord) after 14-day treatment with placebo, baclofen 60 mg/day, diazepam, and baclofen 75 mg/day. The slope of the regression line decreases, and the threshold velocity increases, dependent on the drug and dosage. (From Jones et al., 1970, with permission.)

in quadriceps, triceps, and biceps and in the static component of TSR in biceps and triceps. However, a lessening of clinical disability in patients treated with L-DOPA was combined with an increase in "dynamic sensitivity," at least in the quadriceps, triceps, and biceps muscle (Fig. 13). We made similar observations: a certain increase or reappearance of velocity sensitivity during sinusoidal stimulation was induced by antiparkinsonian drug therapy. In addition to reduction of TSR and increase in velocity sensitivity, the drug also improved the disturbed antagonistic interaction or shortening reaction. It is striking that the same combination of effects on TSR is often found during unspecific maneuvers that facilitate MNs, e.g., the Jendrassik maneuver. An example is shown in Fig. 10, lower half, where the shortening reaction is abolished by the MN facilitation. In addition, Jendrassik facilitation often increased the velocity sensitivity in spastic, rigid, and

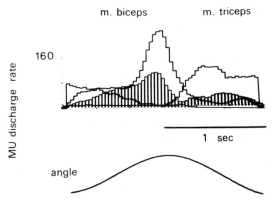

FIG. 12. Effect of intravenously injected biperiden (5 mg) on the TSR in the rigid biceps and triceps muscles of a parkinsonian patient: 30 min after application (bars), the size of the mainly position-sensitive reflex EMG of the biceps muscle is greatly reduced, and the peak of the multi-unit activity shows a slight increase of the small phase lead. The position-sensitive component of the triceps TSR is removed. Total excursion of the forearm, 90°. Maximum angular velocity of the sinusoidal movement, 80°/sec. (From Meyer and Adorjáni, 1980, with permission.)

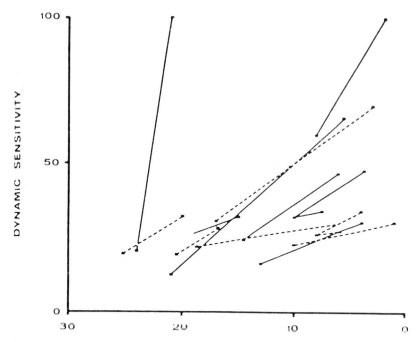

FIG. 13. The relationship of the "dynamic sensitivity" of the quadriceps TSR to clinical improvement (Webster rating scale) during L-DOPA therapy. The clinical improvement is associated with an increase in dynamic sensitivity of the stretch reflex. *Broken line:* patients with previous contralateral thalamotomy. (From Andrews and Burke, 1973, with permission.)

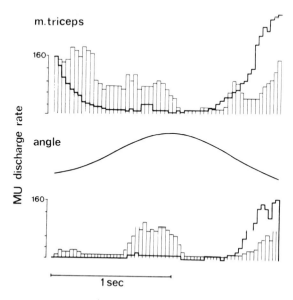

FIG. 14. Restoration of antagonistic inhibition. Top: Mixed spastic–rigid triceps muscle of a patient with spinal tetraparesis. Activity of a group of motor units evoked by sinusoidal movements before *(bars)* and 30 min after *(curve)* injection of a benzodiazepine (motrazepam). Bottom: Rigid triceps muscle of a parkinsonian patient. Activity of a group of motor units evoked by sinusoidal movements before *(bars)* and during *(curve)* a Jendrassik maneuver. Maximum angular velocity, 80°/sec. (Adapted from Meyer and Adorjáni, 1980, with permission.)

rigidospastic muscles (Meyer and Adorjáni, 1980).

DRUG EFFECTS ON TSR IN MIXED FORMS OF ELEVATED MUSCLE TONE

The reaction of muscles with mixed forms of elevated muscle tone (e.g., rigidospastic states in dystonic syndromes, certain paraspastic and tetraspastic motor disorders of spinal, bulbar, or cerebral origin) to drug therapy has not yet been systematically evaluated. Andrews et al. (1973) investigated phenoxybenzamine in athetotic patients and found a reduction in static stretch reflexes and in dynamic shortening reaction. In mixed spastic–rigid forms of elevated muscle tone, we found that a muscle-relaxing therapy led to a reduction of the disturbed antagonistic interaction (Fig. 14, upper half). This was sometimes combined with a reduction in either or both the velocity- and position-dependent TSR.

THE TSR AND MOTOR DISABILITY

It has never been doubted that an increased passively induced TSR represents a

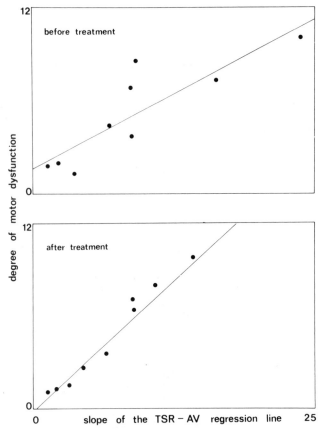

FIG. 15. Relationship of the clinical improvement (maximum disability score, 20) to the slope of the regression line of reflex EMG versus angular velocity (expressed as tangent of the angle between the regression line and the X axis) in nine hemiparetic and tetraspastic patients. **Upper panel:** Before treatment. Correlation coefficient $r = 0.86$. **Lower panel:** After 3-month treatment with a benzodiazepine (motrazepam), 20 to 30 mg/day. Correlation coefficient $r = 0.97$. Mean reduction of the disability score, 16%. Mean reduction of the velocity sensitivity (slope of the regression line of reflex EMG versus angular velocity), 26%.

characteristic part of a movement disorder. Therefore, it is important to know how therapeutic procedures can influence the increased TSR and if decreasing the TSR really improves the motor disability. There exists the time-honored clinical experience that only a certain percentage of patients suffering from an upper motor neuron syndrome may improve with the successful treatment of passive TSR.

As shown earlier, we have compared the disability scores of nine "spastic" patients with the slope of the regression line of reflex EMG versus AV during passive sinusoidal movement (Fig. 15, upper panel). Three months after treatment with an experimental benzodiazepine (motrazepam), a clear increase of the correlation between clinical assessment and slope of regression line was seen (Fig. 15, lower panel). The reasons for this increase in correlation are not yet elucidated; supraspinal drug effects may be responsible. However, a clear mean decrease of velocity sensitivity results from the treatment. The reduction of TSR was accompanied by only slight improvement of motor disability (see increase of the slope of the regression line of disability versus velocity sensitivity). Hence, the pathomechanism reflected by the increased passively induced TSR was associated with only minor disabling features of the underlying motor disorder. It appears that pathological EMG activity to stretch during volition is much more important as a factor of disability.

"Spastic" TSR was investigated under isometric contraction by Neilson and Lance (1978), during sinusoidal volitional movements by McLennan (1977), and during gait and volition by Knutsson and Richards (1979) and Knutsson (1980). McLennan found that the stretch-reflex-suppressing effect of baclofen during passive movement was largely extinguished during voluntary movement.

OUTLOOK

The TSR investigated under different passive and volitional conditions offers a test of considerable informative value in the investigation of muscle-relaxing drugs.

Analysis of Gait and Isokinetic Movements for Evaluation of Antispastic Drugs or Physical Therapies

Evert Knutsson

Department of Clinical Neurophysiology, Karolinska Sjukhuset, Stockholm 104 01, Sweden

Following the introduction of specific antispastic treatments by physical therapy (Levine et al., 1954) and medication (Birkmayer et al., 1967), much interest has been devoted to the study of antispastic effects on motor performance in muscular hypertonias. Many different methods have been designed to evaluate influences on tendon jerks and the H reflexes (Delwaide, 1971, 1973; Knutsson et al., 1971; Pedersen, 1974), tonic vibration reflexes (Hagbarth, 1973), and resistance to passive movements that stretch spastic muscles (Ashby and Burke, 1971; Burke et al., 1970; Herman et al., 1972; Knutsson et al., 1973; Norton et al., 1972; Pedersen et al., 1974). Several forms of therapy have been shown to result in depression of spastic reflexes. This holds true for therapy with baclofen (Birkmayer et al., 1967; Knutsson et al., 1973; Pedersen et al., 1974), diazepam (Hoyt et al., 1972; Peirson et al., 1968; Verrier et al., 1977), dantrolene sodium (Herman et al., 1972; Knutsson and Martensson, 1976; Monster, 1974), and tizanidine (Hassan and McLellan, 1980; Knutsson et al., 1981; Rinne, 1980; Sie and Lakke, 1980). It also holds true for physical therapy by means of cooling (Hartviksen, 1962; Knutsson, 1970), vibration (Hagbarth, 1973), long-term stretch (Odéen, 1981; Odéen and Knutsson, 1982), and the "hold–relax" procedure of the proprioceptive neuromuscular facilitation technique (Knutsson, 1980c).

Except for dantrolene sodium therapy (Knutsson and Martensson, 1976) which interferes with excitation–contraction coupling in muscle fibers (Ellis and Carpenter, 1972; Desmedt and Hainaut, 1977), depression of spastic reflexes in response to therapy does not occur regularly in any of the clinical syndromes with muscular hypertonus. The reason for this is obscure, and information on the features of spasticity responding to therapy is scarce. With baclofen therapy, spasticity caused by spinal lesions is relieved more frequently than spasticity originating from cerebral lesions (Pinto et al., 1972). There seems to be a correlation between the effect induced by baclofen on spastic reflexes and those obtained by local muscle cooling or selective block of thin nerve fibers. These are procedures that share in common a depressive effect on the inflow from muscle spindles. Thus, in patients with spastic reflexes that are depressed by local cooling or selective nerve block, baclofen frequently results in reduced spasticity (Knutsson et al., 1973). In contrast, it usually shows no signficant effect on stretch reflexes and may even enhance them in patients with spastic reflexes that are not reduced by cooling or selective nerve block. These features of variability in effect point to differences in how the muscular hypertonus is sustained or how the inflow from stretch receptors is integrated in the central nervous system. However, at present, neither the mechanisms responsible for muscular hypertonias nor the modes of antispastic action can be defined. Thus, except for the use of local cooling in so-called cryotests (Martensson and Knutsson, 1981) to estimate the antispastic effects to be expected from baclofen medication, there is no rational way to predict the effect of therapy on spastic reflexes.

The rationale for antispastic therapy is the fact that the enhanced stretch reflexes elicited by the stretch imposed on a muscle during movement can give rise to abnormally high restraint and sometimes to a disturbing clonus. When this is so, a reduction of the reflexes can be expected to result in improved motor capacity or relief from some of the effort motion involves. It is, however, not an uncommon finding that spastic reflexes become depressed without any corresponding improvement in motor function. This can have several explanations. First, the spastic reflexes tested in passive limb movements are not identical with those in movements performed by the patient himself. In repeated voluntary sinusoidal knee extensions and flexions, the restraint often is lowered in mild spasticity and increased in severe spasticity (McLellan, 1977). In dynamic, isokinetic movements at maximal effort, the restraint is commonly much larger than in passive movements of identical range and speed (Knutsson and Martensson, 1977, 1980). Only occasionally is it depressed by reciprocal inhibition during maximal voluntary dynamic movements.

In some patients, a low threshold for stretch reflex activation can be found at bedside examination in muscles that are not activated by the stretch imposed on them during walking (Knutsson and Richards, 1979). In consequence, the effects on spastic reflexes observed in tests of passive movements cannot be expected to explain the effects on functional disability that result mainly from the disorganized control of motion. The clinical finding that the degree of spasticity in most patients with spastic paresis appears to mirror the degree of incapacitating restraint in motion seems to derive from the fact that facilitation of stretch reflexes by voluntary effort is more frequent and stronger in severe than in mild spasticity.

Second, to estimate the effect on functional disability resulting from depression of spastic reflexes, the spastic restraint has to be related to the other forces active in motion. A certain restraint, when it opposes strong prime mover muscles, may be of minor importance for motor control. In contrast, when it opposes weak prime-mover muscles, it can have deleterious effects, and its depression by therapy results in a decided improvement in motor capacity. Only when the spastic restraint is crucial to the handicap will its reduction by therapy result in marked improvement in motor capacity. There are many situations in which other components of disorganized motor control are the limiting factors for motor functions.

This argument makes it clear that an understanding of the effects of antispastic therapy on motor capacity cannot be based only on an estimation of the changes of passive restraint to which it may give rise. Information is also needed on the degree of weakness in movements and how this relates to deficient activation of prime mover muscles and to the actual restraint to movements whether or not this is caused by stretch reflex activation, misdirected descending commands, or contracture.

In an attempt to answer these questions, two systems for examination have been developed at our clinic. In one, passive and voluntary movements are tested during speed control by an isokinetic dynamometer, keeping constant a selected speed during motion (Gransberg et al., 1980). In the other, the patterns of movement and muscle activation are determined during walking (Isaksson and Knutsson, 1980). These two systems allow assessments to be made of motor functions not only in simple movements performed with a high degree of volitional drive but also in a type of movement that to a large extent is automatic and normally depends on preprogrammed central commands.

In using these two systems, it is usually possible to understand in the individual case how different components of the disorganized motor control constitute a handicap and to estimate whether or not depression of reflexes can be expected to improve motor capacity. Furthermore, these systems have been used in studies of effects induced by antispastic therapy. These studies have revealed rather complex effects including antispastic as well as antiparetic effects. Depending on how these are related to the residual capacity for adequate activation of the muscles in movements, the result may or may not be an improvement in motor capacity.

In the following sections, descriptions of how isokinetic movements and gait can be used for the analysis of a motor disorder are given with inclusion of some of the basic principles for adequate usage of the methods in analysis. Finally, different types encountered during antispastic therapy are described.

DETERMINATIONS OF TORQUE IN ISOKINETIC MOVEMENTS

The principle for the mode of action of the isokinetic dynamometer (Cybex II, Lumex, Inc.) is shown in Fig. 1. The dynamometer has an electric motor that rotates an axle. The rotation speed can be set at different levels and then kept constant by a servo control unit by feedback from a generator (Perinne, 1968). The rotating axle of the motor is connected to a rotational lever arm by a clutch coupling that allows free motion of the lever arm at all rotational speeds below that of the motor driven axis. When the lever arm is put into rotation by an external force, the coupling becomes engaged as soon as the rotation speed of the lever arm reaches that of the rotating axle, and the lever arm will be inhibited from rotating faster than the motor driven axis. By gear reduction between the motor and the rotational axle connected to the lever arm, not seen in the figure, the braking capacity of the system is sufficient to brake rotation of the lever arm and thus keep the rotation speed almost constant at force application against the lever corresponding to the strongest movement forces in man. The

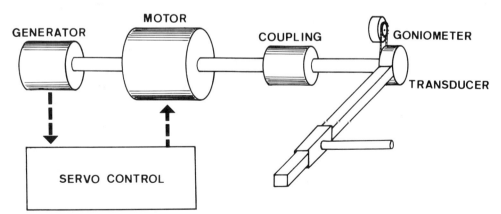

FIG. 1. Schematic representation of isokinetic dynamometer. Full explanation in text. (From Knutsson and Martensson, 1980, with permission.)

braking of movements works in both directions of rotation, although this is not seen in the schematic representation of Fig. 1.

In the original isokinetic dynamometer, the torque of the movement is recorded with a pressure-sensitive potentiometer in a hydraulic system (Nelson et al., 1973; Perinne, 1968). This system allows linear recordings in the range 5 to 500 Nm. In Fig. 1, a transducer is seen between the lever arm and the rotational axis. It has been included in the system to allow linear recordings of torque in the range of 1 to 500 Nm and to enable determinations of torque in passive speed-controlled movements (see below). A goniometer consisting of a rotational potentiometer records angular rotation of the lever arm.

When a force is applied against the lever arm, the arm will start to rotate, and the angular rotation of the arm will accelerate up to the preset speed of the dynamometer. When this speed is reached, the lever arm will start to resist the movement, and this will then continue at the speed set by the dynamometer. Fig. 2 shows records of angular displacement and of torque in voluntary knee extensions at a preset speed of 180°/sec. In A, the extension was made with maximal effort. The preset speed was reached shortly after the start of movement and was then kept constant through the extension to fully extended leg (0°). In B, the extension was made at submaximal effort. The time needed for acceleration up to the preset speed was longer, and the speed of motion was not kept at the preset level during the last part of the extension.

The two records illustrate how torque is recorded only during the part of the movement in which the actual speed of movement is kept at the preset level. When the force is not large enough to maintain this speed, torque will be recorded as zero. In consequence, it must be realized that the torque is not the torque produced in the movement. Only in the phases of movement in which the speed is kept constant at the present level will the record enable determination of the torque produced. The torque recorded is, however, no direct measure of torque produced because of the gravitational force resulting from the weight of the limb segment and the lever arm.

Correction for Torque Resulting from Gravitational Force

In Fig. 3, which shows torque in slow (A) and fast (B) voluntary knee extension in a

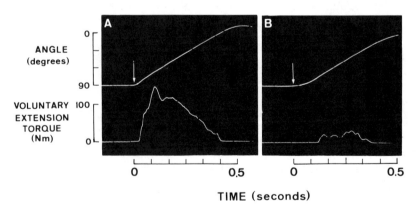

FIG. 2. Angular displacement and torque in voluntary knee extension at maximal (A) and submaximal (B) effort. Movements controlled by the isokinetic dynamometer at a preset speed of 180°/sec. *Arrow* in each record indicates start of movement. (From Knutsson and Martensson, 1980, with permission.)

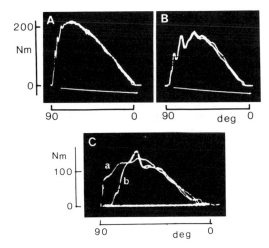

FIG. 3. Torque in isokinetic knee extensions. In A and B, torque in maximum voluntary *(upper curves)* and passive *(lower curves)* movements at preset speeds of 30 and 120°/sec, respectively. *Upper beams* are three superimposed records. In C, torque in maximum voluntary movement with resisted (a) and unresisted (b) acceleration reaching a constant upper speed of 240°/sec.

seated healthy subject, the torque resulting from the weight of the leg and the recording arm is included (lower beams). When the lower leg is extended from a hanging position (90°), thus attaining a horizontal position (0°), the torque caused by weight increases from zero to a maximum level, in this case 20 Nm. The increase, given as negative, is a cosinus function of angle. Thus, it can be calculated from a single determination of gravitational torque at any selected angle. The curve can also be determined by using passive resistance recording (see below) with a relaxed leg or by recording the torque when the relaxed leg is allowed to fall from an extended position during torque measurement with the dynamometer set at low speed. In subjects in whom relaxation is not possible or is uncertain, the torque caused by weight can be estimated from body size.

The torque attributable to muscle action is obtained by correcting the recorded torque for gravitational torque. In movements opposed by gravitation, as, for example, in knee extension in a sitting posture, the torque produced by muscle tension is equal to the sum of recorded and gravitational torque. Thus, the vertical distance between the two curves in Fig. 4A gives the torque produced by muscle action. In movements helped by gravitation, the gravitational force is part of the recorded torque, and the torque produced by muscle action is obtained by subtraction.

In strong movements, the gravitational force is small compared to the torque caused by muscle tension except for the outer ranges of movements. In weak movements, the gravitational force constitutes a relatively large part of the total force active in motion. This may explain why corrections for torque resulting from weight were disregarded in isokinetic determinations until paretic patients were tested (Knutsson and Martensson, 1977). In patients with muscle weakness, the gravitational forces often are of the same order or even larger than the forces produced by muscle action. Thus, the use of uncorrected torque read-

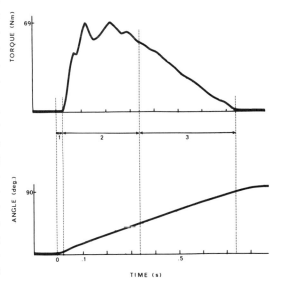

FIG. 4. Phases of movement controlled by isokinetic dynamometer. Phase 1, acceleration. Phase 2, impact and oscillation. Phase 3, constant speed of angular displacement. (From Gransberg et al., 1980, with permission.)

ings for the assessment of degree of disability (Simmons et al., 1982) will, in many cases, result in large errors.

PHASES OF MOVEMENT IN ISOKINETIC TESTS

In records of torque in a movement resisted by the isokinetic dynamometer, three phases of the movement can be distinguished (Gransberg et al., 1980). These phases are indicated in Fig. 4, showing torque and angular displacement in a voluntary knee extension at maximal effort in a healthy subject. The first phase is acceleration. During this, the speed of angular rotation successively increases until the speed has reached the preset level, which was 120°/sec. In this phase, all force produced is used for the acceleration of the lever arm with the attached leg, and torque reading is zero. At the moment the preset speed is reached, the clutch coupling between the rotating axis of the dynamometer and the lever arm starts to resist the movement. In consequence, there will be an impact between the accelerating leg and the braking structure of the dynamometer, and, invariably, the impact is followed by oscillations of the lever arm. The impact force, which is a function of moving mass and acceleration, is included in the first deflection of the torque curve. Thus, the initial part of the torque curve (phase 2) is no measure of momentary muscle torque. It also includes force produced during the preceding acceleration and the oscillations following impact. Only when the oscillations have worn off (phase 3), is the torque curve undisturbed for determination of momentary muscle tension.

In slow movements, the impact will not disturb the torque record for more than a few degrees of angular displacement. In fast movements, the impact force will disturb a larger part of the record. When movements are performed with strong muscle action, the oscillations after impact usually dampen out quickly, leaving a part of the movement for undisturbed reading of torque. In weak movements, the oscillations are not dampened equally fast and may disturb measurements in the full range of motion. In addition to the strength of movement, which is an important factor for the dampening of oscillations in fast movements, the oscillations increase with preset speed, weight, and length of limb segment tested. In practice, this will limit the possibility of testing movements at fast speed.

Recognition of the impact forces requires a certain range of linear frequency response (0–50 Hz) in the torque record. The common use of an inkwriter that cuts responses above 3 Hz (Simmons et al., 1982) will smooth out the torque curve and make identification of disturbances caused by impact and oscillations impossible. In many studies, the peak amplitude of the torque record has been used to estimate peak torque. Actually, the recorded peak torque

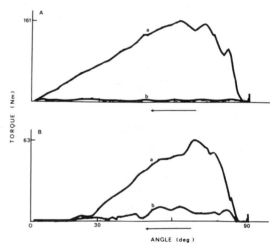

FIG. 5. To illustrate reproducibility in repeated voluntary isokinetic movements at maximal (A) and submaximal (B) effort. Curves give mean torque (a) and difference between mean and maximum torque (b) in 10 trials to repeat identical movements. The movements, knee extensions, start with knee flexed (90°) and continue from right to left in the figure. (From Gransberg et al., 1980, with permission.)

in fast movements is built up by impact force, oscillations of the lever arm, and contractile tension as well as loading and unloading of tissue elasticity. A low-frequency response will also disturb the torque record in the late phase of an isokinetic movement at high speed, since the delay will cause part of the force produced to be read at the wrong angular position. In estimations of dynamic capacity in the disabled, identification of impact forces is important, since the torque recorded in isokinetic movements differs highly in regard to how they are built up by kinetic force and momentary muscle tension. Disregarding this fact will lead to error in the estimation of how tension is produced in the different ranges of motion.

CONTROLLED ACCELERATION TO ELIMINATE IMPACT ARTIFACTS

Elimination of the accelerative artifact following impact can be obtained by giving resistance to the movement during the phase of acceleration (Gransberg and Knutsson, 1980). This is done by modifying the speed setting of the dynamometer during acceleration. At start of movement, the speed of rotation of the dynamometer axle is set at zero. When the muscle contraction starts, the speed of the rotational axle is successively increased up to a final selected speed. By individual selection of the rate of rise of speed to obtain an acceleration as fast as possible but sufficiently slow to match the dynamic capacity, the movement is resisted throughout its full range. The difference in torque record during unresisted and resisted acceleration is seen in Fig. 3C showing torque curves in maximal voluntary knee extensions reaching a constant final speed of 240°/sec. When the movement was resisted during the acceleration (curve a), the torque smoothly reached a maximum at an angular position of 60° and then successively declined as the muscle became shorter. The upper speed as determined

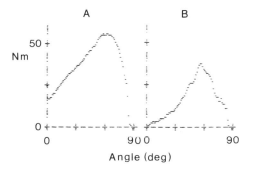

FIG. 6. Mean torque (A) and difference between mean and maximum torque (B) in three repeated voluntary knee extensions in patient with hysteric paresis requested to use maximal effort.

from angular records (not seen) was reached after an angular displacement of 15°. When the movement was unresisted during acceleration (curve b), the acceleration comprised an angular displacement of 8° at zero torque. After this, an abrupt increase in torque, signaling start of resistance and impact, was followed by an overshooting torque curve with a few oscillations.

DETERMINATION OF PASSIVE RESTRAINT

Figure 7 shows a schematic representation of the transducer used to determine torque during passive movements. The central part of the transducer is connected to the rotational axis of the isokinetic dynamometer. From it projects a pair of diagonally positioned steel wings supplied with strain gauges and connected to a bridge carrying the lever arm ("active") for torque recordings. Opposite the lever arm, there is a process with an aperture for inserting another lever arm to enforce passive movements. This lever arm is withdrawn when the transducer records voluntary strength and is used only for the study of passive movements.

To determine torque during passive movements, the limb to be examined is attached to the "active" lever arm after the

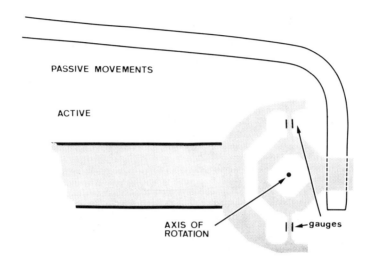

FIG. 7. Schematic representation of transducer used to enable records of passive resistance. Full explanation in text. (From Knutsson and Martensson, 1980, with permission.)

joint axis is aligned with the rotational axis of the transducer. The movement is enforced manually by the examiner through the "passive" lever arm. Speed control of the movement is obtained in the same way as in voluntary movement, by the dynamometer braking movement when the preset speed is reached. The force applied by the examiner to enforce the movement will not be recorded by the strain gauge transducer. It will record only the torque of forces on the "active" lever arm, which are those resulting from the weight of the lever arm itself and the limb examined as well as muscle restraint when present. To determine forces caused by inadvertent muscle activation and contracture in passive movements, the recorded torque is corrected for gravitational force.

An example of a determination of spastic restraint is given in Fig. 8, taken from a study by Knutsson and Martensson (1980). It shows the torque recorded during a passive flexion of the knee in a patient with spastic paraparesis. The torque resulting from weight of the leg and the "active" lever arm, which is a force aiding knee flexion, is indicated by the dotted line. The difference between the two curves gives the spastic restraint for different angular positions of the movement started at maximally extended leg (22°) and continued to 90° of flexion at a speed set at 90°/sec. The spastic restraint opposing the movement derived from activation of the quadriceps muscle is mirrored by the EMG activity in this muscle.

ELECTROMYOGRAPHIC RECORDS IN ISOKINETIC MOVEMENTS

In voluntary dynamic movements, a weakness can be caused by prime mover dysfunction or restraint. In passive movements, restraint can result from muscle activation or contracture. Thus, to distinguish types of restraint and reasons for weakness, EMG records during isokinetic movements are of great value. The problem is to obtain EMG records that are representative for the total muscle activity of the muscles involved in the movements studied. With needle electrodes, the signals from fibers close to the electrode are overrepresented as compared to signals from fibers in other areas. Thus, during isokinetic voluntary movements at different levels of effort, the linearity between force produced and EMG

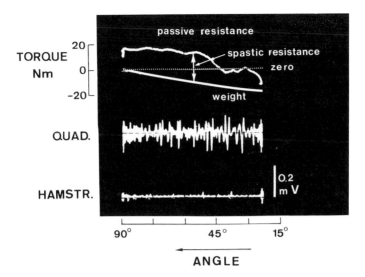

FIG. 8. Record of resistance to passive knee flexion for determination of spastic restraint in patients with paraparesis. Estimated torque caused by the weight of the leg and the recording lever arm given by *solid line*. Difference between recorded torque *(upper beam)* and *solid line* gives spastic restraint during the movement which started at maximally extended leg (22°) and continued to 90° from right to left in the figure. Surface EMG from the quadriceps and the hamstring muscles. (From Knutsson and Martensson, 1980, with permission.)

activity picked up with needle electrodes is poor. With surface electrodes, the linearity between force produced and EMG is quite good, as estimated from knee extensor torque and surface EMG from the quadriceps muscle in healthy subjects. Thus, from a torque level of about 5 Nm to a near maximal level, force and EMG activity are proportional. In patients with motor disorders, proportionality between force and surface EMG is not at all as good, even when patients with restraining muscle activity that complicates the picture are excluded. It seems likely that this results in part from more pronounced differences in activity in deep and superficial parts of the muscles in patients than in healthy subjects. This means that great care has to be taken when force is compared to EMG. There are, however, many situations in which EMG records will help to explain how a motor disorder depends on different components and how this may be altered by therapy.

The practical problems of recording surface EMGs during isokinetic movements without disturbing movement artifacts and pick-up from the speed-controlling system are easily solved by using active electrodes (Medelec, Type AE 15). For averaging, a relatively short time constant is needed for an adequate comparison of activity levels in fast and slow movements. Using active full-wave rectification and bandpass filtering in a three-pole active Paynter filter, 20 to 1,000 Hz, and then low-pass filtering at 100 Hz (Gottlieb and Agarwal, 1970) gives a good adaptation to the frequency spectrum of surface EMG signals (Lindstrom and Magnusson, 1977) and introduces a delay of not more than 10 msec.

Figure 9 illustrates how EMG records during isokinetic movements help to explain the reasons for weakness in voluntary knee extension of a patient with spastic paraparesis. The figure shows torque and EMG activity from the quadriceps (ext) and hamstring (flex) muscles during maximal voluntary and passive knee flexion. In the passive movement, the torque curve shows a restraint starting at an angular position of

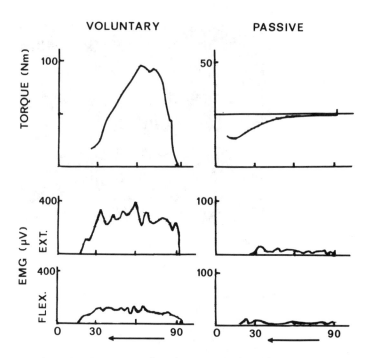

FIG. 9. Voluntary and passive knee extensions in paraparetic patient. Curves give torque corrected for gravitational force and rectified and time-averaged surface EMG from knee extensor and flexor muscles. Each curve is an average of three determinations as processed by a computer. (From Gransberg et al., 1980, with permission.)

60° and successively increasing up to about 20 Nm towards the end of motion. There was no muscle activation as indicated by EMG activity levels corresponding to the noise of the records. Thus, the restraint was most probably caused by contracture. In the voluntary movement, the extension could not be continued to further than an angular position of 25° where the restraint caused by contracture reaches a level of about 15 Nm. In the voluntary extension, the antagonistic hamstring muscles were activated quite strongly, thus imposing some further restraint on the movement.

COMPUTER SYSTEM FOR ANALYSIS OF ISOKINETIC MOVEMENTS

The analysis structure when speed-controlled movements are used to study motor disorders requires collection of a vast amount of data. On one hand, the records from each isokinetic movement include data on functions that vary in the different parts of the movement. On the other hand, motor deficits display many features of speed dependence that demand comparison of components of the disorganized motor control at different speeds of motion. Add to this the fact that each movement has to be examined repeatedly to obtain an adequately reliable analysis and, furthermore, that voluntary and passive movements are to be compared with respect to force and EMG activity from agonist and antagonist muscles.

Hence, it is not an easy task to organize and process these data routinely in clinical studies. Therefore, a computer system was developed (Gransberg et al., 1980). It uses preprogrammed acquisition and interactive computation of data with PDP 11/03 or 11/23 and RT-11 or RSX-11/M as operational system. The program guides the examination by successive requests for signals for reference levels, calibrations, correction for gravitational force, and optional collection of data from a series of voluntary and passive movements of predetermined slow and fast speeds. Data are collected by an A/D converter controlled by a programmable clock and its Schmitt-trigger inputs. The processing includes correction of the torque records for gravitational force, averaging of signals from repeated tests, and organization of the material for display as curves of mean and variability versus angular position and tables of averaged torque, EMG activity, and coactivation index for different angular positions.

The procedure usually applied in the study of the motor performance in one joint in patients with central pareses includes voluntary and passive extension at three speeds, 30°, 60°, and 120°/sec. Each movement is repeated thrice, and the procedure thus includes 18 voluntary and 18 passive movements. This number of voluntary movements can usually be performed without appreciable fatigue and gives an adequate basis for estimation of reproducibility if the patient is allowed a few trials to get used to the test procedure.

GAIT ANALYSIS
Normal Gait

Figure 10 shows a gait record from a healthy individual. It was obtained during free speed walking by collecting and processing signals giving sagittal angular position in three leg segments and surface EMGs from six leg muscles. Angular positions were obtained by the use of rotating plane polarized light for activating photoreceptors attached to leg segments (Polgon, Medelec Ltd.). The receptors are photocells covered with a shield of plane-polarizing material. When polarization of the rotating polarized light fits that of a receptor, the photocell will emit its maximal signal. By converting the phase difference between the signals from two receptors, the difference in angle of their plane polarization is obtained as a DC signal. Hip angle was obtained from a Polgon receptor on the lateral aspect of the thigh and a reference receptor in the walkway. Angular rotation in the knee was obtained from the receptor on the thigh and another on the lower leg. The angular rotation in the ankle was obtained from the receptor on the lower leg and another on the lateral aspect of the foot. Surface EMGs, recorded from three pairs of antagonistic muscle groups acting over the hip, knee, and ankle joints, respectively, were recorded as previously described (Knutsson and Richards, 1979). Signals were rectified and time averaged with a time constant of 0.02 sec.

For processing of the angle and EMG signals in relation to the gait cycle, which was defined as the period from one foot–floor contact to the next with the same foot, signals from foot switches were used to indicate floor contacts with heel, foot sole, and toe. A computer program (Isaksson and Knutsson, 1980) was used to collect and process the signals to obtain curves giving mean and +1 SD (dots) of angular displacements and EMG activity during an averaged gait cycle.

The curves in Fig. 10 were derived from 20 gait cycles in one individual. The small standard deviation indicates a pronounced regularity in patterns of movement and muscle activity. Also, when the gaits in different healthy individuals are compared, the regularity of gait movements and muscle activation becomes apparent. Figure 11 gives the average and ±1 SD (dots) of gait records from a group of healthy subjects in free speed walking derived from the mean curves from 20 to 22 gait cycles in each of eight subjects not including the subject from whom the records of Fig. 10 derive. The standard deviations that give the variability of mean patterns of movement and muscle activity indicate some amplitude variation, but they are most conspicuously

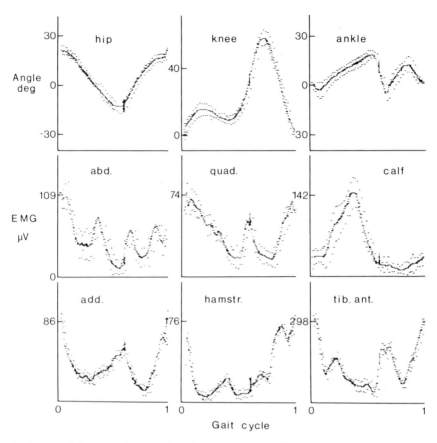

FIG. 10. Gait record from healthy subject in free speed walking. Sagittal angular displacements recorded with Polgon receptors. Zero references are: hip, vertical position of thigh; knee, fully extended leg; ankle, 90° between foot and lower leg. Hip and knee flexions and foot dorsal flexion are given as positive. Rectified and time-averaged (time constant 0.02 sec) surface EMG from hip abductor (abd.) and adductor (add.), quadriceps (quad.), hamstring (hamstr.), calf, and tibialis anterior (tib. ant.) muscles. Curves show mean and ±1 SD *(dots)* from 20 normalized gait cycles. Mean cycle duration was 1.06 sec.

stereotypic. These stereotyped movement and activation patterns of walking became apparent by a time when manual methods of calculations required a longer time constant (0.2 sec) for EMG integration to render amplitude estimation possible (Knutsson and Richards, 1979). When a computer is used for averaging, a shorter time constant (0.02 sec) more suitable for the phasic activations of gait can be used.

The stereotyped patterns of movement and muscle activation in normal walking imply that deviations in patients with motor disorders are quite easy to define. Therefore, the study of gait in patients with disordered motor control presents a means to assess how the functional derangement affects this complex movement. Gait is a type of movement that differs in many respects from the voluntary isokinetic movement. It is highly automatized, requiring a minimum of volitional drive. The force for the movement is largely potential and kinetic energy transformed back and forth by repeatedly letting the body fall forward followed by lifting (Inman, 1966). The muscles are acti-

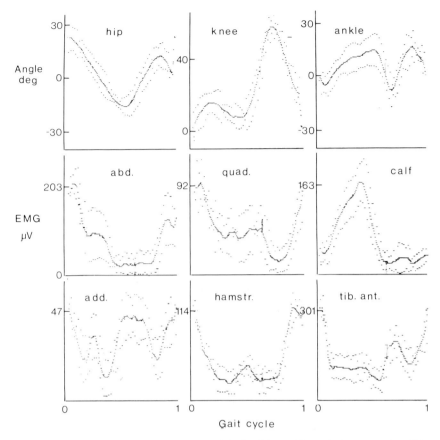

FIG. 11. Gait record as in Fig. 10 averaged from eight healthy subjects in free speed walking; 20 to 22 gait cycles recorded in each subject.

vated in complex sequences in which the muscle function varies among movement production, movement control by restraint, and joint stabilization (Knutsson and Richards, 1979). The control of these actions seems to be highly dependent on preprogrammed commands (Grillner, 1975).

Types of Deranged Gait Control in Spastic Paresis

In spastic hemiparesis, three distinctly different types of disturbances can be distinguished (Knutsson and Richards, 1979; Knutsson, 1981). In these, the crucial disturbances are, respectively, lowered stretch reflex threshold (type 1), low or lacking muscle activation (type 2), and stereotyped coactivation of groups of muscles (type 3). In some hemiparetic patients, the disturbance is more complex, with features of the different types in a mixed pattern. In spastic paraparesis, the same types of disturbances can be seen, although occasionally the disorganization affects only one component in isolation (Knutsson, 1980b). In cerebral palsy, a stereotyped coactivation of several muscle groups is a common feature (Knutsson and Martensson, 1976; Knutsson, 1980a,b), although disturbances caused mainly by low threshold for stretch reflexes and deprivation of capacity to activate the muscles sufficiently have been encountered.

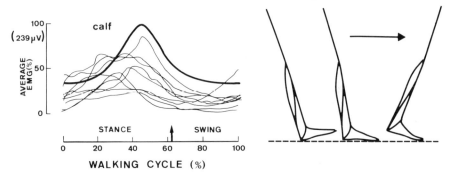

FIG. 12. Averaged EMG activity in the calf muscles in gait of nine patients with spastic hemiparesis *(thin lines)* as compared to the normal averaged activity in these muscles *(heavy line)*, and a schematic representation of the leg and the calf muscles at three stages of the stance phase of normal walking. (From Knutsson and Richards, 1979, with permission.)

Type 1.

Figure 12 illustrates the most common disturbance caused by abnormal stretch reflex activation that occurs in the calf muscles. The left of the figure shows the premature activation of the calf muscles in nine patients (thin lines) as compared to the normal activation (heavy line). To facilitate description of the mechanisms involved, a schematic representation of three stages of the stance phase of walking is given to the right in the figure. Normally, the foot–floor contact is made with the heel, with the toes still elevated from the floor. After this contact, the foot is plantar flexed by the weight load on the heel so that the foot sole makes contact with the floor. The lower leg then starts to pivot forward over the foot. The force for this movement is kinetic energy of the body obtained in the preceding phase of walking. The calf muscles are stretched during this movement, and after a short period of stretch, the EMG starts to rise.

The timing of these events in a healthy subject appears in the records of ankle angle and calf EMG activity in Fig. 10, where the mean gait cycle duration was 1.06 sec, and the mean latency between start of stretch and rise in EMG activity of the calf muscles was about 40 msec. During the pivot of the leg over the foot, the calf muscle activity successively increases. When the body has advanced ahead of the foot, the load on the calf muscles decreases, and the tension built up in the calf becomes sufficient for muscle shortening, resulting in a push-off that propels the body forward. When the calf muscles are activated too early, it will result in a pull backward of the lower leg. Since the body moves forward by inertia, the result will be knee hyperextension. The building up of tension for push-off is ruined in this backward thrust.

In patients with strong stretch reflex activation of the calf muscles, forward pivoting of the leg over the foot is restricted, as exemplified in Fig. 16. This defect will have consequences for the movements of the knee as well as hip joints and result in short steps.

When muscle length changes are compared to EMG to determine those phases of gait in which muscle activation appeared during muscle stretch and thus where abnormal stretch reflex activation could influence walking capacity, four such activations were found in the six leg muscles studied (Knutsson and Richards, 1979). Only in two of these, the calf and the hip adductor muscles, have enhanced or misplaced activations been encountered in spastic patients so far studied.

Type 2.

Figure 13 illustrates the type of disorganized gait control characterized by lack of activity. It was recorded from a patient with spastic paraparesis as a result of multiple sclerosis. As seen in the figure, which shows the means for movements and EMG activity relative to peak levels for the different muscles in a group of healthy subjects, the EMG is extremely low in all muscles examined. This patient had a relatively well-preserved motor capacity in voluntary isokinetic movements. The maximal strength in the knee extensors and flexors was above 50% of the mean strength in comparable healthy individuals. In normal subjects, only about 25% of the maximal voluntary capacity is used in walking on the level, as judged from surface EMG activity (Richards, 1980). Thus, the strength displayed in isokinetic movements would be adequate for normal gait. The problem seems to be that the system for command of gait activations was paralyzed, leaving a capacity for volitionally driven simple movements relatively intact.

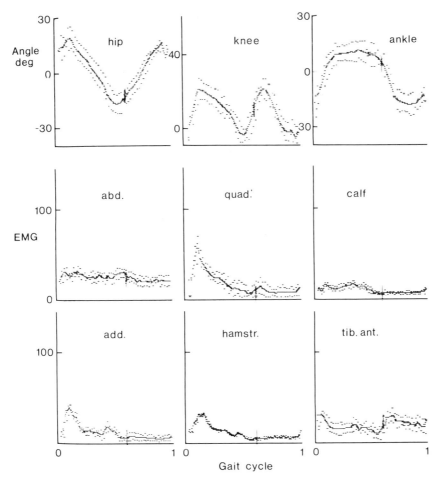

FIG. 13. Gait record in patient with spastic paraparesis with type 2 disorganization of gait control. Records as in Fig. 10, but EMG activity given relative to peak levels of EMG activity for the different muscle groups in normal gait. Averaging derived from 20 gait cycles.

Low or lacking muscle activation can be seen in from two to all six of the examined muscles. Usually, the low degree of activity is mirrored by a low capacity in voluntary isokinetic movements. A marked discrepancy between activation in gait and voluntary simple movements, as in the patient of Fig. 13, indicating a more selective disturbance, is occasionally seen. A few patients lack the normal activation in weight-supporting muscles but activate these muscles during load. This is a feature of activation that probably depends on stretch reflex activation induced by muscle load, which is considered further below (see Fig. 18).

Type 3.

Figure 14 shows an example of stereotyped muscle coactivation in walking. It was recorded from a patient with cerebral palsy and shows how the pattern of EMG is similar in all six muscle groups examined. The normal sequenced activation is completely disrupted, and antagonistic muscles are activated against each other.

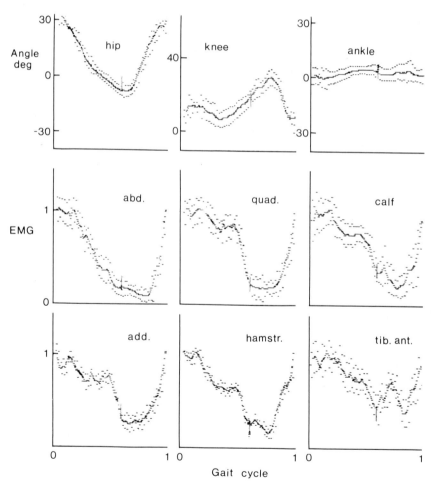

FIG. 14. Gait record of patient with cerebral palsy displaying a type 3 activation pattern in walking. Mean curves derived from 20 gait cycles. The EMG activity is displayed with peak levels for all muscles at unit amplitude. Peak levels (μV) were: Abd., 56; add., 101; quad., 140; hamstr., 60; calf, 66; tib. ant., 123.

EFFECTS OF ANTISPASTIC THERAPY ON DIFFERENT COMPONENTS OF DISORGANIZED MOTOR CONTROL

Effects on Spastic Restraint

Figure 15 illustrates a marked depressive effect on the spastic restraint to foot dorsal flexion induced by tizanidine. It was seen at a dose level of 32 mg/day in a patient with spastic paraparesis after a spinal trauma. The figure shows mean torque corrected for gravitational force from three repeated dorsal flexions starting at an angular position of 120° between foot and leg continuing to a position of 60 to 70° at three different speeds of movement, 30°, 60°, and 120°/sec. In the control records, the passive restraint, in a typical way, increased successively during the flexions. The restraint started earlier during the movement and reached higher levels the faster the speed of motion. At the fastest speed, the restraint reached a maximum of −28 Nm at an angle of about 80° between foot and leg and then fell during the continuation of the movement. During tizanidine medication, the restraint was lowered, the relative changes being largest at the fastest speed.

The effect on the passive restraint to dorsal flexion of the foot was accompanied by a marked improvement in gait capacity with an increase in stride length and walking speed of about 100%. The mechanism of this improvement is elucidated by the partial gait records in Fig. 16 showing angular displacement in the ankle, and the EMG activity in the calf and tibialis anterior muscles averaged from 10 gait cycles during the control and medication periods. For comparison, corresponding records from a healthy individual are included in the figure. In the control period, the foot was kept more plantar flexed than in normal gait throughout the gait cycle. The dorsal flexion was limited to a few degrees from the zero reference level, corresponding to 90° between foot and leg. The EMG record indicates a premature activation of the calf muscles. This was actually a clonic bursting, although this is not seen in the averaged curve other than by the small oscillations of the mean curve.

During the main part of the stance phase, which is the first part of the gait cycle, ending at the vertical bar in the records, the tension in the calf muscles pulls the lower leg backwards, thus inhibiting its normal

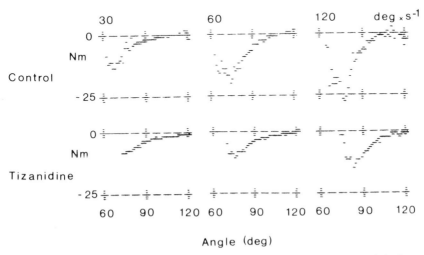

FIG. 15. Passive restraint to foot dorsal flexion in patients with spastic paraparesis before and during tizanidine medication (32 mg/day). Curves give mean torque corrected for gravitational force as derived from three determinations at each speed of motion, 30°, 60°, and 120°/sec.

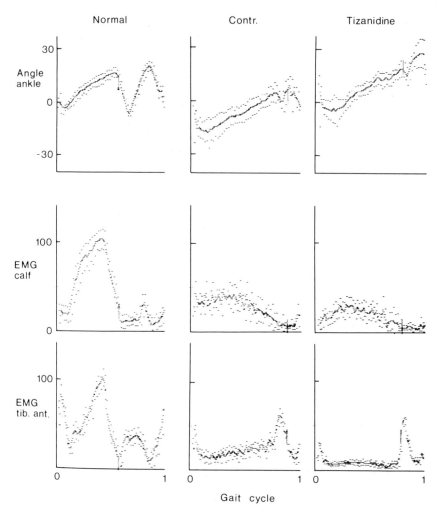

FIG. 16. Angular displacements in ankle and EMG activity in the calf and tibialis anterior muscles in gait of normal subject and paraparetic patient before and during tizanidine medication (32 mg/day). Same patient as in Fig. 15. Records as in Fig. 10 derived from 20 gait cycles in the normal subject and 10 gait cycles in each of the examinations of the patient. The EMG activity is given relative to peak levels of averaged activity for the studied muscles in normal material.

forward pivoting over the foot. Not until the angle between foot and lower leg has passed the perpendicular position (zero reference in the record) can the tension be used for forward thrust. As is typical for this type of abnormal activation of the calf muscles, the EMG activity never reached more than a fraction of the intensity seen in normal gait. Still, the restraint to forward pivoting of the leg can be expected to have been abnormally high. This is because the tension in normal gait first reaches high level when the body has advanced ahead of the foot and the muscle action can be used for stabilization of the leg, control of forward pivoting, and, finally, push-off.

During tizanidine medication, the range of ankle motion was shifted to a more dorsal flexed position, which appears to have been the result of a more delayed and lower calf muscle activation, as indicated by the EMG record. Since the activation of the calf muscles in gait is an activation appearing during a stretch imposed on the muscles by a movement enforced in a preceding gait phase, the antispastic effects illustrated in Figs. 15 and 16 can both be regarded as depressions of stretch reflexes.

Spastic restraint to knee joint extension and flexion is usually low compared to the forces active in motions (Knutsson and Martensson, 1980). Passive restraint to foot dorsal flexion more often constitutes an incapacitating component, as illustrated in Figs. 15 and 16. The restraint to foot dorsal flexion can reach peak levels corresponding to tensions in the Achilles tendon of up to 700 N. Antispastic therapy seldom abolishes high spastic restraint, but the more marked reductions, reaching about 50%, can lead to a decided improvement in motor capacity.

In evaluation of the effects on clonus, the recording of torque is preferably made together with angular displacement on a time base (Fig. 17). The elicitation of clonus in isokinetic movements depends on speed of motion. Thus, clonus elicited at a certain speed of movement may be completely abolished by antispastic therapy, as seen in Fig. 17, showing angular displacement, torque, and EMG records in control and during antispastic therapy with tizanidine in a patient with ankle clonus. Still, clonus can be elicited at a higher speed of motion, and its threshold has to be defined both with regard to speed of passive motion and the angular position at which it starts.

Depressive effects on spastic reflexes are in some patients combined with muscular weakness. This is a feature of several forms of antispastic therapy. With dantrolene sodium therapy, it seems to be related to an inability to recruit more motor unit activity to compensate for the depression of contractile muscle force induced by the drug. Thus, it is mostly seen in patients who reach their upper limits of motor capacity in the movements of daily living (Knutsson and Martensson, 1976).

Muscular weakness is, however, also reported by some patients responding to therapy with depression of stretch reflexes, but without any signs of weakness in tests for voluntary strength. An example of this was the patient whose gait records are shown in Fig. 18B, C. In B is shown the averaged EMG activity from the quadriceps muscle during a gait cycle. By comparison with the normal activation of this muscle in gait (A), it becomes clear that the activation of this muscle at the end of the swing to prepare

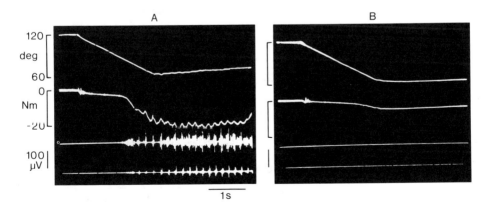

FIG. 17. Records to determine ankle clonus before (A) and during (B) antispastic therapy. Beams are, from above downward, angular displacement, torque, and calf and tibialis anterior surface EMGs. Movements were passive at a preset speed of 30°/sec starting with foot plantar flexed (120°). Record B taken during medication with tizanidine (32 mg/day).

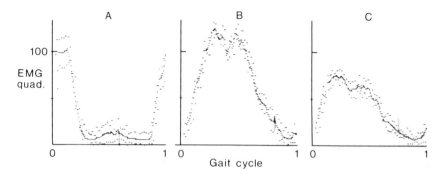

FIG. 18. Averaged EMG activity in the quadriceps muscle in a healthy subject (A) and a paraparetic patient before (B) and during (C) medication with tizanidine (20 mg/day). Mean curves derived from 20 gait cycles in each record. The EMG activity is given relative to peak activity in normal material.

the leg for weight acceptance was lacking. The activation of the muscle was first initiated after the beginning of weight load on the leg at the start of the gait cycle. Thus, the command for normal gait activation was disturbed. Apparently, however, there was a capacity to activate the muscle during load to supply a delayed stability to the leg. During antispastic therapy with tizanidine that resulted in depression of stretch reflexes, the activation of the quadriceps muscle in walking lessened (c), and the gait deteriorated because of instability. It seems likely that abnormal activations of weight-supporting muscles of the type described at least in part depend on activation of stretch reflexes. In consequence, depression of these takes away some of the body support, commonly recognized as the "spastic crutch."

Depression of Antagonist Coactivation

Figure 19 shows torque and EMG in maximal voluntary (A) and passive (B) knee flexions at the same speed of motion, 30°/sec, in a patient with spastic paraparesis. The spastic restraint, indicated by the difference between gravitational torque (dotted line) and recorded torque in B, increased successively during the passive flexion, reaching a maximal level of about 20 Nm at the end of motion. In the volun-

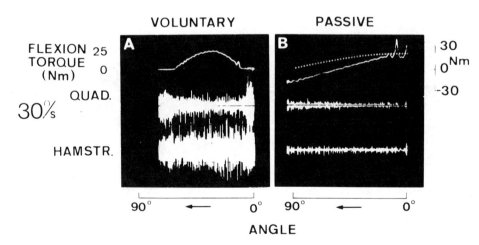

FIG. 19. Torque and surface EMG in voluntary and passive knee flexion at preset speed of 30°/sec in patient with spastic paraparesis. *Dotted line* in B indicates gravitational torque. (From Knutsson and Martensson, 1980, with permission.)

tary motion (A), the restraint to the movement given by the antagonistic quadriceps muscle cannot be determined, but from the EMG records it becomes apparent that restraint in the voluntary movement was much larger. Enhanced restraint in voluntary movement as compared to passive is a common feature in spastic patients (McLellan, 1977; Knutsson and Martensson, 1977, 1980). Thus, the depressive effects on antagonist coactivation form an important subject for study. So far, information on this matter is scarce. With baclofen, a few patients have been reported to respond with depression of antagonist cocontraction (McLellan, 1977). With tizanidine, it was seen only occasionally in a group of patients with significant depressive effects on passive restraint (Knutsson et al., 1981).

A method to assess effects on antagonist coactivation is illustrated in Fig. 20, showing torque and EMG activity in voluntary and passive knee extensions before and after local cooling which has been shown to depress antagonist coactivation in some patients (Martensson and Knutsson, 1981). Cooling of the hamstring muscles was obtained with a chilling pad (Cryomatic Mod CR-1, Chattanooga Corp.) controlled by feedback from a thermistor on the skin surface. The degree of cooling was sufficient to depress the stretch reflexes as seen by the effects on restraint and hamstring EMG in passive knee extensions. Before cooling, the extension could not be performed at the preset speed, 60°/sec, at the end of movement (a), and the movement was completely stopped before the leg was fully extended (b). After cooling, the total range of movement as well as the range of movement at preset speed had become increased, and the torque curve showed a slight increase in voluntary strength. As seen from the averaged EMG, the increased range and strength of movement were associated with a depression of the antagonistic activation of the hamstring muscles. Both before and after cooling, the antagonist activation was larger in voluntary than in passive movements.

Antiparetic Effects

Antiparetic effects induced by antispastic therapy have only recently been described. There were found in response to medication

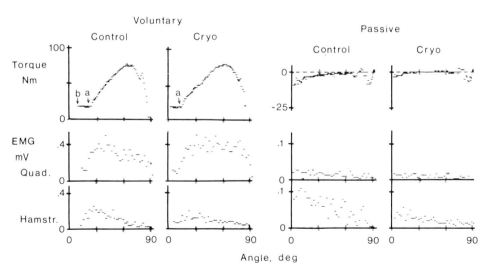

FIG. 20. Torque and surface EMG in voluntary and passive knee extensions in patient with spastic paraparesis before and after local cooling of the hamstring muscles. Each curve averaged from three tests, and torque corrected for gravitational force.

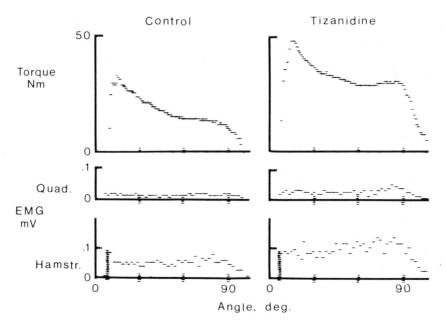

FIG. 21. Torque and surface EMG in voluntary knee flexion before and after a single dose of tizanidine (12 mg) to illustrate an antiparetic effect. Each curve was averaged from three tests, and torque curves are corrected for gravitational force.

with tizanidine (Knutsson et al., 1981) but do not appear to be restricted to this drug since they have also been observed in response to single-dose administration of baclofen and after local cooling. An example of an antiparetic effect induced by tizanidine is given in Fig. 21. It shows torque and EMG activity in voluntary knee flexion before and 1 hr after a single oral dose (12 mg) of tizanidine. As seen in the figure, the strength of the flexion was markedly increased by medication. The increase in strength was accompanied by a marked increase in the hamstring EMG. It could not be ascribed to a diminished antagonist activation. Actually, this was also increased.

The reason for the antiparetic effect is by no means clear. It has always been associated with a depression of passive restraint, indicating a stretch reflex dampening, but whether or not this association is a regular feature has not yet been established. A regular association with stretch reflex depression in the antagonistic muscles would suggest that it depends on a diminished reciprocal inhibition from these. However, this intriguing question calls for more basic information for a reasonable answer.

Electromyographic Analysis of Bicycling on an Ergometer for Evaluation of Spasticity of Lower Limbs in Man

*R. Benecke, B. Conrad, H. M. Meinck, and J. Höhne

Department of Clinical Neurophysiology, University of Göttingen, Göttingen 3400, Federal Republic of Germany

Quantification of muscle tone is necessary for an objective evaluation of therapeutic effects of antispastic drugs. Up until now, most studies have dealt with reflex phenomena in resting muscles, such as the H reflex, tonic and dynamic stretch reflexes, and the tonic vibration reflex (TVR) (Angel and Hofmann, 1963; Olsen and Diamantopoulos, 1967; Burke et al., 1970, 1971; Delwaide et al., 1978). The evaluation of reflex mechanisms alone, however, does not adequately reflect a patient's muscle tone. For example, patients often show exaggerated muscle tone and brisk monosynaptic reflexes but weak TVRs (Hagbarth, 1973; Desmedt and Godaux, 1977c).

One disadvantage of most reflex studies is that they are performed while the muscles are in a resting state: Neilson and Andrews (1973) demonstrated in patients with various motor disturbances that tonic stretch reflexes measured during sustained voluntary contraction and at rest differ in magnitude, duration, wave form, and timing. It has, therefore, been suggested that an abnormal tone, experienced by a cerebral palsied patient as an exaggerated resistance in voluntary movements, is different in character from one assessed by the clinician during passive movement of the relaxed limb (Neilson and Lance, 1978). Furthermore, in addition to exaggerated reflex mechanisms elicited by voluntary contraction of antagonistic muscles or by external influences, a disturbed voluntary program might itself play an important role (see R. J. Grimm, *this volume*). Up to now, this aspect has been neglected (Neilson, 1972).

For these reasons, functionally relevant measurements of spastic motor disorders should include an analysis of natural complex movements involving a large spectrum of tonically and phasically active descending and ascending pathways having actions on the segmental level (Neilson and Lance, 1978).

If spasticity is to be analyzed during a complex natural movement, the following methodological criteria should be fulfilled: (1) the movement chosen for a detailed analysis should have small inter- and intra-individual variance; (2) it should be characterized, like most natural movements, by a combination of active innervation and passive lengthening of antagonistic muscle pairs; and (3) the movement should be one that can be tested on a large number of pa-

---

*To whom correspondence should be addressed: Department of Neurophysiology, University of Göttingen, Robert-Kochtrasse 40, Göttingen, 3400, Federal Republic of Germany.

tients, even those dependent on wheelchairs. Gait analyses have extensive clinical relevance, but only a small number of slightly affected patients can be tested, and large interindividual variations complicate a quantitative approach to spastic gait disturbance (Carlsoo et al., 1974; Bogardh and Richards, 1974). Furthermore, specific spasticity-induced changes of gait might be superimposed on nonspecific, apparently "protective," gait mechanisms (B. Conrad, R. Benecke, J. Carnehl, J. Höhne, and H. M. Meinck, *this volume*).

Analyses of bicycling at different speeds and loads have been introduced as an alternative method for the quantitative evaluation of both spasticity and effectiveness of therapeutic efforts (Benecke and Conrad, 1980; Benecke et al., 1980). Bicycling is a highly stereotyped and reproducible bipedal movement and offers the possibility of differentiating between interacting effects of active voluntary innervation and passive lengthening. Gandy et al. (1980) used a similar movement exercise for the upper extremity: the subjects were instructed to turn a cranked wheel with the arm supported horizontally in front of them.

MATERIAL AND METHODS

A total of 22 women and 17 men in good health volunteered as normal subjects; their ages ranged from 15 to 64 years. The patient group (40 women and 22 men, age range 18 to 68 years) comprised 42 patients with multiple sclerosis (MS), eight patients with amyotrophic lateral sclerosis (ALS), four patients with familial spastic paralysis, and eight patients with a cervical myelopathy. All patients were suffering predominantly from a symmetric paraspastic motor disturbance in the legs. On the basis of clinical criteria, the patients were divided into three groups. Twelve patients showed a positive Babinski response only in conjunction with exaggerated dynamic stretch reflexes in extensor muscles of lower limbs (grade 1); 38 patients displayed sustained clonus and a moderate spastic muscle tone with passive muscle lengthening (grade 2); the remaining 12 patients exhibited marked spastic paraparesis (grade 3).

Bipolar recordings of the EMG of four muscles (tibialis anterior, medial gastrocnemius, rectus femoris, and biceps femoris) were performed in both legs of each subject with surface electrodes placed 3 to 5 cm apart over the belly of each muscle. Muscle activity was studied during active bicycling on a common ergometer machine (Siemens Ergometry System 380B) at various loads (1–10 kpm) and rotation rates (10–60 cpm). Prior to each recording session, the height of the saddle was adjusted to the length of the subject's legs: when the knee joint was maximally extended, the heel just reached the pedal. Visual feedback from a tachometer enabled the subjects to maintain a given rotation rate. The lowest position of the right pedal was defined as 0°. The surface EMGs were fed through an amplifier with a frequency response down 3 db at 60 Hz and 700 Hz and registered on a 16-channel EEG machine (Siemens Mingograf) together with pedal position, torque, and the angles of the knee and ankle joints (Fig. 1).

MECHANICAL EVENTS AND MUSCLE ACTIVATION PATTERNS IN NORMAL BICYCLING

The inter- and intraindividual variations in normal subjects were systematically tested under various experimental conditions (1–10 kpm load; 10–60 cpm). Since a comparison of absolute EMG amplitudes is difficult (Lynn et al, 1978), emphasis was placed on the shape and recruitment ranges of muscle activities. The activity profiles of all muscles investigated markedly changed with increasing rotation rate and load. The range of recruitment was independent of these parameters in rectus femoris muscle, whereas the recruitment range of the other

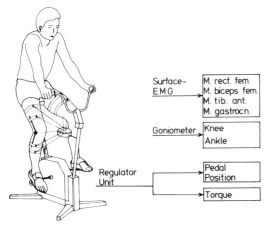

FIG. 1. Experimental set-up for studying dynamic movements of the legs. Activities of various muscles were studied during bicycling on an ergometer. The work load was varied from 1 to 10 kpm.

muscles progressively increased from 1 to 5 kpm load but remained constant at higher loads (5–10 kpm). At lower rotation rates (10–30 cpm), the shape of rectus femoris activity corresponded to an asymmetric spindle form with a short increase and a longer decrease in activity (Fig. 2). Beyond a rotation rate of 40 cpm, the spindle shape appeared symmetric.

Regarding a single experimental task at a given rotation rate and load, the normal activity pattern may be characterized as follows (Fig. 2): (1) at consecutive rotations, each of the investigated muscles is recruited at a distinct pedal position and corresponding angles of the joints; (2) the activities of each muscle show similar shapes at consecutive rotations; (3) in rectus femoris and gastrocnemius muscles, an alternating recruitment on both sides occurs, whereas the activities of biceps femoris and tibialis anterior muscles may overlap. The above characteristics are valid for all loads and rotation rates chosen. They are independent of training effects as well as of individual leg lengths. The angle of the knee joint shows a sinusoidal course, whereas

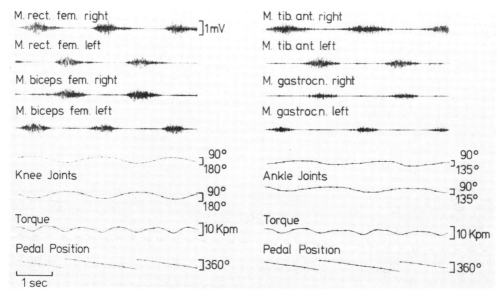

FIG. 2. The EMG patterns of a normal subject during bicycling. Rotation rate: 30 cpm. Work load: 4 kpm. **Left:** EMG of thigh muscles and corresponding mechanical events. The upper curve of knee joint excursions belongs to the right, the lower curve to the left leg. The angle at which the right pedal is in lowest position is defined as 0° (top of sawtooth signal). **Right:** EMG of shank muscles and corresponding mechanical events. Upper curve of ankle joint excursion belongs to the right, lower curve to the left leg (dorsiflexion up).

the ankle joint exhibits a nearly trapezoid excursion.

The curve second to the last in Fig. 2 represents the torque during rotations. The torque is the result of several forces acting on the pedals and is dependent on the interaction of all muscle forces, their anatomic levers, the angle of application of forces to pedal, and the ankle angle. For efficient cycling, it is important that greater forces are not applied in directions that produce small turning effects and vice versa. Quantitative measurements of torque during a complete rotation make it possible to elucidate coordinative deficits by analysis of torque curve profiles, whereas evaluations of "torque integrals" over certain rotation angles can indicate asymmetry of force production between the legs.

QUALITATIVE ASPECTS OF INNERVATION PATTERNS IN SPASTICITY

A comparison of the pathological phenomena in the three groups of patients with increasing spastic disorder shows quantitative as well as qualitative differences. In the first group (increased monosynaptic reflexes and positive Babinski response), a flattening of spindle-shaped activity was observed (see Fig. 6, IB), especially in rectus femoris muscle. Additionally, some patients showed a readily discernible pathological overlapping of activity in the bilateral rectus femoris muscles. The widening of muscle activity patterns in this patient group is caused mainly by delayed termination rather than by premature recruitment. The course of the joint angles and the torque in this patient group showed no pathological changes. The maximal power of the patients was reduced (5–7 kpm load at 40 cpm), indicating paresis, which could often not be detected in a conventional neurological examination.

The dominant finding in the second and third group of patients was a widening of activity patterns in all muscles with both premature recruitment and delayed relaxation (Fig. 3). During passive muscle length-

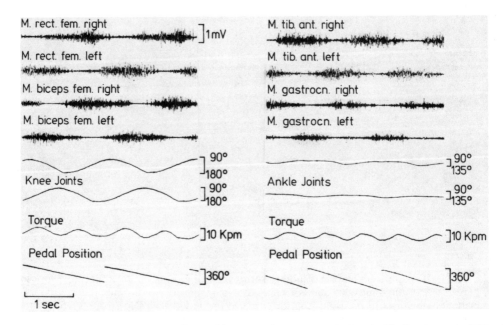

FIG. 3. The EMG patterns of a patient with a spastic paraparesis (grade 2). See legend of Fig. 2. Note enlargement of EMG activities recruited over an extended range of pedal rotation. Homonymous muscles show overlapping activities.

ening, especially of rectus femoris, velocity-dependent synchronized (clonic) and/or desynchronized EMG activity appears. The difference between the two groups was solely quantitative in nature. The pathological muscle recruitment culminated in a sustained activity throughout the entire rotation in patients with severe spastic muscle tone. At consecutive rotations, the pathological patterns appeared in a stereotyped manner like those observed in normal subjects. The courses of knee joint angle and torque were changed remarkably when clonic activity was present, indicating the mechanical consequences of rhythmic muscle activity (Fig. 4). The ankle joint excursions generally showed a decrease in amplitudes, especially a deficit of dorsiflexion. This phenomenon was present not only in patients with paresis of the pretibial flexors but, surprisingly, also in patients with strong sustained activity in tibialis anterior and a relaxed gastrocnemius.

QUANTITATIVE EVALUATION OF SPASTICITY

The question now arises as to how the pathological widening of EMG activity patterns can be expressed in quantitative terms. The first step of the computerized analysis was the automatic determination of onset and end of muscle recruitment in normal subjects. The EMGs of eight rotations were rectified and averaged to obtain data more representative of the activity pattern. The average of integrated EMGs was related to pedal angle and not to time, since variations of rotation rate appeared although subjects were requested to maintain a constant rotation rate with the aid of a tachometer. The ergometry system provided digital signals that allowed determinations of angle position of the pedal (112 pulses/rotation). The EMG data were recorded with a sampling rate of 3°. The averaging procedure was performed with respect to a trigger pulse which was initiated whenever the right pedal reached its lowest position.

Prior to evaluation of onset and end of an activity, the averaged data were smoothed using a symmetric seven-point window with 3 db cutoff at $f_c = 0.36$ degree^{-1}. For a data point to qualify as the onset of an activity, it has to fulfill two conditions. (1) Given the maximum value MAX and the minimum value MIN of the set of all data points, the mean of 10 points prior to onset (P/b) had to be less than $0.18 \times$

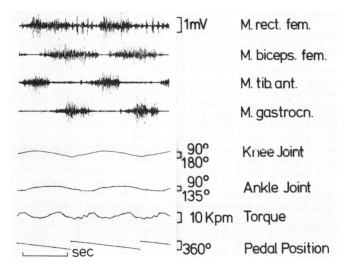

FIG. 4. Clonus activity in rectus femoris during bicycling. Rotation rate: 50 cpm. Work load: 5 kpm. All muscles and joint excursions belong to the right leg.

(MAX − MIN), whereas the corresponding standard deviation had to be less than 1.38 × MIN. (2) From the subset of points obeying condition 1, point P/b was selected in the following way: the slope of a straight line resulting from a least-squares fit involving 20 points following P/b had to be maximal. The end of activity was determined similarly.

Figure 5 shows an example of such a computerized analysis and the resulting mean values and standard deviations of recruitment obtained in normals. On the basis of data in normals, it was decided whether

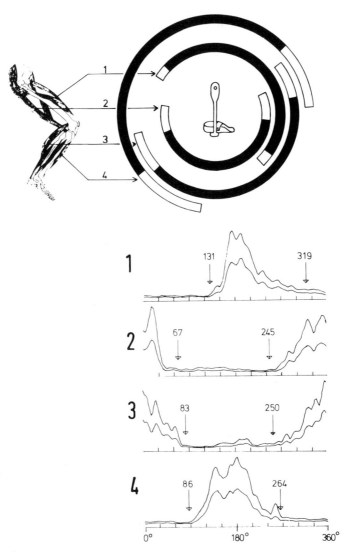

FIG. 5. Onset and end of muscle recruitment in 20 normals while bicycling. Right leg. **Upper part:** Mean angle range *(black)* of four routinely examined muscles and standard deviation *(white)* presented as concentric arcs. The measurements were performed at 30 cpm rotation and 4 kpm work load. **Lower part:** Computerized analysis of muscle recruitment in a normal subject. 1, rectus femoris; 2, biceps femoris; 3, gastrocnemius; 4, tibialis anterior. *Lower curve* shows mean values (eight rotations), *upper curve* standard deviation.

or not onset and end of activity in patients were pathological (>2 SD). This type of computer analysis showed significantly pathological values in the majority (62%) of patients in group 1 and in all patients in groups 2 and 3. The application of these measurements with determination of onset and end of activity was restricted to those muscle activity patterns in which a recruitment period could clearly be differentiated from a period lacking activity. Difficulties occurred when the inactive state was interrupted by intermediate activity (i.e., clonus).

For the above reasons, a more sophisticated quantitative analysis was introduced that could be applied to all pathological EMG patterns. In the range of the rotation in which normally—but not in the patients—a silent state of muscle occurs, the integrals of the EMG were calculated (e.g., activity in the range of 300° to 120° in right rectus femoris muscle). Because comparison of absolute EMG amplitudes is difficult, this pathological activity (I_0) was correlated to the activity in the normal recruitment period (I_1). The use of a quotient (R = I_1/I_0) has the advantage that sources of deviation introduced by the recording conditions can largely be excluded. The calculation of quotient R is demonstrated for a control subject and three patients in Fig. 6. Under identical experimental conditions (30 cpm; 3 kpm), there was a good correlation between the R values and the clinically defined severity of spasticity Fig. 7).

Quantitative measurements of spasticity by quotient R make an objective evaluation of the effectiveness of an antispastic drugs possible. Figure 8 gives an example of a drug-induced effect with diazepam, 10 mg intravenously. It can be seen that R is partly normalized. Parallel to this effect, the power of the patient increased: at a rotation rate of 30 cpm, the maximal applicable work load was 5 kpm prior to and 7 kpm during the maximal effect on R. When the dose was increased to 15 mg diazepam, the quotient R improved further, but the maximal tolerable work load decreased to 3 kpm. Similar results could be obtained with tizanidine or baclofen in increasing doses. It can be assumed that the decrease of patients' maximal power at higher dosages is caused by an additional effect on the motor endplate. These pharmacological experiments were helpful in establishing the optimal dose for improvement of both R and maximal power.

A further method was the determination of the clonus threshold. In Fig. 9, a clonus appeared at a work load of 7 kpm when the rotation rate was 20 cpm; it was also present at a work load of 1 kpm with a rotation rate of 60 cpm. Clonus, once elicited, could be enhanced in magnitude by increasing work load (Fig. 9, right). After adequate doses of antispastic drugs, in addition to the amelioration of the patients' power and R, the clonus threshold increases.

THE COMPUTERIZED ANALYSIS OF THE TORQUE

Figure 10 schematically illustrates the computerized analysis of torque. The following parameters of the torque curve were systematically analyzed: minima and maxima with their corresponding pedal positions (W, X, Y, Z, and A_1, A_2, A_3, A_4, respectively) and torque versus pedal position integrals (I_1, I_2, I_3, I_4). Summation of I_1 and I_2 or I_3 and I_4 leads to a value that roughly represents the power produced by the left or the right leg. By means of the latter analysis, an asymmetry of power between the two legs can be objectively established.

The first question was: What is the difference in amount of work between the two legs? It was surprising that, especially at smaller work loads, remarkable asymmetries could occur, although the riders assumed they were pedaling evenly. The amount of asymmetry in patients with hemiparesis or paraparesis, however, clear-

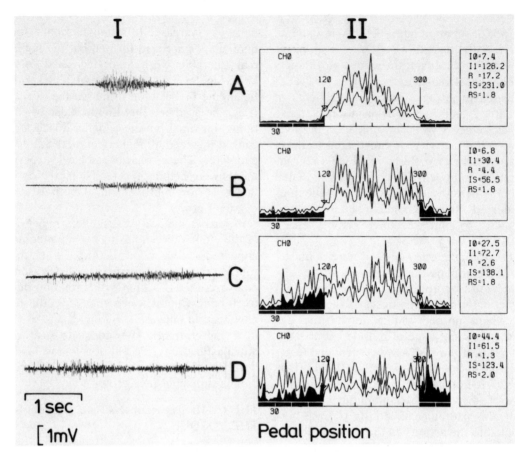

FIG. 6. Quantitative measurements of spasticity by quotient R. **I:** Representative EMG recordings of right rectus femoris muscle in a normal subject (**A**) and in grades 1, 2, and 3 patients (**B, C, D**); 20 cpm, 3 kpm. **II:** Corresponding average (eight rotations) of EMG in a normal subject and three patients with increasingly severe spastic paraparesis. The amplitudes are normalized to peak of standard deviation curves. The *horizontal axis* shows position of pedal with the origin corresponding to lowest position of the right pedal. Computer-evaluated EMG integral (I_1) between 120° and 300° (mean values of muscle recruitment evaluated in normal subjects; *white area*) and between 300° and 120° (I_0; *black area*), where in normals the muscle is resting. $R = I_1/I_0$. IS, integral of the standard deviation curve; RS = IS/I_1. Note flattening of activity in **B** (grade 1 patient) and enhanced duration of muscle recruitment in **C** and **D**. The grade 3 patient (**D**) shows phasic muscle activity intermingled with tonic background. R is inversely related to severity of spastic paraparesis.

ly exceeds that in normal subjects (Fig. 11). In normal subjects, the asymmetry in power production decreases with increasing work load. In pathological states, just the opposite can be observed. The more paretic leg in paraparesis decompensates if greater work loads are applied. This analysis is helpful in the objective quantification of slight paresis in one leg.

Another point of interest was the analysis of the shape of the torque curve, especially of the positions of maxima and minima as indicators of coordinative performances. Despite the remarkable disintegration of EMG patterns in spastic patients, the shape of torque curves remained surprisingly unaffected. A shift of torque maxima (A_2, A_4, see Fig. 10) was observed only in a number

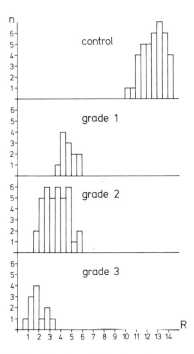

FIG. 7. Distribution of R values in normals and three patient groups with increasing severity of spastic paraparesis. All values established under same experimental situation (30 cpm; 3 kpm).

of grade 3 patients. Since the above phenomenon was regularly observed in patients with a deficit of dorsiflexion in the ankle joint, it is assumed that the resulting unfavorable angle of application of forces to the pedal plays an important role.

DISCUSSION

The experiments on normal subjects and patients with spastic paraparesis provide evidence that study of movement during bicycling is a suitable method for the qualitative and quantitative analysis of negative and positive symptoms. This kind of locomotion is subject to only small intra- and interindividual differences in normal subjects. In patients with spasticity, reproducible changes of the innervation patterns are observed, which can be expressed in quantitative terms. The present method has the advantage over gait analysis that it can also be employed with patients unable to walk as a result of postural decompensation. The specific spasticity-induced changes of innervation patterns are, in contrast to gait analyses, not afflicted by nonspecific "protective" effects, which appear in imminent decompensation of postural functions (B. Conrad, R. Benecke, J. Carnehl, J. Höhne, and H. M. Meinck, *this volume*). It is a further advantage of the present method that during bicycling, passive lengthening of the muscle is performed more physiologically in comparison to passive movements induced either manually (Burke et al., 1970) or by an electromechanical device (Herman, 1970). The amount of clonic activity not only depends on velocity of stretch but also on the activity level of the antagonist (Fig. 9). This might be the reason for the clinical observation of low spastic muscle tone established on passive muscle lengthening in the reclining patient, which contrasts to a pronounced spasticity during natural movements such as walking and standing.

In addition to being a new approach to objective measurement of spasticity, the quantitative and qualitative analysis of bicycling elucidates many pathophysiological

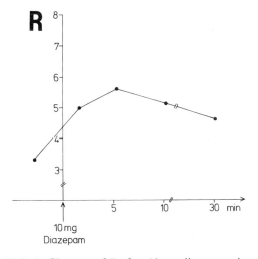

FIG. 8. Changes of R after 10 mg diazepam, i.v.

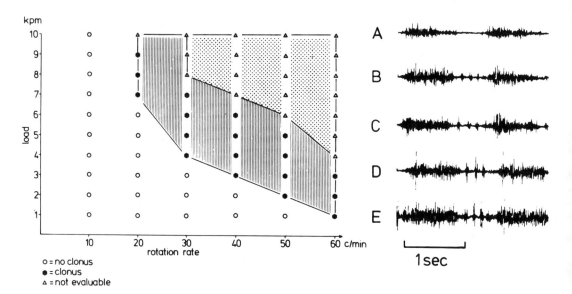

FIG. 9. Evaluation of clonus threshold and maximally obtained work load in a single patient. At constant rotation rate, clonus appears at a distinct work load. The higher the rotation rate, the smaller is the clonus threshold at stepwise enhancement of work loads. If the work load is further raised, depending on rotation rate, a level can be reached at which the patient is no longer able to maintain a constant rotation rate (prolongation of rotation time by at least 10% of expected time in three consecutive rotations). A–E: Raw EMG of right rectus femoris at constant rotation rate (50 cpm) but increasing work loads (1–5 kpm).

aspects of spasticity and paresis in upper motoneuron lesions. By analogy to the clinical symptoms in upper motoneuron lesions, the EMG findings described can be divided into positive and negative signs: (1) lengthened recruitment period of all muscles, especially in extensor muscles, pathological synchronized (clonic) and/or desynchronized reflex activity during passive muscle lengthening (positive); (2) amplitude reduction with flattening of activity shapes (negative). The broadening of the re-

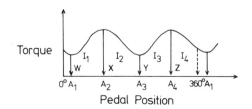

FIG. 10. Scheme of computerized analysis of torque. The curve represents a combination of pedaling forces from both legs.

cruitment periods was the most sensitive phenomenon and appeared when only a positive Babinski sign and exaggerated tendon reflexes, without spastic muscle tone, were clinically present. It might be argued that the prolonged activity represents an early adaptive function of the CNS for compensating paresis, as was assumed for the prolonged ballistic innervation in patients with upper motoneuron (MN) lesions (Hallett, 1979). This explanation seems unlikely, since the surplus activity falls in ranges of pedal position where only small or even negative torques are induced. It might be assumed that the extended recruitment is a result of disfacilitation of inhibitory interneurons that are normally coactivated with the agonist α MNs. The Renshaw cells with their focusing actions on the MN pool are interesting candidates (see E. Pierrot-Deseilligny, R. Katz, and H. Hultborn, *this volume*).

The pathological activity in rectus fem-

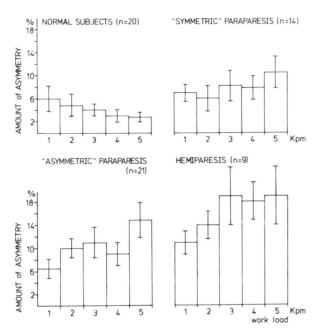

FIG. 11. Asymmetry of force production in normals, spastic paraparesis, and spastic hemiparesis. All values are obtained at rotation rate of 40 cpm. Force symmetry would be present when the sums $I_1 + I_2$ and $I_3 + I_4$ are identical (50%). *Columns* represent mean value of asymmetry in force production between the two legs at increasing work load. *Vertical axis* is amount of force dominance of left ($I_1 + I_2$) or right ($I_3 + I_4$) leg expressed as percentage exceeding 50% level.

oris muscles during muscle lengthening with innervation of both the antagonist and the homonymous muscle of the other leg apparently represents an exaggerated reflex response: it increases in direct proportion to velocity (rotation rate) and muscle effort (work load). In man, the dynamic behavior of the exaggerated stretch reflex in spastic muscle has been attributed to increased activity of the Ia afferents of the muscle spindle through release of dynamic fusimotoneurons (Herman et al., 1973; Burke et al, 1971). Microneurographic recordings, however, have demonstrated that spindle Ia activity is not enhanced in spastic patients (Hagbarth et al., 1973; see D. Burke, *this volume*).

The observation of increased stretch reflexes while bicycling at augmented work load contrasts to the results of Neilson and Lance (1978) who found that the hypersensitivity of tonic stretch reflexes in spasticity was independent of contraction level. The latter authors used isometric contraction, whereas, in our experiments, the reflex activities were studied during different amounts of effort exerted by the entire extremity including the antagonist.

The majority of patients showed a flattening of the activity shape with loss of the dynamic peak. This phenomenon was most pronounced in muscles with clinical paresis. In normal subjects, the dynamic peak of rectus femoris activity is followed by a sudden rise of torque. Since the principle of an orderly recruitment of single motor units (Henneman et al., 1965a; see D. Kernell, *this volume*) is also valid for rapid (ballistic) contractions (Desmedt and Godaux, 1977a, 1981), it can be assumed that the dynamic peak corresponds to augmented recruitment of MNs including large units with high input resistance; these are also the first that cannot be recruited in an upper MN lesion.

The flattening of activity patterns might be the result of such a lack of large MN recruitment.

After a long period of hyperflexion, spastic muscles, especially the triceps, develop a steady retraction. This alteration is associated with an increased resting tension (Herman, 1970) and may limit dorsiflexion in ankle joints in spite of strong activity in tibialis anterior (Dietz et al., 1981). Since all patients investigated suffered almost exclusively from a symmetric spastic paraparesis of the legs, it was caused by a spinal lesion in most cases. Analysis of bicycling in hemiparetic patients showed similar changes in innervation patterns. A sustained activity in flexor muscles as observed in spinal spasticity is, however, a rare finding in hemiparesis, presumably because of the integrity of the dorsal reticulospinal system in cerebral lesions (Lance, 1980).

ACKNOWLEDGMENTS

The authors are indebted to Ms. Karin Thäter and Ms. Sabine Harder for technical assistance, to Mr. Peter Wenig for design of electronic equipment, and to Mr. Randolph Krebs for revision of the English manuscript. This work was supported by the Deutsche Forschungsgemeinschaft (SFB 33).

Motor Control Mechanisms in Health and Disease,
edited by J. E. Desmedt.
Raven Press, New York © 1983.

Neurobionomics of Adaptive Plasticity: Integrating Sensorimotor Function with Environmental Demands

*G. Melvill Jones and G. Mandl

Aviation Medical Research Unit, Department of Physiology, McGill University, Montréal, Quebec H3G 1Y6, Canada

It is a familiar fact that surgical or pathological disruption within the adult nervous system is capable of bringing powerful plastic forces into play in both peripheral and central neural pathways (e.g., Barlow and Gaze, 1977; Flohr and Precht, 1981; Jeannerod and Hecaen, 1979; Schaefer and Meyer, 1973, 1974; Stein et al., 1974; Teuber, 1975). Less well known is the more recent discovery that a vigorous plastic response can equally well be invoked as a result of rearranging sensory–motor relations by means of alterations in the external environment, that is to say, without invasive insult to the nervous system itself.

Thus, to anticipate a subject discussed in later sections of this chapter, a maintained optical modification of vision [e.g., when first wearing new spectacle lenses (Gauthier and Robinson, 1975; Miles and Fuller, 1974) or when entering an underwater visual environment (Gauthier, 1976)] may produce alterations in the normal geometric relationship between head rotation and relative movement of the seen external world. The resulting visual–vestibular mismatch tends to resolve itself by invoking adaptive changes within the CNS, which in turn appropriately alter stimulus–response characteristics in relevant brainstem reflexes (Gonshor and Melvill Jones, 1971, 1973, 1976a,b; Ito et al., 1974, 1979; Keller and Precht, 1979a; Mandl et al., 1981; Melvill Jones and Davies, 1976; Melvill Jones and Gonshor, 1981; Melvill Jones and Mandl, 1979; Miles et al., 1980; Robinson, 1976; see also reviews by Miles and Lisberger, 1981; Wilson and Melvill Jones, 1979).

Similarly, exposure to maintained modification of the gravitational environment (e.g., in space) calls for, and probably achieves, adaptive changes in vestibulospinal and allied sensory–motor systems controlling movement in the lower limbs. As we shall see, numerous other examples of this kind of internal adaptive response to external environmental pressures are currently emerging in the literature (Berthoz and Melvill Jones, 1981; Collewijn and Grootendorst, 1979; Judge and Miles, 1980; Optican and Robinson, 1980; Schultheis and Robinson, 1981).

The general question therefore arises: To what extent can central neural mechanisms, honed by million of years of evolution to precision performance within the boundaries of "normal" everyday environmental

---

*To whom correspondence should be addressed: Aviation Medical Research Unit, Department of Physiology, McIntyre Medical Sciences Building, McGill University, 3655 Drummond Street, Montréal, PQ H3G 1Y6 Canada.

constraints, be adaptively readjusted or rematched when faced with newly emerging environmental exigencies such as those predicated by man-made technological advances? Considered from this viewpoint, the problem area can be subsumed under the newly coined term *neurobionomics,* derived from the more general term bionomics, the latter having been defined as "the branch of biology which deals with the adaptation of living things to their environment" (*Webster's Dictionary,* 2nd ed.).

Thus, neurobionomics is largely concerned with the description and measurement of reflex behavior; the mapping of adaptive behavioral strategies; and the overall scope of those goal-seeking biological processes that come into play during attempted matching of the nervous system's performance to prevailing environmental demands. The present chapter explores the role of adaptive plasticity within the framework of these concepts, using illustrations chosen where appropriate from the authors' experimental studies.

ERROR-ACTIVATED REFLEX CONTROL

It is true that in many response elements of the nervous system there would at first sight appear to be little or no need for plastic modifiability because error-activated feedback supplies the necessary information for automatic response correction and can substantially mask inadequacies of internal neural parameters. Consider, for example, the ocular acquisition and subsequent following of a visual target with the head fixed. Retinal stimulation caused by an offset image is neurally reflected on a retinotopic map in superficial layers of the colliculus (e.g., Robinson, 1972; Schiller and Stryker, 1972). Given appropriate conditions (Goldberg and Wurtz, 1972), this may then cause an enhanced efferent discharge from deeper collicular layers, activating in turn an appropriately coded (Robinson, 1970; van Gisbergen et al., 1981) oculomotor discharge for saccadic repositioning of the eye. If the residual difference between image and foveal position remains unacceptable, negative feedback comes into play and corrects the error by another saccade, as illustrated in some of the records of saccadic eye movement in Fig. 1A.

If the target begins to move, then within the latency of oculomotor response, the image begins to slip across the retina, thus stimulating another error-activated feedback system responsible for the generation of following eye movements (e.g., Rashbass, 1961; Robinson, 1965; Schalén, 1980; Westheimer, 1954). As in the saccadic system, any residual error (in this latter case the vector of retinal image slip velocity) stands to be corrected, after a further reflex response time, by an appropriate change of eye velocity. The basic informational feedback loop describing such systems is outlined in Fig. 2A, in which, as noted above, the head is assumed fixed, so that E represents eye angle (or velocity) relative to fixed spatial coordinates. Target position (or velocity) T is compared with eye position (or velocity) E to produce an error signal $(T - E)$, which drives the eye in a manner tending to resolve the error.

Note that the negative feedback line ("geometric feedback") and summing junction of Fig. 2A depict an external informational pathway and process whereby target and eye position (or velocity) are added to yield the difference (retinal error, $T - E$). This error signal is then translated by the retina into neurophysiological afferent signals feeding into the internal biological system which, in this example, comprises visuo-oculomotor mechanisms. The dash–dot–dash enclosure thus contains the complete, integrated external–internal system. Note particularly that the response characteristics (i.e., input–output dynamics) of the complete system will be altered by modification of any of its component parts, inde-

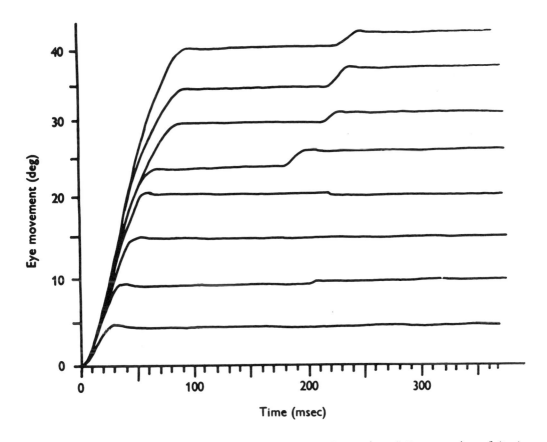

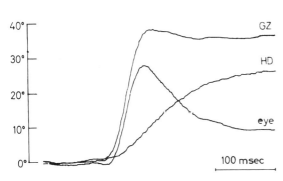

FIG. 1. Examples of the operation of (top) a closed-loop (saccadic) system and (bottom) an open-loop (vestibuloocular) reflex. Top: Patterns of human saccadic eye movement (with head stationary) in response to visual targets offset from 5° to 40°. Note that when the target-eye error is unacceptable, visual feedback leads to its adequate correction by means of secondary saccades. Bottom: Visual acquisition of a target offset by approximately 40° with both eye and head free to move. Note that after target acquisition (see gaze curve, GZ), the traces of eye and head (HD) angles exactly oppose one another (unity gain). Since this gain is retained even when the target light is extinguished just before saccade initiation (i.e., no visual feedback), the system must be capable of operating at unity gain in the open-loop state. Consequently, in this case, VOR gain must be maintained by parametric, rather than closed-loop, feedback control. GZ, gaze angle (head + eye angle with respect to space); HD, head angle with respect to space; eye, eye angle with respect to head. (Top: From Robinson, 1964, with permission. Bottom: From Morasso et al., 1973, with permission.)

pendently of whether they are internal or external to the living organism.

This brings us to the question of what happens when the head is free to move and both the eye and the head turn rapidly towards an offset target. Figure 1B illustrates how, in these circumstances, while the eye first moves saccadically towards the target, the head (HD) follows more sluggishly in the same direction. This inevitably leads to adequate stimulation of the semicircular canals and activation of the vestibuloocular reflex (VOR) in a direction opposed to the initial ocular saccade. As shown in the trace of gaze direction (GZ = eye + HD), the net result is the rapid acquisition of the target by means of a saccadic eye movement, followed by maintained gaze stabilization relative to space through VOR-generated compensatory eye velocity which is equal in magnitude, but opposite in direction, to that of the head. Note that for successful stabilization of gaze, there must consistently be a 1 : 1 relationship between head velocity and the compensatory eye velocity. In other words, after target acquisition, the gain of the VOR (eye vel/head vel) must closely approximate unity throughout this rapid maneuver.

Yet, the latency of visual–oculomotor feedback may prove too long (up to 250 msec delay, Robinson, 1965) to provide effective visual feedback in these circumstances. Indeed, the insignificance of visual feedback in this case is evidenced by the fact that turning out the light just before the sequence in Fig. 1B begins does not alter the compensatory ocular movement (Dichgans et al., 1973). Theoretically, this might be because of the existence of a pre-programed combination of ocular and neck muscle activities. But this does not normally appear to be the case, since unexpected mechanical blockage of head rotation during the movement completely abolishes the compensatory eye movement (Morasso et al., 1973). Another possibility is that head rotation relative to the body activates a neck afferent–oculomotor reflex; but in normal circumstances, this too appears to be almost negligible during rapid movements (Barnes and Forbat, 1979; Morasso et al., 1973). Evidently, in a maneuver such as that in Fig. 1B, the VOR normally operates virtually on its own and at approximately unity gain independently of any immediately available error-activated feedback or other reflex aids. A similar conclusion derives from observations of ocular stabilization during head oscillation at frequencies that lie above the upper limit for successful visual tracking (Baarsma and Collewijn, 1974; Benson and Barnes, 1978; Melvill Jones and Gonshor, 1981; Steinman and Collewijn, 1980).

PARAMETRIC FEEDBACK CONTROL

The fundamental question arises: How is this consistently optimal input (canal rotation)–output (eye rotation) relationship achieved and maintained in situations utilizing the apparently unaided and open-ended VOR?

The generalized scheme of Fig. 2B illustrates a possible approach to this question. The stimulus (S) produces a reflex response (R) without the aid of immediate on-line feedback. Yet, for a proper dynamic relationship between S and R, there must be appropriate numerical settings of parameters controlling signal processing in both internal (e.g., synaptic efficacy) and external components of the system, all of which are assumed to be contained in the dash–dot–dash enclosure as in Fig. 2A. In natural life, however, there will be an undetermined number of potential perturbations acting to modify these parameters: internal influences such as loss of neurons from aging, intrusion of pathological changes from trauma or disease, circadian rhythms, and hormonal cycles; and external influences caused by changes of size and mass associated with normal growth or abnormal environments introduced by man-made vehi-

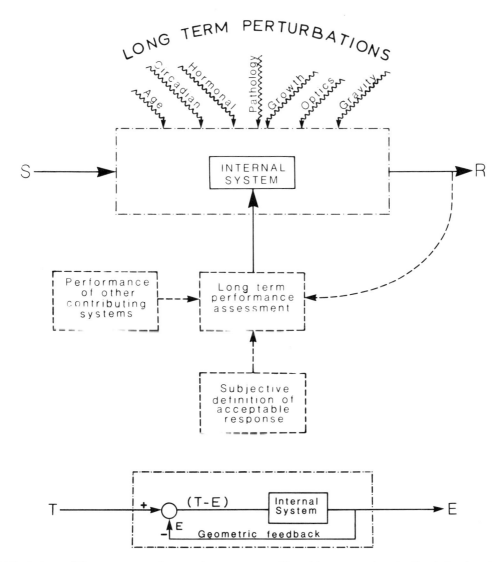

FIG. 2. Two different types of control systems. **A:** Closed-loop negative feedback in the visual control of eye movement. T, angle (or angular velocity) of target relative to space; E, angle (or angular velocity) of the eye relative to space with stationary head; T–E, error representing retinal image displacement (or velocity). The dash–dot–dash rectangle encloses both internal and external components of a fixed system. **B:** Schematic representation of parametric feedback control. S,R, stimulus (input) and response (output), respectively. Both internal and external components of the fixed system are enclosed in the dash–dot–dash rectangle, as in Fig. 2A. Parameters controlling the fixed system are liable to long-term perturbations which may produce sustained changes in S–R relationships. Such sustained rearrangement of stimulus–response relationships are subject to long-term performance assessment, depending on both a presumed subjective definition of response acceptability and the coincident performance of other contributing systems. As an outcome, long-term influences (adaptive plasticity) are brought to bear on internal parameters, causing goal-directed changes in their values.

cles such as those of flight and space (Melvill Jones, 1968), and (of particular relevance to the present theme) the donning of spectacle lenses for the correction of defective visual optics (Gauthier and Robinson, 1975).

How can such an open-ended system as the VOR maintain proper parametric settings along its pathway when faced with the constant intrusion of perturbations such as these? One answer could be by means of "parametric feedback," whereby long-term performance assessment based on a subjective appraisal of response acceptability leads to adaptive updating of parametric settings according to need, with the proviso that that the required modification in any specific element will inherently depend on the simultaneous modification of other contributing systems (Fig. 2B). If parametric modification of this kind does indeed occur, then it should prove possible to provoke alterations of the relevant internal parametric settings by artificially changing external parametric settings. A series of experimental studies was initiated in the mid-1960s to test this hypothesis by using optical reversal of vision, which ultimately calls for overt reversal of the VOR.

EXPERIMENTS WITH REVERSING PRISMS

Methods

The basis of the experimental approach originally adopted is illustrated in the highly schematic diagram of Fig. 3. More detailed aspects of neural pathways contributing to the VOR will be discussed later, and the reader is also referred to more extensive reviews on the system as a whole (Henn et al., 1980; Miles and Lisberger, 1981; Wilson and Melvill Jones, 1979). The semicircular canal feeds its primary afferent input into the vestibular nucleus (VN), from which inhibitory and excitatory interneurons pass on to appropriate nuclei of the oculomotor system (OMN). From OMN, motor nerves leave the brainstem to innervate appropriate antagonistic pairs of extraocular muscles. When the head is rotated in a direction that is excitatory to the primary afferent innervation of the canal, reciprocally organized neural signals will drive the eyes to turn relative to the skull in a direction opposite to that of the skull relative to space. As a result, compensatory eye movement suitable for visual fixation of the surrounding world is induced during head rotation.

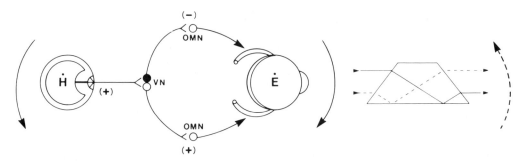

FIG. 3. Schematic diagram of the experimental paradigm based on vision reversal. The open-ended vestibuloocular reflex produces compensatory eye movement opposite to that of the head (*solid curved arrows* depicting head and eye movement). The dove prism produces mirror reversal of vision in the horizontal plane while permitting forward vision in the direct line of sight. When attached to the head, therefore, the dove prism calls for compensatory eye movement in the same direction as that of the head (*dashed curved arrow*). $\dot{H}$, head angular velocity; $\dot{E}$, eye angular velocity; VN, vestibular nucleus; OMN, nuclei of the oculomotor system.

In normal circumstances the VOR induced by slow head rotation operates with the aid of vision and hence in coordination with optical influences. Thus, following Fig. 3, when the head turns to the left relative to a fixed external visual world, visual stimuli will normally cooperate with the VOR in driving the eyes to the right relative to the skull. The significant feature of the present experimental approach is that by using reversing optics, it is possible in effect to "cross innervate" the visual sensory input, but of course in this case without the introduction of invasive surgery. Thus, when the head turns to the left with the reversing optics in place, instead of the visual stimulus providing a right-going optokinetic drive, the new (mirror-reversed) visual stimulus drives the eyes to the left (dashed arrow), which is the same direction as that of movement of the head relative to space. As a result, a direct conflict is established between the vestibular and optokinetic drives to the oculomotor system. If, as adduced above, there is an option to produce gradual alteration of parameters within the VOR according to need, then persistence of this new situation should induce radical changes in those parameters.

To test this presumption, human subjects were first exposed to short periods (16 min) of plane mirror vision during sinusoidal rotation of the whole body (Gonshor, 1970; Gonshor and Melvill Jones, 1976a) and later to long periods (up to 27 days) of maintained vision reversal during normal movement using dove prism goggles mounted on the skull as diagrammed in Fig. 3 (Gonshor and Melvill Jones, 1971, 1976b). During and after the periods of vision reversal, the parameters controlling oculomotor performance were tested daily by sinusoidal head rotation at both low (LF) and high (HF) frequencies. For LF testing, the whole body was passively rotated in the dark at 0.17 Hz and 60°/sec peak velocity. This test stimulus was chosen since it does not of itself produce habituation of the VOR (Gonshor and Melvill Jones, 1969, 1976a; Jager and Henn, 1981), presumably because it lies within the natural range of semicircular canal stimulation (Melvill Jones, 1974).

Subjects were maintained alert by mental arithmetic or, in animals, by sounds of food preparation. Eye movements were recorded either by conventional EOG or by the search coil method of Fuchs and Robinson (1966). It is important to note that during both human and cat experiments, vision reversal was maintained throughout the subject's entire waking life but that the state of the VOR itself was usually tested in the dark. The duration of continuous vision reversal ranged from 2 to 27 days in the human experiments and on the order of 1 year in corresponding experiments with cats.

Low-frequency tests.

Figure 4 illustrates the manner in which VOR parameters were estimated at low frequency. The lower curve gives the angular position of the turntable for one cycle of sinusoidal rotation. Since the skull was fixed to the turntable, this also quantitatively indicates the angular position of the skull relative to space. The middle trace shows the compensatory nystagmoid eye movements induced in the dark by this head rotation. It consists of compensatory smooth pursuit phases of eye rotation interspersed with saccadic repositioning flicks constituting normal quick-phase movements. The top trace shows a replica of the smooth pursuit components of nystagmoid response, aligned mathematically after removal of quick phases into a continuous, approximately sinusoidal curve of cumulative eye position (CEP). Thus, the bottom and top curves, respectively, represent quantitative characteristics of the vestibular (canal) stimulus and oculomotor response of the VOR. Since the stimulus was strictly sinusoidal, and the response was usually well correlated with a sinusoidal least-square fit ($r > 0.97$ in most records), the

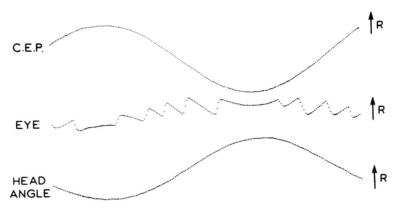

FIG. 4. Method for measurement of VOR parameters during sinusoidal rotation of the head. The curve of compensatory ocular nystagmus is converted by computation (Davies, 1979) into a curve of cumulative eye position (CEP) by alignment of consecutive slow phase segments of nystagmus. The VOR gain and phase are obtained from the parameters defining sine waves fitted to the CEP and head angle curves, respectively. (From Melvill Jones and Davies, 1976, with permission.)

input–output characteristics of the reflex could be assessed in numerical terms as VOR or ocular gain, i.e., the amplitude ratio of response to stimulus, and the phase relationship between the two sinusoidal curves.

Figure 5 shows the time course of changes in gain and phase measured in the dark throughout a set of long-term human experiments. The upper curve shows normalized gain, 1.0 representing the mean control response (absolute value approximately 0.65 gain at 0.17 Hz obtained during the 2 days preceding commencement of vision reversal). The lower curve shows the phase lag of the ocular response (CEP) relative to ideal compensation.

Clearly, a dramatic change of internal parameters controlling gain and phase of the VOR occurred very rapidly after commencement of maintained vision reversal. Initial changes could always be detected after the first few minutes, with 50% attenuation occurring over the first 6 hr (Gonshor and Melvill Jones, 1976b) of the first day. During the first 3 to 4 days of vision reversal, response gain was reduced by approximately 80% to values that often made interpretation of original records of eye movement difficult or impossible. Subsequently, VOR gain began to rise once more towards a plateau value of 0.45 times that of the original control, with the striking additional feature that over this period there were large changes of phase which eventually approached a plateau value of about 125° phase lag relative to normal compensation. Additional important features in the results of Fig. 5 are the following. (1) After return to normal, nonreversed vision, the gain fell to an all-time low within the first hour and subsequently recovered gradually over the next 2 to 3 weeks. (2) Phase lag, which had developed progressively over a period of 2 weeks after the first wearing of the prisms, disappeared within the first few hours after prism removal, followed by a small overshoot and subsequent return to normal phase over the next few days. Although highly significant in their own right (Davies and Melvill Jones, 1976), these latter two features will not be further considered in the present chapter.

High-frequency tests.

At this point, recall that the results of Fig. 5 were obtained from tests performed in the dark at a low rotational frequency of 0.17 Hz, whereas the rationale for formulation of the present experimental approach was

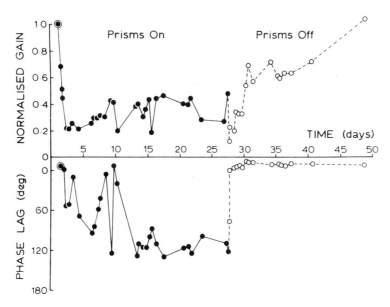

FIG. 5. Change in VOR gain associated with prism reversed vision as tested at low frequency of head oscillation. Normalized gain *(upper curve)* and phase *(lower curve)* of the VOR obtained by the method of Fig. 4 from a human subject during (●) and after (○) wearing reversing prism goggles continuously for 27 days. ◉, control values. Phase is recorded relative to that of perfect compensation with normal vision. The test stimulus was 60°/sec velocity and 0.17 Hz. Each pair of points (gain and phase) was derived from 10 to 20 cycles of sinusoidal stimulation in the dark. (From Melvill Jones, 1977, with permission.)

based on the premise that it is during rapid head movements that the system is denied useful corrective influences originating with immediate visual feedback. Therefore, it becomes important to learn whether parametric changes such as those reflected in Fig. 5 are also evident at high frequencies of head rotation in the prism-adapted subject.

To this effect, tests were also performed at 3 Hz (peak amplitude 60°/sec) head oscillations, since this frequency is too high for effective visual tracking (Eckmiller and Mackeben, 1978; Fuchs, 1967; Melvill Jones and Drazin, 1962; Stark et al., 1962). It should be noted that for reasons detailed elsewhere (Melvill Jones and Gonshor, 1981), these high-frequency tests employed active head rotation relative to the body with eyes open in the light and prisms temporarily removed, anticipating the likely complication of anticompensatory eye movements during quick voluntary head movements in the dark (Barnes, 1979; Melvill Jones, 1964) and the presumed ineffectiveness of vision at this frequency. Of course, this procedure introduces the potential complication of neck afferent influence on oculomotor control, since the head then turns relative to the body. However, from recent experiments, it seems unlikely that this influence would be marked in the present experimental paradigm (Barnes and Forbat, 1979), even though more drastic (surgical) measures (notably bilateral labyrinthectomy) have been shown eventually to induce augmentation of effective neck afferents (Dichgans et al., 1973).

Figure 6 illustrates the time course of changes in ocular gain obtained in this way at both 3.0 Hz and 1.75 Hz head oscillation. Similar to the low-frequency data of Fig. 5, gain was progressively attenuated after donning the reversing prisms (first arrow on abscissa). After prism removal (second ar-

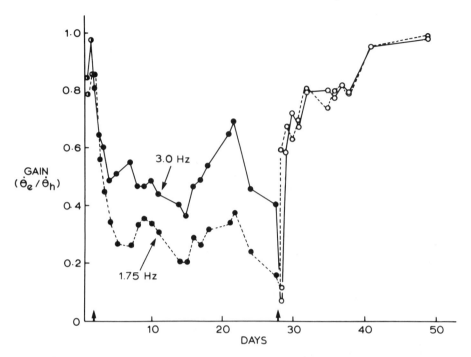

FIG. 6. Change in VOR gain associated with prism-reversed vision as tested at high frequencies of head oscillation. Absolute gain ($\dot{\theta}_e/\dot{\theta}_h$) of responses to voluntary head oscillation in the light and with prisms temporarily removed at 3.0 Hz and 1.75 Hz from the same subject and over the same duration as in Fig. 5. Arrows on the abscissa indicate the times of donning and removal of the prism goggles. ◐, control gains. Mean phase (not shown) was +12° at 3.0 Hz and +30° at 1.75. (From Melvill Jones and Gonshor, 1981, with permission.)

row), there was first a further decrease of gain and then a prolonged period of gradual gain recovery, with the whole process following a time course similar to the one illustrated in Fig. 5. Interestingly, despite these substantial changes of gain, there was never any significant change of phase associated with the HF tests conducted at 3 Hz, although there were marked phase changes at intermediate frequencies (Melvill Jones and Gonshor, 1981).

In the present context, the importance of results such as those in Figs. 5 and 6 rests on the clear demonstration that changing the external relationship between the visual and vestibular stimuli did indeed gradually but substantially change primary parameters of internal reflex pathways linking sensory to motor components of the oculomotor system.

The finding that internal parameters controlling the VOR are adaptively modifiable is not restricted to human subjects and has been described in several other species subjected to a variety of different patterns of optically or mechanically modified vision (see references cited in the introduction). Thus, substantial parametric adjustments can apparently be induced universally in mammalian species, although the precise nature of such changes depends on both the species and the optical devices employed.

SYMBIOSIS OF VOR AND OPTOKINETIC RESPONSE

Since at high frequencies the VOR acts as an open-ended reflex, it is easy to understand the adaptive rationale for parametric changes such as those demonstrated at high frequency as in Fig. 6: the changes always act to reduce retinal image slippage during head movement in the vision-reversed state, even in the absence of effective visual feedback. Indeed, it transpires that in the long-term adapted cat, the changes incurred (acting together with other associated phenomena) produce good retinal image fixation during whole-body rotation with reversing optics in place (Davies and Melvill Jones, 1977). However, as demonstrated by Fig. 5, corresponding parametric changes are also manifest at frequencies of head oscillation sufficiently low to enable closed-loop visual feedback to contribute to compensatory eye movements along the lines depicted in Fig. 2A. In view of this, one may well ask: What function could the VOR be performing at these low frequencies, especially since the normal dark-tested VOR gain is known to be well below unity?

To answer this question, it is essential to note two important features, the functional significance of which has recently been emphasized to good effect by Robinson (1977). First, in normal life, the VOR is generally only activated within a functionally useful context when the visual feedback loop has been closed by having the eyes open in the light. This is unlike the open-loop condition in the dark environment usually employed for experimental testing of the VOR. Second, in normal circumstances, the visual image on the retina is comprised of a wide-field view of a stationary outside world, which is unlike the usual optokinetic test stimulus in which a patterned drum rotates relative to a stationary head in the absence of concurrent vestibular stimulation. It is only when these two features are borne in mind that a sensible interpretation can be made of the apparently incongruous results obtained when either system is tested alone by conventional methods.

Consider first the patterns of neural and oculomotor response to the angular velocity profile of rotational vestibular stimulation in the dark, illustrated in Fig. 7A. The stimulus commences with an angular acceleration up to a steady angular velocity, which is continued for about 50 sec and then decelerated to a stationary condition. After a further 50 sec, the sequence is reversed. During this procedure, compensatory ocular nystagmus is generated by the VOR, with maximum initial slow-phase angular velocity established at the moment of reaching maximum head velocity (reflected in the intensity of nystagmus seen in Fig. 7A). This intense oculomotor activity is paralleled by the corresponding neural activity recorded at the level of the VN. Thereafter, both compensatory eye velocity and neural activity decay exponentially towards the initial condition. On arrest of rotation, the direction of nystagmus reverses, as does the neural signal, because of reversal of the mechanical canal response (Wilson and Melvill Jones, 1979). These reversed postrotational responses again decay exponentially toward base-line conditions.

Next, consider the corresponding responses to wide-field optokinetic stimulation (Fig. 7B) with head fixed. The stimulus trace indicates the commencement and duration of the optokinetic drum stimulus, with the dashed horizontal lines indicating periods of darkness. Typically, with a stimulus such as this, the slow-phase eye velocity commences at a level too slow for accurate following of the rotating drum. Subsequently, eye velocity progressively increases towards a plateau value along a quasiexponential time course. Then, on removal of the stimulus by switching off the light (first arrow), the response continues in the same direction as before, decaying "exponentially" towards initial conditions as a "stored velocity" signal runs out its time course. The progressive augmentation of the optokinetic response (OKR) during steady optokinetic stimulation (optokinetic nystagmus, OKN, reflected in the increasing intensity of nystagmus in Fig. 7B) followed by its exponential decline in darkness

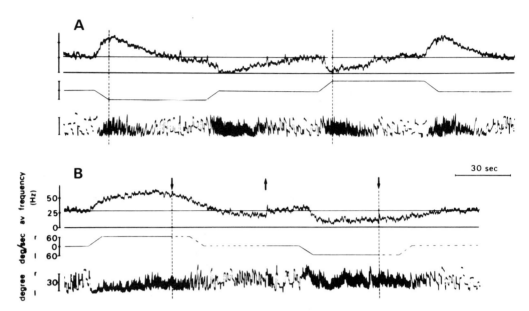

FIG. 7. Patterns of monkey response to (A) vestibular rotational stimulation in the dark and (B) full-field optokinetic stimulation with head still in the light. In A and B, top trace represents running average of single-cell activity in vestibular nucleus; bottom trace ocular nystagmus. Middle trace in A represents table (head) velocity, and in B optokinetic drum velocity. *Downward arrows* indicate light off, *upward arrow* light on. (From Waespe and Henn, 1979, with permission.)

(optokinetic after nystagmus, OKAN) are apparently common to all mammals although manifest in different degrees from species to species (Cohen et al., 1977; Collewijn, 1969; Komatsuzaki et al., 1969; Raphan et al., 1979).

At first sight, the response patterns shown in Figs. 7A and B would both seem to be quite inappropriate for the evolutionarily significant task of stabilizing the retinal image during slow-phase sweeps of ocular nystagmus. Thus, the moment of the most effective vestibular response is restricted to the commencement of head rotation. Conversely, on initiation of optokinetic stimulation, the early part of the response would be associated with substantial image slip on the retina, whereas the maintained response in darkness (OKAN) would seem to serve no sensible purpose at all. A clue to the interpretation of these apparently anomalous individual patterns of oculomotor behavior is contained in the neural records of Figs. 7A and B which were obtained from the same neuron in the VN. Since such neural signals tend to be well correlated with corresponding oculomotor activity (Dichgans et al., 1972; Henn et al., 1974; Waespe and Henn, 1977, 1979), this would suggest that pathways common to both VOR and OKR pass through VN.

The question arises: What is the functional significance of this arrangement? A reasonable answer might be that over the broad sweep of evolutionary experience, vestibular and wide-field optokinetic stimuli must have been encountered simultaneously as complementary inputs describing angular movement of the head relative to the stationary world. Seen in this light, it would therefore seem entirely appropriate that the two subsystems should be closely integrated with one another in the formulation of a common functionally meaningful (oculomotor) response.

However, the precise nature of the interaction at the level of the VN is not intuitively obvious. As indicated by Robinson (1977), simple superposition of the separately determined neural responses, such as those shown in Figs. 7A and B, does not apply here, because the open-loop system in the dark (vestibular testing) is different from the closed-loop system in the light (optokinetic testing). Furthermore, there is the complicating influence of an internally closed loop feeding a component of oculomotor output back to VN via the "optokinetic" (VIS–OK) pathway represented schematically in Fig. 13. As a result, the combined vestibular and optokinetic response shown by the RL curve in Fig. 8A is certainly not the outcome of simple superposition of the vestibular (RD; cf. Fig. 7A top curve) and optokinetic (OK; cf. Fig. 7B top curve) curves.

In order to arrive at a plausible interpretation of the neural convergence observed in VN, it is necessary to consider the integrated closed-loop nature of the combined vestibulovisual system outlined in Fig. 13. It is quite clear from this figure that the experimentally determined VN response (RD, Fig. 8A) is the outcome of an interaction between the primary afferent vestibular signal (H_c) and its processed version fed back on itself via the optokinetic feedback circuit (VIS–OK; the latter feedback pathway is inferred from the fact that purely oculomotor signals can be recorded in VN, as, for instance, during periods of OKAN in Fig. 7B). Consequently, it is H_c, rather than RD, that constitutes one of the two elements summated at VN. A simulated version of H_c is shown as one of the dashed curves in Fig. 8B. Similarly, a simulated estimate of the processed signal (H_v) ultimately fed back into VN during combined vestibular and visual stimulation (Fig. 13) is shown by the second dashed curve in Fig. 8B. Note that the theoretically determind H_v differs from the experimentally determined pure optokinetic response (OK) because of the feedback influence of the concurrent H_c input.

The functional advantage of this linear interaction in the brainstem becomes obvious from a comparison of the inadequate individual vestibular (first-order lead; H_c) and optokinetic (first-order lag, H_v) response

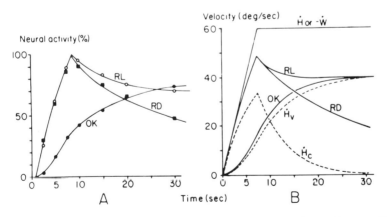

FIG. 8. A: Averaged time course of the discharge rates of six neurons in monkey vestibular nucleus (VN) responding to head rotation in the dark (RD), to optokinetic stimulation with head fixed (OK), and to head rotation in the light (RL). Note that RD and OK do not add up to RL. B: Computer simulation of head movement in the light. $\dot{H}_c$ (cupula) and $\dot{H}_v$ (visual–optokinetic) components, contributing by linear addition at VN to the overall RL curve, are shown as the two *dashed curves*. $\dot{H}$: Vestibular stimulus (head velocity). $-\dot{W}$: Optokinetic stimulus (drum velocity). See text and Fig. 13. (A: From Waespe and Henn, 1977, with permission. B: From Robinson, 1977, with permission.)

curves, with the required wide-band frequency response reflected in the stimulus profile (H or $-W$, Fig. 8B). The combined response curve RL presents a reasonable replica of the original stimulus wave form and thus approaches the ideal oculomotor dynamics required for adequate retinal image stabilization over a wide range of head frequencies.

It should be stressed that the subject of voluntary visual tracking, which is primarily concerned with the foveal pursuit of moving targets rather than reflex ocular stabilization, has been excluded from the present discussion, since pursuit probably enters the system downstream relative to the point of symbiotic VOR–OKR interaction (Lisberger et al., 1981). For further information on this subject, the reader is referred to two contemporary reviews (Henn et al., 1980; Miles and Lisberger, 1981).

ADAPTIVE CHANGES IN THE OKR

The "symbiotic" relationship between vestibular and optokinetic signals, described in the previous section, suggests a sharing of pathways between these two neural channels. One might expect, therefore, a modification in the operating parameters of one of the two subsystems to be reflected in the operating characteristics of the other. Thus, one could ask specifically how a reduction in the gain of, say, the VOR might affect the gain of optokinetic pursuit?

To investigate this question, a normal cat was exposed to wide-field optokinetic monocular stimulation in the absence of a vestibular stimulus (Melvill Jones et al., 1980). The alert behaving animal was trained, as in corresponding vestibular experiments (Melvill Jones and Davies, 1976), to accept head fixation on a servo-driven turntable after implantation of an acrylic fixation structure bolted to the skull in a conventional manner. A search coil (Fuchs and Robinson, 1966) was implanted on the right eye for measurement of horizontal eye movement. Rather than using a concentric drum for optokinetic stimulation, the animal was rotationally accelerated in the dark at below vestibular threshold to a steady velocity of 20°/sec.

To insure complete absence of vestibular stimulation, the 20°/sec rotation was continued for a further 2 min in complete darkness. The room lights were then switched on for 50 sec so that the animal received 20°/sec wide-field optokinetic stimulation from a normal visual environment. This method has the advantage of providing an optokinetic stimulus comprising the natural visual scene, without the complicating factor of dominant spatial frequencies of the kind presented by, for example, a striped optokinetic drum. This was considered particularly important in view of further experiments conducted in stroboscopic light, where undesirable interactions would almost inevitably occur between temporal (flashes) and spatial (stripes) frequencies.

After 50 sec of such optokinetic stimulation, the room lights were extinguished, and eye movement recording continued for a further 50 sec while the turntable continued rotating at constant velocity. Finally, the animal was decelerated in the dark, again at less than vestibular threshold, to a stationary condition and left there for at least a further 3 min to insure complete elimination of any undetected residual vestibular signal.

In accordance with the response patterns shown for monkey in Fig. 7B, eye velocity began at a level well below that of the stimulus and climbed progressively to a plateau value that, in the normal animal, reached an average of 16°/sec, i.e., well below the 20°/sec stimulus velocity. When the light was extinguished, a pattern of OKAN developed with an initially high eye velocity which subsequently declined "exponentially" towards zero. After a series of experiments on the normal cat in normal light to obtain control data, the animal was then fitted with reversing dove prisms (horizontal) as detailed elsewhere (Davies and Melvill Jones, 1976; Melvill Jones, 1977) and diagramed in Fig. 3. Between 5 and 10 days after donning the prisms the animal's VOR was estimated by sinusoidal rotation in the dark as in Fig. 4. As reported elsewhere (Mandl et al., 1981), VOR gain was then on average attenuated by about 70%. In this VOR-adapted state, the animal (with prisms temporarily removed) was exposed to the same form of vestibular-free optokinetic stimulation as described above for control conditions. Under these circumstances, the monocular OKR was found to be substantially attenuated, although the percentage attenuation was only about one-third that of the VOR. Subsequent VOR recovery resulted in simultaneous recovery of the OKR (Melvill Jones et al., 1981).

The above observations support the notion of pathways shared between VOR and OKR. Within this context, it is important to realize that the optokinetic testing was necessarily done in the presence of active visual feedback, leading to closure of a feedback loop (Fig. 2A), which would be expected to mask the full extent of the internal parametric changes and hence could result in the apparently smaller range of optokinetic gain attenuation. In addition, the matter is not without somewhat puzzling conflicts in the literature. Thus, Keller and Precht (1981) did not observe OKN attenuation in their VOR-attenuated cats, whereas Lisberger et al. (1981), working with monkeys, obtained parallel OKR and VOR changes similar to those described above. However, it should be noted that different methods were used to produce VOR attenuation in the different sets of experiments. For example, the experiments of Keller and Precht employed short-term adaptive forcing (4–5 hr), whereas both of the experiments described above and those of Lisberger et al. (1981) employed long-term (>1 week) adaptation. Since it has been shown that VOR adaptation exhibits two distinct time constants (rapid, $\tau \cong 10$ hrs; slow, $\tau \cong 1$ week; Mandl et al., 1981), it is possible that two distinct neural processes are involved, with the long-term one being predominantly responsible for OKR attenuation.

Are these experimentally observed changes in optokinetic performance simply an incidental consequence of changes in the VOR gain, or do they in themselves constitute an adaptive response of the optokinetic subsystem to optical manipulation of the visual input?

Lisberger et al. (1981) concluded from their experiments with monkeys that any form of optical modification of vision should always call for an enhanced optokinetic gain, since this would always tend to increase retinal image stability. According to this view, optokinetic gain reduction coincident with VOR gain attenuation would be construed as maladaptive. This view finds some support from the experimental findings of Collewijn and Grootendorst (1979) who found that in rabbits the integrated visuovestibular response always tended toward a reduction of retinal image slip; i.e., optical manipulation of the visual input always caused an increase in the gain of the optokinetic response regardless of whether the adaptation process called for an increase or a decrease in the VOR gain.

Alternatively, one might argue that a reduction in the (first-order lead) VOR component of the combined oculomotor response to head rotation should be accompanied by a simultaneous reduction in the (first-order lag) optokinetic component if the adequate frequency response of the oculomotor output is to be retained. Following this latter view, the observed parallel adjustments of VOR and OKN might fall into the category of adaptive adjustments.

Clearly, the definition of an adaptive response within the above context is fraught with semantic pitfalls and awaits further clarification from insightful experimental and simulation studies.

SEARCH FOR THE "ERROR" SIGNAL: EXPERIMENTS WITH STROBE ILLUMINATION

Thus far it has been argued that normal sensory–motor integration may be driven by goal-seeking parametric modifications within central neural pathways. It would therefore seem relevant to ascertain the precise nature of the guiding sensory "error" stimulus responsible for molding the observed plastic changes of behavior. Since it can be argued that most of the demonstrated adaptive phenomena occurred in a way that tended to improve retinal image stabilization, it appears likely that the error signal is represented by the vector of smooth retinal image slip. Thus, it would seem important to establish the extent to which such image slip in fact may have been responsible for the observed modifications of VOR and OKR gain in response to vision reversal.

To this end, human subjects wearing reversing prisms were exposed to stroboscopic illumination (4Hz) of sufficiently short flash duration (5 μsec) to prevent any significant smooth retinal image slip during eye movements (Melvill Jones and Mandl, 1979). During a 6- to 8-hr adaptation period, vision-reversed subjects moved freely about a strobe-illuminated lab environment. Subsequent tests of VOR gain, performed in the dark at ⅙ Hz and 60°/sec velocity amplitude, revealed consistent and significant decreases of VOR gain, indicating that the absence of smooth retinal image slip presented no serious impediment for the process of goal-seeking parametric readjustments in the vestibuloocular response system.

An additional quite surprising result emerged from these studies. Control subjects who were exposed to the strobe environment without vision reversal for 6 to 8 hr developed an augmented VOR gain of about 1.2, as tested in the dark. Presumably, this was a result of parametric adjustments in response to the extenuating task of maintaining adequate visual tracking performance during head movements in the dark periods (in this case 250 msec) between consecutive brief strobe flashes.

It is worth noting that additional findings from a related set of strobe studies (Mandl and Melvill Jones, 1979) tend to confirm the idea that environmental exigencies may call into play an oculomotor tracking capability driven by discrete, discontinuous retinal image displacements. Naive human subjects wearing reversing optics for the first time (i.e., unadapted to vision reversal) were rotated on a turntable alternately to the right and to the left at a uniform speed of 10°/sec in either direction. Initially, subjects were in complete darkness, and their eye movement records showed a pattern of normal vestibularly driven compensatory eye movements. Subsequently, at a time unknown to the subject, strobe illumination at 2 Hz was switched on. It required no more than two consecutive flashes to initiate and maintain "reversed" (i.e., opposed to the prevailing vestibular stimulus) oculomotor tracking during the 500-msec dark periods between flashes. Here then, a tracking and short-term velocity storage capability seemed at work, making it possible to generate (visually) appropriate compensatory oculomotor activity, even in a direction opposite to that imposed by the concurrent vestibular drive, and to maintain it even during the relatively prolonged dark interflash intervals. Possibly this capability for short-term prediction in strobe light may be akin to that described by Eckmiller and Mackeben (1978) for monkeys in normal light.

Although discontinuous, rather than smooth, retinal image displacement seemed to have been the key element in shaping the adaptive sensory–motor adjustments to vision reversal described above, the requirement for directional information (i.e., the retinal image velocity vector) would still seem essential for the system's ability to move toward its optimum new performance goal. With this latter requirement in mind, another study was initiated, using strobe-reared cats as subjects. In these animals, a marked deficiency of direction-sensitive visual cells in both cortex and superior colliculus has been demonstrated (Cynader and Chernenko, 1976; Flandrin et al., 1976). Would such animals, whose capability for detecting the direction of visual motion might be assumed to be seriously impaired, be deficient in their adaptation to vision reversal? Three related sets of observations were made in a series of experiments with such developmentally deprived animals.

First, strobe-reared cats were shown to possess both a normally functioning VOR and a vigorous optokinetic response capability (Mandl et al., 1981). There also tended to be retention of a "normal" symbiotic relationship between VOR and OKR, in that exposure to vision reversal (in both normal and strobe light) led to simultaneous attenuation of VOR, OKN, and OKAN (and subsequent return to normal levels after return to nonreversed vision), a situation quite similar to that documented in the normally raised animal (Melvill Jones et al., 1981). Thus, it appears that directionally sensitive cells in the geniculocortical and/or collicular visual systems do not normally play a significant role in mediating directional retinal error information associated with the process of parametric oculomotor readjustments.

Second, the initial phase of the (unadapted) optokinetic response, which in normally reared animals consists of a gradual buildup of pursuit toward stimulus velocity, rose in the strobe-reared animals unusually rapidly toward the final and maintained plateau of pursuit velocity, even though the tests were performed in 8 Hz strobe light in which the environment (because of the brevity of individual flashes) was completely dark for more than 99% of the total testing time (Mandl and Melvill Jones, 1981). The fact that appropriate oculomotor pursuit occurred through almost the entire 50-sec response period in the strobe light—and particularly the initial rapid attainment of virtually full stimulus velocity—again attests to the emergence of some form of quick-acting central rate determination and storage capability brought into operation by intermittent sampling of changes in retinal image position. Since this optokinetic performance was much more adequate than that of the normal cat in strobe light, it would appear that some form of favorable developmental adaptation to the abnormal rearing environment had taken place in the strobe-reared animals.

Interestingly, this enhanced optokinetic capability of the strobe-reared animals was particularly susceptible to suppression by prism reversal, especially with regard to the initial rapid rise of eye velocity toward stimulus velocity. It is important to emphasize that the enhancement of response referred to here was observed using an 8 Hz rearing and testing strobe frequency and was most characteristically expressed during the first 5 sec after introduction of the optokinetic stimulus. Within this context, it would be interesting to know whether the comparably much more depressed optokinetic performance recently reported by Amblard et al. (1981) for their strobe-reared cats was caused by their animals being tested in normal light or, for that matter, by the relatively low 2 Hz stroboscopic rearing frequency used by the latter authors.

The implications of the enhanced optokinetic following in the strobe-reared animals and its loss in the prism-adapted state will be discussed in terms of the operation of potential central mechanisms in a later section. Before we do so, attention should be drawn to a third relevant characteristic of strobe-reared animals, namely, the presence of spontaneous eye oscillations occurring in the dark at a frequency close to that of the strobe light in which the animals were reared (Melvill Jones et al., 1981). Figure 9 reproduces sample records of such oscillations obtained from two strobe-reared animals. These oscillations are of small magnitude ($<1°$) and close to the rearing frequency of 8 Hz. Their numerical characteristics were determined by power spectral analysis but are not relevant to the present theme.

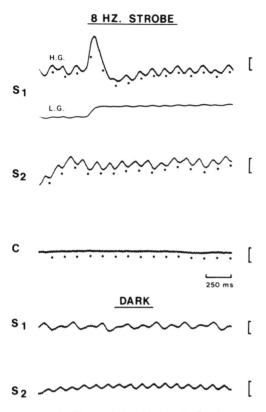

FIG. 9. Oscillatory instability in the oculomotor system of two cats reared from birth in 8 Hz stroboscopic light. S_1,S_2: Two strobe-reared animals. C: Normally reared control animal. L.G., low-gain DC-coupled amplification. All other traces high gain (H.G.) amplification, with high-pass filtering (0.1 Hz cut-off) to retain the trace on the recording paper. Eye calibrations 1° for H.G. traces, 7.4° for L.G. trace. All records from right eye. Down indicates nasal displacement throughout. Note that both strobe-reared animals produced miniature eye oscillations which, in strobe light, were strictly synchronized with the strobe flashes (*dots*) and, in the dark, retained fundamental frequencies close to that of the strobe flashes experienced during rearing. (From Melvill Jones et al., 1981, with permission.)

More important is the effect of adaptation to vision reversal on these spontaneous oscillatory eye movements in the dark.

Note first in the top records of Fig. 10 (Unadapted Dark) two peaks of energy at 4 Hz and 8 Hz, respectively, reflecting the bimodal oscillatory pattern found in this strobe-reared, prism-unadapted animal (see also S_1 DARK trace in Fig. 9). The animal was then exposed in the usual way to maintained vision reversal, after which the oscillations were no longer present (middle records, Adapted Dark). However, as seen in the bottom trace of Fig. 10, intermittent flashes delivered at infrequent intervals did produce transient eye movements with similar oscillatory characteristics to those of the unadapted condition. Apparently, the oscillatory condition was still potentially present, although covert, and could be excited transiently by each individual flashed "stimulus"; this never occurred in the normally reared control animal.

In the present context, the interesting feature is that some aspect of the prism-adapting process has been brought to bear on the central neural network responsible for generating the abnormal oscillatory eye movements of the strobe-reared animal. It is clear that the oscillatory eye movements, having been established as a result of parametric changes in the strobe-reared CNS, are particularly susceptible to adaptive suppression by prism reversal. The fact that such suppression occurred simultaneously with suppression of VOR and OKR perhaps implies that common elements in the CNS engage in all three phenomena (Mandl and Melvill Jones, 1981).

In summary, results from experiments with strobe light indicate that neither smooth retinal image slip nor the functional integrity of direction-sensitive visual cells in cortex and/or superior colliculus seem crucial for driving parametric adjustments in response to vision reversal. In addition, strobe rearing introduces oculomotor abnormalities that may reflect important properties of some of the central neural pathways implicated in both VOR and OKR operation.

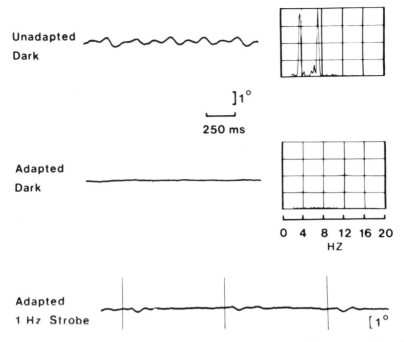

FIG. 10. Effects on oscillatory eye movements of VOR adaptation (60% attenuated) to vision reversal. Power spectra of spontaneous eye movements in the dark obtained from the unadapted and adapted animal (S_1, see Fig. 9) are shown on the right. Note disappearance of spontaneous oscillations after 5 days of adaptation to vision reversal. Bottom trace shows residual oscillatory propensity in response to transient 1 Hz strobe flashes (vertical lines). Eye calibration, 1°. (From Melvill Jones et al., 1981, with permission.)

MOTION SICKNESS: COROLLARY TO ADAPTATION?

As previously reported (Gonshor and Melvill Jones, 1980), the experience of wearing distorting optics (such as reversing prisms) is normally accompanied in human subjects by severe subjective discomfort and nausea. The time course of onset and gradual amelioration of these motion sickness-like symptoms closely parallel the time course of adaptive changes in the visual vestibular–oculomotor system. It was, therefore, of particular interest to discover that human subjects undergoing VOR adaptation to vision reversal in strobe light never experienced nausea or other associated autonomic symptoms. A number of additional observations regarding the nature of adaptive VOR changes in a stroboscopically-illuminated environment have suggested that the noted absence of motion sickness may be related to the decreased rate, and highly selective frequency range, of VOR adaptation under these adverse visual conditions (Melvill Jones and Mandl, 1981).

Thus, although marked adaptation in normal light always occurred over a broad range of frequencies of head oscillation (Gonshor and Melvill Jones, 1976b; Melvill Jones and Gonshor, 1981), equivalent adaptation in 4 Hz stroboscopic illumination was restricted to low head frequencies. The restricted frequency dependence of the adaptive effect seen in strobe light is illustrated in Table 1, which compares the average adaptive attenuation of VOR gain in human subjects examined in both normal light (top line) and strobe light (bottom line). It can be seen that, whereas low-frequency testing

TABLE 1. *Percent decrease in VOR gain as tested at three different frequencies of head oscillation in human subjects after about 6 hr of vision reversal in both normal and 4 Hz strobe light[a]*

	Head frequency		
	0.17 Hz	1.75 Hz	3.0 Hz
Normal light	42% (4)	27% (3)	24% (3)
Strobe light	25% (7)	None (4)	None (4)

[a] At higher head frequencies, there was no evidence for any adaptive effect in strobe light. Numbers in parentheses indicate the number of subjects.

(0.17 Hz) revealed substantial adaptive changes in both data sets, high-frequency testing (1.75 Hz and 3.0 Hz) showed marked gain attenuation in normal light but none in the strobe light. Furthermore, even at the low test frequency of 0.17 Hz (first column, Table 1) VOR gain attenuation resulting from vision reversal in strobe light was less marked than that achieved in normal illumination (Melvill Jones and Mandl, 1979).

As detailed elsewhere (Melvill Jones and Mandl, 1981), the absence of an adaptive effect within the higher frequency range might be ascribed to the fact that, with prisms in place, stroboscopic presentation of the visual world during rapid head movement would invariably be accompanied by relatively large discontinuous shifts of the retinal image with concomitant spatiotemporal pattern dispersion devoid of any meaningful velocity information. In other words, the retinal "error" signal under these environmental conditions is jumbled, with the result that the goal-seeking rematching process lacks the appropriate adaptive drive. If it be assumed that motion sickness and allied autonomic symptoms are a corollary to the nervous system's effort in resolving the conflict inherent in the opposing visual and vestibular drives, the absence of nausea in the strobe experiments might be related to a reduced rate and range of adaptation. Perhaps stroboscopic manipulation of the visual input may offer a convenient method for reducing, or even eliminating, the unpleasant and potentially hazardous autonomic side effects accompanying some aspects of this goal-seeking process of sensory–motor reintegration.

CENTRAL MECHANISMS

It has been shown that the simple procedure of optical modification of vision induces an extensive range of changed behavior which includes attenuation of oculomotor response to both low- and high-frequency head oscillation, attenuation of OKR manifest as changes in both OKN and OKAN, attenuation of the enhanced initial OKN found in strobe-reared cats, and suppression of the spontaneous eye oscillations seen in these latter animals. What kind of central processes might be responsible for these combined changes? The matter is highly controversial at the time of writing, and the following comments must be considered correspondingly speculative.

Historically, the above-described adaptive attenuation of VOR caused by modified vision has been attributed to a cerebellar mechanism schematized in its simplest form in Fig. 11 (Ito, 1970). Ito's scheme followed a theoretical proposal of Marr (1969), which in turn had evolved from earlier ideas generated by Brindley (1964), and which was later developed into a somewhat modified theory by Albus (1971). It proposed that climbing fiber (cf) projections to cerebellar cortex serve the physiological role of producing synaptic modification of parameters controlling parallel fiber inputs to Purkinje cell dendrites. The relevance of this proposal to the adaptive response in VOR rests on the fact that vestibular afferent fibers project via mossy fibers (mf), granule cells, and parallel fibers to Purkinje cells in the cortex of the flocculus lobe of the cerebellum and that these in turn project their output back to the vestibular nuclei. Since direct vestibular projections to VN are al-

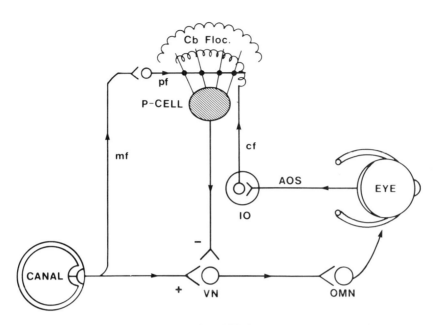

FIG. 11. Elementary diagram to illustrate Ito's (1970) hypothesis of cerebellar floccular lobe (Cb Floc.) participation in parameter control of the VOR. Parallel excitatory and inhibitory "canal" pathways reach VN via direct and transcerebellar (mossy fiber, mf) pathways, respectively, and thence go on to extraocular motor nuclei (OMN). Optic afferents also project to Cb Floc. via the accessory optic system (AOS) and climbing fibers (cf) originating in the inferior olive (IO). The latter projection ("teaching line" of Ito, 1970) is presumed to produce modification of synaptic efficacy at the level of parallel fiber (pf)–Purkinje cell (P-cell) connections appropriate for reduction of retinal image slip (error) signal in the AOS.

ways excitatory, whereas the indirect Purkinje cell projections are always inhibitory, this arrangement allows for the possibility of bringing simultaneous excitatory and inhibitory vestibular influences to bear on the same common neuronal pool in VN (for comprehensive review, see Wilson and Melvill Jones, 1979). Consequently, appropriate changes of neural parameters (e.g., gain) in the two opposing pathways could theoretically bring about not only attenuation of the VOR (for example, by enhancement of inhibitory and/or suppression of excitatory inputs) but even its reversal, by arranging for the inhibitory transcerebellar pathway to predominate over the excitatory one (Davies and Melvill Jones, 1976).

The next important question is: How could these parametric changes be brought about? According to the above scheme, this could be achieved by means of a suitable cf input to these same Purkinje cells, producing appropriate synaptic modification in the transcerebellar (mf) pathway. To be functionally effective, the cf projections should of course carry an appropriate message of the error that has to be corrected. What kind of afferent cf signal would be appropriate? As mentioned earlier, the error signal could well be the vector of retinal image slip velocity, since it is this that should be minimized for clear vision.

Arguing along these lines, Ito and his colleagues initiated a series of experiments to search for a neural connection between optic nerve and cerebellar cortical Purkinje cells in the flocculus (Maekawa and Simpson, 1973; Maekawa and Takeda, 1976; Simpson and Alley, 1974). The outcome of these insightful experiments was the clear identification of a visual pathway containing directionally sensitive cells and running from retina through optic nerve into the accessory optic system (AOS) and thence to the vestibular cerebellar cortex via inferior olive and climbing fibers (Collewijn, 1975; Hoffman et al., 1976; Ito et al., 1977; Simpson et al., 1979). Thus, a proposal emerged whereby any degree of retinal image displacement incurred during head rotation in the light would bring about organized modification of synaptic transmission at the floccular Purkinje cell level by means of changes in synaptic efficacy caused by climbing fiber action via the accessory optic system.

More recently, the matter has been somewhat advanced by the discovery of another important informational pathway projecting to floccular Purkinje cells (Lisberger and Fuchs, 1974, 1978; Miles and Fuller, 1975; Miles et al., 1980). Thus, in the absence of vestibular stimulation, eye movements are accompanied by substantial modulation of Purkinje cell simple spike activity, which appears in monkey to be predominantly related to instantaneous eye angular velocity relative to the skull (Lisberger and Fuchs, 1978). The argument was adduced that since this latter (mossy fiber–simple spike) input to floccular lobe cortex mainly replicates eye angular velocity, and since its phase is such as to make the inhibitory P-cell output supportive of the ongoing eye movement, there would appear to be closure of a positive feedback loop as in Fig. 12 (the -1 box in the direct pathway must be introduced to take into account the fact that normal compensatory eye movements occur in a direction opposite to that of the head). It was therefore proposed that this loop serves primarily to sustain the prevailing eye angular velocity, independently of the manner in which it was initiated (Lisberger and Fuchs, 1978).

In practice, of course, this circuit must be added to those in Fig. 11 to generate a more complex informational system than was at

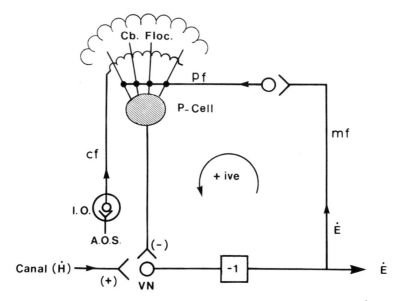

FIG. 12. Oculomotor feedback loop projecting a copy of the neural motor signal ($\dot{E}$) via mossy fibers (mf) to reach P-cells in Cb.Floc. via parallel fibres (pf). Connectivity is such as to establish a positive feedback loop suitable for sustaining ongoing eye movement. The cf projection of optic influence via AOS (see Fig. 11) is also shown. The -1 box in the forward path is necessary to account for the fact that normal compensatory eye movement occurs in the reverse direction to that of head movement. VN, vestibular nucleus.

first envisaged. In particular, the two inputs to floccular Purkinje cells diagramed in Fig. 13 meet in such a manner that, when the normal VOR produces a properly compensatory eye movement, the vestibular and oculomotor mossy fiber inputs arrive with opposite phase at the Purkinje cell and hence cancel one another at that level (Lisberger and Fuchs, 1978; Miles and Fuller, 1974; Miles et al., 1980). This then represents the normal state of affairs when the proper motor response is being produced during head rotation. If now, some systematic optical modification is introduced in the visual input, say telescopic diminution of the visual scene, then because of attempted visual fixation on the stationary but optically scaled-down external world, the oculomotor mf input to cerebellar flocculus (positive feedback loop) will become less active relative to the vestibular one. This in turn should produce reduced eye movement, notably without there having been any parameter changes in the neural networks concerned. However, if this proved unsatisfactory in the sense of being an incomplete adjustment, then there would be stimulation of the accessory optic system (cf) because of image slip, and this would serve to modify internal parameters in both vestibular and oculomotor mf projections passing via the Purkinje cells. Specifically, with diminishing optics, there should be enhancement of the vestibular projection and attenuation of the oculomotor projection. Conversely, magnifying optics would en-

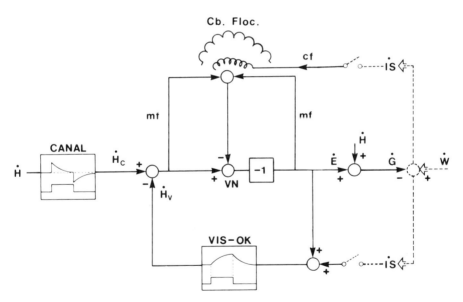

FIG. 13. Composite schematic diagram of elements discussed in the text. The canal response ($\dot{H}_c$) to head velocity ($\dot{H}$) exhibits the response characteristics indicated in box marked CANAL. Both mf projections of Figs. 11 and 12 are shown impinging on P-cell of Cb.Floc., as is the cf projection carrying the optic error signal (image slip velocity, $\dot{IS}$, derived by subtraction of gaze velocity, $\dot{G}$, from world velocity, $\dot{W}$). Also included, but not passing through Cb.Floc. (Keller and Precht, 1979b), is the visual–optokinetic influence with its dynamics (VIS–OK), comprising an efferent copy of eye velocity together with a retinal velocity error signal ($\dot{IS}$), which are eventually fed into VN through pathways that are at present little understood. H_c and H_v are the canal and visual–optokinetic components, respectively, described by computer simulation (Robinson, 1977) as contributing through linear addition to the neural signal at VN (see Fig. 8 and text). The *dashed lines* represent "external" informational pathways. The *switches* indicate presence (closed) or absence (open) of visual input.

hance the oculomotor signal and attenuate the vestibular one. It is important to recall that superimposed on this network would be that shown in the lower pathway of Fig. 13 (VIS–OK). As in Fig. 12, the -1 box is introduced because eye and head movements occur in opposite directions.

With a system such as this, apparently plausible explanations for all of the experimentally demonstrated parametric changes induced by vision reversal could be adduced. First, attenuation of the VOR could be induced as described in connection with Fig. 11, with the added feature that appropriate modifications would take place in the positive feedback loop detailed in Fig. 12. Since this process would lead to reduced output from VN, then OKR output should also be attenuated if the VIS–OK feedback signal projects into the network upstream of VN output (Fig. 13).

Next, it is conceivable that the special conditions normally associated with the strobe-reared animals, namely, enhanced optokinetic following and oscillatory instability in the oculomotor system, could be attributable to manipulation of gain in a closed-loop system akin to that of Fig. 12. Thus, in the original strobe-rearing environment, P-cell activity would be expected to be modulated (via the climbing fibers) by successive brief flash stimuli. The resulting periodic neural signal, once having entered the closed positive feedback loop of Fig. 12, would be fed back to arrive at the P-cell in a manner consistently correlated with the initiating input. Following the original suggestion by Marr (1969), the correlated synaptic activity could result in an increase of synaptic efficacy of the pf pathway forming part of the loop. The resulting increase in loop gain might then account for both the observed oscillatory instability and the enhanced initial optokinetic tracking response. Subsequently, putative changes within that pathway, producing both VOR and OKR attenuation, could theoretically also attenuate both the enhanced initial OKN and the oscillatory instability of the strobe-reared animal.

Clearly, these comments must be considered highly speculative and should be viewed in their proper historical perspective. Indeed, it has recently been reported that, at least in monkey, the adaptive changes in cerebellar flocculus inferred above do not in fact appear to occur in the anticipated manner (Miles et al., 1980). For example, magnifying spectacles were found on average to enhance (rather than attenuate) the transcerebellar vestibular mf signal, whereas reversing prisms tended to produce attenuation (rather than enhancement) in this pathway. Accordingly, Miles and Lisberger (1981) have proposed a "new hypothesis" to account for these intriguing observations, whereby the main modifiable element mediating adaptive gain control in the primate VOR and OKR systems is located in the brainstem pathway rather than floccular neural elements.

CONCLUSIONS

Clearly, the internal mechanism of adaptive plasticity represents a currently controversial issue. However, the main object of this chapter is not to localize specific neural pathways and modifiable elements in the CNS but rather to focus on the general significance of external factors in the shaping of reflex behavior. Thus, the simple device of optical modification of vision serves to induce profound changes in the VOR, in OKN and OKAN, and in the developmentally induced eye oscillations and enhanced OKN of strobe-reared animals. Furthermore, it now transpires that the adaptive influence reaches beyond this to the directional control of eye movement (Berthoz et al., 1981; Schultheis and Robinson, 1981), the reflex coordination of eye and head movement (Berthoz and Melvill Jones, 1981; Guitton and Douglas, 1981), and both visual and vestibular components of postural control (Gonshor and Melvill Jones, 1980).

Evidently, a cascade of integrated reorganization of internal parameters occurs throughout the whole visual–vestibular–oculomotor–postural system. Nor is such reorganization confined to this system alone. Recent studies have shown that quite separate systems, responsible, respectively, for saccadic eye movement (Optican and Robinson, 1980) and accommodation–vergence control (Judge and Miles, 1980), are similarly adaptable. Thus, the phenomenon of adaptive plasticity promises to prove widespread in its distribution throughout the CNS (Schaefer and Meyer, 1973; Llinas and Walton, 1979; Flohr et al., 1981).

From here it is but a short step to conclude that adaptive plasticity, far from being a special attribute confined to the developing organism, in reality represents a common property of the normal adult brain. What, then, might be its normal functional role? Rehabilitation of function after neuroanatomical lesions seems a likely contender, but, returning to the original introductory theme, it now seems certain that an important additional role is to serve the neurobionomic function of integrating sensory–motor activity with changing demands imposed by the external environment on the adult behaving organism. Finally, following the premise that no system in the body is universally free from pathological interference, could it be that some clinical neurological conditions, at present of uncertain etiology, may derive from pathological interference with those basic neural mechanisms that normally are responsible for this form of neurobionomic response to environmental pressures?

ACKNOWLEDGMENTS

The authors wish to thank Dr. J. S. Outerbridge of McGill's Biomedical Engineering Unit for the informative and stimulating discussions regarding various aspects of control theory, Ms. H. Meyer and Mr. M. Sweet for preparing the figures, and Ms. E. Wong for typing the manuscript. Supported by the Canadian Medical Research Council.

Axonal Sprouting and Recovery of Function After Brain Damage

*,**Nakaakira Tsukahara and †Fugio Murakami

**Department of Biophysical Engineering, Osaka University, Osaka, Japan; and †National Institute for Physiological Sciences, Myodaiji, Okasaki, Japan*

It has long been thought that nerve connections in the brain are rigid after their formation at an early developmental stage. However, although this rigidity of the nerve connections probably provides an important basis for some behaviors as the built-in generator of a fixed action pattern, it cannot readily account for various kinds of adaptive behaviors. In this context, it has recently been recognized that some of the nerve connections of the brain have plasticity. This plasticity is thought to be the neural basis for adaptive behaviors.

The most remarkable example of plasticity in the brain is the formation of new synaptic connections, a process called "sprouting" (cf. Edds, 1953), which is now recognized as a widespread phenomenon in the central nervous system. In this chapter, recent studies of sprouting will be dealt with.

In 1958, Liu and Chambers demonstrated light microscopically that dorsal root fibers can sprout in cat spinal cord in which all but one dorsal root or the descending pathways are cut. More recent electron microscopic work of Raisman (1969) and Raisman and Field (1973) in rat septum provided evidence of synaptic formation on septal neurons after interruption of either the fimbrial or medial forebrain bundle fibers. Subsequent works (Moore et al., 1971; Lynch et al., 1974, 1976; Cotman and Lynch, 1976; Bjorklund and Stenevi, 1979; Tsukahara, 1981; Cotman et al., 1981) extended the data base on the nature of axonal sprouting in the CNS.

MORPHOLOGICAL STUDIES

Sprouting During Early Developmental Stages

It is generally agreed that the degree and extent of sprouting are more marked after denervation at the neonatal stage. In some cases, sprouting occurs only after neonatal deafferentiation (Lund and Lund, 1971; Guillery, 1972). In the lateral geniculate body, sprouting of the remaining optic fibers after unilateral enucleation occurs only in kittens partially deafferented before 9 days after birth (Guillery, 1972; Robson et al., 1978; Hickey, 1975). In the superior colliculus, most of the retinotectal fibers terminate contralaterally, and a small portion of the uncrossed retinotectal fibers end in a small area of the ipsilateral superior colliculus. Unilateral lesion of the retinotectal fibers, or eye enucleation, results in the expansion of the projection of the uncrossed, intact retinotectal fibers (Lund, 1972; Lund

*To whom correspondence should be addressed: Department of Biophysical Engineering Science, Osaka University, 1-1 Machikaneyama, Toyanaka, Osaka, Japan.

et al., 1973, 1980; Lund and Lund, 1973, 1976; Lund and Miller, 1975; Jen and Lund, 1981). This change is not observed following lesions of retinal fibers made later than the 12th postnatal day. Similarly, neonatal lesions of corticotectal fibers before the 20th postnatal day result in sprouting of the intact corticotectal fibers from the remaining cortex (Mustari and Lund, 1976).

Other sprouting phenomena also appear to be limited to neonatal life. Hicks and D'Amato (1970) found that unilateral hemispherectomy in neonatal rats resulted in the appearance of an uncrossed corticospinal tract, which is not observed in normal rats. This uncrossed corticospinal tract was not found after ablation of a cerebral hemisphere in adult rats. Similarly, after unilateral cerebral hemispherectomy in neonatal rats, corticotectal fibers crossed the midline and terminated in the contralateral pontine nucleus and superior colliculus (Leong and Lund, 1973; Leong, 1976a), making synaptic contacts on dendrites of the target neurons (Nah and Leong, 1976b; Leong, 1976b, 1977b, 1980; Mihailoff and Castro, 1981). Finally, unilateral lesions of interpositus and dentate nuclei in neonatal rats induced uncrossed cerebellorubral and cerebellothalamic projections from the remaining ipsilateral nuclei (Lim and Leong, 1975; Leong, 1977a). Similar findings were reported in kittens (Kawaguchi et al., 1979; Fujito et al., 1980). Furthermore, unilateral ablation of the sensorimotor cortex in newborn rats resulted in the appearance of a crossed corticorubral projection from the intact contralateral cortex (Nah and Leong, 1976a,b), and the aberrant corticorubral fibers were found to make synaptic contact with the distal dendrites of the red nucleus cells (Nah and Leong, 1976b; Fujito et al., 1980).

Destruction of target neurons during a neonatal stage also results in aberrant retinothalamic (Baisinger et al, 1977; Kalil and Schneider, 1975; Schneider, 1970, 1973) and retinotectal (Miller and Lund, 1975) projections following unilateral lesions of the superior colliculus in neonatal hamsters or neonatal or fetal rats.

Sprouting in the Adult Central Nervous System

Septum.

Raisman (1969) and Raisman and Field (1973) have shown that sprouting occurs in the septal nucleus of adult rats by transecting one of two convergent inputs, either fimbrial afferents to dendrites of septal cells or medial forebrain bundle afferents to dendrites and somata. Following transection of the medial forebrain bundle, axonal sprouts from fimbrial fibers make synaptic contacts on somatic synaptic sites formerly occupied by medial forebrain bundle afferents. Removal of the fimbrial fibers results in multiple synapse formation on dendritic sites formerly occupied by fimbrial synapses. The sprouting of medial forebrain bundle axons causes an increased intensity of staining of adrenergic nerve fibers in the septum following unilateral transection of fimbria (Moore et al., 1971).

Interpeduncular nucleus.

Murray et al. (1979) also found formation of new synaptic contacts, using electron microscopy in interpedunclar nucleus (IPN). In caudal parts of IPN, the fascicular (FR) axons form crest synapses where presynaptic terminals make contact with parallel opposing sites of an attenuated dendritic appendage, one member from ipsilateral FR input and the other from contralateral FR input. Following unilateral destruction of FR, the surviving FR was found to give both members of crest synapses.

Red nucleus.

Sprouting was reported in adult cats (Tsukahara et al., 1975; Hanaway and Smith, 1978; Murakami et al., 1981). After a unilateral lesion of nucleus interpositus of the cerebellum, a second lesion was placed in the frontal cortex. Corticorubral endings normally found on distal dendrites of red nucleus with small cross section were greatly increased on proximal dendrites with large cross section and in the somatic region, indicating corticorubral sprouting on the proximal and somatic region of red nucleus cells (Nakamura et al., 1978; Hanaway and Smith, 1978; Murakami et al., 1981).

Hippocampus.

Cotman and Lynch (1976) and Lynch and Cotman (1975) have utilized the laminar afferent organization of the dentate gyrus to investigate axonal sprouting in adult rats. Because the perforant pathway from the ipsilateral entorhinal cortex terminates on the distal two-thirds of dendritic fields of granule cells, and both associational and commissural fibers end on the proximal one-third, sprouting into restricted synaptic territories can be investigated following interruption of different afferent pathways. After ipsilateral entorhinal lesions in adult rats, the synaptic territory of the associational and commissural fibers expands into the inner one-fourth of the synaptic territory occupied by the perforant fibers (Lynch et al., 1973).

The dentate granule cells also receive a small projection from contralateral cortex (crossed perforant fibers). After unilateral lesions of entorhinal cortex, the projection expands to form synaptic contacts on parts of the dendritic field formerly occupied by the ipsilateral entorhinal fibers (Steward et al., 1974). After bilateral lesions of the entorhinal cortex, the synaptic territory of associational fibers expands up to the inner 35 to 38% of proximal dendrites, contrasted with a distribution restricted to the inner 25% of granule cell dendrites in normal rats (Lynch et al., 1972, 1974, 1976). Thus, removal of major synaptic input to the dentate gyrus results in expansion of intact afferent systems onto deafferented parts of dendritic fields of granule cells.

Sprouting in the dentate gyrus has also been confirmed electron microscopically. After removal of ipsilateral entorhinal cortex in adult rats, 86% of the synapses on the distal two-thirds of granule cell dendrites disappear. However, by 240 days after the lesion, synaptic density returns up to 80% of normal values (Matthews et al., 1976a). This recovery appears to be primarily a result of synapse formation of dendritic spines (Matthews et al., 1976b). Thus, the new synapses appear to occupy deafferented synaptic sites.

Extrinsic afferents to the dentate gyrus from the septum and brainstem consist of relatively few fibers whose zones of termination are not well delimited. Acetylcholine histochemistry has revealed that after entorhinal lesions cholinergic septohippocampal fibers proliferate in the dentate gyrus (Storm-Mathisen, 1974). The response of the brainstem has not been described.

There also is evidence of plasticity of intrinsic hippocampal circuitry following entorhinal lesions. Granule cells in the dentate gyrus receive GABAergic inhibitory inputs from basket cell axons. After an ipsilateral entorhinal lesion, GABAergic interneurons show an increase in both the level of the synthetic enzyme (Nadler et al., 1974) and the amount of GABA they release (Nadler et al., 1977). This may be attributed either to an increase in the effectiveness of synaptic transmission or to sprouting of inhibitory synapses.

The extent of axonal sprouting in the denate gyrus after entorhinal lesions is far less prominent in adults than in neonates. In neonates, the commissural fibers grow out nearly to the superficial margin of the molecular layer, whereas associational fibers expand to the margin of the layer. Septohippocampal fibers also expand to form a dense plexus along the outer margin of the molecular layer. Thus, neonatal lesions of entorhinal cortex induce more marked sprouting than do adult lesion.

Although the mechanisms underlying

these differences between neonates and adults are not clear, it appears that both extent and rate of sprouting vary with age. The rate and extent of the outward expansion of the commissural-associational fibers of dentate gyrus in 3-month-old rats were compared to those of 25-month-old rats (Cotman and Scheff, 1979). Twelve days after lesion, the extent of commissural-associational projection in younger animals (3 months old) showed an increase of approximately 22% compared to normal controls; by contrast, older (25 months old) animals showed only a 10 to 11% increase. Septohippocampal fibers in the younger animals showed significantly less intense staining. The reasons for these differences are not clear.

Spinal cord.

Since Liu and Chambers (1958), sprouting of intact axons in spinal cord has been confirmed electron microscopically (Bernstein and Bernstein, 1971, 1973a,b). New synapses are formed on partially deafferented spinal neurons after partial hemisections in adult rats, monkeys, and humans. Similarly, after spinal hemisection at the midthoracic level in newborn and weanling rats, axons from descending nerve tracts show increased collateral growth rostal to the hemisection (Prendergast and Stelzer, 1976). Pullen and Sears (1978) reported that C synapses, which are one of the seven categories of presynaptic terminals found in spinal MNs, are modified after spinal hemisection. Furthermore, C synapses, which originate from propriospinal neurons, increase in numbers following partial central deafferentation and are found postoperatively in the neuropile on dendrites not normally exhibiting this particular synapse type.

Others.

Potentiality for sprouting in sensory relay nuclei has been reported to be very weak or entirely absent after interruption of the major synaptic input. For example, there is no evidence of sprouting in the spinal trigeminal (Kerr, 1972; Beckerman and Kerr, 1976), lateral cuneate (O'Neal and Westrum, 1973), and dorsal column nuclei (Rustioni and Molenaar, 1975). Similarly, there is no evidence of sprouting in the third-order avian auditory nucleus, nucleus laminaris (Benes et al., 1977). In the lateral geniculate nucleus, sprouting occurs only in kittens partially deafferented before the ninth postnatal day (Guillery, 1972; Robson et al., 1978). However, other studies reveal no evidence of sprouting (Stelzner and Keating, 1977; Hickey, 1975). There is some evidence that suggests that sprouting occurs in the ventrobasal nucleus of the thalamus after cortical ablation (Donoghue and Wells, 1977).

In the motor system, sprouting of bulbospinal noradrenergic fibers (Bjorklund et al., 1971) has been reported, and there is some evidence of sprouting of corticospinal fibers (Kalil and Reh, 1978), but rubrospinal fibers do not appear to sprout (Castro et al., 1977; Castro, 1978).

Evidence of sprouting has been reported in the cerebral cortex (Purpura, 1961; cf. Cotman et al., 1981).

Time Course of Sprouting

The time course of sprouting is characterized by an initial rapid phase followed by a much slower one, lasting for months. The initial signs of sprouting were observed within 2 weeks in dentate gyrus (Steward and Loeshe, 1977; McWilliams and Lynch, 1979), red nucleus (Tsukahara et al., 1975a,b), septum (Moore et al., 1971), and lateral geniculate body (Stenevi et al., 1972). This rapid phase of sprouting is followed by a much slower one that continues well over several months. The time course may be modified by "priming" lesions. This was demonstrated by placing a small priming lesion in the entorhinal cortex, followed by removal of remaining entorhinal cortex.

The reactive sprouting in the dentate gyrus was greatly accelerated after the second entorhinal lesion (Scheff et al., 1978). Thus, whether or not denervation is sufficient to trigger significant growth, it can prime the system for sprouting after additional damage.

Control

In some situations, reinnervation occurs preserving the number of synapses. In both the dentate molecular layer after entorhinal lesions and the septum after fimbrial lesions, the average density of synapses is restored to normal by sprouting (Raisman and Field, 1973; Matthews et al., 1976a). Fimbrial lesions on both sides also lead to recovery of synaptic endings (Field et al., 1980). Similar strict control of the number of synapses has also been reported in the superior cervical sympathetic ganglion of the rat (Raisman, 1977).

Selectivity and Specificity

Sprouting is highly selective in terms of both where it takes place and the types of fibers that display the phenomenon. Even in the same neurons in the same sites, some fiber systems have the potentiality to sprout, and others do not. For example, in the dentate granule cells of hippocampus, commissural, associational, and septohippocampal fibers sprout after ipsilateral entorhinal cortical lesions, whereas the entorhinal inputs do not sprout after lesions of associational and commissural fibers (Cotman and Nadler, 1978).

Pickel et al. (1974) found, 30 days after partial lesion of the superior cerebellar peduncle, enhanced terminal fluorescence indicative of sprouting of noradrenergic fibers from nucleus locus coeruleus not only in the cerebellum but also in the hippocampus. The increased terminal density in hippocampus suggests that sprouting of the hippocampal terminals may be initiated not from denervated postsynaptic cells but from a more remote signal site such as neurons in locus coeruleus or collaterals of the cut axons in the cerebellum. This is consistent with observations of Anden et al. (1966) that lesions of the norepinephrine pathway in the mesencephalon result in increased terminal fluorescence in cerebellum and other brainstem areas after 2 to 4 weeks.

Specificity of reinnervation was reported for the septum. Section of unilateral fimbria and stria terminalis induces innervation of contralateral lateral septal nuclei, which is absent in the normal rat. The new innervation is restricted to septal nuclei, although it coexists with that from bed nuclei in dorsal portion (Field, 1980; Field et al., 1980).

PHYSIOLOGICAL STUDIES

Physiological Effectiveness

Wall and Egger (1971) reported that the ventral posterior lateral nucleus (VPL) in rats is organized somatotopically, with the forelimb represented in the medial two-thirds of the nucleus and the hind limb in the lateral third. Several days after destruction of nucleus gracilis, extracellular unit recordings indicated that the forelimb representation had expanded to occupy two-thirds of the area previously responding to hindlimb stimulation. The expansion was first observed 3 days after the lesion and was complete by 7 days. Wall and Egger suggested that one possible mechanism for this phenomenon is sprouting of the remaining cuneate fibers to the deafferented VPL area.

Lynch et al. (1973) reported that the laminar profile of the negative field potential of hippocampal granule cells to commissural stimulation expanded to span the depth of the molecular layer in rats with a neonatal entorhinal cortical lesion. In normal rats, the maximum negativity to commissural stimulation is found only in the inner molecular layer. This finding suggested that commissural axons that sprout into the outer molecular layer after neonatal entorhinal lesions form functional synaptic connections in the region. However, this may not be a feature of all systems that display sprouting following partial deafferentation. Chow et al. (1973) recorded activity of neurons in the superior colliculus of neonatally enucleated rabbits. Although anatomical investigations showed that the ipsilateral retinal projection had spread into superior colliculus following neonatal enucleation, neurons in the area occupied by new axonal growth do not respond to either stimulation of the intact eye or electric shock stimulation of the optic nerve, a result, they concluded, that brings into question the functional significance of demonstrations of axonal "sprouting" in mammalian systems.

In order to establish the functional effectiveness and the properties of synaptic transmission of newly formed synapses unequivocally, it is necessary to record the synaptic potentials mediated by newly formed synapses. If new synapses are formed at sites different from normal, one could approach the problem by examining the time course of synaptic potentials, which depend on the distance from the synapses to the soma.

The red nucleus (RN) is a good system for examining this problem: RN neurons receive two kinds of synaptic inputs, one from the nucleus interpositus (IP) of the cerebellum on their somata and the other from the sensorimotor cortex (SM) on their distal dendrites. The corticorubral (CR) dendritic EPSPs are characterized by a slowly rising time course, whereas the IP EPSPs are characterized by a rapidly rising time course (Tsukahara and Kosaka, 1968) (Fig. 1A). A change of synaptic location relative to the soma should produce corresponding changes in the electrotonic distortion of the wave form of the EPSPs. Tsukahara et al. (1974, 1975a,b) have shown that a new rapidly rising component is superimposed on the slowly rising corticorubral dendritic EPSPs after IP lesion of adult cats (Fig. 1B–D). Since a slight change of the cable properties (electrotonic length) of dendrites of RN neurons following IP lesions accounts for only a minor portion (less than 5%) of the observed change in the time to peak of the CR EPSPs, Tsukahara et al. concluded that new and active synapses are formed at the proximal portion of the soma–dendritic membrane of RN cells.

An analysis of unitary CR EPSPs before and after lesions of the IP was also consistent with this interpretation (Murakami, 1979; Tsukahara, 1978, 1981). The CR unitary EPSPs in normal cats are characterized by a slowly rising time course and small amplitude (Fig. 2C). Two groups of unitary EPSPs were found in cats with chronic IP lesions, one consisting of CR EPSPs with a shorter time to peak than is found in normal cats (Fig. 2B) and the other consisting of unitary EPSPs within the normal range. The former group was found to be more sensitive to membrane potential displacement than the latter (Fig. 3), which suggests a proximal dendritic or somatic locus of origin. The relationship of the time to peak and amplitude of the CR unitary EPSPs before and after chronic IP lesions can be fitted to a theoretical relationship derived from Rall's compartment model (Rall, 1964) (Fig. 2D). A good concordance between the theoretical curve and the experimental points supports the view that new synapses were formed at the proximal portion of the soma–dendritic membrane of red nucleus cells, close to the soma, after IP lesion.

Properties of Synaptic Transmission

Properties of synaptic transmission of newly formed CR synapses have been investigated in adult cats. Murakami et al. (1977) recorded intracellularly from RN neurons after IP lesions and found that the

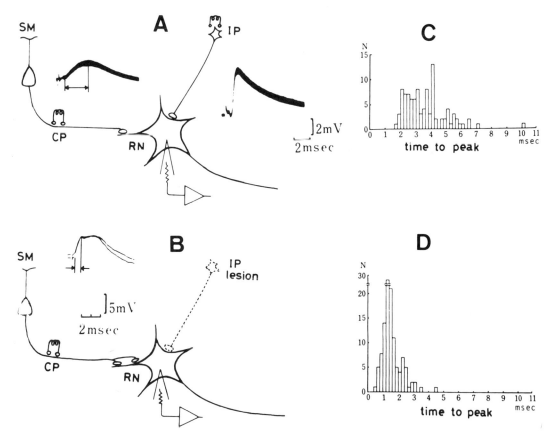

FIG. 1. Lesion-induced sprouting in adult feline red nucleus. A: Synaptic organization of normal red nucleus (RN). Monosynaptic excitatory input from ipsilateral sensorimotor cortex (SM) through the cerebral peduncle (CP) impinges on the distal dendrites, and that from the contralateral nucleus interpositus (IP) of the cerebellum synapses on the soma. Stimulation of CP produces a slowly rising EPSP in a RN cell, and stimulation of IP produces a rapidly rising EPSP as shown in the **inset**. B: After IP lesion, a rapidly rising component appears superimposed on the slowly rising CP EPSPs as shown in the **inset**. C: Frequency distribution of the time to peak of the CP EPSPs of normal cats measured as in the inset of A. D: Same as C but CP EPSPs of cats with chronic IP lesions measured as in the inset of B. (From Tsukahara, 1981, with permission.)

degree and time course of facilitation at newly formed synapses displayed no major differences from those at the normal synapses; however, the mean facilitation decayed more slowly in the newly formed synapses. Posttetanic potentiation, as observed in other part of CNS (e.g., Hughes et al., 1956; Lynch et al., 1976), was observed in the cats with IP lesion and had a similar time course to that in normal cats (Murakami, 1979; Murakami et al., 1976).

Sprouting and Recovery of Function After Lesions

A relationship between collateral sprouting and functional recovery has had some experimental support (McCouch, 1961), but the evidence is circumstantial and fails to identify the primary sites responsible for recovery of junction. Murray and Goldberger (1974) reported that in adult cats intrinsic reflexes elicited by ipsilateral dorsal root in-

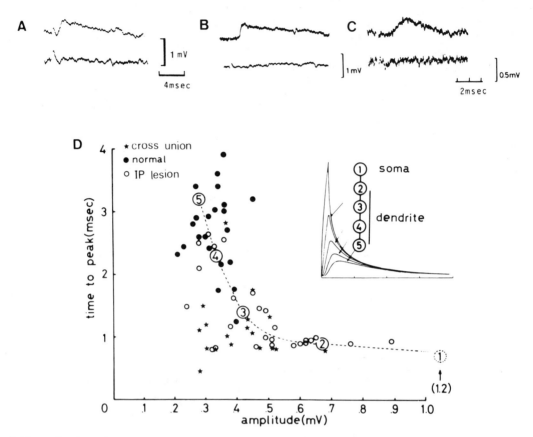

FIG. 2. Corticorubral unitary EPSPs. **A:** Intracellular EPSP evoked by stimulation of cerebral peduncle in a red nucleus cell of a cross-innervated cat. **B:** Same as A but evoked by stimulation of sensorimotor cortex in a cat with lesion of the nucleus interpositus 27 days before acute experiment. **C:** Same as in B but in a normal cat. *Upper traces*, intracellular potentials. *Lower traces*, extracellular field corresponding to the upper traces. **D:** Relationship between time to peak and amplitude of the unitary EPSPs. *Open circles* represent unitary EPSPs of IP-lesioned cats; *stars* represent those of cats cross innervated more than 2 months before recording; *filled circles* represent those of normal cats. *Large open circles* represent time to peak and amplitude of theoretical EPSPs derived by Rall's compartment model initiated at each compartment of a five-compartment chain. The time course of the theoretical EPSPs generated in these compartments is shown in the **inset** of the figure. (From Tsukahara, 1981, with permission.)

put increased after partial hemisection of the spinal cord between T_{12} and L_1. In some cases, reflexes once depressed became hyperactive 2 weeks after the operation. Anatomical examination revealed signs of collateral sprouting from dorsal roots in response to degeneration of descending tracts on the same side. Both degeneration and autoradiographic methods have shown that the site of the sprouting is in Rexed's lamina VI and laterally in lamina VII, where interneurons mediating cutaneous reflexes and stretch reflex facilitation are located. Goldberger (1977) reported that recovery of locomotor patterns after hindlimb deafferentation can be explained by sprouting of ipsilateral descending fibers. But it is not easy to prove a causal relationship between anatomical and behavioral changes.

Dieringer and Precht (1977, 1979) reported a shortening of the time to peak and

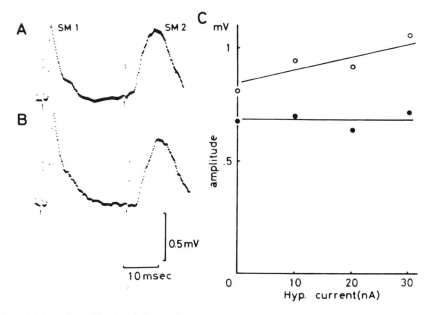

FIG. 3. Sensitivity of amplitude of the EPSPs to membrane polarization. A: Rapidly rising (SM_1) and slowly rising (SM_2) corticorubral EPSP recorded in the same cell by stimulating different points of the sensorimotor cortex in a cat with a lesion of the nucleus interpositus. B: Same as A, but the EPSPs were recorded during hyperpolarization produced by injecting current through a microelectrode. *Upward arrows* indicate the onset of stimulation. C: Relationship between amplitude of EPSP and injected current. *Open circles* correspond to the rapidly rising EPSPs, and *closed circles* to slowly rising ones. (From Murakami et al., 1977, with permission.)

an increase in amplitude of commissural EPSPs in frog vestibular neurons following unilateral chronic labyrinthectomy. The authors suggested that these changes in vestibular EPSPs may be an underlying cellular mechanism for the compensation of postural and locomotor disturbances after unilateral vestibular nerve transection.

Research in rhesus monkeys indicated that changes in unitary EPSPs in MNs produced by group Ia stimulation may also be associated with functional recovery after spinal hemisection. The rise time of unitary EPSPs in MNs produced by Ia stimulation 3 to 6 months postoperatively was slightly shorter on the hemisected side than on the control side (Aoki and Mori, 1978).

Bromberg and Gilman (1978) reported that multiunit activities in the red nucleus show an initial decrease of response and a subsequently slow recovery after lesions of the contralateral interpositus nucleus. To evaluate whether the restoration of activity depends on intact corticofugal fibers, the ipsilateral pericruciate cortex was ablated 6 weeks after the contralateral nucleus interpositus lesion. After the ablation, there was an immediate drop in amplitude of the activity.

Loesche and Steward (1977) reported that recovery of alternation performance in a T maze (for a food reward) occurs following a unilateral entorhinal cortex lesion. They suggested that this recovery results from reinnervation of the denervated granule cells by contralateral entorhinal fibers. Bilateral lesions of entorhinal cortex resulted in a persistent performance deficit. A secondary contralateral entorhinal lesion resulted in a deficit in alternation performance similar to that following one-stage bilateral lesions.

A well-known procedure for testing the recovery and readjustment of behavior is cross innervation or cross connection of muscles. Sperry (1947) reported that motor readjustment occurred after cross innervation of antagonistic muscles in monkeys during certain tasks practiced for a long period of time. This readjustment had never been observed in rats (Sperry, 1942). In cats, cross innervation of forelimb nerves results in some modification of activity when tested in quadripedal locomotion (Tsukahara, 1978). Yumiya et al. (1979) reported that motor readjustment occurred after cross connection of forearm muscles as tested by EMG during walking. Readjustment of the tactile placing reaction was also reported.

The neuronal basis of this behavioral readjustment after cross innervation is not yet clear. Eccles et al. (1972) investigated the reorganization of monosynaptic excitatory input to MNs from muscle afferents but failed to detect a major change. Tsukahara and Fujito (1976) examined the possible synaptic reorganization in red nucleus after cross innervation of antagonistc muscles in cats and found a change in the rise time of corticorubral EPSPs after cross innervation, suggesting that new synapses were formed on proximal parts of red nucleus neurons. Yumiya et al. (1979) examined the input–output relationship in cortical efferent zones after cross connection of forearm muscles in cats but failed to detect any significant changes.

Sprouting Without Degeneration

Most studies of sprouting have been concerned with that induced after a lesion of a major synaptic input of central neurons. These studies even led to a simplified hypothesis that degeneration is essential for sprouting to occur and that vacated synaptic sites are reoccupied by new synaptic connections (Raisman and Field, 1973). But sympathohippocampal sprouting caused by sequential lesion of entorhinal cortex and septal lesion cannot be explained by replacement of synapses (Crutcher and Davis, 1981; Crutcher et al., 1979). In addition, recent reports indicate that the number of vacated postsynaptic sites also changes after functional denervation without degeneration (Cotman and Nadler, 1978). Aguilar et al. (1973) reported that in the peripheral nervous system of salamanders interruption of axoplasmic transport in the peripheral nerve by colchicine application induced sprouting of adjacent peripheral nerves. This sprouting occurred in the absence of nerve degeneration. Thus, the question arises of whether the presence of nerve degeneration is necessary for sprouting, and, consequently, to what extent sprouting occurs in circumstances other than denervation.

Tsukahara and Fujito (1976) found that a new, rapidly rising component in the time to peak of CR EPSPs appeared in cats whose flexor and extensor forelimb nerves were cross innervated (Fig. 4). Since the electrotonic length as well as the membrane time constant did not change after cross innervation, appearance of the new rapidly rising component was not attributable to changes in cable properties of dendrites. The new component appeared after postoperative periods varying from 3 to 10 months and was found in cells innervating upper spinal segments. Red nucleus cells innervating the unoperated lower spinal segments showed a less prominent change in the shape of CR EPSPs (Fig. 5). The possibility that neruons in nucleus interpositus degenerated after cross innervation was excluded because there was no appreciable change in the physiological estimates of the number of cerebellorubral fibers converging on red nucleus cells. Tsukahara and Fujito concluded that sprouting occurs without the presence of nerve degeneration at central synapses.

Other studies in the peripheral nervous system also suggest that degeneration is not necessary for induction of sprouting. Brown and Ironton (1977) and Holland and

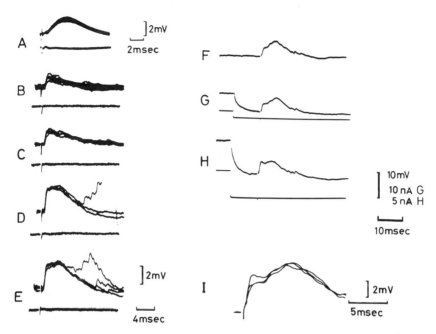

FIG. 4. Corticorubral EPSPs in cross-innervated cats. A: Cerebral peduncle (CP) EPSP induced in a red nucleus (RN) cell in a normal cat. B–H: CP EPSPs in RN cells in cross-innervated cats. B–E: CP EPSPs in a RN cell innervating upper spinal segments with increasing stimulus intensities from B to E from a cat cross innervated 176 days before acute experiment. *Upper traces,* intracellular potentials. *Lower traces,* extracellular records corresponding to upper traces. F–H: CP EPSPs during membrane hyperpolarization by application of current steps as indicated in the *lower traces* in G and H. F: Control EP EPSP. G,H: During membrane hyperpolarization. J: Superimposed tracing of CP EPSPs as indicated in F–H. Voltage and time calibration in E apply to B–D, and voltage and time calibration of H,I also apply to F,G. Records of F–H were from a cat cross innervated 147 days before intracellular recording. (From Tsukahara, 1981, with permission.)

Brown (1980) found that sprouting occurred after a few days of muscle inactivity caused by topical application of tetrodotoxin (TTX) to the motor nerve or of α-bungarotoxin to the muscle. Other studies by Rotshenker and Reichert (1980) showed induction of sprouting in frog by a nerve innervating an intact pectoris muscle when the contralateral nerve innervating the opposite pectoris muscle was crushed; this effect may have been communicated within the spinal cord from the altered MNs (on the lesion side) to the intact MNs.

Signals of Sprouting

The nature of the signals that initiate sprouting and how the signals are transmitted to neuronal somata are not known; however, two possible mechanisms have been proposed. (1) Some signals may be produced by target neurons and transmitted retrogradely to presynaptic fibers. The inducing signal might be released from degenerating tissue (Raisman and Field, 1973) or from inactive muscle (Brown and Ironton, 1977; Holland and Brown, 1980; Eldriges et al., 1981) or from nerve growth factor in sympathetic ganglion (Purves, 1975). (2) Nerve terminals may normally release a factor that prevents sprouting, so that when these nerves are injured or the concentration of this factor decreases, inhibition of sprouting no longer takes place (Diamond et al., 1976).

In the hypoglossal nucleus, there is evidence suggesting that signals transmitted

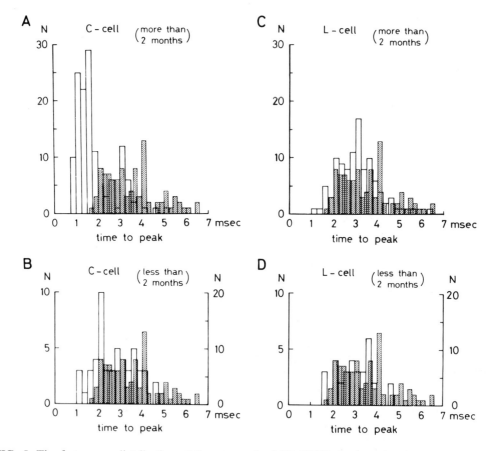

FIG. 5. The frequency distribution of time to peak of CP EPSPs in cross-innervated cats. **A,B:** C cells (RN cells innervating upper spinal segments). **C,D:** L cells (RN cells innervating lower spinal segments). Number of cells is shown on the *ordinate*, and the time to peak of CP EPSPs (msec) on the *abscissa*. The shaded columns in **A–D** illustrate the frequency distribution of time to peak in normal cats, drawn from data of Tsukahara et al. (1975a). The right-hand ordinate scale in **B** and **D** applies to normal cats and is common to **A–D**. The left-hand scale is for cross-innervated cats. **A,C:** Data from cats cross innervated more than 2 months before acute experiments. **B,D:** Data from cats cross innervated less than 2 months before acute experiments. (From Tsukahara, 1981, with permission.)

retrogradely from the target organ to the cell soma influence the number of synaptic connections the target cell receives. Nuerons lose synaptic connections when their axons are severed; the number of synapses returns to normal when the axons regenerate and make synaptic contacts on their targets (Sumner, 1977). Interruption of neuromuscular contact by blockage of transmitter release by botulinum toxin (Sumner, 1977) also produces presynaptic bouton loss (Sumner, 1977; Cull, 1975).

In summary, sprouting is now acknowledged to occur widely in the CNS. To what extent sprouting and formation of functional synapses account for recovery of function after brain damage remains to be explored extensively.

Rehabilitation Versus Passive Recovery of Motor Control Following Central Nervous System Lesions

*Paul Bach-y-Rita

Department of Rehabilitation Medicine, University of Wisconsin Medical School, Madison, Wisconsin

After a CNS lesion, part of the recovery of function relates to resolution of local factors (edema, tissue debris), and another part may occur through neural repair mechanisms which may continue for months or years. Some recovery after CNS lesions occurs spontaneously (see Cotman, 1978; Finger, 1978; Jeannerod and Hecaen, 1979), but specific rehabilitation procedures are important for obtaining a maximum recovery. This chapter deals primarily with the role of rehabilitation in enhancing recovery.

REHABILITATION STUDIES IN THE RAT

The main rehabilitation procedure used is manipulation of environment (Walsh and Greenough, 1976). Schwartz (1964) showed that allowing rats to experience enriched environmental conditions (EC) for 3 months after occipital cortical lesions significantly improved their test scores in maze running (see also Will and Rosenzweig, 1976). Nonlesioned rats are markedly affected in their brain anatomy and chemistry by staying in either enriched (EC) or impoverished (IC) conditions, and these effects (predominating in occipital cortex) were also noted in rats blinded at

---
*Department of Rehabilitation Medicine, University of Wisconsin Medical School, Madison, Wisconsin 53705.

birth (Rosenzweig et al., 1969). A period of 2 hr/day of EC was as effective as 24 hr. The neuroanatomical effects of environmental complexity have been reviewed by Greenough et al. (1979) and Walsh (1981). Optimal recovery is not obtained by simple restoration of general health or by simply providing socialization but through complex stimulation and opportunities for experience (Bach-y-Rita, 1980). We have recently observed that 5 days restriction in a box (5 × 10 × 10 cm, open at top, with food and water available) aggravates rats rendered hemiparetic by an electrolytic lesion in the internal capsule. Control hemiparetic rats placed in normal cages exhibited more motor activities and showed some recovery.

PRIMATE REHABILITATION STUDIES

Black et al. (1975) trained 27 adolescent rhesus monkeys on two motor tasks involving each arm separately, either for flexing the elbow against a load to bring food to the mouth or for squeezing bars in one hand. The precentral forelimb cortical area was then ablated on one side. Active postoperative training of the weak hand led to recovery of 82% of the preoperative function in 6 months when the therapy was initiated immediately. The recovery in 6 months was only to 67% of preoperative control if the training was delayed 4 months after surgery. Only a moderate recovery occurred in the absence of the rehabilitation training. Thus, physical retraining facilitated motor recovery in both rate and magnitude, and it was important to start it soon after the cerebral insult.

MOTOR REHABILITATION IN MAN

Many of the data on human rehabilitation are of little value. Research has been limited by difficulties in assembling sufficient numbers of patients with similar CNS lesions for matched group studies and in quantifying both extent of damage and progress in function. Other difficulties result from the long time scale of recovery and the high cost of CNS rehabilitation programs.

Oculomotor Control in Cerebral Palsied Children

An intensive visuooculomotor training program was applied 8 hr/week for 6 to 10 weeks in a group of cerebral palsy children (Gauthier et al., 1978). To avoid boredom and maintain interest, the program involved projection of movies that were moved horizontally on a screen (beam from projector reflected onto a galvanometer-driven mirror) with separate adjustments of target size and velocity of displacement (Fig. 1). Simultaneous arm tracking and, in some cases, visioacoustic tracking (by means of stereo headphones with the balance between the two channels modulated in synchrony with screen target displacement) was included. Smooth pursuit maximum velocity and frequency increased by 100% or more. Precision of eye movements increased as reaction time decreased. Associated arm movements accelerated and improved the training of eye movement control. Children tested 1 month after completing training had maintained their gains. Reading ability also improved. The data document significant improvement through these specific rehabilitation procedures.

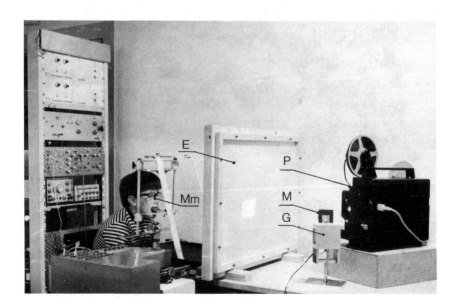

FIG. 1. Apparatus for eye movement training of children with cerebral palsy. The subject is positioned on a bite bar and forehead rest. A Super-8 film projector (P), mirror galvanometer (M,G), and translucent screen are used to project childrens' movies onto the screen and to move the image for eye movement training. A photoelectric technique is used to monitor eye movements (Mm). (Courtesy of G. Gauthier.)

Sensory Feedback Therapy for Hemiparesis

Brudny et al. (1976) applied EMG feedback to 114 patients with hemiparesis, torticollis, or dystonia, with favorable results in most. They displayed the integrated EMG of chosen muscles on an oscilloscope screen and used an auditory display reflecting EMG activities by proportional changes in click rate or tone intensity. In treating a hemiplegic limb, the therapeutic effort was directed towards learning volitional control of muscles, starting proximally and progressing distally. The main goals were decrease of spasticity during movement or stretch and increase of contraction strength in paretic muscles. In torticollis, the goals were a decrease in involuntary spasmodic activity and strengthening of contralateral muscles. A brief initial demonstration on a normal muscle allowed the patient to understand the relationship between muscle contraction or relaxation and the feedback signals. The method is considered to substitute for deficits in proprioception following CNS lesions and to provide an external feedback loop in the disrupted system, thereby augmenting or restoring sensorimotor interactions in voluntary movements. These procedures were useful, even when the program was begun some time after the stroke.

Rehabilitation of Facial Paralysis After VII–XII Anastomosis

The patients with a complete section of the facial nerve (Bell's palsy) in whom the peripheral part of the facial nerve was surgically anastomosed to the central branch of the XI or XII nerve provide an unusual opportunity to evaluate recovery after a known damage (Balliet et al., 1982). Rehabilitation was initiated 18 months to 4 years after surgery and involved EMG feedback therapy, relaxation, training of movement using a mirror, and other methods of specific awareness exercises. In two patients with XII-to-VII anastomosis, individual facial muscle control developed, and facial symmetry was regained (Fig. 2). The patients were able to accomplish eye closure, smile, lip pucker, and frown. Recovery is not spontaneous but requires appropriate rehabilitation procedures. It also appears that the XII motor nucleus is able to mediate rather selective control over facial muscles.

Rehabilitation After a Pontine Infarct

A rather unique case of pontine infarct revealed at autopsy a $10 \times 10 \times 5$ mm lesion in the left half of rostral basis pontis extending from near midline, and the pyramidal tract was severely atrophic in the left medulla (distal to the lesion) and showed nearly total loss of myelin and axonal degeneration (Fig. 3). Below the decussation of pyramid, the right lateral and left ventral corticospinal tracts showed severe degeneration, with only about 3% of myelinated axons remaining at the level of the cervical spinal cord. These changes were limited to the corticospinal pathways of the lower brainstem and spinal cord, and there were no changes rostral to the pontine lesion (Aguilar, 1969). The lesion occurred in a 65-year-old college professor (my father) with no previous history of neurological disease. The application of an intensive rehabilitation therapy program in conjunction with a high degree of motivation and family involvement led to return to full-time work 3 years after the accident. He led an active life until his death at 72 years from an unrelated myocardial infarction precipitated during a mountain hiking trip at 3,000 m altitude.

The remarkable recovery in this case proceeded over a period of 5 years during which the patient carried out an active home therapy program with initially a 3 hr/week physical therapy treatment. Fine movements such as button closing and fast

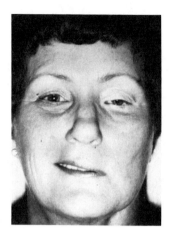 

FIG. 2. Rehabilitation of a facial paralysis patient. J.B. underwent surgery for removal of a left acoustic neuroma, at which time the VII nerve was sectioned. Two years later, a VII–XII cranial nerve anastomosis was performed. **Left:** An attempt to smile 2 years following the anastomosis. **Right:** Smile following 7 months of rehabilitation.

FIG. 3. Micrograph of the lesion in the left basis pontis in a patient who recovered from hemiparesis and survived for 7 years post-lesion. The focus of cystic encephalomalacia measured approximately 1 × 1 × 0.5 cm. (Courtesy of M. J. Aguilar.)

typewriting were regained. The patient was very active and willing to work hard for several hours a day over many years. He designed some of the tasks himself and decided, for example, that he would wash the dinner dishes daily. He would keep his paretic right hand in the warm water and had it participate more and more, over the next few months, in the washing which initially took hours to complete. For typewriting, he forced himself to use his right hand: he visually placed his middle finger above the desired key and allowed his right arm to slightly drop for that finger to strike the key, using gross movements. Through intense and prolonged practice, he could progressively involve individual wrist and, subsequently, finger movements, thereby increasing typing speed. This was indeed an unusually intensive rehabilitation program carried out with high motivation and continuity. The recovery thus achieved suggests that considerable neural plasticity exists even in the aged human brain (see also Bach-y-Rita, 1980).

Recovery of Championship Athletic Performance

Jokl (1964) analyzed factors leading to superior performance in Olympic champions with major motor deficits acquired either before or after reaching their championship level. For example, Harold Connolly (1956 hammer-throw champion) had a combined injury of upper and lower left brachial plexus (Erb–Duchenne–Klumpke–Dejérine type) with marked growth defect and pareses of left arm muscles as a result of a birth trauma (Jokl, 1957). Bill Nieder suffered, 1 year prior to his second Olympic games, spontaneous tearing in right pectoral muscle resulting in reduced force of the main muscle involved in propulsion of the right arm (he is right-handed) during the shotput. Yet his gold medal at the second games actually improved by 5 feet his previous Olympic performance (Jokl, 1964). Among factors in such effective rehabilitation, Jokl considers conceptual originality and creative awareness, appropriate habit formation, intelligence for redesigning motor control strategies, and intensive exercise. The handicapped person is compelled to become aware of sensorimotor features to integrate for action every part of the motor system that remains available.

SENSORY REHABILITATION

The capacity to recover sensory function after CNS damage also suggests potentialities that can be employed in therapy. In view of the motor emphasis of this volume, only one example, related to the rehabilitation of experimental amblyopia resulting from monocular occlusion, will be discussed. Hubel and Wiesel (1970) described in the kitten a "critical period" (fourth to 12th week postnatal) during which even short monocular occlusion by eyelid suture resulted in amblyopia that persisted after the occlusion was removed. The animals did not use the deprived eye in visually guided behavior, and many visual cortical neurons did not respond to photic stimulation of the deprived eye. The changes were reported to be permanent. Although the cats could regain some vision, allowing them to learn discrimination of dark versus light and differences in depth, they never regained pattern vision (Dews and Hubel, 1970).

Chow and Stewart (1972) noted evidence of recovery after monocular occlusion for even longer periods (up to 2 years) provided an appropriate rehabilitation program was carried out. The following factors might have been responsible for their results: (1) initial use of tasks easy to learn; (2) reintroduction of previously mastered problems at various points of the training; (3) use of a go–no–go paradigm that eliminated complexities of a left–right choice point (this factor avoided problems with the disrupted spatial orientation in these animals); (4) training that was intensive, with about 1,000 trials per cat; (5) a great deal of handling and gentling between trials, as food rewards alone were insufficient to maintain progress; (6) adjustment of the difficulty of the task at each point on the basis of the behavioral response. When the errors increased at a given stage of training (with more difficult cues), the cat would be hissing and fighting; then, the remainder of the session was used with easier, previously mastered stimuli. These features are pertinent to the ways human patients should be treated during training on difficult tasks.

CHANGING MOTOR CONTROL

The capacity of the CNS to reorganize function can be documented by studies in which it proved possible to modify motor control systems that were previously considered rather fixed.

Voluntary Ocular Torsion

Balliet and Nakayama (1978a) developed a training procedure to enable a normal man to make large conjugate cyclorotary eye movements at will. The subject was seated in a dark room, and torsional head movements were prevented by a full-mouth bite plate held in the mouth. Fixating a target with the right eye, the subject activated a flash leading to a clear vertical afterimage. He then imaged the afterimage parallel to a vertical luminous real line of equal visual angle which was viewed at 100 cm distance in primary position. The subject had to keep his afterimage matched parallel only by cyclorotating his eye to the real line which he progressively rotated more and more (left or right) from the vertical as the training progressed. Usually 20' of arc increments were used. No instruction was given as to what type of eye movement should be used (slow pursuit, saccade) to achieve this position. Matches that could be held for 5 sec were defined as "cyclofixations."

Subjects usually trained 1 hr a day and could increase torsional eye movements at an average rate of 0.8°/hr of training, with increasing facility. At end of training, the subject could make cyclofixations at any amplitude within the trained range (Fig. 4). One subject acquired a range of 26.5° (12° on one side and 14.5° on the other) in about 30 hr total training. These cyclorotary eye movements can be made in the absence of any tilted or rotary stimulus. Vision alone does not drive eye torsion, which requires voluntary effort. This represents an unsuspected plasticity of the human oculomotor system. At the initial stage of training only, subjects could experience impressions that their head and body were "rolling laterally" (sometimes with visual illusion), and there was intermittent stomach nausea, headache, and fatigue. Subjects felt powerful sensations of body flotation or rapid

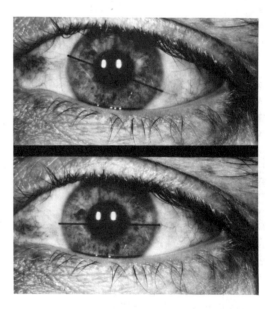

FIG. 4. A tonic voluntary cyclotorsion of 20°. *Bars* drawn across the irises are similar to those used for photographic analysis. (Courtesy of R. Balliet.)

downward falling in the direction of voluntary cyclotorsion. These sensations show similarities to those experienced by pilots in conditions in which conflicting visual and orientational informations occur (Balliet and Nakayama, 1978b). These phenomena were no longer apparent at later stages of training.

Plasticity of the Vestibuloocular Reflex

Reversal of vision by prisms can elicit a complete reversal of the vestibuloocular reflex in man (see G. Melvill Jones and G. Mandl, *this volume*).

NEURAL MECHANISMS

The capacity of the brain to modify motor control and overcome deficits following CNS damage can be enhanced by specific rehabilitation training when the program is functionally oriented and related to the patient's interests and daily living. The mechanisms underlying such capacity are not known. The actual damage to the brain is not repaired (virtual absence of central regeneration), and other mechanisms must be involved (see Finger, 1978; Cotman, 1978; Bach-y-Rita, 1972, 1980, 1981a,b,c), among which the following appear relevant.

Collateral Sprouting

It is well known that intact neurons can produce collateral sprouts of their axon to occupy vacated synaptic spaces at other cells whose normal input has been partly or totally destroyed (Edds, 1953; Raisman, 1977). The newly formed synapses can be functional in CNS (Lynch et al., 1973; N. Tsukahara and F. Murakami, *this volume*), but this is not necessarily beneficial and may result in disturbances. Specificity of information transmission may be upset when neurons are activated or inihibited by other pathways than in the normal condition. Collateral sprouting is thus advantageous in restoring function at denervated muscle fibers (thereby expanding the size of the motor unit and increasing its force; see Brown et al., 1981; for data on man, see Desmedt, 1981b). In the CNS, equivalence of function of neighboring axons is rare, and each neuron receives a variety of different inputs: if one or more such inputs are lost by a lesion, collateral reinnervation by other pathways may actually result in abnormal function.

There is evidence of morphological plasticity in the human cerebral cortex up to at least 80 years, with continuing development of apical dendrites of pyramidal cells in parahippocampal cortex (Buell and Coleman, 1979). There is also evidence that the early components of the somatosensory evoked potential to hand stimulation increase significantly in voltage in healthy octogenarians as compared to young normal adults (Desmedt and Cheron, 1981).

Unmasking

Anatomically established synapses in CNS may be called on when the usually dominating system fails ("unmasking," Wall, 1980). This would compensate for the lost input if the remaining input has appropriate effects. The derepression of function thus implied supposes that connections laid out in the embryo are more diffuse than those actually used in the adult brain. A few remaining fibers might perform the function previously accomplished by the intact pathway, thereby accounting for some recovery after a focal lesion. The unmasking process may involve the strengthening of synaptic access of remaining fibers to the cells that have lost their major input for a particular function. The case of recovery from hemiplegia after a 97% destruction of the pyramidal tract on one side (see above) can be viewed in this light.

There are experimental data suggesting the appearance of increasing reactivity to residual inputs after partial cortical deafferentation in the cat (Franken and Desmedt, 1957; Desmedt and Franken, 1963). The focal lesions used were cortical incisions cutting short corticocortical connections or sections of corpus callosum eliminating contralateral connections. These effects were typically slow to develop and appeared only after 2 to 3 months in the cortical area thus partially deafferented. Unmasking may also lead to an increase in reactivity at spinal level: Bremer (1928) showed that unilateral section of lower lumbar dorsal roots in cats resulted in an exaggerated vestibulospinal reflex on the lesioned side only (when the chronic cat was moved horizontally). In this case, the loss of segmental inputs may have unmasked vestibulospinal connections that supported a potentiated response. In phenomena of sensory substitution, some pathways might even acquire a new functional role through training (see Bach-y-Rita, 1972).

RELEVANT FACTORS

Repetition is important in obtaining functional gains (Gauthier et al., 1978; Chow and Stewart, 1972), and the context of pleasant real-life tasks is helpful. Crossman (1959) showed that the peak performance rate in manual cigar making by normal young women employees in a cigar factory was achieved after about 3 million repetitions. However, repetition alone is not enough (Jokl, 1964). Mind–body interactions and motivation can modify the ultimate results through poorly understood mechanisms. Cannon (1942) discussed Voodoo death as the ultimate example of mind's capacity to modify function, presumably through frontal cortex control of reticular systems in thalamus and brainstem and of autonomic function (see Skinner and Yingling, 1977). Studies of the placebo effect indicate that the state of mind can influence the effects of treatment (Bach-y-Rita, 1980). Effective rehabilitation programs should also take into consideration the factors known to be relevant in the learning process. Miller (1956) discussed the process of "chunking," whereby more and more bits of information can be dealt with together as learning progresses. Also, normal learning occurs with plateaus of consolidation separating phases in which progress is being made (Bryan and Harter, 1899). These mechanisms must be considered when designing scientifically based rehabilitation programs.

In summary, rehabilitation has been shown to increase the rate of recovery of function following CNS lesions as well as to improve the final functional outcome. Furthermore, rehabilitation assists in the substitution (motor and/or sensory) of mechanisms not normally involved in the damaged function.

Bioelectric Control of Powered Limbs for Amputees

*R. B. Stein, D. Charles, and M. Walley

Department of Physiology and Occupational Therapy, University of Alberta, Edmonton, Alberta T6G 2H7, Canada

The dream of replacing a lost limb with artificial devices that function and look like the lost limb is an ancient one. However, for most of recorded history, the artificial limb, where available, consisted of a "peg-leg" or a "hook" controlled by residual body movements. Attempts have been made in each generation to apply existing technology. One example is shown in Fig. 1, which comes from a book published 171 years ago in Berlin (Graefe, 1812). Existing technology consisted of the ability to make fine suits of armor, and, as shown in Fig. 1, cables from the "armor hand" permitted either a precision grip between the thumb and first two fingers (left) or a full fist (right).

The large numbers of amputees created in the two World Wars created an added incentive for the development of improved artificial limbs, and an early attempt to produce a powered upper limb was published soon after World War I (Schlesinger, 1919). However, not much progress was made until after World War II (Reiter, 1948; Battye et al., 1955; Kobrinski et al., 1961). These laboratory studies in Germany, England, and Russia led to several industrial prototypes. Now, 20 years later, perhaps 15,000 electric hands have been produced worldwide (J. Hendrickson Jr., Otto Bock Inc., Minneapolis, Minn., *personal communication*). However, this number represents only a few percent of the more than 100,000 arm amputees estimated to live in the United States alone (Committee on Prosthetics Research and Development, 1971). Only a small percent of the amputees fitted with an electric hand have also received an electric wrist, based on the limited total production of these units (J. Hendrickson Jr., *personal communication*).

In this chapter we review recent progress that promises to extend the applicability of powered prostheses significantly. However, we also consider the reasons for the relatively slow acceptance of these devices and the problems that must be overcome before their full potential can be reached. Even today, for example, the separate control of the digits shown in Fig. 1 remains an elusive goal.

MYOELECTRIC CONTROL OF SINGLE MOVEMENTS

Any artificial device is only as good as the ability of the patient to control it. A variety of control methods have been suggested, but the most widely used involves surface EMG, which is referred to as myoelectric control. The electrical signals from stumps of muscles that remain after amputation are a particularly valuable source of signals. These muscles remain under vol-

---
*To whom correspondence should be addressed.

FIG. 1. An early hand of armor (Graefe, 1812) which permitted control of either a precision grip between the thumb and first two fingers (left) or a full fist (right), a choice that is not regularly found in prostheses even today.

untary control and were often involved in the required movements prior to amputation of the limb. For example, a below-elbow amputee will generally have stumps of the long wrist extensors and flexors which normally assist in opening and closing the hand, respectively (Fig. 2). Thus, it is quite natural for these muscles to be utilized for controlling a powered hand.

However, even though the natural muscles are being used, the mode of control is somewhat different. This is illustrated in Fig. 3A in which a desired movement is made (e.g., opening the hand). Immediately below the movement is the surface EMG produced in such a movement. In many powered limbs, the EMG is then amplified, rectified, and filtered to obtain a smooth signal for controlling the motor (hand, wrist, or elbow). Then, the processed signal is compared with a threshold level (Fig. 3B), and the motor is either on or off depending on the level of the filtered EMG signal with respect to this threshold. In the example shown in Fig. 3B, the processed EMG remains above threshold, and the motor continues at constant velocity until the limit of movement is reached.

Thus, the EMG serves as a simple on–off switch and controls the presence or absence of movement. In the natural movement, the level of EMG was graded with a high level initially to generate an angular acceleration, a lower level during deceleration, and then a maintained EMG to hold the new position. In effect, an amputee must learn instead to vary the duration of an EMG burst; i.e., the extent of the movement is determined by the total period of time that the processed EMG remains above the threshold level.

An alternative scheme is shown in Fig. 3C in which the signal is not filtered but is fed directly to a threshold device (Childress, 1973). The control signal then switches on and off, and the hand receives power in proportion to the time that individual peaks of the rectified EMG exceed

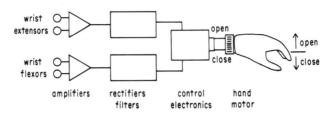

FIG. 2. Schematic diagram of the operation of a myoelectric hand. Muscle potentials are detected on the skin surface overlying stumps of muscles remaining after an amputation. These signals are amplified, rectified, and filtered to provide a smooth signal for controlling the movement of the electric hand motor in two directions (open and close).

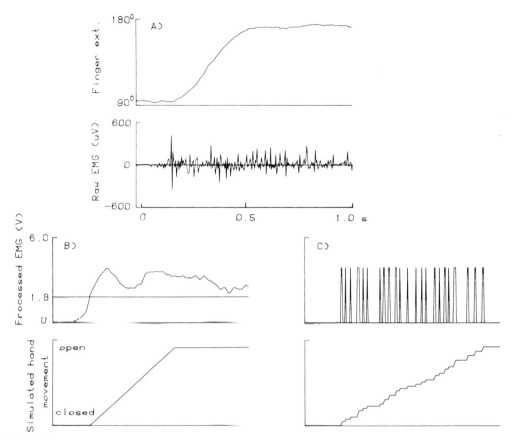

FIG. 3. Stages in processing EMG for control of two types of myoelectric hands. A: To open a natural hand (increasing extension of the first metacarpal joint is indicated as an upward deflection), the surface EMG shown in the second trace is produced. B: The EMG signal has been amplified (30,000x), rectified, filtered (10 Hz low-pass Paynter filter), and compared with a threshold level (1.8 V). When the EMG level is above the threshold level, a control signal is supplied to open the artificial hand at a constant velocity, as shown by the simulated movement. The actual movement would be delayed by the filtering and the inertia of the motor. The opening of the hand continues up to its maximum value as long as the filtered EMG remains above the threshold level. C: An alternative approach is to eliminate the filtering of the EMG. Then, the control signal switches on and off, and the movement of the hand will vary in speed depending on the exact pattern of EMG, but the movement will be less smooth.

threshold. This alternative scheme has a number of advantages. (1) It eliminates the delays inherent in any analog filtering. Thus, the movement can be more responsive to the voluntary commands of the patient. (2) A variable control results, since the hand will move faster the larger the signals are relative to the threshold. A truly proportional control is not achieved, as the input-output curve is nonlinear (see Fig. 4).

Note also that the average speed will remain below that achieved with the filtered signal, since the unfiltered signal will remain below the threshold level a proportion of the time, even with a high mean level. The calculated curve assumes that the EMG can be approximated by a Gaussian signal with a root mean square (RMS) value proportional to the desired force level (Bigland and Lippold, 1954a; Milner-Brown and

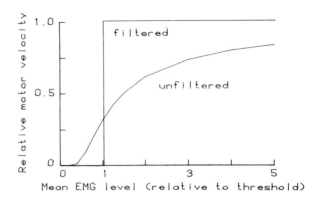

FIG. 4. Mean output (e.g., the mean angular velocity for opening an electric hand) as a function of the mean input (e.g., the root mean square level of the surface EMG). The motor will be on whenever the filtered or unfiltered rectified EMG exceeds a certain threshold level. With the unfiltered signal, variation in speed is possible according to the nonlinear input–output relationship shown.

Stein, 1975). Then, the fraction of time f that the EMG signal will be above the threshold level θ will be

$$f(\theta,\sigma) = \frac{1}{\sigma} \left(\frac{2}{\pi}\right)^{1/2} \int_\theta^\infty e^{-x^2/2\sigma^2} dx$$

The filtered EMG is illustrated for comparison as a simple switch, since the output switches abruptly from off to on as the input level exceeds the threshold. To the extent that the EMG is not perfectly filtered, the step in Fig. 4 will be rounded and give a curve intermediate between the filtered and unfiltered functions shown. The filtered EMG has the advantage that it is more reliable in the presence of noise or variability (for example, as a result of tremor). Any unintentional EMG below the threshold level is less likely to trigger an inadvertent movement. Secondly, it is more efficient in that the motor is not continually being switched on and off during the movement. Rapid, repeated magnetization and demagnetization of the motor's core and frictional losses in starting and stopping the movements waste a certain percentage of the energy available in the battery powering the motor. Since battery power is a major limiting factor in operation of powered limbs, this loss in efficiency is important.

A third method of processing, not shown in Fig. 4, is a true pulse-width modulation, in which a processed EMG signal controls the bidirectional duty cycle of square-wave oscillators. Different amounts of filtering can be introduced prior to the pulse-width modulation with the advantages and disadvantages discussed above. Available prostheses use the various methods described above.

More complex types of processing and input–output curves have been considered (Kwatney et al., 1970; Hogan, 1976; Graupe et al., 1978). Further processing becomes more important the greater the variety of movements required. Suppose, for example, that a below-elbow amputee wishes to control not only a powered electric hand but also a powered wrist. Wrist units become essential for a bilateral amputee, as the reader will realize if he tries to drink a cup of coffee without rotating his wrist. The cup can be easily brought to the mouth, but without pronation, the cup will not be tipped to empty the liquid into the mouth.

TRISTATE CONTROL

Most below-elbow amputees retain at least a portion of the pronator and supinator muscles. However, these muscles are deep lying, so the signals at the surface are small and are not readily detected. Therefore, these muscle remnants cannot be used to obtain the extra movements with conventional signal processing. However, a method of processing EMG that makes these extra movements possible is referred to as tristate control and is illustrated in Fig. 5 (see also Paciaga and Scott, 1978). Rather than having just one threshold level and two states (on and off), a second threshold level is inserted. Then, as show in Fig. 5B, a strong EMG signal that surpasses the second threshold will open the hand, as before. However, an intermediate level can be used to supinate the wrist.

There are several advantages to using the scheme illustrated in Fig. 5 (Stein et al., 1980):

1. Attempts by the patient to activate the supinator muscle (by trying to do the movement he formerly used in turning his wrist) may generate enough EMG signal to produce a supination; i.e., it remains quite a natural movement.
2. If a patient has been using a large, brisk contraction of finger and wrist extensors to open a powered hand, an electric wrist movement can be added at a later date without the patient having to unlearn anything. He simply grades the contraction more carefully so that he can reliably activate the wrist motor with an intermediate level of contraction.
3. A tristate system can also be used with the wrist flexors so that a weak contraction produces a pronation of

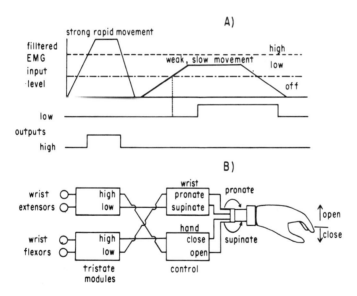

FIG. 5. A: Operation of a tristate module containing high, low, and off states. When the filtered EMG from a strong rapid contraction (left) exceeds the second threshold level, an output appears on the high-level output. A weaker, slow contraction (right) only exceeds the first threshold level and generates a signal on the low-level output after a delay. Note that because of the delay the strong, rapid movement can cross the second threshold before the low-level output is activated. The high-level output internally suppresses the low level, so that conflicting signals are not delivered to a hand or wrist motor. B: Use of two tristate modules to control four movements (pronation and supination of the wrist and opening and closing the hand) from two EMG sources. High-level outputs from wrist extensor and flexor muscles open and close the hand as in Fig. 2. Low-level outputs from the same EMG sources are used to turn the wrist in the two directions. See discussion in the text.

the wrist. The pronators are anatomically close to the wrist flexors, so again the movement may be quite a natural one. Thus, it is possible for a below-elbow amputee with two available muscle sites (wrist extensors and flexors) to control four different movements (opening and closing a hand; pronation and supination of a wrist).

Tristate electronics, which are fully compatible with Otto Bock hands and wrists, are available from at least two sources (Leaf Electronics, Edmonton, Canada; University of New Brunswick Bioengineering Institute, Fredericton, Canada). Earlier units (Dorcas and Scott, 1966; Schmidl, 1973) were bulky enough that two modules could not readily fit in the space available in the forearm shell of the prostheses. They were, however, useful for patients who had only a single EMG site because of either more extensive surgery or a congenital deformity. With only one available EMG site, a weak contraction is used, for example, to close a hand, whereas a stronger contraction is used to open a hand. Patients born without a hand and wrist can often be successfully trained to operate an electric hand and often a wrist satisfactorily. Sorbye (1977) suggests that the success of these efforts is improved if training is started as early as possible, even in the preschool period (2–5 years of age).

In attaining the high (third) state, the processed EMG signal must pass through the intermediate state, which normally would produce a different movement (low output, Fig. 5A). However, the tristate electronics are designed with specific time lags so that the low output is not activated during a rapidly rising contraction. Similarly, the patient relaxes quickly from the high state to prevent activation of the low output. Conversely, the patient increases the contraction slowly when he wants to produce the low output.

Even more states have been attempted (up to 5; Scott et al., 1978), but reliability may decline because of the difficulty of matching the high variability of the EMG signal to the narrower "windows" between one threshold and the next. Alternatively, a larger number of electrodes can be used, and pattern recognition techniques applied to discriminate a variety of functions. These techniques have been applied to both below-elbow (Kato, 1970; Herberts et al., 1978) and above-elbow (Jacobsen et al., 1975) amputees. A commercial multifunctional below-elbow prosthesis (Systemtechnik, Lidingö, Sweden) using these principles is undergoing clinical tests, but the sophisticated methods of computer analysis required to determine the best parameters will probably limit the widespread application of these techniques for some time to come.

Surface EMG electrodes have some inherent limitations in that the location of electrodes built into a prosthesis will vary somewhat from day to day, depending on exactly how the patient dons the limb. In addition, the signals will depend on the moisture of the skin, which determines the skin resistance. The voltage one records externally depends on the extracellular currents generated by the muscle and the effective resistance across the skin. In a country such as Canada with extremely cold winters and mild summers, it is remarkable that any surface-EMG-controlled device works both at 30°C, when the patient is sweating profusely, and at −30°C, when he is shivering in the cold. Other types of control that might reduce some of these difficulties are now discussed.

SURGICAL IMPLANTS

One method for overcoming the problems of surface sites and the limited numbers of muscles that generate surface EMGs involves the use of surgically implanted electrodes. Figure 6 shows a device implanted in 1977 in a short below-elbow amputee (Stein et al., 1980). This amputee had such a short stump (Fig. 7) and weak flexor signals that control of opening and closing a hand was not completely reliable. With external electrodes, there seemed little chance of controlling a wrist (which the amputee wished to do because of his hobbies of carpentry and furniture refinishing). Four EMG electrodes were implanted on the wrist extensors and flexors, supinator and pronator. In addition, a cuff (see below) was placed around the ulnar nerve distal to the exit of the last functional motor branch. All signals were led via insulated wires to a 12-pin socket in a skin interface made of hard carbon (General Atomic, Los Angeles).

Separate amplifiers (with adjustable gain) were used for processing the signals from the four muscles (Fig. 7C), and the amputee inserted a plug into the 12-pin socket to connect the prosthesis each day. Improved control compared to surface electrodes was attained with this implantable system, and greater reliability from day to day. However, in this patient, the skin connector began to work its way out of the skin after a year and had to be removed to prevent infection. Thus, the skin interface probably does not provide a long-term solution at the present time.

TELEMETRY

An alternative that greatly reduces the chances of infection is to use an implanted telemetry system to transmit the EMG signals out of the body as radio waves. The requirements of such a system are stringent:

1. It should operate without a battery, since batteries need replacement from time to time. A cardiac patient who depends on a pacemaker for life undergoes further surgery every few years to change batteries, but an amputee will not repeatedly elect surgery in order to obtain somewhat better

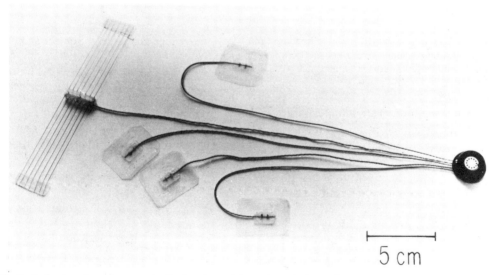

FIG. 6. A device that was surgically implanted in a short below-elbow amputee. It contains EMG probes for four muscle groups (wrist extensors, flexors, pronator, and supinator) as well as a nerve cuff and a percutaneous connector. (From Stein et al., 1980, with permission.)

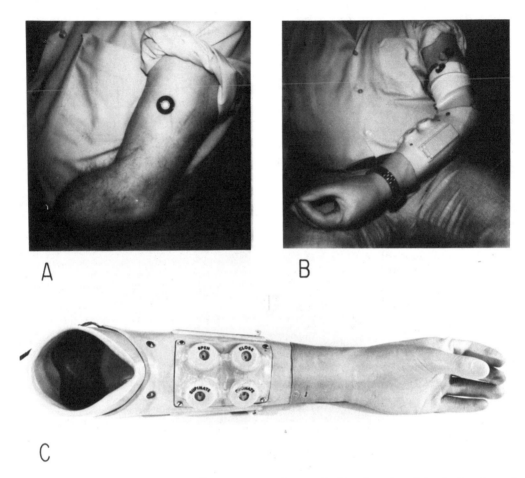

FIG. 7. A: The arm of a short below-elbow amputee after surgical implanation of the device shown in Fig. 6. Note the presence of the percutaneous connnector in the upper arm. B: Prosthesis and cable leading to the percutaneous connector. C: A closer view of the prosthesis shown in B.

function. To operate an internal transmitter as shown in Fig. 8, a source of energy must be provided in the body. A radio signal from an external transmitter may be used for this power. This energy must be rectified and filtered to serve as a power supply for operation of the implanted electronics.

2. The EMG signal from muscles must be amplified and used to frequency-modulate a carrier to transmit the EMG to an external receiver. Often a second set of coils and a distinct frequency is used, although Ko et al. (1979) eliminated the need for separate coils by switching a single coil between transmitting energy and receiving the signals of the electrocardiogram. A simple circuit was implanted by Herberts et al. (1968), and a somewhat more complex one was implanted by Tucker and Peteleski (1977). The devices functioned satisfactorily up to 15 months but eventually were removed because of lead breakage and moisture seepage, which prevented further transmissions. Modern methods have largely

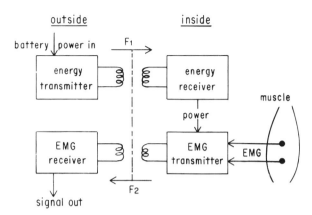

FIG. 8. Schematic diagram for a telemetry device to obtain signals from deeper lying muscles and/or nerves for improved control of electric limbs. Energy is transmitted electromagnetically with frequency F_1 through the skin to a receiver F which rectifies and filters it to supply power to the EMG transmitter. This transmitter amplifies the EMG signals and transmits them, usually with a second frequency F_2, back out through the skin to be received and processed by electronics built into the prosthesis. Note that there is no battery or other electrochemical power source inside the body.

overcome these problems in related applications (Hambrecht, 1979), although obtaining a good seal in implantable devices is not a trivial problem and requires a substantial investment in special tools and materials.

3. Ideally, several channels of EMG should be transmitted if the amputee is to obtain increased function compared to surface electrodes. One method of accomplishing this is to multiplex the input of the EMG transmitter from several sources (Gshwend et al., 1979).

4. Finally, there are severe restraints on size and power consumption. The external devices (energy transmitter and EMG receiver) must be small enough that they fit within the prosthesis of the amputee. They should not add appreciable weight and should not add substantially to the drain on the rechargeable batteries used for operating the prosthesis (e.g., power consumption less than 2 mA at 6 V). Similarly, the internal components must be small and rugged enough to withstand daily use almost indefinitely. The combined noise levels of power source, signal transmitter, and receiver must be low (e.g., 10 μV) compared to the size of the EMG signals.

The constraints on a telemetry system listed above are severe enough that a fully practical system has not yet been developed for widespread use with patients. However, with the developments in microelectronics, we are confident that prosthetic control via telemetry will become practical in the next few years.

NEURAL CONTROL

The higher the level of amputation, the fewer the available sources of EMG signals, and the greater the requirements of the powered prosthesis. For example, an above-elbow amputee needs to control an elbow, a hand, and often a wrist, while having stumps of only two functional muscle groups (elbow flexors such as biceps brachii and elbow extensors such as triceps brachii). Yet, many motoneurons survive, perhaps indefinitely, to distal muscles that have been amputated, and continue to generate action potentials along the stumps of the peripheral nerves (Gordon et al., 1980). Hagbarth and Vallbo developed methods for single unit recordings from human peripheral nerves (Vallbo et al., 1979), but the stability of these recordings over time or with movement is extremely limited (see A. Prochazka and M. Hulliger, *this volume*). More stable methods of recording from single units in motor cortex (Schmidt et al., 1976) and from motoneurons and sensory neurons in the spinal cord and dorsal root ganglion (Loeb et al., 1977; Hoffer et al., 1980) have been developed. However, the stability of recordings from single fibers is presently measured in days or weeks rather than months and years. The inherent variability in the discharges of single neurons is also sufficient that it limits their usefulness in transmitting information for controlling a prosthesis (Stein, 1967; Schmidt et al., 1978; Thomas et al., 1978).

An alternative approach is to record from whole peripheral nerves using cuffs placed around the nerves (Stein et al., 1975, 1977; DeLuca and Gilmore, 1976). The use of cuffs offers two advantages: (1) they provide a high-resistance path for current flow so that the small currents produced by single nerve axons generate a large enough voltage to be detected; (2) they provide some insulation against the much larger currents generated by surrounding muscles. For neural recording, the EMG signal is, in effect, unwanted noise.

The methods of recording from nerves using cuffs are shown schematically in Fig. 9A. Three equally spaced uninsulated electrodes are sewn on the inside of the cuff. The cuff has been slit longitudinally so that it can be opened and placed around the nerve during a sterile surgical procedure. The cuff is then carefully sutured shut to minimize current leakage through the slit. The two end electrodes are connected together so that the two ends of the cuff remain at the same voltage. Hence, currents from external sources including surrounding muscles will flow around and not through the cuff (the flow of a current I across the resistance R of the nerve and the extracellular fluid within the cuff requires a voltage difference $V = IR$); EMG rejection with this method can be exceptionally good. Thus, the small currents generated by the nerve within the cuff can be recorded using the central electrode with respect to the end electrodes (see Fig. 9A) with little EMG contamination.

The signals recorded depend on both the length (l) and diameter (d) of the cuff (Fig. 9B). As the length of the cuff is increased, the voltage recorded increases as the square of the length up to a distance of 1 to 2 cm. Above this length, the signals approach a constant value. The entire curve is shifted down as the square of the cuff diameter. Thus, using a cuff three times as large will give signals only one-ninth as large. Note that the amplitudes are only a few microvolts in size, even for fairly long cuffs. However, by use of low-input-impedance preamplifiers (e.g., QT5-B, Leaf Electronics, Edmonton, Canada) matched to the relatively low input impedance of these cuff electrodes (1–10 kΩ), these small signals can be recorded reliably for long periods of time (Gordon et al., 1980). This is true even during behavioral situations in which considerable amounts of electrical interference are present in the environment (e.g., walking on a motor-driven treadmill).

Our experience to date is largely with cat nerves using cuffs with internal diameters of 1.4 to 3.4 mm. In our one experience using cuffs chronically in a human patient (see Figs. 6 and 7), we were only able to record compound neural potentials. The nerve had been cut following the amputation in this patient over 30 years ago, and sufficient atrophy had presumably occurred that the small volitional signals were not measurable. In a few acute trials with other patients in the operating room, sensory nerve signals could be recorded from human nerves, although they were smaller

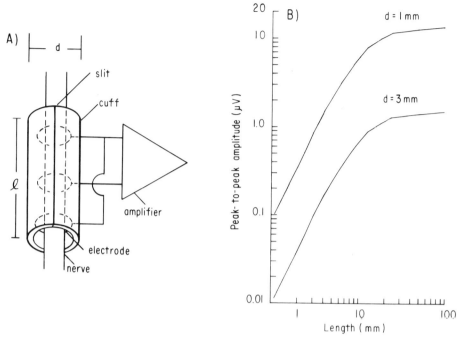

FIG. 9. A: To record from a whole nerve during normal behavior, a cuff is placed around a nerve, and the longitudinal slit in the cuff is sutured closed. Three electrodes in the form of the letter C are sewn on the inside of the cuff. The two end electrodes are shorted together, and the potentials generated by the nerve at the center of the cuff are amplified differentially with respect to the potentials at the two ends of the cuff. Further discussion in the text. B: The voltage recorded from single nerve axons depends on the length and diameter of the cuff relative to the wave length of the nerve impulse and the diameter of the nerve fiber. The amplitude reaches an asymptotic value for cuffs a few centimeters in length and varies inversely with the square of the internal diameter (d) of the cuff. Values shown were computed for a nerve axon with a diameter of 10 μm. (Modified from Stein et al., 1975, with permission.)

because of the large size of human nerves compared to those in the cat (See Fig. 9B).

SENSORY FEEDBACK

Even though we were unsuccessful in recording volitional nerve signals from the amputee shown in Fig. 7, we were able to stimulate through the cuff and produce a clear, conscious sensation. Furthermore, the amputee was able to localize the stimulation to the lateral surface of the "phantom hand" that had been amputated over 30 years previously. This is the normal innervation field of the ulnar nerve, which indicates that the central connections of nerve axons in an adult human are exceedingly stable.

The provision of useful sensory feedback to an amputee has been a long sought after but somewhat elusive goal. Experiments have been reported using direct stimulation of nerves through implanted electrodes (Clippinger et al., 1974) or stimulation of skin (Rohland, 1975; Schmidl, 1977; Brittain et al., 1979; Scott et al., 1980). These studies have been reported on individual cases, but there has not yet been widespread application of these techniques. An amputee does get some feedback from vision (of hand position) and audition (operation of the motor), so most attempts have concentrated on providing force feedback by using strain gauges that sense tension of the fingers of the prosthesis. The aim is to permit the amputee to assess force well enough that he can pick up a delicate object such as an egg on one trial and hold an object tightly in another.

Although it is a desirable objective, we feel that sensory feedback is not yet as high a priority as provision of greatly improved motor control, particularly for high-level amputees. Experiments with monkeys whose arms have been completely deafferented (Polit and Bizzi, 1979; Goodwin and Luschei, 1974; E. Bizzi and W. Abend, *this volume*) have shown that these animals can perform trained movements very satisfactorily against perturbing forces. We still have a long way to go before powered prosthetic limbs can match the performance of even these deafferented limbs. Finally, the amount of information that can be fed back by gross electrical stimuli to the skin or a whole nerve is limited (Solomonow et al., 1978; Walker, et al., 1977; Anani et al., 1977).

OTHER METHODS OF CONTROL

Even with an amputee who has lost his entire arm (shoulder disarticulation), movements about the shoulder girdle remain. Furthermore, movements of the limb normally require contraction of the shoulder muscles for positioning and stabilization of the limb. Thus, one can relate the EMG from shoulder muscles (Wirta et al., 1978) or the position of the shoulder (Simpson, 1971) to desired movements of the limb. The use of shoulder position as a control signal to specify the position of a powered prosthesis has been referred to as "extended physiological proprioception," since the proprioception one has about shoulder position can now be extended to give information about the position of the limb in space. In principle, pressure or positional information from the shoulder could be used for proportional control of arm position. However, the extra complexity and additional battery drain (because of frequent movements resulting from small changes in shoulder position) have limited the practical application of this approach.

We have developed an additional approach to assisting high-level amputees, which we refer to as "touch control." Well-isolated metal contacts are built into the prosthesis at points where the amputee can touch them by movements of the shoulder. Touching the skin to the contact closes an electrical circuit (module TC-2; Leaf Electronics, Edmonton, Canada) and activates a particular movement (Fig. 10). More than one contact can be placed at a given site by using a dome-shaped central contact and an annulus electrically isolated from it. Movement of the shoulder will first touch the dome and produce one movement, whereas further movement of the shoulder will also touch the annulus. Simple logic circuitry (PCL-1; Leaf Electronics) overrides the first movement when this occurs and activates the second movement. If the amputee can produce three distinct movements (e.g., forward flexion, retraction, and elevation of the shoulder), three separate motors can be controlled for movements of the elbow, wrist, and hand in each of two directions (Stein et al., 1980). The success of this complex a procedure depends strongly on the skill of the prosthetist in fabricating a structure that is light, stays in place during body movements, and has sites that are easily but not inadvertently touched. It also depends on the motivation of the patient and the training program for occupational therapy (Stein et al., 1980).

COMPARISON OF POWERED PROSTHESES WITH NORMAL MUSCLES

Up to this point, we have concentrated on the control of powered prostheses. Although reliable, efficient control is essential to the success of powered prostheses, the characteristics of the motors and gears

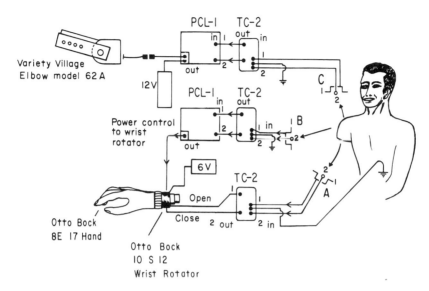

FIG. 10. Components used for a very short above-elbow amputee to control a powered elbow, wrist, and hand. A: Protraction (forward movement) of the humerus brings the skin in contact with the central, nonpriority input of the touch control (TC-2) module, which produces a closing of the hand. Further protraction brings the skin also in contact with the outer, priority input, which cancels the previous input and produces an opening of the hand. B: Abduction of the humerus (away from the body) is used for pronation (nonpriority) and supination (priority) of the wrist using another TC-2 module. C: Retraction (backward movement) of the humerus produces flexion (nonpriority) and extension (priority) of an electric elbow using a third TC-2 module. Power control and logic (PCL-1) modules are used for wrist and elbow, since they do not have control logic built in, as the hand does. (From Stein et al., 1980, with permission.)

compared to normal muscles are equally important. Rather than the superhuman bionic capabilities suggested by some television programs, commercially available powered limbs are still very weak and slow and have a low power-to-weight ratio compared to the natural limbs they replace.

The ability of muscles to respond to neural inputs containing a wide range of frequency components can be assessed using sinusoidally or randomly varying input patterns (Rosenthal et al., 1970; Mannard and Stein, 1973). Figure 11 shows the response of human soleus muscle to a randomly varying pattern of stimuli (Bawa and Stein, 1976). Note that the muscle responds well to frequency components below 2 Hz but that the response declines as the square of frequency at higher frequencies. The response also lags increasingly behind the input at higher frequencies. Thus, muscles behave as "low-pass filters" and reject higher-frequency components such as the 10 Hz component of tremor while responding faithfully to the lower-frequency components that are contained in voluntary behavior. The same pattern of response has been described in a variety of muscles in frogs, cats, and humans. The "break frequency" at which the response begins to fall off sharply is higher in fast limb muscles (5 Hz, cat plantaris muscle; Mannard and Stein, 1973) and eye muscles (8–16 Hz in the cat; Vilis and Lennerstand, 1979) than in slow muscles such as soleus.

The fact that the response falls off as the square of frequency at high frequencies implies that there are two processes limiting the contraction and relaxation of muscles. The range of "break frequencies" from 2 to

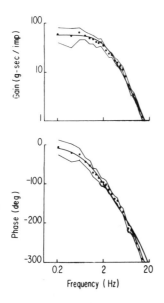

FIG. 11. Values for the gain and phase of responses for human soleus muscle with random stimulation at rest with the ankle held isometrically. The *thick continuous lines* are the curves expected for a linear, second-order system (see text). The *thin lines* show 95% confidence intervals for the data. Scales for gain and frequency are logarithmic. (From Bawa and Stein, 1976, with permission.)

16 Hz implies that the time constants limiting muscle responses are of the order of 10 to 80 msec. There is not space here to discuss the nature of the molecular processes in muscles that provide these limits, but the response characteristics are surprisingly robust. For example, Fig. 12A shows the frequency response of elbow flexors (e.g., biceps brachii) contracting against loads of 0 to 2 kg held in the hand (Aaron and Stein, 1976). The low-frequency behavior is similar, and the peak in the response (near 2 Hz) with the extra loads may indicate a tendency for oscillations to occur because of the interaction of the mass with the elasticity of the muscles.

In comparison, a powered artificial elbow (Liberty Mutual Insurance Co., Boston) shows a response (Fig. 12B) that decreases continuously with frequency and shows phase lags greater than about 90° at all frequencies. This extra phase lag results mainly from the fact that the input controls the velocity and not the position of the prosthesis (see also caption to Fig. 12). Nonetheless, since the response is more than an order of magnitude poorer at 2 Hz than at 0.2 Hz, the prosthesis is not useful for rapid tracking or positioning tasks. Commercially available elbows have other limitations (Liberty Mutual, Boston; Variety Village, Toronto; Fidelity Electronics, Chicago) in that they can lift only 1 to 3 kg placed in the terminal device (hook or hand). The natural human elbow muscles can lift an order of magnitude greater load placed in the hand.

The limitations result to a substantial degree from the most commonly available electrical power sources, which are rechargeable nickel–cadmium batteries. These batteries presently suffice for a day of normal use, and employing faster or more powerful motors would mean that the batteries would run down more quickly. Improvements are long overdue in the capabilities of batteries and the efficiency of the motor and gear trains used. One quickly gains an appreciation of the remarkable abilities of human muscles when they are compared to the weak and inefficient artificial appliances presently available. The limitations are even more severe in considering powered prostheses for the lower limb. Small nickel–cadmium batteries would be completely inadequate for lifting the weight of the body during locomotion, but they may have a role in controlling the damping of the knee joint, which must be small during the swing phase and high during the stance phase of locomotion (Myers and Moskowitz, 1980).

APPLICATION TO UPPER EXTREMITY AMPUTEES

Despite the substantial numbers of powered hands fitted, relatively few follow-up studies have been published. Acceptance of powered hands has generally been

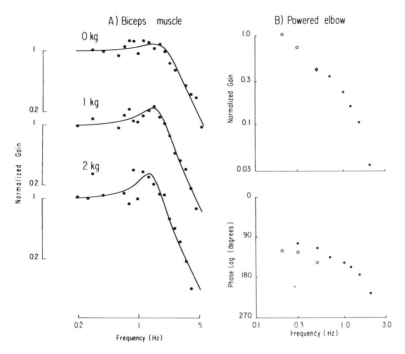

FIG. 12. A: Normalized gain for human biceps brachii muscle during voluntary oscillatory movements with added loads of 0, 1, and 2 kg held in the hand. The gain is remarkably independent of load except for the peak that develops with increasing load near 2 Hz. This peak probably represents the tendency of the mass of the load (and the forearm) to oscillate against the elasticity of the muscle. The scale and data for different loads have been shifted with respect to one another for clarity. B: The gain of a powered elbow (Liberty Mutual, Boston) decreases continuously with frequency from the lowest frequencies studied. The phase of the powered elbow always lags more than 90° behind the input (because the EMG input is used to control the velocity, not the position, of the elbow), and this phase lag increases with frequency. *Open* and *closed symbols* in B represent two different conditions of measurement. See Aaron and Stein (1976) for more details.

good, and claims have been made that over 90% of the below elbow amputees fitted wear the powered prosthesis in preference to their former prosthesis (Soerjanto, 1971; Northmore-Ball et al., 1980). For comparison, rejection rates of conventional below-elbow prostheses have been claimed to be as high as two out of three (Vitali et al., 1978).

Acceptance of powered elbows has been lower, reflecting the severe technical limitations of the elbows mentioned above. Our results on a series of 35 amputees fitted in Edmonton agree with the studies quoted above. In addition to simple measures of acceptance or rejection and subjective responses to questionnaires, we measured performance on a number of objective hand function tests (modified from Jebsen et al., 1969). Wherever possible, function was compared for amputees performing the same task with a myoelectrically controlled hand, with a conventional prosthesis (cable-controlled hook), and with their normal hand in the case of unilateral amputees.

The time required to perform these tasks with the normal hand increased with age, as was expected. The time to perform the same tasks with the conventional hook prosthesis was approximately twice as great and increased with age in a similar fashion. Interestingly, performance on the task with a myoelectric hand was another factor of two slower at any age despite the fact that these patients were generally pleased with their myoelectric prostheses. Most of the amputees wore the powered prosthesis in preference to a conventional prosthesis. Performance on most tasks followed the same pattern, which suggests that patients wear a myoelectric hand for a mixture of cosmetic and functional reasons. They appreciate the improved appearance of the hand compared to the hook (e.g., people do not stare at them), even though the movement is not as fast. Nonetheless, function with a myoelectric hand is improved compared to a passive hand or the so-called "working hand" which has conventional cable control and is in fact extremely difficult to use. It is essential that the speed and accuracy of myoelectric hands be increased in the next few years to the point that they match a hook or perhaps even natural hand movements. However, control must also be improved in parallel with speed, since there will be less time for corrections to be made.

FUTURE DEVELOPMENTS

In conclusion, progress in this field will require developments in three major areas:

1. Improvements must be made in methods for controlling powered artificial limbs, so that control becomes possible not only of the major joints (elbow, wrist, and hand) but also for independent movements of the fingers and powered movements in more than one direction at higher joints (e.g., internal and external rotation of the elbow as well as flexion and extension).
2. More complex prostheses with up to 12 motors have been described (Herberts et al., 1978; Wirta et al., 1978; Kato, 1978; Funakubo et al., 1980), and developments of new, improved, and more versatile gears and motors must proceed in parallel with development of control sites. Many of these are already available for industrial purposes, but their application to prosthetics requires transfer of information across disciplines.
3. Health care delivery methods must also be improved. Although prosthetists may be superbly skilled in fitting artificial limbs, they have little or no experience with electronics. At the same time, as training programs are developed to increase prosthetists' ability to fit and service more complex prostheses, care must be used in designing components that are interchangeable and as compatible with one another as possible.

Coordinated action among scientists and engineers developing new devices, manufacturers producing them, prosthetists fitting them, and governments providing the needed financing is as important as the purely technical developments cited earlier. Above all, the needs and wishes of individual amputees must be kept in mind. The diversity of these requirements means that mass production on a scale found for many consumer goods is unlikely. However, despite the substantial costs and individual attention required, substantial progress has been made since powered limbs were introduced commercially 20 years ago. We hope that at least some benefits of 20th century technology will become available to ever increasing numbers of amputees in the final decades of the century.

ACKNOWLEDGMENTS

The research described in this chapter was supported in part by the Medical Research Council of Canada, the University of Alberta Hospital, the Medical Services Research Foundation of Alberta, and associated private foundations.

References

Aaron, S. L., and Stein, R. B. (1976): Comparison of an EMG-controlled prosthesis and the normal human biceps brachii muscle. *Am. J. Phys. Med.,* 55:1–14.

Abbruzzese, G., Abbruzzese, M., Favale, E., et al. (1980): The effect of hand muscle vibration on the somatosensory evoked potential in man: an interaction between lemniscal and spinocerebellar inputs? *J. Neurol. Neurosurg. Psychiatry,* 43:433–437.

Abdelmoumene, M. (1979): Bases neurophysiologiques du traitement de la douleur dans la douleur. Données sur son origine et son traitement. *Rapports 42° Congrès Français de Médecine,* pp. 29–44. Masson, Paris.

Abend, W. K., Bizzi, E., and Morasso, P. (1982): Human arm trajectory formation. *Brain,* 105:331–348.

Abrahams, V. C. (1972): Neck muscle proprioceptors and vestibulospinal outflow at lumbosacral levels. *Can. J. Physiol. Pharmacol.,* 50:17–21.

Abzug, C., Maeda, M., Peterson, B. W., and Wilson, V. J. (1974): Cervical branching of lumbar vestibulospinal axons. *J. Physiol. (Lond.),* 243:499–522.

Adal, M. N., and Barker, D. (1962): Intramuscular branching of fusimotor fibres. *J. Physiol. (Lond.),* 177:288–299.

Adey, W. R., Porter, R., and Carter, I. D. (1954): Temporal dispersion in cortical afferent volleys as a factor in perception; an evoked potential study of deep somatic sensibility in the monkey. *Brain,* 77:325–344.

Adrian, E. D. (1943): Discharges from vestibular receptors in the cat. *J. Physiol. (Lond.),* 101:389–407.

Adrian, E. D., and Bronk, D. W. (1929): The discharge of impulses in motor nerve fibres. The frequency of discharge in reflex and voluntary contractions. *J. Physiol. (Lond.),* 67:119–151.

Agarwal, G. C., and Gottlieb, G. L. (1972): The muscle silent period and reciprocal inhibition in man. *J. Neurol. Neurosurg. Psychiatry,* 35:72–76.

Agarwal, G. C., and Gottlieb, G. L. (1980): Effect of vibration on the ankle stretch reflex in man. *Electroencephalogr. Clin. Neurophysiol.,* 49:81–92.

Aguayo, A. J., Nair, C. P. V., and Bray, G. M. (1971): Peripheral abnormalities in the Riley-Day syndrome. *Arch. Neurol.,* 24:106–116.

Aguilar, M. J. (1969): Recovery of motor function after unilateral infarction of the basis pontis. *Am. J. Phys. Med.,* 48:279–288.

Aguilar, V. E., Bisby, M. A., Cooper, E., and Diamond, J. (1973): Evidence that axoplasmic transport of trophic factors is involved in the regulation of peripheral nerve fields in salamanders. *J. Physiol. (Lond.),* 234:449–464.

Ahlgren, J. G. A. (1969): The silent period in the EMG of the jaw muscles during mastication and its relationship to tooth contact. *Acta Odontol. Scand.,* 27:219–227.

Akimoto, H., and Saito, Y. (1966): Synchronizing and desynchronizing influences and their interactions on cortical and thalamic neurons. *Prog. Brain Res.,* 21:323–351.

Albe-Fessard, D., and Besson, J. M. (1973): Convergent thalamic and cortical projections. The nonspecific system. In: *Handbook of Sensory Physiology,* edited by A. Iggo, pp. 489–560. Springer, Berlin.

Albe-Fessard, D., Levante, A., and Lamour, Y. (1974): Origin of spinothalamic tract in monkeys. *Brain Res.,* 65:503–509.

Albe-Fessard, D., and Liebeskind, J. (1966): Origines des messages somatosensitifs activant les cellules du cortex motor chez le singe. *Exp. Brain Res.,* 1:127–146.

Albuquerque, E. X., Warnick, J. E., Tasse, J. R., and Sansone, F. M. (1972): Effects of vinblastine and colchicine on neural regulation of the fast and slow skeletal muscles of the rat. *Exp. Neurol.,* 37:607–614.

Albus, J. S. (1971): A theory of cerebellar function. *Math. Biosci.,* 10:25–61.

Allen, W. F. (1919): The application of the Marchi method to the study of radix mesencephalica trigemini in guinea pig. *J. Comp. Neurol.,* 10:169–283.

Allen, G. I., Gilbert, P. F. C., Marini, R., et al. (1977): Integration of cerebral and peripheral inputs by interpositus neurons in monkey. *Exp. Brain Res.,* 27:81–99.

Allen, G. I., and Tsukahara, N. (1978): Cerebrocerebellar communication systems. *Physiol. Rev.,* 54:957–1006.

Allum, J. H. J. (1975): Responses to load disturbances in human shoulder muscles: the hypothesis that one component is a pulse test information signal. *Exp. Brain Res.,* 22:307–326.

Allum, J. H. J., and Budingen, H. Y. (1979): Coupled stretch reflexes in ankle muscles and human posture stability. *Prog. Brain Res.,* 50:185–196.

Allum, J. H. J., Graf, W., Dichgans, J., and Schmidt, C. L. (1976): Visual vestibular interactions in the vestibular nuclei of the goldfish. *Exp. Brain Res.,* 26:463–485.

Alnaes, E. (1967): Static and dynamic properties of Golgi tendon organs in the anterior tibial and soleus muscles of the cat. *Acta Physiol. Scand.,* 70:176–187.

Alstermark, B., Lundberg, A., Norrsell, U., and Sybirska, E. (1981): Integration in descending motor pathways controlling the forelimb in the cat. *Exp. Brain Res.,* 42:299–318.

Alston, W., Angel, R. W., Fink, F. S., and Hofmann, W. W. (1967): Motor activity following the silent period in human muscle. *J. Physiol. (Lond.),* 190:189–202.

Amassian, V. E. (1953): Evoked single cortical unit activity in the somatic sensory areas. *Electroencephalogr. Clin. Neurophysiol.,* 5:415–438.

Amassian, V. E., Patton, H. D., Woodbury, J. W., et

al. (1955): An interpretation of the surface response in somatosensory cortex to peripheral and interareal afferent stimulation. *Electroencephalogr. Clin. Neurophysiol.,* 7:480–483.
Amblard, B., and Cremieux, J. (1976): Role of visual motion information in the maintenance of postural equilibrium in man. *Agressologie,* 17:25–36.
Anani, A. B., Ikeda, K., and Korner, L. (1977): Human ability to discriminate various parameters in afferent electrical nerve stimulation with particular reference to prostheses sensory feedback. *Med. Biol. Eng. Comput.,* 15:363–373.
Andersen, P., Eccles, J. C., and Sears, T. A. (1964): Cortically evoked depolarization of primary afferent fibers in the spinal cord. *J. Neurophysiol.,* 27:63–77.
Andersen, P., Hagan, P. J., Phillips, C. G., and Powell, T. P. S. (1975): Mapping by microstimulation of projections from area 4 to motor units of the baboon's hand. *Proc. R. Soc. Lond. (Biol.),* 188:31–60.
Anderson, M. E., Yoshida, M., and Wilson, V. J. (1972): Tectal and tegmental influences on cat forelimb and hindlimb motoneurons. *J. Neurophysiol.,* 35:462–470.
Andersson, B. F., Lennerstrand, G., and Thoden, U. (1968): Fusimotor effects on position and velocity sensitivity of spindle endings in the external intercostal muscle of the cat. *Acta Physiol. Scand.,* 74:285–300.
Andersson, S. A., and Holmgren, E. (1976): Pain threshold effects of peripheral conditioning. In: *Advances in Pain Research and Therapy.* Vol. 3, edited by J. J. Bonica, J. C. Liebeskind, and D. G. Albe-Fessard, pp. 761–767. Raven Press, New York.
Andrews, C. J., and Burke, D. (1973): A quantitative study of the effect of L-dopa and phenoxybenzamine on the rigidity of Parkinson's disease. *J. Neurol. Neurosurg. Psychiatry,* 36:321–328.
Andrews, C. J., Burke, D., and Lance, J. W. (1972): The response to muscle stretch and shortening in parkinsonian rigidity. *Brain,* 95:795–812.
Andrews, C. J., Neilson, P. D., and Lance, J. W. (1973): Comparison of stretch reflexes and shortening reactions in activated normal subjects with those in Parkinson's disease. *J. Neurol. Neurosurg. Psychiatry,* 36:329–333.
Angel, R. W. (1974): Electromyography during voluntary movement: the two-burst pattern. *Electroencephalogr. Clin. Neurophysiol.,* 36:493–498.
Angel, R. W. (1980): Barognosis in a patient with hemitaxia. *Ann. Neurol.,* 7:73–77.
Angel, R. W. (1981): Electromyographic patterns during ballistic movements in normals and in hemiplegic patients. In: *Progress in Clinical Neurophysiology, Vol. 9, Motor Unit Types, Recruitment and Plasticity in Health and Disease,* edited by J. E. Desmedt, pp. 347–357. Karger, Basel.
Angel, R. W. (1982a): Shortening reaction in patients with cerebellar ataxia. *Ann. Neurol.,* 32:272–278.
Angel, R. W. (1982b): Shortening reaction in normal and Parkinsonian subjects. *Neurology (Minneap.),* 32:246–251.
Angel, R. W., and Alston, W. (1964): Spindle afferent conduction velocity. *Neurology (Minneap.),* 14:674–676.

Angel, R. W., Alston, W., and Higgins, J. R. (1970): Control of movement in Parkinson's disease. *Brain,* 93:1–14.
Angel, R. W., and Hofmann, W. W. (1963): The H reflex in normal spastic and rigid subjects. *Arch. Neurol.* 8:591–596.
Angel, R. W., and Lemon, R. N. (1975): Sensorimotor cortical representation in the rat and the role of the cortex in the production of sensory myoclonic jerks. *J. Physiol. (Lond.),* 248:465–488.
Annett, J., Golby, C. W., and Kay, H. (1958): The measurement of elements in an assembly task. The information output of the human motor system. *Q. J. Exp. Psychol.,* 10:1–11.
Aoki, M., Mori, S., and Fujimori, B. (1976): Exaggeration of knee-jerk following spinal hemisection in monkeys. *Brain Res.,* 107:471–485.
Appelberg, B. (1962): The effect of electrical stimulation in nucleus ruber on response to stretch in primary and secondary spindle afferents. *Acta Physiol. Scand.,* 56:140–151.
Appelberg, B. (1981): Selective central control of dynamic gamma-motoneurones utilised for the functional classification of gamma cells. In: *Muscle Receptors and Movements,* edited by A. Taylor and A. Prochazka, pp. 97–108. Macmillan, London.
Appelberg, B., Hulliger, M., Johansson, H., and Sojka, P. (1981): Reflex activation of dynamic fusimotor neurons by natural stimulation of muscle and joint receptor afferent units. In: *Muscle Receptors and Movement,* edited by A. Taylor and A. Prochazka, pp. 149–161. Macmillan, London.
Appelberg, B., Johansson, H., and Kalistratov, G. (1977): The influence of group II muscle efferents and low-threshold skin afferents on dynamic fusimotor neurones to the triceps surae of the cat. *Brain Res.,* 132:153–158.
Appenteng, K., Lund, J. P., and Seguin, J. J. (1982): The behavior of cutaneous mecanoreceptors recorded in the mandibular division of the gasserian ganglion of the rabbit during movements of the lower jaw. *J. Neurophysiol.,* 47:151–166.
Appenteng, K., Morimoto, T., and Taylor, A. (1980): Fusimotor activity in masseter nerve of the cat during reflex jaw movements. *J. Physiol. (Lond.),* 305:415–431.
Appenteng, K., O'Donovan, M. J., Somjen, G., et al. (1978): The projection of jaw elevator muscle spindle afferents to fifth nerve motoneurones in the cat. *J. Physiol. (Lond.),* 279:409–423.
Appenzeller, O., Kornfeld, M., and Albuquerque, N. M. (1972): Indifference to pain: a chronic peripheral neuropathy with mosaic Schwann cells. *Arch. Neurol.,* 27:322–339.
Araki, T., Eccles, J. C., and Ito, M. (1960): Correlation of inhibitory postsynaptic potential of motoneurones with the latency and time course of inhibition of monosynaptic reflexes. *J. Physiol. (Lond.),* 154:354–377.
Arbib, M. (1981): Perceptual structures and distributed motor control. In: *Handbook of Physiology, Section I, Vol. II, Motor Control,* edited by V. B. Brooks, pp. 1449–1480. American Physiological Society, Washington, D.C.

Armstrong, D. M. (1965): Synaptic excitation and inhibition of Betz cells by antidromic pyramidal volleys. *J. Physiol. (Lond.),* 178:37–38P.

Aronski, A., Kubler, A., Majda, A., and Jakubaska, J. (1975): Comparative study of the action of naloxone and nalorphine in man. *Resusc. Intens. Ther.,* 3:3–12.

Asanuma, C., Thach, W. T., and Jones E. G. (1983): Distribution of cerebellar terminations and their relation to other afferent terminations in the thalamic ventral lateral region of the monkey. *Brain Res. Rev.* 5:237–266.

Asanuma, H. (1975): Recent developments in the study of the columnar arrangement of neurons within the motor cortex. *Physiol. Rev.,* 55:143–156.

Asanuma, H. (1981): The pyramidal tract. In: *Handbook of Physiology, Section I. Vol. II, Motor Control,* edited by V. B. Brooks, pp. 703–733. American Physiological Society, Bethesda.

Asanuma, H., Arnold, A., and Zarzecki, P. (1976): Further study on the excitation of pyramidal tract cells by intracortical microstimulation. *Exp. Brain Res.,* 26:443–461.

Asanuma, H., Larsen, K. D., and Yumiya, H. (1979): Receptive fields of thalamic neurons projecting to the motor cortex in the cat. *Brain Res.,* 172:217–228.

Asanuma, H., and Rosen, I. (1972): Topographical organization of cortical efferent zones projecting to distal forelimb muscles in the monkey. *Exp. Brain Res.,* 14:243–256.

Asanuma, H., and Sakata, H. (1967): Functional organization of a cortical efferent system examined with focal depth stimulation in cats. *J. Neurophysiol.,* 30:35–54.

Asatryan, D. G., and Feldman, A. G. (1965): Functional tuning of nervous system with control of movement or maintenance of a steady posture. *Biophysics,* 10:925–935.

Asbury, A. K., Victor, M., and Adams, R. D. (1963): Uremic polyneuropathy. *Arch. Neurol.,* 8:413–428.

Ashby, P., and Burke, D. (1971): Stretch reflexes in the upper limb of spastic man. *J. Neurol. Neurosurg. Psychiatry,* 34:765–771.

Ashby, P., Burke, D., Rao, S., and Jones, R. F. (1972) The assessment of cyclobenzaprine in the treatment of spasticity. *J. Neurol. Neurosurg. Psychiatry,* 35:599–605.

Ashby, P., Carlen, P., Macleod, S., and Sinha, P. (1977): Ethanol and spinal presynaptic inhibition in man. *Ann. Neurol.,* 1:478–480.

Ashby, P., and Labelle, K. (1977): Effects of extensor and flexor group I afferent volleys on the excitability of individual soleus motoneurones in man. *J. Neurol. Neurosurg. Psychiatry,* 40:910–919.

Ashby, P., and Verrier, M. (1975): Neurophysiological changes following spinal cord lesions in man. *Can. J. Neurol. Sci.,* 2:91–100.

Ashby, P., and Verrier, M. (1976): Neurophysiological changes in hemiplegia: possible explanation for the disparity between muscle tone and tendon reflexes. *Neurology (Minneap.),* 26:1145–1151.

Ashby, P., Verrier, M., and Carleron, S. (1980): Vibratory inhibition of the monosynaptic reflex. In: *Progress in Clinical Neurophysiology, Vol. 8, Spinal and Supraspinal Mechanisms of Voluntary Motor Control and Locomotion,* edited by J. E. Desmedt, pp. 254–262. Karger, Basel.

Ashby, P., Verrier, M., and Lightfoot, E. (1974): Segmental reflex pathway in spinal shock and spinal spasticity in man. *J. Neurol. Neurosurg. Psychiatry,* 37:1352–1360.

Ashby, P., and White, D. F. (1973): Presynaptic inhibition in spasticity and the effect of beta-(4-chlorophenyl)Gaba. *J. Neurol. Sci.,* 20:329–338.

Ashby, P., and Zilm, D. (1978): Synaptic connections to individual tibialis anterior motoneurones in man. *J. Neurol. Neurosurg. Psychiatry,* 41:684–689.

Azarbal, M. (1977): Comparison of Myo-Monitor centric position to centric relation occlusion. *J. Prosthet. Dent.,* 38:331–337.

Azzena, G. B., Tolu, E., and Mameli, O. (1978): Responses of vestibular units to visual input. *Arch. Ital. Biol.,* 116:120–129.

Baarsma, E. A., and Collewijn, J. (1974): Vestibuloocular and optokinetic reactions to rotation and their interaction in the rabbit. *J. Physiol. (Lond.),* 238:603–625.

Bach-y-Rita, P. (1964): Convergent and long latency unit responses in the reticular formation of the cat. *Exp. Neurol.,* 9:327–344.

Bach-y-Rita, P. (1972): *Brain Mechanisms in Sensory Substitution.* Academic Press, New York.

Bach-y-Rita, P. (1980): Brain plasticity as a basis for therapeutic procedures. In: *Recovery of Function: Theoretical Considerations for Brain Injury Rehabilitation,* edited by P. Bach-y-Rita, pp. 225–263. University Park Press, Baltimore.

Bach-y-Rita, P. (1981a): Brain plasticity as a basis for the development of rehabilitation procedures for hemiplegia. *Scand. J. Rehabil. Med.,* 13:73–83.

Bach-y-Rita, P. (1981b): Sensory substitution in rehabilitation. In: *Rehabilitation of the Neurological Patient,* edited by L. Illis, M. Sedgwick, and H. Granville, pp. 361–383. Blackwell, Oxford.

Bach-y-Rita, P. (1981c): Central nervous system: "sprouting" and "unmasking" in rehabilitation. *Arch. Phys. Med. Rehabil.,* 62:413–417.

Bailey, J. O., McCall, W. D., and Ash, M. M. (1977a): The influence of mechanical input parameters on the duration of the EMG silent period in man. *Arch. Oral Biol.,* 22:619–623.

Bailey, J. O., McCall, W. D., and Ash, M. M. (1977b): Electromyographic silent periods and jaw motion parameters: quantitative measures of temporomandibular joint dysfunction. *J. Dent. Res.,* 56:249–253.

Balsinger, J., Lund, R. D., and Miller, B. (1977): Aberrant retinothalamic projections resulting from unilateral tectal lesions made in foetal and neonatal rats. *Exp. Neurol.,* 54:369–382.

Baker, J., Gibson, A., Glickstein, M., and Stein, J. (1976): Visual cells in the pontine nuclei of the cat. *J. Physiol. (Lond.),* 225:415–433.

Baker, M. A., Tyner, C. F., and Towe, A. L. (1971): Observations on single neurons recorded in the sigmoid gyri of awake, nonparalysed cats. *Exp. Neurol.,* 32:388–403.

Baker, R., and Berthoz, A. (1974): Organization of

vestibular nystagmus in oblique oculomotor system. *J. Neurophysiol.*, 37:195–217.
Baker, R., Gresty, M., and Berthoz, A. (1976): Neuronal activity in the prepositus hypoglossi nucleus corelated with vertical and horizontal eye movement in the cat. *Brain Res.*, 101:366–371.
Baker, R., and Llinas, R. (1971): Electrotonic coupling between neurones in the cat mesencephalic nucleus. *J. Physiol. (Lond.)*, 212:45–63.
Baker, R. G., Mano, N., and Shimazu, H. (1969): Postsynaptic potentials in abducens motoneurones induced by vestibular stimulation. *Brain Res.*, 15:577–580.
Baldissera, F., and Gustafsson, B. (1971): Supraspinal control of the discharge evoked by constant current in the alpha-motoneurones. *Brain Res.*, 25:642–644.
Baldissera, F., and Gustafsson, B. (1974): Firing behaviour of a neurone model based on the afterhyperpolarization conductance time course and algebraical summation. *Acta Physiol. Scand.*, 92:27–47.
Baldissera, F., Gustafsson, B., and Parmiggiani, F. (1978): Saturating summation of the afterhyperpolarization conductance in spinal motoneurones: a mechanism for 'secondary range' repetitive firing. *Brain Res.*, 146:69–82.
Baldissera, F., and Parmiggiani, F. (1975): Relevance of motoneural firing adaptation to tension development in the motor unit. *Brain Res.*, 91:315–320.
Balint, R. (1909): Seelenhemmung des "Schauens," optische Ataxie, raümlische Störung des Aufmersamkeit. *Mschr. Psychiatr. Neurol.*, 25:51–81.
Balliet, R., and Nakayama, K. (1978a): Training of voluntary torsion. *Invest. Ophthalmol. Vis. Sci.*, 17:303–314.
Balliet, R., and Nakayama, K. (1978b): Egocentric orientation is influenced by trained voluntary cyclorotary eye movements. *Nature*, 275:214–216.
Balliet, R., Shinn, J. B., and Bach-y-Rita, P. (1982): Facial paralysis rehabilitation: retraining selective muscle control. *Int. Rehabil. Med.* 4:67–74.
Banks, R. W., Barker, D., Bessou, P., et al. (1978): Histological analysis of cat muscle spindles following direct observation of the effects of stimulating dynamic and static motor axons. *J. Physiol. (Lond.)*, 283:605–619.
Banks, R. W., Barker, D., and Stacey, M. J. (1981): Structural aspects of fusimotor effects on spindle sensitivity. In: *Muscle Receptors and Movement*, edited by A. Taylor and A. Prochazka, pp. 5–16. Macmillan, London.
Banks, R. W., Harker, D. W., and Stacey, M. J. (1977): A study of mammalian intrafusal muscle fibres using a combined histochemical and ultrastructural technique. *J. Anat.*, 123:783–796.
Barbeau, A. (1980): High-level levodopa therapy in severely akinetic Parkinsonian patients: twelve years later. In: *Parkinson's Disease: Current Progress, Problems and Management*, edited by U. K. Rinne, J. Klinger, and J. S. Stamm, pp. 229–239. Elsevier, Amsterdam.
Barker, D. (1962): The structure and distribution of muscle receptors. In: *Symposium on Muscle Receptors*, edited by D. Barker, pp. 227–240. Hong Kong University Press, Hong Kong.

Barker, D. (1974): The morphology of muscle receptors. In: *Handbook of Sensory Physiology*, Vol. 3/2, edited by C. C. Hunt, pp. 1–190. Springer, Berlin.
Barker, D., Banks, R. W., Harker, D. W., et al. (1976): Studies of the histochemistry, ultrastructure, motor innervation, and regeneration of mammalian intrafusal muscle fibres. *Prog. Brain Res.*, 44:67–88.
Barker, D., Bessou, P., Jankowska, E., et al. (1978): Identification of intrafusal muscle fibres activated by single fusimotor axons and injected with fluorescent dye in cat tenuissimus spindles. *J. Physiol. (Lond.)*, 275:149–166.
Barker, D., Emonet-Denand, F., Laporte, Y., et al. (1973): Morphological identification and intrafusal distribution of the endings of static fusimotor axons in the cat. *J. Physiol. (Lond.)*, 230:405–427.
Barker, D., Emonet-Denand, F., Laporte, Y., and Stacey, M. J. (1980): Identification of the intrafusal endings of skeletofusimotor axons in the cat. *Brain Res.*, 185:227–237.
Barker, D., Stacey, M. J., and Adal, M. N. (1970): Fusimotor innervation in the cat. *Philos. Trans. R. Soc. Lond. (Biol.)*, 258:315–346.
Barlow, H. B., and Gaze, R. M. (1977): A discussion on structural and functional aspects of plasticity in the nervous system. *Philos. Trans. R. Soc. Lond. (Biol.)*, 278:243–244.
Barmack, N. M., Henkel, C. K., and Pettorossi, V. E. (1978): A subparafascicular projection to the medial vestibular nucleus of the rabbit. *Brain Res.*, 172:339–343.
Barnes, G. R. (1979): Vestibulo-ocular function during coordinated head and eye movements to acquire visual targets. *J. Physiol. (Lond.)*, 287:127–147.
Barnes, G. R., and Forbat, L. N. (1979): Cervical and vestibular afferent control of oculomotor response in man. *Acta Otolaryngol.*, 88:79–87.
Barnes, C. D., and Pompeiano, O. (1970): Presynaptic and postsynaptic effects in the monosynaptic reflex pathway to extensor motoneurones following vibration of synergic muscles. *Arch. Ital. Biol.*, 108:259–294.
Barnett, C. H., and Harding, D. (1955): The activity of antagonist muscles during voluntary movement. *Ann. Phys. Med.*, 2:290–293.
Barrett, E. F., and Barrett, J. N. (1976): Separation of two voltage-sensitive potassium currents, and demonstration of a tetrodotoxin-resistant calcium current in frog motoneurones. *J. Physiol. (Lond.)*, 255:737–774.
Barrett, E. F., Barrett, J. N., and Crill, W. E. (1980): Voltage-sensitive outward currents in cat motoneurones. *J. Physiol. (Lond.)*, 304:251–276.
Barrett, J. N., and Crill, W. E. (1974): Specific membrane properties of cat motoneurones. *J. Physiol. (Lond.)*, 239:301–324.
Bartz, A. E. (1962): Eye-movement latency, duration and response time as a function of angular displacement. *J. Exp. Psychol.*, 64:318–324.
Basbaum, A. I., Clanton, C. H., and Fields, H. L. (1976): Opiate and stimulus produced analgesia: functional anatomy of a medullospinal pathway. *Proc. Natl. Acad. Sci. USA*, 73:4685–4688.
Basmajian, J. V. (1978): *Muscles Alive. Their Func-*

tions Revealed by Electromyography, 4th ed. Williams & Wilkins, Baltimore.

Bastron, R. D., Gengis, S. D., Hoyt, J. L., and Sokoll, M. D. (1972): H reflex during ketamine anesthesia. *Acta Anaesthesiol. Scand.*, 16:96–98.

Bates, J. A. V. (1953): Stimulation of the medial surface of the human cerebral hemisphere after hemispherectomy. *Brain*, 76:405–446.

Bathien, N. (1971): Réflexes spinaux chez l'homme et niveaux d'attention. *Electroencephalogr. Clin. Neurophysiol.*, 30:32–37.

Bathien, N., and Bourdarias, H. (1972): Lower limb cutaneous reflexes in hemiplegia. *Brain*, 95:447–456.

Bathien, N., and Guiheneuc, P. (1974): L'exploration des polynévrites chroniques par la technique du réflexe H. *Rev. Electroencephalogr. Neurophysiol. Clin.*, 4:587–595.

Bathien, N., and Rondot, P. (1977): Reciprocal continuous inhibition in rigidity of Parkinsonism. *J. Neurol. Neurosurg. Psychiatry*, 40:20–24.

Bathien, N., and Willer, J. C. (1971): Effets de différents types d'antalgiques sur le réflexe cutané de flexion chez l'homme. *Thérapie*, 29:709–717.

Battye, C. K., Nightingale, A., and Whillis, J. (1955): The use of myoelectric currents in the operation of prostheses. *J. Bone Joint Surg.*, 37B:506–510.

Bauer, J. A., Wood, G. D., and Held, R. (1969): A device for rapid recording of positioning responses in two dimensions. *Behav. Res. Meth. Instr.*, 1:157–159.

Baumann, T. K., and Hulliger, M. (1981): The high sensitivity of primary spindle afferents to small stretches is not preserved during larger movements of physiological amplitude, unless they are very slow. *Experientia*, 37:606.

Bawa, P. (1981): Interaction of excitatory and inhibitory inputs to spinal motoneurones in man. In: *Progress in Clinical Neurophysiology, Vol. 9, Motor Unit Types, Recruitment and Plasticity in Health and Disease*, edited by J. E. Desmedt, pp. 212–219. Karger, Basel.

Bawa, P., and Stein, R. B. (1976): The frequency response of human soleus muscle. *J. Neurophysiol.*, 39:788–793.

Bawa, P., and Tatton, W. G. (1979): Motor unit responses in muscles stretched by imposed displacements of the monkey wrist. *Exp. Brain Res.*, 37:417–437.

Baxter, D. W., and Olszewski, J. (1960): Congenital universal insensitivity to pain. *Brain*, 83:381–393.

Beck, C. H., and Chambers, W. W. (1970): Speed, accuracy and strength of forelimb movement after unilateral pyramidotomy in rhesus monkeys. *J. Comp. Physiol. Psychol. Monogr.*, 2:1–22.

Beckerman, S. B., and Kerr, F. W. L. (1976): Electrophysiologic evidence that neither sprouting nor neuronal hyperactivity occur following long-term trigeminal or cervical primary deafferentation. *Exp. Neurol.*, 50:427–438.

Begbie, G. H. (1967): Some problems of postural sway. In: *Myotatic, Kinesthetic, and Vestibular Mechanisms*, edited by A. V. S. de Reuck and J. Knight, pp. 80–92. Churchill, London.

Beggs, W. D. A., and Howarth, C. L. (1972): The movement of the hand towards a target. *Q. J. Exp. Psychol.*, 24:448–453.

Behse, F., and Buchthal, F. (1978): Sensory action potentials and nerve biopsy of the sural nerve in neuropathy. *Brain*, 101:473–493.

Belke, R. (1948): Uber die Möglichkeit der isolierten Reizung der hinteren Rückenmarkswurzeln beim Menschen. *Pfluegers Arch.*, 250:733–736.

Bender, L. F., Maynard, F. M., and Hastings, S. V. (1969): The blink reflex as a diagnostic procedure. *Arch. Phys. Med. Rehabil.*, 50:27–31.

Benecke, R., and Conrad, B. (1980): Evaluation of motor deficits in patients suffering from multiple sclerosis. In: *Progress in Multiple Sclerosis*, edited by J. J. Bauer, C. M. Poser, and W. Ritter, pp. 590–595. Springer, Berlin.

Benecke, R., Conrad, B., and Thater, K. (1980): Mechanical and muscular analysis of bicycling in normal subjects and patients with motor deficits. In: *The Skillfulness in Movement: Theory and Application*, edited by H. Nadeau, pp. 604–613. Plenum, New York.

Benes, F. M., Parks, T. N., and Rubel, E. W. (1977): Rapid dendritic atrophy following deafferentation. *Brain Res.*, 122:1–13.

Benita, M., Condé, H., Dormont, J. F., and Schmied, A. (1979): Effects of ventrolateral thalamic nucleus cooling on initiation of forelimb ballistic flexion movements by conditioned cats. *Exp. Brain Res.*, 34:435–452.

Bennett, P. B. (1975): The high nervous pressure syndrome: Man. In: *The Physiology and Medicine of Diving and Compressed Air Work*, edited by P. B. Bennett, and D. H. Elliott. Ballière, London.

Bennett, P. B., Blenkarn, G. D., Roby, J., and Youngblood, D. (1974): Suppression of high pressure nervous syndrome in human deep dives by He-N$_2$-O$_2$. *Undersea Biomed. Res.*, 1:221–237.

Bennett, P. B., Coggin, R., Roby, J., and Willer, J. N. (1980): Rapid compression with Trimix. *VIIth Symposium on Underwater Physiology*, Athens, pp. 66–67.

Bensel, C. R., and Dzendolet, E. (1968): Power spectral density analysis of the standing sway of males. *Percept. Psychophysiol.*, 4:285–288.

Benson, A. J., and Barnes, G. R. (1978): Vision during angular oscillation: the dynamic interaction of visual and vestibular mechanisms. *Aviat. Space Environ. Med.*, 49:340–345.

Berardelli, A., Hallett, M., Kaufman, C., et al. (1982): Stretch reflexes of triceps surae in normal man. *J. Neurol. Neurosurg. Psychiatry*, 6:513–525.

Bergego, C., Pierrot-Deseilligny, E., and Mazieres, L. (1981): Facilitation of transmission in Ib pathways by cutaneous afferents from the controlateral foot sole in man. *Neurosci. Lett.*, 27:297–301.

Berghage, T. E., Lash, L. E., Braithwaite, W. R., and Thalmann, E. D. (1975): Intentional tremor on a helium-oxygen chamber dive to 49,5 ATA. *Undersea Biomed. Res.*, 2:215–222.

Bergmans, J. (1973): Physiological observations on single human nerve fibres. In: *New Developments in*

Electromyography and Clinical Neurophysiology, Vol. 2, edited by J. E. Desmedt, pp. 89–127. Karger, Basel.

Bernhard, C. G., and Bohm, E. (1954): Cortical representation and functional significance of the corticomotoneuronal system. *Arch. Neurol. Psychiatry*, 72:473–502.

Bernstein, J. J., and Bernstein, M. E. (1971): Axonal regeneration and formation of synapses proximal to the site of lesion following hemisection of the rat spinal cord. *Exp. Neurol.*, 30:336–351.

Bernstein, J. J., and Bernstein, M. E. (1973): Neuronal alteration and reinnervation following axonal regeneration and sprouting in mammalian spinal cord. *Brain Behav. Evol.*, 8:135–161.

Bernstein, N. (1967): *The Coordination and Regulation of Movements*. Pergamon, Oxford.

Berthoz, A. (1971): Le corps humain et les vibrations. *La Recherche*, 2:121–129.

Berthoz, A., Lacour, M., Soechting, J. F., and Vidal, P. P. (1979): The role of vision in the control of posture during linear motion. *Prog. Brain Res.*, 50:197–209.

Berthoz, A., Melvill Jones, G., and Begue, A. E. (1981): Differential visual adaptation of vertical canal-dependent vestibulo-ocular reflexes. *Exp. Brain Res.*, 44:19–26.

Berthoz, A., Pavard, B., and Young, L. R. (1975): Perception of linear horizontal self-motion induced by peripheral vision. *Exp. Brain Res.*, 23:471–489.

Bertrand, G. (1956): Spinal efferent pathways from the supplementary motor area. *Brain*, 79:461–473.

Bessette, R. W., Bishop, B., and Mohl, N. (1971): Duration of masseteric silent period in patients with TMJ syndrome. *J. Appl. Physiol.*, 30:864–869.

Bessette, R. W., Duda, L., Mohl, N. D., and Bishop, B. (1973): Effect of biting force on the duration of the masseteric silent period. *J. Dent. Res.*, 52:426–430.

Bessette, R. W., Mohl, N. D., and Di Cosimo, C. J. (1974): Comparison of electromyographic and radiographic examinations in patients with myofascial pain-dysfunction syndrome. *J. Am. Dent. Assoc.*, 89:1358–1364.

Bessette, R. W., and Quinlivan, J. T. (1973): Electromyographic evaluation of the Myo-Monitor. *J. Prosthet. Dent.*, 30:19–24.

Bessette, R. W., and Shatkin, S. S. (1979): Predicting by electromyography the results of nonsurgical treatment of temporomandibular joint syndrome. *Plast. Reconstr. Surg.*, 64:232–238.

Besson, J. M., Wyon-Maillard, M. C., Benoist, J. M., et al. (1973): Effects of phenoperidine on Lamina V cells in the cat dorsal horn. *J. Pharmacol. Exp. Ther.*, 187:239–245.

Bessou, P., and Laporte, Y. (1962): Responses from primary and secondary endings of the same neuromuscular spindle of the tenuissimus muscle of the cat. In: *Symposium on Muscle Receptors*, edited by D. Baker, pp. 105–119. Hong Kong University Press, Hong Kong.

Bessou, P., and Pages, B. (1975): Cinematographic analysis of contractile events produced in intrafusal muscle fibres by stimulation of static and dynamic fusimotor axons. *J. Physiol. (Lond.)*, 252:397–427.

Bianconi, R., and Vandermeulen, J. P. (1963): The response to vibration of the end-organs of mammalian muscle spindles. *J. Neurophysiol.*, 26:177–190.

Biber, M. P., Kneisley, L. W., and Lavail, J. H. (1978): Cortical neurons projecting to the cervical and lumbar enlargement of the spinal cord in young and adult rhesus monkeys. *Exp. Neurol.*, 59:492–508.

Bieber, I., and Fulton, J. F. (1938): The relation of the cerebral cortex to the grasp reflex and to the postural and righting reflexes. *Arch. Neurol. Psychiatry*, 39:435–454.

Bienfang, D. C. (1978): The course of direct projections from the abducens nucleus to the contralateral medial rectus subdivision of the oculomotor nucleus in the cat. *Brain Res.*, 145:277–289.

Bigland, B., and Lippold, O. C. J. (1954a): The relation between force, velocity and integrated electrical activity in human muscles. *J. Physiol. (Lond.)*, 123:214–224.

Bigland, B., and Lippold, O. C. J. (1954b): Motor unit activity in the voluntary contraction of human muscle. *J. Physiol. (Lond.)*, 125:322–335.

Bigland-Ritchie, B. (1981):EMG and fatigue of human voluntary and stimulated contractions. In: *Human Muscle Fatigue: Physiological Mechanisms, Ciba Foundation Symposium*, 82:130–156. Pitman, London.

Bigland-Ritchie, B., Jones, D. A., and Woods, J. J. (1979): Excitation frequency and muscle fatigue: electrical responses during human voluntary and stimulated contractions. *Exp. Neurol.*, 64:414–427.

Biguer, B., Jeannerod, M., and Prablanc, C. (1982): The coordination of eye, head and hand movements during reaching at a single visual target. *Exp. Brain Res.*, 46:301–304.

Binder, M. D., Kroin, J. S., Moore, G. P., et al. (1976): Correlation analysis of muscle spindle responses to single motor unit contractions. *J. Physiol. (Lond.)*, 257:325–336.

Binder, M. D., and Stuart, D. G. (1980): Motor unit-muscle receptors interactions: Design features of the neuromuscular control system. In: *Progress in Clinical Neurophysiology, Vol. 8, Spinal and Supraspinal Mechanisms of Voluntary Motor Control and Locomotion*, edited by J. E. Desmedt, pp. 72–98, Karger, Basel.

Bioulac, B., and Lamarre, Y. (1979): Activity of postcentral cortical neurons of the monkey during conditioned movements of a deafferented limb. *Brain Res.*, 172:427–437.

Birkmayer, W., Danielczyk, W., and Weiler, G. (1967): Zur Objektivierbarkeit des myotonolytisches Effectes eines Aminobuttersäurederivates (CIBA 34647-Ba). *Wien. Med. Wochenschr.*, 117:7–9.

Bishop, B., Machover, S., Johnston, R., and Anderson, M. (1968): A quantitative assessment of gamma-motoneuron contribution to the Achilles tendon reflex in normal subjects. *Arch. Phys. Med. Rehabil.*, 49:145–154.

Bishop, G. H., and Clare, M. H. (1952): Sites of origin of electrical potentials in striate cortex. *J. Neurophysiol.*, 15:201–220.

Bisti, S., Maffei, L., and Piccolino, M. (1974): Visuovestibular interaction in the cat superior colliculus. *J. Neurophysiol.*, 37:146–155.

Bizzi, E. (1968): Discharge of frontal eye field neurons during saccadic and following eye movements in unanaesthetized monkeys. *Exp. Brain Res.*, 10:151–158.

Bizzi, E., Accornero, N., Chapple, W., and Hogan, N. (1981): Central and peripheral mechanisms in motor control. In: *New Perspectives in Cerebral Localization*, edited by R. A. Thompson, pp. 23–34. Raven Press, New York.

Bizzi, E., Accornero, N., Chapple, W., and Hogan, N. (1982): Arm trajectory formation in monkeys. *Exp. Brain Res.*, 46:139–143.

Bizzi, E., Dev, P., Morasso, P., and Polit, A. (1978a): Effect of load disturbances during centrally initiated movements. *J. Neurophysiol.*, 41:542–556.

Bizzi, E., Dev, P., Morasso, P., and Polit, A. (1978b): Role of neck proprioceptors during visually triggered head movements. In: *Progress in Clinical Neurophysiology, Vol. 4, Cerebral Motor Control in Man: Long Loop Mechanisms*, edited by J. E. Desmedt, pp. 141–152, Karger, Basel.

Bizzi, E., and Evarts, E. D. (1971): Translational mechanisms between input and output. *Neurosci. Res. Prog. Bull.*, 9:31–59.

Bizzi, E., Polit, A., and Morasso, P. (1976): Mechanisms underlying achievement of final head position. *J. Neurophysiol.*, 39:435–444.

Bjorklund, A., Katzman, R., Stenevi, U., and West, K. A. (1971): Development and growth of axonal sprouts from noradrenaline and 5-hydroxytryptamine neurons in the rat spinal cord. *Brain Res.*, 13:21–33.

Bjorklund, A., and Stenevi, U. (1979): Regeneration of mono-aminergic and cholinergic neurons in the mammalian central nervous system. *Physiol. Rev.*, 59:62–100.

Black, P., Markowitz, R. S., and Cianci, S. N. (1975): Recovery of motor function after lesions in motor cortex of monkey. In: *Outcome of Severe Damage to the Central Nervous System, Ciba Foundation Symposium*, 34:65–83, Elsevier, Amsterdam.

Blackstock, E., Rushworth, G., and Gath, D. (1972): Electrophysiological studies in alcoholism. *J. Neurol. Neurosurg. Psychiatry*, 35:326–334.

Blass, J. P., and Gibson, G. E. (1978): Studies of pyruvate dehydrogenase deficiency. In: *Advances in Neurology*, Vol. 21, edited by R. Kark, R. Rosenberg, and L. Schut, pp. 181–194. Raven Press, New York.

Boczko, M., and Mumenthaler, M. (1958): Modified pendulousness test to assess tonus of thigh muscles in spasticity. *Neurology (Minneap.)*, 8:846–851.

Bodo, R. C., and Brooks, C. M. (1937): The effect of morphine on blood sugar and reflex activity in the chronic spinal cat. *J. Pharmacol. Exp. Ther.*, 61:82–88.

Boehme, D. H., Marks, N., and Fordice, M. W. (1976): Glycine levels in the degenerated human spinal cord. *J. Neurol. Sci.*, 27:347–352.

Boelhouwer, A. J. W., and Brunia, C. H. M. (1977): Blink reflexes and the state of arousal. *J. Neurol. Neurosurg. Psychiatry*, 40:58–63.

Bogardh, E., and Richards, C. (1974): Gait analysis and re-learning of gait control in hemiplegic patients. *Proceedings: Seventh International Congress World Confederation Physical Therapy*, London, pp. 435–443.

Bonin, von, G. (1949): Architecture of the precentral motor cortex and some adjacent areas. In: *The Precentral Motor Cortex*, edited by P. C. Bucy, pp. 8–82, University of Illinois Press, Urbana.

Bonnet, M., Chouteau, J., Hugon, M., et al. (1973): Activités motrices réflexes spontanées chez le singe sous Heliox (99 ATA). *Forsvarmedicin* 9:314–317.

Boshes, B., Wachs, H., Brumlik, J., et al. (1960): Studies of tone, tremor and speech in normal persons and in parkinsonian patients. *Neurology (Minneap.)*, 10:805–813.

Bossom, J. (1974): Movement without proprioception. *Brain Res.*, 71:285–296.

Botterman, B. R., Eldred, E., and Edgerton, V. R. (1981): Spindle discharge in glucocorticoid-induced muscle atrophy. *Exp. Neurol.*, 72:25–40.

Bouaziz, Z., Bouaziz, M., and Hugon, M. (1975): Modulation of soleus electromyogram by electrical stimulation of medial gastrocnemius nerve in man. *Electromyography*, 15:31–42.

Bouisset, S. (1973): EMG muscle force in normal motor activities. In: *New Developments in Electromyography and Clinical Neurophysiology*, Vol. 1, edited by J. E. Desmedt, pp. 547–583. Karger, Basel.

Bouisset, S., and Maton, B. (1973): Comparison between surface and intramuscular EMG during voluntary movement. In: *New Developments in Electromyography and Clinical Neurophysiology*, Vol. 1, edited by J. E. Desmedt, pp. 533–539. Karger, Basel.

Bouquet, F., and Cioffi, F. (1969): Sull interesse della riposta H di Hoffmann nelle lesioni della prima radice sacrale da ernia discale. *Rass. Int. Clin. Ter.*, 49:925–932.

Bourbonnais, D., Smith, A. M., and Blanchette, G. (1979): Disturbances of both controlled prehension and grasp release following unilateral ablation of the supplementary motor area. *Am. Soc. Neurosci. (Abstr.)*, 5:363.

Boureau, F., Sebille, A., Willer, J. C., et al. (1978): Effets d'une stimulation électrique hétéro-segmentaire percutanée (électro-acupuncture) sur le réflexe nociceptif de flexion chez l'homme. *Ann. Anesthesiol. Fr.*, 5:422–426.

Boureau, F., Willer, J. C., and Yamaguchi, Y. (1979): Abolition par la naloxone de l'effet inhibiteur d'une stimulation electrique peripherique sur la composante tardive du réflexe de clignement. *Electroencephalogr. Clin. Neurophysiol.*, 47:322–328.

Bowery, N. G., Hill, D. R., Hudson, A. L., et al. (1980): Baclofen decreases neurotransmitter release in the mammalian CNS by an action at a novel GABA receptor. *Nature (New Biol.)*, 283:92–94.

Boyd, I. A. (1981): The muscle spindle controversy. *Sci. Prog.*, 67:205–221.

Boyd, I. A., Gladden, M. H., McWilliam, P. N., and Ward, J. (1977): Control of dynamic and static nuclear bag fibres and nuclear chain fibres by gamma- and beta-axons in isolated cat muscle spindles. *J. Physiol. (Lond.)*, 265:133–162.

Braddom, R. I., and Johnson, E. W. (1974): Standardization of H reflex and disgnostic use in S1 radiculopathy. *Arch. Phys. Med. Rehabil.*, 55:161–166.

Brandell, B. R. (1977): Functional roles of the calf and vastus muscles in locomotion. *Am. J. Phys. Med.*, 56:59–74.

Brandstater, M. E., and Lambert, E. H. (1973): Motor unit anatomy. Type and spatial arrangement of muscle fibers. In: *New Developments in Electromyography and Clinical Neurophysiology* Vol. 1, edited by J. E. Desmedt, pp. 14–22. Karger, Basel.

Bratzlavsky, M., De Boever, J., and Vandereecken, H. (1976): Tooth pulpal reflexes in jaw musculature in man. *Arch. Oral Biol.*, 21:491–493.

Brauer, R. W. (1975): The high nervous pressure syndrome: Animals. In: *The Physiology and Medicine of Diving and Compressed Air Work*, edited by P. B. Bennett and D. H. Elliott, pp. 231–247. Bailliere, Tindall, London.

Brauer, R. W., Beaver, R. W., Gillen, H. W., et al. (1980): Diffentiation of the two components of the convulsion stage of the HPNS in vertebrates. *Proceedings: 7th Symposium Underwater Physiology*, Athens, pp. 69–70.

Brauer, R. W., Beaver, R. W., Lasher, S., et al. (1977): Time, rate and temperature factors in the onset of high pressure convulsions. *J. Appl. Physiol.*, 43:173–182.

Brauer, R. W., Beaver, R. W., Mansfield, W. M., et al. (1975): Rate factors in development of the high pressure neurological syndrome. *J. Appl. Physiol.*, 38:220–227.

Brauer, R. W., Goldman, S. M., Beaver, R. W., and Sheeman, M. E. (1974): N_2, H_2 and N_2O antagonism of high pressure neurological syndrome in mice. *Undersea Biomed. Res.*, 1:59–72.

Brauer, R. W., Mansfield, W. M., Beaver, R. W., et al. (1979): Stages in the development of the HPNS of the mouse. *J. Appl. Physiol. Resp.*, 46:128–135.

Brebner, J. (1968): Continuing and reversing the direction of responding movements: some exceptions to the so-called "psychological refractory period." *J. Exp. Psychol.*, 78:120–127.

Bremer, F. (1928): De l'exagération des réflexes, consécutive a la section des racines postérieures. *Ann. Physiol. (Paris)*, 4:730–733.

Bremer, F. (1935): Cerveau "isolé" et physiologie du sommeil. *C. R. Soc. Biol. (Paris)*, 118:1235–1241.

Brenowitz, G. L. and Pubols, L. M. (1981): Increased receptive field size of dorsal horn neurons following chronic spinal cord hemisection. *Brain Res.*, 216:45–59.

Brindley, G. S. (1972): Electrode arrays for making long-lasting electrical connection to connection to spinal roots. *J. Physiol. (Lond.)*, 222:135–136P.

Brindley, G. S., and Craggs, M. D. (1972): The electrical activity in the motor cortex that accompanies voluntary movement. *J. Physiol. (Lond.)*, 223:28–29P.

Brindley, G. S., and Merton, P. A. (1960): The absence of position sense in the human eye. *J. Physiol. (Lond.)*, 153:127–130.

Brinkman, C. (1981): Sensory inputs into the cortical motor areas in the conscious behaving monkey. In: *Brain Mechanisms of Sensation*, edited by Y. Katsuki, L. Norgren, and A. Sata, pp. 71–88. Wiley, London.

Brinkman, C., and Porter, R. (1979): Supplementary motor area in the monkey: activity of neurons during performance of a learned motor task. *J. Neurophysiol.*, 42:681–709.

Brinkman, J., Bush, B. M., and Porter, R. (1978): Deficient influences of peripheral stimuli on precentral neurones in monkeys with dorsal column lesions. *J. Physiol. (Lond.)*, 276:27–48.

Brinkman, J., and Kuypers, H. (1972): Splitbrain monkeys: control of ipsilateral and contralateral arm, hand and finger movements. *Science*, 176:536–539.

Brinkman, J., and Kuypers, H. (1973): Cerebral control of contralateral and ipsilateral arm, hand and finger movements in split-brain rhesus monkey. *Brain*, 96:653–674.

Brodal, A. (1957): *The Reticular Formation of the Brain Stem. Anatomical Aspects and Functional Correlation*, pp. 1–87. Oliver and Boyd, Edinburgh.

Brodal, P. (1978): The cortico-pontine projections in the rhesus monkey. *Brain*, 101:251–284.

Brodal, P., Marsala, J., and Brodal, A. (1967): The cerebral cortical projection to the lateral reticular nucleus in the cat, with special reference to the sensorimotor cortical areas. *Brain Res.*, 6:252–274.

Brodmann, K. (1909): *Vergeichende Lokalisations lehre der Grosshirnrinde in ihren prinzipien Dargestellt auf Grunde des Zellenbaues*, edited by J. A. Barth, pp. 324, Thieme, Leipzig.

Bromberg, M. B., and Gilman, S. (1978): Changes in rubral multiunit activity after lesions in the interpositus nucleus of the cat. *Brain Res.*, 152:353–357.

Brooks, V. B. (1979): Control of intended limb movements by the lateral and intermediate cerebellum. In: *Integration in the Nervous System*, edited by H. Asanuma and V. J. Wilson, pp. 321–357. Igaku Shoin, Tokyo.

Brooks, V. B., Kozlovskaya, I. B., Atkin, A., et al. (1973): Effects of cooling the dentate nucleus on tracking-task performance in monkeys. *J. Neurophysiol.*, 36:974–995.

Brooks, V. B., Rudomin, P., and Slayman, C. L. (1961): Peripheral receptive fields of neurons in the cat's cerebral cortex. *J. Neurophysiol.*, 24:302–325.

Brown, A. G., Hamann, W. C., and Martin, H. F. (1973): Descending influences on spino-cervical tract cell discharges evoked by non-myelinated cutaneous afferent nerve fibres. *Brain Res.*, 53:218–221.

Brown, G. L., and Burns, B. D. (1949): Fatigue and neuromuscular block in mammalian skeletal muscle. *Proc. R. Soc. Lond. (Biol.)*, 136:182–195.

Brown, J. S., and Slater-Hammel, A. T. (1949): Discrete movements in the horizontal plane as a function of their length and direction. *J. Exp. Psychol.*, 38:84–95.

Brown, M. C., Crowe, A., and Matthews, P. B. C. (1965): Observations on the fusimotor fibres of the tibialis posterior muscle of the cat. *J. Physiol. (Lond.)*, 177:140–159.

Brown, M. C., Engberg, I., and Matthews, P. B. C.

(1967a): Fusimotor stimulation and the dynamic sensitivity of the secondary ending on the muscle spindle. *J. Physiol. (Lond.)*, 189:545–550.

Brown, M. C., Engberg, I., and Matthews, P. B. C. (1967b): The relative sensitivity to vibration of muscle receptors of the cat. *J. Physiol. (Lond.)*, 192:773–800.

Brown, M. C., Goodwin, G. M., and Matthews, P. B. C. (1969): After-effects of fusimotor stimulation on the response of muscle spindle primary afferent endings. *J. Physiol. (Lond.)*, 205:677–694.

Brown, M. C., Holland, R. L., and Hopkins, W. G. (1981): Motor nerve sprouting. *Ann. Rev. Neurosci.*, 4:17–42.

Brown, M. C., and Ironton, R. (1977): Motor neurone sprouting induced by prolonged tetrodotoxin block of nerve action potentials. *Nature*, 265:459–461.

Brown, M. C., Lawrence, D. G., and Matthews, P. B. C. (1969): Static fusimotor fibres and the position sensitivity of muscle spindle receptors. *Brain Res.*, 14:173–187.

Brown, S. H., and Cooke, J. D. (1981): Responses to force perturbations preceding voluntary human arm movements. *Brain Res.*, 220:350–355.

Brown, T. G. (1911): The intrinsic factors in the act of progression in the mammal. *Proc. R. Soc. Lond (Biol.)*, 84:308–319.

Brown, W. F., and Rushworth, G. (1973): Reflex latency fluctuations in human single motor units. In: *New Developments in Electromyography and Clinical Neurophysiology*, Vol 3, edited by J. E. Desmedt, pp. 660–665. Karger, Basel.

Brown, W. F., Milner-Brown, H. S., Ball, M., and Girvin, J. P. (1978): Control of the motor cortex on spinal motoneurons in man. In: *Progress in Clinical Neurophysiology, Vol. 4, Cerebral Motor Control in Man: Long Loop Mechanisms*, edited by J. E. Desmedt, pp. 246–262. Karger, Basel.

Browne, J. S. (1975): The responses of muscle spindles in sheep extraocular muscles. *J. Physiol. (Lond.)*, 251:483–496.

Brudny, J., Korein, J., Grynbaum, B. B., et al. (1976): EMG feedback therapy: review of treatment of 114 patients. *Arch. Phys. Med. Rehabil.*, 57:55–61.

Brumlik, J., and Boshes, B. (1961): Muscle tone in normals and in parkinsonism. *Arch. Neurol.*, 4:399–406.

Brune, H. F. (1955): Verstärkung fremdreflektorischer Hemmungsphänomene beim Menschen durch wiederholte Auslösung. *Pfluegers Arch.*, 261:518–526.

Brune, H. F., Dammann, R., and Schenck, E. (1960): Chemische Aktivierung der Muskelspindeln beim Menschen. *Pfluegers Arch.*, 271:397–404.

Brunia, C. H. M. (1973): The influence of diazepam and chlorpromazine on the Achilles tendon and H-reflexes. In: *New Deveopments in Electromyography and Clinical Neurophysiology*, Vol. 3, edited by J. E. Desmedt, pp. 367–370. Karger, Basel.

Bryan, W. L., and Harter, N. (1899): Studies in the telegraphic language: the acquisition of a hierarchy of language. *Psychol. Rev.*, 6:345–375.

Buchthal, F. (1961): The general concept of the motor unit. *Res. Publ. Assoc. Res. Nerv. Ment. Dis.*, 38:1–30.

Buchthal, F., and Rosenfalck, A. (1966): Evoked action potentials and conduction velocity in human sensory nerves. *Brain Res.*, 3:1–122.

Buchthal, F., and Schmalbruch, H. (1980): Motor unit of mammalian muscle. *Physiol. Rev.*, 60:90–142.

Bucy, P. C. (1957): Is there a pyramidal tract? *Brain*, 80:376–392.

Bucy, P. C., Keplinger, J. E., and Sequeira, E. B. (1964): Destruction of the 'pyramidal tract' in man. *J. Neurosurg.*, 21:385–398.

Bucy, P. C., Ladpli, R., and Ehrlich, A. (1966): Destruction of the pyramidal tract in the monkey. The effects of bilateral section of the cerebral peduncles. *J. Neurosurg.*, 25:1–23.

Budingen, H. J., and Freund, H. J. (1976): The relationship between the rate of rise of isometric tension and motor unit recruitment in a human forearm muscle. *Pfluegers Arch.*, 362:61–67.

Buell, S., and Coleman, P. (1979): Dendritic growth in aged human brain and failure of growth in senile dementia. *Science*, 206:854–856.

Buller, A. J. (1957): The ankle-jerk in early hemiplegia. *Lancet*, 2:1262–1263.

Buller, A., and Dornhorst, S. C. (1957): The reinforcement of tendon reflexes. *Lancet*, 2:1260–1262.

Buller, A. J., Eccles, J. C., and Eccles, R. M. (1960): Interactions between motoneurones and muscles in respect of the characteristic speeds of their responses. *J. Physiol. (Lond.)*, 150:417–439.

Buller, A. J., and Lewis, D. M. (1965): The rate of tension development in isometric tetanic contractions of mammalian fast and slow skeletal muscle. *J. Physiol. (Lond.)*, 176:337–354.

Burchfiel, J. L., and Duffy, F. H. (1972): Muscle afferent input to single cells in primate somatosensory cortex. *Brain Res.*, 45:241–249.

Burg, D., Szumski, A. J., Struppler, A., and Velho, F. (1973): Afferent and efferent activation of human muscle receptors involved in reflex and voluntary contraction. *Exp. Neurol.*, 41:754–768.

Burg, D., Szumski, A. J., Struppler, A., and Velho, F. (1974): Assessment of fusimotor contribution to reflex reinforcement in humans. *J. Neurol. Neurosurg. Psychiatry*, 37:1012–1021.

Burg, D., Szumski, A. J., Struppler, A., and Velho, F. (1976): Influence of a voluntary innervation on human muscle spindle sensitivity. In: *The Motor System*, edited by M. Shahani, pp. 95–110. Elsevier, Amsterdam.

Burgess, P. R., and Clark, F. J. (1969): Characteristics of knee joint receptors in the cat. *J. Physiol. (Lond.)*, 203:317–335.

Burgess, P. R., and Perl, E. R. (1973): Cutaneous mechanoreceptors and nociceptors. In: *Handbook of Sensory Physiology*, Vol. 2, edited by A. Iggo, pp. 29–78, Springer, Berlin.

Burke, D. (1980): Muscle spindle function during movement. *Trends Neurosci.*, 3:251–253.

Burke, D. (1981): The activity of human muscle spindle endings in normal motor behaviour. In: *International Review of Neurophysiology*, Vol IV, edited by R. Porter, pp. 91–126. University Park, Baltimore.

Burke, D., Andrews, C. J., and Ashby, P. (1971): Auto-

genic effects of static muscle stretch in spastic man. *Arch. Neurol.*, 25:367–372.
Burke, D., Andrews, C. J., and Knowles, L. (1971): The action of a GABA derivative in human spasticity. *J. Neurol. Sci.*, 14:199–208.
Burke, D., Andrews, C. J., and Lance, J. W. (1972): The tonic vibration reflex in spasticity, Parkinson's disease and normal man. *J. Neurol. Neurosurg. Psychiatry*, 35:477–486.
Burke, D., and Ashby, P. (1972): Are spinal 'presynaptic' inhibitory mechanisms suppressed in spasticity? *J. Neurol. Sci.*, 15:321–326.
Burke, D., and Eklund, G. (1977): Muscle spindle activity in man during standing. *Acta Physiol. Scand.*, 100:187–199.
Burke, D., Gillies, J. D., and Lance, J. W. (1970): The quadriceps stretch reflex in human spasticity. *J. Neurol. Neurosurg. Psychiatry*, 33:216–223.
Burke, D., Hagbarth, K. E., and Lofstedt, L. (1978a): Muscle spindle responses in man to changes in load during accurate position maintenance. *J. Physiol. (Lond.)*, 276:159–164.
Burke, D., Hagbarth, K. E., and Lofstedt, L. (1978b): Muscle spindle activity in man shortening and lengthening contractions. *J. Physiol. (Lond.)*, 277:131–142.
Burke, D., Hagbarth, K. E., Lofstedt, L., and Wallin, B. G. (1976a): The responses of human muscle spindle endings to vibration of non-contracting muscles. *J. Physiol. (Lond.)*, 261:673–693.
Burke, D., Hagbarth, K. E., Lofstedt, L., and Wallin, B. G. (1976b): The responses of human muscle spindle endings to vibration during isometric contraction. *J. Physiol. (Lond.)*, 261:695–711.
Burke, D., Hagbarth, K. E., and Skuse, N. F. (1978): Recruitment order of human spindle endings in isometric voluntary contractions. *J. Physiol. (Lond.)*, 285:101–112.
Burke, D., Hagbarth, K. E., and Skuse, N. F. (1979a): Voluntary activation of spindle endings in human muscles temporarily paralysed by nerve pressure. *J. Physiol. (Lond.)*, 287:329–336.
Burke, D., Hagbarth, K. E., and Wallin, B. G. (1977): Reflex mechanisms in Parkinsonian rigidity. *Scand. J. Rehabil. Med.*, 9:15–23.
Burke, D., Hagbarth, K. E., and Wallin, B. G. (1980a): Alpha-gamma linkage and the mechanisms of reflex reinforcement. In: *Progress in Clinical Neurophysiology, Vol. 8, Spinal and Supraspinal Mechanisms of Voluntary Motor Control and Locomotion*, edited by J. E. Desmedt, pp. 170–180. Karger, Basel.
Burke, D., Hagbarth, K. E., Wallin, B. G., and Lofstedt, L. (1980b): Muscle spindle activity induced by vibration in man: implications for the tonic stretch reflex. In: *Progress in Clinical Neurophysiology, Vol. 8, Spinal and Supraspinal Mechanisms of Voluntary Motor Control and Locomotion*, edited by J. E. Desmedt, pp. 243–253. Karger, Basel.
Burke, D., Knowles, L., Andrews, C. J., and Ashby, P. (1972): Spasticity decerebrate rigidity and the clasp-knife phenomenon: An experimental study in the cat. *Brain*, 95:31–48.
Burke, D., and Lance, J. W. (1973): Studies of the reflex effects of primary and secondary spindle endings in spasticity. In: *New Developments in Electromyography and Clinical Neurophysiology*, Vol. 3, edited by J. E. Desmedt, pp. 475–495. Karger, Basel.
Burke, D., McKeon, B., and Skuse, N. F. (1981a): Dependence of the Achilles tendon reflex on the excitability of spinal reflex pathways. *Ann. Neurol.*, 10:551–556.
Burke, D., McKeon, B., and Skuse, N. F. (1981b): The irrelevance of fusimotor activity to the Achilles tendon jerk of relaxed humans. *Ann. Neurol.*, 10:549–550.
Burke, D., McKeon, B., Skuse, N. F., and Westerman, R. A. (1980a): Anticipation and fusimotor activity in preparation for a voluntary contraction. *J. Physiol. (Lond.)*, 306:337–348.
Burke, D., McKeon, B., and Westerman, R. A. (1980b): Induced changes in the thresholds for voluntary activation of human spindle endings. *J. Physiol. (Lond.)*, 302:171–182.
Burke, D., Skuse, N., and Stuart, D. G. (1979b): The regularity of muscle spindle discharge in man. *J. Physiol. (Lond.)*, 291:277–290.
Burke, R. E. (1967): Motor unit types of cat triceps surae muscle. *J. Physiol. (Lond.)*, 193:141–160.
Burke, R. E. (1968): Group Ia synaptic input to fast and slow twitch motor units of cat triceps surae. *J. Physiol. (Lond.)*, 196:605–630.
Burke, R. E. (1973): On the central nervous system control of fast and slow twitch motor units. In: *New Developments in Electromyography and Clinical Neurophysiology*, Vol. 3, edited by J. E. Desmedt, pp. 69–94, Karger, Basel.
Burke, R. E. (1979): The role of synaptic organization in the control of motor unit activity during movement. *Prog. Brain Res.*, 50:61–67.
Burke, R. E. (1981): Motor unit recruitment: What are the critical factors? In: *Progress in Clinical Neurophysiology, Vol. 9, Motor Unit Types, Recruitment and Plasticity in Health and Disease*, edited by J. E. Desmedt, pp. 61–84. Karger, Basel.
Burke, R. E., and Edgerton, V. R. (1975): Motor unit properties and selective involvement in movement. *Exer. Sports Sci. Rev.*, 3:31–69.
Burke, R. E., Jankowska, E., and Burggencate, G. Ten (1970): A comparison of peripheral and rubrospinal synaptic input to slow and fast twitch motor units of triceps surae. *J. Physiol. (Lond.)*, 207:709–732.
Burke, R. E., Levine, D. N., Tsairis, P., and Zajac, F. E. (1973): Physiological types and histochemical profiles in motor units of the cat gastrocnemius. *J. Physiol. (Lond.)*, 234:723–748.
Burke, R. E., Rudomin, P., and Zajac, F. E. (1976): The effect of activation history on tension production by individual muscle units. *Brain Res.*, 109:515–529.
Burke, R. E., Strick, P. L., Kanda, K., et al. (1977): Anatomy of medial gastrocnemius and soleus motor nuclei in cat spinal cord. *J. Neurophysiol.*, 40:667–680.
Burke, R. E., and Tsairis, P. (1973): Anatomy and innervation ratios in motor units of cat gastrocnemius. *J. Physiol. (Lond.)*, 234:749–765.
Burton, H., and Jones, E. G. (1976): The posterior

thalamic region and its cortical projection in New World and Old World monkeys. *J. Comp. Neurol.*, 168:249–302.

Buser, P., and Imbert, M. (1961): Sensory projections to the motor cortex in cats: a microelectrode study. In: *Sensory Communications*, edited by W. A. Rosenbluth, p. 607–626. MIT Press, Cambridge, Massachusetts.

Bushnell, M. C., Goldberg, M. E., and Robinson, D. L. (1981): Behavioral enhancement of visual responses in monkey cerebral cortex. I. Modulation in posterior parietal cortex related to selective visual attention. *J. Neurophysiol.*, 46:755–772.

Bussel, B., Morin, C., and Pierrot-Deseilligny, E. (1978): Mechanism of monosynaptic reflex reinforcement during Jendrassik manoeuvre in man. *J. Neurol. Neurosurg. Psychiatry*, 41:40–44.

Bussel, B., Katz, R., Pierrot-Deseilligny, E., et al. (1980): Vestibular and proprioceptive influences on the postural reactions to a sudden body displacement in man. In: *Progress in Clinical Neurophysiology, Vol. 8, Spinal and Supraspinal Mechanisms of Voluntary Motor Control and Locomotion*, edited by J. E. Desmedt, pp. 310–322, Karger, Basel.

Bussel, B., and Pierrot-Deseilligny, E. (1977): Inhibition of human motoneurones, probably of Renshaw origin, elicited by an orthodromic motor discharge. *J. Physiol. (Lond.)*, 269:319–339.

Buttner-Ennever, J. A., and Henn, V. (1976): An autoradiographic study of the pathways from the pontine reticular formation involved in horizontal eye movements. *Brain Res.*, 108:155–164.

Buttner, V., and Buettner, V. W. (1978): Parietal cortex neuronal activity in the alert monkey during natural vestibular and optokinetic stimulation. *Brain Res.*, 153:392–397.

Butz, P., Kaufmann, W., and Wiesendanger, M. (1970): Analyse einer raschen Willkurbewegung bei Parkinsonpatienten vor und nach stereotaktischem Eingriff am Thalamus. *Z. Neurol.*, 198:105–119.

Cabelguen, J. M. (1981): Static and dynamic fusimotor controls in various hindlimb muscles during locomotor activity in the decorticate cat. *Brain Res.*, 213:83–98.

Caccia, M. R., McComas, A. J., Upton, A. R. M., and Blogg, T. (1973): Cutaneous reflexes in small muscles of the hand. *J. Neurol. Neurosurg. Psychiatry*, 36:960–977.

Cajal, S. R. (1911): *Histologie du Système Nerveux de l'Homme et des Vertébrés*. Maloine, Paris.

Calvin, W. H. (1972): Steps in production of motoneuron spikes during rhythmic firing. *J. Neurophysiol.*, 35:297–310.

Calvin, W. H. (1974): Three modes of repetitive firing and the role of threshold time course between spikes. *Brain Res.* 69:341–346.

Calvin, W. H., and Sypert, G. W. (1976): Fast and slow pyramidal tract neurons: an intracellular analysis of their contrasting repetitive firing properties in the cat. *J. Neurophysiol.*, 39:420–434.

Campbell, E. J. M., Gandevia, S. C., Killian, K. J., et al. (1980): Changes in the perception of inspiratory resistive loads during partial curarization. *J. Physiol. (Lond.)*, 309:93–100.

Cannon, W. (1942): "Voodoo Death." *Am. Anthropol.*, 44:169–181.

Capaday, C., and Cooke, J. D. (1981): The effects of muscle vibration on the attainment of intended final position during voluntary human arm movements. *Exp. Brain Res.*, 42:228–230.

Carlsoo, S., Dahllof, A. G., and Holm, J. (1974): Kinetic analysis of the gait in patients with hemiparesis and in patients with intermittent claudication. *Scand. J. Rehabil. Med.*, 6:166–179.

Carlsoo, S., Fohlin, L., and Skoglund, G. (1973): Studies of co-contraction of knee muscles. In: *New Developments in Electromyography and Clinical Neurophysiology*, Vol. 1, edited by J. E. Desmedt, pp. 648–655. Karger, Basel.

Carpenter, M. B. (1950): Athetosis and the basal ganglia. *Arch. Neurol. Psychiatry*, 63:875–901.

Carrea, R. M., and Mettler, F. A. (1947): Physiologic consequences of extensive removal of the cerebellar cortex and deep cerebellar nuclei. *J. Comp. Neurol.*, 87:169–288.

Cassel, K., Shaw, K., and Stern, G. (1973): A computerised tracking technique for the assessment of parkinsonian disabilities. *Brain*, 96:815–826.

Casey, K. L. (1966): Unit analysis of nociceptive mechanisms in the thalamus of the awake squirrel monkey. *J. Neurophysiol.*, 29:727–750.

Casey, E. G., Jellife, A. M., Lequesne, P. M., and Millett, Y. (1973): Vincristine neuropathy. *Brain*, 96:69–86.

Cassinari, V., and Pagni, C. A. (1969): *Central Pain*. Harvard University Press, Cambridge, Massachusetts.

Castaigne, P., Held, J. P., Laplane, D., et al. (1973): Etude de l'effet du (R) Lioresal dans la spasticité. *Rev. Neurol.*, 128:245–250.

Castaigne, P., Held, J. P., Pierrot-Deseilligny, E., et al. (1978): Modifications de l'inhibition récurrente de Renshaw induites par la station debout chez le sujet normal et le spastique. *Rev. Neurol.*, 134:85–92.

Castaigne, P., Laplane, D., and Degos, J. D. (1972): Trois cas de négligence motrice par lésion frontale pré-rolandique. *Rev. Neurol.*, 126:5–15.

Castro, A. J. (1978): Analysis of corticospinal and rubrospinal projections after neonatal pyramidotomy in rats. *Brain Res.*, 144:155–158.

Castro, A. J., Clegg, D. A., and McClung, J. R. (1977): The effect of large unilateral cortical lesions on rubrospinal tract sprouting in newborn rats. *Am. J. Anat.*, 149:39–46.

Catsman-Berrevoets, C. E., Kuypers, H. G. J. M., and Lemon, R. N. (1979): Cells of origin of the frontal projections to magnocellular and parvocellular red nucleus and superior colliculus in cynomolgus monkey. An HRP study. *Neurosci. Lett.*, 12:41–46.

Cazin, L., Precht, W., and Lannou, J. (1980): Pathways mediating optokinetic response in vestibular nucleus neurons in the rat. *Pfluegers Arch.*, 384:19–29.

Cervero, F., and Iggo, A. (1980): The substantia gelatinosa of the spinal cord. *Brain*, 103:717–772.

Chadwick, D., Hallett, M., Harris, R., et al. (1977): Clinical, biochemical and physiological factors distinguishing myoclonus responsive to 5-hydroxytryp-

tophan, tryptophan plus a monoamine oxidase inhibitor and clonazepam. *Brain*, 100:455–487.
Chan, C. W. Y., and Kearney, R. E. (1977): Are the "late" EMG responses to limb displacement servo-controlled or "trigger" released? *Am. Soc. Neurosci. Abstr.*, 3:269.
Chan, C. W. Y., and Kearney, R. E. (1982a): Influence of static head tilt upon soleus motoneuron excitability in man. *Neurosci. Lett.*, 33:333–338.
Chan, C. W. Y., and Kearney, R. E. (1982b): Is the functional stretch response servo-controlled or preprogrammed? *Electroencephalogr. Clin. Neurophysiol.*, 53:310–324.
Chan, C. W. Y., Kearney, R. E., and Melvill Jones, G. (1979a): Tibialis anterior response to sudden ankle displacements in normal and Parkinsonian subjects. *Brain Res.*, 173:303–314.
Chan, C. W. Y., Melvill Jones, G., and Catchlove, R. F. H. (1979b): The "late" electromyographic response to limb displacement in man. II. Sensory origin. *Electroencephalogr. Clin. Neurophysiol.*, 46:182–188.
Chan, C. W. Y., Melvill Jones, G., Kearney, R. E., and Watt, D. G. D. (1979c): The "late" electromyographic response to limb displacement in man. *Electroencephalogr. Clin. Neurophysiol.*, 46:173–181.
Chapman, C. R., Wilson, M. E., and Gehrig, J. D. (1976): Comparative effects of acupuncture and transcutaneous stimulation on the perception of painful dental stimuli. *Pain*, 2:265–283.
Chauvel, P., Louvel, J., and Lamarche, M. (1978): Transcortical reflexes and focal motor epilepsy. *Electroencephalogr. Clin. Neurophysiol.*, 45:309–318.
Chavis, D. A., and Pandya, D. N. (1976): Further observations on corticofrontal connections in the rhesus monkey. *Brain Res.*, 117:369–386.
Chen, W. J., and Poppele, R. E. (1978): Small-signal analysis of response of mammalian muscle spindles with large-signal responses. *J. Neurophysiol.*, 41:15–27.
Cheney, P. D., and Fetz, E. E. (1980): Functional classes of primate corticomotoneuronal cells and their relation to active force. *J. Neurophysiol.*, 44:773–791.
Cheney, P. D., and Preston, J. B. (1976): Classification of fusimotor fibres in the primate. *J. Neurophysiol.*, 39:9–19.
Childress, D. S. (1973): An approach to powered grasp. *Proceedings: 4th International Symposium on External Control of Human Extremities*, Belgrade, pp. 159–167.
Chofflon, M., Lachat, J. J., and Ruegg, D. G. (1982): A transcortical loop demonstrated by stimulation of low-threshold muscle afferents in the awake monkey. *J. Physiol. (Lond.)*, 323:393–402.
Chong, K. C., Vojnic, C. D., Quanbu, Y. A., et al. (1978): The assessment of the internal rotation gait in cerebral palsy. *Clin. Orthop.*, 132:145–150.
Chow, K. L., Mathers, L. H., and Spear, P. D. (1973): Spreading of uncrossed retinal projection in superior colliculus of neonatally enucleated rabbits. *J. Comp. Neurol.*, 151:307–322.

Chow, K. L., and Stewart, D. L. (1972): Reversal of structural and functional effects of long-term visual deprivation in cats. *Exp. Neurol.*, 34:409–433.
Christensen, B. N., and Perl, E. R. (1970): Spinal neurons specifically excited by noxious or thermal stimuli. *J. Neurophysiol.*, 33:293–307.
Clamann, H. P. (1981): The influence of different inputs on the recruitment order of muscles and their motor units. In: *Progress in Clinical Neurophysiology, Vol. 9, Motor Unit Types, Recruitment and Plasticity in Health and Disease*, edited by J. E. Desmedt, pp. 176–183. Karger, Basel.
Clark, B. (1964): Thresholds for the perception of angular acceleration in man. *Aerospace Med.*, 38:443–450.
Clarke, A. M. (1967): Effect of the Jendrassik manoeuver on a phasic stretch reflex in normal human subjects during experimental control over supraspinal influences. *J. Neurol. Neurosurg. Psychiatry*, 30:34–42.
Clay, S. A., and Ramseyer, J. C. (1976): The orbicularis oculi reflex in infancy and childhood. Establishment of normal values. *Neurology (Minneap.)*, 26:521–524.
Clippinger, F. W., Avery, R., and Titus, B. R. (1974): A sensory feedback system for an upper-limb amputation prosthesis. *Bull. Prosthet. Res.*, 22:247–258.
Close, R. I. (1972): Dynamic properties of mammalian skeletal muscles. *Physiol. Rev.*, 52:129–197.
Clough, J. F. M., Kernell, D., and Phillips, C. G. (1968): The distribution of monosynaptic excitation from the pyramidal tract and from primary spindle afferents to motoneurones of the baboon's hand and forearm. *J. Physiol. (Lond.)*, 198:145–166.
Clough, J. F. M., Phillips, C. G., and Sheridan, J. D. (1971): The short-latency projection from the baboon's motor cortex to fusimotor neurones of the forearm and hand. *J. Physiol. (Lond.)*, 216:257–279.
Cody, F. W. J., Harrison, L. M., and Taylor, A. (1975): Analysis of activity of muscle spindles of the jaw-closing muscles during normal movements in the cat. *J. Physiol. (Lond.)*, 253:565–582.
Cody, F. W. J., Lee, R. W. H., and Taylor, A. (1972): A functional analysis of the mesencephalic nucleus of the fifth nerve in the cat. *J. Physiol. (Lond.)*, 226:249–261.
Cody, F. W. J., and Taylor, A. (1973): The behaviour of spindles in the jaw-closing muscles during eating and drinking in the cat. *J. Physiol. (Lond.)*, 231:49–50P.
Coggeshall, R. E. (1980): Law of separation of function of the spinal roots. *Physiol. Rev.*, 60:716–755.
Cohen, B., and Henn, V. (1972a): The origin of quick phases of nystagmus in the horizontal plane. *Bibl. Ophthalmol.*, 82:36–55.
Cohen, B., and Henn, V. (1972b): Unit activity in the pontine reticular formation associated with eye movements. *Brain Res.*, 46:403–410.
Cohen, B., Matsuo, V., and Raphan, T. (1977): Quantitative analysis of the velocity characteristics of optokinetic nystagmus and optokinetic after-nystagmus. *J. Physiol. (Lond.)*, 270:321–344.
Colburn, T. R., and Evarts, E. V. (1978): Use of Brush-

less DC torque motors in studies of neuromuscular function. In: *Progress in Clinical Neurophysiology, Vol. 4, Cerebral Motor Control in Man: Long Loop Mechanisms,* edited by J. E. Desmedt, pp. 153–166. Karger, Basel.

Collatos, T. C., Edgerton, V. R., Smith, J. L., and Botterman. B. R. (1977): Contractile properties and fiber type compositions of flexors and extensors of elbow joint in cat: implications for motor control. *J. Neurophysiol.,* 40:1292–1300.

Collewijn, H. (1969): Optokinetic eye movements in the rabbit: input-output relations. *Vision Res.,* 9:117–132.

Collewijn, H. (1975): Direction-selective units in the rabbit's nucleus of the optic tract. *Brain Res.,* 100:489–508.

Collewijn, H., and Grootendorst, A. F. (1979): Adaptation of optokinetic and vestibulo-ocular to modified visual input in the rabbit. *Prog. Brain Res.,* 50: 771–781.

Collins, W. F., Nulsen, F. E., and Randt, C. T. (1960): Relation of peripheral nerve size and sensation in man. *Arch. Neurol.,* 3:381–385.

Collins, W. F., Nulsen, F. E., and Shealy, C. N. (1966): Electrophysiological studies of peripheral nerve fiber size and sensation in man. In: *Pain,* edited by R. S. Knighton and P. R. Dumke, pp. 33–46. Little, Brown, Boston.

Comings, D. E., and Amromin, G. D. (1974): Autosomal dominant insensitivity to pain with hyperplasic myelinopathy and autosomal dominant indifference to pain. *Neurology (Minneap.),* 24:838–848.

Conrad, B. (1978): The motor cortex as a primary device for fast adjustment of programmed motor patterns to afferent signals. In: *Progress in Clinical Neurophysiology, Vol. 4, Cerebral Motor Control in Man: Long Loop Mechanisms,* edited by J. E. Desmedt, pp. 123–140. Karger, Basel.

Conrad, B., and Aschoff, J. C. (1977): Effects of voluntary isometric and isotonic activity on late transcortical reflex components in normal subjects and hemiparetic patients. *Electroencephalogr. Clin. Neurophysiol.,* 42:107–116.

Conrad, B., and Brooks, V. B. (1974): Effects of dentate cooling on rapid alternating arm movements. *J. Neurophysiol.,* 37:792–804.

Conrad, B., Matsunami, C., Meyer-Lohmann, J., et al. (1974): Cortical load compensation during voluntary elbow movements. *Brain Res.,* 81:507–514.

Conrad, B., Meyer-Lohmann, J., Matsunami, K., and Brooks. V. B. (1975): Precentral unit activity following torque pulse injections into elbow movements. *Brain Res.,* 94:219–236.

Cook, J. B., and Nathan, P. W. (1967): On the site of action of diazepam in spasticity in man. *J. Neurol. Sci.,* 5:33–37.

Cooke, J. D. (1979): Dependence of human arm movements on limb mechanical properties. *Brain Res.,* 165:366–369.

Cooke, J. D. (1980): The role of stretch reflexes during active movements. *Brain Res.,* 181:493–497.

Cooker, H. S., Larson, C. R., and Luschei. E. S. (1980): Evidence that the human jaw stretch reflex increases the resistance of the mandible to small displacements *J. Physiol. (Lond.),* 308:61–78.

Coombs, J. S., Eccles, J. C., and Fatt, T. (1955): Excitatory synaptic action in motoneurones. *J. Physiol. (Lond.),* 130:374–395.

Cooper, S. (1960): Muscle spindles and other muscle receptors. In: *The Structure and Function of Muscle,* Vol. 1, edited by G. H. Bourne, pp. 381–420. Academic Press, New York.

Cooper, S., and Daniel, P. M. (1963): Muscle spindles in man. *Brain,* 86:563–586.

Cooper, S., and Eccles, J. C. (1930): The isometric responses of mammalian muscles. *J. Physiol. (Lond.),* 69:377–385.

Coquery, J., and Coulmance, M. (1971): Variations d'amplitude des réflexes monosynaptiques avant un mouvement volontaire. *Physiol. Behav.,* 6:65–69.

Corbin, K. B., and Harrison, F. (1940): Function of the mesencephalic root of fifth cranial nerve. *J. Neurophysiol.,* 3:423–435.

Cordo, P. J., and Nashner, L. M. (1982): Properties of postural adjustments associated with rapid arm movements. *J. Neurophysiol.,* 47:287–302.

Costa, E., and Guidotti, A. (1979): Molecular mechanisms in the receptor action of benzodiazepines. *Annu. Rev. Pharmacol. Toxicol.,* 19:531 545.

Cotman, C. W. (editor) (1978): *Neural Plasticity.* Raven Press, New York.

Cotman, C. W., and Lynch, G. S. (1976): Reactive synaptogenesis in the adult nervous system. In: *Neuronal Recognition,* edited by S. Barondes, pp. 69–108. Plenum, New York.

Cotman, C. W., Nieto-Sampedro, M., and Harris, E. W. (1981): Synapse replacement in the nervous system of adult vertebrates. *Physiol. Rev.,* 61:684–784.

Cotman, C. W., and Scheff, S. W. (1979): Compensatory synapse growth in aged animals after neuronal death. *Mech. Ageing Dev.,* 9:103–117.

Coulter, J. D., Ewing, L., and Carter, C. (1975): Origin of primary sensorimotor cortical projections to lumbar spinal cord of cat and monkey. *Brain Res.,* 103:366–372.

Coulter, J. D., and Jones, E. G. (1977): Differential distribution of corticospinal projections from individual cytoarchitectonic fields in the monkey. *Brain Res.,* 129:335–340.

Courjon, J. H., Jeannerod, M., Ossuzio, I., and Schmid, R. (1977): The role of vision in compensation of vestibulo-ocular reflex after hemilabyrinthectomy in the cat. *Exp. Brain Res.,* 28:235–248.

Courtney, K. R., and Fetz, E. E. (1972): Unit responses recorded from cervical spinal cord of awake monkey. *Brain Res.,* 53:445–450.

Courtois, A., and Cordeau, J. P. (1969): Changes in cortical responsiveness during transition from sleep to wakefulness. *Brain Res.,* 14:199–214.

Courville, J. (1966): The nucleus of the facial nerve: the relation between cellular groups and peripheral branches of the nerve. *Brain Res.,* 1:338–354.

Cowan, W. M., Gottlieb, D. I., Hendrickson, A. O., et al. (1972): The autoradiographic demonstration of axonal connections in the central nervous system. *Brain Res.,* 37:21–51.

Coxe, W. S., and Landau, W. M. (1975): Observation upon the effect of supplementary motor cortex ablation in the monkey. *Brain,* 88:763–773.

Cragg, B. G., and Temperley, H. N. V. (1954): The organisation of neurones: a cooperative analogy. *Electroencephalogr. Clin. Neurophysiol.,* 6:85–92.

Craggs, M. D. (1975): Cortical control of motor prostheses, using the cord-transected baboon as the primate model for human paraplegia. *Adv. Neurol.,* 10:91–101.

Crago, P. E., Houk, J. C., and Hasan, Z. (1976): Regulatory actions of human stretch reflex. *J. Neurophysiol.,* 39:925–935.

Craske, B. (1977): Perception of impossible limb positions induced by tendon vibration. *Science,* 196:71–73.

Crayton, J. M., Smith, R. C., Klass, D., et al. (1977): Electrophysiological (H reflex) studies of patients with tardive dyskinesia. *Am. J. Psychiatry,* 134:775–781.

Crayton, J. W., and Tamminga, C. A. (1978): Spinal-reflex studies as test of dopamine neurotransmission. *Lancet* I:215–216.

Creed, R. S., Denny-Brown, D., Eccles, J. C., et al. (1932): *Reflex Activity in the Spinal Cord.* Oxford University Press, London.

Crossman, E. R. F. W. (1959): A theory of the acquisition of speed-skill. *Ergonomics,* 2:153–166.

Crutcher, K. A., Brothers, L., and Davis, J. N. (1979): Sprouting of sympathetic nerves in the absence of afferent input. *Exp. Neurol.,* 66:778–783.

Crutcher, K. A., and Davis, J. N. (1981): Sympatho-hippocampal sprouting is directed by a target trophic factor. *Brain Res.,* 20:410–414.

Cuello, A. C., Polak, J. M., and Pearse, A. G. (1976): Substance P: a naturally occurring transmitter in human spinal cord. *Lancet,* II:1054–1056.

Cull, R. E. (1975): Role of axonal transport in maintaining central synaptic connections. *Exp. Brain Res.,* 24:97–101.

Cullheim, S. (1978): Relations between cell body size, axon diameter and axon conduction velocity of cat sciatic a-motoneurons stained with horseradish peroxidase. *Neurosci. Lett.,* 8:17–20.

Currier, R. D. (1969): Syndromes of the medulla oblongata. In: *Handbook of Clinical Neurology,* Vol. 2, edited by P. J. Vinken and G. W. Bruyn, pp. 217–237. North-Holland, Amsterdam.

Curtis, D. R., and Ryall, R. W. (1964): Nicotinic and muscarinic receptors of Renshaw cells. *Nature,* 203:652–654.

Cussons, P. D., Hulliger, M., and Matthews, P. B. C. (1977): Effects of fusimotor stimulation on the response of the secondary ending of the muscle spindle to sinusoidal stretching. *J. Physiol. (Lond.),* 270:835–850.

Cynader, M., and Chernenko, G. (1976): Abolition of direction selectivity in the visual cortex of the cat. *Science,* 193:504–505.

Czeh, G., Gallego, R., Kudo, N., and Kuno, M. (1978): Evidence for the maintenance of motoneurone properties by muscle activity. *J. Physiol. (Lond.),* 281:239–252.

Daunton, N., and Melvill Jones, G. (1973): Directional representation of horizontal and vertical acceleration in the neural activity of cat vestibular nuclei. *Proceedings: Annual Science Meeting of Aerospace Medical Association,* p. 144. Pacific Grove, California.

Daunton, N., and Thomsen, D. (1979): Visual modulation of otolith-dependent units in cat vestibular nuclei. *Exp. Brain Res.,* 37:173–176.

Davidoff, R. A. (1973): Alcohol and presynaptic inhibition in an isolated spinal cord preparation. *Arch, Neurol.,* 28:60–63.

Davidoff, R. A. (1978): Pharmacology of spasticity. *Neurology (Minneap.),* 28:46–51.

Davies, P. R. T., and Melvill Jones, G. (1976): An adaptive neural model compatible with plastic changes induced in the human vestibulo-ocular reflex by prolonged optical reversal of vision. *Brain Res.,* 103:546–550.

Davis, C. J. F., and Montgomery, A. (1977): The effect of prolonged inactivity upon the contraction characteristics of fast and slow mammalian twitch muscle. *J. Physiol. (Lond.),* 270:581–594.

Dawson, G. D. (1947): Investigations on a patient subject to myoclonic seizures after sensory stimulation. *J. Neurol. Neurosurg. Psychiatry,* 70:141–149.

Dawson, G. D. (1956): The relative excitability and conduction velocity of sensory and motor nerve fibres in man. *J. Physiol. (Lond.),* 131:436–451.

Dawson, G. D., and Merton, P. E. (1956): "Recurrent" discharges from motoneurones. *Abstracts XXth International Congress of Physiology,* Brussels, pp. 221–222.

Day, B. L., Marsden, C. D., Obeso, J. A., and Rothwell, J. C. (1981): Reciprocal inhibition in the human forearm. *J. Physiol. (Lond.),* 317:59–60P.

Deecke, L., Englitz, H. G., Kornhuber, H. H., and Schmitt, G. (1977): Cerebral potentials preceding voluntary movement in patients with bilateral or unilateral Parkinson akinesia. In: *Progress in Clinical Neurophysiology, Vol. 1, Attention, Voluntary Contraction and Event-Related Cerebral Potentials,* edited by J. E. Desmedt, pp. 151–163. Karger, Basel.

Deecke, L., and Kornhuber, H. H. (1977): Cerebral potentials and the initiation of voluntary movement. In: *Progress in Clinical Neurophysiology, Vol. 1, Attention, Voluntary Contraction and Event-Related Cerebral Potentials,* edited by J. E. Desmedt, pp. 132–150. Karger, Basel.

Deecke, L., and Kornhuber, H. H. (1978): An electrical sign of participation of the mesial 'supplementary' motor cortex in human voluntary finger movement. *Brain Res.,* 159:473–476.

Deecke, L., Scheid, P., and Kornhuber, H. H. (1969): Distribution of readiness potential, pre-motion positivity and motor potential of the human cerebral cortex preceding voluntary finger movements. *Exp. Brain Res.,* 7:158–168.

Degail, P., Lance, J. W., and Neilson, P. D. (1966): Differential effects on tonic and phasic reflex mechanisms produced by vibration of muscles in man. *J. Neurol. Neurosurg. Psychiatry,* 29:1–11.

Dehen, H., Willer, J. C., Bathien, N., and Cambier, J. (1976): Blink reflex in hemiplegia. *Electroencephalogr. Clin. Neurophysiol.,* 40:393–400.

Dehen, H., Willer, J. C., Boureau, F., and Cambier, J. (1977): Congenital insensitivity to pain and the endogenous morphine-like substances. *Lancet*, 2:293–294.

Dehen, H., Willer, J. C., Prier, S., et al. (1978): Congenital insensitivity to pain and the "morphine-like" analgesic system. *Pain*, 5:351–358.

DeJong, R. H., Hershey, W. N., and Wagman, I. H. (1967): Measurement of a spinal reflex response (H-reflex) during general anesthesia in man. *Anesthesiology*, 28:382–389.

DeJong, J. D., and Melvill Jones, G. (1971): Akinesia, hypokinesia, and bradykinesia in the oculomotor system of patients with Parkinson's disease. *Exp. Neurol.*, 32:58–68.

Delacour, J., Liboudan, S., and McNeil, M. (1972): Premotor cortex and instrumental behaviour in monkeys. *Physiol. Behav.*, 8:299–305.

Delgado-Garcia, J., Baker, R., and Highstein, S. M. (1977): The activity of internuclear neurons identified within the abducens nucleus of the alert cat. In: *Control of Gaze by Brain Stem Neurons*, edited by R. Baker and A. Berthoz, pp. 291–300. Elsevier, Amsterdam.

DeLong, M. R., and Georgopoulos, A. P. (1979): Motor functions of the basal ganglia as revealed by studies of single cell activity in the behaving primate. In: *Advances in Neurology*, Vol. 24, pp. 131–140. Raven Press, New York.

DeLuca, C. J., and Gilmore, L. D. (1976): Voluntary nerve signals from severed mammalian nerves: long-term recordings. *Science*, 191:193–195.

Delwaide, P. J. (1971): *Etude Expérimentale de l'Hyperréflexia Tendineuse en Clinique Neurologique*. Editions Arscia, Brussels.

Delwaide, P. J. (1973): Human monosynaptic reflexes and presynaptic inhibition. In: *New Developments in Electromyography and Clinical Neurophysiology*, edited by J. E. Desmedt, Vol. 3, pp. 508–522. Karger, Basel.

Delwaide, P. J. (1977): Excitability of lower limb myotatic arcs under the influence of caloric labyrinthine stimulation. *J. Neurol. Neurosurg. Psychiatry*, 40:970–974.

Delwaide, P. J., Cordonnier, M., and Gadea-Ciria, M. (1978): Excitability relationships between lower limb myotatic arcs in spasticity. *J. Neurol. Neurosurg. Psychiatry*, 41:636–641.

Delwaide, P. J., Dessalle, M., and Juprelle, M. (1975): Etude réflexologique du mode d'action d'un myorelaxant: le LCB 29. *Rev. Electroencephalogr. Neurophysiol. Clin.*, 5:432–438.

Delwaide, P. J., Figiel, C., and Richelle, C. (1977): The effects of postural changes of the upper limb on reflex transmission in the lower limb. Cervico-lumbar reflex interactions in man. *J. Neurol. Neurosurg. Psychiatry*, 40:616–621.

Delwaide, P. J., Martinelli, P., and Crenna, P. (1980): Clinical neurophysiological measurement of spinal reflex activity. In: *Spasticity: Disordered Motor Control*, edited by R. G. Feldman, R. R. Young, and W. P. Koella, pp. 345–371. Year Book, Chicago.

Delwaide, P. J., Schwab, R. S., and Young, R. R. (1974): Polysynaptic spinal reflexes in Parkinson's disease. *Neurology (Minneap.)*, 24:820–827.

Delwaide, P. J., and Toulouse, P. (1981): Facilitation of monosynaptic reflexes by voluntary contraction of muscles in remote parts of the body: Mechanisms involved in the Jendrassik manoeuvre. *Brain*, 104:701–721.

Delwaide, P. J., and Young, R. R. (1973): Electrophysiologic study of polysynaptic reflexes in Parkinsonism. In: *Parkinson's Disease*, Vol. 2, edited by J. Siegfried, pp. 185–189.

Denny-Brown, D. (1929): On the nature of postural reflexes. *Proc. R. Soc. Lond. (Biol.)*, 14:237–252.

Denny-Brown, D. (1951): Hereditary sensory radicular neuropathy. *J. Neurol. Neurosurg. Psychiatry*, 14:237–252.

Denny-Brown, D. (1960): Disease of the basal ganglia. Their relationship to disorders of movement. *Lancet*, 2:1099–1105; 1155–1162.

Denny-Brown, D. (1966): *The Cerebral Control of Movement*. Liverpool University Press, Liverpool.

Denny-Brown, D. (1968): Clinical symptomatology of diseases of the basal ganglia. In: *Handbook of Clinical Neurology*, Vol. 6, *Diseases of the Basal Ganglia*, edited by P. J. Vinken and G. W. Bruyn, pp. 133–172. North-Holland, Amsterdam.

Denny-Brown, D., and Botterell, E. H. (1948): The motor functions of the agranular frontal cortex. *Res. Publ. Assoc. Res. Nerv. Ment. Dis.*, 27:257–346.

Derouesne, C. (1973): Le syndrome pré-moteur. *Rev. Neurol.*, 128:353–363.

Deschesnes, M., and Hammond, C. (1980): Physiological and morphological identification of ventrolateral fibers relaying cerebellar information to the cat motor cortex. *Neuroscience*, 5:1137–1141.

Deschuytere, J., Dekeyser, C., Rosselle, N., and Deschuytere, M. (1981): Monosynaptic reflexes in the flexor carpi ulnaris muscle in man. *Electromyogr. Clin. Neurophysiol.*, 21:213–222.

Deschuytere, J., and Rosselle, N. (1970): EMG and neurophysiological investigation in root compression syndromes in man. *Electromyography*, 10:339–340.

Deschuytere, J., and Rosselle, N. (1971): Electrophysiological study of discharges from spinal origin the anterolateral muscles of the leg in normal adults. *Electromyography*, 11:331–363.

Deschuytere, J., and Rosselle, N. (1973): Diagnostic use of monosynaptic reflexes in L_5 and S_1 root compression. In: *New Developments in Electromyography and Clinical Neurophysiology*, Vol. 3, edited by J. E. Desmedt, pp. 360–366. Karger, Basel.

Deschuytere, J., and Rosselle, N. (1974): Identification of certain EMG patterns of the spinal cord reflexive activity in man. Electrophysiological study of discharges from spinal origin in the forearm flexors in normal adults. *Electromyogr. Clin. Neurophysiol.*, 14:497–511.

Deschuytere, J., and Rosselle, N. (1975): The clinical significance of the reflexive activity derived from the superficial forearm flexors. *Electromyography*, 15:157–158.

Deschuytere, J., Rosselle, N., and Dekeyser, C. (1976): Monosynaptic reflexes in the superficial forearm flexors

in man and their clinical significance. *J. Neurol. Neurosurg. Psychiatry,* 39:555–565.
Descuns, P., Collet, M., Resche, F., et al. (1973): Intérêt du réflexe H dans l'exploration des lésions radiculaires d'origine discale. *Neurochirurgie,* 19:627–640.
Desmedt, J. E. (1950): Etude expérimentale de la dégénérescence wallérienne et de la réinnervation du muscle squelettique. *Arch. Int. Physiol. Biochim.,* 58:23–68;125–156.
Desmedt, J. E. (1958): Méthodes d'étude de la fonction neuromusculaire chez l'homme. *Acta Neurol. Belg.,* 58:977–1017.
Desmedt, J. E. (editor) (1973a): Human reflexes, pathophysiology of motor systems and methodology of human reflexes. In: *New Developments in Electromyography and Clinical Neurophysiology,* Vol. 3. Karger, Basel.
Desmedt, J. E. (1973b): The neuromuscular disorder in myasthenia gravis. In: *New Developments in Electromyography and Clinical Neurophysiology,* Vol. 1, edited by J. E. Desmedt, pp. 241–304. Karger, Basel.
Desmedt, J. E. (1977): Active touch exploration of extrapersonal space elicits specific electrogenesis in the right cerebral hemisphere of intact right-handed man. *Proc. Natl. Acad. Sci. USA,* 74:4037–4040.
Desmedt, J. E. (editor) (1978a): Cerebral motor control in man: long loop mechanisms. In: *Progress in Clinical Neurophysiology,* Vol. 4. Karger, Basel.
Desmedt, J. E. (editor) (1978b): Physiological tremor, pathological tremors and clonus. In: *Progress in Clinical Neurophysiology,* Vol. 5. Karger, Basel.
Desmedt, J. E. (1980a): Patterns of motor commands during various types of voluntary movement in man. *Trends Neurosci.,* 3:265–268.
Desmedt, J. E. (editor) (1980b): Spinal and supraspinal mechanisms of voluntary motor control and locomotion. In: *Progress in Clinical Neurophysiology,* Vol. 8. Karger, Basel.
Desmedt, J. E. (1981a): The size principle of motoneuron recruitment in ballistic or ramp voluntary contractions in man. In: *Progress in Clinical Neurophysiology, Vol. 9, Motor Unit Types, Recruitment and Plasticity in Health and Disease,* edited by J. E. Desmedt, pp. 97–136. Karger, Basel.
Desmedt, J. E. (1981b): Plasticity of motor unit organization studied by coherent electromyography in patients with nerve lesions or with myopathic or neuropathic diseases. In: *Progress in Clinical Neurophysiology, Vol. 9, Motor Unit Types, Recruitment and Plasticity in Health and Disease,* edited by J. E. Desmedt, pp. 250–304. Karger, Basel.
Desmedt, J. E. (1982): Cerebral evoked potentials. In: *Peripheral Neuropathy,* 2nd ed., edited by P. J. Dyck, P. K. Thomas, and E. H. Lambert. Saunders, Philadelphia.
Desmedt, J. E., Brunko, E., and Debecker, J. (1976): Maturation of the somatosensory evoked potentials in normal infants and children with special reference to early N_1 component. *Electroencephalogr. Clin. Neurophysiol.,* 40:43–58.
Desmedt, J. E., Brunko, E., and Debecker, J. (1980): Maturation and sleep correlates of the somatosensory evoked potentials. In: *Progress in Clinical Neurophysiology* Vol. 7, edited by J. E. Desmedt, pp. 146–161. Karger, Basel.
Desmedt, J. E., and Cheron, G. (1981): Non-cephalic reference recording of early somatosensory potentials to finger stimulation in adult or aging normal man: Differentiation of widespread N18 and contralateral N20 from the prerolandic P22 and N30 components. *Electroencephalogr. Clin. Neurophysiol.,* 52:553–570.
Desmedt, J. E., and Franken, L. (1963): Long-term physiological changes in auditory cortex following partial deafferentation. In: *The Effect of Use and Disuse on Neuromuscular Functions,* edited by E. Gutmann and P. Hnik, pp. 264–276. Czeck. Acad. Sci., Prague.
Desmedt, J. E., Franken, L., Borenstein, S., et al. (1966): Le diagnostic des ralentissements de la conduction afferente dans les affections des nerfs périphérique. Intérêt de l'extraction du potentiel évoqué cerebral. *Rev. Neurol.,* 115:155–162.
Desmedt, J. E., and Godaux, E. (1975): Vibration-induced discharge patterns of single motor units in the masseter muscle in man. *J. Physiol. (Lond.),* 253:429–442.
Desmedt, J. E., and Godaux, E. (1976): Habituation of exteroceptive suppression and of exteroceptive reflexes in man as influenced by voluntary contraction. *Brain Res.,* 106:21–29.
Desmedt, J. E., and Godaux, E. (1977a): Ballistic contractions in man: characteristic recruitment pattern of single motor units of the tibialis anterior muscle. *J. Physiol. (Lond.),* 264:673–693.
Desmedt, J. E., and Godaux, E. (1977b): Fast motor units are not preferentially activated in rapid voluntary contractions in man. *Nature,* 267:717–719.
Desmedt, J. E., and Godaux, E. (1978a): Ballistic skilled movements: Load compensation and patterning of the motor commands. In: *Progress in Clinical Neurophysiology, Vol. 4, Cerebral Motor Control in Man: Long Loop Mechanisms,* edited by J. E. Desmedt, pp. 21–55. Karger, Basel.
Desmedt, J. E., and Godaux, E. (1978b): Ballistic contractions in fast or slow human muscles: discharge patterns of single motor units. *J. Physiol. Lond.,* 285:185–196.
Desmedt, J. E., and Godaux, E. (1978c): Mechanism of the vibration paradox: excitatory and inhibitory effects of tendon vibration on single soleus motor units in man. *J. Physiol. (Lond.),* 285:197–207.
Desmedt, J. E., and Godaux, E. (1979): Recruitment patterns of single motor units in the human masseter muscle during brisk jaw clenching. *Arch. Oral Biol.,* 24:171–178.
Desmedt, J. E., and Godaux, E. (1980): The tonic vibration reflex and the vibration paradox in limb and jaw muscles in man. In: *Progress in Clinical Neurophysiology, Vol. 8, Spinal and Supraspinal Mechanisms of Voluntary Motor Control and Locomotion,* edited by J. E. Desmedt, pp. 215–242. Karger, Basel.
Desmedt, J. E. and Godaux, E. (1981): Spinal motoneuron recruitment in man: deordering with direction, but not with speed of voluntary movement. *Science,* 214:933–936.

Desmedt, J. E., and Hainaut, K. (1968): Kinetics of myofilament activation in potentiated contraction: staircase phenomenon in human skeletal muscle. *Nature*, 217:529–532.

Desmedt, J. E., and Hainaut, K. (1977): Inhibition of the intracellular release of calcium by Dantrolene in barnacle giant muscle fibres. *J. Physiol. (Lond.)*, 265:565–585.

Desmedt, J. E., and Hainaut, K. (1978): Excitation-contraction coupling in single muscle fibres and the calcium channel in sarcoplasmic reticulum. *Ann. NY Acad. Sci.*, 307:433–435.

Desmedt, J. E., and Hainaut, K. (1979): Dantrolene and A23187 ionophore: specific action on calcium channels revealed by the aequorin method. *Biochem. Pharmacol.*, 28:957–964.

Desmedt, J. E., and Noel, P. (1973): Average cerebral evoked potentials in the evaluation of lesions. In: *New Developments in Electromyography and Clinical Neurophysiology*, Vol. 2, edited by J. E. Desmedt, pp. 352–371. Karger, Basel.

Desmedt, J. E., and Robertson, D. (1977): Differential enhancement of early and late components of the cerebral somatosensory evoked potentials during fast sequential cognitive tasks in man. *J. Physiol. (Lond.)*, 271:761–782.

Deuel, R. K. (1977): Loss of motor habits after cortical lesions. *Neuropsychologia*, 15:205–215.

Devanandan, M. S., Eccles, R. M., Lewis, D. M., and Stenhouse, D. (1969): Responses of flexor alpha-motoneurones in cats anaesthetised with chloralose. *Exp. Brain Res.*, 8:163–176.

DeVito, J. L., and Smith, O. A. (1959): Projections from the mesial frontal cortex (supplementary motor area) to the cerebral hemispheres and brainstem of the Macaca mulatta. *J. Comp. Neurol.*, 111:261–277.

Dhanarajan, P., Ruegg, D. G., and Wiesendanger, M. (1977): Anatomical investigation of corticopontine projection in primate. *Neuroscience*, 2:913–922.

Diamantopoulos, E., and Zander Olson, P. (1967): Excitability of motor neurones in spinal shock in man. *J. Neurol. Neurosurg. Psychiatry*, 30:427–431.

Diamond, J., Cooper, E., Turner, C., and McIntyre, L. (1976): Trophic regulation of nerve sprouting-neuron-target interactions and spatial relations central sensory nerve fields in salamander skin. *Science*, 193:371–377.

Dichgans, J., Bizzi, E., Morasso, P., and Tagliasco, V. (1973): Mechanisms underlying recovery of head-eye coordination following bilateral labyrinthectomy in monkeys. *Exp. Brain Res.*, 18:548–562.

Dichgans, J., and Brandt, T. (1974): The psychophysics of visually induced perception of self-motion and tilt. In: *The Neurosciences III*, pp. 123–129. M.I.T. Press, Cambridge, Massachusetts.

Dichgans, J., Brandt, T., and Held, R. (1975): The role of vision in gravitational orientation. *Fortschr. Zool.*, 23:255–263.

Dichgans, J., Held, R., Young, L., and Brandt, T. (1972): Moving visual scenes influence the apparent direction of gravity. *Science*, 178:1217–1219.

Dichgans, J., Mauritz, K. H., Allum, J. H., et al. (1976): Postural sway in normals and atactic patients. *Aggressologie*, 17C:15–20.

Dichgans, J., Schmidt, C., and Graf, W. (1973): Visual input improves the speedometer function of the vestibular nuclei in the goldfish. *Exp. Brain Res.*, 18:319–322.

Didday, R. L. A. (1976): A model of visuomotor mechanisms in the frog optic tectum. *Math. Biosci.*, 30:169–180.

Dieringer, N., and Precht, W. (1977): Modification of synaptic input following unilateral labyrinthectomy. *Nature*, 269:431–433.

Dieringer, N. and Precht, W. (1979): Mechanisms of compensation for vestibular deficits in the frog. *Exp. Brain Res.*, 36:311–328.

Dietrichson, P. (1971): Phasic ankle reflex in spasticity and Parkinsonian rigidity. *Acta Neurol. Scand.*, 47:22–51.

Dietrichson, P. (1973): The role of the fusimotor system in spasticity and Parkinsonian rigidity. In: *New Developments in Electromyography and Clinical Neurophysiology*, Vol. 3, edited by J. E. Desmedt, pp. 496–507. Karger, Basel.

Dietrichson, P., Engebretsen, O., Fonstelien, E., and Howland, J. (1978): Quantitation of tremor in man. In: *Progress in Clinical Neurophysiology, Vol. 5, Physiological Tremor, Pathological Tremors and Clonus*, edited by J. E. Desmedt, pp. 90–94. Karger, Basel.

Dietz, V. (1978): Analysis of the electrical muscle activity during maximal contraction and the influence of ischemia. *J. Neurol. Sci.*, 37:187–197.

Dietz, V., Mauritz, K. H., and Dichgans, J. (1980): Body oscillation in balancing due to the segmental stretch reflex activity. *Exp. Brain Res.*, 40:89–95.

Dietz, V., and Noth, J. (1978): Pre-innervation and stretch responses of triceps brachii in man falling with and without visual control. *Brain Res.*, 142:576–579.

Dietz, V., Quintern, J., and Berger, W. (1981): Electrophysiological studies of gait in spasticity and rigidity. Evidence that altered mechanical properties of muscle contribute to hypertonia. *Brain*, 104:431–449.

Dietz, V., Schildbleicher, D., and Noth, J. (1979): Neuronal mechanisms of human locomotion. *J. Neurophysiol.*, 42:1212–1222.

Dimitrijevic, M. R. (1973): Withdrawal reflexes. In: *New Developments in Electromyography and Clinical Neurophysiology*, Vol. 3, edited by J. E. Desmedt, pp. 744–750. Karger, Basel.

Dimitrijevic, M. R., Faganel, J., Gregoric, M., et al. (1972): Habituation: Effects of regular and stochastic stimulation. *J. Neurol. Neurosurg. Psychiatry*, 35:234–242.

Dimitrijevic, M. R., and Nathan, P. W. (1967): Studies of spasticity in man. *Brain*, 90:1–30; 333–358.

Dimitrijevic, M. R., and Nathan, P. W. (1968): Studies of spasticity in man III. Analysis of reflex activity evoked by noxious cutaneous stimulation. *Brain*, 91:349–368.

Dimitrijevic, M. R., and Nathan, P. W. (1971): Studies of spasticity in man V. Dishabituation of the flexion reflex in spinal man. *Brain*, 94:77–90.

Dimitrijevic, M. R., and Nathan, P. W. (1973): Studies of spasticity in man VI. Habituation, dishabituation

and sensitization of tendon reflexes in spinal man. *Brain*, 96:337–354.
Dimitrijevic, M. R., Sherwood, A. M., and Nathan, P. W. (1978): Clonus: Peripheral and central mechanisms. In: *Progress in Clinical Neurophysiology, Vol. 5, Physiological Tremor, Pathological Tremors and Clonus;* edited by J. E. Desmedt, pp. 173–182. Karger, Basel.
Dimitrijevic, M. R., Spencer, W. A., Trontelj, J. V., and Dimitrijevic, M. (1977): Reflex effects of vibration in patients with spinal cord lesions. *Neurology, (Minneap.)*, 27:1078–1086.
Dindar, F., and Verrier, M. (1975): Studies on the receptor responsible for vibration induced inhibition of monosynaptic reflexes in man. *J. Neurol. Neurosurg. Psychiatry*, 38:155–160.
Dinn, J. J., and Crane, D. L. (1970): Schwann cell dysfunction in uremia. *J. Neurol. Neurosurg. Psychiatry*, 33:605–608.
Doetsch, G. S., and Gardner, E. B. (1972): Relationship between afferent input and motor output in sensorimotor cortex of the monkey. *Exp. Neurol.*, 35:78–97.
Doetsch, G. S., and Towe, A. L. (1976): Response properties of distinct neuronal subsets in hindlimb sensorimotor cerebral cortex of the domestic cat. *Exp. Neurol.*, 53:520–547.
Donoghue, J. P., and Wells, J. (1977): Synaptic rearrangement in the mouse following partial cortical deafferentation. *Brain Res.*, 125:351–355.
Dorcas, D. S., and Scott, R. N. (1966): A three-state myoelectric control. *J. Med. Biol. Eng.*, 4:367–370.
Dow, D. S., and Moruzzi, G. (1958): *The Physiology and Pathology of the Cerebellum*. University of Minnesota Press, Minneapolis.
Dreschler, B., Lastovka, M., and Kalvodova, E. (1966): Electrophysiological study of patients with herniated intervertebral disks. *Electromyography*, 6:187.
Dreyer, D. A., Metz, C. B., Schneider, R. J., and Whitsel, B. L. (1974): Differential contributions of spinal pathways to body representation in postcentral gyrus of Macaca mulatta. *J. Neurophysiol.*, 37:119–145.
Dubner, R., Sessle, B. J., and Storey, A. T. (1978): *The Neural Basis of Oral and Facial Function*, pp. 160–165. Plenum, New York.
Dubrovsky, B., and Garcia-Rill, E. (1971): Convergence of tectal and visual cortex input to pericruciate neurons. *Exp. Neurol.*, 33:475–484.
Duensing, F., and Schaefer, K. P. (1957): Die Neuronenaktivität in der Formatio reticularis des Rhombencephalons beim vestibulären Nystagmus. *Arch. Psychiatr. Nervenkr.*, 196:265–290.
Duensing, F., and Schaefer, K. P. (1958): Die Aktivität einzelner Neurone im Bereich der Vestibulariskerne bei Horizontalbeschleunigungen unter besonderer Berücksichtigung des vestibulären Nystagmus. *Arch. Psychiatr. Nervenkr.*, 198:225–252.
Duensing, F., and Schaefer, K. P. (1959): Uber die konvergenz verschiedener labyrintharer Afferenzen auf einzelne Neurone des Vestibulariskerngebietes. *Arch. Psychiatr. Nervenkr.*, 199:345–371.
Duffy, F. H., and Burchfield, J. L. (1971): Somatosensory system: organizational hierarchy from single units in monkey area 5. *Science*, 172:273–275.
DuFresne, J. R., Gurfinkel, V. S., Soechting, J. F., and Terzuolo, C. A. (1978): Response to transient disturbances during intentional forearm flexion in man. *Brain Res.*, 150:103–115.
DuFresne, J. R., Soechting, J. F., and Terzuolo, C. A. (1980): Modulation of the myotatic reflex gain in man during intentional movements. *Brain Res.*, 193:67–84.
Duggan, A. W. (1974): The differential sensitivity to L-Glutamate and L-Aspartate of spinal interneurones and Renshaw cells. *Exp. Brain Res.*, 19:522–528.
Duggan, A. W., Hall, J. G., and Headley, P. M. (1976): Morphine, enkephalin and the substantia gelatinosa. *Nature*, 264:456–458.
Dum, R. P., and Kennedy, T. T. (1980): Physiological and histochemical characteristics of motor units in cat tibialis anterior and extensor digitorum longus muscles. *J. Neurophysiol.*, 43:1615–1630.
Duncan, G. W., Shahani, B. T., and Young, R. R. (1976): An evaluation of baclofen treatment for certain symptoms in patients with spinal cord lesions. *Neurology, (Minneap.)*, 26:441–446.
Duncan, J. (1977): Response selection rules in spatial choice reaction time. In: *Attention and Performance VI*, edited by S. Dornic, pp. 49–61. Wiley, New York.
Dutia, M. B. (1980): Activation of cat muscle spindle primary, secondary and intermediate sensory endings by suxamethonium. *J. Physiol. (Lond.)*, 304:315–330.
Dutia, M. B., and Ferrell, W. R. (1980): The effect of suxamethonium on the response to stretch of Golgi tendon organs in the cat. *J. Physiol. (Lond.)*, 306:511–518.
Dutia, M. B., Lindsay, K. W., and Rosenberg, J. R. (1981): The effect of cerebellectomy on the tonic labyrinth and neck reflexes in the decerebrate cat. *J. Physiol. (Lond.)*, 312:115–123.
Duysens, J., and Pearson, K. G. (1976): The role of cutaneous afferents from the distal hindlimb in the regulation of the step cycle in thalamic cats. *Exp. Brain Res.*, 24:245–255.
Dyck, P. J., Johnson, W. J., Lambert, E. A., and O'Brien, P. C. (1971): Segmental demyelination secondary to axonal degeneration in uremic neuropathy. *Mayo Clinic Proc.*, 46:400–431.
Dyck, P., Thomas, P. K., and Lambert, E. H. (1974): *Peripheral Neuropathy*. Saunders, Philadelphia.
Dyhre-Poulsen, P., Laursen, A. M., Jahnsen, H., and Djorup, A. (1980): Programmed and reflex muscular activity in monkeys landing from a leap. In: *Progress in Clinical Neurophysiology, Vol. 8, Spinal and Supraspinal Mechanisms of Voluntary Motor Control and Locomotion*, edited by J. E. Desmedt, pp. 323–329. Karger, Basel.
Eberhard, H. D., Inman, V. T., and Burchfield, B. (1968): The principal elements in human locomotion. In: *Human Limbs and their Substitutes*, edited by P. E. Klopsteg and P. D. Wilson. Hafner, New York.

Eccles, J. C. (1953): *The Neurophysiological Basis of Mind.* Clarendon Press, Oxford.
Eccles, J. C. (1964): Presynaptic inhibition in the spinal cord. *Prog. Brain Res.,* 12:65–91.
Eccles, J. C. (1966): Long-loop reflexes from the spinal cord to the brainstem and cerebellum. *Atti Accad. Med. Lomb.,* 21:158–176.
Eccles, J. C. (1978): An instruction-selection hypothesis of cerebral learning. In: *Cerebral Correlates of Conscious Experience,* edited by P. A. Buser and A. Rougeul-Buser, pp. 155–175. Elsevier, Amsterdam.
Eccles, J. C. (1980): *The Human Psyche.* Springer, Berlin.
Eccles, J. C., Eccles, R. M., and Lundberg, A. (1957): The convergence of monosynaptic excitatory afferents onto many different species of alpha motoneurones. *J. Physiol. (Lond.),* 137:22–50.
Eccles, J. C., Eccles, R. M., and Lundberg, A. (1958): The action potentials of the alpha motoneurones supplying fast and slow muscles. *J. Physiol. (Lond.),* 142:275–291.
Eccles, J. C., Fatt, P., and Koketsu, K. (1954): Cholinergic and inhibitory synapses in a pathway from motor-axon collaterals to motoneurones. *J. Physiol. (Lond.),* 126:524–562.
Eccles, J. C., Sabah, N. H., and Taborikova, H. (1974): The pathways responsible for excitation and inhibition of fastigial neurones. *Exp. Brain Res.,* 19:78–99.
Eccles, J. C., Schmidt, R. F., and Willis, W. D. (1962): Presynaptic inhibition of the spinal monosynaptic reflex pathway. *J. Physiol. (Lond.),* 161:282–297.
Eccles, J. C., and Sherrington, C. S. (1930): Numbers and contraction-values of individual motor units in some muscles of the limb. *Proc. R. Soc. Lond. (Biol.),* 106:326–357.
Eccles, R. M., and Lundberg, A. (1958): Integrative pattern of Ia synaptic actions on motoneurones of hip and knee muscles. *J. Physiol. (Lond.),* 144:271–298.
Eccles, R. M., and Lundberg, A. (1959): Supraspinal control of interneurones mediating spinal reflexes. *J. Physiol. (Lond.),* 147:565–584.
Echlin, F., and Fessard, A. (1938): Synchronized impulse discharges from receptors in the deep tissues in response to a vibrating stimulus. *J. Physiol. (Lond.),* 93:321–334.
Eckmiller, R., and Mackeben, M. (1978): Pursuit eye movements and their neuronal control in the monkey. *Pfluegers Arch.,* 377:15–23.
Edds, M. V. (1953): Collateral nerve regeneration. *Q. Rev. Biol.,* 28:260–276.
Edgerton, V. R., and Cremer, S. (1981): Motor unit plasticity and possible mechanisms. In: *Progress in Clinical Neurophysiology, Vol. 9, Motor Unit Types, Recruitment and Plasticity in Health and Disease,* edited by J. E. Desmedt, pp. 220–240. Karger, Basel.
Edstrom, L. (1970): Selective changes in the sizes of red and white muscle fibers in upper motor lesions and Parkinsonism. *J. Neurol. Sci.,* 11:537–550.
Edwards, R. H. T., Hill, D. K., Jones, D. A., and Merton, P. A. (1977): Fatigue of long duration in human skeletal muscle after exercise. *J. Physiol. (Lond.),* 272:769–778.
Ehrhardt, K. J., and Wagner, A. (1970): Labyrinthine and neck reflexes recorded from spinal single motoneurons in the cat. *Brain Res.,* 19:87–104.
Eidelberg, E., and Davis, F. (1976): Role of proprioceptive data in performance of a complex visuomotor tracking task. *Brain Res.* 105:580–588.
Eisen, A., and Danon, J. (1974): The orbicularis oculi reflex in acoustic neuromas: A clinical and electrodiagnostic evaluation. *Neurology (Minneap.),* 24:306–311.
Eisen, A., and Humphreys, P. (1974): The Guillain-Barré syndrome. *Arch. Neurol.,* 30:438–443.
Eisen, A., Schomer, D., and Melmed, C. (1977): The application of F-wave in the differentiation of proximal and distal upper limb entrapments. *Neurology, (Minneap.),* 27:662–668.
Eklund, G. (1972): Position sense and state of contraction: the effect of vibration. *J. Neurol. Neurosurg. Psychiatry,* 35:606–611.
Eklund, G., Hagbarth, K. E., and Torebjork, E. (1978): Exteroceptive vibration induced finger flexor reflex in man. *J. Neurol. Neurosurg. Psychiatry,* 41:433–443.
Elble, R. J., and Randall, J. E. (1976): Motor unit activity responsible for 8 to 12 Hz component of human physiological finger tremor. *J. Neurophysiol.,* 39:370–383.
Eldred, E., Granit, R., and Merton, P. A. (1953): Supraspinal control of the muscle spindles and its significance. *J. Physiol. (Lond.),* 122:498–523.
Eldridge, L., Liebhold, M., and Steinbach, J. H. (1981): Alterations in cat skeletal neuromuscular junctions following prolonged inactivity. *J. Physiol. (Lond.),* 313:529–545.
Ellaway, P. H. (1971): Recurrent inhibition of fusimotor neurones exhibiting background discharges in the decerebrate and the spinal cat. *J. Physiol. (Lond.),* 216:419–439.
Ellaway, P. H. (1978): Cumulative sum technique and its application to the analysis of peri-stimulus time histograms. *Electroencephalogr. Clin. Neurophysiol.,* 45:302–304.
Ellaway, P., Emonet-Denand, F., Joffroy, M., and Laporte, Y. (1972): Lack of exclusively fusimotor α axons in flexor and extensor leg muscles of the cat. *J. Neurophysiol.,* 35:149–153.
Ellis, K. O., and Carpenter, J. F. (1972): Studies of the mechanism of action of dantrolene sodium, a skeletal muscle relaxant. *Naunyn Schmiedebergs Arch. Pharmacol.,* 275:83–94.
Elvner, A. M. (1973): Possibility of correcting the urgent voluntary movement and the associated postural actions of human muscles. *Biophysika,* 18:907–911.
Emonet-Denand, F., Jami, L., and Laporte, Y. (1975): Skeleto-fusimotor axons in hind-limb muscles of the cat. *J. Physiol. (Lond.),* 249:153–166.
Emonet-Denand, F., Jami, L., and Laporte, Y. (1980): Histophysiological observations on the skeleto-fusimotor innervation of mammalian spindles. In: *Progress in Clinical Neurophysiology, Vol. 8, Spinal and Supraspinal Mechanisms of Voluntary Motor Control and Locomotion,* edited by J. E. Desmedt, pp. 1–11. Karger, Basel.
Emonet-Denand, F., and Laporte, Y. (1981): Muscle

stretch as a way of detecting brief activation of bag fibres by dynamic axons. In: *Muscle Receptors and Movement,* edited by A. Taylor and A. Prochazka, pp. 67–76. Macmillan, London.

Emonet-Denand, F., Laporte, Y., Matthews, P. B. C., and Petit, J. (1977): On the subdivision of static and dynamic fusimotor actions on the primary ending of the cat muscle spindle. *J. Physiol. (Lond.),* 268:827–861.

Endo, K., Araki, T., and Yagi, N. (1973): The distribution and pattern of axon branching of pyramidal tract cells. *Brain Res.,* 57:484–491.

Engberg, I. (1964): Reflexes to foot muscles in the cat. *Acta Physiol. Scand.,* 62 (Suppl. 235):1–64.

Engberg, I., and Lundberg, A. (1969): An electromyographic analysis of muscular activity in the hindlimb of the cat during unrestrained locomotion. *Acta Physiol. Scand.,* 75:614–630.

Engberg, I., Lundberg, A., and Ryall, R. W. (1968): Reticulospinal inhibition of transmission in reflex pathways. *J. Physiol. (Lond.),* 194:201–223.

Erdman, W. J., and Heather, A. J. (1964): Objective measurement of spasticity by light reflection. *Clin. Pharmacol. Ther.,* 5:883–886.

Erickson, T. C., and Woolsey, C. N. (1951): Observations on the supplementary motor area of man. *Trans. Am. Neurol. Assoc.,* 76:50–56.

Ertekin, C. (1973): Human evoked electrospinogram. In: *New Developments in Electromyography and Clinical Neurophysiology,* Vol. 2, edited by J. E. Desmedt, pp. 344–351. Karger, Basel.

Esslen, E. (1968): Objective kinesiologic and electrotonomyographic observations on spasticity and rigidity. *Biomechanics,* I:339–343.

Evans, J. M., Hogg, M. I. J., Lunn, J. N., and Rosen, M. (1974): A comparative study of the narcotic agonist activity of naloxone and Levallorphan. *Anaesthesia,* 19:721–727.

Evarts, E. V. (1964): Temporal pattern of discharge of pyramidal tract neurons during sleep and waking in the monkey. *J. Neurophysiol.,* 27:152–171.

Evarts, E. V. (1965): Relation of cell size to effects of sleep in pyramidal tract neurons. *Prog. Brain Res.,* 18:81–91.

Evarts, E. V. (1966): Pyramidal tract activity associated with a conditioned hand movement in the monkey. *J. Neurophysiol.,* 19:1011–1027.

Evarts, E. V. (1968): Relation of pyramidal tract activity to force exerted during voluntary movement. *J. Neurophysiol.,* 31:14–27.

Evarts, E. V. (1969): Activity of pyramidal tract neurons during postural fixation. *J. Neurophysiol.,* 32:375–385.

Evarts, E. V. (1972): Contrasts between activity of precentral and postcentral neurons of cerebral cortex during movement in the monkey. *Brain Res.,* 60:25–31.

Evarts, E. V. (1973): Motor cortex reflexes associated with learned movement. *Science,* 179:501–503.

Evarts, E. V. (1974): Precentral and postcentral cortical activity in association with visually triggered movement. *J. Neurophysiol.,* 37:373–381.

Evarts, E. V., and Fromm, C. (1977): Sensory responses in motor cortex neurons during precise motor control. *Neurosci. Lett.,* 5:267–272.

Evarts, E. V., and Fromm, C. (1978): The pyramidal tract neuron as summing point in a closed-loop control system in the monkey. In: *Progress in Clinical Neurophysiology, Vol. 4, Cerebral Motor Control in Man: Long Loop Mechanisms,* edited by J. E. Desmedt, pp. 56–69. Karger, Basel.

Evarts, E. V., and Fromm, C. (1981): Transcortical reflexes and servo control of movement. *Can. J. Physiol. Pharmacol.,* 59:757–775.

Evarts, E. V., and Granit, R. (1976): Relations of reflexes and intended movements. *Prog. Brain Res.,* 44:1–14.

Evarts, E. V., and Tanji, J. (1974): Gating of motor cortex reflexes by prior instruction. *Brain Res.,* 71:479-494.

Evarts, E. V., and Tanji, J. (1976): Reflex and intended responses in motor cortex pyramidal tract neurons of monkey. *J. Neurophysiol.,* 39:1069–1080.

Evarts, E. V., Teravainen, H., and Calne, D. B. (1981): Reaction time in Parkinson's disease. *Brain,* 104:167–186.

Evarts, E. V., and Vaughn, W. J. (1978): Intended arm movements in response to externally produced arm displacements in man. In: *Progress in Clinical Neurophysiology, Vol. 4, Cerebral Motor Control in Man: Long Loop Mechanisms,* edited by J. E. Desmedt, pp. 178–192. Karger, Basel.

Eviatar, L., and Eviatar, A. (1978): Neurovestibular examination of infants and children. *Adv. Otorhinolaryngol.,* 23:529–532.

Ewert, J. P. (1976): The visual system of the toad: behavioral and physiological studies on a pattern recognition system. In: *The Amphibian Visual System,* edited by K. V. Fite, pp. 141–202. Academic Press, New York.

Ezure, K., Inubushi, S., Kibayashi, T., et al. (1980): Mixed type responses on EEG arousal in cat motor cortical neurons. In: *Integrative Control Functions of the Brain,* Vol. 3, edited by M. Ito, pp. 140–141. Elsevier, Amsterdam.

Ezure, K., and Oshima, T. (1981a): Dual activity patterns of fast pyramidal tract cells and their family neurones during EEG arousal in the cat. *Jpn. J. Physiol.,* 31:717–736.

Ezure, K., and Oshima, T. (1981b): Excitation of slow pyramidal tract cells and their family neurones during phasic and tonic phases of EEG arousal. *Jpn. J. Physiol.,* 31:737–748.

Ezure, K., and Oshima, T. (1981c): The structure of EEG arousal as a dynamic ensemble of neuronal activities in cat motor cortex. *Jpn. J. Physiol.,* 31:809–829.

Faganel, J. (1973): Electromyographic analysis of human flexion reflex components. In: *New Developments in Electromyography and Clinical Neurophysiology,* Vol. 3, edited by J. E. Desmedt, pp. 730–733. Karger, Basel.

Fagni, L., Weiss, M., Pellet, J., and Hugon, M. (1981): Tremors in spinal and cerebellar cat at depth (81 and 95 ATA, normoxic heliox). *Undersea Biomed Res.,* 8:68–69.

Feldman, A. G. (1974): Control of muscle length. *Biofizika*, 19:749–751.
Ferguson, I. T. (1978): Electrical study of jaw and orbicularis oculi reflexes after trigeminal nerve surgery. *J. Neurol. Neurosurg. Psychiatry*, 41:819–823.
Ferguson, J. T., Lenman, J. A. R., and Johnson, B. B. (1978): Habituation of the orbicularis oculi reflex in dementia and dyskinetic states. *J. Neurol. Neurosurg. Psychiatry*, 41:824–828.
Fernandez, C., and Goldberg, J. M. (1976): Physiology of peripheral neurons innervating otolith organs of the squirrel monkey. III. Response dynamics. *J. Neurophysiol.*, 39:996–1007.
Fernandez, C., Goldberg, J. M., and Abend, W. K. (1972): Response to static tilts of peripheral neurons innervating otolith organs of the squirrel monkey. *J. Neurophysiol.*, 35:978–997.
Ferrari, E., and Messina, C. (1968): Blink reflexes during sleep and wakefulness in man. *Electroencephalogr. Clin. Neurophysiol.* 32:55–62.
Ferrier, D. (1875): Experiments on the brain of monkeys. *Proc. R. Soc. Lond.*, 23:409–430.
Ferro, J. M., and Calhau, E. S. (1977): Treatment of familial essential myoclonus with propanolol. *Lancet*, 2:143.
Festinger, M. L., and Canon, L. K. (1965): Information about spatial location based on knowledge about efference. *Psychol. Rev.*, 72:373–384.
Fetz, P. (1968): Pyramidal tract effects on interneurons in the cat lumbar dorsal horn. *J. Neurophysiol.*, 31:69–80.
Fetz, E. E., and Baker, M. A. (1969): Response properties of precentral neurons in awake monkeys. *Physiologist*, 12:223.
Fetz, E. E., and Baker, M. A. (1973): Operantly conditioned patterns of precentral unit activity and correlated responses in adjacent cells and contralateral muscles. *J. Neurophysiol.*, 36:179–204.
Fetz, E. E., and Cheney, P. D. (1980): Post spike facilitation of forelimb muscle activity by primate corticomotoneuronal cells. *J. Neurophysiol.*, 44:751–772.
Fetz, E. E., Cheney, P. D., and German, D. C. (1976): Corticomotoneuronal connections of precentral cells detected by post-spike averages of EMG activity in behaving monkeys. *Brain Res.*, 114:505–510.
Fetz, E. E., and Finocchio, D. V. (1971): Operant conditioning of specific patterns of neural and muscular activity. *Science*, 174:431–435.
Fetz, E. E., and Finocchio, D. V. (1972): Operant conditioning of isolated activity in specific muscles and precentral cells. *Brain Res*, 40:19–24.
Fetz, E. E., and Finocchio, D. V. (1975): Correlations between activity of motor cortex cells and arm muscles during operantly conditioned response patterns. *Exp. Brain Res.*, 23:217–240.
Fetz, E. E., Finocchio, D. V., Baker, M. A., and Soso, M. J. (1980): Sensory and motor responses of precentral cortex cells during comparable passive and active joint movements. *J. Neurophysiol.*, 43:1070–1089.
Fetz, E. E., Jankowska, E., Johannisson, T., and Lipski, J. (1979): Autogenic inhibition of motoneurones by impulses in group Ia muscle spindle afferents. *J. Physiol. (Lond.)*, 293:173–195.
Field, P. M., Coldman, D. E., and Raisman, G. (1980): Synaptic formation after injury in the adult rat brain: preferential reinnervation of denervated fimbrial sites by axons of the contralateral fimbria. *Brain Res.*, 189:103–113.
Fields, H. L., Basbaum, A. I., Clanton, C. H., and Anderson, S. D. (1977): Nucleus raphe magnus inhibition of spinal cord dorsal horn neurons. *Brain Res.*, 126:441–453.
Finck, A. D., Ngai, S. H., and Berkowitz, B. A. (1977): Antagonism of general anesthesia by naloxone in the rat. *Anesthesiology*, 46:241–245.
Finger, S. (1978): *Recovery from Brain Damage*. Plenum, New York.
Fink, M. (1978): EEG in psychopharmacology. *Electroencephalogr. Clin. Neurophysiol.*, 45:51–56.
Fiorica, V., Semba, T., and Steggerda, F. (1962): Vestibular responses of the unanesthetized cat recorded during free-fall. *Aerospace Med.*, 33:475–481.
Fish, E., Shankaran, R., and Hsia, J. C. (1979): High neurological syndrome: antagonistic effects of helium pressure and inhalation anesthetics on the dopamine-sensitive cyclic AMP response. *Undersea Biomed. Res.*, 6:189–196.
Fisher, M. A., Shahani, B. T., and Young, R. R. (1979a): Electrophysiologic analysis of the motor system after stroke: the flexor reflex. *Arch. Phys. Med. Rehabil.*, 60:7–11.
Fisher, M. A., Shahani, B. T., and Young, R. R. (1979b): Assessing segmental excitability after acute rostral lesions. II. The blink reflex. *Neurology (Minneap.)*, 29:45–50.
Fitts, P. M., and Deininger, R. L. (1954): S-R compatibility: correspondence among paired elements within stimulus and response codes. *J. Exp. Psychol.*, 48:483–492.
Flandrin, J. M., Kennedy, H., and Amblard, B. (1976): Effects of stroboscopic rearing on the binocularity of cat superior colliculus neurons. *Brain Res.*, 101:576–581.
Floeter, M. K., and Greenough, W. T. (1979): Cerebellar plasticity: modification of Purkinje cell structure by differential rearing in monkeys. *Science*, 206:227–229.
Flohr, H., and Precht, W. (1981): *Lesion-induced Neuronal Plasticity in Sensorimotor Systems*. Springer, Berlin.
Flowers, K. (1975): Ballistic and corrective movements on an aiming task. *Neurology (Minneap.)*, 25:413–421.
Flowers, K. (1976): Visual 'closed-loop' and 'open-loop' characteristics of voluntary movement in patients with Parkinsonism and intention tremor. *Brain*, 99:269–310.
Foerster, O. (1921): Zur Analyse und Pathophysiologie der striaten Bewegunstörungen. *Z. Neurol. Psychiatry*, 73:1–169.
Foerster, O. (1933): The dermatomes of man. *Brain*, 56:1–39.
Foley, J. (1961): The stiffness of spastic muscle. *J. Neurol. Neurosurg. Psychiatry*, 24:125–131.

Foley, J. M., and Held, R. (1972): Visually directed pointing as a function of target distance, direction and available cues. *Percept. Psychophysiol.*, 12:263–268.

Forno, C., and Alston, W. (1967): Uremic polyneuropathy. *Acta Neurol. Scand.*, 43:640–654.

Forssberg, H., Grillner, S., and Rossignol, S. (1977). Phasic gain control of reflexes from the dorsum of the paw during spinal locomotion. *Brain Res.*, 132:121–139.

Forssberg, H., and Nashner, L. M. (1982): Ontogenetic development of postural control in man: adaptation to altered support and visual conditions during stance. *J. Neurosci.*, 2:37–39.

Fra, L., and Brignolio, F. (1968): F and H responses elicited from muscles of the lower limb in normal subjects. *J. Neurol. Sci.*, 7:251–261.

Franken, L., and Desmedt, J. E. (1957): Sensibilisation progressive d'une aire corticale partiellement dénervée à ses afférences résiduelles. *C. R. Soc. Biol. (Paris)*, 151:2204–2208.

Freedman, W., Minassian, S., and Herman, R. (1976): Functional stretch reflex—a cortical reflex? *Prog. Brain Res.*, 44:487–490.

Freund, H. J., and Budingen, H. J. (1978): The relationship between speed and amplitude of the fastest voluntary contraction of human arm muscles. *Exp. Brain Res.*, 131:1–12.

Freund, H. J., Budingen, H. J., and Dietz, V. (1975): Activity of single motor units from human forearm muscles during voluntary isometric contractions. *J. Neurophysiol.*, 38:933–946.

Freund, H. H., and Dietz, V. (1978): The relationship between physiological and pathological tremor. In: *Progress in Clinical Neurophysiology, Vol. 5, Physiological Tremor, Pathological Tremors and Clonus*, edited by J. E. Desmedt, pp. 66–89. Karger, Basel.

Friedman, D. P., and Jones, E. G. (1981): Thalamic input to areas 3a and 2 in monkeys. *J. Neurophysiol.*, 45:59–85.

Friess, S. L., Durant, R. C., Hudak, W. V., and Boyer, R. D. (1975): Effects of moderate pressure He-O$_2$ saturation and response modifiers on neuromuscular function. *Undersea Biomed. Res.*, 2:35–41.

Fritsch, G., and Hitzig, E. (1870): Uber die elektrische Erregbarkeit des Grosshirns. *Arch. Anat. Physiol. Wiss. Med.*, 37:300–332.

Fromm, C., and Evarts, E. V. (1977): Relation of motor cortex neurons to precisely controlled and ballistic movements. *Neurosci. Lett.*, 5:259–265.

Fromm, C., and Evarts, E. V. (1982): Pyramidal tract neurons in somatosensory cortex: central and peripheral inputs during voluntary movement. *Brain Res.*, 238:186–191.

Fromm, C., Evarts, E. V., Kroller, J., and Shinoda, Y. (1981): Activity of motor cortex and red nucleus neurons during voluntary movement. In: *Brain Mechanisms of Perceptual Awareness and Purposeful Behavior*, edited by O. Pompeiano and C. Ajmone Marsan, pp. 269–294. Raven Press, New York.

Fuchs, A. F. (1967): Saccadic and smooth pursuit eye movements in the monkey. *J. Physiol. (Lond.)*, 191:609–631.

Fuchs, A. F., and Kimm, J. (1975): Unit activity in vestibular nucleus of the alert monkey during horizontal angular acceleration and eye movement. *J. Neurophysiol.*, 38:1140–1161.

Fuchs, A. F., and Robinson, D. A. (1966): A method for measuring horizontal and vertical eye movements chronically in the monkey. *J. Appl. Physiol.*, 21:1068–1070.

Fujimori, B., Shimamura, M., Kato, M., et al. (1966): Studies on the mechanism of spasticity following spinal hemisection in the cat. In: *Muscular Afferent and Motor Control*, edited by R. Granit, pp. 397–413. Almqvist and Wiksell, Stockholm.

Fukson, O. I., Berkinblit, M. B., and Feldman, A. G. (1980): The spinal frog takes into account the scheme of its body during the wiping reflex. *Science*, 209:1261–1263.

Fulton, J. F. (1935): A note on the definition of the motor and the premotor areas. *Brain*, 58:311–316.

Fulton, J. F., and Kennard, M. A. (1934): A study of the flaccid and spastic paralysis produced by lesions of the cerebral cortex in primates. *Res. Publ. Assoc. Res. Nerv. Ment. Dis.*, 13:158–210.

Fulton, J. F., and Pi-Suner, J. (1928): A note concerning the possible function of various afferent end-organs in skeletal muscle. *Am. J. Physiol.*, 83:554–562.

Funakubo, H., Isomura, T., Itoh, H., et al. (1980): Total arm prosthesis driven by 12 micromotors, pocketable microcomputer and voice and look-sight microcommanding system. *Proceedings: International Conference on Rehabilitation Engineering*, Toronto, pp. 39–42.

Fung, D. T., Hwang, J. C., and Chung, S. H. (1978): Correlative study of pain perception and masticatory muscle reflexes in man. *Oral Surg.*, 45:44–50.

Furuya, N., Saito, A., Ishikawa, M., and Suzuki, J. (1979): Synaptic events in cat medial rectus motoneurons during vestibular nystagmus. In: *Integrative Control Functions of the Brain*, edited by M. Ito, N. Tsukahara, K. Kubota, and K. Yagi, pp. 197–199. Kodansha, Tokyo.

Futami, T., Shinoda, Y., and Yokota, J. (1979): Spinal axon collaterals of corticospinal neurons identified by intracellular injection of horseradish peroxidase. *Brain Res.*, 164:279–284.

Gabriel, M., Foster, K., and Orona, E. (1980): Interaction of laminae of the cingulate cortex with the anteroventral thalamus during behavioral learning. *Science*, 208:1050–1052.

Gacek, R. R. (1971): Anatomical demonstration of the vestibulo-ocular projections in the cat. *Acta Otolaryngol. (Suppl.)*, 293:1–63.

Gacek, R. R. (1975): The innervation of the vestibular labyrinth. In: *The Vestibular System*, edited by R. F. Naunton, pp. 21–29. Academic Press, New York.

Gacek, R. R. (1977): Location of brainstem neurons projecting to the oculomotor nucleus in the cat. *Exp. Neurol.*, 57:725–749.

Galla, H. J., and Trudell, J. R. (1980): The effect of high pressure on cooperative lipid-protein interaction. Abstr. *Proceedings: 7th Symposium on Underwater Physiology*, Athens, pp. 67–68.

Gallistel, R. (1980): *The Organization of Action: A New Synthesis*. Erlbaum, New York.

Gandevia, S. C. (1982): The perception of motor commands or effort during muscular paralysis. *Brain,* 105:151–160.

Gandevia, S. C., and McCloskey, D. I. (1976): Perceived heaviness of lifted objects and effects of sensory inputs from related non-lifting parts. *Brain Res.,* 109:399–401.

Gandevia, S. C., and McCloskey, D. I. (1977a): Sensations of heaviness. *Brain,* 100:345–354.

Gandevia, S. C., and McCloskey, D. I. (1977b): Effects of related sensory inputs on motor performances in man studied through changes in perceived heaviness. *J. Physiol. (Lond.),* 272:653–672.

Gandevia, S. C., and McCloskey, D. I. (1977c): Changes in motor commands as shown by changes in perceived heaviness, during partial curarization and peripheral anaesthesia in man. *J. Physiol. (Lond.),* 272:673–689.

Gandevia, S. C., and McCloskey, D. I. (1978): Interpretation of perceived motor commands by reference to afferent signals. *J. Physiol. (Lond.),* 283:493–499.

Gandevia, S. C., McCloskey, D. I., and Potter, E. K. (1980): Alterations in perceived heaviness during digital anaesthesia. *J. Physiol. (Lond.),* 306:365–375.

Gandiglio, G., and Fra, L. (1967): Further observations on facial reflexes. *J. Neurol. Sci.,* 5:273–285.

Gandy, M., Johnson, S. W., Lynn, P. A., et al. (1980): Acquisition and analysis of electromyographic data associated with dynamic movements of the arm. *Med. Biol. Eng. Comput.,* 18:57–64.

Garcia, H. A., Fisher, M. A., and Gilai, A. (1979): H reflex analysis of segmental reflex excitability in flexor and extensor muscles. *Neurology (Minneap.),* 29:984–991.

Garcia-Mullin, R., and Mayer, R. F. (1972): H reflexes in acute and chronic hemiplegia. *Brain,* 95:559–572.

Gardette, B., and Rostain, J. C. (1981): Compression technique in deep sea dives, for man in a He-N_2-O_2 mixture. *Undersea Biomed. Res.,* 8:15.

Garnett, R., and Stephens, J. A. (1980): The reflex responses of single motor units in human first dorsal interosseous muscle following cutaneous afferent stimulation. *J. Physiol. (Lond.),* 303:351–364.

Garnett, R., and Stephens, J. A. (1981): Changes in the recruitment threshold of motor units produced by cutaneous stimulation in man. *J. Physiol. (Lond.),* 311:463–473.

Garol, H. W. (1942): The motor cortex of the cat. *J. Neuropathol. Exp. Neurol.,* 1:139–145.

Gassell, M. M. (1970): A critical review of evidence concerning long-loop reflexes excited by muscle afferents in man. *J. Neurol. Neurosurg. Psychiatry,* 33:358–362.

Gassel, M. M. (1973): An objective technique for the analysis of the clinical effectiveness and physiology of action of drugs in man. In: *New Developments in Electromyography and Clinical Neurophysiology,* Vol. 3, edited by J. E. Desmedt, pp. 342–359. Karger, Basel.

Gassel, M. M., and Diamantopoulos, E. (1964a): The effect of procaine nerve block on neuromuscular reflex regulation in man. *Brain,* 87:729–742.

Gassell, M. M., and Diamantopoulos, E. (1964b): The Jendrassik maneuver. *Neurology (Minneap.),* 14:555–560.

Gassell, M. M., and Diamantopoulos, E. (1966): Mechanically and electrically elicited monosynaptic reflexes in man. *J. Appl. Physiol.,* 21:1053–1058.

Gassell, M. M., and Diamantopoulos, E. (1970): Motoneuron excitability in man. *Electroencephalogr. Clin. Neurophysiol.,* 29:190–195.

Gassel, M. M., and Ott, K. H. (1970): Local sign and late effects on motoneuron excitability of cutaneous stimulation in man. *Brain,* 93:95–107.

Gassel, M. M., and Ott, K. H. (1973): An electrophysiological study of the organization of innervation of the tendon jerk in humans. In: *New Developments in Electromyography and Clinical Neurophysiology,* Vol. 3, edited by J. E. Desmedt, pp. 308–316. Karger, Basel.

Gauthier, G. M. (1976): Alterations of the human vestibulo-ocular reflex in a simulated dive at 62 ATA. *Undersea Biomed Res.,* 3:103–112.

Gauthier, G. M., Hofferer, J. M., and Martin, B. A. (1978): A film projecting system as an eye movement diagnostic and training technique for cerebral palsied children. *Electroencephalogr. Clin. Neurophysiol.,* 45:122–127.

Gauthier, G. M., and Robinson, D. A. (1975): Adaptation of human vestibuloocular reflex to magnifying lenses. *Brain Res.,* 92:331–335.

Gauthier, G. M., Roll, J. P., Martin, B., and Harlay, F. (1981): Effects of whole-body vibrations on sensory motor system performance in man. *Aviat. Space Environ. Med.,* 52:473–479.

Geber, J., and Dupelj, M. (1977): The effects of L-Dopa on the spinal monosynaptic mass reflex. *Experientia,* 33:1074–1075.

Gelfan, S., and Rapisarda, A. F. (1964): Synaptic density on spinal neurons of normal dogs and dogs with experimental hind-limb rigidity. *J. Comp. Neurol.,* 123:73–96.

Gelfand, J. M., Gurfinkel, V. S., Tseltlin, M. L., and Shik, M. L. (1971): Problems in the analysis of movements. In: *Models of the Structural-Functional Organization of Certain Biological Systems,* edited by V. S. Gurfinkel, R. Fomin, and V. Tseltlin, pp. 330–345. M.I.T. Press, Cambridge, Massachusetts.

Georgopoulos, A. P., Kalaska, J. F., and Massey, J. J. (1981): Spatial trajectories and reaction times of aimed movements: effects of practice, uncertainty, and change in target location. *J. Neurophysiol.,* 46:725–743.

Gerbrandt, L. K. (1977): Analysis of movement potential components. In: *Progress in Clinical Neurophysiology, Vol. 1, Attention, Voluntary Contraction and Event-Related Cerebral Potentials,* edited by J. E. Desmedt, pp. 174–188. Karger, Basel.

Gernandt, B. E., and Shimamura, M. (1961): Mechanisms of interlimb reflexes in cat. *J. Neurophysiol.,* 24:665–676.

Ghez, C. (1975): Input-output relations of the red nucleus in the cat. *Brain Res.,* 98:93–108.

Ghez, C. (1979): Contributions of central programs to rapid limb movement in the cat. In: *Integration in the Nervous System,* edited by H. Asanuma and V. J. Wilson, pp. 305–320. Igaku-Shoin, Tokyo.

Ghez, C., and Kubota, K. (1977): Activity of red nucleus neurons associated with a skilled forelimb movement in the cat. *Brain Res.*, 129:383–388.

Ghez, C., and Martin, J. H. (1982): Control of rapid limb movement in the cat. III. Agonist-antagonist coupling. *Exp. Brain Res.*, 45:115–125.

Ghez, C., and Pisa, M. (1972): Inhibition of afferent transmission in cuneate nucleus during voluntary movement in the cat. *Brain Res.*, 40:145–151.

Ghez, C., and Shinoda, Y. (1978): Spinal mechanisms of the functional stretch reflex. *Exp. Brain Res.*, 32:55–68.

Ghez, C., and Vicario, D. (1978): The control of rapid limb movement in the cat. I. Response latency. *Exp. Brain Res.*, 33:173–190; 191–203.

Ghez, C., Vicario, D., Martin, J., and Yumiya, H. (1982): Role of the motor cortex in the initiation of motor responses in the cat. *Electroencephalogr. Clin. Neurophysiol.*, 55 (in press).

Gilbert, P. F., and Thach, W. T. (1977): Purkinje cell activity during motor learning. *Brain Res.*, 128:309–328.

Gilden, L., Vaughan, H. G., and Costa, L. D. (1966): Summated human EEG potentials with voluntary movement. *Electroencephalogr. Clin. Neurophysiol.*, 20:433–438.

Gillies, J. D., Burke, D. J., and Lance, J. W. (1971): Tonic vibration reflex in the cat. *J. Neurophysiol.*, 34:252–262.

Gillies, J. D., Lance, J. W., Neilson, P. D., and Tassinari, C. A. (1969): Presynaptic inhibition of the monosynaptic reflex by vibration. *J. Physiol. (Lond.)*, 205:329–339.

Gillings, B. R., and Klineberg, I. J. (1975): Latency and inhibition of human masticatory muscles following stimuli. *J. Dent. Res.*, 54:260–279.

Gilman, S. (1969): The mechanism of cerebellar hypotonia. *Brain*, 92:621–638.

Gilman, S. (1973): Significance of muscle receptor control systems in the pathophysiology of experimental postural abnormalities. In: *New Developments in Electromyography and Clinical Neurophysiology*, Vol. 3, edited by J. E. Desmedt, pp. 175–193. Karger, Basel.

Gilman, S., Carr, D., and Hollanberg, J. (1976): Kinematic effects of deafferentation and cerebellar ablation. *Brain*, 99:311–330.

Gilman, S., and Ebel, H. C. (1970): Fusimotor neuron responses to natural stimuli as a function of prestimulus fusimotor activity in decerebellate cats. *Brain Res.*, 21:367–384.

Gilman, S., Lieberman, J. S., and Marco, L. A. (1974): Spinal mechanisms underlying the effects of unilateral ablation of areas 4 and 6 in monkeys. *Brain*, 97:49–64.

Gilman, S., Marco, L. A., and Ebel, H. C. (1971): Effects of medullary pyramidotomy in the monkey. II. Abnormalities of spindle afferent responses. *Brain*, 94:515–530.

Gilman, S., and Vandermeulen, J. P. (1966): Muscle spindle activity in dystonic and spastic monkeys. *J. Neurophysiol.*, 14:553–563.

Gladden, M. H. (1981): The activity of intrafusal fibres during central stimulation in the cat. In: *Muscle Receptors and Movements*, edited by A. Taylor and A. Prochazka, pp. 109–122. Macmillan, London.

Glencross, D. J. (1977): Control of skilled movements. *Psychol. Bull.*, 84:14–29.

Godaux, E., and Desmedt, J. E. (1975a): Human masseter muscle: H- and tendon reflexes, their paradoxical potentiation by muscle vibration. *Arch. Neurol.*, 32:229–234.

Godaux, E., and Desmedt, J. E. (1975b): Exteroceptive suppression and motor control of the masseter and temporalis muscles in normal man. *Brain Res.*, 85:447–458.

Goldberg, J. M. (1979): Vestibular receptors in mammals: afferent discharge characteristic and efferent control. *Prog. Brain Res.*, 50:355–367.

Goldberg, L. J. (1972): The effect of jaw position on the excitability of two reflexes involving the masseter muscle in man. *Arch. Oral Biol.*, 17:565–576.

Goldberg, L. J. (1981): Masseter muscle excitation induced by stimulation of periodontal and gingival receptors in man. *Brain Res.*, 32:369–381.

Goldberg, L. J., and Derfler, B. (1977): Recruitment order, spike amplitude and twitch tension of single motor units in human masseter muscle. *J. Neurophysiol.*, 40:879–890.

Goldberg, M. E. (1972): Restitution of function in CNS: pathologic grasp in Macaca mulatta. *Exp. Brain Res.*, 15:76–96.

Goldberg, M. E. (1977): Locomotor recovery after unilateral hindlimb deafferentation in cats. *Brain Res.*, 123:59–74.

Goldberg, M. E., and Bushnell, M. C. (1981): Behavioral enhancement of visual responses in monkey cerebral cortex. II. Modulation in frontal eye fields specifically related to saccades. *J. Neurophysiol.*, 46:773–787.

Goldberg, M. E., and Wurtz, R. H. (1972): Activity of superior colliculus in behaving monkey. II. Effect of attention on neuronal responses. *J. Neurophysiol.*, 35:560–574.

Goldfarb, J., and Hu, J. W. (1976): Enhancement of reflexes by naloxone in spinal cats. *Neuropharmacology*, 15:785–792.

Goldring, S., and Ratcheson, R. (1972): Human motor cortex: sensory input from single unit recordings. *Science*, 175:1493–1495.

Goldstein, A. (1976): Opioid peptides (endorphins) in pituitary and brain. *Science*, 193:1081–1086.

Golla, F., and Hettwer, J. (1924): A study of the electromyograms of voluntary movement. *Brain*, 47:57–69.

Gonshor, A., and Melvill Jones, G. (1969): Investigation of habituation to rotational stimulation within the range of natural movement. *Proceedings: Aerospace Medical Association Meeting*, San Francisco, pp. 94–95.

Gonshor, A., and Melvill Jones, G. (1973): Changes of human vestibulo-ocular response induced by vision-reversal during head rotation. *J. Physiol. (Lond.)*, 234:104P.

Gonshor, A., and Melvill Jones, G. (1976a): Short-term adaptative changes in the human vestibulo-ocular reflex arc. *J. Physiol. (Lond.)*, 256:361–379.

Gonshor, A., and Melvill Jones, G. (1976b): Extreme vestibulo-ocular adaptation induced by prolonged optical reversal of vision. *J. Physiol. (Lond.),* 256:381–414.

Gonshor, A., and Melvill Jones, G. (1980): Postural adaptation to prolonged optical reversal of vision in man. *Brain Res.,* 192:239–248.

Goodwill, C. J. (1968): The normal jaw reflex: measurement of the action potential in the masseter muscle. *Ann. Phys. Med.,* 9:183–188.

Goodwin, G. M., Hulliger, M., and Matthews, P. B. C. (1975): The effects of fusimotor stimulation during small amplitude stretching on the frequency-response of the primary ending of the mammalian muscle spindle. *J. Physiol. (Lond.),* 253:175–206.

Goodwin, G. M., and Luschei, E. S. (1974): Effects of destroying spindle afferents from jaw muscles on mastication in monkeys. *J. Neurophysiol.,* 37:967–981.

Goodwin, G. M., and Luschei, E. S. (1975): Discharge of spindle afferents from jaw-closing muscles during chewing in alert monkeys. *J. Neurophysiol.,* 38:560–571.

Goodwin, G. M., McCloskey, D. I., and Matthews, P. B. C. (1972): The contribution of muscle afferents to kinaesthesia shown by vibration induced illusions of movement and by the effects of paralysing joint afferents. *Brain,* 95:705–748.

Goodwin, G. M., McGrath, G. L., and Matthews, P. B. C. (1973): The tonic vibration reflex seen in the acute spinal cat after treatment with DOPA. *Brain Res.,* 49:463–466.

Goor, C., and Ongerboer De Visser, B. W. (1976): Jaw and blink reflexes in trigeminal nerve lesions: an electrodiagnostic study. *Neurology (Minneap.),* 26:95–97.

Gordon, T., Hoffer, J. A., Jhamandas, J., and Stein, R. B. (1980): Long-term effects of axotomy on neural activity during locomotion. *J. Physiol. (Lond.),* 303:243–263.

Gorska, T. (1974): Functional organization of cortical motor areas in adult dogs and puppies. *Acta Neurobiol. Exp.,* 34:171–203.

Goslow, G. E., Stauffer, E. K., Nemeth, W. C., and Stuart, D. G. (1973): The cat step cycle: responses of muscle spindles and tendon organs to passive stretch within the locomotor range. *Brain Res.,* 60:35–54.

Gottlieb, G. L., and Agarwal, G. C. (1970): Filtering of electromyographic signals. *Am. J. Phys. Med.,* 49:142–146.

Gottlieb, G. L., and Agarwal, G. C. (1973): Coordination of posture and movement. In: *New Developments in Electromyography and Clinical Neurophysiology,* Vol. 3, edited by J. E. Desmedt, pp. 418–427. Karger, Basel.

Gottlieb, G. L., and Agarwal, G. C. (1979): Response to sudden torques about ankle in man. Myotatic reflex. *J. Neurol. Neurosurg. Psychiatry,* 42:91–106.

Gottlieb, G. L., Agarwal, G. C., and Stark, L. (1970): Interactions between voluntary and postural mechanisms of the human motor system. *J. Neurophysiol.,* 33:365–381.

Graefe, C. F. (1912): Normen für die Ablösung grösserer Gliedmassen. Berlin.

Granit, R. (1950): Reflex self-regulation of muscle contraction and autogenetic inhibition. *J. Neurophysiol.,* 13:351–372.

Granit, R. (1955): *Receptors and Sensory Perception.* Yale University Press, New Haven.

Granit, R. (1970): *The Basis of Motor Control.* Academic Press, New York.

Granit, R. (1972): Constant errors in the excution and appreciation of movement. *Brain,* 95:649–660.

Granit, R. (1975): The functional role of the muscle spindles. *Brain,* 98:531–556.

Granit, R. (1979): Interpretation of supraspinal effects on the gamma system. *Prog. Brain Res.,* 50:147–154.

Granit, R., and Job, C. (1952): Electromyographic and monosynaptic definition of reflex excitability during muscle stretch. *J. Neurophysiol.,* 15:409–420.

Granit, R., Kernell, D., and Lamarre, Y. (1966): Algebraical summation in synaptic activation of motoneurones firing within the primary range to injected currents. *J. Physiol. (Lond.),* 187:379–399.

Granit, R., Kernell, D., and Shortess, G. K. (1963): The behaviour of mammalian motoneurones during long-lasting orthodromic, antidromic and transmembrane stimulation. *J. Physiol. (Lond.),* 169:743–754.

Granit, R., Philipps, C. G., Skoglund, S., et al. (1957): Differentiation of tonic from phasic ventral horn cells. *J. Neurophysiol.,* 20:470–481.

Gransberg, L., Knutsson, E., and Litton, J. E. (1980): A computer programmed system for the analysis of active and passive isokinetic movements. In: *IEEE 1980 Frontiers of Engineering in Health Care,* pp. 292–295.

Grantyn, R., Baker, R., and Grantyn, A. (1980): Morphological and physiological identification of excitatory pontine reticular neurons projecting to the cat abducens nucleus and spinal cord. *Brain Res.,* 198:221–228.

Grantyn, A. A., and Grantyn, R. (1976): Synaptic actions of tectofugal pathways on abducens motoneurons in the cat. *Brain Res.,* 105:269–285.

Graupe, D., Magnusen, T., and Beese, A. A. (1978): A microprocessor system for myoelectric signal identification. *IEEE Trans. Autom. Control,* AC23:538–544.

Graybiel, A. M. (1977): Direct and indirect pre-oculomotor pathways of the brainstem: an autoradiographic study of the pontine reticular formation in the cat. *J. Comp. Neurol.,* 175:37–78.

Greenough, W. T., Juraska, J. M., and Volkmar, F. R. (1979): Maze training effects on dendritic branching in occipital cortex of adult rats. *Behav. Neural Biol.,* 26:287–297.

Greenwood, R. J., and Hopkins, A. P. (1974): Muscle activity in falling man. *J. Physiol. (Lond.),* 241:26–27P.

Greenwood, R. J., and Hopkins, A. P. (1976a): Muscle responses during sudden falls in man. *J. Physiol. (Lond.),* 254:507–518.

Greenwood, R. J., and Hopkins, A. P. (1976b): Land-

ing from an unexpected fall and a voluntary step. *Brain*, 99:375–386.
Greenwood, R. and Hopkins, A. (1977): Monosynaptic reflexes in falling man. *J. Neurol. Neurosurg. Psychiatry*, 40:448–454.
Greenwood, R., and Hopkins, A. (1980): Motor control during stepping and falling in man. In: *Progress in Clinical Neurophysiology, Vol. 8, Spinal and Supraspinal Mechanisms of Voluntary Motor Control and Locomotion*, edited by J. E. Desmedt, pp. 294–309. Karger, Basel.
Gregoric, M. (1973): Habituation of the blink reflex. In: *New Developments in Electromyography and Clinical Neurophysiology*, Vol. 3, edited by J. E. Desmedt, pp. 673–677. Karger, Basel.
Grigg, P., and Greenspan, B. J. (1977): Response of primate joint afferent neurons to mechanical stimulation of knee joint. *J. Neurophysiol.*, 40:1–8.
Grillner, S. (1972): The role of muscle stiffness in meeting the changing postural and locomotor requirements for force development of the ankle extensors. *Acta Physiol. Scand.*, 86:92–108.
Grillner, S. (1975): Locomotion in vertebrates: central mechanisms and reflex interaction. *Physiol. Rev.*, 55:247–304.
Grillner, S. (1981): Control of locomotion in bipeds, tetrapods and fish. In: *Handbook of Physiology, Section I, Vol. II, Motor Control*, edited by V. B. Brooks, pp. 1179–1236. American Physiological Society, Bethesda.
Grillner, S., and Udo, M. (1970): Is the tonic stretch reflex dependent on suppression of autogenic inhibitory reflexes? *Acta Physiol. Scand.*, 79:13A–14A.
Grimby, L. (1963): Pathological plantar response. *J. Neurol. Neurosurg. Psychiatry*, 26:314–321.
Grimby, L. (1965): Pathological plantar response. I. Flexor and extensor components in early and late reflex parts. II. Loss of significance of stimulus site. *J. Neurol. Neurosurg. Psychiatry*, 28:469–475; 476–481.
Grimby, L., and Hannerz, J. (1974): Differences in recruitment order and discharge pattern of motor units in the early and late flexion reflex components in man. *Acta Physiol. Scand.*, 90:555–564.
Grimby, L., and Hannerz, J. (1977): Firing rate and recruitment order of toe extensor motor units in different modes of voluntary contraction. *J. Physiol. (Lond.)*, 264:865–879.
Grimby, L., and Hannerz, J. (1981): Flexibility of recruitment order of continuously and intermittently discharging motor units in voluntary contraction. In: *Progress in Clinical Neurophysiology, Vol. 9, Motor Unit Types, Recruitment and Plasticity in Health and Disease*, edited by J. E. Desmedt, pp. 201–211. Karger, Basel.
Grimby, L., Hannerz, J., and Hedman, B. (1979): Contraction time and voluntary discharge properties of individual short toe extensor motor units in man. *J. Physiol. (Lond.)*, 289:191–201.
Grimby, L., Hannerz, J., and Hedman, B. (1981): The fatigue and voluntary discharge properties of single motor units in man. *J. Physiol. (Lond.)*, 316:545–554.
Grimm, R. J., and Nashner, L. M. (1978): Long loop dyscontrol. In: *Progress in Clinical Neurophysiology, Vol. 4, Cerebral Motor Control in Man: Long Loop Mechanisms*, edited by J. E. Desmedt, pp. 70–84. Karger, Basel.
Grinker, R. R., Serota, H., and Stein, S. I. (1938): Myoclonic epilepsy. *Arch. Neurol.*, 40:968–981.
Groll-Knapp, E., Ganglberger, J. A., and Haider, M. (1977): Voluntary movement-related slow potentials in cortex and thalamus in man. In: *Progress in Clinical Neurophysiology, Vol. 1, Attention, Voluntary Contraction and Event-Related Cerebral Potentials*, edited by J. E. Desmedt, pp. 164–173. Karger, Basel.
Grossman, W., Jurna, I., and Nell, T. (1975): The effect of reserpine and DOPA on reflex activity in the rat spinal cord. *Exp. Brain Res.*, 22:351–361.
Grusser, O. J., and Thiele, B. (1968): Reaktionen primärer und sekundärer Muskelspindel Afferenzen auf sinusformige Reizung. *Pfluegers Arch.*, 300:161–184.
Gschwend, J. G., Knutti, J. W., Allen, H. V., and Meindl, J. D. (1979): A general purpose implantable multichannel telemetry system for physiological research. *Biotelem. Patient Monit.*, 6:107–117.
Gualtierotti, T., and Paterson, A. S. (1954): Electrical stimulation of the unexposed cortex. *J. Physiol. (Lond.)*, 125:278–291.
Guiheneuc, P. (1974): Mesure de la vitesse de conduction des fibres afférentes proprioceptives chez l'homme normal. *C. R. Soc. Biol.*, 168:589–595.
Guiheneuc, P., and Bathien, N. (1976): Two patterns of results in peripheral neuropathies explored by reflexological methods. Comparison between proximal and distal conduction velocities. *J. Neurol. Sci.*, 30:83–94.
Guiheneuc, P., and Ginet, J. (1971): Le réflexe de Hoffmann—Signification de l'intervalle entre les réponses H et M et intérêt de sa mesure. *C. R. Soc. Biol.*, 165:1763–1766.
Guiheneuc, P., and Ginet, J. (1974): Etude du réflexe de Hoffmann obtenu au niveau du muscle quadriceps de sujets humains normaux. *Electroencephalogr. Clin. Neurophysiol.*, 36:225–231.
Guiheneuc, P., Ginet, J., Descuns, P., et al. (1973): Hoffmann reflex in lumbo-sacral roots injuries. *Electroencephalogr. Clin. Neurophysiol.*, 34:814–815.
Guiheneuc, P., Ginet, J., Groleau, J. Y., and Rojouan, J. (1980): Early phase of Vincristine neuropathy in man: electrophysiological evidence for a dying-back phenomenon with transitory enhancement of spinal transmission of the monosynaptic reflex. *J. Neurol. Sci.*, 45:355–366.
Guiheneuc, P., Ginet, J., and Guenel, J. (1972): Intérêt du réflexe de Hoffmann pour la surveillance des neuropathies au long cours. *Rev. Electroencephalogr. Neurophysiol. Clin.*, 2:125–130.
Guilbaud, G., Besson, J. M., Oliveras, J. L., and Liebeskind, J. C. (1973): Suppression by LSD of the inhibitory effect exerted by dorsal raphe stimulation on certain spinal cord interneurons in the cat. *Brain Res.*, 61:417–422.
Guillain, G., Barré, J. A., and Strohl, A. (1916): Sur un syndrome de radiculo-névrite avec hyperalbuminose du liquide céphalo-rachidien sans réaction cellulaire. Remarques sur les caractères des réflexes tendineux. *Bull. Mem. Soc. Hôp. (Paris)*, 40:1462–1470.

Guillery, R. W. (1972): Experiments to determine whether retinogeniculate axons can form translaminar collateral sprouts in the dorsal lateral geniculate nucleus of the cat. *J. Comp. Neurol.,* 146:407–420.

Gurfinkel, V. S., and Elner, A. M. (1973): On two types of static disturbances in patients with local lesions of the brain. *Aggressologie,* 14D:65–72.

Gurfinkel, V. S., Lipshits, M. I., Mauritz, K. H., and Popov, K. E. (1981): Quantitative analysis of anticipatory postural components in a gross voluntary movement. *Fiziol. Cheloveka (in press).*

Gurfinkel, V. S., Lipshits, M. I., Mori, S., and Popov, K. E. (1976): The stretch reflex during quiet standing in man. *Prog. Brain Res.,* 44:473–486.

Gustafsson, B. (1974): Afterpolarization and the control of repetitive firing in spinal neurones of the cat. *Acta Physiol. Scand.,* Suppl. 46:1–47.

Guttman, L. (1952): Studies on reflex activity of the isolated cord in the spinal man. *J. Nerv. Ment. Dis.,* 116:957–972.

Gydikov, A., and Kosarov, D. (1974): Some features of different motor units in human biceps brachii. *Pfluegers Arch.,* 347:75–88.

Gydikov, A., Kossev, A., Radicheva, N., and Tankow, N. (1981): Interaction between reflexes and voluntary motor activity in man revealed by discharges of separate motor units. *Exp. Neurol.,* 73:331–334.

Haase, J., Cleveland, S., and Ross, H. G. (1975): Problems of postsynaptic autogenous and recurrent inhibition in the mammalian spinal cord. *Rev. Physiol. Biochem. Pharmacol.,* 73:73–129.

Haaxma, R., and Kuypers, H. (1975): Intrahemispheric cortical connections and visual guidance of hand and finger movements. *Brain,* 98:239–260.

Hagbarth, K. E. (1952): Excitatory and inhibitory skin areas for flexor and extensor motoneurones. *Acta Physiol. Scand.* Suppl. 26:1–58.

Hagbarth, K. E. (1960): Spinal withdrawal reflexes in the human lower limbs. *J. Neurol. Neurosurg. Psychiatry,* 23:222–227.

Hagbarth, K. E. (1962): Post-tetanic potentiation of myotatic reflexes in man. *J. Neurol. Neurosurg. Psychiatry,* 25:1–10.

Hagbarth, K. E. (1973): The effect of muscle vibration in normal man and in patients with motor disorders. In: *New Developments in Electromyography and Clinical Neurophysiology,* Vol. 3, edited by J. E. Desmedt, pp. 428–443. Karger, Basel.

Hagbarth, K. E. (1979): Exteroceptive, proprioceptive, and sympathetic activity recorded with microelectrodes from human peripheral nerves. *Mayo Clinic Proc.,* 54:353–365.

Hagbarth, K. E. (1981): Fusimotor and stretch reflex functions studied in recordings from muscle spindle afferents in man. In: *Muscle Receptors and Movement,* edited by A. Taylor and A. Prochazka, pp. 277–285. Macmillan, London.

Hagbarth, K. E., Burke, D., Wallin, G., and Lofstedt, L. (1976): Single unit response to muscle vibration in man. *Prog. Brain Res.,* 44:281–290.

Hagbarth, K. E., and Eklund, G. (1966): Motor effects of vibratory muscle stimuli in man. In: *Muscular Afferents and Motor Control,* edited by R. Granit, pp. 177–186. Almqvist and Wiksell, Stockholm.

Hagbarth, K. E., and Eklund, G. (1968): The effects of muscle vibration in spasticity, rigidity and cerebellar disorders. *J. Neurol. Neurosurg. Psychiatry,* 31:207–213.

Hagbarth, K. E., and Finer, B. (1963): The plasticity of human withdrawal reflexes to noxious skin stimuli in lower limbs. *Prog. Brain Res.,* 1:65–78.

Hagbarth, K. E., Hagglund, J. V., Wallin, E. U., and Young, R. R. (1981): Grouped spindle and electromyographic responses to abrupt wrist extension movement in man. *J. Physiol. (Lond.),* 312:81–96.

Hagbarth, K. E., and Vallbo, A. B. (1968): Discharge characteristics of human muscle afferents during muscle stretch and contraction. *Exp. Neurol.,* 22:674–694.

Hagbarth, K. E., and Vallbo, A. B. (1969): Single unit recordings from muscle nerves in human subjects. *Acta Physiol. Scand.,* 76:321–334.

Hagbarth, K. E., Wallin, G., Burke, D., and Lofstedt, L. (1975a): Effects of the Jendrassik manoeuvre on muscle spindle activity in man. *J. Neurol. Neurosurg. Psychiatry,* 38:1143–1153.

Hagbarth, K. E., Wallin, G., and Lofstedt, L. (1973): Muscle spindle responses to stretch in normal and spastic subjects. *Scand. J. Rehabil. Med.,* 5:156–159.

Hagbarth, K. E., Wallin, G., and Loftstedt, L. (1975b): Muscle spindle activity in man during voluntary fast alternating movements. *J. Neurol. Neurosurg. Psychiatry,* 38:625–635.

Hagbarth, K. E., Wallin, G., Lofstedt, L., and Aquilonius, S. M. (1975c): Muscle spindle activity in alternating tremor of Parkinsonism and in clonus. *J. Neurol. Neurosurg. Psychiatry,* 38:636–641.

Hagbarth, K. E., and Young, R. R. (1979): Participation of the stretch reflex in human physiological tremor. *Brain,* 102:509–526.

Hagbarth, K. E., Young, R. R., Hagglund, J. V., and Wallin, E. U. (1980): Segmentation of human spindle and EMG responses to sudden muscle stretch. *Neurosci. Lett.,* 19:213–217.

Haight, J. R. (1972): The general organization of somatotopic projections to SII cerebral neocortex in the cat. *Brain Res.,* 44:483–502.

Hainaut, K., Duchateau, J., and Desmedt, J. E. (1981): Differential effects on slow and fast motor units of daily muscle training in man. In: *Progress in Clinical Neurophysiology, Vol. 9, Motor Unit Types, Recruitment and Plasticity,* edited by J. E. Desmedt, pp. 241–249. Karger, Basel.

Hall, J. I. (1967): Studies on demyelinated peripheral nerves in guinea-pigs with experimental allergic neuritis. *Brain,* 90:313–332.

Hall, P. V., Smith, J. E., Lane, J., et al. (1979): Glycine and experimental spinal spasticity. *Neurology (Minneap.),* 29:262–267.

Hallett, M. (1979): Ballistic elbow flexion movements in patients with amyotrophic lateral sclerosis. *J. Neurol. Neurosurg. Psychiatry,* 42:232–237.

Hallett, M., Bielawski, M., and Marsden, C. D. (1981): Behavior of the long-latency stretch reflex prior to voluntary movement. *Brain Res.,* 219:178–185.

Hallett, M., Chadwick, D., Adam, J., and Marsden, C. D. (1977a): Reticular reflex myoclonus: a physiological type of human post-hypoxic myoclonus. *J. Neurol. Neurosurg. Psychiatry,* 40:253–312.

Hallett, M., Chadwick, D., and Marsden, C. D. (1977b): Ballistic movement overflow myoclonus. *Brain,* 100.299–312.

Hallett, M., Chadwick, D., and Marsden, C. D. (1979): Cortical reflex myoclonus. *Neurology,* 29:1107–1125.

Hallett, M., and Kaufman, C. (1981): Physiological observations in Sydenham's chorea. *J. Neurol. Neurosurg. Psychiatry,* 44:829–832.

Hallett, M., and Khoshbin, S. (1980): A physiological mechanism of bradykinesia. *Brain,* 103:301–314.

Hallett, M., and Marsden, C. D. (1979): Ballistic flexion movements of the human thumb. *J. Physiol. (Lond.),* 294:33–50.

Hallett, M., and Marsden, C. D. (1981): Physiology and pathophysiology of the ballistic movement pattern. In: *Progress in Clinical Neurophysiology, Vol. 9, Motor Unit Types, Recruitment and Plasticity in Health and Disease,* edited by J. E. Desmedt, pp. 331–346. Karger, Basel.

Hallett, M., Shahani, B. T., and Young, R. R. (1975a): EMG analysis of stereotyped voluntary movements in man. *J. Neurol. Neurosurg. Psychiatry,* 38:1154–1162.

Hallett, M., Shahani, B. T., and Young, R. R. (1975b): EMG analysis of patients with cerebellar deficits. *J. Neurol. Neurosurg. Psychiatry,* 38:1163–1169.

Halliday, A. M. (1967): The electrophysiological study of myoclonus in man. *Brain,* 90:241–284.

Halliday, A. M., and Halliday, E. (1980): Cerebral somatosensory and visual evoked potentials in different clinical forms of myoclonus. In: *Progress in Clinical Neurophysiology, Vol. 7,* edited by J. E. Desmedt, pp. 292–310. Karger, Basel.

Hallin, R. G., and Torebjork, H. E. (1973): Electrically induced A and C fibers responses in intact human skin nerves. *Exp. Brain Res.,* 16:309–320.

Hallin, R. G., and Torebjork, H. E. (1976): Studies on cutaneous A and C fibre afferents, skin nerve blocks and perception. In: *Sensory Functions of the Skin in Primates,* edited by Y. Zotterman, pp. 137–149. Pergamon Press, Oxford.

Hamada, I., Sakai, M., and Kubota, K. (1981): Morphological differences between fast and slow pyramidal tract neurons in the monkey motor cortex as revealed by intracellular injection of horseradish peroxidase by pressure. *Neurosci. Lett.,* 22:233–238.

Hambrecht, F. T. (1979): Neural prostheses. *Ann. Rev. Biophys. Bioeng.,* 8:239–267.

Hammond, P. H. (1954): Involuntary activity in biceps following the sudden application of velocity to the abducted forearm. *J. Physiol. (Lond.),* 127:23P–25P.

Hammond, P. H. (1956): The influences of prior instruction to the subject on an apparently involuntary neuro-muscular response. *J. Physiol. (Lond.),* 132:17P–18P.

Hammond, P. H. (1960): An experimental study of servo action in human muscular control. *Proceedings: IIIrd International Conference on Medical Electronics,* pp. 190–199. Institution of Electrical Engineers, London.

Hammond, P. H., Merton, P. A., and Sutton, G. G. (1956): Nervous gradation of muscular control. *Br. Med. Bull.,* 12:214–218.

Hannam, A. G. (1972): Effect of voluntary contraction of the masseter and other muscles upon the masseteric reflex in man. *J. Neurol. Neurosurg. Psychiatry,* 35:66–71.

Hannam, A. G. (1979): Mastication in man. In: *Oral Motor Behavior: Impact on Oral Conditions and Dental Treatment,* edited by P. Bryant, E. Gale, and J. Rugl, pp. 87–118. National Institutes of Health, Bethesda.

Hannam, A. G., Matthews, B., and Yemm, R. (1968): The unloading reflex in masticatory muscles of man. *Arch. Oral Biol.,* 13:361–364.

Hannam, A. G., Matthews, B., and Yemm, R. (1969): Changes in the activity of the masseter muscle following tooth contact in man. *Arch. Oral Biol.,* 14:1401–1406.

Hannam, A. G., Matthews, B., and Yemm, R. (1970): Receptors involved in the response of the masseter muscle to tooth contact in man. *Arch. Oral Biol.,* 15:17–24.

Hardy, J. D., Wolff, H. G., and Goodell, H. (1948): Studies on pain: an investigation of some quantitative aspects of the dol scale of pain intensity. *J. Clin. Invest.,* 27:380–386.

Harker, D. W., Jami, L., Laporte, Y. and Petit, J. (1977): Fast-conducing skeletofusimotor axons supplying intrafusal chain fibers in the cat peroneus tertius muscle. *J. Neurophysiol.,* 40:791–799.

Harris, D. J. (1979): Hyperbaric hyperreflexia: tendon-jerk and Hoffmann reflexes in man at 43 bars. *Electroencephalogr. Clin. Neurophysiol.,* 47:680–692.

Harris, D. J., and Bennett, P. B. (1981): Effect on tri-mix on reflex excitability in man at pressures up to 66 Bar (Atlantis II). *Undersea Biomed. Res.,* 8:16.

Hartmann-von Monakow, K., Akert, K., and Kunzle, H. (1979): Projections of precentral and premotor cortex to the red nucleus and other midbrain areas in macaca fascicularis. *Exp. Brain Res.,* 34:91–105.

Hartviksen, K. (1962): Ice therapy in spasticity. *Acta Neurol. Scand.,* 38 (Suppl. 3):79–84.

Harvey, R. J., and Matthews, P. B. C. (1961): The response of de-efferented muscle spindle endings in the cat's soleus to slow extension of the muscle. *J. Physiol. (Lond.),* 157:170–192.

Harvey, R. J., Porter, R., and Rawson, J. A. (1977): The natural discharges of Purkinje cells in paravermal regions of lobules V and VI. *J. Physiol. (Lond.),* 271:515–536.

Harvey, R. J., Porter, R., and Rawson, J. A. (1979): Discharges of intracerebellar nuclear cells in monkeys. *J. Physiol. (Lond.),* 297:559–580.

Hasan, Z., and Houk, J. C. (1975): Analysis of response properties of de-efferented mammalian spindle receptors based on frequency response. *J. Neurophysiol.,* 38:663–672.

Hashimoto, S., Gemba, H., and Sasaki, K. (1979): Analysis of slow cortical potentials preceding self-paced hand movements in the monkey. *Exp. Neurol.,* 65:218–229.

Hassan, N., and McLellan, D. L. (1980): Double-blind comparison of single doses of DS 103–282, baclofen

and placebo for suppression of spasticity. *J. Neurol. Neurosurg. Psychiatry,* 43:1132–1136.
Hayashi, H., Suga, M., Satake, M., and Tsubaki, T. (1981): Reduced glycine receptor in the spinal cord in amyotrophic lateral sclerosis. *Ann. Neurol.,* 9:292–293.
Heilman, K. M., Bowers, D., Watson, R. T., and Greer, M. (1976): Reaction time in Parkinson disease. *Arch. Neurol.,* 33:139–140.
Heinbecker, P., Bishop, G. H., and O'Leary, J. L. (1933): Pain and touch fibers in peripheral nerves. *Arch. Neurol.,* 29:771–789.
Held, R., and Gottlieb, N. (1958): Technique for studying adaptation to disarranged hand eye coordination. *Percept. Motor Skills,* 8:83–86.
Hellsing, G. (1977): A tonic vibration reflex evoked in the jaw opening muscles in man. *Arch. Oral Biol.,* 22:175–180.
Hempleman, H. V. (1980): Human physiological studies at 43 bar. *Admiralty Marine Technology Establishment, Physiol. Lab.,* Report R80-402, Alverstoke, United Kingdom.
Henderson, J. V., Lowenhaupt, M. T., and Gilbert, D. L. (1977): Helium pressure alteration of function in squid giant synapse. *Undersea Biomed. Res.,* 4:19–26.
Henderson, W. R., and Smyth, G. E. (1948): Phantom limbs. *J. Neurol. Neurosurg. Psychiatry,* 11:88–112.
Hendrie, A., and Lee, R. G. (1978): Selective effects of vibration on human spinal and long loop reflexes. *Brain Res.,* 157:369–375.
Hendry, S. H. C., Jones, E. G., and Graham, J. (1979): Thalamic relay nuclei for cerebellar and certain related fiber systems in the cat. *J. Comp. Neurol.,* 185:679–713.
Henn, V., Cohen, B., and Young, L. R. (1980): Visual-vestibular interaction in motion perception and the generation of nystagmus. *Neurosci. Res. Prog. Bull.,* 18:4. MIT Press, Cambridge.
Henn, V., Young, L. R., and Finley, C. (1974): Vestibular nucleus units in alert monkey are also influenced by moving visual fields. *Brain Res.,* 71:144–149.
Henneman, E. (1981): Recruitment of motoneurons: the size principle. In: *Progress in Clinical Neurophysiology, Vol. 9, Motor Unit Types, Recruitment and Plasticity in Health and Disease,* edited by J. E. Desmedt, pp. 26–60. Karger, Basel.
Henneman, E., Clamman, H. P., Gillies, J. D., and Skinner, R. D. (1974): Rank order of motoneurones within a pool: Law of combination. *J. Neurophysiol.,* 37:1338–1349.
Henneman, E., and Olson, C. B. (1965): Relations between structure and function in the design of skeletal muscle. *J. Neurophysiol.,* 28:581–598.
Henneman, E., Somjen, G., and Carpenter, D. O. (1965a): Functional significance of cell size in spinal motoneurons. *J. Neurophysiol.,* 28:560–580.
Henneman, E., Somjen, G., and Carpenter, D. O. (1965b): Excitability and inhibitibility of motoneurons of different sizes. *J. Neurophysiol.,* 28:599–620.
Henneman, E., Somjen, G., and Carpenter, D. O. (1969): Functional significance of cell size in spinal motoneurons. *J. Neurophysiol.,* 28:560–580.
Henry, J. L. (1977): Substance P and pain: a possible relation in afferent transmission. In: *Substance P,* edited by U. S. Von Euler and B. Pernow, pp. 231–240. Raven Press, New York.
Henry, J. L. (1980): Pharmacologic studies on baclofen in the spinal cord of the cat. In: *Spasticity: Disordered Motor Control,* edited by R. G. Feldman, R. R. Young and W. P. Koella, pp. 437–452. Year Book, Chicago.
Hepp-Reymond, M. C., Trouche, E., and Weisendanger, M. (1974): Effects of unilateral and bilateral pyramidotomy on a conditioned rapid precision grip in monkeys. *Exp. Brain Res.,* 21:519–527.
Hepp-Reymond, M. C., and Wiesendanger, M. (1971): Unilateral pyramidotomy in monkeys: Effect on force and speed of a conditioned precision grip. *Brain Res.,* 36:117–131.
Hepp-Reymond, M. C., Wyss, U. R., and Anner, R. (1978): Neuronal coding of static force in the primate motor cortex. *J. Physiol. (Paris),* 74:287–291.
Herberts, P., Almstrom, C., and Caine, K. (1978): Clinical application study of multifunctional prosthetic hands. *J. Bone Joint Surg.,* 60B:552–560.
Herberts, P., Kadefors, R., Kaiser, E., and Petersen, I. (1968): Implantation of micro-circuits for myoelectric control of prostheses. *J. Bone Joint Surg.,* 50B:780–791.
Herbison, G. J. (1973): H-reflex in patients with Parkinsonism: effect of levodopa. *Arch. Phys. Med. Rehabil.,* 54:291–295.
Herman, R. (1969): Relationship between the H-reflex and the tendon jerk response. *Electromyography,* 9:359–370.
Herman, R. (1970): The myotatic reflex: Clinico-physiological aspects of spasticity and contracture. *Brain,* 93:273–312.
Herman, R., Freedman, W., Monster, A. W., and Tamai, Y. (1973a): A systematic analysis of myotatic reflex activity in human spastic muscle. In: *New Developments in Electromyography and Clinical Neurophysiology,* Vol. 3, edited by J. E. Desmedt, pp. 556–578. Karger, Basel.
Herman, R., Freedman, W., and Meeks, S. M. (1973b): Physiological aspects of hemiplegic and paraplegic spasticity. In: *New Developments in Electromyography and Clinical Neurophysiology,* Vol. 3, edited by J. E. Desmedt, pp. 579–588. Karger, Basel.
Herman, R., Grillner, S., Stein, P., and Stuart, D. G. (1976): *Neural Control of Locomotion.* Plenum, New York.
Herman, R., Mayer, N., and Mecomber, S. A. (1972): Clinical pharmaco-physiology of Dantrolene sodium. *Am. J. Phys. Med.,* 51:296–311.
Herman, R., Schaumburg, H., and Reiner, S. (1967): A rotational joint apparatus: a device for study of tension length relations of human muscle. *Med. Res. Eng.* 6:18–20.
Herz, E. (1944): Dystonia I. Historical review; analysis of dystonic symptoms and physiological mechanisms. *Arch. Neurol. Psychiatry,* 51:305–318.
Hickey, T. L. (1975): Translaminar growth of axons in the kitten dorsal lateral geniculate nucleus following removal of one eye. *J. Comp. Neurol.,* 161:359–382.
Hicks, S. P., and D'Amato, C. J. (1970): Motorsensory

and visual behavior after hemispherectomy in newborn and mature rats. *Exp. Neurol.*, 29:416–438.
Highstein, S. M. (1973): Synaptic linkage in the vestibuloocular and cerebello-vestibular pathways to the VIth nucleus in the rabbit. *Exp. Brain Res.*, 17:301–314.
Highstein, S. M., and Baker, R. (1978): Excitatory termination of abducens internuclear neurons on medial rectus motoneurons: relationship to syndrome of internuclear ophthalmoplegia. *J. Neurophysiol.*, 41:1647–1661.
Highstein, S. M., Maekawa, K., Steinacker, A., and Cohen, B. (1976): Synaptic input from the pontine reticular nuclei to abducens motoneurons and internuclear neurons in the cat. *Brain Res.*, 112:162–167.
Hikosaka, O., and Igusa, Y. (1980): Axonal projection of prepositus hypoglossi and reticular neurons in the brainstem of the cat. *Exp. Brain Res.*, 39:441–451.
Hikosaka, O., Igusa, Y., Nakao, S., and Shimazu, H. (1978): Direct inhibitory synaptic linkage of pontomedullary reticular burst neurons with abducens motoneurons in the cat. *Exp. Brain Res.*, 33:337–352.
Hikosaka, O., and Kawakami, T. (1977): Inhibitory reticular neurons related to the quick phase of vestibular nystagmus—their location and projection. *Exp. Brain Res.*, 27:377–396.
Hikosaka, O., Maeda, M., Nakao, S., et al. (1977): Presynaptic impulses in the abducens nucleus and their relation to post-synaptic potentials in motoneurons during vestibular nystagmus. *Exp. Brain Res.*, 27:355–376.
Hikosaka, O., Nakao, S., and Shimazu, H. (1980): Postsynaptic inhibition underlying spike suppression of secondary vestibular neurons during quick phases of vestibular nystagmus. *Neurosci. Lett.*, 16:21–26.
Hill, D. K., McDonnell, M. J., and Merton, P. A. (1980): Direct stimulation of the adductor pollicis in man. *J. Physiol. (Lond.)*, 300:2–3P.
Hillyard, S. A., and Picton, T. W. (1979): Event-related brain potentials and selective information processing in man. Cognitive components in cerebral event-related potentials and selective attention. In: *Progress in Clinical Neurophysiology*, Vol. 6, edited by J. E. Desmedt, pp. 1–52. Karger, Basel.
Hines, M. (1943): Control of movements by the cerebral cortex in primates. *Biol. Rev.*, 18:1–31.
Hiraoka, M., and Shimamura, M. (1977): Neural mechanisms of the corneal blinking reflex in cats. *Brain Res.*, 125:265–275.
Hobson, J. A., and Scheibel, A. B. (1980): The brainstem core: Sensorimotor integration and behavioral state control. *Neurosci. Res. Prog. Bull.*, 18:1–173.
Hodes R., and Dement, W. C. (1964): Depression of electrically induced reflexes (H-reflexes) in man during low voltage EEG sleep. *Electrocephalogr. Clin. Neurophysiol.*, 17:617–629.
Hoefer, P. F. A., and Putnam, T. J. (1940): Action potentials of muscles in athetosis and Sydenham chorea. *Arch. Neurol.*, 44:517–531.
Hoehler, F. K., McCann, M. A., and Bernick, D. L. (1981): Habituation of the Hoffman reflex. *Brain Res.*, 220:299–307.

Hoff, E. C. (1935): Corticospinal fibers arising in the premotor area of the monkey. *Arch. Neurol. Psychiatry*, 33:687–698.
Hoffer, J. A., and Andreassen, S. (1981): Regulation of soleus muscle stiffness in premammillary cats: Intrinsic and reflex components. *J. Neurophysiol.*, 45:267–285.
Hoffer, J. A., O'Donovan, M. J., Pratt, C. A., and Loeb, G. E. (1981): Discharge patterns of hindlimb motoneurones during normal cat locomotion. *Science*, 213:466–468.
Hoffmann, K. P., Behrend, K., and Schoppmann, A. (1976): A direct afferent visual pathway from the nucleus of the optic tract to the inferior olive in the cat. *Brain Res.*, 115:150–153.
Hoffmann, P. (1918): Uber die Beziehungen der Sehnenreflexe zur willkürlichen Bewegung und zum Tonus. *Z. Biol.*, 68:351–370.
Hoffmann, P. (1919): Demonstration eines Hemmungreflexes im menschlichen Ruckenmark. *Z. Biol.*, 70:515–524.
Hoffmann, P. (1922): *Untersuchung über die Eigenreflexe (Sehnenreflexe) menschlicher Muskeln.* Springer, Berlin.
Hoffmann, P. (1924): Untersüchungen über die refraktäre Periode des menschlichen Rückenmarkes. *Z. Biol.*, 81:37–48.
Hoffmann, P. (1934): Die physiologischen Eigenschaften der Eigenreflexe. *Ergeb. Physiol.*, 36:15–108.
Hoffmann, P., and Tonnies, J. F. (1948): Nachweis des völligkonstanten Vorkommens des Zungen-Kieferreflexes beim menschen. *Pfluegers Archiv.*, 250:103–108.
Hofsten, C. von (1979): Development of visually directed reaching: the approach phase. *J. Hum. Movemt. Stud.*, 5:160–178.
Hogan, N. (1976): A review of the methods of processing EMG for use as a proportional control signal. *Biomed. Eng.*, 11:81–86.
Hogan, N. (1980): Tuning muscle stiffness can simplify control of natural movement. In: *1980 Advances in Bioengineering*, edited by R. Mow, pp. 279–282. Van C. Asme, New York.
Hohmann, T. C., and Goodgold, J. (1960): A study of abnormal reflex patterns in spasticity. *Am. J. Phys. Med.*, 40:52–55.
Hokfelt, T., Lundberg, J. M., Schultzberg, M., et al. (1980): Cellular localization of peptides in neural structures. *Proc. R. Soc. Lond. (Biol.)*, 210:63–77.
Holland, R. L., and Brown, M. C. (1980): Postsynaptic transmission block can cause terminal sprouting of a motor nerve. *Science*, 207:641–651.
Holloszy, J. O., and Booth, F. W. (1976): Biochemical adaptations to endurance exercise in muscle. *Annu. Rev. Physiol.*, 38:273–291.
Holmes, G. (1917): The symptoms of acute cerebellar injuries due to gunshot injuries. *Brain*, 40:461–535.
Holmes, G. (1922): The Croonian Lectures on the clinical symptoms of cerebellar disease and their interpretation. *Lancet*, i:1177–1182, 1231–1237; ii:59–65, 111–115.
Holst, E. von (1954): Active functions of human visual

perception. English translation in: *Selected Papers of Erich von Holst.* Vol. 1, pp. 192–219. Methuen, London.

Holstege, G., and Kuypers, H. G. J. M. (1977): Propriobulbar fibre connections to trigeminal facial and hypoglossal motor nuclei. An anterograde degeneration study in cat. *Brain,* 100:239–264.

Homma, S. (1973): A survey of Japanese research on muscle vibration. In: *New Developments in Electromyography and Clinical Neurophysiology,* Vol. 3, edited by J. E. Desmedt, pp. 463–468. Karger, Basel.

Homma, S., Kanda, K., and Watanabe, S. (1971): Monosynaptic coding of group Ia afferent discharges during vibratory stimulation of muscle. *Jpn. J. Physiol.,* 21:405–417.

Homma, S., Kanda, K., and Watanabe, S. (1972): Preferred spike intervals in the vibration reflex. *Jpn. J. Physiol.,* 22:421–432.

Homma, S., and Kano, M. (1962): Electrical properties of the tonic reflex arc in the human proprioceptive reflex. In: *Symposium on Muscle Receptors,* edited by D. Barker, pp. 167–174. Hong Kong University Press, Hong Kong.

Hongo, T., Jankowska, E., and Lundberg, A. (1969): The rubrospinal tract. II. Facilitation of interneuronal transmission in reflex paths to motoneurones. *Exp. Brain Res.,* 7:365–391.

Hopf, H. C., Bier, J., Breuer, B., and Scheerer, W. (1973): The blink reflex induced by photic stimuli. In: *New Developments in Electromyography and Clinical Neurophysiology,* Vol. 3, edited by J. E. Desmedt, pp. 666–672. Karger, Basel.

Hopf, H. C., Herbort, R. L., Gnass, M. et al. (1974): Fast and slow contraction times associated with fast and slow spike conduction of skeletal muscle fibres in normal subjects and in spastic hemiparesis. *Z. Neurol.,* 206:193–202.

Hopf, H. C., Hufschmidt, H. J., and Stroder, J. (1965): Development of the trigemino-facial reflex in infants and children. *Ann. Paediatr.,* 204:52–64.

Hopf, H. C., Lowitzsch, K., and Schlegel, H. J. (1973): Central versus proprioceptive influences in brisk voluntary movements. In: *New Developments in Electromyography and Clinical Neurophysiology,* Vol. 3, edited by J. E. Desmedt, pp. 273–276. Karger, Basel.

Hore, J., Preston, J. B., Durkovic, R. G., and Cheney, P. D. (1976): Responses of cortical neurons (areas 3a and 4) to ramp stretch of hindlimb muscles in the baboon. *J. Neurophysiol.,* 39:484–500.

Horn, B. K. P., and Raibert, M. H. (1977): *Configuration Space Control.* Memo No. 458, Artificial Intelligence Lab., MIT Press, Cambridge, Massachusetts.

Horne, M. K., and Porter, R. (1980): The discharges during movement of cells in the ventrolateral thalamus in the monkey. *J. Physiol. (Lond.),* 304:349–372.

Horne, M. K., and Tracey, D. J. (1979): The afferents and projections of the ventroposterolateral thalamus in the monkey. *Exp. Brain Res.,* 36:129–141.

Hornick, R. J. (1973): Vibration. In: *Bioastronomics Data Book.* NASA SP 3006, 2nd ed., Washington, D.C.

Hornykiewicz, O. (1966): Dopamine (3-hydroxytramine) and brain function. *Pharmacol. Rev.,* 18:925–964.

Hornykiewicz, O. (1972): Biochemical and pharmacological aspects of akinesia. In: *Parkinson's Disease: Rigidity, Akinesia, Behaviour,* Vol. 1, edited by K. Siegfried, pp. 127–149. Hans Huber, Berne.

Horsley, V., and Schaefer, E. A. (1888): A record of experiments upon the functions of the cerebral cortex. *Philos. Trans.,* 179:1–45.

Houk, J. C. (1967): A viscoelastic interaction which produces one component of adaptation in responses of Golgi tendon organs. *J. Neurophysiol.,* 30:1482–1493.

Houk, J. C. (1978): Participation of reflex mechanisms and reaction-time processes in the compensatory adjustments to mechanical distrubances. In: *Progress in Clinical Neurophysiology, Vol. 4, Cerebral Motor Control in Man: Long Loop Mechanisms,* edited by J. E. Desmedt, pp. 193–215. Karger, Basel.

Houk, J. C. (1979): Regulation of stiffness by skeletomotor reflexes. *Annu. Rev. Physiol.,* 41:99–114.

Houk, J. C., Crago, P. E., and Rymer, W. Z. (1980): Functional properties of the Golgi tendon organs. In: *Progress in Clinical Neurophysiology, Vol. 8, Spinal and Supraspinal Mechanisms of Voluntary Motor Control and Locomotion,* edited by J. E. Desmedt, pp. 33–43. Karger, Basel.

Houk, J., and Henneman, E. (1967): Responses of Golgi tendon organs to active contractions of the soleus muscle of the cat. *J. Neurophysiol.,* 30:466–481.

Houk, J. C., Rymer, W. Z., and Crago, P. E. (1981): Dependence of the dynamic response of spindle receptors on muscle length and velocity. *J. Neurophysiol.,* 46:143–166.

Houk, J. C., and Simon, W. (1967): Responses of Golgi tendon organs to forces applied to muscle tendon. *J. Neurophysiol.,* 30:1466–1481.

Howarth, C. I., Beggs, W. D. A., and Bowden, J. M. (1971): The relationship between speed and accuracy of movement aimed at a target. *Acta Psychol.,* 35:207–218.

Howe, J. F., Calvin, W. H., and Loeser, J. D. (1976): Impulses reflected from dorsal root ganglia and from focal nerve injuries. *Brain Res.,* 116:139–144.

Hoyt, J. L., Gergis, S. D., and Sokoll, M. D. (1972): Studies on muscle rigidity: Properidol, diazepam and promethazine. *Anesth. Analg.,* 51:188–191.

Hubel, D. H., and Wiesel, T. N. (1970): The period of susceptibility to the physiological effects of unilateral eye closure in kittens. *J. Physiol. (Lond.),* 206:419–436.

Hufschmidt, H. J. (1960): Wird die Silent period nach direkter Muskelreizung durch die Golgi-Sehnenorgane ausgelost? *Pfluegers Arch.,* 271:35–39.

Hufschmidt, A., Dichgans, J., Mauritz, K. H., and Hufschmidt, M. (1980): Some methods and parameters of body sway quantification. *Arch. Psychiatr. Nervenkr.,* 228:135–150.

Hufschmidt, H. J., and Linke, D. (1976): A damping

factor in human voluntary contraction. *J. Neurol. Neurosurg. Psychiatry*, 39:536–537.
Hufschmidt, H. J., and Spuler, H. (1962): Mono- and polysynaptic reflexes of the trigeminal muscles in human beings. *J. Neurol. Neurosurg. Psychiatry*, 25:332–335.
Hugelin, A. (1972): Bodily changes during arousal, attention and emotion. In: *Limbic System, Mechanisms and Autonomic Function,* edited by C. H. Hockman, pp. 202–218. Charles Thomas, Springfield, Illinois.
Hugelin, A., and Bonvallet, M. (1956): Mise en évidence d'un noyau inhibiteur dans le système myotatique masticateur. *C. R. Soc. Biol. (Paris)*, 150:2164–2166.
Hughes, J. (1975): Isolation of an endogenous compound from the brain with pharmacological properties similar to morphine. *Brain Res.*, 88:295–308.
Hughes, J. R., Evarts, E. V., and Marshall, W. H. (1956): Posttetanic potentiation in the visual system of cats. *Am. J. Physiol.*, 186:438–487.
Hugon, M. (1969): Inhibition présynaptique évoquée par stimulation naturelle chez le chat. *Arch. Int. Physiol.*, 77:130–132.
Hugon, M. (1973a): Methodology of the Hoffmann reflex in man. In: *New Developments in Electromyography and Clinical Neurophysiology,* Vol. 3, edited by J. E. Desmedt, pp. 277–293. Karger, Basel.
Hugon, M. (1973b): Exteroceptive reflexes to stimulation of the sural nerve in normal man. In: *New Developments in Electromyography and Clinical Neurophysiology,* Vol. 3, edited by J. E. Desmedt, pp. 713–729. Karger, Basel.
Hugon, M. (1974): La réflectivity monosynaptique et ses déterminants: étude de neurophysiologie animale et humaine. *Rev. Electroencephalogr. Neurophysiol. Clin.*, 4:509–524.
Hugon, M. (1981): La physiologie de la plongée. *La Recherche*, 127:1224–1234.
Hugon, M., Blanc-Garin, J., Gillard, J., and Rostain, J. C. (1979): Perception visuelle, perception somesthésique et signes électroencéphalographiques en plongées à saturation. *Méd. Aéronaut., Spatiale Subaquat. Hyperbare*, 18:307–312.
Hugon, M., Delwaide, P., Pierrot-Deseilligny, E., and Desmedt, J. E. (1973): A discussion of the methodology of the triceps surae T- and H-reflexes. In: *New Developments in Electromyography and Clinical Neurophysiology,* Vol. 3, edited by J. E. Desmedt, pp. 773–780. Karger, Basel.
Hugon, M., and Fagni, L. (1981): Neurophysiological difficulties elicited by high gas pressures. *Jpn. J. EEG-EMG*, Suppl. 11:85–92.
Hugon, M., Fagni, L., Rostain, J. C., and Seki, K. (1980): Somatic evoked potentials and reflexes in monkey during saturation dives in dry chamber. *Proceedings: 7th Symposium on Underwater Physiology,* Athens, p. 69.
Hugon, M., and Lemaire, C. (1975): Cycle d'excitabilité de la fibre nerveuse motrice étudiée chez l'homme normal en hyperbarie à l'helium. *Bull. Med. Sub. Hyp.*, 11:9–17.
Hugon, M., and Seki, K. (1979): Preliminary report on experimental human dives at 31 ATA. Seadragon IV. Jamstec, Yokosuka, Japan.
Huizar, P., Kuno, M., Kudo, N., and Miyata, Y. (1977): Reaction of intact spinal motoneurones to partial denervation of the muscle. *J. Physiol. (Lond.)*, 265:175–191.
Hulliger, M. (1979): The responses of primary spindle afferents to fusimotor stimulation at constant and abruptly changing rates. *J. Physiol. (Lond.)*, 294:461–482.
Hulliger, M. (1981): Muscle spindle afferent units. Functional properties with possible significance in spasticity. In: *Therapie der Spastik—eine Bilanz,* edited by H. J. Bauer, W. P. Koella, and A. Struppler. Verlag für angewandte Wissenschaften, Munich.
Hulliger, M., Matthews, P. B. C., and Noth, J. (1977a): Static and dynamic fusimotor stimulation on the response of Ia fibres to low frequency sinusoidal stretching of widely ranging amplitudes. *J. Physiol. (Lond.)*, 267:811–838.
Hulliger, M., Matthews, P. B. C., and Noth, J. (1977b): Effects of combining static and dynamic fusimotor stimulation on the response of the muscle spindle primary ending to sinusoidal stretching. *J. Physiol. (Lond.)*, 267:839–856.
Hulliger, M., Nordh, E., Thelin, A. E., and Vallbo, A. B. (1979): The responses of afferent fibres from the glabrous skin of the hand during voluntary finger movements in man. *J. Physiol. (Lond.)*, 291:233–249.
Hulliger, M., Nordh, E., and Vallbo, A. B. (1981): The absence of position response in spindle afferent units from human finger muscles during accurate position-holding. *J. Physiol. (Lond.)*, 322:167–180.
Hulliger, M., and Noth, J. (1979): Static and dynamic fusimotor interaction and the possibility of multiple pace-makers operating in the cat muscle spindle. *Brain Res.*, 173:21–28.
Hulliger, M., and Vallbo, A. B. (1979): The responses of muscle spindle afferents during voluntary tracking movements in man. Load dependent servo assistance? *Brain Res.*, 166:401–404.
Hulliger, M., and Vallbo, A. B. (1982): The dependence of discharge rate of spindle afferent units on the size of the load during isotonic position-holding in man. *Brain Res.*, 237:297–309.
Hultborn, H. (1976): Transmission in the pathway of reciprocal Ia inhibition to motoneurones and its control during the tonic stretch reflex. *Progr. Brain Res.*, 44:235–255.
Hultborn, H., Illert, M., and Santini, M. (1976): Convergence on interneurones mediating the reciprocal Ia inhibition of motoneurones. I. Disynaptic inhibition of Ia inhibitory interneurones. *Acta Physiol. Scand.*, 96:193–201.
Hultborn, H., Jankowska, E., and Lindstrom, S. (1971): Recurrent inhibition from motor axon collaterals of transmission in the Ia inhibitory pathway to motoneurones. *J. Physiol. (Lond.)*, 215:591–612.
Hultborn, H., and Pierrot-Deseilligny, E. (1979): Changes in recurrent inhibition during voluntary soleus contractions in man studied by an H-reflex technique. *J. Physiol. (Lond.)*, 297:229–251.

Hultborn, H., Pierrot-Deseilligny, E., and Wigstrom, H. (1979): Recurrent inhibition and afterhyperpolarization following motoneuronal discharges in the cat. *J. Physiol. (Lond.)*, 297:253–266.

Hultborn, H., and Wigstrom, H. (1980): Motor response with long latency and maintained duration evoked by activity in Ia afferents. In: *Progress in Clinical Neurophysiology, Vol. 8, Spinal and Supraspinal Mechanisms of Voluntary Motor Control and Locomotion*, edited by J. E. Desmedt, pp. 99–116. Karger, Basel.

Humphrey, D. R. (1972): Relating motor cortex spike trains to measures of motor performance. *Brain Res.*, 40:7–18.

Humphrey, D. R., and Corrie, W. S. (1978): Properties of pyramidal tract neurons within a functionally defined subregion of primate motor cortex. *J. Neurophysiol.*, 41:216–243.

Humphrey, D. R., and Rietz, R. R. (1976): Cells of origin of corticorubral projections from the arm area of primate motor cortex and their synaptic actions in the red nucleus. *Brain Res.*, 110:162–169.

Humphrey, D. R., Schmidt, E. M., and Thompson, W. D. (1970): Predicting measures of motor performance from multiple cortical spike trains. *Science*, 179:758–762.

Hunt, C. C., and Kuffler, S. W. (1951): Stretch receptor discharges during muscle contraction. *J. Physiol. (Lond.)*, 113:298–315.

Hunt, C. C., and McIntyre, A. K. (1960): Characteristics of responses from receptors from the flexor longus digitorum muscle and the adjoining interosseus region of the cat. *J. Physiol. (Lond.)*, 153:74–87.

Hunt, C. C., and Ottoson, D. (1977): Responses of primary and secondary endings of isolated mammalian muscle spindles to sinusoidal length changes. *J. Neurophysiol.*, 40:1113–1120.

Hunt, C. C., and Wilkinson, R. S. (1980): An analysis of receptor potential and tension of isolated cat muscle spindles in response to sinusoidal stretch. *J. Physiol. (Lond.)*, 302:241–262.

Hunt, R. S., Meltzer, G. E., and Landau, W. M. (1963): Fusimotor function. Part. I. Spinal shock of the cat and the monkey. *Arch. Neurol.*, 9:120–126.

Hunter, W. L., Jr., and Bennett, P. B. (1974): The causes, mechanisms and prevention of the high pressure nervous syndrome. *Undersea Biomed. Res.*, 1:1–28.

Iansek, R., and Porter, R. (1980): The monkey globus pallidus: neuronal discharge properties in relation to movement. *J. Physiol. (Lond.)*, 301:439–455.

Igusa, Y., Sasaki, S., and Shimazu, H. (1980): Excitatory premotor burst neurons in the cat pontine reticular formation related to the quick phase of vestibular nystagmus. *Brain Res.*, 182:451–456.

Illert, M., Lundberg, A., Padel, Y., and Tanaka, R. (1978): Integration in descending motor pathways controlling the forelimb in the cat. 5. Properties of and monosynaptic excitatory convergence on C3-C4 propriospinal neurones. *Exp. Brain Res.*, 33:101–130.

Illert, M., Lundberg, A., and Tanaka, R. (1976): Integration in descending motor pathways controlling the forelimb in the cat. *Exp. Brain Res.*, 26:509–519; 521–540.

Illert, M., Lundberg, A., and Tanaka, R. (1977): Integration in descending motor pathways controlling the forelimb in the cat. III. Convergence on propriospinal neurones transmitting disynaptic excitation from the corticospinal tract and other descending tracts. *Exp. Brain Res.*, 29:323–346.

Ingle, D. (1976): Spatial vision in anurans. In: *The Amphibian Visual System: a Multidisciplinary Approach*, edited by K. V. Fite, pp. 119–141. Academic Press, New York.

Inman, V. T. (1966): Human locomotion. *Can. Med. Assoc. J.*, 94:1047–1054.

Inubushi, S., Kobayashi, T., Oshima, T., and Torii, S. (1978): Intracellular recordings from the motor cortex during EEG arousal in unanaesthetized brain preparations of the cat. *Jpn. J. Physiol.*, 28:669–688.

Isaksson, A. I., and Knutsson, E. (1980): Microcomputer implementation of gait examination in clinical routine. In: *IEEE 1980 Frontiers of Engineering in Health Care*, pp. 42–45.

Ishizuka, N., Mannen, H., Sasaki, S., and Shimazu, H. (1980): Axonal branches and terminations in the cat abducens nucleus of secondary vestibular neurons in the horizontal canal system. *Neurosci. Lett.*, 16:143–148.

Ito, F. (1974): Tonic vibration reflex. *Bull. Tokyo Med. Dent. Univ. (Suppl.)*, 21:37–40.

Ito, M. (1970): Neurophysiological aspects of the cerebellar motor control system. *Int. J. Neurol.*, 7:162–176.

Ito, M., Nisimaru, N., and Yamamoto, M. (1977): Specific patterns of neuronal connections involved in the control of the rabbit's vestibulo-ocular reflexes by the cerebellar flocculus. *J. Physiol. (Lond.)*, 265:833–854.

Ito, M., and Oshima, T. (1962): Temporal summation of after-hyperpolarization following a motoneurone spike. *Nature*, 195:910–911.

Ito, M., and Oshima, T. (1965): Electrical behaviour of the motoneurone membrane during intracellularly applied current steps. *J. Physiol. (Lond.)*, 180:607–635.

Ito, M., Shiida, T., Yagi, N., and Yamamoto, M. (1974): Visual influence on rabbit horizontal vestibulo-ocular reflex presumably effected via the cerebellar flocculus. *Brain Res.*, 65:170–174.

Iyer, V., and Fenichel, G. M. (1976): Normal median nerve proximal latency in carpal tunnel syndrome: a clue to coexisting Martin-Gruber anastomosis. *J. Neurol. Neurosurg. Psychiatry*, 39:449–452.

Jabre, J. (1981): Surface recordings of the H reflex of the flexor carpi radialis. *Muscle Nerve*, 4:435–438.

Jack, J. J. B., Noble, D., and Tsien, R. W. (1975): *Electric Current Flow in Excitable Cells*. Oxford University Press, London.

Jackson, J. H. (1931): *Selected Writing of John Hughlings Jackson*, edited by J. Taylor. Hodder & Stoughton, London.

Jacob, J. J., and Michaud, J. M. (1976): Production par la naloxone d'effets inverses de ceux de la morphine chez le chien éveillé. *Arch. Int. Pharmacodyn. Ther.,* 222:332–340.

Jacob, J. J., Tremblay, E. D., and Colombel, M. C. (1974): Facilitations de réactions nociceptives par la naloxone chez la souris et le rat. *Psychopharmacologia,* 37:217–223.

Jager, J., and Henn, V. (1981). Habituation of the vestibulo-ocular reflex in the monkey during sinusoidal rotation in the dark. *Exp. Brain Res.,* 41:108–114.

Jameson, H. D., Armugasamy, N., and Hardin, W. B. (1968): The supplementary motor area of the raccoon. *Brain Res.,* 11:628–637.

Jami, L., Lan-Couton, D., Malmgren, K., and Petit, J. (1978): Fast and slow skeleto-fusimotor innervation in cat tenuissimus spindles; a study with the glycogen depletion method. *Acta Physiol. Scand.,* 103:284–298.

Jami, L., Lan-Couton, D., Malmgren, K., and Petit, J. (1979): Histophysiological observations on fast skeleto-fusimotor axons. *Brain Res.,* 164:53–59.

Jami, L., and Petit, J. (1975): Correlation between axonal conduction velocity and tetanic tension of motor units in four muscles of the cat hindlimb. *Brain Res.,* 96:114–118.

Jami, L., and Petit, J. (1978): Fusimotor actions on sensitivity of spindle secondary endings to slow muscle stretch in cat peroneus tertius. *J. Neurophysiol.,* 41:860–869.

Jami, L., and Petit, J. (1981): Fusimotor actions on the sensitivity of spindle secondary endings. In: *Muscle Receptors and Movement,* edited by A. Taylor, and A. Prochazka, pp. 51–66. Macmillan, London.

Jane, J. A., Yashon, D., Becker, D. P., et al. (1968): The effect of destruction of the corticospinal tract in the human cerebral peduncle upon motor function and involuntary movements. *J. Neurosurg.,* 29:581–585.

Jankelson, B., Sparks, S., Crane, P. F., and Radke, J. C. (1975): Neural conduction of the Myo-Monitor stimulus: A quantitative analysis. *J. Prosthet. Dent.,* 34:245–253.

Jankowska, E. (1978): Some problems of projections and actions of cortico- and rubrospinal fibres. *J. Physiol. (Paris),* 74:209–214.

Jankowska, E., Padel, Y., and Tanaka, R. (1975): Projections of pyramidal tract cells to α-motoneurones innervating hind-limb muscles in the monkey. *J. Physiol. (Lond.),* 249:637–667.

Jankowska, E., Padel, Y., and Tanaka, R. (1976): Disynaptic inhibition of spinal motoneurones from the motor cortex in the monkey. *J. Physiol. (Lond.),* 258:467–487.

Jankowska, E., and Roberts, W. J. (1972): Synaptic actions of single interneurones mediating reciprocal Ia inhibition of motoneurones. *J. Physiol. (Lond.),* 222:623–642.

Jansen, J. K. S. (1962): Spasticity—functional aspects. *Acta Neurol. Scand.,* 38 (Suppl. 3):41–51.

Jansen, J. K. S., and Rudjord, T. (1964): On the silent period and Golgi tendon organs of the soleus muscle of the cat. *Acta Physiol. Scand.,* 62:364–379.

Jarvilehto, T. (1977): Neural basis of cutaneous sensations analysed by human microelectrodes measurements from human peripheral nerves: a review. *Scand. J. Psychol.,* 18:348–359.

Jasper, H. H. (1958): Recent advances in our understanding of ascending activities of the reticular system. In: *Reticular Formation of the Brain,* edited by H. H. Jasper, L. D. Proctor, R. S. Knighton, W. C. Noshay, and R. T. Costello, pp. 319–331. Little, Brown, Boston.

Jasper, H., and Penfield, W. (1949): Electrocorticograms in man: effect of voluntary movement upon the electrical activity of the precentral gyrus. *Arch. Psychiatry, Nervenkh.,* 183:163–174.

Jeannerod, M. (1981): Intersegmental coordination during reaching at natural visual objects. In: *Attention and Performance IX,* edited by A. Baddeley and J. Long, pp. 153–169. Erlbaum, Hillsdale, New Jersey.

Jeannerod, M., and Biguer, B. (1982): Visuomotor mechanisms in reaching within extrapersonal space. In: *Advances in the Analysis of Visual Behavior,* edited by D. Ingle, M. Goodale, and R. Mansfield, pp. 387–409. MIT Press, Cambridge, Massachusetts.

Jeannerod, M., and Hécaen, H. (1979): *Adaptation et Restauration des Fonctions Nerveuses.* SIMEP, Villeurbanne.

Jeannerod, M., and Prablanc, C. (1978): Résolution et plasticité de la coordination oeil-main. In: *Du Contrôle Moteur à l'Organisation du Geste,* edited by H. Hécaen and M. Jeannerod, pp. 261–289. Masson, Paris.

Jeannerod, M., Prablanc, C., and Perenin, M. T. (1980): Do oculomotor signals contribute in eye-hand coordination? In: *Tutorials in Motor Behaviour,* edited by G. Stelmach and J. Requin, pp. 297–303. Elsevier North-Holland, Amsterdam.

Jebsen, R. H., Taylor, N., Trieschmann, R. B., et al. (1969): An objective and standardized test of hand function. *Arch. Phys. Med. Rehabil.,* 50:311–319.

Jen, L. S., and Lund, R. D. (1981): Experimentally induced enlargement of the uncrossed retinotectal pathways in rats. *Brain Res.,* 211:37–57.

Jendrassik, E. (1885): Zur untersuchungsmethode des Kniephänomens. *Neurol. Centralbl.,* 4:412–415.

Jenner, J. R., and Stephens, J. A. (1979): Evidence for a transcortical cutaneous reflex response in man. *J. Physiol. (Lond.),* 293:39P.

Jerge, C. R. (1963): Organization and function of the trigeminal mesencephalic nucleus. *J. Neurophysiol.,* 26:379–392.

Jessell, T. M., and Iversen, L. L. (1977): Opiate analgesics inhibit substance P release from rat trigeminal nucleus. *Nature,* 268:549–551.

Jex, H. R., and Magdaleno, R. E. (1978): Biomechanical models for vibration feedthrough to hands and head for a semisupine pilot. *Aviat. Space Environ. Med.,* 49:304–316.

Johansson, R. S., and Vallbo, A. B. (1979): Detection of tactile stimuli: Thresholds of afferent units related to psychophysical thresholds in the human hand. *J. Physiol. (Lond.),* 297:405–422.

Johansson, R. S. and Vallbo, A. B. (1980): Spatial properties of the population of mechanoreceptive

units in the glabrous skin of the human hand. *Brain Res.*, 184:353–366.
Johansson, R. S., Vallbo, A. B., and Westling, G. (1980): Thresholds of mechanosensitive afferents in the human hand as measured with von Frey hairs. *Brain Res.*, 184:343–351.
Johansson, R. S., and Westling, G. (1981): Coordination between grip force and vertical lifting force when lifting objects between index and thumb. *Am. Soc. Neurosci. (Abstr.)*, 7:247.
Johns, R. J., Grob, D., and Harvey, A. M. (1957): An electromyographic study of a spinal cord reflex in the normal human arm. *Bull. Johns Hopkins Hosp.*, 101:232–239.
Johnson, R. H., and Spalding, J. M. K. (1964): Progressive sensory neuropathy in children. *J. Neurol. Neurosurg. Psychiatry*, 27:125–130.
Jokl, E. (1957): Neurological case histories of two olympic champions. *JAMA*, 165:129–131.
Jokl, E. (1964): *The Scope of Exercise in Rehabilitation.* Charles C Thomas, Springfield, Illinois.
Jones, D. A., Bigland-Ritchie, B., and Edwards, R. H. T. (1979): Excitation frequency and muscle fatigue: mechanical responses during voluntary and stimulated contractions. *Exp. Neurol.*, 64:401–413.
Jones, E. G. (1972): The development of the "muscular sense" concept in the nineteenth century and the work of H. Charlton Bastian. *J. Hist. Med.*, 27:298–311.
Jones E. G. (1975a): Lamination and differential distribution of thalamic afferents in the sensory-motor cortex of the squirrel monkey. *J. Comp. Neurol.*, 160:167–204.
Jones, E. G. (1975b): Varieties and distribution of non-pyramidal cells in the somatic sensory cortex of the squirrel monkey. *J. Comp. Neurol.*, 160:205–268.
Jones, E. G. (1980): Anatomy of cerebral cortex: columnar input output organization. In: *The Organization of the Cerebral Cortex*, edited by J. O. Schmitt, F. G. Worden, G. Adelman and S. G. Dennis, MIT Press, Cambridge, Massachusetts.
Jones, E. G., Coulter, J. D., Burton, H., and Porter, R. (1977): Cells of origin and terminal distribution of cortico-striatal fibers arising in the sensory-motor cortex of monkeys. *J. Comp. Neurol.*, 173:53–80.
Jones, E. G., Coulter, J. D., and Hendry, S. H. C. (1978): Intracortical connectivity of architectonic fields in the somatic sensory, motor and parietal cortex of monkeys. *J. Comp. Neurol.*, 181:291–349.
Jones, E. G., Coulter, J. D., and Wise, S. P. (1979): Commissural columns in the sensory-motor cortex of monkeys. *J. Comp. Neurol.*, 188:113–136.
Jones, E. G., and Porter, R. (1980): What is area 3a? *Brain Res. Rev.*, 2:1–43.
Jones, E. G., and Powell, T. P. S. (1968): The ipsilateral cortical connexions of the somatic sensory areas in the cat. *Brain Res.*, 9:71–94.
Jones, E. G., and Powell, T. P. S. (1969): Connections of the somatic sensory cortex of the rhesus monkey. I. Ipsilateral cortical connections. *Brain*, 92:477–502.
Jones, E. G., and Powell, T. P. S. (1970a): An electronmicroscopic study of the laminar pattern and mode of termination of the afferent fibre pathways to the somatic sensory cortex. *Philos. Trans. R. Soc. Lond. (Biol.)*, 257:45–62.
Jones, E. G., and Powell, T. P. S. (1970b): Connections of the somatic sensory cortex of the rhesus monkey. III. Thalamic connections. *Brain*, 93:37–56.
Jones, E. G., and Powell, T. P. S. (1970c): An anatomical study of converging sensory pathways within the cerebral cortex of the monkey. *Brain*, 93:793–820.
Jones, E. G., and Wise, S. P. (1977): Size, laminar and columnar distribution of efferent cells in the sensory-motor cortex of monkeys. *J. Comp. Neurol.*, 175:391–438.
Jones, R. F., Burke, D., Marosszeky, J. F., and Gillies, J. D. (1970): A new agent for the control of spasticity. *J. Neurol. Neurosurg. Psychiatry*, 33:464–468.
Joseph, J. (1973): Sequential contraction of muscles producing the same movement at a joint. In: *New Developments in Electromyography and Clinical Neurophysiology*, Vol. 1, edited by J. E. Desmedt, pp. 665–674. Karger, Basel.
Judge, S. J., Wurtz, R. H., and Richmond, B. J. (1980): Vision during saccadic eye movements. I. Visual interactions in striate cortex. *J. Neurophysiol.*, 43:1133–1155.
Kaeser, H. E., and Lambert, E. H. (1962): Nerve function studies in experimental polyneuritis. *Electroencephalogr. Clin. Neurophysiol.* (Suppl.), 22:29–35.
Kalil, K. (1981): Projections of the cerebellar and dorsal column nuclei upon the thalamus of the rhesus monkey. *J. Comp. Neurol.*, 195:25–50.
Kalil, R. E. and Schneider, G. E. (1975): Abnormally synaptic connections of the optic tract in the thalamus after midbrain lesions in newborn hamsters. *Brain Res.*, 100:690–698.
Kanda, K. (1972): Contribution of polysynaptic pathways to the tonic vibration reflex. *Jpn. J. Physiol.*, 22:367–377.
Kanda, K., Burke, R. E., and Walmsley, B. (1977): Differential control of fast and slow twitch motor units in the decerebrate cat. *Exp. Brain Res.*, 29:57–74.
Kanda, K., Homma, S., and Watanabe, S. (1973): Vibration reflex in spastic patients. In: *New Developments in Electromyography and Clinical Neurophysiology*, Vol. 3, edited by J. E. Desmedt, pp. 469–474. Karger, Basel.
Karczmar, A. G., and Dun, J. D. (1978): Cholinergic synapses: physiological, pharmacological and behavioral considerations. In: *Psychopharmacology: A Generation of Progress*, edited by M. A. Lipton, A. DiMascio, and K. F. Killam, pp. 293–305. Raven Press, New York.
Kark, R. A. P., Rodriguez-Budelli, M., and Blass, J. P. (1978): Evidence for a primary defect of lipoamide dehydrogenase in Friedreich's ataxia. In: *Advances in Neurology*, Vol. 21, pp. 163–180. Raven Press, New York.
Karol, E. A., and Pandya, D. N. (1971): The distribution of the corpus callosum in the rhesus monkey. *Brain*, 94:471–486.
Katchalsky, A. K., Rowland, V., and Blumenthal, R.

(1974): Dynamic patterns of brain cell assemblies. *Neurosci. Res. Program Bull.,* 12:1–187.

Kato, I. (1978): Trends in powered upper limb prostheses. *Prosthet. Orthot. Int.,* 2:64–68.

Katz, R. and Pierrot-Deseilligny, E. (1982): Recurrent inhibition in patients with upper motor neuron lesions. *Brain,* 105:103–124.

Kaufmann, P. G., Bennett, P. B., and Farmer, J. C. Jr. (1980): Effect of cerebellar ablation on the high pressure nervous syndrome in rats. *Undersea Biomed. Res.,* 5:63–70.

Kawaguchi, S., Yamamoto, T., and Samejima, A. (1979): Electrophysiological evidence for axonal sprouting of cerebello-thalamic neurons in kittens after neonatal hemicerebellectomy. *Exp. Brain Res.,* 36:21–39.

Kawamura, K., Brodal, A., and Hoddevik, G. (1974): The projection of the superior colliculus onto the reticular formation of the brain stem. An experimental anatomical study in the cat. *Exp. Brain Res.,* 19:1–19.

Kearney, R. E. (1978): Simulation of the human neuromuscular response to ankle rotation with a segmental reflex model. *Comput. Biol. Med.,* 8:329–341.

Kearney, R. E., and Chan, C. W. Y. (1979): Reflex response of human arm muscles to cutaneous stimulation of the foot. *Brain Res.,* 170:214–217.

Keele, C. A. (1962): Sensations aroused by chemical stimulation of the skin. In: *The Assessment of Pain in Man and Animals,* edited by C. A. Keele and R. Smith, pp. 28–40. Livingstone, Edinburgh.

Keele, S. W. (1968): Movement control in skilled motor performance. *Psychol. Bull.,* 70:387–404.

Keele, S. W., and Posner, M. I. (1968): Processing and visual feedback in rapid movements. *J. Exp. Psychol.,* 77:155–158.

Keller, E. L. (1974): Participation of medial pontine reticular formation in eye movement generation in monkey. *J. Neurophysiol.,* 37:316–332.

Keller, E. L. (1978): Gain of the vestibulo-ocular reflex in monkey at high rotational frequencies. *Vision Res.,* 18:311–315.

Keller, E. L., and Daniels, P. (1975): Oculomotor related interaction of visual and vestibular stimulation in vestibular nucleus cells in alert monkey. *Exp. Neurol.,* 46:187–198.

Keller, E. L., and Kamath, B. Y. (1975): Characteristics of head rotation and eye movement related neurons in alert monkey vestibular nucleus. *Brain Res.,* 100:182–187.

Keller, E. L., and Precht, W. (1979): Adaptive modification of central vestibular neurons in response to visual stimulation through reversing prisms. *J. Neurophysiol.,* 42:896–911.

Keller, E. L., and Precht, A. (1981): Adaptive modification in brainstem pathways during vestibulo-ocular reflex recalibration. In: *Neuronal Plasticity in Sensory-Motor Systems,* edited by B. Flohr and R. Precht, pp. 284–294. Springer, Berlin.

Kelso, J. A., and Holt, K. G. (1980): Exploring a vibratory systems analysis of human movement production. *J. Neurophysiol.,* 43:1183–1196.

Kelso, J. A., Holt, K. G., and Flatt, A. E. (1980): The role of proprioceptive in the perception and control of human movement, toward a theoretical reassessment. *Percept. Phychophysiol.,* 28:45–52.

Kelso, J. A., and Stelmach, G. E. (1976): Central and peripheral mechanisms in motor control. In: *Motor Control; Issues and Trends,* edited by G. E. Stelmach, pp. 1–40. Academic Press, New York.

Kemp, J. M., and Powell, T. P. S. (1971): The connections of the striatum and globus pallidus: synthesis and speculation. *Philos. Trans. R. Soc. Lond. (Biol.),* 262:411–457.

Kendig, J. J., and Erickson, N. (1981): H.P.N.S.—Nerve instability at hyperbaric pressures. *Undersea Biomed. Res.,* 8:22.

Kennard, M. A. (1954): The course of ascending fibers in the spinal cord of the cat essential to the recognition of painful stimuli. *J. Comp. Neurol.,* 100:511–524.

Kennedy, W. R. (1970): Innervation of normal human muscle spindles. *Neurology (Minneap.),* 20:463–475.

Kernell, D. (1965a): The adaptation and the relation between discharge frequency and current strength of cat lumbosacral motoneurones stimulated by long-lasting injected currents. *Acta Physiol. Scand.,* 65:65–73.

Kernell, D. (1965b): High frequency repetitive firing of cat lumbosacral motoneurones stimulated by long-lasting injected currents. *Acta Physiol. Scand.,* 65:74–86.

Kernell, D. (1966a): Input resistance, electrical excitability and size of ventral horn cells in cat spinal cord. *Science,* 152:1637–1640.

Kernell, D. (1966b): The repetitive discharge of motoneurones. In: *Muscular Afferents and Motor Control,* Nobel Symposium, edited by R. Granit, pp. 351–362. Almqvist and Wiksell, Stockholm.

Kernell, D. (1968): The repetitive impulse discharge of a simple neurone model compared to that of spinal motoneurones. *Brain Res.,* 11:685–687.

Kernell, D. (1969): Synaptic conductance changes and the repetitive impulse discharge of spinal motoneurones. *Brain Res.,* 15:291–294.

Kernell, D. (1972): The early phase of adaptation in repetitive impulse discharges of cat spinal motoneurones. *Brain Res.,* 41:184–186.

Kernell, D. (1976): Recruitment, rate modulation and the tonic stretch reflex. *Prog. Brain Res.,* 44:257–265.

Kernell, D. (1979): Rhythmic properties of motoneurones innervating muscle fibres of different speed in m. gastrocnemius medialis of the cat. *Brain Res.,* 160:159–162.

Kernell, D., Ducati, A., and Sjoholm, H. (1975): Properties of motor units in the first deep lumbrical muscle of the cat's foot. *Brain Res.,* 98:37–55.

Kernell, D., and Monster, A. W. (1981): Threshold current for repetitive impulse firing in motoneurones innervating muscle fibres of different fatigue sensitivity in the cat. *Brain Res.,* 229:193–196.

Kernell, D., and Monster, A. W. (1982a): Time course and properties of late adaptation in spinal motoneurones of the cat. *Exp. Brain Res.,* 46:191–196.

Kernell, D. and Monster, A. W. (1982b): Motoneurones properties and motor fatigue. An intracellular study of gastrocnemius motoneurones of the cat. *Exp. Brain Res.*, 46:197–204.

Kernell, D., and Sjoholm, H. (1973): Repetitive impulse firing: comparisons between neurone models based on "voltage clamp equations" and spinal motoneurones. *Acta Physiol. Scand.*, 87:40–56.

Kernell, D., and Sjoholm, H. (1975): Recruitment and firing rate modulation of motor unit tension in a small muscle of the cat's foot. *Brain Res.*, 98:57–72.

Kernell, D., and Zwaagstra, B. (1981): Input conductance, axonal conduction velocity and cell size among hindlimb motoneurones of the cat. *Brain Res.*, 204:311–326.

Kerr, F. W. L. (1972): The potential of cervical primary afferents to sprout in the spinal nucleus of V following long term trigeminal denervation. *Brain Res.*, 43:547–560.

Kidokoro, Y., Kubota, K., Shuto, S., and Sumino, R. (1968): Reflex organization of cat masticatory muscles. *J. Neurophysiol.*, 31:695–708.

Kievit, J., and Kuypers, H. G. J. M. (1977): Organization of the thalamo-cortical connexions to the frontal lobe in the rhesus monkey. *Exp. Brain Res.*, 29:299–322.

Kim, R., Nakano, K., Jayaraman, A., and Carpentier, M. (1976): Projections of the globus pallidus and adjacent structures in monkey. *J. Comp. Neurol.*, 169:263–290.

Kimura, J. (1970): Alteration of the orbicularis oculi reflex by pontine lesions: study in multiple sclerosis. *Arch. Neurol.*, 22:156–161.

Kimura, J. (1971a): An evaluation of the facial and trigeminal nerves in polyneuropathy: Electrodiagnostic study in Charcot-Marie-Tooth disease, Guillain-Barré syndrome, and diabetic neuropathy. *Neurology (Minneap.)*, 21:745–752.

Kimura, J. (1971b): Electrodiagnostic study of brainstem strokes. *Stroke*, 2:576–586.

Kimura, J. (1973): The blink reflex as a test for brainstem and higher central nervous system function. In: *New Developments in Electromyography and Clinical Neurophysiology*, Vol. 3, edited by J. E. Desmedt, pp. 682–691. Karger, Basel.

Kimura, J. (1974): Effect of hemispheral lesions on the contralateral blink reflex. *Neurology (Minneap.)*, 24:168–174.

Kimura, J. (1975): Electrically elicited blink reflex in diagnosis of multiple sclerosis. *Brain*, 98:413–426.

Kimura, J. (1976): A method for estimating the refractory period of motor fibers in the human peripheral nerve. *J. Neurol. Sci.*, 28:485–490.

Kimura, J. (1977): Electrical activity in voluntarily contracting muscle. *Arch. Neurol.*, 34:85–88.

Kimura, J. (1978): Proximal versus distal slowing of motor nerve conduction velocity in the Guillain-Barré syndrome. *Ann. Neurol.*, 3:344–350.

Kimura, J., Bodensteiner, J., and Yamada, T. (1977): Electrically elicited blink reflex in normal neonates. *Arch. Neurol.*, 34:246–249.

Kimura, J., and Butzer, J. F. (1975): F wave conduction velocity in Guillain-Barré syndrome. Assessment of nerve segment between axilla and spinal cord. *Arch. Neurol.*, 32:524–529.

Kimura, J., Giron, L. T., and Young, S. M. (1976): Electrophysiological study of Bell palsy. Electrically elicited blink reflex in assessment of prognosis. *Arch. Otolaryngol.*, 102:140–143.

Kimura, J., and Harada, O. (1972): Excitability of the orbicularis oculi reflex in all night sleep: Its suppression in non-rapid eye movement and recovery in rapid eye movement sleep. *Electroencephalogr., Clin. Neurophysiol.*, 33:369–377.

Kimura, J., and Harada, O. (1976): Recovery curves of the blink reflex during wakefulness and sleep. *J. Neurol.*, 213:189–198.

Kimura, J., and Lyon, L. W. (1972): Orbicularis oculi reflex in Wallenberg syndrome: alteration of the late reflex by lesions of the spinal tract and nucleus of the trigeminal nerve. *J. Neurol. Neurosurg. Psychiatry*, 35:228–233.

Kimura, J., and Lyon, L. W. (1973): Alteration of orbicularis oculi reflex by posterior fossa tumors. *J. Neurosurg.*, 38:10–16.

Kimura, J., Powers, J. M., and Van Allen, M. W. (1969): Reflex response of orbicularis oculi muscle to supraorbital nerve stimulation. Study in normal subjects and in peripheral facial paresis. *Arch. Neurol.*, 21:193–199.

Kimura, J., Rodnitzky, R. L., and Okawara, S. (1975): Electrophysiologic analysis of aberrant regeneration after facial nerve paralysis. *Neurology (Minneap.)*, 25:989–993.

Kimura, J., Rodnitzky, R. L., and Van Allen, M. W. (1970): Electrodiagnostic study of trigeminal nerve. *Neurology (Minneap.)*, 20:574–583.

Kimura, J., and Yamada, T. (1980): Electrophysiological assessment of facial hypoesthesia. *Trans. Am. Neurol. Assoc.*, 105:92–94.

Kirkwood, P. A., and Sears, T. A. (1974): Monosynaptic excitation of motoneurons from secondary endings of muscle spindles. *Nature*, 252:243–244.

Kirkwood, P. A., and Sears, T. A. (1975): Monosynaptic excitation of motoneurons from muscle spindle secondary endings of intercostal and triceps surae muscles in the cat. *J. Physiol. (Lond.)*, 245:64P–66P.

Kirkwood, P. A., and Sears, T. A. (1980): The measurement of synaptic connections in the mammalian central nervous system by means of spike triggered averaging. In: *Progress in Clinical Neurophysiology, Vol. 8, Spinal and Supraspinal Mechanisms of Voluntary Motor Control and Locomotion*, edited by J. E. Desmedt, pp. 44–71. Karger, Basel.

Kitai, S. T., Kocsis, J. D., Preston, R. J., and Sugimori, M. (1976): Monosynaptic inputs to caudate neurons identified by intracellular injection of horseradish peroxidase. *Brain Res.*, 109:601–606.

Klee, M. R. (1966): Different effects of the membrane potential of motor cortex units after thalamic and reticular stimulation. In: *The Thalamus*, edited by D. P. Purpura and M. D. Yahr, pp. 287–322. Columbia University Press, New York.

Klee, M. R., Lux, H. D., and Offenloch, K. (1964): Veränderungen der Membranpolarization und der Erregbarkeit von Zellen der motorischen Rinde während hochfrequenter Reizung der Formatio reticularis. *Arch. Psychiatr. Nervenkr.*, 205:237–261.

Kleijn, A. de, and Magnus, R. (1921): Ueber die Funktion der Otolithen. *Pfluegers Arch.*, 186:6–81.

Klein, R., and Posner, M. I. (1974): Attention to visual and kinesthetic components of skills. *Brain Res.*, 71:401–411.

Knox, C. K. (1974): Cross-correlation analysis of stimulus-evoked changes in excitability of spontaneously firing neurones. *J. Neurophysiol.*, 40:616–625.

Knutsson, B. (1961): Comparative value of electromyographic, myelographic and clinical-neurological examinations in diagnosis of lumbar root compression syndrome. *Acta Orthop. Scand. (Suppl.)*, 49:65.

Knutsson, E. (1970): Topical cryotherapy in spasticity. *Scand. J. Rehabil. Med.*, 2:159–163.

Knutsson, E. (1980): Muscle activation patterns of gait in spastic hemiparesis, paraparesis and cerebral palsy. *Scand. J. Rehabil. Med.*, 12 (Suppl. 7): 47–52.

Knutsson, E., Lindblom, U., and Martensson, A. (1973): Differences in effects in gamma and alpha spasticity induced by the GABA derivative baclofen (Lioresal). *Brain*, 96:26–46.

Knutsson, E., and Martensson, A. (1976): Action of Dantrolene sodium in spasticity with a low dependence of fusimotor drive. *J. Neurol. Sci.*, 29:195–212.

Knutsson, E., and Martensson, A. (1977): Activation patterns in spastic muscles stretched by passive and active movements. *Electroencephalogr. Clin. Neurophysiol.*, 43:605.

Knutsson, E., and Martensson, A. (1980): Dynamic motor capacity in spastic paresis and its relation to prime mover dysfunction spastic restraint and antagonist co-activation. *Scand. J. Rehabil. Med.*, 12:93–106.

Knutsson, E., Martensson, A., and Gransberg, L. (1982): Antiparetic and antispastic effects induced by tizanidine in patients with spastic paresis. *J. Neurol. Sci.*, 53:187–204.

Knutsson, E., and Richards, C. (1979): Different types of disturbed motor control in gait of hemiparetic patients. *Brain*, 102:405–430.

Ko, W. H., Hynecek, J., and Homa, J. (1979): Single frequency RF powered ECG telemetry system. *IEEE Trans. Biomed. Eng.*, 26:105–109.

Kobrinski, A. E., Bolkhovitin, S. V., Voskoboinikova, L. M. Ioffe, D. M., Polyan, E. P., Slavutski, Y. L., Sysin, A. Y., and Yakobson, Y. S. (1961): *Problems of Bioelectric Control in Automatic and Remote Control*, Vol. 2, pp. 619–623. Butterworths, London.

Koehler, W., Windhorst, U., Schmidt, J. et al. (1978): Diverging influences on Renshaw cell responses and monosynaptic reflexes from stimulation of capsula interna. *Neurosci. Lett.*, 8:35–39.

Koella, W. P. (1980): Baclofen. In: *Spasticity: Disordered Motor Control*, edited by R. Feldman, R. R. Young, and W. P. Koella, pp. 383–396. Year Book, Chicago.

Koeze, T. H. (1968): The independence of corticomotoneuronal and fusimotor pathways in the production of muscle contraction by motor cortex stimulation. *J. Physiol. (Lond.)*, 197:87–105.

Koeze, T. H. (1973): Muscle spindle afferent studies in the baboon. *J. Physiol. (Lond.)*, 229:297–317.

Koeze, T. H., Phillips, C. G., and Sheridan, J. D. (1968): Thresholds of cortical activation of muscle spindles and α-motoneurones of the baboon's hand. *J. Physiol. (Lond.)*, 195:419–449.

Kolb, E. (1955): Untersuchungen über zentrale kompensation und Kompensationbewegungen einseitig enstateter Frösche. *Z. Vergl. Physiol.*, 37:136–160.

Koll, W., Haase, J., Block, G., and Muslberg, B. (1963): The predilective action of small doses of morphine on nociceptive spinal reflexes of low spinal cats. *Int. J. Neuropharmacol.*, 2:57–65.

Komatsuzaki, A., Harris, H. E., Alpert, J., and Cohen, B. (1969): Horizontal nystagmus of rhesus monkeys. *Acta Otolaryngol.*, 67:535–551.

Kots, Y. M. (1969): Supraspinal control of the segmental centres of muscle antagonists in man. *Biophysics*, 14:176–183.

Kots, Y. M. (1977): *The Organization of Voluntary Movement*. Plenum Press, New York.

Koychabhakdi, N., and Walberg, F. (1977): Cerebellar afferents from neurons in motor nuclei of cranial nerves demonstrated by retrograde axonal transport of horseradish peroxidase. *Brain Res.*, 137:158–163.

Kranz, H. (1981): Control of motoneuron firing during maintained voluntary contraction in normals and in patients with central lesions. In: *Progress in Clinical Neurophysiology, Vol. 9, Motor Unit Types, Recruitment and Plasticity in Health and Disease*, edited by J. E. Desmedt, pp. 358–367. Karger, Basel.

Kranz, H., Adorjani, C., and Baumgartner, G. (1973): The effect of nociceptive cutaneous stimuli on human motoneurons. *Brain*, 96:571–590.

Krassoievitch, M., and Tissot, R. (1971): Etude des réflexes H et T dans la maladie de Parkinson. Modification de ces réflexes par la L-dopa. *Encephale*, 1:1–22.

Krnjevic, K. (1974): Chemical nature of synaptic transmission in vertebrates. *Physiol. Rev.*, 54: 418–540.

Krnjevic, K. (1980): Mechanisms of drug action on spinal and supraspinal reflexes, with special reference to the action. In: *Spasticity: Disordered Motor Control*, edited by R. Feldman, R. R. Young, W. P. Koella, pp. 397–416. Year Book, Chicago.

Krnjevic, K., Puil, E., and Werman, R. (1978): EGTA and motoneuronal after-potentials. *J. Physiol. (Lond.)*, 275:199–223.

Kruger, L. (1956): Characteristics of the somatic afferent projection to the precentral cortex in the monkey. *Am. J. Physiol.*, 186:475–482.

Kubo, T., Matsunaga, T., and Igarashi, M. (1979): Vestibular unitary responses to visual stimulation in the rabbit. *Acta Otolaryngol.*, 88:117–121.

Kubota, K., and Hamada, I. (1978): Visual tracking and neuron activity in the post-arcuate area in monkeys. *J. Physiol. (Paris)*, 74:297–312.

Kubota, K., Tanaka, R., and Tsuzuki, N. (1967): Muscle spindle activity and natural sleep in the cat. *Jpn. J. Physiol.*, 17:613–626.

Kucera, J. (1980): Histochemical study of long nuclear chain fibers in the cat muscle spindle. *Anat. Rec.*, 198:567–580.

Kucera, J., and Dorovini-Zis, K. (1979): Types of human infrafusal muscle fibers. *Muscle Nerve*, 2:437–451.

Kudina, L. P. (1980): Reflex effects of muscle afferents on antagonist studied of single firing motor units in man. *Electroencephalogr. Clin. Neurophysiol.*, 50:214–221.

Kuffler, S. W., and Hunt, C. C. (1952): The mammalian small-nerve fibres; a system for efferent nervous regulation of muscle spindle discharge. *Res. Publ. Assoc. Res. Nerv. Ment. Dis.*, 30:24–37.

Kugelberg, E. (1947): Electromyograms in muscular disorders. *J. Neurol. Neurosurg. Psychiatry*, 10:122-128.

Kugelberg, E. (1948): Demonstration of A and C fibres components in the Babinski plantar response and the pathological flexion reflex. *Brain*, 71:304–319.

Kugelberg, E. (1952): Facial reflexes. *Brain*, 75:385–396.

Kugelberg, E. (1962): Polysynaptic reflexes of clinical importance. *Electroencephagr. Clin. Neurophysiol.*, 22:103–111.

Kugelberg, E. (1973): Properties of the rat hind limb motor units. In: *New Developments in Electromyography and Clinical Neurophysiology, Vol. 1*, edited by J. E. Desmedt, pp. 2–13. Karger, Basel.

Kugelberg, E. (1981): The Motor unit: morphology and function. In: *Progress in Clinical Neurophysiology, Vol. 9, Motor Unit Types, Recruitment and Plasticity in Health and Disease*, edited by J. E. Desmedt, pp. 1–16. Karger, Basel.

Kugelberg, E., Eklund, K., and Grimby, L. (1960): An electromyography study of the nociceptive reflexes of the lower limb. Mechanism of the plantar responses. *Brain*, 83:394–410.

Kugelberg, E., and Widen, L. (1954): Epilepsia partialis continua. *Electroencephalogr. Clin. Neurophysiol.*, 6:503–506.

Kuhn, R. (1950): Functional capacity of the isolated human spinal cord. *Brain*, 1:1–51.

Kuno, M., Miyata, Y., and Munoz-Martinez, E. J. (1974): Properties of fast and slow alpha motoneurones following motor reinnervation. *J. Physiol. (Lond)*, 242:273–288.

Kunzle, H. (1978): An autoradiographic analysis of the efferent connections from premotor and adjacent prefrontal regions (area 6 and 9) in Macaca fascicularis. *Brain Behav. Evol.*, 15:185–234.

Kurizke, J. F., and Gylfe, J. (1962): A new muscle relaxant in spasticity. *Neurology (Minneap.)*, 12:343–350.

Kutas, M., and Donchin, E. (1977): The effect of handedness, of responding hand, and of response force on the contralateral dominance of the readiness potential. In: *Progress in Clinical Neurophysiology, Vol. 1, Attention, Voluntary Contraction and Event-Related Cerebral Potentials*, edited by J. E. Desmedt, pp. 189–210. Karger, Basel.

Kuypers, H. (1958): Some projections from peri-central cortex to the pons and lower brain stem in monkey and chimpanzee. *J. Comp. Neurol.*, 110:221–255.

Kuypers, H. (1973): The anatomical organization of the descending pathways and their contributions to motor control especially in primates. In: *New Developments in Electromyography and Clinical Neurophysiology, Vol. 3*, edited by J. E. Desmedt, pp. 38–68. Karger, Basel.

Kuypers, H. (1978): The general organization of the thalamo-frontal connections in the rhesus monkey. In: *Progress in Clinical Neurophysiology, Vol. 4, Cerebral Motor Control in Man: Long Loop Mechanisms*, edited by J. E. Desmedt, pp. 10–20. Karger, Basel.

Kuypers, H., and Brinkman, J. (1970): Precentral projections to different parts of the spinal intermediate zone in the rhesus monkey. *Brain Res.*, 24:29–48.

Kuypers, H., and Lawrence, D. G. (1967): Cortical projections to the red nucleus and the brain stem in the rhesus monkey. *Brain Res.*, 4:151–188.

Kuypers, H., Swarcbart, M. K., Mishkin, M. and Rosvold, H. E. (1965): Occipito-temporal cortico-cortical connections in the rhesus monkey. *Exp. Neurol.*, 11:245–262.

Kwan, H. C., Mackay, W. A., Murphy, J. T., and Wong, Y. C. (1978): Spatial organization of precentral cortex in awake primates. II. Motor outputs. *J. Neurophysiol.*, 41:1120–1131.

Kwan, H. C., Murphy, J. T., and Repeck, M. W. (1979): Control of stiffness by the medium latency electromyographic response to limb perturbation. *Can. J. Physiol. Pharmacol.*, 57:277–285.

Kwatny, E., Thomas, D. H., and Kwatny, G. (1970): An application of signal processing techniques to the study of myoelectric signals. *IEEE Trans. Biomed. Eng.*, 17:303–313.

Lachat, J. M., Ruegg, D. C., and Wiesendanger, M. (1977): Transcortical facilitation of the H-reflex in monkeys. *Experientia*, 33:781.

Lachman, T., Shahani, B. T., and Young, R. R. (1980): Late responses as aids to diagnosis in peripheral neuropathy. *J. Neurol. Neurosurg. Psychiatry*, 43:156–162.

Lackner, J. M., and Levine, M. S. (1979): Changes in apparent body orientation and sensory localization induced by vibration. *Aviat. Space Environ. Med.*, 50:346–354.

Lacour, M., Vidal, P. P., and Xerri, C. (1981): Early directional influence of visual motion cues on postural control in the falling monkey. *Ann. NY Acad. Sci.*, 374:403–411.

Lacour M., and Xerri, C. (1980): Compensation of postural reactions to fall in the vestibular neurectomized monkey: role of the visual motion cues. *Exp. Brain Res.*, 40:103–110.

Lacour, M., Xerri, C., and Hugon, M. (1978): Muscle responses and monosynaptic reflexes in falling monkey. Role of the vestibular system. *J. Physiol. (Paris)*, 74:427–438.

Lamarre, Y., and Lund, J. P. (1975): Load compensation in human masseter muscles. *J. Physiol. (Lond.)*, 253:21–35.

Lamarre, Y., Spidalieri, G., Bushy, L., and Lund, J. P. (1980): Programming of initiation and execution of ballistic arm movements in the monkey. *Prog. Brain Res.*, 54:157–169.

Lamarre, Y., Spidalieri, G., and Lund, J. P. (1981): Patterns of muscular and motor cortical activity during a simple arm movement in monkey. *Can. J. Physiol. Pharmacol.,* 59:748–756.

Lambertsen, C. J., Gelfand, R., and Clark, J. M. (1978): Predictive studies IV. Work capabilities and physiological effects in He-O₂ excursions to pressures of 400–800–1200 and 1600 feet of sea water. A collaborative investigation from Institute for Environmental Medicine. University of Pennsylvania Medical Center, Philadelphia.

La Motte, C., Pert, C. B., and Synder, S. H. (1976): Opiate receptor binding in primate spinal cord: distribution and changes after dorsal root section. *Brain Res.,* 112:407–412.

Lance, J. W. (1980): The control of muscle tone, reflexes and movement. *Neurology (Minneap.),* 30:1303–1313.

Lance, J. W., Burke, D., and Andrews, C. J. (1973): The reflex effects of muscle vibration. In: *New Developments in Electromyography and Clinical Neurophysiology,* Vol. 3, edited by J. E. Desmedt, pp. 444–462. Karger, Basel.

Lance, J. W., Degail, P., and Neilson, P. D. (1966): Tonic and phasic spinal cord mechanisms in man. *J. Neurol. Neurosurg. Psychiatry,* 29:535–544.

Landau, W. M. (1974): Spasticity: the fable of the neurological demon and the emperor's new therapy. *Arch. Neurol.,* 31:217–219.

Landau, W. M., and Clare, M. H. (1959): The plantar reflex in man, with special reference to some conditions where the extensor response is unexpectedly absent. *Brain,* 82:321–355.

Landau, W. M., and Care, M. H. (1964): Fusimotor function. Part VI. H-reflex, tendon jerk and reinforcement in hemiplegics. *Arch. Neurol.,* 10:128–134.

Landau, W. M., Weave, R. A., and Hornbein, T. F. (1960): Fusimotor nerve function in man. Differential nerve block studies in normal subjects and in spasticity and rigidity. *Arch. Neurol.,* 3:10–23.

Landgren, S., Phillips, C. G., and Porter, R. (1962): Cortical fields of origin of the monosynaptic pyramidal pathways to some alpha motoneurones of the baboon's hand and forearm. *J. Physiol. (Lond.),* 161:112–125.

Landgren, S., and Sifvenius, H. (1969): Projection to cerebral cortex of group I muscle afferents from the cat's hind limb. *J. Physiol. (Lond.),* 200:353–372.

Landsmeer, J. M. F. (1962): Power grip and precision handling. *Ann. Rheum. Dis.,* 21:164–170.

Lannou, J., Cazin, L., and Hamann, K. F. (1980): Response of central vestibular neurons to horizontal linear acceleration in the rat. *Pfluegers Arch.,* 385:123–129.

Laplane, D., Talairach, S., Meininger, V., et al. (1977): Clinical consequences of corticectomies involving the supplementary motor area in man. *J. Neurol. Sci.,* 34:301–314.

Laporte, Y., and Emonet-Denand, F. (1976): The skeleto-fusimotor innervation of cat muscle spindle. *Prog. Brain Res.,* 44:99–106.

Laporte, Y., and Lloyd, D. P. C. (1952): Nature and significance of the reflex connection established by large afferent fibers of muscular origin. *Am. J. Physiol.,* 169:609–621.

Larsen, K. D., and Asanuma, H. (1979): Thalamic projections to the feline motor cortex studied with horseradish peroxidase. *Brain Res.,* 172:209–215.

Larsen, K. D., and Yumiya, H. (1979): Organization of the convergence in the intermediate cerebellar nuclei of somatosensory receptive fields with motor cortical-evoked responses. *Exp. Brain Res.,* 36:477–489.

Larsen, K. D., and Yumiya, H. (1980): Motor cortical modulation of feline red nucleus output: corticorubral and cerebellar-mediated responses. *Exp. Brain Res.,* 38:321–331.

Lasek, R. J., Joseph, B. S., and Whitlock, D. G. (1968): Evaluation of a radioautographic neuroanatomical tracing method. *Brain Res.,* 8:319–336.

Lashley, K. S. (1917): The accuracy of movement in the absence of excitation from a moving organ. *Am. J. Physiol.,* 43:169.

Lawrence, D. G., and Hopkins, D. A. (1976): The development of motor control in the rhesus monkey: evidence concerning the role of cortico-motoneuronal connections. *Brain,* 99:235–254.

Lawrence, D. G., and Kuypers, H. (1968): The functional organization of the motor system in the monkey. I. The effects of bilateral pyramidal lesions. II. The effects of lesions of the descending brain-stem pathway. *Brain,* 91:1–14; 15–36.

Layne, J. M., and Rugh, J. D. (1981): Order effects in recording silent period duration. *J. Dent. Res.,* 60:347.

Leavitt, L. A., and Beasley, W. C. (1964): Clinical application of quantitative methods in the study of spasticity. *Clin. Pharmacol. Ther.,* 5:918–941.

Le Bars, D., Guilbaud, G., Jurna, I., and Besson, J. M. (1976): Differential effects of morphine on responses of dorsal horn lamina V type cells elicited by A and C fibre stimulation in the spinal cat. *Brain Res.,* 115:518–524.

Le Bars, D., Menetrey, D., Conseiller, C., and Besson, J. M. (1975): Depressive effects of morphine upon lamina V cell activities in the dorsal horn of the spinal cat. *Brain Res.,* 98:261–277.

Lee, R. G., and Lishman, J. R. (1975): Visual proprioceptive control of stance. *J. Human Movt. Study,* 1:87–95.

Lee, R. G., Rohs, G. L., and White, D. G. (1979): Long latency EMG responses to load perturbations in hemiplegic patients. *Can. J. Neurol. Sci.,* 6:384.

Lee, R. G., and Stein, R. B. (1981): Resetting of tremor by mechanical perturbations: a comparison of essential tremor and Parkinsonian tremor. *Ann. Neurol.,* 10:523–531.

Lee, R. G., and Tatton, W. G. (1975): Motor responses to sudden limb displacements in primates with specific CNS lesions and in human patients with motor system disorders. *Can. J. Neurol. Sci.,* 2:285–293.

Lee, R. G., and Tatton, W. G. (1978): Long loop reflexes in man: clinical applications. In: *Progress in Clinical Neurophysiology, Vol. 4. Cerebral Motor Control in Man: Long Loop Mechanisms,* edited by J. E. Desmedt, pp. 320–333. Karger, Basel.

Lee, R. G., and Tatton, W. G. (1982): Long latency reflexes to imposed displacements of the human

wrist: dependence on duration of movement. *Exp. Brain Res.*, 45:207–216.
Le Gros-Clark, W. E. (1959): *The Antecedents of Man.* Harper & Row, New York.
Lehmann, H. J., and Ule, G. (1964): Electrophysiological findings and structural changes in circumspect inflammation of peripheral nerves. *Prog. Brain Res.*, 6:169–173.
Lehmann, P. A. (1978): Stereoselective molecular recognition in biology. In: *Receptors and Recognition*, Series A, Vol. 5, edited by P. Cuatrecasas and M. F. Greaves pp. 3–88. Chapman and Hall, London.
Leigh, P. N., Rothwell, J. C., Traub, M. M., and Marsden, C. D. (1980): A patient with reflex myoclonus and muscle rigidity: "jerking stiff man syndrome." *J. Neurol. Neurosurg. Psychiatry*, 43:1125–1131.
Leksell, L. (1945): The action potentials and excitatory effects of the small ventral root fibres to skeletal muscle. *Acta Physiol. Scand.*, 10 (Suppl. 31): 1–84.
Lele, P. P., and Weddell, G. (1959): Sensory nerves of the cornea and cutaneous sensibility. *Exp. Neurol.*, 1:334–359.
Lemaire, C. (1979): Capacité de travail psycho-sensoriel en ambiance hyperbare. *Le Travail Humain*, 42:13 28.
Lemon, R. N. (1981): Variety of functional organization within the monkey motor cortex. *J. Physiol. (Lond.)*, 311:521–540.
Lemon, R. N., Hanby, J. A., and Porter, R. (1976): Relationship between the activity of precentral neurones during active and passive movements in conscious monkeys. *Proc. R. Soc. Lond.*, 194:341–373.
Lemon, R. N., and Porter, R. (1976): Afferent input to movement-related precentral neurones in conscious monkeys. *Proc. R. Soc. Lond. (Biol.)*, 194:313–339.
Lemon, R. N., and van der Burg, J. (1979): Short-latency peripheral inputs to thalamic neurones projecting to the motor cortex in the monkey. *Exp. Brain Res.*, 36:445–462.
Lenman, R. A., and Woolsey, C. N. (1956): Sensory and motor localization in cerebral cortex of the porcupine. *J. Neurophysiol.*, 19:544–563.
Lenman, J. A. R., and Ritchie, A. E. (1970): *Clinical Electromyography.* Pitman, London.
Lennerstrand, G. (1968): Position and velocity sensitivity of muscle spindles in the cat. I. Primary and secondary endings deprived of fusimotor activation. *Acta Physiol. Scand.*, 73:281–299.
Lennerstrand, G., and Thoden U. (1968): Position and velocity sensitivity of muscle spindles in the cat. II. Dynamic fusimotor single-fibre activation of primary endings. III. Static fusimotor single-fibre activation of primary and secondary endings. *Acta Physiol. Scand.*, 74:16–29; 39–49.
Leong, S. K. (1976): An experimental study of the corticofugal system following cerebral lesions in the albino rat. *Exp. Brain Res.*, 26:235–247.
Leong, S. K. (1977): Sprouting of the corticopontine fibers after neonatal cerebellar lesion in the albino rat. *Brain Res.*, 123:164–169.
Leong, S. K. (1980): A qualitative electron microscopic study of the cortico-pontine projections after neonatal cerebellar hemispherectomy. *Brain Res.*, 194:299–310.
Leong, S. K., and Lund, R. D. (1973): Anomalous bilateral corticofugal pathways in albino rats after neonatal lesions. *Brain Res.*, 62:218–221.
Lestienne, F. (1979): Effects of inertial load and velocity on the braking process of voluntary limb movements. *Exp. Brain Res.*, 35:407–418.
Lestienne, F., Polit, A., and Bizzi, E. (1981): Functional organization of the motor process underlying the transition from movement to posture. *Brain Res.*, 230:121–131.
Lestienne, F., Soechting, J., and Berthoz, A. (1977): Postural readjustments induced by linear motion of visual scenes. *Exp. Brain Res.*, 28:363–384.
Levante, A., and Albe-Fessard, D. (1972): Localisation dans les couches VII et VIII de Rexed des cellules d'origine d'un faisceau spino-réticulaire croisé. *C. R. Acad. Sci. (Paris)*, 274:3007–3010.
Levine, J. M., Jossman, P. B., Friend, D. G., et al. (1969): Diazepam in the treatment of spasticity. *J. Chron. Dis.*, 22:57–62.
Levine, M. G., and Kabat, H. (1952): Co-contraction and reciprocal innervation in voluntary movement in man. *Science* 116:115–118.
Levine, M. G., Kabat, H., Knott, M., and Voss, D. E. (1954): Relaxation of spasticity by physiological technics. *Arch. Phys. Med. Rehabil.*, 35:214–223.
Levitt, J., and Levitt, M. (1968): Sensory hind-limb representation in cortex of the cat. *J. Exp. Neurol.*, 22:259–275.
Levy, R. (1963): The relative importance of the gastrocnemius and soleus muscles in the ankle jerk of man. *J. Neurol. Neurosurg. Psychiatry*, 26:148–150.
Lewis, C. H. and Griffin, M. J. (1978): Predicting the effects of dual-frequency vertical vibration on continuous manual control performance. *Ergonomics*, 21:637–650.
Lewis, D., and Porter, R. (1974): Pyramidal tract discharge in relation to movement performance in monkeys with partial anaesthesia of the moving hand. *Brain Res.*, 71:245–251.
Lewis, D. and Proske, U. (1972): The effect of muscle length and rate of fusimotor stimulation on the frequency of discharge in primary endings from muscle spindles in the cat. *J. Physiol. (Lond.)*, 222:511–535.
Lewis, F., Derouesne, J., and Signoret, J. L. (1972): Analyse neuropsychologique du syndrome frontal. *Rev. Neurol. (Paris)*, 127:415–440.
Li, C. L., Cullen, C., and Jasper, H. H. (1956): Laminar microelectrode studies of specific somatosensory cortical potentials. *J. Neurophysiol.*, 19:111–130.
Li, C. L., and Tew, J. M. (1964): Reciprocal activation and inhibition of cortical neurones and voluntary movements in man: cortical cell activity and muscle movement. *Nature*, 203:264–265.
Liberson, W. T. (1962): Monosynaptic reflexes and their clinical significance. *Electroencephalogr. Clin. Neurophysiol. (Suppl.)*, 22:79–89.
Liberson, W. T. (1963): Sensory conduction velocities in normal individuals and in patients with peripheral neuropathies. *Arch. Phys. Med. Rehabil.*, 44:313–310.
Liddell, E. G., and Sherrington, C. (1924): Reflexes in

response to stretch (myotatic reflexes). *Proc. R. Soc. Lond. (Biol.)*, 96:212–242.
Lieberman, J. S., Bailey, C. S., Kitchell, R., and Taylor, R. G. (1979): Muscle spindle afferent discharges following spinal cord transection in the cat. *Arch. Phys. Med. Rehabil.* 60:544–554.
Lim, R. K. (1968): Neuropharmacology of pain and analgesia. In: *Pharmacology of Pain*, edited by D. Armstrong and E. G. Pardo, pp. 169–217. Pergamon, Oxford.
Lim, K. H., and Leong, S. K. (1975): Aberrant bilateral projections from dentate and interposed nuclei in albino rats after neonatal lesions. *Brain Res.*, 96:306–309.
Lindermann, H. H. (1970): Anatomy of the otolith organs. *Adv. Otolaryngol.*, 20:405–433.
Lindqvist, C., and Martensson, A. (1970): Mechanisms involved in cat's blink reflex. *Acta Physiol. Scand.*, 80:149–159.
Lindsay, K. W., Roberts, T. D., and Rosenberg., J. (1976): Asymmetric tonic labyrinth reflexes and neck reflexes in the decerebrate cat. *J. Physiol. (Lond.)*, 261:483–601.
Lisberger, S. G., and Fuchs, A. F. (1974): Response of flocculus Purkinje cells to adequate vestibular stimulation in alert monkey: fixation vs. compensatory eye movement. *Brain Res.*, 69:347–353.
Lisberger, S. G., and Fuchs, A. F. (1978): Role of primate flocculus during rapid behavioral modification of vestibuloocular reflex. I. Purkinje cell activity during visually guided horizontal smooth-pursuit eye movements and passive head rotation. *J. Neurophysiol.*, 41:733–763.
Lisberger, S. G., Miles, F. A., Optican, L. M., and Eighmy, B. B. (1981): Optokinetic responses in monkey and long-term adaptive changes in vestibulo-ocular reflex. *J. Neurophysiol.*, 45:869–890.
Liu, C. N., and Chambers, W. W. (1958): Intraspinal sprouting of dorsal root axons. *Arch. Neurol. Psychiatry,* 79:46–61.
Liu, C. N., and Chambers, W. W. (1964): An experimental study of the corticospinal system in the monkey. The spinal pathway and preterminal distribution of degenerating fibres following discrete lesions of the pre- and postcentral gyri and bulbar pyramid. *J. Comp. Neurol.*, 123:257–284.
Lloyd, D. P. (1943): Conduction and synaptic transmission of the reflex response to stretch in spinal cats. *J. Neurophysiol.*, 6:317–326.
Lloyd, D. P. (1946): Facilitation and inhibition of spinal motoneurons. *J. Neurophysiol.*, 9:421–438.
Lloyd, D. P., and Chang, H. T. (1948): Afferent fibres in muscle nerves. *J. Neurophysiol.*, 11:199–208.
Loe, P. R., Whitsel, B. L., Dreyer, D. A., and Metz, C. B. (1977): Body representation in the ventrobasal thalamus of the macaque: a single unit analysis. *J. Neurophysiol.*, 40:1339–1355.
Loeb, G. E. (1980): Somatosensory unit input to the spinal cord during normal walking. *Can. J. Physiol. Pharmacol.*, 59:627–635.
Loeb, G. A., Bak, M. J., and Duysens, J. (1977): Long-term unit recording from somatosensory neurons in the spinal ganglia of the freely walking cat. *Science*, 197:1192–1194.
Loeb, G. E., and Duysens, J. (1979): Activity patterns in individual hindlimb primary and secondary muscle spindle afferents during normal movements in unrestrained cats. *J. Neurophysiol.*, 42:420–440.
Loeb, G. E., and Hoffer, J. A. (1981): Muscle spindle function during normal and perturbed locomotion in cats. In: *Muscle Receptors and Movement*, edited by A. Taylor and A. Prochazka, pp. 219–228. Macmillan, London.
Loeb, G. E., Walmsley, B., and Duysens, J. (1980): Obtaining proprioceptive information from natural limbs: implantable transducers vs. somatosensory neuron recordings. In: *Physical Sensors for Biomedical Applications*, edited by M. R. Newman et al. CRC Press, Cleveland.
Loesche, J., and Stewart, O. (1977): Behavioral correlates of denervation and reinnervation of the hippocampal formation of the rat: recovery of alternation performance following unilateral entorhinal lesions. *Brain Res. Bull.*, 2:21–39.
Lomo, T., Westgaard, R. H., and Engebretsen, L. (1980): Different stimulation patterns affect contractile properties of denervated rat soleus muscles. In: *Plasticity of Muscle*, edited by D. Pette, pp. 297–309. De Gruyter, Berlin.
Long, C., and Brown, M. E. (1964): Electromyographic kinesiology of the hand:muscle moving the long fingers. *J. Bone Joint Surg.*, 46:1683–1706.
Lorente De No, R. (1933): Vestibulo-ocular reflex arc. *Arch. Neurol. Psychiatry,* 30:245–291.
Lowitzsch, K., Kuhnt, U., Sakmann, C., et al. (1976): Visual pattern evoked responses and blink reflexes in assessment of MS diagnosis. *J. Neurol. Neurosurg. Psychiatry,* 213:17–32.
Lucier, G. E., Ruegg, D. C., and Wiesendanger, M. (1975): Responses of neurones in motor cortex and in area 3a to controlled stretches of forelimb muscles in cebus monkeys. *J. Physiol. (Lond.),* 251: 833–853.
Lund, E. D., Cunningham, T. J., and Lund, J. S. (1973): Modified optic projections after unilateral eye removal in young rat. *Brain Behav. Evol.*, 8:51–72.
Lund, J. P., and Marakami, T. (1981): Interactions between the jaw-opening reflex and mastication. *Can. J. Physiol. Pharmacol.*, 59:683–690.
Lund, J. P., Smith, A., and Lamarre, Y. (1978): Sensory control of mandibular movements and the modulation by set and circumstances. In: *Oral Physiology and Occlusion*, edited by J. H. Perryman, pp. 115–135. Pergamon Press, Elmsford.
Lund, J. P., Smith, A. M., Sessle, B. J., and Murakami, T. (1979): Activity of trigeminal alpha and gamma motoneurons and muscle afferents during performance of a biting task. *J. Neurophysiol.*, 42: 710–725.
Lund, R. D. (1972): Anatomic studies of the superior colliculus. *Invest. Ophthalmol.* 11:434–441.
Lund, R. D., Land, P. W., and Boles, J. (1980): Normal and abnormal uncrossed retinotectal pathways in rats: an HRP study. *J. Comp. Neurol.*, 189:51–72.
Lund, R. D., and Lund, J. S. (1971): Synaptic adjustment after deafferentation of the superior colliculus of the rat. *Science,* 171:804–807.

Lund, R. D., and Lund, J. S. (1973): Reorganization of the retino-tectal pathways in rats after neonatal and retinal lesions. *Exp. Neurol.*, 40:377–390.

Lund, R. D., and Lund, J. S. (1976): Plasticity in the developing visual system: the effects of retinal lesions made in young rats. *J. Comp. Neurol.*, 169:133–154.

Lund, R. D., and Miller, B. F. (1975): Secondary effects of foetal eye damage in rats on intact central optic projections. *Brain Res.*, 92:279–289.

Lundberg, A. (1969): Reflex control of stepping. *The Nansen Memorial Lecture V*, pp. 1–42. Universitetsforlaget, Oslo.

Lundberg, A. (1970): The excitatory control of the Ia inhibitory pathway. In: *Excitatory Synaptic Mechanisms*, edited by P. Andersen and J. K. Jansen, pp. 333–340. Universitetsforlaget, Oslo.

Lundberg, A. (1975): Control of spinal mechanisms from the brain. In: *The Nervous System*, Vol. 1, edited by D. B. Tower, pp. 253–265. Raven Press, New York.

Lundberg, A. (1979): Multisensory control of spinal reflex pathways. *Prog. Brain Res.*, 50:11–28.

Lundberg, A., Malmgren, K., and Schomburg, E. D. (1977): Cutaneous facilitation of transmission in reflex pathways from Ib afferents to motoneurones. *J. Physiol. (Lond.)*, 265:763–780.

Lundberg, A., Malmgren, K., and Schomburg, E. D. (1978): Role of joint afferents in motor control exemplified by effects on reflex pathways from Ib afferents. *J. Physiol. (Lond.)*, 284:327–343.

Lundberg, A., and Winsbury, G. (1960): Selective adequate activation of large afferents from muscle spindles and Golgi tendon organs. *Acta Physiol. Scand.*, 49:155–164.

Lundervold, A. (1957): Electromyographic investigations during typewriting. *Ergonomics*, 1:226–233.

Luschei, E. S., and Fuchs, A. F. (1972): Activity of brain stem neurons during eye movements of alert monkeys. *J. Neurophysiol.*, 35:445–461.

Luscher, H. R., Ruenzel, P., and Henneman, E. (1979): How the size of motoneurones determines their susceptibility to discharge. *Nature*, 282:859–861.

Luttgau, H. C. (1965): The effect of metabolic inhibitors on the fatigue of the action potential in single muscle fibres. *J. Physiol. (Lond.)*, 178:45–67.

Lynch, G., and Cotman, C. W. (1975): The hippocampus as a model for studying anatomical plasticity in the adult brain. In: *Hippocampus*, Vol. 1, edited by R. L. Isaacson and K. H. Pribram, pp. 123–155. Plenum Press, New York.

Lynch, G., Deadwyler, S., and Cotman, C. (1973): Post-lesion axonal growth produces permanent functional connections. *Science*, 180:1364–1366.

Lynch, G., Gall, C., Rose, G., and Cotman, C. W. (1976): Changes in the distribution of the dentate gyrus associational system following unilateral or bilateral entorhinal lesions in the adult rat. *Brain Res.*, 110:57–71.

Lynch, G., Matthews, D. A., Mosko, S., et al. (1972): Induced acetylcholine-esterase-rich layer in rat dentate gyrus following entorhinal lesions. *Brain Res.*, 42:311–318.

Lynn, P. A., Bettles, N. D., Hughes, A. D., and Johnson, S. W. (1978): Influences of electrode geometry on bipolar recordings of the surface electrogram. *Med. Biol. Eng. Comput.*, 16:651–660.

Lyon, M. F. (1951): Hereditary absence of otoliths in the house mouse. *J. Physiol. (Lond.)*, 114:410–418.

Lyon, L. W., Kimura, J., and McCormick, W. F. (1972): Orbicularis oculi reflex in coma: clinical, electrophysiological, and pathological correlations. *J. Neurol. Neurosurg. Psychiatry*, 35:582–588.

Lyon, L. W., and Van Allen, M. W. (1972): Orbicularis oculi reflex. Studies in internuclear ophthalmoplegia and pseudointernuclear ophthalmoplegia. *Arch. Ophthalmol.* 87:148–154.

MacDonald, A. G. (1975): Hydrostatic pressure physiology. In: *The Physiology and Medicine of Diving and Compressed Air*, edited by P. B. Bennett and D. H. Elliott, pp. 78–101. Baillière Tindall, London.

MacDonald, A. G. (1980): Molecular and cellular effects of hydrostatic pressure: a physiologist's view. *Proceedings: 7th Symposium in Underwater Physiology*, Athens, pp. 56–57.

MacDonald, A. G., and Wann, K. T. (1978): *Physiological Aspects of Anaesthetics and Inert Gases*. Academic Press, London.

Mach, E. (1886): *The Analysis of Sensations*. Dover, New York.

Maciewicz, R. J., Eagen, K., Kaneko, C. R. S., and Highstein, S. M. (1977): Vestibular and medullary brain stem afferents to the abducens nucleus in the cat. *Brain Res.*, 123:229–240.

Mackenzie, R. A., Burke, D., Skuse, N. F., and Lethlean, A. K. (1975): Fibre function and perception during cutaneous nerve block. *J. Neurol. Neurosurg. Psychiatry*, 38:865–873.

McClane, T. K., and Martin, W. R. (1967): Effects of morphine, nalorphine, cyclazocine and naloxone on the flexor reflex. *Int. J. Neuropharmacol.*, 6:89–98.

McCloskey, D. I. (1973): Differences between the senses of movement and position shown by the effects of loading and vibration of muscles in man. *Brain Res.*, 61:119–131.

McCloskey, D. I. (1978): Kinaesthetic sensibility. *Physiol. Rev.*, 58:763–820.

McCloskey, D. I. (1980): Kinaesthetic sensations and motor commands in man. In: *Progress in Clinical Neurophysiology, Vol. 8, Spinal and Supraspinal Mechanisms of Voluntary Motor Control and Locomotion*, edited by J. E. Desmedt, pp. 203–214. Karger, Basel.

McCloskey, D. I., Ebeling, P., and Goodwin, G. M. (1974): Estimation of weights and tensions and apparent involvement of a 'sense of effort.' *Exp. Neurol.*, 42:220–232.

McCloskey, D. I., and Torda, T. A. G. (1975): Corollary motor discharges and kinaesthesia. *Brain Res.*, 100:467–470.

McComas, A. J., Sica, R. E. P., and Upton, A. R. M. (1970): Excitability of human motoneurons during effort. *J. Physiol. (Lond.)*, 210:145P–146P.

McCough, G. P. (1961): Factors in the transition to spasticity. In: *The Spinal Cord*, edited by G. Austin, pp. 256–261. Charles C Thomas, Springfield, Illinois.

McCough, G. P., Deering, I. D., and Ling, T. H. (1951): Location and receptors for tonic neck reflexes. *J. Neurophysiol.*, 14:191–195.

McCrea, R. A., Baker, R., and Delgado-Garcia, J. (1970): Afferent and efferent organization of the prepositus hypoglossi nucleus. *Prog. Brain Res.*, 50:653–665.

McCrea, R. A., Yoshida, K., Berthoz, A., and Baker, R. (1980): Eye movement related activity and morphology of second order vestibular neurons terminating in the cat abducens nucleus. *Exp. Brain Res.*, 40:468–473.

McDonagh, J. C., Binder, M. D., Reinking, R. M., and Stuart, D. G. (1980): Tetrapartite classification of motor units of cat tibialis posterior. *J. Neurophysiol.*, 44:696–712.

McDonald, W. I. (1963): The effects of experimental demyelination on conduction in peripheral nerve. *Brain*, 86:481–500.

McDonald, W. I., and Sears, T. A. (1970): The effects of experimental demyelination on conduction in the central nervous system. *Brain*, 93:538–598.

McGeer, P. J., McGeer, E. G., Singh, V. K., and Chase, W. H. (1974): Choline acetyltransferase localization in the central nervous system by immunohistochemistry. *Brain Res.*, 81:373–379.

McGrath, P. A., Sharav, Y., Dubner, R., and Gracely, R. H. (1981): Masseter inhibitory periods and sensations evoked by electrical tooth pulp stimulation. *Pain*, 10:1–17.

McIntyre, A. K. (1951): Afferent limb of the myotatic reflex arc. *Nature*, 168:168–169.

McIntyre, A. K., Proske, U., and Tracey, D. J. (1978): Afferent fibres from muscle receptors in the posterior nerve of the cat's knee joint. *Exp. Brain Res.*, 33:415–424.

McIntyre, A. K., and Robinson, R. G. (1959): Pathway for the jaw-jerk in man. *Brain*, 82:468–474.

McKenna, T. M., Whitsel, B. L., Dreyer, D. A., and Metz, C. B. (1981): Organization of cat anterior parietal cortex: relations among cytoarchitecture, single neuron functional properties and interhemispheric connectivity. *J. Neurophysiol.*, 45:667–697.

McKeon, B., and Burke, D. (1980): Identification of muscle spindle afferents during *in vivo* recordings in man. *Electroencephalogr. Clin. Neurophysiol.*, 48:606–608.

McKinley, J. C., and Berkwitz, N. J. (1928): Quantitative studies on human muscle tonus. *Arch. Neurol. Psychiatry*, 19:1036–1055.

McLellan, D. L. (1973): Effect of baclofen upon monosynaptic and tonic vibration reflexes in patients with spasticity. *J. Neurol. Neurosurg. Psychiatry*, 36:555–560.

McLellan, D. L. (1977): Co-contraction and stretch reflexes in spasticity during treatment with baclofen. *J. Neurol. Neurosurg. Psychiatry*, 40:30–38.

McLeod, J. G., and Penny, R. (1969): Vincristine neuropathy. An electrophysiological and histological study. *J. Neurol. Neurosurg. Psychiatry*, 32:297–304.

McLeod, J. G., and Wray, S. H. (1966): An experimental study of the F wave in the baboon. *J. Neurol. Neurosurg. Psychiatry*, 29:196–200.

McNamara, D. C. (1976): Electrodiagnosis at median occlusal position for human subjects with mandibular joint syndrome. *Arch. Oral Biol.*, 21:325–328.

McNamara, D. C. (1977): Occlusal adjustment for a physiologically balanced occlusion. *J. Prosthet. Dent.*, 38:284–293.

McWilliam, P. N. (1975): The incidence and properties of beta-axons to muscle spindles in the cat hind limb. *Q. J. Exp. Physiol.*, 60:25–36.

McWilliams, R., and Lynch, G. (1979): Terminal proliferation in the partially deafferented dentate gyrus: time course for the appearance and removal of degeneration and the replacement of last terminals. *J. Comp. Neurol.*, 187:191–198.

Maeda, M., Magherini, P. C., and Precht, W. (1977): Functional organization of vestibular and visual inputs to neck and forelimb motoneurons in the frog. *J. Neurophysiol.*, 40:225–243.

Maeda, M., Shimazu, H., and Shinoda, Y. (1971): Rhythmic activities of secondary vestibular efferent fibers recorded within the abducens nucleus during vestibular nystagmus. *Brain Res.*, 34:361–365.

Maeda, M., Shimazu, H., and Shinoda, Y. (1972): Nature of synaptic events in cat abducens motoneurons at slow and quick phase of vestibular nystagmus. *J. Neurophysiol.*, 35:279–296.

Maekawa, K., and Takeda, T. (1976): Electrophysiological identification of the climbing and mossy fiber pathways from the rabbit's retina to the contralateral cerebellar flocculus. *Brain Res.*, 109:169–174.

Magherini, P. C., Pompeiano, O., and Seguin, J. J. (1973): The effect of stimulation of Golgi tendon organs and spindle receptors from hindlimb extensor muscles on supraspinal descending inhibitory mechanisms. *Arch. Ital. Biol.*, 111:24–57.

Magherini, P. C., Thoden, U., and Pompeiano, O. (1971): Spinal-bulbo-spinal reflex inhibition of monosynaptic extensor reflexes of hindlimb in cats. *Arch. Ital. Biol.*, 109:110–129.

Magladery, J. W. (1955): Some observations on spinal reflexes in man. *Pfluegers Arch.*, 261:302–321.

Magladery, J. W., and McDougal, D. B. (1950): Electrophysiological studies of nerve and reflex activity in normal man. I. Identification of certain reflexes in the electromyogram and the conduction velocity of peripheral nerve fibres. *Bull. Johns Hopkins Hosp.*, 86:265–290.

Magladery, J. W., McDougal, D. B., and Stoll, J. (1950): Electrophysiological studies of nerve and reflex activity in normal man. The effects of peripheral ischemia. *Bull. Johns Hopkins Hosp.*, 86:291–317.

Magladery, J. W., Porter, W. E., Park, A. M., and Teasdall, R. D. (1951): Electrophysiological studies of nerve and reflex activity in normal man. IV. The Twoneurone reflex and identification of certain action potentials from spinal roots and cord. *Bull. Johns Hopkins Hosp.*, 88:499–519.

Magladery, J. W., and Teasdall, R. D. (1961): Corneal reflexes: an electromyographic study in man. *Arch. Neurol.*, 5:269–274.

Magladery, J. W., Teasdall, R. D., Park, A. M., and Languth, H. W. (1952): Electrophysiological studies of reflex activity in patients with lesions of the nervous system. I. A comparison of spinal motoneurone excitability following afferent nerve volleys in

normal persons and patients with upper motor neurone lesions. *Bull. Johns Hopkins Hosp.*, 91:219–244.
Magnus, R. (1924): *Körperstellung.* Springer, Berlin.
Magnus, R. (1926): Cameron prize lectures on some results of studies in the physiology of posture. *Lancet*, 211:531–536;585–588.
Magnus, R., and de Kleijn, A. (1912): Die Abhängigkeit des Tonus der Extremitätenmuskeln von der Kopfstellung. *Pfluegers Arch.*, 145:455–548.
Magoun, H. W. (1963): *The Waking Brain,* Charles C Thomas, Springfield, Illinois.
Mai, J. (1978): Depression of spasticity by alpha-adrenergic blockade. *Acta Neurol. Scand.*, 57:65–76.
Mai, J., and Pedersen, E. (1976): Clonus depression by propanolol. *Acta Neurol. Scand.*, 53:395–398.
Malcolm, D. S. (1951): Methods of measuring reflex times applied in sciatica and other conditions due to nerve root compression. *J. Neurol. Neurosurg. Psychiatry*, 14:15–24.
Malcolm, R., and Melvill Jones, G. (1973): Erroneous perception of vertical motion by humans seated in the upright position. *Acta Otolaryngol.*, 77:274–283.
Malis, L. I., Pribram, K. H., and Kruger, L. (1953): Action potentials in motor cortex evoked by peripheral nerve stimulation. *J. Neurophysiol.*, 16:161–167.
Mancall, E. L. (1975): Late (acquired) cortical cerebellar atrophy. In: *Handbook of Clinical Neurology,* Vol. 21, edited by P. J. Vinken and G. W. Bruyn, pp. 477–508. North-Holland, Amsterdam.
Mandl, G., and Melvill Jones, G. (1979): Rapid visual vestibular interaction during visual tracking in strobe light. *Brain Res.*, 165:133–138.
Mandl, G., Melvill Jones, G., and Cynader, M. (1981): Adaptability of the vestibuloocular reflex to vision reversal in strobe reared cats. *Brain Res.*, 209:35–45.
Mann, M. D. (1979): Sets of neurons in somatic cerebral cortex of the cat and their ontogeny. *Brain Res.*, 1:3–45.
Mannard, A., and Stein, R. B. (1973): Determination of the frequency response of isometric soleus muscle in the cat using random nerve stimulation. *J. Physiol.*, 229:275–296.
Mano, N., Oshima, T., and Shimazu, H. (1968): Inhibitory commissural fibers interconnecting the bilateral vestibular nuclei. *Brain Res.*, 8:378–382.
Mano, N., and Yamamoto, K. I. (1980): Simple spike activity of cerebellar cortex related to visually guided wrist tracking movement in the monkey. *J. Neurophysiol.*, 47:713–728.
Mano, T., Takagi, S., and Mitarai, G. (1976): Caractéristiques des décharges unitaires afférentes des fuseaux musculaires chez l'homme. *C. R. Soc. Biol.*, 170:500–503.
Mano, T., Yamazaki, Y., and Takagi, S. (1979): Muscle spindle activity in Parkinsonian rigidity. *Acta Neurol. Scand.*, 60(Suppl. 73):176.
Marczynski, A. T. J. (1969): Invited discussion: Postreinforcement synchronization and the cholinergic system. *Fed. Proc.*, 28:132–134.
Marey, E. J. (1901): La locomotion animale. *Traité de Physiologie Biologique,* Vol. 1. Ballière, Paris.

Marinacci, A. A. (1968): *Applied Electromyography.* Lea & Febiger, Philadelphia.
Mark, R. F. (1963): Tonic stretch reflexes in calf muscles of normal human subjects. *Nature*, 199:50–52.
Mark, R. F., Coquery, J. M., and Paillard, J. (1968): Autogenetic reflex effects of slow or steady stretch of the calf muscles in man. *Exp. Brain Res.*, 6:130–145.
Marks, A. F. (1969): Bullfrog nerve regeneration into porous implants. *Anat. Rec.*, 163:226.
Marr, D. (1969): A theory of cerebellar cortex. *J. Physiol. (Lond.),* 202:437–470.
Marsden, C. D. (1978): The mechanisms of physiological tremor and their significance for pathological tremors. In: *Progress in Clinical Neurophysiology, Vol. 5, Physiological Tremor, Pathological Tremors and Clonus,* edited by J. E. Desmedt, pp. 1–16. Karger, Basel.
Marsden, C. D. (1980): The physiology of myoclonus and its relation to epilepsy. In: *Research and Clinical Forums,* Vol. 2, pp. 31–45. Pergamon, Oxford.
Marsden, C. D. (1981): Movement disorders. In: *Oxford Textbook of Medicine,* Chapter 20. Pergamon, Oxford.
Marsden, C. D. (1982): The mysterious motor function of the basal ganglia. *Neurology, (Minneap.),* 32:514–539.
Marsden, C. D., and Hallett, M. (1979): Ballistic flexion movements of the human thumb. *J. Physiol. (Lond.),* 294:33–50.
Marsden, C. D., Hallett, M., and Fahn, S. (1982): Nosology and pathophysiology of myoclonus. In: *Movement Disorders,* edited by C. D. Marsden and S. Fahn, pp. 196–248. Butterworth, London.
Marsden, C. D., and Harrison, M. J. G. (1974): Idiopathic torsion dystonia (dystonia musculorum deformans). A review of forty-two patients. *Brain*, 97:793–810.
Marsden, C. D., Meadows, J. C., and Merton, P. A. (1969): Muscular wisdom. *J. Physiol. (Lond.),* 200:15P.
Marsden, C. D., Meadows, J. C., and Merton, P. A. (1976): Fatigue in human muscle in relation to the number and frequency of motor impulses. *J. Physiol. (Lond.),* 258:94–95P.
Marsden, C. D., Merton, P. A., and Morton, H. B. (1972): Servo action in human voluntary movement. *Nature*, 238:140–143.
Marsden, W., Merton, P. A., and Morton, H. B. (1973): Is the human stretch reflex cortical rather than spinal? *Lancet*, 1:759–761.
Marsden, C. D., Merton, P. A., and Morton, H. B. (1976a): Servo action in the human thumb. *J. Physiol. (Lond.),* 257:1–44.
Marsden, C. D., Merton, P. A., and Morton, H. B. (1976b): Stretch reflex and servo action in a variety of human muscles. *J. Physiol. (Lond.),* 259:531–560.
Marsden, C. D., Merton, P. A., and Morton, H. B. (1976c): Servo action in human posture. *J. Physiol. (Lond.),* 263:187–188P.
Marsden, C. D., Merton, P. A., and Morton, H. B. (1977a): The sensory mechanism of servo action. *J. Physiol. (Lond.),* 265:531–560.
Marsden, C. D., Merton, P. A., and Morton, H. B.

(1978a): Anticipatory postural responses in the human subject. *J. Physiol. (Lond.),* 275:47–48P.

Marsden, C. D., Merton, P. A., and Morton, H. B. (1979a): Sensitivity, efficacy and disappearance of servo action in human muscle. *J. Physiol. (Lond.),* 292:56P.

Marsden, C. D., Merton, P. A., and Morton, H. P. (1981a): Maximal twitches from stimulation of the motor cortex in man. *J. Physiol. (Lond.),* 312:5P.

Marsden, C. D., Merton, P. A., and Morton, H. B. (1981b): Human postural responses. *Brain,* 104:513–534.

Marsden, C. D., Merton, P. A., Morton, H. B., and Adam, J. (1977b): The effect of posterior column lesions on servo responses from the human long thumb flexor. *Brain,* 100:185–200.

Marsden, C. D., Merton, P. A., Morton, H. B., and Adam, J. (1977c): The effect of lesions of the sensorimotor cortex and capsular pathways on servo responses from human long thumb flexor. *Brain,* 100:503–526.

Marsden, C. D., Merton, P. A., Morton, H. B., and Adam, J. (1978b): The effect of lesions of the central nervous system on long-latency stretch reflexes in the human thumb. In: *Progress in Clinical Neurophysiology, Vol. 4, Cerebral Motor Control in Man: Long Loop Mechanisms,* edited by J. E. Desmedt, pp. 334–341. Karger, Basel.

Marsden, C. D., Merton, P. A., Morton, H. B., et al. (1978c): Automatic and voluntary responses to muscle stretch in man. In: *Progress in Clinical Neurophysiology, Vol. 4, Cerebral Motor Control in Man: Long Loop Mechanisms,* edited by J. E. Desmedt, pp. 167–177. Karger, Basel.

Marsden, C. D., Merton, P. A., Morton, H. B., et al. (1981c): Reliability and efficacity of the long latency stretch reflex in the human thumb. *J. Physiol. (Lond.),* 316:47–60.

Marsden, C. D., Rothwell, J. C., and Traub, M. M. (1979b): Effect of thumb anaesthesia on weight perception, muscle activity and the stretch reflex in man. *J. Physiol. (Lond.),* 294:303–315.

Marshall, W. H., Talbot, S. A., and Ades, H. W. (1943): Cortical response of the anesthetized cat to gross photic and electrical afferent stimulation. *J. Neurophysiol.,* 6:1–15.

Martensson, A., and Knutsson, E. (1981): Effects of baclofen on different components in the disorganized motor control in spastic paresis. In: *Therapie der Spastik,* edited by H. J. Bauer, W. P. Koella, and A. Struppler, pp. 193–199. Verlag Wissenchaften, Munich.

Martin, B., Gauthier, G. M., Roll, J. P., et al. (1980): Effects of whole-body vibration on standing posture in man. *Aviat. Space Environ. Med.,* 51:778–787.

Martin, H. L., Durant, R. C., Hudak, W. V., and Friess, S. L. (1972): Alterations in chemical blockade of cat gastrocnemius soleus tissues under hyperbaric He-O_2 atmospheres. *Toxicol. Appl. Pharmacol.,* 23:82–90.

Martin, J. P. (1965): Tilting reactions and disorders of the basal ganglia. *Brain,* 88:855–874.

Martinelli, P., De Pasqua, V., and Delwaide, P. J. (1979): Réflexes du muscle soléaire évoqués par la stimulation du nerf tibial postérieur versus réflexe tendineux chez l'homme. *Rev. Electroencephalogr. Neurophysiol.,* 9:72–76.

Martinelli, P., and Montagna, R. (1979): Conditioning of the H reflex by stimulation of the posterior tibial nerve in Parkinson's disease. *J. Neurol. Neurosurg. Psychiatry,* 42:701–704.

Martinis, A. S., Sankelson, B., Radke, J., and Adib, F. (1980): Effects of the Myomonitor on cardiac pacemakers. *J. Am. Dent. Assoc.,* 100:203–205.

Masdeu, J. C., Schoene, W. C., and Funkenstein, H. (1978): Aphasia following infarction of the left supplementary motor area. *Neurology (Minneap.),* 20:1220–1223.

Massion, J., and Sasaki, K. (1979): Cerebro-cerebellar interaction. In: *Cerebro-cerebellar Interactions,* edited by J. Massion and K. Sasaki, pp. 261–287. Elsevier, New York.

Matsumura, M. (1979): Intracellular synaptic potentials of primate motor cortex neurons during voluntary movement. *Brain Res.,* 163:33–48.

Matsumura, M., and Kubota, K. (1979): Cortical projection of hand-arm motor area from post-arcuate area in macaque monkey: a histological study of retrograde transport of horseradish peroxidase. *Neurosci. Lett.,* 11:241–246.

Matsumura, K., and Hamada, I. (1981): Characteristics of the ipsilateral movement related neuron in the motor cortex of the monkey. *Brain Res.,* 204:29–42.

Matsunami, K., and Kubota, K. (1972): Muscle afferents of trigeminal mesencephalic tract nucleus and mastication in chronic monkeys. *Jpn. J. Physiol.,* 22:545–555.

Matsuoka, S., Waltz, J. M., Terada, C., Ikeda, T., and Cooper, I. S. (1966): A computer technique for evaluation of recovery cycle of the H reflex in the abnormal movement disorders. *Electroencephalogr. Clin. Neurophysiol.,* 21:496–500.

Matsushita, K., Goto, Y., Okamoto, T., et al. (1974): An EMG study of sprint running. *Res. J. Phys. Educ.,* 19:147–156.

Matthews, B. (1975): Mastication. In: *Applied Physiology of the Mouth,* edited by C. Lavelle, pp. 199–242. Wright, Bristol.

Matthews, B. (1976): Reflexes elicitable from the jaw muscles in man. In: *Mastication,* edited by D. J. Anderson and B. Matthews, pp. 139–146. Wright, Bristol.

Matthews, B., and Whiteside, T. C. D. (1960): Tendon reflexes in free-fall. *Proc. R. Soc. Lond. (Biol.),* 153:195–204.

Matthews, D. A., Cotman, C. W., and Lynch, G. (1976): An electron microscopic study of lesion-induced synaptogenesis in the dentate gyrus of the adult rat. I. Magnitude and time course of degeneration. II. Reappearance of morphologically normal synaptic contacts. *Brain Res.,* 115:1–21; 23–41.

Matthews, P. B. (1962): The differentiation of two types of fusimotor fibre by their effects on the dynamic response of muscle primary endings. *Q. J. Exp. Physiol.,* 47:324–333.

Matthews, P. B. (1963): The response of de-efferented

muscle spindle receptors to stretching at different velocities. *J. Physiol. (Lond.),* 168:660–678.

Matthews, P. B. (1966): The reflex excitation of the soleus muscle of the decerebrate cat, caused by vibration applied to its tendon. *J. Physiol. Lond.,* 184:450–472.

Matthews, P. B. (1972): *Mammalian Muscle Receptors and Their Central Actions.* Williams & Wilkins, Baltimore.

Matthews, P. B. (1980): Developing views on the muscle spindle. In: *Progress in Clinical Neurophysiology, Vol. 8, Spinal and Supraspinal Mechanisms of Voluntary Motor Control and Locomotion,* edited by J. E. Desmedt, pp. 12–27, Karger, Basel.

Matthews, P. B., and Rushworth, G. (1957): The relative sensitivity of muscle nerve fibres to procaine. *J. Physiol. (Lond.),* 135:263–269.

Matthews, P. B., and Stein, R. B. (1969): The sensitivity of muscle spindle afferents to small sinusoidal changes of length. *J. Physiol. (Lond.),* 202:59–82.

Matthews, W. B. (1965): The action of chlorproethazine on spasticity. *Brain,* 88:1057–1064.

Matthews, W. B. (1966): Ratio of maximum H reflex to maximum M response as a measure of spasticity. *J. Neurol. Neurosurg. Psychiatry,* 29:201–204.

Matthews, W. B., Rushworth, G., and Wakefield, G. S. (1972): Dimethothiazine in spasticity. *Acta Neurol. Scand.,* 48:635–644.

Mauguiere, F., Desmedt, J. E., and Courjon, J. (1983): Astereognosis and dissociated loss of frontal or parietal components of somatosensory evoked potentials in hemispheric lesions: detailed correlations with CT scan and clinical signs. *Brain,* 106:271–311.

Mauritz, K. H., Dichgans, J., and Hufschmidt, A. (1977): The angle of visual roll motion determines displacement of subjective visual vertical. *Percept. Psychophysiol.* 22:557–562.

Mauritz, K. H., Dichgans, J., and Hufschmidt, A. (1979): Quantitative analysis of stance in late cortical cerebellar atrophy of the anterior lobe and other forms of cerebellar ataxia. *Brain,* 102:462–482.

Mauritz, K. H., and Dietz, V. (1980): Characteristics of postural instability induced by ischemic blocking of leg afferents. *Exp. Brain Res.,* 38:117–119.

Mauritz, K. H., Dietz, V., and Haller, M. (1980): Balancing as a clinical test in the differential diagnosis of sensorymotor disorders. *J. Neurol. Neurosurg. Psychiatry,* 43:407–412.

Mauritz, K. H., Schmitt, C., and Dichgans, J. (1981): Delayed and enhancement of long latency reflexes as the possible cause of postural tremor in late cerebellar atrophy. *Brain,* 104:97–116.

Mawdsley, C., and Mayer, R. F. (1965): Nerve conduction in alcoholic neuropathy. *Brain,* 88:335–356.

Mayer, D. J. (1975): Pain inhibition by electrical brain stimulation: comparison to morphine. *Neurosci. Res. Program Bull.,* 13:94–99.

Mayer, D. J., and Liebeskind, J. C. (1974): Pain reduction by focal electrical stimulation of the brain: an anatomical and behavioral analysis. *Brain Res.,* 68:73–93.

Mayer, D. J., Price, D. D., and Rafii, A. (1977): Antagonism of acupuncture analgesia in man by the narcotic antagonist naloxone. *Brain Res.,* 121:368–372.

Mayer, R. F., and Denny-Brown, D. (1964): Conduction velocity in peripheral nerve during experimental demyelination in the cat. *Neurology (Minneap.),* 14:714–726.

Mayer, R. F., and Feldman, R. G. (1967): Observations on the nature of the F wave in man. *Neurology (Minneap.),* 17:147–156.

Mayer, R. F., and Mawdsley, C. (1965): Studies in man and cat of the significance of the H wave. *J. Neurol. Neurosurg. Psychiatry,* 28:201–211.

Mayer, R. F., and Mosser, R. S. (1969): Excitability of motoneurones in infants. *Neurology (Minneap.),* 19:932–945.

Mayer, R. F., and Mosser, R. S. (1973): Maturation of human reflexes. In: *New Developments in Electromyography and Clinical Neurophysiology,* Vol. 3, edited by J. E. Desmedt, pp. 294–307. Karger, Basel.

Mayer, R. F., and Young, J. L. (1979): The effects of hemiplegia on single motor units in man. *Acta Neurol. Scand.,* 60(Suppl.73):166–167.

Megaw, E. D. (1972): Direction and extent uncertainty in step input tracking. *J. Motor Behavior,* 4:171–186.

Megaw, E. D. (1974): Possible modification to a rapid on going programmed manual response. *Brain Res.,* 71:425–441.

Mehler, W. R., and Nauta, W. J. (1974): Connections of the basal ganglia and of the cerebellum. *Confin. Neurol.,* 36:205–222.

Meier-Ewert, K., Gleitsman, K., and Reiter, F. (1974): Acoustic jaw reflex in man: its relationship to other brain-stem and microreflexes. *Electroencephalogr. Clin. Neurophysiol.,* 36:629–637.

Meier-Ewert, K., Humme, U., and Dahn, J. (1972): New evidence favouring long-loop reflexes in man. *Arch. Psychiatr. Nervenkr.,* 215:121–128.

Meier-Ewert, K., Schmidt, C., Nordmann, G., et al. (1973): Averaged muscle-responses to repetitive sensory stimuli. In: *New Developments in Electromyography and Clinical Neurophysiology,* Vol. 3, edited by J. E. Desmedt, pp. 767–772. Karger, Basel.

Meinck, H. M. (1976): Deszendierende long-loop Reflexe im menschlichen Rückenmark. 1. Förderung des Triceps surae H-Reflexes durch Reizung von Afferenzen der oberen Extremität. *Z. EEG-EMG,* 7:146–150.

Meinck, H. M. (1980): Facilitation and inhibition of the human H reflex as a function of the amplitude of the control reflex. *Electroencephalogr. Clin. Neurophysiol.,* 48:203–211.

Meinck, H. M., Piesiur-Strehlow, B., and Koehler, W. (1981): Some principles of flexor reflex generation in human leg muscles. *Electroencephalogr. Clin. Neurophysiol.,* 52:140–150.

Meinck, H. M., and Piesiur-Strehlow, B. (1981): Reflexes evoked in leg muscles from arm afferents: a propriospinal pathway in man? *Exp. Brain Res.,* 43:78–86.

Meltzer, G. E., Hunt, R. S., and Landau, W. M. (1963): Fusimotor function. Part III. The spastic monkey. *Arch. Neurol.,* 9:133–136.

Melvill Jones, G. (1964): Predominance of anticompensatory oculomotor response during rapid head rotation. *Aerospace Med.*, 35:965–988.

Melvill Jones, G. (1968): From land to space in a generation: An evolutionary challenge. *Aerospace Med.*, 39:1271–1283.

Melvill Jones, G. (1974): The functional significance of semicircular canal size. In: *Handbook of Sensory Physiology*, Vol. VI/1, edited by H. H. Kornhuber, pp. 171–184. Springer, Berlin.

Melvill Jones, G. (1977): Plasticity in the adult vestibulo-ocular reflex arc. *Philos. Trans. R. Soc. Lond. (Biol.)*, 278:319–334.

Melvill Jones, G., and Davies, P. R. (1976): Adaptation of cat vestibuloocular reflex to 200 days of optically reversed vision. *Brain Res.*, 103:551–554.

Melvill Jones, G., and Drazin, D. H. (1962): Oscillatory motion in flight. In: *Human Problems of Supersonic and Hypersonic Flight*, edited by A. H. Barbour and D. Whittingham, pp. 134–151. Pergamon Press, London.

Melvill Jones, G., and Dejong, J. D. (1971): Dynamic characteristics of saccadic eye movements in Parkinson's disease. *Exp. Neurol.*, 31:17–31.

Melvill Jones, G., and Gonshor, A. (1975): Goal-directed flexibility in the vestibulo-ocular reflex arc, In: *Basic Mechanisms of Ocular Motility and Their Clinical Implications*, edited by G. Lennerstrand and P. Bach-y-Rita, pp. 227–245. Pergamon Press, Oxford.

Melvill Jones, G., and Gonshor, A. (1982): Oculomotor response to rapid head oscillation (0.5–5.0 Hz) after prolonged adaptation to vision-reversal: 'simple' and 'complex' effects. *Exp. Brain Res.*, 45:45–58.

Melvill Jones, G., and Mandl, G. (1979): Effects of strobe light on adaptation of vestibulo-ocular reflex to vision reversal. *Brain Res.*, 164:300–303.

Melvill Jones, G., and Mandl, G. (1981): Motion sickness due to vision reversal: its absence in stroboscopic light. *Ann. NY Acad. Sci.*, 374:303–311.

Melvill Jones, G., Mandl, G., Cynader, M., and Outerbridge, J. S. (1981): Eye oscillations in strobe reared cats. *Brain Res.*, 209:47–60.

Melvill Jones, G., and Watt, D. G. D. (1971a): Observations on the control of stepping and hopping movements in man. *J. Physiol. (Lond.)*, 219:709–729.

Melvill Jones, G., and Watt, D. G. D. (1971b): Muscular control of landing from unexpected falls in man. *J. Physiol. (Lond.)*, 219:729–737.

Melvill Jones, G., and Young, L. R. (1978): Subjective detection of vertical acceleration: a velocity-dependent response? *Acta Otolaryngol.*, 85:45–53.

Melzack, R., and Bromage, P. R. (1973): Experimental phantom limbs. *Exp. Neurol.*, 39:261–269.

Melzack, R., and Wall, P. D. (1965): Pain mechanisms: a new theory. *Science*, 150:971–979.

Mendell, L. M., and Henneman, E. (1971): Terminals of single Ia fibers: location, density and distribution within a pool of 300 homonymous motoneurons. *J. Neurophysiol.*, 34:171–187.

Menetrey, D., Giesler, G. J., Jr., and Besson, J. M. (1977): An analysis of response properties of spinal cord dorsal horn neurones to non-noxious and noxious stimuli in the spinal rat. *Exp. Brain Res.*, 27:15–33.

Merton, P. A. (1951): The silent period in a muscle of the human hand. *J. Physiol. (Lond.)*, 114:183–198.

Merton, P. A. (1953): Speculations in the servo-control of movement. In: *The Spinal Cord*, Ciba Foundation Symposium, edited by J. L. Malcolm and J. A. B. Gray, pp. 84–91. Churchill, London.

Merton, P. A. (1954): Voluntary strength and fatigue. *J. Physiol. (Lond.)*, 123:553–564.

Merton, P. A. (1957): The derivation of the human muscle action potential and its relationship to contraction and fatigue. *International Congress Series*, No. 11, p. 13. Excerpta Medica, Amsterdam.

Merton, P. A. (1972): How we control the contraction of our muscles. *Sci. Am.*, 226:30–37.

Merton, P. A. (1981): Neurophysiology on man. *J. Neurol. Neurosurg. Psychiatry*, 44:861–870.

Merton, P. A., Hill, D. K., and Morton, H. B. (1981): Indirect and direct stimulation of fatigued human muscle. In: *Human Muscle Fatigue: Physiological Mechanisms*, Ciba Foundation Symposium 82, pp. 120–129. Pitman, London.

Merton, P. A., and Morton, H. B. (1980): Stimulation of the cerebral cortex in the intact human subject. *Nature (Lond.)*, 285:227.

Messina, C. (1975): On the nature and meaning of the blink reflex early response. *Electroencephalogr. Clin. Neurophysiol.*, 15:119–124.

Messina, C., and Micalizzi, V. (1970): I riflessi trigemino-facciali nel corso del coma insulinico. *Acta Neurol.*, 25:357–361.

Messina, C., and Quattrone, A. (1973): Comportamento dei riflessi trigemino-facciali in soggetti con lesioni emisferiche. *Riv. Neurol.*, 43:379–386.

Messina, C. D., Rosa, A. E., and Tomasello, F. (1972): Habituation of blink reflexes in parkinsonian patients under levodopa and Amantadine treatment. *J. Neurol. Sci.*, 17:141–148.

Mesulam, M. M., and Mufson, E. J. (1980): The rapid anterograde transport of horseradish peroxidase. *Neuroscience*, 5:1277–1286.

Meyer, M., and Adorjáni, C. (1980): Tonic stretch reflex for quantification of pathological muscle tone. In: *Spasticity: Disordered Motor Control*, edited by R. Feldman, R. Young, and W. Koella. Year Book, Chicago.

Meyer-Lohmann, J., Conrad, B., Matsunami, K., and Brooks, V. B. (1975): Effects of dentate cooling on precentral unit activity following torque pulse injections into elbow movements. *Brain Res.*, 94:237–251.

Meyer-Lohmann, J., Hore, J., and Brooks, V. B. (1977): Cerebellar participation on generation of prompt arm movements. *J. Neurophysiol.*, 40:1038–1050.

Meyer-Lohmann, J., Riebold, W., and Robrecht, D. (1974): Mechanical influence of extrafusal muscle on the static behaviour of de-efferented primary muscle spindle endings in cat. *Pfluegers Arch.*, 352:267–278.

Miglietta, O. E. (1973): The F response after transverse myelotomy. In: *New Developments in Electromyography and Clinical Neurophysiology*, Vol. 3, edited by J. E. Desmedt, pp. 323–327. Karger, Basel.

Mihaloff, G. A., and Castro, A. J. (1981): Autoradiographic and electromicroscopic degeneration evidence for axonal sprouting in the rat corticopontine system. *Neurosci. Lett.* 21:263–273.

Miles, F. A. (1974): Single unit firing patterns in the vestibular nuclei related to voluntary eye movements and passive body rotation in conscious monkeys. *Brain Res.,* 71:215–224.

Miles, F. A., Braitman, D. J., and Dow, B. M. (1980): Long-term adaptive changes in primate vestibuloocular reflex. IV. Electrophysiological observations in flocculus of adapted reflex. *J. Neurophysiol.,* 43:1406–1425.

Miles, F. A., and Evarts, E. V. (1979): Concepts of motor organization. *Ann. Rev. Physiol.,* 30:327–362.

Miles, F. A., and Fuller, J. H. (1974): Adaptive plasticity in the vestibulo-ocular responses of the rhesus monkey. *Brain Res.,* 80:512–516.

Miles, F. A., and Fuller, J. H. (1975): Visual tracking and the primate flocculus. *Science,* 189:1000–1002.

Miles, F. A., and Lisberger, S. G. (1981): Plasticity in the vestibulo-ocular reflex: A new hypothesis. *Ann. Rev. Neurosci.,* 4:273–299.

Millar, J. (1973): Joint afferent fibres responding to muscle stretch, vibration and contraction. *Brain Res.,* 63:380–383.

Miller, A. D., and Brooks, V. B. (1981): Late muscular responses to arm perturbations persist during supraspinal dysfunctions in monkeys. *Exp. Brain Res.,* 41:146–158.

Miller, B., and Lund, R. D. (1975): The pattern of retinotectal connections in albino rats can be modified by fetal surgery. *Brain Res.,* 91:119–125.

Miller, G. A. (1956): The magical number seven, plus or minus two: Some limit on our capacity for processing information. *Psychol. Rev.,* 63:81–97.

Miller, K. W. (1972): Intert gas narcosis and animals under high pressure. *Symp. Soc. Exp. Biol.,* 26:363–378.

Miller, K. W., Patton, W. D., Smith, R. A., and Smith, E. B. (1973): The pressure reversal of general anaesthesia and the critical volume hypothesis. *Mol. Pharmacol.,* 9:131–143.

Miller, S., and Scott, P. D. (1980): Spinal generation of movement in a single limb: Functional implications of a based on the cat. In: *Progress in Clinical Neurophysiology, Vol. 8, Spinal and Supraspinal Mechanisms of Voluntary Motor Control and Locomotion,* edited by J. E. Desmedt, pp. 263–281. Karger, Basel.

Milner-Brown, H. S., Girvin, J. P., and Brown, W. F. (1975a): The effects of motor cortical stimulation on the excitability of spinal motoneurones in man. *Can. J. Neurol. Sci.,* 2:245–253.

Milner-Brown, H. S., and Stein, R. B. (1975): The relation between the surface electromyogram and muscular force. *J. Physiol. (Lond.),* 246:549–569.

Milner-Brown, H. S., Stein, R. B., and Lee, R. G. (1975b): Synchronization of human motor units: possible role of exercise and supraspinal reflexes. *Electroencephalogr. Clin. Neurophysiol.,* 38:245–254.

Milner-Brown, H. S., Stein, R. B., Lee, R. G., and Brown, W. F. (1981): Motor unit recruitment in patients with neuromuscular disorders. In: *Progress in Clinical Neurophysiology, Vol. 9, Motor Unit Types, Recruitment and Plasticity in Health and Disease,* edited by J. E. Desmedt, pp. 305–318. Karger, Basel.

Milner-Brown, H. S., Stein, R. B., and Yemm, R. (1973a): The contractile properties of human motor units during voluntary isometric contractions. *J. Physiol. (Lond.),* 228:285–306.

Milner-Brown, H. S., Stein, R. B., and Yemm, R. (1973b): The orderly recruitment of human motor units during voluntary isometric contractions. *J. Physiol. (Lond.),* 230:359–370.

Milner-Brown, H. S., Stein, R. B., and Yemm, R. (1973c): Changes in firing rate of human motor units during linearly changing voluntary contractions. *J. Physiol. (Lond.),* 230:371–390.

Mishelevich, D. J. (1969): Repetitive firing to current in cat motoneurons as a function of muscle unit twitch type. *Exp. Neurol.,* 25:401–409.

Mizuno, Y., Tanaka, R., and Yanagisawa, N. (1971): Reciprocal group I inhibition on triceps surae motoneurons in man. *J. Neurophysiol.,* 34:1010–1017.

Moffie, D. (1971): Late results of bulbar trigeminal tractomy. *J. Neurol. Neurosurg. Psychiatry.,* 34:270–274.

Moffie, D., Ongerboer De Visser, B. W., and Van Der Sande, J. J. (1979): Pure motor hemiplegia; localization of the pyramidal tract in the internal capsule. *Ned. Tijdschr. Geneeskd.,* 123:822–825.

Mohler, C. W., Goldberg, V. B., and Wurtz, R. H. (1973): Visual receptive fields of frontal eye field neurons. *Brain Res.,* 61:385–389.

Moldaver, J. (1973): Some comments on blink reflexes. In: *New Developments in Electromyography and Clin. Neurophysiology,* Vol. 3, edited by J. E. Desmedt, pp. 658–659. Karger, Basel.

Moll, L., and Kuypers, H. (1977): Premotor cortical ablations in monkeys: contralateral changes in visually guided reaching behavior. *Science,* 198:317–319.

Money, K. E., and Scott, J. W. (1962): Functions of separate sensory receptors of nonauditory labyrinth of the cat. *Am. J. Physiol.,* 202:1211–1220.

Monster, A. W. (1974): Spasticity and the effects of Dantrolene sodium. *Arch. Phys. Med.,* 55:373–383.

Monster, A. W. (1979): Firing rate behavior of human motor units during isometric voluntary contraction: relation to unit size. *Brain Res.,* 171:349–354.

Monster, A. W., and Chan, H. (1977): Isometric force production by motor units of extensor digitorum communis muscle in man. *J. Neurophysiol.,* 40:1432–1443.

Monster, A. W., Herman, R., and Altland, N. R. (1973): Effect of the peripheral and central 'sensory' component in the calibration of position. In: *New Developments in Electromyography and Clinical Neurophysiology,* Vol. 3, edited by J. E. Desmedt, pp. 383–403. Karger, Basel.

Moore, R. Y., Bjorklund, A., and Stenevi, U. (1971): Plastic changes in the adrenergic innervation of the rat septal area in response to denervation. *Brain Res.,* 33:13–35.

Mor, J., and Carmon, A. (1975): Laser emitted radiant heat for pain research. *Pain,* 1:233–237.

Morasso, P. (1981): Spatial control of arm movements. *Exp. Brain Res.,* 42:223–227.

Morasso, P., Bizzi, E., and Dichgans, J. (1973): Adjustment of saccade characteristics during head movements. *Exp. Brain Res.,* 16:492–500.

Morgan-Hughes, J. A. (1968): Experimental diphteritic neuropathy, a pathological and electrophysiological study. *J. Neurol. Sci.*, 7:157–175.

Mori, S., and Ishida, A. (1978): High frequency postural tremor and underlying neural mechanism. In: *Progress in Clinical Neurophysiology, Vol. 5, Physiological Tremors, Pathological Tremors and Clonus*, edited by J. E. Desmedt, pp. 51–65. Karger, Basel.

Mortimer, J. A., and Webster, D. D. (1978): Relationships between quantitative measures of rigidity and tremor and the electromyographic responses to load perturbations in unselected normal subjects and Parkinson patients. In: *Progress in Clinical Neurophysiology, Vol. 4, Cerebral Motor Control in Man: Long Loop Mechanisms*, edited by J. E. Desmedt, pp. 342–360. Karger, Basel.

Mortimer, J. A., and Webster, D. D. (1979): Evidence for a quantitative association between EMG stretch responses and Parkinsonian rigidity. *Brain Res.*, 162:169–173.

Mortimer, J. A., Webster, D. D., and Dukich, T. G. (1981): Changes in short and long latency stretch responses during the transition from posture to movement. *Brain Res.*, 229:337–351.

Moruzzi, G. (1972): The sleep waking cycle. *Ergeb. Physiol.*, 64:1–165.

Moruzzi, G., and Magoun, H. W. (1949): Brain stem reticular formation and activation of the EEG. *Electroencephalogr. Clin. Neurophysiol.*, 1:455–473.

Motter, B. C., and Mountcastle, V. B. (1981): The functional properties of the light sensitive neurons of the posterior parietal cortex studied in waking monkeys: focal sparing and opponent vector organization. *J. Neurosci.*, 1:3–26.

Mountcastle, V. B. (1957): Modality and topographic properties of single neurons of cat's somatic sensory cortex. *J. Neurophysiol.*, 20:408–434.

Mountcastle, V. B. (1979): An organizing principle for cerebral function: The unit module and the distributed system. In: *The Neurosciences. Fourth Study Program*, edited by F. O. Schmitt and F. G. Worden, pp. 21–42. MIT Press, Cambridge, Massachusetts.

Mountcastle, V. B., Lynch, J. C., Georgopoulos, A., et al. (1975): Posterior parietal association cortex of the monkey: command functions for operations within extrapersonal space. *J. Neurophysiol.*, 38:871–908.

Mountcastle, V. B., and Powell, T. P. S. (1959): Central nervous mechanisms subserving position sense and kinesthesis. *Bull. Johns Hopkins Hosp.*, 105:173–200.

Mowbray, G. H., and Rhoades, M. V. (1959): On the reduction of choice reaction times with practice. *Q. J. Exp. Psychol.*, 11:16–23.

Muakkassa, K. F., and Strick, P. L. (1979): Frontal lobe inputs to primate motor cortex: evidence for four somatotopically organized 'premotor' areas. *Brain Res.*, 177:176–182.

Muir, R. B., and Porter, R. (1973): The effect of a preceding stimulus on temporal facilitation at corticomotoneuronal synapses. *J. Physiol. (Lond.)*, 228:749–763.

Murakami, F., Fujito, Y., and Tsukahara, N. (1976): Physiological properties of the newly formed cortico-rubral synapses of red nucleus neurons due to collateral sprouting. *Brain Res.*, 103:147–151.

Murakami, F., Katsumaru, H., Saito, K., and Tsukahara, N. (1981): An electromicroscopic study of corticorubral projections to red nucleus neurons identified by intracellular injection of HRP. *Brain Res.*, 242:41–54.

Murphy, J. T., Kwan, H. C., MacKay, W. A., and Wong, Y. C. (1978): Spatial organization of precentral cortex in awake primates. *J. Neurophysiol.*, 41:1132–1139.

Murphy, J. T., Kwan, H. C., and Repeck, M. W. (1979): Functional significance of long loop reflex responses to limb perturbation. In: *Advances in Neurology*, Vol. 24, edited by L. J. Poirier, T. L. Sourkes, and P. Bedard, pp. 123–129. Raven Press, New York.

Murphy, J. T., Wong, Y. C., and Kwan, H. C. (1974): Distributed feedback systems for muscle control. *Brain Res.*, 71:495–505.

Murphy, J. T., Wong, Y. C., and Kwan, H. C. (1975): Afferent-efferent linkages in motor cortex for single forelimb muscles. *J. Neurophysiol.*, 38:990–1014.

Murphy, P. R. (1981): The recruitment order of α-motoneurones in the decerebrate rabbit. *J. Physiol. (Lond.)*, 315:59–67.

Murray, E. A., and Coulter, J. D. (1981): Organization of corticospinal neurons in the monkey. *J. Comp. Neurol.*, 195:339–365.

Murray, J. C., and Thompson, J. W. (1957): The occurrence and function of collateral sprouting the sympathetic nervous system of the cat. *J. Physiol. (Lond.)*, 135:133–162.

Murray, M., and Goldberger, M. E. (1974): Restitution of function and collateral sprouting in the cat spinal cord: The partially hemisected animal. *J. Comp. Neurol.*, 158:19–36.

Murray, M., Zimmer, J., and Raisman, G. (1979): Quantitative electron microscopic evidence for reinnervation in the adult rat interpeduncular nucleus after lesion of the fascicules retroflexus. *J. Comp. Neurol.*, 187:147–168.

Murray, M. P. (1967) Gait as a total pattern of movement. *Am. J. Phys. Med.*, 46:290–330.

Murray, M. P., Sepic, S. B., Garoner, G. M., and Downs, W. J. (1978): Walking patterns of men with Parkinsonism. *Am. J. Phys. Med.*, 57:278–294.

Murthy, K. S. K. (1978): Vertebrate fusimotor neurones and their influences on motor behaviour. *Prog. Neurobiol.*, 11:249–307.

Mustari, M. J., and Lund, R. D. (1976): An aberrant crossed visual corticotectal pathway in albino rats. *Brain Res.*, 112:37–44.

Muybridge, E. (1901): *The Human Figure in Motion*. Chapman and Hall, London.

Myers, D. R., and Moskowitz, G. D. (1980): Myoelectric knee controller for A/K prostheses. *Proceedings: International Conference on Rehabilitation, Engineering*, Toronto, pp. 51–54.

Nadler, J., Cotman, C. W., and Lynch, G. S. (1974): Biochemical plasticity of short-axon interneurons: increased glutamate decarboxylase activity in the denervated area of rat dentate gyrus following entorhinal lesion. *Exp. Neurol.*, 45:403–413.

Nadler, J. V., White, W. F., Vaca, K. W., and Cotman, C. W. (1977): Calcium-dependent-aminobutyrate release by interneurons of rat hippocampal lesions: lesion-induced plasticity. *Brain Res.*, 131:231–258.

Naess, K., and Storm-Mathisen, A. (1955): Fatigue of sustained tetanic contractions. *Acta Physiol. Scand.*, 34:351–366.

Nah, S. H., and Leong, S. K. (1976): Bilateral cortifugal projection to the red nucleus after neonatal lesions in the albino rat. *Brain Res.*, 107:433–436.

Nakamura, Y. (1980): Brainstem neuronal mechanisms controlling the trigeminal motoneuron activity. In: *Progress in Clinical Neurophysiology, Vol. 8, Spinal and Supraspinal Mechanisms of Voluntary Motor Control and Locomotion*, edited by J. E. Desmedt, pp. 181–202. Karger, Basel.

Nakamura, Y., Mizuno, N., and Konishi, A. (1978): A quantitative electron microscope study of cerebellar axon terminals on the magnocellular red nucleus neurons in the cat. *Brain Res.*, 147:17–27.

Nakao, S., Curthoys, I. S., and Markham, C. H. (1980): Direct inhibitory projection of pause neurons to nystagmus-related pontomedullary reticular burst neurons in the cat. *Exp. Brain Res.*, 40:283–293.

Nakao, S., and Sasaki, S. (1980): Excitatory input from interneurons in the abducens nucleus to medial rectus motoneurons mediating conjugate horizontal nystagmus in the cat. *Exp. Brain Res.*, 39:23–32.

Nakao, S., Sasaki, S., Schor, R. H., and Shimazu, H. (1981): Functional organization of premotor neurons in the cat medial vestibular nucleus related to slow and fast phases of nystagmus. *Exp. Brain Res.*, 45:371–385.

Nakao, S., Sasaki, S., and Shimazu, H. (1977): Nuclear delay of impulse transmission in abducens motoneurons during fast eye movements of visual and vestibular origin in alert cats. *J. Neurophysiol.*, 40:1415–1423.

Namerow, N. S. (1973): Observations of the blink reflex in multiple sclerosis. In: *New Developments in Electromyography and Clinical Neurophysiology*, Vol. 3, edited by J. E. Desmedt, pp. 692–696. Karger, Basel.

Namerow, N. S., and Etemadi, A. (1970): The orbicularis oculi reflex in multiple sclerosis. *Neurology (Minneap.)*, 20:1200–1203.

Napier, J. R. (1956): The prehensile movements of the human hand. *J. Bone Joint Surg.*, 38b:902–913.

Napier, J. R. (1960): Studies of the hands of living primates. *Proc. Zool. Soc. Lond.*, 134:647–657.

Napier, J. R. (1962): The evolution of the hand. *Sci. Am.*, 207:56–62.

Napier, J. R. (1976): *The Human Hand*. Carolina Biological Supply, Burlington, North Carolina.

Napier, J. R., and Napier, P. H. (1967): *A Handbook of Living Primates*. Academic Press, London.

Narabayashi, H., and Ohye, C. (1978): Parkinsonian tremor and nucleus ventralis intermedius of the human thalamus. In: *Progress in Clinical Neurophysiology, Vol. 5, Physiological Tremor, Pathological Tremors and Clonus*, edited by J. E. Desmedt, pp. 165–172. Karger, Basel.

Nashner, L. M. (1971): A model describing the vestibular detection of body sway motion. *Acta Otolaryngol.*, 72:429–436.

Nashner, L. M. (1976): Adapting reflexes controlling the human posture. *Exp. Brain Res.*, 26:59–72.

Nashner, L. M. (1977): Fixed patterns of rapid postural responses among leg muscles during stance. *Exp. Brain Res.*, 30:13–24.

Nashner, L. M. (1980): Balance adjustment of humans perturbed while walking. *J. Neurophysiol.*, 44:650–664.

Nashner, L., and Berthoz, A. (1978): Visual contribution to rapid motor responses during postural control. *Brain Res.*, 150:403–407.

Nashner, L. M., Black, F. O., and Wall, C. (1982): Adaptation to altered support and visual conditions during stance: patients with vestibular deficits. *J. Neurosci.*, 2:536–544.

Nashner, L. M., and Cordo, P. J. (1981): Relation of postural responses and reaction-time voluntary movements in human leg muscles. *Exp. Brain Res.*, 43:395–405.

Nashner, L. M., and Grimm, R. J. (1978): Analysis of multiloop dyscontrols in standing cerebellar patients. In: *Progress in Clinical Neurophysiology, Vol. 4, Cerebral Motor Control in Man: Long Loop Mechanisms*, edited by J. E. Desmedt, pp. 300–319. Karger, Basel.

Nashner, L. M., Woollacott, M., and Tuma, G. (1979): Organization of rapid response to postural and locomotor-like perturbations of standing man. *Exp. Brain Res.*, 36:463–476.

Nashold, B. S. (1966): An electronic method of measuring and recording resistance to passive muscle stretch. *J. Neurosurg.*, 24:310–314.

Nathan, P. W. (1976): The gate control theory of pain. A critical review. *Brain*, 99:123–158.

Nathan, P. W., and Smith, M. C. (1955a): Long descending tracts in man. Review of present knowledge. *Brain*, 78:248–303.

Nathan, P. W., and Smith, M. C. (1955b): The Babinski response: a review and new observations. *J. Neurol. Neurosurg. Psychiatry*, 18:250–259.

Nathan, P. W., and Smith, M. C. (1959): Fasciculi proprii of the spinal cord in man: Review of present knowledge. *Brain*, 82:610–668.

Navas, F., and Stark, L. (1968): Sampling on intermittency in hand control system dynamics. *Biophys. J.*, 8:252–302.

Neilson, P. D. (1972): Interaction between voluntary contraction and tonic stretch reflex transmission in normal and spastic patients. *J. Neurol. Neurosurg. Psychiatry*, 35:853–860.

Neilson, P. D. (1974): Measurement of involuntary arm movement in athetotic patients. *J. Neurol. Neurosurg. Psychiatry*, 37:171–177.

Neilson, P. D., and Andrews, C. J. (1973): Comparison of the tonic stretch reflex in athetotic patients during rest and voluntary activity. *J. Neurol. Neurosurg. Psychiatry*, 36:547–554.

Neilson, P. D., and Lance, J. W. (1978): Reflex transmission characteristics during voluntary activity in normal man and patients with movement disorders. In: *Progress in Clinical Neurophysiology, Vol. 4, Cerebral Motor Control in Man: Long Loop Mecha-*

nisms, edited by J. E. Desmedt, pp. 263–299. Karger, Basel.
Neilson, P. D., and Neilson, M. G. (1978): The role of action reflexes in the damping of mechanical oscillations. *Brain Res.,* 142:439–453.
Nelson, A. J., Moffroid, M., and Whipple, R. (1973): The relationship of integrated electromyographic discharge to isokinetic contractions. In: *New Developments in Electromyography and Clinical Neurophysiology,* Vol. 1, edited by J. E. Desmedt, pp. 584–595. Karger, Basel.
Newsom Davis, J. (1973): Pathological interoceptive responses in respiratory muscles and the mechanism of hiccup. In: *New Developments in Electrophysiology and Clinical Neurophysiology,* Vol. 3, edited by J. E. Desmedt, pp. 751–760. Karger, Basel.
Newsom Davis, J. (1975): The response to stretch of human intercostal muscle spindles studied *in vitro. J. Physiol. (Lond.),* 249:561-579.
Newton, R. A., and Price, D. D. (1975): Modulation of cortical and pyramidal tract induced motor responses by electrical stimulation of the basal ganglia. *Brain Res.,* 85:403-422.
Nichols, T. R., and Houk, J. C. (1976): Improvement in linearity and regulation of stiffness that results from actions of stretch reflex. *J. Neurophysiol.,* 39:119–142.
Nicolis, G., and Prigogine, I. (1977): *Self-Organization in Nonequilibrium Systems—From Dissipative Structures to Order Through Fluctuations.* Wiley, New York.
Nielsen, V. K. (1973): The peripheral nerve function in chronic renal failure. V. Sensory and motor conduction velocity. *Acta Med. Scand.,* 194:445–454.
Nieoullon, A., and Rispal-Padel, L. (1976): Somatotopic localization in cat motor cortex. *Brain Res.,* 105:405–422.
Nobin, A., and Bjorklund, A. (1973): Topography of the monoamine neuron systems in the human brain as revealed in fetuses. *Acta Physiol. Scand.,* 388:1–40.
Nogushi, T., Homma, S., and Nakajima, Y. (1979): Measurements of excitatory postsynaptic potentials in the stretch reflex of normal subjects and spastic patients. *J. Neurol. Neurosurg. Psychiatry,* 42:1100–1105.
Noordenbos, W., and Wall, P. D. (1976): Diverse sensory functions with an almost divided spinal cord. A case of spinal cord transection with preservation of part of one anterolateral quadrant. *Pain,* 2:185–195.
Norris, F. H., and Gasteiger, E. L. (1955): Action potentials of single motor units in normal muscle. *Electroencephalogr. Clin. Neurophysiol,* 7:115–126.
Northmore-Ball, M. P., Heger, H., and Hunter, G. A. (1980): The below-elbow myoelectric prosthesis. A comparison of the Otto Bock myo-electric prosthesis with the hook and functional hand. *J. Bone Joint Surg.,* 62B:363–367.
Norton, B. J., Bomze, H. A., and Chaplin, H. (1972): An approach to the objective measurement of spasticity. *Phys. Ther.,* 52:15–23.
Notermans, S. L. (1966): Measurement of pain threshold determined by electrical stimulation and its clinical application. *Neurology (Minneap.),* 16:1071–1086.

Notermans, S. L., and Vingerhoets, H. M. (1974): The importance of the Hoffmann reflex in the diagnosis of lumbar root lesions. *Clin. Neurol. Neurosurg.,* 1:54–65.
Nutt, J. G., Miroz, E. A., Leeman, S. E., et al. (1980): Substance P in human cerebrospinal fluid: reductions in peripheral neuropathy and autonomic dysfunction. *Neurology (Minneap.),* 30:1280-1285.
Obeso, J. A., Rothwell, J. C., and Marsden, C. D. (1981): Simple tics in Gilles de la Tourette's syndrome are not prefaced by a normal premovement potential. *J. Neurol. Neurosurg. Psychiatry,* 44:735–738.
Odeen, I. (1981). Reduction of muscular hypertonus by long-term muscle stretch. *Scan. J. Rehabil. Med.,* 13:93–99.
Odeen, I., and Knutsson, E. (1982): Evaluation of the effects of muscle stretch and weight load in patients with spastic paraplegia. *Scand. J. Rehabil. Med. (in press).*
Odgen, R., and Franz, S. I. (1917): On cerebral motor control: the recovery from experimentally produced hemiplegia. *Psychobiology,* 1:33–50.
Oliver, L. C. (1952): The supranuclear arc of the corneal reflex. *Acta Psychiatr. Scand.,* 27:329–333.
Oliveras, J. L., Besson, J. M., Guilbaud, G., and Liebeskind., J. C. (1974): Behavioral and electrophysiological evidence of pain inhibition from midbrain stimulation in the cat. *Exp. Brain Res.,* 20:32–44.
Oliveras, J. L., Hosobuchi, Y., Redjemi, F., et al. (1977): Opiate antagonist naloxone, strongly reduces analgesia induced by stimulation of the raphe nucleus (centralis inferior). *Brain Res.,* 120:221–229.
Olsen, P. Z., and Diamantopoulos, E. (1967): Excitability of spinal motor neurones in normal subjects and patients with spasticity, parkinsonian rigidity and cerebellar hypotonia. *J. Neurol. Neurosurg. Psychiatry,* 30:325–331.
Olszewski, J. (1952): *The Thalamus of the Macaca Mulatta. An Atlas for Use with the Stereotaxic Instrument.* Karger, Basel.
O'Neal, J. T., and Westrum, L. E. (1973): The fine structural synaptic organization of the cat lateral cuneate nucleus. A study of sequential alterations in degeneration. *Brain Res.,* 51:97–124.
Ongerboer De Visser, B. W. (1980): The corneal reflex: electrophysiological and anatomical data in man. *Prog. Neurobiol.,* 15:71–83.
Ongerboer De Visser, B. W. (1981): Corneal reflex latency in lesions of the lower postcentral region. *Neurology (Minneap.),* 31:701–707.
Ongerboer De Visser, B. W. (1982): Afferent limb of the human jaw reflex: electrophysiological and anatomical study. *Neurology (Minneap.),* 32:563–566.
Ongerboer De Visser, B. W., and Goor, C. (1974): Electromyographic and reflex study in idiopathic and symptomatic trigeminal neuralgias, latency of the jaw and blink reflexes. *J. Neurol. Neurosurg. Psychiatry,* 37:1225–1230.
Ongerboer De Visser, B. W., and Goor, C. (1976a): Jaw reflexes and masseter myograms in mesencephalic and pontine lesions. *J. Neurol. Neurosurg. Psychiatry,* 39:90–92.

Ongerboer De Visser, B. W., and Goor, C. (1976b): Cutaneous silent period in masseter muscles. Electrophysiological and anatomical data. *J. Neurol. Neurosurg. Psychiatry*, 39:674–679.

Ongerboer De Visser, B. W., and Kuypers, H. (1978): Late blink reflex changes in lateral medullary lesions. An electrophysiological and neuro-anatomical study of Wallenberg's syndrome. *Brain*, 101:285–294.

Ongerboer De Visser, B. W., Mechelse, K., and Megens, P. H. A. (1977): Corneal reflex latency in trigeminal nerve lesions. *Neurology (Minneap.)*, 27:1164–1167.

Ongerboer De Visser, B. W., and Moffie, D. (1979): Effects of brainstem and thalamic lesions on the corneal reflex. *Brain*, 102:595–608.

Oppenheimer, H. (1908): *Lerhbach der Nervenkrankheiten*. Karger, Berlin.

Optican, L. M., and Robinson, D. A. (1980): Cerebellar-dependent adaptive control of primate saccadic system. *J. Neurophysiol.*, 44:1058–1076.

Orgogozo, J. M., and Larsen, B. (1979): Activation of the supplementary motor area during voluntary movement in man suggests it works as a supramotor area. *Science*, 206:847–850.

Ornhagen, H. C. H. (1979): Rate of pressure change and hyperbaric bradycardia in the mouse sinus node. *Undersea Biomed. Res.*, 6:241–249.

Oscarsson, O. (1965): Functional organization of spino- and cuneocerebellar tracts. *Physiol. Rev.*, 45:495–522.

Oscarsson, O. (1973): Functional organization of spino-cerebellar paths. In: *Handbook of Sensory Physiology. II. Somatosensory System*, edited by A. Iggo, pp. 339–380. Springer, Berlin.

Oscarsson, O., and Rosen, I. (1963): Projection to the cerebral cortex of large muscle spindle afferents in forelimb nerves of the cat. *J. Physiol. (Lond.)*, 169:924–945.

Oscarsson, O., and Uddenberg, N. (1965): Properties of afferent connections to the rostral spinocerebellar tract in the cat. *Acta Physiol. Scand.*, 64:143–153.

Oshima, T. (1969): Studies of pyramidal tract cells. In: *Basic Mechanisms of the Epilepsies*, edited by H. H. Jasper, A. A. Ward, and A. Pope, pp. 253–261. Little, Brown, Boston.

Oshima, T. (1978): Intracellular activities of cortical laminae I-III neurones during EEG arousal. *Behav. Brain Sci.*, 3:500–501.

Oshima, T. (1981): Cortical neurones in arousal. *Jpn. J. EEG-EMG*, Suppl. 4:73–78.

Ott, K. H., and Gassel, M. M. (1969): Methods of tendon jerk reinforcement. *J. Neurol. Neurosurg. Psychiatry*, 32:541–547.

Ovalle, W. K., and Smith, R. S. (1972): Histochemical identification of three types of intrafusal muscle fibres in the cat and monkey based on the myosin ATPase reaction. *Can. J. Physiol. Pharmacol.*, 50:195–202.

Overend, W. (1896): Preliminary note on a new cranial reflex. *Lancet*, 1:619.

Pagni, C. A., Ettore, G., Infuso, L., and Marossere, F. (1964): EMG responses to capsular stimulation in the human. *Experientia*, 20:691–692.

Paillard, J. (1955): *Réflexes et Régulations d'Origine Proprioceptive Chez l'Homme*. Arnette, Paris.

Paillard, J. (1980): The multichanneling of visual cues and the organization of a visually guided response. In: *Tutorials in Motor Behavior*, edited by G. E. Stelmach and J. Requin, pp. 259–279. Elsevier, Amsterdam.

Paillard, J., Bert, J., Zwingelstein, J., and Giudicelli, P. (1961): Recherche d'une méthode d'approche de l'action neurophysiologique de diverses drogues. *Rev. Neurol. (Paris)*, 104:227–228.

Paintal, A. S. (1973): Conduction in mammalian nerve fibres. In: *New Developments in Electromyography and Clinical Neurophysiology*, Vol. 2, edited by J. E. Desmedt, pp. 19–41. Karger, Basel.

Palmer, C., Schmidt, E. M., and McIntosh, J. S. (1981): Corticospinal and corticorubral projections from the supplementary motor area in monkey. *Brain Res.*, 209:305–314.

Pal'Tsev, Y. I., and El'Ner, A. N. (1967): Preparation and compensatory period during voluntary movement in patients with involvement of the brain of different localization. *Biophysics*, 12:161–168.

Panayiotopoulos, C. P., and Lagos, G. (1980): Tibial nerve H reflex and F wave studies in patients with uremic neuropathy. *Muscle Nerve*, 3:423–426.

Panayiotopoulos, C. P., and Scarpalezos, S. (1977): F wave studies on the deep peroneal nerve. *J. Neurol. Sci.*, 31:331–341.

Pandya, D. N., and Kuypers, H. G. J. M. (1969): Cortico-cortical connections in the rhesus monkey. *Brain Res.*, 13:13–36.

Pandya, D. N., and Vignolo, L. A. (1971): Intra- and interhemispheric projections of the precentral, premotor and arcuate areas in the rhesus monkey. *Brain Res.* 26:217–233.

Pappas, C. L., and Strick, P. L. (1981): Physiological demonstration of multiple representation in the forelimb region of the cat motor cortex. *J. Comp. Neurol.*, 200:481–490; 491–500.

Parmentier, J. L., Shaivastan, B. B., and Bennett, P. B. (1981): Hydrostatic pressure-reduce synaptic efferency by inhibiting transmitter release. *Undersea Biomed. Res.* 8:175–183.

Partanen, V. S. (1978): Double discharges in neuromuscular disease. *J. Neurol. Sci.*, 36:377–382.

Partridge, L. D. (1965): Modifications of neural output signals by muscles: a frequency response study. *J. Appl. Physiol.*, 20:150–156.

Patton, N. J., and Mortensen, O. A. (1971): An electromyographic study of reciprocal activity of muscles. *Anat. Rec.*, 170:255–268.

Paty, D. W., Blume, W. T., Brown, W. F., et al. (1979): Chronic progressive myelopathy. *Ann. Neurol.*, 6:419–424.

Paul, R. L., Merzenich, M., and Goodman, H. (1972): Representation of slowly and rapidly adapting cutaneous mechanoreceptors of the hand in Brodmann's areas 3 and 1 of Macaca mulatta. *Brain Res.*, 36:229–249.

Pavlov, I. P. (1927): *Conditioned Reflexes. An Investigation of the Physiological Activity of the Cerebral Cortex*. Dover, New York.

Payne, W. H. (1967): Visual reaction times on a circle about the fovea. *Science*, 155:481–482.
Pearson, K.G., and Duysens, J. (1976): Function of segmental reflexes in the control of stepping in cockroaches and cats. In: *Neural Control of Locomotion*, edited by R. Herman, S. Griller, P. Stein, and D. G. Stuart, pp. 519–538. Plenum, New York.
Peat, M., Dubo, H. I. Winter, D. A. et al. (1976): Electromyographic temporal analysis of gait: hemiplegic locomotion. *Arch. Phys. Med. Rehabil.*, 57:421–425.
Pedersen, E., Arlien-Soborg, P., and Mai, J. (1974): The mode of action of the GABA derivative baclofen in human spasticity. *Acta Neurol. Scand.*, 50:665–680.
Pedersen, E., Dietrichson, P., Gormsen, J., and Arlien-Soborg, P. (1974): Measurement of phasic and tonic stretch reflexes in antispastic and antiparkinsonian therapy. *Scand. J. Rehabil. Med. (Suppl.)*, 3:51–60.
Pedersen, E., and Mai, J. (1978): Clonus and stretch reflex in spastic patients. In: *Progress in Clinical Neurophysiology, Vol. 5, Physiological Tremor, Pathological Tremors and Clonus*, edited by J. E. Desmedt, pp. 183–191. Karger Basel.
Pedotti, A. (1977): A study of motor coordination and neuromuscular activities in human locomotion. *Biol. Cybern.*, 26:53–62.
Peirson, G. A., Fowlks, E. W., and King, P. S. (1968): Long-term follow-up on the use of diazepam in the treatment of spasticity. *Am. J. Phys. Ther.*, 47:143–149.
Penders, C., and Boniver, R. (1972): Exploration electrophysiologique du reflexe de clignement dans la paralysie faciale a frigore. *ORL*, 34:17–26.
Penders, C. A., and Delwaide, P. J. (1971): Blink reflex de clignement chez l'homme. Particularités electrophysiologiques de la reponse precoce. *Arch. Int. Physiol. Biochem.*, 77:351–354.
Penders, C.A.;, and Delwaide, P. J. (1971): Blink reflex studies in patients with Parkinsonism before and during therapy. *J. Neurol. Neurosurg. Psychiatry*, 34:674–678.
Penders, C. A., and Delwaide, P. J. (1973): Physiologic approach to the human blink reflex. In: *New Developments in Electromyography and Clinical Neurophysiology*, Vol. 3, edited by J. E. Desmedt, pp. 649–657. Karger, Basel.
Penfield, W. (1958): *The Excitable Cortex in Conscious Man*. Liverpool University Press, Liverpool.
Penfield, W., and Boldrey, E. (1937): Somatic motor and sensory representation in the cerebral cortex of man as studied by electrical stimulation. *Brain*, 60:389–443.
Penfield W., and Jasper, H. (1954): *Epilepsy and the Functional Anatomy of the Human Brain*. Little, Brown, Boston.
Penfield, W., and Rasmussen, T. (1950): *The Cerebral Cortex in Man*. Macmillan, New York.
Penfield, W., and Welch, K. (1951): The supplementary motor area of the cerebral cortex. *Arch. Neurol. Psychiatry*, 66:289–317.
Perl, E. R. (1971): Is pain a specific sensation? *J. Psychiatry. Res.*, 8:273–287.
Perret, C., and Buser, P. (1972): Static and dynamic fusimotor activity during locomotor movements in the cat. *Brain Res.*, 40:165–169.
Person, R. S. (1958): An electromyographic investigation on coordination of the activity of antagonistic muscles in man during the development of a motor habit. *Pavlov J. Higher Nerv. Activ.*, 8:13–23.
Pert, C. B., Kuhar, M. J., and Snyder, S. H. (1976): Opiate receptor: autoradiographic localization in rat brain. *Proc. Natl. Acad. Sci. USA*, 73:3729–3733.
Petajan, J. H. (1979): Motor unit control in Huntington's disease. In: *Advances in Neurology*, Vol. 23, edited by T. Chase, N. Wexler, and A. Barbeau, pp. 163–176. Raven Press, New York.
Petajan, J. H. (1981): Motor unit frequency control in normal man. In: *Progress in Clinical Neurophysiology, Vol. 9, Motor Unit Types, Recruitment and Plasticity in Health and Disease*, edited by J. E. Desmedt, pp. 184–200. Karger, Basel.
Petajan, J. H., and Philip, P. A. (1969): Frequency control of motor unit action potentials. *Electroencephalogr. Clin. Neurophysiol.*, 27:66–72.
Petersen, I. (1973): EMG study of behaviour of sphincter and related muscles during orgasm and ejaculation. In: *New Developments in Electromyography and Clinical Neurophysiology*, Vol. 2, edited by J. E. Desmedt, pp. 439–446. Karger, Basel.
Perterson, B. W. (1970): Distribution of neural responses to tilting within vestibular nuclei of the cat. *J. Neurophysiol.*, 33:750–767.
Peterson, B. W., and Coulter, J. D. (1977): A new long spinal projection from the vestibular nuclei in the cat. *Brain Res.*, 122:351–356.
Phillips, C. G. (1959): Actions of antidromic pyramidal volleys on single Betz cells in the cat. *J. Exp. Physiol.*, 44:1–25.
Phillips, C. G. (1969): Motor apparatus of the baboon's hand. *Proc. R. Soc. Lond. (Biol.)*, 173:141–174.
Phillips, C. G., and Porter, R. (1964): The pyramidal projection to motoneurones of some muscle groups of the baboon's forelimb. *Prog. Brain Res.*, 12:222–242.
Phillips, C. G., and Porter, R. (1977): *Corticospinal Neurones*. Academic Press, London.
Phillips, C. G., Powell, T. P. S., and Wiesendanger, M. (1971): Projection from low threshold muscle afferents of hand and forearm to area 3a of baboon's cortex. *J. Physiol. (Lond.)*, 217:419–446.
Phillips, D. S., Denny, D. D., Robertson, R. T., et al. (1972): Cortical projections of ascending nonspecific systems. *Physiol. Behav.*, 8:269–277.
Phillips, S. J., Richens, A., and Shand, D. G. (1973): Adrenergic control of tendon jerk reflexes in man. *Br. J. Pharmacol.*, 47:595–605.
Pickel, V., Segal, M., and Bloom, F. E. (1974):Axonal proliferation following lesions of cerebellar peduncles. *J. Comp. Neurol.*, 155:43–60.
Pierrot-Deseilligny, E., Bergego, C., and Katz, R. (1982): Reversal in cutaneous control of Ib pathways during voluntary contraction. *Brain Res.*, 233:400–403.
Pierrot-Deseilligny, E., Bergego, C., Katz, R., and Morin, C. (1981a): Cutaneous depression of Ib reflex pathways to motoneurones in man. *Exp. Brain Res.*, 42:351–361.

Pierrot-Deseilligny, E., and Bussel, B. (1975): Evidence for recurrent inhibition by motoneurones in human subjects. *Brain Res.*, 88:105–108.

Pierrot-Deseilligny, E., Bussel, B., Held, J. P., and Katz, R. (1976): Excitability of human motoneurones after discharge in a conditioning reflex. *Electroencephalogr. Clin. Neurophysiol.*, 40:279–287.

Pierrot-Deseilligny, E., Bussel, B., and Morin, C. (1973): Supraspinal control of the changes induced in H-reflex by cutaneous stimulation, as studied in normal and spastic man. In: *New Developments in Electromyography and Clinical Neurophysiology,* Vol. 3, edited by J. E. Desmedt, pp. 550–555. Karger, Basel.

Pierrot-Deseilligny, E., and Lacert, P. (1973): Amplitude and variability of monosynaptic reflexes prior to various voluntary movements in normal and spastic man. In: *New Developments in Electromyography and Clinical Neurophysiology,* Vol. 3, edited by J. E. Desmedt, pp. 538–549. Karger, Basel.

Pierrot-Deseilligny, E., Katz, R., and Morin, C. (1979): Evidence for Ib inhibition in human subjects. *Brain Res.*, 166:176–179.

Pierrot-Deseilligny, E., and Morin, C. (1980): Evidence for supraspinal influences on Renshaw inhibition during motor activity in man. In: *Progress in Clinical Neurophysiology,* Vol. 8, *Spinal and Supraspinal Mechanisms of Voluntary Motor Control and Locomotion,* edited by J. E. Desmedt, pp. 142–169. Karger, Basel.

Pierrot-Deseilligny, E., Morin, C., Bergego, C., and Tankov, N. (1981b): Pattern of group I fibre projections from ankle flexor and extensor muscles in man. *Exp. Brain Res.*, 42:337–350.

Pierrot-Deseilligny, E., Morin, C., Katz, R., and Bussel, B. (1977): Influence of posture and voluntary movement on recurrent inhibition in human subjects. *Brain Res.*, 124:427–436.

Piesiur-Strehlow, B., and Meinck, H. M. (1980): Response patterns of human lumbosacral motoneurone pools to distant somatosensory stimuli. *Electroencephalogr. Clin. Neurophysiol.*, 48:673–682.

Pinto, O. de S., Polikar, M., and Debono, G. (1972): Results of international clinical trials with Lioresal. *Postgrad. Med. J.,* 48 (Suppl. 5):18–23.

Piper, H. (1912): *Elektrophysiologie Menschlicher Muskeln.* Springer, Berlin.

Polit, A., and Bizzi, E. (1979): Characteristics of motor programs underlying arm movements in monkey. *J. Neurophysiol.,* 42:183–194.

Pollock, L. J., Boshes, B., Finkelman, I., et al. (1951): Spasticity. Pseudospontaneous spasms and other reflex activities. *Arch. Neurol.,* 66:537–560.

Pomeranz, B., and Chiu, D. (1976): Naloxone blockage of acupuncture analgesia: endorphin implicated. *Life Sci.,* 19:1757–1762.

Pompeiano, O., and Barnes, C. D. (1971). Effect of sinusoidal muscle stretch on neurons in medial and descending vestibular nuclei. *J. Neurophysiol.,* 34:725–734.

Poppele, R., and Bowman, R. J. (1970): Quantitative description of linear behavior of mammalian muscle spindles. *J. Neurophysiol.,* 33:59–72.

Poppele, R. E., and Kennedy, W. R. (1974): Comparison between behaviour of human and cat muscle spindle recorded *in vitro. Brain Res.,* 75:316–319.

Porter, R. (1970): Early facilitation in corticomotoneuronal synapses. *J. Physiol. (Lond.),* 207:733–745.

Porter, R. (1972): Relationship of discharges of cortical neurons to movement in free-to-move monkeys. *Brain Res.,* 40:39–43.

Porter, R., and Hore, J. (1969): The time course of minimal corticomotoneuronal excitatory postsynaptic potentials in lumber motoneurons of the monkey. *J. Neurophysiol.,* 32:443–451.

Porter R., and Lewis, M. (1975): Relationship of neuronal discharges in the precentral gyrus of monkeys to the performance of arm movements. *Brain Res.,* 98:21–36.

Porter, R. Lewis, M., and Linklater, G. F. (1971): A headpiece for recording neurones in unrestrained monkeys. *Electroencephalogr. Clin. Neurophysiol.,* 30:91–93.

Porter, R., and Rack, P. M. H. (1976): Timing of the responses in the motor cortex of monkeys to an unexpected disturbance of finger position. *Brain Res.,* 103:201–213.

Posner, M. I., Snyner, C. R. R., and Davison, B. J. (1980): Attention and the detection of signals. *J. Exp. Psychol.,* 109:160–174

Poulton, E. C. (1974): *Tracking Skill and Manual Control.* Academic Press, New York.

Powell, T. P. S., and Mountcastle, V. B. (1959): Some aspects of the functional organization of the cortex of the postcentral gyrus of the monkey. *Bull. Johns Hopkins Hosp.,* 105:133–162.

Prablanc, C., Echallier, J. F., Komilis, E., and Jeannerod, M. (1979): Optimal response of eye and hand motor systems in pointing at a visual target. *Biol. Cybern.,* 35:113–124.

Prablanc, C., and Jeannerod, M. (1973): Continuous recording of hand position in the study of complex visuomotor tasks. *Neuropsychologia,* 11:123–125.

Prablanc, C., and Jeannerod, M. (1974): Latence et précision des saccades en fonction de l'intensité, de la durée, et de la position rétinienne d'un stimulus. *Rev. Electroencephalogr. Neurophysiol. Clin.,* 4:484–488.

Precht, W., and Shimazu, H. (1965): Functional connections of tonic and kinetic vestibular neurons with primary vestibular afferents. *J. Neurophysiol.,* 28:1014–1028.

Precht, W., Schwindt, P. C., and Magherini, P. C. (1974): Tectal influences on cat ocular motoneurons. *Brain Res.,* 82:27–40.

Precht, W., and Strata, P. (1980): On the pathway mediating optikinetic responses in vestibular nuclear neurons. *Neuroscience,* 5:777–789.

Prendergast, J., and Stelzner, D. J. (1976): Increases in collateral axonal growth rostral to thoracic hemisection in neonatal and weanling rat. *J. Comp. Neurol.,* 166:145–162.

Preston, J. B., Shende, M. C., and Uemura, K. (1967): The motor cortex-pyramidal system: patterns of facilitation and inhibition on motoneurons innervating limb musculature of cat and baboon. In: *Neurophysiological Basis of Normal and Abnormal Motor Ac-*

tivities, edited by M. D. Yahr and D. P. Purpura, pp. 61–72. Raven Press, New York.
Preston, J. B., and Whitlock, D. G. (1960): Precentral facilitation and inhibition of spinal motoneurons. *J. Neurophysiol.*, 23:154–170.
Price, D. D., and Dubner, E. (1977): Neurons that subserve the sensory discriminative aspect of pain. *Pain*, 3:307–338.
Prochazka, A. (1980): Muscle spindle activity during walking and during free fall. In: *Spinal and Supraspinal Mechanisms of Voluntary Motor Control and Locomotion. Progress in Clinical Neurophyiology*, Vol. 8, edited by J. E. Desmedt, pp. 282–293. Karger, Basel.
Prochazka, A., Schofield, P., Westerman, R. A., and Ziccone, S. P. (1977): Reflexes in cat ankle muscles after landing from falls. *J. Physiol. (Lond.)*, 272:705–719.
Prochazka, A., Stephens, J. A., and Wand, P. (1979): Muscle spindle discharge in normal and obstructed movements. *J. Physiol. (Lond.)*, 287:57–66.
Prochazka, A., and Wand, P. (1980): Tendon organ discharge during voluntary movements in cats. *J. Physiol. (Lond.)*, 303:385–390.
Prochazka, A., and Wand, P. (1981a): Independence of fusimotor and skeletomotor systems during voluntary movement. In: *Muscle Receptors and Movement*, edited by A. Taylor and A Prochazka, pp. 229–243. Macmillan, London.
Prochazka, A., and Wand, P. (1981b): Muscle spindle responses to rapid stretching in normal cats. In: *Muscle Receptors and Movement*, edited by A. Taylor and A. Prochazka, pp. 157–161. Macmillan, London.
Prochazka, A., Westerman, R. A., and Ziccone, S. P. (1976): Discharges of single hindlimb afferents in the freely-moving cat. *J. Neurophysiol.*, 39:1090–1104.
Prochazka, A., Westerman, R. A., and Ziccone, S. P. (1977): Ia afferent activity during a variety of voluntary movements in the cat. *J. Physiol. (Lond.)*, 268:423–448.
Proske, U. (1981): The Golgi tendon organ: properties of the receptor and reflex action of impulses arising from tendon organs. *Int. Rev. Physiol.*, 25:47–90.
Proske, U., and Gregory, J. E. (1977): The time-course of recovery of the initial burst of primary endings of muscle spindles. *Brain Res.*, 121:358–361.
Proske, U., and Waite, P. M. E. (1976): The relation between tension and axonal conduction velocity for motor units in the medial gastrocnemius muscle of the cat. *Exp. Brain Res.*, 26:325–328.
Pullen, A. H., and Sears, T. A. (1978): Modification of 'C' synapses following partial central deafferentation of thoracic motoneurons. *Brain Res.*, 145:141–146.
Purpura, D. P. (1961): Analysis of axodendritic synaptic organizations in immature cerebral cortex. *Ann. NY Acad. Sci.*, 94:604–654.
Purves, D. (1975): Functional and structural changes in mammalian sympathetic neurones following interruption of their axons. *J. Physiol. (Lond.)*, 252:429–463.
Purves-Stuart, J. (1937): *The Diagnosis of Nervous Diseases*. Arnold, London.

Putkonen, P. T. S., Courjon, J. H., and Jeannerod, M. (1977): Compensation of postural effects of hemilabyrinthectomy in the cat. A sensory substitution process? *Exp. Brain Res.*, 28:249–257.
Rack, P. M. H. (1978): Mechanical and reflex factors in human tremor. In: *Progress in Clinical Neurophysiology, Vol. 5, Physiological Tremor, Pathological Tremors and Clonus*, edited by J. E. Desmedt, pp. 17–27. Karger, Basel.
Rack, P. M. H., Ross, H. F., and Brown, T. I. N. (1978): Reflex responses during sinusoidal movement of human limbs. In: *Progress in Clinical Neurophysiology, Vol. 4, Cerebral Motor Control in Man: Long Loop Mechanisms*, edited by J. E. Desmedt, pp. 216–228. Karger, Basel.
Rack, P. M. H., and Westbury, D. R. (1966): The effects of suxamethonium and acetylcholine on the behavior of cat muscle spindles during dynamic stretching, and during fusimotor stimulation. *J. Physiol. (Lond.)*, 186:698–713.
Rack, P. M. H., and Westbury, D. R. (1969): The effects of length and stimulus rate on tension in the isometric cat soleus muscle. *J. Physiol. (Lond.)*, 204:443–460.
Rademaker, G. (1935): *Reaction Labyrinthiques et Equilibre*. Masson, Paris.
Raibert, M. (1976): A state space model for sensorimotor control and learning. *AIM 351, Artificial Intelligence Lab.*, MIT Press, Cambridge, Massachusetts.
Raisman, G. (1969): Neuronal plasticity in the septal nuclei of the adult rat. *Brain Res.*, 14:25–48.
Raisman, G. (1977): Formation of synapses in the adult rat after injury: Similarities and differences between a peripheral and a central nervous site. *Philos. Trans. R. Soc. Lond.*, 278:349–359.
Rall, W. (1964): Theoretical significance of dendritic trees for neuronal input-output relations. In: *Neural Theory and Modeling*, edited by R. F. Reiss, pp. 73–87. Standford University Press, Stanford.
Ralston, H. (1976): Energetics in human walking. In: *Neural Control of Locomotion* Vol. 18, edited by R. Herman, S. Grillner, P. Stein, and D. Stuart. Plenum Press, New York.
Ralston, H. (1980): Fine structure of the spinal cord. In: *The Spinal Cord and its Reaction to Traumatic Injury*, edited by W. E. Windle, pp. 81–94. Dekker, Basel.
Raphan, T., and Cohen, B. (1978): Brainstem mechanisms for rapid and slow eye movements. *Annu. Rev. Physiol.*, 40:527–552.
Raphan, T., Matsuo, V., and Cohen, B. (1979): Velocity storage in the vestibuloocular reflex arc. *Exp. Brain Res.*, 35:229–248.
Rashbass, C. (1961): The relationship between saccadic and smooth tracking eye movements. *J. Physiol. (Lond.)*, 159:326–338.
Rasminsky, M. (1978): Physiology of conduction in demyelinated axons. In: *Physiology and Pathobiology of Axons*, edited by S. G. Waxman, pp. 361–376. Raven Press, New York.
Reimen, J. C., and Ash, M. A. (1974): Myo-Monitor centric: an evaluation. *J. Prosthet. Dent.*, 31:137–145.
Reinking, R. M., Stephens, J. A., and Stuart, D. G.

(1975): The tendon organs of cat medial gastrocnemius: significance of motor unit size and type for the activation of Ib afferents. *J. Physiol. (Lond.)*, 250:491–512.

Reiss, R. F. (1962): A theory and stimulation of rhythmic behavior due to reciprocal inhibition in small nerve nets. *Proceedings: AFIPS Spring Joint Computer Conference*, 21:171–194.

Reiter, R. (1948): Eine neue Elektrokunsthand. *Grenzgzbeite der Medizin*, 1:133–135.

Renou, G., Rondot, P., and Bathien, N. (1973): Influence of peripheral stimulation on the silent period between burst of Parkinsonian tremor. In: *New Developments in Electromyography and Clinical Neurophysiology*, Vol. 3, edited by J. E. Desmedt, pp. 635–640. Karger, Basel.

Renshaw, B. (1941): Influence of discharge of motoneurones upon excitation of neighboring motoneurones. *J. Neurophysiol.*, 4:167–183.

Renshaw, B. (1946): Central effects of centripetal impulses in axons of spinal ventral roots. *J. Neurophysiol.*, 9:191–204.

Rethelyi, M and Szenthagothai, J. (1973): Distribution and connections of afferent fibres in the spinal cord. In: *Handbook of Sensory Physiology*, Vol. 2, edited by A. Iggo, pp. 207–252. Springer, Berlin.

Rexed, B. (1954): A cytoarchitectonic atlas of the spinal cord in the cat. *J. Comp. Neurol.*, 100:297–379.

Reynold, D. V. (1969): Surgery in the rat during electrical analgesia induced by focal brain stimulation. *Science*, 164:444–445.

Reynolds, S. F., and Blass, J. P. (1976): Selective cerebellar damage in partial pyruvate dehydrogenase deficiency. *Neurology (Minneap.)*, 26:625–628.

Richwien, R. (1966): Zur diagnose der Zentralen Störungen der Corneareflexes in afferenten Schenkel. *Dtsch. Z. Nervenheilk.*, 189:181–187.

Riddoch, G. (1917): The reflex functions of the completely divided spinal cord in man compared with those associated with less severe lesions. *Brain*, 40:264–402.

Riley, C. M., Day, R. L., Greeley, D. M., and Langford, W. S. (1949): Central autonomic dysfunction with defective lacrimation. *Pediatrics*, 3:468–478.

Rinne, U. K. (1980): Tizanidine treatment of spasticity in multiple sclerosis and chronic myelopathy. *Curr. Ther. Res.*, 28:827–836.

Rispal-Padel, L., and Latreille, J. (1974): The organization of projection from the cerebellar nuclei to the contralateral motor cortex in the cat. *Exp. Brain Res.*, 19:36–60.

Rizzoli, A. A. (1968): Distribution of glutamic acid, aspartic acid, gamma-aminobutyric acid and glycine in six areas of cat spinal cord before and after transection. *Brain*, 91:11–18.

Robert, B. L., and Witowsky, P. (1975): A functional analysis of the mesencephalic nucleus of the fifth nerve in the selachian brain. *Proc. R. Soc. Lond. (Biol.)*, 190:473–495.

Roberts, T. D. (1968): Labyrinthine control of the postural muscles. *NASA*, SP-152:149–168.

Roberts, T. D. (1973): Reflex balance. *Nature*, 244:156–158.

Roberts, T. D. (1975): The behavioral vertical. *Fortschr. Zool.*, 23:192–198.

Roberts, T. D. (1976): The role of vestibular and neck receptors in locomotion. In: *Neural Control of Locomotion*, edited by R. Herman, S. Grillner, P. Stein, and D. Stuart. Plenum, New York.

Roberts, T. D. (1979): *Neurophysiology of Postural Mechanisms*. Butterworth, London.

Robinson, D. A. (1965): The mechanics of smooth pursuit eye movement. *J. Physiol. (Lond.)*, 180:569–591.

Robinson, D. A. (1969): The mechanics of human saccadic eye movement. *J. Physiol. (Lond.)*, 204:443–460.

Robinson, D. A. (1970): Oculomotor unit behavior in the monkey. *J. Neurophysiol.*, 33:393–404.

Robinson, D. A. (1972): Eye movements evoked by collicular stimulation in the alert monkey. *Vision Res.*, 12:1795–1808.

Robinson, D. A. (1976): Adaptive gain control of the vestibulo-ocular reflex by the cerebellum. *J. Neurophysiol.*, 39:954–969.

Robinson, D. A. (1977): Linear addition of optokinetic and vestibular signals in the vestibular nucleus. *Exp. Brain Res.*, 30:447–450.

Robinson, D. A. (1981): The use of control systems analysis in the neurophysiology of eye movements. *Annu. Rev. Neurosci.*, 4:463–503.

Robinson, D. L., Goldberg, M. E., and Stanton, G. B. (1978): Parietal association cortex in the primate: sensory mechanisms and behavioral modulations. *J. Neurophysiol.*, 41:910–932.

Robson, J. A., Mason, C. A., and Guillery, R. W. (1978): Terminal arbors of axons that have formed abnormal connections. *Science*, 201:635–637.

Rockel, A. J., Hiorns, R. W., and Powell, T. P. S. (1980): The basic uniformity in structure of the neocortex. *Brain*, 103:221–244.

Rohland, T. A. (1975): Sensory feedback for powered limb prosthesis. *Med. Biol. Eng.*, 13:300–301.

Roland, P. E., Larsen, B., Lassen, N. A., and Shihoj, E. (1980): Supplementary and other cortical areas in organization of voluntary movements in man. *J. Neurophysiol.*, 43:118–136.

Roll, J. P., Bonnet, M., and Hugon, M. (1973): The baboon as a model for the study of spinal reflexes. In: *New Developments in Electromyography and Clinical Neurophysiology*, Vol. 3, edited by J. E. Desmedt, pp. 194–208. Karger, Basel.

Roll, J. P., Gilhodes, J. C., and Tardy-Gervet, M. F. (1980a): Effets perceptifs et moteurs des vibrations musculaires chez l'homme normal. *Arch. Ital. Biol.*, 118:51–71.

Roll, J. P., Lacour, M., Hugon, M. and Bonnet, M. (1978): Spinal reflex activity in man under hyperbaric heliox conditions (31 and 62 ATA). In: *Underwater Physiology VI*, edited by C. W. Shilling and M. W. Beckett, pp. 21–28. FASEB, Bethesda, Maryland.

Roll, J. P., Martin, B., Gauthier, G. M., and Mussa-Ivaldi, F. (1980b): Effects of whole-body vibration on spinal reflexes in man. *Aviat. Space Environ. Med.*, 51:1227–1233.

Roll, J. P. and Vedel, J. P. (1981): Kinaesthetic role of

muscle afferents in man studied by tendon vibration and microneurography. *Exp. Brain Res.*, 47:177–190.
Ron, S., Robinson, D. A., and Skavenski, A. A. (1972): Saccades and the quick phase of nystagmus. *Vision Res.*, 12:2015–2022.
Rondot, P., and Bathien, N. (1978): Pathophysiology of Parkinsonian tremor. In: *Progress in Clinical Neurophysiology, Vol. 5, Physiological Tremor, Pathological Tremors and Clonus*, edited by J. E. Desmedt, pp. 138–149. Karger, Basel.
Rondot, P., Bathien, N., and Ribadeau-Dumas, J. L. (1975): Indications of Piribedil in L-Dopa-treated parkinsonian patients. In: *Advances in Neurology*, Vol. 9, edited by D. B. Calne, T. N. Chase, and A. Barbeau, pp. 373–382. Raven Press, New York.
Rondot, P., Jedynak, C. P., and Ferrey, G. (1978): Pathological tremors: Nosological correlates. In: *Progress in Clinical Neurophysiology, Vol. 5, Physiological Tremor, Pathological Tremors and Clonus*, edited by J. E. Desmedt, pp. 95–113. Karger, Basel.
Rondot, P., and Metral, S. (1973): Analysis of the shortening reaction in man. In: *New Developments in Electromyography and Clinical Neurophysiology*, Vol. 3, edited by J. E. Desmedt, pp. 629–634. Karger, Basel.
Rondot, P., and Scherrer, J. (1966): Contraction réflexe provoquée par le raccourcissement passif du muscle dans l'athétose et les dystonies d'attitude. *Rev. Neurol. (Paris)*, 114:329–337.
Rosen, I., and Asanuma, H. (1972): Peripheral afferent inputs to the forelimb area of the monkey motor cortex: input-output relations. *Exp. Brain Res.*, 14:257–273.
Rosen, I., Fehling, C., Sedwick, M., and Elmquist, D. (1977): Focal reflex epilepsy with myoclonus: electrophysiological investigation and therapeutic implications. *Electroencephalogr. Clin. Neurophysiol.*, 42:95–106.
Rosenthal, N. P., McKean, T. A., Roberts, W. J., and Terzuolo, C. A. (1970): Frequency analysis of stretch reflex and its main subsystems in triceps surae muscles of the cat. *J. Neurophysiol.*, 33:713–749.
Rosenzweig, M., Bennett, E. L., Diamond, M. C., et al. (1969): Influences of environmental complexity and visual stimulation on development of occipital cortex in rats. *Brain Res.*, 14:427–445.
Ross, R. T. (1972): Corneal reflex in hemisphere disease. *J. Neurol. Neurosurg. Psychiatry*, 35:877–880.
Rossignol, S., and Gauther, L. (1980): An analysis of mechanisms controlling the reversal of crossed spinal reflexes. *Brain Res.*, 182:31–45.
Rossignol, S., and Melvill Jones, G. (1976): Audiospinal influence in man studied by the H-reflex and its possible role on rhythmic movements synchronized to sound. *Electroencephalogr. Clin. Neurophysiol.*, 41:83–92.
Rostain, J. C., Gardette-Chaufour, M. C., Doucet, J., and Gardette, B. (1981): Problems posed by compression at depths greater than 800 m in Papio Papio. *Undersea Biomed. Res.*, 8:17.
Rotshenker, S., and Reichert, J. (1980): Motor axon sprouting and site of synaptic formation in intact innervated skeletal muscle of the frog. *J. Comp. Neurol.*, 193:413–422.
Rothwell, J. C., Traub, M. M., and Marsden, C. D. (1980): Influence of voluntary intent on the human long-latency stretch reflex. *Nature*, 286:496–498.
Rothwell, J. C., Traub, M. M., and Marsden, C. D. (1982): Automatic and "voluntary" responses compensating for disturbances of human thumb movements. *Brain Res.*, 248:33–41.
Routtenberg, A. (1968): The two arousal hypothesis: reticular formation and limbic system. *Psychol. Rev.*, 75:51–79.
Rowbotham, G. F. (1939): Observations of the effects of trigeminal denervation. *Brain*, 62:364–380.
Ruffini, A. (1905): Les dispositifs anatomiques de la sensibilité cutanée sur les expansions nerveuses de la peau chez l'homme et quelques autres mammifères. *Rev. Gen. Histol.*, 1:421–538.
Rushton, D. N., Rothwell, J. C., and Craggs, M. D. (1981): Gating of somatosensory evoked potentials during different kinds of movement in man. *Brain*, 104:465–491.
Rushworth, G. (1960): Spasticity and rigidity: an experimental study and review. *J. Neurol. Neurosurg. Psychiatry*, 23:99–117.
Rushworth, G. (1962): Observation on blink reflexes. *J. Neurol. Neurosurg. Psychiatry*, 25:93–108.
Rushworth, G. (1964): Some aspects of the pathophysiology of spasticity and rigidity. *Clin. Pharmacol. Ther.*, 5:828–836.
Rushworth, G., Lisman, W. A., Hughes, J. T., and Oppenheimer, D. R. (1961): Intense rigidity of the arms due to isolation of motoneurones by a spinal tumour. *J. Neurol., Neurosurg. Psychiatry*, 24:132–142.
Russel, W. R., and Demyer, W. (1961): The quantitative cortical origin of pyramidal axons of Macaca rhesus, with some remarks on the slow rate of axolysis. *Neurology, (Minneap.)*, 11:96–108.
Russel, W. J., and Tate, M. A. (1975): A device for applying nociceptive stimulation by pressure. *J. Physiol. (Lond.)*, 248:5–7.
Rustioni, A., and Kaufman, A. B. (1977): Identification of cells of origin of nonprimary afferents to the dorsal column nuclei of the cat. *Exp. Brain Res.*, 27:1–14.
Rustioni, A., and Molenaar, I. (1975): Dorsal column nuclei afferents in the lateral funiculus of the cat: distribution pattern and absence of sprouting after chronic deafferentation. *Exp. Brain Res.*, 23:1–12.
Ryall, R. W. (1970): Renshaw cell mediated inhibition of Renshaw cells: patterns of excitation and inhibition from impulses in motor axon collaterals. *J. Neurophysiol.*, 33:257–270.
Ryall, R. W. (1979): *Mechanisms of Drug Action on the Nervous System*. Cambridge University Press, Cambridge.
Ryall, R. W., Piercey, M. F., Polosa, C., and Goldfarb, J. (1972): Excitation of Renshaw cells in relation to orthodromic and antidromic excitation of motoneurons. *J. Neurophysiol.*, 35:137–148.
Rymer, W. Z., Houk, J. C., and Crago, P. E. (1979): Mechanisms of the clasp-knife reflex studied in an animal model. *Exp. Brain Res.*, 37:93–113.
Sahrmann, S. A., and Norton, B. J. (1977): The relationship of voluntary movement to spasticity in the

upper motor neuron syndrome. *Arch. Neurol.,* 2:460–465.
Said, G., and Landrieu, P. (1978): Etude quantitative des fibres nerveuses isolées dans les polynévrites alcooliques. *J. Neurol. Sci.,* 35:317–330.
Saito, M., Tomonaga, M., Hirayama, K., and Narabayashi, H. (1977): Histochemical study of normal human muscle spindle. *J. Neurol. Neurosurg. Psychiatry,* 216:79–89.
Sakai, M. (1978): Single unit activity in a border area between the dorsal prefrontal and premotor regions in motor task of monkeys. *Brain Res.,* 147:377–383.
Sakai, T., and Preston, J. B. (1978): Evidence for a transcortical reflex: primate corticospinal tract neuron responses to ramp stretch of muscle. *Brain Res.,* 159:463–467.
Sakata, H., Shibutani, H., and Kawano, K. (1980): Spatial properties of visual fixation neurons in posterior parietal association cortex of monkey. *J. Neurophysiol.,* 43:1654–1673.
Sakitt, B. (1980): A spring model and equivalent neural network for arm posture control. *Biol. Cybern.,* 37:227–234.
Saletu, B. (1977): Cerebral evoked potentials in psychopharmacology. In: *Progress in Clinical Neurophysiology, Vol. 2, Psychopharmacology Correlates of Evoked Potentials,* edited by J. E. Desmedt, pp. 175–207. Karger, Basel.
Saltzman, E. (1979): Levels of sensorimotor representation. *J. Math. Psychol.,* 20:91–163.
Sanides, F. (1968): The architecture of the cortical taste nerve areas in squirrel monkey, and their relationships to insular, sensorimotor and prefrontal regions. *Brain Res.,* 8:97–124.
Sasaki, K., Matsuda, Y., and Mizuno, N. (1973): Distribution of cerebellar-induced responses in the cerebral cortex. *Exp. Neurol.,* 39:342–354.
Sasaki, S., and Shimazu, H. (1981): Reticulo-vestibular organization participating in generation of horizontal fast eye movement. *Ann. NY Acad. Sci.,* 374:130–145.
Satoh, M., and Takagi, H. (1971): Enhancement by morphine of the central descending inhibitory influence on spinal sensory transmission. *Eur. J. Pharmacol.,* 14:60–65.
Sax, D. S., and Johnson, T. L. (1980): Spinal reflex activity in man. Measurement in relation to spasticity. In: *Spasticity: Disordered Motor Control,* edited by R. G. Feldman, R. R. Young, and W. P. Koella, pp. 301–314. Yearbook, Chicago.
Sax, D. S., Johnson, T. L., and Feldman, R. G. (1977): L-dopa effects on H reflex recovery in Parkinson's disease. *Ann. Neurol.,* 2:120–124.
Schaefer, K. P., and Meyer, D. L. (1973): Compensatory mechanisms following labyrinthine lesions in the guinea pig. In: *Memory and Transfer of Information,* edited by H. P. Zippel, pp. 203–232. Plenum, New York.
Schaefer, K. P., and Meyer, D. L. (1974): Compensation of vestibular lesions. In: *Handbook of Sensory Physiology,* Vol. VI/2, edited by H. Kornhuber, pp. 463–490. Springer, Berlin.
Schäfer, S. S., and Schäfer, S. (1973): The behaviour of the proprioceptors of the muscle and the innervation of the fusimotor system during cold shivering. *Exp. Brain Res.,* 17:364–380.
Schalen, L. (1980): Quantification of tracking eye movements in normal subjects. *Acta Otolaryngol.,* 90:404–413.
Schaltenbrand, G. (1937): Myographische Untersuchungen in der Klinik. *Dtsch. Z. Nervenheilk.,* 142:1–17; 143:1–38.
Schaltenbrand, G. (1958): Myographische Untersuchungen reflektorischer Störungen des Muskeltonus. *Arch. Phys. Ther.* 4:264–273.
Scheff, S. W., Bernado, L. S., and Cotman, C. W. (1978): Effect of serial lesions on sprouting in the dentate gyrus. *Brain Res.,* 150:45–53.
Schenck, E., and Dietz, V. (1975): Alkoholische polyneuropathie. *Arch. Psychiatr. Nervenkr.,* 220:159–170.
Schenck, E., and Lauck-Koehler, B. (1950): Fremdreflektorische Beeinflüssung der Erregbarkeit des menschlichen Rückenmarks für Eigenreflexe. *Pfluegers Arch.,* 252:423–444.
Schenck, E., and Lauck-Koehler, B. (1954): Der Entladungsrhythmus der Motoneurone in phasischer fremd Reflex en beim Menschen. *Pfluegers Arch.,* 260:61–73.
Schenck, E., and Manz, F. (1973): The blink reflex in Bell's palsy. In: *New Developments in Electromyography and Clinical Neurophysiology,* Vol. 3, edited by J. E. Desmedt, pp. 678–681. Karger, Basel.
Schieber, M. H., and Thach, W. T. (1980): Alpha-gamma dissociation during slow tracking movements of the monkey's wrist: preliminary evidence from spinal ganglion recording. *Brain Res.,* 202:213–216.
Schiller, P. H., and Stalberg, E. (1978): F responses studied with single fibre EMG in normal subjects and spastic patients. *J. Neurol. Neurosurg. Psychiatry,* 41:45–53.
Schiller, P. H., and Stryker, M. (1972): Single-unit recording and stimulation in superior colliculus of the alert rhesus monkey. *J. Neurophysiol.,* 35:915–924.
Schlesinger, G. (1919): Der mechanische Aufbau der kunstlichen Glieder. In: *Ersatzglieder und Arbeitshilfen,* pp. 407–410. Springer, Berlin.
Schmidt, E. M., Bak, M. J., and McIntosh, J. S. (1976): Long-term recording from cortical neurons. *Exp. Neurol.,* 52:496–506.
Schmidt, E. M., Jost, R. G., and Davis, K. K. (1975): Re-examination of the force relationship of cortical cell discharge patterns with conditioned wrist movements. *Brain Res.,* 83:213–223.
Schmidt, E. M., McIntosh, J. S., Durelli, L., and Bak, M. J. (1978): Fine control of operantly conditioned firing patterns of cortical neurons. *Exp. Neurol.,* 61:349–369.
Schmidt, E. M., and Thomas, J. S. (1981): Motor unit recruitment order: Modification under volitional control. In: *Progress in Clinical Neurophysiology, Vol. 9, Motor Unit Types, Recruitment and Plasticity in Health and Disease,* edited by J. E. Desmedt, pp. 145–148. Karger, Basel.
Schmidl, H. (1973): The INAIL-CECA Prostheses. *Orthot. Prosthet.,* 37:6–12.
Schmidl, H. (1977): The importance of information feedback in prostheses for the upper limbs. *Prosthet. Orthot. Int.,* 1:21–24.

Schmidt, R. F. (1971): Presynaptic inhibition in the vertebrate nervous system. *Ergeb. Physiol.*, 63:20–101.
Schmidt, R. A., Zelanik, H., Hawkins, B., et al. (1979): Motor output variability: a theory for the accuracy of rapid motor acts. *Psychol. Rev.*, 86:415–451.
Schmied, A., Benita, M., Conde, H., and Dormont, J. F. (1979): Activity of ventrolateral thalamic neurons in relation to a simple reaction time task in the cat. *Exp. Brain Res.*, 36:285–300.
Schmitz, M. A., Simons, A. K., and Boettcher, C. A. (1960): The effect of low frequency, high amplitude, whole-body vertical vibration on human performance. *Contract DA 49007-MD-797, Report 130.* Bostrom Laboratory, Wisconsin.
Schneider, P. (1968): Quantitative Analyse und Mechanismen der Bradykinesie bei Parkinsonpatienten. *Dtsch. Z. Nervenheilk.*, 194:89–102.
Schneider, G. E. (1970): Mechanisms of functional recovery following lesions of visual cortex or superior colliculus in neonatal and adult hamster. *Brain Behav. Evol.*, 3:295–323.
Schneider, G. E. (1973): Early lesions of superior colliculus. Factors affecting the formation of abnormal retinal projections. *Brain Behav. Evol.*, 8:73–109.
Schoen, J. H. (1964): Comparative aspects of the descending fibre systems in the spinal cord. *Prog. Brain Res.*, 11:203–222.
Schoenen, J. (1973): Organisation cytoarchitectonique de la moelle épinière de différents mammifères et de l'homme. *Acta Neurol. Belg.*, 73:348–358.
Schoenen, J. (1981): *Organisation Cytoarchitectonique de la Moelle Epinière de l'Homme.* Sciences et Lettres, Liège.
Schomburg, E. D., Meinck, H. M., Haustein, J., and Roesler, J. (1978): Functional organization of the spinal reflex pathway from forelimb afferents to hindlimb motoneurones in the cat. *Brain Res.*, 139:21–33.
Schomburg, E. D., Roesler, J., and Meinck, H. M. (1977): Phase-dependent transmission in the excitatory propriospinal reflex pathway from forelimb afferents to lumbar motoneurones during fictive locomotion. *Neurosci. Lett.*, 4:249–252.
Schultheis, L. W., and Robinson, D. A. (1981): Directional plasticity of the vestibulo-ocular reflex in the cat. *Ann. NY Acad. Sci.*, 374:504–512.
Schultz, A. H. (1949): Age changes and variability in gibbons. *Am. J. Phys. Anthropol.*, 2:1–129.
Schwartz, S. (1964): Effect of neonatal cortical lesions and early environmental factors on adult rat behavior. *J. Comp. Physiol. Psychol.*, 57:72–77.
Schwartz, D. W. F., Deecke, L., and Frederickson, J. M. (1973): Cortical projection of group I muscle afferents to areas 2, 3a, and the vestibular field in the rhesus monkey. *Exp. Brain Res.*, 17:516–526.
Schwindt, P. C. (1973): Membrane-potential trajectories underlying motoneuron rhythmic firing at high rates. *J. Neurophysiol.*, 36:434–449.
Schwindt, P. C., and Calvin, W. H. (1972): Membrane-potential trajectories between spikes underlying motoneurone firing rates. *J. Neurophysiol.*, 35:311–325.
Schwindt, P., and Crill, W. (1980): Role of a persistent inward current in motoneuron bursting during spinal seizures. *J. Neurophysiol.*, 43:1296–1318.
Scott, R. N., Brittain, R. H., Caldwell, R. R., et al. (1980): Sensory feedback system compatible with myoelectric control. *Med. Biol. Eng. Comput.*, 18:65–69.
Scott, R. N., Paciaga, J. E., and Parker, P. A. (1978): Operator error in multistate myoelectric control system. *Med. Biol. Eng. Comput.*, 16:296–301.
Sears, T. A. (1973): Servo control of the intercostal muscles. In: *New Developments in Electromyography and Clinical Neurophysiology,* Vol. 3, edited by J. E. Desmedt, pp. 404–417. Karger, Basel.
Sebille, A. (1978): Etude electrophysiologique des neuropathies latentes induites par la perhexiline. *Electroencephalogr. Clin. Neurophysiol.*, 45:666–670.
Sebille, A. (1980): The Hoffmann reflex of the soleus muscle. A study in leprosy. *J. Neurol. Sci.*, 45:373–378.
Seeman, P. (1972): The membrane actions of anaesthetics and tranquillisers. *Pharmacol. Rev.*, 24:583–655.
Seki, K., and Hugon, M. (1976): Critical flicker frequency (CFF) and subjective fatigue during an oxyhelium saturation dive at 62 ATA. *Undersea Biomed. Res.*, 3:235–247.
Sessle, B. J. (1977): Identification of alpha and gamma trigeminal motoneurons and effects of stimulation of amygdala, cerebellum and cerebral cortex. *Exp. Neurol.*, 54:303–322.
Sessle, B. S., and Schmitt, A. (1972): Effects of controlled tooth stimulations on jaw muscle activity in man. *Arch. Oral Biol.*, 17:1597–1607.
Severin, F. V., Orlovski, G. N., and Shik, M. L. (1967): Work of the muscle receptors during controlled locomotion. *Biophysics*, 12:575–586.
Shahani, B. T. (1968): Effects of sleep on human reflexes with a double component. *J. Neurol. Neurosurg. Psychiatry*, 31:574–579.
Shahani, B. T. (1970): The human blink reflex. *J. Neurol. Neurosurg. Psychiatry*, 33:792–800.
Shahani, B. T., and Young, R. R. (1978): Action tremreflexes. *J. Neurol. Neurosurg. Psychiatry*, 34:616–627.
Shahani, B. T., and Young, R. R. (1972): Human orbicularis oculi reflexes. *Neurology (Minneap.)*, 22:149–154.
Shahani, B. T., and Young, R. R. (1973a): Studies of the normal human silent period. In: *New Developments in Electromyography and Clinical Neurophysiology,* Vol. 3, edited by J. E. Desmedt, pp. 589–602. Karger, Basel.
Shahani, B. T., and Young, R. R. (1973b): Blink reflexes in orbicularis oculi. In: *New Developments in Electromyography and Clinical Neurophysiology,* Vol. 3, edited by J. E. Desmedt, pp. 641–648. Karger, Basel.
Shahani, B. T., and Young, R. R. (1978): Action tremor spasms. In: *New Developments in Electromyography and Clinical Neurophysiology,* Vol. 3, edited by J. E. Desmedt, pp. 734–743. Karger, Basel.
Shahani, B. T., and Young, R. R. (1978): Action Tremors. In: *Progress in Clinical Neurophysiology, Vol. 5, Physiological Tremor, Pathological Tremors and*

Clonus, edited by J. E. Desmedt, pp. 129–137. Karger, Basel.
Shapiro, A. K., Shapiro, E. S., Brunn, R. D., and Sweet, R. D. (editors) (1978): *Gilles de la Tourette's syndrome*. Raven Press, New York.
Shapovalov, A. I. (1973): Extrapyramidal control of primate motoneurons. In: *New Developments in Electromyography and Clinical Neurophysiology*, Vol. 3, edited by J. E. Desmedt, pp. 145–158. Karger, Basel.
Shapovalov, A. I. (1975): Neuronal organization and synaptic mechanisms of supraspinal motor control in vertebrates. *Rev. Physiol. Biochem. Pharmacol.*, 72:1–54.
Sharma, A., K., and Thomas, P. K. (1974): Peripheral nerve structure and function in experimental diabetes. *J. Neurol. Sci.*, 23:1–15.
Sharpless, S., and Jasper, H. (1956): Habituation of the arousal reaction. *Brain*, 79:655–681.
Sharrard, W. J. W. (1955): The distribution of the permanent paralysis in the lower limb in poliomyelitis. *J. Bone Joint Surg.*, 37B:540–548.
Sherrington, C. S. (1906): *The Integrative Action of the Nervous System*. Yale University Press, New Haven.
Sherrington, C. S. (1910): Flexion of the limb crossed extension reflex, and reflex stepping and standing. *J. Physiol. (Lond.)*, 40:28–121.
Shibasaki, H., Motomura, S., Yamashita, Y., et al. (1981): Periodic synchronous discharge and myoclonus in Creutzfeld-Jacob disease. *Ann. Neurol.*, 9:150–156.
Shibasaki, H., Yamashita, Y., and Kuroiwa, Y. (1978): Electroencephalographic studies of myoclonus. Myoclonus-related cortical spikes and high amplitude somatosensory evoked potentials. *Brain*, 101:447–460.
Shimamura, J., and Aoki, M. (1969): Effects of spinobulbo-spinal reflex volleys on flexor motoneurons of hindlimb in the cat. *Brain Res.*, 16:333–349.
Shimamura, M. (1973a): Spinal-bulbo-spinal and propriospinal reflexes in various vertebrates. *Brain Res.*, 64:141–165.
Shimamura, M. (1973b): Neural mechanisms of the startle reflex in cerebral palsy, with special reference to its relationship with spino-bulbo-spinal reflexes. In: *New Developments in Electromyography and Clinical Neurophysiology*, Vol. 3, edited by J. E. Desmedt, pp. 761–766. Karger, Basel.
Shimamura, M., and Akert, K. (1965): Peripheral nervous relations of propriospinal and spino-bulbo-spinal reflex systems. *Jpn. J. Physiol.*, 15:638–647.
Shimamura, M., and Livingston, R. B. (1963): Longitudinal conduction systems serving spinal and brainstem co-ordination. *J. Neurophysiol.*, 26:208–272.
Shimamura, M., Logure, I., and Igusa, Y. (1976): Ascending spinal tracts on the spino-bulbo-spinal reflex in cats. *Jpn. J. Physiol.*, 26:577–589.
Shimamura, M., Mori, S., Matsushima, S., and Fujimori, B. (1964): On the spino-bulbo-spinal reflex in dogs, monkeys and man. *Jpn. J. Physiol.*, 14:411–421.
Shimamura, M., Mori, S., and Yamauchi, T. (1967): Effects of spino-bulbo-spinal reflex volleys on extensor motoneurons of hindlimb in cats. *J. Neurophysiol.*, 30:319–332.
Shimazu, H., Hongo, T., Kubota, K., and Narabayashi, H. (1962): Rigidity and spasticity in man. Electromyographic analysis with reference to the role of the globus pallidus. *Arch. Neurol.*, 6:10–17.
Shimazu, H., and Precht, W. (1965): Tonic and kinetic responses of cat's vestibular neurons to horizontal angular acceleration. *J. Neurophysiol.*, 28:991–1013.
Shimazu, H., and Precht, W. (1966): Inhibition of central vestibular neurons from the contralateral labyrinth and its mediating pathway. *J. Neurophysiol.*, 29:467–492.
Shinoda, Y., Arnold, A. P., and Asanuma, H. (1976): Spinal branching of corticospinal axons in the cat. *Exp. Brain Res.*, 26:215–234.
Shinoda, Y., Yokota, J. I., and Futami, T. (1981): Divergent projections of individual corticospinal axons to motoneurons of multiple muscles in the monkey. *Neurosci. Lett.*, 23:7–12.
Sidowski, J. B. (1966): *Experimental Methods and Instrumentation in Psychology*. McGraw-Hill, New York.
Sie, O. G., and Lakke, J. P. (1980): The spasmolytic properties of 5-chloro-4-(2-imidozolin-2-yl-amino)-2, 1, 3-benzothiadiazolehydrochloride (DS 103-282) *Clin. Neurol. Neurosurg.*, 80:273–279.
Silfverskiold, B. P. (1969): Romberg's test in the cerebellar syndrome occurring in chronic alcoholism. *Acta Neurol. Scand.*, 45:292–302.
Silfverskiold, B. P. (1977): Cortical cerebellar degeneration associated with a specific disorder of standing and locomotion. *Acta Neurol. Scand.*, 55:257–272.
Simantov, R., Kuhar, M. J. Pasternak, G. W., and Snyder, S. H. (1976): The regional distribution of a morphine-like factor enkephalin in monkey brain. *Brain Res.*, 106:189–197.
Simard, T. G., Ladd, H. W., and Hamonet, C. (1973): Myoelectric control of ortheses. In: *New Developments in Electromyography and Clinical Neurophysiology*, Vol. 1, edited by J. E. Desmedt, pp. 683–691. Karger, Basel.
Simmons, J. W., Rath, D. and Merta, R. (1982): Calculation of disability using the Cybex II system. *Orthopedics*, 5:181–185.
Simon, E. J., and Hiller, J. M. (1978): The opiate receptors. *Annu. Rev. Pharmacol. Toxicol.*, 18:371–394.
Simons, D. G., and Dimitrijevic, M. R. (1972): Quantitative variations in the force of quadriceps responses to serial patellar tendon taps in normal man. *Am. J. Phys. Med.*, 51:240–263.
Simons, D. G., and Lamonte, R. J. (1971): An automated system to measure reflex responses to patellar tendon tap in man. *Am. J. Phys. Med.*, 50:72–79
Simoyama, M., and Tanaka, R. (1974): Reciprocal Ia inhibition at the onset of voluntary movements in man. *Brain Res.*, 82:334–337.
Simpson, D. C. (1971): Extended physiological proprioception. *International Symposium for Control of Upper-Extremity Protheses and Ortheses*. Göteborg, Sweden.
Simpson, J. I., and Alley, K. E. (1974): Visual climbing fiber input to rabbit vestibulo-cerebellum: a source

of direction-specific information. *Brain Res.*, 82: 302–308.

Simpson, J. I., Soodak, R. E., and Hess, R. (1979): The accessory optic system and its relation to the vestibulo-cerebellum. *Prog. Brain Res.*, 50:715–724.

Singer, W. (1979): Central-core control of visual-cortex functions. In: *The Neurosciences: Fourth Study Program*, edited by F. O. Schmitt and F. G. Worden, pp. 1093–1110. MIT Press, Cambridge, Massachusetts.

Sjolund, B., Terenius, L., and Eriksson, M. (1977): Increased cerebrospinal fluid levels of endorphins after electro-acupuncture. *Acta Physiol. Scand.*, 100:382–384.

Sjoquist, O. (1938): Studies on pain conduction in the trigeminal nerve. *Acta Psychiatr. Scand. (Suppl.)*, 17:1–139.

Skiba, T. J., and Laskin, D. M. (1981): Masticatory muscle silent periods in patients with MPD syndrome before and after treatment. *J. Dent. Res.*, 60:699–706.

Skinner, J. E., and Yingling, C. D. (1977): Central gating mechanisms that regulate event-related potentials and behavior. In: *Progress in Clinical Neurophysiology, Vol. 1, Attention, Voluntary Contraction and Event-Related Cerebral Potentials*, edited by J. E. Desmedt, pp. 175–207. Karger, Basel.

Sloper, J. J. (1973): An electron microscope study of the termination of afferent connections of the primary somatic sensory cortex. *J. Neurocytol.*, 2:361–368.

Sloper, J. J., Hiorns, R. W., and Powell, T. P. S. (1979): A qualitative and quantitative electron microscopic study of the neurons in the primate motor and somatic sensory cortices. *Philos. Trans. R. Soc. Lond.*, 285:141–171.

Sloper, J. J., and Powell, T. P. S. (1979): Ultrastructure features of the sensorimotor cortex of the primate. *Philos. Trans. R. Soc. Lond.*, 285:123–139.

Smith, A. M. (1979): The activity of supplementary motor area neurons during a maintained precision grip. *Brain Res.*, 172:315–327.

Smith, A. M. (1981): The co-activation of antagonist muscles. *Can. J. Physiol. Pharmacol.*, 7:733–747.

Smith, A. M., and Bourbonnais, D. (1981): Neuronal activity in cerebellar cortex related to control of prehensile force. *J. Neurophysiol.*, 45:286–303.

Smith, A. M., Hepp-Reymond, M. C., and Wyss, U. R. (1975): Relation to activity in precentral cortical neurons to force and rate of force change during isometric contractions of finger muscles. *Exp. Brain Res.*, 23:315–332.

Smith, G. A. (1980): Models of choice reaction time. In: *Reaction Times*, edited by A. T. Welford, pp. 173–214. Academic Press, New York.

Smith, J. E., Hall, P. V., Campbell, R. L. et al. (1976): Levels of gamma-amino-butyric acid in the dorsal grey lumbar spinal cord during the development of experimental spinal spasticity. *Life Sci.*, 19:1525–1530.

Smith, J. L., Betts, B., Edgerton, V. R., and Zernicke, R. F. (1980): Rapid ankle extension during paw shakes. *J. Neurophysiol.*, 43:612–619.

Smith, J. L., Edgerton, V. R., Betts, B., and Collatos, T. C. (1977): EMG of slow and fast ankle extensors of cat during posture, locomotion, and jumping. *J. Neurophysiol.*, 40:503–513.

Smith, J. L., Roberts, E. M., and Atkins, E. (1972): Fusimotor neuron block and voluntary arm movement in man. *Am. J. Phys. Med.*, 51:225–239.

Smith, J. L., and Spector, S. A. (1981): Unique contributions of slow and fast extensor muscles to the control of limb movements. In: *Progress in Clinical Neurophysiology, Vol. 9, Motor Unit Types, Recruitment and Plasticity in Health and Disease*, edited by J. E. Desmedt, pp. 161–175. Karger, Basel.

Smith, R. D., Marcarian, H. Q., and Niemer, W. T. (1969): Direct projections from the masseteric nerve to the mesencephalic nucleus. *J. Comp. Neurol.*, 133:495–502.

Smyth, G. E. (1939): The systematization and central connections of the spinal tract and nucleus of the trigeminal nerve. *Brain*, 62:41–87.

Snow, P. J., Rose, P. K., and Brown, A. G. (1976): Tracing axons and axon collaterals of spinal neurons using intracellular injection of horseradish peroxydase. *Science*, 191:312–313.

Soechting, J. F., Dufresne, J. R., and Lacquaniti, F. (1981): Time-varying properties of the myotatic response in man during some simple motor tasks. *J. Neurophysiol.*, 46:1226–1243.

Soechting, J. F., and Lacquaniti, F. (1981): Invariant characteristics of a pointing movement in man. *J. Neurosci.*, 1:710–720.

Soerjanto, R. (1971): *On the Application of the Myoelectric Hand-Prosthesis in the Netherlands*. Institute of Medical Physics, TNO Rqt. 1.1.59–3. Utrecht, Netherlands.

Solomonov, M., Raplee, L., and Lyman, J. (1978): Electrotactile two point discrimination as a function of frequency, pulse and pulse time delay. *Ann. Biomed. Eng.*, 6:117–125.

Sommer, J. (1940): Periphere Bahnung von Muskeleigenreflexen als Wesen des Jendrassikschen Phänomens. *Dtsch. Z. Nervenheilk.*, 150:249–262.

Sorbye, R. (1977): Myoelectrically controlled hand prosthesis in children. *Int. J. Rehabil. Res.*, 1:15–25.

Soso, M. J., and Fetz, E. E. (1980): Responses of identified cells in postcentral cortex of awake monkeys during comparable active and passive joint movements. *J. Neurophysiol.*, 43:1090–1110.

Sparks, D. L., and Travis, R. P. (1971): Firing patterns of reticular formation neurons during horizontal eye movements. *Brain Res.*, 33:477–481.

Sperry, R. W. (1942): Transplantation of motor nerves and muscles in the forelimb of the rat. *J. Comp. Neurol.*, 76:283–321.

Sperry, R. W. (1947): Effect of crossing nerves to antagonistic limb muscles in the monkey. *Arch. Neurol. Psychiatry*, 58:452–473.

Sperry, R. W. (1950): Neural basis of the spontaneous optokinetic response produced by visual inversion. *J. Comp. Physiol. Psychol.*, 43:482–489.

Spiegel, E. A., Wycis, H. T., Baird, H. W. et al. (1956): Pallidum and muscle tone. *Neurology (Minneap.)*, 6:350–356.

Spoendlin, H. H. (1966): The ultrastructure of the vestibular sense organ. In: *The Vestibular System and its Diseases*, edited by R. J. Wolfson, pp. 39–68. University of Pennsylvania Press, Philadelphia.

Sprague, J. M., and Chambers, W. W. (1953): Regulation of posture in intact and decerebrate cat. *J. Neurophysiol.*, 16:451–463.

Stalberg, E., and Thiele, B. (1973): Discharge pattern of motoneurones in humans. In: *New Developments in Electromyography and Clinical Neurophysiology*, Vol. 3, edited by J. E. Desmedt, pp. 234–241. Karger, Basel.

Stalberg, E., and Trontelj, J. V. (1979): *Single Fibre Electromyography* Mirvalle Press, Old Woking, U.K.

Stanley, E. F. (1978): Reflexes evoked in human thenar muscles during voluntary activity and their conduction pathways. *J. Neurol. Neurosurg. Psychiatry*, 41:1016–1023.

Stark, L. (1968): *Neurological Control Systems*. Plenum, New York.

Stark, L., Vossius, G., and Young, L. (1962): Predictive control of eye tracking movements. *IRE Trans. Human Factors Electron*, HFE-3:52–57.

Stauffer, E. K., Watt, D. G. D., Taylor, A., et al. (1976): Analysis of muscle receptor connections by spike triggered averaging. *J. Neurophysiol.*, 39:1393–1402.

Stefanis, C., and Jasper, H. (1964): Recurrent collateral inhibition in pyramidal tract neurons. *J. Neurophysiol.*, 27:855–877.

Steg, G. (1966): Efferent muscle control in rigidity. In: *Afferent and Motor Control*, edited by R. Granit, pp. 437–443. Almqvist, Stockholm.

Steiger, H. J., and Buttner-Ennever, J. (1978): Relationship between motoneurons and interneurons in the abducens nucleus. *Brain Res.*, 148:181–188.

Stein, G., Rosen, J. J., and Butters, N. (1974): *Plasticity and Recovery of Function in the Central Nervous System*. Academic Press, New York.

Stein, J. (1978): Long loop motor control in monkeys. In: *Progress in Clinical Neurophysiology, Vol. 4, Cerebral Motor Control in Man: Long Loop Mechanisms*, edited by J. E. Desmedt, pp. 107–122. Karger, Basel.

Stein, R. B. (1967): The information capacity of nerve cells using a frequency code. *Biophys. J.*, 7:797–826.

Stein, R. B. (1974): Peripheral control movement. *Physiol. Rev.*, 54:215–243.

Stein, R. B., and Bertoldi, R. (1981): The size principle. In: *Progress in Clinical Neurophysiology. Vol. 9, Motor Unit Types, Recruitment and Plasticity in Health and Disease*, edited by J. E. Desmedt, pp. 85–96. Karger, Basel.

Stein, R. B., Charles, D., Davis, L., et al. (1975): Principles underlying new methods for chronic neural recording. *Can. J. Neurol. Sci.*, 2:235–244.

Stein, R. B., Charles, D., Hoffer, J. A., et al. (1980): New approaches for the control of powered prostheses, particularly by high-level arm amputees. *Bull. Prosthet. Res.*, 17:51–62.

Stein, R. B., French, A. S., Mannard, A., and Yemm, R. (1972): New methods for analysing motor function in man and animals. *Brain Res.*, 40:187–192.

Stein, R. B., Nichols, T. R., Jhamandas, J., et al. (1977): Stable long-term recordings from cat peripheral nerves. *Brain Res.*, 128:21–38.

Stein, R. B., and Oguztorelli, M. N. (1978): Reflex involvement in the generation and control of tremor and clonus. In: *Progress in Clinical Neurophysiology, Vol. 5, Physiological Tremor, Pathological Tremors and Clonus*, edited by J. E. Desmedt, pp. 28–50. Karger, Basel.

Steinman, R. M., and Collewijn, H. (1980): Binocular retinal image motion during active head rotation. *Vision Res.*, 20:415–429.

Stelzner, D. J., and Keating, E. G. (1977): Lack of intralaminar sprouting of retinal axons in monkeys LGN. *Brain Res.*, 126:201–210.

Stenevi, U., Bjorklund, A., and Moore, R. Y. (1972): Growth of intact central adrenergic axons in the denervated lateral geniculate body. *Exp. Neurol.*, 35:290–299.

Stephens, J. A., and Taylor, A. (1972): Fatigue of maintained voluntary muscle contraction in man. *J. Physiol. (Lond.)*, 220:1–18.

Stephens, J. A., and Usherwood, T. P. (1977): The mechanical properties of human motor units with special reference to their fatigability and recruitment threshold. *Brain Res.*, 125:91–97.

Steriade, M. (1970): Ascending control of thalamic and cortical responsiveness. *Int. Rev. Neurobiol.*, 12:87–144.

Steriade, M. (1978): Cortical long axoned cells and putative interneurons during the sleep-waking cycle. *Behav. Brain Res. Sci.*, 3:465–514.

Steriade, M., Deschenes, M., and Oakson, G. (1974): Inhibitory processes and interneuronal apparatus in motor cortex during sleep and waking. *J. Neurophysiol.*, 37:1065–1092.

Steriade, M., and Hobson, J. A. (1976): Neuronal activity during the sleep-waking cycle. *Prog. Neurobiol.*, 6:155–376.

Stern, J., Mendell, J., and Clark, K. (1968): H reflex suppression by thalamic stimulation and drug administration. *J. Neurosurg.*, 29:393–396.

Stern, J., and Ward, A. A. (1960): Inhibition of the muscle spindle discharge by ventrolateral thalamic stimulation. Its relation to Parkinsonism. *Arch. Neurol.*, 3:193–204.

Stern, J., and Ward, A. A. (1962): Supraspinal and drug modulation of the alpha motor system. *Arch. Neurol.*, 6:404–413.

Stern, P., and Bokonjic, R. (1964): Glycine therapy in seven cases of spasticity. *Pharmacology*, 12:117–119.

Steward, O., Cotman, C. W., and Lynch, G. S. (1974): Growth of a new fiber projection in the brain of adult rats. *Exp. Brain Res.*, 20:45–66.

Steward, O., and Loesche, J. (1977): Quantitative autoradiographic analysis of the time course of proliferation of contralateral entorhinal efferents in the dentate gyrus denervated by ipsilateral entorhinal lesions. *Brain Res.*, 125:11–21.

Storm-Mathisen, J. (1974): Choline acetyltransferase and acetylcholinesterase in fascia dentata following lesion of the entorhinal afferents. *Brain Res.*, 80:181–197.

Strick, P. L. (1976): Anatomical analysis of ventrolateral thalamic input in primate motor cortex. Activity of ventrolateral thalamic neurons during arm movement. *J. Neurophysiol.*, 39:1020–1031; 1032–1044.

Strick, P. L. (1978): Cerebellar involvement in 'volitional' muscle responses to load changes. In: *Progress in Clinical Neurophysiology, Vol. 4, Cerebral Motor Control in Man: Long Loop Mechanisms,* edited by J. E. Desmedt, pp. 85–93. Karger, Basel.

Strick, P. L., and Preston, J. B. (1978a): Multiple representation in the primate motor cortex. *Brain Res.*, 154:366–370.

Strick, P. L., and Preston, J. B. (1978b): Sorting of somatosensory afferent information in primate motor cortex. *Brain Res.*, 156:364–368.

Strick, P. L., and Preston, J. B. (1982): Two representations of the hand in area 4 of a primate. *J. Neurophysiol.*, 48:139–149; 150–159.

Strick, P. L., and Sterling, P. (1974): Synaptic termination of afferents from the ventrolateral nucleus of the thalamus in the cat motor cortex. *J. Comp. Neurol.*, 153:77–106.

Struppler, A. (1960): Neuere Ausblicke in der konservativen Therapie der Spastik. *Nervenarzt,* 31:369–372.

Struppler, A. (1981): A critique of the papers by Hagbarth and Vallbo. In: *Muscle Receptors and Movement,* edited by A. Taylor and A. Prochazka, pp. 291–294. Macmillan, London.

Struppler, A., Burg, D., and Erbel, F. (1973): The unloading reflex under normal and pathological conditions in man. In: *New Developments in Electromyography and Clinical Neurophysiology, Vol. 3,* edited by J. E. Desmedt, pp. 603–617. Karger, Basel.

Struppler, A., Erbel, F., and Velho, F. (1978): An overview on the pathophysiology of Parkinsonian and other pathological tremors. In: *Progress in Clinical Neurophysiology, Vol. 5, Physiological Tremor, Pathological Tremors and Clonus,* edited by J. E. Desmedt, pp. 114–128. Karger, Basel.

Struppler, A., Kessel, F. K., and Weidenbach, W. (1960): Electrophysiologische Untersuchungen an de-afferentierten menschlichen Muskeln: Masseter-Studien. *Med. Monatsschr.* 14:25–26.

Struppler, A., Landau, W. M., and Mehls, O. (1969): Analyse des Entlastungsreflexes am Menschen. *Pfluegers Arch.*, 313:155–167.

Struppler, A., and Preuss, R. (1959): Untersuchungen über periphere und zentrale faktoren der eigenreflexregbarkeit am menschen mit hilfe des Jendrassikschen Hangriffes. *Pfluegers Arch.*, 268:425–434.

Struppler, A., and Velho, F. (1976): Single muscle spindle afferent recordings in human flexor reflex. In: *The Motor System,* edited by M. Shahani, pp. 197–207. Elsevier, Amsterdam.

Stuart, D. G., Goslow, G. E., Mosher, C. G. and Reinking, R. M. (1970): Stretch responsiveness of Golgi tendon organs. *Exp. Brain Res.*, 10:463–476.

Stuart, D. G., Moscher, C. G., Gerlach, R. L., and Reinking, R. M. (1972): Mechanical arrangement and transducing properties of golgi tendon organs. *Exp. Brain Res.*, 14:274–292.

Stuart, D. G., Willis, W. D., and Reinking, R. M. (1971): Stretch-evoked excitatory postsynaptic potentials in motoneurons. *Brain Res.*, 33:115–125.

Sumino, R. (1971): Central neural pathways involved in the jaw opening reflex in the cat. In: *Oral-Facial Sensory and Motor Mechanisms,* edited by R. Dubner and Y. Kawamura, pp. 315–322. Appleton-Century-Crofts, New York.

Sumner, B. R. (1977): Ultrastructural responses of the hypoglossal nucleus to the presence in the tongue of botulinum toxin. *Exp. Brain Res.*, 30:313–321.

Sunderland, S. (1945): The intraneural topography of the radial, median and ulnar nerves. *Brain,* 68:243–299.

Sutherland, D. H. (1966): An electromyographic study of the planter flexors of the ankle in normal walking on the level. *J. Bone Joint Surg.*, 48:66–71.

Sutherland, D. H., Cooper, L., and Daniel, D. (1980): The role of the ankle plantar flexors in normal walking. *J. Bone Joint Surg.*, 62:354–363.

Sutton, G. G. (1975): Receptors in focal reflex myoclonus. *J. Neurol. Neurosurg. Psychiatry,* 38:505–507.

Sutton, G. G., and Mayer, R. F. (1974): Focal reflex myoclonus. *J. Neurol. Neurosurg. Psychiatry,* 37:207–217.

Swanson, A. G. Buchan, G. C., and Alvord, E. C. (1965): Anatomic changes in congenital insensitivity to pain. *Arch. Neurol.*, 12:12–18.

Swash, M., and Fox, K. P. (1972): Muscle spindle innervation in man. *J. Anat. (Lond.),* 112:61–80.

Szentagothai, J. (1949): Functional representation in the motor trigeminal nucleus. *J. Comp. Neurol.*, 90:111–120.

Szentagothai, J. (1975): The 'module-concept' in cerebral cortex architecture. *Brain Res.*, 95:475–496.

Szentagothai, J. (1978): The neuron network of the cerebral cortex: a functional interpretation. *Proc. R. Soc. Lond. (Biol.),* 201:219–248.

Szentagothai, J., and Rethelyi, M. (1973): Cyto- and neuropil architecture of the spinal cord. In: *New Developments in Electromyography and Clinical Neurophysiology, Vol. 3,* edited by J. E. Desmedt, pp. 20–37. Karger, Basel.

Szumski, J., Burg, D., Struppler, A., and Velho, F. (1974): Activity of muscle spindles during muscle twitch and clonus in normal and spastic human subjects. *Electroencephalogr. Clin. Neuophysiol.*, 37:589–597.

Taborikova, H. (1963): Supraspinal influences on H-reflexes. In: *New Developments in Electromyography and Clinical Neurophysiology, Vol. 3,* edited by J. E. Desmedt, pp. 328–335. Karger, Basel.

Taborikova, H., Provini, L., and Decandia, M. (1966): Evidence that muscle stretch evokes long-loop reflexes from higher centres. *Brain Res.*, 2:192–194.

Taborikova, H., and Sax, D. S. (1969): Conditioning of H-reflexes by a preceding subthreshold H-reflex stimulus. *Brain,* 92:203–212.

Tackmann, W., Minkenberg, R., and Strenge, H. (1977): Correlation of electrophysiological and quantitative histological findings in the sural nerve of man. Studies on alcoholic neuropathy. *J. Neurol. Neurosurg. Psychiatry,* 216:289–299.

Takahashi, K. (1965): Slow and fast groups of pyramidal tract cells and their respective membrane properties. *J. Neurophysiol.*, 28:908–924.
Takahashi, T., and Otsuka, M. (1975): Regional distribution of substance P in the spinal cord and nerve roots of the cat and the effect of dorsal root section. *Brain Res.*, 87:1–11.
Takagi, H., Matsunara, M., Yanai, A., and Ogiu, K. (1955): The effect of analgesics on the spinal reflex activity of the cat. *Jpn. J. Pharmacol.*, 4:176–187.
Takamori, M. (1967): H reflex study in upper motoneuron diseases. *Neurology (Minneap.)*, 17:32–40.
Talairach, J., and Bancaud, J. (1966): The supplementary motor area in man. *Int. J. Neurol.*, 5:330–347.
Talbott, R. E. (1974): Modification of postural response of the normal dog by blindfolding. *J. Physiol. (Lond.)*, 243:309–320.
Talboth, W. H., Darian-Smith, I., Kornhuber, H. H., and Mountcastle, V. B. (1968): The sense of flutter-vibration: comparison of the human capacity with response patterns of mechano-receptive afferents from the monkey hand. *J. Neurophysiol.*, 31:301–334.
Tanaka, R. (1974): Reciprocal Ia inhibition during voluntary movements in man. *Exp. Brain Res.*, 21:529–540.
Tanaka, R. (1976): Reciprocal Ia inhibition and voluntary movements in man. *Prog. Brain Res.*, 44:291–302.
Tanaka, R. (1980): Inhibitory mechanism in reciprocal innervation in voluntary movements. In: *Progress in Clinical Neurophysiology, Vol. 8. Spinal and Supraspinal Mechanisms of Voluntary Motor Control and Locomotion*, edited by J. E. Desmedt, pp. 117–128. Karger Basel.
Tanji, J. (1975): Activity of neurons in cortical area 3a during maintenance of steady postures by the monkey. *Brain Res.*, 88:549–553.
Tanji, J., and Evarts, E. V. (1976): Anticipatory activity of motor cortex neurons in relation to direction of an intended movement. *J. Neurophysiol.*, 34:1062–1066.
Tanji, J., and Kato, M. (1973): Recruitment of motor units in voluntary contraction of a finger muscle in man. *Exp. Neurol.*, 40:759–770.
Tanji, J., and Kato, M. (1981): Activity of low and high threshold motor units of abductor digiti quinti in slow and fast voluntary contractions. In: *Progress in Clinical Neurophysiology, Vol. 9, Motor Unit Types, Recruitment and Plasticity in Health and Disease*, edited by J. E. Desmedt, pp. 137–144. Karger, Basel.
Tanji, J., and Kurata, K. (1979): Neuronal activity in the cortical supplementary motor area related with distal and proximal forelimb movements. *Neurosci. Lett.*, 12:201–206.
Tanji, J., and Kurata, K. (1982): Comparison of movement related activity in two cortical motor areas of primates. *J. Neurophysiol.*, 48:633–653.
Tanji, J., Taniguchi, K., and Saga, T. (1980): The supplementary motor area: Neuronal responses to motor instructions. *J. Neurophysiol.*, 43:60–68.
Tanji, J., and Wise, S. P. (1981): Submodality distribution in the sensorimotor cortex of the unanesthetized monkey. *J. Neurophysiol.*, 45:467–481.
Tatcher, D. B., and Van Allen, M. W. (1971): Corneal reflex latency. *Neurology (Minneap.)*, 21:735–737.
Tatton, W. G., and Bawa, P. (1979): Input-output properties of motor unit responses in muscle stretched by imposed displacements of the monkey wrist. *Exp. Brain Res.*, 37:439–457.
Tatton, W. G., Bawa, P., and Bruce, I. C. (1979): Altered motor cortical activity in extrapyramidal rigidity. In: *Advances in Neurology, Vol. 24, The Extrapyramidal System and its Disorders*, edited by L. J. Poirier, T. L. Sourkes, and P. Bedard, pp. 141–160. Raven Press, New York.
Tatton, W. G., Bawa, P., Bruce, I. C., and Lee, R. G. (1978): Long loop reflexes in monkeys: an interpretative base for human reflexes. In: *Progress in Clinical Neurophysiology, Vol. 4, Cerebral Motor Control in Man: Long Loop Mechanisms*, edited by J. E. Desmedt, pp. 229–245. Karger, Basel.
Tatton, W. G., and Bedingham, W. (1981): Characteristic long latency reflex abnormalities in dystonia and rigidity. *Can. J. Neurol. Sci.*, 8:200.
Tatton, W. G., Forner, S. D., Gerstein, G. L., et al. (1975): The effect of postcentral cortical lesions on motor responses to sudden upper limb displacements in monkeys. *Brain Res.*, 96:108–113.
Tatton, W. G., and Lee, R. G. (1975): Evidence for abnormal long-loop reflexes in rigid Parkinsonian patients. *Brain Res.*, 100:671–676.
Taub, E. (1976): Movements in nonhuman primates deprived of somatosensory feedback. *Exerc. Sports Sci. Rev.* 4:335–374.
Taub, E. (1981): Somatosensory deafferentation research with monkeys. In: *Behavioral Psychology and Rehabilitation Medicine*. Williams & Wilkins, Baltimore.
Taub, E., Bacon, R. C., and Berman, A. J. (1965): Acquisition of a trace-conditioned avoidance response after deafferentation of the responding limb. *J. Comp. Physiol. Psychol.*, 59:275–279.
Taub, E., and Berman, A. J. (1968): Movement and learning in the absence of sensory feedback. In: *The Neuropsychology of Spatially Oriented Behavior*, edited by S. J. Freedman, pp. 174–192. Dorsey Press, Springfield, Illinois.
Taub, E., Ellman, S. J., and Berman, A. J. (1966): Deafferentation in monkeys: effect on conditioned grasp response. *Science*, 151:593–594.
Taub, E., Goldberg, I. A., and Taub, P. (1975): Deafferentation in monkeys: pointing at a target without visual feedback. *Exp. Neurol.*, 46:178–186.
Taylor, A. (1981): A critique of the papers by Vallbo and Hagbarth. In: *Muscle Receptors and Movements*, edited by A. Taylor and A. Prochazka, pp. 287–290, Macmillan, London.
Taylor, A., and Appenteng, K. (1981): Distinctive modes of static and dynamic fusimotor drive in jaw muscles. In: *Muscles Receptors and Movement*, edited by A. Taylor and A. Prochazka, pp. 179–192. Macmillan, London.
Taylor, A., and Cody, F. W. J. (1974): Jaw muscle spin-

dle activity in the cat during normal movements of eating and drinking. *Brain Res.,* 71:523–530.

Taylor, A., and Davey, M. R. (1968): Behaviour of jaw muscle stretch receptors during active and passive movements in the cat. *Nature,* 220:301–302.

Taylor, A., and Prochachazka, A. (editors) (1981): *Muscle Receptors and Movements.* Macmillan, London.

Taylor, A., Stephens, J. A., Somjen, G., et al. (1978): Extracellular spike triggered averaging for plotting synaptic projections. *Brain Res.,* 140:344–348.

Taylor, F. V., and Birmingham, H. P. (1948): Studies of tracking behavior. II. The acceleration pattern of quick manual corrective responses. *J. Exp. Psychol.,* 38:783–795.

Teasdall, R. D., Languth, H. W., and Magladery, J. W. (1952): Electrophysiological studies of reflex activity in patients with lesions of the nervous system. *Bull. Johns Hopkins Hosp.,* 91:245–256; 267–275.

Teichner, W. H. (1954): Recent studies of simple reaction time. *Psychol. Bull.,* 51:128–149.

Telford, C. W. (1931): The refractory phase of voluntary and associative responses. *J. Exp. Psychol.,* 14:1–36.

Terenius, L. (1978): Endogenous peptides and analgesia. *Annu. Rev. Pharmacol. Toxicol.,* 18:189–204.

Terzuolo, C. A., Soechting, J. F., and Viviani, P. (1973): Studies on the control of some simple motor tasks. I. Relations between parameters of movements and EMG activities. *Brain Res.,* 58:212–216.

Terzuolo, C. A., and Viviani, P. (1973): Parameters of motion and EMG during simple motor tasks in normal subjects and cerebellar patients. In: *Cerebellum, Epilepsy and Behavior,* edited by I. S. Cooper, M. Rikland, and R. S. Snider. pp. Plenum, New York.

Teuber, H. L. (1975): Recovery of function after brain injury in man; in outcome of severe damage to the central nervous system. *Ciba Foundation Symposium,* 34:159–186. Elsevier, Amsterdam.

Thach, W. T. (1970): Discharge of cerebellar neurons related to two maintained postures and two prompt movements. *J. Neurophysiol.,* 33:527–536; 537–547.

Thach, W. T. (1978a): Single unit studies of long loops involving the motor cortex and cerebellum during limb movements in monkeys. In: *Progress in Clinical Neurophysiology, Vol. 4, Cerebral Motor Control in Man: Long Loop Mechanisms,* edited by J. E. Desmedt, pp 94–106. Karger, Basel.

Thach, W. T. (1978b): Correlation of neural discharge with pattern and force of muscular activity, joint position and direction of intended next movement in motor cortex and cerebellum. *J. Neurophysiol.,* 41:654–676.

Thach, W. T., and Jones, E. G. (1979): The cerebellar dentatothalamic connection: terminal field, lamellae, rods and somatotopy. *Brain Res.,* 169:168–172.

Thiebaut, F., and Isch, F. (1958): Etude sémiologique de mouvement choreique. *Rev. Prat. (Paris),* 8:127–130.

Thoden, U., Dichgans, J., and Savidis, T. (1977): Direction-specific optokinetic modulation of monosynaptic hind limb reflexes in cats. *Exp. Brain Res.,* 30:155–160.

Thoden, U., Magherini, P. C., and Pompeiano, O. (1971): Proprioceptive influence on supraspinal descending inhibitory mechanisms. *Arch. Ital. Biol.,* 109:130–151.

Thom, R. (1975): *Structural Stability and Morphogenesis.* Benjamine, Reading, Pennsylvania.

Thomas, J. S., Brown, J., and Lucier, G. E. (1977): Influence of task set on muscular responses to arm perturbations in normal subjects and Parkinson patients. *Exp. Neurol.,* 55:618–628.

Thomas, J.E., and Lambert, E. H. (1960): Ulnar nerve conduction velocity and H reflex in infants and children. *J. Appl. Physiol.,* 15:1–9.

Thomas, J. E., Regan, T. J., and Klass, D. W. (1977): Epilepsia partialis continua. *Arch. Neurol.,* 34:266–275.

Thomas, J. S., Schmidt, E. M., and Hambrecht, F. T. (1978): Facility of motor unit control during tasks defined directly in terms of unit behaviors. *Exp. Neurol.,* 59:384–395.

Thomas, P. K., Calne, D. B., and Stewart, G. (1974): Hereditary motor and sensory polyneuropathy (peroneal muscular atrophy). *Ann. Hum. Genet.,* 38:111–153.

Thompson, F. J., and Fernandez, J. J. (1975): Patterns of cortical projection to hindlimb muscle motoneurone pools. *Brain Res.,* 97:33–46.

Thompson, J. A. (1980): How do we use visual information to control locomotion. *Trends Neurosci.,* 3:247–250.

Thorne, J. (1965): Central responses to electrical activation of the peripheral nerves supplying the intrinsic hand muscles. *J. Neurol. Neurosurg. Psychiatry,* 28:482–495.

Thrush, D. C. (1973): Congenital insensitivity to pain. *Brain,* 96:369–386.

Thulin, C. A., and Blom, S. (1974): Rigidity and tremor induced in feline muscles deprived of their fusimotor control. *J. Neurol. Sci.,* 21:299–308.

Tilney, F., and Pike, F. H. (1925): Muscular coordination experimentally studied in its relation to the cerebellum. *Arch. Neurol. Psychiatry,* 12:289–334.

Tobin, W. E., and Sandler, S. G. (1966): Depression of muscle spindle function with vincristine. A preliminary report. *Nature,* 212:90–91.

Torebjork, H. E., and Hallin, R. G. (1973): Perceptual changes accompanying controlled preferential blocking of A and C fibre responses in intact human skin nerves. *Exp. Brain Res.,* 16:321–332.

Torebjork, H. E., Ochoa, J. L., and Cann, F. V. (1979): Paresthesia—Abnormal impulse generation in sensory nerve fibres in man. *Acta Physiol. Scand.,* 105:518–520.

Torring, J., Pedersen, R., and Klemar, B. (1981): Standardisation of the electrical elicitation of the human flexor reflex. *J. Neurol. Neurosurg. Psychiatry,* 44:129–132.

Toulouse, P., and Delwaide, P. J. (1980): Topographical aspects of reflex facilitation by remote contraction. *Arch. Phys. Med. Rehabil.,* 61:511–516.

Towe, A. L., and Harding, G. (1970): Extracellular microelectrode sampling bias. *Exp. Neurol.,* 29:366–381.

Towe, A. L., Whitehorn, D., and Nyquist, J. K. (1968):

Differential activity among wide-field neurons of the cat postcruciate cerebral cortex. *Exp. Neurol.*, 20:497–521.
Tower, S. S. (1940a): Unit for sensory reception in cornea. *J. Neurophysiol.*, 3:486–500.
Tower, S. S. (1940b): Pyramidal lesion in the monkey. *Brain*, 63:36–90.
Toyama, K., Maekawa, K., and Takeda, T. (1977): Convergence of retinal inputs onto visual cortical cells. I. A study of the cells monosynaptically excited from the lateral geniculate body. II. A study of the cells disynaptically excited from the lateral geniculate body. *Brain Res.*, 137:207–220; 221–231.
Tracey, D. J. (1979): Characteristics of wrist joint receptors in the cat. *Exp. Brain. Res.*, 34:165–176.
Tracey, D. J., Asanuma, C., Jones, E. G., and Porter, R. (1980a): Thalamic relay to motor cortex: afferent pathways from brain stem, cerebellum and spinal cord in monkeys. *J. Neurophysiol.*, 44:532–554.
Tracey, D. J., Walmsley, B., and Brinkman, J. (1980b): Long loop reflexes can be obtained in spinal monkeys. *Neurosci. Lett.*, 18:59–65.
Traub, M. M., Rothwell, J. C., and Marsden, C. D. (1980): Anticipatory postural reflexes in Parkinson's disease and other akinetic-rigid syndromes and in cerebellar ataxia. *Brain*, 103:393–412.
Traub, M. M., Rothwell, J. C., and Marsden, C. D. (1981): A grab reflex in the human hand. *Brain*, 103:869–884.
Travis, A. M. (1955): Neurological deficiencies following supplementary motor area lesions in Macaca mulatta. *Brain*, 78:174–201.
Trevino, D. L., Coulter, J. D., and Willis, W. D. (1973): Location of cells of origin of spinothalamic tract in lumbar enlargement of the monkey. *J. Neurophysiol.*, 36:750–761.
Troni, W. (1981): Analysis of conduction velocity in the H pathway. *J. Neurol. Sci.*, 51:223–233; 235–246.
Trontelj, J. V. (1973): A study of the F response by single fibre electromyography. In: *New Developments in Electromyography and Clinical Neurophysiology*, Vol, 3, edited by J. E. Desmedt, pp. 318–322. Karger, Basel.
Trontelj, M. A., and Trontelj, J. V. (1978): Reflex arc of the first component of the human blink: a single motoneurone study. *J. Neurol. Neurosurg. Psychiatry*, 41:538–547.
Tsukahara, N. (1978): Synaptic plasticity in the red nucleus. In: *Neuronal Plasticity*, edited by C. W. Cotman, pp. 113–130. Raven Press, New York.
Tsukahara, N. (1981): Synaptic plasticity in the mammalian central nervous system. *Annu. Rev. Neurosci.*, 4:351–381.
Tsukahara, N., and Fujito, Y. (1976): Physiological evidence of formation of new synapses from cerebellum in the red nucleus neurons following crossunion of forelimb nerves. *Brain Res.*, 106:184–188.
Tsukahara, N., Fuller, D. R., and Brooks, V. B. (1968): Collateral pyramidal influences on the cortico-rubrospinal system. *J. Neurophysiol.*, 31:467–484.
Tsukahara, N., Hultborn, H., Urakami, F., and Fujito, Y. (1975): Electrophysiological study of formation of new synapses and collateral sprouting in red nucleus neurons after partial denervation. *J. Neurophysiol.*, 38:1359–1372.
Tsukahara, N., and Kosaka, K. (1968): The mode of cerebral excitation of red nucleus neurons. *Exp. Brain Res.*, 5:102–117.
Tucker, F. R., and Peteleski, N. (1977): Myoelectronic telemetry implant for myo-electric control of a powered prosthesis. *Can. Elec. Eng. J.*, 2:3–7.
Tuttle, R. H. (1967): Knuckle-walking and the evolution of the hominoid hand. *Am. J. Phys. Anthropol.*, 26:171–206.
Tyler, D. B., and Bard, P. (1949): Motion sickness. *Physiol. Rev.*, 29:311–319.
Tyner, C. F. (1975): The naming of neurons. Application of taxonomic theory to the study of cellular populations. *Brain Behav. Evol.*, 12:75–96.
Upton, A. R., McComas, A. J., and Sica, R. E. (1971): Potentiation of late responses evoked in muscles during effort. *J. Neurol. Neurosurg. Psychiatry*, 34:699–711.
Vallbo, A. B. (1970a): Slowly adapting muscle receptors in man. *Acta Physiol. Scand.*, 78:315–333.
Vallbo, A. B. (1970b): Discharge patterns in human muscle spindle afferents during isometric voluntary contractions. *Acta Physiol. Scand.*, 80:552–566.
Vallbo, A. B. (1971): Muscle spindle response at the onset of isometric voluntary contractions in man. Time difference between fusimotor and skeletomotor effects. *J. Physiol. (Lond.)*, 318:405–431.
Vallbo, A. B. (1973): Muscle spindle afferent discharge from resting and contracting muscles in normal human subjects. In: *New Developments in Electromyography and Clinical Neurophysiology*, Vol. 3, edited by J. E. Desmedt, pp. 251–262. Karger, Basel.
Vallbo, A. B. (1974): Human muscle spindle discharge during isometric voluntary contractions. Amplitude relations between spindle frequency and torque. *Acta Physiol. Scand.*, 90:319–336.
Vallbo, A. B. (1981): Independence of skeletomotor and fusimotor activity in man. *Brain Res.*, 223:176–180.
Vallbo, A. B., and Hagbarth, K. E. (1968): Activity from skin mechanoreceptors recorded percutaneously in awake human subjects. *Exp. Neurol.*, 21:270–289.
Vallbo, A. B., and Hagbarth, K. E. (1973): Microelectrode recordings from human peripheral nerves. In: *New Developments in Electromyography and Clinical Neurophysiology*, Vol. 2, edited by J. E. Desmedt, pp. 67–84, Karger, Basel.
Vallbo, A. B., Hagbarth, K. E., Torebjork, H. E., and Wallin, B. G. (1979): Somatosensory proprioceptive and sympathetic activity in human peripheral nerves. *Physiol. Rev.*, 59:919–957.
Vallbo, A. B., and Hulliger, M. (1979): Flexible balance between skeletomotor and fusimotor activity during voluntary movements in man. *Neurosci. Lett.* (Suppl.), 3:103.
Vallbo, A. B., and Hulliger, M. (1981): Independence of skeletomotor and fusimotor activity in man? *Brain Res.*, 223:176–180.
Vallbo, A. B., Hulliger, M., and Nordh, E. (1981): Do spindle afferents monitor joint position in man? A study with active position-holding. *Brain Res.*, 204:209–213.

Van Bogaert, L., Radermaker, J., and Titeca, J. (1950): Les syndromes myocloniques. *Folia Psychiatr. Neurol. (Neerl.),* 53:650–690.

Van Boxtel, A. (1979): Selective effects of vibration on monosynaptic and late EMG responses in human soleus muscle after stimulation of the posterior tibial nerves or a tendon tap. *J. Neurol. Neurosurg. Psychiatry,* 42:995–1004.

Van Gisbergen, J. A. M., Robinson, D. A., and Gielen, S. (1981): A quantitative analysis of generation of saccadic eye movements by burst neurons. *J. Neurophysiol.,* 45:417–442.

Van Hees, J., and Gybels, J. M. (1972): Pain related to single afferent C fibers from human skin. *Brain Res.,* 48:397–400.

Van Ingen Schenau, G. J. (1980): Some fundamental aspects of the biomechanics of overground versus treadmill locomotion. *Med. Sci. Sports Exerc.,* 12: 257–261.

Vazuka, F. A. (1958): Comparative effects of relaxant drugs on human skeletal muscle hyperactivity. *Neurology (Minneap.),* 8:446–454.

Veale, J. L., Rees, S., and Mark, R. F. (1973): Reshaw cell activity in normal and spastic man. In: *New Developments in Electromyography and Clinical Neurophysiology,* Vol. 3, edited by J. E. Desmedt, pp. 523–537. Karger, Basel.

Vecchierini, M. F., and Guiheneuc, P. (1979): Electrophysiological study of the peripheral nervous system in children: changes in proximal and distal conduction velocity from birth to age five. *J. Neurol. Neurosurg. Psychiatry,* 42:753–759.

Vecchierini, M. F., and Guiheneuc, P. (1981): Excitability of the monosynaptic reflex pathway in the child from birth to four years of age. *J. Neurol. Neurosurg. Psychiatry,* 44:309–314.

Verrier, M., Ashby, P., and McLeod, S. (1977): Diazepam effect on reflex activity in patients with complete spinal cord lesions and in those with other causes of spasticity. *Arch. Phys. Med.,* 58:148–153.

Victor, M. (1975): Polyneuropathy due to nutritional deficiency and alcoholism. In: *Peripheral Neuropathy,* Vol. 2, edited by P. J. Dyck, P. J. Thomas, and E. H. Lambert, pp. 1031–1066. Saunders, Philadelphia.

Vidal, P. P., Gouny, M., and Berthoz, A. (1978): Role de la vision dans le déclenchement de réactions posturales rapides. *Arch. Ital. Biol.,* 116:281–291.

Vidal, P. P., Lacour, M., and Berthoz, A. (1979): Contribution of vision to muscle responses in monkey during free-fall: visual stabilization decreases vestibular dependent response. *Exp. Brain Res.,* 37: 241–252.

Vierck, C. J., Hamilton, D. M., and Thornby, J. L. (1971): Pain reactivity of monkeys after lesions to the dorsal lateral columns of the spinal cord. *Exp. Brain Res.,* 13:140–158.

Vighetto, A., and Perenin, M. T. (1981): Ataxie optique. *Rev. Neurol. (Paris),* 137:357–372.

Vigouret, J., Teschemaker, H., Albus, K., and Herz, A. (1973): Differentiation between spinal and supraspinal sites of action of morphine when inhibiting the hindleg flexor reflex in rabbits. *Neuropharmacology,* 12:111–121.

Vilis, T., and Cooke, J. (1976): Modulation of the functional stretch reflex by the segmental reflex pathway. *Exp. Brain Res.,* 25:247–254.

Vilis, T., and Hore, J. (1980): Central neural mechanisms contributing to cerebellar tremor produced by limb perturbations. *J. Neurophysiol.,* 43:279–291.

Vilis, T., Hore, J., Meyer-Lohmann, J., and Brooks, V. B. (1976): Dual nature of the precentral responses to limb perturbations revealed by cerebellar cooling. *Brain Res.,* 117:336–340.

Vilis, T., and Lennerstrand, G. (1979): Dynamics of isometric tension in single motor units in the inferior oblique muscle of the cat. *Acta Physiol. Scand.,* 105:390–392.

Vince, M. A. (1948): The intermittency of control movements and the psychophysical refractory period. *Br. J. Psychol.,* 38:149–157.

Vince, M. A., and Welford, A. T. (1967): Time taken to change the speed of a response. *Nature,* 213: 532–533.

Visser, S. L. (1965): The significance of the Hoffmann reflex in the EMG examination of patients with herniation of the nucleus pulposus. *Psychiatr. Neurol. Neurochir. (Amsterdam),* 68:300–305.

Vitali, M., Robinson, K. P., Andrews, B. G., and Harris, E. E. (1978): *Amputations and Prostheses.* Bailliere Tindall, London.

Viviani, P., and Terzuolo, C. (1980): Space-time invariance in learned motor skills. In: *Tutorials in Motor Behavior,* edited by G. E. Stelmach and J. Requin, pp. 525–533. North-Holland, Amsterdam.

Vogt, B. A., and Pandya, D. N. (1978): Cortico-cortical connections of somatic sensory cortex (areas 3, 1 and 2) in the rhesus monkey. *J. Comp. Neurol.,* 177:179–192.

Vogt, C., and Vogt, O. (1919): Allegemeinere Ergebnisse unserer Hirnforschung. *J. Psychol. Neurol. (Leipzig),* 25:277–462.

Von Holst, E. (1954): Relations between the central nervous system and the peripheral organs. *Br. J. Anim. Behav.,* 2:89–94.

Von Holst, E., and Mittelstadt, H. (1950): Das Reafferenzprinzip. *Naturwissenschaften,* 37:464–476.

Vrbova, G., Gordon, T., and Jones, R. (1978): *Nerve-Muscle Interaction.* Chapman and Hall, London.

Vredenbregt, J. and Rau, G. (1973): Surface electromyography in relation to force, muscle length and endurance. In: *New Developments in Electromyography and Clinical Neurophysiology,* Vol. 1, edited by J. E. Desmedt, pp. 607–622. Karger, Basel.

Wachholder, K. (1928): Willkurliche Haltung und Bewegung. *Erbeg. Physiol.,* 26:568–775.

Wachholder, K., und Altenburger, H. (1926): Beiträge zur Physiologie der Willkürlichen Bewegung. IX. Mitteilung Fortlaufende Hin- und Herlewegungen. X. Einzelbewegungen. *Pfluegers Arch.,* 214:625–641; 642–661.

Waespe, W., and Henn, V. (1977): Neuronal activity in the vestibular nuclei of the alert monkey during vestibular and optokinetic stimulation. *Exp. Brain Res.,* 27:523–538.

Waespe, W., and Henn, V. (1978): Conflicting visual vestibular stimulation and vestibular nucleus activity in alert monkey. *Exp. Brain Res.,* 33:203–211.

Waespe, W., and Henn, V. (1979): The velocity response of vestibular nucleus neurons during vestibular, visual and combined angular acceleration. *Exp. Brain Res.*, 37:337–347.

Wager, E. W., and Buerger, A. A. (1974): A linear relationship between H reflex latency and sensory conduction velocity in diabetic neuropathy. *Neurology (Minneap.)*, 24:711–714.

Walker, A. E. (1949): Afferent connections. In: *The Precentral Motor Cortex*, 2nd edition, edited by P. C. Bucy, pp. 112–132. University of Illinois Press, Urbana.

Walker, C. F., Lockhead, G. R., Marble, D. R., and McElhaney, J. H. (1977): Parameters of stimulation and perception in an artificial sensory feedback system. *J. Bioeng.*, 1:251–260.

Walker, D. D., and Kimura, J. (1978): A fast-recovery electrode amplifier for electrophysiology. *Electroencephalogr. Clin. Neurophysiol.*, 45:789–792.

Wall, P. D. (1967): The laminar organization of dorsal and effects of descending impulses. *J. Physiol. (Lond.)*, 188:403–423.

Wall, P. D. (1970): The sensory and motor role of impulses traveling in the dorsal columns towards cerebral cortex. *Brain*, 93:505–524.

Wall, P. D. (1978a): The gate control theory of pain mechanisms. A re-examination and a re-statement. *Brain*, 101:1–18.

Wall, P. D. (1978b): Mechanisms of plasticity of connection following damage in adult mammalian systems. In: *Recovery of Function: Theoretical Considerations for Brain Injury Rehabilitation*, edited by P. Bach-y-Rita, pp. 91–105. University Park, Baltimore.

Wall, P. D., and Dubner, R. (1972): Somatosensory pathways. *Annu. Rev. Physiol.*, 34:315–336.

Wall, P. D., and Egger, M. D. (1971): Formation of new connexions in adult rat brains after partial deafferentation. *Nature*, 232:542–545.

Wall, P. D., Freeman, J., and Major, D. (1967): Dorsal horn cells in spinal and in freely moving rats. *Exp. Neurol.*, 19:519–529.

Wall, P. D., Waxman, S., and Basbaum, A. I. (1974): Ongoing activity in peripheral nerve: injury discharge. *Exp. Neurol.*, 45:576–589.

Wallin, B. G., Hongell, A., and Hagbarth, K. E. (1973): Recordings from muscle afferents in Parkinsonian rigidity. In: *New Developments in Electromyography and Clinical Neurophysiology*, Vol. 3, edited by J. E. Desmedt, pp. 263–272. Karger, Basel.

Wallin, G., and Hagbarth, K. E. (1978): Muscle spindle activity in man during voluntary alternating movements, Parkinsonian tremor and clonus. In: *Progress in Clinical Neurophysiology, Vol. 5, Physiological Tremor, Pathological Tremors and Clonus*, edited by J. E. Desmedt, pp. 150–159. Karger, Basel.

Walmsley, B., Hodgson, J. A., and Burke, R. E. (1978): Forces produced by medial gastrocnemius and soleus muscles during locomotion in freely moving cats. *J. Neurophysiol.*, 41:1203–1216.

Walsh, E. G. (1962): The perception of rhythmically repeated linear motion in the horizontal plane. *Br. J. Psychol.*, 53:439–445.

Walsh, E. G. (1964): The perception of rhythmically repeated linear motion in the vertical plane. *Q. J. Exp. Physiol.*, 49:58–65.

Walsh, J. C., and Mc Leod, J. G. (1970): Alcoholic neuropathy. *J. Neurol. Sci.*, 10:457–469.

Walsh, J. T., and Cordeau, J. P. (1965): Responsiveness in the visual system during various phases of sleep and waking. *Exp. Neurol.*, 11:80–103.

Walsh, R. N. (1981): Effects of environmental complexity and deprivation on brain anatomy and histology: a review. *Int. J. Neurosci.*, 12:33–51.

Walsh, R. N., and Greenough, W. T. (editors) (1976): *Environments as Therapy for Brain Disfunction*. Plenum, New York.

Walshe, F. M. (1919): On the genesis and physiological significance of spasticity and other disorders of motor innervation with a consideration of the functional relationships of the pyramidal tract. *Brain*, 42:1–28.

Walshe, F. M. (1935): Syndrome of the premotor cortex and the definition of the terms "premotor" and "motor." *Brain*, 58:49–80.

Wand, P., Prochazka, A., and Sontag, K. H. (1980): Neuromuscular responses to gait perturbations in freely moving cats. *Exp. Brain Res.*, 38:109–114.

Wann, K. T., MacDonald, A. G., Harper, A. A., and Ashford, R. (1980): Transient versus state effects of high hydrostatic pressure. *Proceedings: 7th Symposium in Underwater Physiology (A satellite of the XXVIII Int. Congr. Physiol. Sci.)*, Athens, pp. 61–62.

Watson, C. W., and Denny-Brown, D. (1953): Myoclonus epilepsy as a symptom of diffuse neuronal disease. *Arch. Neurol. Psychiatry*, 70:151–168.

Watt, D. G. D. (1976): Responses of cats to sudden falls: an otolith originating reflex assisting landing. *J. Physiol. (Lond.)*, 39:257–265.

Watt, D. G. D., Stauffer, E. K., Taylor, A., et al. (1976): Analysis of muscle receptor connections by spike-triggered averaging. I. Spindle primary and tendon organ afferents. *J. Neurophysiol.*, 39:1374–1392.

Watt, D. G. D., and Zucker, H. A. (1980): Adaptation of an otolith-spinal reflex during prolonged exposure to altered gravity. *Am. Soc. Neurosci. (Abstr.)*, 6:558.

Waxman, S. G. (1980): The structural basis for axonal conduction abnormalities in demyelinating diseases. In: *Progress in Clinical Neurophysiology, Vol. 7, Clinical Uses of Somatosensory Evoked Potentials*, edited by J. E. Desmedt, pp. 170–189. Karger, Basel.

Weaver, R. A., Landau, W. M., and Higgins, J. F. (1963): Fusimotor function. Part II. Evidence of fusimotor depression in human spinal shock. *Arch. Neurol.*, 9:127–132.

Webster, D. D. (1964): The dynamic quantitation of spasticity with automated integrals of passive motion resistance. *Clin. Pharmacol. Ther.*, 5:900–908.

Webster, D. D. (1966): Rigidity in extrapyramidal disease. *J. Neurosurg.*, 24:299–307.

Weddell, G. (1941): The pattern of cutaneous innervation in relation to cutaneous sensibility. *J. Anat.*, 75:346–367.

Weddell, G. (1945): The anatomy of cutaneous sensibility. *Br. Med. Bull.*, 3:167–172.

Weddell, G., and Sinclair, D. C. (1953): The anatomy

of pain sensibility. *Acta Neuroveg. (Wien)*, 7:135–146.
Welch, W. K., and Kennard, M. A. (1944): Relation of cerebral cortex to spasticity and flaccidity. *J. Neurophysiol.*, 7:256–269.
Welford, A. T. (1952): The psychological refractory period and the timing of high-speed performance. *Br. J. Psychol.*, 43:2–19.
Welford, A. T. (1968): *Fundamentals of Skills*. Methuen, London.
Welford, A. T. (1980): Choice reaction time: basis concepts. In: *Reaction Times*, edited by A. T. Welford, pp. 73–128. Academic Press, New York.
Welford, A. T., Norris, A. H., and Shock, N. W. (1969): Speed and accuracy of movement and their changes with age. In: *Attention and Performance II*, edited by W. G. Koster. *Acta Psychol.*, 30:3–15.
Welker, W. I., Benjamin, R. M., Miles, R. C., and Woolsey, C. N. (1957): Motor effects of stimulation of cerebral cortex of squirrel monkey. *J. Neurophysiol.*, 20:347–364.
Westbury, D. (1972): A study of stretch and vibration reflexes of the cat by intracellular recording from motoneurones. *J. Physiol. (Lond.)*, 226:37–56.
Westheimer, G. (1954): Eye movement responses to a horizontally moving visual stimulus. *Arch. Ophthalmol.*, 52:932.
Westling, K. G., and Johansson, R. S. (1980): Factors setting the grip force when lifting an object with index and thumb. *Neurosci. Lett. (Suppl.)*, 5:113.
Westphal, C. (1880): Ueber eine Art paradoxer Muskelcontraction. *Arch. Psychiatr. Nervenkr.*, 10:243–248.
White, C., and Rickens, A. (1974): Thymoxamine and spasticity. *Lancet*, I:686–687.
Wiesendanger, M. (1973a): Some aspects of pyramidal tract functions in primates. In: *New Developments in Electromyography and Clinical Neurophysiology*, Vol. 3, edited by J. E. Desmedt, pp. 159–174. Karger, Basel.
Wiesendanger, M. (1973b): Input from muscle and cutaneous nerves of the hand and forearm to neurones of the precentral gyrus of baboons and monkeys. *J. Physiol. (Lond.)*, 228:203–219.
Wiesendanger, M. (1981): Organization of secondary motor areas of cerebral cortex. In: *Handbook of Physiology*, Section 1, Vol. 2, pp. 1121–1147. Elsevier, Amsterdam.
Wiesendanger, M., Ruegg, D. G., and Lucier, G. E. (1975): Why transcortical reflexes? *Can. J. Neurol. Sci.*, 2:295–301.
Wiesendanger, M., Ruegg, D. G., and Wiesendanger, R. (1979): The corticopontine system in primates: anatomical and functional considerations, In: *Cerebro-cerebellar Interactions*, edited by J. Massion and H. Sasaki, pp. 45–65. Elsevier, Amsterdam.
Wiesendanger, M., Seguin, J. J., and Künzle, H. (1973): The supplementary motor area—a control system for posture? In: *Control of Posture and Locomotion*, edited by R. B. Stein, pp. 331–346. Plenum, New York.
Wieser, H. G., Graf, H. P., Bernoulli, C., and Siegfried, J. (1978): Quantitative analysis of intracerebral recordings in epilepsia partialis continua. *Electroencephalogr. Clin. Neurophysiol.*, 44:14–22.
Wikler, A., and Frank, K. (1944): Hindlimb reflexes of chronic spinal dog during cycles of addiction to morphine and methadone. *J. Pharmacol., Exp. Ther.*, 80:176–187.
Willems, E. (1911): Localisation motrice et kinesthesique. Les noyaux masticateur et mésencéphalique du trijumeau chez le lapin. *Le Névraxe*, 12:1–12.
Willer, J. C. (1975): Influence de l'anticipation de la douleur sur les fréquences cardiaque et respiratoire et sur le réflexe nociceptif chez l'homme. *Physiol. Behav.*, 15:411–415.
Willer, J. C. (1977): Comparative study of perceived pain and nociceptive flexion reflex in man. *Pain*, 3:69–80.
Willer, J. C., and Bathien, N. (1975): Determination of an indication of pain by the sapheno-bicipital reflex method. *Electomyography*, 15:127–135.
Willer, J. C., and Bathien, N. (1977): Pharmacological modulations on the nociceptive flexion reflex in man. *Pain*, 3:111–119.
Willer, J. C., Bathien, N., and Hugelin, A. (1976): Evaluation of central and peripheral pain mechanisms by nociceptive flexion reflex in man. In: *Advances in Pain Research and Therapy*, Vol. 1, edited by J. J. Bonica and D. Albe-Fessard, pp. 131–135. Raven Press, New York.
Willer, J. C., Boureau, F., and Albe-Fessard, D. (1978a): Role of large diameter cutaneous afferents in transmission of nociceptive messages: electrophysiological study in man. *Brain Res.*, 152:358–364.
Willer, J. C., Boureau, F., and Berny, J. (1979a): Nociceptive flexion reflexes elicited by noxious laser radiant heat in man. *Pain*, 7:15–20.
Willer, J. C., Boureau, F., Dauthier, C., and Bonora, M. (1979b): Study of naloxone in normal awake man: effects on heart rate and respiration. *Neuropharmacology*, 18:469–472.
Willer, J. C., and Bussel, B. (1980): Evidence for a direct spinal mechanism in morphine-induced inhibition of nociceptive reflexes in humans. *Brain Res.*, 187:212–215.
Willer, J. C., and Dehen, H. (1977): Respective importance of different electrophysiological parameters in alcoholic neuropathy. *J. Neurol. Sci.*, 33:387–396.
Willer, J. C., Dehen, H., Boureau, F., and Cambier, J. (1978b): Congenital insensitivity to pain and naloxone. *Lancet*, 1:739.
Willer, J. C., Dehen, H., Boureau, F., and Calbier, H. (1978c): Further observations on the endogenous morphine-like system in relation to congenital insensitivity to pain. *J. Med.*, 9:269–272.
Willer, J. C., Roby, A., Boulu, P., and Boureau, F. (1982): Comparative effects of electroacupuncture and transcutaneous nerve stimulation on the human blink reflex. *Pain*, 14:267–278.
Williams, R. G. (1980): Sensitivity changes shown by spindle receptors in chronically immobilized skeletal muscle. *J. Physiol. (Lond.)*, 306:2P.
Willis, W. D., and Coggeshall, R. E. (1978): *Sensory Mechanisms of the Spinal Cord*. Wiley, New York.
Willis, W. D., Trevino, D. L., Coulter, J. D., and Maunz, R. A. (1974): Responses of primate spino-

thalamic tract neurons to natural stimulation of the hindlimb. *J. Neurophysiol.*, 37:358–372.
Wilson, V. J., and Melvill Jones, G. (1979): *Mammalian Vestibular Physiology.* Plenum, New York.
Wilson, V. J., and Peterson, B. W. (1981): Vestibulospinal and reticulospinal systems. In: *Handbook of Physiology,* Section 1, Vol. 2, pp. 667–702. American Physiological Society, Washington, D. C.
Windhorst, U., and Meyer-Lohmann, J. (1977): The influence of extrafusal muscle activity on discharge patterns of primary muscle spindle endings. *Pfluegers Arch.*, 372:131–138.
Windle, W. F. (1926): The distribution and probable significance of unmyelinated nerve fibres in the trigeminal nerve of the cat. *J. Comp. Neurol.*, 41:453–477.
Winkelman, R. K. (1960): *Nerve Endings in Normal and Pathologic Skin.* Charles C Thomas, Springfield, Illinois.
Winter, D. A. (1976): The locomotion laboratory as a clinical assessment system. *Med. Prog. Technol.*, 4:95–106.
Wirta, R. W., Taylor, D. R., and Finlay, F. R. (1978): Pattern-recognition arm prosthesis. *Bull. Prosthet. Res.*, 30:8–35.
Wise, S. P., and Tanji, J. (1981a): Supplementary and precentral motor cortex. Contrast in responsiveness to peripheral input in the hindlimb area of the unanesthetized monkey. *J. Comp. Neurol.*, 195:433–451.
Wise, S. P., and Tanji, J. (1981b): Neuronal responses in sensorimotor cortex to ramp displacements and maintained positions imposed on the hindlimb of the unanesthetized monkey. *J. Neurophysiol.*, 45:482–500.
Wolff, R. (1913): Die Areflexie der Cornea bei organischen Nervenkrankheiten. *Arch. Psychiatr.*, 52:716–721.
Wolpaw, J. R. (1980): Correlations between task-related activity and responses to perturbations in primate sensorimotor cortex. *J. Neurophysiol.*, 44:1122–1138.
Wong, Y. C., Kwan, H. C., Mackay, W. A., and Murphy, J. T. (1978): Spatial organization of precentral cortex in awake primates. I. Somatosensory inputs. *J. Neurophysiol.*, 41:1107–1120.
Woodworth, R. S. (1899): The accuracy of voluntary movements. *Psychol. Rev. Monogr.*, 3:1–114.
Woolsey, C. N. (1958): Organization of somatic sensory and motor areas of the cerebral cortex. In: *Biological and Biochemical Basis of Behaviour,* edited by H. Harlow and C. N. Woolsey, pp. 63–81. University of Wisconsin Press, Madison.
Woolsey, C. N. (1964): Cortical localization as defined by evoked potential and electrical stimulation studies. In: *Cerebral Localization and Organization,* edited by G. Schaltenbrand and C. N. Woolsey. University of Wisconsin Press, Madison.
Woolsey, C. N. (1975): Cortical motor map of Macaca mulatta after chronic section of medullary pyramid. In: *Cerebral Localization,* edited by R. Zulch, N. Creutzfeldt, and B. Galbraith, pp. 19–31. Springer, Berlin.
Woolsey, C. N., Settlage, P. H., Meyer, D. R., et al. (1952): Patterns of localization in precentral and "supplementary" motor areas and their relation to the concept of a premotor area. *Res. Publ. Assoc. Res. Nerv. Ment. Dis.*, 30:238–264.
Wuerker, R. B., McPhedran, A. M., and Henneman, E. (1965): Properties of motor units in the heterogenous pale muscle (m. gastrocnemius) of the cat. *J. Neurophysiol.*, 28:85–99.
Wulff, C. H., and Gilliatt, R. W. (1979): F waves in patients with hand wasting caused by a cervical rib and band. *Muscle Nerve,* 2:452–457.
Wurtz, R. H., and Mohler, C. W. (1976): Enhancement of visual responses in monkey striata cortex and frontal eye fields. *J. Neurophysiol.*, 39:766–772.
Wyman, R. J., Waldron, I., and Wachtel, G. M. (1974): Lack of fixed order of recruitment in cat motoneuron pools. *Exp. Brain Res.*, 29:101–114.
Yanagisawa, N. (1980): Reciprocal reflex connections in motor disorders in man. In: *Progress in Clinical Neurophysiology, Vol. 8, Spinal and Supraspinal Mechanisms of Voluntary Motor Control and Locomotion,* edited by J. E. Desmedt, pp. 129–141. Karger, Basel.
Yanagisawa, N., and Goto, A. (1971): Dystonia musculorum deformans. *J. Neurol. Sci.*, 13:39–65.
Yanagisawa, N., and Tanaka, R. (1978): Reciprocal Ia inhibition in spastic paralysis in man. *Electroencephalogr. Clin. Neurophysiol. (Suppl.),* 34:521–526.
Yanagisawa, N., Tanaka, R., and Ito, Z. (1976): Reciprocal Ia inhibition in spastic hemiplegia of man. *Brain,* 99:555–574.
Yanagisawa, N., Tsukagoshi, H., Toyokura, Y., and Narabayashi, H. (1975): Huntington's chorea. *Shinkei Naika (Neurol. Med.),* 2:459–471.
Yap, C. B. (1967): Spinal segmental and long-loop reflexes on spinal motoneurone excitability in spasticity and rigidity. *Brain,* 90:887–896.
Yates, S. K., and Brown, W. F. (1981): The human jaw jerk: electrophysiologic methods to measure the latency, normal values and changes in multiple sclerosis. *Neurology (Minneap.),* 31:632–634.
Yemm, R. (1972): The responses of the masseter and temporal muscles following electrical stimulation of oral nervous membrane in man. *Arch. Oral. Biol.*, 17:23–33.
Yin, T. C. T., and Mountcasle, V. B. (1977): Visual input to the visuomotor mechanism of the monkey's parietal lobe. *Science,* 197:1381–1383.
Yoshida, K., McCrea, R., Berthoz, A., and Vidal, P. P. (1981): Properties of immediate premotor inhibitory burst neurons controlling horizontal rapid eye movements in the cat. In: *Progress in Oculomotor Research,* edited by A. Fuchs and W. Becker. Elsevier, Amsterdam.
Young, J. L., and Mayer, R. F. (1981): Physiological properties and classification of single motor units activated by intramuscular microstimulation in the first dorsal interosseous muscle in man. In: *Progress in Clinical Neurophysiology, Vol. 9, Motor Unit Types, Recruitment and Plasticity in Health and Disease,* edited by J. E. Desmedt, pp. 17–25. Karger, Basel.
Young, L. R. (1970): On visual-vestibular interaction. In: *The Role of the Vestibular Organs in Space Ex-*

ploration, edited by A. Proctor. NASA SP-314: 205–210.

Young, L., Dichgans, J., Murphy, R., and Brandt, T. (1973): Interaction of optokinetic and vestibular stimuli in motion perception. *Acta Otolaryngol.,* 76:24–31.

Young, L. R., and Jacob, L. M. (1968): A revised dynamic otolith model. *Aerospace Med.,* 39:606–608.

Young, L. R., and Oman, C. M. (1969): Model for vestibular adaptation to horizontal rotation. *Aerospace Med.,* 40:1076–1080.

Young, R. R. (1973): The clinical significance of exteroceptive reflexes. In: *New Developments in Electromyography and Clinical Neurophysiology,* Vol. 3. edited by J. E. Desmedt, pp. 697–712. Karger, Basel.

Young, R. R., and Delwaide, P. J. (1981): Drug therapy. Spasticity. *N. Engl. J. Med.,* 304:28–33; 96–99.

Young, R. R., and Hagbarth, K. E. (1980): Physiological tremor enhanced by manoeuvres affecting the segmental stretch reflex. *J. Neurol. Neurosurg. Psychiatry,* 43:248–256.

Young, R. R., and Shahani, B. T. (1979): Clinical neurophysiological aspects of post-hypoxic intentional myoclonus. In: *Advances in Neurology,* Vol. 26, edited by S. Fahn, J. N. Davis, and L. P. Rowland, pp. 85–105. Raven Press, New York.

Yu, S. K., Schmitt, A., and Sessle, B. J. (1973): Inhibitory effects on jaw muscle activity of innocuous and noxius stimulation of facial and intraoral sites in man. *Arch. Oral. Biol.,* 18:861–870.

Yumiya, H., Larsen, D. D., and Asanuma, H. (1979): Motor readjustment and input-output relationship of motor cortex following cross-connection of forearm muscles in cats. *Brain Res.,* 177:566–570.

Zajac, F. E. (1981): Recruitment and rate modulation of motor units during locomotion. In: *Progress in Clinical Neurophysiology, Vol. 9, Motor Unit Types Recruitment and Plasticity in Health and Disease,* edited by J. E. Desmedt, pp. 149–160. Karger, Basel.

Zalkind, V. I. (1971): Method for an adequate stimulation of receptors of the cat carpo-radialis joint. *Sechenov Physiol. J. USSR,* 57:1123–1127.

Zander-Olsen, P., and Diamantopoulos, E. (1967): Excitability of spinal motor neurones in normal subjects and patients with spasticity, Parkinsonian rigidity and cerebellar hypotonia. *J. Neurol. Neurosurg. Psychiatry,* 30:325–331.

Zeeman, W., and Dyken, P. (1968): Dystonia musculorum deformans. In: *Handbook of Clinical Neurology, Vol. 6, Diseases of the Basal Ganglia,* edited by P. J. Vinken and G. W. Bruyn. pp. 517–543. North-Holland, Amsterdam.

Zucker, R. A. (1973): Theoretical implications of the size principle of motoneurone recruitment. *J. Theor. Biol.,* 38:587–596.

Zuckerman, E. G., and Glaser, G. H. (1972): Urea-induced myoclonic seizures. *Arch. Neurol.,* 27: 14–28.

Zwaagstra, B., and Kernell, D. (1981): Sizes of soma and stem dendrites in intracellular labelled α-motoneurones of cat. *Brain Res.,* 204:295–309.

Subject Index

A, see Alpha entries
Abducens motor activity, 566
Abducens nucleus, premotor interneurons in, 584–585
Abduction threshold, 246
Abductor hallucis F wave, 975
Acceleration responses, 99
Acetylcholine, 982
Achilles tendon jerk, 135
Achilles tendon vibration, 678–679
Acoustic neuroma, 778
Actin-myosin mechanical interaction, 242
Action, translating sensation into, 61
Action potentials
 failure of, 183–185
 synchronous, 388
Activation failure, 180–181
Acupuncture, 824–826
Aγ fibers, 809,817–818
Adaptation phases, 214–215
 late, 223–224
Adaptive equilibrium controls, 618–619
Adaptive plasticity, neurobionomics of, 1047–1071
Adduction of thumb, 389
Adductor pollicis, human, 207
AFB (audiovisual feedback), 897
Afferent discharge
 during voluntary movement, 94–97
 fusimotor activity and, 123–129
Afferent drive sources on task-related cortical cells, 369–370
Afferent identification
 in human neurography, 100–105
 reliability of animal, 105–106
Afferent information, gait pattern and, 721,723
Afferent input
 importance of, 47
 kinesthetic, 162–165
 peripheral
  into PM neurons, 415,418
  into SMA, 406
  into SMA neurons, 401–402
Afferent pathways
 distribution of, 266–267
 of short-latency inputs, 263–285
Afferent projections to precentral neurons, 316–319

Afferent proprioceptive CV, 934
Afferent volley dispersion, 140–142
Afferent volleys, pain and, 814–818
Afferents
 conduction velocity of, 97,98
 identification of, 97–106
 proprioceptive, see Proprioceptive afferents
 spindle, see Spindle afferents
Afterdischarge, 923–924
After-hyperpolarization (AHP), 215
Aging, 954–955
Aging effects on competence, 2
Agonist-antagonist innervation, 35
AHP (after-hyperpolarization), 215
Akinesia, parkinsonian, 480
Alcoholics, chronic, 936–937
Alpha-axon conduction velocity, 234
Alpha fibers, 97
Alpha-gamma coactivation, 103,114–116
 term, 129
Alpha-gamma cosilence, 114
Alpha-gamma independence or dissociation, 129–130
Alpha-gamma linkage, 93,129
Alpha-motoneurons (MN), 34–36,93
Alpha motor axon velocity, 234
Amino acids, 982
Amplitude, movement, 15
Amputees
 powered limbs for, 1093–1108
 upper extremity, 1106–1108
Anesthesia
 on crossed triceps tulips, 656
 local, spasticity and, 134
 long-latency stretch reflexes and, 535
Angiotensin II, 983
Ankle clonus, 916
Ankle jerk, 135
Ankle joint angle, 139–140
Ankle muscles, group I fibers from, 705,707
Antagonist coactivation, depression of, 1032–1033
Antagonist co-contraction, 347
 joint stiffness and, 364–365
Antagonist muscles, cortical control of, 347–372
Antagonist vibration response, 687
Anterior-posterior partitioning of somatosensory inputs, 366
Antidromic activity, MN inhibition and, 464

1181

Antidromic collision, 199–207
Antiparetic effects, 1033–1034
Antiparkinsonian drugs, 1007–1010
Antispastic therapies
  gait analysis and, 1013–1034
  motor control components and, 1029–1034
Apical dendrites, 304
Area 3a, cytoarchitecture of, 265–266
Area 4, 303,393
  input-output organization of, 321–327
Area 6, 402
Arm, cutaneomuscular reflex of, 791
Arm trajectory formation, 31–45
  multijoint, 39–45
  single-degree-of-freedom, 36–39
Arousal, behavioral, 287–300
Arousal circuit models
  formulation of, 294–297
  functional implications of, 297–299
Artificial muscular wisdom, 181–183
Ascending-size-order of muscle unit recruitment, 218,219–220
Aspartate, 983
Ataxia
  bilateral, 562
  cerebellar, see Cerebellar ataxia
  optic, see Optic ataxia
  "sensory," 905
  stance, 634–637
  unilateral, 561
Athetosis, 851
  EMG analysis of, 913
Athetotic cerebral palsy, 440
Audiovisual feedback (AFB), 897
Automatic actions
  equilibrium and, 610–612
  equilibrium control and, 615–616
  locomotion and, 612–613
  postural support and, 613–615
Automatic behavior, term, 607
Automatic equilibrium controls, 618–619
Axon collaterals, 308
Axon diameters of gamma fibers, 123
Axonal sprouting, see Sprouting
Axons, pyramidal tract, 310–311

B, see Beta entries
Baclofen, 990–991,1013,1033
  spasticity and, 7–8
Balance at high pressure, 833
Ballistic, term, 13
Ballistic contractions, 232–234
  isometric and isotonic, 243
  mechanical design of, 238–240
Ballistic double discharge, 240–243
Ballistic force threshold, 234

Ballistic mode of voluntary motor control, 230–232
Ballistic movement, 230–231
  dystonia and, 853–861
  myotatic responses and, 544–546
Ballistic movement pattern, 907–909
BAU (burst after unloading), 557
Behavior, automatic, term, 607
Behavioral arousal, 287–300
Behaviors, simple, 607
Bell's palsy, 776–778
Bends, compression and decompression, 831
Beta axons, 113
Beta fibers, 97
Beta innervation, skeletofusimotor, 112–113
Beta-motoneurons, 117
Biceps, long-latency reflexes in, 510,514–515,516
Bicycling
  normal, 1036–1038
  spasticity and, 1035–1046
Bilateral ataxia, 562
Bilaterality
  of PM neurons, 415,417
  of SMA neurons, 403–405
Bioelectric control of powered limbs for amputees, 1093–1108
Bipedal gait, reflex control of, 699–716
Blink reflex, 730–732,773
  acoustic neuroma and, 778
  Bell's palsy and, 776–778
  cerebral lesions and, 768–769
  Charcot-Marie-Tooth disease (CMT), 779
  early, 731
  electrically elicited, 773–786
  facial hypesthesia and, 780–782
  Guillain-Barré syndrome (GBS), 779
  late, 731–732
  multiple sclerosis and, 778–780
  normal values in adults and infants, 774–776
  parasellar meningioma and, 783
  paratrigeminal syndrome and, 776
  polyneuropathy and, 778
  R_1 and R_2 of, 784
  trigeminal nerve lesions and, 763
  trigeminal neuralgia and, 776
  Wallenberg syndrome and, 763–767,780
Blink reflex latencies, 747–772
Block
  ischemic, 195
  neuromuscular, 209
  pressure and lidocaine, 171
Bradykinesia, 6
  in parkinsonism, 884
Brain damage, sprouting after, 1073–1084
Brainstem control, endogenous opiate, 811
"Break frequency," 1105
Bricolage, 300
Burst activities, synchronism of, 578

Burst after unloading (BAU), 557
Burst neurons, 575
  excitatory (EBNs), 577–584
  inhibitory (IBNs), 576–577
  medium-lead, 583
Burst termination, 66
"Busy line" effect, 674,675

C fibers, 809
Carcinomatous radiculopathy, 946–947
Carpal tunnel syndrome (CTS), 972,974
Cascade transmissions, 296–299
Catch phenomenon, 242
Caudal compartments of area, 4,76
Caudal motor cortex, projections to, 84–89
Cell physiology under pressure, 837–838
Cell S, 164
Cell size, 249–250
Cellular response types, 289
Central nervous system (CNS)
  hierarchical design of, 3,11
  lesions, motor control recovery following, 1085–1092
  sprouting in, 1073–1084
Centripetal interaction torque, 40
CEP (cumulative eye position), 1053–1054
Cerebellar ataxia, 8–10
  physostigmine on, 9–10
  SR in, 561–562
  visuomotor tracking in, 889–895
Cerebellar lesions, postural instability with, 633–643
Cerebellar patients, postural activity with, 637–638
Cerebellar Purkinje cells, 383–385
Cerebellopontine angle tumor, 778
Cerebellum
  movement of wrist and, 381,383–385
  role of, 475
Cerebral blood flow, regional, 407–408
Cerebral cortex, monkey, SMA and PM of, 393–408
Cerebral lesions, 759–761
  corneal and blink reflexes and, 768–769
Cerebral palsy
  athetotic, 440
  gait with, 1028
Ceruletide, 994–995
CFF (visual critical fusion frequency), 833
Chain fibers, types of, 109
Charcot-Marie-Tooth disease (CMT), 779
  F waves and, 966–967,970,972,973
Chin vibration, 679
Cholecystokinin, 979,983
Chorea
  EMG analysis of, 914
  motor unit control in, 900–902
  pathophysiology of, 872–874,875
  senile, 914
  Sydenham's, see Sydenham's chorea
  tracking movements in, 886–888
Circular vection, 600
"Click" test, 898,902
Clonus, 486
  ankle, 916
  after spinal injury, 918
Closed-loop condition, 15
  error in, 24–25
  static and dynamic, 18
CMT, see Charcot-Marie-Tooth disease
CNS, see Central nervous system
Coactivation
  alpha-gamma, 103
  antagonist, depression of, 1032–1033
Coactivation system, 371
Coactivation zone, discharge pattern in, 363
"Coal hammer grip," 375
Co-contraction, 441
  antagonist, 347
Co-contraction cells, 369
Co-contraction mode of control, 371–372
Collateral sprouting, 1091
Collaterals
  axon, 308
  recurrent, 304–305
Collision, antidromic, 199–207
Compensation to peripheral disturbances, 47–59
Competence, aging effects on, 2
Compression bends, 831
Conditioning reflex, test reflex and, 444–447
Conductance, input, 220–221
Conduction paths, 106
Conduction velocity (CV)
  of afferent, 97,98
  afferent proprioceptive, 934
  alpha-axon, 234
  axonal, 222
  distal nerve, 942–943
  F-wave, 965–968
  proximal nerve, 930–936
Congenital insensitivity to pain, 821–824
Contraction
  ballistic, see Ballistic contractions
  muscular, see Muscular contractions
  ramp, 232
  vibration-induced, 155–156,163
Contraction times, twitch (CT), 228–229
Control signals at motoneuron pool, 365–366
Convergent facilitation concept, 498
Cooled muscle, 176
Coriolis torque, 40
Corneal reflex, 732–735
  cerebral lesions and, 768–769
  trigeminal nerve lesions and, 763
  Wallenberg syndrome and, 763–767

Corneal reflex latencies, 757–772
Corollary discharge, 151,344
   kinesthetic afferent input and, 162–165
Cortical atrophy, late, 634–636
Cortical cells, task-related, afferent drive sources on, 369–370
Cortical control of antagonist muscles, 347–372
Cortical interneurons, 306
Cortical myoclonus, 865
   spontaneous, 866
Cortical neuronal activity, movement and, 302–303
Cortical neurons
   laminar distribution, 293–294
   response patterns of, 290–294
   taxonomy of, 299–300
Cortical unit recordings, 474
Corticocortical afferents, 305
Corticocortical connections, 85–86
   of sensory-motor cortex, 277–279
Corticofugal neurons, 329,342
Corticomotoneuronal cells, 337
Corticomotoneuronal synapses, 309
Corticorubral system, 342
   pyramidal tract neurons versus, 329–345
Corticospinal axons, 308
Corticospinal discharge, influences of, 306–309
Corticospinal fibers, 282
Corticospinal neurons, 303–304
   discharges during movement of, 309–313
   discharges of, 301
Corticospinal pathways, stimulation of, 387–391
Cortico-subcortical myoclonus, 866
Corticothalamic labeling, 270–271
Critical fusion frequency (CFF), visual, 833
Crossed triceps tulips, 654–655, 659
   anesthesia on, 656
Cryotests, 1014
CT (twitch contraction times), 228–229
CTS (carpal tunnel syndrome), 972,974
Cumulative eye position (CEP), 1053–1054
Current-threshold maps in precentral motor cortex, 367–368
Cutaneomuscular reflex organization, 787–796
   eliciting, 789
   of hand and arm, 791
   interlimb, 794–795
   of leg, 791–794
   methodological considerations, 787–791
Cutaneous afferents, Ib reflex pathways and, 710–711
Cutaneous control of Ib reflex pathways, 708–712
Cutaneous depression of Ib effects, 709–710
Cutaneous facilitation of large motor units, 253–261
Cutaneous input
   MN recruitment and, 257–259
   proprioceptive inputs and, 797
Cutaneous receptive fields, 358
Cutaneous receptors, 253
Cutaneous stimulation site, changing, 711–712
Cutaneous stimulus strength, varying, 709
CV, *see* Conduction velocity

Dantrolene sodium therapy, 1013
Deafferentiation, 56
Deceleration response, 99
Decerebrate cat, 133–134
Decompression bends, 831
Deep sea diving, 829–849
Degeneration, sprouting without, 1082–1083
Deiters' nucleus, 622–623
Delayed feedback, 70–75
Dendrites, apical, 304
Dendritic delayed depolarization, 242
Depolarization
   dendritic delayed, 242
   hyperpolarization and, 288–289
Dermoepidermal plexus, superficial, 812
DF (disfacilitation), 289,566
DI (disinhibition), 289,566
Diabetic neuropathy, 972
Diabetic polyneuropathy (DPN), 974
Digastric jaw opening reflex, 749
Discharge, ballistic doublet, 240–243
Discharge frequency
   coefficient of variation of, 100
   steady-state force and, 335–340
Discharge patterns
   in coactivation zone, 363
   in jaw closer and limb muscles, 680–683
Disfacilitation (DF), 289,566
Disinhibition (DI), 289,566
Disk prolapse, intervertebral, 945–946
Displacement
   duration of, long-latency reflexes and, 493–497
   mechanical, rapid postural reactions to, 645–659
   passive, responses to, 340–341
   ramp-and-hold, 341
Display polarity, 64
Distal forelimb, motor representation of, 321–324
Distal nerve CV, 942–943
Disynaptic Ia inhibitory pathway, 435–437
Disynaptic pathway, 142
Diving, deep sea, 829–849
Dopamine, 982
Dorsal column lesion, 472–473
Dorsal column pathway, 810
Dorsal root recordings
   in cats and monkeys, 96–97
   in monkeys, 123
Double facilitation, 800
Double stimulation, 69–70
Doublet discharge, ballistic, 240–243

Doublet onset firing, 241
DPN (diabetic polyneuropathy), 974
Drug effects on TSR, 1010
Drug value in clinical pharmacology, 995–996
Drugs
    antiparkinsonian, 1007–1010
    muscle relaxant, *see* Muscle relaxant drugs
    neurotropic, 977–996
Dynamic force changes, long-latency stretch reflexes and, 534–537
Dynamic index, 98–99
    suxamethonium and, 99–100
Dynamic ratio, 98,99
Dynamic sensitivity, 1004
"Dynamic TSR," 999
Dynamically sensitive spindle afferents, 111
Dyscontrol, motor, 685–697
Dyskinesia, EMG analysis of, 913
Dystonia
    ballistic movements and, 853–861
    EMG analysis of, 912–913
    EMG patterns in, 852–858
    long-latency stretch reflexes and, 855
    myoclonic, 853
    pathophysiology of, 851–863
    reciprocal inhibition and, 855–859
    voluntary movement and, 853

E (postsynaptic excitation), 289
E + DF response, 293–294
E + E response, 291–293
EBNs (excitatory burst neurons), 577–584
Eccentricity, target, 22
EEG (electroencephalograph) arousal, 287–300
Efference copy, 151,344
Electrical tooth pulp stimulation, 749
Electrically elicited blink reflex, 773–786
Electrically induced reflexes, 473–474
Electro-acupuncture, 824–826
Electrode, recording, 94
Electroencephalograph (EEG) arousal, 287–300
Electromyography, *see* EMG entries
Electrophysiological studies in man, 315–316
Electrophysiological tests, 981,984–985
EMG activity, 94–95
EMG activity changes, excitability curves and, 805–806
EMG method, needle, 227
"EMG modulation curve" technique, 788
EMG recordings, 101
    ergometer, 1035–1046
    surface, analysis of movements with, 907–914
Enhanced vision (EV condition), 600–604
Enkephalins, 979,983
Environmental demands, sensorimotor function and, 1047–1071
Epilepsia partialis continua, 865–866

Epinephrine, 982
EPSPs, *see* Excitatory postsynaptic potentials
Equilibrium
    automatic actions and, 610–612
    posture and, 694–696
    sensory inputs and, 608
Equilibrium control, 615–619
Equilibrium position, 37,38–39
Ergometer EMG, 1035–1046
Error, static, 51
Error-activated reflex control, 1048–1050
Error-feedback system, 499
EV condition (enhanced vision), 600–604
Evoked step, simulated, 126–128
Exafferent signals, 151
Excitability, reflex, 58–59
Excitability curves, EMG activity changes and, 805–806
Excitatory burst neurons (EBNs), 577–584
Excitatory fusimotor bias, 58
Excitatory nonspecific input, 295
Excitatory postsynaptic potentials (EPSPs)
    increases of, 289
    monosynaptic, 219
Excitatory vestibular neurons, 567–570
Extension, "mass" term, 915
Exteroceptive afferents, long loops and, 802–803
Exteroceptive influences on lower limb, 797–807
Exteroceptive input effect on recruitment patterns, 255–257
Exteroceptive reflexes in clinical neuropharmacology, 993–994
Exteroceptive suppression
    adequate stimulus for, 749–750
    in masseter muscles, 735–738
    periods of, 750–752
Extrafusal muscle fibers, 109
Eye-hand sequence in visuomotor task, 22–25
Eye latency, 23
Eye movement, 15
    horizontal conjugate, 565–588
Eye movement-related neurons, 565
Eye position, cumulative (CEP), 1053–1054

f-I relation, 214
F ratio, 965–968
F waves, 191,200–206,952,960
    abductor hallucis and, 975
    central latency of, 962–965
    Charcot-Marie-Tooth disease and, 966–967, 970,972,973
    clinical uses of, 961–975
    conduction velocity of, 965–968
    determination of, 961–962
    Guillain-Barré syndrome and, 963–967,970–972
    in normal subjects and patients with polyneuro-

F waves *(contd.)*
    pathy, 968–975
    SP and, 461,463
Facial hypesthesia, 775
    blink reflex and, 780–782
Facilatatory effects, convergent, on MNs, 497–499
Facilitation
    double, 800
    intensity of, 665–666
Fall, postural reactions to, 592–597
Familial ataxia, 903–904
Fascicles
    dorsal root, 96
    muscle, 94
Fast-twitch fatigue-resistant (FR) muscle unit, 213
Fast-twitch fatigue-sensitive (FF) muscle unit, 213
Fatigue
    under hight pressure, 831
    ischemic, 179
    motoneuron discharge during, 169–211
    motor impulses and, 180–183
    muscular, 152–153
    muscular wisdom and, 169–211
    redundancy and, 4
    voluntary, 390
Fatigue curves, tetanic, 183
FCR (flexor carpi radialis) muscle, 956–960
FCU (flexor carpi ulnaris) muscle, 957–960
Feedback
    continuous, tracking performance and, 65–75
    delayed, 70–75
    "geometric," 1048
    visual, 13
Feedback conditions on hand movements, 18, 20–22
Feedback control, parametric, 1050–1052
Fentanyl citrate, 823
FF (fast-twitch fatigue-sensitive) muscle unit, 213
Finger grip movement, 18
Firing duration, muscle unit type and, 222–223
Firing frequencies
    muscle cooling and, 193–199
    of single motor units, 185–199
Firing patterns
    doublet onset, 241
    of wrist flexor and extensor motor units, 351–357
Firing ranges, 214,242
Firing rate
    muscle length and, 121
    muscle unit type and, 223–224
    percent modulation in (PMFR), 360–361
Fitt's law, 14–15
"Flat-foot" phase of gait, 723,724
Flexion, "mass" term, 915
"Flexion-extension wave," 721
Flexion reflex, 804–805
    nociceptive, pain research and, 809–827
    pain sensations and, 811–827
Flexion threshold, 246
Flexor carpi radialis (FCR) muscle, 956–960
Flexor carpi ulnaris (FCU) muscle, 957–960
Flexor pollicis longus (FPL), long-latency stretch reflexes of, 512–514,516
Flexor reflex, 787
Flexor reflex afferents (FRA), 253, 257
Flexor reflex organization, *see* Cutaneomuscular reflex organization
Foot control, vibration and, 693
Foot tracking, vibration and, 692–693
Force records, 101
Force-speed control, size principle and, 227–251
Forearm movement, 34–35
Forearm positioning, vibration and, 691–692
Forelimb, distal, motor representation of, 321–324
Foveal vision condition, 18
Foveation, 29
FPL, *see* Flexor pollicis longus
FR (fast-twitch fatigue-resistant) muscle unit, 213
FRA (flexor reflex afferents), 253,257
Frequency
    "break," 1105
    peak, 186
    recruitment, 898
Frequency-current relation of spinal MNs, 213–216
Friedreich's ataxia, 903–904
"Frozen gait," 883
Functional stretch response (FSR), 7,468–472, 474,518
Fusimotor activity
    afferent discharge and, 123–129
    phasic, 120
    skeletomotor activity and, 118–119
    spontaneous, in resting muscles, 112
    tendon jerk and, 139–143
Fusimotor bias, excitatory, 58
Fusimotor control concepts, 129–132
Fusimotor drive
    muscle spindle responsiveness and, 138
    tendon jerk and, 138
Fusimotor efferents, 93
Fusimotor neurons
    recruitment order of, 116–118
    static and dynamic, 113–114
Fusimotor role in muscle tone disorders, 133–150
Fusimotor system
    selective blockade of, 134–136
    skeletomotor system and, 130–131

GABA (gamma aminobutyric acid), 7,980,982
Gait
    with cerebral palsy, 1028
    deranged, 1025–1028

"frozen," 883
normal, 1023–1025
paraspastic, 718–721
Gait analysis, 1023–1028
  antispastic therapies and, 1013–1034
  in neurological diseases, 718
Gait mechanisms, protective, 723–726
Gait pattern, afferent information and, 721,723
Gamma-activation, tonic, 130,131
Gamma aminobutyric acid (GABA), 7,980,982
Gamma excitation, 380
Gamma fibers, 97
  axon diameters of, 123
Gamma-fusimotor fibers, 113
Gamma loop, 47
Gamma-motoneurons, 117
Ganglia, dorsal root, 96
Gastrocnemius muscle, 223–224
Gastrocnemius-soleus (GS) spinal stretch reflex, 699
Gate control theory of pain, 810–811
GBS, see Guillain-Barré syndrome
"Geometric feedback," 1048
Gilles de la Tourette's syndrome, 874–881
Glomeruli, functioning, 2
Glutamate, 980,983
Glycine, 982
Golgi tendon organs, 237–238
Granule cell attenuation zone, 266
Group I discharges, proprioceptive, 699
Group I fiber projections in lower limb, 700–708
Group I fibers from ankle muscles, 705,707
Group I reflex pathways, functional significance, 712–716
Grouping, response, 70
GS (gastrocnemius-soleus) spinal stretch reflex, 699
Guillain-Barré syndrome (GBS), 779
  F wave and, 963–967,970–972

H index, 927,930–945
  diminution of, 940–942
H/M ratio, 930–947
  increase of, 944
H reflex, 136,137–143,665
  in clinical neuropharmacology, 987–988
  under high pressure, 838,842–845
  in lower limb, 951–956,959–960
  in man, 433–440
  masseteric, 743–744
  normal values for, 931
  recovery curve after stimulation of tibial nerve, 992
  recovery curve of, 991–992
  tendon jerk and, 137–143
  tibialis anterior, 798
  in upper limb, 951,956–960

H reflex amplitude, 628–630
H' test reflex, 444–454
H1 reflex discharge, 444–451
Hand
  cutaneomuscular reflex of, 791
  primate, 373–374
Hand latency, 23
Hand movement
  duration, 23–24
  feedback conditions on, 18,20–22
Head position, preprogrammed, 33
Head-trunk vibration (HTV), 687
Head vibration, (HV), 687
Heaviness, 156–160
Helium effects, pressure and, 836–837
Hemiparesis
  sensory feedback therapy for, 1087
  spastic, 1025–1026
Hemiplegia
  long-latency reflexes in, 504,506–508
  spastic, 439–440
Hierarchical design of CNS, 3,11
High pressure
  cell physiology under, 837–838
  clinical neurophysiology under, 838–840
  fatigue under, 831
  H reflex under, 838,842–845
  helium effects and, 836–837
  human performance under, 830–835
  myoclonus under, 842
  neurological syndrome (HPNS), 840–841
  neurophysiology at, 835–849
  reflexes under, 841–842
  sensorimotor abilities under, 832–833
  somatic evoked potential under, 846–849
  tremor under, 845–846
Hippocampus, sprouting in, 1075–1076
Hodgkin's disease, 937
Hoffmann reflex, see H reflex
Homotopic input-output relations, 61
HPNS (high-pressure neurological syndrome), 840–841
HTV (head-trunk vibration), 687
Hung-up tendon jerk, 873,876
Huntington's chorea, see Chorea
HV (head vibration), 687
Hyperbaric conditions, motor control in, 829–849; see also High pressure entries
Hyperpolarization, depolarization and, 288–289
Hyper-reflexia, 471
Hypertonia, disorders associated with, 147–150
Hypesthesia, facial, see Facial hypesthesia
Hypotonia, 146
  disorders associated with, 144–147

I (postsynaptic inhibition), 289
I + DI response, 293,295

I + E response, 293–294
Ia afferents, 140
Ia fibers, 703
Ia inhibitory pathway
  in central motor diseases, 439–440
    athetotic cerebral palsy, 440
    spastic hemiplegia, 439–440
  disynaptic, 435–437
  in normal subjects, 437–439
    at rest, 437–438
    at voluntary contraction, 438–439
  reciprocal, 433–441
Ia interneuron, 991
IAPP (isopropylamino-2-pyridine phosphate), 947–948
Ib afferents, 140
Ib effects
  cutaneous depression of, 709–710
  tonic depression of, 712
Ib inhibition, 699
  depression of, 708–709
Ib inhibitory pathways, 261
Ib reflex pathways
  cutaneous afferents and, 710–711
  cutaneous control of, 708–712
IBNs (inhibitory burst neurons), 576–577
Illumination, strobe, 1061–1065
Impact artifacts, 1019
Impedance, mechanical, of wrist, 349–351
Implants, surgical, 1098–1099
Inactivation, 214
Inertial-gravitational input, 617
Inhibition
  recurrent, *see* Recurrent inhibition
  vibration-induced, *see* Vibration-induced inhibition
Inhibitory burst neurons (IBNs), 576–577
Inhibitory nonspecific input, 295
Inhibitory postsynaptic potentials (IPSPs)
  increases of, 289
  late, 256
Inhibitory vestibular neurons, 570–572
Initial burst, 99
Initial peak velocity (IPV), 890–893
Innervation, agonist-antagonist, 35
Input conductance, 220–221
Input-output organization
  of area 4, 321–327
Input-output relations, homotopic, 61
Insensitivity, congenital, to pain, 821–824
Interaction torques, 40
Intercostal spindle, human, 110
Interlimb cutaneomuscular reflexes, 794–795
Intermediate equilibrium position, 38–39
Interneuronal transmission, 991
Interneurons, 257; *see also* Renshaw cells
  cortical, 306
  premotor, in abducens nucleus, 584–585

Interpeduncular nucleus, sprouting in, 1074
Interpositus nucleus, 383
Intervertebral disk prolapse, 945–946
Intrafusal muscle fibers, 108
Intramuscular EMG recordings, 101
Involuntary movements, 3–4
  analysis with surface EMG, 909–914
IPSPs, *see* Inhibitory postsynaptic potentials
IPV (initial peak velocity), 890–893
Ischemic block, 195
Ischemic fatigue, 179
Isokinetic movements
  computer system for analysis of, 1022–1023
  electromyographic records in, 1020–1022
  torque in, 1015–1018
Isokinetic tests, movement in, 1018–1019
Isometric ballistic contractions, 243
Isometric tensions, 153–154
Isomorphic map, 83
Isopropylamino-2-pyridine phosphate (IAPP), 947–948
Isotonic ballistic contractions, 243
Isotonic ramp threshold, 245
Isotonic twitch response, 97–98

Jaw closer muscles, 679
  discharge patterns in, 680–683
  loading, 739–743
  unloading, 745
Jaw reflex, 729–730
  human, 739–755
  myotatic, 731
  silent period after, 753–755
Jaw-opening muscles, loading, 744
Jaw-opening reflex, 748
  digastric, 749
Jendrassik maneuver, 136–137, 661
  analysis of, 661–669
  factors influencing facilitation, 664–666
  functional role, 667–668
  mechanisms of conditioning, 666–667
  methodological considerations, 662–664
  after spinal injury, 918, 919
Jitter, 142, 681
Joint movement, control of, 347–372
Joint reversals, 44
Joint stiffness
  antagonist co-contraction and, 364–365
  control of, 347–372

Ketamine, 96
Kinesia paradoxica phenomena, 5
Kinesiological analysis, 718
Kinesthetic afferent inputs, 162–165
Kinesthetic perceptive defects, 694
Knee extension, 714
Knee flexion, 715

Labyrinthectomy, unilateral, 627
Labyrinthine reflex control, tonic, of limb posture, 623–631
Lag cells, 76
Landing
 muscle responses after, 595–597
 muscle responses before, 593–595
Latency, response, see Response latency
Lateral pinch, 376
Lead cells, 76
 classes of, 81–82
 in rostral motor cortex, 77–80
 stimulus-response relations in, 80–83
Leg
 cutaneomuscular reflex of, 791–794
 stretch reflexes in, 518
Leg tracking, visuomotor, 883–888
Leg vibration (LV), 687
Lemniscal fibers, 282
 terminal distribution of, 268
Length-tension curves, 35–36, 364–365
Lesion
 central nervous system, motor control recovery following, 1085–1092
 cerebral, see Cerebral lesions
 of dorsal column, 472–473
 root, 945–947
 of SMA, see Supplementary motor area, lesions of
 sprouting after, 1079–1082
 suprasegmental, 792
 trigeminal nerve, 758–759
Lidocaine block, 171
Limb
 lower
  assessing spasticity of, 1035–1046
  exteroceptive influences on, 797–807
  group I fiber projections in, 700–708
  H reflexes in, 951–956, 959–960
  spinal reflexes in, 687–689
 phantom, 152
 powered, 1093–1108
 rigidity in, 5
 upper, H reflexes in, 951, 956–960
Limb muscles, discharge patterns in, 680–683
Limb posture, tonic labyrinthine reflex control of, 623–631
Limb stiffness concept, 499, 500
Linear vection, 600
Loading
 jaw-closing muscles, 739–743
 jaw-opening muscles, 744
Loads, viscous, 66
Locomotion
 automatic actions and, 612–613
 pathophysiology of human, 717–726
 sensorimotor actions and, 717–718
 treadmill, 718

Locomotor patterns, vestibular system and, 723
Long-latency myotatic reflexes
 changes in, 541–554
  functional interpretation of, 548, 550–551
  in normal subjects, 542–546
  in Parkinson's disease patients and karate experts, 546–548, 549
 clinical applications, 502–508
 dependence on duration of displacement, 493–497
 functional significance of, 499–502
 in man, 489–508
 mechanisms underlying, 490–499
Long-latency response, 7
Long-latency stretch responses
 anesthesia and, 535
 in biceps, 510, 514–515, 516
 dynamic force changes and, 534–537
 dystonia and, 855
 flexor pollicis longus, 512–514, 516
 functions of, 524–537
 history, 509–512
 origin and function, 509–539
 physiological nature of, 478–482
 produced by large disturbances, 526–528
 produced by small disturbances, 525–526
 segmental versus suprasegmental contributions to, 467–487
 spindle volley and, 518, 520–523
 subsequent responses to perturbation and, 531–533
 supraspinal contribution to, 482–487
 in thumb flexor, 514
 thumb flexor and, 524–533
 transcerebral concept of, 523–524
 transcortical concept of, 516–524
Long-latency suprasegmental responses, 640–643
Long-loop concept, 489
Long loops, exteroceptive afferents and, 802–803
LV (leg vibration), 687

M, see Motor response
M-class cells, 369
M wave, 135–136, 200
 multiples of threshold strength for (xMT), 434
M1 component, 489
M2-3 component, 490
Map
 isomorphic, 83
 proprioceptive, 27
 visual, 13–14
Martin-Gruber anastomosis, 186
"Mass flexion and extension," terms, 915
Masseter motor unit, 681
Masseter muscles
 exteroceptive suppression in, 735–738
 M responses of, 743

Masseter muscles *(contd.)*
  SP in human, 735–738
Masseter reflex, 729–730
  vibration-induced potentiation of, 677–678
Masseter tonic vibration reflex, 745–748
Masseteric H reflex, 743–744
Maturation of sensory mechanisms, 618–619
Maximal shocks, 204–205
Medium-lead burst neurons, 583
Membrane potential changes, 288
"Microchorea," 901
Microneurography in humans, 94–95
Microswitch shoes, 718
Midbrain recordings in cats and monkeys, 95–96
MN, *see* Alpha-motoneurons
Monoamines, 982
Monosynaptic excitatory postsynaptic potentials, 219
Monosynaptic reflex, 142
  in polyneuropathy, 927–949
Monosynaptic stretch reflex (MSR), 468
Morphine, 821
  mechanisms of, 820–821
α-Motoneuron, *see* Alpha-motoneurons
Motoneuron discharge, 188
  during fatigue, 169–211
Motoneuron inhibition, antidromic activity and, 464
Motoneuron pool, 213
  control signals at, 365–366
  excitability of, 986–988
  rank ordering in, 237–238
Motoneuron recruitment
  cutaneous input and, 257–259
  size principle of, 227–251
Motoneurons, 227
  convergent facilitatory effects on, 497–499
  cutaneous control of Ib reflex pathways to, 708–712
  spinal, *see* Spinal motoneurons
Motor act, complex, 607
Motor activity, preprogrammed, 479–482
Motor area, supplementary, *see* Supplementary motor area
Motor commands
  movement sensations and, 152
  muscle sense and, 151–167
"Motor" compartment, 978,980
Motor control
  changing, 1090–1091
  higher-order, SMA role in, 430–431
  in hyperbaric conditions, 829–849
  muscle afferents and, 93–132
  PM function in, 418–419
  in precision grip, 253–261
  of prehension, 373–385
  recovery after CNS lesions, 1085–1092
  SMA function in, 406–408
  after spinal injury, 915–926
  of voluntary contractions, 47–59
Motor control components, antispastic therapy effects on, 1029–1034
Motor cortex, 378–380
  caudal, *see* Caudal motor cortex
  functional organization of, 301–319
  in motor responses, 75–84
  precentral, *see* Precentral motor cortex
  primate, connectivities of, 263–285
  rostral, *see* Rostral motor cortex
  sensory motor processing in, 61–92
  structure of, 303–306
  thalamic relay to, 267–276
Motor cortical arousal, structure of, 287–300
Motor disorders
  TSR and, 1010–1011
  vibration-induced inhibition in, 673–674
Motor dyscontrol, 685–697
Motor effects, vibration, 686–687
Motor impulses, fatigue and, 180–183
Motor neurons, *see* Motoneurons
Motor performance variables, 349
"Motor point," 135
Motor representation of distal forelimb, 321–324
Motor response (M), 929,962
  of masseter, 743
  motor cortex in, 75–84
Motor strategy, term, 351
Motor stroke, 155
Motor system, hierarchically organized, 3
Motor threshold, 55–56
Motor unit, 227
  large, cutaneous facilitation of, 253–261
  masseter, 681
  phasic, 243–245
  rate gradation of, 240–243
  recruitment, 680–683
  single (SMU)
    in chorea, 900–902
    in familial ataxia, 903–904
    firing frequencies of, 185–199
    in Friedreich's ataxia, 903–904
    in movement disorders, 897–905
    in parkinsonism, 899–900
    responses to mechanical perturbances, 492–493
    in spastic paraparesis, 903
  soleus, 676–677
  wrist flexor and extensor, 351–357
  wrist flexor and extensor activity summary, 356
Motor unit spikes, single, 53
Motosensory areas, 316
Movable platform, movement control on, 607–619
Movable platform techniques, 607–610
Movement
  active
    responses to stop of, 340–341

SUBJECT INDEX

temporal relation to, 332–335
assumptions, 3
axioms, 3–4
ballistic, *see* Ballistic movement
control, open-loop hypothesis of, 534,535
cortical neuronal activity and, 302–303
discharges during, of corticospinal neurons, 309–313
eye, 15
finger grip, 18
forearm, 34–35
hand, *see* Hand movement
involuntary, *see* Involuntary movement
isokinetic, *see* Isokinetic movements
in isokinetic tests, 1018–1019
joint, *see* Joint movement
learned, neurons during, 393–420
multijoint, 45
myotatic gain preceding, 546
neuronal activity and, 405,415,417
peripheral versus central control of, 56
PM in control of, 409–410
preprogramming of, by SMA, 407–408
program disorders of, *see* Program disorders of movement
reaching, *see* Reaching movement
single, myoelectric control of, 1093–1096
sinusoidal, 999
small versus large, 56
theory of servo-assistance of, 344
voluntary, *see* Voluntary movement
wrist, cerebellum and, 381,383–385
Movement amplitude, 15
Movement control
on movable platform, 607–619
servo-assistance mechanism of, 479
Movement disorders, motor unit control in, 897–905
Movement-related neurons
in PM, 411–414
in SMA, 398–400
Movement sensations, motor commands and, 152
Movement speed, 14–15
Movement time (MT), 15,48,52
Ms (motosensory area), 316
MSR (monosynaptic stretch reflex), 468
MT (movement time), 15,48,52
Multifocal muscle jerks, 869
Multijoint arm movements, 39–45
Multijoint movement, 45
Multiple sclerosis, 778–780
Multiunit recordings, 101–102
Muscle afferents, motor control and, 93–132
Muscle contraction, recruitment and, 224–226
Muscle cooling, firing frequency and, 193–199
Muscle fascicle, 94
Muscle fibers, 108,109
Muscle force, gradation of, 213–226

Muscle jerks
multifocal, 869
pathophysiology of, 865–881
Muscle length
firing rate and, 121
variations in, 131
zero, 148
Muscle relaxant drugs, 987
in spasticity, 1006–1007
tonic stretch reflex and, 1005–1011
Muscle relaxation, degree of, 101
Muscle responses, 593–597
Muscle sense, motor commands and, 151–167
Muscle shortening, 58
Muscle spindle afferent activity, 143–150
Muscle spindle afferents, 93
Muscle spindle discharge during voluntary movement, 108–129
Muscle spindle responsiveness, fusimotor drive and, 138
Muscle temperature, 170
Muscle tone, 146
Muscle tone disorders, fusimotor role in, 133–150
Muscle twitch, 97–98,102–104
Muscle unit, 213
endurance, recruitment order and, 221–222
firing rate stability and, 223–224
force, rate modulation of, 216–219
grouping, 213
recruitment, ascending-size-order of, 218,219–220
tension-frequency relation of, 216
type, firing duration and, 222–223
Muscle warming-up effect, 173–174
Muscles
antagonist, *see* Antagonist muscles
cooled, 176
normal, powered prostheses versus, 1104–1106
resting, spontaneous fusimotor activity in, 112
slow versus fast, 217–218
synergic, recruitment patterns in, 253–255
Muscular contractions
judgments about, 151–167
passive shortening and, 555–563
pyramidal tract discharge and, 313–315
Muscular fatigue, 152–153
Muscular force, sensations of, 152–162
Muscular wisdom
artificial, 181–183
fatigue and, 169–211
Myo-monitor, 743–744
"Myocentric occlusal position," 744
Myoclonic dystonia, 853
Myoclonic jerk, 473
Myoclonus
categories of, 865
EEG-EMG correlation in, 868–870
EMG analysis of, 909–912

Myoclonus *(contd.)*
  EMG correlates of, 866–868
  under high pressure, 842
  pathophysiology of, 865–872
  somatosensory evoked potentials in, 870–872
Myoelectric control of single movements, 1093–1096
Myofilaments, 111
Myotatic arc, excitability of, 986
Myotatic gain preceding movement, 546
Myotatic jaw reflex, 731
Myotatic reflex gain, 542
Myotatic reflexes
  ballistic movements and, 544–546
  long-latency, *see* Long-latency myotatic reflexes
Myotatic stretch reflexes, 7

Naloxone, 821
Neocerebellar lesions, 636,637
Neural control of prostheses, 1101–1103
Neural mechanisms, 1091–1092
Neurobionomics
  of adaptive plasticity, 1047–1071
  term, 1048
Neurography, human, afferent identification in, 100–105
Neurological diseases, gait analysis in, 718
Neurological syndrome, high-pressure (HPNS), 840–841
Neuromuscular block, 209
Neuronal activity, movement and, 405,415,417
Neuronal discharge patterns in wrist control area, 357–363
Neuronal function, contextual pluralism of, 300
Neurons
  motor, *see* Motoneurons
  single, during learned movement, 393–420
Neuropathy, diabetic, 972
Neuropeptides in spinal cord, 981
Neuropharmacology, clinical, 977–996
Neurophysiology, high-pressure, 835–849
Neuropil, 303
Neurotensin, 983
Neurotransmission in spinal circuits, 978–981
Neurotransmitters, 977
  currently known, 982–983
Nociceptive flexion reflexes, pain research and, 809–827
Nociceptive input, 809
Nociceptive reflexes, methodology for, 812–813
Nonprogram disorders, 1,2
Norepinephrine, 982
Nuclear bag 1 and 2, 108–109
Nuclear chain fibers, 108
Nucleus X, 267
Nystagmus
  horizontal, premotor mechanisms and, 585–588
  vestibular, 565–588
Nystagmus-related type II neurons, 573

Ocular torsion, voluntary, 1090–1091
Oculomotor function, 565
OKR, *see* Optokinetic response
Open-loop condition, 15
  error in, 24–25
  of final position, 31–36
Open-loop hypothesis of movement control, 534, 535
Opiate brainstem control, endogenous, 811
Opiates, spinal effect of, 811
Opposability, index of, 373
Optic ataxia, reaching errors in, 27–29
Optimal T-f range, 216
Optokinetic pathway (VIS-OK), 1059
Optokinetic response (OKR), 1056–1064
  adaptive changes in, 1060–1061
Oral ventral lateral nucleus, 267
Otolith organs, 621
Otolith-spinal reflexes, 628
Otolith system, 589
Otolith units, second-order, 590
Otolithic receptors, 621–622
Overshooting, 24,26
Oxytocin, 979,983

p1 plates, 108
p2 plates, 108
Pain
  afferent volleys, 814–818
  congenital insensitivity to, 821–824
  gate control theory of, 810–811
Pain input control, 810–811
Pain mechanisms, 809–810
Pain research, nociceptive flexion reflexes and, 809–827
Pain sensations, flexion reflexes and, 811–827
Pain transmission, 809–810
Paleocerebellum, 634
Palmaris longus (PL) muscle, 956–958,960
Parallel S-R processor, 90–91
Parametric feedback control, 1050–1052
Paraparesis, spastic, *see* Spastic paraparesis
Paraplegia, 915
Parasellar meningioma, 783
Paraspastic gait, 718–721
Paratrigeminal syndrome, 776
Parkinsonian akinesia, 480
Parkinsonian rigidity, 147,149–150
Parkinson's disease
  bradykinesia in, 884
  long-latency reflexes in, 502–504
  motor unit control in, 899–900

program disorder in, 5–6
quick-tracking movement in, 884–885
slow-tracking movement in, 885–886
SR in, 559–561
treatment, 4–6
Pathophysiology
  of chorea, 872–874,875
  of dystonia, 851–863
  of human locomotion, 717–726
  of muscle jerks, 865–881
  of myoclonus, 865–872
  of tics, 874–881
Peak frequency, 186
Pectoralis tulip, 647
Pendulum test after spinal injury, 918
Pentobarbital, 96
Peptides, 983
Percent modulation in firing rate (PMFR), 360–361
Perceptive defects, kinesthetic, 694
Perceptual effects, 151
  vibration, 686
Perceptual threshold (PT), 709
Performance, relative errors in, 48
Performance context, 4
Peripheral disturbances, compensation to, 47–59
Peripheral vision condition, 20
Perirolandic cortex, right, 302
Perturbations, delivery of, 57
Phantom limb, 152
Pharmacological correlations with electrophysiological tests, 984–995
Pharmacology, clinical, drug value in, 995–996
Phasic activation, 53–55
Phasic control signal, 65–67
Phasic fusimotor activity, 120
Phasic motor units, 243–245
Phasic response pattern, 222
Phasic soleus contractions, 451–454
Phasic stretch reflex, conditioning, 920–923
Phenothiazine, 504–506
Phosphene in visual field, 391
Physostigmine in cerebral ataxia, 9–10
Picrotoxin, 675
Pinch
  lateral, 376
  skin, 256–257
Piper rhythm and frequency, 208
PL (palmaris longus) muscle, 956–958,960
Plantar stimulation, 922
Plasticity
  adaptive, neurobionomics of, 1047–1071
  of vestibuloocular reflex, 1091
Platform studies of spasticity, 6–7
PMFR (percent modulation in firing rate), 360–361
Polarity, display, 64
Polyneuropathy, 778

diabetic (DPN), 974
  F waves in, 968–975
  monosynaptic reflexes in, 927–949
Polysynaptic stretch reflex (PSR), 480–481
Pontine infarct, rehabilitation after, 1087–1089
Position sensitivity, 1003,1004
Positional control, term, 524
Postcentral gyrus, 301
Postcruciate gyrus, 89
Postspike facilitation, 308–309
Postsynaptic excitation (E), 289
Postsynaptic inhibition (I), 289
Postural activity in cerebellar patients, 637–638
Postural adjustments, rapid, vestibular and visual control of, 589–605
Postural control, vision in, 598
Postural instability
  with cerebellar lesions, 633–643
  quantitative analysis of, 633–634
Postural perturbations, visual cues and, 598–604
Postural preactivation, 637
Postural reactions, rapid, to mechanical displacement, 645–659
Postural reactions to fall, 592–597
Postural support, automatic actions and, 613–615
Postural tremor, 638–643
"Postural tulip," 647–659
Posture
  equilibrium and, 694–696
  limb, see Limb posture
  SMA and, 395
  SMA function in, 406–408
Posture-movement transitions, 542–543
Potentiation, vibration-induced, 671–683
Power grips, 374–375
Precentral cortical neurons, 474
Precentral gyrus, 301
  primary motor area of, 393
  wrist control areas of, 357
Precentral motor area, SMA versus, 428–430
Precentral motor cortex, current-threshold maps in, 367–368
Precentral neurons
  in abducens nucleus, 584–585
  afferent projections to, 316–319
  in medial vestibular nucleus, 566–572
Precision grips, 374–375
  motor control of, 253–261
"Precision handling," term, 375
Prehension, 375
  form and control of, 374–378
  somatosensory afferents in, 373–385
Premotor area
  historical background, 408–409
  identification of, 410–411
  of monkey cerebral cortex, 393–408
  in motor control, 418–419
  in movement control, 409–410

Premotor area *(contd.)*
  neuronal organization of, 565–588
  pyramidal tract neurons in, 414–415,418
  somatotopy in, 415
Premotor area neurons
  bilaterality of, 415,417
  movement-related, 411–414
  natural activity of, 410
  peripheral afferent input into, 415,418
Premotor burst neurons in reticular formation, 575–584
Premotor cortex of monkeys, 283–284
Premotor mechanisms, horizontal nystagmus and, 585–588
Pressure, high, *see* High pressure *entries*
Pressure block, 171
Primary firing range, 214
Primary motor area
  gating of response by SMA, 406–407
  of precentral gyrus, 393
Primate hands, 373–374
Prisms, reversing, experiments with, 1052–1056
Progabide, 990
Program disorders of movement, 1–11
  development of, 1–2
  future work, 10–11
  list, 2
  in parkinsonism, 5–6
Propanidid, 96
Proprioceptive afferent function concept, 129–132
Proprioceptive afferents, characteristics of, 97
Proprioceptive group I discharges, 699
Proprioceptive inputs
  cutaneous inputs and, 797
  regulatory role of, 47–59
Proprioceptive map, 27
Propriospinal interneuron system, 924–926
Prostheses
  neural control of, 1101–1103
  powered, normal muscles versus, 1104–1106
  sensory feedback of, 1103–1104
  tristate, 1096–1098
Protective gait mechanisms, 723–726
Proximal nerve CV, 930–936
  distal CV changes versus, 942–943
PSR (polysynaptic stretch reflex), 480–481
PT (perceptual threshold), 709
PT (pyramidal tract), *see* Pyramidal tract *entries*
PTNs (pyramidal tract neurons), 253,257
Pulse step command, 39
Pulse step control, 65–67
Purkinje cells, cerebellar, 383–385
Pyramidal cells, 303
Pyramidal neurons, 90
Pyramidal tract antidromic identification, 331
Pyramidal tract axons, 310–311
Pyramidal tract (PT) cells, fast and slow, 287–293

Pyramidal tract discharge, muscular contraction and, 313–315
Pyramidal tract neurons (PTNs), 253,257
  corticorubral neurons versus, 329–345
  in PM, 414–415,418
  in SMA, 400–401

Quadriceps (Q) contraction, 699–700

RA-III reflex, 813–821
  supraspinal influences on, 818–820
Radiculopathy
  carcinomatous, 946–947
  infectious or inflammatory, 947
Ramp-and-hold displacement, 341
Ramp-and-hold stretches, 99,104
Ramp contraction, 232
Ramp mode of voluntary motor control, 230–232
Ramp threshold, isotonic, 245
Rank deordering, 245–248
Rank ordering in motoneuron pool, 237–238
Rate gradation, 213
  of motor units, 240–243
Rate modulation of muscle unit force, 216–219
Reaching movement
  dynamic aspect of, 17
  at natural objects, 15–18,19
  in optic ataxia, 27–29
  visual control of, 13–29
Reaction torque, 40
Receptive fields, simple and complex, 75–76
Reciprocal Ia inhibitory pathway, 433–441
Reciprocal mode of control, 371–372
Recording electrode, 94
Recruitment deordering, 254
Recruitment frequency, 898
Recruitment gradation, 213
  of muscle contraction, 224–226
Recruitment order
  of fusiform neurons, 116–118
  muscle unit, 218,219–220
  muscle unit endurance and, 221–222
  size principle and, 219–221
Recruitment patterns
  exteroceptive input effects on, 255–257
  in synergic muscles, 253–255
Recruitment thresholds
  mean, 246–248
  twitch force and, 229–251
Recurrent collaterals, 304–305
Recurrent inhibition
  dystonia and, 855–859
  evidence for, 443–447
  phasic soleus contractions and, 451–454
  role of, 446–447,448–449

serving as control of reciprocal Ia inhibition, 455–457
serving as variable gain regulator for motor output, 455
during silent period, 459–465
tonic soleus voluntary contractions and, 449–451
during triceps contractions, 447–449
Red nucleus
  antidromic identification, 331
  input-output plan, 61
  sprouting in, 1074–1075
Reduced vision (RV condition), 601–604
Redundancy, 2,4
Reference trajectory, 45
Reflex control
  of bipedal gait, 699–716
  error-activated, 1048–1050
Reflex data, patterns of, 935–940
Reflex excitability, 58–59
Reflex gain, 144
Reflex reinforcement by remote voluntary muscle contraction, 661–669
Reflex responses in spinocerebellar atrophy, 638–643
Reflex responsiveness, 499,501–502
Reflex threshold of soleus, 143
"Reflex" tulip, 648
Reflexes
  blink, see Blink reflex
  clinical neuropharmacology and, 977–996
  corneal, see Corneal reflex
  electrically induced, 473–474
  H, see H reflex
  under high pressure, 841–842
  jaw, see Jaw reflex
  monosynaptic, see Monosynaptic reflex
  myotatic, see Myotatic reflexes
  segmental, 475–478
Refractoriness, transient, 70
Rehabilitation, 1085–1092
  in man, 1086–1089
  after pontine infarct, 1087–1089
  primate, 1085–1086
  rat, 1085
  sensory, 1089–1090
Relative errors in performance, 48
Renal insufficiency, chronic, 935–936
Renshaw cell activity, 142
Renshaw cells, 443–457,991; see also Interneurons
  during voluntary movements, 454–457
Response
  skilled, 232
  triggered, 74
Response grouping, 70
Response latency, 67–68
  variance in, 83

Response topographies, 90
Restraint, passive, determination of, 1019–1020
Restricted vision condition, 20
Reticular activating system, ascending, concept of, 287
Reticular formation, premotor burst neurons in, 575–584
Reticulospinal tract, 623
Reversals, joint, 44
Rexed layer IX, 3
Rhythmic threshold, 214
Rigidity
  antiparkinsonian drugs in, 1007–1010
  decerebrate, 133–134
  in limb, 5
  parkinsonian, 147,149–150
  "resting," 1007
  TSR in, 1003–1005
Root lesions, 945–947
Rostral compartments of area 4, 76
Rostral motor cortex
  lead cells in, 77–80
  projections to, 84–89
RV condition (reduced vision), 601–604

S motor unit, 228
S (slow-twitch fatigue-resistant) muscle unit, 213
S-R processors (stimulus-response transformations), 90–91
Saccular afferents, 590
Saccular maculae, 589
SBS (spino-bulbo-spinal) pathways, 467
SC (selective contraction), 662–663
Scalp, stimulation of corticospinal pathways through, 387–391
Sciatic nerve block, 145
SCT (spinocervical tract), 810
Secondary firing range, 214,242
Segmental contributions to long-latency stretch responses, 467–487
Segmental reflexes, 475–478
Segmental stretch reflex responses, short-latency, 639–640
Selective contraction (SC), 662–663
Semicircular canals, 621
Semilunar ganglion, 105
Senile chorea, 914
Sensation
  of muscular force, 152–162
  translating, into action, 61
Sensitivity
  dynamic, 1004
  position, 1003,1004
  small-amplitude, 100
  velocity, 1003
Sensorimotor area, 316

Sensorimotor function
  environmental demands and, 1047–1071
  under high pressure, 832–833
  locomotion and, 717–718
"Sensory ataxia," 905
"Sensory" compartment, 980
Sensory conditions
  altered, equilibrium control under, 615–616
  vestibular inputs and, 616–618
Sensory convergence area, SMA as, 395
Sensory cortex, thalamic relay to, 276–277
Sensory effects, vibration, 685–686
Sensory feedback of prostheses, 1103–1104
Sensory feedback therapy for hemiparesis, 1087
Sensory fields, peripheral, 157–160
Sensory inputs, equilibrium and, 608
Sensory mechanisms, maturation of, 618–619
Sensory-motor cortex, corticocortical connectivity of, 277–279
Sensory-motor processing, 61–92
Sensory-motor tasks, 691–694
Sensory pathways, distribution of, 266–267
Sensory rehabilitation, 1089–1090
SEP, see Somatosensory evoked potential
Septal nucleus, sprouting in, 1074
Serotonin, 982
Servo-assistance mechanism of movement control, 479
Servo-assistance of movement, theory of, 344
Servo hypothesis, 499
Servo-information hypothesis, 499
Shock
  spinal, 144–146
  supramaximal, 204–205
Shock strength, tetanic tension and, 172
Shoes, microswitch, 718
Short-latency inputs, afferent pathways of, 263–285
Short-latency projections, 319
Short-latency segmental stretch reflex responses, 639–640
Shortening
  muscle, 58
  passive, muscular contractions and, 555–563
Shortening reaction (SR), 555
  in cerebellar ataxia, 561–562
  family, 555–557
  mechanism and significance of, 562–563
  in parkinsonism, 559–561
  paradoxical, 504
  in triceps brachii, 557–559
Signs, 2–3
Silent period (SP)
  F wave and, 461,463
  in human masseter muscles, 735–738
  after jaw reflex, 753–755
  motoneurons during, 459–465
Simulated evoked step, 126–128

Single-channel formulation, 68
Single-degree-of-freedom arm movements, 36–39
Single-unit recordings, 102
Sinusoidal movements, 999
Size principle
  force-speed control and, 227–251
  of motoneuron recruitment, 227–251
  recruitment order and, 219–221
Skeletofusimotor beta innervation, 112–113
Skeletomotor activity, fusimotor activity and, 118–119
Skeletomotor system, fusimotor system and, 130–131
Skilled response, 232
Skin pinch, 256–257
Skin stimulation, 813–814
Slow-twitch fatigue-resistant (S) muscle unit, 213
Sm (sensorimotor) area, 316
SMA, see Supplementary motor area
Small-amplitude sensitivity, 100
SMU, see Motor unit, single
Soleus
  excitability of motoneuron pools, 798–806
  phasic contractions, 451–454
  reflex threshold of, 143
  tonic voluntary contractions, 449–451
Soleus motor unit, 676–677
Somatic sensory areas, subcortical projections of, 279–283
Somatosensory afferents in prehension, 373–385
Somatosensory evoked potential (SEP), 253
  under high pressure, 846–849
  myoclonus and, 870–872
Somatosensory inputs, anterior-posterior partitioning of, 366
Somatostatin, 979,983
Somatotopy
  in PM, 415
  in SMA, 402–403
Somesthetic information, 691
SP (silent period), see Silent period
SP (substance P), 811,979,983
Spastic hemiparesis, 1025–1026
Spastic hemiplegia, 439–440
Spastic paraparesis, 899,1027
  motor unit control in, 903
Spastic restraint, 1029–1032
Spasticity, 6–8,148–149
  assessing, of lower limb, 1035–1046
  baclofen and, 7–8
  bicycling and, 1035–1046
  innervation patterns in, 1038–1039
  local anesthetic and, 134
  muscle relaxants in, 1006–1007
  platform studies of, 6–7
  quantitative evaluation of, 1039–1041
  TSR in, 999–1003

Speed, movement, 14–15
Spike suppression of secondary type I MVN neurons, 572–575
Spike-triggered averaging (STA), 228
Spikes, single motor unit, 53
Spinal circuits, neurotransmission in, 978–981
Spinal cord
    neuropeptides in, 981
    sprouting in, 1076
    vestibulospinal influences to, 622–623
Spinal cord chamber, 96
Spinal cord dorsal horn mechanisms, 994–995
Spinal effect of opiates, 811
Spinal gray matter, 978
Spinal injury, motor control after, 915–926
Spinal motoneurons
    frequency-current relation of, 213–216
    functional properties of, 213–226
Spinal myoclonus, 865
Spinal reflexes
    in lower limbs, 687–689
    suprasegmental activation of, 919–920
Spinal shock, 144–146
Spindle afferent discharge, 93
Spindle afferents, 96
    discharge rates, 110–111
    dynamically sensitive, 111
    position sensitivity, 111
    primary and secondary, 104–105
Spindle firing rate, 58
Spindle properties in man, 108–119
Spindle responses to rapid stretch, 121–123
Spindle volley, long-latency stretch reflexes and, 518, 520–523
Spindles, studies of human, 109–110
Spino-bulbo-spinal (SBS) pathways, 467
Spinocerebellar atrophy, reflex responses in, 638–643
Spinocervical tract (SCT), 810
Spinothalamic terminations, 269
Spinothalamic tract (STT), 810
Spinovestibular pathway, 271
Sprouting
    after brain damage, 1073–1084
    collateral, 1091
    without degeneration, 1082–1083
    after lesions, 1079–1082
    morphological studies, 1073–1077
    physiological studies, 1077–1084
    signals of, 1083–1084
    term, 1073
    time course of, 1076–1077
SR, see Shortening reaction
STA (spike-triggered averaging), 228
Stabilization method, visual, 598
Stance ataxia, 634–637
Startle reaction, 594
Static error, 51

"Static TSR," 999
Steady-state equilibrium position, 37
Steady-state force, discharge frequency and, 335–340
Stellate cells, 305–306
Stiffness
    joint, see Joint stiffness
    wrist, 351
Stimulation
    double, 69–70
    skin, 813–814
    sural nerve, 798–806, 813
Stimulus-response compatibility, 74
Stimulus-response relations in lead cells, 80–83
Stimulus-response transformations (S-R processors), 90–91
Stimulus strength, 700
Stimulus uncertainty, 67–68
Stop effect, 340
Stretch, rapid, spindle responses to, 121–123
Stretch reflex
    in leg, 518
    monosynaptic, 468
    myotatic, 7
    transcortical, 479
    vibration and, 490–492
"Stretch reflex" gain, 144
Stretch reflex studies after spinal injury, 917–923
Stretch responses, long-latency, see Long-latency stretch responses
Strobe illumination, 1061–1065
Stroke, motor, 155
STT (spinothalamic tract), 810
Subcortical myoclonus, 865
Subcortical projections of somatic sensory areas, 279–283
Substance P (SP), 811, 979, 983
Summation, temporal, 953–954
Supination movements, unopposed, 49–50
Supplementary motor area (SMA), 284–285, 380–381, 382
    anatomical connections, 421–427
    electrical stimulation series, 422
    fore- and hindlimb representation areas in, 422–423
    functional organization of, 421–431
    gating of primary motor area responses by, 406–407
    historical background, 394
    identification of, 397–398
    lesions of, 408
        deficits following, 422
    of monkey cerebral cortex, 393–408
    in motor control, 406–408
    peripheral afferent input into, 406
    possible function of, 394–396
    posture and, 395
    posture control and, 406–408

Supplementary motor area (SMA) *(contd.)*
  precentral motor area versus, 428–430
  preprogamming of movement by, 407–408
  pyramidal tract neurons in, 400–401
  role in higher-order motor control, 430–431
  as sensory convergence area, 395
  somatotopy in, 402–403
Supplementary motor area neurons
  activity, 396–397
  afferent input into, 401–402
  bilaterality of, 403–405
  movement-related, 398–400
  related to distal and proximal forelimb movements, 425–428
  related to fore- and hindlimb movements, 423–425
Suppressed vision (SV condition), 601–604
Supramaximal shocks, 204–205
Suprasegmental activation of spinal reflexes, 919–920
Suprasegmental contributions to long-latency stretch responses, 467–487
Suprasegmental lesions, 792
Suprasegmental responses, long-latency, 640–643
Supraspinal circuit, 803–804
Supraspinal contribution to long-latency stretch responses, 482–487
Supraspinal control loss, 793
Supraspinal influences on RA-III reflex, 818–820
Supraspinal signal, excitatory, 66
Suprathreshold intensity, 215
Sural nerve stimulation, 798–806, 813
Surface EMG recording, 101
Surgical implants, 1098–1099
Suxamethonium, dynamic index and, 99–100
SV condition (suppressed vision), 601–604
Sway motions, 608
Sway rotations, 609
Sydenham's chorea, EMG analysis of, 913–914
Symptoms, 2–3
Synaptic transmission, properties of, 1078–1079
System redundancy, 2

T cells (target cells), 810
T-f relation, 216
T/H ratio, 930–947
  increase of, 944–945
T reflex (tendon reflex), 688–689
T responses, 928
Target-acquisition phase, 27
Target cells (T cells), 810
Target eccentricity, 22
Target steps, double, 69–70
Targeted movements, sensory motor processing of, 61–92
Taurine, 983
"Tea-cup" tulip, 659

Telemetry, 1099–1101
Temperature, muscle, 170
Temporal relation to active movement, 332–335
Temporal summation, 953–954
Temporomandibular joint pain dysfunction (TMJ), 753–754
Tendon jerk
  conditioning, 917
  fusimotor activity and, 139–143
  fusimotor drive and, 138
  H reflex and, 137–143
  hung-up, 873, 876
Tendon organ discharge during voluntary movement, 106–108
Tendon organs, 106
Tendon reflex (T reflex), 688–689
Tendon taps after spinal injury, 917–918
Tendon vibration, 687
Tension
  isometric, 153–154
  peak, 172
Tension-frequency relation of muscle units, 216
Tensor fasciae latae (TFL), 649–650
Tetanic fatigue curves, 183
Tetanic tension, shock strength and, 172
TFL (tensor fasciae latae), 649–650
Thalamic afferents, 285
Thalamic relay
  to motor cortex, 267–276
  to sensory cortex, 276–277
Thalamocortical projections, 86–87, 305–306
Thalamocortical relay cells, 269
Threshold strength, multiples of, for M wave, (xMT), 434
Threshold velocity, 1003
Thumb, adduction of, 389
Thumb flexor, long-latency stretch reflexes and, 524–533
Thumb tulip, 647
Thymoxamine, 988, 990
Tib_1 response, 639–643
Tib_2 response, 641–643
Tibialis anterior H reflex, 798
Tibialis anterior motoneuron pools, excitability of, 798–806
Tics
  EEG-EMG correlation in simple, 876–881
  EMG analysis of, 914
  EMG correlates of, 875–876
  pathophysiology of, 874–881
Time, movement (TM), 15, 48, 52
Time-to-peak, minimum, 239
Timing, subjective, of voluntary
  muscular contractions, 165–167
Tizanidine, 990–991, 1033
TMJ (temporomandibular joint pain dysfunction), 753–754
Tonic activation, 55

Tonic command, 65–67
Tonic depression of Ib effects, 712
Tonic gamma-activation, 130,131
Tonic labyrinthine reflex control of limb posture, 623–631
Tonic labyrinthine reflex interaction, 631–632
Tonic neck reflex interaction, 631–632
Tonic soleus voluntary contractions, 449–451
Tonic stretch reflex (TSR)
  drug effects on, 1010
  "dynamic," 999
  electromyograph approach to, 998
  mechanomyographic approach to, 997–998
  motor disability and, 1010–1011
  muscle relaxant drugs and, 1005–1011
  passive characteristics of, 998–1005
  in rigidity, 1003–1005
  in spasticity, 999–1003
  "static," 999
Tonic vibration reflex (TVR), 147,624–625,686
  exteroceptive, 687
  in man, 671–683
  masseter, 745–748
Tooth pulp stimulation, electrical, 749
Torque
  computerized analysis of, 1041–1043
  interaction, 40
  in isokinetic movements, 1015–1018
Torsion dystonia, 851
Tracking performance, continuous feedback system and, 65–75
Trail endings, 108
Trajectory
  reference, 45
  term, 39
Trajectory formation, arm, see Arm trajectory formation
Transcerebellar loop, 475
Transcerebral concept of long-latency stretch reflexes, 523–524
Transcortical concept of long-latency stretch reflexes, 516–524
Transcortical stretch reflex, 479
Transcortical stretch reflex mechanism, 509–510
Transcutaneous nerve stimulation, 825–826
Transient refractoriness, 70
Translational mechanisms, 62
Treadmill locomotion, 718
Tremor
  high-pressure, 845–846
  postural, 638–643
TRH, 983
Triamcinolone, 138
Triceps brachii, SR in, 557–559
Triceps contractions, recurrent inhibition during, 447–449
Triceps surae motoneurons, 135
Triceps tulips, crossed, see Crossed triceps tulips

Trigeminal nerve, 727–728
Trigeminal nerve lesions, 758–759
  corneal and blink reflexes and, 763
Trigeminal neuralgia, 776
Trigeminal nucleus, 827
Trigeminal reflexes in man, 727–738
Triggered responses, 74
Triphasic pattern, 556
Tristate prostheses, 1096–1098
Trisynaptic pathway, 142
TSR, see Tonic stretch reflex
"Tulip" records, 647–659
TVR, see Tonic vibration reflex
Twitch, muscle, 97–98,102–104
Twitch amplitude, twitch duration and, 353
Twitch contraction times (CT), 228–229
Twitch contractions, maximum, 103
Twitch duration, twitch amplitude and, 353
Twitch force, recruitment threshold and, 229–251
Twitch response, isotonic, 97–98

Uncertainty, stimulus, 67–68
Undershooting, 24,26
Unilateral ataxia, 561
Unilateral labyrinthectomy, 627
Unit activity, 77–80
Unloading
  burst after (BAU), 557
  jaw-closing muscles, 745
Unloading reflex, 745
Unmasking, 1091–1092
Utricular afferents, 590
Utricular maculae, 589

Variation of discharge frequency, coefficient of, 100
Vasopressin, 983
Velocity sensitivity, 1003
Ventral posterior lateral nucleus (VPL), 266–277, 317–319
Ventral posterior medial nucleus, 276
Ventral root recordings, 106
Vestibular control of rapid postural adjustments, 589–597
Vestibular inputs, sensory conditions and, 616–618
Vestibular neurons
  excitatory, 567–570
  inhibitory, 570–572
Vestibular nucleus, medial
  neurons, secondary type I, 572–575
  premotor neurons in, 566–572
Vestibular nystagmus, 565–588
Vestibular organs, peripheral, 621
Vestibular system, locomotor patterns and, 723
Vestibulocerebellar lesions, 636,637

Vestibuloocular reflex (VOR), 73, 589, 1050–1058
　plasticity of, 1091
Vestibuloreticular connection, 587
Vestibulospinal influences to spinal cord, 622–623
Vestibulospinal reflexes, 589
Vestibulospinal tract, 622
Vibration
　Achilles tendon, 678–679
　chin, 679
　externally applied, 98
　foot control and, 693
　foot tracking and, 692–693
　forearm positioning and, 691–692
　high-intensity, 685
　motor effects, 686–687
　perceptual effects, 686
　sensory effects, 685–686
　sensory-motor tasks and, 691–694
　stretch reflex and, 490–492
　tendon, 687
　whole body (WBV), 685–697
Vibration effects, 671
Vibration-induced contraction, 155–156, 163
Vibration-induced inhibition
　in patients with motor disorders, 673–674
　possible mechanisms underlying, 678–680
Vibration-induced potentiation of masseter reflex, 677–678
Vibration paradox
　features of, 673
　in man, 671–683
　mechanisms involved in, 674–677
Vibration parameters, 672–673
Vibration reflex
　after spinal injury, 918–919
　tonic, see Tonic vibration reflex
Vibration response, antagonist, 687
Vibratory inhibition in clinical neuropharmacology, 989–991
VIS-OK (optokinetic pathway), 1059
Viscous loads, 66
Vision in postural control, 598
Visual control
　of rapid postural adjustments, 598–605
　of reaching movements, 13–29
Visual critical fusion frequency (CFF), 833
Visual cues, postural perturbations and, 598–604
Visual feedback, 13
Visual field, phosphene in, 391
Visual "map," 13–14
Visual motion cues, 603–605
Visual stabilization method, 598
Visual-vestibular interactions, 604–605
Visuomotor cells, 419

Visuomotor leg tracking, 883–888
Visuomotor mechanisms, two-level organization of, 14
Visuomotor task, eye-hand sequence in, 22–25
Visuomotor tracking in cerebellar ataxia, 889–895
Voluntary contractions
　motor control of, 47–59
　tonic soleus, 449–451
Voluntary fatigue, 390
Voluntary motor control, ballistic versus ramp modes of, 230–232
Voluntary movement, 3–4
　afferent activity during, 94–97
　analysis with surface EMG, 907–909
　changes with, 443–457
　dystonia and, 853
　muscle afferent function and, 93–132
　muscle spindle discharge during, 108–129
　Renshaw cell control during, 454–457
　tendon organ discharge during, 106–108
Voluntary muscular contractions
　remote, reflex reinforcement by, 661–669
　subjective timing of, 165–167
Voluntary ocular torsion, 1090–1091
Voluntary potential (VP), 459–465
Voluntary response, later (VOL), 529
VOR, see Vestibuloocular reflex
VP (voluntary potential), 459–465
VPL (ventral posterior lateral nucleus), 266–277, 317–319

Wallenberg's syndrome, 759
　blink reflex and, 780
　corneal and blink reflexes with, 763–767
Warming-up effect, muscle, 173–174
WBV (whole body vibration), 685–697
Weight heaviness, estimating, 156–157
Westphal phenomenon, 504, 851
Whole body vibration (WBV), 685–697
Withdrawal reflex, 920
Wrist
　mechanical impedance of, 349–351
　movement of, cerebellum and, 381, 383–385
Wrist control area
　neuronal discharge patterns in, 357–363
　of precentral gyrus, 357
Wrist flexor and extensor motor unit activity, 356
Wrist flexor and extensor motor units, 351–357
Wrist stiffness, 351

Zero muscle length, 148